Get the most out of
MyMathLab®

MyMathLab creates personalized experiences to help each student achieve success and provides powerful tools so instructors can create the perfect learning experiences for their courses.

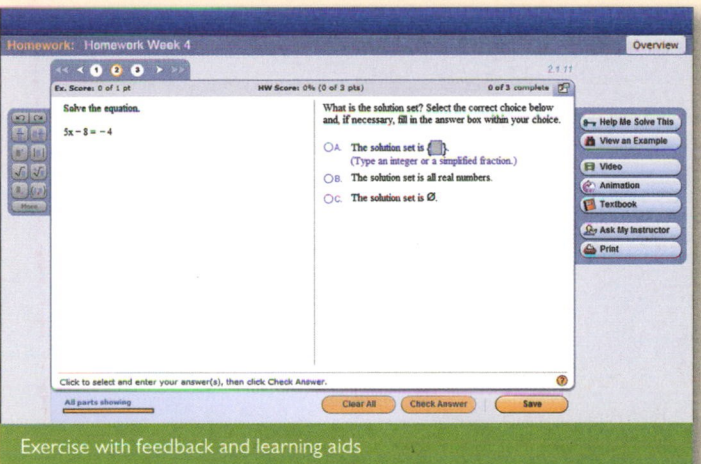

Exercise with feedback and learning aids

Personalized Support
for Students

- MyMathLab comes with many learning resources—eText, animations, videos, and more— all designed to support you as you complete your assignments.

- Whether you're doing homework or working from the adaptive study plan, you'll receive immediate feedback, so you'll know exactly where you need help.

Data-Driven Reporting
for Instructors

- MyMathLab's comprehensive online gradebook automatically tracks students' results on tests, quizzes, homework, and in the study plan.

- The Reporting Dashboard makes it easier than ever to identify topics where students are struggling, or specific students who may need extra help.

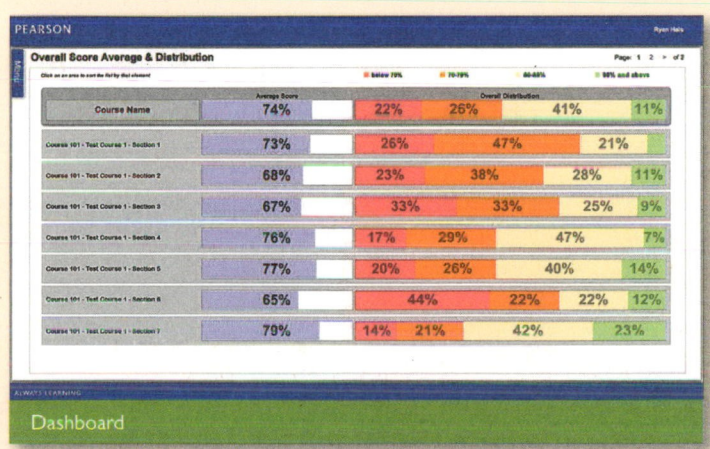

Dashboard

EDITION
6

Beginning and Intermediate Algebra

EDITION 6

Beginning and Intermediate Algebra

MARGARET L. LIAL
American River College

JOHN HORNSBY
University of New Orleans

TERRY MCGINNIS

PEARSON

Boston   Columbus   Indianapolis   New York   San Francisco   Hoboken   Amsterdam
Cape Town   Dubai   London   Madrid   Milan   Munich   Paris   Montreal   Toronto   Delhi
Mexico City   São Paulo   Sydney   Hong Kong   Seoul   Singapore   Taipei   Tokyo

Editorial Director: Chris Hoag
Editor in Chief: Michael Hirsch
Editorial Assistant: Chase Hammond
Program Manager: Beth Kaufman
Project Manager: Christine Whitlock
Program Management Team Lead: Karen Wernholm
Project Management Team Lead: Peter Silvia
Media Producer: Shana Siegmund
TestGen Sr. Content Developer: John R. Flanagan
MathXL Executive Content Manager: Rebecca Williams
Marketing Manager: Rachel Ross
Marketing Assistant: Kelly Cross
Senior Author Support/Technology Specialist: Joe Vetere
Rights and Permissions Project Manager: Diahanne Lucas Dowridge
Procurement Specialist: Carol Melville
Associate Director of Design: Andrea Nix
Program Design Lead: Beth Paquin
Text and Cover Design, Production Coordination, Composition, and Illustrations: Cenveo Publisher Services
Cover Image: Boyan Dimitrov/Shutterstock

Acknowledgments of third party content appear on page C-1, which constitutes an extension of this copyright page.

Library of Congress Cataloging-in-Publication Data
Lial, Margaret L.
  Beginning & intermediate algebra/Margaret Lial, American River College, John Hornsby, University of New Orleans, Terry McGinnis.—6th edition.
    pages cm
  ISBN 978-0-321-96916-3
1. Algebra–Textbooks. I. Hornsby, John. II. McGinnis, Terry. III. Title. IV. Title: Beginning and intermediate algebra.
  QA152.3.L52 2016
  512.9–dc23

                                                    2013049387

2 3 4 5 6 7 8 9 10—CRK—19 18 17 16 15

www.pearsonhighered.com

ISBN 13: 978-0-321-96916-3
ISBN 10: 0-321-96916-2

To Margaret L. Lial

On March 16, 2012, the mathematics education community lost one of its most influential members with the passing of our beloved mentor, colleague, and friend Marge Lial. On that day, Marge lost her long battle with ALS. Throughout her illness, Marge showed the remarkable strength and courage that characterized her entire life.

Margaret L. Lial

We would like to share a few comments from among the many messages we received from friends, colleagues, and others whose lives were touched by our beloved Marge:

"What a lady"

"A remarkable person"

"Gracious to everyone"

"One of a kind"

"Truly someone special"

"A loss in the mathematical world"

"A great friend"

"Sorely missed but so fondly remembered"

"Even though our crossed path was narrow, she made an impact and I will never forget her."

"There is talent and there is Greatness. Marge was truly Great."

"Her true impact is almost more than we can imagine."

In the world of college mathematics publishing, Marge Lial was a rock star. People flocked to her, and she had a way of making everyone feel like they truly mattered. And to Marge, they did. She and Chuck Miller began writing for Scott Foresman in 1970. As her illness progressed, she told us that she could no longer continue because "just getting from point A to point B" had become too challenging. That's our Marge—she even gave a geometric interpretation to her illness.

It has truly been an honor and a privilege to work with Marge Lial these past twenty years. While we no longer have her wit, charm, and loving presence to guide us, so much of who we are as mathematics educators has been shaped by her influence. We will continue doing our part to make sure that the Lial name represents excellence in mathematics education. And we remember daily so many of the little ways she impacted us, including her special expressions, "Margisms" as we like to call them. She often ended emails with one of them— the single word "Onward."

We conclude with a poem penned by another of Marge's coauthors, Callie Daniels.

Your courage inspires me
Your strength…impressive
Your wit humors me
Your vision…progressive

Your determination motivates me
Your accomplishments pave my way
Your vision sketches images for me
Your influence will forever stay.

Thank you, dearest Marge.
Knowing you and working with you has been a divine gift.

John Hornsby
Terry McGinnis

CONTENTS

Additional topics available in MyMathLab®: Sets; Introduction to Calculators; Solving Systems of Linear Equations by Matrix Methods; Piecewise Linear Functions; Conic Sections; Nonlinear Systems of Equations; Second-Degree Inequalities and Systems of Inequalities; Linear Programming; Sequences and Series; The Binomial Theorem

PREFACE

It is with great pleasure that we offer the sixth edition of *Beginning and Intermediate Algebra*. We have remained true to the original goal that has guided us over the years—to provide the best possible text and supplements package to help students succeed and instructors teach. This edition faithfully continues that process through enhanced explanations of concepts, new and updated examples and exercises, student-oriented features like Vocabulary Lists, Pointers, Cautions, Problem-Solving Hints, and Now Try Exercises, as well as an extensive package of helpful supplements and study aids.

This text is part of a series that includes the following books, all by Lial, Hornsby, and McGinnis:

- *Beginning Algebra*, Twelfth Edition,
- *Intermediate Algebra*, Twelfth Edition,
- *Algebra for College Students*, Eighth Edition.

WHAT'S NEW IN THIS EDITION?

We are pleased to offer the following new features:

VOCABULARY LISTS New vocabulary is now given at the beginning of appropriate sections. The list format allows students to preview vocabulary that is introduced in the section and also to review and check-off words they are able to correctly define upon completing a section.

CONCEPT CHECK EXERCISES Each exercise set begins with a set of Concept Check problems that facilitate students' mathematical thinking and conceptual understanding. Problem types include multiple-choice, true/false, matching, completion, and *What Went Wrong?* exercises. Many emphasize new vocabulary.

EXTENDING SKILLS EXERCISES These problems, scattered throughout selected exercise sets, expand on section objectives. Some are challenging in nature.

MIXED REVIEW EXERCISES Each chapter review has been expanded to include a dedicated set of Mixed Review Exercises to help students further synthesize concepts.

MARGIN ANSWERS TO REVIEW COMPONENTS To provide immediate reference and enable students to get the most out of review opportunities, answers are included in the margins for Summary, Chapter Review, and Mixed Review Exercises, plus Chapter Tests and Cumulative Review Exercises.

SPECIFIC CONTENT CHANGES include the following:

- **New Chapter R** provides a thorough review of fractions (Section R.1) as well as decimals and percents (all new Section R.2).

- **Application Sections 2.4, 2.7, 5.6, and 7.7** include new and/or updated problem-solving examples, exercises, and hints.

- **Section 2.5** now covers solving a linear equation in two variables x and y for y as preparation for Chapter 3 on forms of linear equations.

- **The presentation of linear equations in two variables in Chapter 3** includes three new examples of graphing and writing equations of lines, along with several groups of new exercises that make connections between tables, equations, and graphs.

- **Slope-intercept form and point-slope form** are now covered in separate **Sections 3.4 and 3.5**.

- **Expanded Summary Exercises** in **Chapter 2** continue our emphasis on the difference between simplifying expressions and solving equations. A new example in the **Chapter 5** Summary Exercises illustrates applying general factoring strategies.

- **Systems of linear equations** are covered in **Chapter 7** immediately following the review sections on linear equations in two variables.

- **Chapter 8** now covers inequalities and absolute value and begins with a **new review section on linear inequalities in one variable. Section 8.4** includes new material on **solving systems of linear inequalities.**

- **The introduction to relations and functions in Section 9.1** has a new example and enhanced discussion.

- **Section 10.3** now includes a new objective and discussion, examples, and exercises on equations and graphs of **circles.**

- **Presentations of the following topics have been enhanced and expanded:**

  Applying operations on real numbers (Sections 1.4, 1.5)
  Solving linear equations in one variable (Section 2.1)
  Solving linear inequalities in one variable with fractional coefficients (Section 2.8)
  Understanding polynomial vocabulary and adding and subtracting polynomials (Section 4.4)
  Factoring trinomials (Sections 5.3, 5.4)
  Solving equations with rational expressions (Section 6.6)
  Applying composition of functions (Section 9.3)
  Solving quadratic inequalities (Section 11.8)
  Finding inverse functions (Section 12.1)
  Approximating exponentials and logarithms (Sections 12.2, 12.3)

ACKNOWLEDGMENTS

The comments, criticisms, and suggestions of users, nonusers, instructors, and students have positively shaped this textbook over the years, and we are most grateful for the many responses we have received. The feedback gathered for this edition was particularly helpful.

We especially wish to thank the following individuals who provided invaluable suggestions.

Barbara Aaker, *Community College of Denver*
Kim Bennekin, *Georgia Perimeter College*
Dixie Blackinton, *Weber State University*
Callie Daniels, *St. Charles Community College*
Cheryl Davids, *Central Carolina Technical College*
Robert Diaz, *Fullerton College*
Chris Diorietes, *Fayetteville Technical Community College*
Sylvia Dreyfus, *Meridian Community College*
Sabine Eggleston, *Edison State College*
LaTonya Ellis, *Bishop State Community College*
Beverly Hall, *Fayetteville Technical Community College*
Sandee House, *Georgia Perimeter College*
Joe Howe, *St. Charles Community College*
Donna Kessler, *Moberly Area Community College*
Lynette King, *Gadsden State Community College*
Linda Kodama, *Windward Community College*
Carlea McAvoy, *South Puget Sound Community College*
Rick McBride, *Santa Fe Community College*
James Metz, *Kapi'olani Community College*
Jean Millen, *Georgia Perimeter College*
Molly Misko, *Gadsden State Community College*
Jane Roads, *Moberly Area Community College*
Yvonne Sandoval, *Pima Community College, West Campus*
Melanie Smith, *Bishop State Community College*
Erik Stubsten, *Chattanooga State Technical Community College*
Tong Wagner, *Greenville Technical College*
Rick Woodmansee, *Sacramento City College*
Sessia Wyche, *University of Texas at Brownsville*

Over the years, we have come to rely on an extensive team of experienced professionals. Our sincere thanks go to these dedicated individuals at Pearson Arts & Sciences, who worked long and hard to make this revision a success: Chris Hoag, Maureen O'Connor, Michael Hirsch, Rachel Ross, Beth Kaufman, Christine Whitlock, Chase Hammond, Kelly Cross, and Shana Siegmund.

Additionally, Shannon d'Hemecourt, Rachel Haskell, Eddie Herring, and Abby Tanenbaum did a great job helping us update real data applications. We are also grateful to Kathy Diamond and Marilyn Dwyer of Cenveo, Inc., for their excellent production work; Bonnie Boehme for supplying her copyediting expertise; Aptara for their photo research; and Lucie Haskins for producing another accurate, useful index. Callie Daniels, Perian Herring, Jack Hornsby, Paul Lorczak, and Sarah Sponholz did a thorough, timely job accuracy checking manuscript and page proofs and Lisa Collette checked the index.

We particularly thank the many students and instructors who have used this textbook over the years. You are the reason we do what we do. It is our hope that we have positively impacted your mathematics journey. We would welcome any comments or suggestions you might have via email to math@pearson.com.

John Hornsby
Terry McGinnis

STUDENT SUPPLEMENTS

STUDENT'S SOLUTIONS MANUAL

- Provides detailed solutions to the odd-numbered, section-level exercises and to all Now Try Exercises, Relating Concepts, Summary, Chapter Review, Mixed Review, Chapter Test, and Cumulative Review Exercises

ISBNs: 0-321-96986-3/978-0-321-96986-6

LIAL VIDEO LIBRARY

The **Lial Video Library,** available in MyMathLab, provides students with a wealth of video resources to help them navigate the road to success. All video resources in the library include optional subtitles in English. The **Lial Video Library** includes the following resources:

- **Section Lecture Videos** offer a new navigation menu that allows students to easily focus on the key examples and exercises that they need to review in each section. Optional Spanish subtitles are available.

- **Solutions Clips** show an instructor working through the complete solutions to selected exercises from the text. Exercises with a solution clip are marked in the text and e-book with a Play Button icon ▶.

- **Quick Review Lectures** provide a short summary lecture of each key concept from the Quick Reviews at the end of every chapter in the text.

- **Chapter Test Prep Videos** provide step-by-step solutions to all exercises from the Chapter Tests. These videos provide guidance and support when students need it the most—the night before an exam. The Chapter Test Prep Videos are also available on YouTube (searchable using author name and book title).

LIAL VIDEO LIBRARY WORKBOOK

NEW! The **Lial Video Library Workbook** is an unbound, three-hole-punched workbook/note-taking guide that students can use in conjunction with the Lial Video Library.

The notebook helps students develop organized notes as they work along with the videos. The notebook includes:

- Guided Examples to be used in conjunction with the Lial Video Library Section Lecture Videos, plus corresponding Now Try Exercises for each text objective.
- Extra practice exercises for every section of the text with ample space for students to show their work.
- Learning objectives and key vocabulary terms for every text section, along with vocabulary practice problems.

ISBNs: 0-321-96972-3 / 978-0-321-96972-9

INSTRUCTOR SUPPLEMENTS

ANNOTATED INSTRUCTOR'S EDITION

- Provides "on-page" answers to all text exercises in an easy-to-read margin format, along with helpful Teaching Tips and extensive Classroom Examples

ISBNs: 0-321-96942-1 / 978-0-321-96942-2

INSTRUCTOR'S SOLUTIONS MANUAL (Download only)

- Provides complete answers to all text exercises, including all Classroom Examples and Now Try Exercises

ISBNs: 0-321-96974-X / 978-0-321-96974-3

INSTRUCTOR'S RESOURCE MANUAL WITH TESTS
(Download only)

- Contains two diagnostic pretests, four free-response and two multiple-choice test forms per chapter, and two final exams
- Includes a mini-lecture for each section of the text with objectives, key examples, and teaching tips
- Provides a correlation guide from the fifth to the sixth edition

ISBNs: 0-321-96975-8 / 978-0-321-96975-0

AVAILABLE FOR STUDENTS AND INSTRUCTORS

MYMATHLAB® ONLINE COURSE (access code required)

MyMathLab from Pearson is the world's leading online resource in mathematics, integrating interactive homework, assessment, and media in a flexible, easy-to-use format. It provides **engaging experiences** that personalize, stimulate, and measure learning for each student. And it comes from an **experienced partner** with educational expertise and an eye on the future.

To learn more about how MyMathLab combines proven learning applications with powerful assessment, visit www.mymathlab.com or contact your Pearson representative.

MYMATHLAB® READY TO GO COURSE (access code required)

These new "Ready to Go" courses provide students with all the same great MyMathLab features, but make it easier for instructors to get started. Each course includes pre-assigned homework and quizzes to make creating a course even simpler. Ask your Pearson representative about the details for this particular course or to see a copy of this course.

MATHXL® ONLINE COURSE (access code required)

MathXL® is the homework and assessment engine that runs MyMathLab. (MyMathLab is MathXL plus a learning management system.)

With MathXL, instructors can:

- Create, edit, and assign online homework and tests using algorithmically generated exercises correlated at the objective level to the textbook.
- Create and assign their own online exercises and import TestGen tests for added flexibility.
- Maintain records of all student work tracked in MathXL's online gradebook.

With MathXL, students can:

- Take chapter tests in MathXL and receive personalized study plans and/or personalized homework assignments based on their test results.
- Use the study plan and/or the homework to link directly to tutorial exercises for the objectives they need to study.
- Access supplemental animations and video clips directly from selected exercises.

MathXL is available to qualified adopters. For more information, visit our website at www.mathxl.com, or contact your Pearson representative.

TESTGEN®

TestGen® (www.pearsoned.com/testgen) enables instructors to build, edit, print, and administer tests using a computerized bank of questions developed to cover all the objectives of the text. TestGen is algorithmically based, allowing instructors to create multiple but equivalent versions of the same question or test with the click of a button. Instructors can also modify test bank questions or add new questions. The software and testbank are available for download from Pearson Education's online catalog.

POWERPOINT® LECTURE SLIDES

- Present key concepts and definitions from the text
- Available for download at www.pearsonhighered.com

ISBNs: 0-321-96976-6 / 978-0-321-96976-7

STUDY SKILLS

Using Your Math Textbook

Your textbook is a valuable resource. You will learn more if you fully make use of the features it offers.

General Features

Locate each general feature and complete any blanks.

- **Table of Contents** Find this at the front of the text. *Mark the chapters and sections you will cover, as noted on your course syllabus.*

- **Answer Section** *Tab this section at the back of the book* so you can refer to it frequently when doing homework. Answers to odd-numbered section exercises are provided.

- **Glossary** Find this feature after the answer section at the back of the text. It provides an alphabetical list of the key terms found in the text, with definitions and section references. *Using the glossary, an equation is a statement that _____.*

- **List of Formulas** Inside the back cover of the text is a helpful list of geometric formulas, along with review information on triangles and angles. Use these for reference throughout the course. *The formula for finding the volume of a cube is _____.*

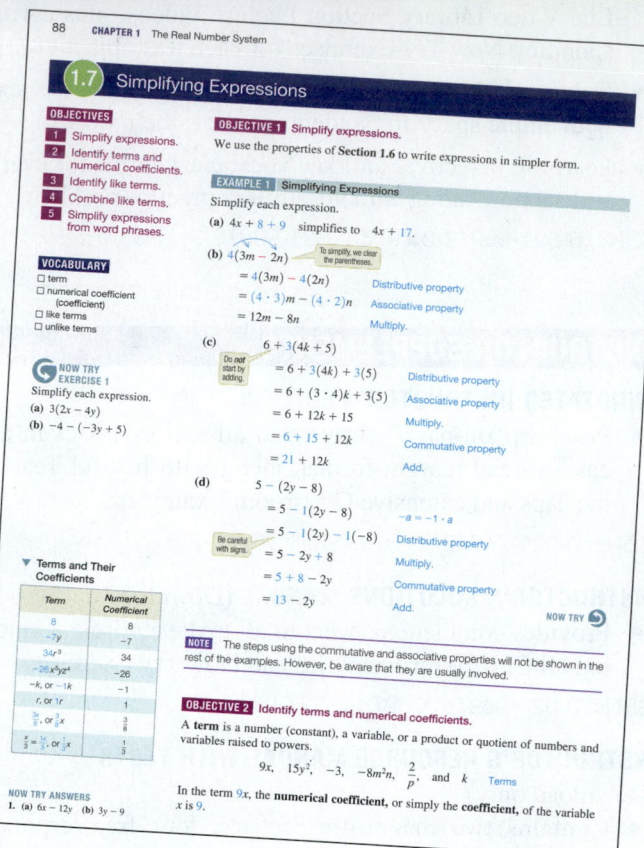

Specific Features

Look through Chapter 2 and give the number of a page that includes an example of each of the following specific features.

- **Objectives** The objectives are listed at the beginning of each section and again within the section as the corresponding material is presented. Once you finish a section, ask yourself if you have accomplished them. *See page _____.*

- **Vocabulary List** Important vocabulary is listed at the beginning of each section. You should be able to define these terms when you finish a section. *See page _____.*

- **Now Try Exercises** These margin exercises allow you to immediately practice the material covered in the examples and prepare you for the exercises. Check your results using the answers at the bottom of the page. *See page _____.*

- **Pointers** These small shaded balloons provide on-the-spot warnings and reminders, point out key steps, and give other helpful tips. *See page _____.*

- **Cautions** These provide warnings about common errors that students often make or trouble spots to avoid. *See page _____.*

- **Notes** These provide additional explanations or emphasize other important ideas. *See page _____.*

- **Problem-Solving Hints** These boxes give helpful tips or strategies to use when you work applications. *See page _____.*

Prealgebra Review

R.1 Fractions

R.2 Decimals and Percents

Study Skills *Reading Your Math Textbook*

R.1 Fractions

VOCABULARY

☐ natural (counting) numbers
☐ whole numbers
☐ fractions
☐ numerator
☐ denominator
☐ proper fraction
☐ improper fraction
☐ factors
☐ product
☐ prime number
☐ composite number
☐ lowest terms
☐ greatest common factor (GCF)
☐ mixed number
☐ reciprocals
☐ quotient
☐ dividend
☐ divisor
☐ sum
(continued)

In the study of elementary mathematics, the numbers used most often are the **natural (counting) numbers,**

$$1, 2, 3, 4, \ldots,$$

the **whole numbers,**

$$0, 1, 2, 3, 4, \ldots,$$

and **fractions,** such as

$$\frac{1}{2}, \quad \frac{2}{3}, \quad \text{and} \quad \frac{11}{12}.$$

The three dots, or *ellipsis points,* indicate that each list of numbers continues in the same way indefinitely.

The parts of a fraction are named as shown.

$$\text{Fraction bar} \rightarrow \frac{3}{8} \begin{array}{l} \leftarrow \text{Numerator} \\ \leftarrow \text{Denominator} \end{array}$$

NOTE Fractions are a way to represent parts of a whole. In a fraction, the **numerator** gives the number of parts being represented. The **denominator** gives the total number of equal parts in the whole. See **FIGURE 1**.

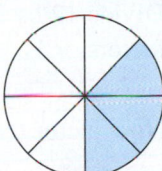

The shaded region represents $\frac{3}{8}$ of the circle.

FIGURE 1

A fraction is classified as being either a **proper fraction** or an **improper fraction.**

Proper fractions	$\dfrac{1}{5}, \ \dfrac{2}{7}, \ \dfrac{9}{10}, \ \dfrac{23}{25}$	Numerator is **less than** denominator. Value is less than 1.
Improper fractions	$\dfrac{3}{2}, \ \dfrac{5}{5}, \ \dfrac{11}{7}, \ \dfrac{28}{4}$	Numerator is **greater than or equal to** denominator. Value is greater than or equal to 1.

VOCABULARY (continued)
☐ least common denominator (LCD)
☐ difference
☐ circle graph (pie chart)

OBJECTIVE 1 Learn the definition of *factor*.

In the statement $3 \times 6 = 18$, the numbers 3 and 6 are **factors** of 18. Other factors of 18 include 1, 2, 9, and 18. The result of the multiplication, 18, is the **product.** We can represent the product of two numbers, such as 3 and 6, in several ways.

$$3 \times 6, \quad 3 \cdot 6, \quad (3)(6), \quad (3)6, \quad 3(6) \qquad \text{Products}$$

We *factor* a number by writing it as the product of two or more numbers.

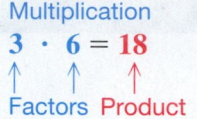

Multiplication
$3 \cdot 6 = 18$
Factors Product

Factoring
$18 = 3 \cdot 6$
Product Factors

Factoring is the reverse of multiplying two numbers to get the product.

NOTE In algebra, a raised dot · is often used instead of the × symbol to indicate multiplication because × may be confused with the letter *x*.

A natural number greater than 1 is **prime** if it has only itself and 1 as factors. "Factors" are understood here to mean natural number factors.

$$2, 3, 5, 7, 11, 13, 17, 19, 23, 29, 31, 37 \qquad \text{First dozen prime numbers}$$

A natural number greater than 1 that is not prime is a **composite number.**

$$4, 6, 8, 9, 10, 12, 14, 15, 16, 18, 20, 21 \qquad \text{First dozen composite numbers}$$

By agreement, the number 1 is neither prime nor composite.

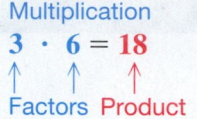

NOW TRY EXERCISE 1

Write 60 as a product of prime factors.

EXAMPLE 1 Factoring Numbers

Write each number as a product of prime factors.

(a) 35   Write 35 as the product of the prime factors 5 and 7.

$$35 = 5 \cdot 7$$

(b) 24   We show a factor tree on the right, with prime factors circled.

Divide by the least prime factor of 24, which is 2. $24 = 2 \cdot 12$

Divide 12 by 2 to find two factors of 12. $24 = 2 \cdot 2 \cdot 6$

Now factor 6 as $2 \cdot 3$. $24 = \underline{2 \cdot 2 \cdot 2 \cdot 3}$

All factors are prime.

NOW TRY

NOTE When factoring, we need not start with the least prime factor. No matter which prime factor we start with, we will *always* obtain the same prime factorization. We verify this in **Example 1(b)** by starting with 3 instead of 2.

Divide 24 by 3. $24 = 3 \cdot 8$

Divide 8 by 2. $24 = 3 \cdot 2 \cdot 4$

Divide 4 by 2. $24 = \underline{3 \cdot 2 \cdot 2 \cdot 2}$

The same prime factors result.

NOW TRY ANSWER
1. $2 \cdot 2 \cdot 3 \cdot 5$

OBJECTIVE 2 Write fractions in lowest terms.

The **basic principle of fractions** is used to write a fraction in *lowest terms*.

Basic Principle of Fractions

If the numerator and denominator of a fraction are multiplied or divided by the same nonzero number, the value of the fraction is not changed.

A fraction is in **lowest terms** when the numerator and denominator have no factors in common (other than 1).

Writing a Fraction in Lowest Terms

Step 1 Write the numerator and the denominator in factored form.

Step 2 Divide the numerator and the denominator by the **greatest common factor (GCF)**, the product of all factors common to both.

🔄 **NOW TRY EXERCISE 2**

Write each fraction in lowest terms.

(a) $\dfrac{30}{42}$ (b) $\dfrac{10}{70}$ (c) $\dfrac{72}{120}$

EXAMPLE 2 Writing Fractions in Lowest Terms

Write each fraction in lowest terms.

(a) $\dfrac{10}{15} = \dfrac{2 \cdot 5}{3 \cdot 5} = \dfrac{2 \cdot 1}{3 \cdot 1} = \dfrac{2}{3}$

The factored form shows that 5 is the greatest common factor of 10 and 15. Dividing both numerator and denominator by 5 gives $\dfrac{10}{15}$ in lowest terms as $\dfrac{2}{3}$.

(b) $\dfrac{15}{45}$

By inspection, the greatest common factor of 15 and 45 is 15.

$$\dfrac{15}{45} = \dfrac{15}{3 \cdot 15} = \dfrac{1}{3 \cdot 1} = \dfrac{1}{3}$$

> Remember to write 1 in the numerator.

If the GCF is not obvious, factor the numerator and denominator into prime factors.

$$\dfrac{15}{45} = \dfrac{3 \cdot 5}{3 \cdot 3 \cdot 5} = \dfrac{1 \cdot 1}{3 \cdot 1 \cdot 1} = \dfrac{1}{3} \qquad \text{The same answer results.}$$

(c) $\dfrac{150}{200} = \dfrac{3 \cdot 50}{4 \cdot 50} = \dfrac{3 \cdot 1}{4 \cdot 1} = \dfrac{3}{4}$ 50 is the greatest common factor of 150 and 200.

Another strategy is to choose *any* common factor and work in stages.

$$\dfrac{150}{200} = \dfrac{15 \cdot 10}{20 \cdot 10} = \dfrac{3 \cdot 5 \cdot 10}{4 \cdot 5 \cdot 10} = \dfrac{3 \cdot 1 \cdot 1}{4 \cdot 1 \cdot 1} = \dfrac{3}{4} \qquad \text{The same answer results.}$$

NOW TRY 🔄

OBJECTIVE 3 Convert between improper fractions and mixed numbers.

A **mixed number** is a single number that represents the sum of a natural number and a proper fraction.

NOW TRY ANSWERS

2. (a) $\dfrac{5}{7}$ (b) $\dfrac{1}{7}$ (c) $\dfrac{3}{5}$

Mixed number $\rightarrow 5\dfrac{3}{4} = 5 + \dfrac{3}{4}$

 **NOW TRY EXERCISE 3**

Write $\frac{92}{5}$ as a mixed number.

EXAMPLE 3 Converting an Improper Fraction to a Mixed Number

Write $\frac{59}{8}$ as a mixed number.

The fraction bar represents division. $\left(\frac{a}{b} \text{ means } a \div b.\right)$ Divide the numerator of the improper fraction by the denominator.

$$
\begin{array}{r}
7 \leftarrow \text{Quotient} \\
\text{Denominator of fraction} \rightarrow 8)\overline{59} \leftarrow \text{Numerator of fraction} \\
\underline{56} \\
3 \leftarrow \text{Remainder}
\end{array}
\qquad \frac{59}{8} = 7\frac{3}{8}
$$

NOW TRY

 **NOW TRY EXERCISE 4**

Write $11\frac{2}{3}$ as an improper fraction.

EXAMPLE 4 Converting a Mixed Number to an Improper Fraction

Write $6\frac{4}{7}$ as an improper fraction.

Multiply the denominator of the fraction by the natural number and then add the numerator to obtain the numerator of the improper fraction.

$$7 \cdot 6 = 42 \quad \text{and} \quad 42 + 4 = 46$$

The denominator of the improper fraction is the same as the denominator in the mixed number, which is 7 here. Thus, $6\frac{4}{7} = \frac{46}{7}$.

NOW TRY

Multiplying Fractions

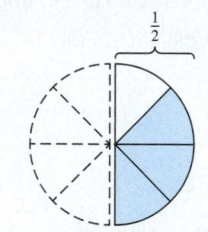

$\frac{3}{4}$ of $\frac{1}{2}$ is equivalent to $\frac{3}{4} \cdot \frac{1}{2}$, which equals $\frac{3}{8}$ of the circle.

FIGURE 2

OBJECTIVE 4 Multiply and divide fractions.

FIGURE 2 illustrates multiplying fractions.

> **Multiplying Fractions**
>
> If $\frac{a}{b}$ and $\frac{c}{d}$ are fractions, then $\frac{a}{b} \cdot \frac{c}{d} = \frac{a \cdot c}{b \cdot d}$.

That is, to multiply two fractions, multiply their numerators and then multiply their denominators.

EXAMPLE 5 Multiplying Fractions

Find each product, and write it in lowest terms.

(a) $\quad \dfrac{3}{8} \cdot \dfrac{4}{9}$

$= \dfrac{3 \cdot 4}{8 \cdot 9}$ Multiply numerators.
           Multiply denominators.

$= \dfrac{3 \cdot 4}{2 \cdot 4 \cdot 3 \cdot 3}$ Factor the denominator.

$= \dfrac{\overset{1}{\cancel{}}}{2 \cdot 3}$ Divide numerator and denominator by 3 and 4, or by 12.

Remember to write 1 in the numerator.

$= \dfrac{1}{6}$ *Make sure the product is in lowest terms.*

NOW TRY ANSWERS

3. $18\frac{2}{5}$

4. $\frac{35}{3}$

NOW TRY
EXERCISE 5

Find each product, and write
it in lowest terms.

(a) $\dfrac{4}{7} \cdot \dfrac{5}{8}$ **(b)** $3\dfrac{2}{5} \cdot 6\dfrac{2}{3}$

(b)

$$2\dfrac{1}{3} \cdot 5\dfrac{1}{4}$$

$$= \dfrac{7}{3} \cdot \dfrac{21}{4} \qquad \text{Write each mixed number as an improper fraction.}$$

$$= \dfrac{7 \cdot 21}{3 \cdot 4} \qquad \begin{array}{l}\text{Multiply numerators.}\\ \text{Multiply denominators.}\end{array}$$

$$= \dfrac{7 \cdot 3 \cdot 7}{3 \cdot 4} \qquad \text{Factor the numerator.}$$

Think: $\frac{49}{4}$ means $49 \div 4$.

$$= \dfrac{49}{4}, \quad \text{or} \quad 12\dfrac{1}{4} \qquad \begin{array}{l}\text{Write in lowest terms}\\ \text{and as a mixed number.}\end{array}$$

NOW TRY

NOTE Some students prefer to factor and divide out any common factors *before* multiplying.

$$\dfrac{3}{8} \cdot \dfrac{4}{9} \qquad \text{Example 5(a)}$$

$$= \dfrac{3}{2 \cdot 4} \cdot \dfrac{4}{3 \cdot 3} \qquad \text{Factor.}$$

$$= \dfrac{1}{2 \cdot 3} \qquad \text{Divide out common factors. Multiply.}$$

$$= \dfrac{1}{6} \qquad \text{The same answer results.}$$

▼ Reciprocals

Number	Reciprocal
$\frac{3}{4}$	$\frac{4}{3}$
$\frac{11}{7}$	$\frac{7}{11}$
$\frac{1}{5}$	5, or $\frac{5}{1}$
10, or $\frac{10}{1}$	$\frac{1}{10}$

A number and its reciprocal have a product of 1. For example,
$\frac{3}{4} \cdot \frac{4}{3} = \frac{12}{12}$, or 1.

Two numbers are **reciprocals** of each other if their product is 1. Because division is the inverse or opposite of multiplication, we use reciprocals to divide fractions. **FIGURE 3** illustrates dividing fractions.

Dividing Fractions

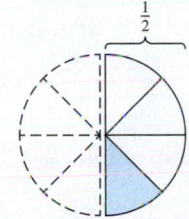

$\frac{1}{2} \div 4$ is equivalent to $\frac{1}{2} \cdot \frac{1}{4}$, which equals $\frac{1}{8}$ of the circle.

FIGURE 3

Dividing Fractions

If $\dfrac{a}{b}$ and $\dfrac{c}{d}$ are fractions, then $\quad \dfrac{a}{b} \div \dfrac{c}{d} = \dfrac{a}{b} \cdot \dfrac{d}{c}.$

Multiply by the reciprocal.

That is, to divide by a fraction, multiply by its reciprocal.

As an example of why this procedure works, we know that

$$20 \div 10 = 2 \quad \text{and also that} \quad 20 \cdot \dfrac{1}{10} = 2.$$

The answer to a division problem is a **quotient.** In $\frac{a}{b} \div \frac{c}{d}$, the first fraction $\frac{a}{b}$ is the **dividend,** and the second fraction $\frac{c}{d}$ is the **divisor.**

NOW TRY ANSWERS
5. (a) $\frac{5}{14}$ (b) $\frac{68}{3}$, or $22\frac{2}{3}$

**NOW TRY
EXERCISE 6**

Find each quotient, and write it in lowest terms.

(a) $\dfrac{2}{7} \div \dfrac{8}{9}$ (b) $3\dfrac{3}{4} \div 4\dfrac{2}{7}$

EXAMPLE 6   Dividing Fractions

Find each quotient, and write it in lowest terms.

(a) $\dfrac{3}{4} \div \dfrac{8}{5}$

$\quad = \dfrac{3}{4} \cdot \dfrac{5}{8}$ Multiply by the reciprocal of the divisor.

$\quad = \dfrac{3 \cdot 5}{4 \cdot 8}$ Multiply numerators.
Multiply denominators.

$\quad = \dfrac{15}{32}$ ◁ Make sure the quotient is in lowest terms.

(b) $\dfrac{3}{4} \div \dfrac{5}{8}$

$\quad = \dfrac{3}{4} \cdot \dfrac{8}{5}$ Multiply by the reciprocal.

$\quad = \dfrac{3 \cdot 4 \cdot 2}{4 \cdot 5}$ Multiply and factor.

$\quad = \dfrac{6}{5}, \quad \text{or} \quad 1\dfrac{1}{5}$

(c) $\dfrac{5}{8} \div 10$ ◁ Think of 10 as $\frac{10}{1}$ here.

$\quad = \dfrac{5}{8} \cdot \dfrac{1}{10}$ Multiply by the reciprocal.

$\quad = \dfrac{5 \cdot 1}{8 \cdot 2 \cdot 5}$ Multiply and factor.

$\quad = \dfrac{1}{16}$ ◁ Remember to write 1 in the numerator.

(d) $1\dfrac{2}{3} \div 4\dfrac{1}{2}$

$\quad = \dfrac{5}{3} \div \dfrac{9}{2}$ Write each mixed number as an improper fraction.

$\quad = \dfrac{5}{3} \cdot \dfrac{2}{9}$ Multiply by the reciprocal of the divisor.

$\quad = \dfrac{5 \cdot 2}{3 \cdot 9}$ Multiply numerators.
Multiply denominators.

$\quad = \dfrac{10}{27}$ The quotient is in lowest terms.   **NOW TRY** ↺

Adding Fractions

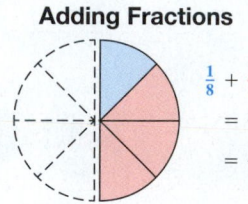

$\dfrac{1}{8} + \dfrac{3}{8}$

$= \dfrac{4}{8}$

$= \dfrac{1}{2}$

FIGURE 4

OBJECTIVE 5 Add and subtract fractions.

The result of adding two numbers is the **sum** of the numbers. For example, $2 + 3 = 5$, so 5 is the sum of 2 and 3.

FIGURE 4 illustrates adding fractions.

Adding Fractions

If $\dfrac{a}{b}$ and $\dfrac{c}{b}$ are fractions, then $\dfrac{a}{b} + \dfrac{c}{b} = \dfrac{a + c}{b}$.

That is, to find the sum of two fractions having the *same* denominator, add the numerators and *keep the same denominator.*

NOW TRY ANSWERS

6. (a) $\dfrac{9}{28}$ (b) $\dfrac{7}{8}$

NOW TRY
EXERCISE 7

Find the sum, and write it in lowest terms.

$$\frac{1}{8} + \frac{3}{8}$$

EXAMPLE 7 Adding Fractions (Same Denominator)

Find each sum, and write it in lowest terms.

(a) $\dfrac{3}{7} + \dfrac{2}{7}$

$= \dfrac{3+2}{7}$ Add numerators. Keep the same denominator.

$= \dfrac{5}{7}$

(b) $\dfrac{2}{10} + \dfrac{3}{10}$

$= \dfrac{2+3}{10}$ Add numerators. Keep the same denominator.

$= \dfrac{5}{10}$

$= \dfrac{1}{2}$ Write in lowest terms.

NOW TRY

If the fractions to be added do *not* have the same denominator, we must first rewrite them with a common denominator. For example, to rewrite $\dfrac{3}{4}$ as an equivalent fraction with denominator 12, think as follows.

$$\frac{3}{4} = \frac{?}{12}$$

We must find the number that can be multiplied by 4 to give 12. Because $4 \cdot 3 = 12$, we multiply the numerator and the denominator by 3.

$$\frac{3}{4} = \frac{3 \cdot 3}{4 \cdot 3} = \frac{9}{12}$$

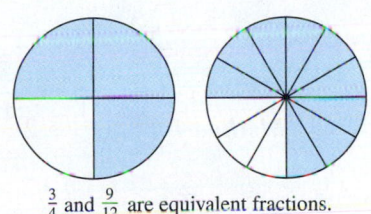

$\frac{3}{4}$ is equivalent to $\frac{9}{12}$.
See **FIGURE 5.**

$\frac{3}{4}$ and $\frac{9}{12}$ are equivalent fractions.

FIGURE 5

NOTE The process of writing an equivalent fraction is the reverse of writing a fraction in lowest terms.

Finding the Least Common Denominator (LCD)

To add or subtract fractions with different denominators, find the **least common denominator (LCD)** as follows.

Step 1 Factor each denominator using prime factors.

Step 2 The LCD is the product of every (different) factor that appears in any of the factored denominators. If a factor is repeated, use the greatest number of repeats as factors of the LCD.

Step 3 Write each fraction with the LCD as the denominator.

NOW TRY ANSWER

7. $\frac{1}{2}$

**NOW TRY
EXERCISE 8**

Find each sum, and write it in lowest terms.

(a) $\dfrac{5}{12} + \dfrac{3}{8}$ (b) $3\dfrac{1}{4} + 5\dfrac{5}{8}$

EXAMPLE 8   Adding Fractions (Different Denominators)

Find each sum, and write it in lowest terms.

(a) $\dfrac{4}{15} + \dfrac{5}{9}$

Step 1   To find the LCD, factor each denominator using prime factors.

$$15 = 5 \cdot 3 \quad \text{and} \quad 9 = 3 \cdot 3$$

3 is a factor of both denominators.

Step 2 $$\text{LCD} = 5 \cdot 3 \cdot 3 = 45$$

In this example, the LCD needs one factor of 5 and two factors of 3 because the second denominator has two factors of 3.

Step 3   Write each fraction with 45 as denominator.

$$\frac{4}{15} = \frac{4 \cdot 3}{15 \cdot 3} = \frac{12}{45} \quad \text{and} \quad \frac{5}{9} = \frac{5 \cdot 5}{9 \cdot 5} = \frac{25}{45}$$

At this stage, the fractions are not in lowest terms.

$$\frac{4}{15} + \frac{5}{9}$$

$$= \frac{12}{45} + \frac{25}{45} \qquad \text{Use the equivalent fractions with the common denominator.}$$

Make sure the sum is in lowest terms.

$$= \frac{37}{45} \qquad \text{Add numerators. Keep the same denominator.}$$

(b) $3\dfrac{1}{2} + 2\dfrac{3}{4}$

Method 1 $3\dfrac{1}{2} + 2\dfrac{3}{4}$

$$= \frac{7}{2} + \frac{11}{4} \qquad \text{Write each mixed number as an improper fraction.}$$

Think: $\dfrac{7 \cdot 2}{2 \cdot 2} = \dfrac{14}{4}$

$$= \frac{14}{4} + \frac{11}{4} \qquad \text{Find a common denominator. The LCD is 4.}$$

$$= \frac{25}{4}, \quad \text{or} \quad 6\frac{1}{4} \qquad \text{Add. Write as a mixed number.}$$

Method 2
$$\begin{array}{l} 3\dfrac{1}{2} = 3\dfrac{2}{4} \\[2mm] +\, 2\dfrac{3}{4} = 2\dfrac{3}{4} \\ \hline \end{array}$$

Write $3\dfrac{1}{2}$ as $3\dfrac{2}{4}$. Then add vertically. Add the whole numbers and the fractions separately.

$$5\frac{5}{4} = 5 + 1\frac{1}{4} = 6\frac{1}{4}, \quad \text{or} \quad \frac{25}{4} \qquad \text{The same answer results.}$$

NOW TRY

NOW TRY ANSWERS
8. (a) $\dfrac{19}{24}$ (b) $\dfrac{71}{8}$, or $8\dfrac{7}{8}$

The result of subtracting one number from another number is the **difference** of the numbers. For example, $9 - 5 = 4$, so 4 is the difference of 9 and 5.

FIGURE 6 illustrates subtracting fractions.

Subtracting Fractions

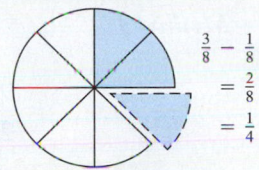

$$\frac{3}{8} - \frac{1}{8}$$
$$= \frac{2}{8}$$
$$= \frac{1}{4}$$

FIGURE 6

Subtracting Fractions

If $\dfrac{a}{b}$ and $\dfrac{c}{b}$ are fractions, then $\qquad \dfrac{a}{b} - \dfrac{c}{b} = \dfrac{a-c}{b}.$

That is, to find the difference of two fractions having the *same* denominator, subtract the numerators and ***keep the same denominator.***

EXAMPLE 9 Subtracting Fractions

Find each difference, and write it in lowest terms.

(a) $\dfrac{15}{8} - \dfrac{3}{8}$

$= \dfrac{15-3}{8}$ Subtract numerators.
            Keep the same denominator.

$= \dfrac{12}{8}$

$= \dfrac{3}{2},$ or $1\dfrac{1}{2}$ Write in lowest terms and as a mixed number.

(b) $\dfrac{7}{18} - \dfrac{4}{15}$

$= \dfrac{7 \cdot 5}{2 \cdot 3 \cdot 3 \cdot 5} - \dfrac{4 \cdot 2 \cdot 3}{2 \cdot 3 \cdot 3 \cdot 5}$ $18 = 2 \cdot 3 \cdot 3$ and $15 = 3 \cdot 5$, so the LCD is $2 \cdot 3 \cdot 3 \cdot 5 = 90$.

$= \dfrac{35}{90} - \dfrac{24}{90}$ Write equivalent fractions.

$= \dfrac{11}{90}$ Subtract. The answer is in lowest terms.

(c) $\dfrac{15}{32} - \dfrac{11}{45}$

     Because $32 = 2 \cdot 2 \cdot 2 \cdot 2 \cdot 2$ and $45 = 3 \cdot 3 \cdot 5$, there are no common factors except 1. The LCD is $32 \cdot 45 = 1440$.

$= \dfrac{15 \cdot 45}{32 \cdot 45} - \dfrac{11 \cdot 32}{45 \cdot 32}$

$= \dfrac{675}{1440} - \dfrac{352}{1440}$ Write equivalent fractions.

$= \dfrac{323}{1440}$ Subtract numerators.
            Keep the common denominator.

NOW TRY
EXERCISE 9

Find each difference, and write it in lowest terms.

(a) $\dfrac{5}{11} - \dfrac{2}{9}$ **(b)** $4\dfrac{1}{3} - 2\dfrac{5}{6}$

(d) $4\dfrac{1}{2} - 1\dfrac{3}{4}$

Method 1 $4\dfrac{1}{2} - 1\dfrac{3}{4}$

$= \dfrac{9}{2} - \dfrac{7}{4}$ Write each mixed number as an improper fraction.

Think: $\dfrac{9 \cdot 2}{2 \cdot 2} = \dfrac{18}{4}$ $= \dfrac{18}{4} - \dfrac{7}{4}$ Find a common denominator. The LCD is 4.

$= \dfrac{11}{4},$ or $2\dfrac{3}{4}$ Subtract. Write as a mixed number.

Method 2 $4\dfrac{1}{2} = 4\dfrac{2}{4} = 3\dfrac{6}{4}$ The LCD is 4.
$4\dfrac{2}{4} = 3 + 1 + \dfrac{2}{4} = 3 + \dfrac{4}{4} + \dfrac{2}{4} = 3\dfrac{6}{4}$

$-1\dfrac{3}{4} = 1\dfrac{3}{4} = 1\dfrac{3}{4}$
_____

$2\dfrac{3}{4},$ or $\dfrac{11}{4}$ The same answer results.

NOW TRY

OBJECTIVE 6   Solve applied problems that involve fractions.

EXAMPLE 10   Adding Fractions to Solve an Applied Problem

NOW TRY
EXERCISE 10

A board is $10\dfrac{1}{2}$ ft long. If it must be divided into four pieces of equal length for shelves, how long must each piece be?

The diagram in **FIGURE 7** appears in the book *Woodworker's 39 Sure-Fire Projects*. Find the height of the desk to the top of the writing surface.

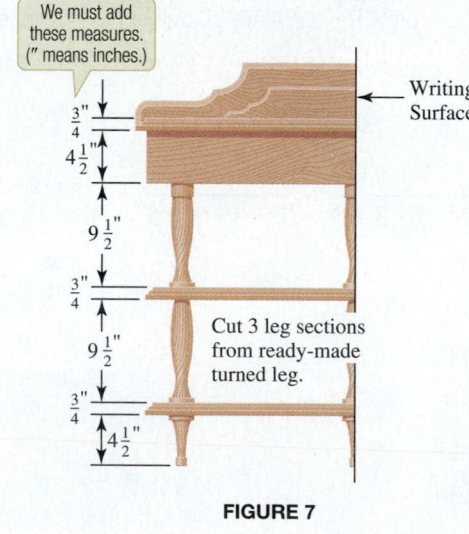

We must add these measures. (" means inches.)

$\dfrac{3}{4}$"
$4\dfrac{1}{2}$"
$9\dfrac{1}{2}$"
$\dfrac{3}{4}$"
$9\dfrac{1}{2}$"
$\dfrac{3}{4}$"
$4\dfrac{1}{2}$"

← Writing Surface

Cut 3 leg sections from ready-made turned leg.

FIGURE 7

$\dfrac{3}{4} \rightarrow \dfrac{3}{4}$ Use Method 2 from **Example 8(b).** The common denominator is 4.

$4\dfrac{1}{2} = 4\dfrac{2}{4}$

$9\dfrac{1}{2} = 9\dfrac{2}{4}$

$\dfrac{3}{4} \rightarrow \dfrac{3}{4}$

$9\dfrac{1}{2} = 9\dfrac{2}{4}$

$\dfrac{3}{4} \rightarrow \dfrac{3}{4}$

$+ 4\dfrac{1}{2} = 4\dfrac{2}{4}$
_____

$26\dfrac{17}{4}$ Because $\dfrac{17}{4}$ is an improper fraction, this is *not* the final form of the answer.

Think: $\dfrac{17}{4}$ means $17 \div 4$.

NOW TRY ANSWERS
9. (a) $\dfrac{23}{99}$ **(b)** $\dfrac{3}{2}$, or $1\dfrac{1}{2}$
10. $2\dfrac{5}{8}$ ft

Because $\dfrac{17}{4} = 4\dfrac{1}{4}$, we have $26\dfrac{17}{4} = 26 + 4\dfrac{1}{4} = 30\dfrac{1}{4}$. The height is $30\dfrac{1}{4}$ in.

NOW TRY

OBJECTIVE 7 Interpret data in a circle graph.

In a **circle graph,** or **pie chart,** a circle is used to indicate the total of all the data categories represented. The circle is divided into *sectors,* or wedges, whose sizes show the relative magnitudes of the categories. The sum of all the fractional parts must be 1 (for 1 whole circle).

**NOW TRY
EXERCISE 11**

Refer to the circle graph in **FIGURE 8.**

(a) Which region had the least number of Internet users?

(b) Estimate the number of Internet users in Asia.

(c) How many Internet users were there in Asia?

EXAMPLE 11 Using a Circle Graph to Interpret Information

In a recent year, there were about 2100 million (that is, 2.1 billion) Internet users worldwide. The circle graph in **FIGURE 8** shows the fractions of these users living in various regions of the world.

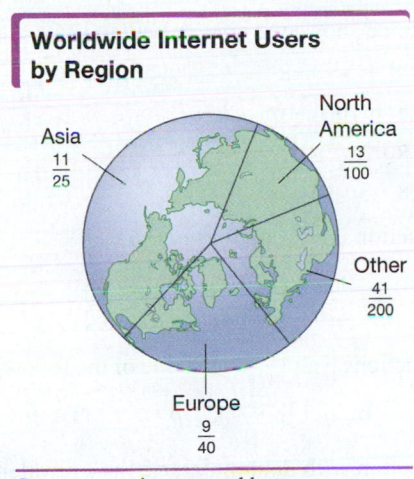

Worldwide Internet Users by Region

Asia $\frac{11}{25}$

North America $\frac{13}{100}$

Other $\frac{41}{200}$

Europe $\frac{9}{40}$

Source: www.internetworldstats.com

FIGURE 8

(a) Which region had the largest share of Internet users? What was that share?

The sector for Asia is the largest. Asia had the largest share of Internet users, $\frac{11}{25}$.

(b) Estimate the number of Internet users in North America.

A share of $\frac{13}{100}$ can be rounded to $\frac{10}{100}$, or $\frac{1}{10}$, and the total number of Internet users, 2100 million, can be rounded to 2000 million (or 2 billion). We multiply $\frac{1}{10}$ by 2000.

$$\frac{1}{10} \cdot 2000 = 200 \text{ million}$$ Approximate number of Internet users in North America

(c) How many Internet users were there in North America?

$$\frac{13}{100} \cdot 2100$$ Multiply the actual fraction from the graph for North America by the number of users.

$$= \frac{13}{100} \cdot \frac{2100}{1}$$ $a = \frac{a}{1}$, for all a.

$$= \frac{27{,}300}{100}$$ Multiply numerators.
Multiply denominators.

$$= 273$$ Divide.

Thus, 273 million, or 273,000,000 people in North America used the Internet.

NOW TRY

NOW TRY ANSWERS

11. (a) North America

(b) 1000 million, or 1 billion
$\left(\frac{11}{25} \text{ is about } \frac{1}{2}.\right)$

(c) 924 million, or 924,000,000

R.1 Exercises

FOR EXTRA HELP   MyMathLab®

▶ *Complete solution available in MyMathLab*

Concept Check   *Decide whether each statement is* true *or* false. *If it is false, explain why.*

1. In the fraction $\frac{5}{8}$, 5 is the numerator and 8 is the denominator.

2. The mixed number equivalent of the improper fraction $\frac{31}{5}$ is $6\frac{1}{5}$.

3. The fraction $\frac{7}{7}$ is proper.

4. The number 1 is prime.

5. The fraction $\frac{13}{39}$ is in lowest terms.

6. The reciprocal of $\frac{6}{2}$ is $\frac{3}{1}$.

7. The product of 10 and 2 is 12.

8. The difference of 10 and 2 is 5.

Concept Check   *Choose the letter of the correct response.*

9. Which choice shows the correct way to write $\frac{16}{24}$ in lowest terms?

A. $\frac{16}{24} = \frac{8+8}{8+16} = \frac{8}{16} = \frac{1}{2}$

B. $\frac{16}{24} = \frac{4 \cdot 4}{4 \cdot 6} = \frac{4}{6}$

C. $\frac{16}{24} = \frac{8 \cdot 2}{8 \cdot 3} = \frac{2}{3}$

D. $\frac{16}{24} = \frac{14+2}{21+3} = \frac{2}{3}$

10. Which fraction is *not* equal to $\frac{5}{9}$?

A. $\frac{15}{27}$ **B.** $\frac{30}{54}$ **C.** $\frac{40}{74}$ **D.** $\frac{55}{99}$

11. For the fractions $\frac{p}{q}$ and $\frac{r}{s}$, which one of the following can serve as a common denominator?

A. $q \cdot s$ **B.** $q + s$ **C.** $p \cdot r$ **D.** $p + r$

12. Which fraction with denominator 24 is equivalent to $\frac{5}{8}$?

A. $\frac{21}{24}$ **B.** $\frac{15}{24}$ **C.** $\frac{5}{24}$ **D.** $\frac{10}{24}$

Identify each number as prime, composite, *or* neither. *If the number is composite, write it as a product of prime factors.* **See Example 1.**

13. 19   **14.** 31   **15.** 30   **16.** 50

▶ **17.** 64   **18.** 81   **19.** 1   **20.** 0

21. 57   **22.** 51   **23.** 79   **24.** 83   **25.** 124

26. 138   **27.** 500   **28.** 700   **29.** 3458   **30.** 1025

Write each fraction in lowest terms. **See Example 2.**

31. $\frac{8}{16}$ **32.** $\frac{4}{12}$ ▶ **33.** $\frac{15}{18}$ **34.** $\frac{16}{20}$ **35.** $\frac{64}{100}$

36. $\frac{55}{200}$ **37.** $\frac{18}{90}$ **38.** $\frac{16}{64}$ **39.** $\frac{144}{120}$ **40.** $\frac{132}{77}$

Write each improper fraction as a mixed number. **See Example 3.**

41. $\frac{12}{7}$ **42.** $\frac{16}{9}$ **43.** $\frac{77}{12}$ **44.** $\frac{101}{15}$ **45.** $\frac{83}{11}$ **46.** $\frac{67}{13}$

Write each mixed number as an improper fraction. **See Example 4.**

47. $2\frac{3}{5}$ **48.** $5\frac{6}{7}$ **49.** $10\frac{3}{8}$ **50.** $12\frac{2}{3}$ **51.** $10\frac{1}{5}$ **52.** $18\frac{1}{6}$

Find each product or quotient, and write it in lowest terms. See Examples 5 and 6.

53. $\dfrac{4}{5} \cdot \dfrac{6}{7}$ **54.** $\dfrac{5}{9} \cdot \dfrac{2}{7}$ **55.** $\dfrac{2}{3} \cdot \dfrac{15}{16}$ **56.** $\dfrac{3}{5} \cdot \dfrac{20}{21}$

▶ **57.** $\dfrac{1}{10} \cdot \dfrac{12}{5}$ **58.** $\dfrac{1}{8} \cdot \dfrac{10}{7}$ **59.** $\dfrac{15}{4} \cdot \dfrac{8}{25}$ **60.** $\dfrac{21}{8} \cdot \dfrac{4}{7}$

61. $21 \cdot \dfrac{3}{7}$ **62.** $36 \cdot \dfrac{4}{9}$ **63.** $3\dfrac{1}{4} \cdot 1\dfrac{2}{3}$ **64.** $2\dfrac{2}{3} \cdot 1\dfrac{3}{5}$

65. $2\dfrac{3}{8} \cdot 3\dfrac{1}{5}$ **66.** $3\dfrac{3}{5} \cdot 7\dfrac{1}{6}$ ▶ **67.** $\dfrac{5}{4} \div \dfrac{3}{8}$ **68.** $\dfrac{7}{5} \div \dfrac{3}{10}$

69. $\dfrac{32}{5} \div \dfrac{8}{15}$ **70.** $\dfrac{24}{7} \div \dfrac{6}{21}$ **71.** $\dfrac{3}{4} \div 12$ **72.** $\dfrac{2}{5} \div 30$

73. $6 \div \dfrac{3}{5}$ **74.** $8 \div \dfrac{4}{9}$ **75.** $6\dfrac{3}{4} \div \dfrac{3}{8}$ **76.** $5\dfrac{3}{5} \div \dfrac{7}{10}$

77. $2\dfrac{1}{2} \div 1\dfrac{5}{7}$ **78.** $2\dfrac{2}{9} \div 1\dfrac{2}{5}$ **79.** $2\dfrac{5}{8} \div 1\dfrac{15}{32}$ **80.** $2\dfrac{3}{10} \div 1\dfrac{4}{5}$

Find each sum or difference, and write it in lowest terms. See Examples 7–9.

81. $\dfrac{7}{15} + \dfrac{4}{15}$ **82.** $\dfrac{2}{9} + \dfrac{5}{9}$ ▶ **83.** $\dfrac{7}{12} + \dfrac{1}{12}$ **84.** $\dfrac{3}{16} + \dfrac{5}{16}$

▶ **85.** $\dfrac{5}{9} + \dfrac{1}{3}$ **86.** $\dfrac{4}{15} + \dfrac{1}{5}$ **87.** $\dfrac{3}{8} + \dfrac{5}{6}$ **88.** $\dfrac{5}{6} + \dfrac{2}{9}$

89. $3\dfrac{1}{8} + 2\dfrac{1}{4}$ **90.** $4\dfrac{2}{3} + 2\dfrac{1}{6}$ **91.** $3\dfrac{1}{4} + 1\dfrac{4}{5}$ **92.** $5\dfrac{3}{4} + 1\dfrac{1}{3}$

93. $\dfrac{7}{9} - \dfrac{2}{9}$ **94.** $\dfrac{8}{11} - \dfrac{3}{11}$ ▶ **95.** $\dfrac{13}{15} - \dfrac{3}{15}$ **96.** $\dfrac{11}{12} - \dfrac{3}{12}$

97. $\dfrac{7}{12} - \dfrac{1}{3}$ **98.** $\dfrac{5}{6} - \dfrac{1}{2}$ **99.** $\dfrac{7}{12} - \dfrac{1}{9}$ **100.** $\dfrac{11}{16} - \dfrac{1}{12}$

101. $4\dfrac{3}{4} - 1\dfrac{2}{5}$ **102.** $3\dfrac{4}{5} - 1\dfrac{4}{9}$ **103.** $6\dfrac{1}{4} - 5\dfrac{1}{3}$ **104.** $5\dfrac{1}{3} - 4\dfrac{1}{2}$

Work each problem involving fractions.

105. For each description, write a fraction in lowest terms that represents the region described.

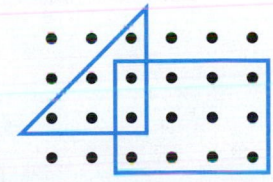

   (a) The dots in the rectangle as a part of the dots in the entire figure

   (b) The dots in the triangle as a part of the dots in the entire figure

   (c) The dots in the overlapping region of the triangle and the rectangle as a part of the dots in the triangle alone

   (d) The dots in the overlapping region of the triangle and the rectangle as a part of the dots in the rectangle alone

106. At the conclusion of the Pearson Learning softball league season, batting statistics for five players were as shown in the table.

Player	At-Bats	Hits	Home Runs
Maureen	36	12	3
Christine	40	9	2
Chase	11	5	1
Joe	16	8	0
Greg	20	10	2

Use this information to answer each question. Estimate as necessary.

(a) Which player got a hit in exactly $\frac{1}{3}$ of his or her at-bats?

(b) Which player got a hit in just less than $\frac{1}{2}$ of his or her at-bats?

(c) Which player got a home run in just less than $\frac{1}{10}$ of his or her at-bats?

(d) Which player got a hit in just less than $\frac{1}{4}$ of his or her at-bats?

(e) Which two players got hits in exactly the same fractional parts of their at-bats? What was the fractional part, expressed in lowest terms?

Use the table to answer Exercises 107 and 108.

107. How many cups of water would be needed for eight microwave servings?

108. How many teaspoons of salt would be needed for five stove-top servings? (*Hint:* 5 servings is halfway between 4 and 6 servings.)

	Microwave	Stove Top		
Servings	**1**	**1**	**4**	**6**
Water	$\frac{3}{4}$ cup	1 cup	3 cups	4 cups
Grits	3 Tbsp	3 Tbsp	$\frac{3}{4}$ cup	1 cup
Salt (optional)	Dash	Dash	$\frac{1}{4}$ tsp	$\frac{1}{2}$ tsp

Source: Package of Quaker Quick Grits.

The Pride Golf Tee Company, the only U.S. manufacturer of wooden golf tees, has created the Professional Tee System, shown in the figure. Use the information given to work Exercises 109 and 110. (Source: The Gazette.)

109. Find the difference in length between the ProLength Plus and the once-standard Shortee.

110. The ProLength Max tee is the longest tee allowed by the U.S. Golf Association's *Rules of Golf*. How much longer is the ProLength Max than the Shortee?

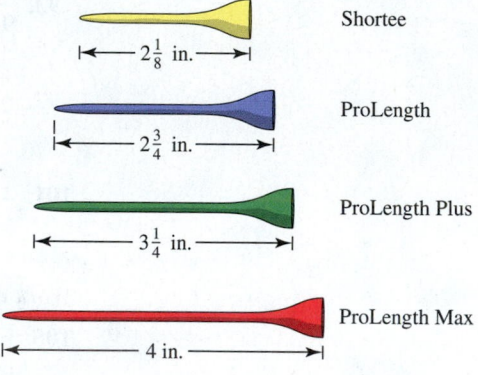

Solve each problem. See Example 10.

111. A hardware store sells a 40-piece socket wrench set. The measure of the largest socket is $\frac{3}{4}$ in., while the measure of the smallest is $\frac{3}{16}$ in. What is the difference between these measures?

112. Two sockets in a socket wrench set have measures of $\frac{9}{16}$ in. and $\frac{3}{8}$ in. What is the difference between these two measures?

113. A piece of property has an irregular shape, with five sides, as shown in the figure. Find the total distance around the piece of property. (This distance is the **perimeter** of the figure.)

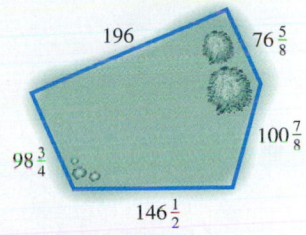

Measurements are in feet.

114. Find the perimeter of the triangle in the figure.

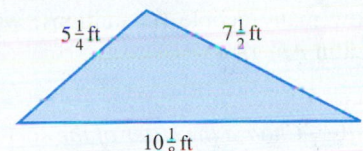

115. A board is $15\frac{5}{8}$ in. long. If it must be divided into three pieces of equal length, how long must each piece be?

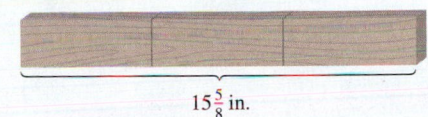

116. Paul's favorite recipe for barbecue sauce calls for $2\frac{1}{3}$ cups of tomato sauce. The recipe makes enough barbecue sauce to serve seven people. How much tomato sauce is needed for one serving?

117. A cake recipe calls for $1\frac{3}{4}$ cups of sugar. A caterer has $15\frac{1}{2}$ cups of sugar on hand. How many cakes can he make?

118. Kyla needs $2\frac{1}{4}$ yd of fabric to cover a chair. How many chairs can she cover with $23\frac{2}{3}$ yd of fabric?

119. It takes $2\frac{3}{8}$ yd of fabric to make a costume for a school play. How much fabric would be needed for seven costumes?

120. A cookie recipe calls for $2\frac{2}{3}$ cups of sugar. How much sugar would be needed to make four batches of cookies?

121. First published in 1953, the digest-sized *TV Guide* has changed to a full-sized magazine. The full-sized magazine is 3 in. wider than the old guide. What is the difference in their heights? (*Source: TV Guide.*)

122. Under existing standards, most of the holes in Swiss cheese must have diameters between $\frac{11}{16}$ and $\frac{13}{16}$ in. To accommodate new high-speed slicing machines, the U.S. Department of Agriculture wants to reduce the minimum size to $\frac{3}{8}$ in. How much smaller is $\frac{3}{8}$ in. than $\frac{11}{16}$ in.? (*Source: U.S. Department of Agriculture.*)

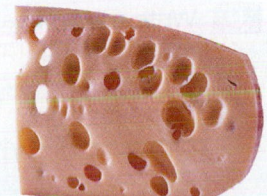

Approximately 40 million people living in the United States were born in other countries. The circle graph gives the fractional number from each region of birth for these people. Use the graph to answer each question. **See Example 11.**

123. What fractional part of the foreign-born population was from other regions?

124. What fractional part of the foreign-born population was from Latin America or Asia?

125. About how many people (in millions) were born in Europe?

126. About how many people (in millions) were born in Latin America?

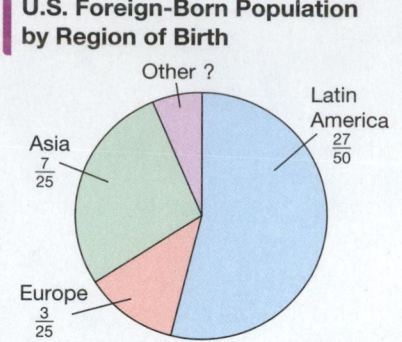

U.S. Foreign-Born Population by Region of Birth

Source: U.S. Census Bureau.

Extending Skills   Choose the letter of the correct response.

127. Estimate the best approximation for the sum.

$$\frac{14}{26} + \frac{98}{99} + \frac{100}{51} + \frac{90}{31} + \frac{13}{27}$$

   A. 5   **B.** 6   **C.** 7   **D.** 8

128. Estimate the best approximation for the product.

$$\frac{202}{50} \cdot \frac{99}{100} \cdot \frac{21}{40} \cdot \frac{75}{36}$$

   A. 3   **B.** 4   **C.** 8   **D.** 16

R.2   Decimals and Percents

OBJECTIVES

1   Write decimals as fractions.

2   Add and subtract decimals.

3   Multiply and divide decimals.

4   Write fractions as decimals.

5   Write percents as decimals and decimals as percents.

6   Write percents as fractions and fractions as percents.

7   Solve applied problems that involve percents.

Fractions are one way to represent parts of a whole. Another way is with a decimal fraction or **decimal,** a number written with a decimal point.

$$9.4, \quad 14.001, \quad 0.25 \qquad \text{Decimal numbers}$$

Each digit in a decimal number has a place value, as shown below.

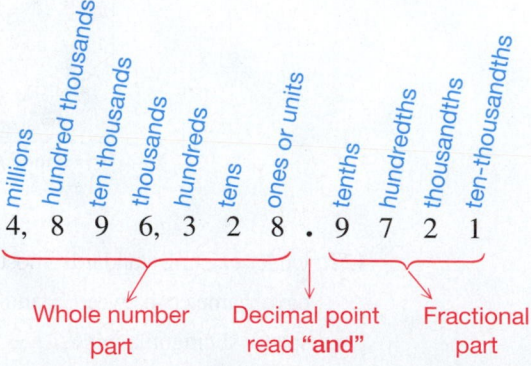

Each successive place value is ten times greater than the place value to its right and one-tenth as great as the place value to its left.

OBJECTIVE 1 Write decimals as fractions.

Place value is used to write a decimal number as a fraction.

Converting a Decimal to a Fraction

Read the decimal using the correct place value. Write it in fractional form just as it is read.

- The numerator will be the digits to the right of the decimal point.
- The denominator will be a power of 10—that is, 10 for tenths, 100 for hundredths, and so on.

NOW TRY EXERCISE 1

Write each decimal as a fraction. (Do not write in lowest terms.)

(a) 0.8   **(b)** 0.431   **(c)** 2.58

EXAMPLE 1 Writing Decimals as Fractions

Write each decimal as a fraction. (Do not write in lowest terms.)

(a) 0.95

We read 0.95 as "ninety-five **hundredths**." The equivalent fractional form is

$$0.95 = \frac{95}{100}. \leftarrow \text{For hundredths}$$

(b) 0.056

We read 0.056 as "fifty-six **thousandths**." The equivalent fractional form is

$$0.056 = \frac{56}{1000}.$$

For thousandths

> Do not confuse 0.056 with 0.56, read "fifty-six *hundredths*," which is the fraction $\frac{56}{100}$.

(c) 4.2095

We read this decimal number, which is greater than 1, as "Four *and* two thousand ninety-five ten-thousandths."

$$4.2095 = 4\frac{2095}{10,000} \qquad \text{Write the decimal number as a mixed number.}$$

$$= \frac{42,095}{10,000} \qquad \text{Write the mixed number as an improper fraction.}$$

NOW TRY

OBJECTIVE 2 Add and subtract decimals.

EXAMPLE 2 Adding and Subtracting Decimals

Add or subtract as indicated.

(a) 6.92 + 14.8 + 3.217

Place the digits of the decimal numbers in columns by place value, so that tenths are in one column, hundredths in another column, and so on.

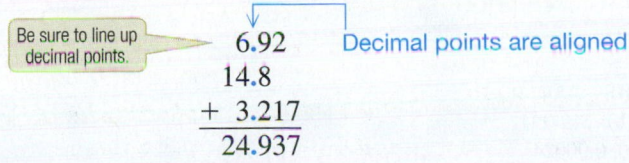

Be sure to line up decimal points.
$$\begin{array}{r} 6.92 \\ 14.8 \\ + \ 3.217 \\ \hline 24.937 \end{array}$$
Decimal points are aligned.

**NOW TRY
EXERCISE 2**

Add or subtract as indicated.

(a) $68.9 + 42.72 + 8.973$

(b) $351.8 - 2.706$

To avoid errors, attach zeros as placeholders so that all the decimal numbers are the same length.

$$
\begin{array}{r}
6.92 \\
14.8 \\
+\ \ 3.217 \\
\end{array}
\quad \text{becomes} \quad
\begin{array}{r}
6.920 \\
14.800 \\
+\ \ 3.217 \\
\hline
24.937 \\
\end{array}
$$

Attach 0s.

6.92 is equivalent to 6.920.
14.8 is equivalent to 14.800.

(b) $47.6 - 32.509$

$$
\begin{array}{r}
47.6 \\
-\ 32.509 \\
\end{array}
\quad \text{becomes} \quad
\begin{array}{r}
47.600 \\
-\ 32.509 \\
\hline
15.091 \\
\end{array}
$$

Write the decimal numbers in columns, attaching 0s to 47.6.

(c) $3 - 0.253$

$$
\begin{array}{r}
3.000 \\
-\ 0.253 \\
\hline
2.747 \\
\end{array}
$$

A whole number is assumed to have the decimal point at the right of the number. Write 3 as 3.000.

NOW TRY

OBJECTIVE 3   Multiply and divide decimals.

Multiplying Decimals

Step 1   Ignore the decimal points, and multiply as if the numbers were whole numbers.

Step 2   Add the number of **decimal places** (digits to the *right* of the decimal point) in each factor. Place the decimal point that many digits from the right in the product.

**NOW TRY
EXERCISE 3**

Multiply.

(a) 9.32×1.4

(b) 0.06×0.004

EXAMPLE 3   Multiplying Decimals

Multiply.

(a) 29.3×4.52

$$
\begin{array}{r}
29.3 \\
\times\ \ \ 4.52 \\
\hline
586 \\
1465 \\
1172 \\
\hline
132.436 \\
\end{array}
$$

1 decimal place
2 decimal places
$1 + 2 = 3$

3 decimal places

(b) 31.42×65

$$
\begin{array}{r}
31.42 \\
\times\ \ \ \ \ 65 \\
\hline
15710 \\
18852 \\
\hline
2042.30 \\
\end{array}
$$

2 decimal places
0 decimal places
$2 + 0 = 2$

2 decimal places

The final 0 can be dropped, and the product can be written 2042.3.

(c) 0.05×0.3

Here $5 \times 3 = 15$. Be careful placing the decimal point.

2 decimal places   1 decimal place

$$0.05 \qquad \times \qquad 0.3$$

$$= 0.015$$ Do *not* write 0.150.

$2 + 1 = 3$ decimal places
Attach 0 as a placeholder in the tenths place.

NOW TRY

NOW TRY ANSWERS
2. **(a)** 120.593   **(b)** 349.094
3. **(a)** 13.048   **(b)** 0.00024

$$5 \leftarrow \text{Quotient}$$
$$\text{Divisor} \rightarrow 25\overline{)125}$$
$$\uparrow$$
$$\text{Dividend}$$

Remember this terminology for the parts of a division problem.

Dividing Decimals

Step 1 Change the **divisor** (the number we are dividing *by*) into a whole number by moving the decimal point as many places as necessary to the right.

Step 2 Move the decimal point in the **dividend** (the number we are dividing *into*) to the right by the same number of places.

Step 3 Move the decimal point straight up, and then divide as with whole numbers to find the **quotient.**

NOW TRY EXERCISE 4

Divide.

(a) $451.47 \div 14.9$

(b) $7.334 \div 1.3$

(Round the quotient to two decimal places.)

EXAMPLE 4 Dividing Decimals

Divide.

(a) $233.45 \div 11.5$

Write the problem as follows. $11.5\overline{)233.45}$

$$11.5\,)\overline{233.4\,5}$$
To change the divisor 11.5 into a whole number, move each decimal point one place to the right.

To see why this works, write the division in fractional form and multiply by $\frac{10}{10}$. The result is the same as when we moved the decimal point one place to the right in the divisor and the dividend.

$$\frac{233.45}{11.5} \cdot \frac{10}{10} = \frac{2334.5}{115}$$
Multiplying by $\frac{10}{10}$ is equivalent to multiplying by 1.

Move the decimal point straight up and divide as with whole numbers.

$$
\begin{array}{r}
20.3 \\
115\overline{)2334.5} \\
\underline{230} \\
345 \\
\underline{345} \\
0
\end{array}
$$
Move the decimal point straight up.

In the second step of the division, 115 does not divide into 34, so we used zero as a placeholder in the quotient.

(b) $8.949 \div 1.25$ (Round the quotient to two decimal places.)

$$1.25\,)\overline{8.949}$$
Move each decimal point two places to the right.

$$
\begin{array}{r}
7.159 \\
125\overline{)894.900} \\
\underline{875} \\
199 \\
\underline{125} \\
740 \\
\underline{625} \\
1150 \\
\underline{1125} \\
25
\end{array}
$$
Move the decimal point straight up, and divide as with whole numbers. Attach 0s as placeholders.

We carried out the division to three decimal places so that we could round to two decimal places, obtaining the quotient 7.16.

NOW TRY ANSWERS
4. (a) 30.3 (b) 5.64

NOW TRY

NOTE To round 7.159 in **Example 4(b)** to two decimal places (that is, to the nearest hundredth), we look at the digit to the *right* of the hundredths place. If this digit is 5 or greater, we round up. If it is less than 5, we drop the digit(s) beyond the desired place.

Hundredths place
↓

7.159   9, the digit to the right of the hundredths place, is 5 or greater.

≈ 7.16   Round 5 up to 6. ≈ means "is approximately equal to."

Multiplying or Dividing by Powers of 10 (Shortcuts)

- To *multiply* by a power of 10, *move the decimal point to the right* as many places as the number of zeros.

- To *divide* by a power of 10, *move the decimal point to the left* as many places as the number of zeros.

In both cases, insert 0s as placeholders if necessary.

**NOW TRY
EXERCISE 5**

Multiply or divide as indicated.

(a) 294.72 × 10

(b) 4.793 ÷ 100

EXAMPLE 5   Multiplying and Dividing by Powers of 10

Multiply or divide as indicated.

(a) 48.731 × 100

$= 48.73\,1$ or   4873.1   Move the decimal point two places to the right because 100 has two 0s.

(b) 48.7 ÷ 1000

$= 048.7$ or   0.0487   Move the decimal point three places to the left because 1000 has three 0s. Insert a 0 in front of the 4 to do this.   NOW TRY

OBJECTIVE 4 Write fractions as decimals.

Converting a Fraction to a Decimal

Because a fraction bar indicates division, write a fraction as a decimal by dividing the numerator by the denominator.

EXAMPLE 6   Writing Fractions as Decimals

Write each fraction as a decimal.

(a) $\dfrac{19}{8}$

$$\begin{array}{r} 2.375 \\ 8\overline{)19.000} \\ \underline{16} \\ 30 \\ \underline{24} \\ 60 \\ \underline{56} \\ 40 \\ \underline{40} \\ 0 \end{array}$$

Divide 19 by 8. Add a decimal point and as many 0s as necessary.

(b) $\dfrac{2}{3}$

$$\begin{array}{r} 0.6666\ldots \\ 3\overline{)2.0000\ldots} \\ \underline{18} \\ 20 \\ \underline{18} \\ 20 \\ \underline{18} \\ 20 \\ \underline{18} \\ 20 \end{array}$$

$\dfrac{2}{3} = 0.6666\ldots$ ← Repeating decimal

$\dfrac{19}{8} = 2.375$ ← Terminating decimal

**NOW TRY
EXERCISE 6**

Write each fraction as a decimal. For repeating decimals, write the answer by first using bar notation and then rounding to the nearest thousandth.

(a) $\dfrac{17}{20}$ **(b)** $\dfrac{2}{9}$

Because the remainder in the division in part (a) is 0, this quotient is a **terminating decimal.** The remainder in the division in part (b) is never 0. Because a number, in this case 2, is always left after the subtraction, this quotient is a **repeating decimal.** A convenient notation for a repeating decimal is a bar over the digit (or digits) that repeats.

$$\frac{2}{3} = 0.6666\ldots, \quad \text{or} \quad 0.\overline{6}$$

We often round repeating decimals to as many places as needed.

$$\frac{2}{3} \approx 0.667 \qquad \text{An approximation to the nearest thousandth}$$

NOW TRY

OBJECTIVE 5 Write percents as decimals and decimals as percents.

The word **percent** means **"per 100."** Percent is written with the symbol **%. *One percent means "one per one hundred," or "one one-hundredth."***

Percent, Decimal, and Fraction Equivalents
$$\mathbf{1\% = 0.01, \quad or \quad 1\% = \frac{1}{100}}$$

**NOW TRY
EXERCISE 7**

Write each percent as a decimal.

(a) 23% **(b)** 350%

EXAMPLE 7 Writing Percents as Decimals

Write each percent as a decimal.

(a) 73%

We use the fact that $1\% = 0.01$ and convert as follows.

$$73\% = 73 \cdot 1\% = 73 \cdot 0.01 = 0.73$$

(b) $125\% = 125 \cdot 1\% = 125 \cdot 0.01 = 1.25$ $1\% = 0.01$

> A percent greater than 100 represents a number greater than 1.

(c) $3\frac{1}{2}\%$

First write the fractional part as a decimal.

$$3\frac{1}{2}\% = (3 + 0.5)\% = 3.5\%$$

Now write the percent in decimal form.

$$3.5\% = 3.5 \cdot 1\% = 3.5 \cdot 0.01 = 0.035 \qquad 1\% = 0.01 \qquad \text{NOW TRY}$$

**NOW TRY
EXERCISE 8**

Write each decimal as a percent.

(a) 0.71 **(b)** 1.32

EXAMPLE 8 Writing Decimals as Percents

Write each decimal as a percent.

(a) 0.32

This conversion is the opposite of what we did in **Example 7** when we wrote percents as decimals. We use $1\% = 0.01$ in reverse.

$$0.32 = 32 \cdot 0.01 = 32 \cdot 1\% = 32\% \qquad 0.01 = 1\%$$

(b) $0.05 = 5 \cdot 0.01 = 5 \cdot 1\% = 5\% \qquad 0.01 = 1\%$

(c) $2.63 = 263 \cdot 0.01 = 263 \cdot 1\% = 263\%$

> A number greater than 1 is more than 100%.

NOW TRY

NOW TRY ANSWERS
6. (a) 0.85 **(b)** $0.\overline{2}$, 0.222
7. (a) 0.23 **(b)** 3.50, or 3.5
8. (a) 71% **(b)** 132%

Converting Percents and Decimals (Shortcuts)

- To convert a percent to a decimal, move the decimal point two places to the *left* and drop the % symbol.

- To convert a decimal to a percent, move the decimal point two places to the *right* and attach a % symbol.

Divide by 100.

Drop %. Move 2 places left.

Decimal **Percent**

Multiply by 100. Attach %.

Move 2 places right.

NOW TRY
EXERCISE 9

Convert each percent to a decimal and each decimal to a percent.

(a) 52% **(b)** 2%

(c) 0.45 **(d)** 3.5

EXAMPLE 9 Converting Percents and Decimals by Moving the Decimal Point

Convert each percent to a decimal and each decimal to a percent.

(a) $45\% = 0.45$ **(b)** $250\% = 2.50$, or 2.5 **(c)** $9\% = 09\% = 0.09$

(d) $0.57 = 57\%$ **(e)** $1.5 = 1.50 = 150\%$ **(f)** $0.327 = 32.7\%$

NOW TRY

OBJECTIVE 6 Write percents as fractions and fractions as percents.

NOW TRY
EXERCISE 10

Write each percent as a fraction. Give answers in lowest terms.

(a) 20% **(b)** 160%

EXAMPLE 10 Writing Percents as Fractions

Write each percent as a fraction. Give answers in lowest terms.

(a) 8%

We use the fact that $1\% = \frac{1}{100}$, and convert as follows.

$$8\% = 8 \cdot 1\% = 8 \cdot \frac{1}{100} = \frac{8}{100}$$

In lowest terms, $\qquad \frac{8}{100} = \frac{2 \cdot 4}{25 \cdot 4} = \frac{2}{25}.$

Thus, $8\% = \frac{2}{25}.$

(b) $175\% = 175 \cdot 1\% = 175 \cdot \frac{1}{100} = \frac{175}{100}$

In lowest terms,

$$\frac{175}{100} = \frac{7 \cdot 25}{4 \cdot 25} = \frac{7}{4}, \quad \text{or} \quad 1\frac{3}{4}.$$

> A number greater than 1 is more than 100%.

(c) 13.5%

$$= 13\frac{1}{2} \cdot 1\% \qquad \text{Write 13.5 as a mixed number.}$$

$$= \frac{27}{2} \cdot \frac{1}{100} \qquad \text{Write } 13\frac{1}{2} \text{ as an improper fraction. Use the fact that } 1\% = \frac{1}{100}.$$

$$= \frac{27}{200} \qquad \text{Multiply the fractions.}$$

NOW TRY

NOW TRY ANSWERS
9. **(a)** 0.52 **(b)** 0.02
   (c) 45% **(d)** 350%
10. **(a)** $\frac{1}{5}$ **(b)** $\frac{8}{5}$, or $1\frac{3}{5}$

We know that 100% of something is the whole thing. One way to convert a fraction to a percent is to multiply by 100%, which is equivalent to 1.

NOW TRY
EXERCISE 11
Write each fraction as a percent.

(a) $\dfrac{6}{25}$ (b) $\dfrac{7}{2}$

EXAMPLE 11 Writing Fractions as Percents

Write each fraction as a percent.

(a) $\dfrac{2}{5}$ (b) $\dfrac{1}{6}$

$= \dfrac{2}{5} \cdot 100\%$ Multiply by 1 in the form 100%.

$= \dfrac{2}{5} \cdot \dfrac{100}{1}\%$

$= \dfrac{2 \cdot 5 \cdot 20}{5 \cdot 1}\%$ Multiply and factor.

$= \dfrac{2 \cdot 20}{1}\%$ Divide out the common factor.

$= 40\%$ Simplify.

(b)

$= \dfrac{1}{6} \cdot 100\%$

$= \dfrac{1}{6} \cdot \dfrac{100}{1}\%$

$= \dfrac{1 \cdot 2 \cdot 50}{2 \cdot 3 \cdot 1}\%$

$= \dfrac{50}{3}\%$

$= 16\dfrac{2}{3}\%$, or $16.\overline{6}\%$

NOW TRY

OBJECTIVE 7 Solve applied problems that involve percents.

The decimal form of a percent is generally used in calculations.

NOW TRY
EXERCISE 12
A winter coat is on sale for 60% off. The regular price is $120. Find the amount of the discount and the sale price of the coat.

EXAMPLE 12 Using Percent to Solve an Applied Problem

A DVD with a regular price of $18 is on sale this week at 22% off. Find the amount of the discount and the sale price of the DVD.

The discount is 22% *of* 18. The word *of* here means multiply.

$$22\% \quad of \quad 18$$
$$\downarrow \qquad \downarrow \qquad \downarrow$$
$$0.22 \quad \cdot \quad 18 \quad \text{Write 22\% as a decimal.}$$
$$= 3.96 \qquad \text{Multiply.}$$

The discount is $3.96. The sale price is found by subtracting.

$$\$18.00 - \$3.96 = \$14.04 \qquad \text{Original price} - \text{discount} = \text{sale price}$$

NOW TRY

NOW TRY ANSWERS
11. (a) 24% (b) 350%
12. $72; $48

R.2 Exercises

FOR EXTRA HELP   MyMathLab®

● *Complete solution available in MyMathLab*

Concept Check   *In Exercises 1–4, provide the correct response.*

1. In the decimal number 367.9412, name the digit that has each place value.

   (a) tens   (b) tenths   (c) thousandths   (d) ones or units   (e) hundredths

2. Write a decimal number that has 5 in the thousands place, 0 in the tenths place, and 4 in the ten-thousandths place.

3. For the decimal number 46.249, round to the place value indicated.

   (a) hundredths   (b) tenths   (c) ones or units   (d) tens

4. Round each decimal to the nearest thousandth.

(a) $0.\overline{8}$ (b) $0.\overline{5}$ (c) 0.9762 (d) 0.8642

Write each decimal as a fraction. (Do not write in lowest terms.) ***See Example 1.***

5. 0.4	**6.** 0.6	**7.** 0.64	**8.** 0.82	**9.** 0.138
10. 0.104	**11.** 0.043	**12.** 0.087	**13.** 3.805	**14.** 5.166

Add or subtract as indicated. ***See Example 2.***

15. $25.32 + 109.2 + 8.574$ **16.** $90.527 + 32.43 + 589.8$ **17.** $28.73 - 3.12$

18. $46.88 - 13.45$ **19.** $43.5 - 28.17$ **20.** $345.1 - 56.31$

21. $3.87 + 15 + 2.9$ **22.** $8.2 + 1.09 + 12$ **23.** $32.56 + 47.356 + 1.8$

24. $75.2 + 123.96 + 3.897$ ▶ **25.** $18 - 2.789$ **26.** $29 - 8.582$

Multiply or divide as indicated. ***See Examples 3–5.***

27. 12.8×9.1 **28.** 34.04×0.56 **29.** 0.2×0.03 **30.** 0.07×0.004

▶ **31.** $78.65 \div 11$ **32.** $73.36 \div 14$ **33.** $19.967 \div 9.74$ **34.** $44.4788 \div 5.27$

35. 57.116×100 **36.** 82.053×100 **37.** 0.094×1000 **38.** 0.025×1000

39. $1.62 \div 10$ **40.** $8.04 \div 10$ **41.** $124.03 \div 100$ **42.** $490.35 \div 100$

Concept Check   *Complete the following table of fraction, decimal, and percent equivalents.*

	Fraction in Lowest Terms (or Whole Number)	Decimal	Percent
43.	$\frac{1}{100}$	0.01	
44.	$\frac{1}{50}$		2%
45.		0.05	5%
46.	$\frac{1}{10}$		
47.	$\frac{1}{8}$	0.125	
48.			20%
49.	$\frac{1}{4}$		
50.	$\frac{1}{3}$		
51.			50%
52.	$\frac{2}{3}$		$66\frac{2}{3}\%$, or $66.\overline{6}\%$
53.		0.75	
54.	1	1.0	

Write each fraction as a decimal. For repeating decimals, write the answer by first using bar notation and then rounding to the nearest thousandth. ***See Example 6.***

55. $\frac{3}{8}$ **56.** $\frac{7}{8}$ **57.** $\frac{5}{4}$ **58.** $\frac{9}{5}$

▶ **59.** $\frac{5}{9}$ **60.** $\frac{8}{9}$ **61.** $\frac{1}{6}$ **62.** $\frac{5}{6}$

Write each percent as a decimal. See Examples 7 and 9(a)–9(c).

63. 54%   **64.** 39%   **65.** 7%   **66.** 4%

67. 117%   **68.** 189%   **69.** 2.4%   **70.** 3.1%

71. $6\frac{1}{4}\%$ **72.** $5\frac{1}{2}\%$ **73.** 0.8%   **74.** 0.9%

Write each decimal as a percent. See Examples 8 and 9(d)–9(f).

75. 0.79   **76.** 0.83   **77.** 0.02   **78.** 0.08

79. 0.004   **80.** 0.005   **81.** 1.28   **82.** 2.35

83. 0.4   **84.** 0.6   **85.** 6   **86.** 10

Write each percent as a fraction. Give answers in lowest terms. See Example 10.

87. 51%   **88.** 47%   **89.** 15%   **90.** 35%   **91.** 2%

92. 8%   **93.** 140%   **94.** 180%   **95.** 7.5%   **96.** 2.5%

Write each fraction as a percent. See Example 11.

97. $\dfrac{4}{5}$ **98.** $\dfrac{3}{25}$ **99.** $\dfrac{7}{50}$ **100.** $\dfrac{9}{20}$ **101.** $\dfrac{2}{11}$

102. $\dfrac{4}{9}$ **103.** $\dfrac{9}{4}$ **104.** $\dfrac{8}{5}$ **105.** $\dfrac{13}{6}$ **106.** $\dfrac{31}{9}$

Solve each problem. See Example 12.

107. What is 50% of 320?   **108.** What is 25% of 120?

109. What is 6% of 80?   **110.** What is 5% of 70?

111. What is 14% of 780?   **112.** What is 26% of 480?

Solve each problem. See Example 12.

113. Elwyn's bill for dinner at a restaurant was $89. He wants to leave a 20% tip. How much should he leave for the tip? What is his total bill for dinner and tip?

114. Gary earns $15 per hour at his job. He recently received a 7% raise. How much per hour was his raise? What is his new hourly rate?

115. Find the discount on a leather recliner with a regular price of $795 if the recliner is on sale at 15% off. What is the sale price of the recliner?

116. A laptop computer with a regular price of $597 is on sale at 20% off. Find the amount of the discount and the sale price of the computer.

In a recent year, approximately 60 million people from other countries visited the United States. The circle graph shows the distribution of these international visitors by country or region. Use the graph to work each problem.

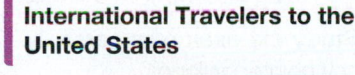

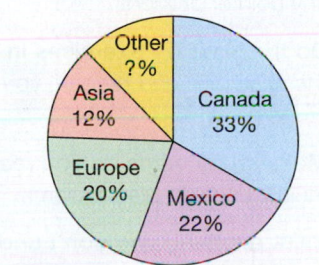

Source: U.S. Department of Commerce.

117. About how many travelers visited the United States from Canada?

118. About how many travelers visited the United States from Mexico?

119. What percent of travelers visited the United States from places other than Canada, Mexico, Europe, or Asia? (*Hint:* The sum of the parts of the graph must equal 1 whole, that is, 100%.)

120. Use the answer from **Exercise 119** to find about how many travelers visited the United States from places other than Canada, Mexico, Europe, or Asia.

Reading Your Math Textbook

Take time to read each section and its examples before doing your homework. You will learn more and be better prepared to work the exercises your instructor assigns.

Approaches to Reading Your Math Textbook

Student A learns best by listening to her teacher explain things. She "gets it" when she sees the instructor work problems. She previews the section before the lecture, so she knows generally what to expect. **Student A carefully reads the section in her text** *AFTER* **she hears the classroom lecture on the topic.**

Student B learns best by reading on his own. He reads the section and works through the examples before coming to class. That way, he knows what the teacher is going to talk about and what questions he wants to ask. **Student B carefully reads the section in his text** *BEFORE* **he hears the classroom lecture on the topic.**

Which of these reading approaches works best for you—that of Student A or Student B?

Tips for Reading Your Math Textbook

- **Turn off your cell phone.** You will be able to concentrate more fully on what you are reading.

- **Survey the material.** Glance over the assigned material to get an idea of the "big picture." Look at the list of objectives to see what you will be learning.

- **Read slowly.** Read only one section—or even part of a section—at a sitting, with paper and pencil in hand.

- **Pay special attention to important information given in colored boxes or set in bold-face type.**

- **Study the examples carefully.** Pay particular attention to the blue side comments and any pointer balloons.

- **Do the Now Try exercises in the margin on separate paper as you go.** These mirror the examples and prepare you for the exercise set. The answers are given at the bottom of the page.

- **Make study cards as you read.** Make cards for new vocabulary, rules, procedures, formulas, and sample problems.

- **Mark anything you don't understand.** *ASK QUESTIONS* in class—everyone will benefit. Follow up with your instructor, as needed.

Think through and answer each question.

1. Which two or three reading tips will you try this week?

2. Did the tips you selected improve your ability to read and understand the material? Explain.

1

The Real Number System

Positive and *negative numbers,* shown here to indicate gains and losses, are examples of *real numbers,* the subject of this chapter.

+1.44% 100
+1.49% 200
+.71% 199
-2.09% 300
+2.13% 300
+2.62% 5000
+.54% 800
+3.35%
-1.19% 542
+.53% 100
100
0%
-1.89% 0
0% 350

27

1.1 Exponents, Order of Operations, and Inequality

VOCABULARY

☐ exponent (power)
☐ base
☐ exponential expression
☐ inequality

NOW TRY EXERCISE 1

Find the value of each exponential expression.

(a) 6^2 **(b)** $\left(\dfrac{4}{5}\right)^3$

OBJECTIVE 1 Use exponents.

In **Chapter R,** we factored a number as the product of its prime factors.

>   81   can be written as $3 \cdot 3 \cdot 3 \cdot 3$.   The factor 3 appears four times.

In algebra, repeated factors are written with an *exponent,* so the product $3 \cdot 3 \cdot 3 \cdot 3$ is written as 3^4 and read as "3 to the fourth power."

$$\underbrace{3 \cdot 3 \cdot 3 \cdot 3}_{4 \text{ factors of } 3} = 3^4 \leftarrow \text{Exponent}$$

The number 4 is the **exponent,** or **power,** and 3 is the **base** in the **exponential expression** 3^4. A natural number exponent, then, tells how many times the base is used as a factor. *A number raised to the first power is simply that number.* For example,

$$5^1 = 5 \quad \text{and} \quad \left(\frac{1}{2}\right)^1 = \frac{1}{2}. \qquad \text{In general, } a^1 = a.$$

EXAMPLE 1 Evaluating Exponential Expressions

Find the value of each exponential expression.

(a) 5^2 means $\underbrace{5 \cdot 5}$,   which equals   25.

> 5 is used as a factor 2 times.

Read 5^2 as "5 to the second power" or, more commonly, "5 squared."

(b) 6^3 means $\underbrace{6 \cdot 6 \cdot 6}$,   which equals   216.

> 6 is used as a factor 3 times.

Read 6^3 as "6 to the third power" or, more commonly, "6 cubed."

(c) 2^5 means $2 \cdot 2 \cdot 2 \cdot 2 \cdot 2$,   which equals   32.   2 is used as a factor 5 times.

Read 2^5 as "2 to the fifth power."

(d) $\left(\dfrac{2}{3}\right)^3$ means $\dfrac{2}{3} \cdot \dfrac{2}{3} \cdot \dfrac{2}{3}$,   which equals $\dfrac{8}{27}$. $\frac{2}{3}$ is used as a factor 3 times.

(e) $(0.3)^2$ means $0.3(0.3)$,   which equals   0.09.   0.3 is used as a factor 2 times.

> **NOW TRY**

⚠ **CAUTION** *Squaring, or raising a number to the second power, is not the same as doubling the number.* For example,

$$3^2 \text{ means } 3 \cdot 3, \textit{ not } 2 \cdot 3.$$

Thus $3^2 = 9$, *not* 6. Similarly, cubing, or raising a number to the third power, does *not* mean tripling the number.

OBJECTIVE 2 Use the rules for order of operations.

When an expression involves more than one operation, we often use **grouping symbols,** such as parentheses ( ), to indicate the order in which the operations should be performed.

NOW TRY ANSWERS
1. (a) 36 **(b)** $\frac{64}{125}$

Consider the following expression.

$$5 + 2 \cdot 3$$

To show that the multiplication should be performed before the addition, we could use parentheses to group $2 \cdot 3$.

$$5 + (2 \cdot 3) \quad \text{equals} \quad 5 + 6, \quad \text{which equals} \quad 11.$$

If the addition is to be performed first, the parentheses should group $5 + 2$.

$$(5 + 2) \cdot 3 \quad \text{equals} \quad 7 \cdot 3, \quad \text{which equals} \quad 21.$$

Other grouping symbols are brackets $[\;\;]$, braces $\{\;\;\}$, and fraction bars. (For example, in $\frac{8 - 2}{3}$, the expression $8 - 2$ is "grouped" in the numerator.)

To simplify an expression that involves more than one operation, we use the following rules for **order of operations.** This order is used by most calculators and computers.

Order of Operations

If grouping symbols are present, work within them, innermost first (and above and below fraction bars separately), in the following order.

Step 1   Apply all **exponents.**

Step 2   Do any **multiplications** or **divisions** in order from left to right.

Step 3   Do any **additions** or **subtractions** in order from left to right.

If no grouping symbols are present, start with Step 1.

NOTE   Multiplication is understood in expressions with parentheses such as

$$3(7), \quad (6)2, \quad (-5)(-4), \quad \text{or} \quad 3(4 + 1).$$

EXAMPLE 2   Using the Rules for Order of Operations

Find the value of each expression.

(a) $4 + 5 \cdot 6$ A helpful strategy is to label the order in which the
$\quad$ ② ①   operations should be performed.

$= 4 + 30$ Multiply.

$= 34$ Add.

(b) $9(6 + 11)$
$\quad$ ② ①

$= 9(17)$ Add inside the parentheses.

$= 153$ Multiply.

(c) $6 \cdot 8 + 5 \cdot 2$
$\quad$ ① ③ ②

$= 48 + 10$ Multiply, working from left to right.

$= 58$ Add.

**NOW TRY
EXERCISE 2**

Find the value of each expression.

(a) $15 - 2 \cdot 6$

(b) $8 + 2(5 - 1)$

(c) $6(2 + 4) - 7 \cdot 5$

(d) $8 \cdot 10 \div 4 - 2^3 + 3 \cdot 4^2$

(d) $16 - 3(2 + 3)$ ◁ ⟨Do *not* subtract $16 - 3$ first.⟩

③ ② ①

$= 16 - 3(5)$ Add inside the parentheses.

$= 16 - 15$ Multiply.

$= 1$ Subtract.

(e) $2(5 + 6) + 7 \cdot 3$

② ① ④ ③

$= 2(11) + 7 \cdot 3$ Add inside the parentheses.

$= 22 + 21$ Multiply, working from left to right.

$= 43$ Add.

⟨$2^3 = 2 \cdot 2 \cdot 2$, not $2 \cdot 3$.⟩

(f) $9 - 2^3 + 5$

$= 9 - 2 \cdot 2 \cdot 2 + 5$ Apply the exponent.

$= 9 - 8 + 5$ Multiply.

$= 1 + 5$ Subtract.

$= 6$ Add.

(g) $72 \div 2 \cdot 3 + 4 \cdot 2^3 - 3^3$ ◁ ⟨Think: $3^3 = 3 \cdot 3 \cdot 3$⟩

$= 72 \div 2 \cdot 3 + 4 \cdot 8 - 27$ Apply the exponents.

$= 36 \cdot 3 + 4 \cdot 8 - 27$ Divide.

$= 108 + 32 - 27$ Multiply, working from left to right.

$= 140 - 27$ Add.

$= 113$ Subtract.

Multiplications and divisions are done from left to right as they appear. Then additions and subtractions are done from left to right as they appear.

NOW TRY ↻

OBJECTIVE 3 Use more than one grouping symbol.

In an expression such as $2(8 + 3(6 + 5))$, we often use brackets in place of the outer pair of parentheses.

EXAMPLE 3 Using Brackets and Fraction Bars as Grouping Symbols

Simplify each expression.

(a) $2[8 + 3(6 + 5)]$

$= 2[8 + 3(11)]$ Add inside the parentheses.

$= 2[8 + 33]$ Multiply inside the brackets.

$= 2[41]$ Add inside the brackets.

$= 82$ Multiply.

NOW TRY ANSWERS
2. (a) 3 **(b)** 16
   (c) 1 **(d)** 60

**NOW TRY
EXERCISE 3**

Simplify each expression.

(a) $7[(3^2 - 1) + 4]$

(b) $\dfrac{9(14 - 4) - 2}{4 + 3 \cdot 6}$

(b) $\dfrac{4(5 + 3) + 3}{2(3) - 1}$ Simplify the numerator and denominator separately.

$= \dfrac{4(8) + 3}{2(3) - 1}$ Work inside the parentheses in the numerator.

$= \dfrac{32 + 3}{6 - 1}$ Multiply.

$= \dfrac{35}{5}$ Add and subtract.

$= 7$ Divide.

NOW TRY

NOTE   The expression $\dfrac{4(5 + 3) + 3}{2(3) - 1}$ in **Example 3(b)** can be written as the quotient

$$[4(5 + 3) + 3] \div [2(3) - 1].$$

This shows that the fraction bar "groups" the numerator and denominator separately.

OBJECTIVE 4 Know the meanings of $\neq$, $<$, $>$, $\leq$, and $\geq$.

So far, we have used the equality symbol $=$. The symbols

$$\neq, \quad <, \quad >, \quad \leq, \quad \text{and} \quad \geq \qquad \text{Inequality symbols}$$

are used to express an **inequality,** a statement that two expressions may not be equal. The equality symbol with a slash through it, $\neq$, means "is not equal to."

$$7 \neq 8 \qquad \text{7 is not equal to 8.}$$

If two numbers are not equal, then one of the numbers must be less than the other. The symbol $<$ represents "is less than."

$$7 < 8 \qquad \text{7 is less than 8.}$$

The symbol $>$ means "is greater than."

$$8 > 2 \qquad \text{8 is greater than 2.}$$

To keep the meanings of the symbols $<$ and $>$ clear, remember that the symbol always points to the lesser number.

$$\text{Lesser number} \rightarrow \quad 8 < 15$$

$$15 > 8 \quad \leftarrow \text{Lesser number}$$

The symbol $\leq$ means "is less than or equal to."

$$5 \leq 9 \qquad \text{5 is less than or equal to 9.}$$

If either the $<$ part or the $=$ part is true, then the inequality $\leq$ is true. The statement $5 \leq 9$ is true because $5 < 9$ is true.

The symbol $\geq$ means "is greater than or equal to."

$$9 \geq 5 \qquad \text{9 is greater than or equal to 5.}$$

NOW TRY ANSWERS
3. (a) 84 (b) 4

**NOW TRY
EXERCISE 4**

Determine whether each
statement is *true* or *false*.

(a) $12 \neq 10 - 2$

(b) $5 > 4 \cdot 2$

(c) $7 \leq 7$

(d) $\dfrac{5}{9} > \dfrac{7}{11}$

EXAMPLE 4 Using Inequality Symbols

Determine whether each statement is *true* or *false*.

(a) $6 \neq 5 + 1$ This statement is false because $6 = 5 + 1$.

(b) $5 + 3 < 19$ The statement $5 + 3 < 19$ is true because $8 < 19$.

(c) $15 \leq 20 \cdot 2$ The statement $15 \leq 20 \cdot 2$ is true because $15 < 40$.

(d) $25 \geq 30$ Both $25 > 30$ and $25 = 30$ are false, so $25 \geq 30$ is false.

(e) $12 \geq 12$ Because $12 = 12$, this statement is true.

(f) $9 < 9$ Because $9 = 9$, this statement is false.

(g) $\dfrac{6}{15} \geq \dfrac{2}{3}$ Find a common denominator.

$\dfrac{6}{15} \geq \dfrac{10}{15}$ Both $\dfrac{6}{15} > \dfrac{10}{15}$ and $\dfrac{6}{15} = \dfrac{10}{15}$ are false, so $\dfrac{6}{15} \geq \dfrac{2}{3}$ is false.   **NOW TRY**

OBJECTIVE 5 Translate word statements to symbols.

**NOW TRY
EXERCISE 5**

Write each word statement in
symbols.

(a) Ten is not equal to eight
minus two.

(b) Fifty is greater than fifteen.

(c) Eleven is less than or equal
to twenty.

EXAMPLE 5 Translating from Words to Symbols

Write each word statement in symbols.

(a) Twelve equals ten plus two.

$$12 = 10 + 2$$

(b) Nine is less than ten.

$$9 < 10$$

(c) Fifteen is not equal to eighteen.

$$15 \neq 18$$

(d) Seven is greater than four.

$$7 > 4$$

(e) Thirteen is less than or equal to
forty.

$$13 \leq 40$$

(f) Eleven is greater than or equal to
eleven.

$$11 \geq 11$$

NOW TRY

OBJECTIVE 6 Write statements that change the direction of inequality symbols.

Any statement involving $<$ can be converted to one with $>$, and any statement
involving $>$ can be converted to one with $<$. *We do this by reversing the order of
the numbers and the direction of the symbol.*

Interchange numbers.

$$6 < 10 \qquad \text{becomes} \qquad 10 > 6$$

Reverse symbol.

**NOW TRY
EXERCISE 6**

Write the statement as another
true statement with the
inequality symbol reversed.

$$8 < 9$$

NOW TRY ANSWERS

4. (a) true   (b) false
   (c) true   (d) false

5. (a) $10 \neq 8 - 2$ (b) $50 > 15$
   (c) $11 \leq 20$

6. $9 > 8$

EXAMPLE 6 Converting between Inequality Symbols

Parts (a)–(c) each show a statement written in two equally correct ways. In each
inequality, the inequality symbol points toward the lesser number.

(a) $5 > 2$, $2 < 5$ (b) $\dfrac{1}{2} \leq \dfrac{3}{4}$, $\dfrac{3}{4} \geq \dfrac{1}{2}$ (c) $1.2 \geq 0.5$, $0.5 \leq 1.2$

NOW TRY

▼ Summary of Equality and Inequality Symbols

Symbol	Meaning	Example
=	Is equal to	$0.5 = \frac{1}{2}$ means 0.5 is equal to $\frac{1}{2}$.
≠	Is not equal to	$3 \neq 7$ means 3 is not equal to 7.
<	Is less than	$6 < 10$ means 6 is less than 10.
>	Is greater than	$15 > 14$ means 15 is greater than 14.
≤	Is less than or equal to	$4 \leq 8$ means 4 is less than or equal to 8.
≥	Is greater than or equal to	$1 \geq 0$ means 1 is greater than or equal to 0.

⚠ **CAUTION** Equality and inequality symbols are used to write mathematical **sentences.** Operation symbols $(+, -, \cdot, \text{ and } \div)$ are used to write mathematical **expressions** that represent a number. Compare the following.

Sentence: $4 < 10 \leftarrow$ Gives the relationship between 4 and 10

Expression: $4 + 10 \leftarrow$ Tells how to operate on 4 and 10 to get 14

1.1 Exercises

FOR EXTRA HELP   ▶   MyMathLab®

▶ *Complete solution available in MyMathLab*

Concept Check *Decide whether each statement is* true *or* false. *If it is false, explain why.*

1. $3^2 = 6$ **2.** $1^3 = 3$ **3.** $3^1 = 1$

4. The expression 6^2 means that 2 is used as a factor 6 times.

5. When evaluated, $4 + 3(8 - 2)$ is equal to 42.

6. When evaluated, $12 \div 2 \cdot 3$ is equal to 2.

Concept Check *For each expression, label the order in which the operations should be performed. Do not actually perform them.*

7. $18 - 2 + 3$ ◯ ◯ **8.** $28 - 6 \div 2$ ◯ ◯ **9.** $2 \cdot 8 - 6 \div 3$ ◯ ◯ ◯

10. $40 + 6(3 - 1)$ ◯ ◯ ◯ **11.** $3 \cdot 5 - 2(4 + 2)$ ◯ ◯ ◯ ◯ **12.** $9 - 2^3 + 3 \cdot 4$ ◯ ◯ ◯ ◯

Find the value of each exponential expression. See Example 1.

13. 3^2 **14.** 8^2 ▶ **15.** 7^2 **16.** 4^2 **17.** 12^2

18. 14^2 **19.** 4^3 **20.** 5^3 **21.** 10^3 **22.** 11^3

23. 3^4 **24** 6^4 **25.** 4^5 **26.** 3^5 **27.** $\left(\dfrac{1}{6}\right)^2$

28. $\left(\dfrac{1}{3}\right)^2$ **29.** $\left(\dfrac{2}{3}\right)^4$ **30.** $\left(\dfrac{3}{4}\right)^3$ **31.** $(0.4)^3$ **32.** $(0.5)^4$

Find the value of each expression. See Examples 2 and 3.

33. $64 \div 4 \cdot 2$ **34.** $250 \div 5 \cdot 2$ ▶ **35.** $13 + 9 \cdot 5$

36. $11 + 7 \cdot 6$ **37.** $25.2 - 12.6 \div 4.2$ **38.** $12.4 - 9.3 \div 3.1$

39. $\dfrac{1}{4} \cdot \dfrac{2}{3} + \dfrac{2}{5} \cdot \dfrac{11}{3}$ **40.** $\dfrac{9}{4} \cdot \dfrac{2}{3} + \dfrac{4}{5} \cdot \dfrac{5}{3}$ **41.** $9 \cdot 4 - 8 \cdot 3$

42. $11 \cdot 4 + 10 \cdot 3$

43. $20 - 4 \cdot 3 + 5$

44. $18 - 7 \cdot 2 + 6$

45. $10 + 40 \div 5 \cdot 2$

46. $12 + 64 \div 8 - 4$

47. $18 - 2(3 + 4)$

48. $30 - 3(4 + 2)$

49. $3(4 + 2) + 8 \cdot 3$

50. $9(1 + 7) + 2 \cdot 5$

51. $18 - 4^2 + 3$

52. $22 - 2^3 + 9$

53. $2 + 3[5 + 4(2)]$

54. $5 + 4[1 + 7(3)]$

55. $5[3 + 4(2^2)]$

56. $6[2 + 8(3^3)]$

▶ 57. $3^2[(11 + 3) - 4]$

58. $4^2[(13 + 4) - 8]$

59. $\dfrac{6(3^2 - 1) + 8}{8 - 2^2}$

60. $\dfrac{2(8^2 - 4) + 8}{29 - 3^3}$

61. $\dfrac{4(6 + 2) + 8(8 - 3)}{6(4 - 2) - 2^2}$

62. $\dfrac{6(5 + 1) - 9(1 + 1)}{5(8 - 6) - 2^3}$

Extending Skills   *Insert one pair of parentheses in each expression so that the given value results when the operations are performed.*

63. $3 \cdot 6 + 4 \cdot 2$
   $= 60$

64. $2 \cdot 8 - 1 \cdot 3$
   $= 42$

65. $10 - 7 - 3$
   $= 6$

66. $15 - 10 - 2$
   $= 7$

67. $8 + 2^2$
   $= 100$

68. $4 + 2^2$
   $= 36$

First simplify both sides of each inequality. Then determine whether the given statement is true or false. **See Examples 2–4.**

▶ 69. $9 \cdot 3 - 11 \le 16$

70. $6 \cdot 5 - 12 \le 18$

71. $5 \cdot 11 + 2 \cdot 3 \le 60$

72. $9 \cdot 3 + 4 \cdot 5 \ge 48$

73. $0 \ge 12 \cdot 3 - 6 \cdot 6$

74. $10 \le 13 \cdot 2 - 15 \cdot 1$

75. $45 \ge 2[2 + 3(2 + 5)]$

76. $55 \ge 3[4 + 3(4 + 1)]$

77. $[3 \cdot 4 + 5(2)] \cdot 3 > 72$

78. $2 \cdot [7 \cdot 5 - 3(2)] \le 58$

79. $\dfrac{3 + 5(4 - 1)}{2 \cdot 4 + 1} \ge 3$

80. $\dfrac{7(3 + 1) - 2}{3 + 5 \cdot 2} \le 2$

81. $3 \ge \dfrac{2(5 + 1) - 3(1 + 1)}{5(8 - 6) - 4 \cdot 2}$

82. $7 \le \dfrac{3(8 - 3) + 2(4 - 1)}{9(6 - 2) - 11(5 - 2)}$

Write each statement in words, and decide whether it is true or false. **See Examples 4 and 5.**

83. $5 < 17$

84. $8 < 12$

85. $5 \ne 8$

86. $6 \ne 9$

87. $7 \ge 14$

88. $6 \ge 12$

89. $15 \le 15$

90. $21 \le 21$

91. $\dfrac{1}{3} = \dfrac{3}{10}$

92. $\dfrac{10}{6} = \dfrac{3}{2}$

93. $2.5 > 2.50$

94. $1.80 > 1.8$

Write each word statement in symbols. **See Example 5.**

95. Fifteen is equal to five plus ten.

96. Twelve is equal to twenty minus eight.

▶ 97. Nine is greater than five minus four.

98. Ten is greater than six plus one.

99. Sixteen is not equal to nineteen.

100. Three is not equal to four.

101. One-half is less than or equal to two-fourths.

102. One-third is less than or equal to three-ninths.

Write each statement with the inequality symbol reversed while keeping the same meaning. See Example 6.

103. $5 < 20$

104. $30 > 9$

105. $\dfrac{4}{5} > \dfrac{3}{4}$

106. $\dfrac{5}{4} < \dfrac{3}{2}$

107. $2.5 \geq 1.3$

108. $4.1 \leq 5.3$

One way to measure a person's cardiofitness is to calculate how many METs, or metabolic units, he or she can reach at peak exertion. One MET is the amount of energy used when sitting quietly. To calculate ideal METs, we can use the following expressions.

$14.7 - \text{age} \cdot 0.13$ For women

$14.7 - \text{age} \cdot 0.11$ For men

(Source: New England Journal of Medicine.)

109. A 40-yr-old woman wishes to calculate her ideal MET.

  (a) Write the expression, using her age.

  (b) Calculate her ideal MET. (*Hint:* Use the rules for order of operations.)

  (c) Researchers recommend that a person reach approximately 85% of his or her MET when exercising. Calculate 85% of the ideal MET from part (b). Then refer to the following table. What activity listed in the table can the woman do that is approximately this value?

Activity	METs	Activity	METs
Golf (with cart)	2.5	Skiing (water or downhill)	6.8
Walking (3 mph)	3.3	Swimming	7.0
Mowing lawn (power mower)	4.5	Walking (5 mph)	8.0
Ballroom or square dancing	5.5	Jogging	10.2
Cycling	5.7	Skipping rope	12.0

Source: Harvard School of Public Health.

  (d) Repeat parts (a)–(c) for a 55-yr-old man.

110. Repeat parts (a)–(c) of **Exercise 109** using your age and gender.

The table shows the number of pupils per teacher in U.S. public schools in selected states.

111. Which states had a number greater than 13.8?

112. Which states had a number that was at most 14.7?

113. Which states had a number not less than 13.8?

114. Which states had a number less than 13.0?

State	Pupils per Teacher
Alaska	16.2
Texas	14.7
California	24.1
Wyoming	12.5
Maine	12.3
Idaho	17.6
Missouri	13.8

Source: National Center for Education Statistics.

STUDY SKILLS

Taking Lecture Notes

Study the set of sample math notes given here.

- **Include the date and title** of the day's lecture topic.

- **Include definitions,** written here in parentheses—don't trust your memory.

- **Skip lines and write neatly** to make reading easier.

- **Emphasize direction words** (like *simplify*) with their explanations.

- **Mark important concepts with stars, underlining, etc.**

- **Use two columns,** which allows an example and its explanation to be close together.

- **Use brackets and arrows** to clearly show steps, related material, etc.

With a partner or in a small group, compare lecture notes. Answer each question.

1. What are you doing to show main points in your notes (such as boxing, using stars, etc.)?

2. In what ways do you set off explanations from worked problems and subpoints (such as indenting, using arrows, circling, etc.)?

3. What new ideas did you learn by examining your class-mates' notes?

4. What new techniques will you try in your notes?

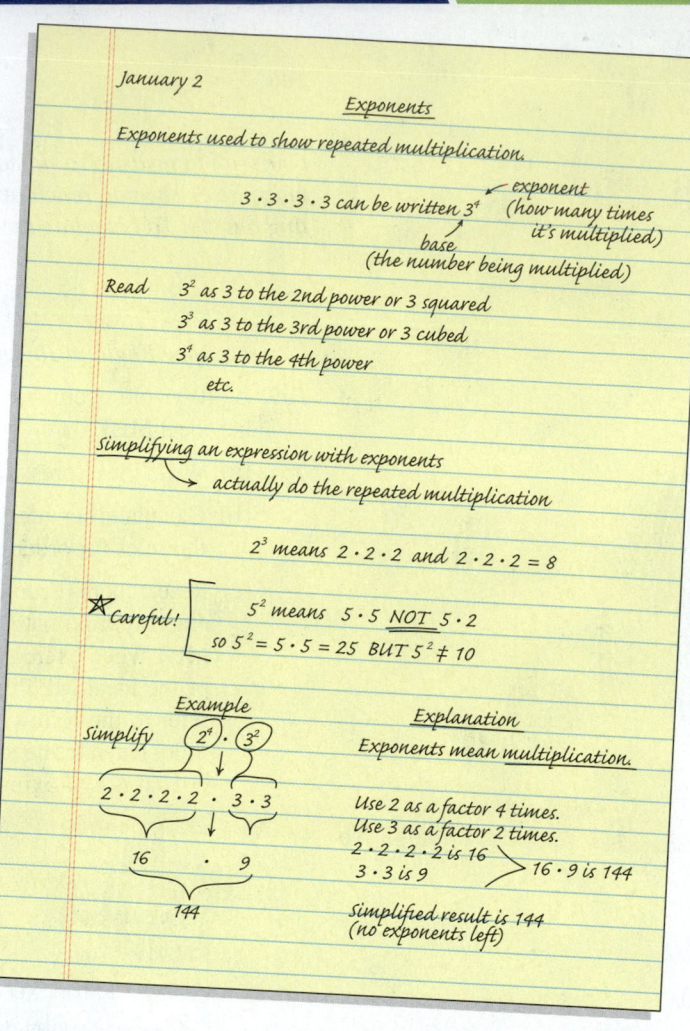

1.2 Variables, Expressions, and Equations

A **constant** is a fixed, unchanging number. A **variable** is a symbol, usually a letter, used to represent an unknown number.

$$5, \quad \frac{3}{4}, \quad 8\frac{1}{2}, \quad 10.8 \qquad \text{Constants} \qquad \Big| \qquad a, \quad x, \quad y, \quad z \qquad \text{Variables}$$

An **algebraic expression** is a sequence of constants, variables, operation symbols, and/or grouping symbols formed according to the rules of algebra.

$$x + 5, \quad 2m - 9, \quad 8p^2 + 6(p - 2) \qquad \text{Algebraic expressions}$$

$2m$ means $2 \cdot m$, the product of 2 and m.

$6(p - 2)$ means the product of 6 and $p - 2$.

VOCABULARY
- ☐ constant
- ☐ variable
- ☐ algebraic expression
- ☐ equation
- ☐ solution
- ☐ set
- ☐ element

 **NOW TRY EXERCISE 1**

Find the value of each expression for $x = 6$.

(a) $9x$ **(b)** $4x^2$

 **NOW TRY EXERCISE 2**

Find the value of each expression for $x = 4$ and $y = 7$.

(a) $3x + 4y$ **(b)** $\dfrac{6x - 2y}{2y - 9}$

(c) $4x^2 - y^2$

OBJECTIVE 1   Evaluate algebraic expressions, given values for the variables.

To *evaluate* an expression means to find its *value*. An algebraic expression can have different numerical values for different values of the variables.

EXAMPLE 1   Evaluating Algebraic Expressions

Find the value of each expression for $x = 5$.

(a) $8x$ ⟵ Multiplication is understood.

$= 8 \cdot x$

$= 8 \cdot 5$ Let $x = 5$.

$= 40$ Multiply.

(b) $3x^2$

$= 3 \cdot x^2$ $\boxed{5^2 = 5 \cdot 5}$

$= 3 \cdot 5^2$ Let $x = 5$.

$= 3 \cdot 25$ Square 5.

$= 75$ Multiply.   **NOW TRY**

⚠ **CAUTION**

$3x^2$ means $3 \cdot x^2$,   **not** $3x \cdot 3x$.   See **Example 1(b)**.

Unless parentheses are used, the exponent refers only to the variable or constant just before it. We use parentheses to write $3x \cdot 3x$ with exponents as $(3x)^2$.

EXAMPLE 2   Evaluating Algebraic Expressions

Find the value of each expression for $x = 5$ and $y = 3$.

(a) $2x + 7y$ ⟵ We could use parentheses and write $2(5) + 7(3)$.

 Follow the rules for order of operations.

$= 2 \cdot 5 + 7 \cdot 3$ Let $x = 5$ and $y = 3$.

$= 10 + 21$ Multiply.

$= 31$ Add.

(b) $\dfrac{9x - 8y}{2x - y}$

$= \dfrac{9 \cdot 5 - 8 \cdot 3}{2 \cdot 5 - 3}$ Let $x = 5$ and $y = 3$.

$= \dfrac{45 - 24}{10 - 3}$ Multiply.

$= \dfrac{21}{7}$ Subtract.

$= 3$ Divide.

(c) $x^2 - 2y^2$ $\boxed{3^2 = 3 \cdot 3}$

$= 5^2 - 2 \cdot 3^2$ Let $x = 5$ and $y = 3$.

$\boxed{5^2 = 5 \cdot 5}$

$= 25 - 2 \cdot 9$ Apply the exponents.

$= 25 - 18$ Multiply.

$= 7$ Subtract.   **NOW TRY**

NOW TRY ANSWERS

1. **(a)** 54   **(b)** 144
2. **(a)** 40   **(b)** 2   **(c)** 15

**NOW TRY
EXERCISE 3**

Write each word phrase as an algebraic expression, using x as the variable.

(a) The sum of a number and 10

(b) A number divided by 7

(c) The product of 3 and the difference of 9 and a number

OBJECTIVE 2 Translate word phrases to algebraic expressions.

EXAMPLE 3   Using Variables to Write Word Phrases as Algebraic Expressions

Write each word phrase as an algebraic expression, using x as the variable.

(a) The sum of a number and 9

$$x + 9, \quad \text{or} \quad 9 + x \qquad \text{"Sum" is the answer to an addition problem.}$$

(b) 7 minus a number

$$7 - x \qquad \text{"Minus" indicates subtraction.}$$

The expression $x - 7$ is incorrect. We cannot subtract in either order and obtain the same result.

(c) A number subtracted from 12

$$12 - x \quad \boxed{\text{Be careful with order.}}$$

Compare this result with "12 subtracted from a number," which is $x - 12$.

(d) The product of 11 and a number

$$11 \cdot x, \quad \text{or} \quad 11x$$

(e) 5 divided by a number

$$5 \div x, \quad \text{or} \quad \frac{5}{x} \quad \boxed{\tfrac{x}{5} \text{ is \textit{not} correct here.}}$$

(f) The product of 2 and the difference of a number and 8

We are multiplying 2 times "something." This "something" is the difference of a number and 8, written $x - 8$. We use parentheses around this difference.

$$2 \cdot (x - 8), \quad \text{or} \quad 2(x - 8) \quad \boxed{\begin{array}{l} 8 - x, \text{ which means the difference} \\ \text{of 8 and a number, is \textit{not} correct.} \end{array}}$$ **NOW TRY**

OBJECTIVE 3 Identify solutions of equations.

An **equation** is a statement that two algebraic expressions are equal. *An equation always includes the equality symbol, =.*

$$\left. \begin{array}{lll} x + 4 = 11, & 2y = 16, & 4p + 1 = 25 - p, \\[2mm] \dfrac{3}{4}x + \dfrac{1}{2} = 0, & z^2 = 4, & 4(m - 0.5) = 2m \end{array} \right\} \text{Equations}$$

To **solve an equation** means to find the value of the variable that makes the equation true. Such a value of the variable is a **solution** of the equation.

EXAMPLE 4   Deciding Whether a Number Is a Solution of an Equation

Decide whether each equation has the given number as a solution.

(a) $5p + 1 = 36$; 7

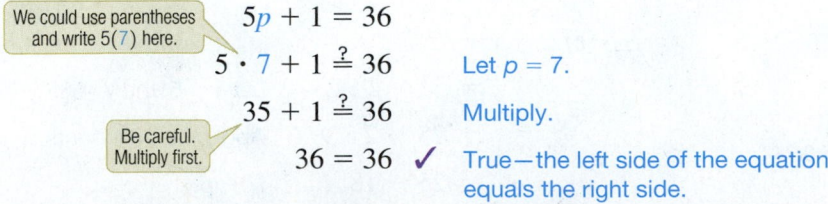

$$5p + 1 = 36$$
$$5 \cdot 7 + 1 \overset{?}{=} 36 \qquad \text{Let } p = 7.$$
$$35 + 1 \overset{?}{=} 36 \qquad \text{Multiply.}$$
$$36 = 36 \quad \checkmark \qquad \text{True—the left side of the equation equals the right side.}$$

$\boxed{\text{We could use parentheses and write } 5(7) \text{ here.}}$

$\boxed{\text{Be careful. Multiply first.}}$

NOW TRY ANSWERS
3. (a) $x + 10$, or $10 + x$ **(b)** $\frac{x}{7}$
(c) $3(9 - x)$

The number 7 is a solution of the equation.

NOW TRY
EXERCISE 4
Decide whether the equation has the given number as a solution.

$8k + 5 = 61; \quad 7$

(b) $9m - 6 = 32; \quad 4$

$$9m - 6 = 32$$
$$9 \cdot 4 - 6 \stackrel{?}{=} 32 \qquad \text{Let } m = 4.$$
$$36 - 6 \stackrel{?}{=} 32 \qquad \text{Multiply.}$$
$$30 = 32 \qquad \text{False—the left side does } not \text{ equal the right side.}$$

The number 4 is not a solution of the equation.

NOW TRY

OBJECTIVE 4 Identify solutions of equations from a set of numbers.

A **set** is a collection of objects. In mathematics, these objects are usually numbers. The objects that belong to a set are its **elements.** They are written between **braces { }.**

$$\{1, 2, 3, 4, 5\} \leftarrow \text{The set containing the elements 1, 2, 3, 4, and 5}$$

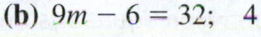

NOW TRY
EXERCISE 5
Write the word statement as an equation. Use x as the variable. Then find the solution of the equation from the set $\{0, 2, 4, 6, 8, 10\}$.

The sum of a number and nine is equal to the difference of 25 and the number.

EXAMPLE 5 Finding a Solution from a Given Set

Write each word statement as an equation. Use x as the variable. Then find the solution of the equation from the following set.

$$\{0, 2, 4, 6, 8, 10\}$$

(a) The sum of a number and four is six.

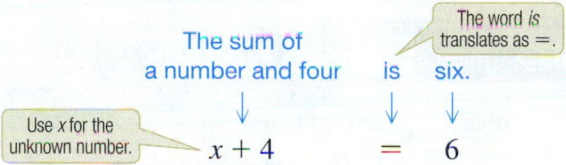

One by one, mentally substitute each number from the given set $\{0, 2, 4, 6, 8, 10\}$ in $x + 4 = 6$. Because $2 + 4 = 6$ is true, 2 is the only solution.

(b) Nine more than five times a number is 49.

Start with $5x$, and then add 9 to it.
The word *is* translates as $=$.

$$5x + 9 \qquad = \qquad 49 \qquad 5 \cdot x = 5x$$

Substitute each of the given numbers. The solution is 8 because

$$5 \cdot 8 + 9 = 49 \text{ is true.}$$

(c) The sum of a number and 12 is equal to four times the number.

The sum of a number and 12
is equal to
four times the number.

$$x + 12 \qquad = \qquad 4x \qquad 4 \cdot x = 4x$$

Substituting each of the given numbers leads to a true statement only for $x = 4$ because

NOW TRY ANSWERS
4. yes
5. $x + 9 = 25 - x; \ 8$

$$4 + 12 = 4(4) \text{ is true.}$$

NOW TRY

OBJECTIVE 5 Distinguish between *equations* and *expressions*.

To distinguish between equations and expressions, remember the following.

An equation is a sentence—it has something on the left side, an = symbol, and something on the right side.

An expression is a phrase that represents a number.

$$\underbrace{4x + 5}_{\text{Left side}} \overset{\uparrow}{=} \underbrace{9}_{\text{Right side}} \qquad\qquad 4x + 5$$

Equation (to solve)   Expression (to simplify or evaluate)

 NOW TRY EXERCISE 6

Decide whether each of the following is an *equation* or an *expression*.

(a) $2x + 5 = 6$

(b) $2x + 5 - 6$

EXAMPLE 6   Distinguishing between Equations and Expressions

Decide whether each of the following is an *equation* or an *expression*.

(a) $2x - 3$ — Ask, "Is there an equality symbol?" The answer is no, so this is an expression.

(b) $2x - 3 = 8$ — Because there is an equality symbol with something on either side of it, this is an equation.

(c) $5x^2 + 2y^2$ — There is no equality symbol. This is an expression.

NOW TRY

NOW TRY ANSWERS

6. (a) equation   **(b)** expression

1.2 Exercises

FOR EXTRA HELP   ▶   MyMathLab®

▶ *Complete solution available in MyMathLab*

Concept Check   *Choose the letter(s) of the correct response.*

1. The expression $8x^2$ means _____.

   A. $8 \cdot x \cdot 2$ **B.** $8 \cdot x \cdot x$ **C.** $8 + x^2$ **D.** $8x \cdot 8x$

2. If $x = 2$ and $y = 1$, then the value of xy is _____.

   A. $\dfrac{1}{2}$ **B.** 1 **C.** 2 **D.** 3

3. The sum of 15 and a number x is represented by _____.

   A. $15 + x$ **B.** $15 - x$ **C.** $x - 15$ **D.** $15x$

4. Which of the following are expressions? Which are equations?

   A. $6x = 7$ **B.** $6x + 7$ **C.** $6x - 7$ **D.** $6x - 7 = 0$

Give a short explanation.

5. Explain why $2x^3$ is not the same as $2x \cdot 2x \cdot 2x$.

6. Why are "7 less than a number" and "7 is less than a number" translated differently?

7. When evaluating the expression $5x^2$ for $x = 4$, explain why 4 must be squared *before* multiplying by 5.

8. Suppose that the directions on a test read "*Solve each expression.*" How could we politely correct the person who wrote these directions?

Find the value of each expression for **(a)** $x = 4$ *and* **(b)** $x = 6$. *See Example 1.*

9. $x + 7$ **10.** $x - 3$ **11.** $4x$ **12.** $6x$ ▶ **13.** $4x^2$

14. $5x^2$ **15.** $\dfrac{x + 1}{3}$ **16.** $\dfrac{x - 2}{5}$ **17.** $\dfrac{3x - 5}{2x}$ **18.** $\dfrac{4x - 1}{3x}$

19. $3x^2 + x$ **20.** $2x + x^2$ **21.** $6.459x$ **22.** $3.275x$

Find the value of each expression for **(a)** $x = 2$ *and* $y = 1$ *and* **(b)** $x = 1$ *and* $y = 5$. *See Example 2.*

▶ **23.** $8x + 3y + 5$ **24.** $4x + 2y + 7$ **25.** $3(x + 2y)$ **26.** $2(2x + y)$

27. $x + \dfrac{4}{y}$ **28.** $y + \dfrac{8}{x}$ **29.** $\dfrac{x}{2} + \dfrac{y}{3}$ **30.** $\dfrac{x}{5} + \dfrac{y}{4}$

31. $\dfrac{2x + 4y - 6}{5y + 2}$ **32.** $\dfrac{4x + 3y - 1}{x}$ **33.** $2y^2 + 5x$ **34.** $6x^2 + 4y$

35. $\dfrac{3x + y^2}{2x + 3y}$ **36.** $\dfrac{x^2 + 1}{4x + 5y}$ **37.** $0.841x^2 + 0.32y^2$ **38.** $0.941x^2 + 0.25y^2$

Write each word phrase as an algebraic expression, using x as the variable. **See Example 3.**

▶ **39.** Twelve times a number   **40.** Fifteen times a number

41. Nine added to a number   **42.** Six added to a number

43. Four subtracted from a number   **44.** Seven subtracted from a number

45. A number subtracted from seven   **46.** A number subtracted from four

47. The difference of a number and 8   **48.** The difference of 8 and a number

49. 18 divided by a number   **50.** A number divided by 18

51. The product of 6 and four less than a number   **52.** The product of 9 and five more than a number

Decide whether each equation has the given number as a solution. **See Example 4.**

53. $4m + 2 = 6$;   1   **54.** $2r + 6 = 8$;   1

▶ **55.** $2y + 3(y - 2) = 14$;   3   **56.** $6x + 2(x + 3) = 14$;   2

57. $6p + 4p + 9 = 11$; $\dfrac{1}{5}$ **58.** $2x + 3x + 8 = 20$; $\dfrac{12}{5}$

59. $3r^2 - 2 = 46$;   4   **60.** $2x^2 + 1 = 19$;   3

61. $\dfrac{3}{8}x + \dfrac{1}{4} = 1$;   2   **62.** $\dfrac{7}{10}x + \dfrac{1}{2} = 4$;   5

63. $0.5(x - 4) = 80$;   20   **64.** $0.2(x - 5) = 70$;   40

Write each word statement as an equation. Use x as the variable. Then find the solution of the equation from the set $\{2, 4, 6, 8, 10\}$. *See Example 5.*

▶ **65.** The sum of a number and 8 is 18.   **66.** A number minus three equals 1.

67. One more than twice a number is 5.   **68.** The product of a number and 3 is 6.

69. Sixteen minus three-fourths of a number is 13.

70. The sum of six-fifths of a number and 2 is 14.

71. Three times a number is equal to 8 more than twice the number.

72. Twelve divided by a number equals $\dfrac{1}{3}$ times that number.

Decide whether each of the following is an equation *or an* expression. ***See Example 6.***

 73. $3x + 2(x - 4)$ **74.** $8y - (3y + 5)$ **75.** $7t + 2(t + 1) = 4$

76. $9r + 3(r - 4) = 2$ **77.** $x + y = 9$ **78.** $x + y - 9$

A ***mathematical model*** *is an equation that describes the relationship between two quantities. For example, the life expectancy at birth of Americans can be approximated by the equation*

$$y = 0.180x - 283,$$

where x is a year between 1960 and 2010 and y is age in years. (Source: Centers for Disease Control and Prevention.)

Use this model to approximate life expectancy (to the nearest year) in each of the following years.

79. 1960   **80.** 1975

81. 1995   **82.** 2010

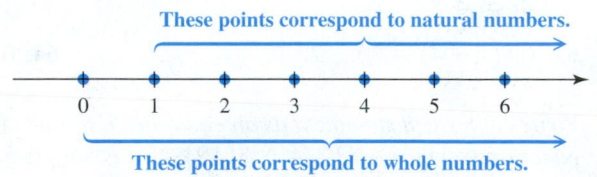

1.3   Real Numbers and the Number Line

OBJECTIVES

1. Classify numbers and graph them on number lines.
2. Tell which of two real numbers is less than the other.
3. Find the additive inverse of a real number.
4. Find the absolute value of a real number.
5. Interpret meanings of real numbers from a table of data.

VOCABULARY

☐ natural (counting) numbers
☐ whole numbers
☐ number line
☐ integers
☐ signed numbers
☐ rational numbers
☐ graph
☐ coordinate
☐ irrational numbers
☐ real numbers
☐ additive inverse (opposite)
☐ absolute value

OBJECTIVE 1 Classify numbers and graph them on number lines.

The set of numbers used for counting is the *natural numbers*. The set of *whole numbers* includes 0 with the natural numbers.

Natural Numbers and Whole Numbers

$\{1, 2, 3, 4, 5, \ldots\}$ is the set of **natural numbers** (or **counting numbers**).

$\{0, 1, 2, 3, 4, 5 \ldots\}$ is the set of **whole numbers.**

We can represent numbers on a **number line** like the one in **FIGURE 1**.

These points correspond to natural numbers.

| 0 | 1 | 2 | 3 | 4 | 5 | 6 |

These points correspond to whole numbers.

FIGURE 1

To draw a number line, choose any point on the line and label it 0. Then choose any point to the right of 0 and label it 1. Use the distance between 0 and 1 as the scale to locate, and then label, other points.

The natural numbers are located to the right of 0 on the number line. For each natural number, we can place a corresponding number to the left of 0, labeling the points -1, -2, -3, and so on, as shown in **FIGURE 2** on the next page. Each is the **opposite,** or **negative,** of a natural number. The natural numbers, their opposites, and 0 form the set of *integers*.

Integers

$\{\ldots, -3, -2, -1, 0, 1, 2, 3, \ldots\}$ is the set of **integers.**

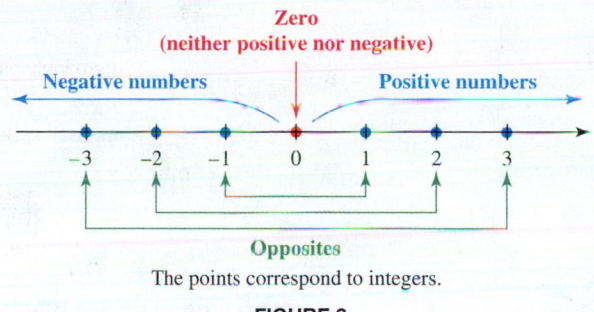

The points correspond to integers.

FIGURE 2

Positive numbers and *negative numbers* are **signed numbers.**

NOW TRY
EXERCISE 1

Use an integer to express the number in boldface italics in the following statement.

At its deepest point, the floor of West Okoboji Lake sits **136** ft below the water's surface. (*Source:* www.watersafetycouncil.org)

EXAMPLE 1   Using Negative Numbers

Use an integer to express the number in boldface italics in each statement.

(a) The lowest Fahrenheit temperature ever recorded was **129°** below zero at Vostok, Antarctica, on July 21, 1983. (*Source: World Almanac and Book of Facts.*)

Use -129 because "below zero" indicates a negative number.

(b) General Motors had a loss of about $**399** billion in 2013. (*Source: The Wall Street Journal.*)

A loss indicates a negative "profit," here -399.   **NOW TRY**

Fractions, reviewed in **Section R.1,** are examples of *rational numbers.*

Rational Numbers

$\{x \mid x$ is a quotient of two integers, with denominator not $0\}$ is the set of **rational numbers.**

(Read the part in the braces as "the set of all numbers x such that x is a quotient of two integers, with denominator not 0.")

NOTE   The set symbolism used in the definition of rational numbers,

$\{x \mid x$ has a certain property$\}$,

is **set-builder notation.** We use this notation when it is not possible to list all the elements of a set.

Because any number that can be written as the quotient of two integers (that is, as a fraction) is a rational number, *all integers, mixed numbers, terminating (or ending) decimals, and repeating decimals are rational.*

▼ Rational Numbers

Rational Number	Equivalent Quotient of Two Integers
-5	$\frac{-5}{1}$ (means $-5 \div 1$)
$1\frac{3}{4}$	$\frac{7}{4}$ (means $7 \div 4$)
0.23 (terminating decimal)	$\frac{23}{100}$ (means $23 \div 100$)
$0.3333\ldots$, or $0.\overline{3}$ (repeating decimal)	$\frac{1}{3}$ (means $1 \div 3$)
4.7	$\frac{47}{10}$ (means $47 \div 10$)

To **graph** a number, we place a dot on a number line at the point that corresponds to the number. The number is the **coordinate** of the point.

NOW TRY EXERCISE 2

Graph each rational number on a number line.

$-3, \quad \dfrac{17}{8}, \quad -2.75, \quad 1\dfrac{1}{2}, \quad -\dfrac{3}{4}$

EXAMPLE 2   Graphing Rational Numbers

Graph each rational number on a number line.

$$-\frac{3}{2}, \quad -\frac{2}{3}, \quad 0.5, \quad 1\frac{1}{3}, \quad \frac{23}{8}, \quad 3.25, \quad 4$$

To locate the improper fractions on a number line, write them as mixed numbers or decimals. The graph is shown in **FIGURE 3**.

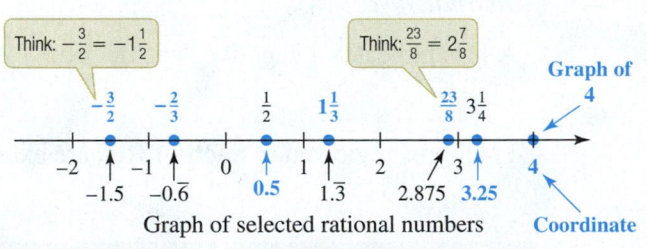

Graph of selected rational numbers

FIGURE 3

Think of the graph of a set of numbers as a picture of the set.

NOW TRY

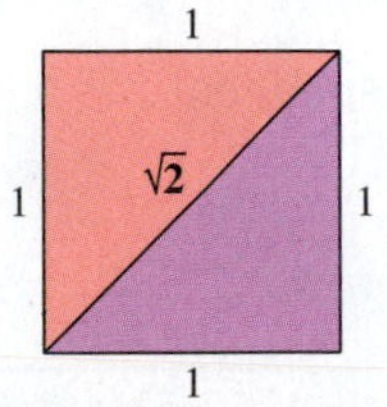

Not all numbers are rational. For example, the square root of 2, written $\sqrt{2}$, cannot be written as a quotient of two integers. Because of this, $\sqrt{2}$ is an *irrational number*. (See **FIGURE 4**.)

This square has diagonal of length $\sqrt{2}$. The number $\sqrt{2}$ is an irrational number.

FIGURE 4

Irrational Numbers

$\{x \,|\, x$ is a nonrational number represented by a point on a number line$\}$ is the set of **irrational numbers.**

The decimal form of an irrational number neither terminates nor repeats.

Both rational and irrational numbers can be represented by points on a number line and together form the set of *real numbers*. See **FIGURE 5** on the next page.

NOW TRY ANSWER

2.

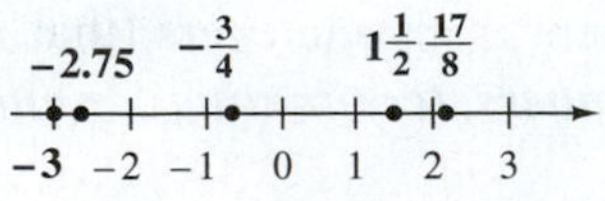

Real Numbers

$\{x \mid x$ is a rational or an irrational number$\}$ is the set of **real numbers.***

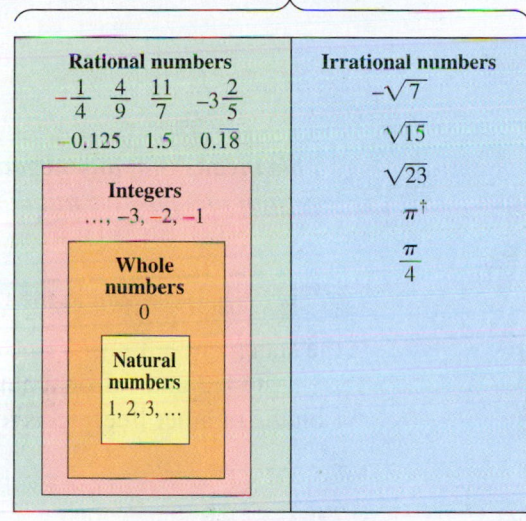

FIGURE 5

**NOW TRY
EXERCISE 3**

List the numbers in the following set that belong to each set of numbers.

$$\left\{-7, -\tfrac{4}{5}, 0, \sqrt{3}, 2.7, \pi, 13\right\}$$

(a) Whole numbers

(b) Integers

(c) Rational numbers

(d) Irrational numbers

EXAMPLE 3   Determining Whether a Number Belongs to a Set

List the numbers in the following set that belong to each set of numbers.

$$\left\{-5, -\tfrac{2}{3}, 0, 0.\overline{6}, \sqrt{2}, 3\tfrac{1}{4}, 5, 5.8\right\}$$

(a) Natural numbers: 5

(b) Whole numbers: 0 and 5
The whole numbers consist of the natural (counting) numbers and 0.

(c) Integers: -5, 0, and 5

(d) Rational numbers: $-5, -\tfrac{2}{3}, 0, 0.\overline{6} \left(\text{or } \tfrac{2}{3}\right), 3\tfrac{1}{4} \left(\text{or } \tfrac{13}{4}\right), 5,$ and $5.8 \left(\text{or } \tfrac{58}{10}\right)$
Each of these numbers can be written as the quotient of two integers.

(e) Irrational numbers: $\sqrt{2}$

(f) Real numbers: All the numbers in the set are real numbers.   NOW TRY

OBJECTIVE 2   Tell which of two real numbers is less than the other.

Given any two different positive integers, we can determine which number is less than the other. Positive numbers decrease as the corresponding points on a number line go to the left. For example,

$8 < 12$ because 8 is to the left of 12 on a number line.

This ordering is extended to all real numbers by definition.

NOW TRY ANSWERS

3. (a) 0, 13
   (b) -7, 0, 13
   (c) $-7, -\tfrac{4}{5}, 0, 2.7, 13$
   (d) $\sqrt{3}, \pi$

---

*An example of a number that is not a real number is the square root of a negative number, such as $\sqrt{-5}$.

†The value of the irrational number π (pi) is approximately 3.141592654. The decimal digits continue forever with no repeated pattern.

Ordering of Real Numbers

For any two real numbers a and b, **a is less than b** if a lies to the left of b on a number line. See **FIGURE 6**.

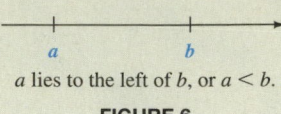

a lies to the left of b, or $a < b$.
FIGURE 6

This means that any negative number is less than 0, and any negative number is less than any positive number. Also, 0 is less than any positive number.

NOW TRY
EXERCISE 4
Determine whether the statement is *true* or *false*.

$$-8 \leq -9$$

EXAMPLE 4   Determining the Order of Real Numbers

Is the statement $-3 < -1$ *true* or *false*?

Locate -3 and -1 on a number line. See **FIGURE 7**. Because -3 lies to the left of -1 on the number line, -3 is less than -1. The statement $-3 < -1$ is true.

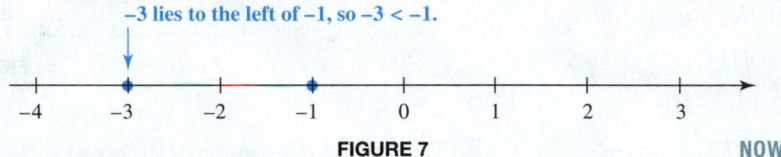

−3 lies to the left of −1, so −3 < −1.

FIGURE 7   NOW TRY

Ordering of Real Numbers

For any two real numbers a and b, **a is greater than b** if a lies to the right of b on a number line. See **FIGURE 8**.

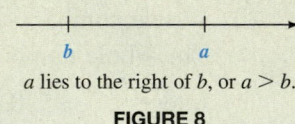

a lies to the right of b, or $a > b$.
FIGURE 8

In **FIGURE 7** above, $-1 > -3$ because -1 lies to the right of -3 on the number line.

OBJECTIVE 3   Find the additive inverse of a real number.

By a property of the real numbers, for any real number x (except 0), there is exactly one number on a number line the same distance from 0 as x, but on the *opposite* side of 0. See **FIGURE 9**. Such pairs of numbers are *additive inverses,* or *opposites,* of each other.

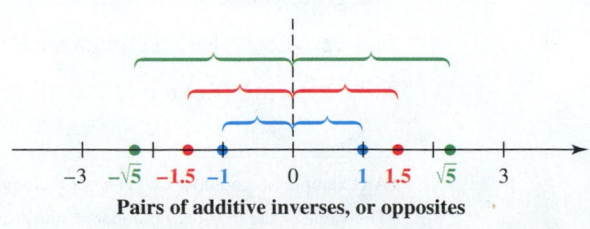

Pairs of additive inverses, or opposites

FIGURE 9

NOW TRY ANSWER
4. false

Additive Inverse

The **additive inverse** of a number x is the number that is the same distance from 0 on a number line as x, but on the *opposite* side of 0.

▼ **Additive Inverses**

Number	Additive Inverse
7	-7
-3	$-(-3)$, or 3
0	0
19	-19
$-\frac{2}{3}$	$\frac{2}{3}$
0.52	-0.52

The additive inverse of a nonzero number is found by changing the sign of the number.

A nonzero number and its additive inverse have opposite signs.

We indicate the additive inverse of a number by writing the symbol $-$ in front of the number. For example, the additive inverse of 7 is written -7 (read "negative 7"). We could write the additive inverse of -3 as $-(-3)$, but we know that 3 is the opposite of -3. Because a number can have only one additive inverse, 3 and $-(-3)$ must represent the same number.

$$-(-3) = 3$$

This idea can be generalized.

Double Negative Rule

For any real number x, the following holds.

$$-(-x) = x$$

OBJECTIVE 4   **Find the absolute value of a real number.**

Because additive inverses are the same distance from 0 on a number line, a number and its additive inverse have the same *absolute value*. The **absolute value** of a real number x, written $|x|$ and read **"the absolute value of x,"** can be defined as the distance between 0 and the number on a number line.

$|2| = 2$ The distance between 2 and 0 on a number line is 2 units.

$|-2| = 2$ The distance between -2 and 0 on a number line is also 2 units.

Distance is a physical measurement, which is never negative. *Therefore, the absolute value of a number is never negative.*

Absolute Value

For any real number x, the absolute value of x is defined as follows.

$$|x| = \begin{cases} x & \text{if } x \geq 0 \\ -x & \text{if } x < 0 \end{cases}$$

By this definition, if x is a positive number or 0, then its absolute value is x itself. For example, since 8 is a positive number,

$$|8| = 8.$$

If x is a negative number, then its absolute value is the additive inverse of x.

$|-8| = -(-8) = 8$ The additive inverse of -8 is 8.

⚠ CAUTION

$$|x| = \begin{cases} x & \text{if } x \geq 0 \\ -x & \text{if } x < 0 \end{cases}$$ Definition of absolute value

The "$-x$" in the second part of the definition of absolute value does *not* represent a negative number. Because x is negative in the second part, it follows that $-x$ represents the *opposite* of a negative number—that is, a positive number. ***The absolute value of a number is never negative.***

NOW TRY
EXERCISE 5

Find each absolute value.

(a) $|4|$ **(b)** $|-4|$

(c) $-|-4|$

EXAMPLE 5   Finding Absolute Values

Find each absolute value.

(a) $|0| = 0$

(b) $|5| = 5$

(c) $|-5| = -(-5) = 5$

(d) $-|5| = -(5) = -5$

(e) $-|-5| = -(5) = -5$

(f) $|8 - 2| = |6| = 6$

(g) $-|8 - 2| = -|6| = -6$

Absolute value bars are grouping symbols. In parts (f) and (g), we perform any operations inside the absolute value bars *before* finding the absolute value.

NOW TRY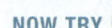

OBJECTIVE 5   Interpret meanings of real numbers from a table of data.

NOW TRY
EXERCISE 6

In the table for **Example 6,** which category represents a decrease for both years?

EXAMPLE 6   Interpreting Data

The Consumer Price Index (CPI) measures the average change in prices of goods and services purchased by urban consumers in the United States. The table shows the percent change in the CPI for selected categories of goods and services from 2009 to 2010 and from 2010 to 2011. Use the table to answer each question.

Category	Change from 2009 to 2010	Change from 2010 to 2011
Appliances	−4.5	−1.2
Education	4.4	4.2
Gasoline	18.4	26.4
Housing	−0.4	1.3
Medical care	3.4	3.0

Source: U.S. Bureau of Labor Statistics.

(a) Which category in which year represents the greatest percent decrease?

We must find the negative number with the greatest absolute value. The number that satisfies this condition is -4.5, so the greatest percent decrease was shown by appliances from 2009 to 2010.

(b) Which category in which year represents the least change?

We must find the number (either positive, negative, or zero) with the least absolute value. From 2009 to 2010, housing showed the least change, a decrease of 0.4%.

NOW TRY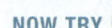

NOW TRY ANSWERS

5. (a) 4   **(b)** 4   **(c)** −4

6. appliances

1.3 Exercises

FOR EXTRA HELP

  MyMathLab®

▶ *Complete solution available in MyMathLab*

Concept Check   *Complete each statement.*

1. The number _____ is a whole number, but not a natural number.

2. The natural numbers, their additive inverses, and 0 form the set of _____ .

3. The additive inverse of every negative number is a (*negative / positive*) number.

4. If x and y are real numbers with $x > y$, then x lies to the (*left / right*) of y on a number line.

5. A rational number is the _____ of two integers, with the _____ not equal to 0.

6. Decimal numbers that neither terminate nor repeat are _____ numbers.

7. *Concept Check*   Match each expression in Column I with its value in Column II. Choices in Column II may be used once, more than once, or not at all.

I	II
(a) $\lvert -9 \rvert$	**A.** 9
(b) $-(-9)$	**B.** -9
(c) $-\lvert -9 \rvert$	**C.** Neither A nor B
(d) $-\lvert -(-9) \rvert$	**D.** Both A and B

8. *Concept Check*   Fill in each blank with the correct values: The opposite of -5 is _____ , while the absolute value of -5 is _____ . The additive inverse of -5 is _____ , while the additive inverse of the absolute value of -5 is _____ .

9. Students often say "Absolute value is always positive." Is this true? Explain.

10. *Concept Check*   *True* or *false*: If a is negative, then $\lvert a \rvert = -a$.

Concept Check   *Exercises 11–28 check understanding of the various sets of numbers.*

In Exercises 11–16, give a number that satisfies the given condition.

11. An integer between 3.6 and 4.6

12. A rational number between 2.8 and 2.9

13. A whole number that is not positive and is less than 1

14. A whole number greater than 3.5

15. An irrational number that is between $\sqrt{12}$ and $\sqrt{14}$

16. A real number that is neither negative nor positive

In Exercises 17–22, decide whether each statement is true *or* false.

17. Every natural number is positive.

18. Every whole number is positive.

19. Every integer is a rational number.

20. Every rational number is a real number.

21. Some numbers are both rational and irrational.

22. Every terminating decimal is a rational number.

In Exercises 23–28, give three numbers between -6 *and 6 that satisfy each given condition.*

23. Positive real numbers but not integers

24. Real numbers but not positive numbers

25. Real numbers but not whole numbers

26. Rational numbers but not integers

27. Real numbers but not rational numbers

28. Rational numbers but not negative numbers

In Exercises 29–32, use an integer to express each number in boldface italics representing a change. In Exercises 33 and 34, use a rational number. **See Example 1.**

29. Between July 1, 2011, and July 1, 2012, the population of the United States increased by approximately *2,259,105.* (*Source:* U.S. Census Bureau.)

30. Between 2011 and 2012, the number of movie screens in the United States increased by *207.* (*Source:* Motion Picture Association of America.)

31. From 2011 to 2012, attendance at the first game of the World Series went from 46,406 to 42,982, a decrease of **3424**. (*Source:* Major League Baseball.)

32. In 1935, there were 15,295 banks in the United States. By 2012, the number was 7083, a decrease of **8212** banks. (*Source:* Federal Deposit Insurance Corporation.)

33. On Friday, August 23, 2013, the Dow Jones Industrial Average (DJIA) closed at 15,010.51. On the previous day it had closed at 14,963.74. Thus, on Friday, it closed up **46.77** points. (*Source: The Washington Post.*)

34. On Tuesday, August 27, 2013, the NASDAQ closed at 3578.52. On the previous day, it had closed at 3657.57. Thus, on Tuesday, it closed down **79.05** points. (*Source: The Washington Post.*)

*Graph each number on a number line. **See Example 2**.*

35. $0, 3, -5, -6$

36. $2, 6, -2, -1$

37. $-2, -6, -4, 3, 4$

38. $-5, -3, -2, 0, 4$

39. $\frac{1}{4}, 2\frac{1}{2}, -3.8, -4, -1\frac{5}{8}$

40. $5.25, 4\frac{5}{9}, -2\frac{1}{3}, 0, -3\frac{2}{5}$

*For Exercises 41 and 42, **see Example 3**. List all numbers from each set that are the following.*

   (a) natural numbers   **(b)** whole numbers   **(c)** integers
   (d) rational numbers   **(e)** irrational numbers   **(f)** real numbers

▶ **41.** $\left\{ -9, -\sqrt{7}, -1\frac{1}{4}, -\frac{3}{5}, 0, 0.\overline{1}, \sqrt{5}, 3, 5.9, 7 \right\}$

42. $\left\{ -5.3, -5, -\sqrt{3}, -1, -\frac{1}{9}, 0, 0.\overline{27}, 1.2, 3, \sqrt{11} \right\}$

*For each number, find **(a)** the additive inverse and **(b)** the absolute value. **See Objective 3 and Example 5**.*

43. -7 **44.** -4 **45.** 8 **46.** 10

47. $-\frac{3}{4}$ **48.** $-\frac{2}{5}$ **49.** 5.6 **50.** 8.1

*Find each absolute value. **See Example 5**.*

▶ **51.** $|-6|$ **52.** $|-14|$ **53.** $-|12|$ **54.** $-|19|$ **55.** $-\left|-\frac{2}{3}\right|$

56. $-\left|-\frac{4}{5}\right|$ **57.** $|6-3|$ **58.** $|9-4|$ **59.** $-|6-3|$ **60.** $-|9-4|$

*Select the lesser of the two given numbers. **See Examples 4 and 5**.*

61. $-11, -3$ **62.** $-8, -13$ **63.** $-7, -6$

64. $-16, -17$ **65.** $4, |-5|$ **66.** $4, |-3|$

67. $|-3.5|, |-4.5|$ **68.** $|-8.9|, |-9.8|$ **69.** $-|-6|, -|-4|$

70. $-|-2|, -|-3|$ **71.** $|5-3|, |6-2|$ **72.** $|7-2|, |8-1|$

Decide whether each statement is true *or* false. ***See Examples 4 and 5**.*

▶ **73.** $-5 < -2$ **74.** $-8 > -2$ **75.** $-4 \leq -(-5)$

76. $-6 \leq -(-3)$ **77.** $|-6| < |-9|$ **78.** $|-12| < |-20|$

79. $-|8| > |-9|$ **80.** $-|12| > |-15|$ **81.** $-|-5| \geq -|-9|$

82. $-|-12| \leq -|-15|$ **83.** $|6-5| \geq |6-2|$ **84.** $|13-8| \leq |7-4|$

The table shows the percent change in the Consumer Price Index (CPI) for selected categories of goods and services from 2009 to 2010 and from 2010 to 2011. Use the table to answer each question. **See Example 6.**

85. Which category in which year represents the greatest percentage increase?

86. Which category in which year represents the greatest percentage decrease?

87. Which category in which year represents the least change?

88. Which categories represent a decrease for both years?

Category	Change from 2009 to 2010	Change from 2010 to 2011
Communication	−0.3	−1.6
Fuel and other utilities	1.7	2.9
Medical care	3.4	3.0
Public transportation	6.3	7.2
Shelter	−0.4	1.3

Source: U.S. Bureau of Labor Statistics.

STUDY SKILLS

Completing Your Homework

You are ready to do your homework **AFTER** you have read the corresponding textbook section and worked through the examples and Now Try exercises.

Homework Tips

- **Survey the exercise set.** Glance over the problems that your instructor has assigned to get a general idea of the types of exercises you will be working. Skim directions, and note any references to section examples.

- **Work problems neatly.** Use pencil and write legibly, so others can read your work. Skip lines between steps. Clearly separate problems from each other.

- **Show all your work.** It is tempting to take shortcuts. Include ALL steps.

- **Check your work frequently to make sure you are on the right track.** It is hard to unlearn a mistake. For all odd-numbered problems, answers are given in the back of the book.

- **If you have trouble with a problem, refer to the corresponding worked example in the section.** The exercise directions will often reference specific examples to review. Pay attention to every line of the worked example to see how to get from step to step.

- **If you are having trouble with an even-numbered problem, work the corresponding odd-numbered problem.** Check your answer in the back of the book, and apply the same steps to work the even-numbered problem.

- **Do some homework problems every day.** This is a good habit, even if your math class does not meet each day.

- **Mark any problems you don't understand.** Ask your instructor about them.

Think through and answer each question.

1. What is your instructor's policy regarding homework?

2. Think about your current approach to doing homework. Be honest in your assessment.

   (a) What are you doing that is working well?

   (b) What improvements could you make?

3. Which one or two homework tips will you try this week? Why?

1.4   Adding and Subtracting Real Numbers

VOCABULARY

☐ sum
☐ addends
☐ difference
☐ minuend
☐ subtrahend

NOW TRY
EXERCISE 1

Use a number line to find each sum.

(a) $3 + 5$ **(b)** $-1 + (-3)$

OBJECTIVE 1 Add two numbers with the same sign.

Recall that the answer to an addition problem is a **sum.** The numbers being added are the **addends.**

$$x + y = z \leftarrow \text{Sum}$$
$$\underset{\text{Addends}}{}$$

EXAMPLE 1   Adding Numbers (Same Sign) on a Number Line

Use a number line to find each sum.

(a) $2 + 3$

Step 1   Start at 0 and draw an arrow 2 units to the *right*. See **FIGURE 10.**

Step 2   From the right end of that arrow, draw another arrow 3 units to the *right*.

The number below the end of the second arrow is 5, so $2 + 3 = 5$.

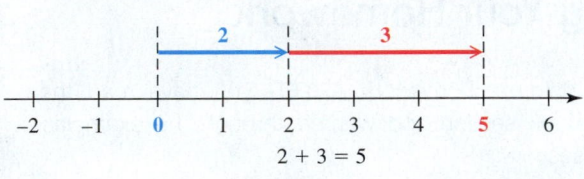

$$2 + 3 = 5$$

FIGURE 10

(b) $-2 + (-4)$

(We put parentheses around -4 due to the $+$ and $-$ symbols next to each other.)

Step 1   Start at 0 and draw an arrow 2 units to the *left*. See **FIGURE 11.**

Step 2   From the left end of the first arrow, draw a second arrow 4 units to the *left* to represent the addition of a *negative* number.

The number below the end of the second arrow is -6, so $-2 + (-4) = -6$.

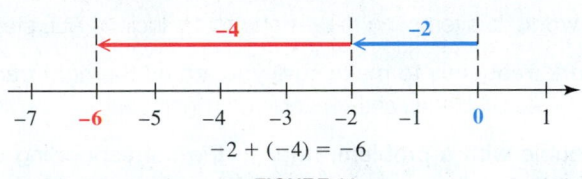

$$-2 + (-4) = -6$$

FIGURE 11

NOW TRY

In **Example 1(b),** the sum of the two negative numbers -2 and -4 is a negative number whose distance from 0 is the sum of the distance of -2 from 0 and the distance of -4 from 0. *That is, the sum of two negative numbers is the negative of the sum of their absolute values.*

Adding Signed Numbers (Same Sign)

To add two numbers with the *same* sign, add their absolute values. The sum has the same sign as the addends.

Examples: $2 + 4 = 6$ and $-2 + (-4) = -6$

**NOW TRY
EXERCISE 2**

Find each sum.

(a) $-6 + (-11)$

(b) $-\dfrac{2}{5} + \left(-\dfrac{1}{2}\right)$

EXAMPLE 2 Adding Two Negative Numbers

Find each sum.

(a) $-9 + (-2)$ Both addends are negative.

$= -(|-9| + |-2|)$ Add the absolute values of the addends.

       Sign of each addend

$= -(9 + 2)$ Take the absolute values.

$= -11$ The sum of two negative numbers is negative.

(b) $-\dfrac{1}{4} + \left(-\dfrac{2}{3}\right)$ Both addends are negative.

Think: $\left|-\dfrac{3}{12}\right| = \dfrac{3}{12}$ and $\left|-\dfrac{8}{12}\right| = \dfrac{8}{12}$ $= -\dfrac{3}{12} + \left(-\dfrac{8}{12}\right)$ Write equivalent fractions using the LCD, 12.

$= -\dfrac{11}{12}$ Add the absolute values of the addends. Use the common negative sign.

(c) $-2.6 + (-4.7)$ Both addends are negative.

$= -7.3$ Add the absolute values of the addends. Use the common negative sign.

     NOW TRY

OBJECTIVE 2 Add two numbers with different signs.

**NOW TRY
EXERCISE 3**

Use a number line to find the sum.

$4 + (-8)$

EXAMPLE 3 Adding Numbers (Different Signs) on a Number Line

Use a number line to find the sum $-2 + 5$.

Step 1   Start at 0 and draw an arrow 2 units to the *left*. See **FIGURE 12**.

Step 2   From the left end of this arrow, draw a second arrow 5 units to the *right* to represent the addition of a *positive* number.

The number below the end of the second arrow is 3, so $-2 + 5 = 3$.

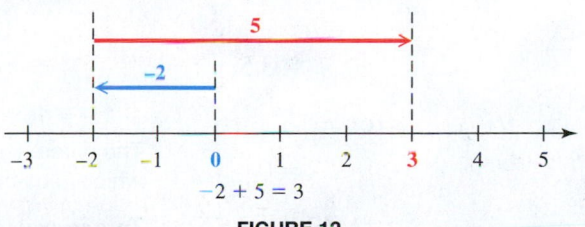

$-2 + 5 = 3$

FIGURE 12   **NOW TRY**

Adding Signed Numbers (Different Signs)

To add two numbers with *different* signs, find their absolute values and subtract the lesser absolute value from the greater. The sum has the same sign as the addend with greater absolute value.

Examples: $-2 + 6 = 4$ and $2 + (-6) = -4$

NOW TRY EXERCISE 4

Find each sum.

(a) $7 + (-4)$

(b) $\dfrac{2}{3} + \left(-2\dfrac{1}{9}\right)$

(c) $-5.7 + 3.7$

(d) $-10 + 10$

EXAMPLE 4   Adding Signed Numbers (Different Signs)

Find each sum.

(a) $-12 + 5$

$\overset{|-12|\quad|5|}{}$ Find the absolute value of each addend, and subtract the lesser from the greater.

$= -(12 - 5)$

Use the sign of the addend with greater absolute value.

$= -7$

(b) $-8 + 12$ Find the absolute value of each addend, and subtract the lesser from the greater.

$= +(12 - 8)$

Use the sign of the addend with greater absolute value.

$= 4$ The $+$ symbol is understood.

(c) $\dfrac{5}{6} + \left(-1\dfrac{1}{3}\right)$

$= \dfrac{5}{6} + \left(-\dfrac{4}{3}\right)$ Write the mixed number as an improper fraction.

$= \dfrac{5}{6} + \left(-\dfrac{8}{6}\right)$ Find a common denominator.

$= -\left(\dfrac{8}{6} - \dfrac{5}{6}\right)$ $\left|\dfrac{5}{6}\right| = \dfrac{5}{6}$ and $\left|-\dfrac{8}{6}\right| = \dfrac{8}{6}$;
Subtract the lesser absolute value from the greater.

Use a $-$ symbol because $\left|-\dfrac{8}{6}\right| > \left|\dfrac{5}{6}\right|$.

$= -\dfrac{3}{6}$ Subtract the fractions.

$= -\dfrac{1}{2}$ Write in lowest terms.

(d) $8.1 + (-4.6)$ $|8.1| = 8.1$ and $|-4.6| = |4.6|$;
Subtract the lesser absolute value from the greater.

$= +(8.1 - 4.6)$

$|8.1| > |-4.6|$

$= 3.5$

(e) $-16 + 16$ $|-16| = 16$ and $|16| = 16$;

$= 0$ The difference of the absolute values is 0, which is neither positive nor negative.

(f) $42 + (-42)$ ***In general, when additive inverses are added, the sum is 0.***

$= 0$

NOW TRY

OBJECTIVE 3 Use the definition of subtraction.

Recall that the answer to a subtraction problem is a **difference.** In the subtraction $x - y$, x is the **minuend** and y is the **subtrahend.**

$$x \quad - \quad y \quad = \quad z$$

Minuend   Subtrahend   Difference

NOW TRY EXERCISE 5

Use a number line to find the difference.

$6 - 2$

EXAMPLE 5 Subtracting Numbers on a Number Line

Use a number line to find the difference $7 - 4$.

Step 1 Start at 0 and draw an arrow 7 units to the *right*. See **FIGURE 13**.

Step 2 From the right end of the first arrow, draw a second arrow 4 units to the *left* to represent the *subtraction*.

The number below the end of the second arrow is 3, so $7 - 4 = 3$.

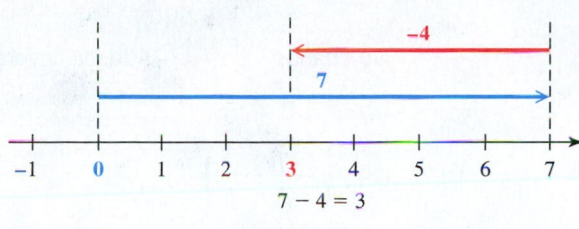

$7 - 4 = 3$

FIGURE 13

NOW TRY

The procedure used in **Example 5** to find the difference $7 - 4$ is exactly the same procedure that would be used to find the sum $7 + (-4)$.

$$7 - 4 \quad \text{is equal to} \quad 7 + (-4).$$

This suggests that *subtracting* a positive number from a greater positive number is the same as *adding* the additive inverse of the lesser number to the greater.

Definition of Subtraction

For any real numbers x and y, the following holds.

$$x - y = x + (-y)$$

To subtract y from x, add the additive inverse (or opposite) of y to x. That is, change the subtrahend to its opposite and add.

Example: $4 - 9$
$$= 4 + (-9)$$
$$= -5$$

EXAMPLE 6 Subtracting Signed Numbers

Find each difference.

(a) $12 - 3$

    ⌐ Change − to +.

$= 12 + (-3)$

No change ⌐   ⌐ Additive inverse of 3

$= 9$ 12 has the greater absolute value, so the sum is positive.

(b) $5 - 7$

    ⌐ Change − to +.

$= 5 + (-7)$

No change ⌐   ⌐ Additive inverse of 7

$= -2$ −7 has the greater absolute value, so the sum is negative.

NOW TRY ANSWER
5. 4

**NOW TRY
EXERCISE 6**

Find each difference.

(a) $-5 - (-11)$

(b) $4 - 15$

(c) $-\dfrac{5}{7} - \dfrac{1}{3}$

(d) $5.25 - (-3.24)$

(c) $-8 - 15$

Change $-$ to $+$.

$= -8 + (-15)$

No change ⬏   ⬑ Additive inverse of 15

$= -23$ The sum of two negative numbers is negative.

(d) $-3 - (-5)$

Change $-$ to $+$.

$= -3 + 5$

No change ⬏   ⬑ Additive inverse of -5

$= 2$ 5 has the greater absolute value, so the sum is positive.

(e) $\dfrac{3}{8} - \left(-\dfrac{4}{5}\right)$

$= \dfrac{15}{40} - \left(-\dfrac{32}{40}\right)$ Write equivalent fractions using the LCD, 40.

$= \dfrac{15}{40} + \dfrac{32}{40}$ Definition of subtraction

$= \dfrac{47}{40}$,   or $1\dfrac{7}{40}$ Add the fractions. Write as a mixed number.

(f) $-8.75 - (-2.41)$

$= -8.75 + 2.41$ Definition of subtraction

$= -6.34$ Add the decimals.   **NOW TRY**

Uses of the Symbol $-$

We use the symbol $-$ for three purposes.

1. *It can represent subtraction,* as in $9 - 5 = 4$.

2. *It can represent negative numbers,* such as -10, -2, and -3.

3. *It can represent the additive inverse (or opposite) of a number,* as in "the additive inverse (or opposite) of 8 is -8."

We may see more than one use of $-$ in the same expression, such as $-6 - (-9)$, where -9 is subtracted from -6. The meaning of the symbol $-$ depends on its position in the algebraic expression.

OBJECTIVE 4   Use the rules for order of operations when adding and subtracting signed numbers.

EXAMPLE 7   Using the Rules for Order of Operations

Perform each indicated operation.

(a) $-6 - [2 - (8 + 3)]$ Work from the inside out.

$= -6 - [2 - 11]$ Add inside the parentheses.

$= -6 - [2 + (-11)]$ Definition of subtraction

$= -6 - [-9]$ Add inside the brackets.

$= -6 + 9$ Definition of subtraction

$= 3$ Add.

NOW TRY ANSWERS

6. **(a)** 6   **(b)** -11

   (c) $-\dfrac{22}{21}$, or $-1\dfrac{1}{21}$

   (d) 8.49

**NOW TRY
EXERCISE 7**

Perform each indicated operation.

(a) $8 - [(-3 + 7) - (3 - 9)]$

(b) $3|6 - 9| - |4 - 12|$

(b) $5 + [(-3 - 2) - (4 - 1)]$ Work within each set of parentheses inside the brackets.

$= 5 + [(-3 + (-2)) - 3]$

$= 5 + [(-5) - 3]$

$= 5 + [(-5) + (-3)]$ *Show all steps to avoid sign errors.*

$= 5 + [-8]$

$= -3$

(c) $\dfrac{2}{3} - \left[\dfrac{1}{12} - \left(-\dfrac{1}{4} \right) \right]$

$= \dfrac{8}{12} - \left[\dfrac{1}{12} - \left(-\dfrac{3}{12} \right) \right]$ Write equivalent fractions using the LCD, 12.

$= \dfrac{8}{12} - \left[\dfrac{1}{12} + \dfrac{3}{12} \right]$ Work inside the brackets.

$= \dfrac{8}{12} - \dfrac{4}{12}$ Add inside the brackets.

$= \dfrac{4}{12}$ Subtract.

$= \dfrac{1}{3}$ Write in lowest terms.

(d) $|4 - 7| + 2|6 - 3|$ *$2|6 - 3|$ means $2 \cdot |6 - 3|$.*

$= |-3| + 2|3|$ Work within the absolute value bars.

$= 3 + 2 \cdot 3$ Evaluate the absolute values.

$= 3 + 6$ Multiply.   *Be careful. Multiply first.*

$= 9$ Add.   **NOW TRY**

OBJECTIVE 5 Translate words and phrases involving addition and subtraction.

▼ **Words and Phrases That Indicate Addition**

Word or Phrase	Example	Numerical Expression and Simplification
Sum of	The *sum of* −3 and 4	−3 + 4, which equals 1
Added to	5 *added to* −8	−8 + 5, which equals −3
More than	12 *more than* −5	−5 + 12, which equals 7
Increased by	−6 *increased by* 13	−6 + 13, which equals 7
Plus	3 *plus* 14	3 + 14, which equals 17

EXAMPLE 8 Translating Words and Phrases (Addition)

Write a numerical expression for each phrase, and simplify the expression.

(a) The *sum of* −8 and 4 and 6

$-8 + 4 + 6$ simplifies to $-4 + 6$, which equals 2.

Add in order from left to right.

NOW TRY ANSWERS
7. **(a)** −2 **(b)** 1

**NOW TRY
EXERCISE 8**

Write a numerical expression for the phrase, and simplify the expression.

The sum of −3 and 7, increased by 10

(b) 3 more than −5, increased by 12

$$(-5 + 3) + 12 \quad \text{simplifies to} \quad -2 + 12, \quad \text{which equals} \quad 10.$$

Here we *simplified* each expression by performing the operations.   **NOW TRY**

▼ **Words and Phrases That Indicate Subtraction**

Word, Phrase, or Sentence	Example	Numerical Expression and Simplification
Difference of	The *difference of* −3 and −8	−3 − (−8) simplifies to −3 + 8, which equals 5
Subtracted from*	12 *subtracted from* 18	18 − 12, which equals 6
From . . . , subtract	From 12, subtract 8.	12 − 8 simplifies to 12 + (−8), which equals 4
Less	6 *less* 5	6 − 5, which equals 1
Less than*	6 *less than* 5	5 − 6 simplifies to 5 + (−6), which equals −1
Decreased by	9 *decreased by* −4	9 − (−4) simplifies to 9 + 4, which equals 13
Minus	8 *minus* 5	8 − 5, which equals 3

*Be careful with order when translating.

⚠ **CAUTION**   When subtracting two numbers, be careful to write them in the correct order. In general,

$$x - y \neq y - x. \qquad \text{For example, } 5 - 3 \neq 3 - 5.$$

Think carefully before interpreting an expression involving subtraction.

**NOW TRY
EXERCISE 9**

Write a numerical expression for each phrase, and simplify the expression.

(a) The difference of 5 and −8, decreased by 4

(b) 7 less than −2

| **EXAMPLE 9** | **Translating Words and Phrases (Subtraction)** |

Write a numerical expression for each phrase, and simplify the expression.

(a) The difference of −8 and 5

When "difference of" is used, write the numbers in the order given.

$$-8 - 5 \quad \text{simplifies to} \quad -8 + (-5), \quad \text{which equals} \quad -13.$$

(b) 4 subtracted from the sum of 8 and −3

Here the operation of addition is also used, as indicated by the words *sum of*. First, add 8 and −3. Next, subtract 4 from this sum.

$$[8 + (-3)] - 4 \quad \text{simplifies to} \quad 5 - 4, \quad \text{which equals 1.}$$

(c) 4 less than −6

Here, 4 must be taken *from* −6, so write −6 first.

Be careful with order. $-6 - 4 \quad \text{simplifies to} \quad -6 + (-4), \quad \text{which equals} \quad -10.$

Notice that "4 less than −6" differs from "4 *is less than* −6." The second of these is symbolized 4 < −6 (which is a false statement).

(d) 8, decreased by 5 less than 12

First, write "5 less than 12" as 12 − 5. Next, subtract 12 − 5 from 8.

$$8 - (12 - 5) \quad \text{simplifies to} \quad 8 - 7, \quad \text{which equals} \quad 1.$$

NOW TRY

NOW TRY ANSWERS
8. (−3 + 7) + 10; 14
9. (a) [5 − (−8)] − 4; 9
    (b) −2 − 7; −9

**NOW TRY
EXERCISE 10**

Find the difference between a gain of 226 yd on the football field by the Chesterfield Bears and a loss of 7 yd by the New London Wildcats.

EXAMPLE 10 Solving an Application Involving Subtraction

The record-high temperature in the United States is 134°F, recorded at Death Valley, California, in 1913. The record low is −80°F, at Prospect Creek, Alaska, in 1971. See **FIGURE 14**. What is the difference between these highest and lowest temperatures? (*Source: National Climatic Data Center.*)

We must subtract the lowest temperature from the highest temperature.

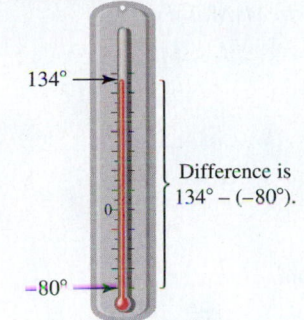

$$134 - (-80)$$

Order of numbers matters in subtraction.

$$= 134 + 80 \quad \text{Definition of subtraction}$$

$$= 214 \quad \text{Add.}$$

The difference between the two temperatures is 214°F.

134° →

Difference is 134° − (−80°).

0

−80° →

FIGURE 14

NOW TRY

OBJECTIVE 6 Use signed numbers to interpret data.

**NOW TRY
EXERCISE 11**

Refer to **FIGURE 15** and use a signed number to represent the change in enrollment from 1985 to 1990.

EXAMPLE 11 Using a Signed Number to Interpret Data

The bar graph in **FIGURE 15** shows public high school (grades 9–12) enrollment in millions of students for selected years from 1980 to 2010.

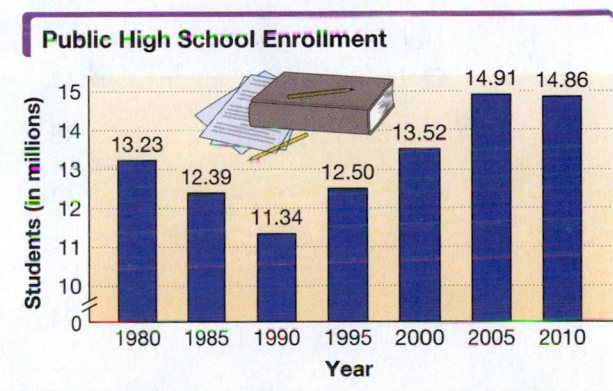

Public High School Enrollment

13.23   12.39   11.34   12.50   13.52   14.91   14.86

Students (in millions)

1980 1985 1990 1995 2000 2005 2010

Year

Source: U.S. National Center for Education Statistics.

FIGURE 15

(a) Use a signed number to represent the change in enrollment in millions from 2000 to 2005.
Start with the number for 2005. Subtract from it the number for 2000.

2005   2000
↓   ↓

$$14.91 - 13.52 = +1.39 \text{ million students}$$ ← *increase.* A positive number indicates an *increase.* The bar for 2005 is "higher" than the bar for 2000.

(b) Use a signed number to represent the change in enrollment in millions from 2005 to 2010.
Start with the number for 2010. Subtract from it the number for 2005.

2010   2005
↓   ↓

$$14.86 - 14.91 = -0.05 \text{ million students}$$ ← *decrease.* A negative number indicates a *decrease.* The bar for 2010 is "lower" than the bar for 2005.

NOW TRY

NOW TRY ANSWERS
10. 233 yd
11. −1.05 million students

1.4 Exercises

FOR
EXTRA
HELP

 MyMathLab®

▶ *Complete solution available*
in MyMathLab

Concept Check   *Complete each of the following.*

▶ **1.** The sum of two negative numbers will always be a (*positive*/*negative*) number. Give a number-line illustration using the sum $-2 + (-3)$.

2. The sum of a number and its opposite will always be _____.

▶ **3.** When adding a positive number and a negative number, where the negative number has the greater absolute value, the sum will be a (*positive*/*negative*) number. Give a number-line illustration using the sum $-4 + 2$.

4. To simplify the expression $8 + [-2 + (-3 + 5)]$, one should begin by adding _____ and _____, according to the rules for order of operations.

5. By the definition of subtraction, in order to perform the subtraction $-6 - (-8)$, we must add the opposite of _____ to _____ to obtain _____.

6. "The difference of 7 and 12" translates as _____, while "the difference of 12 and 7" translates as _____.

Concept Check   *Suppose that x represents a positive number and y represents a negative number. Determine whether the given expression must represent a positive number or a negative number.*

7. $x - y$ **8.** $y - x$ **9.** $y - |x|$ **10.** $x + |y|$

*Find each sum. **See Examples 1–7.***

▶ **11.** $-6 + (-2)$ **12.** $-9 + (-2)$ **13.** $-5 + (-7)$

14. $-11 + (-5)$ **15.** $6 + (-4)$ **16.** $11 + (-8)$

17. $4 + (-6)$ **18.** $3 + (-7)$ **19.** $-16 + 7$

20. $-13 + 6$ **21.** $6 + (-6)$ **22.** $-11 + 11$

23. $-\dfrac{1}{3} + \left(-\dfrac{4}{15}\right)$ **24.** $-\dfrac{1}{4} + \left(-\dfrac{5}{12}\right)$ ▶ **25.** $-\dfrac{1}{6} + \dfrac{2}{3}$

26. $-\dfrac{6}{25} + \dfrac{19}{20}$ **27.** $\dfrac{5}{8} + \left(-\dfrac{17}{12}\right)$ **28.** $\dfrac{9}{10} + \left(-\dfrac{11}{8}\right)$

29. $2\dfrac{1}{2} + \left(-3\dfrac{1}{4}\right)$ **30.** $1\dfrac{3}{8} + \left(-2\dfrac{1}{4}\right)$ **31.** $-3.5 + 12.4$

32. $-12.5 + 21.3$ **33.** $-2.34 + (-3.67)$ **34.** $-1.25 + (-6.88)$

35. $4 + [13 + (-5)]$ **36.** $6 + [12 + (-3)]$ **37.** $8 + [-2 + (-1)]$

38. $12 + [-3 + (-4)]$ **39.** $-2 + [5 + (-1)]$ **40.** $-8 + [9 + (-2)]$

41. $-6 + [6 + (-9)]$ **42.** $-3 + [3 + (-8)]$ **43.** $[(-9) + (-3)] + 12$

44. $[(-8) + (-6)] + 14$ **45.** $-6.1 + [3.2 + (-4.8)]$ **46.** $-9.4 + [5.8 + (-7.9)]$

47. $[-3 + (-4)] + [5 + (-6)]$ **48.** $[-8 + (-3)] + [4 + (-6)]$

49. $[-4 + (-3)] + [8 + (-1)]$ **50.** $[-5 + (-9)] + [16 + (-2)]$

51. $[-4 + (-6)] + [-3 + (-8)] + [12 + (-11)]$

52. $[-2 + (-11)] + [-12 + (-2)] + [18 + (-6)]$

Find each difference. See Examples 1–7.

53. $4 - 7$ **54.** $8 - 13$ ▶ **55.** $5 - 9$ **56.** $6 - 11$

57. $-7 - 1$ **58.** $-9 - 4$ **59.** $-8 - 6$ **60.** $-9 - 5$

61. $7 - (-2)$ **62.** $9 - (-2)$ **63.** $-6 - (-2)$ **64.** $-7 - (-5)$

65. $2 - (3 - 5)$ **66.** $-3 - (4 - 11)$ **67.** $\dfrac{1}{2} - \left(-\dfrac{1}{4}\right)$

68. $\dfrac{1}{3} - \left(-\dfrac{1}{12}\right)$ **69.** $-\dfrac{3}{4} - \dfrac{5}{8}$ **70.** $-\dfrac{5}{6} - \dfrac{1}{2}$

71. $\dfrac{5}{8} - \left(-\dfrac{1}{2} - \dfrac{3}{4}\right)$ **72.** $\dfrac{9}{10} - \left(\dfrac{1}{8} - \dfrac{3}{10}\right)$ **73.** $3.4 - (-8.2)$

74. $5.7 - (-11.6)$ **75.** $-6.4 - 3.5$ **76.** $-4.4 - 8.6$

Perform each indicated operation. See Examples 1–7.

▶ **77.** $(4 - 6) + 12$ **78.** $(3 - 7) + 4$ **79.** $(8 - 1) - 12$

80. $(9 - 3) - 15$ **81.** $6 - (-8 + 3)$ **82.** $8 - (-9 + 5)$

83. $2 + (-4 - 8)$ **84.** $6 + (-9 - 2)$ **85.** $|-5 - 6| + |9 + 2|$

86. $|-4 + 8| + |6 - 1|$ **87.** $|-8 - 2| - |-9 - 3|$ **88.** $|-4 - 2| - |-8 - 1|$

89. $\left(-\dfrac{3}{4} - \dfrac{5}{2}\right) - \left(-\dfrac{1}{8} - 1\right)$ **90.** $\left(-\dfrac{3}{8} - \dfrac{2}{3}\right) - \left(-\dfrac{9}{8} - 3\right)$

91. $\left(-\dfrac{1}{2} + 0.25\right) - \left(-\dfrac{3}{4} + 0.75\right)$ **92.** $\left(-\dfrac{3}{2} - 0.75\right) - \left(0.5 - \dfrac{1}{2}\right)$

93. $-9 + [(3 - 2) - (-4 + 2)]$ **94.** $-8 - [(-4 - 1) + (9 - 2)]$

95. $-3 + [(-5 - 8) - (-6 + 2)]$ **96.** $-4 + [(-12 + 1) - (-1 - 9)]$

97. $-9.12 + [(-4.8 - 3.25) + 11.279]$ **98.** $-7.62 - [(-3.99 + 1.427) - (-2.8)]$

Write a numerical expression for each phrase, and simplify the expression. See Examples 8 and 9.

99. The sum of -5 and 12 and 6   **100.** The sum of -3 and 5 and -12

101. 14 added to the sum of -19 and -4 **102.** -2 added to the sum of -18 and 11

▶ **103.** The sum of -4 and -10, increased by 12   **104.** The sum of -7 and -13, increased by 14

105. $\dfrac{2}{7}$ more than the sum of $\dfrac{5}{7}$ and $-\dfrac{9}{7}$ **106.** 1.85 more than the sum of -1.25 and -4.75

▶ **107.** The difference of 4 and -8 **108.** The difference of 7 and -14

109. 8 less than -2 **110.** 9 less than -13

111. The sum of 9 and -4, decreased by 7   **112.** The sum of 12 and -7, decreased by 14

113. 12 less than the difference of 8 and -5 **114.** 19 less than the difference of 9 and -2

The table gives scores (above or below par—that is, above or below the score "standard") for selected golfers during the 2013 PGA Tour Championship. Write a signed number that represents the total score (above or below par) for the four rounds for each golfer.

	Golfer	Round 1	Round 2	Round 3	Round 4
115.	Steve Stricker	−4	+1*	−2	−5
116.	Phil Mickelson	+1	−3	0	−2
117.	Charl Schwartzel	−2	+9	+7	−4
118.	Kevin Streelman	−1	+2	+4	−3

*Golf scoring commonly includes a + symbol with a score over par.
Source: *The Gazette.*

Solve each problem. ***See Example 10.***

119. Based on 2020 population projections, New York will lose 5 seats in the U.S. House of Representatives, Pennsylvania will lose 4 seats, and Ohio will lose 3. Write a signed number that represents the total number of seats these three states are projected to lose. (*Source:* Population Reference Bureau.)

120. Michigan is projected to lose 3 seats in the U.S. House of Representatives and Illinois 2 in 2020. The states projected to gain the most seats are California with 9, Texas with 5, Florida with 3, Georgia with 2, and Arizona with 2. Write a signed number that represents the algebraic sum of these changes. (*Source:* Population Reference Bureau.)

121. The surface, or rim, of a canyon is at altitude 0. On a hike down into the canyon, a party of hikers stops for a rest at 130 m below the surface. The hikers then descend another 54 m. Write the new altitude as a signed number.

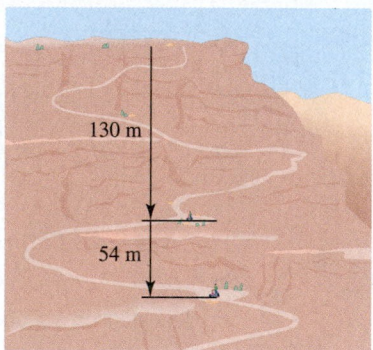

130 m

54 m

122. A pilot announces to the passengers that the current altitude of their plane is 34,000 ft. Because of turbulence, the pilot is forced to descend 2100 ft. Write the new altitude as a signed number.

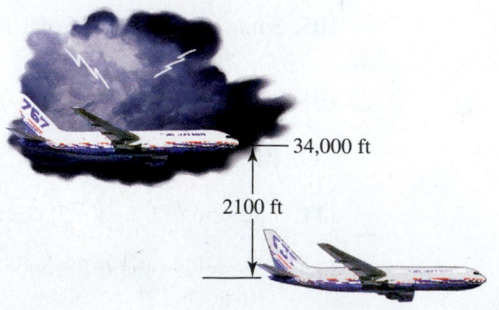

34,000 ft

2100 ft

123. The lowest temperature ever recorded in Arkansas was −29°F. The highest temperature ever recorded there was 149°F more than the lowest. What was this highest temperature? (*Source: National Climatic Data Center.*)

124. On January 23, 1943, the temperature rose 49°F in two minutes in Spearfish, South Dakota. If the starting temperature was −4°F, what was the temperature two minutes later?

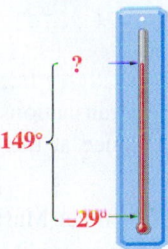 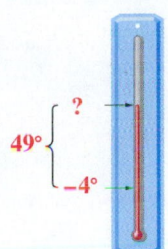

125. The lowest temperature ever recorded in Illinois was −36°F on January 5, 1999. The lowest temperature ever recorded in Utah was on February 1, 1985, and was 33°F lower than Illinois's record low. What is the record low temperature for Utah? (*Source:* National Climatic Data Center.)

126. The lowest temperature ever recorded in South Carolina was −19°F. The lowest temperature ever recorded in Wisconsin was 36° lower than South Carolina's record low. What is the record low temperature for Wisconsin? (*Source:* National Climatic Data Center.)

127. Nadine enjoys playing Triominoes every Wednesday night. Last Wednesday, on four successive turns, her scores were

$$-19, \quad 28, \quad -5, \quad \text{and} \quad 13.$$

What was her final score for the four turns?

128. Bruce also enjoys playing Triominoes. On five successive turns, his scores were

$$-13, \quad 15, \quad -12, \quad 24, \quad \text{and} \quad 14.$$

What was his total score for the five turns?

129. In 2005, Americans saved −0.5% of their after-tax incomes. In July 2013, they saved 4.4%. (*Source:* U.S. Bureau of Economic Analysis.)

(a) Express the difference between these amounts as a positive number.

(b) How could Americans have a negative personal savings rate in 2005?

130. In 2000, the U.S. federal budget had a surplus of $236 billion. In 2015, the federal budget projected a deficit of $576 billion. Express the difference between these amounts as a positive number. (*Source:* U.S. Treasury Department.)

131. In 2005, undergraduate college students had an average of $4906 in student loans. This average increased $788 by 2010, then dropped $154 by 2012. What was the average amount in student loans in 2012? (*Source:* The College Board.)

132. Among entertainment expenditures, the average annual spending per U.S. household on fees and admissions was $526 in 2001. This amount increased $80 by 2006 and then decreased $12 by 2011. What was the average household expenditure for fees and admissions in 2011? (*Source:* U.S. Bureau of Labor Statistics.)

133. In August, Susan began with a checking account balance of $904.89. Her withdrawals and deposits for August are as follows:

Withdrawals	Deposits
$35.84	$85.00
$26.14	$120.76
$3.12	

Assuming no other transactions, what was her account balance at the end of August?

134. In September, Jeffery began with a checking account balance of $904.89. His withdrawals and deposits for September are as follows:

Withdrawals	Deposits
$41.29	$80.59
$13.66	$276.13
$84.40	

Assuming no other transactions, what was his account balance at the end of September?

135. Linda owes $870.00 on her MasterCard account. She returns two items costing $35.90 and $150.00 and receives credit for these on the account. Next, she makes a purchase of $82.50 and then two more purchases of $10.00 each. She makes a payment of $500.00. She then incurs a finance charge of $37.23. How much does she still owe?

136. Marcial owes $679.00 on his Visa account. He returns three items costing $36.89, $29.40, and $113.55 and receives credit for these on the account. Next, he makes purchases of $135.78 and $412.88 and two purchases of $20.00 each. He makes a payment of $400. He then incurs a finance charge of $24.57. How much does he still owe?

The graph shows annual returns in percent for Class A shares of the Invesco S&P 500 Index Fund from 2009 to 2013. Use a signed number to represent the change in percent return for each period. **See Example 11.**

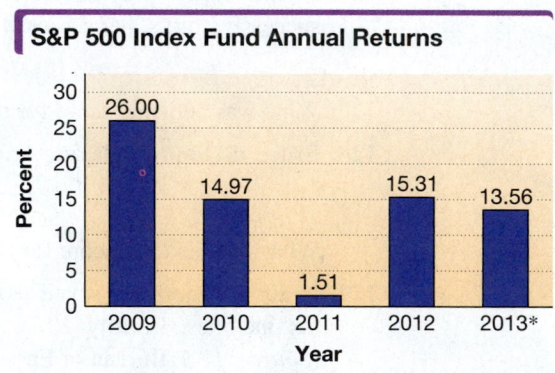

S&P 500 Index Fund Annual Returns

*Through second quarter
Source: Invesco.

137. 2009 to 2010

139. 2011 to 2012

138. 2010 to 2011

140. 2009 to 2013

The two tables show the heights of some selected mountains and the depths of some selected trenches. Use the information given to answer Exercises 141–146 on the next page.

Mountain	Height (in feet)
Foraker	17,400
Wilson	14,246
Pikes Peak	14,110

Trench	Depth (in feet, as a negative number)
Philippine	−32,995
Cayman	−24,721
Java	−23,376

Source: World Almanac and Book of Facts.

141. What is the difference between the height of Mt. Foraker and the depth of the Philippine Trench?

142. What is the difference between the height of Pikes Peak and the depth of the Java Trench?

143. How much deeper is the Cayman Trench than the Java Trench?

144. How much deeper is the Philippine Trench than the Cayman Trench?

145. How much higher is Mt. Wilson than Pikes Peak?

146. If Mt. Wilson and Pikes Peak were stacked one on top of the other, how much higher would they be than Mt. Foraker?

STUDY SKILLS

Using Study Cards

You may have used "flash cards" in other classes. In math, "study cards" can help you remember terms and definitions, procedures, and concepts. Use study cards to

- Help you understand and learn the material;
- Quickly review when you have a few minutes;
- Review before a quiz or test.

One of the advantages of study cards is that you learn while you are making them.

Vocabulary Cards

Put the word and a page reference on the front of the card. On the back, write the definition, an example, any related words, and a sample problem (if appropriate).

Procedure ("Steps") Cards

Write the name of the procedure on the front of the card. Then write each step in words. On the back of the card, put an example showing each step.

Make a vocabulary card and a procedure card for material you are learning now.

Front of Card

Back of Card

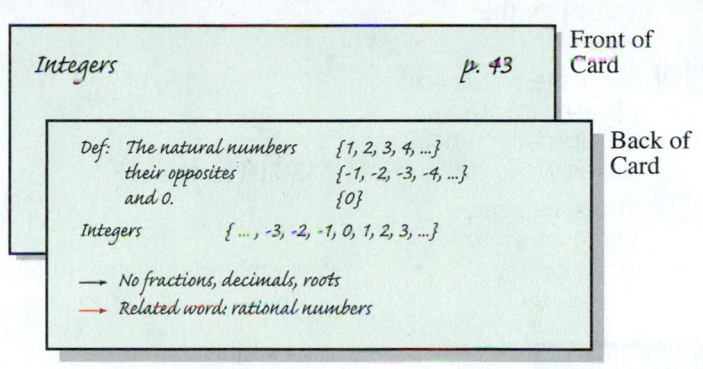

Front of Card

Back of Card

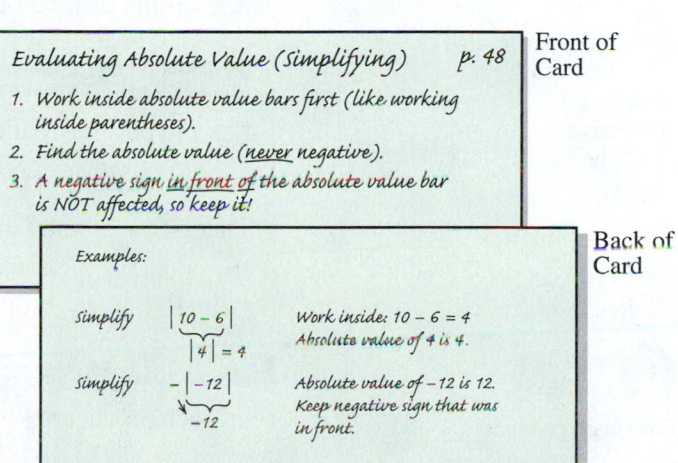

1.5 Multiplying and Dividing Real Numbers

VOCABULARY

☐ product
☐ factor
☐ multiplicative inverse (reciprocal)
☐ quotient
☐ dividend
☐ divisor

The result of multiplication is a **product.** We know that the product of two positive numbers is positive. We also know that the product of 0 and any positive number is 0, so we extend that property to all real numbers.

> **Multiplication Property of 0**
>
> For any real number x, the following hold.
>
> $$x \cdot 0 = 0 \quad \text{and} \quad 0 \cdot x = 0$$

OBJECTIVE 1 Find the product of a positive number and a negative number.

Look at the following pattern.

$$3 \cdot 5 = 15$$
$$3 \cdot 4 = 12$$
$$3 \cdot 3 = 9$$
$$3 \cdot 2 = 6$$
$$3 \cdot 1 = 3$$
$$3 \cdot 0 = 0$$
$$3 \cdot (-1) = ?$$

The products decrease by 3.

What should $3 \cdot (-1)$ equal? The product $3 \cdot (-1)$ represents the sum

$$-1 + (-1) + (-1), \quad \text{which equals} \quad -3,$$

so the product should be -3. Also, $3 \cdot (-2)$ and $3 \cdot (-3)$ represent the sums

$$-2 + (-2) + (-2), \quad \text{which equals} \quad -6$$

and

$$-3 + (-3) + (-3), \quad \text{which equals} \quad -9.$$

These results maintain the pattern in the list.

> **Multiplying Signed Numbers (Different Signs)**
>
> For any positive real numbers x and y, the following hold.
>
> $$x(-y) = -(xy) \quad \text{and} \quad (-x)y = -(xy)$$
>
> That is, the product of two numbers with different signs is negative.
>
> *Examples:* $6(-3) = -18 \quad \text{and} \quad (-6)3 = -18$

NOW TRY EXERCISE 1

Find each product.
(a) $-11(9)$ (b) $3.1(-2.5)$

EXAMPLE 1 Multiplying Signed Numbers (Different Signs)

Find each product.

(a) $8(-5)$
$= -(8 \cdot 5)$
$= -40$

The product of two numbers with *different* signs is *negative.*

(b) $-9\left(\dfrac{1}{3}\right)$
$= -\left(9 \cdot \dfrac{1}{3}\right)$
$= -3$

(c) $-6.2(4.1)$
$= -(6.2 \cdot 4.1)$
$= -25.42$

NOW TRY

NOW TRY ANSWERS
1. (a) -99 (b) -7.75

OBJECTIVE 2 Find the product of two negative numbers.

Look at another pattern.

$$-5(4) = -20$$
$$-5(3) = -15$$
$$-5(2) = -10$$
$$-5(1) = -5$$
$$-5(0) = 0$$
$$-5(-1) = \,?$$

The products increase by 5.

The numbers in color on the left side of the equality symbols decrease by 1 for each step down the list. The products on the right increase by 5 for each step down the list. To maintain this pattern, $-5(-1)$ should be 5 more than $-5(0)$, or 5 more than 0, so

$$-5(-1) = 5.$$

The pattern continues with

$$-5(-2) = 10$$
$$-5(-3) = 15$$
$$-5(-4) = 20$$
$$-5(-5) = 25, \quad \text{and so on.}$$

These results suggest the next rule.

Multiplying Two Negative Numbers

For any positive real numbers x and y, the following holds.

$$-x(-y) = xy$$

That is, the product of two negative numbers is positive.

Example: $-5(-4) = 20$

NOW TRY EXERCISE 2

Find each product.

(a) $-8(-11)$ **(b)** $-\dfrac{1}{7}\left(-\dfrac{5}{2}\right)$

EXAMPLE 2   Multiplying Two Negative Numbers

Find each product.

(a) $-9(-2)$

$\quad = 18$

(b) $-\dfrac{2}{3}\left(-\dfrac{3}{2}\right)$

$\quad = 1$

(c) $-0.5(-1.25)$

$\quad = 0.625$

The product of two numbers with the *same* sign is *positive.*   **NOW TRY**

The following box summarizes multiplying signed numbers.

Multiplying Signed Numbers

The product of two numbers with the *same* sign is *positive.*

The product of two numbers with *different* signs is *negative.*

Examples: $7(3) = 21$, $-7(-3) = 21$, $-7(3) = -21$,   and $7(-3) = -21$

NOW TRY ANSWERS

2. (a) 88   **(b)** $\dfrac{5}{14}$

OBJECTIVE 3 Identify factors of integers.

The definition of **factor** from **Section R.1** can be extended to integers. If the product of two integers is a third integer, then each of the two integers is a *factor* of the third.

▼ Integer Factors

Integer	18	20	15	7	1
Pairs of Factors	1, 18	1, 20	1, 15	1, 7	1, 1
	2, 9	2, 10	3, 5	−1, −7	−1, −1
	3, 6	4, 5	−1, −15		
	−1, −18	−1, −20	−3, −5		
	−2, −9	−2, −10			
	−3, −6	−4, −5			

▼ Reciprocals

Number	*Reciprocal (Multiplicative Inverse)*
4	$\frac{1}{4}$
0.3, or $\frac{3}{10}$	$\frac{10}{3}$
−5	$\frac{1}{-5}$, or $-\frac{1}{5}$
$-\frac{5}{8}$	$-\frac{8}{5}$

A number and its reciprocal have a product of **1.** For example,

$$4 \cdot \frac{1}{4} = \frac{4}{4}, \text{ or } 1.$$

0 has no reciprocal because the product of 0 and any number is 0, not 1.

OBJECTIVE 4 Use the reciprocal of a number to apply the definition of division.

The definition of division depends on the idea of a *reciprocal,* or *multiplicative inverse,* of a number.

Reciprocal, or Multiplicative Inverse

Pairs of numbers whose product is 1 are **reciprocals,** or **multiplicative inverses,** of each other.

Recall that the answer to a division problem is a **quotient.** For example, we can write the quotient of 15 and 3 as $15 \div 3$, which equals 5. We obtain the same answer if we multiply $15 \cdot \frac{1}{3}$, the reciprocal of 3. This suggests the next definition.

Definition of Division

For any real numbers x and y, where $y \neq 0$, the following holds.

$$x \div y = x \cdot \frac{1}{y}$$

That is, to divide two numbers, multiply the first number (the **dividend**) by the reciprocal, or multiplicative inverse, of the second number (the **divisor**).

Example: $15 \div 3$

$$= 15 \cdot \frac{1}{3}$$

$$= 5$$

Recall that an equivalent form of $x \div y$ is $\frac{x}{y}$, where the fraction bar represents division. ***In algebra, quotients are usually represented with a fraction bar.*** For example,

$$15 \div 3 \quad \text{is equivalent to} \quad \frac{15}{3}.$$

NOTE The following forms all represent division, where $y \neq 0$.

$$x \div y, \quad \frac{x}{y}, \quad x/y, \quad \text{and} \quad y\overline{)x}$$

Because division is defined in terms of multiplication, the rules for multiplying signed numbers also apply to dividing them.

Dividing Signed Numbers

The quotient of two numbers with the *same* sign is *positive*.

The quotient of two numbers with *different* signs is *negative*.

Examples: $\quad \dfrac{15}{3} = 5, \quad \dfrac{-15}{-3} = 5, \quad \dfrac{15}{-3} = -5, \quad \text{and} \quad \dfrac{-15}{3} = -5$

**NOW TRY
EXERCISE 3**

Find each quotient.

(a) $\dfrac{-10}{5}$ (b) $\dfrac{-1.44}{-0.12}$

(c) $-\dfrac{3}{8} \div \dfrac{7}{10}$

EXAMPLE 3 Dividing Signed Numbers

Find each quotient.

(a) $\dfrac{8}{-2} = -4$ (b) $\dfrac{-100}{5} = -20$ (c) $\dfrac{-4.5}{-0.09} = 50$

(d) $-\dfrac{1}{8} \div \left(-\dfrac{3}{4}\right)$

$\quad = -\dfrac{1}{8} \cdot \left(-\dfrac{4}{3}\right)$ Multiply by the reciprocal of the divisor.

$\quad = \dfrac{1}{6}$ Multiply the fractions. Write in lowest terms.

NOW TRY

Consider the quotient $\dfrac{12}{3}$.

$$\frac{12}{3} = 4 \qquad \text{because} \qquad 4 \cdot 3 = 12.$$ Multiply to check a division problem.

Using this relationship between multiplication and division, we investigate division by 0. Consider the quotient $\dfrac{0}{3}$.

$$\frac{0}{3} = 0 \qquad \text{because} \qquad 0 \cdot 3 = 0.$$

Now consider $\dfrac{3}{0}$.

$$\frac{3}{0} = ?$$

We need to find a number that when multiplied by 0 will equal 3, that is, $? \cdot 0 = 3$. *No* real number satisfies this equation because the product of any real number and 0 must be 0. Thus,

$\dfrac{x}{0}$ **is not a number, and** *division by 0 is undefined.* **If a division problem involves division by 0, write "undefined."**

Division Involving 0

For any real number x, where $x \neq 0$, the following hold.

$$\frac{0}{x} = 0 \quad \text{and} \quad \frac{x}{0} \text{ is undefined.}$$

Examples: $\quad \dfrac{0}{-10} = 0 \quad$ and $\quad \dfrac{-10}{0}$ is undefined.

From the definitions of multiplication and division of real numbers,

$$\frac{-40}{8} = -5 \quad \text{and} \quad \frac{40}{-8} = -5, \quad \text{so} \quad \frac{-40}{8} = \frac{40}{-8}.$$

Based on this example, the quotient of a positive number and a negative number can be expressed in three equivalent forms.

Equivalent Forms

For any positive real numbers x and y, the following are equivalent.

$$\frac{-x}{y}, \quad \frac{x}{-y}, \quad \text{and} \quad -\frac{x}{y}$$

Similarly, the quotient of two negative numbers can be expressed as a quotient of two positive numbers.

Equivalent Forms

For any positive real numbers x and y, the following are equivalent.

$$\frac{-x}{-y} \quad \text{and} \quad \frac{x}{y}$$

NOTE   Although we use the forms $\frac{-x}{y}$, $\frac{x}{-y}$, and $\frac{-x}{-y}$ in our work with algebraic expressions, we generally give final answers in the form $-\frac{x}{y}$ or $\frac{x}{y}$.

OBJECTIVE 5   Use the rules for order of operations when multiplying and dividing signed numbers.

EXAMPLE 4   Using the Rules for Order of Operations

Perform each indicated operation.

(a) $-9(2) - (-3)(2)$

$\quad = -18 - (-6) \qquad$ Multiply.

$\quad = -18 + 6 \qquad$ Definition of subtraction

$\quad = -12 \qquad$ Add.

(b) $-5(-2 - 3)$

$\quad = -5(-5) \qquad$ Subtract inside the parentheses.

$\quad = 25 \qquad$ Multiply.

 **NOW TRY
EXERCISE 4**

Perform each indicated operation.

(a) $-4(6) - (-5)(5)$

(b) $\dfrac{12(-4) - 6(-3)}{-4(7 - 16)}$

(c) $-6 + 2(3 - 5)$ Begin inside the parentheses.

Do *not* add first.

$= -6 + 2(-2)$ Subtract inside the parentheses.

$= -6 + (-4)$ Multiply.

$= -10$ Add.

(d) $\dfrac{5(-2) - 3(4)}{2(1 - 6)}$ Simplify the numerator and denominator separately.

$= \dfrac{-10 - 12}{2(-5)}$ Multiply in the numerator.
Subtract inside the parentheses in the denominator.

$= \dfrac{-22}{-10}$ Subtract in the numerator.
Multiply in the denominator.

$= \dfrac{11}{5}$ Write in lowest terms.   **NOW TRY**

OBJECTIVE 6   Evaluate algebraic expressions given values for the variables.

EXAMPLE 5   Evaluating Algebraic Expressions

 **NOW TRY
EXERCISE 5**

Evaluate $\dfrac{3x^2 - 12}{y}$ for $x = -4$ and $y = -3$.

Evaluate each expression for $x = -1$, $y = -2$, and $m = -3$.

(a) $(3x + 4y)(-2m)$

Use parentheses around substituted negative values to avoid errors.

$= [3(-1) + 4(-2)][-2(-3)]$ Substitute the given values for the variables.

$= [-3 + (-8)][6]$ Multiply.

$= [-11]6$ Add inside the brackets.

$= -66$ Multiply.

(b) $2x^2 - 3y^2$

Think: $(-2)^2 = -2(-2) = 4$

$= 2(-1)^2 - 3(-2)^2$ Substitute -1 for x and -2 for y.

Think: $(-1)^2 = -1(-1) = 1$

$= 2(1) - 3(4)$ Apply the exponents.

$= 2 - 12$ Multiply.

$= -10$ Subtract.

(c) $\dfrac{4y^2 + x}{m}$

$= \dfrac{4(-2)^2 + (-1)}{-3}$ Substitute -2 for y, -1 for x, and -3 for m.

$= \dfrac{4(4) + (-1)}{-3}$ Apply the exponent.

$= \dfrac{16 + (-1)}{-3}$ Multiply.

$= \dfrac{15}{-3}$ Add.

$= -5$ Divide.   **NOW TRY**

NOW TRY ANSWERS

4. (a) 1   **(b)** $-\dfrac{5}{6}$

5. -12

OBJECTIVE 7 Translate words and phrases involving multiplication and division.

▼ **Words and Phrases That Indicate Multiplication**

Word or Phrase	Example	Numerical Expression and Simplification
Product of	The *product of* −5 and −2	−5(−2), which equals 10
Times	13 *times* −4	13(−4), which equals −52
Twice (meaning "2 times")	*Twice* 6	2(6), which equals 12
Triple (meaning "3 times")	*Triple* 4	3(4), which equals 12
Of (used with fractions)	$\frac{1}{2}$ *of* 10	$\frac{1}{2}$(10), which equals 5
Percent of	12*% of* −16	0.12(−16), which equals −1.92
As much as	$\frac{2}{3}$ *as much as* 30	$\frac{2}{3}$(30), which equals 20

NOW TRY
EXERCISE 6

Write a numerical expression for each phrase, and simplify the expression.

(a) Twice the sum of −10 and 7

(b) 40% of the difference of 45 and 15

EXAMPLE 6 Translating Words and Phrases (Multiplication)

Write a numerical expression for each phrase, and simplify the expression.

(a) The product of 12 and the sum of 3 and −6

$$12[3 + (−6)] \quad \text{simplifies to} \quad 12[−3], \quad \text{which equals} \quad −36.$$

(b) Twice the difference of 8 and −4

$$2[8 − (−4)] \quad \text{simplifies to} \quad 2[12], \quad \text{which equals} \quad 24.$$

> The "difference of *a* and *b*" means *a* − *b*.

(c) Two-thirds of the sum of −5 and −3

$$\frac{2}{3}[−5 + (−3)] \quad \text{simplifies to} \quad \frac{2}{3}[−8], \quad \text{which equals} \quad −\frac{16}{3}.$$

(d) 15% of the difference of 14 and −2

> Remember that 15% = 0.15.

$$0.15[14 − (−2)] \quad \text{simplifies to} \quad 0.15[16], \quad \text{which equals} \quad 2.4.$$

(e) Double the product of 3 and 4

> Double means "2 times."

$$2 \cdot (3 \cdot 4) \quad \text{simplifies to} \quad 2(12), \quad \text{which equals} \quad 24.$$

NOW TRY

▼ **Phrases That Indicate Division**

Phrase	Example	Numerical Expression and Simplification
Quotient of	The *quotient of* −24 and 3	$\frac{−24}{3}$, which equals −8
Divided by	−16 *divided by* −4	$\frac{−16}{−4}$, which equals 4
Ratio of	The *ratio of* 2 to 3	$\frac{2}{3}$

NOW TRY ANSWERS
6. (a) 2(−10 + 7); −6
  (b) 0.40(45 − 15); 12

When translating a phrase involving division into a fraction, we write the first number named as the numerator and the second as the denominator.

**NOW TRY
EXERCISE 7**

Write a numerical expression for the phrase, and simplify the expression.

The quotient of 21 and the sum of 10 and −7

EXAMPLE 7 Interpreting Words and Phrases (Division)

Write a numerical expression for each phrase, and simplify the expression.

(a) The quotient of 14 and the sum of −9 and 2

"Quotient" indicates division. $\dfrac{14}{-9+2}$ simplifies to $\dfrac{14}{-7}$, which equals −2.

(b) The product of 5 and −6, divided by the difference of −7 and 8

$\dfrac{5(-6)}{-7-8}$ simplifies to $\dfrac{-30}{-15}$, which equals 2.   **NOW TRY**

OBJECTIVE 8 Translate simple sentences into equations.

EXAMPLE 8 Translating Sentences into Equations

Write each sentence as an equation, using x as the variable. Then find the solution of the equation from the set of integers between −12 and 12, inclusive.

(a) Three times a number is −18.

The word *times* indicates multiplication. The word *is* translates as =.

$3 \cdot x = -18,$ or $3x = -18$ $3 \cdot x = 3x$

The integer between −12 and 12, inclusive, that makes this statement true is −6 because $3(-6) = -18$. The solution of the equation is −6.

(b) The sum of a number and 9 is 12.

$$x + 9 = 12$$

Because $3 + 9 = 12$, the solution of this equation is 3.

(c) The difference of a number and 5 is 0.

$$x - 5 = 0$$

Because $5 - 5 = 0$, the solution of this equation is 5.

(d) The quotient of 24 and a number is −2.

$$\frac{24}{x} = -2$$

Here, x must be a negative number because the numerator is positive and the quotient is negative. Because $\dfrac{24}{-12} = -2$, the solution is −12.   **NOW TRY**

**NOW TRY
EXERCISE 8**

Write each sentence as an equation, using x as the variable. Then find the solution of the equation from the set of integers between −12 and 12, inclusive.

(a) The sum of a number and −4 is 7.

(b) The difference of −8 and a number is −11.

⚠ **CAUTION** In **Examples 6 and 7** above, the *phrases* translate as *expressions*, while in **Example 8,** the *sentences* translate as *equations*. *An expression is a phrase. An equation is a sentence with something on the left side, an = symbol, and something on the right side.*

$\dfrac{5(-6)}{-7-8}$ $3x = -18$

↑   ↑

Expression   Equation

1.5 Exercises

FOR EXTRA HELP   **MyMathLab®**

● *Complete solution available in MyMathLab*

Concept Check Fill in each blank with one of the following.

greater than 0   less than 0   equal to 0

1. The product or the quotient of two numbers with the same sign is _____.

2. The product or the quotient of two numbers with different signs is _____.

3. If three negative numbers are multiplied, the product is _____.

4. If two negative numbers are multiplied and then their product is divided by a negative number, the result is _____.

5. If a negative number is squared and the result is added to a positive number, the result is _____.

6. The reciprocal of a negative number is _____.

7. If three positive numbers, five negative numbers, and zero are multiplied, the product is _____.

8. The cube of a negative number is _____.

● **9.** *Concept Check* Complete this statement: The quotient formed by any nonzero number divided by 0 is _____, and the quotient formed by 0 divided by any nonzero number is _____. Give an example of each quotient.

10. *Concept Check* Which expression is undefined?

A. $\dfrac{4+4}{4+4}$ **B.** $\dfrac{4-4}{4+4}$ **C.** $\dfrac{4-4}{4-4}$ **D.** $\dfrac{4-4}{4}$

Find each product. See Examples 1 and 2.

11. $5(-6)$ **12.** $3(-4)$ **13.** $-5(-6)$ **14.** $-3(-4)$ **15.** $-10(-12)$

16. $-9(-5)$ **17.** $3(-11)$ **18.** $3(-15)$ **19.** $-0.5(0)$ **20.** $-0.3(0)$

21. $-6.8(0.35)$ **22.** $-4.6(0.24)$ **23.** $-\dfrac{3}{8} \cdot \left(-\dfrac{10}{9}\right)$ **24.** $-\dfrac{5}{6} \cdot \left(-\dfrac{16}{15}\right)$

25. $\dfrac{2}{15}\left(-1\dfrac{1}{4}\right)$ **26.** $\dfrac{3}{7}\left(-1\dfrac{5}{9}\right)$ **27.** $-8\left(-\dfrac{3}{4}\right)$ **28.** $-6\left(-\dfrac{2}{3}\right)$

Find all integer factors of each number. See Objective 3.

29. 32   **30.** 36   ● **31.** 40   **32.** 50   **33.** 31   **34.** 17

Find each quotient. See Example 3 and the discussion of division involving 0.

● **35.** $\dfrac{15}{5}$ **36.** $\dfrac{35}{5}$ **37.** $\dfrac{-42}{6}$ **38.** $\dfrac{-28}{7}$ **39.** $\dfrac{-32}{-4}$

40. $\dfrac{-35}{-5}$ ● **41.** $\dfrac{96}{-16}$ **42.** $\dfrac{38}{-19}$ **43.** $\dfrac{-8.8}{2.2}$ **44.** $\dfrac{-4.6}{0.23}$

45. $-\dfrac{4}{3} \div \left(-\dfrac{1}{8}\right)$ **46.** $-\dfrac{6}{5} \div \left(-\dfrac{1}{3}\right)$ **47.** $-\dfrac{5}{6} \div \dfrac{8}{9}$ **48.** $-\dfrac{7}{10} \div \dfrac{3}{4}$

49. $\dfrac{0}{-5}$ **50.** $\dfrac{0}{-9}$ **51.** $\dfrac{11.5}{0}$ **52.** $\dfrac{15.2}{0}$

Perform each indicated operation. See Example 4.

53. $7 - 3 \cdot 6$

54. $8 - 2 \cdot 5$

55. $-10 - (-4)(2)$

56. $-11 - (-3)(6)$

▶ **57.** $-7(3 - 8)$

58. $-5(4 - 7)$

59. $7 + 2(4 - 1)$

60. $5 + 3(6 - 4)$

61. $-4 + 3(2 - 8)$

62. $-8 + 4(5 - 7)$

63. $(12 - 14)(1 - 4)$

64. $(8 - 9)(4 - 12)$

65. $(7 - 10)(10 - 4)$

66. $(5 - 12)(19 - 4)$

67. $(-2 - 8)(-6) + 7$

68. $(-9 - 4)(-2) + 10$

69. $3(-5) + |3 - 10|$

70. $4(-8) + |4 - 15|$

71. $\dfrac{-5(-6)}{9 - (-1)}$

72. $\dfrac{-12(-5)}{7 - (-5)}$

73. $\dfrac{-21(3)}{-3 - 6}$

74. $\dfrac{-40(3)}{-2 - 3}$

75. $\dfrac{-10(2) + 6(2)}{-3 - (-1)}$

76. $\dfrac{-12(4) + 5(3)}{-14 - (-3)}$

77. $\dfrac{3^2 - 4^2}{7(-8 + 9)}$

78. $\dfrac{5^2 - 7^2}{2(3 + 3)}$

79. $\dfrac{8(-1) - |(-4)(-3)|}{-6 - (-1)}$

80. $\dfrac{-27(-2) - |6 \cdot 4|}{-2(3) - 2(2)}$

81. $\dfrac{-13(-4) - (-8)(-2)}{(-10)(2) - 4(-2)}$

82. $\dfrac{-5(2) + [3(-2) - 4]}{-3 - (-1)}$

A few years ago, the following question and expression appeared on boxes of Swiss Miss Hot Cocoa Mix:

On average, how many mini-marshmallows are in one serving?

$$3 + 2 \times 4 \div 2 - 3 \times 7 - 4 + 47$$

83. The box gave 92 as the answer. What is the *correct* answer?

84. **WHAT WENT WRONG?** Explain the algebraic error that somebody at the company made in calculating the answer.

Evaluate each expression for $x = 6$, $y = -4$, and $a = 3$. See Example 5.

85. $5x - 2y + 3a$

86. $6x - 5y + 4a$

▶ **87.** $(2x + y)(3a)$

88. $(5x - 2y)(-2a)$

89. $\left(\dfrac{1}{3}x - \dfrac{4}{5}y\right)\left(-\dfrac{1}{5}a\right)$

90. $\left(\dfrac{5}{6}x + \dfrac{3}{2}y\right)\left(-\dfrac{1}{3}a\right)$

91. $(-5 + x)(-3 + y)(3 - a)$

92. $(6 - x)(5 + y)(3 + a)$

93. $-2y^2 + 3a$

94. $5x - 4a^2$

95. $\dfrac{2y - x}{a - 3}$

96. $\dfrac{xy + 8a}{x - 6}$

Write a numerical expression for each phrase, and simplify the expression. See Examples 6 and 7.

97. The product of -9 and 2, added to 9

98. The product of 4 and -7, added to -12

▶ **99.** Twice the product of -1 and 6, subtracted from -4

100. Twice the product of -8 and 2, subtracted from -1

101. Nine subtracted from the product of 1.5 and -3.2

102. Three subtracted from the product of 4.2 and -8.5

103. The product of 12 and the difference of 9 and -8

104. The product of -3 and the difference of 3 and -7

▶ **105.** The quotient of -12 and the sum of -5 and -1

106. The quotient of -20 and the sum of -8 and -2

107. The sum of 15 and -3, divided by the product of 4 and -3

108. The sum of -18 and -6, divided by the product of 2 and -4

109. Two-thirds of the difference of 8 and -1

110. Three-fourths of the sum of -8 and 12

111. 20% of the product of -5 and 6

112. 30% of the product of -8 and 5

113. The sum of $\frac{1}{2}$ and $\frac{5}{8}$, times the difference of $\frac{3}{5}$ and $\frac{1}{3}$

114. The sum of $\frac{3}{4}$ and $\frac{1}{2}$, times the difference of $\frac{2}{3}$ and $\frac{1}{6}$

115. The product of $-\frac{1}{2}$ and $\frac{3}{4}$, divided by $-\frac{2}{3}$

116. The product of $-\frac{2}{3}$ and $-\frac{1}{5}$, divided by $\frac{1}{7}$

*Write each sentence as an equation, using x as the variable. Then find the solution of the equation from the set of integers between -12 and 12, inclusive. **See Example 8.***

▶ **117.** The quotient of a number and 3 is -3.

118. The quotient of a number and 4 is -1.

119. 6 less than a number is 4.

120. 7 less than a number is 2.

121. When 5 is added to a number, the result is -5.

122. When 6 is added to a number, the result is -3.

*To find the **average (mean)** of a group of numbers, we add the numbers and then divide the sum by the number of values added. For example, we find the average of 14, 8, 3, 9, and 1, as follows.*

$$\frac{14 + 8 + 3 + 9 + 1}{5} \quad \begin{matrix} \leftarrow \text{ Given numbers} \\ \leftarrow \text{ Number of values} \end{matrix}$$

$$= \frac{35}{5} \qquad \text{Add.}$$

$$\text{Average} \rightarrow \; = 7 \qquad \text{Divide.}$$

Find the average of each group of numbers.

123. 23, 18, 13, -4, and -8

124. 18, 12, 0, -4, and -10

125. -15, 29, 8, -6

126. -17, 34, 9, -2

127. All integers between -10 and 14, inclusive

128. All even integers between -18 and 4, inclusive

*The operation of division is used in **divisibility tests.** A divisibility test allows us to determine whether a given number is divisible (without remainder) by another number.*

129. An integer is divisible by 2 if its last digit is divisible by 2, and not otherwise. Show that

    (a) 3,473,986 is divisible by 2   and   **(b)** 4,336,879 is not divisible by 2.

130. An integer is divisible by 3 if the sum of its digits is divisible by 3, and not otherwise. Show that

    (a) 4,799,232 is divisible by 3   and   **(b)** 2,443,871 is not divisible by 3.

131. An integer is divisible by 4 if its last two digits form a number divisible by 4, and not otherwise. Show that

(a) 6,221,464 is divisible by 4   and   (b) 2,876,335 is not divisible by 4.

132. An integer is divisible by 5 if its last digit is divisible by 5, and not otherwise. Show that

(a) 3,774,595 is divisible by 5   and   (b) 9,332,123 is not divisible by 5.

133. An integer is divisible by 6 if it is divisible by both 2 and 3, and not otherwise. Show that

(a) 1,524,822 is divisible by 6   and   (b) 2,873,590 is not divisible by 6.

134. An integer is divisible by 8 if its last three digits form a number divisible by 8, and not otherwise. Show that

(a) 2,923,296 is divisible by 8   and   (b) 7,291,623 is not divisible by 8.

135. An integer is divisible by 9 if the sum of its digits is divisible by 9, and not otherwise. Show that

(a) 4,114,107 is divisible by 9   and   (b) 2,287,321 is not divisible by 9.

136. An integer is divisible by 12 if it is divisible by both 3 and 4, and not otherwise. Show that

(a) 4,253,520 is divisible by 12   and   (b) 4,249,474 is not divisible by 12.

SUMMARY EXERCISES Performing Operations with Real Numbers

Operations with Signed Numbers

Addition

Same sign   Add the absolute values of the numbers. The sum has the same sign as the addends.

Different signs   Find the absolute values of the numbers, and subtract the lesser absolute value from the greater. The sum has the same sign as the addend with greater absolute value.

Subtraction

Add the additive inverse (or opposite) of the subtrahend to the minuend.

Multiplication and Division

Same sign   The product or quotient of two numbers with the same sign is positive.

Different signs   The product or quotient of two numbers with different signs is negative.

Division by 0 is undefined.

Perform each indicated operation.

1. $14 - 3 \cdot 10$

2. $-3(8) - 4(-7)$

3. $(3 - 8)(-2) - 10$

4. $-6(7 - 3)$

5. $7 + 3(2 - 10)$

6. $-4[(-2)(6) - 7]$

7. $(-4)(7) - (-5)(2)$

8. $-5[-4 - (-2)(-7)]$

9. $40 - (-2)[8 - 9]$

1. −16   **2.** 4
3. 0   **4.** −24
5. −17   **6.** 76
7. −18   **8.** 90
9. 38

10. 4

11. -5 **12.** 5

13. $-\dfrac{7}{2}$, or $-3\dfrac{1}{2}$

14. 4

15. 13   **16.** $\dfrac{5}{4}$, or $1\dfrac{1}{4}$

17. 9   **18.** $\dfrac{37}{10}$, or $3\dfrac{7}{10}$

19. 0   **20.** 25

21. 14   **22.** undefined

23. -4 **24.** $\dfrac{6}{5}$, or $1\dfrac{1}{5}$

25. -1 **26.** $\dfrac{52}{37}$, or $1\dfrac{15}{37}$

27. $\dfrac{17}{16}$, or $1\dfrac{1}{16}$ **28.** $-\dfrac{2}{3}$

29. 3.33   **30.** 1.02

31. 0   **32.** 24

33. -7 **34.** -3

35. -1 **36.** $\dfrac{1}{2}$

37. $-\dfrac{5}{13}$ **38.** 5

39. $-\dfrac{8}{27}$ **40.** 4

10. $\dfrac{5(-4)}{-7-(-2)}$

11. $\dfrac{-3-(-9+1)}{-7-(-6)}$

12. $\dfrac{5(-8+3)}{13(-2)+(-7)(-3)}$

13. $\dfrac{6^2-8}{-2(2)+4(-1)}$

14. $\dfrac{16(-8+5)}{15(-3)+(-7-4)(-3)}$

15. $\dfrac{9(-6)-3(8)}{4(-7)+(-2)(-11)}$

16. $\dfrac{2^2+4^2}{5^2-3^2}$

17. $\dfrac{(2+4)^2}{(5-3)^2}$

18. $\dfrac{4^3-3^3}{-5(-4+2)}$

19. $\dfrac{-9(-6)+(-2)(27)}{3(8-9)}$

20. $|-4(9)|-|-11|$

21. $\dfrac{6(-10+3)}{15(-2)-3(-9)}$

22. $\dfrac{3^2-5^2}{(-9)^2-9^2}$

23. $\dfrac{(-10)^2+10^2}{-10(5)}$

24. $-\dfrac{3}{4}\div\left(-\dfrac{5}{8}\right)$

25. $\dfrac{1}{2}\div\left(-\dfrac{1}{2}\right)$

26. $\dfrac{8^2-12}{(-5)^2+2(6)}$

27. $\left[\dfrac{5}{8}-\left(-\dfrac{1}{16}\right)\right]+\dfrac{3}{8}$

28. $\left(\dfrac{1}{2}-\dfrac{1}{3}\right)-\dfrac{5}{6}$

29. $-0.9(-3.7)$

30. $-5.1(-0.2)$

31. $|-2(3)+4|-|-2|$

32. $40+2(-5-3)$

Evaluate each expression for $x=-2$, $y=3$, *and* $a=4$.

33. $-x+y-3a$

34. $(x-y)-(a-2y)$

35. $\left(\dfrac{1}{2}x+\dfrac{2}{3}y\right)\left(-\dfrac{1}{4}a\right)$

36. $\dfrac{2x+3y}{a-xy}$

37. $\dfrac{x^2-y^2}{x^2+y^2}$

38. $-x^2+3y$

39. $\left(\dfrac{x}{y}\right)^3$

40. $\left(\dfrac{a}{x}\right)^2$

1.6 Properties of Real Numbers

OBJECTIVES

1. Use the commutative properties.
2. Use the associative properties.
3. Use the identity properties.
4. Use the inverse properties.
5. Use the distributive property.

In the basic properties covered in this section, a, b, and c represent real numbers.

OBJECTIVE 1 Use the commutative properties.

The word *commute* means to go back and forth. We might commute to work or to school. If we travel from home to work and follow the same route from work to home, we travel the same distance each time. The **commutative properties** say that if two numbers are added or multiplied in either order, the result is the same.

Commutative Properties

$$a+b=b+a \qquad \text{Addition}$$

$$ab=ba \qquad \text{Multiplication}$$

**NOW TRY
EXERCISE 1**

Use a commutative property
to complete each statement.

(a) $7 + (-3) = -3 +$ _____

(b) $(-5)4 = 4 \cdot$ _____

**NOW TRY
EXERCISE 2**

Use an associative property to
complete each statement.

(a) $-9 + (3 + 7) =$ _____

(b) $5[(-4) \cdot 9] =$ _____

EXAMPLE 1 Using the Commutative Properties

Use a commutative property to complete each statement.

(a) $-8 + 5 = 5 + \underline{\ ?\ }$ Notice that the "order" changed.

$-8 + 5 = 5 + (-8)$ Commutative property of addition

(b) $(-2)7 = \underline{\ ?\ } (-2)$

$-2(7) = 7(-2)$ Commutative property of multiplication   **NOW TRY**

OBJECTIVE 2 Use the associative properties.

When we *associate* one object with another, we think of those objects as being grouped together. The **associative properties** say that when we add or multiply three numbers, we can group the first two together or the last two together and obtain the same answer.

> **Associative Properties**
>
> $$(a + b) + c = a + (b + c) \quad \text{Addition}$$
> $$(ab)c = a(bc) \quad \text{Multiplication}$$

EXAMPLE 2 Using the Associative Properties

Use an associative property to complete each statement.

(a) $-8 + (1 + 4) = (-8 + \underline{\ ?\ }) + 4$ The "order" is the same.
The "grouping" changed.

$-8 + (1 + 4) = (-8 + 1) + 4$ Associative property of addition

(b) $[2 \cdot (-7)] \cdot 6 = 2 \cdot \underline{\ ?\ }$

$[2 \cdot (-7)] \cdot 6 = 2 \cdot [(-7) \cdot 6]$ Associative property of multiplication

NOW TRY

By the associative property, the sum (or product) of three numbers will be the same no matter how the numbers are "associated" in groups. Parentheses can be left out if a problem contains only addition (or multiplication). For example,

$$(-1 + 2) + 3 \quad \text{and} \quad -1 + (2 + 3) \quad \text{can be written as} \quad -1 + 2 + 3.$$

EXAMPLE 3 Distinguishing between Properties

Decide whether each statement is an example of a commutative property, an associative property, or both.

(a) $(2 + 4) + 5 = 2 + (4 + 5)$

The order of the three numbers is the same on both sides of the equality symbol. The only change is in the *grouping,* or association, of the numbers. This is an example of the associative property.

(b) $6 \cdot (3 \cdot 10) = 6 \cdot (10 \cdot 3)$

The same numbers, 3 and 10, are grouped on each side. On the left, the 3 appears first, but on the right, the 10 appears first. The only change involves the *order* of the numbers, so this is an example of the commutative property.

NOW TRY ANSWERS
1. **(a)** 7 **(b)** -5
2. **(a)** $(-9 + 3) + 7$
   (b) $[5 \cdot (-4)] \cdot 9$

Is $5 + (7 + 6) = 5 + (6 + 7)$ an example of a commutative property or an associative property?

(c) $(8 + 1) + 7 = 8 + (7 + 1)$

Both the order and the grouping are changed. On the left, the order of the three numbers is 8, 1, and 7. On the right, it is 8, 7, and 1. On the left, the 8 and 1 are grouped. On the right, the 7 and 1 are grouped. Therefore, *both* properties are used.

NOW TRY

Find each sum or product.

(a) $8 + 54 + 7 + 6 + 32$

(b) $5(37)(20)$

EXAMPLE 4 Using the Commutative and Associative Properties

Find each sum or product.

(a) $23 + 41 + 2 + 9 + 25$

$= (41 + 9) + (23 + 2) + 25$

$= 50 + 25 + 25$ Use the commutative and associative properties.

$= 100$

(b) $25(69)(4)$

$= 25(4)(69)$

$= 100(69)$

$= 6900$ **NOW TRY**

OBJECTIVE 3 Use the identity properties.

If a child wears a costume on Halloween, the child's appearance is changed, but his or her *identity* is unchanged. The identity of a real number is unchanged when identity properties are applied.

The **identity properties** say that the sum of 0 and any number equals that number, and the product of 1 and any number equals that number.

Identity Properties

$$a + 0 = a \quad \text{and} \quad 0 + a = a \quad \text{Addition}$$

$$a \cdot 1 = a \quad \text{and} \quad 1 \cdot a = a \quad \text{Multiplication}$$

The number 0 leaves the identity, or value, of any real number unchanged by addition, so 0 is the **identity element for addition,** or the **additive identity.** Because multiplication by 1 leaves any real number unchanged, 1 is the **identity element for multiplication,** or the **multiplicative identity.**

Use an identity property to complete each statement.

(a) $\dfrac{2}{5} \cdot \underline{} = \dfrac{2}{5}$

(b) $8 + \underline{} = 8$

EXAMPLE 5 Using the Identity Properties

Use an identity property to complete each statement.

(a) $-3 + \underline{\,?\,} = -3$

$-3 + 0 = -3$

Identity property of addition

(b) $\underline{\,?\,} \cdot \dfrac{1}{2} = \dfrac{1}{2}$

$1 \cdot \dfrac{1}{2} = \dfrac{1}{2}$

Identity property of multiplication

NOW TRY

NOW TRY ANSWERS

3. commutative

4. (a) 107 **(b)** 3700

5. (a) 1 **(b)** 0

NOW TRY
EXERCISE 6

Simplify.

(a) $\dfrac{16}{20}$ (b) $\dfrac{2}{5} + \dfrac{3}{20}$

EXAMPLE 6 **Using the Identity Property to Simplify Expressions**

Simplify. In part (a), write in lowest terms. In part (b), perform the operation.

(a) $\dfrac{49}{35}$

$= \dfrac{7 \cdot 7}{5 \cdot 7}$ Factor.

$= \dfrac{7}{5} \cdot \dfrac{7}{7}$ Write as a product.

$= \dfrac{7}{5} \cdot 1$ Divide.

$= \dfrac{7}{5}$ Identity property

(b) $\dfrac{3}{4} + \dfrac{5}{24}$

$= \dfrac{3}{4} \cdot 1 + \dfrac{5}{24}$ Identity property

$= \dfrac{3}{4} \cdot \dfrac{6}{6} + \dfrac{5}{24}$ Use $1 = \dfrac{6}{6}$ to obtain a common denominator.

$= \dfrac{18}{24} + \dfrac{5}{24}$ Multiply.

$= \dfrac{23}{24}$ Add.   **NOW TRY**

OBJECTIVE 4 Use the inverse properties.

Each day before we go to work or school, we probably put on our shoes. Before we go to sleep at night, we probably take them off, and this leads to the same situation that existed before we put them on. These operations from everyday life are examples of *inverse* operations.

The **inverse properties** of addition and multiplication lead to the additive and multiplicative identities, respectively. Recall that $-a$ is the **additive inverse,** or **opposite,** of a and $\dfrac{1}{a}$ is the **multiplicative inverse,** or **reciprocal,** of the nonzero number a.

Inverse Properties

$a + (-a) = 0$ and $-a + a = 0$ Addition

$a \cdot \dfrac{1}{a} = 1$ and $\dfrac{1}{a} \cdot a = 1$ $(a \neq 0)$ Multiplication

---

NOW TRY
EXERCISE 7

Use an inverse property to complete each statement.

(a) $10 + \underline{\qquad} = 0$

(b) $-9 \cdot \underline{\qquad} = 1$

EXAMPLE 7 **Using the Inverse Properties**

Use an inverse property to complete each statement.

(a) $\underline{\ ?\ } + \dfrac{1}{2} = 0$

$-\dfrac{1}{2} + \dfrac{1}{2} = 0$

(b) $4 + \underline{\ ?\ } = 0$

$4 + (-4) = 0$

(c) $-0.75 + \dfrac{3}{4} = \underline{\ ?\ }$

$-0.75 + \dfrac{3}{4} = 0$

The inverse property of addition is used in parts (a)–(c).

(d) $\underline{\ ?\ } \cdot \dfrac{5}{2} = 1$

$\dfrac{2}{5} \cdot \dfrac{5}{2} = 1$

(e) $-5(\underline{\ ?\ }) = 1$

$-5\left(-\dfrac{1}{5}\right) = 1$

(f) $4(0.25) = \underline{\ ?\ }$

$4(0.25) = 1$

The inverse property of multiplication is used in parts (d)–(f).   **NOW TRY**

NOW TRY ANSWERS

6. (a) $\dfrac{4}{5}$ (b) $\dfrac{11}{20}$

7. (a) -10 (b) $-\dfrac{1}{9}$

**NOW TRY
EXERCISE 8**

Simplify.

$$-\frac{1}{3}x + 7 + \frac{1}{3}x$$

EXAMPLE 8   Using Properties to Simplify an Expression

Simplify.

$$-2x + 10 + 2x$$

$= (-2x + 10) + 2x$	Order of operations
$= [10 + (-2x)] + 2x$	Commutative property
$= 10 + [(-2x) + 2x]$	Associative property
$= 10 + 0$	Inverse property
$= 10$	Identity property **NOW TRY**

For *any* value of *x*, $-2x$ and $2x$ are additive inverses.

NOTE   The steps of **Example 8** may be skipped when we actually do the simplification.

OBJECTIVE 5   Use the distributive property.

The word *distribute* means "to give out from one to several." Consider the following expressions.

$$2(5 + 8) \quad \text{equals} \quad 2(13), \quad \text{which equals} \quad 26.$$

$$2(5) + 2(8) \quad \text{equals} \quad 10 + 16, \quad \text{which equals} \quad 26.$$

Because both expressions equal 26,

$$2(5 + 8) = 2(5) + 2(8).$$

This is an example of the *distributive property of multiplication with respect to addition,* the only property involving *both* addition and multiplication. With this property, a product can be changed to a sum or difference. This idea is illustrated in **FIGURE 16**.

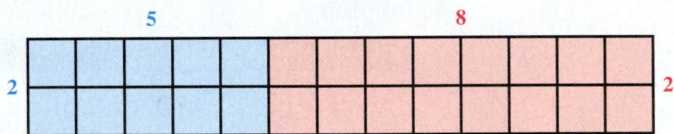

The area of the left part is 2(5) = 10.
The area of the right part is 2(8) = 16.
The total area is 2(5 + 8) = 2(13) = 26,
or the total area is 2(5) + 2(8) = 10 + 16 = 26.
Thus, 2(5 + 8) = 2(5) + 2(8).

FIGURE 16

The **distributive property** says that multiplying a number a by a sum of numbers $b + c$ gives the same result as multiplying a by b and a by c and then adding the two products.

Distributive Property

$$a(b + c) = ab + ac \qquad \text{and} \qquad (b + c)a = ba + ca$$

As the arrows show, the a outside the parentheses is "distributed" over the b and c inside. The distributive property is also valid for multiplication over subtraction.

$$a(b - c) = ab - ac \qquad \text{and} \qquad (b - c)a = ba - ca$$

NOW TRY ANSWER
8. 7

The distributive property can be extended to more than two numbers.

$$a(b + c + d) = ab + ac + ad$$

NOW TRY EXERCISE 9

Use the distributive property to rewrite each expression.

(a) $2(p + 5)$

(b) $-5(4x + 1)$

(c) $6(2r + t - 5z)$

EXAMPLE 9 Using the Distributive Property

Use the distributive property to rewrite each expression.

(a)
$$5(9 + 6)$$

> We could write $5(9) + 5(6)$ here.

$$= 5 \cdot 9 + 5 \cdot 6 \qquad \text{The factor 5 is "distributed" to the numbers 9 and 6.}$$

> Multiply first.

$$= 45 + 30 \qquad \text{Multiply.}$$

$$= 75 \qquad \text{Add.}$$

(b) $4(x + 5 + y)$

$$= 4x + 4 \cdot 5 + 4y \qquad \text{Distributive property}$$

$$= 4x + 20 + 4y \qquad \text{Multiply.}$$

(c)
$$-\frac{1}{2}(4x + 3)$$

> Think: $-\frac{1}{2}(4x) = \left(-\frac{1}{2} \cdot 4\right)x = \left(-\frac{1}{2} \cdot \frac{4}{1}\right)x$

$$= -\frac{1}{2}(4x) + \left(-\frac{1}{2}\right)(3) \qquad \text{Distributive property}$$

> This step is often omitted.

$$= -2x + \left(-\frac{3}{2}\right) \qquad \text{Multiply.}$$

$$= -2x - \frac{3}{2} \qquad \text{Definition of subtraction}$$

(d) $3(k - 9)$

> Be careful here.

$$= 3[k + (-9)] \qquad \text{Definition of subtraction}$$

$$= 3k + 3(-9) \qquad \text{Distributive property}$$

$$= 3k - 27 \qquad \text{Multiply; definition of subtraction}$$

(e) $-2(3x - 4)$

$$= -2[3x + (-4)] \qquad \text{Definition of subtraction}$$

$$= -2(3x) + (-2)(-4) \qquad \text{Distributive property}$$

$$= (-2 \cdot 3)x + (-2)(-4) \qquad \text{Associative property}$$

$$= -6x + 8 \qquad \text{Multiply.}$$

(f)
$$8(3r + 11t + 5z)$$

$$= 8(3r) + 8(11t) + 8(5z) \qquad \text{Distributive property}$$

> This step is often omitted.

$$= (8 \cdot 3)r + (8 \cdot 11)t + (8 \cdot 5)z \qquad \text{Associative property}$$

$$= 24r + 88t + 40z \qquad \text{Multiply.} \qquad \text{NOW TRY}$$

⚠ **CAUTION** In practice, we often omit the first step in **Examples 9(d) and 9(e),** where we rewrote the subtraction as addition of the additive inverse.

$$3(k - 9) \qquad \text{Example 9(d)}$$

$$= 3k - 3(9) \qquad \text{Be careful not to make a sign error.}$$

$$= 3k - 27 \qquad \text{Multiply.}$$

NOW TRY ANSWERS

9. (a) $2p + 10$ **(b)** $-20x - 5$
(c) $12r + 6t - 30z$

The expression $-a$ may be interpreted as $-1 \cdot a$. Using this result and the distributive property, we can *remove* (or *clear*) *parentheses* from some expressions.

NOW TRY
EXERCISE 10

Write each expression without parentheses.

(a) $-(2 - r)$

(b) $-(2x - 5y - 7)$

EXAMPLE 10   Using the Distributive Property to Remove (Clear) Parentheses

Write each expression without parentheses.

(a)
$$-(2y + 3)$$

> The − symbol indicates a factor of −1.

$$= -1 \cdot (2y + 3) \qquad -a = -1 \cdot a$$
$$= -1 \cdot 2y + (-1) \cdot 3 \qquad \text{Distributive property}$$
$$= -2y - 3 \qquad \text{Multiply; definition of subtraction}$$

(b) $-(-9w - 2)$

$$= -1(-9w - 2)$$
$$= -1(-9w) - 1(-2)$$
$$= 9w + 2$$

We can also interpret the negative sign in front of the parentheses to mean the *opposite* of each of the terms within the parentheses.

$$-1(-9w - 2)$$
$$\downarrow \qquad \downarrow$$
$$= +9w + 2$$

(c)
$$-(-x - 3y + 6z)$$
$$= -1(-1x - 3y + 6z) \qquad \boxed{\text{Be careful with signs.}}$$
$$= -1(-1x) - 1(-3y) - 1(6z) \qquad \text{Distributive property}$$

> $-1(-1x) = 1x = x$

$$= x + 3y - 6z \qquad \text{Multiply.} \qquad \text{NOW TRY}$$

Here is a summary of the basic properties of real numbers.

Properties of Addition and Multiplication

For any real numbers a, b, and c, the following properties hold.

Commutative Properties $a + b = b + a \qquad ab = ba$

Associative Properties $(a + b) + c = a + (b + c)$
$$(ab)c = a(bc)$$

Identity Properties There is a real number 0 such that
$$a + 0 = a \qquad \text{and} \qquad 0 + a = a.$$

There is a real number 1 such that
$$a \cdot 1 = a \qquad \text{and} \qquad 1 \cdot a = a.$$

Inverse Properties For each real number a, there is a single real number $-a$ such that
$$a + (-a) = 0 \qquad \text{and} \qquad (-a) + a = 0.$$

For each nonzero real number a, there is a single real number $\frac{1}{a}$ such that
$$a \cdot \frac{1}{a} = 1 \qquad \text{and} \qquad \frac{1}{a} \cdot a = 1.$$

Distributive Properties $a(b + c) = ab + ac \qquad (b + c)a = ba + ca$

NOW TRY ANSWERS
10. (a) $-2 + r$
    (b) $-2x + 5y + 7$

1.6 Exercises

  MyMathLab®

▶ *Complete solution available in MyMathLab*

1. *Concept Check* Match each item in Column I with the correct choice(s) from Column II. Choices may be used once, more than once, or not at all.

I

(a) Identity element for addition

(b) Identity element for multiplication

(c) Additive inverse of a

(d) Multiplicative inverse, or reciprocal, of the nonzero number a

(e) The number that is its own additive inverse

(f) The two numbers that are their own multiplicative inverses

(g) The only number that has no multiplicative inverse

(h) An example of the associative property

(i) An example of the commutative property

(j) An example of the distributive property

II

A. $(5 \cdot 4) \cdot 3 = 5 \cdot (4 \cdot 3)$

B. 0

C. $-a$

D. -1

E. $5 \cdot 4 \cdot 3 = 60$

F. 1

G. $(5 \cdot 4) \cdot 3 = 3 \cdot (5 \cdot 4)$

H. $5(4 + 3) = 5 \cdot 4 + 5 \cdot 3$

I. $\dfrac{1}{a}$

2. *Concept Check* Complete each statement.

The commutative property allows us to change the (*order / grouping*) of the addends in a sum or the factors in a product.

The associative property allows us to change the (*order / grouping*) of the addends in a sum or the factors in a product.

Concept Check Tell whether or not the following everyday activities are commutative.

3. Washing your face and brushing your teeth

4. Putting on your left sock and putting on your right sock

5. Preparing a meal and eating a meal

6. Starting a car and driving away in a car

7. Putting on your socks and putting on your shoes

8. Getting undressed and taking a shower

Concept Check Work each problem involving the properties of real numbers.

9. Use parentheses to show how the associative property can be used to give two different meanings to the phrase "foreign sales clerk."

10. Use parentheses to show how the associative property can be used to give two different meanings to the phrase "defective merchandise counter."

11. Evaluate the following expressions.

$$25 - (6 - 2) \quad \text{and} \quad (25 - 6) - 2.$$

Do you think subtraction is associative?

12. Evaluate the following expressions.

$$180 \div (15 \div 3) \quad \text{and} \quad (180 \div 15) \div 3.$$

Do you think division is associative?

13. Complete the table and each statement beside it.

Number	Additive Inverse	Multiplicative Inverse
5		
-10		
$-\frac{1}{2}$		
$\frac{3}{8}$		
x		$(x \neq 0)$
$-y$		$(y \neq 0)$

In general, a number and its additive inverse have (*the same*/*opposite*) signs.

A number and its multiplicative inverse have (*the same*/*opposite*) signs.

14. The following conversation took place between one of the authors of this book and his son, Jack, when Jack was 4 years old.

DADDY: "Jack, what is 3 + 0?"

JACK: "3."

DADDY: "Jack, what is 4 + 0?"

JACK: "4. And Daddy, *string* plus zero equals *string*!"

What property of addition did Jack recognize?

Use a commutative or an associative property to complete each statement. State which property is used. **See Examples 1 and 2.**

15. $-15 + 9 = 9 +$ _____

16. $6 + (-2) = -2 +$ _____

17. $-8 \cdot 3 =$ _____ $\cdot (-8)$

18. $-12 \cdot 4 = 4 \cdot$ _____

19. $(3 + 6) + 7 = 3 + ($ _____ $+ 7)$

20. $(-2 + 3) + 6 = -2 + ($ _____ $+ 6)$

21. $7 \cdot (2 \cdot 5) = ($ _____ $\cdot 2) \cdot 5$

22. $8 \cdot (6 \cdot 4) = (8 \cdot$ _____ $) \cdot 4$

Decide whether each statement is an example of a commutative, *an* associative, *an* identity, *an* inverse, *or the* distributive property. **See Examples 1, 2, 3, 5, 6, 7, and 9.**

23. $4 + 15 = 15 + 4$

24. $3 + 12 = 12 + 3$

25. $5 \cdot (13 \cdot 7) = (5 \cdot 13) \cdot 7$

26. $-4 \cdot (2 \cdot 6) = (-4 \cdot 2) \cdot 6$

▶ **27.** $-6 + (12 + 7) = (-6 + 12) + 7$

28. $(-8 + 13) + 2 = -8 + (13 + 2)$

29. $-9 + 9 = 0$

30. $1 + (-1) = 0$

▶ **31.** $\frac{2}{3}\left(\frac{3}{2}\right) = 1$

32. $\frac{5}{8}\left(\frac{8}{5}\right) = 1$

33. $1.75 + 0 = 1.75$

34. $-8.45 + 0 = -8.45$

35. $(4 + 17) + 3 = 3 + (4 + 17)$

36. $(-8 + 4) + 12 = 12 + (-8 + 4)$

37. $2(x + y) = 2x + 2y$

38. $9(t + s) = 9t + 9s$

▶ **39.** $-\frac{5}{9} = -\frac{5}{9} \cdot \frac{3}{3} = -\frac{15}{27}$

40. $-\frac{7}{12} = -\frac{7}{12} \cdot \frac{7}{7} = -\frac{49}{84}$

41. $4(2x) + 4(3y) = 4(2x + 3y)$

42. $6(5t) - 6(7r) = 6(5t - 7r)$

Find each sum or product. **See Example 4.**

43. $97 + 13 + 3 + 37$

44. $49 + 199 + 1 + 1$

▶ **45.** $1999 + 2 + 1 + 8$

46. $2998 + 3 + 2 + 17$

47. $159 + 12 + 141 + 88$

48. $106 + 8 + (-6) + (-8)$

49. $843 + 627 + (-43) + (-27)$ **50.** $1846 + 1293 + (-46) + (-93)$

51. $5(47)(2)$ **52.** $2(79)5$

53. $-4 \cdot 5 \cdot 93 \cdot 5$ **54.** $2 \cdot 25 \cdot 67 \cdot (-2)$

Simplify each expression. ***See Examples 7 and 8.***

▶ **55.** $6t + 8 - 6t + 3$ **56.** $9r + 12 - 9r + 1$ **57.** $\dfrac{2}{3}x - 11 + 11 - \dfrac{2}{3}x$

58. $\dfrac{1}{5}y + 4 - 4 - \dfrac{1}{5}y$ **59.** $\left(\dfrac{9}{7}\right)(-0.38)\left(\dfrac{7}{9}\right)$ **60.** $\left(\dfrac{4}{5}\right)(-0.73)\left(\dfrac{5}{4}\right)$

61. $t + (-t) + \dfrac{1}{2}(2)$ **62.** $w + (-w) + \dfrac{1}{4}(4)$

63. *Concept Check* A student used the distributive property to rewrite the expression $-3(4 - 6)$ as shown.

$$-3(4 - 6)$$
$$= -3(4) - 3(6)$$
$$= -12 - 18$$
$$= -30$$

This answer is incorrect. **WHAT WENT WRONG?** Rewrite the given expression correctly.

64. *Concept Check* A student wrote the expression $-(3x + 4)$ without parentheses as shown.

$$-(3x + 4)$$
$$= -1(3x + 4)$$
$$= -1(3x) + 4$$
$$= -3x + 4$$

This answer is incorrect. **WHAT WENT WRONG?** Rewrite the given expression correctly.

65. Explain how the procedure for changing $\dfrac{3}{4}$ to $\dfrac{9}{12}$ requires the use of the multiplicative identity element, 1.

66. Explain how the procedure for changing $\dfrac{9}{12}$ to $\dfrac{3}{4}$ requires the use of the multiplicative identity element, 1.

Use the distributive property to rewrite each expression. ***See Example 9.***

67. $5(9 + 8)$ **68.** $6(11 + 8)$ ▶ **69.** $4(t + 3)$

70. $5(w + 4)$ **71.** $7(z - 8)$ **72.** $8(x - 6)$

73. $-8(r + 3)$ **74.** $-11(x + 4)$ **75.** $-\dfrac{1}{4}(8x + 3)$

76. $-\dfrac{1}{3}(9x + 5)$ **77.** $-5(y - 4)$ **78.** $-9(g - 4)$

79. $2(6x + 5)$ **80.** $3(3x + 4)$ **81.** $-3(2x - 5)$

82. $-4(3x - 2)$ **83.** $-6(8x + 1)$ **84.** $-5(4x + 1)$

85. $-\dfrac{4}{3}(12y + 15z)$ **86.** $-\dfrac{2}{5}(10b + 20a)$ **87.** $8(3r + 4s - 5y)$

88. $2(5u - 3v + 7w)$ **89.** $-3(8x + 3y + 4z)$ **90.** $-5(2x - 5y + 6z)$

Write each expression without parentheses. ***See Example 10.***

▶ **91.** $-(4t + 3m)$ **92.** $-(9x + 12y)$ **93.** $-(-5c - 4d)$

94. $-(-13x - 15y)$ **95.** $-(-q + 5r - 8s)$ **96.** $-(-z + 5w - 9y)$

1.7 Simplifying Expressions

VOCABULARY

☐ term
☐ numerical coefficient (coefficient)
☐ like terms
☐ unlike terms

NOW TRY EXERCISE 1

Simplify each expression.

(a) $3(2x - 4y)$

(b) $-4 - (-3y + 5)$

▼ Terms and Their Coefficients

Term	Numerical Coefficient
8	8
$-7y$	-7
$34r^3$	34
$-26x^5yz^4$	-26
$-k$, or $-1k$	-1
r, or $1r$	1
$\frac{3x}{8}$, or $\frac{3}{8}x$	$\frac{3}{8}$
$\frac{x}{3} = \frac{1x}{3}$, or $\frac{1}{3}x$	$\frac{1}{3}$

NOW TRY ANSWERS
1. (a) $6x - 12y$ (b) $3y - 9$

OBJECTIVE 1 Simplify expressions.

We use the properties of **Section 1.6** to write expressions in simpler form.

EXAMPLE 1 Simplifying Expressions

Simplify each expression.

(a) $4x + 8 + 9$ simplifies to $4x + 17$.

(b) $4(3m - 2n)$ To simplify, we clear the parentheses.

$$= 4(3m) - 4(2n) \qquad \text{Distributive property}$$
$$= (4 \cdot 3)m - (4 \cdot 2)n \qquad \text{Associative property}$$
$$= 12m - 8n \qquad \text{Multiply.}$$

(c) $\qquad 6 + 3(4k + 5)$ Do *not* start by adding.

$$= 6 + 3(4k) + 3(5) \qquad \text{Distributive property}$$
$$= 6 + (3 \cdot 4)k + 3(5) \qquad \text{Associative property}$$
$$= 6 + 12k + 15 \qquad \text{Multiply.}$$
$$= 6 + 15 + 12k \qquad \text{Commutative property}$$
$$= 21 + 12k \qquad \text{Add.}$$

(d) $\qquad 5 - (2y - 8)$

$$= 5 - 1(2y - 8) \qquad -a = -1 \cdot a$$
$$= 5 - 1(2y) - 1(-8) \qquad \text{Distributive property}$$
$$= 5 - 2y + 8 \qquad \text{Multiply.}$$ Be careful with signs.
$$= 5 + 8 - 2y \qquad \text{Commutative property}$$
$$= 13 - 2y \qquad \text{Add.}$$ NOW TRY

NOTE The steps using the commutative and associative properties will not be shown in the rest of the examples. However, be aware that they are usually involved.

OBJECTIVE 2 Identify terms and numerical coefficients.

A **term** is a number (constant), a variable, or a product or quotient of numbers and variables raised to powers.

$$9x, \quad 15y^2, \quad -3, \quad -8m^2n, \quad \frac{2}{p}, \quad \text{and} \quad k \qquad \text{Terms}$$

In the term $9x$, the **numerical coefficient,** or simply the **coefficient,** of the variable x is 9.

> ⚠️ **CAUTION** It is important to be able to distinguish between *terms* and *factors*. Consider the following expressions.

$$8x^3 + 12x^2$$ This expression has **two terms**, $8x^3$ and $12x^2$. Terms are separated by a $+$ or $-$ symbol.

$$(8x^3)(12x^2)$$ This is a **one-term** expression. The **factors** $8x^3$ and $12x^2$ are multiplied.

OBJECTIVE 3 Identify like terms.

Terms with exactly the same variables that have the same exponents on the variables are **like terms**.

Like Terms	**Unlike Terms**	
$9t$ and $4t$	$4y$ and $7t$	Different variables
$6x^2$ and $-5x^2$	$17x$ and $-8x^2$	Different exponents
$-2pq$ and $11pq$	$4xy^2$ and $4xy$	Different exponents
$3x^2y$ and $5x^2y$	$-7wz^3$ and $2xz^3$	Different variables

OBJECTIVE 4 Combine like terms.

The distributive property

$$a(b + c) = ab + ac \quad \text{can be written "in reverse" as} \quad ab + ac = a(b + c).$$

This last form, which may be used to find the sum or difference of like terms, provides justification for **combining like terms**.

NOW TRY EXERCISE 2

Simplify each expression.
(a) $4x + 6x - 7x$ (b) $z + z$
(c) $4p^2 - 3p^2$

EXAMPLE 2 Combining Like Terms

Simplify each expression.

(a) $-9m + 5m$
$= (-9 + 5)m$ Distributive property in reverse
$= -4m$

(b) $6r + 3r + 2r$
$= (6 + 3 + 2)r$
$= 11r$

(c) $4x + x$
$= 4x + 1x$ $x = 1x$
$= (4 + 1)x$
$= 5x$

(d) $16y^2 - 9y^2$
$= (16 - 9)y^2$
$= 7y^2$

(e) $32y + 10y^2$ These unlike terms cannot be combined.   **NOW TRY**

> ⚠️ **CAUTION** Only like terms may be combined.

NOW TRY ANSWERS
2. (a) $3x$ (b) $2z$ (c) p^2

Simplifying an Expression

An expression has been simplified when the following conditions have been met.

- All grouping symbols have been removed.

- All like terms have been combined.

- Operations have been performed, when possible.

EXAMPLE 3   Simplifying Expressions

Simplify each expression.

(a) $14y + 2(6 + 3y)$ *Start by distributing the 2.*

$= 14y + 2(6) + 2(3y)$ Distributive property

$14y + 6y$
$= (14 + 6)y$
$= 20y$

$= 14y + 12 + 6y$ Multiply.

$= 20y + 12$ Combine like terms.

(b) $9k - 6 - 3(2 - 5k)$ *Be careful with signs.*

$= 9k - 6 - 3(2) - 3(-5k)$ Distributive property

$= 9k - 6 - 6 + 15k$ Multiply.

$= 24k - 12$ Combine like terms.

(c) $-(2 - r) + 10r$

$= -1(2 - r) + 10r$ $-a = -1 \cdot a$, or $-1(a)$

$= -1(2) - 1(-r) + 10r$ Distributive property

Be careful with signs. $= -2 + 1r + 10r$ Multiply.

$= -2 + 11r$ Combine like terms.

Alternatively, $-(2 - r)$ can be thought of as the *opposite* of $(2 - r)$ —that is, $-2 + r$—which can then be added to $10r$ to obtain $-2 + 11r$.

(d) $100[0.03(x + 4)]$

$= [(100)(0.03)](x + 4)$ Associative property

$= 3(x + 4)$ Multiply.

$= 3x + 3(4)$ Distributive property

$= 3x + 12$ Multiply.

(e) $5(2a - 6) - 3(4a - 9)$

$= 5(2a) + 5(-6) - 3(4a) - 3(-9)$ Distributive property twice

$= 10a - 30 - 12a + 27$ Multiply.

$= -2a - 3$ Combine like terms.

NOW TRY
EXERCISE 3

Simplify each expression.

(a) $5k - 6 - (3 - 4k)$

(b) $\dfrac{1}{4}x - \dfrac{2}{3}(x - 9)$

(f) $-\dfrac{2}{3}(x - 6) - \dfrac{1}{6}x$

$$= -\dfrac{2}{3}x - \dfrac{2}{3}(-6) - \dfrac{1}{6}x \qquad \text{Distributive property}$$

$$= -\dfrac{2}{3}x + 4 - \dfrac{1}{6}x \qquad \text{Multiply.}$$

$$= -\dfrac{4}{6}x + 4 - \dfrac{1}{6}x \qquad \text{Find a common denominator.}$$

$$= -\dfrac{5}{6}x + 4 \qquad \text{Combine like terms.} \qquad \text{NOW TRY} \ \text{\textcircled{↻}}$$

NOTE Examples 2 and 3 suggest that like terms may be combined by adding or subtracting the coefficients of the terms and keeping the same variable factors.

OBJECTIVE 5 Simplify expressions from word phrases.

EXAMPLE 4 Translating Words into a Mathematical Expression

NOW TRY
EXERCISE 4

Translate the phrase into a mathematical expression using x as the variable, and simplify.

Twice a number, subtracted from the sum of the number and 5

Translate the phrase into a mathematical expression using x as the variable, and simplify.

The sum of 9, five times a number,
four times the number, and
six times the number

The word "sum" indicates that the terms should be added. Use x for the number.

$$9 + 5x + 4x + 6x \quad \text{simplifies to} \quad 9 + 15x. \qquad \text{Combine like terms.}$$

> This is an expression to be simplified, *not* an equation to be solved.

NOW TRY ANSWERS

3. **(a)** $9k - 9$ **(b)** $-\dfrac{5}{12}x + 6$

4. $(x + 5) - 2x; \ -x + 5$

NOW TRY $\text{\textcircled{↻}}$

1.7 Exercises

FOR EXTRA HELP   ▶   MyMathLab®

▶ *Complete solution available in MyMathLab*

Concept Check *Choose the letter of the correct response.*

1. Which expression is a simplified form of $-(6x - 3)$?

    A. $-6x - 3$ **B.** $-6x + 3$ **C.** $6x - 3$ **D.** $6x + 3$

2. Which is an example of a term with numerical coefficient 5?

    A. $5x^3y^7$ **B.** x^5 **C.** $\dfrac{x}{5}$ **D.** 5^2xy^3

3. Which is an example of a pair of like terms?

    A. $6t, 6w$ **B.** $-8x^2y, 9xy^2$ **C.** $5ry, 6yr$ **D.** $-5x^2, 2x^3$

4. Which is a correct translation for "six times a number, subtracted from the product of eleven and the number" (if x represents the number)?

    A. $6x - 11x$ **B.** $11x - 6x$ **C.** $(11 + x) - 6x$ **D.** $6x - (11 + x)$

5. *Concept Check* A student simplified the expression $7x - 2(3 - 2x)$ as shown.

$$7x - 2(3 - 2x)$$
$$= 7x - 2(3) - 2(2x)$$
$$= 7x - 6 - 4x$$
$$= 3x - 6$$

WHAT WENT WRONG? Find the correct simplified answer.

6. *Concept Check* A student simplified the expression $3 + 2(4x - 5)$ as shown.

$$3 + 2(4x - 5)$$
$$= 5(4x - 5)$$
$$= 5(4x) + 5(-5)$$
$$= 20x - 25$$

WHAT WENT WRONG? Find the correct simplified answer.

Simplify each expression. See Example 1.

7. $4r + 19 - 8$

8. $7t + 18 - 4$

9. $7(3x - 4y)$

10. $8(2p - 9q)$

▶ **11.** $5 + 2(x - 3y)$

12. $8 + 3(s - 6t)$

13. $-2 - (5 - 3p)$

14. $-10 - (7 - 14r)$

15. $6 + (4 - 3x) - 8$

16. $-12 + (7 - 8x) + 6$

In each term, give the numerical coefficient of the variable(s). See Objective 2.

▶ **17.** $-12k$

18. $-11y$

19. $3m^2$

20. $9n^6$

21. xw

22. pq

23. $-x$

24. $-t$

25. $\dfrac{x}{2}$

26. $\dfrac{x}{6}$

27. $\dfrac{2x}{5}$

28. $\dfrac{8x}{9}$

29. $-0.5x^3$

30. $-1.75x^2$

31. 10

32. 15

Identify each group of terms as like *or* unlike*. See Objective 3.*

▶ **33.** $8r, -13r$

34. $-7x, 12x$

35. $5z^4, 9z^3$

36. $8x^5, -10x^3$

37. $4, 9, -24$

38. $7, 17, -83$

39. x, y

40. t, s

Simplify each expression. See Examples 1–3.

41. $7y + 6y$

42. $5m + 2m$

43. $-6x - 3x$

44. $-4z - 8z$

▶ **45.** $12b + b$

46. $19x + x$

47. $3k + 8 + 4k + 7$

48. $15z + 1 + 4z + 2$

49. $-5y + 3 - 1 + 5 + y - 7$

50. $2k - 7 - 5k + 6 - 1 + 2$

51. $-2x + 3 + 4x - 17 + 20$

52. $r - 6 - 12r - 4 + 16$

53. $16 - 5m - 4m - 2 + 2m$

54. $6 - 3z - 2z - 5 - 2z$

55. $-10 + x + 4x - 7 - 4x$

56. $-p + 10p - 3p - 4 - 5p$

57. $1 + 7x + 11x - 1 + 5x$

58. $-r + 2 - 5r - 2 + 4r$

59. $-\dfrac{4}{3} + 2t + \dfrac{1}{3}t - 8 - \dfrac{8}{3}t$

60. $-\dfrac{5}{6} + 8x + \dfrac{1}{6}x - 7 - \dfrac{7}{6}$

61. $6y^2 + 11y^2 - 8y^2$

62. $-9m^3 + 3m^3 - 7m^3$

63. $2p^2 + 3p^2 - 8p^3 - 6p^3$

64. $5y^3 + 6y^3 - 3y^2 - 4y^2$

▶ **65.** $2(4x + 6) + 3$

66. $4(6y + 9) + 7$

67. $-6 - 4(y - 7)$

68. $-4 - 5(t - 13)$

69. $13p + 4(4 - 8p)$

70. $5x + 3(7 - 2x)$

71. $3t - 5 - 2(2t - 4)$

72. $8p + 6 - 3(3p - 1)$

73. $100[0.05(x + 3)]$

74. $100[0.06(x + 5)]$

75. $10[0.3(5 - 3x)]$

76. $10[0.5(8 - 2z)]$

77. $-5(5y - 9) + 3(3y + 6)$

78. $-3(2t + 4) + 8(2t - 4)$

79. $2(5r + 3) - 3(2r - 3)$

80. $3(2y - 5) - 4(5y - 7)$

81. $8(2k - 1) - (4k - 3)$

82. $6(3p - 2) - (5p + 1)$

83. $-\dfrac{4}{3}(y - 12) - \dfrac{1}{6}y$

84. $-\dfrac{7}{5}(t - 15) - \dfrac{1}{2}t$

85. $\dfrac{1}{2}(2x + 4) - \dfrac{1}{3}(9x - 6)$

86. $\dfrac{1}{4}(8x + 16) - \dfrac{1}{5}(20x - 15)$

87. $-\dfrac{2}{3}(5x + 7) - \dfrac{1}{3}(4x + 8)$

88. $-\dfrac{3}{4}(7x + 9) - \dfrac{1}{4}(5x + 7)$

89. $-7.5(2y + 4) - 2.9(3y - 6)$

90. $8.4(6t - 6) + 2.4(9 - 3t)$

91. $-2(-3k + 2) - (5k - 6) - 3k - 5$

92. $-2(3r - 4) - (6 - r) + 2r - 5$

93. $-4(-3x + 3) - (6x - 4) - 2x + 1$

94. $-5(8x + 2) - (5x - 3) - 3x + 17$

Extending Skills   *Write each of the following as a mathematical expression, and simplify.*

95. Add $3x - 2$ to $4x + 8$.

96. Add $8t + 5$ to $10t - 8$.

97. Subtract $x - 7$ from $5x + 1$.

98. Subtract $3x - 5$ from $2x - 3$.

Translate each phrase into a mathematical expression using x as the variable, and simplify.
See Example 4.

▶ **99.** Five times a number, added to the sum of the number and three

100. Six times a number, added to the sum of the number and six

101. A number multiplied by -7, subtracted from the sum of 13 and six times the number

102. A number multiplied by 5, subtracted from the sum of 14 and eight times the number

103. Six times a number added to -4, subtracted from twice the sum of three times the number and 4 (*Hint: Twice* means two times.)

104. Nine times a number added to 6, subtracted from triple the sum of 12 and 8 times the number (*Hint: Triple* means three times.)

RELATING CONCEPTS For Individual or Group Work (Exercises 105–108)

A manufacturer has fixed costs of $1000 to produce gizmos. Each gizmo costs $5 to make. The fixed cost to produce gadgets is $750, and each gadget costs $3 to make.
Work Exercises 105–108 in order.

105. Write an expression for the cost to make x gizmos. (*Hint:* The cost will be the sum of the fixed cost and the cost per item times the number of items.)

106. Write an expression for the cost to make y gadgets.

107. Write an expression for the total cost to make x gizmos and y gadgets.

108. Simplify the expression from **Exercise 107.**

STUDY SKILLS

Reviewing a Chapter

Your textbook provides material to help you prepare for quizzes or tests in this course. Refer to the **Chapter 1 Summary** as you read through the following techniques.

Chapter Reviewing Techniques

- **Review the Key Terms and any New Symbols.** Make a study card for each. Include a definition, an example, a sketch (if appropriate), and a section or page reference.

- **Take the Test Your Word Power quiz** to check your understanding of new vocabulary. The answers immediately follow.

- **Read the Quick Review.** Pay special attention to the headings. Study the explanations and examples given for each concept. Try to think about the whole chapter.

- **Reread your lecture notes.** Focus on what your instructor has emphasized in class, and review that material in your text.

- **Look over your homework.** Pay special attention to any trouble spots.

- **Work the Review Exercises.** They are grouped by section. Answers are included in the margin for quick reference.

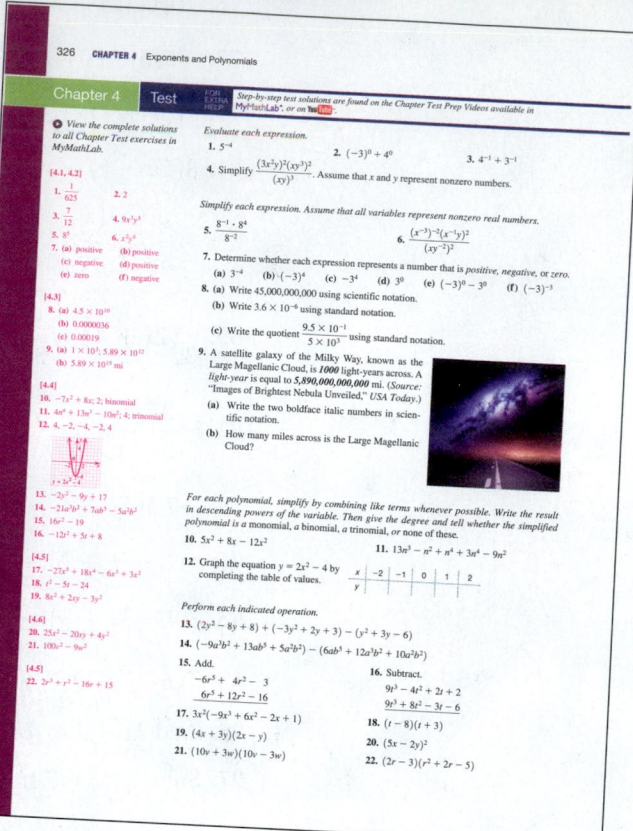

  ▶ Pay attention to direction words, such as *simplify, solve,* and *evaluate*.

  ▶ Are your answers exact and complete? Did you include the correct labels, such as $, cm^2, ft, etc.?

  ▶ Make study cards for difficult problems.

- **Work the Mixed Review Exercises.** They are in mixed-up order. Check your answers in the margin.

- **Take the Chapter Test under test conditions.**

  ▶ Time yourself.

  ▶ Use a calculator or notes (if your instructor permits them on tests).

  ▶ Take the test in one sitting.

  ▶ Show all your work.

  ▶ Check your answers in the margin. Section references are provided.

Reviewing a chapter will take some time. Avoid rushing through your review in one night. Use the suggestions over a few days or evenings to better understand the material and remember it longer.

Follow these reviewing techniques to prepare for your next test.

1. How much time did you spend reviewing for your test? Was it enough?

2. How did the reviewing techniques work for you?

3. What will you do differently when reviewing for your next test?

Chapter 1	Summary

Key Terms

1.1
exponent (power)
base
exponential expression
inequality

1.2
constant
variable
algebraic expression
equation
solution
set
element

1.3
natural (counting) numbers
whole numbers
number line
integers
signed numbers
rational numbers
graph
coordinate
irrational numbers
real numbers
additive inverse (opposite)
absolute value

1.4
sum
addends
difference
minuend
subtrahend

1.5
product
factor
multiplicative inverse
  (reciprocal)
quotient
dividend
divisor

1.6
identity element for addition
  (additive identity)
identity element for
  multiplication
  (multiplicative identity)

1.7
term
numerical coefficient
  (coefficient)
like terms
unlike terms

New Symbols

a^n n factors of a

$[\ ]$ brackets

$=$ is equal to

$\neq$ is not equal to

$<$ is less than

$>$ is greater than

$\leq$ is less than or equal to

$\geq$ is greater than or
     equal to

$\{\ \}$ set braces

$\{x \mid x \text{ has a certain}$
     $\textbf{property}\}$
     set-builder notation

$-x$ additive inverse, or
     opposite, of x

$|x|$ absolute value of x

$\dfrac{1}{x}$ multiplicative inverse,
     or reciprocal, of x
     (where $x \neq 0$)

$a(b), (a)b, (a)(b), a \cdot b,$
   or ab a times b

$a \div b, \dfrac{a}{b}, a/b, \text{ or } b\overline{)a}$
     a divided by b

Test Your Word Power

See how well you have learned the vocabulary in this chapter.

1. An **exponent** is
   A. a symbol that tells how many numbers are being multiplied
   B. a number raised to a power
   C. a number that tells how many times a factor is repeated
   D. one of two or more numbers that are multiplied.

2. A **variable** is
   A. a symbol used to represent an unknown number
   B. a value that makes an equation true
   C. a solution of an equation
   D. the answer in a division problem.

3. An **integer** is
   A. a positive or negative number
   B. a natural number, its opposite, or zero
   C. any number that can be graphed on a number line
   D. the quotient of two numbers.

4. The **absolute value** of a number is
   A. the graph of the number
   B. the reciprocal of the number
   C. the opposite of the number
   D. the distance between 0 and the number on a number line.

5. A **term** is
   A. a numerical factor

   B. a number, a variable, or a product or quotient of numbers and variables raised to powers
   C. one of several variables with the same exponents
   D. a sum of numbers and variables raised to powers.

6. A **numerical coefficient** is
   A. the numerical factor of the variable(s) in a term
   B. the number of terms in an expression
   C. a variable raised to a power
   D. the variable factor in a term.

ANSWERS

1. C; *Example:* In 2^3, the number 3 is the exponent (or power), so 2 is a factor three times, and $2^3 = 2 \cdot 2 \cdot 2 = 8$. **2.** A; *Examples: a, b, c*
3. B; *Examples:* $-9, 0, 6$ **4.** D; *Examples:* $|2| = 2$ and $|-2| = 2$ **5.** B; *Examples:* $6, \frac{x}{2}, -4ab^2$ **6.** A; *Examples:* The term 3 has numerical coefficient 3, $8z$ has numerical coefficient 8, and $-10x^4y$ has numerical coefficient -10.

Quick Review

CONCEPTS | **EXAMPLES**

1.1 Exponents, Order of Operations, and Inequality

Order of Operations
Work within any parentheses or brackets and above and below fraction bars first. Always follow this order.

Step 1   Apply all exponents.

Step 2   Do any multiplications or divisions in order from left to right.

Step 3   Do any additions or subtractions in order from left to right.

Simplify $36 - 4(2^2 + 3)$.

$$36 - 4(2^2 + 3)$$
$$= 36 - 4(4 + 3) \qquad \text{Apply the exponent.}$$
$$= 36 - 4(7) \qquad \text{Add inside the parentheses.}$$
$$= 36 - 28 \qquad \text{Multiply.}$$
$$= 8 \qquad \text{Subtract.}$$

1.2 Variables, Expressions, and Equations

To *evaluate* an expression means to find its *value*. Evaluate an expression with a variable by substituting a given number for the variable.

Find the value of $2x + y^2$ for $x = 3$ and $y = -4$.

$$2x + y^2$$
$$= 2(3) + (-4)^2 \qquad \text{Substitute.}$$
$$= 6 + 16 \qquad \text{Multiply. Apply the exponent.}$$
$$= 22 \qquad \text{Add.}$$

Values of a variable that make an equation true are solutions of the equation.

Is 2 a solution of $5x + 3 = 18$?

$$5(2) + 3 \stackrel{?}{=} 18 \qquad \text{Let } x = 2.$$
$$13 = 18 \qquad \text{False}$$

2 is not a solution.

1.3 Real Numbers and the Number Line

Ordering Real Numbers
a is less than b if a lies to the left of b on a number line.

a is greater than b if a lies to the right of b on a number line.

The additive inverse, or opposite, of x is $-x$.

The absolute value of x, written $|x|$, is the distance between x and 0 on a number line.

Graph -2, 0, and 3.

$$-2 < 3 \qquad 3 > 0 \qquad 0 < 3$$

$$-(5) = -5 \qquad -(-7) = 7 \qquad -0 = 0$$

$$|13| = 13 \qquad |0| = 0 \qquad |-5| = 5$$

1.4 Adding and Subtracting Real Numbers

Adding Two Signed Numbers
Same sign   Add their absolute values. The sum has the same sign as the addends.

Different signs   Subtract their absolute values. The sum has the sign of the addend with greater absolute value.

Definition of Subtraction

$$x - y = x + (-y)$$

Add.

$$9 + 4 = 13$$
$$-8 + (-5) = -13$$
$$7 + (-12) = -5$$
$$-5 + 13 = 8$$

Subtract.

$5 - (-2)$	$-3 - 4$	$-2 - (-6)$
$= 5 + 2$	$= -3 + (-4)$	$= -2 + 6$
$= 7$	$= -7$	$= 4$

CONCEPTS	EXAMPLES

1.5 Multiplying and Dividing Real Numbers

Multiplying and Dividing Two Signed Numbers

Same sign   The product (or quotient) is *positive*.

Different signs   The product (or quotient) is *negative*.

Definition of Division

$$x \div y = x \cdot \frac{1}{y} \quad (\text{where } y \neq 0)$$

0 divided by a nonzero number equals 0.
Division by 0 is undefined.

Multiply or divide.

$$6 \cdot 5 = 30 \qquad -7(-8) = 56 \qquad \frac{-24}{-6} = 4$$

$$-6(5) = -30 \qquad \frac{-18}{9} = -2 \qquad \frac{49}{-7} = -7$$

$$\frac{10}{2} = 10 \cdot \frac{1}{2} = 5$$

$$\frac{0}{5} = 0 \qquad \frac{5}{0} \text{ is undefined.}$$

1.6 Properties of Real Numbers

Commutative Properties

$$a + b = b + a$$
$$ab = ba$$

$$7 + (-1) = -1 + 7$$
$$5(-3) = (-3)5$$

Associative Properties

$$(a + b) + c = a + (b + c)$$
$$(ab)c = a(bc)$$

$$(3 + 4) + 8 = 3 + (4 + 8)$$
$$[-2(6)]4 = -2[(6)4]$$

Identity Properties

$$a + 0 = a \qquad 0 + a = a$$
$$a \cdot 1 = a \qquad 1 \cdot a = a$$

$$-7 + 0 = -7 \qquad 0 + (-7) = -7$$
$$9 \cdot 1 = 9 \qquad 1 \cdot 9 = 9$$

Inverse Properties

$$a + (-a) = 0 \qquad -a + a = 0$$
$$a \cdot \frac{1}{a} = 1 \qquad \frac{1}{a} \cdot a = 1 \quad (\text{where } a \neq 0)$$

$$7 + (-7) = 0 \qquad -7 + 7 = 0$$
$$-2\left(-\frac{1}{2}\right) = 1 \qquad -\frac{1}{2}(-2) = 1$$

Distributive Properties

$$a(b + c) = ab + ac$$
$$(b + c)a = ba + ca$$
$$a(b - c) = ab - ac$$

$$5(4 + 2) = 5(4) + 5(2)$$
$$(4 + 2)5 = 4(5) + 2(5)$$
$$9(5 - 4) = 9(5) - 9(4)$$

1.7 Simplifying Expressions

Only like terms may be combined. We use the distributive property.

Simplify each expression.

$$-3y^2 + 6y^2 + 14y^2$$
$$= (-3 + 6 + 14)y^2$$
$$= 17y^2$$

$$4(3 + 2x) - 6(5 - x)$$
$$= 4(3) + 4(2x) - 6(5) - 6(-x)$$
$$= 12 + 8x - 30 + 6x$$
$$= 14x - 18$$

Chapter 1 | Review Exercises

1.1 *Find the value of each exponential expression.*

1. 5^4

2. $\left(\dfrac{3}{5}\right)^3$

3. $\left(\dfrac{1}{8}\right)^2$

4. $(0.1)^3$

Find the value of each expression.

5. $8 \cdot 5 - 13$

6. $16 + 12 \div 4 - 2$

7. $20 - 2(5 + 3)$

8. $7[3 + 6(3^2)]$

9. $\dfrac{9(4^2 - 3)}{4 \cdot 5 - 17}$

10. $\dfrac{6(5 - 4) + 2(4 - 2)}{3^2 - (4 + 3)}$

Decide whether each statement is true or false.

11. $12 \cdot 3 - 6 \cdot 6 \leq 0$

12. $3[5(2) - 3] > 20$

13. $9 \leq 4^2 - 8$

Write each word statement in symbols.

14. Thirteen is less than seventeen.

15. Five plus two is not equal to ten.

16. Two-thirds is greater than or equal to four-sixths.

1.2 *Find the value of each expression for $x = 6$ and $y = 3$.*

17. $2x + 6y$

18. $4(3x - y)$

19. $\dfrac{x}{3} + 4y$

20. $\dfrac{x^2 + 3}{3y - x}$

Write each word phrase as an algebraic expression, using x as the variable.

21. Six added to a number

22. A number subtracted from eight

23. Nine subtracted from six times a number

24. Three-fifths of a number added to 12

Decide whether each equation has the given number as a solution.

25. $5x + 3(x + 2) = 22$;   2

26. $\dfrac{t + 5}{3t} = 1$;   6

Write each word statement as an equation. Use x as the variable. Then find the solution of the equation from the set $\{0, 2, 4, 6, 8, 10\}$.

27. Six less than twice a number is 10.

28. The product of a number and 4 is 8.

1.3 *Graph each number on a number line.*

29. $-4, -\dfrac{1}{2}, 0, 2.5, 5$

30. $-3, -1\dfrac{1}{2}, \dfrac{2}{3}, 2.25, 3$

Classify each number, using the sets natural numbers, whole numbers, integers, rational numbers, irrational numbers, and real numbers.

31. $\dfrac{4}{3}$

32. $0.\overline{63}$

33. 19

34. $\sqrt{6}$

Select the lesser of the two given numbers.

35. $-10, 5$

36. $-8, -9$

37. $-\dfrac{2}{3}, -\dfrac{3}{4}$

38. $0, -|23|$

Decide whether each statement is true or false.

39. $12 > -13$

40. $0 > -5$

41. $-9 < -7$

42. $-13 \geq -13$

---

1. 625

2. $\dfrac{27}{125}$

3. $\dfrac{1}{64}$

4. 0.001

5. 27

6. 17

7. 4

8. 399

9. 39

10. 5

11. true

12. true

13. false

14. $13 < 17$

15. $5 + 2 \neq 10$

16. $\dfrac{2}{3} \geq \dfrac{4}{6}$

17. 30

18. 60

19. 14

20. 13

21. $x + 6$

22. $8 - x$

23. $6x - 9$

24. $12 + \dfrac{3}{5}x$

25. yes

26. no

27. $2x - 6 = 10$; 8

28. $4x = 8$; 2

29.

30.

31. rational numbers, real numbers

32. rational numbers, real numbers

33. natural numbers, whole numbers, integers, rational numbers, real numbers

34. irrational numbers, real numbers

35. -10

36. -9

37. $-\dfrac{3}{4}$

38. $-|23|$

39. true

40. true

41. true

42. true

43. (a) 9 (b) 9
44. (a) 0 (b) 0
45. (a) −6 (b) 6
46. (a) $\frac{5}{7}$ (b) $\frac{5}{7}$
47. 12 48. −3
49. −19 50. −7
51. −6 52. −4
53. −17 54. $-\frac{29}{36}$
55. −21.8 56. −14
57. −10 58. −19
59. −11 60. −1
61. 7
62. $-\frac{43}{35}$, or $-1\frac{8}{35}$
63. 10.31 64. −12
65. 2 66. −3
67. (−31 + 12) + 19; 0
68. [−4 + (−8)] + 13; 1
69. −4 − (−6); 2
70. [4 + (−8)] − 5; −9
71. $26.25 72. −10°F
73. −$29 74. −10°
75. 38 76. 14,840.95
77. 36 78. −105
79. $\frac{1}{2}$ 80. 10.08
81. −20 82. −10
83. −24 84. −35
85. 4 86. −20
87. $-\frac{3}{4}$ 88. 11.3
89. −1

*For each number, find **(a)** the additive inverse and **(b)** the absolute value.*

43. −9   **44.** 0   **45.** 6   **46.** $-\frac{5}{7}$

Find each absolute value.

47. $|-12|$ **48.** $-|3|$ **49.** $-|-19|$ **50.** $-|9-2|$

1.4 *Perform each indicated operation.*

51. −10 + 4   **52.** 14 + (−18)   **53.** −8 + (−9)

54. $\frac{4}{9} + \left(-\frac{5}{4}\right)$ **55.** −13.5 + (−8.3)   **56.** (−10 + 7) + (−11)

57. [−6 + (−8) + 8] + [9 + (−13)]   **58.** (−4 + 7) + (−11 + 3) + (−15 + 1)

59. −7 − 4   **60.** −12 − (−11)

61. 5 − (−2)   **62.** $-\frac{3}{7} - \frac{4}{5}$

63. 2.56 − (−7.75)   **64.** (−10 − 4) − (−2)

65. (−3 + 4) − (−1)   **66.** −(−5 + 6) − 2

Write a numerical expression for each phrase, and simplify the expression.

67. 19 added to the sum of −31 and 12   **68.** 13 more than the sum of −4 and −8

69. The difference of −4 and −6   **70.** Five less than the sum of 4 and −8

Solve each problem.

71. George found that his checkbook balance was −$23.75, so he deposited $50.00. What is his new balance?

72. The low temperature in Yellowknife, in the Canadian Northwest Territories, one January day was −26°F. It rose 16° that day. What was the high temperature?

73. Reginald owed a friend $28. He repaid $13, but then borrowed another $14. What positive or negative amount represents his present financial status?

74. If the temperature drops 7° below its previous level of −3°, what is the new temperature?

75. A quarterback passed for a gain of 8 yd, was sacked for a loss of 12 yd, and then threw a 42 yd touchdown pass. What positive or negative number represents the total net yardage for the plays?

76. On Friday, August 30, 2013, the Dow Jones Industrial Average closed at 14,810.31, down 30.64 from the previous day. What was the closing value the previous day? (*Source: The Washington Post.*)

1.5 *Perform each indicated operation.*

77. −12(−3)   **78.** 15(−7)   **79.** $-\frac{4}{3}\left(-\frac{3}{8}\right)$ **80.** −4.8(−2.1)

81. 5(8 − 12)   **82.** (5 − 7)(8 − 3)   **83.** 2(−6) − (−4)(−3)

84. 3(−10) − 5   **85.** $\frac{-36}{-9}$ **86.** $\frac{220}{-11}$

87. $-\frac{1}{2} \div \frac{2}{3}$ **88.** $\frac{-33.9}{-3}$ **89.** $\frac{-5(3) - 1}{8 - 4(-2)}$

90. 2
91. 1   **92.** 0.5
93. −18   **94.** −18
95. 125   **96.** −423
97. −4(5) − 9; −29
98. $\frac{5}{6}[12 + (-6)]$; 5
99. $\frac{12}{8 + (-4)}$; 3
100. $\frac{-20(12)}{15 - (-15)}$; −8
101. 8x = −24; −3
102. $\frac{x}{3}$ = −2; −6

103. identity property
104. identity property
105. inverse property
106. inverse property
107. associative property
108. associative property
109. distributive property
110. commutative property
111. 7y + 14   **112.** −48 + 12t
113. 6s + 15y   **114.** 4r − 5s

115. 11m   **116.** $16p^2$
117. $16p^2 + 2p$ **118.** −4k + 12
119. −2m + 29 **120.** −5k − 1
121. −2(3x) − 7x; −13x
122. (5 + 4x) + 8x; 5 + 12x

90. $\dfrac{5(-2) - 3(4)}{-2[3 - (-2)] - 1}$ **91.** $\dfrac{10^2 - 5^2}{8^2 + 3^2 - (-2)}$ **92.** $\dfrac{(0.6)^2 + (0.8)^2}{(-1.2)^2 - (-0.56)}$

Evaluate each expression for x = −5, y = 4, and z = −3.

93. $6x - 4z$ **94.** $5x + y - z$ **95.** $5x^2$ **96.** $z^2(3x - 8y)$

Write a numerical expression for each phrase, and simplify the expression.

97. Nine less than the product of −4 and 5

98. Five-sixths of the sum of 12 and −6

99. The quotient of 12 and the sum of 8 and −4

100. The product of −20 and 12, divided by the difference of 15 and −15

Write each sentence as an equation, using x as the variable. Then find the solution from the set of integers between −12 and 12, inclusive.

101. 8 times a number is −24.   **102.** The quotient of a number and 3 is −2.

1.6 *Decide whether each statement is an example of a* commutative, *an* associative, *an* identity, *an* inverse, *or the* distributive *property.*

103. 6 + 0 = 6   **104.** 5 · 1 = 5

105. $-\dfrac{2}{3}\left(-\dfrac{3}{2}\right) = 1$ **106.** 17 + (−17) = 0

107. 5 + (−9 + 2) = [5 + (−9)] + 2   **108.** w(xy) = (wx)y

109. 3(x + y) = 3x + 3y   **110.** (1 + 2) + 3 = 3 + (1 + 2)

Use the distributive property to rewrite each expression. Simplify if possible.

111. 7(y + 2)   **112.** −12(4 − t)   **113.** 3(2s + 5y)   **114.** −(−4r + 5s)

1.7 *Simplify each expression.*

115. 2m + 9m   **116.** $15p^2 - 7p^2 + 8p^2$

117. $5p^2 - 4p + 6p + 11p^2$ **118.** −2(3k − 5) + 2(k + 1)

119. 7(2m + 3) − 2(8m − 4)   **120.** −(2k + 8) − (3k − 7)

Translate each phrase into a mathematical expression using x as the variable, and simplify.

121. Seven times a number, subtracted from the product of −2 and three times the number

122. A number multiplied by 8, added to the sum of 5 and four times the number

Chapter 1 Mixed Review Exercises

Complete the table.

1. 3; 3; $-\dfrac{1}{3}$ **2.** 12; −12; $\dfrac{1}{12}$

3. $-\dfrac{2}{3}; \dfrac{2}{3}; \dfrac{2}{3}$ **4.** 0.2; 0.2; 5

	Number	Absolute Value	Additive Inverse	Multiplicative Inverse
1.	−3			
2.	12			
3.				$-\dfrac{3}{2}$
4.			−0.2	

5. rational numbers, real numbers
6. 37
7. $\frac{8}{3}$, or $2\frac{2}{3}$ 8. $-\frac{1}{24}$
9. 2
10. $-\frac{28}{15}$, or $-1\frac{13}{15}$
11. $-\frac{3}{2}$, or $-1\frac{1}{2}$
12. $\frac{25}{36}$
13. 16   14. 77.6
15. 11   16. $16t - 36$
17. $8x^2 - 21y^2$ 18. 24
19. $-47°F$ 20. 14,776 ft

5. To which of the following sets does $0.\overline{6}$ belong: natural numbers, whole numbers, integers, rational numbers, irrational numbers, real numbers?

6. Evaluate $(x + 6)^3 - y^3$ for $x = -2$ and $y = 3$.

Perform each indicated operation.

7. $\dfrac{6(-4) + 2(-12)}{5(-3) + (-3)}$

8. $\dfrac{3}{8} - \dfrac{5}{12}$

9. $\dfrac{8^2 + 6^2}{7^2 + 1^2}$

10. $-\dfrac{12}{5} \div \dfrac{9}{7}$

11. $2\dfrac{5}{6} - 4\dfrac{1}{3}$

12. $\left(\dfrac{5}{6}\right)^2$

13. $[(-2) + 7 - (-5)] + [-4 - (-10)]$

14. $-16(-3.5) - 7.2(-3)$

15. $-8 + [(-4 + 17) - (-3 - 3)]$

16. $-4(2t + 1) - 8(-3t + 4)$

17. $5x^2 - 12y^2 + 3x^2 - 9y^2$

18. $(-8 - 3) - 5(2 - 9)$

Solve each problem.

19. The highest temperature ever recorded in Iowa was 118°F. The lowest temperature ever recorded in the state was 165° lower than the highest temperature. What is the record low temperature for Iowa? (*Source:* National Climatic Data Center.)

20. The top of Mt. Whitney, visible from Death Valley, has an altitude of 14,494 ft above sea level. The bottom of Death Valley is 282 ft below sea level. Using 0 as sea level, find the difference between these two elevations. (*Source:* World Almanac and Book of Facts.)

Chapter 1 — Test

FOR EXTRA HELP MyMathLab® or on YouTube.

Step-by-step test solutions are found on the Chapter Test Prep Videos available in

▶ *View the complete solutions to all Chapter Test exercises in MyMathLab.*

[1.1]
1. true   2. false

[1.3]
3. rational numbers, real numbers
4. $-|-8|$, or -8

[1.5]
5. $\dfrac{-6}{2 + (-8)}$; 1   6. negative

[1.4, 1.5]
7. 4
8. $-\dfrac{17}{6}$, or $-2\dfrac{5}{6}$
9. 2   10. 6
11. 108   12. $\dfrac{30}{7}$, or $4\dfrac{2}{7}$

[1.5]
13. -70 14. 3

Decide whether each statement is true *or* false.

1. $4[-20 + 7(-2)] \leq 135$

2. $\left(\dfrac{1}{2}\right)^2 + \left(\dfrac{2}{3}\right)^2 = \left(\dfrac{1}{2} + \dfrac{2}{3}\right)^2$

3. To which of the following sets does $-\dfrac{2}{3}$ belong: natural numbers, whole numbers, integers, rational numbers, irrational numbers, real numbers?

4. Select the lesser number of 6 and $-|-8|$.

5. Write a numerical expression for the phrase, and simplify the expression.

The quotient of -6 and the sum of 2 and -8

6. If a and b are both negative, is $\dfrac{a + b}{a \cdot b}$ positive or negative?

Perform each indicated operation.

7. $-2 - (5 - 17) + (-6)$

8. $-5\dfrac{1}{2} + 2\dfrac{2}{3}$

9. $-6.2 - [-7.1 + (2.0 - 3.1)]$

10. $4^2 + (-8) - (2^3 - 6)$

11. $(-5)(-12) + 4(-4) + (-8)^2$

12. $\dfrac{30(-1 - 2)}{-9[3 - (-2)] - 12(-2)}$

Evaluate each expression for $x = -2$ and $y = 4$.

13. $3x - 4y^2$

14. $\dfrac{5x + 7y}{3(x + y)}$

Solve each problem.

15. The highest elevation in Argentina is Mt. Aconcagua, which is 6960 m above sea level. The lowest point in Argentina is the Valdés Peninsula, 40 m below sea level. Find the difference between the highest and lowest elevations.

16. For a certain system of rating relief pitchers, 3 points are awarded for a save, 3 points are awarded for a win, 2 points are subtracted for a loss, and 2 points are subtracted for a blown save. If Craig Kimbrel of the Atlanta Braves has 4 saves, 3 wins, 2 losses, and 1 blown save, how many points does he have?

17. For 2012, the U.S. federal government collected $2.45 trillion in revenues, but spent $3.54 trillion. Write the federal budget deficit as a signed number. (*Source*: U.S. Department of Management and Budget.)

Match each statement in Column I with the property it illustrates in Column II.

I	II
18. $3x + 0 = 3x$	**A.** Commutative property
19. $(5 + 2) + 8 = 8 + (5 + 2)$	**B.** Associative property
20. $-3(x + y) = -3x + (-3y)$	**C.** Inverse property
21. $-5 + (3 + 2) = (-5 + 3) + 2$	**D.** Identity property
22. $-\dfrac{5}{3}\left(-\dfrac{3}{5}\right) = 1$	**E.** Distributive property

23. What property is used to clear parentheses and write $3(x + 1)$ as $3x + 3$?

24. Consider the expression $-6[5 + (-2)]$.

 (a) Evaluate it by first working within the brackets.

 (b) Evaluate it by using the distributive property.

 (c) Why must the answers in parts (a) and (b) be the same?

Simplify each expression.

25. $8x + 4x - 6x + x + 14x$ **26.** $5(2x - 1) - (x - 12) + 2(3x - 5)$

2

Linear Equations and Inequalities in One Variable

Solving *linear equations,* the subject of this chapter, can be thought of in terms of the concept of balance.

103

2.1 The Addition Property of Equality

VOCABULARY

☐ equation
☐ linear equation in one variable
☐ solution
☐ solution set
☐ equivalent equations

An **equation** is a statement asserting that two algebraic expressions are equal.

⚠ CAUTION *Remember that an equation includes an equality symbol.*

Equation
↓
Left side → $x - 5 = 2$ ← Right side

An equation can be solved.

Expression
↓
$x - 5$

An expression **cannot** be solved. (It can be *evaluated* for a given value, or *simplified*.)

OBJECTIVE 1 Identify linear equations.

Linear Equation in One Variable

A **linear equation in one variable** (here x) can be written in the form

$$Ax + B = C,$$

where A, B, and C are real numbers and $A \neq 0$.

Examples: $4x + 9 = 0$, $2x - 3 = 5$, and $x = 7$ Linear equations

$x^2 + 2x = 5$, $\quad x^3 = -1$, $\quad \dfrac{1}{x} = 6$, and $\quad |2x + 6| = 0$ *Non*linear equations

A **solution** of an equation is a number that makes the equation true when it replaces the variable. An equation is solved by finding its **solution set,** the set of all solutions. Equations with exactly the same solution sets are **equivalent equations.**

A linear equation in x is *solved* by using a series of steps to produce a simpler equivalent equation of the form

$$x = \textbf{a number} \quad \text{or} \quad \textbf{a number} = x.$$

OBJECTIVE 2 Use the addition property of equality.

In the linear equation $x - 5 = 2$, both $x - 5$ and 2 represent the same number because that is the meaning of the equality symbol. To solve the equation, we change the left side from $x - 5$ to just x, as follows.

$x - 5 = 2$	Given equation
$x - 5 + 5 = 2 + 5$	Add 5 to *each* side to keep them equal.
$x + 0 = 7$	Additive inverse property
$x = 7$	Additive identity property

Add 5. It is the opposite (additive inverse) of -5, and $-5 + 5 = 0$.

To check that 7 is the solution, we replace x with 7 in the original equation.

CHECK		
	$x - 5 = 2$	Original equation
	$7 - 5 \overset{?}{=} 2$	Let $x = 7$.
	$2 = 2$ ✓	True

The left side equals the right side.

We write a solution set using set braces.

The final equation is true, so 7 is the solution and $\{7\}$ is the solution set.

To solve the equation $x - 5 = 2$, we used the **addition property of equality.**

> **Addition Property of Equality**
>
> If A, B, and C represent real numbers, then the equations
>
> $$A = B \quad \text{and} \quad A + C = B + C \quad \text{are equivalent.}$$
>
> *That is, the same number may be added to each side of an equation without changing the solution set.*

$$x - 5 \quad = \quad 2$$

$$x - 5 + 5 \quad = \quad 2 + 5$$

FIGURE 1

In this property, any quantity that represents a real number C can be added to each side of an equation to obtain an equivalent equation.

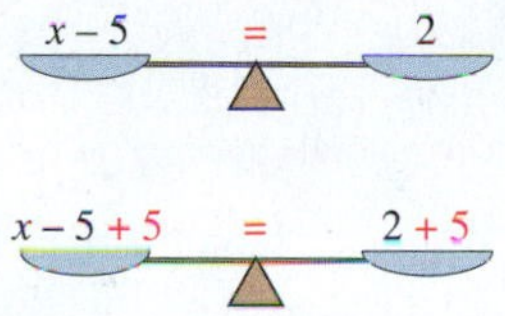

NOTE   Equations can be thought of in terms of a balance. Thus, adding the *same* quantity to each side does not affect the balance. See **FIGURE 1**.

NOW TRY EXERCISE 1

Solve $x - 13 = 4$.

EXAMPLE 1   Applying the Addition Property of Equality

Solve $x - 16 = 7$.

Our goal is to get an equivalent equation of the form $x = $ a number.

$$x - 16 = 7$$

$$x - 16 + 16 = 7 + 16 \qquad \text{Add 16 to each side.}$$

$$x = 23 \qquad \text{Combine like terms.}$$

CHECK   Substitute 23 for x in the *original* equation.

$$x - 16 = 7 \qquad \text{Original equation}$$

$$23 - 16 \overset{?}{=} 7 \qquad \text{Let } x = 23.$$

7 is *not* the solution.

$$7 = 7 \checkmark \qquad \text{True}$$

A true statement results, so 23 is the solution and $\{23\}$ is the solution set.

NOW TRY

> ❗ **CAUTION**   *The final line of the check does not give the solution to the problem.* It confirms that the value found is actually a solution.

NOW TRY EXERCISE 2

Solve $t - 5.7 = -7.2$.

EXAMPLE 2   Applying the Addition Property of Equality

Solve $x - 2.9 = -6.4$.

Our goal is to isolate x.

$$x - 2.9 = -6.4$$

$$x - 2.9 + 2.9 = -6.4 + 2.9 \qquad \text{Add 2.9 to each side.}$$

$$x = -3.5$$

CHECK

$$x - 2.9 = -6.4 \qquad \text{Original equation}$$

$$-3.5 - 2.9 \overset{?}{=} -6.4 \qquad \text{Let } x = -3.5.$$

$$-6.4 = -6.4 \checkmark \qquad \text{True}$$

NOW TRY ANSWERS
1. $\{17\}$
2. $\{-1.5\}$

A true statement results, so the solution set is $\{-3.5\}$.

NOW TRY

In **Section 1.4,** subtraction was defined as addition of the opposite. Thus, we can also use the following when solving an equation.

> ### Addition Property of Equality Extended to Subtraction
>
> The same number may be *subtracted* from each side of an equation without changing the solution set.

NOW TRY
EXERCISE 3

Solve $-15 = x + 12$.

EXAMPLE 3   Applying the Addition Property of Equality

Solve $-7 = x + 22$.

Here, the variable x is on the right side of the equation.

$$-7 = x + 22 \qquad \text{The variable can be isolated on } \textit{either} \text{ side.}$$

$$-7 - 22 = x + 22 - 22 \qquad \text{Subtract 22 from each side.}$$

$$-29 = x, \quad \text{or} \quad x = -29 \qquad \text{Rewrite; a number } = x, \text{ or } x = \text{a number.}$$

CHECK
$$-7 = x + 22 \qquad \text{Original equation}$$
$$-7 \stackrel{?}{=} -29 + 22 \qquad \text{Let } x = -29.$$
$$-7 = -7 \;\checkmark \qquad \text{True}$$

The check confirms that the solution set is $\{-29\}$.   **NOW TRY**

> **NOTE**   In **Example 3,** what happens if we subtract $-7 - 22$ incorrectly, obtaining $x = -15$ (instead of $x = -29$) as the last line of the solution? A check should indicate an error.
>
> **CHECK**
> $$-7 = x + 22 \qquad \text{Original equation from } \textbf{Example 3}$$
> $$-7 \stackrel{?}{=} -15 + 22 \qquad \text{Let } x = -15.$$
> $$\underset{\text{The left side does } \textit{not} \text{ equal the right side.}}{-7 = 7} \qquad \text{False}$$
>
> The false statement indicates that -15 is *not* a solution of the equation. If this happens, rework the problem.

NOW TRY
EXERCISE 4

Solve $x - 5 = 2x$.

EXAMPLE 4   Subtracting a Variable Term

Solve $6x - 8 = 7x$.

$$6x - 8 = 7x$$
$$6x - 8 - 6x = 7x - 6x \qquad \text{Subtract } 6x \text{ from each side.}$$
$$-8 = x \qquad \text{Combine like terms.}$$

CHECK
$$6x - 8 = 7x \qquad \text{Original equation}$$
$$6(-8) - 8 \stackrel{?}{=} 7(-8) \qquad \text{Let } x = -8.$$
$$-48 - 8 \stackrel{?}{=} -56 \qquad \text{Multiply.}$$
$$-56 = -56 \;\checkmark \qquad \text{True}$$

Use parentheses when substituting to avoid errors.

A true statement results, so the solution set is $\{-8\}$.   **NOW TRY**

NOW TRY ANSWERS
3. $\{-27\}$
4. $\{-5\}$

What happens in **Example 4** if we start by subtracting $7x$ from each side?

$$6x - 8 = 7x \qquad \text{Original equation from \textbf{Example 4}}$$
$$6x - 8 - 7x = 7x - 7x \qquad \text{Subtract } 7x \text{ from each side.}$$
$$-8 - x = 0 \qquad \text{Combine like terms.}$$
$$-8 - x + 8 = 0 + 8 \qquad \text{Add 8 to each side.}$$
$$-x = 8 \qquad \text{Combine like terms.}$$

This result gives the value of $-x$, but not of x itself. However, it does say that the additive inverse of x is 8, which means that x must be -8.

$$x = -8 \qquad \text{Same result as in \textbf{Example 4}}$$

(This result can also be justified by the multiplication property of equality, covered in **Section 2.2**.) We can make the following generalization.

If a is a number and $-x = a$, then $x = -a$.

NOW TRY
EXERCISE 5

Solve $\frac{2}{3}x + 4 = \frac{5}{3}x$.

EXAMPLE 5 Subtracting a Variable Term (Fractional Coefficients)

Solve $\frac{3}{5}x + 15 = \frac{8}{5}x$.

$$\frac{3}{5}x + 15 = \frac{8}{5}x \qquad \text{Original equation}$$

$$\frac{3}{5}x + 15 - \frac{3}{5}x = \frac{8}{5}x - \frac{3}{5}x \qquad \text{Subtract } \frac{3}{5}x \text{ from each side.}$$

> From now on we will skip this step.

$$15 = 1x \qquad \qquad \frac{3}{5}x - \frac{3}{5}x = 0; \frac{8}{5}x - \frac{3}{5}x = \frac{5}{5}x = 1x$$

$$15 = x \qquad \text{Multiplicative identity property}$$

Check by replacing x with 15 in the original equation. The solution set is $\{15\}$.

NOW TRY

NOW TRY
EXERCISE 6

Solve $6x - 8 = 12 + 5x$.

EXAMPLE 6 Applying the Addition Property of Equality Twice

Solve $8 - 6p = -7p + 5$.

$$8 - 6p = -7p + 5$$
$$8 - 6p + 7p = -7p + 5 + 7p \qquad \text{Add } 7p \text{ to each side.}$$
$$8 + p = 5 \qquad \text{Combine like terms.}$$
$$8 + p - 8 = 5 - 8 \qquad \text{Subtract 8 from each side.}$$
$$p = -3 \qquad \text{Combine like terms.}$$

CHECK
$$8 - 6p = -7p + 5 \qquad \text{Original equation}$$
$$8 - 6(-3) \stackrel{?}{=} -7(-3) + 5 \qquad \text{Let } p = -3.$$
$$8 + 18 \stackrel{?}{=} 21 + 5 \qquad \text{Multiply.}$$
$$26 = 26 \ \checkmark \qquad \text{True}$$

NOW TRY ANSWERS
5. $\{4\}$
6. $\{20\}$

The check results in a true statement, so the solution set is $\{-3\}$. **NOW TRY**

NOTE *There are often several correct ways to solve an equation.* In the equation

$$8 - 6p = -7p + 5, \quad \text{See Example 6.}$$

we could begin by adding $6p$ (instead of $7p$) to each side. Combining like terms and subtracting 5 from each side gives $3 = -p$. (Try this.) If $3 = -p$, then $-3 = p$, and the variable has been isolated on the right side of the equation. The same solution results.

OBJECTIVE 3 Simplify, and then use the addition property of equality.

EXAMPLE 7 Combining Like Terms When Solving

NOW TRY EXERCISE 7

Solve.

$$5x - 10 - 12x$$
$$= 4 - 8x - 9$$

Solve $3t - 12 + t + 2 = 5 + 3t + 2$.

$3t - 12 + t + 2 = 5 + 3t + 2$	
$4t - 10 = 7 + 3t$	Combine like terms.
$4t - 10 - 3t = 7 + 3t - 3t$	Subtract $3t$ from each side.
$t - 10 = 7$	Combine like terms.
$t - 10 + 10 = 7 + 10$	Add 10 to each side.
$t = 17$	Combine like terms.

CHECK	$3t - 12 + t + 2 = 5 + 3t + 2$	Original equation
	$3(17) - 12 + 17 + 2 \overset{?}{=} 5 + 3(17) + 2$	Let $t = 17$.
	$51 - 12 + 17 + 2 \overset{?}{=} 5 + 51 + 2$	Multiply.
	$58 = 58$ ✓	True

The check results in a true statement, so the solution set is $\{17\}$. **NOW TRY**

EXAMPLE 8 Using the Distributive Property When Solving

NOW TRY EXERCISE 8

Solve.

$$4(3x - 2) - (11x - 4) = 3$$

Solve $3(2 + 5x) - (1 + 14x) = 6$.

$$3(2 + 5x) - (1 + 14x) = 6$$

Be sure to distribute to *all* terms within the parentheses.

$$3(2 + 5x) - 1(1 + 14x) = 6 \qquad -(1 + 14x) = -1(1 + 14x)$$

$3(2) + 3(5x) - 1(1) - 1(14x) = 6$	Distributive property
Be careful here. $6 + 15x - 1 - 14x = 6$	Multiply.
$x + 5 = 6$	Combine like terms.
$x + 5 - 5 = 6 - 5$	Subtract 5 from each side.
$x = 1$	Combine like terms.

Check by substituting 1 for x in the original equation. The solution set is $\{1\}$.

NOW TRY

NOW TRY ANSWERS
7. $\{5\}$
8. $\{7\}$

⚠ CAUTION *Be careful to apply the distributive property correctly* in a problem like that in **Example 8**, or a sign error may result.

2.1 Exercises

  MyMathLab®

 Complete solution available in MyMathLab

Concept Check   *Complete each statement with the correct response. The following terms may be used once, more than once, or not at all.*

linear	expression	solution set	multiplication
equation	addition	equivalent equations	variable

1. A(n) _____ includes an equality symbol, while a(n) _____ does not.

2. A(n) _____ equation in one _____ (here x) can be written in the form $Ax + B (= / \neq) C$.

3. Equations that have exactly the same solution set are _____.

4. The _____ property of equality states that the same expression may be added to or subtracted from each side of an equation without changing the _____.

5. *Concept Check*   Decide whether each of the following is an *equation* or an *expression*. If it is an equation, solve it. If it is an expression, simplify it.

(a) $5x + 8 - 4x + 7$ **(b)** $-6y + 12 + 7y - 5$

(c) $5x + 8 - 4x = 7$ **(d)** $-6y + 12 + 7y = -5$

6. *Concept Check*   Which pairs of equations are equivalent equations?

A. $x + 2 = 6$ and $x = 4$ **B.** $10 - x = 5$ and $x = -5$

C. $x + 3 = 9$ and $x = 6$ **D.** $4 + x = 8$ and $x = -4$

7. *Concept Check*   Which of the following are *not* linear equations in one variable?

A. $x^2 - 5x + 6 = 0$ **B.** $x^3 = x$

C. $3x - 4 = 0$ **D.** $7x - 6x = 3 + 9x$

8. Explain how to check a solution of an equation.

Solve each equation, and check the solution. See Examples 1–6.

9. $x - 3 = 9$ **10.** $x - 9 = 8$ **11.** $x - 12 = 19$

12. $x - 18 = 22$ **13.** $x - 6 = -9$ **14.** $x - 5 = -7$

15. $r + 8 = 12$ **16.** $x + 7 = 11$ **17.** $x + 28 = 19$

18. $x + 47 = 26$ **19.** $x + \dfrac{1}{4} = -\dfrac{1}{2}$ **20.** $x + \dfrac{2}{3} = -\dfrac{1}{6}$

21. $7 + r = -3$ **22.** $8 + k = -4$ **23.** $2 = p + 15$

24. $5 = z + 19$ **25.** $-4 = x - 14$ **26.** $-7 = x - 22$

27. $-\dfrac{1}{3} = x - \dfrac{3}{5}$ **28.** $-\dfrac{1}{4} = x - \dfrac{2}{3}$ **29.** $x - 8.4 = -2.1$

30. $x - 15.5 = -5.1$ **31.** $t + 12.3 = -4.6$ **32.** $x + 21.5 = -13.4$

33. $3x = 2x + 7$ **34.** $5x = 4x + 9$ **35.** $10x + 4 = 9x$

36. $8t + 5 = 7t$ **37.** $8x - 3 = 9x$ **38.** $6x - 4 = 7x$

39. $6t - 2 = 5t$ **40.** $4z - 6 = 3z$ **41.** $\dfrac{2}{5}w - 6 = \dfrac{7}{5}w$

42. $\dfrac{2}{7}z - 2 = \dfrac{9}{7}z$ **43.** $\dfrac{1}{2}x + 5 = -\dfrac{1}{2}x$ **44.** $\dfrac{1}{5}x + 7 = -\dfrac{4}{5}x$

45. $5.6x + 2 = 4.6x$ **46.** $9.1x + 5 = 8.1x$ **47.** $1.4x - 3 = 0.4x$

48. $1.9t - 6 = 0.9t$ **49.** $5p = 4p$ **50.** $8z = 7z$

51. $3x + 7 - 2x = 0$ **52.** $5x + 4 - 4x = 0$ **53.** $3x + 7 = 2x + 4$

54. $9x + 5 = 8x + 4$ **55.** $8t + 6 = 7t + 6$ **56.** $13t + 9 = 12t + 9$

57. $-4x + 7 = -5x + 9$ **58.** $-6x + 3 = -7x + 10$ **59.** $5 - x = -2x - 11$

60. $3 - 8x = -9x - 1$ **61.** $1.2y - 4 = 0.2y - 4$ **62.** $7.7r - 6 = 6.7r - 6$

Solve each equation, and check the solution. **See Examples 7 and 8.**

63. $3x + 6 - 10 = 2x - 2$ **64.** $8x + 4 - 8 = 7x - 1$

65. $5t + 3 + 2t - 6t = 4 + 12$ **66.** $4x - 6 + 3x - 6x = 3 + 10$

▶ **67.** $6x + 5 + 7x + 3 = 12x + 4$ **68.** $4x + 3 + 8x + 1 = 11x + 2$

69. $5.2q - 4.6 - 7.1q = -0.9q - 4.6$ **70.** $4.0x + 2.7 - 9.6x = -4.6x + 2.7$

71. $\dfrac{5}{7}x + \dfrac{1}{3} = \dfrac{2}{5} - \dfrac{2}{7}x + \dfrac{2}{5}$ **72.** $\dfrac{6}{7}s - \dfrac{3}{4} = \dfrac{4}{5} - \dfrac{1}{7}s + \dfrac{1}{6}$

73. $(5y + 6) - (3 + 4y) = 10$ **74.** $(8r + 3) - (1 + 7r) = 6$

▶ **75.** $2(p + 5) - (9 + p) = -3$ **76.** $4(k + 6) - (8 + 3k) = -5$

77. $-6(2b + 1) + (13b - 7) = 0$ **78.** $-5(3w - 3) + (16w + 1) = 0$

79. $10(-2x + 1) = -19(x + 1)$ **80.** $2(-3r + 2) = -5(r - 3)$

Extending Skills   *Solve each equation, and check the solution.* **See Example 8.**

81. $-2(8p + 2) - 3(2 - 7p) - 2(4 + 2p) = 0$

82. $-5(1 - 2z) + 4(3 - z) - 7(3 + z) = 0$

83. $4(7x - 1) + 3(2 - 5x) - 4(3x + 5) = -6$

84. $9(2m - 3) - 4(5 + 3m) - 5(4 + m) = -3$

Concept Check   *Work each problem.*

85. Write an equation that requires the use of the addition property of equality, where 6 must be added to each side and the solution is a negative number.

86. Write an equation that requires the use of the addition property of equality, where $\frac{1}{2}$ must be subtracted from each side and the solution is a positive number.

Write an equation using the information given in the problem. Use x as the variable. Then solve the equation.

87. Three times a number is 17 more than twice the number. Find the number.

88. One added to three times a number is three less than four times the number. Find the number.

89. If six times a number is subtracted from seven times the number, the result is -9. Find the number.

90. If five times a number is added to three times the number, the result is the sum of seven times the number and 9. Find the number.

STUDY SKILLS

Managing Your Time

Many college students juggle a difficult schedule and multiple responsibilities, including school, work, and family demands.

Time Management Tips

- **Read the syllabus for each class.** Understand class policies, such as attendance, late homework, and make-up tests. Find out how you are graded.

- **Make a semester or quarter calendar.** Put test dates and major due dates for *all* your classes on the *same* calendar. Try using a different color pen for each class.

- **Make a weekly schedule.** After you fill in your classes and other regular responsibilities, block off some study periods. Aim for 2 hours of study for each 1 hour in class.

- **Choose a regular study time and place** (such as the campus library). Routine helps.

- **Keep distractions to a minimum.** Get the most out of the time you have set aside for studying by limiting interruptions. Turn off your cell phone. Take a break from social media. Avoid studying in front of the TV.

- **Make "to-do" lists.** Number tasks in order of importance. Cross off tasks as you complete them.

- **Break big assignments into smaller chunks.** Make deadlines for each smaller chunk so that you stay on schedule.

- **Take breaks when studying.** Do not try to study for hours at a time. Take a 10-minute break each hour or so.

- **Ask for help when you need it.** Talk with your instructor during office hours. Make use of the learning center, tutoring center, counseling office, or other resources available at your school.

Think through and answer each question.

1. Evaluate when and where you are currently studying. Are the places you named quiet and comfortable? Are you studying when you are most alert?

2. How many hours do you have available for studying this week?

3. Which two or three of the above suggestions will you try this week to improve your time management?

4. Once the week is over, evaluate how these suggestions worked. What will you do differently next week?

2.2   The Multiplication Property of Equality

OBJECTIVES

1   Use the multiplication property of equality.
2   Simplify, and then use the multiplication property of equality.

OBJECTIVE 1   Use the multiplication property of equality.

The addition property of equality is not enough to solve some equations. Consider the following.

$$3x + 2 = 17$$
$$3x + 2 - 2 = 17 - 2 \qquad \text{Subtract 2 from each side.}$$
$$3x = 15 \qquad \text{Combine like terms.}$$

The coefficient of x is 3, not 1 as desired. The **multiplication property of equality** is needed to change $3x = 15$ to an equation of the form

$$x = \text{a number}.$$

Because $3x = 15$, both $3x$ and 15 must represent the same number. Multiplying both $3x$ and 15 by the same number will result in an equivalent equation.

Multiplication Property of Equality

If A, B, and C represent real numbers, where $C \neq 0$, then the equations

$$A = B \quad \text{and} \quad AC = BC \quad \text{are equivalent.}$$

That is, each side of an equation may be multiplied by the same nonzero number without changing the solution set.

In $3x = 15$, we must change $3x$ to $1x$, or x. To do this, we multiply each side of the equation by $\frac{1}{3}$, the *reciprocal* of 3, because $\frac{1}{3} \cdot 3 = \frac{3}{3} = 1$.

$$3x = 15$$
$$\frac{1}{3} \cdot 3x = \frac{1}{3} \cdot 15 \qquad \text{Multiply each side by } \tfrac{1}{3}, \text{ the reciprocal of 3.}$$
$$\left(\frac{1}{3} \cdot 3 \right) x = \frac{1}{3} \cdot 15 \qquad \text{Associative property}$$

The product of a number and its reciprocal is 1.

$$1x = 5 \qquad \text{Multiplicative inverse property}$$
$$x = 5 \qquad \text{Multiplicative identity property}$$

The solution is 5. We can check this result in the original equation.

Just as the addition property of equality permits *subtracting* the same number from each side of an equation, the multiplication property of equality permits *dividing* each side of an equation by the same nonzero number.

$$3x = 15$$
$$\frac{3x}{3} = \frac{15}{3} \qquad \text{Divide each side by 3.}$$
$$x = 5 \qquad \text{Same result as above}$$

> **Multiplication Property of Equality Extended to Division**
>
> We can divide each side of an equation by the same nonzero number without changing the solution. ***Do not, however, divide each side by a variable, because the variable might be equal to 0.***

NOTE It is usually easier to multiply on each side of an equation if the coefficient of the variable is a fraction, and divide on each side if the coefficient is an integer.

To solve $\frac{3}{4}x = 12$,   it is easier to multiply by $\frac{4}{3}$ than to divide by $\frac{3}{4}$.

To solve $5x = 20$,   it is easier to divide by 5 than to multiply by $\frac{1}{5}$.

 **NOW TRY EXERCISE 1**

Solve $8x = 80$.

EXAMPLE 1 Applying the Multiplication Property of Equality

Solve $5x = 60$.

$$5x = 60 \quad \text{— Our goal is to isolate } x.$$

$$\frac{5x}{5} = \frac{60}{5} \qquad \text{Divide each side by 5, the coefficient of } x.$$

Dividing by 5 is the same as multiplying by $\frac{1}{5}$.

$$x = 12 \qquad \frac{5x}{5} = \frac{5}{5}x = 1x = x$$

CHECK Substitute 12 for x in the original equation.

$$5x = 60 \qquad \text{Original equation}$$

$$5(12) \overset{?}{=} 60 \qquad \text{Let } x = 12.$$

$$60 = 60 \checkmark \qquad \text{True}$$

A true statement results, so the solution set is $\{12\}$.   **NOW TRY**

NOW TRY EXERCISE 2

Solve $10x = -24$.

EXAMPLE 2 Applying the Multiplication Property of Equality

Solve $25x = -30$.

$$25x = -30$$

$$\frac{25x}{25} = \frac{-30}{25} \qquad \text{Divide each side by 25, the coefficient of } x.$$

To avoid errors later, show the division as a separate step.

$$x = -\frac{30}{25} \qquad \frac{-a}{b} = -\frac{a}{b}$$

$$x = -\frac{6}{5} \qquad \text{Write in lowest terms.}$$

CHECK $\qquad\qquad 25x = -30 \qquad \text{Original equation}$

$$\frac{25}{1}\left(-\frac{6}{5}\right) \overset{?}{=} -30 \qquad \text{Let } x = -\frac{6}{5}.$$

$$-30 = -30 \checkmark \qquad \text{True}$$

NOW TRY ANSWERS
1. $\{10\}$
2. $\left\{-\frac{12}{5}\right\}$

The check confirms that the solution set is $\left\{-\frac{6}{5}\right\}$.   **NOW TRY**

**NOW TRY
EXERCISE 3**

Solve $7.02 = -1.3x$.

EXAMPLE 3   Solving a Linear Equation (Decimal Coefficient)

Solve $6.09 = -2.1x$.

$$6.09 = -2.1x \quad \text{\footnotesize Isolate } x \text{ on the right.}$$

$$\frac{6.09}{-2.1} = \frac{-2.1x}{-2.1} \qquad \text{Divide each side by } -2.1.$$

$$-2.9 = x, \quad \text{or} \quad x = -2.9$$

Check by replacing x with -2.9 in the original equation. The solution set is $\{-2.9\}$.

NOW TRY

**NOW TRY
EXERCISE 4**

Solve $\frac{x}{5} = -7$.

EXAMPLE 4   Solving a Linear Equation (Fractional Coefficient)

Solve $\frac{x}{4} = 3$.

$$\frac{x}{4} = 3$$

$$\frac{1}{4}x = 3 \qquad \frac{x}{4} = \frac{1x}{4} = \frac{1}{4}x$$

$$4 \cdot \frac{1}{4}x = 4 \cdot 3 \qquad \text{Multiply each side by 4, the reciprocal of } \frac{1}{4}.$$

$$\boxed{4 \cdot \frac{1}{4}x = 1x = x} \quad x = 12 \qquad \text{Multiplicative inverse property; multiplicative identity property}$$

CHECK

$$\frac{x}{4} = 3 \qquad \text{Original equation}$$

$$\frac{12}{4} \stackrel{?}{=} 3 \qquad \text{Let } x = 12.$$

$$3 = 3 \quad ✓ \qquad \text{True}$$

A true statement results, so the solution set is $\{12\}$.

NOW TRY

**NOW TRY
EXERCISE 5**

Solve $\frac{4}{7}z = -16$.

EXAMPLE 5   Solving a Linear Equation (Fractional Coefficient)

Solve $\frac{3}{4}w = 6$.

$$\frac{3}{4}w = 6$$

$$\frac{4}{3} \cdot \frac{3}{4}w = \frac{4}{3} \cdot 6 \qquad \text{Multiply each side by } \frac{4}{3}, \text{ the reciprocal of } \frac{3}{4}.$$

$$\boxed{\text{Reciprocals have a product of 1.}} \quad 1 \cdot w = \frac{4}{3} \cdot \frac{6}{1} \qquad \text{Multiplicative inverse property}$$

$$w = 8 \qquad \text{Multiplicative identity property; multiply fractions.}$$

Check to confirm that the solution set is $\{8\}$.

NOW TRY

NOW TRY ANSWERS

3. $\{-5.4\}$

4. $\{-35\}$

5. $\{-28\}$

In **Section 2.1,** we obtained $-x = 8$ in our alternative solution to **Example 4.** We reasoned that because the additive inverse (or opposite) of x is 8, then x must equal -8. We can use the multiplication property of equality to obtain the same result.

NOW TRY
EXERCISE 6

Solve $-x = -9$.

EXAMPLE 6   Applying the Multiplication Property of Equality

Solve $-x = 8$.

$$-x = 8$$

$$-1x = 8 \qquad -x = -1x$$

$$-1(-1x) = -1(8) \qquad \text{Multiply each side by } -1.$$

$$[-1(-1)]x = -8 \qquad \text{Associative property; multiply.}$$

These steps are usually omitted.

$$1x = -8 \qquad \text{Multiplicative inverse property}$$

$$x = -8 \qquad \text{Multiplicative identity property}$$

CHECK

$$-x = 8 \qquad \text{Original equation}$$

$$-(-8) \overset{?}{=} 8 \qquad \text{Let } x = -8.$$

$$8 = 8 \ \checkmark \qquad \text{True}$$

A true statement results, so $\{-8\}$ is the solution set.  **NOW TRY**

OBJECTIVE 2   Simplify, and then use the multiplication property of equality.

NOW TRY
EXERCISE 7

Solve $9n - 6n = 21$.

EXAMPLE 7   Combining Like Terms When Solving

Solve $5m + 6m = 33$.

$$5m + 6m = 33$$

$$11m = 33 \qquad \text{Combine like terms.}$$

$$\frac{11m}{11} = \frac{33}{11} \qquad \text{Divide by 11.}$$

$$m = 3 \qquad \text{Multiplicative identity property; divide.}$$

CHECK

$$5m + 6m = 33 \qquad \text{Original equation}$$

$$5(3) + 6(3) \overset{?}{=} 33 \qquad \text{Let } m = 3.$$

$$15 + 18 \overset{?}{=} 33 \qquad \text{Multiply.}$$

$$33 = 33 \ \checkmark \qquad \text{True}$$

NOW TRY ANSWERS
6. $\{9\}$ 7. $\{7\}$

A true statement results, so the solution set is $\{3\}$. **NOW TRY**

2.2 Exercises

FOR EXTRA HELP   ▶ MyMathLab®

▶ *Complete solution available in MyMathLab*

1. *Concept Check*   Tell whether to use the addition or multiplication property of equality to solve each equation. *Do not actually solve.*

(a) $3x = 12$ (b) $3 + x = 12$ (c) $-x = 4$ (d) $-12 = 6 + x$

Concept Check   Choose the letter of the correct response.

2. Which equation does *not* require the use of the multiplication property of equality?

A. $3x - 5x = 6$ **B.** $-\dfrac{1}{4}x = 12$ **C.** $5x - 4x = 7$ **D.** $\dfrac{x}{3} = -2$

3. In the solution of a linear equation, the next-to-the-last step reads "$-x = -\frac{3}{4}$." Which of the following is the solution of this equation?

    A. $-\frac{3}{4}$ **B.** $\frac{3}{4}$ **C.** -1 **D.** $\frac{4}{3}$

4. Which of the following is the solution of the equation $-x = -24$?

    A. 24   **B.** -24 **C.** 1   **D.** -1

Concept Check   *By what number is it necessary to multiply both sides of each equation to isolate x on the left side? Do not actually solve.*

5. $\frac{4}{5}x = 8$ 6. $\frac{2}{3}x = 6$ 7. $\frac{x}{10} = 5$ 8. $\frac{x}{100} = 10$

9. $-\frac{9}{2}x = -4$ 10. $-\frac{8}{3}x = -11$ 11. $-x = 0.75$ 12. $-x = 0.48$

Concept Check   *By what number is it necessary to divide both sides of each equation to isolate x on the left side? Do not actually solve.*

13. $6x = 5$ 14. $7x = 10$ 15. $-4x = 16$ 16. $-13x = 26$

17. $0.12x = 48$ 18. $0.21x = 63$ 19. $-x = 25$ 20. $-x = 50$

Solve each equation, and check the solution. **See Examples 1–6.**

21. $6x = 36$ 22. $8x = 64$ 23. $2m = 15$ 24. $3m = 10$

25. $4x = -20$ 26. $5x = -60$ 27. $-7x = 28$ 28. $-9x = 36$

▶ 29. $10t = -36$ 30. $10s = -54$ 31. $-6x = -72$ 32. $-4x = -64$

33. $4r = 0$ 34. $7x = 0$ ▶ 35. $-x = 12$ 36. $-t = 14$

37. $-x = -\frac{3}{4}$ 38. $-x = -\frac{1}{2}$ 39. $0.2t = 8$ 40. $0.9x = 18$

41. $-0.3x = 9$ 42. $-0.5x = 20$ 43. $0.6x = -1.44$ 44. $0.8x = -2.96$

45. $-9.1 = -2.6x$ 46. $-7.2 = -4.5x$ ▶ 47. $-2.1m = 25.62$ 48. $-3.9x = 32.76$

49. $\frac{1}{4}x = -12$ 50. $\frac{1}{5}p = -3$ ▶ 51. $\frac{z}{6} = 12$ 52. $\frac{x}{5} = 15$

53. $\frac{x}{7} = -5$ 54. $\frac{r}{8} = -3$ ▶ 55. $\frac{2}{7}p = 4$ 56. $\frac{3}{8}x = 9$

57. $-\frac{5}{6}t = -15$ 58. $-\frac{3}{4}z = -21$ 59. $-\frac{7}{9}x = \frac{3}{5}$ 60. $-\frac{5}{6}x = \frac{4}{9}$

Solve each equation, and check the solution. **See Example 7.**

▶ 61. $4x + 3x = 21$ 62. $8x + 3x = 121$ 63. $6r - 8r = 10$

64. $3p - 7p = 24$ 65. $\frac{2}{5}x - \frac{3}{10}x = 2$ 66. $\frac{2}{3}x - \frac{5}{9}x = 4$

67. $7m + 6m - 4m = 63$ 68. $9r + 2r - 7r = 68$ 69. $-6x + 4x - 7x = 0$

70. $-5x + 4x - 8x = 0$ 71. $8w - 4w + w = -3$ 72. $9x - 3x + x = -4$

73. $\frac{1}{3}x - \frac{1}{4}x + \frac{1}{12}x = 3$ 74. $\frac{2}{5}x + \frac{1}{10}x - \frac{1}{20}x = 18$

75. $0.9w - 0.5w + 0.1w = -3$ 76. $0.5x - 0.6x + 0.3x = -1$

Concept Check   *Work each problem.*

77. Write an equation that requires the use of the multiplication property of equality, where each side must be multiplied by $\frac{2}{3}$ and the solution is a negative number.

78. Write an equation that requires the use of the multiplication property of equality, where each side must be divided by 100 and the solution is not an integer.

Write an equation using the information given in the problem. Use x as the variable. Then solve the equation.

79. When a number is multiplied by 4, the result is 6. Find the number.

80. When a number is multiplied by -4, the result is 10. Find the number.

81. When a number is divided by -5, the result is 2. Find the number.

82. If twice a number is divided by 5, the result is 4. Find the number.

2.3 More on Solving Linear Equations

OBJECTIVES

1 Learn and use the four steps for solving a linear equation.

2 Solve equations with no solution or infinitely many solutions.

3 Solve equations with fractions or decimals as coefficients.

4 Write expressions for two related unknown quantities.

VOCABULARY

☐ conditional equation
☐ identity
☐ contradiction
☐ empty (null) set

OBJECTIVE 1 Learn and use the four steps for solving a linear equation.

We now apply *both* properties of equality to solve linear equations.

Solving a Linear Equation in One Variable

Step 1 **Simplify each side separately.** Use the distributive property as needed.
- Clear any parentheses.
- Clear any fractions or decimals.
- Combine like terms.

Step 2 **Isolate the variable terms on one side.** Use the addition property of equality so that all terms with variables are on one side of the equation and all constants (numbers) are on the other side.

Step 3 **Isolate the variable.** Use the multiplication property of equality to obtain an equation that has just the variable with coefficient 1 on one side.

Step 4 **Check.** Substitute the value found into the *original* equation. If a true statement results, write the solution set. If not, rework the problem.

Remember that when we solve an equation, our primary goal is to isolate the variable on one side of the equation.

EXAMPLE 1 Solving a Linear Equation

Solve $-6x + 5 = 17$.

Step 1 There are no parentheses, fractions, or decimals in this equation, so this step is not necessary.

Our goal is to isolate *x*. $\qquad -6x + 5 = 17$

Step 2 $\qquad\qquad -6x + 5 - 5 = 17 - 5$ Subtract 5 from each side.

$\qquad\qquad\qquad\qquad -6x = 12$ Combine like terms.

Step 3 $\qquad\qquad\qquad \dfrac{-6x}{-6} = \dfrac{12}{-6}$ Divide each side by -6.

$\qquad\qquad\qquad\qquad x = -2$

**NOW TRY
EXERCISE 1**

Solve $7 + 2m = -3$.

Step 4   Check by substituting -2 for x in the original equation.

CHECK

$$-6x + 5 = 17 \qquad \text{Original equation}$$

$$-6(-2) + 5 \stackrel{?}{=} 17 \qquad \text{Let } x = -2.$$

$$12 + 5 \stackrel{?}{=} 17 \qquad \text{Multiply.}$$

$$17 = 17 \ \checkmark \qquad \text{True}$$

The check confirms that -2 is the solution. The solution set is $\{-2\}$.   **NOW TRY**

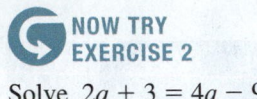

**NOW TRY
EXERCISE 2**

Solve $2q + 3 = 4q - 9$.

| EXAMPLE 2 | Solving a Linear Equation |

Solve $3x + 2 = 5x - 8$.

Step 1   There are no parentheses, fractions, or decimals in the equation.

Our goal is to isolate x. $\longrightarrow \quad 3x + 2 = 5x - 8$

Step 2

$$3x + 2 - 5x = 5x - 8 - 5x \qquad \text{Subtract } 5x \text{ from each side.}$$

$$-2x + 2 = -8 \qquad \text{Combine like terms.}$$

$$-2x + 2 - 2 = -8 - 2 \qquad \text{Subtract 2 from each side.}$$

$$-2x = -10 \qquad \text{Combine like terms.}$$

Step 3

$$\frac{-2x}{-2} = \frac{-10}{-2} \qquad \text{Divide each side by } -2.$$

$$x = 5$$

Step 4   Check by substituting 5 for x in the original equation.

CHECK

$$3x + 2 = 5x - 8 \qquad \text{Original equation}$$

$$3(5) + 2 \stackrel{?}{=} 5(5) - 8 \qquad \text{Let } x = 5.$$

$$15 + 2 \stackrel{?}{=} 25 - 8 \qquad \text{Multiply.}$$

$$17 = 17 \ \checkmark \qquad \text{True}$$

The check confirms that 5 is the solution. The solution set is $\{5\}$.   **NOW TRY**

NOTE   *Remember that the variable can be isolated on either side of the equation.*
In **Example 2,** x will be isolated on the right if we begin by subtracting $3x$.

$$3x + 2 = 5x - 8 \qquad \text{Equation from \textbf{Example 2}}$$

$$3x + 2 - 3x = 5x - 8 - 3x \qquad \text{Subtract } 3x \text{ from each side.}$$

$$2 = 2x - 8 \qquad \text{Combine like terms.}$$

$$2 + 8 = 2x - 8 + 8 \qquad \text{Add 8 to each side.}$$

$$10 = 2x \qquad \text{Combine like terms.}$$

$$\frac{10}{2} = \frac{2x}{2} \qquad \text{Divide each side by 2.}$$

$$5 = x \qquad \text{The same solution results.}$$

NOW TRY ANSWERS

1. $\{-5\}$

2. $\{6\}$

There are often several equally correct ways to solve an equation.

 **NOW TRY
EXERCISE 3**

Solve.

$3(z - 6) - 5z = -7z + 7$

EXAMPLE 3   Solving a Linear Equation

Solve $4(k - 3) - k = k - 6$.

Step 1   Clear parentheses using the distributive property.

$$4(k - 3) - k = k - 6$$

$4(k) + 4(-3) - k = k - 6$	Distributive property
$4k - 12 - k = k - 6$	Multiply.
$3k - 12 = k - 6$	Combine like terms.

Step 2

$3k - 12 - k = k - 6 - k$	Subtract k.
$2k - 12 = -6$	Combine like terms.
$2k - 12 + 12 = -6 + 12$	Add 12.
$2k = 6$	Combine like terms.

Step 3

$\dfrac{2k}{2} = \dfrac{6}{2}$	Divide by 2.
$k = 3$	

Step 4   **CHECK**

$4(k - 3) - k = k - 6$	Original equation
$4(3 - 3) - 3 \overset{?}{=} 3 - 6$	Let $k = 3$.
$4(0) - 3 \overset{?}{=} 3 - 6$	Work inside the parentheses.
$-3 = -3$ ✓	True

The solution set is $\{3\}$.

NOW TRY

 **NOW TRY
EXERCISE 4**

Solve.

$5x - (x + 9) = x - 4$

EXAMPLE 4   Solving a Linear Equation

Solve $8z - (3 + 2z) = 3z + 1$.

Step 1

$8z - (3 + 2z) = 3z + 1$	
$8z - 1(3 + 2z) = 3z + 1$	Multiplicative identity property
$8z - 1(3) - 1(2z) = 3z + 1$	Distributive property
$8z - 3 - 2z = 3z + 1$	Multiply.
$6z - 3 = 3z + 1$	Combine like terms.

> Be careful with signs.

Step 2

$6z - 3 - 3z = 3z + 1 - 3z$	Subtract $3z$.
$3z - 3 = 1$	Combine like terms.
$3z - 3 + 3 = 1 + 3$	Add 3.
$3z = 4$	Combine like terms.

Step 3

$\dfrac{3z}{3} = \dfrac{4}{3}$	Divide by 3.
$z = \dfrac{4}{3}$	

NOW TRY ANSWERS

3. $\{5\}$

4. $\left\{\dfrac{5}{3}\right\}$

Step 4   Check that $\left\{\dfrac{4}{3}\right\}$ is the solution set.

NOW TRY

⚠ **CAUTION**   In an expression such as $8z - (3 + 2z)$ in **Example 4,** the $-$ sign acts like a factor of -1 and affects the sign of *every* term within the parentheses.

$$8z - (3 + 2z) \leftarrow \text{Left side of the equation in \textbf{Example 4}}$$

$$= 8z - 1(3 + 2z)$$

$$= 8z + (-1)(3 + 2z)$$

$$= 8z - 3 - 2z$$

Change to $-$ in *both* terms.

**NOW TRY
EXERCISE 5**

Solve.

$$24 - 4(7 - 2t) = 4(t - 1)$$

EXAMPLE 5   Solving a Linear Equation

Solve $4(4 - 3x) = 32 - 8(x + 2)$.

Step 1

$$4(4 - 3x) = 32 - 8(x + 2) \quad \text{Be careful with signs.}$$

$$16 - 12x = 32 - 8x - 16 \qquad \text{Distributive property}$$

$$16 - 12x = 16 - 8x \qquad\qquad \text{Combine like terms.}$$

Step 2

$$16 - 12x + 8x = 16 - 8x + 8x \qquad \text{Add } 8x.$$

$$16 - 4x = 16 \qquad\qquad \text{Combine like terms.}$$

$$16 - 4x - 16 = 16 - 16 \qquad \text{Subtract 16.}$$

$$-4x = 0 \qquad\qquad \text{Combine like terms.}$$

Step 3

$$\frac{-4x}{-4} = \frac{0}{-4} \qquad\qquad \text{Divide by } -4.$$

$$x = 0$$

Step 4   **CHECK**

$$4(4 - 3x) = 32 - 8(x + 2) \qquad \text{Original equation}$$

$$4[4 - 3(0)] \overset{?}{=} 32 - 8(0 + 2) \qquad \text{Let } x = 0.$$

$$4(4 - 0) \overset{?}{=} 32 - 8(2) \qquad \text{Multiply and add.}$$

$$4(4) \overset{?}{=} 32 - 16 \qquad \text{Subtract and multiply.}$$

$$16 = 16 \; \checkmark \qquad \text{True}$$

Because a true statement results, the solution set is $\{0\}$.   **NOW TRY**

NOTE   As the check in **Example 5** confirms, it is perfectly acceptable for an equation to have solution set $\{0\}$.

OBJECTIVE 2   Solve equations with no solution or infinitely many solutions.

Each equation so far has had exactly one solution. An equation with exactly one solution is a **conditional equation** because it is only true under certain conditions. Some equations may have no solution or infinitely many solutions.

NOW TRY ANSWER
5. $\{0\}$

 **NOW TRY EXERCISE 6**

Solve.

$-3(x - 7) = 2x - 5x + 21$

EXAMPLE 6 Solving an Equation That Has Infinitely Many Solutions

Solve $5x - 15 = 5(x - 3)$.

$5x - 15 = 5(x - 3)$	
$5x - 15 = 5x - 15$	Distributive property
$5x - 15 - 5x = 5x - 15 - 5x$	Subtract $5x$.
$-15 = -15$	Combine like terms.
$-15 + 15 = -15 + 15$	Add 15.
$0 = 0$	True

> Notice that the variable "disappeared."

Solution set: {all real numbers}

Because the last statement $(0 = 0)$ is true, *any* real number is a solution. We could have predicted this from the second line in the solution.

$5x - 15 = 5x - 15$ ← This is true for *any* value of x.

Try several values for x in the original equation to see that they all satisfy it.

An equation with both sides exactly the same, like $0 = 0$, is an **identity.** An identity is true for *all* replacements of the variables. As shown above, we write the solution set as

{all real numbers}.   NOW TRY

⚠ **CAUTION** In **Example 6,** do not write {0} as the solution set. While 0 is a solution, there are infinitely many other solutions. *For the solution set to be {0}, the last line must include a variable, such as x, and read x = 0 (as in Example 5), not 0 = 0.*

 **NOW TRY EXERCISE 7**

Solve.

$-4x + 12 = 3 - 4(x - 3)$

EXAMPLE 7 Solving an Equation That Has No Solution

Solve $2x + 3(x + 1) = 5x + 4$.

$2x + 3(x + 1) = 5x + 4$	
$2x + 3x + 3 = 5x + 4$	Distributive property
$5x + 3 = 5x + 4$	Combine like terms.
$5x + 3 - 5x = 5x + 4 - 5x$	Subtract $5x$.
$3 = 4$	False

> Again, the variable "disappeared."

There is no solution.   Solution set: ∅

A false statement $(3 = 4)$ results. A **contradiction** is an equation that has no solution. Its solution set is the **empty set,** or **null set,** symbolized ∅.

NOW TRY

⚠ **CAUTION** **Do not** write {∅} to represent the empty set.

NOW TRY ANSWERS
6. {all real numbers}
7. ∅

▼ **Solution Sets of Equations**

Type of Equation	Final Equation in Solution	Number of Solutions	Solution Set
Conditional (See Examples 1–5.)	$x =$ a number	One	{a number}
Identity (See Example 6.)	A true statement with no variable, such as 0 = 0	Infinitely many	{all real numbers}
Contradiction (See Example 7.)	A false statement with no variable, such as 3 = 4	None	∅

OBJECTIVE 3 Solve equations with fractions or decimals as coefficients.

To avoid messy computations, we clear an equation of fractions by multiplying each side by the least common denominator (LCD) of all the fractions in the equation. Doing this will give an equation with only *integer* coefficients.

NOW TRY EXERCISE 8

Solve.

$$\frac{1}{2}x + \frac{5}{8}x = \frac{3}{4}x - 6$$

EXAMPLE 8   Solving a Linear Equation (Fractional Coefficients)

Solve $\frac{2}{3}x - \frac{1}{2}x = -\frac{1}{6}x - 2$.

Step 1   The LCD of all the fractions in the equation is 6.

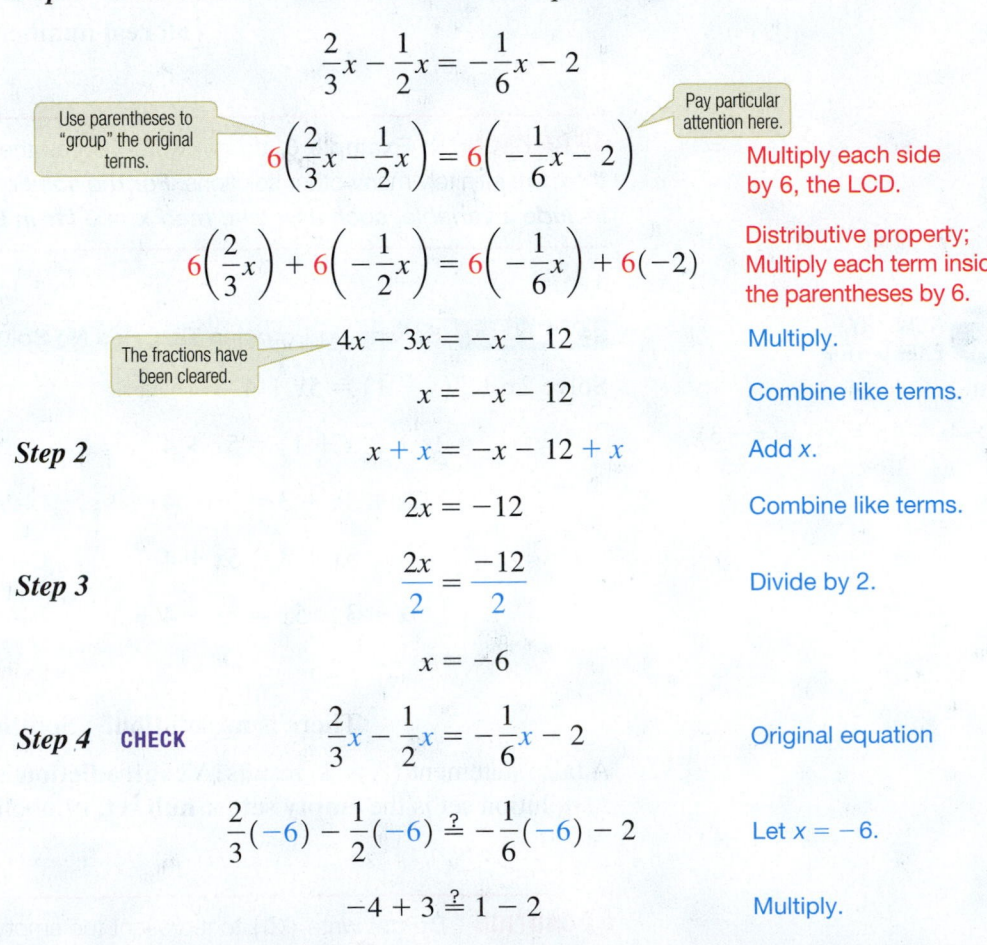

$$\frac{2}{3}x - \frac{1}{2}x = -\frac{1}{6}x - 2$$

Use parentheses to "group" the original terms.

Pay particular attention here.

$$6\left(\frac{2}{3}x - \frac{1}{2}x\right) = 6\left(-\frac{1}{6}x - 2\right)$$ Multiply each side by 6, the LCD.

$$6\left(\frac{2}{3}x\right) + 6\left(-\frac{1}{2}x\right) = 6\left(-\frac{1}{6}x\right) + 6(-2)$$ Distributive property; Multiply each term inside the parentheses by 6.

The fractions have been cleared.

$$4x - 3x = -x - 12$$ Multiply.

$$x = -x - 12$$ Combine like terms.

Step 2 $$x + x = -x - 12 + x$$ Add x.

$$2x = -12$$ Combine like terms.

Step 3 $$\frac{2x}{2} = \frac{-12}{2}$$ Divide by 2.

$$x = -6$$

Step 4   **CHECK** $$\frac{2}{3}x - \frac{1}{2}x = -\frac{1}{6}x - 2$$ Original equation

$$\frac{2}{3}(-6) - \frac{1}{2}(-6) \stackrel{?}{=} -\frac{1}{6}(-6) - 2$$ Let $x = -6$.

$$-4 + 3 \stackrel{?}{=} 1 - 2$$ Multiply.

$$-1 = -1 \checkmark$$ True

NOW TRY ANSWER
8. {−16}

The check confirms that the solution set is {−6}.

NOW TRY

⊘ **CAUTION** *When clearing an equation of fractions, be sure to multiply every term on each side of the equation by the LCD.*

 **NOW TRY EXERCISE 9**

Solve.

$$\frac{2}{3}(x + 2) - \frac{1}{2}(3x + 4) = -4$$

EXAMPLE 9 Solving a Linear Equation (Fractional Coefficients)

Solve $\frac{1}{3}(x + 5) - \frac{3}{5}(x + 2) = 1$.

Step 1 We clear the parentheses first. Then we clear the fractions.

$$\frac{1}{3}(x + 5) - \frac{3}{5}(x + 2) = 1 \quad \text{◁ Study Step 1 carefully.}$$

$$\frac{1}{3}(x) + \frac{1}{3}(5) - \frac{3}{5}(x) - \frac{3}{5}(2) = 1 \qquad \text{Distributive property}$$

$$\frac{1}{3}x + \frac{5}{3} - \frac{3}{5}x - \frac{6}{5} = 1 \qquad \text{Multiply.}$$

$$15\left(\frac{1}{3}x + \frac{5}{3} - \frac{3}{5}x - \frac{6}{5}\right) = 15(1) \qquad \text{Multiply each side by 15, the LCD.}$$

$$15\left(\frac{1}{3}x\right) + 15\left(\frac{5}{3}\right) + 15\left(-\frac{3}{5}x\right) + 15\left(-\frac{6}{5}\right) = 15(1) \qquad \text{Distributive property}$$

Think: $15\left(\frac{1}{3}x\right)$
$= \left(\frac{15}{1} \cdot \frac{1}{3}\right)x$
$= 5x$

$$5x + 25 - 9x - 18 = 15 \qquad \text{Multiply.}$$

$$-4x + 7 = 15 \qquad \text{Combine like terms.}$$

Step 2 $$-4x + 7 - 7 = 15 - 7 \qquad \text{Subtract 7.}$$

$$-4x = 8 \qquad \text{Combine like terms.}$$

Step 3 $$\frac{-4x}{-4} = \frac{8}{-4} \qquad \text{Divide by } -4.$$

$$x = -2$$

Step 4 Check to confirm that $\{-2\}$ is the solution set.   **NOW TRY** ↻

EXAMPLE 10 Solving a Linear Equation (Decimal Coefficients)

Solve $0.1t + 0.05(20 - t) = 0.09(20)$.

Step 1 $$0.1t + 0.05(20 - t) = 0.09(20) \quad \text{◁ Clear the parentheses first.}$$

$$0.1t + 0.05(20) + 0.05(-t) = 0.09(20) \qquad \text{Distributive property}$$

$$0.1t + 1 - 0.05t = 1.8 \qquad \text{(*) Multiply.}$$

The decimals here are expressed as tenths (0.1 and 1.8) and hundredths (0.05). We choose the least exponent on 10 to eliminate the decimal points, which will make all coefficients integers. Here, we multiply by 10^2—that is, 100.

Now clear the decimals. $$100(0.1t + 1 - 0.05t) = 100(1.8) \qquad \text{Multiply by 100.}$$

$$100(0.1t) + 100(1) + 100(-0.05t) = 100(1.8) \qquad \text{Distributive property}$$

$$10t + 100 - 5t = 180 \qquad \text{Multiply.}$$

$$5t + 100 = 180 \qquad \text{Combine like terms.}$$

**NOW TRY
EXERCISE 10**

Solve.

$0.05(13 - t) - 0.2t = 0.08(30)$

Step 2

$$5t + 100 - 100 = 180 - 100 \qquad \text{Subtract 100.}$$

$$5t = 80 \qquad \text{Combine like terms.}$$

Step 3

$$\frac{5t}{5} = \frac{80}{5} \qquad \text{Divide by 5.}$$

$$t = 16$$

Step 4   Check to confirm that $\{16\}$ is the solution set.   NOW TRY

NOTE   In **Example 10**, multiplying by 100 is the same as moving the decimal points two places to the right.

$$0.10t + 1.00 - 0.05t = 1.80 \qquad \text{Equation (*) from \textbf{Example 10}}$$
$$ \qquad \text{with 0s included as placeholders}$$

$$10t + 100 - 5t = 180 \qquad \text{Multiply by 100.}$$

OBJECTIVE 4   Write expressions for two related unknown quantities.

**NOW TRY
EXERCISE 11**

Two numbers have a sum of 18. If one of the numbers is represented by m, find an expression for the other number.

EXAMPLE 11   Translating a Phrase into an Algebraic Expression

Perform each translation.

(a)   Two numbers have a sum of 23. If one of the numbers is represented by x, find an expression for the other number.

First, suppose that the sum of two numbers is 23, and one of the numbers is 10. To find the other number, we would subtract 10 from 23.

$$23 - 10 \leftarrow \text{This gives 13 as the other number.}$$

Instead of using 10 as one of the numbers, we use x. The other number would be obtained in the same way—by subtracting x from 23.

$$23 - x. \qquad \boxed{\begin{array}{l} x - 23 \text{ is not correct.} \\ \text{Subtraction is not} \\ \text{commutative.} \end{array}}$$

To check, find the sum of the two numbers.

$$x + (23 - x)$$

$$= 23, \quad \text{as required.}$$

(b)   Two numbers have a product of 24. If one of the numbers is represented by x, find an expression for the other number.

Suppose that one of the numbers is 4. To find the other number, we would divide 24 by 4.

$$\frac{24}{4} \leftarrow \begin{array}{l} \text{This gives 6 as the other number.} \\ \text{The product } 6 \cdot 4 \text{ is 24.} \end{array}$$

In the same way, if x is one of the numbers, then we divide 24 by x to find the other number.

$$\frac{24}{x} \leftarrow \text{The other number} \qquad \text{NOW TRY} \; \text{}$$

NOW TRY ANSWERS
10. $\{-7\}$
11. $18 - m$

2.3 Exercises

 FOR EXTRA HELP   ▶ **MyMathLab®**

▶ *Complete solution available in MyMathLab*

Concept Check Using the methods of this section, what should we do first when solving each equation? Do not actually solve.

1. $7x + 8 = 1$

2. $7x - 5x + 15 = 8 + x$

3. $3(2t - 4) = 20 - 2t$

4. $\frac{3}{4}z = -15$

5. $\frac{2}{3}x - \frac{1}{6} = \frac{3}{2}x + 1$

6. $0.9x + 0.3(x + 12) = 6$

7. *Concept Check* Suppose that when solving three linear equations, we obtain the final results shown in parts (a)–(c). Fill in the blanks in parts (a)–(c), and then match each result with the solution set in choices A–C for the *original* equation.

(a) $6 = 6$ (The original equation is a(n) _____.)   **A.** $\{0\}$

(b) $x = 0$ (The original equation is a(n) _____ equation.)   **B.** {all real numbers}

(c) $-5 = 0$ (The original equation is a(n) _____.)   **C.** $\varnothing$

8. *Concept Check* Which equation does *not* have {all real numbers} as its solution set?

A. $5x = 4x + x$ **B.** $2(x + 6) = 2x + 12$ **C.** $\frac{1}{2}x = 0.5x$ **D.** $3x = 2x$

Solve each equation, and check the solution. **See Examples 1–7.**

9. $3x + 2 = 14$

10. $4x + 3 = 27$

11. $-5z - 4 = 21$

12. $-7w - 4 = 10$

13. $4p - 5 = 2p$

14. $6q - 2 = 3q$

15. $2x + 9 = 4x + 11$

16. $7p + 8 = 9p - 2$

17. $5m + 8 = 7 + 3m$

18. $4r + 2 = r - 6$

19. $-12x - 5 = 10 - 7x$

20. $-16w - 3 = 13 - 8w$

▶ **21.** $12h - 5 = 11h + 5 - h$

22. $-4x - 1 = -5x + 1 + 3x$

23. $7r - 5r + 2 = 5r + 2 - r$

24. $9p - 4p + 6 = 7p + 6 - 3p$

▶ **25.** $3(4x + 2) + 5x = 30 - x$

26. $5(2m + 3) - 4m = 2m + 25$

▶ **27.** $-2p + 7 = 3 - (5p + 1)$

28. $4x + 9 = 3 - (x - 2)$

▶ **29.** $11x - 5(x + 2) = 6x + 5$

30. $6x - 4(x + 1) = 2x + 4$

▶ **31.** $6(3w + 5) = 2(10w + 10)$

32. $4(2x - 1) = -6(x + 3)$

33. $-(4x + 2) - (-3x - 5) = 3$

34. $-(6k - 5) - (-5k + 8) = -3$

▶ **35.** $3(2x - 4) = 6(x - 2)$

36. $3(6 - 4x) = 2(-6x + 9)$

37. $6(4x - 1) = 12(2x + 3)$

38. $6(2x + 8) = 4(3x - 6)$

Solve each equation, and check the solution. **See Examples 8–10.**

▶ **39.** $\frac{3}{5}t - \frac{1}{10}t = t - \frac{5}{2}$

40. $-\frac{2}{7}r + 2r = \frac{1}{2}r + \frac{17}{2}$

41. $\frac{3}{4}x - \frac{1}{3}x + 5 = \frac{5}{6}x$

42. $\frac{1}{5}x - \frac{2}{3}x - 2 = -\frac{2}{5}x$

43. $\frac{1}{7}(3x + 2) - \frac{1}{5}(x + 4) = 2$

44. $\frac{1}{4}(3x - 1) + \frac{1}{6}(x + 3) = 3$

45. $-\dfrac{1}{4}(x - 12) + \dfrac{1}{2}(x + 2) = x + 4$

46. $\dfrac{1}{9}(p + 18) + \dfrac{1}{3}(2p + 3) = p + 3$

47. $\dfrac{2}{3}k - \left(k - \dfrac{1}{2}\right) = \dfrac{1}{6}(k - 51)$

48. $-\dfrac{5}{6}q - (q - 1) = \dfrac{1}{4}(-q + 80)$

49. $0.75x - 3.2 = 0.55 - 0.5x$

50. $1.35x - 0.6 = 1.65 + 2.1x$

51. $0.8t + 0.15 = 2t - 1.35$

52. $-0.12p + 3.4 = 0.84 + 5p$

53. $0.2(60) + 0.05x = 0.1(60 + x)$

54. $0.3(30) + 0.15x = 0.2(30 + x)$

55. $1.00x + 0.05(12 - x) = 0.10(63)$

56. $0.92x + 0.98(12 - x) = 0.96(12)$

57. $0.6(10{,}000) + 0.8x = 0.72(10{,}000 + x)$

58. $0.2(5000) + 0.3x = 0.25(5000 + x)$

Solve each equation, and check the solution. See Examples 1–10.

59. $10(2x - 1) = 8(2x + 1) + 14$

60. $9(3k - 5) = 12(3k - 1) - 51$

61. $24 - 4(7 - 2t) = 4(t - 1)$

62. $8 - 2(2 - x) = 4(x + 1)$

63. $4(x + 8) = 2(2x + 6) + 20$

64. $4(x + 3) = 2(2x + 8) - 4$

65. $\dfrac{1}{2}(x + 2) + \dfrac{3}{4}(x + 4) = x + 5$

66. $\dfrac{1}{3}(x + 3) + \dfrac{1}{6}(x - 6) = x + 3$

67. $0.1(x + 80) + 0.2x = 14$

68. $0.3(x + 15) + 0.4(x + 25) = 25$

69. $9(v + 1) - 3v = 2(3v + 1) - 8$

70. $8(t - 3) + 4t = 6(2t + 1) - 10$

Perform each translation. See Example 11.

71. Two numbers have a sum of 11. One of the numbers is q. What expression represents the other number?

72. Two numbers have a sum of 34. One of the numbers is r. What expression represents the other number?

73. The product of two numbers is 9. One of the numbers is x. What expression represents the other number?

74. The product of two numbers is -6. One of the numbers is m. What expression represents the other number?

75. A baseball player got 65 hits one season. He got h of the hits in one game. What expression represents the number of hits he got in the rest of the games?

76. A hockey player scored 42 goals in one season. He scored n goals in one game. What expression represents the number of goals he scored in the rest of the games?

77. Monica is x years old. What expression represents her age 15 yr from now? 5 yr ago?

78. Chandler is y years old. What expression represents his age 4 yr ago? 11 yr from now?

79. Cliff has r quarters. Express the value of the quarters in cents.

80. Claire has y dimes. Express the value of the dimes in cents.

81. A clerk has t dollars, all in \$5 bills. What expression represents the number of \$5 bills?

82. A clerk has v dollars, all in \$10 bills. What expression represents the number of \$10 bills?

83. A plane ticket costs x dollars for an adult and y dollars for a child. Find an expression that represents the total cost for 3 adults and 2 children.

84. A concert ticket costs p dollars for an adult and q dollars for a child. Find an expression that represents the total cost for 4 adults and 6 children.

STUDY SKILLS

Using Study Cards Revisited

We introduced study cards previously. Another type of study card follows.

Practice Quiz Cards

Write a problem with direction words (like *solve, simplify*) on the front of the card, and work the problem on the back. Make one for each type of problem you learn.

Make a practice quiz card for material you are learning now.

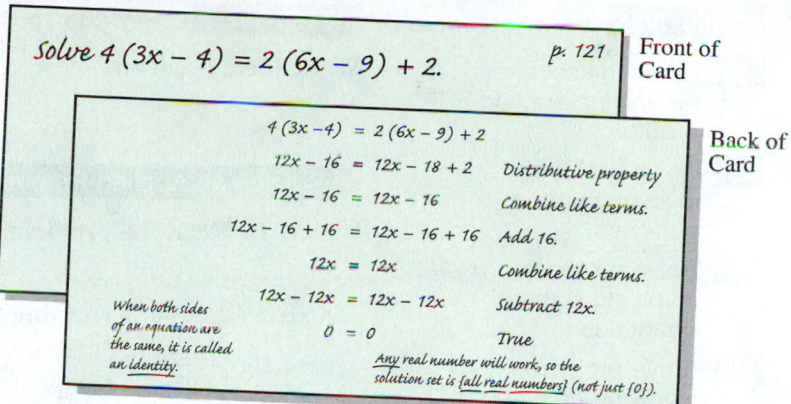

Front of Card

Solve $4(3x - 4) = 2(6x - 9) + 2$. *p. 121*

Back of Card

$$4(3x - 4) = 2(6x - 9) + 2$$
$$12x - 16 = 12x - 18 + 2 \quad \text{Distributive property}$$
$$12x - 16 = 12x - 16 \quad \text{Combine like terms.}$$
$$12x - 16 + 16 = 12x - 16 + 16 \quad \text{Add 16.}$$
$$12x = 12x \quad \text{Combine like terms.}$$
$$12x - 12x = 12x - 12x \quad \text{Subtract 12x.}$$
$$0 = 0 \quad \text{True}$$

When both sides of an equation are the same, it is called an identity.

Any real number will work, so the solution set is {all real numbers} (not just {0}).

SUMMARY EXERCISES Applying Methods for Solving Linear Equations

Concept Check Decide whether each of the following is an equation *or an* expression. *If it is an equation, solve it. If it is an expression, simplify it.*

1. $x + 2 = -3$

2. $4p - 6 + 3p - 8$

3. $-(m - 1) - (3 + 2m)$

4. $6q - 9 = 12 + 3q$

5. $5x - 9 = 3(x - 3)$

6. $\frac{2}{3}x + 8 = \frac{1}{4}x$

7. $2 - 6(z + 1) - 4(z - 2) - 10$

8. $7(p - 2) + p = 2(p + 2)$

9. $\frac{1}{2}(x + 10) - \frac{2}{3}x$

10. $-4(k + 2) + 3(2k - 1)$

Solve each equation, and check the solution.

11. $-6z = -14$

12. $2m + 8 = 16$

13. $12.5x = -63.75$

14. $-x = -12$

15. $\frac{4}{5}x = -20$

16. $7m - 5m = -12$

17. $-x = 6$

18. $\frac{x}{-2} = 8$

19. $4x + 2(3 - 2x) = 6$

20. $x - 16.2 = 7.5$

21. $7m - (2m - 9) = 39$

22. $2 - (m + 4) = 3m - 2$

23. $-3(m - 4) + 2(5 + 2m) = 29$

24. $-0.3x + 2.1(x - 4) = -6.6$

25. $0.08x + 0.06(x + 9) = 1.24$

26. $3(m + 5) - 1 + 2m = 5(m + 2)$

27. $-2t + 5t - 9 = 3(t - 4) - 5$

28. $2.3x + 13.7 = 1.3x + 2.9$

29. $0.2(50) + 0.8r = 0.4(50 + r)$

30. $r + 9 + 7r = 4(3 + 2r) - 3$

31. $2(3 + 7x) - (1 + 15x) = 2$

32. $0.6(100 - x) + 0.4x = 0.5(92)$

33. $\frac{1}{4}x - 4 = \frac{3}{2}x + \frac{3}{4}x$

34. $\frac{3}{4}(z - 2) - \frac{1}{3}(5 - 2z) = -2$

---

1. equation; $\{-5\}$

2. expression; $7p - 14$

3. expression; $-3m - 2$

4. equation; $\{7\}$

5. equation; $\{0\}$

6. equation; $\left\{-\dfrac{96}{5}\right\}$

7. expression; $-10z - 6$

8. equation; $\{3\}$

9. expression; $-\dfrac{1}{6}x + 5$

10. expression; $2k - 11$

11. $\left\{\dfrac{7}{3}\right\}$ 12. $\{4\}$

13. $\{-5.1\}$ 14. $\{12\}$

15. $\{-25\}$ 16. $\{-6\}$

17. $\{-6\}$ 18. $\{-16\}$

19. {all real numbers}

20. $\{23.7\}$

21. $\{6\}$ 22. $\{0\}$

23. $\{7\}$ 24. $\{1\}$

25. $\{5\}$ 26. $\varnothing$

27. $\varnothing$ 28. $\{-10.8\}$

29. $\{25\}$

30. {all real numbers}

31. $\{3\}$ 32. $\{70\}$

33. $\{-2\}$ 34. $\left\{\dfrac{14}{17}\right\}$

2.4 Applications of Linear Equations

VOCABULARY

☐ consecutive integers
☐ consecutive even (or odd) integers
☐ degree
☐ complementary angles
☐ right angle
☐ supplementary angles
☐ straight angle

OBJECTIVE 1 **Learn the six steps for solving applied problems.**

While there is not one specific method, we suggest the following.

Solving an Applied Problem

Step 1 **Read** the problem carefully. *What information is given? What is to be found?*

Step 2 **Assign a variable** to represent the unknown value. Make a sketch, diagram, or table, as needed. If necessary, express any other unknown values in terms of the variable.

Step 3 **Write an equation** using the variable expression(s).

Step 4 **Solve** the equation.

Step 5 **State the answer.** Label it appropriately. *Does it seem reasonable?*

Step 6 **Check** the answer in the words of the *original* problem.

OBJECTIVE 2 **Solve problems involving unknown numbers.**

EXAMPLE 1   **Finding the Value of an Unknown Number**

The product of 4, and a number decreased by 7, is 100. What is the number?

Step 1 **Read** the problem carefully. We are asked to find a number.

Step 2 **Assign a variable** to represent the unknown quantity.

$$\text{Let } x = \text{the number.}$$

Step 3 **Write an equation.**

Writing a "word equation" is often helpful.

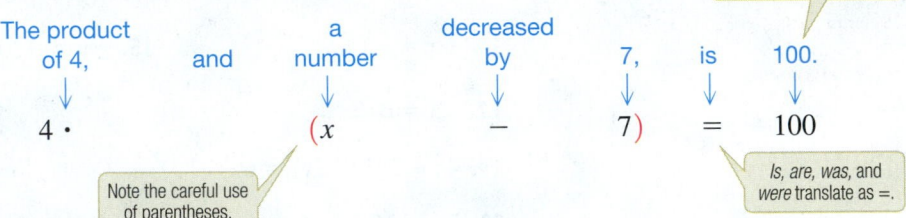

The product of 4,	and	a number	decreased by	7,	is	100.
$4\cdot$		$(x$	$-$	$7)$	$=$	100

Note the careful use of parentheses.

Is, are, was, and were translate as =.

Step 4 **Solve.**

$$4(x - 7) = 100 \qquad \text{Equation from Step 3}$$

$$4x - 28 = 100 \qquad \text{Distributive property}$$

$$4x - 28 + 28 = 100 + 28 \qquad \text{Add 28.}$$

$$4x = 128 \qquad \text{Combine like terms.}$$

$$\frac{4x}{4} = \frac{128}{4} \qquad \text{Divide by 4.}$$

$$x = 32$$

 **NOW TRY EXERCISE 1**

The product of 7, and a number increased by 3, is -63. What is the number?

Step 5 **State the answer.** The number is 32.

Step 6 **Check.** When 32 is decreased by 7, we obtain $32 - 7 = 25$. If 4 is multiplied by 25, we obtain 100, as required. The answer, 32, is correct.

NOW TRY

⚠ **CAUTION** Because of the commas in the problem in **Example 1,** writing the equation as

$$4x - 7 = 100 \text{ is } incorrect.$$

This equation corresponds to the statement "The product of 4 and a number, decreased by 7, is 100."

 **NOW TRY EXERCISE 2**

If 5 is added to a number, the result is 7 less than three times the number. Find the number.

EXAMPLE 2 Finding the Value of an Unknown Number

If 6 is subtracted from five times a number, the result is 9 more than twice the number. Find the number.

Step 1 **Read** the problem. We are asked to find a number.

Step 2 **Assign a variable** to represent the unknown quantity.

$$\text{Let } x = \text{the number.}$$

Step 3 **Write an equation.**

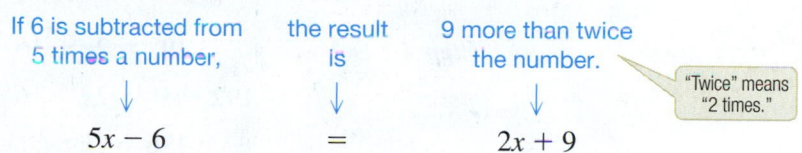

If 6 is subtracted from 5 times a number,	the result is	9 more than twice the number.
↓	↓	↓
$5x - 6$	$=$	$2x + 9$

"Twice" means "2 times."

Step 4 **Solve.**

$$5x - 6 - 2x = 2x + 9 - 2x \qquad \text{Subtract } 2x.$$
$$3x - 6 = 9 \qquad \text{Combine like terms.}$$
$$3x - 6 + 6 = 9 + 6 \qquad \text{Add 6.}$$
$$3x = 15 \qquad \text{Combine like terms.}$$
$$\frac{3x}{3} = \frac{15}{3} \qquad \text{Divide by 3.}$$
$$x = 5$$

Step 5 **State the answer.** The number is 5.

Step 6 **Check.** If 6 is subtracted from 5 times the number, we have

$$5 \cdot 5 - 6, \quad \text{which equals 19.}$$

Nine more than twice the number would be

$$2 \cdot 5 + 9, \quad \text{which also equals 19.}$$

The answer, 5, checks.

NOW TRY

OBJECTIVE 3 Solve problems involving sums of quantities.

PROBLEM-SOLVING HINT To solve problems involving sums of quantities, choose a variable to represent one of the unknowns. ***Then represent the other quantity in terms of the same variable.*** (See Example 11 in **Section 2.3**.)

NOW TRY ANSWERS
1. -12
2. 6

NOW TRY EXERCISE 3

In the 2012 Summer Olympics in London, England, Germany won 21 fewer medals than Great Britain. The two countries won a total of 109 medals. How many medals did each country win? (*Source: World Almanac and Book of Facts.*)

EXAMPLE 3   **Finding Numbers of Olympic Medals**

In the 2012 Summer Olympics in London, England, the United States won 16 more medals than China. The two countries won a total of 192 medals. How many medals did each country win? (*Source: World Almanac and Book of Facts.*)

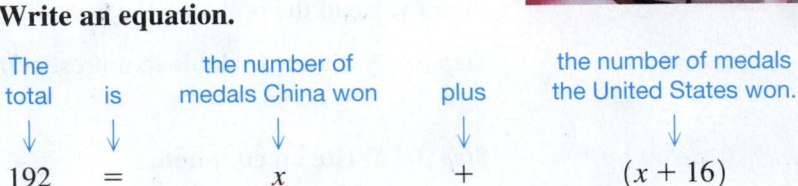

Step 1   **Read** the problem. We are given the total number of medals and asked to find the number each country won.

Step 2   **Assign a variable.**

Let x = the number of medals China won.

Then $x + 16$ = the number of medals the United States won.

Step 3   **Write an equation.**

The total	is	the number of medals China won	plus	the number of medals the United States won.
↓	↓	↓	↓	↓
192	=	x	+	$(x + 16)$

Step 4   **Solve.**

$$192 = 2x + 16 \qquad \text{Combine like terms.}$$
$$192 - 16 = 2x + 16 - 16 \qquad \text{Subtract 16.}$$
$$176 = 2x \qquad \text{Combine like terms.}$$
$$\frac{176}{2} = \frac{2x}{2} \qquad \text{Divide by 2.}$$
$$88 = x, \quad \text{or} \quad x = 88 \leftarrow \text{Medals China won}$$

Step 5   **State the answer.** The variable x represents the number of medals China won, so China won 88 medals.

$$x + 16$$
$$= 88 + 16$$
$$= 104 \leftarrow \text{Medals the United States won}$$

Step 6   **Check.** The United States won 104 medals and China won 88, so the total number of medals was $104 + 88 = 192$. Because $104 - 88 = 16$, the United States won 16 more medals than China. This agrees with the information given in the problem. The answer checks.

NOW TRY

NOTE   The problem in **Example 3** could also be solved by letting x represent the number of medals the United States won. Then $x - 16$ would represent the number of medals China won. The equation would be different.

$$192 = x + (x - 16) \qquad \text{Alternative equation for \textbf{Example 3}}$$

The solution of this equation is 104, which is the number of U.S. medals. The number of Chinese medals would be $104 - 16 = 88$. ***The answers are the same,*** whichever approach is used, even though the equation and its solution are different.

NOW TRY ANSWER

3. Great Britain: 65 medals; Germany: 44 medals

⚠ **CAUTION** The nature of an applied problem may restrict the set of possible solutions. For instance, an answer such as -33 medals or $25\frac{1}{2}$ medals would be inappropriate in **Example 3.** Be sure that an answer is reasonable given the context of the problem.

**NOW TRY
EXERCISE 4**

In one week, the owner of Carly's Coffeehouse found that the number of orders for bagels was $\frac{2}{3}$ the number of orders for chocolate scones. If the total number of orders for the two items was 525, how many orders were placed for bagels?

EXAMPLE 4 Finding the Number of Orders for Tea

The owner of Terry's Coffeehouse found that on one day the number of orders for tea was $\frac{1}{3}$ the number of orders for coffee. If the total number of orders for the two drinks was 76, how many orders were placed for tea?

Step 1 **Read** the problem. It asks for the number of orders for tea.

Step 2 **Assign a variable.** Because of the way the problem is stated, let the variable represent the number of orders for *coffee*.

Let x = the number of orders for coffee.

Then $\frac{1}{3}x$ = the number of orders for tea.

Step 3 **Write an equation.** Use the fact that the total number of orders was 76.

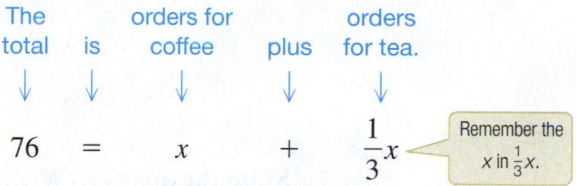

| The total | is | orders for coffee | plus | orders for tea. |
| 76 | = | x | + | $\frac{1}{3}x$ |

Remember the x in $\frac{1}{3}x$.

Step 4 **Solve.** $76 = \frac{4}{3}x$ $\quad x = 1x = \frac{3}{3}x$;
Combine like terms.

$$\frac{3}{4}(76) = \frac{3}{4}\left(\frac{4}{3}x\right) \quad \text{Multiply by } \tfrac{3}{4}, \text{ the reciprocal of } \tfrac{4}{3}.$$

Be careful. This is *not* the answer.

$57 = x \leftarrow$ Number of orders for coffee

Step 5 **State the answer.** In this problem, *x does not represent the quantity that we must find.* The number of orders for tea was $\frac{1}{3}x$.

$$\frac{1}{3}(57) = 19 \leftarrow \text{Number of orders for tea}$$

Step 6 **Check.** The number of orders for tea, 19, is one-third the number of orders for coffee, 57, and

$$19 + 57 = 76, \quad \text{as required.}$$

This agrees with the information given in the problem. The answer checks.

NOW TRY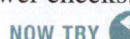

PROBLEM-SOLVING HINT   In **Example 4,** it was easier to let the variable represent the quantity that was *not* specified. This required extra work in Step 5 to find the number of orders for tea. In some cases, this approach is easier.

NOW TRY ANSWER
4. 210 bagel orders

**NOW TRY
EXERCISE 5**

At the Sherwood Estates pool party, each resident brought four guests. If a total of 175 people visited the pool that day, how many were residents and how many were guests?

Tank

| Oil x | Gasoline $16x$ | $= 68$ |

FIGURE 2

EXAMPLE 5 Analyzing a Gasoline-Oil Mixture

A lawn trimmer uses a mixture of gasoline and oil. The mixture contains 16 oz of gasoline for each 1 ounce of oil. If the tank holds 68 oz of the mixture, how many ounces of oil and how many ounces of gasoline does it require when it is full?

Step 1   **Read** the problem. We must find how many ounces of oil and gasoline are needed to fill the tank.

Step 2   **Assign a variable.**

Let x = the number of ounces of oil required.

Then $16x$ = the number of ounces of gasoline required.

A diagram like the one in **FIGURE 2** is helpful.

Step 3   **Write an equation.**

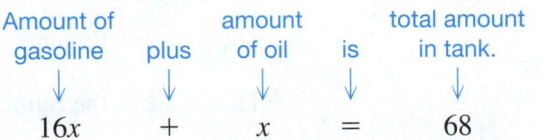

$$16x \quad + \quad x \quad = \quad 68$$

Step 4   **Solve.** $17x = 68$ Combine like terms.

$$\frac{17x}{17} = \frac{68}{17}$$ Divide by 17.

$$x = 4$$

Step 5   **State the answer.** When full, the lawn trimmer requires 4 oz of oil, and

$$16x = 16(4) \text{ oz}$$

$$= 64 \text{ oz of gasoline.}$$

Step 6   **Check.** Because $4 + 64 = 68$, and 64 is 16 times 4, the answer checks.

NOW TRY

PROBLEM-SOLVING HINT   Sometimes we must find three unknown quantities. ***When three unknowns are compared in pairs, let the variable represent the unknown found in both pairs.***

EXAMPLE 6 Dividing a Board into Pieces

A project calls for three pieces of wood. The longest piece must be twice the length of the middle-sized piece. The shortest piece must be 10 in. shorter than the middle-sized piece. If a board 70 in. long is to be used, how long must each piece be?

Step 1   **Read** the problem. There will be three answers.

Step 2   **Assign a variable.** The middle-sized piece appears in both pairs of comparisons, so let x represent the length, in inches, of the middle-sized piece.

Let x = the length of the middle-sized piece.

Then $2x$ = the length of the longest piece,

and $x - 10$ = the length of the shortest piece.

NOW TRY ANSWER
5. 35 residents; 140 guests

NOW TRY
EXERCISE 6

Over a 6-hr period, a basketball player spent twice as much time lifting weights as practicing free throws and 2 hr longer watching game films than practicing free throws. How many hours did he spend on each task?

A sketch is helpful. See **FIGURE 3**.

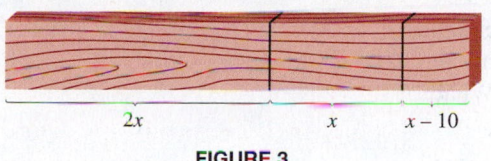

$2x$ x $x - 10$

FIGURE 3

Step 3 **Write an equation.**

Longest	plus	middle-sized	plus	shortest	is	total length.
↓	↓	↓	↓	↓	↓	↓
$2x$	$+$	x	$+$	$(x - 10)$	$=$	70

Step 4 **Solve.**

$$4x - 10 = 70 \qquad \text{Combine like terms.}$$
$$4x - 10 + 10 = 70 + 10 \qquad \text{Add 10.}$$
$$4x = 80 \qquad \text{Combine like terms.}$$
$$\frac{4x}{4} = \frac{80}{4} \qquad \text{Divide by 4.}$$
$$x = 20$$

Step 5 **State the answer.** The middle-sized piece is 20 in. long, the longest piece is $2(20) = 40$ in. long, and the shortest piece is $20 - 10 = 10$ in. long.

Step 6 **Check.** The lengths sum to 70 in. All problem conditions are satisfied.

NOW TRY

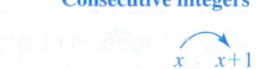

Consecutive integers

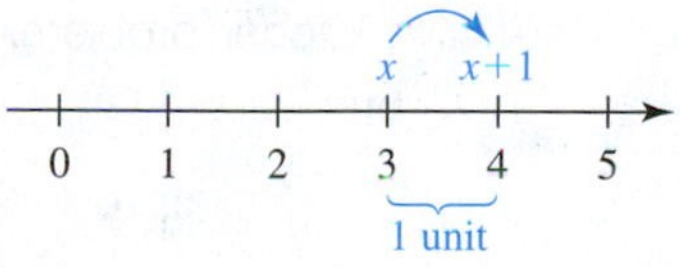

x $x+1$

0 1 2 3 4 5

1 unit

FIGURE 4

OBJECTIVE 4 Solve problems involving consecutive integers.

Two integers that differ by 1 are **consecutive integers.** For example, 3 and 4, 16 and 17, and -2 and -1 are pairs of consecutive integers. See **FIGURE 4**.

> *In general, if x represents an integer, then $x + 1$ represents the next greater consecutive integer.*

EXAMPLE 7 Finding Consecutive Integers

Two pages that face each other in this book have 269 as the sum of their page numbers. What are the page numbers?

Step 1 **Read** the problem. Because the two pages face each other, they must have page numbers that are consecutive integers.

x $x+1$

FIGURE 5

Step 2 **Assign a variable.**

Let $x =$ the lesser page number.

Then $x + 1 =$ the greater page number.

FIGURE 5 illustrates the situation.

NOW TRY ANSWER
6. practicing free throws: 1 hr;
lifting weights: 2 hr;
watching game films: 3 hr

Step 3 **Write an equation.** The sum of the page numbers is 269.

$$x + (x + 1) = 269$$

 **NOW TRY EXERCISE 7**

Two pages that face each other in a book have 593 as the sum of their page numbers. What are the page numbers?

Step 4 **Solve.**

$x + (x + 1) = 269$	Equation from Step 3
$2x + 1 = 269$	Combine like terms.
$2x + 1 - 1 = 269 - 1$	Subtract 1.
$2x = 268$	Combine like terms.
$\dfrac{2x}{2} = \dfrac{268}{2}$	Divide by 2.
$x = 134$	

Step 5 **State the answer.** The lesser page number is 134, and the greater page number is $134 + 1 = 135$. (Your book is opened to these two pages.)

Step 6 **Check.** The sum of 134 and 135 is 269. The answer checks. **NOW TRY**

Consecutive *even* integers, such as 2 and 4, and 8 and 10, differ by 2. Similarly, **consecutive *odd* integers**, such as 1 and 3, and 9 and 11, also differ by 2. See **FIGURE 6**.

Consecutive even integers

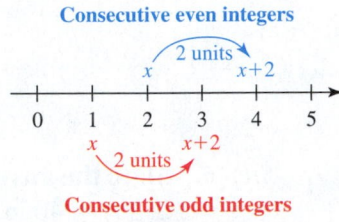

FIGURE 6

In general, if x represents an even (or odd) integer, then x + 2 represents the next greater consecutive even (or odd) integer, respectively.

In this book, we list consecutive integers in increasing order.

PROBLEM-SOLVING HINT　If $x =$ the lesser (least) integer in a consecutive integer problem, then the following apply.

- For two consecutive integers, use　　**x,　x + 1.**
- For two consecutive *even* integers, use　**x,　x + 2.**
- For two consecutive *odd* integers, use　**x,　x + 2.**

EXAMPLE 8　Finding Consecutive Odd Integers

If the lesser of two consecutive odd integers is doubled, the result is 7 more than the greater of the two integers. Find the two integers.

Step 1 **Read** the problem. We must find two consecutive odd integers.

Step 2 **Assign a variable.**

Let　　　$x =$ the lesser consecutive odd integer.

Then　$x + 2 =$ the greater consecutive odd integer.

Step 3 **Write an equation.**

If the lesser is doubled,	the result is	7	more than	the greater.
↓	↓	↓	↓	↓
$2x$	$=$	7	$+$	$(x + 2)$

**NOW TRY
EXERCISE 8**

Find two consecutive odd integers such that the sum of twice the lesser and three times the greater is 191.

Step 4 **Solve.**

$$2x = 9 + x \qquad \text{Combine like terms.}$$
$$2x - x = 9 + x - x \qquad \text{Subtract } x.$$
$$x = 9 \qquad \text{Combine like terms.}$$

Step 5 **State the answer.** The lesser integer is 9. The greater is $9 + 2 = 11$.

Step 6 **Check.** When 9 is doubled, we obtain 18, which is 7 more than the greater odd integer, 11. The answer checks.

NOW TRY

OBJECTIVE 5 Solve problems involving supplementary and complementary angles.

An angle can be measured by a unit called the **degree** (°), which is $\frac{1}{360}$ of a complete rotation. See **FIGURE 7**.

- Two angles whose sum is 90° are **complementary**, or *complements* of each other.
- An angle that measures 90° is a **right angle.**
- Two angles whose sum is 180° are **supplementary**, or *supplements* of each other.
- An angle that measures 180° is a **straight angle.**

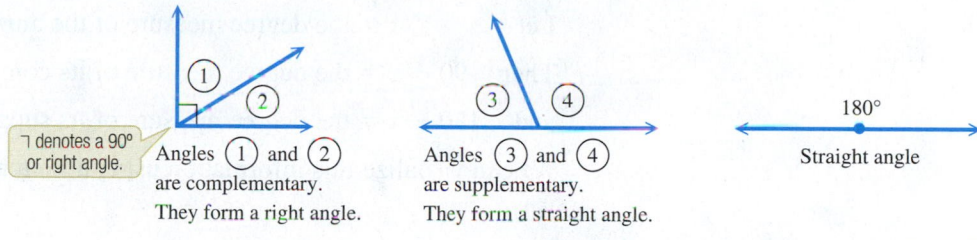

FIGURE 7

PROBLEM-SOLVING HINT Let *x* represent the degree measure of an angle.

90 − *x* represents the degree measure of its complement.

180 − *x* represents the degree measure of its supplement.

EXAMPLE 9 Finding the Measure of an Angle

Find the measure of an angle whose complement is five times its measure.

Step 1 **Read** the problem. We must find the measure of an angle, given information about the measure of its complement.

Step 2 **Assign a variable.**

Let $x =$ the degree measure of the angle.

Then $90 - x =$ the degree measure of its complement.

Step 3 **Write an equation.**

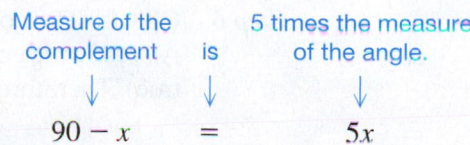

$$90 - x \qquad = \qquad 5x$$

**NOW TRY
EXERCISE 9**

Find the measure of an angle whose complement is twice its measure.

Step 4   **Solve.**

$$90 - x = 5x \qquad \text{Equation from Step 3}$$
$$90 - x + x = 5x + x \qquad \text{Add } x.$$
$$90 = 6x \qquad \text{Combine like terms.}$$
$$\frac{90}{6} = \frac{6x}{6} \qquad \text{Divide by 6.}$$
$$15 = x, \quad \text{or} \quad x = 15$$

Step 5   **State the answer.** The measure of the angle is 15°.

Step 6   **Check.** If the angle measures 15°, then $90° - 15° = 75°$ is the measure of its complement. 75° is equal to five times 15°, as required.   **NOW TRY**

**NOW TRY
EXERCISE 10**

Find the measure of an angle whose supplement is 46° less than three times its complement.

EXAMPLE 10   Finding the Measure of an Angle

Find the measure of an angle whose supplement is 10° more than twice its complement.

Step 1   **Read** the problem. We are to find the measure of an angle, given information about its complement and its supplement.

Step 2   **Assign a variable.**

Let x = the degree measure of the angle.

Then $90 - x$ = the degree measure of its complement,

and $180 - x$ = the degree measure of its supplement.

We can visualize this information using a sketch. See **FIGURE 8**.

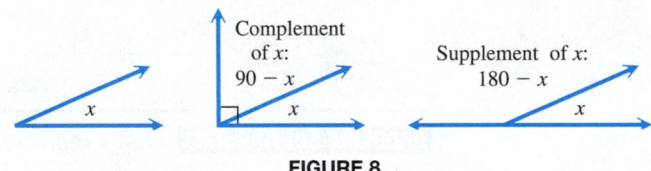

FIGURE 8

Step 3   **Write an equation.**

Supplement	is	10	more than	twice	its complement.
↓	↓	↓	↓	↓	↓
$180 - x$	=	10	+	$2 \cdot$	$(90 - x)$

> Be sure to use parentheses here.

Step 4   **Solve.**

$$180 - x = 10 + 180 - 2x \qquad \text{Distributive property}$$
$$180 - x = 190 - 2x \qquad \text{Combine like terms.}$$
$$180 - x + 2x = 190 - 2x + 2x \qquad \text{Add } 2x.$$
$$180 + x = 190 \qquad \text{Combine like terms.}$$
$$180 + x - 180 = 190 - 180 \qquad \text{Subtract 180.}$$
$$x = 10$$

Step 5   **State the answer.** The measure of the angle is 10°.

Step 6   **Check.** The complement of 10° is 80° and the supplement of 10° is 170°. Also, 170° is equal to 10° more than twice 80° (that is, $170 = 10 + 2(80)$ is true). Therefore, the answer checks.

NOW TRY ANSWERS
9. 30°
10. 22°

NOW TRY

2.4 Exercises

 FOR EXTRA HELP   MyMathLab®

Complete solution available in MyMathLab

Concept Check   *Which choice would **not** be a reasonable answer? Justify your response.*

1. A problem requires finding the number of cars on a dealer's lot.

   A. 0   **B.** 45   **C.** 1   **D.** $6\frac{1}{2}$

2. A problem requires finding the number of hours a light bulb is on during a day.

   A. 0   **B.** 4.5   **C.** 13   **D.** 25

3. A problem requires finding the distance traveled in miles.

   A. −10   **B.** 1.8   **C.** $10\frac{1}{2}$ **D.** 50

4. A problem requires finding the time in minutes.

   A. 0   **B.** 10.5   **C.** −5   **D.** 90

*Solve each problem. In each case, give an equation using x as the variable and give the answer. **See Examples 1 and 2.***

5. The sum of a number and 9 is −26. What is the number?

6. The difference of a number and 11 is −31. What is the number?

7. The product of 8, and a number increased by 6, is 104. What is the number?

8. The product of 5, and 3 more than twice a number, is 85. What is the number?

9. If 2 is added to five times a number, the result is equal to 5 more than four times the number. Find the number.

10. If four times a number is added to 8, the result is equal to three times the number, added to 5. Find the number.

11. If 2 is subtracted from a number and this difference is tripled, the result is 6 more than the number. Find the number.

12. If 3 is added to a number and this sum is doubled, the result is 2 more than the number. Find the number.

13. When 6 is added to $\frac{3}{4}$ of a number, the result is 4 less than the number. Find the number.

14. When $\frac{2}{3}$ of a number is added to 10, the result is 5 more than the number. Find the number.

15. The sum of three times a number and 7 more than the number is the same as the difference of −11 and twice the number. What is the number?

16. If 4 is added to twice a number and this sum is multiplied by 2, the result is the same as if the number is multiplied by 3 and 4 is added to the product. What is the number?

Complete the six suggested problem-solving steps to solve each problem.

17. The 150-member Iowa legislature includes 4 fewer Democrats than Republicans. (No other parties are represented.) How many Democrats and how many Republicans are there in the legislature? (*Source:* www.legis.iowa.gov) **(See Example 3.)**

   Step 1 **Read** the problem carefully.

   We must find the number of Democrats and the number of _____.

   Step 2 **Assign a variable.**

   Let x = the number of Republicans.

   Then _____ = the number of _____.

Step 3 **Write an equation.**

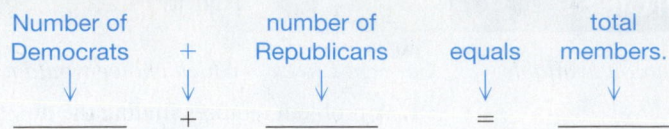

Complete Steps 4–6 to solve the problem.

18. The sum of two consecutive even integers is 254. Find the integers. (**See Example 8.**)

Step 1 **Read** the problem carefully.

We must find two consecutive _____ .

Step 2 **Assign a variable.**

Let $x =$ the lesser of the two _____ even integers.

Then _____ = the greater of the two consecutive even integers.

Step 3 **Write an equation.**

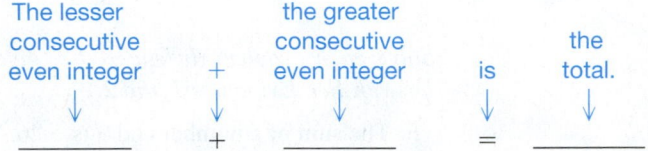

Complete Steps 4–6 to solve the problem.

Solve each problem. See Example 3.

19. New York and Ohio are among the states with the most remaining drive-in movie screens. New York has 2 more screens than Ohio, and there are 56 screens total in the two states. How many drive-in movie screens remain in each state? (*Source:* www.drive-ins.com)

20. Two of the most watched episodes in television were the final episodes of *M*A*S*H* and *Cheers*. The total number of viewers for these two episodes was about 92 million, with 8 million more people watching the *M*A*S*H* episode than the *Cheers* episode. How many people watched each episode? (*Source:* Nielsen Media Research.)

21. During the 113th session, the U.S. Senate had a total of 98 Democrats and Republicans. There were 6 more Democrats than Republicans. How many members of each party were there? (*Source:* www.thegreenpapers.com)

22. During the 113th session, the total number of Democrats and Republicans in the U.S. House of Representatives was 432. There were 32 more Republicans than Democrats. How many members of each party were there? (*Source:* www.thegreenpapers.com)

23. Madonna and Bruce Springsteen had the two top-grossing concert tours for 2012, together generating about $427 million in ticket sales. If Bruce Springsteen took in $29 million less than Madonna, how much did each tour generate? (*Source:* www.billboard.com)

24. The Toyota Camry and the Honda Accord were the top-selling passenger cars in the United States in 2012. Accord sales were 27 thousand less than Camry sales, and 691 thousand of the two cars were sold. How many of each car were sold? (*Source:* www.edmunds.com)

25. In the 2012–2013 NBA regular season, the Miami Heat won two more than four times as many games as they lost. The Heat played 82 games. How many wins and losses did the team have? (*Source:* www.NBA.com)

26. In the 2013 regular baseball season, the Cleveland Indians won 48 fewer than twice as many games as they lost. They played 162 regular-season games. How many wins and losses did the team have? (*Source:* www.MLB.com)

27. A one-cup serving of orange juice contains 3 mg less than four times the amount of vitamin C as a one-cup serving of pineapple juice. Servings of the two juices contain a total of 122 mg of vitamin C. How many milligrams of vitamin C are in a serving of each type of juice? (*Source:* U.S. Agriculture Department.)

28. A one-cup serving of pineapple juice has 9 more than three times as many calories as a one-cup serving of tomato juice. Servings of the two juices contain a total of 173 calories. How many calories are in a serving of each type of juice? (*Source:* U.S. Agriculture Department.)

Solve each problem. ***See Examples 4 and 5.***

▶ 29. In one day, a store sold $\frac{2}{3}$ as many DVDs as Blu-ray discs. The total number of DVDs and Blu-ray discs sold that day was 280. How many DVDs were sold?

30. A workout that combines weight training and aerobics burns a total of 371 calories. If weight training burns $\frac{2}{5}$ as many calories as aerobics, how many calories does weight training burn?

31. The world's largest taco contained approximately 1 kg of onion for every 6.6 kg of grilled steak. The total weight of these two ingredients was 617.6 kg. To the nearest tenth of a kilogram, how many kilograms of each ingredient were used to make the taco? (*Source: Guinness World Records.*)

32. As of 2013, the combined population of China and India was estimated at 2.5 billion. If there were about 0.9 as many people living in India as China, what was the population of each country, to the nearest tenth of a billion? (*Source:* U.S. Census Bureau.)

33. The value of a "Mint State-63" (uncirculated) 1950 Jefferson nickel minted at Denver is twice the value of a 1945 nickel in similar condition minted at Philadelphia. Together, the total value of the two coins is $24.00. What is the value of each coin? (*Source:* Yeoman, R., *A Guide Book of United States Coins.*)

34. U.S. five-cent coins are made from a combination of nickel and copper. For every 1 lb of nickel, 3 lb of copper are used. How many pounds of copper would be needed to make 560 lb of five-cent coins? (*Source:* The United States Mint.)

▶ 35. A recipe for whole-grain bread calls for 1 oz of rye flour for every 4 oz of whole-wheat flour. How many ounces of each kind of flour should be used to make a loaf of bread weighing 32 oz?

36. A medication contains 9 mg of active ingredients for every 1 mg of inert ingredients. How much of each kind of ingredient would be contained in a single 250-mg caplet?

Solve each problem. ***See Example 6.***

37. An office manager booked 55 airline tickets. He booked 7 more tickets on American Airlines than United Airlines. On Southwest Airlines, he booked 4 more than twice as many tickets as on United. How many tickets did he book on each airline?

38. A mathematics textbook editor spent 7.5 hr making telephone calls, writing e-mails, and attending meetings. She spent twice as much time attending meetings as making telephone calls and 0.5 hr longer writing e-mails than making telephone calls. How many hours did she spend on each task?

▶ **39.** A party-length submarine sandwich that is 59 in. long is cut into three pieces. The middle piece is 5 in. longer than the shortest piece, and the shortest piece is 9 in. shorter than the longest piece. How long is each piece?

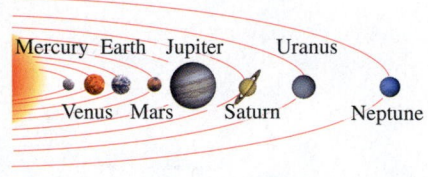

40. A three-foot-long deli sandwich must be split into three pieces so that the middle piece is twice as long as the shortest piece and the shortest piece is 8 in. shorter than the longest piece. How long should the three pieces be?

41. The United States earned 104 medals at the 2012 Summer Olympics. The number of silver medals earned was the same as the number of bronze medals. The number of gold medals was 17 more than the number of silver medals. How many of each kind of medal did the United States earn? (*Source: World Almanac and Book of Facts.*)

42. China earned a total of 88 medals at the 2012 Summer Olympics. The number of gold medals earned was 15 more than the number of bronze medals. The number of bronze medals earned was 4 fewer than the number of silver medals. How many of each kind of medal did China earn? (*Source: World Almanac and Book of Facts.*)

43. Venus is 31.2 million mi farther from the sun than Mercury, while Earth is 57 million mi farther from the sun than Mercury. If the total of the distances from these three planets to the sun is 196.2 million mi, how far away from the sun is Mercury? (All distances given here are *mean* (*average*) distances.) (*Source: The New York Times Almanac.*)

44. Saturn, Jupiter, and Uranus have a total of 156 known satellites (moons). Jupiter has 5 more satellites than Saturn, and Uranus has 35 fewer satellites than Saturn. How many known satellites does Uranus have? (*Source:* http://solarsystem.nasa.gov)

45. The sum of the measures of the angles of any triangle is 180°. In triangle *ABC*, angles *A* and *B* have the same measure, while the measure of angle *C* is 60° greater than each of angles *A* and *B*. What are the measures of the three angles?

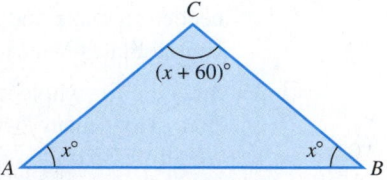

46. In triangle *ABC*, the measure of angle *A* is 141° more than the measure of angle *B*. The measure of angle *B* is the same as the measure of angle *C*. Find the measure of each angle. (*Hint:* **See Exercise 45.**)

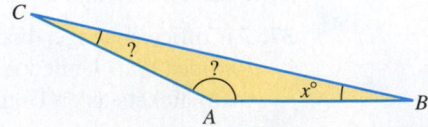

Solve each problem. ***See Examples 7 and 8.***

47. The numbers on two consecutively numbered gym lockers have a sum of 137. What are the locker numbers?

48. The numbers on two consecutive checkbook checks have a sum of 357. What are the numbers?

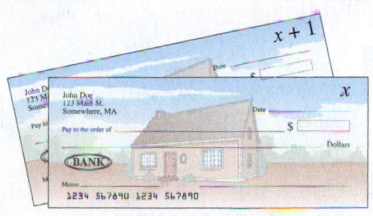

▶ **49.** Two pages that are back-to-back in this book have 203 as the sum of their page numbers. What are the page numbers?

50. Two apartments have numbers that are consecutive integers. The sum of the numbers is 59. What are the two apartment numbers?

▶ **51.** Find two consecutive even integers such that the lesser added to three times the greater gives a sum of 46.

52. Find two consecutive even integers such that six times the lesser added to the greater gives a sum of 86.

53. Find two consecutive odd integers such that 59 more than the lesser is four times the greater.

54. Find two consecutive odd integers such that twice the greater is 17 more than the lesser.

55. When the lesser of two consecutive integers is added to three times the greater, the result is 43. Find the integers.

56. If five times the lesser of two consecutive integers is added to three times the greater, the result is 59. Find the integers.

Extending Skills   *Solve each problem.*

57. If the sum of three consecutive even integers is 60, what is the first of the three even integers? (*Hint:* If x and $x + 2$ represent the first two consecutive even integers, how would we represent the third consecutive even integer?)

58. If the sum of three consecutive odd integers is 69, what is the third of the three odd integers?

59. If 6 is subtracted from the third of three consecutive odd integers and the result is multiplied by 2, the answer is 23 less than the sum of the first and twice the second of the integers. Find the integers.

60. If the first and third of three consecutive even integers are added, the result is 22 less than three times the second integer. Find the integers.

Solve each problem. ***See Examples 9 and 10.***

61. Find the measure of an angle whose complement is four times its measure.

62. Find the measure of an angle whose complement is five times its measure.

63. Find the measure of an angle whose supplement is eight times its measure.

64. Find the measure of an angle whose supplement is three times its measure.

▶ **65.** Find the measure of an angle whose supplement measures 39° more than twice its complement.

66. Find the measure of an angle whose supplement measures 38° less than three times its complement.

67. Find the measure of an angle such that the difference between the measures of its supplement and three times its complement is 10°.

68. Find the measure of an angle such that the sum of the measures of its complement and its supplement is 160°.

2.5   Formulas and Additional Applications from Geometry

A **formula** is an equation in which variables are used to describe a relationship. For example, formulas exist for finding perimeters and areas of geometric figures, calculating money earned on bank savings, and converting among measurements.

$$P = 4s, \quad \mathcal{A}^* = \pi r^2, \quad I = prt, \quad F = \frac{9}{5}C + 32 \qquad \text{Formulas}$$

Many of the formulas used in this book are given inside the back cover.

OBJECTIVE 1   Solve a formula for one variable, given values of the other variables.

The **area** of a plane (two-dimensional) geometric figure is a measure of the surface covered by the figure. Area is measured in square units.

EXAMPLE 1   Using Formulas to Evaluate Variables

Find the value of the remaining variable in each formula.

(a) $\mathcal{A} = LW$; $\mathcal{A} = 64, L = 10$

This formula gives the area of a rectangle. See **FIGURE 9**.

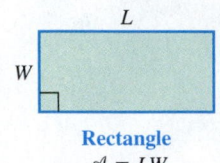

Rectangle
$\mathcal{A} = LW$
FIGURE 9

In this book, $\mathcal{A}$ denotes area.

$\mathcal{A} = LW$ Solve for W.

$64 = 10W$ Let $\mathcal{A} = 64$ and $L = 10$.

$\dfrac{64}{10} = \dfrac{10W}{10}$ Divide by 10.

$6.4 = W$

The width is 6.4. Because $10(6.4) = 64$, the given area, the answer checks.

(b) $\mathcal{A} = \frac{1}{2}h(b + B)$; $\mathcal{A} = 210, B = 27, h = 10$

This formula gives the area of a trapezoid. See **FIGURE 10**.

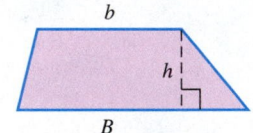

Trapezoid
$\mathcal{A} = \frac{1}{2}h(b + B)$
FIGURE 10

$$\mathcal{A} = \frac{1}{2}h(b + B)$$

Solve for b.

$210 = \dfrac{1}{2}(10)(b + 27)$ Let $\mathcal{A} = 210, h = 10, B = 27$.

$210 = 5(b + 27)$ Multiply $\frac{1}{2}(10)$.

$210 = 5b + 135$ Distributive property

$210 - 135 = 5b + 135 - 135$ Subtract 135.

$75 = 5b$ Combine like terms.

$\dfrac{75}{5} = \dfrac{5b}{5}$ Divide by 5.

$15 = b$

The length of the shorter parallel side, b, is 15. This answer checks because

$$\frac{1}{2}(10)(15 + 27) = 210, \quad \text{as required.}$$

**NOW TRY
EXERCISE 1**

Find the value of the remaining variable.

$P = 2a + 2b;$

$P = 78, a = 12$

NOW TRY ANSWER
1. $b = 27$

NOW TRY

---

* In this book, we use $\mathcal{A}$ to denote area.

OBJECTIVE 2 Use a formula to solve an applied problem.

The **perimeter** of a plane (two-dimensional) geometric figure is the measure of the outer boundary of the figure. For a polygon (e.g., a rectangle, square, or triangle), it is the sum of the lengths of the sides.

NOW TRY EXERCISE 2

Kurt's garden is in the shape of a rectangle. The length is 10 ft less than twice the width, and the perimeter is 160 ft. Find the dimensions of the garden.

EXAMPLE 2 Finding the Dimensions of a Rectangular Yard

Cathleen's backyard is in the shape of a rectangle. The length is 5 m less than twice the width, and the perimeter is 80 m. Find the dimensions of the yard.

Step 1 **Read** the problem. We must find the dimensions of the yard.

Step 2 **Assign a variable.** Let W = the width of the lot, in meters. The length is 5 meters less than twice the width, so the length is $L = 2W - 5$. See **FIGURE 11**.

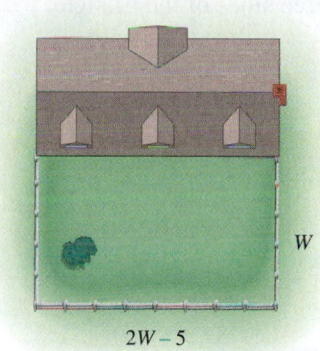

FIGURE 11

Step 3 **Write an equation.** Use the formula for the perimeter of a rectangle.

$$P = 2L + 2W \qquad \text{Perimeter of a rectangle}$$

Perimeter   = 2 · Length   + 2 · Width

$$80 \quad = 2(2W - 5) \quad + \quad 2W \qquad \begin{array}{l}\text{Substitute 80 for perimeter } P \\ \text{and } 2W - 5 \text{ for length } L.\end{array}$$

Step 4 **Solve.**

$$80 = 4W - 10 + 2W \qquad \text{Distributive property}$$
$$80 = 6W - 10 \qquad \text{Combine like terms.}$$
$$80 + 10 = 6W - 10 + 10 \qquad \text{Add 10.}$$
$$90 = 6W \qquad \text{Combine like terms.}$$
$$\frac{90}{6} = \frac{6W}{6} \qquad \text{Divide by 6.}$$

We must also find the length.

$$15 = W$$

Step 5 **State the answer.** The width is 15 m and the length is $2(15) - 5 = 25$ m.

Step 6 **Check.** If the width is 15 m and the length is 25 m, the perimeter is

$$2(25) + 2(15) = 80 \text{ m}, \quad \text{as required.} \qquad \text{NOW TRY}$$

EXAMPLE 3 Finding the Dimensions of a Triangle

The longest side of a triangle is 3 ft longer than the shortest side. The medium side is 1 ft longer than the shortest side. If the perimeter of the triangle is 16 ft, what are the lengths of the three sides?

Step 1 **Read** the problem. We must find the lengths of the sides of a triangle.

Step 2 **Assign a variable.**

Let s = the length of the shortest side, in feet,

$s + 1$ = the length of the medium side, in feet, and,

$s + 3$ = the length of the longest side in feet.

It is a good idea to draw a sketch. See **FIGURE 12**.

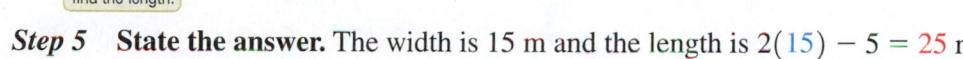

FIGURE 12

NOW TRY ANSWER
2. width: 30 ft; length: 50 ft

NOW TRY
EXERCISE 3

The perimeter of a triangle is 30 ft. The longest side is 1 ft longer than the medium side, and the shortest side is 7 ft shorter than the medium side. What are the lengths of the three sides of the triangle?

Step 3   **Write an equation.** Use the formula for the perimeter of a triangle.

$$P = a + b + c \qquad \text{Perimeter of a triangle}$$

$$16 = s + (s + 1) + (s + 3) \qquad \text{Substitute.}$$

Step 4   **Solve.** $16 = 3s + 4$ Combine like terms.

$$12 = 3s \qquad \text{Subtract 4.}$$

$$4 = s \qquad \text{Divide by 3.}$$

Step 5   **State the answer.** The shortest side, s, has length 4 ft.

$$s + 1 = 4 + 1 = 5 \text{ ft} \qquad \text{Length of medium side}$$

$$s + 3 = 4 + 3 = 7 \text{ ft} \qquad \text{Length of longest side}$$

Step 6   **Check.** The medium side, 5 ft, is 1 ft longer than the shortest side, and the longest side, 7 ft, is 3 ft longer than the shortest side. The perimeter is

$$4 + 5 + 7 = 16 \text{ ft}, \quad \text{as required.} \qquad \text{NOW TRY} \;\; \text{⟳}$$

NOW TRY
EXERCISE 4

The area of a triangle is 77 cm². The base is 14 cm. Find the height of the triangle.

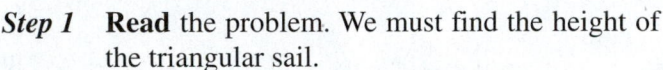

EXAMPLE 4   Finding the Height of a Triangular Sail

The area of a triangular sail of a sailboat is 126 ft^2. (Recall that " ft^2" means "square feet.") The base of the sail is 12 ft. Find the height of the sail.

Step 1   **Read** the problem. We must find the height of the triangular sail.

Step 2   **Assign a variable.** Let $h =$ the height of the sail, in feet. See **FIGURE 13**.

Step 3   **Write an equation.** Use the formula for the area of a triangle.

$$\mathcal{A} = \frac{1}{2}bh \qquad \begin{array}{l}\mathcal{A} \text{ is area, } b \text{ is base,} \\ \text{and } h \text{ is height.}\end{array}$$

FIGURE 13

$$126 = \frac{1}{2}(12)h \qquad \text{Let } \mathcal{A} = 126, \, b = 12.$$

Step 4   **Solve.** $126 = 6h$ Multiply.

$$21 = h \qquad \text{Divide by 6.}$$

Step 5   **State the answer.** The height of the sail is 21 ft.

Step 6   **Check** to see that the values $\mathcal{A} = 126$, $b = 12$, and $h = 21$ satisfy the formula for the area of a triangle.

$$126 = \frac{1}{2}(12)(21) \text{ is true.} \qquad \text{NOW TRY} \;\; \text{⟳}$$

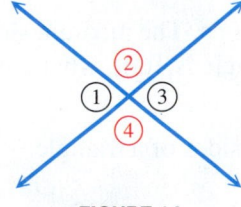

FIGURE 14

NOW TRY ANSWERS
3. 5 ft, 12 ft, 13 ft
4. 11 cm

OBJECTIVE 3   Solve problems involving vertical angles and straight angles.

FIGURE 14 shows two intersecting lines forming angles that are numbered ①, ②, ③, and ④. Angles ① and ③ lie "opposite" each other. They are **vertical angles.** Another pair of vertical angles is ② and ④. *Vertical angles have equal measures.*

Consider angles ① and ②. When their measures are added, we obtain 180°, the measure of a straight angle. There are three other angle pairs that form straight angles: ② and ③, ③ and ④, and ① and ④.

**NOW TRY
EXERCISE 5**

Find the measure of each marked angle in the figure.

EXAMPLE 5 Finding Angle Measures

Refer to the appropriate figure in each part.

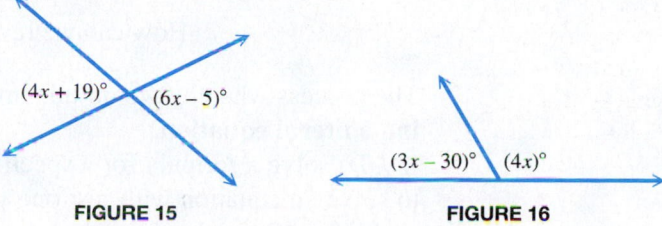

FIGURE 15   **FIGURE 16**

(a) Find the measure of each marked angle in **FIGURE 15**.

The marked angles are vertical angles, so they have equal measures.

$$4x + 19 = 6x - 5 \qquad \text{Set } 4x + 19 \text{ equal to } 6x - 5.$$
$$19 = 2x - 5 \qquad \text{Subtract } 4x.$$
$$24 = 2x \qquad \text{Add 5.}$$

This is *not* the answer. $\longrightarrow \quad 12 = x \qquad \text{Divide by 2.}$

Replace x with 12 in the expression for the measure of each angle.

$4x + 19$		$6x - 5$	
$= 4(12) + 19$	Let $x = 12$.	$= 6(12) - 5$	Let $x = 12$.
$= 48 + 19$	Multiply.	$= 72 - 5$	Multiply.
$= 67$	Add.	$= 67$	Subtract.

The angles have equal measures, as required. Each measures 67°.

(b) Find the measure of each marked angle in **FIGURE 16**.

The measures of the marked angles must add to 180° because together they form a straight angle. (They are also *supplements* of each other.)

$$(3x - 30) + 4x = 180 \qquad \text{Supplementary angles sum to } 180°.$$
$$7x - 30 = 180 \qquad \text{Combine like terms.}$$
$$7x = 210 \qquad \text{Add 30.}$$

Don't stop here. $\longrightarrow \quad x = 30 \qquad \text{Divide by 7.}$

Replace x with 30 in the expression for the measure of each angle.

$3x - 30$		$4x$	
$= 3(30) - 30$	Let $x = 30$.	$= 4(30)$	Let $x = 30$.
$= 90 - 30$	Multiply.	$= 120$	Multiply.
$= 60$	Subtract.		

The two angles measure 60° and 120°, which add to 180°, as required.   **NOW TRY**

⚠ **CAUTION** In **Example 5**, the answer is *not* the value of x. **Remember to substitute the value of the variable into the expression given for each angle.**

NOW TRY ANSWER

5. 32°, 32°

OBJECTIVE 4 Solve a formula for a specified variable.

Sometimes we want to rewrite a formula in terms of a *different* variable in the formula. For example, consider $\mathcal{A} = LW$, the formula for the area of a rectangle.

$$\text{How can we rewrite } \mathcal{A} = LW \text{ in terms of } W?$$

The process whereby we do this involves **solving for a specified variable,** or **solving a literal equation.**

To solve a formula for a specified variable, we use the *same* steps that we used to solve an equation with just one variable. Consider the parallel reasoning to solve each of the following for x.

$3x + 4 = 13$		$ax + b = c$	
$3x + 4 - 4 = 13 - 4$	Subtract 4.	$ax + b - b = c - b$	Subtract b.
$3x = 9$		$ax = c - b$	
$\dfrac{3x}{3} = \dfrac{9}{3}$	Divide by 3.	$\dfrac{ax}{a} = \dfrac{c - b}{a}$	Divide by a.
$x = 3$	Equation solved for x	$x = \dfrac{c - b}{a}$	Formula solved for x

When we solve a formula for a specified variable, we treat the specified variable as if it were the ONLY variable in the equation, and we treat the other variables as if they were numbers.

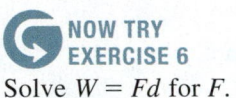
NOW TRY EXERCISE 6
Solve $W = Fd$ for F.

EXAMPLE 6 Solving for a Specified Variable

Solve $\mathcal{A} = LW$ for W.

W is multiplied by L, so we undo the multiplication by dividing each side by L.

$$\mathcal{A} = LW \quad \text{Our goal is to isolate } W.$$

$$\frac{\mathcal{A}}{L} = \frac{LW}{L} \qquad \qquad \text{Divide by } L.$$

$$\frac{\mathcal{A}}{L} = W, \quad \text{or} \quad W = \frac{\mathcal{A}}{L} \qquad \frac{LW}{L} = \frac{L}{L} \cdot W = 1 \cdot W = W \qquad \text{NOW TRY} \circlearrowleft$$

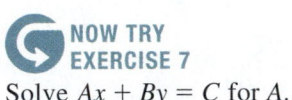
NOW TRY EXERCISE 7
Solve $Ax + By = C$ for A.

EXAMPLE 7 Solving for a Specified Variable

Solve $P = 2L + 2W$ for L.

$$P = 2L + 2W \quad \text{Our goal is to isolate } L.$$

$$P - 2W = 2L + 2W - 2W \qquad \text{Subtract } 2W.$$

$$P - 2W = 2L \qquad \text{Combine like terms.}$$

$$\frac{P - 2W}{2} = \frac{2L}{2} \qquad \text{Divide by 2.}$$

$$\frac{P - 2W}{2} = L, \quad \text{or} \quad L = \frac{P - 2W}{2} \qquad \frac{2L}{2} = \frac{2}{2} \cdot L = 1 \cdot L = L$$

$$\frac{P - 2W}{2} \neq P - W$$

NOW TRY $\circlearrowleft$

NOW TRY ANSWERS

6. $F = \dfrac{W}{d}$

7. $A = \dfrac{C - By}{x}$

**NOW TRY
EXERCISE 8**

Solve $S = \frac{1}{2}(a + b + c)$
for a.

EXAMPLE 8 Solving for a Specified Variable

Solve $A = \frac{1}{2}h(b + B)$ for B.

$$A = \frac{1}{2}h(b + B)$$

> Our goal is to isolate B.

$$A = \frac{1}{2}hb + \frac{1}{2}hB \qquad \text{Clear the parentheses using the distributive property.}$$

$$2 \cdot A = 2\left(\frac{1}{2}hb + \frac{1}{2}hB\right) \qquad \text{Multiply each side by 2 to clear the fractions.}$$

$$2 \cdot A = 2 \cdot \frac{1}{2}hb + 2 \cdot \frac{1}{2}hB \qquad \text{Distributive property}$$

$$2A = hb + hB \qquad \text{Multiply; } 2 \cdot \frac{1}{2} = \frac{2}{2} = 1$$

$$2A - hb = hb + hB - hb \qquad \text{Subtract } hb.$$

$$2A - hb = hB \qquad \text{Combine like terms.}$$

$$\frac{2A - hb}{h} = \frac{hB}{h} \qquad \text{Divide by } h.$$

$$\frac{2A - hb}{h} = B, \quad \text{or} \quad B = \frac{2A - hb}{h} \qquad \text{NOW TRY}$$

**NOW TRY
EXERCISE 9**

Solve each equation for y.

(a) $5x + y = 3$

(b) $x - 2y = 8$

EXAMPLE 9 Solving for a Specified Variable

Solve each equation for y.

(a)
$$2x - y = 7$$

> Our goal is to isolate y.

$$2x - y - 2x = 7 - 2x \qquad \text{Subtract } 2x.$$

$$-y = 7 - 2x \qquad \text{Combine like terms.}$$

$$-1(-y) = -1(7 - 2x) \qquad \text{Multiply by } -1.$$

$$y = -7 + 2x \qquad \text{Multiply; distributive property}$$

$$y = 2x - 7 \qquad -a + b = b - a$$

We could have added y and subtracted 7 from each side of the equation to isolate y on the right, giving $2x - 7 = y$, a different form of the same result. There is often more than one way to solve for a specified variable.

(b)
$$-3x + 2y = 6$$

$$-3x + 2y + 3x = 6 + 3x \qquad \text{Add } 3x.$$

$$2y = 3x + 6 \qquad \text{Combine like terms; commutative property}$$

$$\frac{2y}{2} = \frac{3x + 6}{2} \qquad \text{Divide by 2.}$$

> Be careful here.

$$y = \frac{3x}{2} + \frac{6}{2} \qquad \frac{a + b}{c} = \frac{a}{c} + \frac{b}{c}$$

> $\frac{3x}{2} = \frac{3}{2} \cdot \frac{x}{1} = \frac{3}{2}x$

$$y = \frac{3}{2}x + 3 \qquad \text{Simplify.}$$

NOW TRY ANSWERS

8. $a = 2S - b - c$

9. (a) $y = -5x + 3$

  (b) $y = \frac{1}{2}x - 4$

Although we could have given our answer as $y = \frac{3x + 6}{2}$, we simplified further in preparation for our work in **Chapter 3**.

NOW TRY

2.5 Exercises

  FOR EXTRA HELP   ▶   **MyMathLab®**

▶ *Complete solution available in MyMathLab*

Concept Check *Decide whether perimeter or area would be used to solve a problem concerning the measure of the quantity.*

1. Carpeting for a bedroom

2. Sod for a lawn

3. Fencing for a yard

4. Baseboards for a living room

5. Tile for a bathroom

6. Fertilizer for a garden

7. Determining the cost of replacing a linoleum floor with a wood floor

8. Determining the cost of planting rye grass in a lawn for the winter

A formula is given along with the values of all but one of the variables. Find the value of the variable that is not given. Use 3.14 as an approximation for π (pi). ***See Example 1.***

9. $P = 2L + 2W$ (perimeter of a rectangle); $L = 8, W = 5$

10. $P = 2L + 2W$; $L = 6, W = 4$

11. $A = \dfrac{1}{2}bh$ (area of a triangle); $b = 8, h = 16$

12. $A = \dfrac{1}{2}bh$; $b = 10, h = 14$

13. $P = a + b + c$ (perimeter of a triangle); $P = 12, a = 3, c = 5$

14. $P = a + b + c$; $P = 15, a = 3, b = 7$

▶ **15.** $d = rt$ (distance formula); $d = 252, r = 45$

16. $d = rt$; $d = 100, t = 2.5$

17. $A = \dfrac{1}{2}h(b + B)$ (area of a trapezoid);

$A = 91, h = 7, b = 12$

18. $A = \dfrac{1}{2}h(b + B)$; $A = 75, b = 19, B = 31$

19. $C = 2\pi r$ (circumference of a circle); $C = 16.328$

20. $C = 2\pi r$; $C = 8.164$

21. $C = 2\pi r$; $C = 20\pi$

22. $C = 2\pi r$; $C = 100\pi$

23. $A = \pi r^2$ (area of a circle); $r = 4$

24. $A = \pi r^2$; $r = 12$

25. $S = 2\pi rh$; $S = 120\pi, h = 10$

26. $S = 2\pi rh$; $S = 720\pi, h = 30$

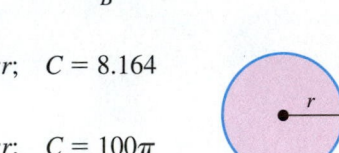

The **volume** *of a three-dimensional geometric figure is a measure of the space occupied by the figure. For example, the volume of a gasoline tank determines how many gallons of gasoline it would take to completely fill the tank. Volume is measured in cubic units.*

In Exercises 27–32, a formula for the volume (V) of a three-dimensional object is given, along with values for the other variables. Evaluate V. (Use 3.14 as an approximation for π.) ***See Example 1.***

27. $V = LWH$ (volume of a rectangular box); $L = 10, W = 5, H = 3$

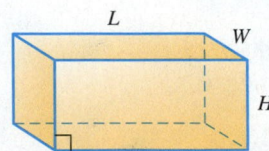

28. $V = LWH$; $L = 12, W = 8, H = 4$

29. $V = \dfrac{1}{3} Bh$ (volume of a pyramid); $\quad B = 12, h = 13$

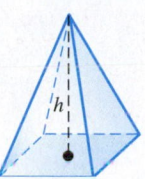

30. $V = \dfrac{1}{3} Bh$; $\quad B = 36, h = 4$

31. $V = \dfrac{4}{3} \pi r^3$ (volume of a sphere); $\quad r = 12$

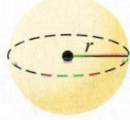

32. $V = \dfrac{4}{3} \pi r^3$; $\quad r = 6$

Simple interest I in dollars is calculated using the formula

$$I = prt, \qquad \text{Simple interest formula}$$

where p represents principal, or amount, in dollars that is invested or borrowed, r represents annual interest rate, expressed as a decimal, and t represents time, in years.
    In Exercises 33–38, find the value of the remaining variable in the simple interest formula. **See Example 1.** *(Hint: Write percents as decimals.* **See Section R.2.)**

33. $p = \$7500, r = 4\%, t = 2$ yr

34. $p = \$3600, r = 3\%, t = 4$ yr

35. $I = \$33, r = 2\%, t = 3$ yr

36. $I = \$270, r = 5\%, t = 6$ yr

37. $I = \$180, p = \$4800, r = 2.5\%$

38. $I = \$162, p = \$2400, r = 1.5\%$

Solve each problem. **See Examples 2 and 3.**

39. The length of a rectangle is 9 in. more than the width. The perimeter is 54 in. Find the length and the width of the rectangle.

40. The width of a rectangle is 3 ft less than the length. The perimeter is 62 ft. Find the length and the width of the rectangle.

▶ **41.** The perimeter of a rectangle is 36 m. The length is 2 m more than three times the width. Find the length and the width of the rectangle.

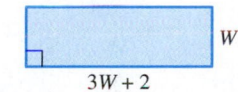

42. The perimeter of a rectangle is 36 yd. The width is 18 yd less than twice the length. Find the length and the width of the rectangle.

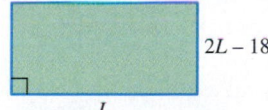

▶ **43.** The longest side of a triangle is 3 in. longer than the shortest side. The medium side is 2 in. longer than the shortest side. If the perimeter of the triangle is 20 in., what are the lengths of the three sides?

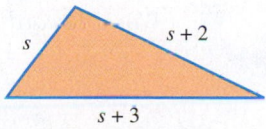

44. The perimeter of a triangle is 28 ft. The medium side is 4 ft longer than the shortest side, while the longest side is twice as long as the shortest side. What are the lengths of the three sides?

45. Two sides of a triangle have the same length. The third side measures 4 m less than twice that length. The perimeter of the triangle is 24 m. Find the lengths of the three sides.

46. A triangle is such that its medium side is twice as long as its shortest side and its longest side is 7 yd less than three times its shortest side. The perimeter of the triangle is 47 yd. What are the lengths of the three sides?

Use a formula to write an equation, and then solve each problem. (Use 3.14 as an approximation for π.) **Formulas are found inside the back cover of this book. See Examples 2–4.**

47. One of the largest fashion catalogues in the world was published in Hamburg, Germany. Each of the 212 pages in the catalogue measured 1.2 m by 1.5 m. What was the perimeter of a page? What was the area? (*Source: Guinness World Records.*)

48. One of the world's largest mandalas (sand paintings) measures 12.24 m by 12.24 m. What is the perimeter of the sand painting? To the nearest hundredth of a square meter, what is the area? (*Source: Guinness World Records.*)

▶ **49.** The area of a triangular road sign is 70 ft². If the base of the sign measures 14 ft, what is the height of the sign?

50. The area of a triangular advertising banner is 96 ft². If the height of the banner measures 12 ft, what is the measure of the base?

51. A prehistoric ceremonial site dating to about 3000 B.C. was discovered in southwestern England. The site is a nearly perfect circle, consisting of nine concentric rings that probably held upright wooden posts. Around this timber temple is a wide, encircling ditch enclosing an area with a diameter of 443 ft. Find this enclosed area to the nearest thousand square feet. (*Hint:* Find the radius. Then use $\mathcal{A} = \pi r^2$.) (*Source: Archaeology.*)

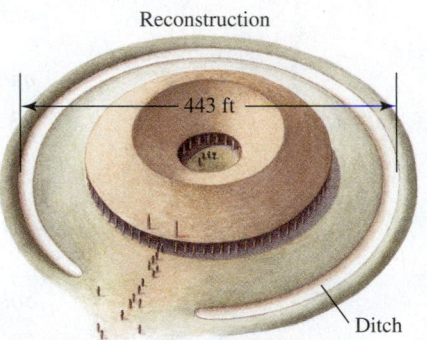

Reconstruction

443 ft

Ditch

52. The Rogers Centre in Toronto, Canada, is the first stadium with a hard-shell, retractable roof. The steel dome is 630 ft in diameter. To the nearest foot, what is the circumference of this dome? (*Source:* www.ballparks.com)

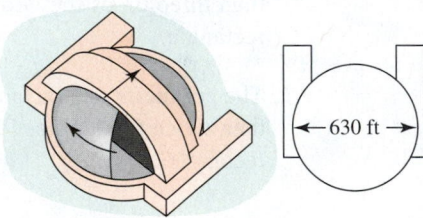

630 ft

53. One of the largest drums ever constructed was made from Japanese cedar and cowhide, with radius 7.87 ft. What was the area of the circular face of the drum? What was the circumference of the drum? Round answers to the nearest hundredth. (*Source: Guinness World Records.*)

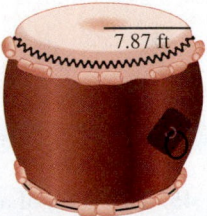

7.87 ft

54. A drum played at the Royal Festival Hall in London had radius 6.5 ft. What was the area of the circular face of the drum? What was the circumference of the drum? (*Source: Guinness World Records.*)

55. The survey plat depicted here shows two lots that form a trapezoid. The measures of the parallel sides are 115.80 ft and 171.00 ft. The height of the trapezoid is 165.97 ft. Find the combined area of the two lots. Round the answer to the nearest hundredth of a square foot.

56. Lot A in the survey plat is in the shape of a trapezoid. The parallel sides measure 26.84 ft and 82.05 ft. The height of the trapezoid is 165.97 ft. Find the area of Lot A. Round the answer to the nearest hundredth of a square foot.

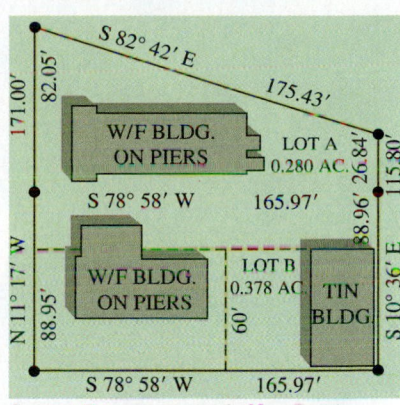

Source: Property survey in New Roads, Louisiana.

57. The U.S. Postal Service requires that any box sent by Priority Mail® have length plus girth (distance around) totaling no more than 108 in. The maximum volume that meets this condition is contained by a box with a square end 18 in. on each side. What is the length of the box? What is the maximum volume? (*Source:* United States Postal Service.)

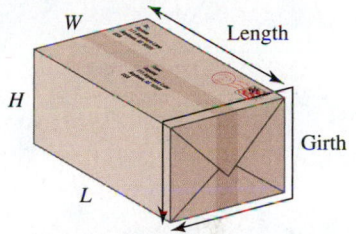

58. One of the world's largest sandwiches, made by Wild Woody's Chill and Grill in Roseville, Michigan, was 12 ft long, 12 ft wide, and $17\frac{1}{2}$ in. $\left(1\frac{11}{24}\text{ ft}\right)$ thick. What was the volume of the sandwich? (*Source: Guinness World Records.*)

Not to scale

Find the measure of each marked angle. See Example 5.

59.

$(x + 1)°$ $(4x - 56)°$

60.
$(10x + 7)°$ $(7x + 3)°$

61.
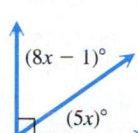
$(8x - 1)°$
$(5x)°$

62.
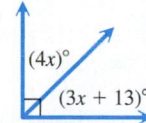
$(4x)°$
$(3x + 13)°$

63.
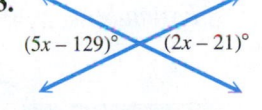
$(5x - 129)°$ $(2x - 21)°$

64.
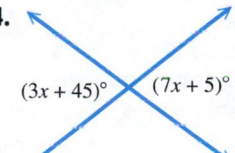
$(3x + 45)°$ $(7x + 5)°$

65.
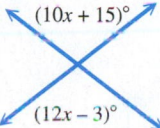
$(10x + 15)°$
$(12x - 3)°$

66.
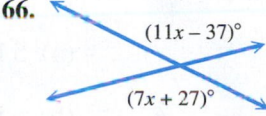
$(11x - 37)°$
$(7x + 27)°$

Solve each formula for the specified variable. See Examples 6–8.

67. $d = rt$ for t

68. $d = rt$ for r

69. $\mathcal{A} = bh$ for b

70. $\mathcal{A} = LW$ for L

71. $C = \pi d$ for d

72. $P = 4s$ for s

73. $V = LWH$ for H

74. $V = LWH$ for W

75. $I = prt$ for r

76. $I = prt$ for p

77. $A = \frac{1}{2}bh$ for h

78. $A = \frac{1}{2}bh$ for b

79. $V = \frac{1}{3}\pi r^2 h$ for h

80. $V = \pi r^2 h$ for h

81. $P = a + b + c$ for b

82. $P = a + b + c$ for a

▶ **83.** $P = 2L + 2W$ for W

84. $A = p + prt$ for r

▶ **85.** $y = mx + b$ for m

86. $y = mx + b$ for x

87. $Ax + By = C$ for y

88. $Ax + By = C$ for x

▶ **89.** $M = C(1 + r)$ for r

90. $A = p(1 + rt)$ for t

91. $P = 2(a + b)$ for a

92. $P = 2(a + b)$ for b

93. $S = \frac{1}{2}(a + b + c)$ for b

94. $S = \frac{1}{2}(a + b + c)$ for c

95. $C = \frac{5}{9}(F - 32)$ for F

96. $A = \frac{1}{2}h(b + B)$ for b

Solve each equation for y. **See Example 9.**

97. $6x + y = 4$

98. $3x + y = 6$

99. $5x - y = 2$

100. $4x - y = 1$

101. $-3x + 5y = -15$

102. $-2x + 3y = -9$

103. $x - 3y = 12$

104. $x - 5y = 10$

2.6 Ratio, Proportion, and Percent

OBJECTIVES

1. Write ratios.
2. Solve proportions.
3. Solve applied problems using proportions.
4. Find percents and percentages.

VOCABULARY

☐ ratio
☐ proportion
☐ terms of a proportion
☐ extremes
☐ means
☐ cross products of a proportion
☐ percent
☐ percentage
☐ base

**NOW TRY
EXERCISE 1**

Write a ratio for each word phrase. Express fractions in lowest terms.

(a) 7 in. to 4 in.

(b) 45 sec to 2 min

OBJECTIVE 1 Write ratios.

A **ratio** is a comparison of two quantities using a quotient.

> **Ratio**
>
> The ratio of a number a to a number b (where $b \neq 0$) is written as follows.
>
> $$a \text{ to } b, \quad a:b, \quad \text{or} \quad \frac{a}{b}$$
>
> Writing a ratio as a quotient $\frac{a}{b}$ is most common in algebra.
>
> *Examples:*   2 to 3,   2:3, $\frac{2}{3}$

EXAMPLE 1   Writing Word Phrases as Ratios

Write a ratio for each word phrase. Express fractions in lowest terms.

(a) 5 hr to 3 hr $\dfrac{5 \text{ hr}}{3 \text{ hr}} = \dfrac{5}{3}$

(b) 6 hr to 3 days

First convert 3 days to hours.

$$3 \text{ days} = 3 \cdot 24 = 72 \text{ hr} \qquad \text{1 day = 24 hr}$$

Now write the ratio using the common unit of measure, hours.

$$\frac{6 \text{ hr}}{3 \text{ days}} = \frac{6 \text{ hr}}{72 \text{ hr}} = \frac{6}{72} = \frac{1}{12} \qquad \text{Write in lowest terms.} \qquad \textbf{NOW TRY}$$

Applications of ratios occur regularly in everyday life. For example, automobile manufacturers report "miles per gallon" (abbreviated mpg) for their vehicles. This is a ratio found by dividing number of miles driven by number of gallons of gasoline used.

Another application of ratios is in *unit pricing,* to see which size of an item offered in different sizes produces the best price per unit.

EXAMPLE 2 Finding Price per Unit

A Jewel-Osco supermarket charges the following prices for a jar of extra crunchy peanut butter.

Peanut Butter

Size	Price
18 oz	$3.49
28 oz	$4.99
40 oz	$6.79

Which size is the best buy? That is, which size has the lowest unit price?

To find the best buy, write ratios comparing the price for each size jar to the number of units (ounces) per jar.

Peanut Butter

Size	Unit Price (dollars per ounce)
18 oz	$\frac{\$3.49}{18} = \$0.194$
28 oz	$\frac{\$4.99}{28} = \$0.178$
40 oz	$\frac{\$6.79}{40} = \$0.170$ ← Best buy

To find the price per ounce, the number of ounces goes in the denominator.

(Results are rounded to the nearest thousandth.)

Because the 40-oz size produces the lowest unit price, it is the best buy. Buying the largest size does not always provide the best buy, although it often does, as in this case.

NOW TRY

OBJECTIVE 2 Solve proportions.

A ratio is used to compare two numbers or amounts. A **proportion** says that two ratios are equal. For example, the proportion

$$\frac{3}{4} = \frac{15}{20}$$

A proportion is a special type of equation.

says that the ratios $\frac{3}{4}$ and $\frac{15}{20}$ are equal. In the proportion

$$\frac{a}{b} = \frac{c}{d} \quad (\text{where } b, d \neq 0),$$

$a, b, c,$ and d are the **terms** of the proportion. The terms a and d are the **extremes,** and the terms b and c are the **means.** We read the proportion $\frac{a}{b} = \frac{c}{d}$ as

"a is to b as c is to d."

NOW TRY EXERCISE 2

A supermarket charges the following prices for a certain brand of liquid detergent.

Size	Price
75 oz	$ 8.94
100 oz	$13.97
150 oz	$19.97

Which size is the best buy? What is the unit price (to the nearest thousandth) for that size?

NOW TRY ANSWERS
1. (a) $\frac{7}{4}$ (b) $\frac{3}{8}$
2. 75 oz; $0.119 per oz

Multiplying each side of this proportion by the common denominator, bd, gives the following.

$$\frac{a}{b} = \frac{c}{d}$$

$$bd \cdot \frac{a}{b} = bd \cdot \frac{c}{d} \qquad \text{Multiply each side by } bd.$$

$$\frac{b}{b}(d \cdot a) = \frac{d}{d}(b \cdot c) \qquad \text{Associative and commutative properties}$$

$$ad = bc \qquad \text{Commutative and identity properties}$$

We can also find the products ad and bc by multiplying diagonally.

$$ad = bc$$

$$\frac{a}{b} = \frac{c}{d}$$

For this reason, ad and bc are the **cross products of the proportion.**

Cross Products of a Proportion

If $\frac{a}{b} = \frac{c}{d}$, then the cross products ad and bc are equal—that is, ***the product of the extremes equals the product of the means.***

Also, if $ad = bc$, then $\frac{a}{b} = \frac{c}{d}$ (where $b, d \neq 0$).

NOTE If $\frac{a}{c} = \frac{b}{d}$, then $ad = cb$, or $ad = bc$. This means that the two proportions are equivalent, and the proportion

$$\frac{a}{b} = \frac{c}{d} \quad \text{can also be written as} \quad \frac{a}{c} = \frac{b}{d} \quad \text{(where } c, d \neq 0\text{)}.$$

Sometimes one form is more convenient to work with than the other.

> **NOW TRY**
> **EXERCISE 3**
> Decide whether each proportion is *true* or *false*.
> (a) $\frac{1}{3} = \frac{33}{100}$ (b) $\frac{4}{13} = \frac{16}{52}$

EXAMPLE 3 Deciding Whether Proportions Are True

Decide whether each proportion is *true* or *false*.

(a) $\dfrac{3}{4} = \dfrac{15}{20}$

Check to see whether the cross products are equal.

$$4 \cdot 15 = 60$$
$$\frac{3}{4} = \frac{15}{20}$$
$$3 \cdot 20 = 60$$

The cross products are equal, so the proportion is true.

(b) $\dfrac{6}{7} = \dfrac{30}{32}$

The cross products, $6 \cdot 32 = 192$ and $7 \cdot 30 = 210$, are not equal, so the proportion is false.

NOW TRY

NOW TRY ANSWERS
3. (a) false (b) true

Four numbers are used in a proportion. If any three of these numbers are known, the fourth can be found.

**NOW TRY
EXERCISE 4**

Solve.

$$\frac{9}{7} = \frac{x}{56}$$

EXAMPLE 4   Finding an Unknown in a Proportion

Solve $\frac{5}{9} = \frac{x}{63}$.

$$\frac{5}{9} = \frac{x}{63} \qquad \boxed{\text{Solve for } x.}$$

$$5 \cdot 63 = 9 \cdot x \qquad \text{Cross products must be equal.}$$

$$315 = 9x \qquad \text{Multiply.}$$

$$35 = x \qquad \text{Divide by 9.}$$

Check by substituting 35 for x in the proportion. The solution set is $\{35\}$.

 **NOW TRY**

**NOW TRY
EXERCISE 5**

Solve.

$$\frac{k-3}{6} = \frac{3k+2}{4}$$

EXAMPLE 5   Solving an Equation Using Cross Products

Solve $\frac{m-2}{5} = \frac{m+1}{3}$.

$$\frac{m-2}{5} = \frac{m+1}{3}$$

$$3(m-2) = 5(m+1) \qquad \boxed{\text{Be sure to use parentheses.}} \qquad (*) \quad \text{Cross products}$$

$$3m - 6 = 5m + 5 \qquad \text{Distributive property}$$

$$-2m - 6 = 5 \qquad \text{Subtract } 5m.$$

$$-2m = 11 \qquad \text{Add 6.}$$

$$m = -\frac{11}{2} \qquad \text{Divide by } -2.$$

The solution set is $\left\{-\frac{11}{2}\right\}$.

 **NOW TRY**

NOTE   When we set cross products equal to each other, we are actually multiplying each ratio in the proportion by a common denominator.

$$\frac{m-2}{5} = \frac{m+1}{3} \qquad \text{See **Example 5.**}$$

$$15\left(\frac{m-2}{5}\right) = 15\left(\frac{m+1}{3}\right) \qquad \text{Multiply each ratio by 15, the LCD.}$$

$$\boxed{\begin{array}{l} 15\left(\frac{m-2}{5}\right) \\ = 15 \cdot \frac{1}{5}(m-2) \\ = 3(m-2) \end{array}} \qquad 3(m-2) = 5(m+1) \qquad \text{This is equation (*) from **Example 5.**}$$

⚠️ **CAUTION**   *The cross-product method cannot be used directly if there is more than one term on either side of the equality symbol.*

$$\underbrace{\frac{m-1}{5} = \frac{m+1}{3} - 4}_{\text{2 terms}}, \qquad \underbrace{\frac{x}{3} + \frac{5}{4} = \frac{1}{2}}_{\text{2 terms}}$$

Do *not* use the cross-product method to solve equations in this form.

NOW TRY ANSWERS
4. $\{72\}$ **5.** $\left\{-\frac{12}{7}\right\}$

**NOW TRY
EXERCISE 6**

Twenty gallons of gasoline costs $69.80. How much would 27 gal of the same gasoline cost?

OBJECTIVE 3   Solve applied problems using proportions.

EXAMPLE 6   Applying Proportions

After Lee Ann pumped 5.0 gal of gasoline, the display showing the price read $18.10. When she finished pumping the gasoline, the price display read $52.49. How many gallons did she pump?

To solve this problem, set up a proportion, with prices in the numerators and gallons in the denominators. Let $x =$ the number of gallons she pumped.

$$\text{Price} \longrightarrow \frac{\$18.10}{5.0} = \frac{\$52.49}{x} \longleftarrow \text{Price}$$
$$\text{Gallons} \longrightarrow \qquad\qquad \longleftarrow \text{Gallons}$$

> Be sure that numerators represent the *same* quantities and denominators represent the *same* quantities.

$$18.10x = 5.0(52.49) \qquad \text{Cross products}$$
$$18.10x = 262.45 \qquad \text{Multiply.}$$
$$x = 14.5 \qquad \text{Divide by 18.10.}$$

She pumped 14.5 gal. Check this answer. Notice that the way the proportion was set up uses the fact that the unit price is the same, no matter how many gallons are purchased.

NOW TRY

OBJECTIVE 4   Find percents and percentages.

A percent is a ratio where the second number is always 100. For example,

50% represents the ratio of 50 to 100, that is, $\frac{50}{100}$, or, as a decimal, 0.50.

27% represents the ratio of 27 to 100, that is, $\frac{27}{100}$, or, as a decimal, 0.27.

The word **percent** means **"per 100."** One percent means "one per 100."

$$1\% = 0.01, \quad \text{or} \quad 1\% = \frac{1}{100} \qquad \begin{array}{l}\text{Percent, decimal, and fraction} \\ \text{equivalents (\textbf{Section R.2})}\end{array}$$

We can solve a percent problem involving $x\%$ by writing it as a proportion. The amount, or **percentage,** is compared to the **base** (the whole amount).

$$\frac{\boldsymbol{amount}}{\boldsymbol{base}} = \frac{\boldsymbol{x}}{\boldsymbol{100}}$$

We can also write this proportion as follows.

$$\frac{\text{amount}}{\text{base}} = \text{percent (as a decimal)} \qquad \begin{array}{l}\frac{x}{100} \text{ or } 0.01x \text{ is equivalent} \\ \text{to } x \text{ percent.}\end{array}$$

$$\boldsymbol{amount} = \boldsymbol{percent\ (as\ a\ decimal)} \cdot \boldsymbol{base} \qquad \text{Basic percent equation}$$

EXAMPLE 7   Solving Percent Equations

Solve each problem.

(a) What is 15% of 600?

Let $n =$ the number. The word *of* indicates multiplication.

What   is   15%   of   600?

> Translate each word or phrase to write the equation.

$$n \quad = \quad 0.15 \quad \cdot \quad 600 \qquad \text{Write the percent equation.}$$
$$n = 90 \qquad \text{Multiply.}$$

> Write 15% as a decimal.

Thus, 90 is 15% of 600.

NOW TRY ANSWER
6. $94.23

NOW TRY EXERCISE 7

Solve each problem.

(a) What is 20% of 70?

(b) 40% of what number is 130?

(c) 121 is what percent of 484?

(b) 32% of what number is 64?

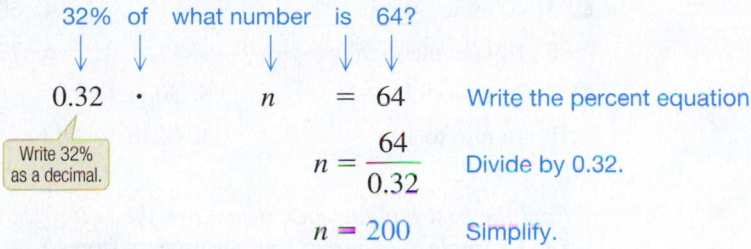

| 32% | of | what number | is | 64? |

$$0.32 \cdot n = 64 \qquad \text{Write the percent equation.}$$

Write 32% as a decimal.

$$n = \frac{64}{0.32} \qquad \text{Divide by 0.32.}$$

$$n = 200 \qquad \text{Simplify.}$$

32% of 200 is 64.

(c) 90 is what percent of 360?

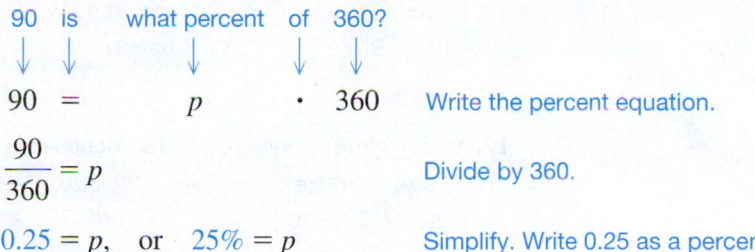

| 90 | is | what percent | of | 360? |

$$90 = p \cdot 360 \qquad \text{Write the percent equation.}$$

$$\frac{90}{360} = p \qquad \text{Divide by 360.}$$

$$0.25 = p, \quad \text{or} \quad 25\% = p \qquad \text{Simplify. Write 0.25 as a percent.}$$

Thus, 90 is 25% of 360.

NOW TRY

NOW TRY EXERCISE 8

A winter coat is on a clearance sale for $48. The regular price is $120. What percent of the regular price is the savings?

EXAMPLE 8 Solving an Applied Percent Problem

A newspaper ad offered a set of tires at a sales price of $258. The regular price was $300. What percent of the regular price was the savings?

The savings on the tires amounted to $300 − $258 = $42. We can now restate the problem: What percent of 300 is 42?

| What percent | of | 300 | is | 42? |

$$p \cdot 300 = 42 \qquad \text{Write the percent equation.}$$

$$p = \frac{42}{300} \qquad \text{Divide by 300.}$$

$$p = 0.14, \quad \text{or} \quad 14\% \qquad \text{Simplify. Write 0.14 as a percent.}$$

NOW TRY ANSWERS

7. **(a)** 14 **(b)** 325 **(c)** 25%

8. 60%

The sale price represents a 14% savings.

NOW TRY

2.6 Exercises

FOR EXTRA HELP

 MyMathLab®

▶ *Complete solution available in MyMathLab*

1. *Concept Check* Match each ratio in Column I with the ratio equivalent to it in Column II.

I	II
(a) 75 to 100	**A.** 80 to 100
(b) 5 to 4	**B.** 50 to 100
(c) $\frac{1}{2}$	**C.** 3 to 4
(d) 4 to 5	**D.** 15 to 12

2. *Concept Check* Which of the following represent a ratio of 4 days to 2 weeks?

A. $\frac{4}{2}$ **B.** $\frac{4}{7}$ **C.** $\frac{4}{14}$

D. $\frac{2}{1}$ **E.** $\frac{2}{7}$ **F.** $\frac{1}{2}$

G. $\frac{2}{4}$ **H.** $\frac{7}{2}$ **I.** 2

Write a ratio for each word phrase. Express fractions in lowest terms. ***See Example 1.***

▶ **3.** 40 mi to 30 mi

4. 60 ft to 70 ft

5. 120 people to 90 people

6. 72 dollars to 220 dollars

▶ **7.** 20 yd to 8 ft

8. 30 in. to 8 ft

9. 24 min to 2 hr

10. 16 min to 1 hr

11. 60 in. to 2 yd

12. 720 sec to 1 hr

Find the best buy for each item. Give the unit price to the nearest thousandth for that size. ***See Example 2.*** (*Source:* Various grocery stores.)

13. Granulated Sugar

Size	Price
4 lb	$3.29
10 lb	$7.49

14. Applesauce

Size	Price
23 oz	$1.99
48 oz	$3.49

15. Orange Juice

Size	Price
64 oz	$2.99
89 oz	$4.79
128 oz	$6.49

16. Salad Dressing

Size	Price
8 oz	$1.69
16 oz	$1.97
36 oz	$5.99

17. Maple Syrup

Size	Price
8.5 oz	$5.79
12.5 oz	$7.99
32 oz	$16.99

18. Mouthwash

Size	Price
16.9 oz	$3.39
33.8 oz	$3.49
50.7 oz	$5.29

19. Tomato Ketchup

Size	Price
32 oz	$1.79
36 oz	$2.69
40 oz	$2.49
64 oz	$4.38

20. Grape Jelly

Size	Price
12 oz	$1.05
18 oz	$1.73
32 oz	$1.84
48 oz	$2.88

21. Laundry Detergent

Size	Price
87 oz	$7.88
131 oz	$10.98
263 oz	$19.96

22. Spaghetti Sauce

Size	Price
14 oz	$1.79
24 oz	$1.77
48 oz	$3.65

Decide whether each proportion is true *or* false. ***See Example 3.***

▶ **23.** $\dfrac{5}{35} = \dfrac{8}{56}$

24. $\dfrac{4}{12} = \dfrac{7}{21}$

25. $\dfrac{120}{82} = \dfrac{7}{10}$

26. $\dfrac{27}{160} = \dfrac{18}{110}$

27. $\dfrac{\frac{1}{2}}{5} = \dfrac{1}{10}$

28. $\dfrac{\frac{1}{3}}{6} = \dfrac{1}{18}$

Solve each equation. ***See Examples 4 and 5.***

▶ **29.** $\dfrac{k}{4} = \dfrac{175}{20}$

30. $\dfrac{x}{6} = \dfrac{18}{4}$

31. $\dfrac{49}{56} = \dfrac{z}{8}$

32. $\dfrac{20}{100} = \dfrac{z}{80}$

33. $\dfrac{x}{24} = \dfrac{15}{16}$

34. $\dfrac{x}{4} = \dfrac{12}{30}$

35. $\dfrac{z}{2} = \dfrac{z+1}{3}$

36. $\dfrac{m}{5} = \dfrac{m-2}{2}$

▶ **37.** $\dfrac{3y-2}{5} = \dfrac{6y-5}{11}$

38. $\dfrac{2r+8}{4} = \dfrac{3r-9}{3}$

39. $\dfrac{5k+1}{6} = \dfrac{3k-2}{3}$

40. $\dfrac{x+4}{6} = \dfrac{x+10}{8}$

41. $\dfrac{2p+7}{3} = \dfrac{p-1}{4}$

42. $\dfrac{3m-2}{5} = \dfrac{4-m}{3}$

43. $\dfrac{2(x-4)}{3} = \dfrac{4(x-3)}{5}$

44. $\dfrac{9(x-3)}{6} = \dfrac{6(x-2)}{2}$

Solve each problem. (Assume that all items are equally priced.) ***See Example 6.***

45. If 16 candy bars cost $20.00, how much do 24 candy bars cost?

46. If 12 ring tones cost $30.00, how much do 8 ring tones cost?

47. Eight quarts of oil cost $14.00. How much do 5 qt of oil cost?

48. Four tires cost $398.00. How much do 7 tires cost?

49. If 9 pairs of jeans cost $121.50, find the cost of 5 pairs.

50. If 7 shirts cost $87.50, find the cost of 11 shirts.

51. If 6 gal of premium unleaded gasoline costs $22.56, how much would it cost to completely fill a 15-gal tank?

52. If sales tax on a $16.00 DVD is $1.32, find the sales tax on a $120.00 DVD player.

Solve each problem. (In Exercises 57–60, round answers to the nearest tenth.) ***See Example 6.***

53. Biologists tagged 500 fish in North Bay. At a later date, they found 7 tagged fish in a sample of 700. Estimate the total number of fish in North Bay to the nearest hundred.

54. Researchers at West Okoboji Lake tagged 840 fish. A later sample of 1000 fish contained 18 that were tagged. Approximate the fish population in West Okoboji Lake to the nearest hundred.

55. The distance between Kansas City, Missouri, and Denver is 600 mi. On a certain wall map, this is represented by a length of 2.4 ft. On the map, how many feet would there be between Memphis and Philadelphia, two cities that are actually 1000 mi apart?

56. The distance between Singapore and Tokyo is 3300 mi. On a certain wall map, this distance is represented by 11 in. The actual distance between Mexico City and Cairo is 7700 mi. How far apart are they on the same map?

57. A wall map of the United States has a distance of 8.5 in. between Memphis and Denver, two cities that are actually 1040 mi apart. The actual distance between St. Louis and Des Moines is 333 mi. How far apart are St. Louis and Des Moines on the map?

58. A wall map of the United States has a distance of 8.0 in. between New Orleans and Chicago, two cities that are actually 912 mi apart. The actual distance between Milwaukee and Seattle is 1940 mi. How far apart are Milwaukee and Seattle on the map?

59. On a world globe, the distance between Capetown and Bangkok, two cities that are actually 10,080 km apart, is 12.4 in. The actual distance between Moscow and Berlin is 1610 km. How far apart are Moscow and Berlin on this globe?

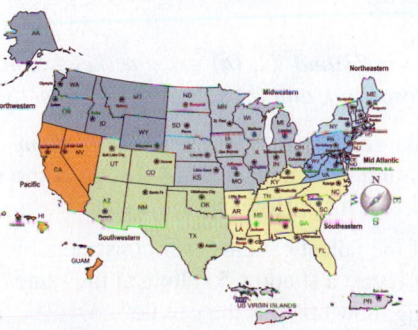

60. On a world globe, the distance between Rio de Janeiro and Hong Kong, two cities that are actually 17,615 km apart, is 21.5 in. The actual distance between Paris and Stockholm is 1605 km. How far apart are Paris and Stockholm on this globe?

61. According to the directions on a bottle of Armstrong® Concentrated Floor Cleaner, for routine cleaning, $\frac{1}{4}$ cup of cleaner should be mixed with 1 gal of warm water. How much cleaner should be mixed with $10\frac{1}{2}$ gal of water?

62. The directions on the bottle mentioned in **Exercise 61** also specify that, for extra-strength cleaning, $\frac{1}{2}$ cup of cleaner should be used for each gallon of water. How much cleaner should be mixed with $15\frac{1}{2}$ gal of water for extra-strength cleaning?

63. On September 23, 2013, the exchange rate between euros and U.S. dollars was 1 euro to $1.3492. Ashley went to Rome and exchanged her U.S. currency for euros, receiving 300 euros. How much in U.S. dollars did she exchange? (*Source:* www.xe.com/ucc)

64. If 8 U.S. dollars can be exchanged for 103.0 Mexican pesos, how many pesos can be obtained for $65? (Round to the nearest tenth.)

*Two triangles are **similar** if they have the same shape (but not necessarily the same size). Similar triangles have sides that are proportional. The figure shows two similar triangles. Notice that the ratios of the corresponding sides all equal $\frac{3}{2}$.*

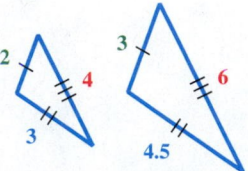

$$\frac{3}{2} = \frac{3}{2}, \quad \frac{4.5}{3} = \frac{3}{2}, \quad \frac{6}{4} = \frac{3}{2}$$

If we know that two triangles are similar, we can set up a proportion to solve for the length of an unknown side.

　　Find the lengths x and y as needed in each pair of similar triangles.

65.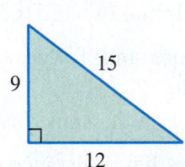

66.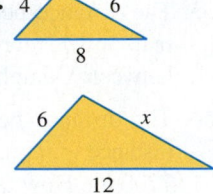

67.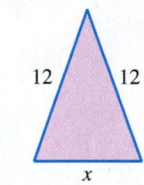

68.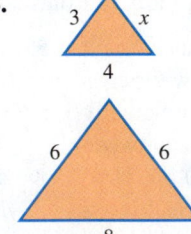

69.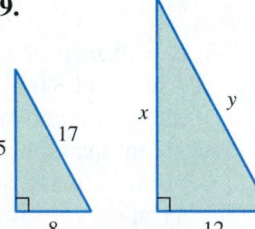

70.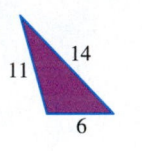

*For Exercises 71 and 72, **(a)** draw a sketch consisting of two right triangles depicting the situation described, and **(b)** solve the problem. (Source: Guinness World Records.)*

71. An enlarged version of the chair used by George Washington at the Constitutional Convention casts a shadow 18 ft long at the same time a vertical pole 12 ft high casts a shadow 4 ft long. How tall is the chair?

72. One of the tallest candles ever constructed was exhibited at the 1897 Stockholm Exhibition. If it cast a shadow 5 ft long at the same time a vertical pole 32 ft high cast a shadow 2 ft long, how tall was the candle?

The Consumer Price Index (CPI) provides a means of determining the purchasing power of the U.S. dollar from one year to the next. Using the period from 1982 to 1984 as a measure of 100.0, the CPI for selected years from 1999 through 2011 is shown in the table. To use the CPI to predict a price in a particular year, we set up a proportion and compare it with a known price in another year.

$$\frac{\text{price in year } A}{\text{index in year } A} = \frac{\text{price in year } B}{\text{index in year } B}$$

Year	Consumer Price Index
1999	166.6
2001	177.1
2003	184.0
2005	195.3
2007	207.3
2009	214.5
2011	224.9

Source: U.S. Bureau of Labor Statistics.

Use the CPI figures in the table to find the amount that would be charged for using the same amount of electricity that cost $225 in 1999. Give answers to the nearest dollar.

73. in 2001   **74.** in 2003   **75.** in 2007   **76.** in 2011

Children are often given antibiotics in liquid form, called an oral suspension. Pharmacists make up these suspensions by mixing medication in powder form with water. They use proportions to calculate the volume of the suspension for the amount of medication that has been prescribed. In Exercises 77 and 78, do each of the following.

(a) Find the total amount of medication in milligrams to be given over the full course of treatment.

(b) Write a proportion that can be solved to find the total volume of the liquid suspension that the pharmacist will prepare. Use x as the variable.

(c) Solve the proportion to determine the total volume of the oral suspension.

77. Logan's pediatric nurse practitioner has prescribed 375 mg of Amoxil a day for 7 days to treat his ear infection. The pharmacist uses 125 mg of Amoxil in each 5 mL of the suspension. (*Source:* www.drugs.com)

78. An Amoxil oral suspension can also be made by using 250 mg for each 5 mL of suspension. Ava's pediatrician prescribed 900 mg a day for 10 days to treat her bronchitis. (*Source:* www.drugs.com)

Solve each problem. **See Examples 7 and 8.**

79. What is 18% of 780?   **80.** What is 23% of 480?

81. 42% of what number is 294?   **82.** 18% of what number is 108?

83. 120% of what number is 510?   **84.** 140% of what number is 315?

85. 4 is what percent of 50?   **86.** 8 is what percent of 64?

87. What percent of 30 is 36?   **88.** What percent of 48 is 96?

89. Clayton earned 48 points on a 60-point geometry project. What percent of the total points did he earn?

90. On a 75-point algebra test, Grady scored 63 points. What percent of the total points did he score?

91. A laptop computer that has a regular price of $700 is on sale for $504. What percent of the regular price is the savings?

92. An all-in-one desktop computer that has a regular price of $980 is on sale for $833. What percent of the regular price is the savings?

93. Tyler has a monthly income of $1500. His rent is $480 per month. What percent of his monthly income is his rent?

94. Lily has a monthly income of $2200. She has budgeted $154 per month for entertainment. What percent of her monthly income did she budget for entertainment?

95. Anna saved $1950, which was 65% of the amount she needed for a used car. What was the total amount she needed for the car?

96. Bryn had $525, which was 70% of the total amount she needed for a deposit on an apartment. What was the total deposit she needed?

2.7 Further Applications of Linear Equations

OBJECTIVES

1   Use percent in solving problems involving rates.

2   Solve problems involving mixtures.

3   Solve problems involving simple interest.

4   Solve problems involving denominations of money.

5   Solve problems involving distance, rate, and time.

NOW TRY EXERCISE 1

(a) How much pure alcohol is in 70 L of a 20% alcohol solution?

(b) Find the annual simple interest if $3200 is invested at 2%.

OBJECTIVE 1   **Use percent in solving problems involving rates.**

Recall from **Sections R.2 and 2.6** that the word "percent" means "per 100." One percent means "one per 100."

$$1\% = 0.01, \quad \text{or} \quad 1\% = \frac{1}{100} \qquad \text{Percent, decimal, and fraction equivalents}$$

PROBLEM-SOLVING HINT   Mixing different concentrations of a substance or different interest rates involves percents. To obtain the amount of pure substance or the interest, we multiply as follows.

Mixture Problems	Interest Problems (annual)
base · rate (%) = percentage	principal · rate (%) = interest
$b \cdot r \quad = \quad p$	$p \quad \cdot \quad r \quad = \quad I$

In an equation, percent is always written as a decimal (or a fraction).

EXAMPLE 1   **Using Percents to Find Percentages**

(a) If a chemist has 40 L of a 35% acid solution, then the amount of pure acid in the solution is found by multiplying.

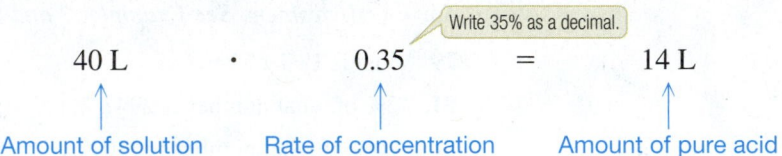

Write 35% as a decimal.

40 L · 0.35 = 14 L

Amount of solution   Rate of concentration   Amount of pure acid

(b) If $1300 is invested for one year at 3% simple interest, the amount of interest earned in the year is calculated as follows.

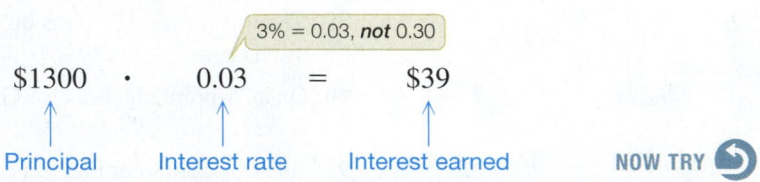

3% = 0.03, **not** 0.30

$1300 · 0.03 = $39

Principal   Interest rate   Interest earned   **NOW TRY**

PROBLEM-SOLVING HINT   In the applications that follow, using a table helps organize the information in a problem and more easily set up an equation, which is usually the most difficult step.

NOW TRY ANSWERS
1. (a) 14 L   **(b)** $64

OBJECTIVE 2 Solve problems involving mixtures.

EXAMPLE 2 Solving a Mixture Problem

A chemist mixes 20 L of a 40% acid solution with some 70% acid solution to obtain a mixture that is 50% acid. How many liters of the 70% acid solution should she use?

Step 1   **Read** the problem. Note the percent of each solution and of the mixture.

Step 2   **Assign a variable** to represent the unknown quantity.

Let x = the number of liters of 70% acid solution needed.

As in **Example 1(a),** the amount of pure acid in this solution is the product of the percent of strength and the number of liters of solution.

$$0.70x \qquad \text{Liters of pure acid in } x \text{ liters of 70\% solution}$$

The amount of pure acid in the 20 L of 40% solution is found similarly.

$$0.40(20) \qquad \text{Liters of pure acid in the 40\% solution}$$

The new solution will contain $(x + 20)$ liters of 50% solution. The amount of pure acid in this solution is again found by multiplying.

$$0.50(x + 20) \qquad \text{Liters of pure acid in the 50\% solution}$$

FIGURE 17 illustrates this information, which is organized in the table.

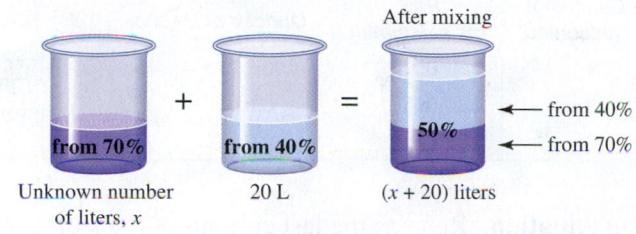

Liters of Solution	Rate (as a decimal)	Liters of Pure Acid
x	0.70	$0.70x$
20	0.40	$0.40(20)$
$x + 20$	0.50	$0.50(x + 20)$

Sum must equal

FIGURE 17

Step 3   **Write an equation.** The number of liters of pure acid in the 70% solution added to the number of liters of pure acid in the 40% solution will equal the number of liters of pure acid in the final mixture.

Pure acid in 70% solution   plus   pure acid in 40% solution   is   pure acid in 50% solution.

$$0.70x \qquad + \qquad 0.40(20) \qquad = \qquad 0.50(x + 20)$$

Refer to the last column of the table.

Step 4   **Solve.** First clear the parentheses. Then clear the decimals.

$0.70x = 0.7x$ and $0.50x = 0.5x.$

$0.7x + 8 = 0.5x + 10$	Multiply; distributive property
$10(0.7x + 8) = 10(0.5x + 10)$	Multiply by 10.
$7x + 80 = 5x + 100$	Distributive property
$2x + 80 = 100$	Subtract 5x.
$2x = 20$	Subtract 80.
$x = 10$	Divide by 2.

Step 5   **State the answer.** The chemist needs to use 10 L of 70% solution.

**NOW TRY
EXERCISE 2**

A certain seasoning is 70% salt. How many ounces of this seasoning must be mixed with 30 oz of dried herbs containing 10% salt to obtain a seasoning that is 50% salt?

Step 6   **Check.** If 10 L of 70% solution are used, the amounts of pure acid are the same.

$$0.70(10) + 0.40(20) \quad \text{Sum of the two solutions}$$
$$= 7 + 8$$
$$= 15$$

$$0.50(10 + 20) \quad \text{Mixture}$$
$$= 0.50(30)$$
$$= 15 \qquad \text{NOW TRY} \; \circlearrowleft$$

⚠️ **CAUTION**   In a mixture problem, the concentration of the final mixture must be **between** the concentrations of the two solutions making up the mixture.

**NOW TRY
EXERCISE 3**

How many liters of a 25% saline solution must be mixed with a 10% saline solution to obtain 15 L of a 15% solution?

| EXAMPLE 3 | Solving a Mixture Problem |

How many ounces of a seasoning that is 15% pepper must be mixed with a version that is 30% pepper to obtain 9 oz of a seasoning that is 20% pepper?

Step 1   **Read** the problem. We are given the *total* amount of the mixture. We must find the amount of the seasoning that is 15% pepper.

Step 2   **Assign a variable.** Use the fact that the total mixture is 9 oz.

Let $x =$ the number of ounces of seasoning that is 15% pepper.

Then $9 - x =$ the number of ounces of seasoning that is 30% pepper.

Ounces of Seasoning	Rate (as a decimal)	Ounces of Pepper
x	0.15	$0.15x$
$9 - x$	0.30	$0.30(9 - x)$
9	0.20	$0.20(9)$

Use a table to organize the given information.

Step 3   **Write an equation.** Refer to the last column of the table.

Pepper in 15% seasoning	plus	pepper in 30% seasoning	is	pepper in 20% seasoning.
↓	↓	↓	↓	↓
$0.15x$	$+$	$0.30(9 - x)$	$=$	$0.20(9)$

Step 4   **Solve.**

$$0.15x + 2.7 - 0.3x = 1.8 \qquad \text{Distributive property; multiply.}$$

> To multiply by 100, move the decimal point in each term 2 places to the right.

$$15x + 270 - 30x = 180 \qquad \text{Multiply by 100.}$$
$$-15x + 270 = 180 \qquad \text{Combine like terms.}$$
$$-15x = -90 \qquad \text{Subtract 270.}$$
$$x = 6 \qquad \text{Divide by } -15.$$

Step 5   **State the answer.** 6 oz of seasoning that is 15% pepper is needed. (This means that $9 - 6 = 3$ oz of the 30% pepper seasoning is needed, although the problem does not specifically ask for this amount.)

Step 6   **Check.** The ounces of pepper before and after mixing are the same.

NOW TRY ANSWERS
2. 60 oz
3. 5 L

$$0.15(6) + 0.30(9 - 6) \quad \text{Sum of the two seasonings}$$
$$= 0.9 + 0.9$$
$$= 1.8$$

$$0.20(9) \quad \text{Mixture}$$
$$= 1.8$$

$$\text{NOW TRY} \; \circlearrowleft$$

OBJECTIVE 3 Solve problems involving simple interest.

The formula for simple interest

$$I = prt \quad \text{becomes} \quad I = pr \quad \text{when time } t = 1 \text{ (for annual interest),}$$

as shown in the Problem-Solving Hint at the beginning of this section. Multiplying the total amount (principal) by the rate (rate of interest) gives the percentage (amount of interest).

NOW TRY EXERCISE 4

A financial advisor invests some money in a municipal bond paying 3% annual interest and $5000 more than that amount in a certificate of deposit paying 4% annual interest. To earn $410 per year in interest, how much should he invest at each rate?

EXAMPLE 4 Solving a Simple Interest Problem

Susan plans to invest some money at 2% and $2000 more than this amount at 4%. To earn $380 per year in interest, how much should she invest at each rate?

Step 1 **Read** the problem. There will be two answers.

Step 2 **Assign a variable.**

Let x = the amount invested at 2% (in dollars).

Then $x + 2000$ = the amount invested at 4% (in dollars).

Amount Invested (in dollars)	Rate (as a decimal)	Interest for One Year (in dollars)
x	0.02	0.02x
$x + 2000$	0.04	0.04(x + 2000)

Use a table to organize the given information.

Step 3 **Write an equation.** Multiply amount by rate to obtain interest earned. The two amounts of interest must total $380.

Interest at 2%	plus	interest at 4%	is	total interest.
↓	↓	↓	↓	↓
0.02x	+	0.04(x + 2000)	=	380

Step 4 **Solve.**

$$0.02x + 0.04x + 80 = 380 \quad \text{Distributive property}$$
$$2x + 4x + 8000 = 38{,}000 \quad \text{Multiply by 100.}$$
$$6x + 8000 = 38{,}000 \quad \text{Combine like terms.}$$
$$6x = 30{,}000 \quad \text{Subtract 8000.}$$
$$x = 5000 \quad \text{Divide by 6.}$$

Step 5 **State the answer.** At 2%, she should invest $5000. At 4%, she should invest

$$\$5000 + \$2000 = \$7000.$$

Step 6 **Check.** The sum of the two interest amounts is

$$0.02(\$5000) + 0.04(\$7000)$$
$$= \$100 + \$280$$
$$= \$380, \quad \text{as required.} \qquad \text{NOW TRY}$$

OBJECTIVE 4 Solve problems involving denominations of money.

PROBLEM-SOLVING HINT   To obtain the total value in problems that involve different denominations of money or items with different monetary values, we multiply as follows.

Money Denominations Problems

number · value of one item = total value

Examples: 30 dimes have a value of 30($0.10) = $3.

15 five-dollar bills have a value of 15($5) = $75.

**NOW TRY
EXERCISE 5**
Clayton has saved $5.65 in
dimes and quarters. He has
10 more quarters than
dimes. How many of each
denomination of coin
does he have?

EXAMPLE 5   Solving a Money Denominations Problem

A bank teller has 25 more $5 bills than $10 bills. The total value of the money is
$200. How many of each denomination of bill does she have?

Step 1   **Read** the problem. We must find the number of each denomination of bill.

Step 2   **Assign a variable.**

Let x = the number of $10 bills.

Then $x + 25$ = the number of $5 bills.

Number of Bills	Denomination (in dollars)	Total Value (in dollars)
x	10	$10x$
$x + 25$	5	$5(x + 25)$

Step 3   **Write an equation.** Multiplying the number of bills by the denomination
gives the monetary value. The value of the tens added to the value of the
fives must be $200.

$$
\begin{array}{ccccc}
\text{Value of} & & \text{value of} & & \\
\text{tens} & \text{plus} & \text{fives} & \text{is} & \$200. \\
\downarrow & \downarrow & \downarrow & \downarrow & \downarrow \\
10x & + & 5(x + 25) & = & 200
\end{array}
$$

Step 4   **Solve.**

$$10x + 5x + 125 = 200 \qquad \text{Distributive property}$$
$$15x + 125 = 200 \qquad \text{Combine like terms.}$$
$$15x = 75 \qquad \text{Subtract 125.}$$
$$x = 5 \qquad \text{Divide by 15.}$$

Step 5   **State the answer.** The teller has 5 tens and 5 + 25 = 30 fives.

Step 6   **Check.** The teller has 30 − 5 = 25 more fives than tens. The value is

$$5(\$10) + 30(\$5) = \$200, \quad \text{as required.} \qquad \text{NOW TRY} \circlearrowleft$$

OBJECTIVE 5 Solve problems involving distance, rate, and time.

If a car travels at an average rate of 50 mph for 2 hr, then it travels

$$50 \times 2 = 100 \text{ mi.}$$

This is an example of the basic relationship between distance, rate, and time.

$$\textbf{distance = rate · time,} \quad \text{or} \quad \textbf{\textit{d = rt}}$$

NOW TRY ANSWER
5. dimes: 9; quarters: 19

By solving, in turn, for r and t in the formula $d = rt$, we obtain two other equivalent forms of the formula.

Forms of the Distance Formula

$$d = rt \qquad r = \frac{d}{t} \qquad t = \frac{d}{r}$$

NOW TRY EXERCISE 6

It took a driver 6 hr to travel from St. Louis to Fort Smith, a distance of 400 mi. What was the driver's rate, to the nearest hundredth?

EXAMPLE 6 Finding Distance, Rate, or Time

Solve each problem using a form of the distance formula.

(a) The speed (rate) of sound is 1088 ft per sec at sea level at 32°F. Find the distance sound travels in 5 sec under these conditions.

We must find distance, given rate and time, using $d = rt$ (or $rt = d$).

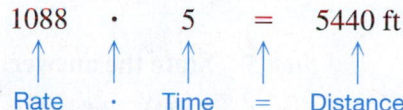

$$1088 \quad \cdot \quad 5 \quad = \quad 5440 \text{ ft}$$

Rate · Time = Distance

(b) The winner of the first Indianapolis 500 race (in 1911) was Ray Harroun, driving a Marmon Wasp at an average rate of 74.59 mph. (*Source: Universal Almanac.*) How long did it take him to complete the 500 mi?

We must find time, given rate and distance, using $t = \frac{d}{r}$ $\left(\text{or } \frac{d}{r} = t\right)$.

Distance → $\dfrac{500}{74.59} = 6.70$ hr (rounded) ← Time
Rate →

To convert 0.70 hr to minutes, we multiply by 60 to obtain $0.70(60) = 42$. It took Harroun about 6 hr, 42 min, to complete the race.

(c) At the 2012 Olympic Games, American swimmer Missy Franklin won a gold medal with a time of 58.33 sec in the women's 100-m backstroke swimming event. (*Source: World Almanac and Book of Facts.*) Find her rate.

We must find rate, given distance and time, using $r = \frac{d}{t}$ $\left(\text{or } \frac{d}{t} = r\right)$.

Distance → $\dfrac{100}{58.33} = 1.71$ m per sec (rounded) ← Rate   **NOW TRY**
Time →

EXAMPLE 7 Solving a Distance-Rate-Time Problem

Two cars leave Iowa City, Iowa, at the same time and travel east on Interstate 80. One travels at a constant rate of 55 mph. The other travels at a constant rate of 63 mph. In how many hours will the distance between them be 24 mi?

Step 1 **Read** the problem carefully.

Step 2 **Assign a variable.** We are looking for time.

Let t = the number of hours until the distance between them is 24 mi.

The sketch in **FIGURE 18** shows what is happening in the problem.

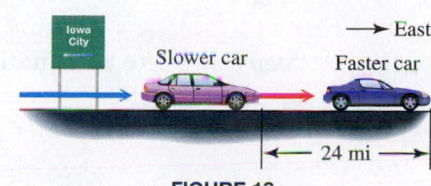

FIGURE 18

NOW TRY ANSWER
6. 66.67 mph

NOW TRY
EXERCISE 7
From a point on a straight road, two bicyclists ride in the same direction. One travels at a rate of 18 mph. The other travels at a rate of 20 mph. In how many hours will they be 5 mi apart?

To construct a table, we fill in the rates given in the problem, using t for the time traveled by each car. Because $d = rt$, or $rt = d$, we multiply rate by time to find expressions for the distances traveled.

	Rate	Time	Distance	
Faster Car	63	t	$63t$	The quantities $63t$ and $55t$
Slower Car	55	t	$55t$	represent the two distances.

Step 3   **Write an equation.**

$$63t - 55t = 24$$ The *difference* between the larger distance and the smaller distance is 24 mi.

Step 4   **Solve.** $8t = 24$ Combine like terms.

$$t = 3$$ Divide by 8.

Step 5   **State the answer.** It will take the cars 3 hr to be 24 mi apart.

Step 6   **Check.** After 3 hr, the faster car will have traveled $63 \cdot 3 = 189$ mi and the slower car will have traveled $55 \cdot 3 = 165$ mi. The difference is

$$189 - 165 = 24, \quad \text{as required.}$$ **NOW TRY**

PROBLEM-SOLVING HINT   In distance-rate-time problems, once we have filled in two pieces of information in each row of a table, we can automatically fill in the third piece of information, using the appropriate form of the distance formula. Then we set up the equation based on a sketch and the information in the table.

EXAMPLE 8   **Solving a Distance-Rate-Time Problem**

Two planes leave Memphis at the same time. One heads south to New Orleans. The other heads north to Chicago. The Chicago plane flies 50 mph faster than the New Orleans plane. In $\frac{1}{2}$ hr, the planes are 275 mi apart. What are their rates?

Step 1   **Read** the problem carefully.

Step 2   **Assign a variable.**

Let $r =$ the rate of the slower plane.

Then $r + 50 =$ the rate of the faster plane.

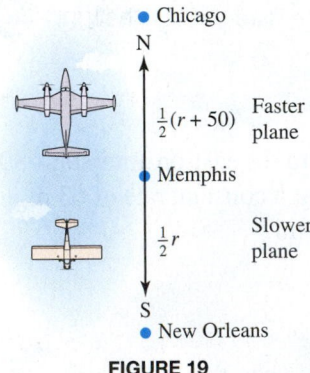
• Chicago
N
$\frac{1}{2}(r + 50)$ Faster plane
• Memphis
$\frac{1}{2}r$ Slower plane
S
• New Orleans
FIGURE 19

	Rate	Time	Distance	
Slower Plane	r	$\frac{1}{2}$	$\frac{1}{2}r$	Sum is 275 mi.
Faster Plane	$r + 50$	$\frac{1}{2}$	$\frac{1}{2}(r + 50)$	

Step 3   **Write an equation.** As **FIGURE 19** shows, the planes are headed in *opposite* directions. The *sum* of their distances equals 275 mi.

$$\frac{1}{2}r + \frac{1}{2}(r + 50) = 275$$

NOW TRY
EXERCISE 8

Two cars leave a parking lot at the same time, one traveling east and the other traveling west. The westbound car travels 6 mph faster than the eastbound car. In $\frac{1}{4}$ hr, they are 35 mi apart. What are their rates?

NOW TRY ANSWER

8. slower car: 67 mph;
   faster car: 73 mph

Step 4   Solve.

$$\frac{1}{2}r + \frac{1}{2}(r + 50) = 275 \qquad \text{Equation from Step 3}$$

$$\frac{1}{2}r + \frac{1}{2}r + 25 = 275 \qquad \text{Distributive property}$$

$$r + 25 = 275 \qquad \text{Combine like terms.}$$

Rate of slower plane → $r = 250$ Subtract 25.

Step 5   State the answer. The slower plane (headed south) has a rate of 250 mph.

$$250 + 50 = 300 \text{ mph} \leftarrow \text{Rate of faster plane}$$

Step 6   Check. Verify that $\frac{1}{2}(250) + \frac{1}{2}(300) = 275$ mi, as required.   NOW TRY

2.7 Exercises

FOR EXTRA HELP   MyMathLab®

Complete solution available in MyMathLab

Concept Check   *In Exercises 1–7, choose the letter of the correct response.*

1. Which expression represents the amount of pure alcohol in x liters of a 75% alcohol solution?

   A. $0.75x$ liters   **B.** $(75 + x)$ liters   **C.** $(75 - x)$ liters   **D.** $75x$ liters

2. Which expression represents the value of x quarters?

   A. $25x$ dollars   **B.** $\dfrac{25}{x}$ dollars   **C.** $0.25x$ dollars   **D.** $(x + 0.25)$ dollars

3. If a minivan travels at 55 mph for t hours, which expression represents the distance traveled?

   A. $(t + 55)$ miles   **B.** $(t - 55)$ miles   **C.** $55t$ miles   **D.** $\dfrac{55}{t}$ miles

4. If a car travels at r mph for 6 hr, which expression represents the distance traveled?

   A. $\dfrac{r}{6}$ miles   **B.** $(r - 6)$ miles   **C.** $(r + 6)$ miles   **D.** $6r$ miles

5. Suppose that a chemist is mixing two acid solutions, one of 20% concentration and the other of 30% concentration. Which concentration could *not* be obtained?

   A. 22%   **B.** 24%   **C.** 28%   **D.** 32%

6. Suppose that water is added to a 24% alcohol mixture. Which concentration could be obtained? (*Hint:* The solution is being diluted.)

   A. 22%   **B.** 26%   **C.** 28%   **D.** 30%

7. Which choice is the best estimate for the average rate of a bus trip of 405 mi that lasted 8.2 hr?

   A. 50 mph   **B.** 30 mph   **C.** 60 mph   **D.** 40 mph

8. **Concept Check**   An automobile averages 45 mph and travels for 30 min. Is the distance traveled $45 \cdot 30 = 1350$ mi? If not, explain why not, and give the correct distance.

Answer each question. **See Example 1 and the Problem-Solving Hint preceding Example 5.**

9. How much pure alcohol is in 150 L of a 30% alcohol solution?

10. How much pure acid is in 250 mL of a 14% acid solution?

11. If $25,000 is invested for 1 yr at 3% simple interest, how much interest is earned?

12. If $10,000 is invested for 1 yr at 3.5% simple interest, how much interest is earned?

13. What is the monetary value of 35 half-dollars?

14. What is the monetary value of 283 nickels?

Solve each problem. ***See Examples 2 and 3.***

15. How many liters of 25% acid solution must a chemist add to 80 L of 40% acid solution to obtain a mixture that is 30% acid?

Liters of Solution	Rate (as a decimal)	Liters of Pure Acid
x	0.25	$0.25x$
80	0.40	$0.40(80)$
$x + 80$	0.30	$0.30(x + 80)$

16. How many gallons of 50% antifreeze must be mixed with 80 gal of 20% antifreeze to obtain a mixture that is 40% antifreeze?

Gallons of Mixture	Rate (as a decimal)	Gallons of Pure Antifreeze
x	0.50	$0.50x$
80	0.20	$0.20(80)$
$x + 80$	0.40	$0.40(x + 80)$

17. A pharmacist has 20 L of a 10% drug solution. How many liters of 5% drug solution must be added to obtain a mixture that is 8%?

Liters of Solution	Rate (as a decimal)	Liters of Pure Drug
20		$20(0.10)$
	0.05	
	0.08	

18. A certain metal is 20% tin. How many kilograms of this metal must be mixed with 80 kg of a metal that is 70% tin to obtain a metal that is 50% tin?

Kilograms of Metal	Rate (as a decimal)	Kilograms of Pure Tin
x	0.20	
	0.70	
	0.50	

19. In a chemistry class, 12 L of a 12% alcohol solution must be mixed with a 20% solution to obtain a 14% solution. How many liters of the 20% solution are needed?

20. How many liters of a 10% alcohol solution must be mixed with 40 L of a 50% solution to obtain a 40% solution?

21. Minoxidil is a drug that has proven to be effective in treating male pattern baldness. Water must be added to 20 mL of a 4% minoxidil solution to dilute it to a 2% solution. How many milliliters of water should be used? (*Hint:* Water is 0% minoxidil.)

22. A pharmacist wishes to mix a solution that is 2% minoxidil. She has on hand 50 mL of a 1% solution, and she wishes to add some 4% solution to it to obtain the desired 2% solution. How much 4% solution should she add?

23. How many liters of a 60% acid solution must be mixed with a 75% acid solution to obtain 20 L of a 72% solution?

24. How many gallons of a fruit drink that is 50% real juice must be mixed with a fruit drink that is 20% real juice to obtain 12 gal of a fruit drink that is 40% real juice?

Solve each problem. ***See Example 4.***

25. Arlene is saving money for her college education. She deposited some money in a savings account paying 5% and $1200 less than that amount in a second account paying 4%. The two accounts produced a total of $141 interest in 1 yr. How much did she invest at each rate?

26. Margaret won a prize for her work. She invested part of the money in a certificate of deposit at 2% and $3000 more than that amount in a bond paying 3%. Her annual interest income was $390. How much did Margaret invest at each rate?

27. An artist invests in a tax-free bond paying 6%, and $6000 more than three times as much in mutual funds paying 5%. Her total annual interest income from the investments is $825. How much does she invest at each rate?

28. With income earned by selling the rights to his life story, an actor invests some of the money at 3% and $30,000 more than twice as much at 4%. The total annual interest earned from the investments is $5600. How much is invested at each rate?

29. Jamal had $2500, some of which he deposited in a mutual fund account paying 8%. The rest he deposited in a money market account paying 2%. How much did he deposit in each account if the total annual interest was $152?

Amount Invested (in dollars)	Rate (as a decimal)	Interest for One Year (in dollars)
x	0.08	
	0.02	

30. Carter invested a total of $9000 in two accounts, one paying 1% and the other paying 4%. If he earned total annual interest of $285, how much did he deposit in each account?

Amount Invested (in dollars)	Rate (as a decimal)	Interest for One Year (in dollars)
x	0.01	
	0.04	

Solve each problem. See Example 5.

31. A coin collector has $1.70 in dimes and nickels. She has two more dimes than nickels. How many nickels does she have?

Number of Coins	Denomination (in dollars)	Total Value (in dollars)
x	0.05	0.05x
	0.10	

32. A bank teller has $725 in $5 bills and $20 bills. The teller has five more twenties than fives. How many $5 bills does the teller have?

Number of Bills	Denomination (in dollars)	Total Value (in dollars)
x	5	
x + 5	20	

33. In January 2013, U.S. first-class mail rates increased to 46 cents for the first ounce, and 20 cents for each additional ounce. If Sabrina spent $15.50 for a total of 45 stamps of these two denominations, how many stamps of each denomination did she buy? (*Source:* U.S. Postal Service.)

34. A movie theater has two ticket prices: $8 for adults and $5 for children. If the box office took in $4116 from the sale of 600 tickets, how many tickets of each kind were sold?

35. Harriet operates a coffee shop. One of her customers wants to buy two kinds of beans: Arabian Mocha and Colombian Decaf. If she wants twice as much Arabian Mocha as Colombian Decaf, how much of each can she buy for a total of $87.50? (Prices are listed on the sign.)

36. See **Exercise 35.** Another one of Harriet's customers wants to buy Italian Espresso beans and Kona Deluxe beans. If he wants four times as much Kona Deluxe as Italian Espresso, how much of each can he buy for a total of $247.50?

Arabian Mocha.........$8.50/lb
Chocolate Mint........$10.50/lb
Colombian Decaf........$8.00/lb
French Roast..........$7.50/lb
Guatemalan Spice......$9.50/lb
Hazelnut Decaf.......$10.00/lb
Italian Espresso........$9.00/lb
Kona Deluxe..........$11.50/lb

*Solve each problem. **See Example 6.***

▶ 37. A driver averaged 53 mph and took 10 hr to travel from Memphis to Chicago. What is the distance between Memphis and Chicago?

38. A small plane traveled from Warsaw to Rome, averaging 164 mph. The trip took 2 hr. What is the distance from Warsaw to Rome?

39. The winner of the 2013 Indianapolis 500 (mile) race was Tony Kanaan, who drove his Dellara-Chevrolet to victory at a rate of 187.433 mph. What was his time (to the nearest thousandth of an hour)? (*Source: USA Today.*)

40. In 2013, Ryan Newman drove his Chevrolet to victory in the Brickyard 400 (mile) race at a rate of 153.485 mph. What was his time (to the nearest thousandth of an hour)? (*Source: World Almanac and Book of Facts.*)

*In Exercises 41–44, find the rate on the basis of the information provided. Round answers to the nearest hundredth. All events were at the 2012 Olympics. (Source: World Almanac and Book of Facts.) **See Example 6.***

	Event	Participant	Distance	Time
41.	200-m run, women	Allyson Felix, USA	200 m	21.88 sec
42.	400-m run, women	Sanya Richards-Ross, USA	400 m	49.55 sec
43.	110-m hurdles, men	Aries Merritt, USA	110 m	12.92 sec
44.	200-m run, men	Usain Bolt, Jamaica	200 m	19.32 sec

*Solve each problem. **See Examples 7 and 8.***

▶ 45. From a point on a straight road, Marco and Celeste ride bicycles in the same direction. Marco rides at 10 mph and Celeste rides at 12 mph. In how many hours will they be 15 mi apart?

46. At a given hour, two steamboats leave a city in the same direction on a straight canal. One travels at 18 mph and the other travels at 24 mph. In how many hours will the boats be 9 mi apart?

47. Atlanta and Cincinnati are 440 mi apart. John leaves Cincinnati, driving toward Atlanta at an average rate of 60 mph. Pat leaves Atlanta at the same time, driving toward Cincinnati in her antique auto, averaging 28 mph. How long will it take them to meet?

	r	t	d
John	60	t	$60t$
Pat	28	t	$28t$

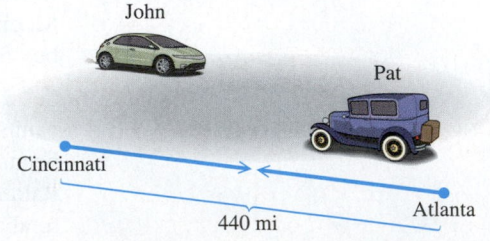

48. St. Louis and Portland are 2060 mi apart. A small plane leaves Portland, traveling toward St. Louis at an average rate of 90 mph. Another plane leaves St. Louis at the same time, traveling toward Portland and averaging 116 mph. How long will it take them to meet?

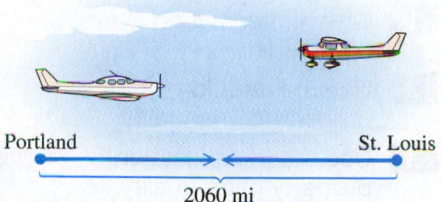

	r	t	d
Plane Leaving Portland	90	t	90t
Plane Leaving St. Louis	116	t	116t

Portland   St. Louis

2060 mi

49. A train leaves Kansas City, Kansas, and travels north at 85 km per hr. Another train leaves at the same time and travels south at 95 km per hour. How long will it take before they are 315 km apart?

50. Two steamers leave a port on a river at the same time, traveling in opposite directions. Each is traveling at 22 mph. How long will it take for them to be 110 mi apart?

51. Two planes leave an airport at the same time, one flying east, the other flying west. The eastbound plane travels 150 mph slower. They are 2250 mi apart after 3 hr. Find the rate of each plane.

52. Two trains leave a city at the same time. One travels north, and the other travels south 20 mph faster. In 2 hr, the trains are 280 mi apart. Find their rates.

	r	t	d
Eastbound	x − 150	3	
Westbound	x	3	

	r	t	d
Northbound	x	2	
Southbound	x + 20	2	

53. Two cars start from towns 400 mi apart and travel toward each other. They meet after 4 hr. Find the rate of each car if one travels 20 mph faster than the other.

54. Two cars leave towns 230 km apart at the same time, traveling directly toward one another. One car travels 15 km per hr slower than the other. They pass one another 2 hr later. What are their rates?

Extending Skills   *Solve each problem.*

55. Kevin is three times as old as Bob. Three years ago the sum of their ages was 22 yr. How old is each now? (*Hint:* Write an expression first for the age of each now and then for the age of each three years ago.)

56. A store has 39 qt of milk, some in pint cartons and some in quart cartons. There are six times as many quart cartons as pint cartons. How many quart cartons are there? (*Hint:* 1 qt = 2 pt)

57. A table is three times as long as it is wide. If it were 3 ft shorter and 3 ft wider, it would be square (with all sides equal). How long and how wide is the table?

58. Elena works for $8 an hour. A total of 25% of her salary is deducted for taxes and insurance. How many hours must she work to take home $450?

59. Paula received a paycheck for $585 for her weekly wages less 10% deductions. How much was she paid before the deductions were made?

60. At the end of a day, the owner of a gift shop had $2394 in the cash register. This amount included sales tax of 5% on all sales. Find the amount of the sales.

2.8 Solving Linear Inequalities

VOCABULARY

☐ inequality
☐ linear inequality in one variable
☐ interval
☐ three-part inequality

An **inequality** relates algebraic expressions using the symbols

< "is less than,"	≤ "is less than or equal to,"
> "is greater than,"	≥ "is greater than or equal to."

Linear Inequality in One Variable

A **linear inequality in one variable** (here x) can be written in the form

$$Ax + B < C, \quad Ax + B \leq C, \quad Ax + B > C, \quad \text{or} \quad Ax + B \geq C,$$

where A, B, and C represent real numbers and $A \neq 0$.

Examples: $x + 5 < 2$, $z - \dfrac{3}{4} \geq 5$,   and $2k + 5 \leq 10$ Linear inequalities

---

We solve a linear inequality by finding all real number solutions of it. For example, the solution set $\{x \mid x \leq 2\}$ includes *all real numbers* that are less than or equal to 2, not just the *integers* less than or equal to 2.

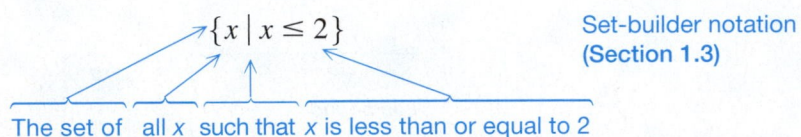

Set-builder notation (Section 1.3)

The set of   all x such that x is less than or equal to 2

OBJECTIVE 1   Graph intervals on a number line.

Graphing is a good way to show the solution set of an inequality. To graph all real numbers belonging to the set

$$\{x \mid x \leq 2\},$$

we place a square bracket at 2 on a number line and draw an arrow extending from the bracket to the left (because all numbers *less than* 2 are also part of the graph). See **FIGURE 20**.

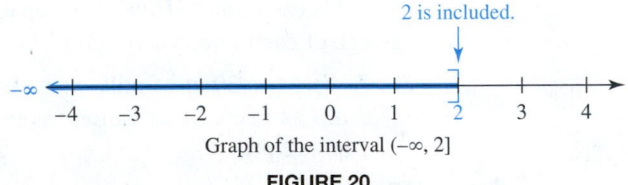

Graph of the interval $(-\infty, 2]$

FIGURE 20

The set of numbers less than or equal to 2 is an example of an **interval** on a number line. We can write this interval using **interval notation** as follows.

$$(-\infty, 2] \quad \text{Interval notation}$$

The **negative infinity symbol** $-\infty$ does not indicate a number, but shows that the interval includes *all* real numbers less than 2. Again, the square bracket indicates that 2 is part of the solution. Intervals that continue indefinitely in the positive direction are written with the **positive infinity symbol** ∞.

NOW TRY EXERCISE 1

Write each inequality in interval notation, and graph the interval.

(a) $x < -1$ (b) $-2 \leq x$

EXAMPLE 1 Graphing Intervals on a Number Line

Write each inequality in interval notation, and graph the interval.

(a) $x > -5$

The statement $x > -5$ says that x can represent any number greater than -5 but cannot equal -5. The interval is written $(-5, \infty)$. We graph this interval by placing a parenthesis at -5 and drawing an arrow to the right, as in **FIGURE 21**. The parenthesis at -5 indicates that -5 is *not* part of the graph.

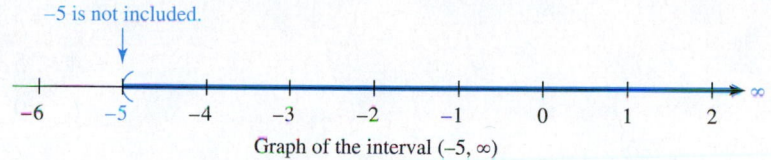

Graph of the interval $(-5, \infty)$

FIGURE 21

(b) $3 > x$

The statement $3 > x$ means the same as $x < 3$. *The inequality symbol continues to point toward the lesser number.* The graph of $x < 3$, written in interval notation as $(-\infty, 3)$, is shown in **FIGURE 22**.

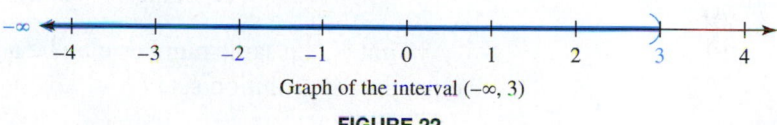

Graph of the interval $(-\infty, 3)$

FIGURE 22

CHECK To confirm that the interval in **FIGURE 22** is graphed in the proper direction, select a value that is part of the graph and substitute it into the given inequality $3 > x$. For example, we select 0 and substitute to obtain $3 > 0$, a true statement. ✓

NOW TRY

Important Concepts Regarding Interval Notation

1. A parenthesis indicates that an endpoint is *not included* in a solution set.
2. A bracket indicates that an endpoint is *included* in a solution set.
3. A parenthesis is *always* used next to an infinity symbol, $-\infty$ or ∞.
4. The set of all real numbers is written in interval notation as $(-\infty, \infty)$.

▼ **Methods of Expressing Solution Sets of Linear Inequalities**

Set-Builder Notation	Interval Notation	Graph
$\{x \mid x < a\}$	$(-\infty, a)$	
$\{x \mid x \leq a\}$	$(-\infty, a]$	
$\{x \mid x > a\}$	(a, ∞)	
$\{x \mid x \geq a\}$	$[a, \infty)$	
$\{x \mid x$ is a real number$\}$	$(-\infty, \infty)$	

NOW TRY ANSWERS
1. (a) $(-\infty, -1)$
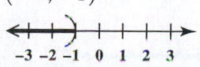
(b) $[-2, \infty)$

> **NOTE**   Some texts use a solid circle ● rather than a square bracket to indicate that an endpoint is included in a number line graph. An open circle ○ is used to indicate noninclusion, rather than a parenthesis.

OBJECTIVE 2 Use the addition property of inequality.

Consider the true inequality $2 < 5$. Add 4 to each side.

$$2 < 5$$
$$2 + 4 < 5 + 4 \qquad \text{Add 4.}$$
$$6 < 9 \qquad \text{True}$$

The result is a true statement. This suggests the **addition property of inequality.**

Addition Property of Inequality

If A, B, and C represent real numbers, then the inequalities

$$A < B \quad \text{and} \quad A + C < B + C \quad \text{are equivalent.}^*$$

That is, the same number may be added to each side of an inequality without changing the solution set.

*This also applies to $A \leq B$, $A > B$, and $A \geq B$.

Consider the inequality $2 < 5$ again. This time subtract 4 from each side.

$$2 < 5$$
$$2 - 4 < 5 - 4 \qquad \text{Subtract 4.}$$
$$-2 < 1 \qquad \text{True}$$

Again, a true statement results. *As with the addition property of equality, the same number may be **subtracted** from each side of an inequality.*

NOW TRY
EXERCISE 2
Solve the inequality, and graph the solution set.

$$5 + 5x \geq 4x + 3$$

EXAMPLE 2 Using the Addition Property of Inequality

Solve $7 + 3x \geq 2x - 5$, and graph the solution set.

$$7 + 3x \geq 2x - 5 \qquad \text{As with equations, our goal is to isolate } x.$$
$$7 + 3x - 2x \geq 2x - 5 - 2x \qquad \text{Subtract } 2x.$$
$$7 + x \geq -5 \qquad \text{Combine like terms.}$$
$$7 + x - 7 \geq -5 - 7 \qquad \text{Subtract 7.}$$
$$x \geq -12 \qquad \text{Combine like terms.}$$

The solution set is $[-12, \infty)$. Its graph is shown in **FIGURE 23**.

NOW TRY ANSWER
2. $[-2, \infty)$

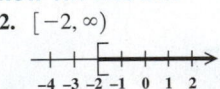

```
    -13 -12 -11 -10 -9  -8  -7  -6  -5  -4  -3  -2  -1   0
```

FIGURE 23

NOW TRY ↻

NOTE Because an inequality has many solutions, we cannot check all of them by substitution as we did with the single solution of an equation. To check the solutions in the interval $[-12, \infty)$ in **Example 2,** we first substitute -12 for x in the related *equation*.

CHECK
$$7 + 3x = 2x - 5 \qquad \text{Related equation}$$
$$7 + 3(-12) \stackrel{?}{=} 2(-12) - 5 \qquad \text{Let } x = -12.$$
$$7 - 36 \stackrel{?}{=} -24 - 5 \qquad \text{Multiply.}$$
$$-29 = -29 \ \checkmark \qquad \text{True}$$

A true statement results, so -12 is indeed the "boundary" point. Next we test a number other than -12 from the interval $[-12, \infty)$. We choose 0.

CHECK
$$7 + 3x \geq 2x - 5 \qquad \text{Original inequality}$$
$$7 + 3(0) \stackrel{?}{\geq} 2(0) - 5 \qquad \text{Let } x = 0.$$

0 is easy to substitute.
$$7 + 0 \stackrel{?}{\geq} 0 - 5 \qquad \text{Multiply.}$$
$$7 \geq -5 \ \checkmark \qquad \text{True}$$

Again, a true statement results. The checks confirm that solutions to the inequality are in the interval $[-12, \infty)$. Any number "outside" the interval $[-12, \infty)$—that is, any number in $(-\infty, -12)$—will give a false statement when tested. (Try this.)

OBJECTIVE 3 Use the multiplication property of inequality.

Consider the true inequality $3 < 7$. Multiply each side by the positive number 2.

$$3 < 7$$
$$2(3) < 2(7) \qquad \text{Multiply each side by 2.}$$
$$6 < 14 \qquad \text{True}$$

The result is a true statement. Now multiply each side of $3 < 7$ by the negative number -5.

$$3 < 7$$
$$-5(3) < -5(7) \qquad \text{Multiply each side by } -5.$$
$$-15 < -35 \qquad \text{False}$$

To obtain a true statement when multiplying each side by -5, *we must reverse the direction of the inequality symbol.*

$$3 < 7$$
$$-5(3) > -5(7) \qquad \text{Multiply by } -5. \text{ Reverse the direction of the symbol.}$$
$$-15 > -35 \qquad \text{True}$$

NOTE The above illustrations began with the inequality $3 < 7$, a true statement involving two positive numbers. Similar results occur when one or both of the numbers is negative. Verify this by multiplying each of the following inequalities first by 2 and then by -5.

$$-3 < 7, \quad 3 > -7, \quad \text{and} \quad -7 < -3$$

These observations suggest the **multiplication property of inequality.**

Multiplication Property of Inequality

Let A, B, and C represent real numbers, where $C \neq 0$.

1. If C is *positive,* then the inequalities

$$A < B \quad \text{and} \quad AC < BC \quad \text{are equivalent.*}$$

2. If C is *negative,* then the inequalities

$$A < B \quad \text{and} \quad AC > BC \quad \text{are equivalent.*}$$

That is, each side of an inequality may be multiplied by the same positive number without changing the direction of the inequality symbol. *If the multiplier is negative, we must reverse the direction of the inequality symbol.*

*This also applies to $A \leq B$, $A > B$, and $A \geq B$.

As with the multiplication property of equality, the same nonzero number may be divided into each side of an inequality.

Note the following differences for positive and negative numbers.

1. When each side of an inequality is multiplied or divided by a *positive number,* the direction of the inequality symbol *does not change.*

2. *Reverse the direction of the inequality symbol ONLY when multiplying or dividing each side of an inequality by a NEGATIVE NUMBER.*

EXAMPLE 3   Using the Multiplication Property of Inequality

Solve each inequality, and graph the solution set.

(a) $3x < -18$

We divide each side by 3, a positive number, so the direction of the inequality symbol *does not* change. *(It does not matter that the number on the right side of the inequality is negative.)*

$$3x < -18$$

> 3 is *positive.* Do NOT reverse the direction of the symbol.

$$\frac{3x}{3} < \frac{-18}{3} \qquad \text{Divide by 3.}$$

$$x < -6$$

The solution set is $(-\infty, -6)$. The graph is shown in **FIGURE 24**.

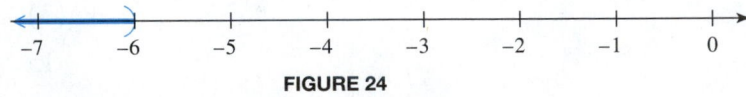

FIGURE 24

(b) $-4x \geq 8$

Here, each side of the inequality must be divided by -4, a negative number, which *does* require changing the direction of the inequality symbol.

$$-4x \geq 8$$

> To avoid errors, show the division as a separate step.

> -4 is *negative.* Change $\geq$ to $\leq$.

$$\frac{-4x}{-4} \leq \frac{8}{-4} \qquad \text{Divide by } -4. \text{ Reverse the symbol.}$$

$$x \leq -2$$

NOW TRY
EXERCISE 3
Solve the inequality, and graph the solution set.

$$-5k \geq 15$$

The solution set $(-\infty, -2]$ is graphed in **FIGURE 25**.

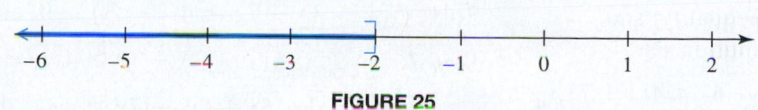

FIGURE 25

NOW TRY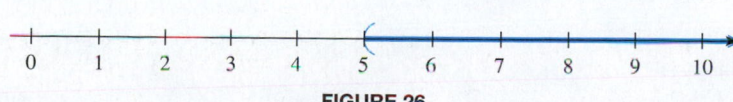

OBJECTIVE 4 Solve linear inequalities using both properties of inequality.

> **Solving a Linear Inequality in One Variable**
>
> **Step 1**   **Simplify each side separately.** Use the distributive property as needed.
> - Clear any parentheses.
> - Clear any fractions or decimals.
> - Combine like terms.
>
> **Step 2**   **Isolate the variable terms on one side.** Use the addition property of inequality so that all terms with variables are on one side of the inequality and all constants (numbers) are on the other side.
>
> **Step 3**   **Isolate the variable.** Use the multiplication property of inequality to obtain an inequality in one of the following forms, where k is a constant (number).
>
> $$\text{variable} < k, \quad \text{variable} \leq k, \quad \text{variable} > k, \quad \text{or} \quad \text{variable} \geq k$$
>
> *Remember: Reverse the direction of the inequality symbol only when multiplying or dividing each side of an inequality by a negative number.*

NOW TRY
EXERCISE 4
Solve the inequality, and graph the solution set.

$$6 - 2t + 5t \leq 8t - 4$$

EXAMPLE 4   Solving a Linear Inequality

Solve $3x + 2 - 5 > -x + 7 + 2x$, and graph the solution set.

Step 1   Combine like terms and simplify.

$$3x + 2 - 5 > -x + 7 + 2x$$
$$3x - 3 > x + 7$$

Step 2   Use the addition property of inequality.

$$3x - 3 - x > x + 7 - x \qquad \text{Subtract } x.$$
$$2x - 3 > 7 \qquad\qquad \text{Combine like terms.}$$
$$2x - 3 + 3 > 7 + 3 \qquad \text{Add 3.}$$
$$2x > 10 \qquad\qquad \text{Combine like terms.}$$

Step 3   Use the multiplication property of inequality.

Because 2 is positive, keep the symbol >.

$$\frac{2x}{2} > \frac{10}{2} \qquad \text{Divide by 2.}$$
$$x > 5$$

NOW TRY ANSWERS
3. $(-\infty, -3]$

4. $[2, \infty)$

The solution set is $(5, \infty)$. Its graph is shown in **FIGURE 26**.

FIGURE 26

NOW TRY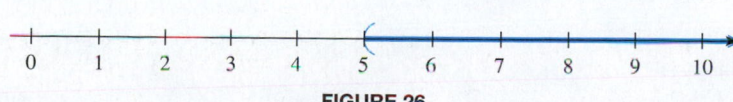

NOW TRY
EXERCISE 5

Solve the inequality, and graph the solution set.

$$2x - 3(x - 6) < 4(x + 7)$$

EXAMPLE 5   Solving a Linear Inequality

Solve $5(x - 3) - 7x \geq 4(x - 3) + 9$, and graph the solution set.

Step 1	$5(x - 3) - 7x \geq 4(x - 3) + 9$	
	$5x - 15 - 7x \geq 4x - 12 + 9$	Distributive property
	$-2x - 15 \geq 4x - 3$	Combine like terms.
Step 2	$-2x - 15 - 4x \geq 4x - 3 - 4x$	Subtract $4x$.
	$-6x - 15 \geq -3$	Combine like terms.
	$-6x - 15 + 15 \geq -3 + 15$	Add 15.
	$-6x \geq 12$	Combine like terms.
Step 3	$\dfrac{-6x}{-6} \leq \dfrac{12}{-6}$	Divide by -6. Reverse the symbol.

Because -6 is negative, change $\geq$ to $\leq$.

$$x \leq -2$$

The solution set is $(-\infty, -2\,]$. Its graph is shown in **FIGURE 27**.

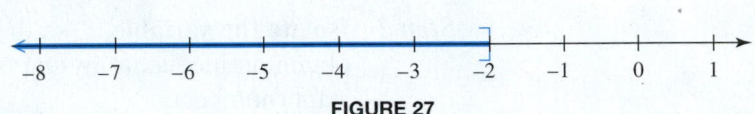

FIGURE 27

NOW TRY

NOW TRY
EXERCISE 6

Solve the inequality, and graph the solution set.

$$\frac{1}{8}(x + 4) \geq \frac{1}{6}(2x + 8)$$

EXAMPLE 6   Solving a Linear Inequality with Fractions

Solve $\frac{3}{4}(x - 6) < \frac{2}{3}(5x + 1)$, and graph the solution set.

Step 1	$\dfrac{3}{4}(x - 6) < \dfrac{2}{3}(5x + 1)$	Clear the parentheses first. Then clear the fractions.
	$\dfrac{3}{4}x - \dfrac{9}{2} < \dfrac{10}{3}x + \dfrac{2}{3}$	Distributive property
	$12\left(\dfrac{3}{4}x - \dfrac{9}{2}\right) < 12\left(\dfrac{10}{3}x + \dfrac{2}{3}\right)$	Multiply each side by the LCD, 12.
	$9x - 54 < 40x + 8$	Distributive property
Step 2	$9x - 54 - 40x < 40x + 8 - 40x$	Subtract $40x$.
	$-31x - 54 < 8$	Combine like terms.
	$-31x - 54 + 54 < 8 + 54$	Add 54.
	$-31x < 62$	Combine like terms.
Step 3	$\dfrac{-31x}{-31} > \dfrac{62}{-31}$	Divide by -31. Reverse the symbol.
	$x > -2$	

The solution set is $(-2, \infty)$. Its graph is shown in **FIGURE 28**.

NOW TRY ANSWERS

5. $(-2, \infty)$

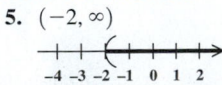

6. $(-\infty, -4\,]$

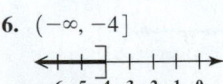

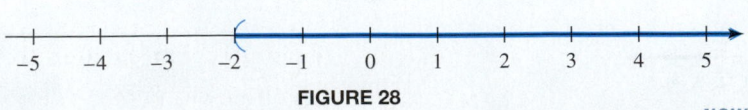

FIGURE 28

NOW TRY

OBJECTIVE 5 Solve applied problems using inequalities.

▼ Words and Phrases That Indicate Inequality

Word or Phrase	Example	Inequality
Is more than	A number *is more than* 4	$x > 4$
Is less than	A number *is less than* −12	$x < -12$
Exceeds	A number *exceeds* 3.5	$x > 3.5$
Is at least	A number *is at least* 6	$x \geq 6$
Is at most	A number *is at most* 8	$x \leq 8$

! CAUTION Do not confuse statements such as "5 is more than a number" with the phrase "5 more than a number." The first of these is expressed as $5 > x$, while the second is expressed as $x + 5$, or $5 + x$.

The next example uses the idea of finding the average of a number of scores. *To find the average of n numbers, add the numbers and divide by n.* We use the six problem-solving steps from **Section 2.4**, changing Step 3 to "Write an inequality."

NOW TRY EXERCISE 7

Will has grades of 98 and 85 on his first two tests in algebra. If he wants an average of at least 90 after his third test, what score must he make on that test?

EXAMPLE 7 Finding an Average Test Score

John has grades of 86, 88, and 78 on his first three tests in geometry. If he wants an average of at least 80 after his fourth test, what are the possible scores he can make on that test?

Step 1 **Read** the problem again.

Step 2 **Assign a variable.** Let x = John's score on his fourth test.

Step 3 **Write an inequality.**

$$\underset{\text{Average}}{\downarrow} \qquad \underset{\substack{\text{is at} \\ \text{least 80.}}}{\downarrow \downarrow}$$

$$\frac{86 + 88 + 78 + x}{4} \geq 80 \qquad \text{To find his average after four tests, add the test scores and divide by 4.}$$

Step 4 **Solve.**

$$\frac{252 + x}{4} \geq 80 \qquad \text{Add in the numerator.}$$

$$4\left(\frac{252 + x}{4}\right) \geq 4(80) \qquad \text{Multiply by 4.}$$

$$252 + x \geq 320$$

$$252 + x - 252 \geq 320 - 252 \qquad \text{Subtract 252.}$$

$$x \geq 68 \qquad \text{Combine like terms.}$$

Step 5 **State the answer.** He must score 68 or more on the fourth test to have an average of *at least* 80.

Step 6 **Check.** $\dfrac{86 + 88 + 78 + 68}{4} = \dfrac{320}{4} = 80$

To complete the check, also show that any number greater than 68 (but less than or equal to 100) makes the average greater than 80.

NOW TRY ANSWER
7. 87 or more

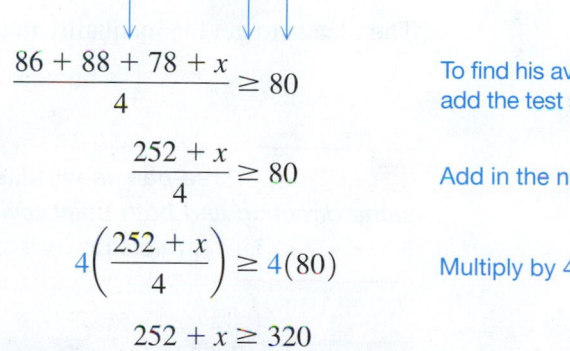

NOW TRY

> ⚠ **CAUTION**   In applied problems, remember that
>
> is at least   translates as   is greater than or equal to
>
> and   is at most   translates as   is less than or equal to.

OBJECTIVE 6 Solve linear inequalities with three parts.

An inequality that says that one number is *between* two other numbers is a **three-part inequality.** For example,

$$-3 < 5 < 7 \quad \text{says that} \quad 5 \text{ is between } -3 \text{ and } 7.$$

NOW TRY
EXERCISE 8

Write the inequality in interval notation, and graph the interval.

$$0 \leq x < 2$$

EXAMPLE 8 Graphing a Three-Part Inequality

Write the inequality in interval notation, and graph the interval.

$$-1 \leq x < 3$$

The statement is read "-1 is less than or equal to x *and* x is less than 3." We want the set of numbers *between* -1 and 3, with -1 included and 3 excluded. In interval notation, we write $[-1, 3)$, using a square bracket at -1 because -1 is part of the graph and a parenthesis at 3 because 3 is not part of the graph. See **FIGURE 29**.

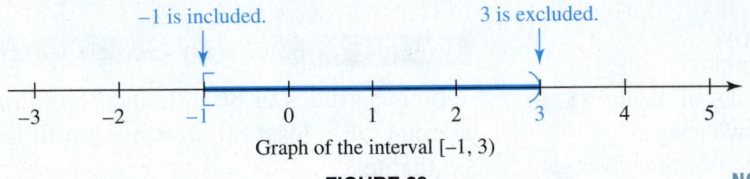

−1 is included.   3 is excluded.

Graph of the interval [−1, 3)

FIGURE 29   **NOW TRY** ↻

The three-part inequality $3 < x + 2 < 8$ says that $x + 2$ is between 3 and 8. We solve this inequality as follows.

$$3 - 2 < x + 2 - 2 < 8 - 2 \qquad \text{Subtract 2 from } each \text{ part.}$$
$$1 < \quad x \quad < 6$$

The idea is to get the inequality in the form

$$\text{a number} < x < \text{another number.}$$

> ⚠ **CAUTION**   *Three-part inequalities are written so that the symbols point in the same direction and both point toward the lesser number.* It would be *wrong* to write $8 < x + 2 < 3$, which would imply that $8 < 3$, a false statement.

EXAMPLE 9 Solving Three-Part Inequalities

Solve each inequality, and graph the solution set.

(a) $4 < \quad 3x - 5 \quad \leq 10$ ⟵ Work with all three parts at the same time.

$$4 + 5 < 3x - 5 + 5 \leq 10 + 5 \qquad \text{Add 5 to each part.}$$
$$9 < \quad 3x \quad \leq 15 \qquad \text{Combine like terms.}$$

Remember to divide all *three* parts by 3.

$$\frac{9}{3} < \quad \frac{3x}{3} \quad \leq \frac{15}{3} \qquad \text{Divide each part by 3.}$$
$$3 < \quad x \quad \leq 5$$

NOW TRY ANSWER
8. [0, 2)

−3 −2 −1 0 1 2 3

**NOW TRY
EXERCISE 9**

Solve the inequality, and graph the solution set.

$$-4 \leq \frac{3}{2}x - 1 \leq 0$$

The solution set is $(3, 5]$. Its graph is shown in **FIGURE 30**.

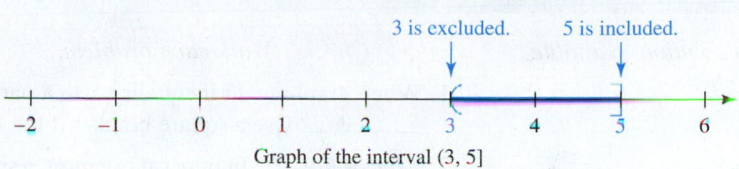

3 is excluded. 5 is included.

Graph of the interval $(3, 5]$

FIGURE 30

(b)

$$-4 \leq \frac{2}{3}m - 1 < 8$$

$3(-4) \leq 3\left(\frac{2}{3}m - 1\right) < 3(8)$	Multiply each part by 3 to clear the fraction.
$-12 \leq 2m - 3 < 24$	Distributive property
$-12 + 3 \leq 2m - 3 + 3 < 24 + 3$	Add 3 to each part.
$-9 \leq 2m < 27$	Combine like terms.
$\dfrac{-9}{2} \leq \dfrac{2m}{2} < \dfrac{27}{2}$	Divide each part by 2.
$-\dfrac{9}{2} \leq m < \dfrac{27}{2}$	

The solution set is $\left[-\dfrac{9}{2}, \dfrac{27}{2}\right)$. Its graph is shown in **FIGURE 31**.

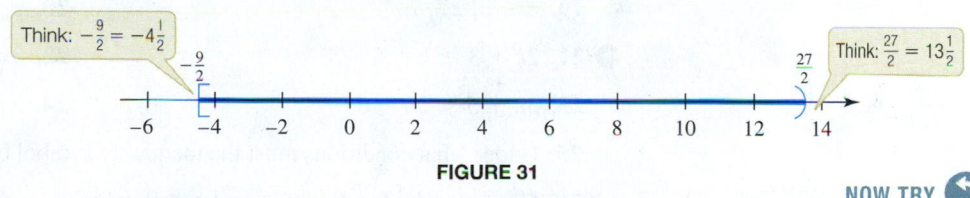

Think: $-\dfrac{9}{2} = -4\dfrac{1}{2}$

Think: $\dfrac{27}{2} = 13\dfrac{1}{2}$

FIGURE 31

NOW TRY

⚠ **CAUTION** Be especially careful of whether to use parentheses or square brackets when writing and graphing solution sets of three-part inequalities. The following table illustrates the four possibilities that may occur.

▼ **Methods of Expressing Solution Sets of Three-Part Inequalities**

Set-Builder Notation	Interval Notation	Graph
$\{x \mid a < x < b\}$	(a, b)	
$\{x \mid a < x \leq b\}$	$(a, b]$	
$\{x \mid a \leq x < b\}$	$[a, b)$	
$\{x \mid a \leq x \leq b\}$	$[a, b]$	

NOW TRY ANSWER

9. $\left[-2, \dfrac{2}{3}\right]$

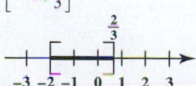

2.8 Exercises

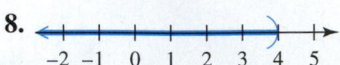

▶ *Complete solution available in MyMathLab*

Concept Check   Work each problem.

1. When graphing an inequality, use a parenthesis if the inequality symbol is _____ or _____ . Use a square bracket if the inequality symbol is _____ or _____ .

2. *True* or *false?*   In interval notation, a square bracket is sometimes used next to an infinity symbol.

3. In interval notation, the set $\{x \mid x > 0\}$ is written _____ .

4. In interval notation, the set of all real numbers is written _____ .

Concept Check   Write an inequality involving the variable x that describes each set of numbers graphed.

5.
```
   ←──(─┼──┼──┼──┼──┼──┼──→
     -4 -3 -2 -1  0  1  2  3
```

6.
```
   ←──[─┼──┼──┼──┼──┼──┼──┼──→
     -4 -3 -2 -1  0  1  2  3  4
```

7.
```
   ←──┼──┼──┼──┼──┼──┼──]──→
     -2 -1  0  1  2  3  4  5
```

8.
```
   ←──┼──┼──┼──┼──┼──┼──)─┼──→
     -2 -1  0  1  2  3  4  5
```

Write each inequality in interval notation, and graph the interval. **See Example 1.**

▶ **9.** $k \le 4$ **10.** $x \le 3$ **11.** $x < -3$ **12.** $r < -11$

13. $t > 4$ **14.** $m > 5$ **15.** $0 \ge x$

16. $1 \ge x$ **17.** $-\dfrac{1}{2} \le x$ **18.** $-\dfrac{3}{4} \le x$

Solve each inequality. Write the solution set in interval notation, and graph it. **See Example 2.**

19. $z - 8 \ge -7$ **20.** $p - 3 \ge -11$

▶ **21.** $2k + 3 \ge k + 8$ **22.** $3x + 7 \ge 2x + 11$

23. $3n + 5 < 2n - 6$ **24.** $5x - 2 < 4x - 5$

25. Under what conditions must the inequality symbol be reversed when solving an inequality?

26. *Concept Check*   If $p < q$ and $r < 0$, which one of the following statements is false?

    A. $pr < qr$ **B.** $pr > qr$ **C.** $p + r < q + r$ **D.** $p - r < q - r$

Solve each inequality. Write the solution set in interval notation, and graph it. **See Example 3.**

27. $3x < 18$ **28.** $5x < 35$ **29.** $2y \ge -20$

30. $6m \ge -24$ ▶ **31.** $-8t > 24$ **32.** $-7x > 49$

33. $-x \ge 0$ **34.** $-k < 0$ **35.** $-\dfrac{3}{4}r < -15$

36. $-\dfrac{7}{8}t < -14$ **37.** $-0.02x \le 0.06$ **38.** $-0.03v \ge -0.12$

Solve each inequality. Write the solution set in interval notation, and graph it. **See Examples 4–6.**

39. $8x + 9 \le -15$ **40.** $6x + 7 \le -17$

41. $-4x - 3 < 1$ **42.** $-5x - 4 < 6$

43. $5r + 1 \ge 3r - 9$ **44.** $6t + 3 < 3t + 12$

45. $5x - 2 \leq -x + 10$

46. $3x - 9 \geq -2x + 6$

47. $-7x + 4 > -3x - 2$

48. $-8x + 1 < -4x + 11$

▶ **49.** $6x + 3 + x < 2 + 4x + 4$

50. $-4w + 12 + 9w \geq w + 9 + w$

51. $-x + 4 + 7x \leq -2 + 3x + 6$

52. $14x - 6 + 7x > 4 + 10x - 10$

53. $5(t - 1) > 3(t - 2)$

54. $7(m - 2) < 4(m - 4)$

▶ **55.** $5(x + 3) - 6x \leq 3(2x + 1) - 4x$

56. $2(x - 5) + 3x < 4(x - 6) + 1$

57. $\dfrac{1}{3}(5x - 4) \geq \dfrac{2}{5}(x + 3)$

58. $\dfrac{5}{12}(5x - 7) < \dfrac{5}{6}(x - 5)$

59. $\dfrac{2}{3}(p + 3) > \dfrac{5}{6}(p - 4)$

60. $\dfrac{7}{9}(x - 4) \leq \dfrac{4}{3}(x + 5)$

61. $\dfrac{4}{5}x - \dfrac{1}{2}(x + 3) \leq \dfrac{3}{10}$

62. $\dfrac{1}{6}x + \dfrac{1}{3}(x - 1) > \dfrac{1}{2}$

63. $4x - (6x + 1) \leq 8x + 2(x - 3)$

64. $2z - (4z + 3) > 6z + 3(z + 4)$

65. $5(2k + 3) - 2(k - 8) > 3(2k + 4) + k - 2$

66. $2(3z - 5) + 4(z + 6) \geq 2(3z + 2) + 3z - 15$

Concept Check *Translate each statement into an inequality. Use x as the variable.*

67. You must be at least 18 yr old to vote.

68. Less than 1 in. of rain fell.

69. Chicago received more than 5 in. of snow.

70. A full-time student must take at least 12 credits.

71. Tracy could spend at most $20 on a gift.

72. The car's speed exceeded 60 mph.

Solve each problem. See Example 7.

▶ **73.** Christy has scores of 76 and 81 on her first two algebra tests. If she wants an average of at least 80 after her third test, what possible scores can she make on that test?

74. Joseph has scores of 96 and 86 on his first two geometry tests. What possible scores can he make on his third test so that his average is at least 90?

75. The average monthly precipitation in Houston, TX, for October, November, and December is 4.6 in. If 5.7 in. falls in October and 4.3 in. falls in November, how many inches must fall in December so that the average monthly precipitation for these months exceeds 4.6 in.? (*Source:* National Climatic Data Center.)

76. The average monthly precipitation in New Orleans, LA, for June, July, and August is 6.7 in. If 8.1 in. falls in June and 5.7 in. falls in July, how many inches must fall in August so that the average monthly precipitation for these months exceeds 6.7 in.? (*Source:* National Climatic Data Center.)

77. When 2 is added to the difference between six times a number and 5, the result is greater than 13 added to five times the number. Find all such numbers.

78. When 8 is subtracted from the sum of three times a number and 6, the result is less than 4 more than the number. Find all such numbers.

79. The formula for converting Fahrenheit temperature to Celsius is

$$C = \frac{5}{9}(F - 32).$$

If the Celsius temperature on a certain winter day in Minneapolis is never less than $-25°$, how would we describe the corresponding Fahrenheit temperatures? (*Source:* National Climatic Data Center.)

80. The formula for converting Celsius temperature to Fahrenheit is

$$F = \frac{9}{5}C + 32.$$

The Fahrenheit temperature of Phoenix has never exceeded $122°$. How would we describe this using Celsius temperature? (*Source:* National Climatic Data Center.)

81. For what values of x would the rectangle have a perimeter of at least 400?

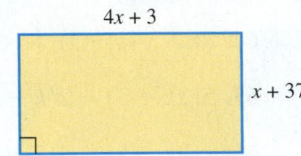

82. For what values of x would the triangle have a perimeter of at least 72?

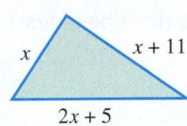

83. An international phone call costs \$2.00, plus \$0.30 per minute or fractional part of a minute. If x represents the number of minutes of the length of the call, then $2 + 0.30x$ represents the cost of the call. If Alan has \$5.60 to spend on a call, what is the maximum total time he can use the phone?

84. At the Speedy Gas'n Go, a car wash costs \$3.00 and gasoline is selling for \$3.60 per gallon. Carla has \$48.00 to spend, and her car is so dirty that she must have it washed. What is the maximum number of gallons of gasoline that she can purchase?

A company that produces DVDs has found that revenue from sales of DVDs is \$5 per DVD, less sales costs of \$100. Production costs are \$125, plus \$4 per DVD. Profit (P) is given by revenue (R) less cost (C), so the company must find the production level x that makes

$$P > 0, \quad \text{that is,} \quad R - C > 0. \qquad P = R - C$$

85. Write an expression for revenue R, letting x represent the production level (number of DVDs to be produced).

86. Write an expression for production costs C in terms of x.

87. Write an expression for profit P, and then solve the inequality $P > 0$.

88. Describe the solution in terms of the problem.

Concept Check Write a three-part inequality involving the variable x that describes each set of numbers graphed.

89.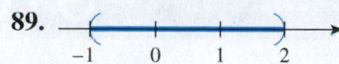

90.

91.

92.

Write each inequality in interval notation, and graph the interval. **See Example 8.**

93. $8 \leq x \leq 10$

94. $3 \leq x \leq 5$

95. $0 < y \leq 10$

96. $-3 \leq x < 0$

97. $4 > x > -3$

98. $6 \geq x \geq -4$

Solve each inequality. Write the solution set in interval notation, and graph it. **See Example 9.**

99. $-8 < 4x \leq 4$

100. $-3 \leq 3x < 12$

▶ **101.** $-5 \leq 2x - 3 \leq 9$

102. $-7 \leq 3x - 4 \leq 8$

103. $10 < 7p + 3 < 24$

104. $-8 \leq 3r - 1 \leq -1$

105. $-4 < -2x < 12$

106. $9 < -3x < 15$

107. $5 < 1 - 6m < 12$

108. $-1 \leq 1 - 5q \leq 16$

109. $6 \leq 3(x - 1) < 18$

110. $-4 < 2(x + 1) \leq 6$

111. $-12 \leq \frac{1}{2}z + 1 \leq 4$

112. $-6 \leq \frac{1}{3}x + 3 \leq 5$

113. $1 \leq 3 + \frac{2}{3}p \leq 7$

114. $2 < 6 + \frac{3}{4}x < 12$

115. $-7 \leq \frac{5}{4}r - 1 \leq -1$

116. $-12 \leq \frac{3}{7}x + 2 \leq -4$

RELATING CONCEPTS For Individual or Group Work (Exercises 117–120)

Work Exercises 117–120 in order, to see the connection between the solution of an equation and the solutions of the corresponding inequalities.

117. Solve the following equation, and graph the solution set on a number line.

$$3x + 2 = 14$$

118. Solve the following inequality, and graph the solution set on a number line.

$$3x + 2 > 14$$

119. Solve the following inequality, and graph the solution set on a number line.

$$3x + 2 < 14$$

120. If we were to graph all the solution sets from **Exercises 117–119** on the same number line, describe the graph that we would obtain. (This is the **union** of all the solution sets.)

Taking Math Tests

Techniques to Improve Your Test Score	Comments
Come prepared with a pencil, eraser, paper, and calculator, if allowed.	Working in pencil lets you erase, keeping your work neat.
Scan the entire test, note the point values of different problems, and plan your time accordingly.	To do 20 problems in 50 minutes, allow $50 \div 20 = 2.5$ minutes per problem. Spend less time on the easier problems.
Do a "knowledge dump" when you get the test. Write important notes, such as formulas, in a corner of the test.	Writing down tips and information that you've learned at the beginning allows you to relax later.
Read directions carefully, and circle any significant words. When you finish a problem, reread the directions. Did you do what was asked?	Pay attention to any announcements written on the board or made by your instructor. Ask if you don't understand something.
Show all your work. Many teachers give partial credit if some steps are correct, even if the final answer is wrong. *Write neatly.*	If your teacher can't read your writing, you won't get credit for it. If you need more space to work, ask to use extra paper.
Write down anything that might help solve a problem: a formula, a diagram, etc. If necessary, circle the problem and come back to it later. Do *not* erase anything you wrote down.	If you know even a little bit about a problem, write it down. The answer may come to you as you work on it, or you may get partial credit. Don't spend too long on any one problem.
If you can't solve a problem, make a guess. Do not change it unless you find an obvious mistake.	Have a good reason for changing an answer. Your first guess is usually your best bet.
Check that the answer to an application problem is reasonable and makes sense. Reread the problem to make sure you've answered the question.	Use common sense. Can the father really be seven years old? Would a month's rent be $32,140? Remember to label your answer if needed: $, years, inches, etc.
Check for careless errors. Rework each problem without looking at your previous work. Then compare the two answers.	Reworking a problem from the beginning forces you to rethink it. If possible, use a different method to solve the problem.

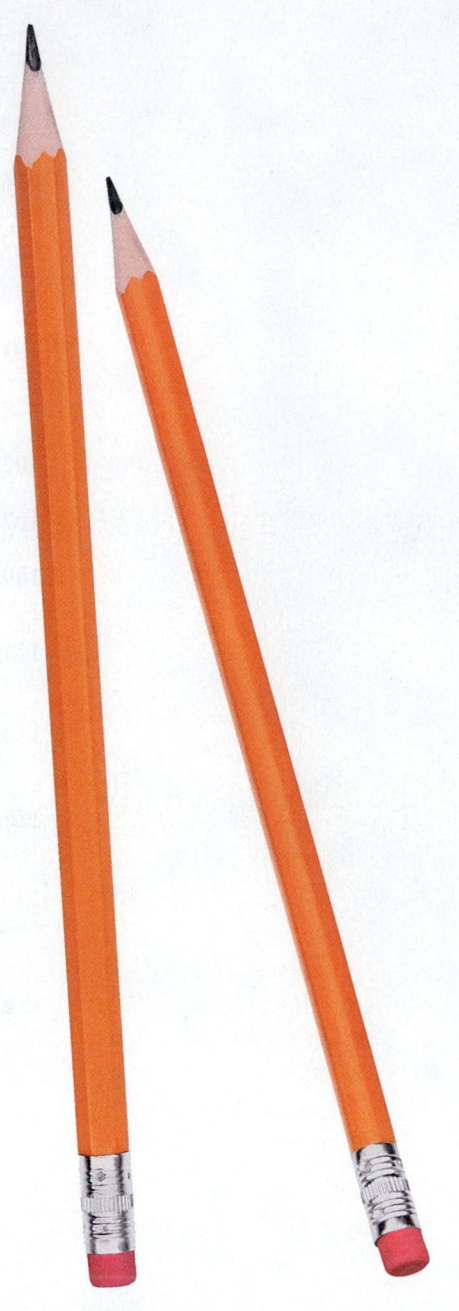

Think through and answer each question.

1. What two or three tips will you try when you take your next math test?

2. How did the tips you selected work for you when you took your math test?

3. What will you do differently when taking your next math test?

Chapter 2 Summary

Key Terms

2.1
equation
linear equation in one
  variable
solution
solution set
equivalent equations

contradiction
empty (null) set

2.4
consecutive integers
consecutive even
  (or odd) integers
degree
complementary angles
right angle
supplementary angles
straight angle

2.3
conditional equation
identity

2.5
formula
area
perimeter
vertical angles

2.6
ratio
proportion
terms of a proportion
extremes
means

cross products of
  a proportion
percent
percentage
base

2.8
inequality
linear inequality in
  one variable
interval
three-part inequality

New Symbols

$\varnothing$ empty set
$1°$ one degree
$\lceil$ right angle

a **to** b, $a : b$, or $\dfrac{a}{b}$
ratio of a to b

∞ infinity
$-\infty$ negative infinity
$(-\infty, \infty)$ set of all real
  numbers

(a, b) interval notation
  for $a < x < b$
$[a, b]$ interval notation
  for $a \le x \le b$

Test Your Word Power

See how well you have learned the vocabulary in this chapter.

1. A **solution set** is the set of numbers that
   A. make an expression undefined
   B. make an equation false
   C. make an equation true
   D. make an expression equal to 0.

2. **Complementary angles** are angles
   A. formed by two parallel lines
   B. whose sum is 90°
   C. whose sum is 180°
   D. formed by perpendicular lines.

3. **Supplementary angles** are angles
   A. formed by two parallel lines
   B. whose sum is 90°
   C. whose sum is 180°
   D. formed by perpendicular lines.

4. A **ratio**
   A. compares two quantities using a quotient
   B. says that two quotients are equal
   C. is a product of two quantities
   D. is a difference of two quantities.

5. A **proportion**
   A. compares two quantities using a quotient
   B. says that two quotients are equal
   C. is a product of two quantities
   D. is a difference of two quantities.

6. An **inequality** is
   A. a statement that two algebraic expressions are equal
   B. a point on a number line
   C. an equation with no solutions
   D. a statement that relates algebraic expressions using $<, \le, >,$ or $\ge$.

ANSWERS

1. C; *Example:* {8} is the solution set of $2x + 5 = 21$. **2.** B; *Example:* Angles with measures 35° and 55° are complementary angles.
3. C; *Example:* Angles with measures 112° and 68° are supplementary angles. **4.** A; *Example:* $\frac{7 \text{ in.}}{12 \text{ in.}}$, or $\frac{7}{12}$ **5.** B; *Example:* $\frac{2}{3} = \frac{8}{12}$
6. D; *Examples:* $x < 5,\ 7 + 2y \ge 11,\ -5 < 2z - 1 \le 3$

Quick Review

CONCEPTS	EXAMPLES

2.1 The Addition Property of Equality

The same number may be added to (or subtracted from) each side of an equation without changing the solution set.

Solve.
$$x - 6 = 12$$
$$x - 6 + 6 = 12 + 6 \qquad \text{Add 6.}$$
$$x = 18 \qquad \text{Combine like terms.}$$

Solution set: $\{18\}$

2.2 The Multiplication Property of Equality

Each side of an equation may be multiplied (or divided) by the same nonzero number without changing the solution set.

Solve.
$$\frac{3}{4}x = -9$$
$$\frac{4}{3} \cdot \frac{3}{4}x = \frac{4}{3} \cdot (-9) \qquad \text{Multiply by } \tfrac{4}{3} \text{, the reciprocal of } \tfrac{3}{4}.$$
$$x = -12$$

Solution set: $\{-12\}$

2.3 More on Solving Linear Equations

Step 1 Simplify each side separately.
- Clear any parentheses.
- Clear any fractions or decimals.
- Combine like terms.

Step 2 Isolate the variable terms on one side.

Step 3 Isolate the variable.

Step 4 Check.

Solve.
$$2x + 2(x + 1) = 14 + x$$
$$2x + 2x + 2 = 14 + x \qquad \text{Distributive property}$$
$$4x + 2 = 14 + x \qquad \text{Combine like terms.}$$
$$4x + 2 - x - 2 = 14 + x - x - 2 \qquad \text{Subtract } x. \text{ Subtract 2.}$$
$$3x = 12 \qquad \text{Combine like terms.}$$
$$\frac{3x}{3} = \frac{12}{3} \qquad \text{Divide by 3.}$$
$$x = 4$$

CHECK $\quad 2(4) + 2(4 + 1) \stackrel{?}{=} 14 + 4 \qquad$ Let $x = 4$.
$$18 = 18 \ \checkmark \qquad \text{True}$$

Solution set: $\{4\}$

2.4 Applications of Linear Equations

Step 1 Read.

Step 2 Assign a variable.

Step 3 Write an equation.

Step 4 Solve the equation.

Step 5 State the answer.

Step 6 Check.

One number is five more than another. Their sum is 21. What are the numbers?

Let $\quad x =$ the lesser number.
Then $\quad x + 5 =$ the greater number.

$$x + (x + 5) = 21$$
$$2x + 5 = 21 \qquad \text{Combine like terms.}$$
$$2x = 16 \qquad \text{Subtract 5.}$$
$$x = 8 \qquad \text{Divide by 2.}$$

The numbers are 8 and 13.

13 is five more than 8, and $8 + 13 = 21$. The answer checks.

CONCEPTS	EXAMPLES

2.5 Formulas and Additional Applications from Geometry

To find the value of one of the variables in a formula, given values for the others, substitute the known values into the formula.

Find L if $\mathcal{A} = LW$, given that $\mathcal{A} = 24$ and $W = 3$.

$$\mathcal{A} = LW$$

$$24 = L \cdot 3 \qquad \mathcal{A} = 24, W = 3$$

$$\frac{24}{3} = \frac{L \cdot 3}{3} \qquad \text{Divide by 3.}$$

$$8 = L$$

To solve a formula for one of the variables, isolate that variable by treating the other variables as numbers and using the steps for solving equations.

Solve $P = 2a + 2b$ for b.

$$P = 2a + 2b$$

$$P - 2a = 2a + 2b - 2a \qquad \text{Subtract } 2a.$$

$$P - 2a = 2b \qquad \text{Combine like terms.}$$

$$\frac{P - 2a}{2} = \frac{2b}{2} \qquad \text{Divide by 2.}$$

$$\frac{P - 2a}{2} = b, \quad \text{or} \quad b = \frac{P - 2a}{2}$$

2.6 Ratio, Proportion, and Percent

To write a ratio, express quantities in the same units.

4 ft to 8 in. can be written 48 in. to 8 in., which is the ratio

$$\frac{48}{8}, \quad \text{or} \quad \frac{6}{1}.$$

To solve a proportion, use the method of cross products.

Solve.

$$\frac{x}{12} = \frac{35}{60}$$

$$60x = 12 \cdot 35 \qquad \text{Cross products}$$

$$60x = 420 \qquad \text{Multiply.}$$

$$x = 7 \qquad \text{Divide by 60.}$$

Solution set: $\{7\}$

To solve a percent problem, use the percent equation.

amount = percent (as a decimal) · base

65   is   what percent   of   325?

$$65 = p \cdot 325$$

$$\frac{65}{325} = p$$

$$0.2 = p, \quad \text{or} \quad 20\% = p$$

65 is 20% of 325.

CONCEPTS	EXAMPLES

2.7 Further Applications of Linear Equations

Step 1   Read.

Step 2   Assign a variable. Make a table and/or draw a sketch to help solve the problem.

The three forms of the formula relating distance, rate, and time are

$$d = rt, \quad r = \frac{d}{t}, \quad \text{and} \quad t = \frac{d}{r}.$$

Two cars leave from the same point, traveling in opposite directions. One travels at 45 mph and the other at 60 mph. How long will it take them to be 210 mi apart?

Let t = time it takes for the two cars to be 210 mi apart.

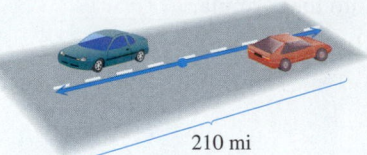

210 mi

	Rate	Time	Distance
One Car	45	t	$45t$
Other Car	60	t	$60t$

The sum of the distances is 210 mi.

Step 3   Write an equation.

Step 4   Solve the equation.

$$45t + 60t = 210$$
$$105t = 210 \quad \text{Combine like terms.}$$
$$t = 2 \quad \text{Divide by 105.}$$

Steps 5 and 6   State the answer and check the solution.

It will take them 2 hr to be 210 mi apart.

2.8 Solving Linear Inequalities

Step 1   Simplify each side separately.

- Clear any parentheses.
- Clear any fractions or decimals.
- Combine like terms.

Step 2   Isolate the variable terms on one side.

Step 3   Isolate the variable.

Be sure to reverse the direction of the inequality symbol when multiplying or dividing by a negative number.

Solve the inequality, and graph the solution set.

$$3(1 - x) + 5 - 2x > 9 - 6$$
$$3 - 3x + 5 - 2x > 9 - 6 \quad \text{Distributive property}$$
$$8 - 5x > 3 \quad \text{Combine like terms.}$$
$$8 - 5x - 8 > 3 - 8 \quad \text{Subtract 8.}$$
$$-5x > -5 \quad \text{Combine like terms.}$$
$$\frac{-5x}{-5} < \frac{-5}{-5} \quad \begin{array}{l}\text{Divide by } -5.\\ \text{Change} > \text{to} <.\end{array}$$
$$x < 1$$

Solution set: $(-\infty, 1)$

To solve a three-part inequality such as

$$4 < 2x + 6 \leq 8,$$

work with all three parts at the same time.

Solve the inequality, and graph the solution set.

$$4 < \quad 2x + 6 \quad \leq 8$$
$$4 - 6 < \quad 2x + 6 - 6 \quad \leq 8 - 6 \quad \text{Subtract 6.}$$
$$-2 < \quad 2x \quad \leq 2 \quad \text{Combine like terms.}$$
$$\frac{-2}{2} < \quad \frac{2x}{2} \quad \leq \frac{2}{2} \quad \text{Divide by 2.}$$
$$-1 < \quad x \quad \leq 1$$

Solution set: $(-1, 1]$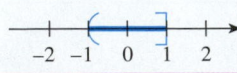

Chapter 2	Review Exercises

Answers (left column):

1. $\{6\}$
2. $\{-12\}$
3. $\{7\}$
4. $\left\{\dfrac{2}{3}\right\}$
5. $\{11\}$
6. $\{17\}$
7. $\{5\}$
8. $\{-4\}$
9. $\{5\}$
10. $\{-12\}$
11. $\left\{\dfrac{64}{5}\right\}$
12. $\{4\}$
13. {all real numbers}
14. $\{-19\}$
15. {all real numbers}
16. $\{20\}$
17. $\varnothing$
18. $\{-1\}$
19. $-\dfrac{7}{2}$
20. Democrats: 71; Republicans: 47
21. Hawaii: 6425 mi^2; Rhode Island: 1212 mi^2
22. Seven Falls: 300 ft; Twin Falls: 120 ft
23. $80°$
24. 11, 13
25. $h = 11$
26. $\mathcal{A} = 28$
27. $r = 4.75$
28. $V = 904.32$
29. $h = \dfrac{\mathcal{A}}{b}$
30. $h = \dfrac{2\mathcal{A}}{b + B}$

For Exercises 31 and 32, there are other correct forms.

31. $y = -x + 11$
32. $y = \dfrac{3}{2}x - 6$
33. $135°$; $45°$
34. $100°$; $100°$

2.1–2.3 *Solve each equation.*

1. $x - 5 = 1$
2. $x + 8 = -4$
3. $3t + 1 = 2t + 8$
4. $5z = 4z + \dfrac{2}{3}$
5. $(4r - 2) - (3r + 1) = 8$
6. $3(2x - 5) = 2 + 5x$
7. $7x = 35$
8. $12r = -48$
9. $2p - 7p + 8p = 15$
10. $\dfrac{x}{12} = -1$
11. $\dfrac{5}{8}q = 8$
12. $12m + 11 = 59$
13. $3(2x + 6) - 5(x + 8) = x - 22$
14. $5x + 9 - (2x - 3) = 2x - 7$
15. $\dfrac{1}{2}r - \dfrac{r}{3} = \dfrac{r}{6}$
16. $0.1(x + 80) + 0.2x = 14$
17. $3x - (-2x + 6) = 4(x - 4) + x$
18. $\dfrac{1}{2}(x + 3) - \dfrac{2}{3}(x - 2) = 3$

2.4 *Solve each problem.*

19. If 7 is added to five times a number, the result is equal to three times the number. Find the number.

20. In 2013, Illinois had 118 members in its House of Representatives, consisting of only Democrats and Republicans. There were 24 more Democrats than Republicans. How many representatives from each party were there? (*Source:* www.ilga.gov)

21. The land area of Hawaii is 5213 mi^2 greater than the area of Rhode Island. Together, the areas total 7637 mi^2. What is the area of each of the two states?

22. The height of Seven Falls in Colorado is $\dfrac{5}{2}$ the height of Twin Falls in Idaho. The sum of the heights is 420 ft. Find the height of each. (*Source: World Almanac and Book of Facts.*)

23. The supplement of an angle measures 10 times the measure of its complement. What is the measure of the angle?

24. Find two consecutive odd integers such that when the lesser is added to twice the greater, the result is 24 more than the greater integer.

2.5 *A formula is given along with the values of all but one of the variables. Find the value of the variable that is not given. Use 3.14 as an approximation for π.*

25. $\mathcal{A} = \dfrac{1}{2}bh$; $\mathcal{A} = 44, b = 8$
26. $\mathcal{A} = \dfrac{1}{2}h(b + B)$; $h = 8, b = 3, B = 4$
27. $C = 2\pi r$; $C = 29.83$
28. $V = \dfrac{4}{3}\pi r^3$; $r = 6$

Solve each formula for the specified variable.

29. $\mathcal{A} = bh$ for h
30. $\mathcal{A} = \dfrac{1}{2}h(b + B)$ for h

Solve each equation for y.

31. $x + y = 11$
32. $3x - 2y = 12$

Find the measure of each marked angle.

33. $(8x - 1)°$ $(3x - 6)°$

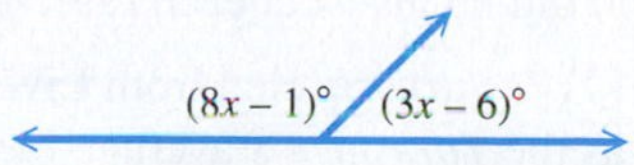

34.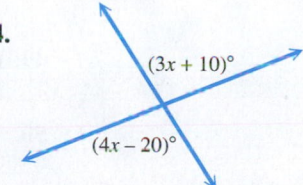
$(3x + 10)°$ $(4x - 20)°$

Solve each problem.

35. 2 cm   **36.** 42.2°; 92.8°

37. $\dfrac{3}{2}$ **38.** $\dfrac{3}{4}$

39. $\left\{\dfrac{7}{2}\right\}$ **40.** $\left\{-\dfrac{8}{3}\right\}$

41. $3.06   **42.** 375 km

43. 8 gold medals

44. 18 oz; $0.249

45. 175%   **46.** 2500

47. 3.75 L

48. $5000 at 5%; $5000 at 3%

49. 8.2 mph   **50.** $2\dfrac{1}{2}$ hr

35. The perimeter of a certain rectangle is 16 times the width. The length is 12 cm more than the width. Find the width of the rectangle.

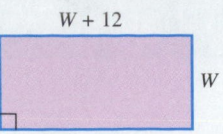

36. A baseball diamond is a square with a side of 90 ft. The pitcher's mound is located 60.5 ft from home plate, as shown in the figure. Find the measures of the angles marked in the figure. (*Hint:* Recall that the sum of the measures of the angles of any triangle is 180°.)

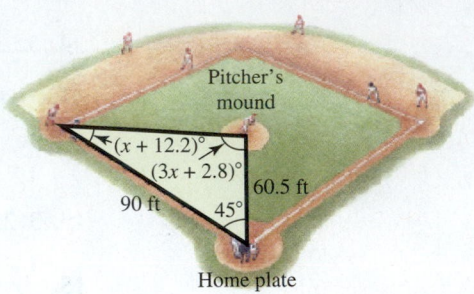

2.6   *Write a ratio for each word phrase. Express fractions in lowest terms.*

37. 60 cm to 40 cm

38. 90 in. to 10 ft

Solve each equation.

39. $\dfrac{p}{21} = \dfrac{5}{30}$

40. $\dfrac{5+x}{3} = \dfrac{2-x}{6}$

Solve each problem.

41. The tax on a $24.00 item is $2.04. How much tax would be paid on a $36.00 item?

42. The distance between two cities on a road map is 32 cm. The two cities are actually 150 km apart. The distance on the map between two other cities is 80 cm. How far apart are these cities?

43. In the 2012 Summer Olympics in London, England, Italian athletes earned 28 medals. Two of every 7 medals were gold. How many gold medals did Italy earn? (*Source: World Almanac and Book of Facts.*)

44. Find the best buy. Give the unit price to the nearest thousandth for that size. (*Source: Jewel-Osco.*)

Cereal

Size	Price
9 oz	$ 3.49
14 oz	$ 3.99
18 oz	$ 4.49

45. What percent of 12 is 21?

46. 36% of what number is 900?

2.7   *Solve each problem.*

47. A nurse must mix 15 L of a 10% solution of a drug with some 60% solution to obtain a 20% mixture. How many liters of the 60% solution will be needed?

48. Robert invested $10,000, from which he earns an annual income of $400 per year. He invested part of the $10,000 at 5% annual interest and the remainder in bonds paying 3% interest. How much did he invest at each rate?

49. In 1846, the vessel *Yorkshire* traveled from Liverpool to New York, a distance of 3150 mi, in 384 hr. What was the *Yorkshire's* average rate? Round the answer to the nearest tenth.

50. Two planes leave St. Louis at the same time. One flies north at 350 mph and the other flies south at 420 mph. In how many hours will they be 1925 mi apart?

51. $[-4, \infty)$ ⊢⊢⊢⊢⊢⊢⊢⊢⊢→
          -4   0

52. $(-\infty, 7)$ ←⊢⊢⊢⊢⊢⊢⊢⊢→
           0   7

53. $[-5, 6)$ ⊢⊢⊢⊢⊢⊢⊢⊢⊢→
          -5   0   6

54. B

55. $[-3, \infty)$ ⊢⊢⊢⊢⊢⊢⊢⊢→
          -3   0

56. $(-\infty, 2)$ ←⊢⊢⊢⊢⊢⊢⊢→
           0   2

57. $[3, \infty)$ ⊢⊢⊢⊢⊢⊢⊢⊢→
          0   3

58. $[46, \infty)$ ⊢⊢⊢⊢⊢⊢⊢⊢→
          0 10   40 46

59. $(-\infty, -5)$ ←⊢⊢⊢⊢⊢⊢→
            -5   0

60. $(-\infty, -4)$ ←⊢⊢⊢⊢⊢→
            -4   0

61. $\left[-2, \frac{3}{2}\right]$ ⊢⊢⊢⊢⊢⊢→
            -2   0 1 2

62. $\left(\frac{4}{3}, 5\right]$ ⊢⊢⊢⊢⊢⊢→
            0 1 2   5

63. 88 or more

64. all numbers less than or equal to $-\frac{1}{3}$

2.8 *Write each inequality in interval notation, and graph the interval.*

51. $x \geq -4$ **52.** $x < 7$ **53.** $-5 \leq x < 6$

54. Which inequality requires reversing the inequality symbol when it is solved?

     A. $4x \geq -36$ **B.** $-4x \leq 36$ **C.** $4x < 36$ **D.** $4x > 36$

Solve each inequality. Write the solution set in interval notation, and graph it.

55. $x + 6 \geq 3$ **56.** $5x < 4x + 2$

57. $-6x \leq -18$ **58.** $8(x - 5) - (2 + 7x) \geq 4$

59. $4x - 3x > 10 - 4x + 7x$ **60.** $3(2x + 5) + 4(8 + 3x) < 5(3x + 7)$

61. $-3 \leq 2x + 1 \leq 4$ **62.** $9 < 3x + 5 \leq 20$

Solve each problem.

63. Awilda has grades of 94 and 88 on her first two calculus tests. What possible scores on a third test will give her an average of at least 90?

64. If nine times a number is added to 6, the result is at most 3. Find all such numbers.

Chapter 2   **Mixed Review Exercises**

Solve.

1. {7}   **2.** $r = \dfrac{I}{pt}$

3. $(-\infty, 2)$ **4.** {-9}

5. {70}   **6.** $\left\{\dfrac{13}{4}\right\}$

7. ∅

8. {all real numbers}

9. 4000 calories

10. DiGiorno: $668.7 million;
    Red Baron: $268.8 million

11. 160 oz, $0.062

12. 24°, 66°

1. $\dfrac{x}{7} = \dfrac{x - 5}{2}$ **2.** $I = prt$ for r

3. $-2x > -4$ **4.** $2k - 5 = 4k + 13$

5. $0.05x + 0.02x = 4.9$ **6.** $2 - 3(x - 5) = 4 + x$

7. $9x - (7x + 2) = 3x + (2 - x)$ **8.** $\dfrac{1}{3}s + \dfrac{1}{2}s + 7 = \dfrac{5}{6}s + 5 + 2$

9. Athletes in vigorous training programs can eat 50 calories per day for every 2.2 lb of body weight. To the nearest hundred, how many calories can a 175-lb athlete consume per day? (*Source: The Gazette.*)

10. In a recent year, the top-selling frozen pizza brands, DiGiorno and Red Baron, together had sales of $937.5 million. Red Baron's sales were $399.9 million less than DiGiorno's. What were sales in millions for each brand? (*Source:* www.aibonline.org)

11. Find the best buy. Give the unit price to the nearest thousandth for that size. (*Source:* Jewel-Osco.)

Laundry Detergent

Size	Price
50 oz	$3.99
100 oz	$7.29
160 oz	$9.99

12. Find the measure of each marked angle.

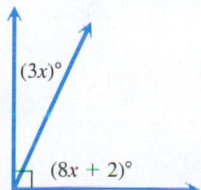

13. 13 hr
14. faster train: 80 mph;
slower train: 50 mph
15. 44 m
16. 50 m or less

13. Janet drove from Louisville to Dallas, a distance of 819 mi, averaging 63 mph. What was her driving time?

14. Two trains are 390 mi apart. They start at the same time and travel toward one another, meeting 3 hr later. If the rate of one train is 30 mph more than the rate of the other train, find the rate of each train.

15. The perimeter of a triangle is 96 m. One side is twice as long as another, and the third side is 30 m long. What is the length of the longest side?

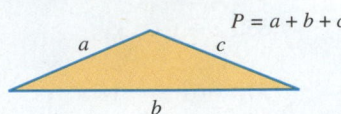

$$P = a + b + c$$

16. The perimeter of a certain square cannot be greater than 200 m. Find the possible values for the length of a side.

Chapter 2   Test

FOR EXTRA HELP *Step-by-step test solutions are found on the Chapter Test Prep Videos available in* MyMathLab® *or on* YouTube.

▶ *View the complete solutions to all Chapter Test exercises in MyMathLab.*

[2.1–2.3]
1. {−6}   **2.** {21}
3. ∅   **4.** {30}
5. {all real numbers}
6. $\left\{\dfrac{13}{4}\right\}$

[2.4]
7. wins: 97; losses: 65
8. Hawaii: 4021 mi²;
Maui: 728 mi²;
Kauai: 551 mi²
9. 50°   **10.** 24, 26

[2.5]
11. (a) $W = \dfrac{P - 2L}{2}$
(b) 18
12. $y = \dfrac{5}{4}x - 2$
(There are other correct forms.)
13. 75°, 75°

Solve each equation.

1. $5x + 9 = 7x + 21$

2. $-\dfrac{4}{7}x = -12$

3. $7 - (x - 4) = -3x + 2(x + 1)$

4. $0.06(x + 20) + 0.08(x - 10) = 4.6$

5. $-8(2x + 4) = -4(4x + 8)$

6. $2 - 3(x - 5) = 3 + (x + 1)$

Solve each problem.

7. In the 2013 baseball season, the St. Louis Cardinals won 33 less than twice as many games as they lost. They played 162 regular-season games. How many wins and losses did the Cardinals have? (*Source:* www.MLB.com)

8. Three islands in the Hawaiian island chain are Hawaii (the Big Island), Maui, and Kauai. Together, their areas total 5300 mi². The island of Hawaii is 3293 mi² larger than the island of Maui, and Maui is 177 mi² larger than Kauai. What is the area of each island?

9. Find the measure of an angle if its supplement measures 10° more than three times its complement.

10. If the lesser of two consecutive even integers is tripled, the result is 20 more than twice the greater integer. Find the two integers.

11. The formula for the perimeter of a rectangle is $P = 2L + 2W$.

(a) Solve for W.

(b) If $P = 116$ and $L = 40$, find the value of W.

12. Solve the equation $5x - 4y = 8$ for y.

13. Find the measure of each marked angle.

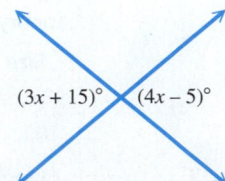

$(3x + 15)°$ $(4x - 5)°$

[2.6]

14. {6}

15. {−29}

16. 40%

17. 16 oz; $0.249

18. 2300 mi

[2.7]

19. $8000 at 3%; $14,000 at 4.5%

20. 4 hr

[2.8]

21. (a) $x < 0$

(b) $-2 < x \le 3$

22. $(-\infty, 11)$

23. $(-2, 6]$

24. $(-\infty, 4]$

25. 83 or more

Solve each equation.

14. $\dfrac{z}{8} = \dfrac{12}{16}$

15. $\dfrac{x+5}{3} = \dfrac{x-3}{4}$

Solve each problem.

16. What percent of 65 is 26?

17. Find the best buy. Give the unit price to the nearest thousandth for that size. (*Source:* Jewel-Osco.)

Cheese Slices

Size	Price
8 oz	$ 2.99
16 oz	$ 3.99
48 oz	$14.69

18. The distance between Milwaukee and Boston is 1050 mi. On a certain map, this distance is represented by 42 in. On the same map, Seattle and Cincinnati are 92 in. apart. What is the actual distance between Seattle and Cincinnati?

19. Carlos invested some money at 3% simple interest and $6000 more than that amount at 4.5% simple interest. After 1 yr, his total interest from the two accounts was $870. How much did he invest at each rate?

20. Two cars leave from the same point, traveling in opposite directions. One travels at a constant rate of 50 mph, while the other travels at a constant rate of 65 mph. How long will it take for them to be 460 mi apart?

21. Write an inequality involving x that describes the numbers graphed.

(a)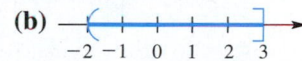

(b)

Solve each inequality. Write the solution set in interval notation, and graph it.

22. $-3x > -33$

23. $-10 < 3x - 4 \le 14$

24. $-4x + 2(x - 3) \ge 4x - (3 + 5x) - 7$

25. Susan has grades of 76 and 81 on her first two algebra tests. If she wants an average of at least 80 after her third test, what score must she make on that test?

Chapters R–2 — Cumulative Review Exercises

[R.1]

1. $\dfrac{37}{60}$

2. $\dfrac{48}{5}$

[R.2]

3. 34.03

[1.2]

4. $\dfrac{1}{2}x - 18$

5. $\dfrac{6}{x + 12} = 2$

[1.3]

6. true

[1.4, 1.5]

7. −8

8. 28

9. 0

Perform each indicated operation.

1. $\dfrac{5}{6} + \dfrac{1}{4} - \dfrac{7}{15}$

2. $\dfrac{9}{8} \cdot \dfrac{16}{3} \div \dfrac{5}{8}$

3. $4.8 + 12.5 + 16.73$

Translate from words to symbols. Use x as the variable.

4. The difference of half a number and 18

5. The quotient of 6 and 12 more than a number is 2.

6. *True* or *false?* $\dfrac{8(7) - 5(6 + 2)}{3 \cdot 5 + 1} \ge 1$

Perform each indicated operation.

7. $\dfrac{-4(9)(-2)}{-3^2}$

8. $(-7 - 1)(-4) + (-4)$

9. $\dfrac{6(-4) - (-2)(12)}{3^2 + 7^2}$

[1.2–1.5]

10. $-\dfrac{19}{3}$

[1.6]
11. distributive property
12. inverse property
13. identity property

[2.1–2.3]
14. $\{-1\}$
15. $\{-1\}$
16. $\{-12\}$

[2.6]
17. $\{26\}$

[2.5]
18. $y = -\dfrac{3}{4}x + 6$

(There are other correct forms.)

19. $c = P - a - b - B$

[2.8]
20. $(-\infty, 1]$

21. $(-1, 2]$

[2.4]
22. 4 cm; 9 cm; 27 cm

[2.5]
23. 12.42 cm

[2.7]
24. slower car: 40 mph;
faster car: 60 mph

[R.2]
25. (a) 532,000
(b) 504,000
(c) 336,000

---

10. Find the value of $\dfrac{3x^2 - y^3}{-4z}$ for $x = -2$, $y = -4$, and $z = 3$.

Name the property illustrated by each equation.

11. $7(p + q) = 7p + 7q$ **12.** $7 + (-7) = 0$ **13.** $3.5(1) = 3.5$

Solve each equation, and check the solution.

14. $2r - 6 = 8r$ **15.** $4 - 5(s + 2) = 3(s + 1) - 1$

16. $\dfrac{2}{3}x + \dfrac{3}{4}x = -17$ **17.** $\dfrac{2x + 3}{5} = \dfrac{x - 4}{2}$

18. Solve $3x + 4y = 24$ for y.   **19.** Solve $P = a + b + c + B$ for c.

Solve each inequality. Write the solution set in interval notation, and graph it.

20. $6(r - 1) + 2(3r - 5) \le -4$ **21.** $-18 \le -9z < 9$

Solve each problem.

22. A 40-cm piece of yarn must be cut into three pieces. The longest piece is to be three times as long as the middle-sized piece, and the shortest piece is to be 5 cm shorter than the middle-sized piece. Find the length of each piece.

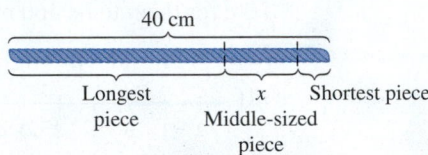

23. A fully inflated professional basketball has a circumference of 78 cm. What is the radius of a circular cross section through the center of the ball? (Use 3.14 as an approximation for π.) Round the answer to the nearest hundredth.

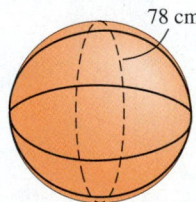

24. Two cars are 400 mi apart. Both start at the same time and travel toward one another. They meet 4 hr later. If the rate of one car is 20 mph faster than the other, what is the rate of each car?

25. The graph shows the breakdown of the colors chosen for new 2012 model-year compact/sports cars sold in the United States. If approximately 2.8 million of these cars were sold, about how many were each color? (*Source:* Ward's Auto Group.)

   (a) White   (b) Silver   (c) Red

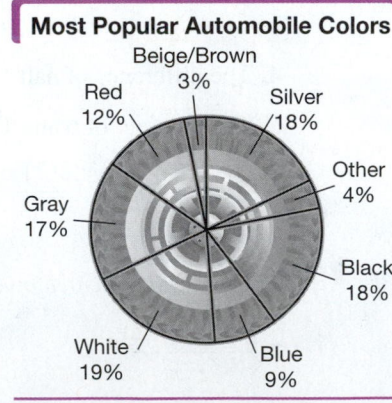

Most Popular Automobile Colors

Source: DuPont Automotive Products.

3

Linear Equations and Inequalities in Two Variables

We determine location on a map using *coordinates,* a concept that is based on a *rectangular coordinate system,* one of the topics of this chapter.

3.1 Linear Equations and Rectangular Coordinates

VOCABULARY

☐ line graph
☐ linear equation in two variables
☐ ordered pair
☐ table of values
☐ x-axis
☐ y-axis
☐ origin
☐ rectangular (Cartesian) coordinate system
☐ quadrant
☐ plane
☐ coordinates
☐ plot
☐ scatter diagram

OBJECTIVE 1   Interpret graphs.

A line graph is used to show changes or trends in data over time. To form a **line graph,** we connect a series of points representing data with line segments.

EXAMPLE 1   Interpreting a Line Graph

The line graph in **FIGURE 1** shows average prices of a gallon of regular unleaded gasoline in the United States for the years 2005 through 2012.

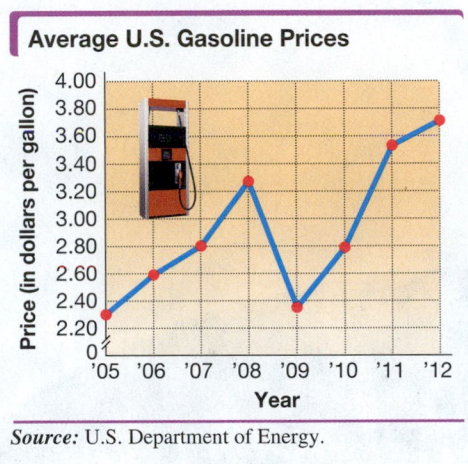

Average U.S. Gasoline Prices

Source: U.S. Department of Energy.

FIGURE 1

(a) Between which years did the average price of a gallon of gasoline decrease?

The line between 2008 and 2009 *falls* from left to right, so the average price of a gallon of gasoline decreased from 2008 to 2009.

(b) What was the general trend in the average price of a gallon of gasoline from 2009 through 2012?

The line graph *rises* from left to right from 2009 to 2012, so the average price of a gallon of gasoline increased over those years.

(c) Estimate the average price of a gallon of gasoline in 2009 and 2012. About how much did the price increase between 2009 and 2012?

Move up from 2009 on the horizontal scale to the point plotted for 2009. This point is about three-fourths of the way between the lines on the vertical scale for $2.20 and $2.40—that is, about $2.35 per gallon.

Locate the point plotted for 2012. Moving across to the vertical scale, this point is about halfway between the lines for $3.60 and $3.80—that is, about $3.70 per gallon.

Between 2009 and 2012, the average price increased about

NOW TRY EXERCISE 1

Refer to the line graph in **FIGURE 1**.

(a) Estimate the average price of a gallon of gasoline in 2010.

(b) About how much did the average price of a gallon of gasoline increase from 2010 to 2012?

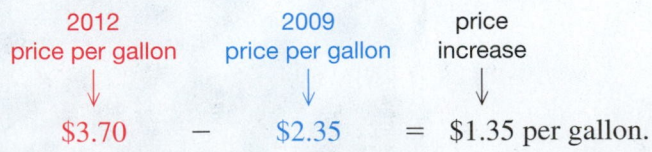

2012	2009	price
price per gallon	price per gallon	increase
↓	↓	↓
$3.70 −	$2.35 =	$1.35 per gallon.

NOW TRY ANSWERS
1. (a) about $2.80
   (b) about $0.90

NOW TRY

Year	Average Price (in dollars per gallon)
2005	2.30
2006	2.59
2007	2.80
2008	3.27
2009	2.35
2010	2.79
2011	3.53
2012	3.71

Source: U.S. Department of Energy.

The line graph in **FIGURE 1** relates years to average prices for a gallon of gasoline. We can also represent these two related quantities using a table of data, as shown in the margin. In table form, we can see more precise data rather than estimating it. Trends in the data are easier to see from the graph, which gives a "picture" of the data.

We can extend these ideas to the subject of this chapter, *linear equations in two variables.* A linear equation in two variables, one for each of the quantities being related, can be used to represent the data in a table or graph. ***The graph of a linear equation in two variables is a line.***

Linear Equation in Two Variables

A **linear equation in two variables** (here x and y) can be written in the form

$$Ax + By = C,$$

where A, B, and C are real numbers and A and B are not both 0. This form is called *standard form.*

Examples: $3x + 4y = 9$, $x - y = 0$,   and $x + 2y = -8$ Linear equations in two variables

---

NOTE Other linear equations in two variables, such as

$$y = 4x + 5 \quad \text{and} \quad 3x = 7 - 2y,$$

are not written in standard form, but could be algebraically rewritten in this form. We discuss the forms of linear equations in more detail in **Sections 3.4 and 3.5.**

OBJECTIVE 2 Write a solution as an ordered pair.

Recall from **Section 1.2** that a *solution* of an equation is a number that makes the equation true when it replaces the variable. For example, the linear equation in *one* variable

$$x - 2 = 5$$

has solution 7 because replacing x with 7 gives a true statement.

A solution of a linear equation in **two** *variables requires* **two** *numbers, one for each variable.* For example, a true statement results when we replace x with 2 and y with 13 in the equation $y = 4x + 5$ because

$$13 = 4(2) + 5. \qquad \text{Let } x = 2 \text{ and } y = 13.$$

The pair of numbers $x = 2$ and $y = 13$ gives a solution of the equation $y = 4x + 5$. The phrase "$x = 2$ and $y = 13$" is abbreviated

$$\underset{\text{Ordered pair}}{\underbrace{(\overset{x\text{-value}}{2}, \overset{y\text{-value}}{13})}}$$

with the x-value, 2, and the y-value, 13, given as a pair of numbers written inside parentheses. ***The x-value is always given first.*** A pair of numbers such as $(2, 13)$ is an **ordered pair.**

> **! CAUTION** The ordered pairs $(2, 13)$ and $(13, 2)$ are *not* the same. In the first pair, $x = 2$ and $y = 13$. In the second pair, $x = 13$ and $y = 2$. ***The order in which the numbers are written in an ordered pair is important.***

OBJECTIVE 3 Decide whether a given ordered pair is a solution of a given equation.

We substitute the x- and y-values of an ordered pair into a linear equation in two variables to see whether the ordered pair is a solution. An ordered pair that is a solution of an equation is said to *satisfy* the equation.

NOW TRY
EXERCISE 2
Decide whether each ordered pair is a solution of the equation.

$$3x - 7y = 19$$

(a) $(3, 4)$ **(b)** $(-3, -4)$

EXAMPLE 2   Deciding Whether Ordered Pairs Are Solutions of an Equation

Decide whether each ordered pair is a solution of the equation $2x + 3y = 12$.

(a) $(3, 2)$

Substitute 3 for x and 2 for y in the given equation.

$$2x + 3y = 12$$
$$2(3) + 3(2) \overset{?}{=} 12 \qquad \text{Let } x = 3 \text{ and } y = 2.$$
$$6 + 6 \overset{?}{=} 12 \qquad \text{Multiply.}$$
$$12 = 12 \ \checkmark \quad \text{True}$$

This result is true, so $(3, 2)$ is a solution of $2x + 3y = 12$.

(b) $(-2, -7)$

$$2x + 3y = 12$$
$$2(-2) + 3(-7) \overset{?}{=} 12 \qquad \text{Let } x = -2 \text{ and } y = -7.$$
$$-4 + (-21) \overset{?}{=} 12 \qquad \text{Multiply.}$$
$$-25 = 12 \qquad \text{False}$$

> Use parentheses to avoid errors.

This result is false, so $(-2, -7)$ is *not* a solution of $2x + 3y = 12$.   **NOW TRY**

OBJECTIVE 4 Complete ordered pairs for a given equation.

Substituting a number for one variable in a linear equation makes it possible to find the value of the other variable.

EXAMPLE 3   Completing Ordered Pairs

Complete each ordered pair for the equation $y = 4x + 5$.

(a) $(7, \underline{\quad})$ > The x-value always comes first.

In this ordered pair, $x = 7$. To find the corresponding value of y, replace x with 7 in the given equation.

$$y = 4x + 5$$
$$y = 4(7) + 5 \qquad \text{Let } x = 7.$$
$$y = 28 + 5 \qquad \text{Multiply.}$$
$$y = 33 \qquad \text{Add.}$$

> Solve for the value of y.

The ordered pair is $(7, 33)$.

NOW TRY ANSWERS
2. (a) no **(b)** yes

**NOW TRY
EXERCISE 3**

Complete each ordered pair
for the equation.

$$y = 3x - 12$$

(a) $(4, \underline{\quad})$ **(b)** $(\underline{\quad}, 3)$

(b) $(\underline{\quad}, -3)$

In this ordered pair, $y = -3$. Find the corresponding value of x by replacing y with -3 in the given equation.

$y = 4x + 5$	*Solve for the value of x.*
$-3 = 4x + 5$	Let $y = -3$.
$-8 = 4x$	Subtract 5 from each side.
$-2 = x$	Divide each side by 4.

The ordered pair is $(-2, -3)$.

NOW TRY

OBJECTIVE 5 Complete a table of values.

Ordered pairs are often displayed in a **table of values.** Although we usually write tables of values vertically, they may be written horizontally.

EXAMPLE 4 Completing Tables of Values

Complete the table of values for each equation. Write the results as ordered pairs.

(a) $x - 2y = 8$

x	y	Ordered Pairs
2		$(2, \underline{\quad})$
10		$(10, \underline{\quad})$
	0	$(\underline{\quad}, 0)$
	-2	$(\underline{\quad}, -2)$

From the first row of the table, let $x = 2$ in the equation. From the second row of the table, let $x = 10$.

If $x = 2$,	If $x = 10$,
then $x - 2y = 8$	then $x - 2y = 8$
becomes $2 - 2y = 8$	becomes $10 - 2y = 8$
$-2y = 6$	$-2y = -2$
$y = -3.$	$y = 1.$

The first two ordered pairs are $(2, -3)$ and $(10, 1)$. From the third and fourth rows of the table, let $y = 0$ and $y = -2$, respectively.

If $y = 0$,	If $y = -2$,
then $x - 2y = 8$	then $x - 2y = 8$
becomes $x - 2(0) = 8$	becomes $x - 2(-2) = 8$
$x - 0 = 8$	$x + 4 = 8$
$x = 8.$	$x = 4.$

The last two ordered pairs are $(8, 0)$ and $(4, -2)$. The completed table of values and corresponding ordered pairs follow.

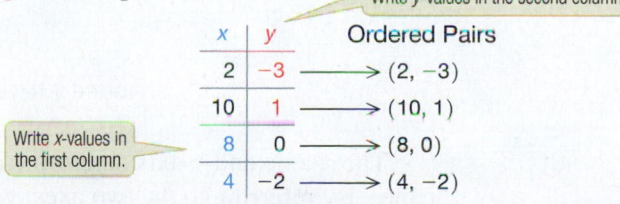

	x	y	Ordered Pairs
	2	-3	$(2, -3)$
	10	1	$(10, 1)$
Write x-values in the first column.	8	0	$(8, 0)$
	4	-2	$(4, -2)$

Write y-values in the second column.

NOW TRY ANSWERS

3. (a) $(4, 0)$ **(b)** $(5, 3)$

Each ordered pair is a solution of the given equation $x - 2y = 8$.

 **NOW TRY EXERCISE 4**

Complete the table of values for the equation. Write the results as ordered pairs.

$$5x - 4y = 20$$

x	y
0	
	0
2	

(b) $x = 5$

x	y
	-2
	6
	3

The given equation is $x = 5$. No matter which value of y is chosen, the value of x is always 5.

x	y	Ordered Pairs
5	-2	⟶ (5, -2)
5	6	⟶ (5, 6)
5	3	⟶ (5, 3)

NOW TRY

NOTE   We can think of $x = 5$ in **Example 4(b)** as an equation in two variables by rewriting $x = 5$ as

$$x + 0y = 5.$$

This form of the equation shows that, for any value of y, the value of x is 5. Similarly, $y = 4$ can be written

$$0x + y = 4.$$

OBJECTIVE 6   **Plot ordered pairs.**

In **Section 2.3,** we saw that linear equations in *one* variable had either one, zero, or an infinite number of real number solutions. These solutions could be graphed on *one* number line. For example, the linear equation in one variable $x - 2 = 5$ has solution 7, which is graphed on the number line in **FIGURE 2.**

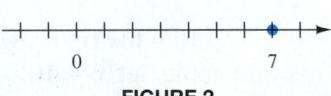

FIGURE 2

Every linear equation in *two* variables has an infinite number of ordered pairs (x, y) as solutions. To graph these solutions, we need *two* number lines, one for each variable, drawn at right angles as in **FIGURE 3.** The horizontal number line is the *x*-axis, and the vertical line is the *y*-axis. The point at which the *x*-axis and *y*-axis intersect is the **origin.** Together, the *x*-axis and *y*-axis form a **rectangular coordinate system.**

The rectangular coordinate system is divided into four regions, or **quadrants.** These quadrants are numbered counterclockwise, as shown in **FIGURE 3.**

René Descartes (1596–1650)

The rectangular coordinate system is also called the **Cartesian coordinate system,** in honor of René Descartes, the French mathematician credited with its invention.

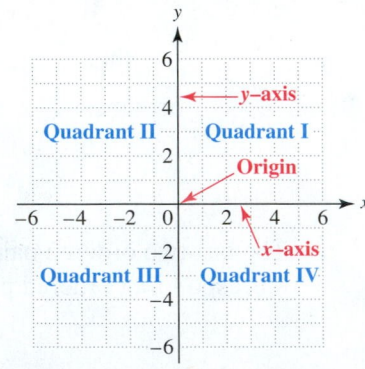

FIGURE 3 Rectangular Coordinate System

The *x*-axis and *y*-axis determine a **plane**—a flat surface illustrated by a sheet of paper. By referring to the two axes, we can associate every point in the plane with an ordered pair. The numbers in the ordered pair are the **coordinates** of the point.

NOW TRY ANSWER

4.

x	y
0	-5
4	0
2	$-\frac{5}{2}$

$(0, -5), (4, 0), \left(2, -\frac{5}{2}\right)$

> **NOTE** In a plane, *both* numbers in the ordered pair are needed to locate a point. The ordered pair is a name for the point.

NOW TRY EXERCISE 5

Plot each point in a rectangular coordinate system.

$(-3, 1), (2, -4), (0, -1),$
$\left(\frac{5}{2}, 3\right), (-4, -3), (-4, 0)$

EXAMPLE 5 Plotting Ordered Pairs

Plot each point in a rectangular coordinate system.

(a) $(2, 3)$ **(b)** $(-1, -4)$ **(c)** $(-2, 3)$ **(d)** $(3, -2)$ **(e)** $\left(\frac{3}{2}, 2\right)$

(f) $(4, -3.75)$ **(g)** $(5, 0)$ **(h)** $(0, -3)$ **(i)** $(0, 0)$

The point $(2, 3)$ from part (a) is **plotted** (graphed) in **FIGURE 4**. The other points are plotted in **FIGURE 5**. In each case, we begin at the origin.

Step 1 Move right or left the number of units that corresponds to the *x*-coordinate in the ordered pair—*right if the x-coordinate is positive or left if the x-coordinate is negative.*

Step 2 Then turn and move up or down the number of units that corresponds to the *y*-coordinate in the ordered pair—*up if the y-coordinate is positive or down if the y-coordinate is negative.*

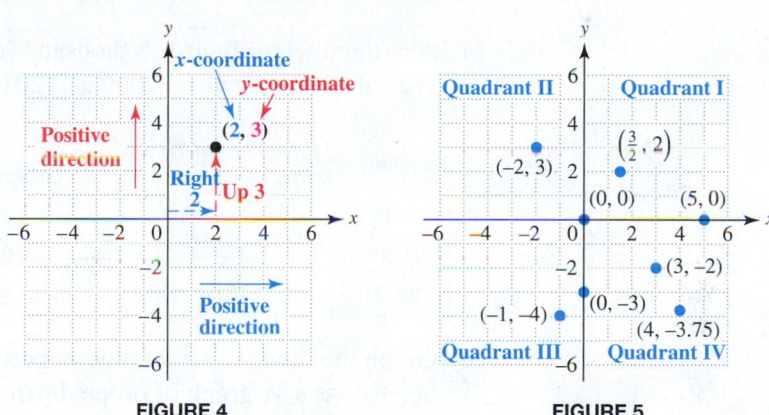

FIGURE 4   **FIGURE 5**

Notice in **FIGURE 5** that the point $(-2, 3)$ is in quadrant II, whereas the point $(3, -2)$ is in quadrant IV. *The order of the coordinates is important. The x-coordinate is always given first in an ordered pair.*

To plot the point $\left(\frac{3}{2}, 2\right)$, think of the improper fraction $\frac{3}{2}$ as the mixed number $1\frac{1}{2}$ and move $\frac{3}{2}$ $\left(\text{or } 1\frac{1}{2}\right)$ units to the right along the *x*-axis. Then turn and go 2 units up, parallel to the *y*-axis. The point $(4, -3.75)$ is plotted similarly, by approximating the location of the decimal *y*-coordinate.

The point $(5, 0)$ lies on the *x*-axis because the *y*-coordinate is 0. The point $(0, -3)$ lies on the *y*-axis because the *x*-coordinate is 0. The point $(0, 0)$ is at the origin.

Points on the axes themselves are not in any quadrant.   NOW TRY

NOW TRY ANSWER

5.

We can use a linear equation in two variables to mathematically describe, or *model,* certain real-life situations.

**NOW TRY
EXERCISE 6**

Use the linear equation in
Example 6 to approximate
the number of twin births, to
the nearest thousand, in 2010.
Interpret the results.

EXAMPLE 6   Using a Linear Equation to Model Twin Births

The annual number of twin births in the United States from 2006 through 2011 can
be approximated by the linear equation

Number of twin births⌐   ⌐Year

$$y = -1.421x + 2989,$$

which relates x, the year, and y, the number of twin births in thousands. (*Source:*
National Center for Health Statistics.)

(a) Complete the table of values for the given linear equation.

x (Year)	y (Number of Twin Births, in thousands)
2006	
2009	
2011	

To find y when $x = 2006$, we substitute into the equation.

$$y = -1.421x + 2989$$

$$y = -1.421(2006) + 2989 \qquad \text{Let } x = 2006.$$

⌐ ≈ means "is
approximately equal to."

$$y \approx 138 \qquad \text{Use a calculator.}$$

In 2006, there were about 138 thousand (or 138,000) twin births.
We substitute the years 2009 and 2011 in the same way to complete the table.

x (Year)	y (Number of Twin Births, in thousands)	Ordered Pairs (x, y)	
2006	138	⟶ (2006, 138)	Here each year x is paired with a number of twin births y (in thousands).
2009	134	⟶ (2009, 134)	
2011	131	⟶ (2011, 131)	

(b) Graph the ordered pairs found in part (a).
See **FIGURE 6**. A graph of ordered pairs of data is a **scatter diagram.**

Number of Twin Births

Notice the axis labels and scales.
Each grid square represents 1 unit in
the horizontal direction and 2 units
in the vertical direction. We show a
break in the y-axis, to indicate the
jump from 0 to 130.

FIGURE 6

NOW TRY ANSWER

6. $y \approx 133$; There were
approximately 133 thousand
(or 133,000) twin births in the
U.S. in 2010.

A scatter diagram enables us to describe how the two quantities are related
to each other. In **FIGURE 6**, the plotted points could be connected to approximate a
straight ***line,*** so the variables x (year) and y (number of twin births) have a ***line***ar
relationship. The decrease in the number of twin births is also reflected.

NOW TRY

> ⚠ **CAUTION**   The equation in **Example 6** is valid only for the years 2006 through 2011. **Do not assume that it would provide reliable data for other years.**

3.1 Exercises

FOR EXTRA HELP   MyMathLab®

▶ *Complete solution available in MyMathLab*

Concept Check   Complete each statement.

1. The symbol (x, y) (*does/does not*) represent an ordered pair, while the symbols $[x, y]$ and $\{x, y\}$ (*do/do not*) represent ordered pairs.

2. The origin is represented by the ordered pair _____ .

3. The point whose graph has coordinates $(-4, 2)$ is in quadrant _____ .

4. The point whose graph has coordinates $(0, 5)$ lies on the _____-axis.

5. The ordered pair $(4, \underline{})$ is a solution of the equation $y = 3$.

6. The ordered pair $(\underline{}, -2)$ is a solution of the equation $x = 6$.

Concept Check   Fill in each blank with the word positive *or the word* negative.

The point with coordinates (x, y) is in

7. quadrant III if x is _____ and y is _____ .

8. quadrant II if x is _____ and y is _____ .

9. quadrant IV if x is _____ and y is _____ .

10. quadrant I if x is _____ and y is _____ .

11. A point (x, y) has the property that $xy < 0$. In which quadrant(s) must the point lie? Explain.

12. A point (x, y) has the property that $xy > 0$. In which quadrant(s) must the point lie? Explain.

The line graph shows the overall unemployment rate in the U.S. civilian labor force for the years 2006 through 2012. Use the graph to work Exercises 13–16. **See Example 1.**

13. Between which pairs of consecutive years did the unemployment rate decrease?

14. What was the general trend in the unemployment rate between 2007 and 2010?

15. Estimate the overall unemployment rate in 2011 and 2012. About how much did the unemployment rate decline between 2011 and 2012?

16. During which year(s)

   (a) was the unemployment rate greater than 9%, but less than 10%?

   (b) did the unemployment rate stay the same?

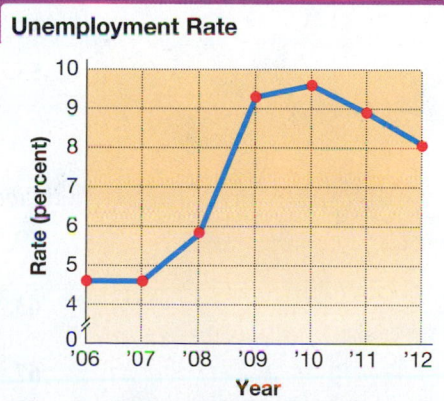

Source: Bureau of Labor Statistics.

Decide whether the given ordered pair is a solution of the given equation. **See Example 2.**

▶ 17. $x + y = 8$; $(0, 8)$ 18. $x + y = 9$; $(0, 9)$ 19. $2x + y = 5$; $(3, -1)$

20. $2x - y = 6$; $(4, 2)$ ▶ 21. $5x - 3y = 15$; $(5, 2)$ 22. $4x - 3y = 6$; $(2, 1)$

23. $x = -4y$; $(-8, 2)$ 24. $y = 3x$; $(2, 6)$ 25. $y = 2$; $(4, 2)$

26. $x = -6$; $(-6, 5)$ 27. $x - 6 = 0$; $(4, 2)$ 28. $x + 4 = 0$; $(-6, 2)$

*Complete each ordered pair for the equation $y = 2x + 7$. **See Example 3.***

29. $(5, \_\_)$ **30.** $(2, \_\_)$ **31.** $(\_\_, -3)$ **32.** $(\_\_, 0)$

*Complete each ordered pair for the equation $y = -4x - 4$. **See Example 3.***

33. $(\_\_, 0)$ **34.** $(0, \_\_)$ **35.** $(\_\_, 24)$ **36.** $(\_\_, 16)$

*Complete the table of values for each equation. Write the results as ordered pairs. **See Example 4.***

37. $4x + 3y = 24$

x	y
0	
	0
	4

38. $2x + 3y = 12$

x	y
0	
	0
	8

39. $4x - 9y = -36$

x	y
0	
	0
	8

40. $3x - 5y = -15$

x	y
0	
	0
	-6

41. $x = 12$

x	y
	3
	8
	0

42. $x = -9$

x	y
	6
	2
	-3

43. $y = -10$

x	y
4	
0	
-4	

44. $y = -6$

x	y
8	
4	
-2	

45. $y + 2 = 0$

x	y
9	
2	
0	

46. $y + 6 = 0$

x	y
6	
3	
0	

47. $x - 4 = 0$

x	y
	4
	0
	-4

48. $x - 8 = 0$

x	y
	8
	3
	0

49. Do $(3, 4)$ and $(4, 3)$ correspond to the same point in the plane? Explain.

50. Do $(4, -1)$ and $(-1, 4)$ represent the same ordered pair? Explain.

*Give ordered pairs for the points labeled A–H in the figure. (Coordinates of the points shown are integers.) Identify the quadrant in which each point is located. **See Example 5.***

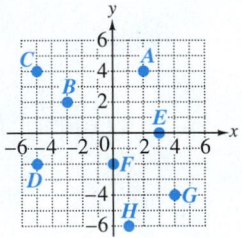

51. A **52.** B **53.** C **54.** D

55. E **56.** F **57.** G **58.** H

*Plot and label each point in a rectangular coordinate system. **See Example 5.***

59. $(6, 2)$ **60.** $(5, 3)$ **61.** $(-4, 2)$ **62.** $(-3, 5)$

63. $\left(-\dfrac{4}{5}, -1\right)$ **64.** $\left(-\dfrac{3}{2}, -4\right)$ **65.** $(3, -1.75)$ **66.** $(5, -4.25)$

67. $(0, 4)$ **68.** $(0, -3)$ **69.** $(4, 0)$ **70.** $(-3, 0)$

*Complete the table of values for each equation. Then plot and label the ordered pairs. **See Examples 4 and 5.***

71. $x - 2y = 6$

x	y
0	
	0
2	
	-1

72. $2x - y = 4$

x	y
0	
	0
1	
	-6

73. $3x - 4y = 12$

x	y
0	
	0
-4	
	-4

74. $2x - 5y = 10$

x	y
0	
	0
−5	
	−3

75. $y + 4 = 0$

x	y
0	
5	
−2	
	−3

76. $x - 5 = 0$

x	y
	1
	0
	6
	−4

77. Look at the graphs of the ordered pairs in **Exercises 71–76.** Describe the pattern indicated by the plotted points.

78. Answer each question.

   (a) A line through the plotted points in **Exercise 75** would be horizontal. What do you notice about the *y*-coordinates of the ordered pairs?

   (b) A line through the plotted points in **Exercise 76** would be vertical. What do you notice about the *x*-coordinates of the ordered pairs?

Solve each problem. See Example 6.

79. Suppose that it costs a flat fee of $20 plus $5 per day to rent a pressure washer. Therefore, the cost *y* in dollars to rent the pressure washer for *x* days is given by the linear equation

$$y = 5x + 20.$$

Express each of the following as an ordered pair.

   (a) When the washer is rented for 5 days, the cost is $45.

   (b) We paid $50 when we returned the washer, so we must have rented it for 6 days.

80. Suppose that it costs $5000 to start up a business selling snow cones. Furthermore, it costs $0.50 per cone in labor, ice, syrup, and overhead. Then the cost *y* in dollars to make *x* snow cones is given by the linear equation

$$y = 0.50x + 5000.$$

Express each of the following as an ordered pair.

   (a) When 100 snow cones are made, the cost is $5050.

   (b) When the cost is $6000, the number of snow cones made is 2000.

81. The table shows the rate (in percent) at which 2-year college students (public) completed a degree within 3 years.

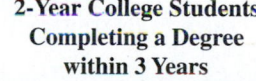

2-Year College Students Completing a Degree within 3 Years

Year	Percent
2008	29.3
2009	28.3
2010	28.0
2011	26.9
2012	25.4
2013	22.5

Source: ACT.

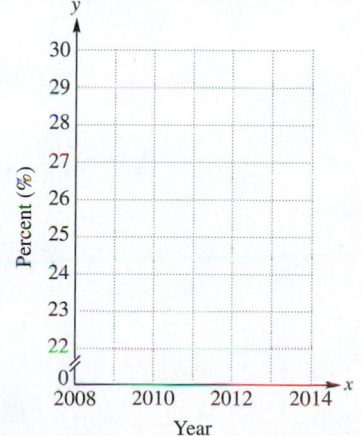

   (a) Write the data from the table as ordered pairs (x, y), where *x* represents the year and *y* represents the percent.

   (b) What does the ordered pair (2013, 22.5) mean in the context of this problem?

   (c) Make a scatter diagram of the data, using the ordered pairs from part (a) and the given grid.

   (d) Describe the pattern indicated by the points on the scatter diagram. What happened to the rates at which 2-year college students complete a degree within 3 years?

82. The table shows the number of U.S. students who studied abroad (in thousands).

Academic Year	Number of Students (in thousands)
2005	224
2006	242
2007	262
2008	260
2009	270
2010	274
2011	283

Source: Institute of International Education.

U.S. Students Studying Abroad

(a) Write the data from the table as ordered pairs (x, y), where x represents the year and y represents the number of U.S. students (in thousands) studying abroad.

(b) What does the ordered pair $(2011, 283)$ mean in the context of this problem?

(c) Make a scatter diagram of the data, using the ordered pairs from part (a) and the given grid.

(d) Describe the pattern indicated by the points on the scatter diagram. What was the trend in the number of U.S. students studying abroad during these years?

83. The maximum benefit for the heart from exercising occurs if the heart rate is in the target heart rate zone. The lower limit of this target zone can be approximated by the linear equation

$$y = -0.65x + 143,$$

where x represents age and y represents heartbeats per minute. (*Source: The Gazette.*)

Age	Heartbeats (per minute)
20	
40	
60	
80	

(a) Complete the table of values for this linear equation.

(b) Write the data from the table of values as ordered pairs.

(c) Make a scatter diagram of the data. Do the points lie in an approximately linear pattern?

84. (See **Exercise 83.**) The upper limit of the target heart rate zone can be approximated by the linear equation

$$y = -0.85x + 187,$$

where x represents age and y represents heartbeats per minute. (*Source: The Gazette.*)

Age	Heartbeats (per minute)
20	
40	
60	
80	

(a) Complete the table of values for this linear equation.

(b) Write the data from the table of values as ordered pairs.

(c) Make a scatter diagram of the data. Describe the pattern indicated by the data.

85. See **Exercises 83 and 84.** What is the target heart rate zone for age 20?   Age 40?

86. See **Exercises 83 and 84.** What is the target heart rate zone for age 60?   Age 80?

STUDY SKILLS

Analyzing Your Test Results

An exam is a learning opportunity—learn from your mistakes. After a test is returned, do the following:

- **Note what you got wrong and why you had points deducted.**

- **Figure out how to solve the problems you missed.** Check your textbook or notes, or ask your instructor. Rework the problems correctly.

- **Keep all quizzes and tests that are returned to you.** Use them to study for future tests and the final exam.

Typical Reasons for Errors on Math Tests

These are test taking errors. They are easy to correct if you read carefully, show all your work, proofread, and double-check units and labels.

1. You read the directions wrong.

2. You read the question wrong or skipped over something.

3. You made a computation error.

4. You made a careless error. (For example, you incorrectly copied a correct answer onto a separate answer sheet.)

5. Your answer was not complete.

6. You labeled your answer wrong. (For example, you labeled an answer "ft" instead of "ft^2.")

7. You didn't show your work.

These are test preparation errors. Be sure to practice all the kinds of problems that you will see on tests.

8. You didn't understand a concept.

9. You were unable to set up the problem (in an application).

10. You were unable to apply a procedure.

Below are sample charts for tracking your test taking progress. Refer to the tests you have taken so far in your course, and use the charts to find out if you tend to make certain kinds of errors. Check the appropriate box when you've made an error in a particular category.

Test Taking Errors

Test	Read directions wrong	Read question wrong	Computation error	Not exact or accurate	Not complete	Labeled wrong	Didn't show work
1							
2							
3							

Test Preparation Errors

Test	Didn't understand concept	Didn't set up problem correctly	Couldn't apply concept to new situation
1			
2			
3			

What will you do to avoid these kinds of errors on your next test?

3.2 Graphing Linear Equations in Two Variables

VOCABULARY

☐ graph, graphing
☐ x-intercept
☐ y-intercept
☐ horizontal line
☐ vertical line

OBJECTIVE 1 Graph linear equations by plotting ordered pairs.

There are infinitely many ordered pairs that satisfy a linear equation in two variables. We find these ordered-pair solutions by choosing as many values of x (or y) as we wish and then completing each ordered pair.

For example, consider the equation

$$x + 2y = 7.$$

If we choose $x = 1$, then we can substitute to find the corresponding value of y.

$$x + 2y = 7 \qquad \text{Given equation.}$$
$$1 + 2y = 7 \qquad \text{Let } x = 1.$$
$$2y = 6 \qquad \text{Subtract 1.}$$
$$y = 3 \qquad \text{Divide by 2.}$$

If $x = 1$, then $y = 3$, so the ordered pair $(1, 3)$ is a solution.

$$1 + 2(3) = 7 \qquad (1, 3) \text{ is a solution.}$$

This ordered pair and other solutions of $x + 2y = 7$ are graphed in **FIGURE 7**.

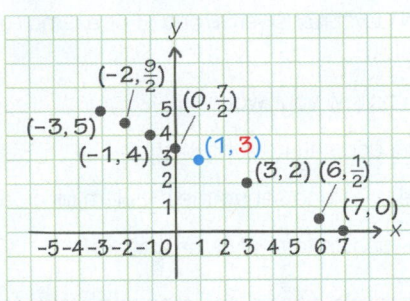

| FIGURE 7 | FIGURE 8 |

Notice that the points plotted in **FIGURE 7** all appear to lie on a straight line, as shown in **FIGURE 8**. In fact, the following is true.

Every point on the line represents a solution of the equation $x + 2y = 7$, and every solution of the equation corresponds to a point on the line.

The line gives a "picture" of all the solutions of the equation

$$x + 2y = 7.$$

The line extends indefinitely in both directions, as suggested by the arrowhead on each end, and is the **graph** of the equation $x + 2y = 7$. **Graphing** is the process of plotting ordered pairs and drawing a line through the corresponding points.

Graph of a Linear Equation

The graph of any linear equation in two variables is a straight line.

Notice that the word ***line*** appears in the name "***line***ar equation."

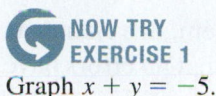

NOW TRY
EXERCISE 1

Graph $x + y = -5$.

EXAMPLE 1 Graphing a Linear Equation

Graph $x - y = -3$.

At least two different ordered pairs are needed to draw the graph. To find them, we arbitrarily choose values for x or y and substitute them into the equation. We choose $x = 0$ to find one ordered pair and $y = 0$ to find another.

$x - y = -3$	$x - y = -3$
$0 - y = -3$ 0 is easy to substitute.	$x - 0 = -3$ 0 is easy to substitute.
$-y = -3$ Subtract.	$x = -3$ Subtract.
$y = 3$ Multiply by -1.	
One ordered pair is $(0, 3)$.	One ordered pair is $(-3, 0)$.

We find a third ordered pair (as a check) by choosing some other value for x or y. We let $x = 2$.

$$x - y = -3$$
$$2 - y = -3 \quad \text{We arbitrarily let } x = 2. \text{ Other numbers could be used for } x, \text{ or for } y, \text{ instead.}$$
$$-y = -5 \quad \text{Subtract 2.}$$
$$y = 5 \quad \text{Multiply by } -1.$$

This gives the ordered pair $(2, 5)$. We plot the three ordered-pair solutions and draw a line through them. See **FIGURE 9**.

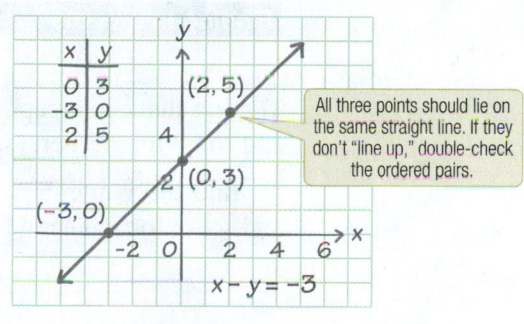

All three points should lie on the same straight line. If they don't "line up," double-check the ordered pairs.

FIGURE 9

NOW TRY

EXAMPLE 2 Graphing a Linear Equation

Graph $4x - 5y = 20$.

To find three ordered pairs that are solutions of $4x - 5y = 20$, we choose three arbitrary values for x or y that we think will be easy to substitute.

Let $x = 0$.	Let $y = 0$.	Let $y = 2$.
$4x - 5y = 20$	$4x - 5y = 20$	$4x - 5y = 20$
$4(0) - 5y = 20$	$4x - 5(0) = 20$	$4x - 5(2) = 20$
$0 - 5y = 20$	$4x - 0 = 20$	$4x - 10 = 20$
$-5y = 20$	$4x = 20$	$4x = 30$
$y = -4$	$x = 5$	$\dfrac{30}{4} = 7\frac{1}{2} \quad x = 7\frac{1}{2}$
Ordered pair: $(0, -4)$	Ordered pair: $(5, 0)$	Ordered pair: $\left(7\frac{1}{2}, 2\right)$

NOW TRY ANSWER

1.

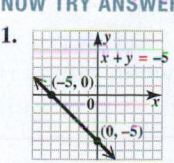

**NOW TRY
EXERCISE 2**
Graph $2x - 4y = 8$.

We plot the three ordered-pair solutions and draw a line through them. See **FIGURE 10.** Two points determine the line, and the third point is used to check that no errors have been made.

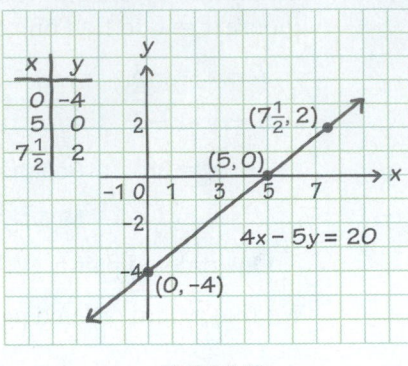

FIGURE 10   **NOW TRY**

OBJECTIVE 2 Find intercepts.

In **FIGURE 10**, the graph intersects (crosses) the x-axis at $(5, 0)$ and the y-axis at $(0, -4)$. For this reason, $(5, 0)$ is the **x-intercept** and $(0, -4)$ is the **y-intercept** of the graph. The intercepts are often convenient points to use when graphing linear equations.

Finding Intercepts

To find the x-intercept, let $y = 0$ in the given equation and solve for x. Then $(x, 0)$ is the x-intercept.

To find the y-intercept, let $x = 0$ in the given equation and solve for y. Then $(0, y)$ is the y-intercept.

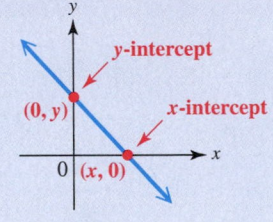

EXAMPLE 3 Graphing a Linear Equation Using Intercepts

Find the intercepts for the graph of $2x + y = 4$. Then draw the graph.

To find the intercepts, we first let $x = 0$ and then let $y = 0$. To find a third point, we arbitrarily let $x = 4$.

Let $x = 0$.	Let $y = 0$.	Let $x = 4$.
$2x + y = 4$	$2x + y = 4$	$2x + y = 4$
$2(0) + y = 4$	$2x + 0 = 4$	$2(4) + y = 4$
$0 + y = 4$	$2x = 4$	$8 + y = 4$
$y = 4$	$x = 2$	$y = -4$
y-intercept: $(0, 4)$	x-intercept: $(2, 0)$	Third point: $(4, -4)$

NOW TRY ANSWER
2.
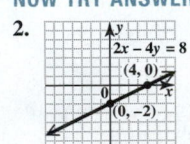

The graph, with the two intercepts in red, and a table of values is shown in **FIGURE 11** on the next page.

NOW TRY
EXERCISE 3
Find the intercepts for the graph of $x + 2y = 2$. Then draw the graph.

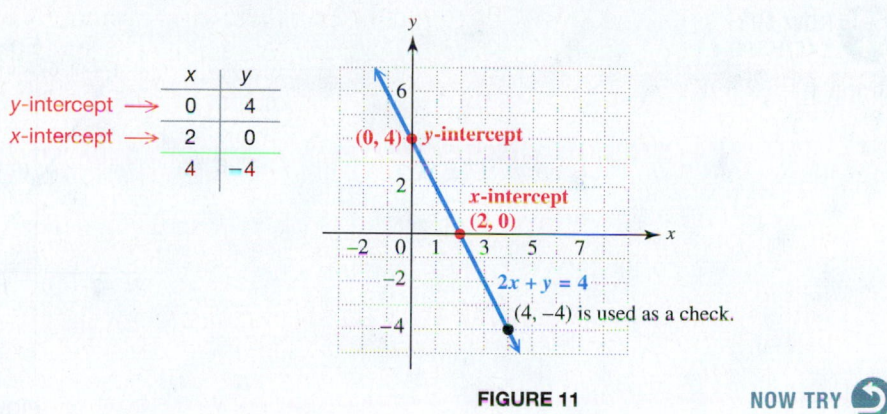

	x	y
y-intercept ⟶	0	4
x-intercept ⟶	2	0
	4	−4

$(0, 4)$ • y-intercept

x-intercept $(2, 0)$

$2x + y = 4$

$(4, −4)$ is used as a check.

FIGURE 11

NOW TRY

⚠ **CAUTION** *When choosing x- or y-values to find ordered pairs to plot, be careful to choose so that the resulting points are not too close together.* For example, using $(−1, −1)$, $(0, 0)$, and $(1, 1)$ to graph $x − y = 0$ may result in an inaccurate line. It is better to choose points whose x-values differ by at least 2.

EXAMPLE 4 Graphing a Linear Equation Using Intercepts

Graph $y = -\dfrac{3}{2}x + 3$.

Although this linear equation is not in standard form $(Ax + By = C)$, it *could* be written in that form. To find the intercepts, we first let $x = 0$ and then let $y = 0$.

$$y = -\frac{3}{2}x + 3$$
$$y = -\frac{3}{2}(0) + 3 \quad \text{Let } x = 0.$$
$$y = 0 + 3 \quad \text{Multiply.}$$
$$y = 3 \quad \text{Add.}$$

y-intercept: $(0, 3)$

$$y = -\frac{3}{2}x + 3$$
$$0 = -\frac{3}{2}x + 3 \quad \text{Let } y = 0.$$
$$\frac{3}{2}x = 3 \quad \text{Add } \frac{3}{2}x.$$
$$x = 2 \quad \text{Multiply by } \frac{2}{3}.$$

x-intercept: $(2, 0)$

To find a third point, we arbitrarily let $x = -2$.

$$y = -\frac{3}{2}x + 3 \qquad \text{Choosing a multiple of 2 makes multiplying by } -\frac{3}{2} \text{ easier.}$$
$$y = -\frac{3}{2}(-2) + 3 \quad \text{Let } x = -2.$$
$$y = 3 + 3 \quad \text{Multiply.}$$
$$y = 6 \quad \text{Add.}$$

Third point: $(-2, 6)$

NOW TRY ANSWER
3. x-intercept: $(2, 0)$;
   y-intercept: $(0, 1)$

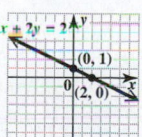

We plot the three ordered-pair solutions and draw a line through them, as shown in **FIGURE 12** on the next page.

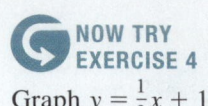

**NOW TRY
EXERCISE 4**

Graph $y = \frac{1}{3}x + 1$.

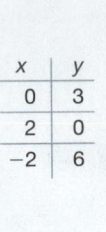

x	y
0	3
2	0
−2	6

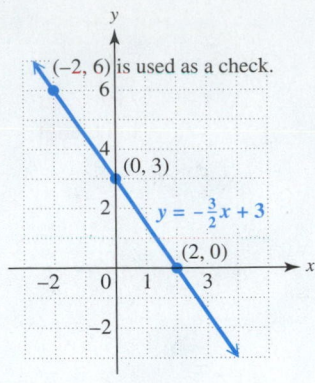

FIGURE 12   NOW TRY

OBJECTIVE 3   Graph linear equations of the form $Ax + By = 0$.

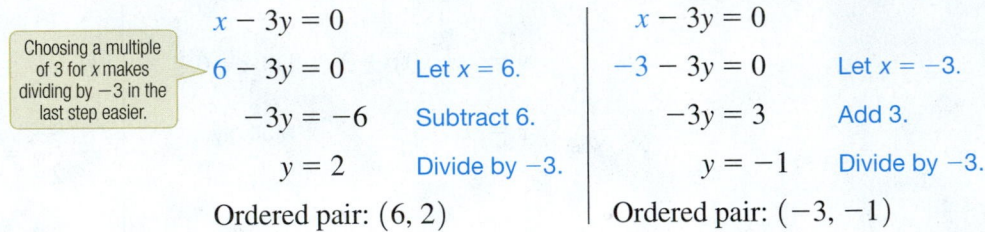

EXAMPLE 5   Graphing an Equation with x- and y-Intercepts (0, 0)

**NOW TRY
EXERCISE 5**

Graph $2x + y = 0$.

Graph $x - 3y = 0$.

To find the *y*-intercept, let $x = 0$.

$$x - 3y = 0$$
$$0 - 3y = 0 \qquad \text{Let } x = 0.$$
$$-3y = 0 \qquad \text{Subtract.}$$
$$y = 0 \qquad \text{Divide by } -3.$$

y-intercept: $(0, 0)$

To find the *x*-intercept, let $y = 0$.

$$x - 3y = 0$$
$$x - 3(0) = 0 \qquad \text{Let } y = 0.$$
$$x - 0 = 0 \qquad \text{Multiply.}$$
$$x = 0 \qquad \text{Subtract.}$$

x-intercept: $(0, 0)$

The *x*- and *y*-intercepts are the *same* point, $(0, 0)$. We must select *two other values* for *x* or *y* to find two other points on the graph. We choose $x = 6$ and $x = -3$.

> Choosing a multiple of 3 for *x* makes dividing by −3 in the last step easier.

$$x - 3y = 0$$
$$6 - 3y = 0 \qquad \text{Let } x = 6.$$
$$-3y = -6 \qquad \text{Subtract 6.}$$
$$y = 2 \qquad \text{Divide by } -3.$$

Ordered pair: $(6, 2)$

$$x - 3y = 0$$
$$-3 - 3y = 0 \qquad \text{Let } x = -3.$$
$$-3y = 3 \qquad \text{Add 3.}$$
$$y = -1 \qquad \text{Divide by } -3.$$

Ordered pair: $(-3, -1)$

We use the three ordered-pair solutions to draw the graph in **FIGURE 13**.

NOW TRY ANSWERS

4.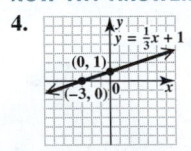

5.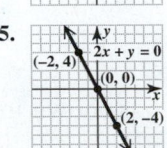

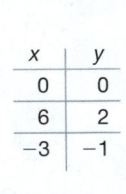

x	y
0	0
6	2
−3	−1

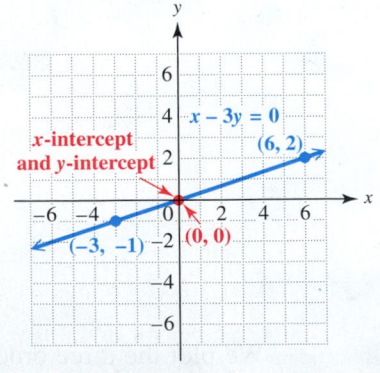

FIGURE 13   NOW TRY

Line through the Origin

The graph of a linear equation of the form

$$Ax + By = 0,$$

where A and B are nonzero real numbers, passes through the origin $(0, 0)$.

OBJECTIVE 4 Graph linear equations of the form $y = b$ or $x = a$.

Consider the following linear equations.

$$y = -4 \quad \text{can be written as} \quad 0x + y = -4.$$

$$x = 3 \quad \text{can be written as} \quad x + 0y = 3.$$

When the coefficient of x or y is 0, the graph is a horizontal or vertical line.

NOW TRY
EXERCISE 6
Graph $y = 2$.

EXAMPLE 6 Graphing a Horizontal Line ($y = b$)

Graph $y = -4$.

For any value of x, the value of y is always -4. Three ordered-pair solutions of the equation are shown in the table of values. Drawing a line through these points gives the **horizontal line** in **FIGURE 14**.

The y-intercept is $(0, -4)$.

There is no x-intercept.

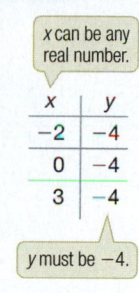

x can be any real number.

x	y
-2	-4
0	-4
3	-4

y must be -4.

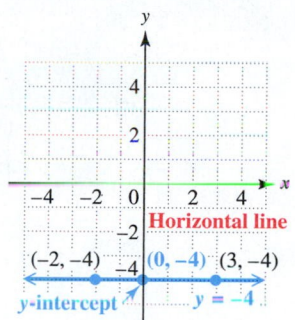

FIGURE 14

NOW TRY

NOW TRY
EXERCISE 7
Graph $x + 4 = 0$.

EXAMPLE 7 Graphing a Vertical Line ($x = a$)

Graph $x - 3 = 0$.

First we add 3 to each side of the equation $x - 3 = 0$ to obtain $x = 3$. All ordered-pair solutions of this equation have x-coordinate 3. Any number can be used for y. Three ordered pairs that satisfy the equation are given in the table of values. The graph is the **vertical line** in **FIGURE 15**.

The x-intercept is $(3, 0)$.

There is no y-intercept.

y can be any real number.

x	y
3	3
3	0
3	-2

x must be 3.

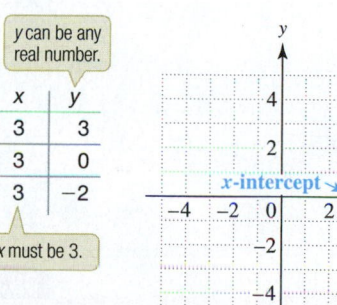

FIGURE 15

NOW TRY

NOW TRY ANSWERS

6.

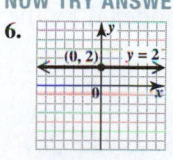

7.

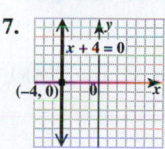

Horizontal and Vertical Lines

The graph of $y = b$, where b is a real number, is a **horizontal line** with y-intercept $(0, b)$ and no x-intercept (unless the horizontal line is the x-axis itself).

The graph of $x = a$, where a is a real number, is a **vertical line** with x-intercept $(a, 0)$ and no y-intercept (unless the vertical line is the y-axis itself).

Keep the following in mind regarding the x- and y-axes.

- **The x-axis is the horizontal line given by the equation $y = 0$.**
- **The y-axis is the vertical line given by the equation $x = 0$.**

> ⚠ **CAUTION**   The equations of horizontal and vertical lines are often confused with each other. The graph of $y = b$ is parallel to the x-axis and the graph of $x = a$ is parallel to the y-axis (for $a \neq 0$ and $b \neq 0$).

▼ **Forms of Linear Equations**

Equation	To Graph	Example
$Ax + By = C$ (where A, B, and $C \neq 0$)	Find any two points on the line. A good choice is to find the intercepts. Let $x = 0$, and find the corresponding value of y. Then let $y = 0$, and find x. As a check, find a third point by choosing a value for x or y that has not yet been used.	
$Ax + By = 0$	The graph passes through the point $(0, 0)$. To find additional points that lie on the graph, choose any values for x or y, except 0.	
$y = b$	Draw a horizontal line, through the point $(0, b)$.	
$x = a$	Draw a vertical line, through the point $(a, 0)$.	

OBJECTIVE 5   Use a linear equation to model data.

EXAMPLE 8   Using a Linear Equation to Model Internet Use

In the United States, the weekly time spent online y in hours can be modeled by the linear equation

$$y = 0.96x + 9.1,$$

where $x = 0$ represents 2000, $x = 1$ represents 2001, and so on. (*Source: The 2013 Digital Future Report*, USC.)

(a) Use the equation to approximate weekly time spent online in the years 2000, 2006, and 2012.

NOW TRY EXERCISE 8

Use **(a)** the graph and **(b)** the equation in **Example 8** to approximate weekly time spent online in 2009. (Round the answer in part (b) to the nearest tenth.)

Substitute the appropriate value for each year x to find weekly time spent online that year.

$$y = 0.96x + 9.1 \qquad \text{Given linear equation}$$

For 2000: $\quad y = 0.96(0) + 9.1 \qquad$ Replace x with 0.

$$y = 9.1 \text{ hr} \qquad \text{Multiply, and then add.}$$

For 2006: $\quad y = 0.96(6) + 9.1 \qquad$ 2006 − 2000 = 6

$$y \approx 14.9 \text{ hr} \qquad \text{Replace } x \text{ with 6.}$$

For 2012: $\quad y = 0.96(12) + 9.1 \qquad$ 2012 − 2000 = 12

$$y \approx 20.6 \text{ hr} \qquad \text{Replace } x \text{ with 12.}$$

(b) Write the information from part (a) as three ordered pairs, and use them to graph the given linear equation.

Because x represents the year and y represents the time, the ordered pairs are

$$(0, 9.1), \quad (6, 14.9), \quad \text{and} \quad (12, 20.6).$$

See **FIGURE 16**. (Arrowheads are not included with the graphed line because the data are for the years 2000 to 2012 only—that is, from $x = 0$ to $x = 12$.)

Internet Use in the U.S.

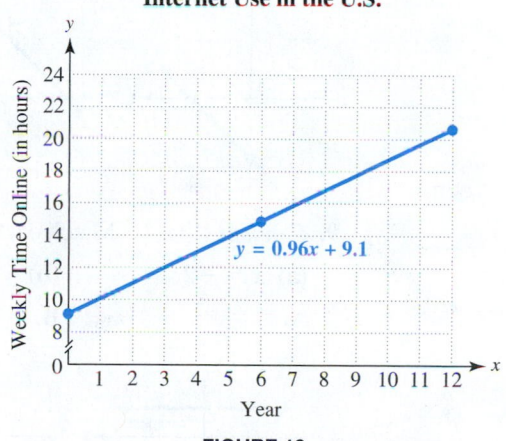

FIGURE 16

(c) Use the graph and then the equation to approximate weekly time spent online in 2010.

For 2010, $x = 10$. On the graph, find 10 on the horizontal axis, move up to the graphed line and then across to the vertical axis. It appears that weekly time spent online in 2010 was about 19 hr. To use the equation, substitute 10 for x.

$$y = 0.96x + 9.1 \qquad \text{Given linear equation}$$

$$y = 0.96(10) + 9.1 \qquad \text{Let } x = 10.$$

$$y = 18.7 \qquad \text{Multiply, and then add.}$$

This result for 2010 is close to our estimate of 19 hr from the graph. **NOW TRY**

NOW TRY ANSWERS
8. (a) about 18 hr
   (b) 17.7 hr

3.2 Exercises

FOR EXTRA HELP   MyMathLab®

▶ *Complete solution available in MyMathLab*

Concept Check *Fill in each blank with the correct response.*

1. A linear equation in two variables x and y can be written in the form $Ax +$ _____ $=$ _____ , where A, B, and C are real numbers and A and B are not both _____ .

2. The graph of any linear equation in two variables is a straight _____ . Every point on the line represents a _____ of the equation.

3. ***Concept Check***   Match the information about each graph in Column I with the correct linear equation in Column II.

I	**II**
(a) The graph of the equation has y-intercept $(0, -4)$.	**A.** $3x + y = -4$
(b) The graph of the equation has $(0, 0)$ as x-intercept and y-intercept.	**B.** $x - 4 = 0$
(c) The graph of the equation does not have an x-intercept.	**C.** $y = 4x$
(d) The graph of the equation has x-intercept $(4, 0)$.	**D.** $y = 4$

4. ***Concept Check***   Which of these equations have a graph with only one intercept?

   A. $x + 8 = 0$ **B.** $x - y = 3$ **C.** $x + y = 0$ **D.** $y = 4$

Concept Check   *Identify the intercepts of each graph. (Coordinates of the points shown are integers.)*

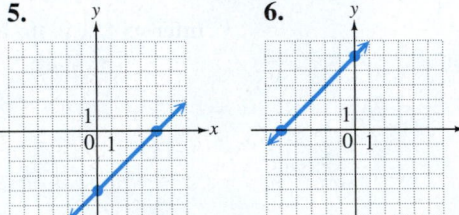

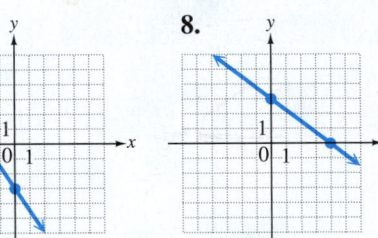

5.   **6.**   **7.**   **8.**

9. ***Concept Check***   Match each equation in (a)–(d) with its graph in A–D.

   (a) $x = -2$ **(b)** $y = -2$ **(c)** $x = 2$ **(d)** $y = 2$

   A.   **B.**   **C.**   **D.**

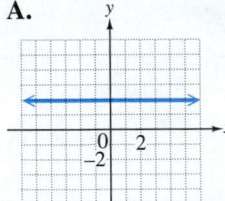

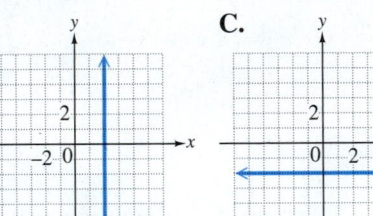

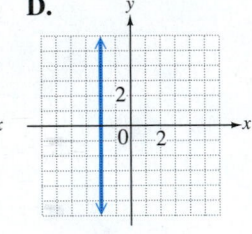

10. ***Concept Check***   What is the equation of the x-axis? What is the equation of the y-axis?

*Complete the given ordered-pair solutions for each equation. Then graph each equation by plotting the points and drawing a line through them. **See Examples 1–4.***

▶ **11.** $x + y = 5$

   $(0, \underline{}), (\underline{}, 0), (2, \underline{})$

12. $x - y = 2$

   $(0, \underline{}), (\underline{}, 0), (5, \underline{})$

▶ **13.** $y = \dfrac{2}{3}x + 1$

   $(0, \underline{}), (3, \underline{}), (-3, \underline{})$

14. $y = -\dfrac{3}{4}x + 2$

   $(0, \underline{}), (4, \underline{}), (-4, \underline{})$

15. $3x = -y - 6$

   $(0, \underline{}), (\underline{}, 0), \left(-\dfrac{1}{3}, \underline{}\right)$

16. $x = 2y + 3$

   $(\underline{}, 0), (0, \underline{}), \left(\underline{}, \dfrac{1}{2}\right)$

Find the x- and y-intercepts for the graph of each equation. See Examples 1–7.

17. $x - y = 8$ **18.** $x - y = 7$ **19.** $5x - 2y = 20$ **20.** $-3x + 2y = 12$

21. $x + 6y = 0$ **22.** $3x + y = 0$ **23.** $y = -2x + 4$ **24.** $y = 3x + 6$

25. $y = \dfrac{1}{3}x - 2$ **26.** $y = \dfrac{1}{4}x - 1$ **27.** $2x - 3y = 0$ **28.** $4x - 5y = 0$

29. $x - 4 = 0$ **30.** $x - 5 = 0$ **31.** $y = 2.5$ **32.** $y = -1.5$

Graph each linear equation. See Examples 1–7.

33. $x - y = 4$ **34.** $x - y = 5$ **35.** $2x + y = 6$

36. $-3x + y = -6$ **37.** $y = 2x - 5$ **38.** $y = 4x + 3$

39. $x = y + 2$ **40.** $x = -y + 6$ **41.** $2x - 5y = 10$

42. $3x + 2y = 6$ **43.** $3x + 7y = 14$ **44.** $6x - 5y = 18$

45. $y = -\dfrac{3}{4}x + 3$ **46.** $y = -\dfrac{2}{3}x - 2$ ▶ **47.** $y - 2x = 0$

48. $y + 3x = 0$ **49.** $y = -6x$ **50.** $y = 4x$

▶ **51.** $y = -1$ **52.** $y = 3$ **53.** $x = 5$

54. $x = -1$ ▶ **55.** $x + 2 = 0$ **56.** $x - 4 = 0$

57. $-3y = 15$ **58.** $-2y = 12$ **59.** $x + 2 = 8$ **60.** $x - 1 = -4$

Concept Check   *Describe what the graph of each linear equation will look like in the coordinate plane. (Hint: Rewrite the equation if necessary so that it is in a more recognizable form.)*

61. $3x = y - 9$ **62.** $2x = y - 4$ **63.** $x - 10 = 1$ **64.** $x + 4 = 3$

65. $3y = -6$ **66.** $5y = -15$ **67.** $2x = 4y$ **68.** $3x = 9y$

Extending Skills   *Plot each set of points, and draw a line through them. Then give an equation of the line.*

69. $(3, 5)$, $(3, 0)$, and $(3, -3)$ **70.** $(1, 3)$, $(1, 0)$, and $(1, -1)$

71. $(-3, -3)$, $(0, -3)$, and $(4, -3)$ **72.** $(-5, 5)$, $(0, 5)$, and $(3, 5)$

Solve each problem. See Example 8.

73. The weight y (in pounds) of a man taller than 60 in. can be approximated by the linear equation

$$y = 5.5x - 220,$$

where x is the height of the man in inches.

(a) Use the equation to approximate the weights of men whose heights are 62 in., 66 in., and 72 in.

(b) Write the information from part (a) as three ordered pairs.

(c) Graph the equation for $x \geq 62$, using the data from part (b).

(d) Use the graph to estimate the height of a man who weighs 155 lb. Then use the equation to find the height of this man to the nearest inch.

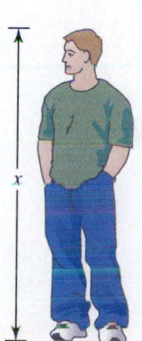

74. The height y (in centimeters) of a woman can be approximated by the linear equation

$$y = 3.9x + 73.5,$$

where x is the length of her radius bone (from the wrist to the elbow) in centimeters.

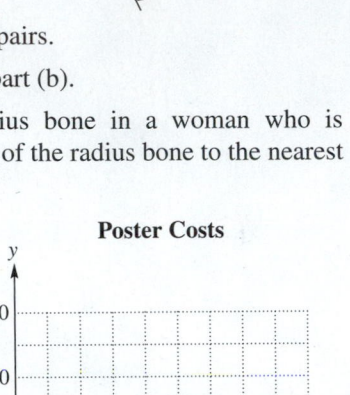

(a) Use the equation to approximate the heights of women with radius bones of lengths 20 cm, 26 cm, and 22 cm.

(b) Write the information from part (a) as three ordered pairs.

(c) Graph the equation for $x \geq 20$, using the data from part (b).

(d) Use the graph to estimate the length of the radius bone in a woman who is 167 cm tall. Then use the equation to find the length of the radius bone to the nearest centimeter.

75. As a fundraiser, a club is selling posters. The printer charges a $25 set-up fee, plus $0.75 for each poster. The cost y in dollars to print x posters is given by the linear equation

$$y = 0.75x + 25.$$

(a) What is the cost y in dollars to print 50 posters?   To print 100 posters?

(b) Find the number of posters x if the printer billed the club for costs of $175.

(c) Write the information from parts (a) and (b) as three ordered pairs.

(d) Use the data from part (c) and the given grid to graph the equation for $x \geq 0$.

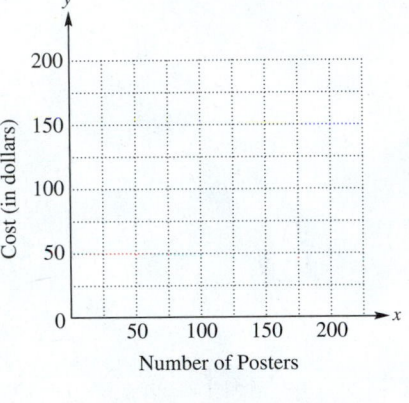

Poster Costs

Cost (in dollars) / Number of Posters

76. A gas station is selling gasoline for $3.50 per gallon and charges $7 for a car wash. The cost y in dollars for x gallons of gasoline and a car wash is given by the linear equation

$$y = 3.50x + 7.$$

(a) What is the cost y in dollars for 9 gal of gasoline and a car wash?   For 4 gal of gasoline and a car wash?

(b) Find the number of gallons of gasoline x if the cost for gasoline and a car wash is $35.

(c) Write the information from parts (a) and (b) as three ordered pairs.

(d) Use the data from part (c) and the given grid to graph the equation for $x \geq 0$.

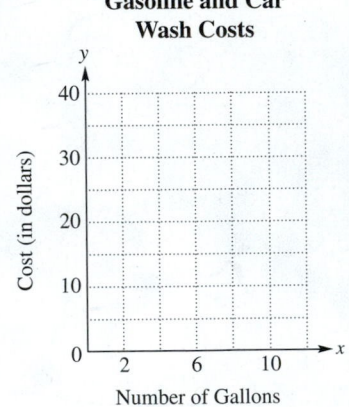

Gasoline and Car Wash Costs

Cost (in dollars) / Number of Gallons

77. The graph shows the value of a sport-utility vehicle (SUV) over the first 5 yr of ownership. Use the graph to do the following.

(a) Determine the initial value of the SUV.

(b) Find the **depreciation** (loss in value) from the original value after the first 3 yr.

(c) What is the annual or yearly depreciation in each of the first 5 yr?

(d) What does the ordered pair (5, 5000) mean in the context of this problem?

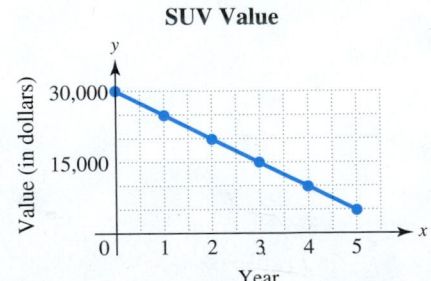

SUV Value

Value (in dollars) / Year

78. Demand for an item is often closely related to its price. As price increases, demand decreases, and as price decreases, demand increases. Suppose demand for a video game is 2000 units when the price is $40, and demand is 2500 units when the price is $30.

(a) Let x be the price and y be the demand for the game. Graph the two given pairs of prices and demands on the given grid.

(b) Assume that the relationship is linear. Draw a line through the two points from part (a). From the graph, estimate the demand if the price drops to $20.

(c) Use the graph to estimate the price if the demand is 3500 units.

(d) Write the prices and demands from parts (b) and (c) as ordered pairs.

Video Game Price/Demand

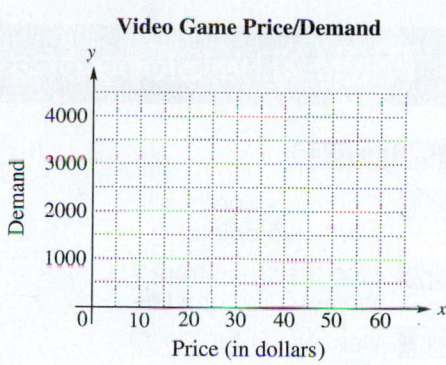

79. U.S. per capita consumption y of cheese in pounds from 2000 through 2012 is shown in the graph and modeled by the linear equation

$$y = 0.307x + 30.1,$$

where $x = 0$ represents 2000, $x = 2$ represents 2002, and so on.

(a) Use the equation to approximate cheese consumption (to the nearest tenth) in 2000, 2008, and 2012.

(b) Use the graph to estimate cheese consumption for the same years.

(c) How do the approximations using the equation compare with the estimates from the graph?

(d) The USDA projects that per capita consumption of cheese in 2022 will be 36.8 lb. Use the equation to approximate per capita cheese consumption (to the nearest tenth) in 2022. How does the approximation using the equation compare to the USDA projection?

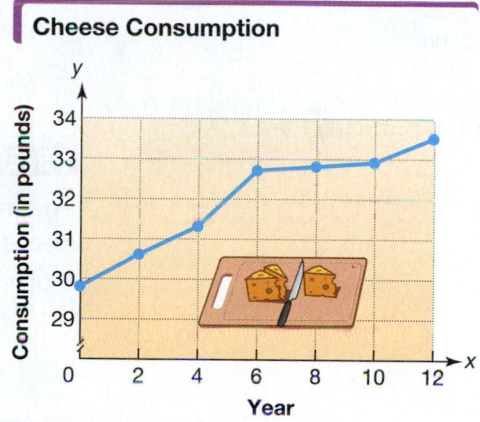

Cheese Consumption

Source: U.S. Department of Agriculture.

80. The number of U.S. marathon finishers y in thousands from 1990 through 2010 are shown in the graph and modeled by the linear equation

$$y = 13.36x + 220.8,$$

where $x = 0$ represents 1990, $x = 5$ represents 1995, and so on.

(a) Use the equation to approximate the number of U.S. marathon finishers in 1990, 2000, and 2010 to the nearest thousand.

(b) Use the graph to estimate the number of U.S. marathon finishers for the same years.

(c) How do the approximations using the equation compare to the estimates from the graph?

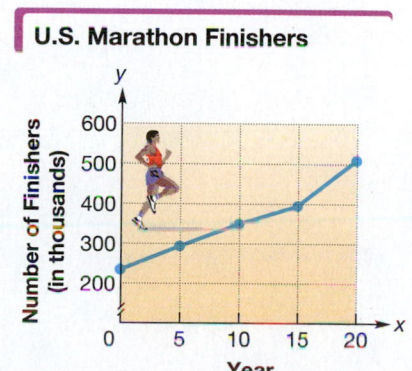

U.S. Marathon Finishers

Source: Running U.S.A.

3.3 The Slope of a Line

An important characteristic of the lines we graphed in **Section 3.2** is their slant, or "steepness" as viewed from *left to right*. See **FIGURE 17**.

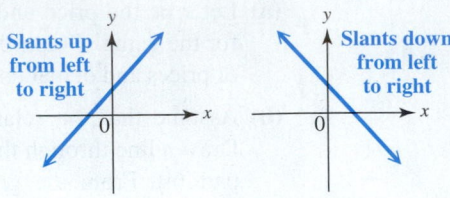

FIGURE 17

One way to measure the steepness of a line is to compare the vertical change in the line with the horizontal change while moving along the line from one fixed point to another. This measure of steepness is the *slope* of the line.

OBJECTIVE 1 **Find the slope of a line given two points.**

To find the steepness, or slope, of the line in **FIGURE 18**, we begin at point Q and move to point P. The vertical change, or **rise,** is the change in the y-values, which is the difference

$$6 - 1 = 5 \text{ units.}$$

The horizontal change, or **run,** is the change in the x-values, which is the difference

$$5 - 2 = 3 \text{ units.}$$

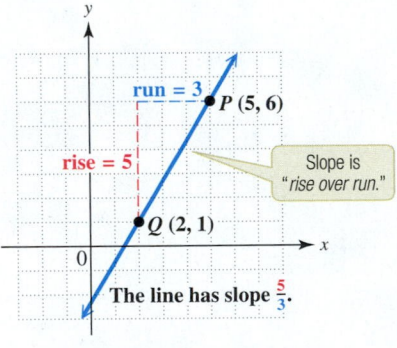

FIGURE 18

Remember from **Section 2.6** that one way to compare two numbers is by using a ratio. **Slope** is a ratio of the vertical change in *y* to the horizontal change in *x*. The line in **FIGURE 18** has

$$\text{slope} = \frac{\text{vertical change in } y \text{ (rise)}}{\text{horizontal change in } x \text{ (run)}} = \frac{5}{3}.$$

To confirm this ratio, we can count grid squares. We start at point Q in **FIGURE 18** and count *up* 5 grid squares to find the vertical change (rise). To find the horizontal change (run) and arrive at point P, we count to the *right* 3 grid squares. The slope is $\frac{5}{3}$, as found above.

We can summarize this discussion as follows.

Slope of a Line

Slope is a single number that allows us to determine the direction in which a line is slanting from left to right, as well as how much slant there is to the line.

**NOW TRY
EXERCISE 1**

Find the slope of the line.

EXAMPLE 1 Finding the Slope of a Line

Find the slope of the line in **FIGURE 19**.

We use the coordinates of the two points shown on the line. The vertical change is the difference in the *y*-values.

$$-1 - 3 = -4$$

The horizontal change is the difference in the *x*-values.

$$6 - 2 = 4$$

Thus, the line has

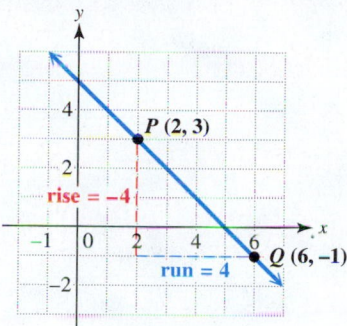

FIGURE 19

$$\text{slope} = \frac{\text{change in } y \text{ (rise)}}{\text{change in } x \text{ (run)}} = \frac{-4}{4}, \text{ or } -1.$$

Counting grid squares, we begin at point *P* and count *down* 4 grid squares. Then we count to the *right* 4 grid squares to reach point *Q*. Because we counted down, we write the vertical change as a negative number, -4 here. The slope is $\frac{-4}{4}$, or -1.

NOW TRY

NOTE *The slope of a line is the same for any two points on the line.* In **FIGURE 19**, find the points $(3, 2)$ and $(5, 0)$ on the line. If we start at $(3, 2)$ and count *down* 2 units and then to the *right* 2 units, we arrive at $(5, 0)$. The slope is $\frac{-2}{2}$, or -1, the same slope we found in **Example 1.**

The concept of slope is used in many everyday situations. See **FIGURE 20**.

• A highway with a 10%, or $\frac{1}{10}$, grade (or slope) rises 1 m for every 10 m horizontally.

• A roof with pitch (or slope) $\frac{5}{12}$ rises 5 ft for every 12 ft that it runs horizontally.

• A stairwell with slope $\frac{8}{12}$ $\left(\text{or } \frac{2}{3}\right)$ indicates a vertical rise of 8 ft for a horizontal run of 12 ft.

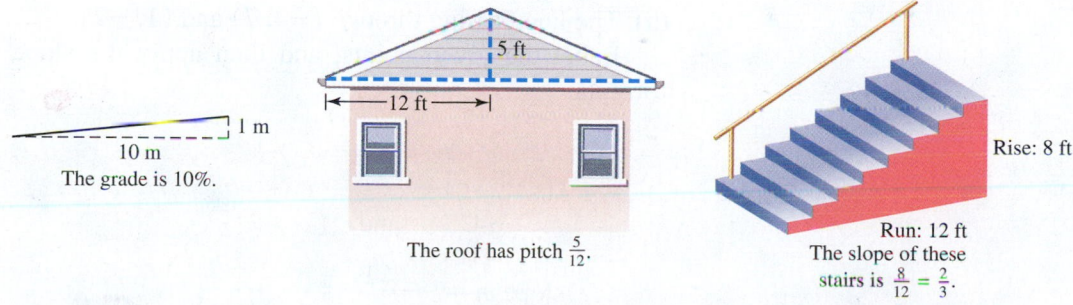

FIGURE 20

We can generalize the preceding discussion and find the slope of a line through two nonspecific points (x_1, y_1) and (x_2, y_2). (This notation is called **subscript notation.** Read x_1 as "*x*-sub-one" and x_2 as "*x*-sub-two.") See **FIGURE 21** on the next page.

NOW TRY ANSWER

1. 3

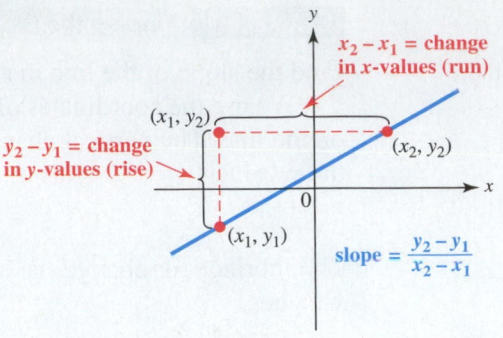

FIGURE 21

Moving along the line from the point (x_1, y_1) to the point (x_2, y_2), we see that y changes by $y_2 - y_1$ units. This is the vertical change (rise). Similarly, x changes by $x_2 - x_1$ units, which is the horizontal change (run). The slope of the line is the ratio of $y_2 - y_1$ to $x_2 - x_1$.

Slope Formula

The **slope** m of the line passing through the points (x_1, y_1) and (x_2, y_2) is defined as follows. (Traditionally, the letter m represents slope.)

$$m = \frac{\text{change in } y}{\text{change in } x} = \frac{y_2 - y_1}{x_2 - x_1} \quad (\text{where } x_1 \neq x_2)$$

The slope gives the change in y for each unit of change in x.

NOTE Subscript notation is used to identify a point. It does *not* indicate an operation. **Note the difference between x_2, which represents a nonspecific value, and x^2, which means $x \cdot x$.** Read x_2 as "x-sub-two," *not* "x squared."

EXAMPLE 2 Finding Slopes of Lines

Find the slope of each line.

(a) The line passing through $(-4, 7)$ and $(1, -2)$

Label the given points, and then apply the slope formula.

$$(x_1, y_1) \qquad (x_2, y_2)$$
$$(-4, 7) \quad \text{and} \quad (1, -2)$$

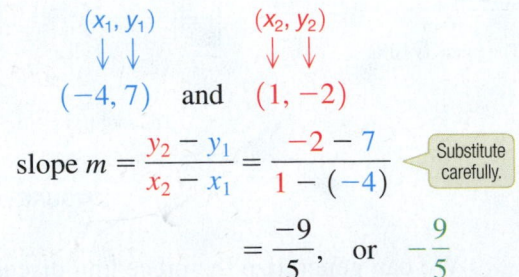

$$\text{slope } m = \frac{y_2 - y_1}{x_2 - x_1} = \frac{-2 - 7}{1 - (-4)} \quad \boxed{\text{Substitute carefully.}}$$

$$= \frac{-9}{5}, \quad \text{or} \quad -\frac{9}{5}$$

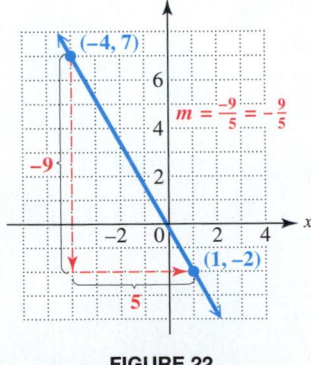

FIGURE 22

Begin at $(-4, 7)$ and count grid squares in **FIGURE 22** to confirm that the slope is $\frac{-9}{5}$, or $-\frac{9}{5}$.

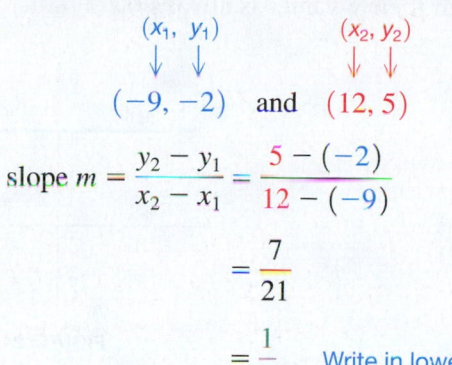

NOW TRY EXERCISE 2

Find the slope of the line passing through $(4, -5)$ and $(-2, -4)$.

(b) The line passing through $(-9, -2)$ and $(12, 5)$
Label the points, and then apply the slope formula.

$$\overset{(x_1,\ y_1)}{(-9, -2)} \quad \text{and} \quad \overset{(x_2,\ y_2)}{(12, 5)}$$

$$\text{slope } m = \frac{y_2 - y_1}{x_2 - x_1} = \frac{5 - (-2)}{12 - (-9)}$$

$$= \frac{7}{21}$$

$$= \frac{1}{3} \qquad \text{Write in lowest terms.}$$

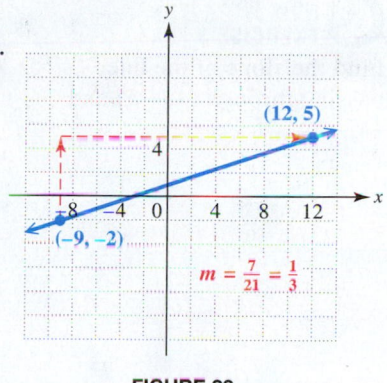

FIGURE 23

Confirm this calculation using **FIGURE 23**. (Note the scale on the x- and y-axes.)

The same slope is obtained if we label the points in reverse order. *It makes no difference which point is identified as* (x_1, y_1) *or* (x_2, y_2).

$$\overset{(x_2,\ y_2)}{(-9, -2)} \quad \text{and} \quad \overset{(x_1,\ y_1)}{(12, 5)}$$

$$\text{slope } m = \frac{y_2 - y_1}{x_2 - x_1} = \frac{-2 - 5}{-9 - 12} \qquad \text{Substitute.}$$

$$= \frac{-7}{-21} \qquad \text{Subtract.}$$

$$= \frac{1}{3} \qquad \text{The same slope results.}$$

NOW TRY

The slopes of the lines in **FIGURES 22** and **23** suggest the following.

Orientation of Lines with Positive and Negative Slopes

A line with positive slope rises (slants up) from left to right.

A line with negative slope falls (slants down) from left to right.

EXAMPLE 3 Finding the Slope of a Horizontal Line

Find the slope of the line passing through $(-5, 4)$ and $(2, 4)$.

$$\overset{(x_1,\ y_1)}{(-5, 4)} \quad \text{and} \quad \overset{(x_2,\ y_2)}{(2, 4)} \qquad \text{Label the points.}$$

$$m = \frac{y_2 - y_1}{x_2 - x_1} = \frac{4 - 4}{2 - (-5)} \qquad \text{Substitute in the slope formula.}$$

$$= \frac{0}{7} \qquad \text{Subtract.}$$

$$= 0 \qquad \text{Slope 0}$$

NOW TRY ANSWER

2. $-\frac{1}{6}$

**NOW TRY
EXERCISE 3**
Find the slope of the line
passing through $(1, -3)$ and
$(4, -3)$.

As shown in **FIGURE 24**, the line passing through the two points $(-5, 4)$ and $(2, 4)$ is horizontal, with equation $y = 4$. *All horizontal lines have slope 0* because the difference in their y-values is always 0.

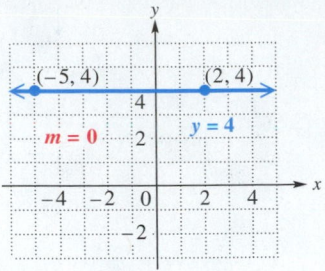

FIGURE 24

NOW TRY

**NOW TRY
EXERCISE 4**
Find the slope of the line
passing through $(-2, 1)$ and
$(-2, -4)$.

EXAMPLE 4   Applying the Slope Concept to a Vertical Line

Find the slope of the line passing through $(6, 2)$ and $(6, -4)$.

$$(x_1, y_1) \qquad (x_2, y_2)$$
$$\downarrow \downarrow \qquad\quad \downarrow \downarrow$$
$$(6, 2) \quad \text{and} \quad (6, -4) \qquad \text{Label the points.}$$

$$m = \frac{y_2 - y_1}{x_2 - x_1} = \frac{-4 - 2}{6 - 6} \qquad \text{Substitute in the slope formula.}$$

$$= \frac{-6}{\mathbf{0}} \qquad \text{Undefined slope}$$

Because division by 0 is undefined, this line has undefined slope. (This is why the slope formula has the restriction $x_1 \neq x_2$.)

The graph in **FIGURE 25** shows that this line is vertical, with equation $x = 6$. All points on a vertical line have the same x-value, so *the slope of any vertical line is undefined.*

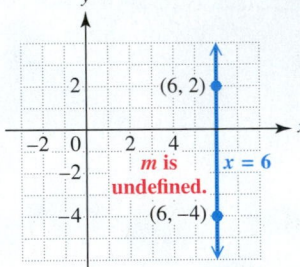

FIGURE 25

NOW TRY

> **Slopes of Horizontal and Vertical Lines**
>
> A **horizontal line,** which has an equation of the form $y = b$ (where b is a constant (number)), has **slope 0.**
>
> A **vertical line,** which has an equation of the form $x = a$ (where a is a constant (number)), has **undefined slope.**

FIGURE 26 summarizes the four cases for slopes of lines.

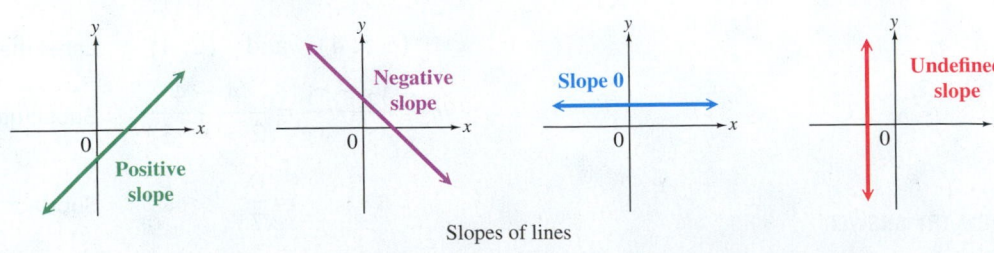

Slopes of lines

FIGURE 26

NOW TRY ANSWERS
3. 0
4. undefined slope

OBJECTIVE 2   Find the slope from the equation of a line.

Consider this linear equation.

$$y = -3x + 5$$

We can find the slope of this line using any two points on the line. Because the equation is solved for y, we find two points by choosing two different values of x and then finding the corresponding values of y. We arbitrarily choose $x = -2$ and $x = 4$.

$y = -3x + 5$		$y = -3x + 5$	
$y = -3(-2) + 5$	Let $x = -2$.	$y = -3(4) + 5$	Let $x = 4$.
$y = 6 + 5$	Multiply.	$y = -12 + 5$	Multiply.
$y = 11$	Add.	$y = -7$	Add.

The ordered pairs are $(-2, 11)$ and $(4, -7)$. Now we apply the slope formula.

$$m = \frac{-7 - 11}{4 - (-2)} = \frac{-18}{6} = -3$$

The slope, -3, is the same number as the coefficient of x in the given equation $y = -3x + 5$. It can be shown that this always happens, *as long as the equation is solved for y.* This fact is used to find the slope of a line from its equation.

Finding the Slope of a Line from Its Equation

Step 1   Solve the equation for y. (See **Section 2.5.**)

Step 2   The slope is given by the coefficient of x.

EXAMPLE 5   Finding Slopes from Equations

Find the slope of each line.

(a) $2x - 5y = 4$

Step 1   Solve the equation for y.

$$2x - 5y = 4 \quad \text{Isolate } y \text{ on one side.}$$

$$-5y = 4 - 2x \qquad \text{Subtract } 2x.$$

$$-5y = -2x + 4 \qquad \text{Commutative property}$$

$$\boxed{\frac{-2x}{-5} = \frac{-2}{-5}x = \frac{2}{5}x} \quad y = \frac{2}{5}x - \frac{4}{5} \qquad \text{Divide } each \text{ term by } -5.$$

$$\uparrow \atop \text{Slope}$$

Step 2   The slope is given by the coefficient of x, so the slope is $\frac{2}{5}$.

(b) $$8x + 4y = 1$$

$$\text{Solve for } y. \quad 4y = 1 - 8x \qquad \text{Subtract } 8x.$$

$$4y = -8x + 1 \qquad \text{Commutative property}$$

$$y = -2x + \frac{1}{4} \qquad \text{Divide } each \text{ term by } 4.$$

The slope is given by the coefficient of x, which is -2.

NOW TRY EXERCISE 5

Find the slope of the line.

$$3x + 5y = -1$$

(c)

$$3y + x = -3$$

> We omit the step showing the commutative property.

$$3y = -x - 3 \qquad \text{Subtract } x.$$

$$y = \frac{-x}{3} - 1 \qquad \text{Divide } \textit{each} \text{ term by 3.}$$

> The slope is $-\frac{1}{3}$, *not* $\frac{-x}{3}$ or $-\frac{x}{3}$.

$$y = -\frac{1}{3}x - 1 \qquad \frac{-x}{3} = \frac{-1x}{3} = -\frac{1}{3}x$$

The coefficient of x is $-\frac{1}{3}$, so the slope of this line is $-\frac{1}{3}$.

NOW TRY

NOTE We can solve the linear equation $Ax + By = C$ (where $B \neq 0$) for y to show that, in general, the slope of a line is $m = -\frac{A}{B}$.

OBJECTIVE 3 **Use slopes to determine whether two lines are parallel, perpendicular, or neither.**

Two lines in a plane that never intersect are **parallel.** We use slopes to tell whether two lines are parallel.

FIGURE 27 on the next page shows the graphs of $x + 2y = 4$ and $x + 2y = -6$. These lines appear to be parallel. We solve each equation for y to find the slope.

$$x + 2y = 4 \qquad\qquad\qquad\qquad x + 2y = -6$$

$$2y = -x + 4 \quad \text{Subtract } x. \qquad\qquad 2y = -x - 6 \quad \text{Subtract } x.$$

$$y = \frac{-x}{2} + 2 \quad \text{Divide by 2.} \qquad\qquad y = \frac{-x}{2} - 3 \quad \text{Divide by 2.}$$

$$y = -\frac{1}{2}x + 2 \quad \frac{-x}{2} = \frac{-1x}{2} = -\frac{1}{2}x \qquad\qquad y = -\frac{1}{2}x - 3 \quad \frac{-x}{2} = \frac{-1x}{2} = -\frac{1}{2}x$$

> The slope is $-\frac{1}{2}$, not $-\frac{x}{2}$.

↑ Slope   ↑ Slope

Both lines have slope $-\frac{1}{2}$. *Nonvertical parallel lines always have equal slopes.*

FIGURE 28 shows the graphs of $x + 2y = 4$ and $2x - y = 6$. These lines appear to be **perpendicular** (that is, they intersect at a 90° angle). As shown above, solving $x + 2y = 4$ for y gives $y = -\frac{1}{2}x + 2$, with slope $-\frac{1}{2}$. We solve $2x - y = 6$ for y.

$$2x - y = 6$$

$$-y = -2x + 6 \qquad \text{Subtract } 2x.$$

$$y = 2x - 6 \qquad \text{Multiply by } -1.$$

↑ Slope

The product of the slopes of the two lines is

$$-\frac{1}{2}(2) = -1.$$

NOW TRY ANSWER

5. $-\frac{3}{5}$

▼ Negative Reciprocals

Number	Negative Reciprocal
$\frac{3}{4}$	$-\frac{4}{3}$
$\frac{1}{2}$	$-\frac{2}{1}$, or -2
-6, or $-\frac{6}{1}$	$\frac{1}{6}$
-0.4, or $-\frac{4}{10}$	$\frac{10}{4}$, or 2.5

The product of each number and its negative reciprocal is **−1.**

It can be proved that the product of the slopes of two perpendicular lines, neither of which is vertical, is always −**1.** This means that the slopes of perpendicular lines are negative (or opposite) reciprocals—if one slope is the nonzero number a, then the other is $-\frac{1}{a}$.

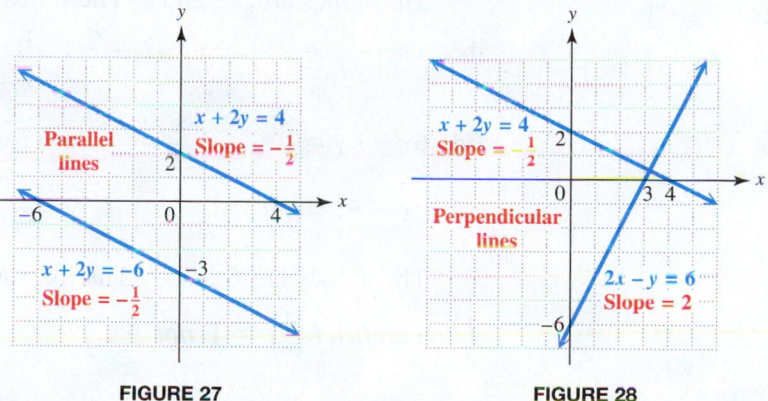

FIGURE 27   FIGURE 28

Slopes of Parallel and Perpendicular Lines

Two lines with the same slope are parallel.

Two lines whose slopes have a product of -1 are perpendicular.

EXAMPLE 6 Deciding Whether Two Lines Are Parallel or Perpendicular

Decide whether each pair of lines is *parallel, perpendicular,* or *neither.*

(a) $x + 3y = 7$

$-3x + y = 3$

Find the slope of each line by first solving each equation for y.

$$x + 3y = 7 \qquad\qquad -3x + y = 3$$
$$3y = -x + 7 \quad \text{Subtract } x. \qquad y = 3x + 3 \quad \text{Add } 3x.$$
$$y = -\frac{1}{3}x + \frac{7}{3} \quad \text{Divide by 3.}$$

The slope is $-\frac{1}{3}$.   The slope is 3.

The slopes are not equal, so the lines are not parallel. Find the product of the slopes.

$$-\frac{1}{3}(3) = -1 \qquad \text{The slopes are negative reciprocals.}$$

The two lines are perpendicular because the product of their slopes is -1.

(b) $4x - y = 4$

$8x - 2y = -12$

Solve each equation for y, and identify the slope.

$$4x - y = 4 \qquad\qquad 8x - 2y = -12$$
$$-y = -4x + 4 \quad \text{Subtract } 4x. \qquad -2y = -8x - 12 \quad \text{Subtract } 8x.$$
$$y = 4x - 4 \quad \text{Multiply by } -1. \qquad y = 4x + 6 \quad \text{Divide by } -2.$$

Both lines have slope 4, so the lines are parallel.

NOW TRY
EXERCISE 6
Decide whether the pair of
lines is *parallel, perpendicular,*
or *neither.*

$$2x - 3y = 1$$
$$4x + 6y = 5$$

(c) $4x + 3y = 6$ $\xrightarrow{\text{Solve for } y.}$ $y = -\dfrac{4}{3}x + 2$

$2x - \ y = 5$ $\xrightarrow{\hspace{1cm}}$ $y = 2x - 5$

The slopes are $-\dfrac{4}{3}$ and 2. These two lines are neither parallel nor perpendicular,

because $-\dfrac{4}{3} \neq 2$ and $-\dfrac{4}{3} \cdot 2 \neq -1$.

(d) $6x - \ y = 1$ $\xrightarrow{\text{Solve for } y.}$ $y = 6x - 1$

$x - 6y = -12$ $\xrightarrow{\hspace{1cm}}$ $y = \dfrac{1}{6}x + 2$

The slopes are 6 and $\dfrac{1}{6}$. The lines are not parallel, nor are they perpendicular.

$\left(Be\ careful.\ 6\left(\dfrac{1}{6}\right) = 1,\ not\ -1.\right)$

NOW TRY

NOW TRY ANSWER
6. neither

3.3 Exercises

FOR
EXTRA
HELP

▶ MyMathLab®

▶ *Complete solution available*
in MyMathLab

Concept Check *Work each problem involving slope.*

1. Slope is a measure of the _____ of a line. Slope is the (*horizontal / vertical*) change compared to the (*horizontal / vertical*) change while moving along the line from one point to another.

2. Slope is the _____ of the vertical change in _____, called the (*rise / run*), to the horizontal change in _____, called the (*rise / run*).

3. Look at the graph at the right.

(a) Start at the point $(-1, -4)$ and count vertically up to the horizontal line that goes through the other plotted point. What is this vertical change? (Remember: "up" means positive, "down" means negative.)

(b) From this new position, count horizontally to the other plotted point. What is this horizontal change? (Remember: "right" means positive, "left" means negative.)

(c) What is the ratio (quotient) of the numbers found in parts (a) and (b)? What do we call this number?

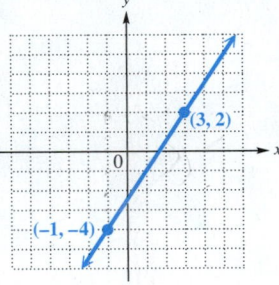

4. See **Exercise 3.** If we were to *start* at the point $(3, 2)$ and *end* at the point $(-1, -4)$ would the answer to **Exercise 3(c)** be the same? Explain.

5. Match the graph of each line in (a)–(d) with its slope in A–D. (Coordinates of the points shown are integers.)

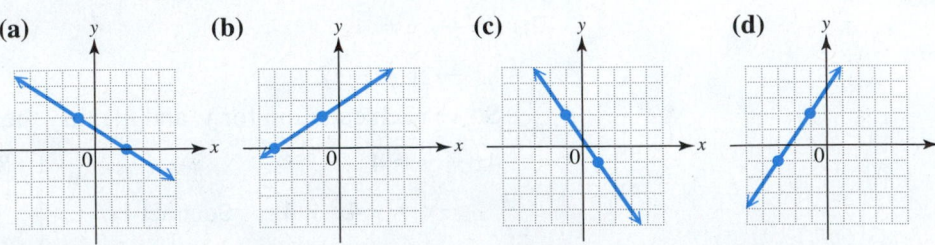

A. $\dfrac{2}{3}$ **B.** $\dfrac{3}{2}$ **C.** $-\dfrac{2}{3}$ **D.** $-\dfrac{3}{2}$

6. Decide whether the line with the given slope rises from left to right, falls from left to right, is horizontal, or is vertical.

(a) $m = -4$ (b) $m = 0$ (c) m is undefined. (d) $m = \dfrac{3}{7}$

Concept Check *On a pair of axes similar to the one shown, sketch the graph of a straight line having the indicated slope.*

7. Negative   **8.** Positive

9. Undefined   **10.** Zero

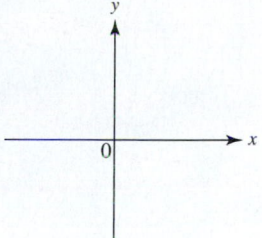

Concept Check *The figure at the right shows a line that has a positive slope (because it rises from left to right) and a positive y-value for the y-intercept (because it intersects the y-axis above the origin).*

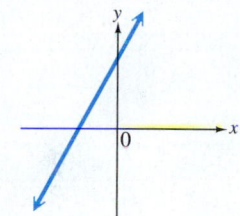

For each line in Exercises 11–16, decide whether

(a) *the slope is positive, negative, or zero and*

(b) *the y-value of the y-intercept is positive, negative, or zero.*

11.

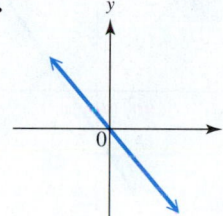

12.

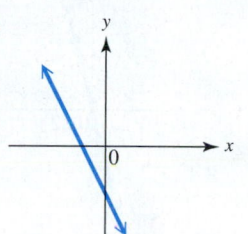

13.

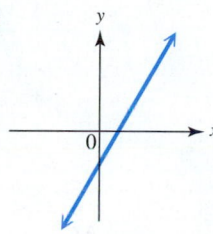

14.

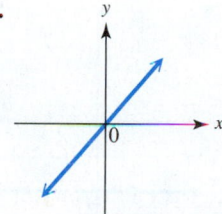

15.

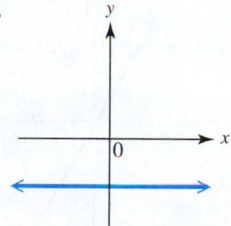

16.

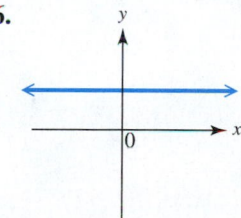

17. *Concept Check* A student was asked to find the slope of the line through the points $(2, 5)$ and $(-1, 3)$. His answer, $-\dfrac{2}{3}$, was incorrect. He showed his work as

$$\frac{3 - 5}{2 - (-1)} = \frac{-2}{3} = -\frac{2}{3}.$$

WHAT WENT WRONG? Give the correct slope.

18. *Concept Check* A student was asked to find the slope of the line through the points $(-2, 4)$ and $(6, -1)$. Her answer, $-\dfrac{8}{5}$, was incorrect. She showed her work as

$$\frac{6 - (-2)}{-1 - 4} = \frac{8}{-5} = -\frac{8}{5}.$$

WHAT WENT WRONG? Give the correct slope.

Concept Check   *Find each slope.*

19. What is the slope (or grade) of this hill?

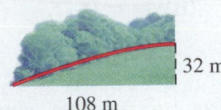

32 m
108 m

20. What is the slope (or pitch) of this roof?

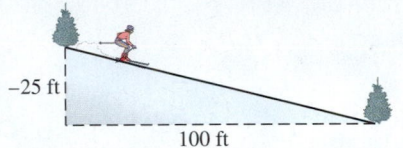

6 ft
20 ft

21. What is the slope of the slide? (*Hint:* The slide *drops* 8 ft vertically as it extends 12 ft horizontally.)

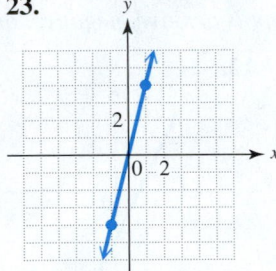

−8 ft
12 ft

22. What is the slope (or grade) of this ski slope? (*Hint:* The ski slope *drops* 25 ft vertically for every 100 horizontal feet.)

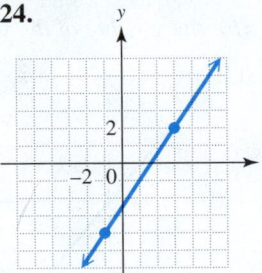

−25 ft
100 ft

Use the coordinates of the indicated points to find the slope of each line. (Coordinates of the points shown are integers.) **See Example 1.**

23.

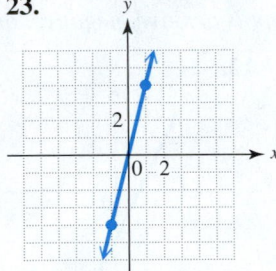

24.

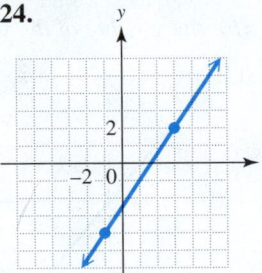

25.

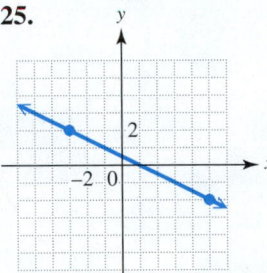

26.

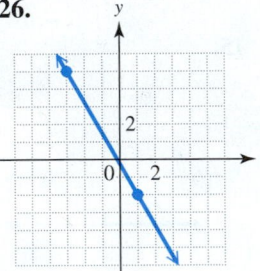

27.

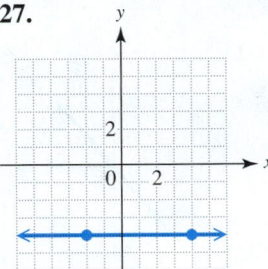

28.

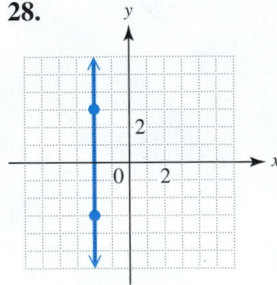

Find the slope of the line passing through each pair of points. **See Examples 2–4.**

29. $(1, -2)$ and $(-3, -7)$ **30.** $(4, -1)$ and $(-2, -8)$ **31.** $(0, 3)$ and $(-2, 0)$

32. $(8, 0)$ and $(0, -5)$ **33.** $(4, 3)$ and $(-6, 3)$ **34.** $(6, 5)$ and $(-12, 5)$

35. $(-2, 4)$ and $(-3, 7)$ **36.** $(-4, 5)$ and $(-5, 8)$

37. $(-12, 3)$ and $(-12, -7)$ **38.** $(-8, 6)$ and $(-8, -1)$

39. $(4.8, 2.5)$ and $(3.6, 2.2)$ **40.** $(3.1, 2.6)$ and $(1.6, 2.1)$

41. $\left(-\dfrac{7}{5}, \dfrac{3}{10}\right)$ and $\left(\dfrac{1}{5}, -\dfrac{1}{2}\right)$ **42.** $\left(-\dfrac{4}{3}, \dfrac{1}{2}\right)$ and $\left(\dfrac{1}{3}, -\dfrac{5}{6}\right)$

Find the slope of each line. See Example 5.

43. $y = 5x + 12$ **44.** $y = 2x + 3$ **45.** $4y = x + 1$

46. $2y = x + 4$ ▶ **47.** $3x - 2y = 3$ **48.** $6x - 4y = 4$

49. $-3x + 2y = 5$ **50.** $-2x + 4y = 5$ **51.** $x + y = -4$ **52.** $x - y = -2$

53. $y = -5$ **54.** $y = 4$ **55.** $x = 6$ **56.** $x = -2$

Find the slope of each line in two ways by doing the following.

(a) Give any two points that lie on the line, and use them to determine the slope.

(b) Solve the equation for y, and identify the slope from the equation.

See Objective 2 and Example 5.

57. $2x + y = 10$ **58.** $-4x + y = -8$ **59.** $5x - 3y = 15$ **60.** $3x + 2y = 12$

Each table of values gives several points that lie on a line.

(a) Use any two of the ordered pairs to find the slope of the line.

(b) What is the x-intercept of the line?   The y-intercept?

(c) Graph the line.

61.

x	y
-4	0
-2	2
0	4
1	5

62.

x	y
-4	3
-1	0
0	-1
2	-3

63.

x	y
3	-3
0	-2
-3	-1
-6	0

64.

x	y
-1	-6
0	-4
2	0
5	6

Concept Check *Answer each question.*

65. What is the slope of a line whose graph is

   (a) parallel to the graph of $3x + y = 7$?

   (b) perpendicular to the graph of $3x + y = 7$?

66. What is the slope of a line whose graph is

   (a) parallel to the graph of $-5x + y = -3$?

   (b) perpendicular to the graph of $-5x + y = -3$?

67. If two lines are both vertical or both horizontal, which of the following are they?

   A. Parallel   **B.** Perpendicular   **C.** Neither parallel nor perpendicular

68. If a line is vertical, what is true of any line that is perpendicular to it?

For each pair of equations, give the slopes of the lines and then determine whether the two lines are parallel, perpendicular, or neither. See Example 6.

▶ **69.** $2x + 5y = 4$
    $4x + 10y = 1$

70. $-4x + 3y = 4$
    $-8x + 6y = 0$

71. $8x - 9y = 6$
    $8x + 6y = -5$

72. $5x - 3y = -2$
    $3x - 5y = -8$

73. $3x - 2y = 6$
    $2x + 3y = 3$

74. $3x - 5y = -1$
    $5x + 3y = 2$

75. $5x - y = 1$
    $x - 5y = -10$

76. $3x - 4y = 12$
    $4x + 3y = 12$

The graph shows album sales (which include CD, vinyl, cassette, and digital albums) and music purchases (which include digital tracks, albums, singles, and music videos) in millions of units from 2004 through 2012. Use the graph to work Exercises 77 and 78.

77. Locate the line on the graph that represents music purchases.

    (a) Write two ordered pairs (x, y), where x is the year and y is purchases in millions of units, to represent the data for the years 2004 and 2012.

    (b) Use the ordered pairs from part (a) to find the slope of the line.

    (c) Interpret the meaning of the slope in the context of this problem.

78. Locate the line on the graph that represents album sales. Repeat parts (a)–(c) of **Exercise 77.** For part (a), x is the year and y is sales in millions of units.

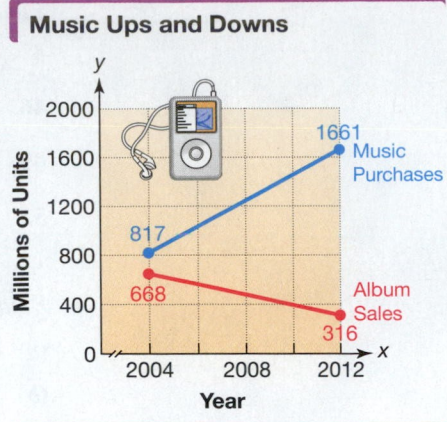

Music Ups and Downs

Source: Nielsen SoundScan.

RELATING CONCEPTS   For Individual or Group Work   (Exercises 79–84)

FIGURE A *gives the percent of freshmen at 4-year colleges and universities who planned to major in the Biological Sciences.* **FIGURE B** *shows the percent of the same group of students who planned to major in Business.* **Work Exercises 79–84 in order.**

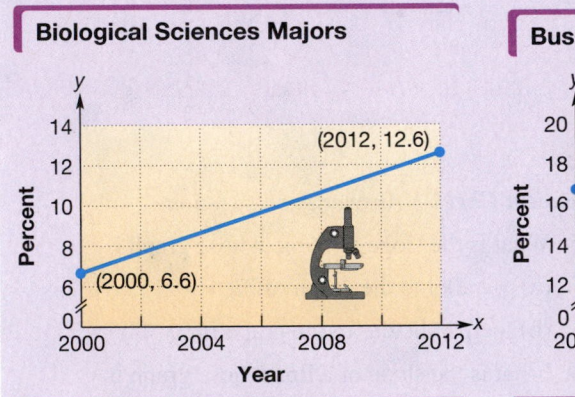

Biological Sciences Majors

Source: Higher Education Research Institute.

FIGURE A

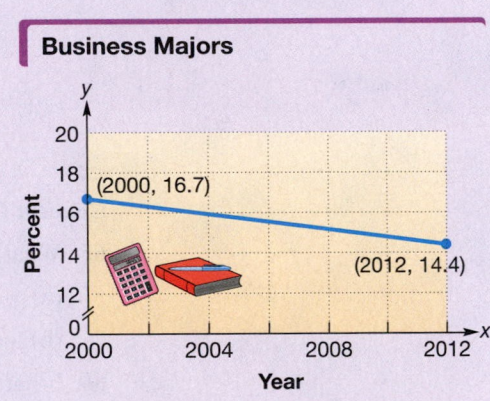

Business Majors

Source: Higher Education Research Institute.

FIGURE B

79. Use the given ordered pairs to find the slope of the line in **FIGURE A**.

80. The slope of the line in **FIGURE A** is (*positive* / *negative*). This means that during the period represented, the percent of freshmen planning to major in the Biological Sciences (*increased* / *decreased*).

81. The slope of a line represents the *rate of change*. Based on **FIGURE A**, what was the increase in the percent of freshmen *per year* who planned to major in the Biological Sciences during the period shown?

82. Use the given ordered pairs to find the slope of the line in **FIGURE B** to the nearest tenth.

83. The slope of the line in **FIGURE B** is (*positive* / *negative*). This means that during the period represented, the percent of freshmen planning to major in Business (*increased* / *decreased*).

84. Based on **FIGURE B**, what was the decrease in the percent of freshmen *per year* who planned to major in Business?

Preparing for Your Math Final Exam

Your math final exam is likely to be a comprehensive exam, which means it will cover material from the entire term. **One way to prepare for it now is by working a set of Cumulative Review Exercises** each time your class finishes a chapter. This continual review will help you remember concepts and procedures as you progress through the course.

Final Exam Preparation Suggestions

1. **Figure out the grade you need to earn on the final exam to get the course grade you want.** Check your course syllabus for grading policies, or ask your instructor if you are not sure.

   How many points do you need to earn on your math final exam to get the grade you want?

2. **Create a final exam week plan.** Set priorities that allow you to spend extra time studying. This may mean making adjustments, in advance, in your work schedule or enlisting extra help with family responsibilities.

   What adjustments do you need to make for final exam week? List two or three here.

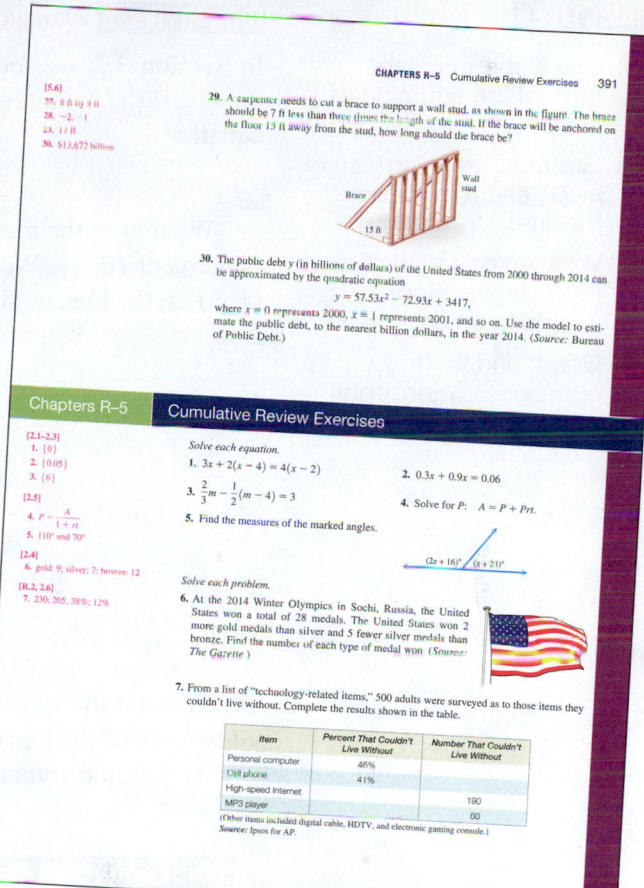

3. **Use the following suggestions to guide your studying.**

   - **Begin reviewing several days before the final exam.** DON'T wait until the last minute.

   - **Know exactly which chapters and sections will be covered on the exam.**

   - **Divide up the chapters.** Decide how much you will review each day.

   - **Keep returned quizzes and tests.** Use them to review.

   - **Practice all types of problems. Use the Cumulative Review Exercises** at the end of each chapter in your textbook beginning in Chapter 2. All answers, with section references, are given in the margins.

   - **Review or rewrite your notes** to create summaries of important information.

   - **Make study cards for all types of problems.** Carry the cards with you, and review them whenever you have a few minutes.

   - **Take plenty of short breaks as you study to reduce physical and mental stress.** Exercising, listening to music, and enjoying a favorite activity are effective stress busters.

   Finally, *DON'T* stay up all night the night before an exam—*get a good night's sleep.*

   Which of these suggestions will you use as you study for your math final exam? List two or three here.

3.4 Slope-Intercept Form of a Linear Equation

OBJECTIVE 1 Use slope-intercept form of the equation of a line.

In **Section 3.3,** we found the slope (steepness) of a line by solving the equation of the line for y. In that form, the slope is the coefficient of x. For example, the line with equation

$$y = 2x + 3 \quad \text{has slope} \quad 2.$$

What does the number 3 represent? To find out, suppose a line has slope m and y-intercept $(0, b)$. We can find an equation of this line by choosing another point (x, y) on the line, as shown in **FIGURE 29.** Then we apply the slope formula.

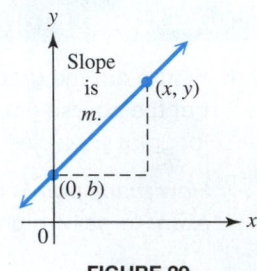

$$m = \frac{y - b}{x - 0} \quad \leftarrow \text{Change in } y\text{-values}$$
$$\phantom{m = \frac{y - b}{x - 0}} \quad \leftarrow \text{Change in } x\text{-values}$$

$$m = \frac{y - b}{x} \quad \text{Subtract in the denominator.}$$

$$mx = y - b \quad \text{Multiply by } x.$$

$$mx + b = y \quad \text{Add } b.$$

$$y = mx + b \quad \text{Rewrite.}$$

FIGURE 29

This result is the *slope-intercept form* of the equation of a line. Both the *slope* and the *y-intercept* of the line can be read directly from this form. For the line with equation $y = 2x + 3$, the number 3 gives the y-intercept $(0, 3)$.

Slope-Intercept Form

The **slope-intercept form** of the equation of a line with slope m and y-intercept $(0, b)$ is

$$y = mx + b.$$

Slope ⤴ ⤴ $(0, b)$ is the y-intercept.

The intercept given is the y-intercept.

EXAMPLE 1 Identifying Slopes and *y*-Intercepts

Identify the slope and y-intercept of the line with each equation.

(a) $y = -4x + 1$

Slope ⤴ ⤴ y-intercept $(0, 1)$

(b) $y = x - 8$ can be written as $y = 1x + (-8)$.

Slope ⤴ ⤴ y-intercept $(0, -8)$

(c) $y = 6x$ can be written as $y = 6x + 0$.

Slope ⤴ ⤴ y-intercept $(0, 0)$

(d) $y = \dfrac{x}{4} - \dfrac{3}{4}$ can be written as $y = \dfrac{1}{4}x + \left(-\dfrac{3}{4}\right)$.

Slope ⤴ ⤴ y-intercept $\left(0, -\dfrac{3}{4}\right)$

NOW TRY

NOTE Slope-intercept form is an especially useful form for a linear equation because of the information we can determine from it. It is also the form used by graphing calculators and the one that describes a *linear function*.

OBJECTIVE 2 Graph a line using its slope and a point on the line.

We can use the slope and the point represented by the *y*-intercept to graph a line.

Graphing a Line Using the Slope and *y*-Intercept

Step 1 Write the equation in slope-intercept form $y = mx + b$, if necessary, by solving for *y*.

Step 2 Identify the *y*-intercept. Plot the point $(0, b)$.

Step 3 Identify the slope *m* of the line. Use the geometric interpretation of slope ("*rise over run*") to find another point on the graph by counting from the *y*-intercept.

Step 4 Join the two points with a line to obtain the graph. (If desired, obtain a third point, such as the *x*-intercept, as a check.)

EXAMPLE 2 Graphing Lines Using Slopes and *y*-Intercepts

Graph the equation of each line using the slope and *y*-intercept.

(a) $y = \dfrac{2}{3}x - 1$

    Step 1 The equation is in slope-intercept form.

$$y = \dfrac{2}{3}x - 1$$

           ↑   ↑

    Slope   Value of *b* in *y*-intercept $(0, b)$

    Step 2 The *y*-intercept is $(0, -1)$. Plot this point. See **FIGURE 30**.

    Step 3 The slope is $\dfrac{2}{3}$. By definition,

$$\text{slope } m = \dfrac{\text{change in } y \text{ (rise)}}{\text{change in } x \text{ (run)}} = \dfrac{2}{3}.$$

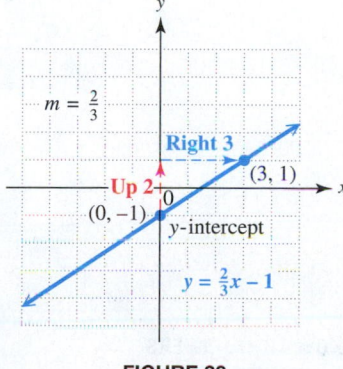

From the *y*-intercept, count up 2 units and to the right 3 units to the point $(3, 1)$.

    Step 4 Draw the line through the points $(0, -1)$ and $(3, 1)$ to obtain the graph in **FIGURE 30**.

FIGURE 30

(b) $3x + 4y = 8$

    Step 1 Solve for *y* to write the equation in slope-intercept form.

$$3x + 4y = 8$$

    Isolate y on one side. $4y = -3x + 8$ Subtract 3*x*.

Slope-intercept form → $y = -\dfrac{3}{4}x + 2$ Divide by 4.

**NOW TRY
EXERCISE 2**

Graph $3x + 2y = 8$ using the slope and y-intercept.

Step 2   The y-intercept in $y = -\frac{3}{4}x + 2$ is $(0, 2)$. Plot this point. See **FIGURE 31**.

Step 3   The slope is $-\frac{3}{4}$, which can be written as either $\frac{-3}{4}$ or $\frac{3}{-4}$. We use $\frac{-3}{4}$ here.

$$m = \frac{\text{change in } y \text{ (rise)}}{\text{change in } x \text{ (run)}} = \frac{-3}{4}$$

From the y-intercept, count *down* 3 units (because of the negative sign) and to the right 4 units to the point $(4, -1)$.

Step 4   Draw the line through the two points $(0, 2)$ and $(4, -1)$ to obtain the graph in **FIGURE 31**.

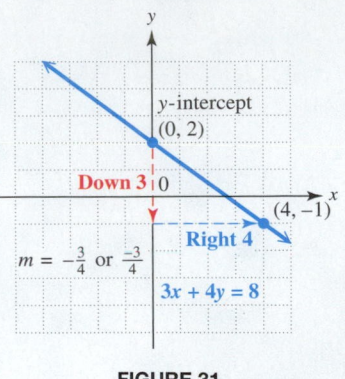

FIGURE 31

NOW TRY

NOTE   In Step 3 of **Example 2(b)**, we could use $\frac{3}{-4}$ for the slope. From the y-intercept $(0, 2)$ in **FIGURE 31**, count up 3 units and to the *left* 4 units (because of the negative sign) to the point $(-4, 5)$. Verify that this produces the same line.

**NOW TRY
EXERCISE 3**

Graph the line passing through the point $(-3, -4)$, with slope $\frac{5}{2}$.

EXAMPLE 3   Graphing a Line Using Its Slope and a Point

Graph the line passing through the point $(-2, 3)$, with slope -4.

First, plot the point $(-2, 3)$. See **FIGURE 32**. Then write the slope -4 as

$$m = \frac{\text{change in } y \text{ (rise)}}{\text{change in } x \text{ (run)}} = \frac{-4}{1}.$$

Locate another point on the line by counting *down* 4 units from $(-2, 3)$ and then to the right 1 unit. Finally, draw the line through this new point P and the given point $(-2, 3)$. See **FIGURE 32**.

We could have written the slope as $\frac{4}{-1}$ instead. In this case, we would move up 4 units from $(-2, 3)$ and then to the *left* 1 unit. Verify that this produces the same line.

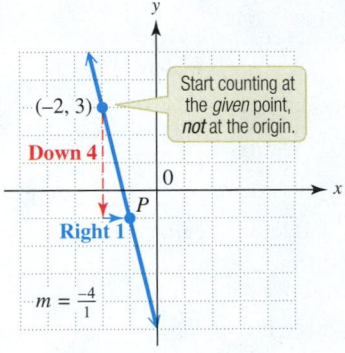

FIGURE 32

NOW TRY

OBJECTIVE 3   Write an equation of a line using its slope and any point on the line.

We can use the slope-intercept form to write the equation of a line if we know the slope and any point on the line.

NOW TRY ANSWERS

2.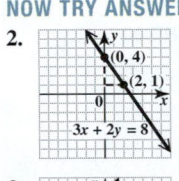

3.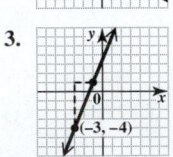

EXAMPLE 4   Using Slope-Intercept Form to Write Equations

Write an equation in slope-intercept form of the line passing through the given point and having the given slope.

(a) $(0, -1)$, $m = \frac{2}{3}$

Because the point $(0, -1)$ is the y-intercept, $b = -1$. We can substitute this value for b and the given slope $m = \frac{2}{3}$ directly into slope-intercept form $y = mx + b$ to write an equation.

**NOW TRY
EXERCISE 4**

Write an equation in slope-intercept form of the line passing through the given point and having the given slope.

(a) $(0, 2)$, $m = -4$

(b) $(-2, 1)$, $m = 3$

Slope $\downarrow$ $\downarrow$ y-intercept is $(0, b)$.

$$y = mx + b \qquad \text{Slope-intercept form}$$

$$y = \frac{2}{3}x + (-1) \qquad \text{Substitute.}$$

$$y = \frac{2}{3}x - 1 \qquad \text{Definition of subtraction}$$

(b) $(2, 5)$, $m = 4$

This line passes through the point $(2, 5)$, which is **not** the y-intercept because the x-coordinate is 2, **not** 0. **We cannot substitute directly as in part (a).** We can find the y-intercept by substituting $x = 2$ and $y = 5$ from the given point and the given slope $m = 4$ into $y = mx + b$ and solving for b.

$$y = mx + b \qquad \text{Slope-intercept form}$$

$$5 = 4(2) + b \qquad \text{Let } x = 2,\ y = 5,\ \text{and } m = 4.$$

$(0, b)$ is the y-intercept. Don't stop here.

$$5 = 8 + b \qquad \text{Multiply.}$$

$$-3 = b \qquad \text{Subtract 8.}$$

Now substitute the values of m and b into slope-intercept form.

$$y = mx + b \qquad \text{Slope-intercept form}$$

$$y = 4x - 3 \qquad \text{Let } m = 4 \text{ and } b = -3. \qquad \text{NOW TRY}$$

OBJECTIVE 4 Graph and write equations of horizontal and vertical lines.

EXAMPLE 5 Graphing Horizontal and Vertical Lines Using Slope and a Point

Graph each line passing through the given point and having the given slope.

(a) $(4, -2)$, $m = 0$

**NOW TRY
EXERCISE 5**

Graph each line passing through the given point and having the given slope.

(a) $(-3, 3)$, undefined slope

(b) $(3, -3)$, slope 0

Recall from **Section 3.3** that horizontal lines have slope 0. To graph this line, plot the point $(4, -2)$ and draw the horizontal line through it. See **FIGURE 33**.

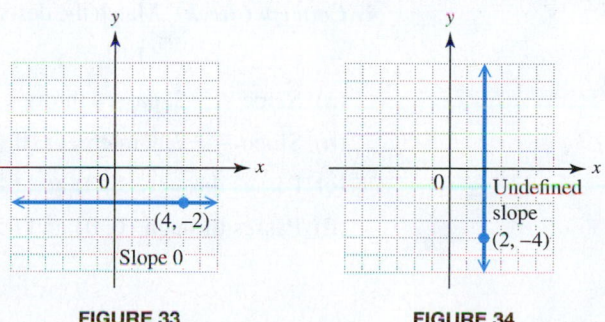

FIGURE 33   **FIGURE 34**

(b) $(2, -4)$, undefined slope

Vertical lines have undefined slope. To graph this line, plot the point $(2, -4)$ and draw the vertical line through it. See **FIGURE 34**.   **NOW TRY**

NOW TRY ANSWERS

4. (a) $y = -4x + 2$

  (b) $y = 3x + 7$

5. (a)

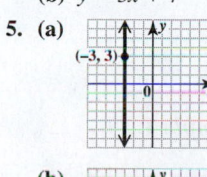

  (b)

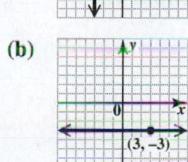

 **NOW TRY
EXERCISE 6**

Write an equation of the line passing through the point $(-1, 1)$ and having the given slope.

(a) Undefined slope

(b) $m = 0$

EXAMPLE 6 **Writing Equations of Horizontal and Vertical Lines**

Write an equation of the line passing through the point $(2, -2)$ and having the given slope.

(a) Slope 0

This line is horizontal because it has slope 0. Recall that a horizontal line through the point (a, b) has equation $y = b$. The y-coordinate of the point $(2, -2)$ is -2, so the equation is $y = -2$. See **FIGURE 35**.

(b) Undefined slope

This line is vertical because it has undefined slope. A vertical line through the point (a, b) has equation $x = a$. The x-coordinate of $(2, -2)$ is 2, so the equation is $x = 2$. See **FIGURE 35**.

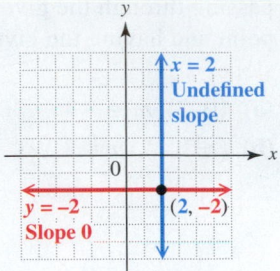

FIGURE 35

NOW TRY

NOW TRY ANSWERS

6. (a) $x = -1$ **(b)** $y = 1$

3.4 Exercises

FOR EXTRA HELP

 MyMathLab®

● *Complete solution available in MyMathLab*

Concept Check *Fill in each blank with the correct response.*

1. In slope-intercept form $y = mx + b$ of the equation of a line, the slope is _____ and the y-intercept is _____ .

2. The line with equation $y = -\dfrac{x}{2} - 3$ has slope _____ and y-intercept _____ .

3. *Concept Check* Match each equation in parts (a)–(d) with the graph in A–D that would most closely resemble its graph.

(a) $y = x + 3$ **(b)** $y = -x + 3$ **(c)** $y = x - 3$ **(d)** $y = -x - 3$

A. 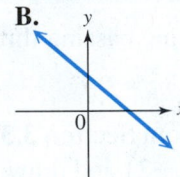   **B.** 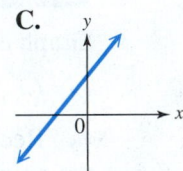   **C.** 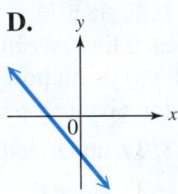   **D.**

4. *Concept Check* Match the description in Column I with the correct equation in Column II.

I	II
(a) Slope $= -2$, passes through $(4, 1)$	**A.** $y = 4x$
(b) Slope $= -2$, y-intercept $(0, 1)$	**B.** $y = \dfrac{1}{4}x$
(c) Passes through $(0, 0)$ and $(4, 1)$	**C.** $y = -4x$
(d) Passes through $(0, 0)$ and $(1, 4)$	**D.** $y = -2x + 1$
	E. $2x + y = 9$

Identify the slope and y-intercept of the line with each equation. **See Example 1.**

5. $y = \dfrac{5}{2}x - 4$

6. $y = \dfrac{7}{3}x - 6$

7. $y = -x + 9$

8. $y = x + 1$

9. $y = \dfrac{x}{5} - \dfrac{3}{10}$

10. $y = \dfrac{x}{7} - \dfrac{5}{14}$

Graph the equation of each line using the slope and y-intercept. **See Example 2.**

11. $y = 3x + 2$ **12.** $y = 4x - 4$ **13.** $y = -\dfrac{1}{3}x + 4$

14. $y = -\dfrac{1}{2}x + 2$ **15.** $2x + y = -5$ **16.** $3x + y = -2$

17. $4x - 5y = 20$ **18.** $6x - 5y = 30$

Graph each line passing through the given point and having the given slope. **See Examples 3 and 5.**

19. $(0, 1), m = 4$ **20.** $(0, -5), m = -2$ **21.** $(1, -5), m = -\dfrac{2}{5}$

22. $(2, -1), m = -\dfrac{1}{3}$ **23.** $(-1, 4), m = \dfrac{2}{5}$ **24.** $(-2, 2), m = \dfrac{3}{2}$

25. $(0, 0), m = -2$ **26.** $(0, 0), m = -3$ **27.** $(-2, 3), m = 0$

28. $(3, 2), m = 0$ **29.** $(2, 4)$, undefined slope   **30.** $(3, -2)$, undefined slope

31. $(5, -5)$, slope 0   **32.** $(-4, 4)$, slope 0

Concept Check *Answer each question.*

33. What is the common name given to a vertical line whose *x*-intercept is the origin?

34. What is the common name given to a line with slope 0 whose *y*-intercept is the origin?

Use the geometric interpretation of slope ("rise over run," from **Section 3.3**) *to find the slope of each line. Then, by identifying the y-intercept from the graph, write the slope-intercept form of the equation of the line. (Coordinates of the points shown are integers.)*

35. 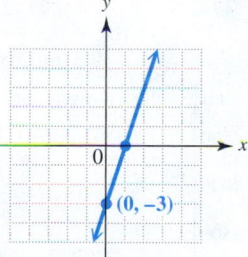   **36.** 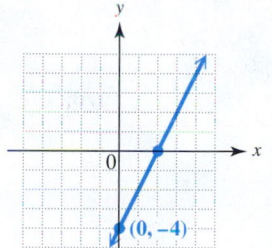   **37.**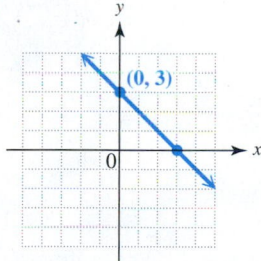

38. 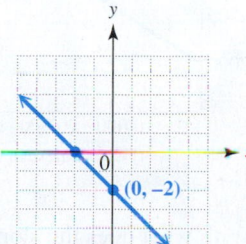   **39.** 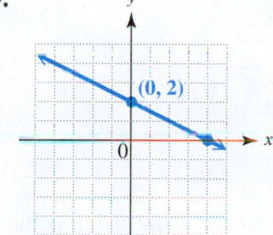   **40.**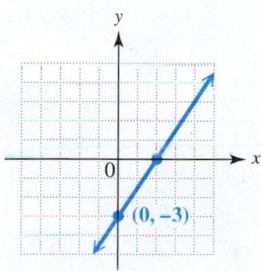

Write an equation in slope-intercept form (if possible) of the line passing through the given point and having the given slope. **See Examples 4 and 6.**

41. $(0, -3), m = 4$ **42.** $(0, 6), m = -5$ **43.** $(0, -7), m = -1$

44. $(0, -9), m = 1$ **45.** $(4, 1), m = 2$ **46.** $(2, 7), m = 3$

47. $(-1, 3)$, $m = -4$ **48.** $(-3, 1)$, $m = -2$ **49.** $(9, 3)$, $m = 1$

50. $(8, 4)$, $m = 1$ **51.** $(-4, 1)$, $m = \dfrac{3}{4}$ **52.** $(2, 1)$, $m = \dfrac{5}{2}$

53. $(0, 3)$, $m = 0$ **54.** $(0, -4)$, $m = 0$ **55.** $(2, -6)$, undefined slope

56. $(-1, 7)$, undefined slope   **57.** $(0, -2)$, undefined slope   **58.** $(0, 5)$, undefined slope

59. $(6, -6)$, slope 0   **60.** $(-3, 3)$, slope 0

Each table of values gives several points that lie on a line.

(a) Use any two of the ordered pairs to find the slope of the line.

(b) Identify the y-intercept of the line.

(c) Use the slope and y-intercept from parts (a) and (b) to write an equation of the line in slope-intercept form.

(d) Graph the equation.

61.

x	y
0	−1
3	5
5	9

62.

x	y
0	4
2	2
4	0

63.

x	y
−9	1
−6	0
0	−2

64.

x	y
−10	−1
0	3
5	5

Extending Skills   Solve each problem.

65. In his job, Andrew earns 5% commission on his sales, plus a base salary of $2000 per month. This is illustrated in the graph and can be modeled by the linear equation

$$y = 0.05x + 2000,$$

where y is his monthly salary in dollars and x is his sales, also in dollars.

Monthly Salary

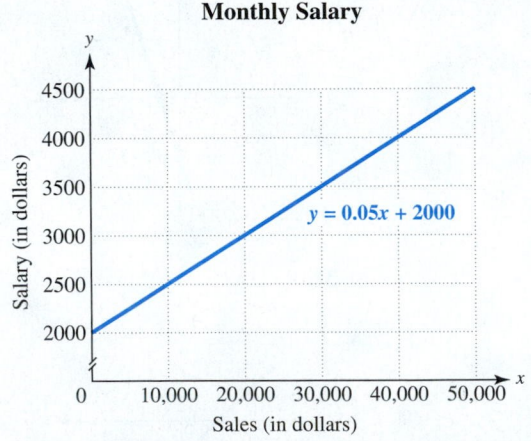

(a) What is the slope? With what does the slope correspond in the problem?

(b) What is the y-intercept? With what does the y-value of the y-intercept correspond in the problem?

(c) Use the equation to determine Andrew's monthly salary if his sales are $10,000. Confirm this using the graph.

(d) Use the graph to determine his sales if he wants to earn a monthly salary of $3500. Confirm this using the equation.

66. The cost to rent a moving van is $0.50 per mile, plus a flat fee of $100. This is illustrated in the graph and can be modeled by the linear equation

$$y = 0.50x + 100,$$

where y is the total rental cost in dollars and x is the number of miles driven.

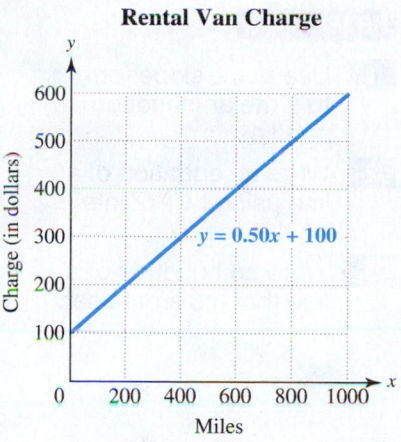

Rental Van Charge

(a) What is the slope? With what does the slope correspond in the problem?

(b) What is the y-intercept? With what does the y-value of the y-intercept correspond in the problem?

(c) Use the equation to determine the total charge if 400 mi are driven. Confirm this using the graph.

(d) Use the graph to determine the number of miles driven if the charge is $500. Confirm this using the equation.

Extending Skills   *The cost y of producing x items is, in some cases, expressed in the form*

$$y = mx + b.$$

The value of b gives the **fixed cost** *(the cost that is the same no matter how many items are produced), and the value of m is the* **variable cost** *(the cost of producing an additional item). Use this information to work Exercises 67 and 68.*

67. It costs $400 to start up a business selling snow cones. Each snow cone costs $0.25 to produce.

(a) What is the fixed cost?   **(b)** What is the variable cost?

(c) Write the cost equation.

(d) What will be the cost of producing 100 snow cones, based on the cost equation?

(e) How many snow cones will be produced if the total cost is $775?

68. It costs $2000 to purchase a copier, and each copy costs $0.02 to make.

(a) What is the fixed cost?   **(b)** What is the variable cost?

(c) Write the cost equation.

(d) What will be the cost of producing 10,000 copies, based on the cost equation?

(e) How many copies will be produced if the total cost is $2600?

RELATING CONCEPTS   For Individual or Group Work   (Exercises 69–72)

A line with equation written in slope-intercept form $y = mx + b$ *has slope m and y-intercept* $(0, b)$. *Recall from* **Section 3.1** *that the standard form of a linear equation in two variables is*

$$Ax + By = C, \qquad \text{Standard form}$$

where A, B, and C are real numbers and A and B are not both 0. **Work Exercises 69–72 in order.**

69. Write the standard form of a linear equation in slope-intercept form—that is, solved for y—to show that, in general, the slope is given by $-\dfrac{A}{B}$ (where $B \neq 0$).

70. Use the fact that $m = -\dfrac{A}{B}$ to find the slope of the line with each equation.

     (a) $2x + 3y = 18$ **(b)** $4x - 2y = -1$ **(c)** $3x - 7y = 21$

71. Refer to the slope-intercept form found in **Exercise 69.** What is the y-intercept?

72. Use the result of **Exercise 71** to find the y-intercept of each line in **Exercise 70.**

3.5   Point-Slope Form of a Linear Equation and Modeling

OBJECTIVE 1   Use point-slope form to write an equation of a line.

There is another form that can be used to write an equation of a line. To develop this form, let m represent the slope of a line and let (x_1, y_1) represent a given point on the line. Let (x, y) represent any other point on the line. See **FIGURE 36**.

$$m = \frac{y - y_1}{x - x_1} \qquad \text{Definition of slope}$$

$$m(x - x_1) = y - y_1 \qquad \text{Multiply each side by } x - x_1.$$

$$y - y_1 = m(x - x_1) \qquad \text{Rewrite.}$$

This result is the *point-slope form* of the equation of a line.

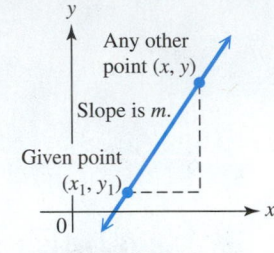

FIGURE 36

Point-Slope Form

The **point-slope form** of the equation of a line with slope m passing through the point (x_1, y_1) is

$$y - y_1 = m(x - x_1).$$

with arrows indicating: Slope (pointing to m), Given point (pointing to x_1 and y_1).

NOW TRY EXERCISE 1

Write an equation of the line passing through $(3, -1)$, with slope $-\frac{2}{5}$. Give the final answer in slope-intercept form.

EXAMPLE 1   Using Point-Slope Form to Write Equations

Write an equation of each line. Give the final answer in slope-intercept form.

(a) The line passing through $(-2, 4)$, with slope -3

The given point is $(-2, 4)$, so $x_1 = -2$ and $y_1 = 4$. Also, $m = -3$. Substitute these values into the point-slope form.

$$y - y_1 = m(x - x_1) \qquad \text{Point-slope form}$$

> Only y_1, m, and x_1 are replaced with numbers.

$$y - 4 = -3[x - (-2)] \qquad \text{Let } y_1 = 4, m = -3, x_1 = -2.$$

$$y - 4 = -3(x + 2) \qquad \text{Definition of subtraction}$$

$$y - 4 = -3x - 6 \qquad \text{Distributive property}$$

> The answer is in $y = mx + b$ form as specified.

$$y = -3x - 2 \qquad \text{Add 4.}$$

(b) The line passing through $(4, 2)$, with slope $\frac{3}{5}$

$$y - y_1 = m(x - x_1) \qquad \text{Point-slope form}$$

$$y - 2 = \frac{3}{5}(x - 4) \qquad \text{Let } y_1 = 2, m = \frac{3}{5}, x_1 = 4.$$

$$y - 2 = \frac{3}{5}x - \frac{12}{5} \qquad \text{Distributive property}$$

> Do not clear fractions here because we want the answer in slope-intercept form—that is, solved for y.

$$y = \frac{3}{5}x - \frac{12}{5} + \frac{10}{5} \qquad \text{Add } 2 = \frac{10}{5} \text{ to each side.}$$

$$y = \frac{3}{5}x - \frac{2}{5} \qquad \text{Combine like terms.} \qquad \text{NOW TRY}$$

NOW TRY ANSWER

1. $y = -\frac{2}{5}x + \frac{1}{5}$

OBJECTIVE 2 Write an equation of a line using two points on the line.

Many of the linear equations in **Sections 3.1–3.4** were given in **standard form**

$$Ax + By = C, \qquad \text{Standard form}$$

where A, B, and C are real numbers and A and B are not both 0. In most cases, A, B, and C are rational numbers. For consistency in this book, we give answers so that A, B, and C are integers with greatest common factor 1 and $A \geq 0$. (If $A = 0$, then we give $B > 0$.)

NOTE The definition of standard form is not the same in all texts. A linear equation can be written in many different, equally correct, ways. For example,

$$3x + 4y = 12, \quad 6x + 8y = 24, \quad \text{and} \quad -9x - 12y = -36$$

all represent the same set of ordered pairs. When giving answers in standard form, let us agree that $3x + 4y = 12$ is preferable to the other forms because the greatest common factor of 3, 4, and 12 is 1 and $A \geq 0$.

NOW TRY EXERCISE 2

Write an equation of the line passing through the points $(4, 1)$ and $(6, -2)$. Give the final answer in

(a) slope-intercept form and

(b) standard form.

EXAMPLE 2 Writing an Equation of a Line Using Two Points

Write an equation of the line passing through the points $(-2, 5)$ and $(3, 4)$. Give the final answer in slope-intercept form and then in standard form.

First, find the slope of the line.

$$
\begin{array}{cc}
(x_1, y_1) & (x_2, y_2) \\
\downarrow\downarrow & \downarrow\downarrow \\
(3, 4) \quad \text{and} & (-2, 5)
\end{array}
\qquad \text{Label the points.}
$$

$$\text{slope } m = \frac{y_2 - y_1}{x_2 - x_1} = \frac{5 - 4}{-2 - 3} \qquad \text{Apply the slope formula.}$$

$$= \frac{1}{-5}, \quad \text{or} \quad -\frac{1}{5} \qquad \text{Simplify the fraction.}$$

Now use $m = -\frac{1}{5}$ and either $(-2, 5)$ or $(3, 4)$ as (x_1, y_1) in the point-slope form.

$$y - y_1 = m(x - x_1) \qquad \text{We choose } (3, 4).$$

$$y - 4 = -\frac{1}{5}(x - 3) \qquad \text{Let } y_1 = 4, m = -\frac{1}{5}, x_1 = 3.$$

$$y - 4 = -\frac{1}{5}x + \frac{3}{5} \qquad \text{Distributive property}$$

$$y = -\frac{1}{5}x + \frac{3}{5} + \frac{20}{5} \qquad \text{Add } 4 = \frac{20}{5} \text{ to each side.}$$

Slope-intercept form $\longrightarrow y = -\frac{1}{5}x + \frac{23}{5} \qquad \text{Combine like terms.}$

$$5y = -x + 23 \qquad \text{Multiply by 5 to clear fractions.}$$

Standard form $\longrightarrow x + 5y = 23 \qquad \text{Add } x.$ **NOW TRY**

NOW TRY ANSWERS

2. (a) $y = -\frac{3}{2}x + 7$

(b) $3x + 2y = 14$

NOTE   In **Example 2,** the same result would be found using $(-2, 5)$ for (x_1, y_1). We could also substitute the slope and either given point in slope-intercept form $y = mx + b$ and then solve for b, as in **Section 3.4, Example 4(b).**

▼ Summary of the Forms of Linear Equations

Equation	Description	Example
$y = mx + b$	**Slope-intercept form** Slope is m. y-intercept is $(0, b)$.	$y = \frac{3}{2}x - 6$
$y - y_1 = m(x - x_1)$	**Point-slope form** Slope is m. Line passes through (x_1, y_1).	$y + 3 = \frac{3}{2}(x - 2)$
$Ax + By = C$ (where A, B, and C are real numbers and A and B are not both 0)	**Standard form** Slope is $-\frac{A}{B}$ $(B \neq 0)$. x-intercept is $\left(\frac{C}{A}, 0\right)$ $(A \neq 0)$. y-intercept is $\left(0, \frac{C}{B}\right)$ $(B \neq 0)$.	$3x - 2y = 12$
$x = a$	**Vertical line** Slope is undefined. x-intercept is $(a, 0)$.	$x = 3$
$y = b$	**Horizontal line** Slope is 0. y-intercept is $(0, b)$.	$y = 3$

OBJECTIVE 3   **Write an equation of a line that fits a data set.**

If a given set of data fits a linear pattern—that is, if its graph consists of points lying close to a straight line—we can write a linear equation that models the data.

EXAMPLE 3   **Writing an Equation of a Line That Models Data**

The table lists the average annual cost (in dollars) of tuition and fees for in-state students at public 4-year colleges and universities for selected years. Year 1 represents 2001, year 3 represents 2003, and so on. Plot the data and write an equation that approximates it.

Letting y represent the cost in year x, we plot the data as shown in **FIGURE 37**.

Year	Cost (in dollars)
1	3766
3	4645
5	5491
7	6185
9	7050
11	8244

Source: The College Board.

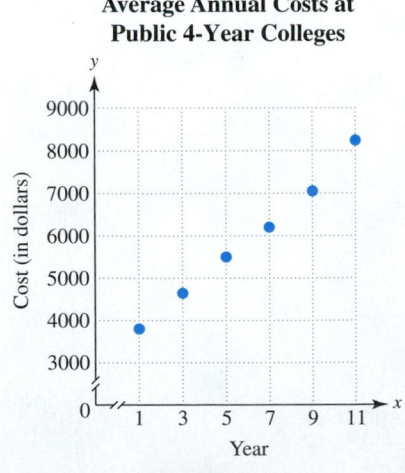

Average Annual Costs at Public 4-Year Colleges

FIGURE 37

NOW TRY EXERCISE 3

Use the points (3, 4645) and (5, 5491) to write an equation in slope-intercept form that approximates the data of **Example 3.** How well does this equation approximate the cost in 2011?

The points appear to lie approximately in a straight line. We choose the ordered pairs (5, 5491) and (7, 6185) from the table and find the slope of the line through these points.

$$m = \frac{y_2 - y_1}{x_2 - x_1} = \frac{6185 - 5491}{7 - 5} = 347 \qquad \text{Let } (7, 6185) = (x_2, y_2) \text{ and } (5, 5491) = (x_1, y_1).$$

The slope, 347, is positive, indicating that tuition and fees *increased* $347 each year. Now substitute this slope and the point (5, 5491) in the point-slope form to find an equation of the line.

$$y - y_1 = m(x - x_1) \qquad \text{Point-slope form}$$

$$y - 5491 = 347(x - 5) \qquad \text{Let } (x_1, y_1) = (5, 5491), m = 347.$$

$$y - 5491 = 347x - 1735 \qquad \text{Distributive property}$$

$$y = 347x + 3756 \qquad \text{Add 5491.}$$

Thus, the equation $y = 347x + 3756$ can be used to model the data.

To see how well this equation approximates the ordered pairs in the data table, let $x = 9$ (for 2009) and find y.

$$y = 347x + 3756 \qquad \text{Equation of the line}$$

$$y = 347(9) + 3756 \qquad \text{Substitute 9 for } x.$$

$$y = 3123 + 3756 \qquad \text{Multiply.}$$

$$y = 6879 \qquad \text{Add.}$$

The corresponding value in the table for $x = 9$ is 7050, so the equation approximates the data reasonably well.

NOW TRY

NOW TRY ANSWER

3. $y = 423x + 3376$; The equation gives $y = 8029$ when $x = 11$, which approximates the data reasonably well.

NOTE In **Example 3**, if we had chosen two different data points, we would have obtained a slightly different equation. See **Now Try Exercise 3**.

Also, we could have used slope-intercept form $y = mx + b$ (instead of point-slope form) to write an equation that models the data.

3.5 Exercises

 FOR EXTRA HELP

 MyMathLab®

▶ *Complete solution available in MyMathLab*

Concept Check *Work each problem.*

1. Match each form or description in Column I with the corresponding equation in Column II.

I	II
(a) Point-slope form	**A.** $x = a$
(b) Horizontal line	**B.** $y = mx + b$
(c) Slope-intercept form	**C.** $y = b$
(d) Standard form	**D.** $y - y_1 = m(x - x_1)$
(e) Vertical line	**E.** $Ax + By = C$

2. Write the equation $y + 1 = -2(x - 5)$ first in slope-intercept form and then in standard form.

3. Which equations are equivalent to $2x - 3y = 6$?

A. $y = \dfrac{2}{3}x - 2$

B. $-2x + 3y = -6$

C. $y = -\dfrac{3}{2}x + 3$

D. $y - 2 = \dfrac{2}{3}(x - 6)$

4. In the summary box following **Example 2,** we give the equations

$$y = \frac{3}{2}x - 6 \quad \text{and} \quad y + 3 = \frac{3}{2}(x - 2)$$

as examples of equations in slope-intercept form and point-slope form, respectively. Write each of these equations in standard form. What do you notice?

*Write an equation of the line passing through the given point and having the given slope. Give the final answer in slope-intercept form. See **Example 1.***

5. $(1, 7)$, $m = 5$

6. $(2, 9)$, $m = 6$

7. $(6, -3)$, $m = 1$

8. $(-4, 4)$, $m = 1$

9. $(1, -7)$, $m = -3$

10. $(1, -5)$, $m = -7$

11. $(3, -2)$, $m = -1$

12. $(-5, 4)$, $m = -1$

▶ **13.** $(-2, 5)$, $m = \dfrac{2}{3}$

14. $(4, 2)$, $m = -\dfrac{1}{3}$

15. $(6, -3)$, $m = -\dfrac{4}{5}$

16. $(7, -2)$, $m = -\dfrac{7}{2}$

*Write an equation of the line passing through the given pair of points. Give the final answer in (a) slope-intercept form and (b) standard form. See **Example 2.***

17. $(4, 10)$ and $(6, 12)$

18. $(8, 5)$ and $(9, 6)$

19. $(-4, 0)$ and $(0, 2)$

20. $(0, -2)$ and $(-3, 0)$

21. $(-2, -1)$ and $(3, -4)$

22. $(-1, -7)$ and $(-8, -2)$

23. $\left(-\dfrac{2}{3}, \dfrac{8}{3}\right)$ and $\left(\dfrac{1}{3}, \dfrac{7}{3}\right)$

24. $\left(\dfrac{1}{2}, \dfrac{3}{2}\right)$ and $\left(-\dfrac{1}{4}, \dfrac{5}{4}\right)$

Write an equation of the given line through the given points. Give the final answer in (a) slope-intercept form and (b) standard form.

25.

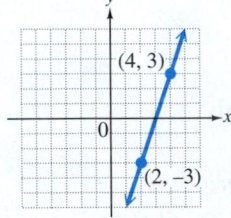

26.

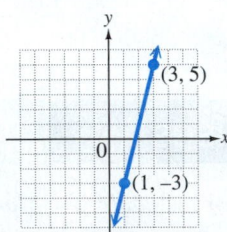

27.

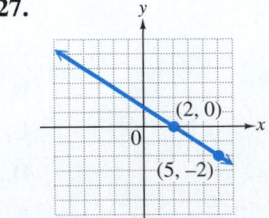

28.

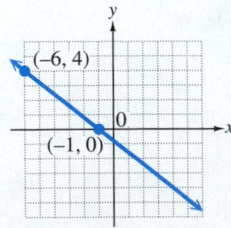

Extending Skills *Write an equation of the line satisfying the given conditions. Give the final answer in slope-intercept form. (Hint: Recall the relationships among slopes of parallel and perpendicular lines in Section 3.3.)*

29. Perpendicular to $x - 2y = 7$; y-intercept $(0, -3)$

30. Parallel to $5x - y = 10$; y-intercept $(0, -2)$

31. Through $(2, 3)$; parallel to $4x - y = -2$

32. Through $(4, 2)$; perpendicular to $x - 3y = 7$

33. Through $(2, -3)$; parallel to $3x = 4y + 5$

34. Through $(-1, 4)$; perpendicular to $2x = -3y + 8$

Solve each problem. See Example 3.

35. The table lists the average annual cost y (in dollars) of tuition and fees at 2-year colleges for selected years x, where year 1 represents 2008, year 2 represents 2009, and so on.

Year	Cost (in dollars)
1	2530
2	2790
3	2940
4	3070
5	3220

Source: The College Board.

(a) Write five ordered pairs (x, y) for the data.

(b) Plot the ordered pairs (x, y). Do the points lie approximately in a straight line?

(c) Use the ordered pairs $(1, 2530)$ and $(4, 3070)$ to write an equation of a line that approximates the data. Give the final equation in slope-intercept form.

(d) Use the equation from part (c) to estimate the average annual cost at 2-year colleges in 2013 to the nearest dollar. (*Hint:* What is the value of x for 2013?)

36. The table gives heavy-metal nuclear waste y (in thousands of metric tons) from spent reactor fuel awaiting permanent storage. (*Source: Scientific American.*)

Year x	Waste y
1995	32
2000	42
2010	61
2020*	76

*Estimate by the U.S. Department of Energy.

Let $x = 0$ represent 1995, $x = 5$ represent 2000 (since $2000 - 1995 = 5$), and so on.

(a) For 1995, the ordered pair is $(0, 32)$. Write ordered pairs (x, y) for the data for the other years given in the table.

(b) Plot the ordered pairs (x, y). Do the points lie approximately in a straight line?

(c) Use the ordered pairs $(0, 32)$ and $(25, 76)$ to write an equation of a line that approximates the data. Give the final equation in slope-intercept form.

(d) Use the equation from part (c) to estimate the amount of nuclear waste in 2015. (*Hint:* What is the value of x for 2015?)

The points on the graph show the number of colleges y that teamed up with banks to issue student ID cards which doubled as debit cards for recent years x. The graph of a linear equation that models the data is also shown.

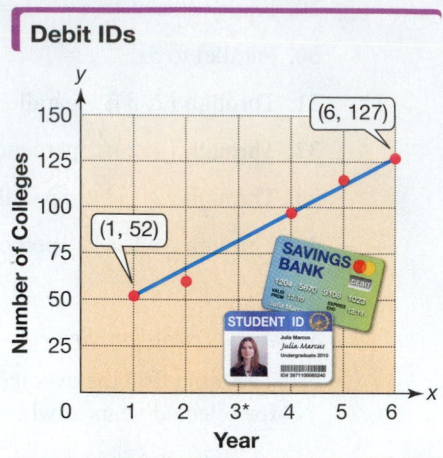

Debit IDs

Source: CR80News.
* Data for year 3 is unavailable.

37. Use the ordered pairs shown on the graph to write an equation of the line that models the data. Give the final equation in slope-intercept form.

38. Use the equation from **Exercise 37** to estimate the number of colleges that teamed up with banks to offer debit IDs in year 3 when data was unavailable.

The points on the graph indicate years of life expected at birth y in the United States for selected years x. The graph of a linear equation that models the data is also shown.

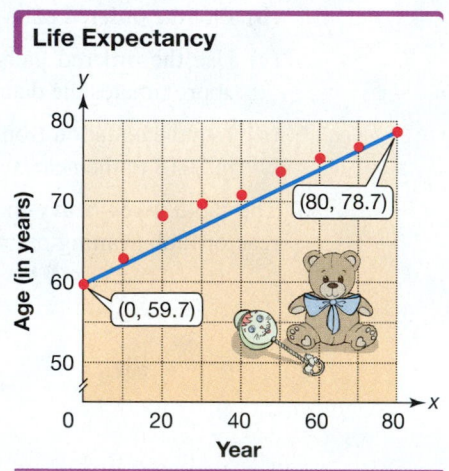

Life Expectancy

Source: National Center for Health Statistics.

Here x = 0 represents 1930, x = 10 represents 1940, and so on.

39. Use the ordered pairs shown on the graph to write an equation of the line that models the data. Give the final equation in slope-intercept form.

40. Use the equation from **Exercise 39** to do the following.

(a) Find years of life expected at birth in 2000. (*Hint:* What is the value of *x* for 2000?) Round the answer to the nearest tenth.

(b) How does the answer in part (a) compare to the actual value of 76.8 yr?

(c) Project years of life expected at birth in 2020. (*Hint:* What is the value of *x* for 2020?) Round the answer to the nearest tenth. Does the answer seem reasonable?

SUMMARY EXERCISES   Applying Graphing and Equation-Writing Techniques for Lines

Answers (left column):

1. (a) B (b) D
   (c) A (d) C
2. A, B

3.
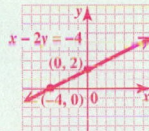

4.
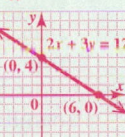

5.
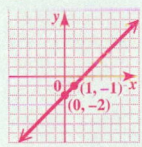

6.
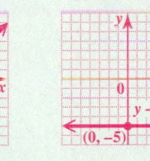

7.
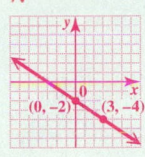

8.
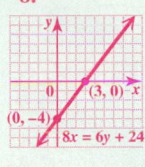

9.
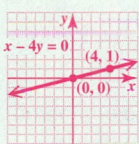

10.
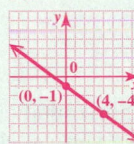

11.
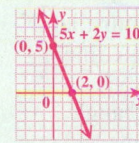

12.
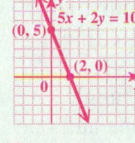

13.
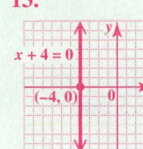

14.
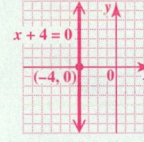

15. $y = -2x + 6$ 16. $y = \frac{4}{3}x + 8$

17. $y = \frac{1}{2}x - 2$

18. $y = -\frac{2}{3}x + 5$

19. $y = -3x - 6$ 20. $y = \frac{3}{2}x + 12$

21. $y = -4x - 3$ 22. $x = 0$

23. $y = \frac{2}{3}x$ 24. $y = -x - 4$

25. $y = x - 5$ 26. $y = 0$

27. $y = \frac{5}{3}x + 5$

28. $y = -5x - 8$

---

1. *Concept Check* Match the description in Column I with the correct equation in Column II.

I

(a) Slope $= -0.5$, $b = -2$

(b) x-intercept $(4, 0)$, y-intercept $(0, 2)$

(c) Passes through $(4, -2)$ and $(0, 0)$

(d) $m = \frac{1}{2}$, passes through $(-2, -2)$

II

A. $y = -\frac{1}{2}x$

B. $y = -\frac{1}{2}x - 2$

C. $x - 2y = 2$

D. $x + 2y = 4$

E. $x = 2y$

2. *Concept Check* Which equations are equivalent to $2x + 5y = 20$?

A. $y = -\frac{2}{5}x + 4$ B. $y - 2 = -\frac{2}{5}(x - 5)$

C. $y = \frac{5}{2}x - 4$ D. $2x = 5y - 20$

Graph each line, using the given information or equation.

3. $x - 2y = -4$ **4.** $2x + 3y = 12$

5. $m = 1$, y-intercept $(0, -2)$ **6.** $y - 4 = -9$

7. $m = -\frac{2}{3}$, passes through $(3, -4)$ **8.** $8x = 6y + 24$

9. $x - 4y = 0$ **10.** $m = -\frac{3}{4}$, passes through $(4, -4)$

11. $5x + 2y = 10$ **12.** $x + 5y = 0$

13. $x + 4 = 0$ **14.** $y = -x + 6$

Write an equation in slope-intercept form of each line represented by the table of ordered pairs or the graph.

15.

x	y
3	0
1	4
-1	8

16.

x	y
-6	0
0	8
3	12

17.
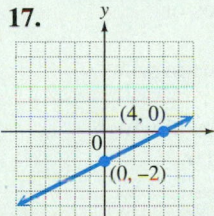

18.
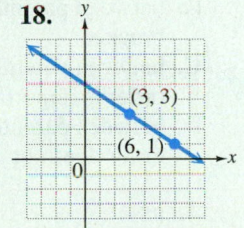

Write an equation of each line. Give the final answer in slope-intercept form if possible.

19. $m = -3$, $b = -6$ **20.** $m = \frac{3}{2}$, through $(-4, 6)$

21. Through $(1, -7)$ and $(-2, 5)$ **22.** Through $(0, 0)$, undefined slope

23. Through $(0, 0)$ and $(3, 2)$ **24.** $m = -1$, $b = -4$

25. Through $(5, 0)$ and $(0, -5)$ **26.** Through $(0, 0)$, $m = 0$

27. $m = \frac{5}{3}$, through $(-3, 0)$ **28.** Through $(1, -13)$ and $(-2, 2)$

Chapter 3	Summary

Key Terms

3.1

line graph
linear equation in two
    variables
ordered pair
table of values
x-axis
y-axis

origin
rectangular (Cartesian)
    coordinate system
quadrant
plane
coordinates
plot
scatter diagram

3.2

graph, graphing
x-intercept
y-intercept
horizontal line
vertical line

3.3

rise
run
slope
parallel lines
perpendicular lines

New Symbols

(x, y) ordered pair

m slope

(x_1, y_1) subscript notation
        (read "*x*-sub-one,
        y-sub-one")

Test Your Word Power

See how well you have learned the vocabulary in this chapter.

1. A **linear equation in two variables**
   is an equation that can be written in
   the form
   A. $Ax + By < C$
   B. $ax = b$
   C. $y = x^2$
   D. $Ax + By = C$.

2. An **ordered pair** is a pair of
   numbers written
   A. in numerical order between
      brackets
   B. between parentheses or
      brackets
   C. between parentheses in which
      order is important
   D. between parentheses in which
      order does not matter.

3. An **intercept** is
   A. the point where the *x*-axis and
      y-axis intersect
   B. a pair of numbers written in
      parentheses in which order matters
   C. one of the four regions
      determined by a rectangular
      coordinate system
   D. the point where a graph intersects
      the *x*-axis or the *y*-axis.

4. The **slope** of a line is
   A. the measure of the run over the
      rise of the line
   B. the distance between two points
      on the line
   C. the ratio of the change in *y* to the
      change in *x* along the line

   D. the horizontal change compared
      to the vertical change of two
      points on the line.

5. Two lines in a plane are **parallel** if
   A. they represent the same line
   B. they never intersect
   C. they intersect at a 90° angle
   D. one has a positive slope and one
      has a negative slope.

6. Two lines in a plane are
   perpendicular if
   A. they represent the same line
   B. they never intersect
   C. they intersect at a 90° angle
   D. one has a positive slope and one
      has a negative slope.

ANSWERS

1. D; *Examples:* $3x + 2y = 6$, $x = y - 7$, $y = 4x$ **2.** C; *Examples:* $(0, 3)$, $(-3, 8)$, $(4, 0)$ **3.** D; *Example:* The graph of the equation $4x - 3y = 12$
has *x*-intercept $(3, 0)$ and *y*-intercept $(0, -4)$.   **4.** C; *Example:* The line through $(3, 6)$ and $(5, 4)$ has slope $\frac{4 - 6}{5 - 3} = \frac{-2}{2} = -1$.   **5.** B; *Example:* See
FIGURE 27 in **Section 3.3**.   **6.** C; *Example:* See **FIGURE 28** in **Section 3.3**.

Quick Review

CONCEPTS

EXAMPLES

3.1   Linear Equations and Rectangular Coordinates

An ordered pair is a solution of an equation if it satisfies the equation.

Decide whether $(2, -5)$ and $(0, -6)$ are solutions of $4x - 3y = 18$.

$$4(2) - 3(-5) \overset{?}{=} 18 \qquad\qquad 4(0) - 3(-6) \overset{?}{=} 18$$
$$8 + 15 \overset{?}{=} 18 \qquad\qquad\qquad 0 + 18 \overset{?}{=} 18$$
$$23 = 18 \quad \text{False} \qquad\qquad 18 = 18 \checkmark \quad \text{True}$$

$(2, -5)$ is not a solution. $\qquad$ $(0, -6)$ is a solution.

If a value of either variable in an equation is given, then the value of the other variable can be found by substitution.

Complete the ordered pair $(0, \underline{\quad})$ for the given equation.

$$3x = y + 4$$
$$3(0) = y + 4 \qquad \text{Let } x = 0.$$
$$0 = y + 4 \qquad \text{Multiply.}$$
$$-4 = y \qquad \text{Subtract 4.}$$

The ordered pair is $(0, -4)$.

To plot an ordered pair, begin at the origin.

Step 1   Move right or left the number of units corresponding to the x-coordinate—right if it is positive or left if it is negative.

Step 2   Then turn and move up or down the number of units corresponding to the y-coordinate—up if it is positive or down if it is negative.

Plot the ordered pair $(-3, 4)$.

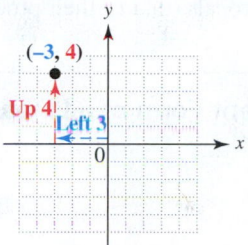

3.2   Graphing Linear Equations in Two Variables

To graph a linear equation, follow these steps.

Step 1   Find at least two ordered pairs that satisfy the equation. (It is good practice to find a third ordered pair as a check.)

Step 2   Plot the corresponding points.

Step 3   Draw a straight line through the points.

Graph $x - 2y = 4$.

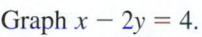

	x	y
y-intercept $\rightarrow$	0	-2
x-intercept $\rightarrow$	4	0
	-2	-3

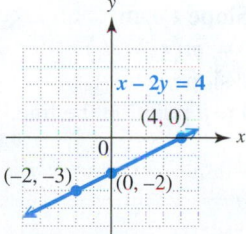

The graph of $Ax + By = 0$ passes through the origin. Find and plot two other points that satisfy the equation. Then draw a straight line through the points.

The graph of $y = b$ is a horizontal line through $(0, b)$.

The graph of $x = a$ is a vertical line through $(a, 0)$.

Graph $2x + 3y = 0$.

x	y
-3	2
0	0
3	-2

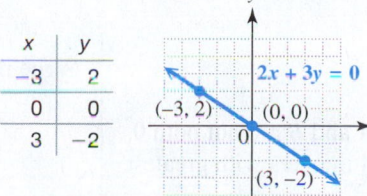

Graph $y = -3$ and $x = -3$.

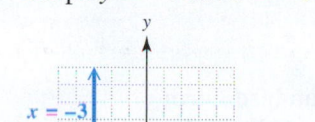
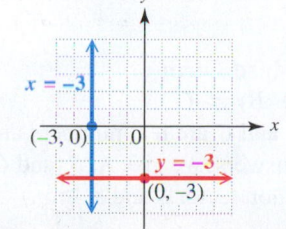

CONCEPTS

EXAMPLES

3.3 The Slope of a Line

The slope m of the line passing through the points (x_1, y_1) and (x_2, y_2) is defined as follows.

$$m = \frac{\text{change in } y}{\text{change in } x} = \frac{y_2 - y_1}{x_2 - x_1} \quad (\text{where } x_1 \neq x_2)$$

Horizontal lines have slope 0.

Vertical lines have undefined slope.

To find the slope of a line from its equation, solve for y. The slope is the coefficient of x.

The line passing through $(-2, 3)$ and $(4, -5)$ has slope

$$m = \frac{-5 - 3}{4 - (-2)} = \frac{-8}{6} = -\frac{4}{3}.$$

The line $y = -2$ has slope 0.

The line $x = 4$ has undefined slope.

Find the slope of the line with the following equation.

$$3x - 4y = 12$$

$$-4y = -3x + 12 \qquad \text{Subtract } 3x.$$

$$y = \frac{3}{4}x - 3 \qquad \text{Divide by } -4.$$

Slope ⬆

Parallel lines have the same slope.

The lines $y = 3x - 1$ and $y = 3x + 4$ are parallel because both have slope 3.

The slopes of perpendicular lines, neither of which is vertical, are negative reciprocals (that is, their product is -1).

The lines $y = -3x - 1$ and $y = \frac{1}{3}x + 4$ are perpendicular because their slopes are -3 and $\frac{1}{3}$, and $-3\left(\frac{1}{3}\right) = -1$.

3.4 Slope-Intercept Form of a Linear Equation

Slope-Intercept Form
$$y = mx + b$$
m is the slope. $(0, b)$ is the y-intercept.

Write an equation of the line with slope 2 and y-intercept $(0, -5)$.

$$y = 2x - 5$$

3.5 Point-Slope Form of a Linear Equation and Modeling

Point-Slope Form
$$y - y_1 = m(x - x_1)$$
m is the slope.
(x_1, y_1) is a point on the line.

Write an equation of the line passing through $(-4, 5)$ with slope $-\frac{1}{2}$.

$$y - 5 = -\frac{1}{2}[x - (-4)] \qquad \begin{array}{l}\text{Substitute for } m \text{ and } (x_1, y_1) \\ \text{in the point-slope form.}\end{array}$$

$$y - 5 = -\frac{1}{2}(x + 4) \qquad \text{Definition of subtraction}$$

$$y - 5 = -\frac{1}{2}x - 2 \qquad \text{Distributive property}$$

$$y = -\frac{1}{2}x + 3 \qquad \text{Add 5.}$$

Standard Form
$$Ax + By = C$$
A, B, and C are real numbers and A and B are not both 0. (In answers, we give A, B, and C as integers with greatest common factor 1 and $A \geq 0$.)

Write the equation $y = -\frac{1}{2}x + 3$ in standard form.

$$y = -\frac{1}{2}x + 3$$

$$-2y = x - 6 \qquad \text{Multiply each term by } -2.$$

$$x + 2y = 6 \qquad A = 1, B = 2, C = 6$$

Chapter 3 | Review Exercises

Answer column (left):

1. $-1; 2; 1$
2. $2; \dfrac{3}{2}; \dfrac{14}{3}$
3. $0; \dfrac{8}{3}; -9$
4. $7; 7; 7$
5. yes
6. no
7. yes
8. no
9. I
10. II
11. none
12. none

Graph for Exercises 9–12

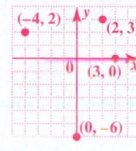

13.
$\left(-\dfrac{5}{2}, 0\right); (0, 5)$
14.
$\left(\dfrac{8}{3}, 0\right); (0, 4)$

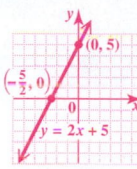 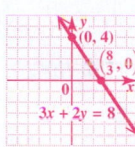

15. $(-4, 0);$
$(0, -2)$
16. $(-6, 0);$
no y-intercept

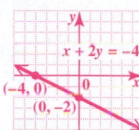 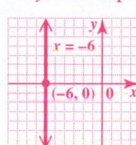

17. $-\dfrac{1}{2}$
18. undefined
19. 3
20. 0
21. undefined
22. $\dfrac{3}{2}$
23. $-\dfrac{1}{3}$
24. $\dfrac{3}{2}$
25. (a) 2 (b) $\dfrac{1}{3}$
26. parallel
27. perpendicular
28. neither
29. $y = -x + \dfrac{2}{3}$
30. $y = -\dfrac{1}{2}x + 4$
31. $y = x - 7$
32. $y = \dfrac{2}{3}x + \dfrac{14}{3}$
33. $y = -\dfrac{3}{4}x - \dfrac{1}{4}$
34. $x = -4$
35. $y = 1$
36. $y = -\dfrac{1}{3}x + 1$

3.1 *Complete the given ordered pairs for each equation.*

1. $y = 3x + 2$; $(-1, \underline{}), (0, \underline{}), (\underline{}, 5)$

2. $4x + 3y = 6$; $(0, \underline{}), (\underline{}, 0), (-2, \underline{})$

3. $x = 3y$; $(0, \underline{}), (8, \underline{}), (\underline{}, -3)$

4. $x - 7 = 0$; $(\underline{}, -3), (\underline{}, 0), (\underline{}, 5)$

Decide whether the given ordered pair is a solution of the given equation.

5. $x + y = 7$; $(2, 5)$ 6. $2x + y = 5$; $(-1, 3)$

7. $3x - y = 4$; $\left(\dfrac{1}{3}, -3\right)$ 8. $x = -1$; $(0, -1)$

Identify the quadrant in which each point is located. Then plot and label each point in a rectangular coordinate system.

9. $(2, 3)$ 10. $(-4, 2)$ 11. $(3, 0)$ 12. $(0, -6)$

3.2 *Find the x- and y-intercepts for the graph of each equation. Then draw the graph.*

13. $y = 2x + 5$ 14. $3x + 2y = 8$ 15. $x + 2y = -4$ 16. $x = -6$

3.3 *Find the slope of each line. (In Exercises 26 and 27, coordinates of the points shown are integers.)*

17. Through $(2, 3)$ and $(-4, 6)$ 18. Through $(2, 5)$ and $(2, 8)$

19. $y = 3x - 4$ 20. $y = 5$ 21. $x = -7$

22. 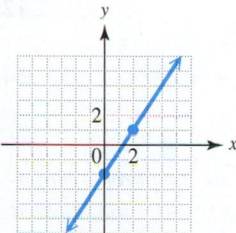   23. 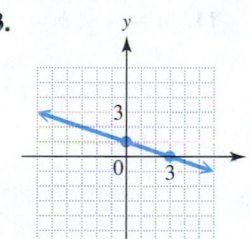   24. The line passing through these points

x	y
0	1
2	4
6	10

25. Find each slope.

(a) A line whose graph is parallel to the graph of $y = 2x + 3$

(b) A line whose graph is perpendicular to the graph of $y = -3x + 3$

Decide whether each pair of lines is **parallel, perpendicular,** *or* **neither.**

26. $3x + 2y = 6$ 27. $x - 3y = 1$ 28. $x - 2y = 8$
    $6x + 4y = 8$ $3x + y = 4$ $x + 2y = 8$

3.4, 3.5 *Write an equation of each line. Give the final answer in slope-intercept form (if possible).*

29. $m = -1, b = \dfrac{2}{3}$ 30. Through $(2, 3)$ and $(-4, 6)$

31. Through $(4, -3), m = 1$ 32. Through $(-1, 4), m = \dfrac{2}{3}$

33. Through $(1, -1), m = -\dfrac{3}{4}$ 34. Through $(-4, 1)$, undefined slope

35. Slope 0, through $(-4, 1)$ 36. The line in **Exercise 23**

Chapter 3   Mixed Review Exercises

1. A
2. C, D
3. A, B, D
4. D
5. C
6. B

7. $\left(-\dfrac{5}{2}, 0\right)$;
$(0, -5)$; -2

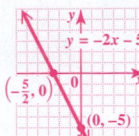

8. $(0, 0)$; $(0, 0)$;
$-\dfrac{1}{3}$

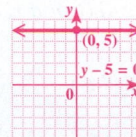

9. no x-intercept;
$(0, 5)$; 0

10. $(-5, 0)$;
no y-intercept;
undefined slope

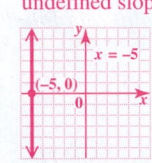

11. (a) $y = -\dfrac{1}{4}x - \dfrac{5}{4}$
 (b) $x + 4y = -5$

12. (a) $y = -3x + 30$
 (b) $3x + y = 30$

13. (a) $y = -\dfrac{4}{7}x - \dfrac{23}{7}$
 (b) $4x + 7y = -23$

14. (a) $y = -5$ **(b)** $y = -5$

15. Because the graph falls from left to right, the slope is negative.

16. $(0, 39.6)$, $(3, 36.0)$

17. $y = -1.2x + 39.6$

18. 37.2% ($x = 2$);
It is a little high, as we might expect. The actual data point lies slightly below the graph of the line.

In Exercises 1–6, match each statement to the appropriate graph or graphs in A–D. Graphs may be used more than once.

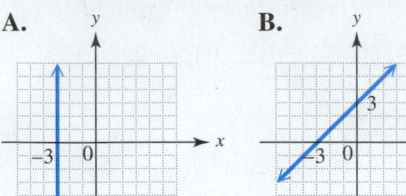

A.   **B.**   **C.**   **D.**

1. The line shown in the graph has undefined slope.

2. The graph of the equation has y-intercept $(0, -3)$.

3. The graph of the equation has x-intercept $(-3, 0)$.

4. The line shown in the graph has negative slope.

5. The graph is that of the equation $y = -3$.

6. The line shown in the graph has slope 1.

Find the x- and y-intercepts and the slope of each line. Then graph the line.

7. $y = -2x - 5$

8. $x + 3y = 0$

9. $y - 5 = 0$

10. $x = -5$

Write an equation of each line. Give the final answer in (a) slope-intercept form and (b) standard form (if possible).

11. $m = -\dfrac{1}{4}$, $b = -\dfrac{5}{4}$

12. Through $(8, 6)$, $m = -3$

13. Through $(3, -5)$ and $(-4, -1)$

14. Slope 0, through $(5, -5)$

The points on the graph indicate the percent y of 4-year college students in public schools who earned a degree within 5 years of entry for selected years x. The graph of a linear equation that models the data is also shown. Here x = 0 represents 2010, x = 1 represents 2011, and so on.

15. Because the points of the graph lie approximately in a linear pattern, a straight line can be used to model the data. Will this line have positive or negative slope? Explain.

16. Write two ordered pairs (x, y) for the data for 2010 and 2013.

17. Use the two ordered pairs from **Exercise 16** to write an equation of a line that models the data. Give the final equation in slope-intercept form.

18. Use the equation from **Exercise 17** to approximate the percent for 2012. (What is the value of x for 2012?) How does the answer compare to the actual value of 36.6%?

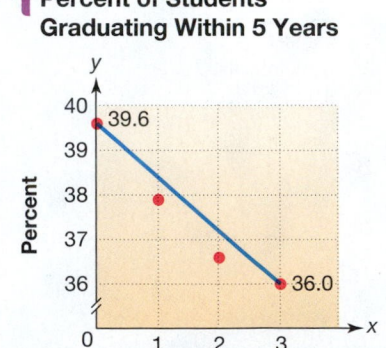

Percent of Students Graduating Within 5 Years

Source: ACT.

<table>
<tr><td>Chapter 3</td><td>Test</td></tr>
</table>

FOR EXTRA HELP

Step-by-step test solutions are found on the Chapter Test Prep Videos available in MyMathLab® *or on* YouTube.

▶ *View the complete solutions to all Chapter Test exercises in MyMathLab.*

[3.1]
1. -6; -10; -5
2. no

[3.2]
3. To find the x-intercept, let $y = 0$, and to find the y-intercept, let $x = 0$.

4. $(2, 0)$; $(0, 6)$ 5. $(0, 0)$; $(0, 0)$

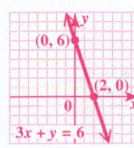 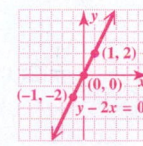
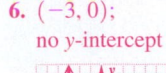 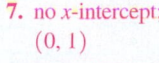

6. $(-3, 0)$;   7. no x-intercept;
   no y-intercept $(0, 1)$

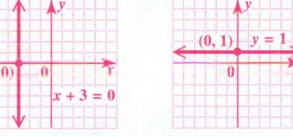

8. $(4, 0)$; $(0, -4)$

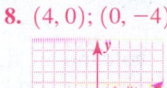
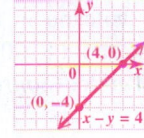

[3.3]
9. $-\dfrac{8}{3}$ 10. -2

11. undefined   12. 0

13. $\dfrac{5}{2}$

[3.4, 3.5]
14. $y = 2x + 6$ 15. $y = \dfrac{5}{2}x - 4$

16. $y = -9x + 12$

[3.1, 3.5]
17. The slope is negative because sales are decreasing.
18. $(0, 209)$, $(13, 145)$;
    (a) -4.9
    (b) $y = -4.9x + 209$
19. 184.5 thousand;
    The equation gives an approximation that is a little high.
20. In 2013, worldwide snowmobile sales were 145 thousand.

---

1. Complete the ordered pairs $(0, \underline{})$, $(\underline{}, 0)$, $(\underline{}, -3)$ for the equation $3x + 5y = -30$.

2. Is $(4, -1)$ a solution of $4x - 7y = 9$?

3. How do we find the x-intercept of the graph of a linear equation in two variables? How do we find the y-intercept?

Graph each linear equation. Give the x- and y-intercepts.

4. $3x + y = 6$ 5. $y - 2x = 0$ 6. $x + 3 = 0$

7. $y = 1$ 8. $x - y = 4$

Find the slope of each line. (In Exercise 13, coordinates of the points shown are integers.)

9. Through $(-4, 6)$ and $(-1, -2)$ 10. $2x + y = 10$

11. $x + 12 = 0$ 12. A line whose graph is parallel to the graph of $y - 4 = 6$

13.

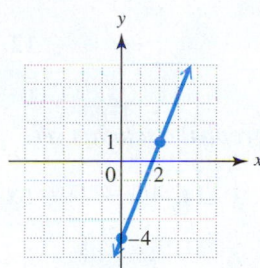

Write an equation of each line. Give the final answer in slope-intercept form.

14. Through $(-1, 4)$, $m = 2$ 15. The line in **Exercise 13**

16. Through $(2, -6)$ and $(1, 3)$

The graph shows worldwide snowmobile sales y for selected years x, where $x = 0$ represents 2000, $x = 1$ represents 2001, and so on. Use the graph to work Exercises 17–20.

17. Is the slope of the line in the graph positive or negative? Explain.

18. Write two ordered pairs (x, y) for the data points shown in the graph.
    (a) Use the ordered pairs to find the slope of the line to the nearest tenth.
    (b) Write an equation of a line that models the data. Give the final equation in slope-intercept form.

19. Use the equation from **Exercise 18(b)** to approximate worldwide snowmobile sales for 2005. How does the answer compare to the actual sales of 173.7 thousand?

20. What does the ordered pair $(13, 145)$ mean in the context of this problem?

Worldwide Snowmobile Sales

Source: www.snowmobile.org

Chapters R–3   Cumulative Review Exercises

[R.1]

1. $\frac{301}{40}$, or $7\frac{21}{40}$ **2.** 6

[1.4]   **[1.5]**

3. 7   **4.** $\frac{73}{18}$, or $4\frac{1}{18}$

[1.1–1.5]

5. true   **6.** −43

[1.6]   **[1.7]**

7. distributive   **8.** $-p + 2$
   property

[2.5]

9. $h = \frac{3V}{\pi r^2}$

[2.3]

10. $\{-1\}$

11. $\{2\}$

[2.6]

12. $\{-13\}$

[2.8]

13. $(-2.6, \infty)$

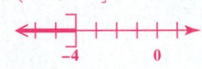
   −2.6 −1 0

14. $(0, \infty)$

   0

15. $(-\infty, -4]$

   −4   0

[2.4]

16. 13 mi

[R.2]

17. (a) \$7000   **(b)** \$10,000

18. about \$30,000

[3.2]   **[3.3]**

19. $(-4, 0)$; $(0, 3)$ **20.** $\frac{3}{4}$

[3.2]

21.

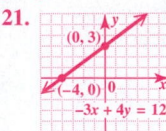
$-3x + 4y = 12$

[3.3]

22. perpendicular

[3.4, 3.5]

23. $y = 3x - 11$ **24.** $y = 4$

Perform each indicated operation.

1. $10\frac{5}{8} - 3\frac{1}{10}$

2. $\frac{3}{4} \div \frac{1}{8}$

3. $5 - (-4) + (-2)$

4. $\dfrac{(-3)^2 - (-4)(2^4)}{5(2) - (-2)^3}$

5. *True* or *false?* $\dfrac{4(3 - 9)}{2 - 6} \geq 6$

6. Find the value of $xz^3 - 5y^2$ for $x = -2$, $y = -3$, and $z = -1$.

7. What property does $3(-2 + x) = -6 + 3x$ illustrate?

8. Simplify $-4p - 6 + 3p + 8$ by combining like terms.

Solve.

9. $V = \frac{1}{3}\pi r^2 h$ for h

10. $6 - 3(1 + x) = 2(x + 5) - 2$

11. $-(m - 3) = 5 - 2m$

12. $\dfrac{x - 2}{3} = \dfrac{2x + 1}{5}$

Solve each inequality, and graph the solution set.

13. $-2.5x < 6.5$

14. $4(x + 3) - 5x < 12$

15. $\frac{2}{3}x - \frac{1}{6}x \leq -2$

Solve each problem.

16. Mount Mayon in the Philippines is the most perfectly shaped conical volcano in the world. Its base is a circle with circumference 80 mi, and it has a height of about 8100 ft. (One mile is 5280 ft.) Find the radius of the circular base to the nearest mile. (*Source:* www.britannica.hk)

17. Over the next 45 yr, baby boomers are expected to inherit \$10.4 trillion from their parents, an average of \$50,000 each. The circle graph shows how they plan to spend their inheritances.

   (a) How much of the \$50,000 is expected to go toward a home purchase?

   (b) How much of the \$50,000 is expected to go toward retirement?

18. Use the answer from **Exercise 17(b)** to estimate the amount expected to go toward paying off debts or funding children's education.

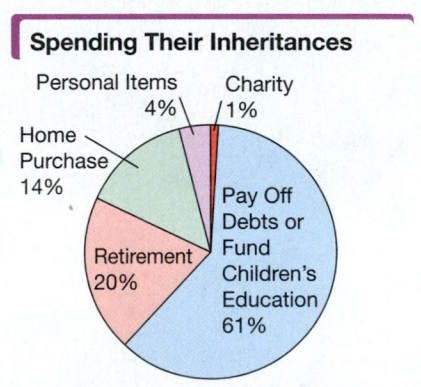

Spending Their Inheritances

Source: First Interstate Bank Trust and Private Banking Group.

Consider the linear equation $-3x + 4y = 12$. Find the following.

19. The *x*- and *y*-intercepts   **20.** The slope   **21.** The graph

22. Are the lines with equations $x + 5y = -6$ and $y = 5x - 8$ *parallel*, *perpendicular*, or *neither*?

Write an equation of each line. Give the final answer in slope-intercept form if possible.

23. Through $(2, -5)$, slope 3

24. Through $(0, 4)$ and $(2, 4)$

4

Exponents and Polynomials

Numbers such as the one shown here are often written using *scientific notation,* a method of expressing them using a power of 10 as a factor.

17,512,083,047,998

$= 1.7512083047998 \times 10^{13}$

261

4.1   The Product Rule and Power Rules for Exponents

OBJECTIVES

1   Use exponents.
2   Use the product rule for exponents.
3   Use the rule $(a^m)^n = a^{mn}$.
4   Use the rule $(ab)^m = a^m b^m$.
5   Use the rule $\left(\frac{a}{b}\right)^m = \frac{a^m}{b^m}$.
6   Use combinations of the rules for exponents.
7   Use the rules for exponents in a geometry application.

VOCABULARY

☐ base
☐ exponent (power)
☐ exponential expression

**NOW TRY
EXERCISE 1**

Write $4 \cdot 4 \cdot 4$ in exponential form and evaluate.

**NOW TRY
EXERCISE 2**

Identify the base and the exponent of each expression. Then evaluate.

(a) $(-3)^4$ **(b)** -3^4

NOW TRY ANSWERS
1. 4^3; 64
2. **(a)** -3; 4; 81   **(b)** 3; 4; -81

OBJECTIVE 1   Use exponents.

Recall from **Section 1.1** that in the expression 5^2, the number 5 is the **base** and 2 is the **exponent,** or **power.** The expression 5^2 is an **exponential expression.** Although we do not usually write the exponent when it is 1, the following holds for any quantity a in general.

$$a^1 = a$$

EXAMPLE 1   Using Exponents

Write $3 \cdot 3 \cdot 3 \cdot 3$ in exponential form and evaluate.

Because 3 occurs as a factor four times, the base is 3 and the exponent is 4. The exponential expression is 3^4, read "3 to the fourth power" or simply "3 to the fourth."

$$\underbrace{3 \cdot 3 \cdot 3 \cdot 3}_{\text{4 factors of 3}} \quad \text{means} \quad 3^4, \quad \text{which equals} \quad 81.$$

NOW TRY

EXAMPLE 2   Evaluating Exponential Expressions

Identify the base and the exponent of each expression. Then evaluate.

Expression	Base	Exponent	Value
(a) 5^4	5	4	$5 \cdot 5 \cdot 5 \cdot 5$, which equals 625
(b) -5^4	5	4	$-1 \cdot (5 \cdot 5 \cdot 5 \cdot 5)$, which equals -625
(c) $(-5)^4$	-5	4	$(-5)(-5)(-5)(-5)$, which equals 625

NOW TRY

> ⚠ **CAUTION**   Compare **Examples 2(b) and 2(c).** In the expression -5^4, the absence of parentheses shows that the exponent 4 applies only to the base 5, not -5. In $(-5)^4$, the parentheses show that the exponent 4 applies to the base -5. In summary, $-a^n$ and $(-a)^n$ are not necessarily the same.

Expression	Base	Exponent	Example
$-a^n$	a	n	$-3^2 = -(3 \cdot 3) = -9$
$(-a)^n$	$-a$	n	$(-3)^2 = (-3)(-3) = 9$

OBJECTIVE 2   Use the product rule for exponents.

To develop the product rule, we use the definition of exponents.

$$2^4 \cdot 2^3$$

$$= (\underbrace{2 \cdot 2 \cdot 2 \cdot 2}_{\text{4 factors}})(\underbrace{2 \cdot 2 \cdot 2}_{\text{3 factors}})$$

$$= \underbrace{2 \cdot 2 \cdot 2 \cdot 2 \cdot 2 \cdot 2 \cdot 2}_{4 + 3 = 7 \text{ factors}}$$

$$= 2^7$$

Also,

$$6^2 \cdot 6^3$$
$$= (6 \cdot 6)(6 \cdot 6 \cdot 6)$$
$$= 6 \cdot 6 \cdot 6 \cdot 6 \cdot 6$$
$$= 6^5.$$

Generalizing from these examples, we have the following.

$2^4 \cdot 2^3$ is equal to 2^{4+3}, which equals 2^7.

$6^2 \cdot 6^3$ is equal to 6^{2+3}, which equals 6^5.

In each case, adding the exponents gives the exponent of the product, suggesting the **product rule for exponents.**

> ### Product Rule for Exponents
>
> For any positive integers m and n, $\quad a^m \cdot a^n = a^{m+n}$.
> (Keep the same base and add the exponents.)
>
> *Example:* $\quad 6^2 \cdot 6^5 = 6^{2+5} = 6^7$

> ⊘ **CAUTION** Do not multiply the bases when using the product rule. ***Keep the same base and add the exponents.*** For example,
>
> $$6^2 \cdot 6^5 = 6^7, \quad \textit{not} \quad 36^7.$$

NOW TRY EXERCISE 3

Use the product rule for exponents to simplify each expression, if possible.

(a) $(-5)^2(-5)^4$

(b) $y^2 \cdot y \cdot y^5$

(c) $(2x^3)(4x^6)$

(d) $2^4 \cdot 5^3$

(e) $3^2 + 3^3$

EXAMPLE 3 Using the Product Rule

Use the product rule for exponents to simplify each expression, if possible.

(a) $6^3 \cdot 6^5$

$\quad = 6^{3+5} \qquad$ Product rule

$\quad = 6^8 \qquad$ Add the exponents.

(b) $(-4)^7(-4)^2$

$\quad = (-4)^{7+2} \qquad$ Product rule

$\quad = (-4)^9 \qquad$ Add the exponents.

(c) $x^2 \cdot x$

$\quad = x^2 \cdot x^1 \qquad a = a^1$, for all a.

$\quad = x^{2+1} \qquad$ Product rule

$\quad = x^3 \qquad$ Add the exponents.

(d) $m^4 m^3 m^5$

$\quad = m^{4+3+5} \qquad$ Product rule

$\quad = m^{12} \qquad$ Add the exponents.

(e) $\boxed{\text{Think: } 2^3 \text{ means } 2 \cdot 2 \cdot 2.}$ $2^3 \cdot 3^2$ $\boxed{\text{Think: } 3^2 \text{ means } 3 \cdot 3.}$ The product rule does not apply. ***The bases are different.***

$\quad = 8 \cdot 9 \qquad$ Evaluate 2^3 and 3^2.

$\quad = 72 \qquad$ Multiply.

(f) $2^3 + 2^4$ The product rule does not apply. ***This is a sum, not a product.***

$\quad = 8 + 16 \qquad$ Evaluate 2^3 and 2^4.

$\quad = 24 \qquad$ Add.

NOW TRY ANSWERS
3. (a) $(-5)^6$ **(b)** y^8 **(c)** $8x^9$
 (d) The product rule does not apply; 2000
 (e) The product rule does not apply; 36

(g) $(2x^3)(3x^7)$ $\boxed{2x^3 \text{ means } 2 \cdot x^3 \text{ and } 3x^7 \text{ means } 3 \cdot x^7.}$

$\quad = (2 \cdot 3) \cdot (x^3 \cdot x^7) \qquad$ Commutative and associative properties

$\quad = 6x^{3+7} \qquad$ Multiply and then use the product rule.

$\quad = 6x^{10} \qquad$ Add the exponents.

NOW TRY ↻

> ⚠️ **CAUTION**   Be sure that you understand the difference between *adding* and *multiplying* exponential expressions.
>
> $8x^3 + 5x^3$ means $(8 + 5)x^3$,   which equals $13x^3$.
>
> $(8x^3)(5x^3)$ means $(8 \cdot 5)x^{3+3}$,   which equals $40x^6$.

OBJECTIVE 3   Use the rule $(a^m)^n = a^{mn}$.

We can simplify an expression such as $(5^2)^4$ with the product rule for exponents, as follows.

$$(5^2)^4$$

$$= 5^2 \cdot 5^2 \cdot 5^2 \cdot 5^2 \quad \text{Definition of exponent}$$

$$= 5^{2+2+2+2} \quad \text{Product rule}$$

$$= 5^8 \quad \text{Add.}$$

Observe that $2 \cdot 4 = 8$. This example suggests **power rule (a) for exponents.**

> ### Power Rule (a) for Exponents
>
> For any positive integers m and n, $(a^m)^n = a^{mn}$.
> (Raise a power to a power by multiplying exponents.)
>
> *Example:* $(3^2)^4 = 3^{2 \cdot 4} = 3^8$

**NOW TRY
EXERCISE 4**

Use power rule (a) for exponents to simplify.

(a) $(4^7)^5$ **(b)** $(y^4)^7$

EXAMPLE 4   Using Power Rule (a)

Use power rule (a) for exponents to simplify.

(a) $(2^5)^3$

$$= 2^{5 \cdot 3}$$

$$= 2^{15}$$

(b) $(5^7)^2$

$$= 5^{7 \cdot 2}$$

$$= 5^{14}$$

(c) $(x^2)^5$

$$= x^{2 \cdot 5} \quad \text{Power rule (a)}$$

$$= x^{10} \quad \text{Multiply.}$$

NOW TRY

OBJECTIVE 4   Use the rule $(ab)^m = a^m b^m$.

Consider the following.

$$(4x)^3$$

$$= (4x)(4x)(4x) \quad \text{Definition of exponent}$$

$$= (4 \cdot 4 \cdot 4)(x \cdot x \cdot x) \quad \text{Commutative and associative properties}$$

$$= 4^3 x^3 \quad \text{Definition of exponent}$$

This example suggests **power rule (b) for exponents.**

> ### Power Rule (b) for Exponents
>
> For any positive integer m, $(ab)^m = a^m b^m$.
> (Raise a product to a power by raising each factor to the power.)
>
> *Example:* $(2p)^5 = 2^5 p^5$

NOW TRY ANSWERS

4. (a) 4^{35} **(b)** y^{28}

NOW TRY EXERCISE 5

Use power (b) for exponents to simplify.

(a) $(-5ab)^3$ **(b)** $(4t^3p^5)^2$

EXAMPLE 5 Using Power Rule (b)

Use power rule (b) for exponents to simplify.

(a) $(3xy)^2$

$\quad = 3^2x^2y^2$ Power rule (b)

$\quad = 9x^2y^2$ $\quad 3^2 = 3 \cdot 3 = 9$

(b) $5(pq)^2$

$\quad = 5(p^2q^2)$ Power rule (b)

$\quad = 5p^2q^2$ Multiply.

(c) $(2m^2p^3)^4$

$\quad = 2^4(m^2)^4(p^3)^4$ Power rule (b)

$\quad = 2^4m^8p^{12}$ Power rule (a)

$\quad = 16m^8p^{12}$ $\quad 2^4 = 2 \cdot 2 \cdot 2 \cdot 2 = 16$

(d) $(-5^6)^3$

$\quad = (-1 \cdot 5^6)^3$ $\quad -a = -1 \cdot a$

$\quad = (-1)^3 \cdot (5^6)^3$ Power rule (b)

Raise -1 to the designated power.

$\quad = -1 \cdot 5^{18}$ Power rule (a)

$\quad = -5^{18}$ Multiply.

NOW TRY

⚠ **CAUTION** *Power rule (b) does not apply to a sum.*

$$(4x)^2 = 4^2x^2, \quad \text{but} \quad (4 + x)^2 \neq 4^2 + x^2.$$

OBJECTIVE 5 Use the rule $\left(\dfrac{a}{b}\right)^m = \dfrac{a^m}{b^m}$.

Because the quotient $\dfrac{a}{b}$ can be written as $a\left(\dfrac{1}{b}\right)$, we use this fact, power rule (b), and some properties of real numbers to obtain **power rule (c) for exponents.**

> **Power Rule (c) for Exponents**
>
> For any positive integer m, $\quad \left(\dfrac{a}{b}\right)^m = \dfrac{a^m}{b^m} \quad$ (where $b \neq 0$).
>
> (Raise a quotient to a power by raising both numerator and denominator to the power.)
>
> *Example:* $\quad \left(\dfrac{5}{3}\right)^2 = \dfrac{5^2}{3^2}$

NOW TRY EXERCISE 6

Use power (c) for exponents to simplify.

(a) $\left(\dfrac{p}{q}\right)^5$ **(b)** $\left(\dfrac{1}{4}\right)^3$

$\quad (q \neq 0)$

NOW TRY ANSWERS

5. (a) $-125a^3b^3$ **(b)** $16t^6p^{10}$

6. (a) $\dfrac{p^5}{q^5}$ **(b)** $\dfrac{1}{64}$

EXAMPLE 6 Using Power Rule (c)

Use power rule (c) for exponents to simplify.

(a) $\left(\dfrac{2}{3}\right)^5$

$\quad = \dfrac{2^5}{3^5}$

$\quad = \dfrac{32}{243}$

(b) $\left(\dfrac{m}{n}\right)^4$

$\quad = \dfrac{m^4}{n^4}$

$\quad (n \neq 0)$

(c) $\left(\dfrac{1}{5}\right)^4$

$\quad = \dfrac{1^4}{5^4}$ Power rule (c)

$\quad = \dfrac{1}{625}$ Simplify.

NOW TRY

> **NOTE** In **Example 6(c),** we used the fact that $1^4 = 1$.
>
> *In general, $1^n = 1$, for any integer n.*

Rules for Exponents

For positive integers m and n, the following hold.

			Examples
Product rule	$a^m \cdot a^n = a^{m+n}$		$6^2 \cdot 6^5 = 6^{2+5} = 6^7$
Power rules (a)	$(a^m)^n = a^{mn}$		$(3^2)^4 = 3^{2 \cdot 4} = 3^8$
(b)	$(ab)^m = a^m b^m$		$(2p)^5 = 2^5 p^5$
(c)	$\left(\dfrac{a}{b}\right)^m = \dfrac{a^m}{b^m}$	(where $b \neq 0$)	$\left(\dfrac{5}{3}\right)^2 = \dfrac{5^2}{3^2}$

OBJECTIVE 6 Use combinations of the rules for exponents.

EXAMPLE 7 Using Combinations of the Rules

Simplify.

(a) $\left(\dfrac{2}{3}\right)^2 \cdot 2^3$

$= \dfrac{2^2}{3^2} \cdot \dfrac{2^3}{1}$ Power rule (c)

$= \dfrac{2^2 \cdot 2^3}{3^2 \cdot 1}$ Multiply fractions.

$= \dfrac{2^{2+3}}{3^2}$ Product rule

$= \dfrac{2^5}{3^2}$ Add.

$= \dfrac{32}{9}$ Apply the exponents.

(b) $(5x)^3 (5x)^4$

$= (5x)^7$ Product rule

$= 5^7 x^7$ Power rule (b)

An equally correct way to simplify this expression follows.

$(5x)^3 (5x)^4$

$= 5^3 x^3 5^4 x^4$ Power rule (b)

$= 5^3 \cdot 5^4 x^3 x^4$ Commutative property

$= 5^7 x^7$ Product rule

(c) $(2x^2 y^3)^4 (3xy^2)^3$

$= 2^4 (x^2)^4 (y^3)^4 \cdot 3^3 x^3 (y^2)^3$ Power rule (b)

$= 2^4 x^8 y^{12} \cdot 3^3 x^3 y^6$ Power rule (a)

$= 2^4 \cdot 3^3 x^8 x^3 y^{12} y^6$ Commutative and associative properties

$= 16 \cdot 27 x^{11} y^{18}$ Apply the exponents; product rule

$= 432 x^{11} y^{18}$ Multiply.

Notice that $(2x^2 y^3)^4$ means $2^4 x^{2 \cdot 4} y^{3 \cdot 4}$, **not** $(2 \cdot 4) x^{2 \cdot 4} y^{3 \cdot 4}$.

**NOW TRY
EXERCISE 7**

Simplify.

(a) $\left(\dfrac{3}{5}\right)^3 \cdot 3^2$ (b) $(8k)^5(8k)^4$

(c) $(x^4y)^5(-2x^2y^5)^3$

(d) $(-x^3y)^2(-x^5y^4)^3$

> Think of the negative sign as a factor of -1.

$= (-1 \cdot x^3y)^2(-1 \cdot x^5y^4)^3$ $-a = -1 \cdot a$

$= (-1)^2(x^3)^2y^2 \cdot (-1)^3(x^5)^3(y^4)^3$ Power rule (b)

$= (-1)^2(x^6)(y^2)(-1)^3(x^{15})(y^{12})$ Power rule (a)

$= (-1)^5(x^{21})(y^{14})$ Product rule

$= -x^{21}y^{14}$ Simplify.   **NOW TRY**

⚠️ **CAUTION**   Be aware of the distinction between $(2y)^3$ and $2y^3$.

$$(2y)^3 = 2y \cdot 2y \cdot 2y = 8y^3,$$ The base is $2y$.

while $$2y^3 = 2 \cdot y \cdot y \cdot y.$$ The base is y.

OBJECTIVE 7   Use the rules for exponents in a geometry application.

**NOW TRY
EXERCISE 8**

Find an expression that represents the area of the figure. Assume $x > 0$.

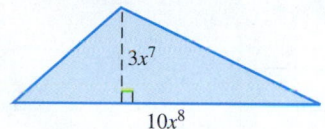

| **EXAMPLE 8**   Using Area Formulas |

Find an expression that represents the area in (a) **FIGURE 1** and (b) **FIGURE 2**.

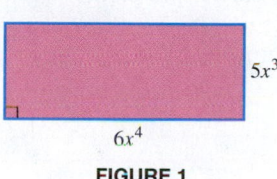

FIGURE 1

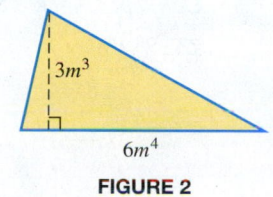

Assume $x > 0$, $m > 0$.

FIGURE 2

(a) For **FIGURE 1**, use the formula for the area of a rectangle.

$\mathcal{A} = LW$ Area formula

$\mathcal{A} = (6x^4)(5x^3)$ Substitute.

$\mathcal{A} = 6 \cdot 5 \cdot x^{4+3}$ Commutative property; product rule

$\mathcal{A} = 30x^7$ Multiply. Add the exponents.

(b) **FIGURE 2** is a triangle with base $6m^4$ and height $3m^3$.

$\mathcal{A} = \dfrac{1}{2}bh$ Area formula

$\mathcal{A} = \dfrac{1}{2}(6m^4)(3m^3)$ Substitute.

$\mathcal{A} = \dfrac{1}{2}(6 \cdot 3 \cdot m^{4+3})$ Properties of real numbers; product rule

$\mathcal{A} = 9m^7$ Multiply. Add the exponents.   **NOW TRY**

NOW TRY ANSWERS

7. (a) $\dfrac{243}{125}$ (b) 8^9k^9
   (c) $-8x^{26}y^{20}$

8. $15x^{15}$

4.1 Exercises

    MyMathLab®

● *Complete solution available in MyMathLab*

Concept Check   Decide whether each statement is true or false. If false, tell why.

1. $3^3 = 9$

2. $(x^2)^3 = x^5$

3. $(-3)^4 = 3^4$

4. $\left(\dfrac{1}{5}\right)^2 = \dfrac{1}{5^2}$

Write each expression in exponential form. **See Example 1.**

5. $w \cdot w \cdot w \cdot w \cdot w \cdot w$

6. $t \cdot t \cdot t \cdot t \cdot t \cdot t \cdot t \cdot t$

● **7.** $\left(\dfrac{1}{2}\right)\left(\dfrac{1}{2}\right)\left(\dfrac{1}{2}\right)\left(\dfrac{1}{2}\right)\left(\dfrac{1}{2}\right)\left(\dfrac{1}{2}\right)$

8. $\left(\dfrac{1}{4}\right)\left(\dfrac{1}{4}\right)\left(\dfrac{1}{4}\right)\left(\dfrac{1}{4}\right)\left(\dfrac{1}{4}\right)$

9. $(-4)(-4)(-4)(-4)$

10. $(-3)(-3)(-3)(-3)(-3)(-3)$

11. $(-7y)(-7y)(-7y)(-7y)$

12. $(-8p)(-8p)(-8p)(-8p)(-8p)$

13. Explain how the expressions $(-3)^4$ and -3^4 are different.

14. Explain how the expressions $(5x)^3$ and $5x^3$ are different.

Identify the base and the exponent of each expression. In Exercises 15–18, also evaluate. **See Example 2.**

● **15.** 3^5

16. 2^7

● **17.** $(-3)^5$

18. $(-2)^7$

19. $(-6x)^4$

20. $(-8x)^4$

21. $-6x^4$

22. $-8x^4$

Concept Check   Simplify each expression.

23. $8^2 \cdot 8^5$

$\quad = 8^{\text{—}+\text{—}}$

$\quad = \underline{}$

24. $5m^2 \cdot 2m^6$

$\quad = (5 \cdot \underline{}) \cdot (m^{\text{—}} \cdot m^{\text{—}})$

$\quad = \underline{} m^{\text{—}+\text{—}}$

$\quad = \underline{}$

Use the product rule for exponents to simplify each expression, if possible. Write each answer in exponential form. **See Example 3.**

● **25.** $5^2 \cdot 5^6$

26. $3^6 \cdot 3^7$

27. $4^2 \cdot 4^7 \cdot 4^3$

28. $5^3 \cdot 5^8 \cdot 5^2$

29. $(-7)^3(-7)^6$

30. $(-9)^8(-9)^5$

● **31.** $t^3 \cdot t^8 \cdot t^{13}$

32. $n^5 \cdot n^6 \cdot n^9$

33. $(-8r^4)(7r^3)$

34. $(10a^7)(-4a^3)$

● **35.** $(-6p^5)(-7p^5)$

36. $(-5w^8)(-9w^8)$

● **37.** $(5x^2)(-2x^3)(3x^4)$

38. $(12y^3)(4y)(-3y^5)$

● **39.** $3^8 + 3^9$

40. $4^{12} + 4^5$

41. $5^8 \cdot 3^9$

42. $6^3 \cdot 8^9$

Use the power rules for exponents to simplify each expression. Assume that variables in denominators are not zero. Write each answer in exponential form. **See Examples 4–6.**

● **43.** $(4^3)^2$

44. $(8^3)^6$

● **45.** $(t^4)^5$

46. $(y^6)^5$

47. $(7r)^3$

48. $(11x)^4$

● **49.** $(5xy)^5$

50. $(9pq)^6$

51. $(-5^2)^6$

52. $(-9^4)^8$

53. $(-8^3)^5$

54. $(-7^5)^7$

55. $8(qr)^3$

56. $4(vw)^5$

57. $\left(\dfrac{9}{5}\right)^8$

58. $\left(\dfrac{12}{7}\right)^3$

▶ 59. $\left(\dfrac{1}{2}\right)^3$ 60. $\left(\dfrac{1}{3}\right)^5$ 61. $\left(\dfrac{a}{b}\right)^3$ 62. $\left(\dfrac{r}{t}\right)^4$

63. $\left(\dfrac{x}{2}\right)^3$ 64. $\left(\dfrac{y}{3}\right)^4$ 65. $\left(-\dfrac{2x}{y}\right)^5$ 66. $\left(-\dfrac{4p}{q}\right)^3$

Simplify each expression. **See Example 7.**

67. $\left(\dfrac{5}{2}\right)^3 \cdot \left(\dfrac{5}{2}\right)^2$ 68. $\left(\dfrac{3}{4}\right)^5 \cdot \left(\dfrac{3}{4}\right)^6$ ▶ 69. $\left(\dfrac{9}{8}\right)^3 \cdot 9^2$

70. $\left(\dfrac{8}{5}\right)^4 \cdot 8^3$ 71. $(2x)^9(2x)^3$ 72. $(6y)^5(6y)^8$

73. $(-6p)^4(-6p)$ 74. $(-13q)^3(-13q)$ 75. $(6x^2y^3)^5$

76. $(5r^5t^6)^7$ 77. $(x^2)^3(x^3)^5$ 78. $(y^4)^5(y^3)^5$

79. $(2w^2x^3y)^2(x^4y)^5$ 80. $(3x^4y^2z)^3(yz^4)^5$ ▶ 81. $(-r^4s)^2(-r^2s^3)^5$

82. $(-ts^6)^4(-t^3s^5)^3$ 83. $\left(\dfrac{5a^2b^5}{c^6}\right)^3$ $(c \neq 0)$ 84. $\left(\dfrac{6x^3y^9}{z^5}\right)^4$ $(z \neq 0)$

85. *Concept Check* A student wrote the following as a simplification of $(10^2)^3$.

$$1000^6$$

WHAT WENT WRONG?

86. *Concept Check* A student wrote the following as a simplification of $(3x^2y^3)^4$.

$$3 \cdot 4x^8y^{12}, \quad \text{or} \quad 12x^8y^{12}$$

WHAT WENT WRONG?

Find an expression that represents the area of each figure. **See Example 8.** *(If necessary, refer to the formulas found inside the back cover of this book. The ⌐ in the figures indicates 90° angles.)*

▶ 87.

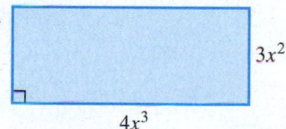

88.

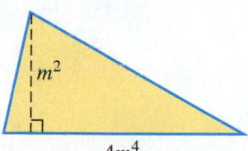

89.

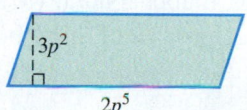

90.

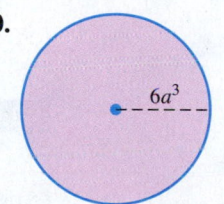

Find an expression that represents the volume of each figure. (If necessary, refer to the formulas found inside the back cover of this book.)

91.

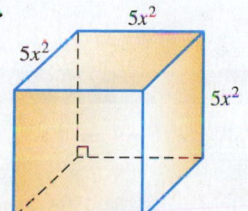

92.

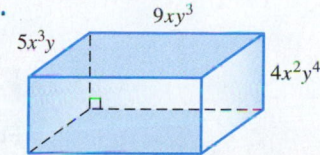

Compound interest is interest paid on both the principal and the interest earned earlier. The formula for compound interest, which involves an exponential expression, is

$$A = P(1 + r)^n,$$

where A is the amount accumulated from a principal of P dollars left untouched for n years with an annual interest rate r (expressed as a decimal).

In Exercises 93–96, use the preceding formula and a calculator to find A to the nearest cent.

93. $P = \$250$, $r = 0.04$, $n = 5$ **94.** $P = \$400$, $r = 0.04$, $n = 3$

95. $P = \$1500$, $r = 0.015$, $n = 6$ **96.** $P = \$2000$, $r = 0.015$, $n = 4$

4.2 Integer Exponents and the Quotient Rule

OBJECTIVES

1. Use 0 as an exponent.
2. Use negative numbers as exponents.
3. Use the quotient rule for exponents.
4. Use combinations of the rules for exponents.

Consider the following list.

$$2^4 = 16$$
$$2^3 = 8$$
$$2^2 = 4$$

As exponents decrease by 1, the results are divided by 2 each time.

Each time we decrease the exponent by 1, the value is divided by 2 (the base). Using this pattern, we can continue the list to lesser and lesser integer exponents.

$$2^1 = 2$$
$$2^0 = 1$$
$$2^{-1} = \frac{1}{2}$$
$$2^{-2} = \frac{1}{4}$$
$$2^{-3} = \frac{1}{8}$$

We continue the pattern here.

From the preceding list, it appears that we should define 2^0 as 1 and bases raised to negative exponents as reciprocals of those bases.

OBJECTIVE 1 Use 0 as an exponent.

The definitions of 0 and negative exponents must be consistent with the rules for exponents from **Section 4.1.** For example, if we define 6^0 to be 1, then

$$6^0 \cdot 6^2 = 1 \cdot 6^2 = 6^2 \quad \text{and} \quad 6^0 \cdot 6^2 = 6^{0+2} = 6^2,$$

and we see that the product rule is satisfied. Check that the power rules are also valid for a 0 exponent. Thus, we define a 0 exponent as follows.

Zero Exponent

For any nonzero real number a, $a^0 = 1$.

Example: $17^0 = 1$

NOW TRY EXERCISE 1

Evaluate.

(a) 6^0

(b) -12^0

(c) $(-12x)^0$ $(x \neq 0)$

(d) $14^0 - 12^0$

EXAMPLE 1 Using Zero Exponents

Evaluate.

(a) $60^0 = 1$

(b) $(-60)^0 = 1$

(c) $-60^0 = -(1) = -1$

(d) $y^0 = 1$ $(y \neq 0)$

(e) $6y^0 = 6(1) = 6$ $(y \neq 0)$

(f) $(6y)^0 = 1$ $(y \neq 0)$

(g) $8^0 + 11^0 = 1 + 1 = 2$

(h) $-8^0 - 11^0 = -1 - 1 = -2$

NOW TRY

> ⚠ **CAUTION** Look again at **Examples 1(b)** and **1(c)**. In $(-60)^0$, the base is -60, and because any nonzero base raised to the 0 exponent is 1, $(-60)^0 = 1$. In -60^0, which can be written $-(60)^0$, the base is 60, so $-60^0 = -1$.

OBJECTIVE 2 Use negative numbers as exponents.

From the lists at the beginning of this section, $2^{-2} = \frac{1}{4}$ and $2^{-3} = \frac{1}{8}$. We can make a conjecture that 2^{-n} should equal $\frac{1}{2^n}$. Is the product rule valid in such cases? For example,

$$6^{-2} \cdot 6^2 = 6^{-2+2} = 6^0 = 1.$$

The expression 6^{-2} behaves as if it were the reciprocal of 6^2, because their product is 1. The reciprocal of 6^2 is also $\frac{1}{6^2}$, leading us to define 6^{-2} as $\frac{1}{6^2}$, and generalize accordingly.

Negative Exponents

For any nonzero real number a and any integer n, $\quad a^{-n} = \dfrac{1}{a^n}$.

Example: $\quad 3^{-2} = \dfrac{1}{3^2}$

By definition, a^{-n} and a^n are reciprocals.

$$a^n \cdot a^{-n} = a^n \cdot \frac{1}{a^n} = 1$$

Because $1^n = 1$, the definition of a^{-n} can also be written as follows.

$$a^{-n} = \frac{1}{a^n} = \frac{1^n}{a^n} = \left(\frac{1}{a}\right)^n$$

For example, $\qquad 6^{-3} = \left(\frac{1}{6}\right)^3 \quad$ and $\quad \left(\frac{1}{3}\right)^{-2} = 3^2.$

NOW TRY ANSWERS

1. **(a)** 1　**(b)** -1　**(c)** 1　**(d)** 0

NOW TRY
EXERCISE 2

Write with positive exponents and simplify.

(a) 2^{-3}

(b) $\left(\dfrac{1}{7}\right)^{-2}$

(c) $\left(\dfrac{3}{2}\right)^{-4}$

(d) $3^{-2} + 4^{-2}$

(e) p^{-4} $(p \neq 0)$

EXAMPLE 2   **Using Negative Exponents**

Write with positive exponents and simplify. Assume that all variables represent nonzero real numbers.

(a) $3^{-2} = \dfrac{1}{3^2} = \dfrac{1}{9}$ $a^{-n} = \dfrac{1}{a^n}$

(b) $5^{-3} = \dfrac{1}{5^3} = \dfrac{1}{125}$ $a^{-n} = \dfrac{1}{a^n}$

(c) $\left(\dfrac{1}{2}\right)^{-3} = 2^3 = 8$ $\dfrac{1}{2}$ and 2 are reciprocals.
(Reciprocals have a product of 1.)

Notice that we can change the base to its reciprocal if we also change the sign of the exponent.

(d) $\left(\dfrac{2}{5}\right)^{-4}$

$= \left(\dfrac{5}{2}\right)^4$ $\dfrac{2}{5}$ and $\dfrac{5}{2}$ are reciprocals.

$= \dfrac{5^4}{2^4}$ Power rule (c)

$= \dfrac{625}{16}$ Apply the exponents.

(e) $\left(\dfrac{4}{3}\right)^{-5}$

$= \left(\dfrac{3}{4}\right)^5$ $\dfrac{4}{3}$ and $\dfrac{3}{4}$ are reciprocals.

$= \dfrac{3^5}{4^5}$ Power rule (c)

$= \dfrac{243}{1024}$ Apply the exponents.

(f) $4^{-1} - 2^{-1}$

$= \dfrac{1}{4} - \dfrac{1}{2}$ Apply the exponents.

$= \dfrac{1}{4} - \dfrac{2}{4}$ Find a common denominator.

$= -\dfrac{1}{4}$ Subtract.

(g) $3p^{-2}$

$= \dfrac{3}{1} \cdot \dfrac{1}{p^2}$ $a^{-n} = \dfrac{1}{a^n}$

$= \dfrac{3}{p^2}$ Multiply.

(h) $\dfrac{1}{x^{-4}}$

$= \dfrac{1^{-4}}{x^{-4}}$ $1^n = 1$, for any integer n.

$= \left(\dfrac{1}{x}\right)^{-4}$ Power rule (c)

$= x^4$ $\dfrac{1}{x}$ and x are reciprocals.

(i) $x^3 y^{-4}$

$= \dfrac{x^3}{1} \cdot \dfrac{1}{y^4}$ $a^{-n} = \dfrac{1}{a^n}$

$= \dfrac{x^3}{y^4}$ Multiply.

In general, $\dfrac{1}{a^{-n}} = a^n$.   **NOW TRY**

⚠ **CAUTION**   *A negative exponent does not indicate a negative number. Negative exponents lead to reciprocals.*

Expression	Example	
a^{-n}	$3^{-2} = \dfrac{1}{3^2} = \dfrac{1}{9}$	Not negative
$-a^{-n}$	$-3^{-2} = -\dfrac{1}{3^2} = -\dfrac{1}{9}$	Negative

---

NOW TRY ANSWERS

2. (a) $\dfrac{1}{8}$ **(b)** 49   **(c)** $\dfrac{16}{81}$

(d) $\dfrac{25}{144}$ **(e)** $\dfrac{1}{p^4}$

Consider the following.

$$\frac{2^{-3}}{3^{-4}}$$

$$= \frac{\frac{1}{2^3}}{\frac{1}{3^4}} \qquad \text{Definition of negative exponent}$$

$$= \frac{1}{2^3} \div \frac{1}{3^4} \qquad \frac{a}{b} \text{ means } a \div b.$$

$$= \frac{1}{2^3} \cdot \frac{3^4}{1} \qquad \text{To divide, multiply by the reciprocal of the divisor.}$$

$$= \frac{3^4}{2^3} \qquad \text{Multiply.}$$

Therefore, $\qquad \dfrac{2^{-3}}{3^{-4}} = \dfrac{3^4}{2^3}.$

Changing from Negative to Positive Exponents

For any nonzero numbers a and b and any integers m and n, the following hold.

$$\frac{a^{-m}}{b^{-n}} = \frac{b^n}{a^m} \quad \text{and} \quad \left(\frac{a}{b}\right)^{-m} = \left(\frac{b}{a}\right)^m$$

Examples: $\quad \dfrac{3^{-5}}{2^{-4}} = \dfrac{2^4}{3^5} \quad$ and $\quad \left(\dfrac{4}{5}\right)^{-3} = \left(\dfrac{5}{4}\right)^3$

**NOW TRY
EXERCISE 3**

Write with positive exponents and simplify. Assume that all variables represent nonzero real numbers.

(a) $\dfrac{5^{-3}}{6^{-2}}$ **(b)** $m^2 n^{-4}$

(c) $\dfrac{x^2 y^{-3}}{5z^{-4}}$

EXAMPLE 3 Changing from Negative to Positive Exponents

Write with positive exponents and simplify. Assume that all variables represent non-zero real numbers.

(a) $\dfrac{4^{-2}}{5^{-3}} = \dfrac{5^3}{4^2} = \dfrac{125}{16}$

(b) $\dfrac{m^{-5}}{p^{-1}} = \dfrac{p^1}{m^5} = \dfrac{p}{m^5}$

(c) $\dfrac{a^{-2}b}{3d^{-3}} = \dfrac{bd^3}{3a^2}$ \qquad Notice that b in the numerator and 3 in the denominator are not affected.

(d) $\left(\dfrac{x}{2y}\right)^{-4}$

$= \left(\dfrac{2y}{x}\right)^4 \qquad$ Negative-to-positive rule

$= \dfrac{2^4 y^4}{x^4} \qquad$ Power rules (b) and (c)

$= \dfrac{16y^4}{x^4} \qquad$ Apply the exponent.

NOW TRY ANSWERS

3. (a) $\dfrac{36}{125}$ **(b)** $\dfrac{m^2}{n^4}$ **(c)** $\dfrac{x^2 z^4}{5y^3}$

NOW TRY

⚠️ **CAUTION**   Be careful. We cannot use the rule $\frac{a^{-m}}{b^{-n}} = \frac{b^n}{a^m}$ to change negative exponents to positive exponents if the exponents occur in a *sum* or *difference* of terms. For example,

$$\frac{5^{-2} + 3^{-1}}{7 - 2^{-3}} \quad \text{would be written with positive exponents as} \quad \frac{\frac{1}{5^2} + \frac{1}{3}}{7 - \frac{1}{2^3}}.$$

OBJECTIVE 3   Use the quotient rule for exponents.

Consider the following.

$$\frac{6^5}{6^3} = \frac{6 \cdot 6 \cdot 6 \cdot 6 \cdot 6}{6 \cdot 6 \cdot 6} = 6^2$$

The difference of the exponents, $5 - 3 = 2$, is the exponent in the quotient.

Also,

$$\frac{6^2}{6^4} = \frac{6 \cdot 6}{6 \cdot 6 \cdot 6 \cdot 6} = \frac{1}{6^2} = 6^{-2}.$$

Here, $2 - 4 = -2$. These examples suggest the **quotient rule for exponents.**

Quotient Rule for Exponents

For any nonzero real number a and any integers m and n,

$$\frac{a^m}{a^n} = a^{m-n}.$$

(Keep the same base and subtract the exponents.)

Example: $\dfrac{5^8}{5^4} = 5^{8-4} = 5^4$

⚠️ **CAUTION**   A common **error** is to write $\frac{5^8}{5^4} = 1^{8-4} = 1^4$. **This is incorrect.** By the quotient rule, the quotient must have the *same base*, 5, just as in the product rule.

$$\frac{5^8}{5^4} = 5^{8-4} = 5^4$$

We can confirm this by writing out the factors.

$$\frac{5^8}{5^4} = \frac{5 \cdot 5 \cdot 5 \cdot 5 \cdot 5 \cdot 5 \cdot 5 \cdot 5}{5 \cdot 5 \cdot 5 \cdot 5} = 5^4$$

EXAMPLE 4   Using the Quotient Rule

Simplify. Assume that all variables represent nonzero real numbers.

(a) $\dfrac{5^8}{5^6} = 5^{8-6} = 5^2 = 25$

           Keep the same base.

(b) $\dfrac{4^2}{4^5} = 4^{2-5} = 4^{-3} = \dfrac{1}{4^3} = \dfrac{1}{64}$

**NOW TRY
EXERCISE 4**

Simplify. Assume that all variables represent nonzero real numbers.

(a) $\dfrac{6^3}{6^4}$ **(b)** $\dfrac{t^4}{t^{-5}}$

(c) $\dfrac{(p+q)^{-3}}{(p+q)^{-7}}$ $(p \neq -q)$

(d) $\dfrac{5^2xy^{-3}}{3^{-1}x^{-2}y^2}$

(c) $\dfrac{5^{-3}}{5^{-7}} = 5^{-3-(-7)} = 5^4 = 625$

> Be careful with signs.

(d) $\dfrac{q^5}{q^{-3}} = q^{5-(-3)} = q^8$

(e) $\dfrac{3^2x^5}{3^4x^3}$

$= \dfrac{3^2}{3^4} \cdot \dfrac{x^5}{x^3}$

$= 3^{2-4} \cdot x^{5-3}$ Quotient rule

$= 3^{-2}x^2$ Subtract.

$= \dfrac{x^2}{3^2}$ Definition of negative exponent

$= \dfrac{x^2}{9}$ Apply the exponent.

(f) $\dfrac{(m+n)^{-2}}{(m+n)^{-4}}$

$= (m+n)^{-2-(-4)}$

$= (m+n)^{-2+4}$

$= (m+n)^2$ $(m \neq -n)$

The restriction $m \neq -n$ is necessary to prevent a denominator of 0 in the original expression. Division by 0 is undefined.

(g) $3x^{-5}$

> Avoid the error of applying -5 to 3.

$= 3 \cdot \dfrac{1}{x^5}$ -5 applies *only* to x.

$= \dfrac{3}{x^5}$ Multiply.

(h) $\dfrac{7x^{-3}y^2}{2^{-1}x^2y^{-5}}$

$= \dfrac{7 \cdot 2^1 y^2 y^5}{x^2 x^3}$ Negative-to-positive rule

$= \dfrac{14y^7}{x^5}$ Multiply; product rule

NOW TRY

The definitions and rules for exponents are summarized here.

Definitions and Rules for Exponents

For any integers m and n, the following hold.   **Examples**

Product rule $a^m \cdot a^n = a^{m+n}$ $7^4 \cdot 7^5 = 7^{4+5} = 7^9$

Zero exponent $a^0 = 1$ (where $a \neq 0$) $(-3)^0 = 1$

Negative exponent $a^{-n} = \dfrac{1}{a^n}$ (where $a \neq 0$) $5^{-3} = \dfrac{1}{5^3}$

Quotient rule $\dfrac{a^m}{a^n} = a^{m-n}$ (where $a \neq 0$) $\dfrac{2^2}{2^5} = 2^{2-5} = 2^{-3} = \dfrac{1}{2^3}$

Power rules (a) $(a^m)^n = a^{mn}$ $(4^2)^3 = 4^{2 \cdot 3} = 4^6$

      (b) $(ab)^m = a^m b^m$ $(3k)^4 = 3^4 k^4$

      (c) $\left(\dfrac{a}{b}\right)^m = \dfrac{a^m}{b^m}$ (where $b \neq 0$) $\left(\dfrac{2}{3}\right)^2 = \dfrac{2^2}{3^2}$

Negative-to-positive rules $\dfrac{a^{-m}}{b^{-n}} = \dfrac{b^n}{a^m}$ (where $a \neq 0$, $b \neq 0$) $\dfrac{2^{-4}}{5^{-3}} = \dfrac{5^3}{2^4}$

     $\left(\dfrac{a}{b}\right)^{-m} = \left(\dfrac{b}{a}\right)^m$ $\left(\dfrac{4}{7}\right)^{-2} = \left(\dfrac{7}{4}\right)^2$

NOW TRY ANSWERS

4. **(a)** $\dfrac{1}{6}$ **(b)** t^9 **(c)** $(p+q)^4$

    (d) $\dfrac{75x^3}{y^5}$

OBJECTIVE 4 Use combinations of the rules for exponents.

 NOW TRY EXERCISE 5

Simplify. Assume that all variables represent nonzero real numbers.

(a) $\dfrac{3^{15}}{(3^3)^4}$ **(b)** $(4t)^5(4t)^{-3}$

(c) $\left(\dfrac{7y^4}{10}\right)^{-3}$ **(d)** $\dfrac{(a^2b^{-2}c)^{-3}}{(2ab^3c^{-4})^5}$

(e) $\dfrac{(5k)^{-6}(5k)^8}{(5k)^7(5k)^{-4}}$

EXAMPLE 5 Using Combinations of Rules

Simplify. Assume that all variables represent nonzero real numbers.

(a) $\dfrac{(4^2)^3}{4^5}$

$= \dfrac{4^6}{4^5}$ Power rule (a)

$= 4^{6-5}$ Quotient rule

$= 4^1$ Subtract.

$= 4$ $a^1 = a$, for all a.

(b) $(2x)^3(2x)^2$

$= (2x)^5$ Product rule

$= 2^5x^5$ Power rule (b)

$= 32x^5$ $2^5 = 32$

(c) $\left(\dfrac{2x^3}{5}\right)^{-4}$

$= \left(\dfrac{5}{2x^3}\right)^4$ Negative-to-positive rule

$= \dfrac{5^4}{2^4x^{12}}$ Power rules (a)–(c)

$= \dfrac{625}{16x^{12}}$ Apply the exponents.

(d) $\left(\dfrac{3x^{-2}}{4^{-1}y^3}\right)^{-3}$

$= \dfrac{3^{-3}x^6}{4^3y^{-9}}$ Power rules (a)–(c)

$= \dfrac{x^6y^9}{4^3 \cdot 3^3}$ Negative-to-positive rule

$= \dfrac{x^6y^9}{1728}$ $4^3 \cdot 3^3 = 64 \cdot 27 = 1728$

(e) $\dfrac{(4m)^{-3}}{(3m)^{-4}}$

$= \dfrac{4^{-3}m^{-3}}{3^{-4}m^{-4}}$ Power rule (b)

$= \dfrac{3^4m^4}{4^3m^3}$ Negative-to-positive rule

$= \dfrac{3^4m^{4-3}}{4^3}$ Quotient rule

$= \dfrac{3^4m^1}{4^3}$ Subtract.

$= \dfrac{81m}{64}$ Apply the exponents.

(f) $\dfrac{(7y)^{-3}(7y)^4}{(7y)^{12}(7y)^{-10}}$

$= \dfrac{(7y)^{-3+4}}{(7y)^{12+(-10)}}$ Product rule

$= \dfrac{(7y)^1}{(7y)^2}$ Add.

$= (7y)^{1-2}$ Quotient rule

$= (7y)^{-1}$ Subtract.

$= \dfrac{1}{7y}$ $a^{-1} = \frac{1}{a}$, for $a \neq 0$.

NOW TRY

NOW TRY ANSWERS

5. (a) 27   **(b)** $16t^2$ **(c)** $\dfrac{1000}{343y^{12}}$

(d) $\dfrac{c^{17}}{32a^{11}b^9}$ **(e)** $\dfrac{1}{5k}$

NOTE Because the steps can be done in several different orders, there are many equally correct ways to simplify expressions like those in **Examples 5(c)–5(f).**

4.2 Exercises

FOR EXTRA HELP   ▶ MyMathLab®

▶ *Complete solution available in MyMathLab*

Concept Check *Decide whether each expression is* positive, negative, *or* 0.

1. $(-2)^{-3}$ **2.** $(-3)^{-2}$ **3.** -2^4 **4.** -3^6

5. $\left(\dfrac{1}{4}\right)^{-2}$ **6.** $\left(\dfrac{1}{5}\right)^{-2}$ **7.** $1 - 5^0$ **8.** $1 - 7^0$

Decide whether each expression is equal to 0, 1, or −1. See Example 1.

▶ 9. 9^0 **10.** 3^0 **11.** $(-2)^0$ **12.** $(-12)^0$

13. -8^0 **14.** -6^0 **15.** $-(-6)^0$ **16.** $-(-13)^0$

17. $(-4)^0 - 4^0$ **18.** $(-11)^0 - 11^0$ **19.** $\dfrac{0^{10}}{12^0}$ **20.** $\dfrac{0^5}{2^0}$

21. $8^0 - 12^0$ **22.** $6^0 - 13^0$ **23.** $\dfrac{0^2}{2^0 + 0^2}$ **24.** $\dfrac{2^0}{0^2 + 2^0}$

Concept Check In Exercises 25 and 26, match each expression in Column I with the equivalent expression in Column II. Choices in Column II may be used once, more than once, or not at all. (In Exercise 25, $x \neq 0$.)

I	II		I	II
25. (a) x^0	**A.** 0		**26. (a)** -2^{-4}	**A.** 8
(b) $-x^0$	**B.** 1		**(b)** $(-2)^{-4}$	**B.** 16
(c) $7x^0$	**C.** -1		**(c)** 2^{-4}	**C.** $-\dfrac{1}{16}$
(d) $(7x)^0$	**D.** 7		**(d)** $\dfrac{1}{2^{-4}}$	**D.** -8
(e) $-7x^0$	**E.** -7		**(e)** $\dfrac{1}{-2^{-4}}$	**E.** -16
(f) $(-7x)^0$	**F.** $\dfrac{1}{7}$		**(f)** $\dfrac{1}{(-2)^{-4}}$	**F.** $\dfrac{1}{16}$

Evaluate each expression. See Examples 1 and 2.

27. $6^0 + 8^0$ **28.** $4^0 + 2^0$ **▶ 29.** 4^{-3} **30.** 5^{-4}

31. $\left(\dfrac{1}{2}\right)^{-4}$ **32.** $\left(\dfrac{1}{3}\right)^{-3}$ **33.** $\left(\dfrac{6}{7}\right)^{-2}$

34. $\left(\dfrac{2}{3}\right)^{-3}$ **35.** $(-3)^{-4}$ **36.** $(-4)^{-3}$

37. $5^{-1} + 3^{-1}$ **38.** $6^{-1} + 2^{-1}$ **39.** $3^{-2} - 2^{-1}$

40. $6^{-2} - 3^{-1}$ **41.** $\left(\dfrac{1}{2}\right)^{-1} + \left(\dfrac{2}{3}\right)^{-1}$ **42.** $\left(\dfrac{1}{3}\right)^{-1} + \left(\dfrac{4}{3}\right)^{-1}$

Concept Check Simplify each expression.

43. $\dfrac{5^{11}}{5^8}$

$= 5^{\underline{\ \ }-\underline{\ \ }}$

$= 5^{\underline{\ \ }}$

$= \underline{\ \ \ \ }$

44. $\dfrac{6^{-5}}{6^{-2}}$

$= 6^{\underline{\ \ }-(\underline{\ \ })}$

$= 6^{\underline{\ \ }}$

$= \dfrac{1}{6^{\underline{\ \ }}}$

$= \underline{\ \ \ \ }$

Simplify each expression. Assume that all variables represent nonzero real numbers and no denominators are zero. See Examples 2–4.

▶ 45. $\dfrac{5^8}{5^5}$ **46.** $\dfrac{11^6}{11^3}$ **▶ 47.** $\dfrac{3^{-2}}{5^{-3}}$ **48.** $\dfrac{4^{-3}}{3^{-2}}$

49. $\dfrac{5}{5^{-1}}$ **50.** $\dfrac{6}{6^{-2}}$ **51.** $\dfrac{x^{12}}{x^{-3}}$ **52.** $\dfrac{y^4}{y^{-6}}$

53. $\dfrac{1}{6^{-3}}$ **54.** $\dfrac{1}{5^{-2}}$ **55.** $\dfrac{2}{r^{-4}}$ **56.** $\dfrac{3}{s^{-8}}$

57. $\dfrac{4^{-3}}{5^{-2}}$ **58.** $\dfrac{6^{-2}}{5^{-4}}$ **59.** $p^5 q^{-8}$ **60.** $x^{-8} y^4$

61. $\dfrac{r^5}{r^{-4}}$ **62.** $\dfrac{a^6}{a^{-4}}$ **63.** $\dfrac{x^{-3} y}{4 z^{-2}}$ **64.** $\dfrac{p^{-5} q^{-4}}{9 r^{-3}}$

65. $\dfrac{(a+b)^{-3}}{(a+b)^{-4}}$ **66.** $\dfrac{(x+y)^{-8}}{(x+y)^{-9}}$ **67.** $\dfrac{(x+2y)^{-3}}{(x+2y)^{-5}}$ **68.** $\dfrac{(p-3q)^{-2}}{(p-3q)^{-4}}$

Simplify each expression. Assume that all variables represent nonzero real numbers. **See Example 5.**

69. $\dfrac{(7^4)^3}{7^9}$ **70.** $\dfrac{(5^3)^2}{5^2}$ **71.** $x^{-3} \cdot x^5 \cdot x^{-4}$ **72.** $y^{-8} \cdot y^5 \cdot y^{-2}$

73. $\dfrac{(3x)^{-2}}{(4x)^{-3}}$ **74.** $\dfrac{(2y)^{-3}}{(5y)^{-4}}$ ▶ **75.** $\left(\dfrac{x^{-1} y}{z^2}\right)^{-2}$ **76.** $\left(\dfrac{p^{-4} q}{r^{-3}}\right)^{-3}$

77. $(6x)^4 (6x)^{-3}$ **78.** $(10y)^9 (10y)^{-8}$ **79.** $\dfrac{(m^7 n)^{-2}}{m^{-4} n^3}$

80. $\dfrac{(m^{-8} n^{-4})^2}{m^2 n^5}$ **81.** $\dfrac{(x^{-1} y^2 z)^{-2}}{(x^{-3} y^3 z)^{-1}}$ **82.** $\dfrac{(a^2 b^3 c^4)^{-4}}{(a^{-2} b^{-3} c^{-4})^{-5}}$

83. $\left(\dfrac{xy^{-2}}{x^2 y}\right)^{-3}$ **84.** $\left(\dfrac{wz^{-5}}{w^{-3} z}\right)^{-2}$ **85.** $\dfrac{(2r)^{-4} (2r)^5}{(2r)^9 (2r)^{-7}}$

86. $\dfrac{(8x)^{-8} (8x)^9}{(8x)^{13} (8x)^{-11}}$ **87.** $\dfrac{(-4y)^8 (-4y)^{-8}}{(-4y)^{-26} (-4y)^{27}}$ **88.** $\dfrac{(-9p)^{16} (-9p)^{-16}}{(-9p)^{-41} (-9p)^{42}}$

89. *Concept Check*   A student simplified $\dfrac{16^3}{2^2}$ as shown.

$$\frac{16^3}{2^2} = \left(\frac{16}{2}\right)^{3-2} = 8^1 = 8$$

WHAT WENT WRONG?   Give the correct answer.

90. *Concept Check*   A student simplified -5^4 as shown.

$$-5^4 = (-5^4) = 625$$

WHAT WENT WRONG?   Give the correct answer.

Extending Skills   *Simplify each expression. Assume that all variables represent nonzero real numbers.*

91. $\dfrac{(4a^2 b^3)^{-2} (2ab^{-1})^3}{(a^3 b)^{-4}}$ **92.** $\dfrac{(m^6 n)^{-2} (m^2 n^{-2})^3}{m^{-1} n^{-2}}$

93. $\dfrac{(2y^{-1} z^2)^2 (3y^{-2} z^{-3})^3}{(y^3 z^2)^{-1}}$ **94.** $\dfrac{(3p^{-2} q^3)^2 (5p^{-1} q^{-4})^{-1}}{(p^2 q^{-2})^{-3}}$

95. $\dfrac{(9^{-1} z^{-2} x)^{-1} (4z^2 x^4)^{-2}}{(5z^{-2} x^{-3})^2}$ **96.** $\dfrac{(4^{-1} a^{-1} b^{-2})^{-2} (5a^{-3} b^4)^{-2}}{(3a^{-3} b^{-5})^2}$

SUMMARY EXERCISES   Applying the Rules for Exponents

Simplify each expression. Use only positive exponents in the answers. Assume that all variables represent nonzero real numbers.

Answers (left column):

1. $105x^7y^{14}$
2. $-128a^{10}b^{15}c^4$
3. $\dfrac{729w^3x^9}{y^{12}}$
4. $\dfrac{x^4y^6}{16}$
5. c^{22}
6. $\dfrac{1}{k^4t^{12}}$
7. $\dfrac{11}{30}$
8. $y^{12}z^3$
9. $\dfrac{x^6}{y^5}$
10. 0
11. $\dfrac{1}{z^2}$
12. $\dfrac{9}{r^2s^2t^{10}}$
13. $\dfrac{300x^3}{y^3}$
14. $\dfrac{3}{5x^6}$
15. x^8
16. $\dfrac{y^{11}}{x^{11}}$
17. $\dfrac{a^6}{b^4}$
18. $6ab$
19. $\dfrac{61}{900}$
20. 1
21. $\dfrac{343a^6b^9}{8}$
22. 1
23. -1
24. 0
25. $\dfrac{27y^{18}}{4x^9}$
26. $\dfrac{1}{a^8b^{12}c^{16}}$
27. $\dfrac{x^{15}}{216z^9}$
28. $\dfrac{q}{8p^6r^3}$
29. x^6y^6
30. 0
31. $\dfrac{343}{x^{15}}$
32. $\dfrac{9}{x^6}$
33. $5p^{10}q^9$
34. $\dfrac{7}{24}$
35. $\dfrac{r^{14}t}{2s^2}$
36. 1
37. $8p^{10}q$
38. $\dfrac{1}{mn^3p^3}$
39. -1
40. (a) D (b) D
   (c) E (d) B
   (e) J (f) F
   (g) I (h) B
   (i) E (j) F

Exercises:

1. $(10x^2y^4)^2(10xy^2)^3$
2. $(-2ab^3c)^4(-2a^2b)^3$
3. $\left(\dfrac{9wx^3}{y^4}\right)^3$

4. $(4x^{-2}y^{-3})^{-2}$
5. $\dfrac{c^{11}(c^2)^4}{(c^3)^3(c^2)^{-6}}$
6. $\left(\dfrac{k^4t^2}{k^2t^{-4}}\right)^{-2}$

7. $5^{-1}+6^{-1}$
8. $\dfrac{(3y^{-1}z^3)^{-1}(3y^2)}{(y^3z^2)^{-3}}$
9. $\dfrac{(2xy^{-1})^3}{2^3x^{-3}y^2}$

10. $-4^0+(-4)^0$
11. $(z^4)^{-3}(z^{-2})^{-5}$
12. $\left(\dfrac{r^2st^5}{3r}\right)^{-2}$

13. $\dfrac{(3^{-1}x^{-3}y)^{-1}(2x^2y^{-3})^2}{(5x^{-2}y^2)^{-2}}$
14. $\left(\dfrac{5x^2}{3x^{-4}}\right)^{-1}$
15. $\left(\dfrac{-9x^{-2}}{9x^2}\right)^{-2}$

16. $\dfrac{(x^{-4}y^2)^3(x^2y)^{-1}}{(xy^2)^{-3}}$
17. $\dfrac{(a^{-2}b^3)^{-4}}{(a^{-3}b^2)^{-2}(ab)^{-4}}$
18. $(2a^{-30}b^{-29})(3a^{31}b^{30})$

19. $5^{-2}+6^{-2}$
20. $\left(\dfrac{(x^{43}y^{23})^2}{x^{-26}y^{-42}}\right)^0$
21. $\left(\dfrac{7a^2b^3}{2}\right)^3$

22. $-(-19^0)$
23. $-(-13)^0$
24. $\dfrac{0^{13}}{13^0}$

25. $\dfrac{(2xy^{-3})^{-2}}{(3x^{-2}y^4)^{-3}}$
26. $\left(\dfrac{a^2b^3c^4}{a^{-2}b^{-3}c^{-4}}\right)^{-2}$
27. $(6x^{-5}z^3)^{-3}$

28. $(2p^{-2}qr^{-3})(2p)^{-4}$
29. $\dfrac{(xy)^{-3}(xy)^5}{(xy)^{-4}}$
30. $52^0-(-8)^0$

31. $\dfrac{(7^{-1}x^{-3})^{-2}(x^4)^{-6}}{7^{-1}x^{-3}}$
32. $\left(\dfrac{3^{-4}x^{-3}}{3^{-3}x^{-6}}\right)^{-2}$
33. $(5p^{-2}q)^{-3}(5pq^3)^4$

34. $8^{-1}+6^{-1}$
35. $\left(\dfrac{4r^{-6}s^{-2}t}{2r^8s^{-4}t^2}\right)^{-1}$
36. $(13x^{-6}y)(13x^{-6}y)^{-1}$

37. $\dfrac{(8pq^{-2})^4}{(8p^{-2}q^{-3})^3}$
38. $\left(\dfrac{mn^{-2}p}{m^2np^4}\right)^{-2}\left(\dfrac{mn^{-2}p}{m^2np^4}\right)^3$
39. $-(-8^0)^0$

40. **Concept Check**   *Match each expression (a)–(j) in Column I with the equivalent expression A–J in Column II. Choices in Column II may be used once, more than once, or not at all.*

I		II	
(a) 2^0+2^0	(b) $2^1\cdot2^0$	A. 0	B. 1
(c) 2^0-2^{-1}	(d) 2^1-2^0	C. -1	D. 2
(e) $2^0\cdot2^{-2}$	(f) $2^1\cdot2^1$	E. $\dfrac{1}{2}$	F. 4
(g) $2^{-2}-2^{-1}$	(h) $2^0\cdot2^0$	G. -2	H. -4
(i) $2^{-2}\div2^{-1}$	(j) $2^0\div2^{-2}$	I. $-\dfrac{1}{4}$	J. $\dfrac{1}{4}$

4.3 Scientific Notation

VOCABULARY
☐ scientific notation
☐ standard notation

OBJECTIVE 1   Express numbers in scientific notation.

Numbers occurring in science are often extremely large (such as the distance from Earth to the sun, 93,000,000 mi) or extremely small (the wavelength of blue light, approximately 0.000000475 m). Because of the difficulty of working with many zeros, scientists often express such numbers with exponents using *scientific notation*.

Scientific Notation

A number is written in **scientific notation** when it is expressed in the form

$$a \times 10^n,$$

where $1 \le |a| < 10$ and n is an integer.

In scientific notation, there is *always* one nonzero digit before the decimal point.

$3.19 \times 10^1 = 3.19 \times 10 = 31.9$	Decimal point moves 1 place to the right.
$3.19 \times 10^2 = 3.19 \times 100 = 319.$	Decimal point moves 2 places to the right.
$3.19 \times 10^3 = 3.19 \times 1000 = 3190.$	Decimal point moves 3 places to the right.
$3.19 \times 10^{-1} = 3.19 \times 0.1 = 0.319$	Decimal point moves 1 place to the left.
$3.19 \times 10^{-2} = 3.19 \times 0.01 = 0.0319$	Decimal point moves 2 places to the left.
$3.19 \times 10^{-3} = 3.19 \times 0.001 = 0.00319$	Decimal point moves 3 places to the left.

NOTE   In scientific notation, the multiplication cross $\times$ is commonly used.

A number in scientific notation is always written with the decimal point after the first nonzero digit and then multiplied by the appropriate power of 10. For example, 56,200 is written 5.62×10^4 because

$$56,200 = 5.62 \times 10,000 = 5.62 \times 10^4.$$

Other examples include the following.

42,000,000	is written	$4.2 \times 10^7,$
0.000586	is written	$5.86 \times 10^{-4},$
and 2,000,000,000	is written	$2 \times 10^9.$

To write a number in scientific notation, follow these steps. For a negative number, follow these steps using the *absolute value* of the number. Then make the result negative.

Converting a Positive Number to Scientific Notation

Step 1   **Position the decimal point.** Place a caret, ^, to the right of the first nonzero digit, where the decimal point will be placed.

Step 2   **Determine the numeral for the exponent.** Count the number of digits from the decimal point to the caret. This number gives the absolute value of the exponent on 10.

Step 3   **Determine the sign for the exponent.** Decide whether multiplying by 10^n should make the result of Step 1 greater or less.

* The exponent should be positive to make the result greater.

* The exponent should be negative to make the result less.

NOW TRY EXERCISE 1

Write each number in scientific notation.

(a) 12,600,000

(b) 0.00027

(c) −0.0000341

EXAMPLE 1   Using Scientific Notation

Write each number in scientific notation.

(a) 93,000,000

   Step 1   Place a caret to the right of the 9 (the first nonzero digit) to mark the new location of the decimal point.

$$9_\wedge 3{,}000{,}000$$

   Step 2   Count from the decimal point, which is understood to be after the last 0, to the caret.

$$9_\wedge 3{,}000{,}000. \leftarrow \text{Decimal point}$$

Count 7 places.

   Step 3   Because 9.3 is to be made greater, the exponent on 10 is positive.

$$93{,}000{,}000 = 9.3 \times 10^7$$

(b) $63{,}200{,}000{,}000 = 6.3200000000 = 6.32 \times 10^{10}$

10 places

(c) 0.00462

   Move the decimal point to the right of the first nonzero digit, and count the number of places the decimal point was moved.

$$0.00462 \quad \text{3 places}$$

Because 0.00462 is *less* than 4.62, the exponent must be *negative*.

$$0.00462 = 4.62 \times 10^{-3}$$

(d) $-0.0000762 = -7.62 \times 10^{-5}$ Remember the negative sign.   **NOW TRY**

5 places

NOTE   When writing a positive number in scientific notation, think as follows.

* If the original number is "large," like 93,000,000, use a *positive* exponent on 10 because positive is greater than negative.

* If the original number is "small," like 0.00462, use a *negative* exponent on 10 because negative is less than positive.

OBJECTIVE 2 Convert numbers in scientific notation to standard notation.

Multiplying a number by a positive power of 10 will make the number greater. Multiplying by a negative power of 10 will make the number less.

We refer to a number such as 475 as the **standard notation** of 4.75×10^2.

NOW TRY EXERCISE 2

Write each number in standard notation.

(a) 5.71×10^4

(b) 2.72×10^{-5}

(c) -8.81×10^{-4}

EXAMPLE 2 Writing Numbers in Standard Notation

Write each number in standard notation.

(a) 6.2×10^3

Because the exponent is positive, we make 6.2 greater by moving the decimal point 3 places to the right. We attach two zeros.

$$6.2 \times 10^3 = 6.200 = 6200$$

(b) $4.283 \times 10^6 = 4.283000 = 4{,}283{,}000$ Move 6 places to the right. Attach zeros as necessary.

(c) $-7.04 \times 10^{-3} = -0.00704$ Move 3 places to the left.

The exponent tells the number of places and the direction in which the decimal point is moved.

NOW TRY

OBJECTIVE 3 Use scientific notation in calculations.

NOW TRY EXERCISE 3

Perform each calculation. Write answers in both scientific and standard notation.

(a) $(6 \times 10^7)(7 \times 10^{-4})$

(b) $\dfrac{18 \times 10^{-3}}{6 \times 10^4}$

EXAMPLE 3 Multiplying and Dividing with Scientific Notation

Perform each calculation.

(a)
$$(7 \times 10^3)(5 \times 10^4)$$
$$= (7 \times 5)(10^3 \times 10^4)$$ Commutative and associative properties
$$= 35 \times 10^7$$ Multiply. Use the product rule.
$$= (3.5 \times 10^1) \times 10^7$$ Write 35 in scientific notation.
$$= 3.5 \times (10^1 \times 10^7)$$ Associative property
$$= 3.5 \times 10^8$$ Product rule
$$= 350{,}000{,}000$$ Write in standard notation.

Don't stop. This number is not in scientific notation because 35 is not between 1 and 10.

(b)
$$\frac{4 \times 10^{-5}}{2 \times 10^3}$$
$$= \frac{4}{2} \times \frac{10^{-5}}{10^3}$$
$$= 2 \times 10^{-8}$$ Divide. Use the quotient rule.
$$= 0.00000002$$ Write in standard notation.

NOW TRY

NOTE Multiplying or dividing numbers written in scientific notation may produce an answer in the form $a \times 10^0$. Because $10^0 = 1$, $a \times 10^0 = a$. For example,

$$(8 \times 10^{-4})(5 \times 10^4) = 40 \times 10^0 = 40. \qquad 10^0 = 1$$

Also, if $a = 1$, then $a \times 10^n = 10^n$. For example, we could write $1{,}000{,}000$ as 10^6 instead of 1×10^6.

NOW TRY ANSWERS

2. (a) 57,100 **(b)** 0.0000272
  (c) −0.000881
3. (a) 4.2×10^4, or 42,000
  (b) 3×10^{-7}, or 0.0000003

 **NOW TRY EXERCISE 4**

See **Example 4.** About how much would 8,000,000 nanometers measure in inches?

EXAMPLE 4 Using Scientific Notation to Solve an Application

A *nanometer* is a very small unit of measure that is equivalent to about 0.00000003937 in. About how much would 700,000 nanometers measure in inches? (*Source:* www.conversion-metric.org)

Write each number in scientific notation, and then multiply.

$$700,000(0.00000003937)$$

$$= (7 \times 10^5)(3.937 \times 10^{-8}) \qquad \text{Write in scientific notation.}$$

$$= (7 \times 3.937)(10^5 \times 10^{-8}) \qquad \text{Properties of real numbers}$$

$$= 27.559 \times 10^{-3} \qquad \text{Multiply. Use the product rule.}$$

Don't stop here. $= (2.7559 \times 10^1) \times 10^{-3} \qquad \text{Write 27.559 in scientific notation.}$

$$= 2.7559 \times 10^{-2} \qquad \text{Product rule}$$

$$= 0.027559 \qquad \text{Write in standard notation.}$$

Thus, 700,000 nanometers would measure

$$2.7559 \times 10^{-2} \text{ in.,} \quad \text{or} \quad 0.027559 \text{ in.} \qquad \text{NOW TRY} \quad \text{}$$

 **NOW TRY EXERCISE 5**

The land area of California is approximately 1.6×10^5 mi², and the 2012 population of California was approximately 3.8×10^7 people. Use this information to estimate the number of square miles per California resident in 2012. (*Source:* U.S. Census Bureau.)

EXAMPLE 5 Using Scientific Notation to Solve an Application

In 2013, the gross federal debt was about $\$1.7218 \times 10^{13}$ (which is more than $17 trillion). The population of the United States was approximately 317 million that year. About how much would each person have had to contribute in order to pay off the federal debt? (*Source:* www.usgovernmentdebt.us; www.census.gov)

Write the population in scientific notation. Then divide to obtain the per person contribution.

$$\frac{1.7218 \times 10^{13}}{317,000,000}$$

$$= \frac{1.7218 \times 10^{13}}{3.17 \times 10^8} \qquad \text{Write 317 million in scientific notation.}$$

$$= \frac{1.7218}{3.17} \times 10^5 \qquad \text{Quotient rule}$$

$$= 0.54315 \times 10^5 \qquad \text{Divide. Round to 5 decimal places.}$$

$$= 54,315 \qquad \text{Write in standard notation.}$$

Each person would have to pay about $54,315. NOW TRY

NOW TRY ANSWERS
4. 3.1496×10^{-1} in., or 0.31496 in.
5. 4.2×10^{-3} mi², or 0.0042 mi²

4.3 Exercises

 FOR EXTRA HELP MyMathLab®

▶ *Complete solution available in MyMathLab*

Concept Check *Match each number written in scientific notation in Column I with the correct choice from Column II. Not all choices in Column II will be used.*

I	II	I	II
1. (a) 4.6×10^{-4}	**A.** 46,000	**2. (a)** 1×10^9	**A.** 1 billion
(b) 4.6×10^4	**B.** 460,000	**(b)** 1×10^6	**B.** 100 million
(c) 4.6×10^5	**C.** 0.00046	**(c)** 1×10^8	**C.** 1 million
(d) 4.6×10^{-5}	**D.** 0.000046	**(d)** 1×10^{10}	**D.** 10 billion
	E. 4600		**E.** 100 billion

Concept Check   *Determine whether or not each number is written in scientific notation as defined in* **Objective 1.** *If it is not, write it as such.*

3. 4.56×10^4 **4.** 7.34×10^6 **5.** $5{,}600{,}000$ **6.** $34{,}000$

7. 0.8×10^2 **8.** 0.9×10^3 **9.** 0.004 **10.** 0.0007

11. *Concept Check*   Write each number in scientific notation.

(a) $63{,}000$

The first nonzero digit is _____. The decimal point should be moved _____ places.
$$63{,}000 = \text{\_\_\_\_\_} \times 10^{\text{\_\_\_}}$$

(b) 0.0571

The first nonzero digit is _____. The decimal point should be moved _____ places.
$$0.0571 = \text{\_\_\_\_\_} \times 10^{\text{\_\_\_}}$$

12. *Concept Check*   Write each number in standard notation.

(a) 4.2×10^3

Move the decimal point _____ places to the _____.
$$4.2 \times 10^3 = \text{\_\_\_\_\_}$$

(b) 6.42×10^{-3}

Move the decimal point _____ places to the _____.
$$6.42 \times 10^{-3} = \text{\_\_\_\_\_}$$

Write each number in scientific notation. See Example 1.

▶ **13.** $5{,}876{,}000{,}000$ **14.** $9{,}994{,}000{,}000$ **15.** $82{,}350$ **16.** $78{,}330$

17. 0.000007 **18.** 0.0000004 **19.** 0.00203 **20.** 0.0000578

21. $-13{,}000{,}000$ **22.** $-25{,}000{,}000{,}000$ **23.** -0.006 **24.** -0.01234

Write each number in standard notation. See Example 2.

▶ **25.** 7.5×10^5 **26.** 8.8×10^6 **27.** 5.677×10^{12} **28.** 8.766×10^9

29. 1×10^{12} **30.** 1×10^7 **31.** 6.21×10^0 **32.** 8.56×10^0

33. 7.8×10^{-4} **34.** 8.9×10^{-5} **35.** 5.134×10^{-9} **36.** 7.123×10^{-10}

37. -4×10^{-3} **38.** -6×10^{-4} **39.** -8.1×10^5 **40.** -9.6×10^6

Perform the indicated operations. Write each answer (a) in scientific notation and (b) in standard notation. See Example 3.

41. $(2 \times 10^8)(3 \times 10^3)$

42. $(4 \times 10^7)(3 \times 10^3)$

▶ **43.** $(5 \times 10^4)(3 \times 10^2)$

44. $(8 \times 10^5)(2 \times 10^3)$

45. $(3 \times 10^{-4})(-2 \times 10^8)$

46. $(4 \times 10^{-3})(-2 \times 10^7)$

47. $(6 \times 10^3)(4 \times 10^{-2})$

48. $(7 \times 10^5)(3 \times 10^{-4})$

49. $(9 \times 10^4)(7 \times 10^{-7})$

50. $(6 \times 10^4)(8 \times 10^{-8})$

51. $\dfrac{9 \times 10^{-5}}{3 \times 10^{-1}}$ **52.** $\dfrac{12 \times 10^{-4}}{4 \times 10^{-3}}$ **53.** $\dfrac{8 \times 10^3}{-2 \times 10^2}$

54. $\dfrac{15 \times 10^4}{-3 \times 10^3}$ **55.** $\dfrac{2.6 \times 10^{-3}}{2 \times 10^2}$ **56.** $\dfrac{9.5 \times 10^{-1}}{5 \times 10^3}$

57. $\dfrac{4 \times 10^5}{8 \times 10^2}$ **58.** $\dfrac{3 \times 10^9}{6 \times 10^5}$ **59.** $\dfrac{-4.5 \times 10^4}{1.5 \times 10^{-2}}$

60. $\dfrac{-7.2 \times 10^3}{6.0 \times 10^{-1}}$ **61.** $\dfrac{-8 \times 10^{-4}}{-4 \times 10^3}$ **62.** $\dfrac{-5 \times 10^{-6}}{-2 \times 10^2}$

Calculators can express numbers in scientific notation. The displays often use notation such as

$$5.4\,\text{E}\,3 \quad \text{to represent} \quad 5.4 \times 10^3.$$

Similarly, 5.4E−3 represents 5.4×10^{-3}. They can also perform operations with numbers entered in scientific notation.

Predict the display the calculator would give for the expression shown in each screen.

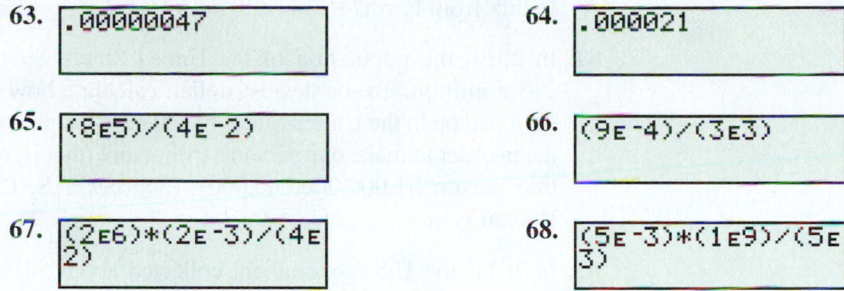

63. `.00000047`

64. `.000021`

65. `(8E5)/(4E-2)`

66. `(9E-4)/(3E3)`

67. `(2E6)*(2E-3)/(4E
2)`

68. `(5E-3)*(1E9)/(5E
3)`

Extending Skills *Use scientific notation to calculate the result in each expression. Write answers in scientific notation.*

69. $\dfrac{650{,}000{,}000(0.0000032)}{0.00002}$

70. $\dfrac{3{,}400{,}000{,}000(0.000075)}{0.00025}$

71. $\dfrac{0.00000072(0.00023)}{0.000000018}$

72. $\dfrac{0.000000081(0.000036)}{0.00000048}$

73. $\dfrac{0.0000016(240{,}000{,}000)}{0.00002(0.0032)}$

74. $\dfrac{0.000015(42{,}000{,}000)}{0.000009(0.000005)}$

Each statement contains a number in boldface italic type. If the number is in scientific notation, write it in standard notation. If the number is not in scientific notation, write it as such.
See Examples 1 and 2.

75. A *muon* is an atomic particle closely related to an electron. According to the Web page *Muon basics*, the half-life of a muon is about 2 millionths ($\boldsymbol{2 \times 10^{-6}}$) of a second. (*Source:* www2.fisica.unlp.edu.ar)

76. There are 13 red balls and 39 black balls in a box. Mix them up and draw 13 out one at a time without returning any ball . . . the probability that the 13 drawings each will produce a red ball is . . . $\boldsymbol{1.6 \times 10^{-12}}$. (*Source:* Weaver, W., *Lady Luck.*)

77. An electron and a positron attract each other in two ways: the electromagnetic attraction of their opposite electric charges, and the gravitational attraction of their two masses. The electromagnetic attraction is

$$\boldsymbol{4{,}200{,}000{,}000{,}000{,}000{,}000{,}000{,}000{,}000{,}000{,}000{,}000{,}000}$$

times as strong as the gravitational. (*Source:* Asimov, I., *Isaac Asimov's Book of Facts.*)

78. The name "googol" applies to the number

$$\boldsymbol{10{,}000{,}000{,}000{,}000{,}000{,}000{,}000{,}000{,}000{,}000{,}000{,}000{,}000{,}000{,}000{,}000{,}000{,}}$$
$$\boldsymbol{000{,}000{,}000{,}000{,}000{,}000{,}000{,}000{,}000{,}000{,}000{,}000{,}000{,}000{,}000{,}000.}$$

It was created by Edward Kasner and his nephew in 1938. The Web search engine Google honors this number. Sergey Brin, president and cofounder of Google, Inc., was a mathematics major. He chose the name Google to describe the vast reach of this search engine. (*Source: The Gazette.*)

Use scientific notation to calculate the answer to each problem. ***See Examples 3–5.***

79. The Double Helix Nebula, a conglomeration of dust and gas stretching across the center of the Milky Way galaxy, is 25,000 light-years from Earth. If one light-year is about 6,000,000,000,000 mi, about how many miles is the Double Helix Nebula from Earth? (*Source:* www.spitzer.caltech.edu)

80. Pollux, one of the brightest stars in the night sky, is 33.7 light-years from Earth. If one light-year is about 6,000,000,000,000 mi (that is, 6 trillion mi), about how many miles is Pollux from Earth? (*Source: World Almanac and Book of Facts.*)

81. In 2012, the population of the United States was about 313.9 million. To the nearest dollar, calculate how much each person in the United States would have had to contribute in order to make one person a trillionaire (that is, to give that person $1,000,000,000,000). (*Source:* U.S. Census Bureau.)

82. In 2011, the U.S. government collected about $4105 per person in individual income taxes. If the population at that time was 310,000,000, how much did the government collect in taxes for 2011?
(*Source:* www.usgovernmentrevenue.com)

83. Before Congress raised the debt limit in 2013, it was 1.67×10^{13}. To the nearest dollar, how much was this for every man, woman, and child in the country? Use 314 million as the population of the United States. (*Source:* www.census.gov)

84. In 2010, the state of Minnesota had about 8.1×10^4 farms with an average of 3.44×10^2 acres per farm. What was the total number of acres devoted to farmland in Minnesota that year? (*Source:* U.S. Department of Agriculture.)

85. Light travels at a speed of 1.86×10^5 mi per sec. When Venus is 6.68×10^7 mi from the sun, how long (in seconds) does it take light to travel from the sun to Venus? (*Source: World Almanac and Book of Facts.*)

86. The distance to Earth from Pluto is 4.58×10^9 km. *Pioneer 10* transmitted radio signals from Pluto to Earth at the speed of light, 3.00×10^5 km per sec. About how long (in seconds) did it take for the signals to reach Earth?

87. During the 2012–2013 season, Broadway shows grossed a total of 1.14×10^9. Total attendance for the season was 1.16×10^7. What was the average ticket price (to the nearest cent) for a Broadway show? (*Source:* The Broadway League.)

88. In 2012, 1.08×10^{10} was spent to attend motion pictures in the United States and Canada. The total number of tickets sold was 1.36 billion. What was the average ticket price (to the nearest cent) for a movie? (*Source:* Motion Picture Association of America.)

89. In 2012, the world's fastest computer could perform 10,000,000,000,000,000 calculations per second. How many could it perform per minute? Per hour? (*Source:* www.japantimes.co.jp)

90. In 2013, it was reported that the world's fastest computer could handle 33.86 quadrillion calculations per second. (*Hint:* 1 quadrillion $= 1 \times 10^{15}$) How many could it perform per minute? Per hour? (*Source*: www.top500.org)

RELATING CONCEPTS For Individual or Group Work (Exercises 91–94)

In 1935, Charles F. Richter devised a scale to compare the intensities of earthquakes. The *intensity* of an earthquake is measured relative to the intensity of a standard *zero-level* earthquake of intensity I_0. The relationship is equivalent to

$$I = I_0 \times 10^R, \quad \text{where } R \text{ is the } \textbf{Richter scale} \text{ measure.}$$

For example, if an earthquake has magnitude 5.0 on the Richter scale, then its intensity is calculated as

$$I = I_0 \times 10^{5.0} = I_0 \times 100{,}000,$$

which is 100,000 times as intense as a zero-level earthquake.

Intensity $I_0 \times 10^0$ $I_0 \times 10^1$ $I_0 \times 10^2$ $I_0 \times 10^3$ $I_0 \times 10^4$ $I_0 \times 10^5$ $I_0 \times 10^6$ $I_0 \times 10^7$ $I_0 \times 10^8$

Richter Scale 0 1 2 3 4 5 6 7 8

To compare two earthquakes, such as one that measures 8.0 to one that measures 5.0, calculate the *ratio* of their intensities.

$$\frac{\text{intensity } 8.0}{\text{intensity } 5.0} = \frac{I_0 \times 10^{8.0}}{I_0 \times 10^{5.0}} = \frac{10^8}{10^5} = 10^{8-5} = 10^3 = 1000$$

An earthquake that measures 8.0 is 1000 time as intense as one that measures 5.0.

The table lists information for selected earthquakes with their years, locations, and magnitudes. **Work Exercises 91–94 in order.**

Year	Earthquake Location	Richter Scale Measurement
1960	Chile	9.5
1952	Kamchatka	9.0
2007	Southern Sumatra, Indonesia	8.5
2013	Obihoro, Japan	6.9
2002	Hindu Kush, Afghanistan	5.9

Source: earthquake.usgs.gov

91. Compare the intensity of the 1960 Chile earthquake with the 2007 Southern Sumatra earthquake.

92. Compare the intensity of the 2013 Obihoro earthquake with the 2002 Hindu Kush earthquake.

93. Compare the intensity of the 1952 Kamchatka earthquake with the 2007 Southern Sumatra earthquake. $\left(\textit{Hint: } 10^{0.5} = \sqrt{10}\right)$

94. Suppose an earthquake measures 7.5 on the Richter scale. How would the intensity of the 1960 Chile earthquake compare to it?

4.4   Adding, Subtracting, and Graphing Polynomials

OBJECTIVES

1 Identify terms and coefficients.
2 Combine like terms.
3 Know the vocabulary for polynomials.
4 Evaluate polynomials.
5 Add and subtract polynomials.
6 Graph equations defined by polynomials of degree 2.

VOCABULARY

☐ term
☐ leading term
☐ numerical coefficient (coefficient)
☐ like terms
☐ unlike terms
☐ polynomial
☐ descending powers
☐ degree of a term
☐ degree of a polynomial
☐ monomial
☐ binomial
☐ trinomial
☐ parabola
☐ vertex
☐ axis of symmetry (axis)

 **NOW TRY EXERCISE 1**

Name the coefficient of each term in the expression. Give the number of terms.

$$t - 10t^2$$

NOW TRY ANSWER

1. $1; -10$; two terms

OBJECTIVE 1   Identify terms and coefficients.

In an expression such as

$$4x^3 + 6x^2 + 5x + 8,$$

the quantities $4x^3$, $6x^2$, $5x$, and 8 are **terms.** (See **Section 1.7.**) In the **leading** (or first) **term** $4x^3$, the number 4 is the **numerical coefficient,** or simply the **coefficient,** of x^3. In the same way, 6 is the coefficient of x^2 in the term $6x^2$, and 5 is the coefficient of x in the term $5x$. The constant term 8 can be thought of as

$$8 \cdot 1 = 8x^0 \quad \text{because} \quad x^0 = 1,$$

so 8 is the coefficient in the term 8.

EXAMPLE 1   Identifying Coefficients

Name the coefficient of each term in each expression. Give the number of terms.

(a) $x - 6x^4 + 3$ can be written as $1x + (-6x^4) + 3x^0$.

The coefficients are 1, −6, and 3.

There are three terms: x, $-6x^4$, and 3.

(b) $5 - v^3$ can be written as $5v^0 + (-1v^3)$.

There are two terms.

The coefficients are 5 and −1.

NOW TRY

OBJECTIVE 2   Combine like terms.

Recall from **Section 1.7** that **like terms** have exactly the same variables, with the same exponents on the variables. *Only the coefficients may differ.*

$19m^5$ and $14m^5$	
$6y^9$, $-37y^9$, and y^9	Examples of **like terms**
$3pq$ and $-2pq$	
$2xy^2$ and $-xy^2$	

$7x$ and $7y$	
z^4 and z	Examples of **unlike terms**
$2pq$ and $2p$	
$-4xy^2$ and $5x^2y$	

Using the distributive property, we combine like terms by adding or subtracting their coefficients.

EXAMPLE 2   Combining Like Terms

Simplify by combining like terms.

(a) $-4x^3 + 6x^3$

$= (-4 + 6)x^3$ $ac + bc$
$\qquad\qquad\qquad = (a + b)c$

$= 2x^3$

(b) $9x^6 - 14x^6 + x^6$ $x^6 = 1x^6$

$= (9 - 14 + 1)x^6$

$= -4x^6$

NOW TRY
EXERCISE 2

Simplify by combining like terms.

(a) $x - \dfrac{2}{5}x$

(b) $3x^2 - x^2 + 2x$

(c) $y + \dfrac{2}{3}y$

$= 1y + \dfrac{2}{3}y$ $y = 1y$

$= \left(\dfrac{3}{3} + \dfrac{2}{3}\right)y$ $1 = \dfrac{3}{3}$; Distributive property

$= \dfrac{5}{3}y$ Add the fractions.

(e) $12m^2 + 5m + 4m^2$

$= (12 + 4)m^2 + 5m$

$= 16m^2 + 5m$ ← Stop here. These are unlike terms.

(d) $8rs - 13rs + 9rs$

$= (8 - 13 + 9)rs$

$= 4rs$

(f) $5u + 11v$

These are unlike terms. They cannot be combined.

NOW TRY

> ⚠ **CAUTION** In **Example 2(e),** we cannot combine $16m^2$ and $5m$ because the exponents on the variables are different. *Unlike terms have different variables or different exponents on the same variables.*

OBJECTIVE 3 Know the vocabulary for polynomials.

> **Polynomial in *x***
>
> A **polynomial in *x*** is a term or the sum of a finite number of terms of the form
>
> $$ax^n,\quad \text{for any real number } a \text{ and any whole number } n.$$

For example,

$$16x^8 - 7x^6 + 5x^4 - 3x^2 + 4$$ Polynomial in *x* (The 4 can be written as $4x^0$.)

is a polynomial in *x*. This polynomial is written in **descending powers** of the variable because the exponents on *x* decrease from left to right. By contrast,

$$2x^3 - x^2 + \dfrac{4}{x}, \quad \text{or} \quad 2x^3 - x^2 + 4x^{-1}, \quad \text{Not a polynomial}$$

is not a polynomial in *x*. A variable appears in a denominator or as a factor to a negative power in a numerator.

> **NOTE** We can define *polynomial* using any variable and not just *x*, as in **Example 2(e).** Polynomials may have terms with more than one variable, as in **Example 2(d).**

The **degree of a term** is the sum of the exponents on the variables. The **degree of a polynomial** is the greatest degree of any nonzero term of the polynomial.

▼ **Degrees of Terms and Polynomials**

Term	Degree	Polynomial	Degree
$3x^4$	4	$3x^4 - 5x^2 + 6$	4
$5x$, or $5x^1$	1	$5x + 7$	1
-7, or $-7x^0$	0	$x^5 + 3x^6 - 7$	6
$2x^2y$, or $2x^2y^1$	$2 + 1 = 3$	$2x^2y + xy - 5y^2$	3

NOW TRY ANSWERS

2. **(a)** $\dfrac{3}{5}x$ **(b)** $2x^2 + 2x$

A polynomial with only one term is a **monomial.** (*Mono-* means "one," as in *mono*rail.) A polynomial with exactly two terms is a **binomial.** (*Bi-* means "two," as in *bi*cycle.) A polynomial with exactly three terms is a **trinomial.** (*Tri-* means "three," as in *tri*angle.)

$$9m, \quad -6y^5, \quad a^2, \quad \text{and} \quad 6 \qquad \text{Monomials}$$

$$-9x^4 + 9x^3, \quad 8m^2 + 6m, \quad \text{and} \quad 3m^5 - 9m^2 \qquad \text{Binomials}$$

$$9m^3 - 4m^2 + 6, \quad \frac{19}{3}y^2 + \frac{8}{3}y + 5, \quad \text{and} \quad -3m^5 - 9m^2 + 2 \qquad \text{Trinomials}$$

NOW TRY EXERCISE 3

Simplify, give the degree, and tell whether the simplified polynomial is a *monomial,* a *binomial,* a *trinomial,* or *none of these.*

(a) $3x^2 + 2x - 4$

(b) $x^3 + 4x^3$

(c) $x^8 - x^7 + 2x^8$

EXAMPLE 3   Classifying Polynomials

For each polynomial, first simplify, if possible. Then give the degree and tell whether the simplified polynomial is a *monomial,* a *binomial,* a *trinomial,* or *none of these.*

(a) $2x^3 + 5$ The polynomial cannot be simplified. It is a binomial of degree 3.

(b) $6x - 8x + 13x$

$= 11x$ Combine like terms to simplify.

The degree is 1 (because $x = x^1$). The simplified polynomial is a monomial.

(c) $4xy - 5xy + 2xy$

$= xy$ Combine like terms to simplify.

The degree is 2 (because $xy = x^1y^1$, and $1 + 1 = 2$). The simplified polynomial is a monomial.   NOW TRY

OBJECTIVE 4   Evaluate polynomials.

A polynomial usually represents different numbers for different values of the variable.

NOW TRY EXERCISE 4

Find the value for $t = -3$.

$$4t^3 - t^2 - t$$

EXAMPLE 4   Evaluating a Polynomial

Find the value of $3x^4 + 5x^3 - 4x - 4$ for **(a)** $x = -2$ and **(b)** $x = 3$.

(a)

$$3x^4 + 5x^3 - 4x - 4$$

Use parentheses to avoid errors.

$= 3(-2)^4 + 5(-2)^3 - 4(-2) - 4$ Substitute -2 for x.

$= 3(16) + 5(-8) - 4(-2) - 4$ Apply the exponents.

$= 48 - 40 + 8 - 4$ Multiply.

$= 12$ Add and subtract.

(b)

$$3x^4 + 5x^3 - 4x - 4$$

Replace x with 3.

$= 3(3)^4 + 5(3)^3 - 4(3) - 4$ Let $x = 3$.

$= 3(81) + 5(27) - 4(3) - 4$ Apply the exponents.

$= 243 + 135 - 12 - 4$ Multiply.

$= 362$ Add and subtract.   NOW TRY

NOW TRY ANSWERS

3. **(a)** The polynomial cannot be simplified; degree 2; trinomial
   (b) $5x^3$; degree 3; monomial
   (c) $3x^8 - x^7$; degree 8; binomial
4. -114

> **! CAUTION**   Use parentheses around the numbers that are substituted for the variable, as in **Example 4.** *Be particularly careful when substituting a negative number for a variable that is raised to a power, or a sign error may result.*

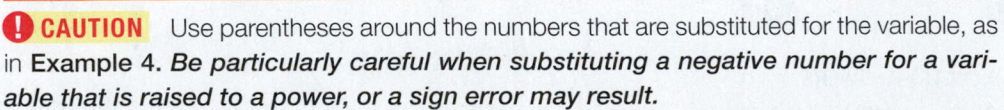

OBJECTIVE 5 Add and subtract polynomials.

Adding Polynomials

To add two polynomials, add like terms.

NOW TRY
EXERCISE 5
Find each sum.
(a) Add $7y^3 - 4y^2 + 2$ and $-6y^3 + 5y^2 - 3$.
(b) Add $-5x^4 - 2x + 3$ and $x^3 - 5x$.

EXAMPLE 5 Adding Polynomials Vertically

Find each sum.

(a) Add $6x^3 - 4x^2 + 3$ and $-2x^3 + 7x^2 - 5$.

$$\begin{array}{l} 6x^3 - 4x^2 + 3 \\ \underline{-2x^3 + 7x^2 - 5} \end{array}$$ Write like terms in columns.

Now add, column by column.

Add the coefficients only. Do *not* add the exponents.

$$\begin{array}{ccc} 6x^3 & -4x^2 & 3 \\ \underline{-2x^3} & \underline{7x^2} & \underline{-5} \\ 4x^3 & 3x^2 & -2 \end{array}$$

Add the three sums together to obtain the answer.

$$4x^3 + 3x^2 + (-2) = 4x^3 + 3x^2 - 2 \leftarrow \text{Final sum}$$

(b) Add $2x^2 - 4x + 3$ and $x^3 + 5x$.

Write like terms in columns and add column by column.

$$\begin{array}{l} 2x^2 - 4x + 3 \\ \underline{x^3 \qquad + 5x} \\ x^3 + 2x^2 + \ x + 3 \end{array}$$ Leave spaces for missing terms.

NOW TRY

The polynomials in **Example 5** also can be added horizontally.

NOW TRY
EXERCISE 6
Add $10x^4 - 3x^2 - x$ and $x^4 - 3x^2 + 5x$ horizontally.

EXAMPLE 6 Adding Polynomials Horizontally

Find each sum.

(a) Add $6x^3 - 4x^2 + 3$ and $-2x^3 + 7x^2 - 5$.

$$(6x^3 - 4x^2 + 3) + (-2x^3 + 7x^2 - 5) = 4x^3 + 3x^2 - 2$$ Same answer as found in **Example 5(a)**

(b) Add $2x^2 - 4x + 3$ and $x^3 + 5x$.

$$(2x^2 - 4x + 3) + (x^3 + 5x)$$
$$= x^3 + 2x^2 - 4x + 5x + 3$$ Commutative property
$$= x^3 + 2x^2 + x + 3$$ See **Example 5(b)**.   NOW TRY

In **Section 1.4**, the difference $x - y$ was defined as $x + (-y)$. (We find the difference $x - y$ by adding x and the opposite of y.)

$$7 - 2 \quad \text{is equivalent to} \quad 7 + (-2), \quad \text{which equals} \quad 5.$$
$$-8 - (-2) \quad \text{is equivalent to} \quad -8 + 2, \quad \text{which equals} \quad -6.$$

A similar method is used to subtract polynomials.

NOW TRY ANSWERS
5. **(a)** $y^3 + y^2 - 1$
  (b) $-5x^4 + x^3 - 7x + 3$
6. $11x^4 - 6x^2 + 4x$

> **Subtracting Polynomials**
>
> To subtract two polynomials, change all the signs of the subtrahend (second polynomial) and add the result to the minuend (first polynomial).

**NOW TRY
EXERCISE 7**

Perform each subtraction.

(a) $(3x - 8) - (5x - 9)$

(b) $(4t^4 - t^2 + 7)$
　$- (5t^4 - 3t^2 + 1)$

EXAMPLE 7　Subtracting Polynomials Horizontally

Perform each subtraction.

(a) $(5x - 2) - (3x - 8)$

$$= (5x - 2) + [-(3x - 8)] \qquad \text{Definition of subtraction}$$

$$= (5x - 2) + [-1(3x - 8)] \qquad -a = -1a$$

$$= (5x - 2) + (-3x + 8) \qquad \text{Distributive property}$$

$$= 2x + 6 \qquad \text{Combine like terms.}$$

CHECK　To check a subtraction problem, use the fact that

$$\text{if} \quad a - b = c, \quad \text{then} \quad a = b + c.$$

Here, add $3x - 8$ and $2x + 6$.

$$(3x - 8) + (2x + 6)$$
$$= 5x - 2 \ \checkmark$$

(b) Subtract $6x^3 - 4x^2 + 2$ from $11x^3 + 2x^2 - 8$.

> Be careful to write the problem in the correct order.

$$(11x^3 + 2x^2 - 8) - (6x^3 - 4x^2 + 2)$$

$$= (11x^3 + 2x^2 - 8) + (-6x^3 + 4x^2 - 2)$$

$$= 5x^3 + 6x^2 - 10 \qquad \text{Combine like terms.}$$

CHECK　Add $6x^3 - 4x^2 + 2$ and $5x^3 + 6x^2 - 10$.

$$(6x^3 - 4x^2 + 2) + (5x^3 + 6x^2 - 10)$$

$$= 11x^3 + 2x^2 - 8 \ \checkmark \qquad \text{NOW TRY} $$

Subtraction can also be done in columns. We use vertical subtraction in **Section 4.7** when we divide polynomials.

**NOW TRY
EXERCISE 8**

Subtract by columns.

$(12x^2 - 9x + 4)$
　$- (-10x^2 - 3x + 7)$

EXAMPLE 8　Subtracting Polynomials Vertically

Subtract by columns: $(14y^3 - 6y^2 + 2y - 5) - (2y^3 - 7y^2 - 4y + 6)$.

$$14y^3 - 6y^2 + 2y - 5$$
$$\underline{2y^3 - 7y^2 - 4y + 6} \qquad \text{Arrange like terms in columns.}$$

Change all signs in the second row (the subtrahend), and then add.

$$14y^3 - 6y^2 + 2y - 5$$
$$\underline{-2y^3 + 7y^2 + 4y - 6} \qquad \text{Change all signs.}$$
$$12y^3 + y^2 + 6y - 11 \qquad \text{Add.}$$

NOW TRY ANSWERS

7. (a) $-2x + 1$
　(b) $-t^4 + 2t^2 + 6$
8. $22x^2 - 6x - 3$

NOW TRY

**NOW TRY
EXERCISE 9**

Perform the indicated
operations.

$(6p^4 - 8p^3 + 2p - 1)$
$\quad - (-7p^4 + 6p^2 - 12)$
$\quad + (p^4 - 3p + 8)$

EXAMPLE 9 Adding and Subtracting More Than Two Polynomials

Perform the indicated operations.

$$(4 - x + 3x^2) - (2 - 3x + 5x^2) + (8 + 2x - 4x^2)$$

Rewrite, using the definition of subtraction.

$$(4 - x + 3x^2) - (2 - 3x + 5x^2) + (8 + 2x - 4x^2)$$

$$= (4 - x + 3x^2) + (-2 + 3x - 5x^2) + (8 + 2x - 4x^2)$$

$$= (2 + 2x - 2x^2) + (8 + 2x - 4x^2) \qquad \text{Combine like terms.}$$

$$= 10 + 4x - 6x^2 \qquad \text{Combine like terms.}$$

NOW TRY

**NOW TRY
EXERCISE 10**

Subtract.

$(4x^2 - 2xy + y^2)$
$\quad - (6x^2 - 7xy + 2y^2)$

EXAMPLE 10 Adding and Subtracting Multivariable Polynomials

Add or subtract as indicated.

(a) $(4a + 2ab - b) + (3a - ab + b)$

$$= 4a + 2ab - b + 3a - ab + b$$

$$= 7a + ab \qquad \text{Combine like terms.}$$

(b) $(2x^2y + 3xy + y^2) - (3x^2y - xy - 2y^2)$

$$= 2x^2y + 3xy + y^2 - 3x^2y + xy + 2y^2$$

$$= -x^2y + 4xy + 3y^2$$

> Be careful with signs. The coefficient of xy is 1.

NOW TRY

OBJECTIVE 6 Graph equations defined by polynomials of degree 2.

In **Chapter 3,** we graphed linear equations (which are actually polynomial equations of degree 1). By plotting points, we can graph polynomial equations of degree 2.

EXAMPLE 11 Graphing Equations Defined by Polynomials of Degree 2

Graph each equation.

(a) $y = x^2$

It is easier to select values for x and find corresponding y-values. Selecting $x = 2$ and substituting in $y = x^2$ gives

$$y = 2^2 = 4.$$

The point $(2, 4)$ is on the graph of $y = x^2$. (Recall that in an ordered pair such as $(2, 4)$, *the x-value comes first and the y-value second.*) We show some ordered pairs that satisfy $y = x^2$ in the table with **FIGURE 3** on the next page. If we plot the ordered pairs from the table on a coordinate system and draw a smooth curve through them, we obtain the graph shown in **FIGURE 3**.

The graph of $y = x^2$ is the graph of a function, because each input x is related to just one output y. The curve in **FIGURE 3** is a **parabola.** The point $(0, 0)$, the *lowest* point on this graph, is the **vertex** of the parabola. The vertical line through the vertex (the y-axis here) is the **axis of symmetry,** or simply the **axis,** of the parabola. This axis is a line of symmetry for the graph. If the graph is folded on this line, the two halves will coincide.

NOW TRY ANSWERS

9. $14p^4 - 8p^3 - 6p^2 - p + 19$
10. $-2x^2 + 5xy - y^2$

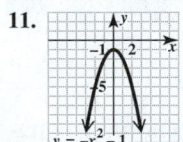

**NOW TRY
EXERCISE 11**

Graph $y = -x^2 - 1$.

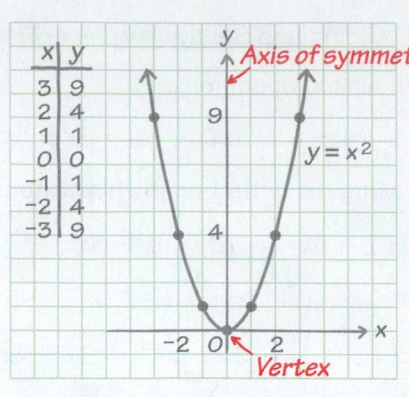

FIGURE 3

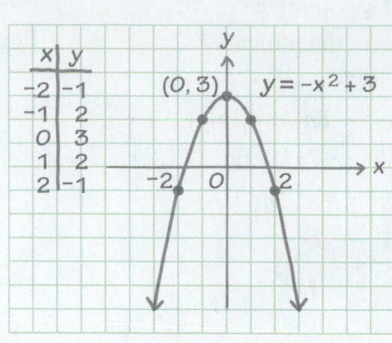

FIGURE 4

(b) $y = -x^2 + 3$

Plot points to obtain the graph. For example, let $x = -2$ and $x = 0$.

$$y = -(-2)^2 + 3 \qquad y = -0^2 + 3$$
$$y = -4 + 3 \qquad\qquad y = 0 + 3$$
$$y = -1 \qquad\qquad\quad y = 3$$

The points $(-2, -1)$, $(0, 3)$, and several others are shown in the table that accompanies the graph in **FIGURE 4**. The vertex $(0, 3)$ is the *highest* point on this graph. The graph opens downward because x^2 has a negative coefficient.

NOW TRY

NOW TRY ANSWER

11.

NOTE *All polynomials of degree 2 have parabolas as their graphs.* When graphing, find points until the vertex and points on either side of it are located. (In this section, all parabolas have their vertices on the *x*-axis or the *y*-axis.)

4.4 Exercises

FOR EXTRA HELP   ▶ MyMathLab®

▶ *Complete solution available in MyMathLab*

Concept Check   *Complete each statement.*

1. In the term $4x^6$, the coefficient of x^6 is _____ and the exponent is _____ .

2. The expression $4x^3 - 5x^2$ has exactly (*one / two / three*) term(s).

3. The degree of the term $-3x^9$ is _____ .

4. The polynomial $4x^2 + y^2$ (*is / is not*) an example of a trinomial.

5. When $x^2 + 10$ is evaluated for $x = 3$, the result is _____ .

6. $5x^{\underline{\quad}} + 3x^3 - 7x$ is a trinomial of degree 6.

7. Combining like terms in $-3xy - 2xy + 5xy$ gives _____ .

8. _____ is an example of a monomial with coefficient 8, in the variable x, having degree 5.

For each expression, determine the number of terms and name the coefficients of the terms. See Example 1.

▶ 9. $6x^4$ **10.** $-9y^5$ **11.** t^4 **12.** s^7

13. $-19r^2 - r$ **14.** $2y^3 - y$ **15.** $x + 8x^2 + 5x^3$ **16.** $v - 2v^3 - v^7$

In each polynomial, simplify by combining like terms whenever possible. In Exercises 17–26, write the result in descending powers of the variable. See Example 2 and Objective 3.

▶ 17. $-3m^5 + 5m^5$ **18.** $-4y^3 + 3y^3$ **19.** $2r^5 + (-3r^5)$

20. $9y^2 + (-19y^2)$ **21.** $0.2m^5 - 0.5m^2$ **22.** $-0.9y + 0.9y^2$

23. $-3x^5 + 3x^5 - 5x^5$ **24.** $6x^3 - 9x^3 + 10x^3$ **25.** $-4p^7 + 8p^7 + 5p^9$

26. $-3a^8 + 4a^8 - 3a^2$ **27.** $-1.5x^2 + 5.3x^2 - 3.8x^2$ **28.** $8.6y^4 - 10.3y^4 + 1.7y^4$

29. $-4xy^2 + 3xy^2 - 2xy^2 + xy^2$ **30.** $3pr^5 - 8pr^5 + pr^5 + 2pr^5$

31. $-\dfrac{1}{3}tu^7 + \dfrac{2}{5}tu^7 + \dfrac{1}{15}tu^7 - \dfrac{8}{5}tu^7$ **32.** $-\dfrac{3}{4}p^2q - \dfrac{1}{3}p^2q + \dfrac{7}{12}p^2q - \dfrac{1}{6}p^2q$

For each polynomial, first simplify, if possible, and write the result in descending powers of the variable. Then give the degree and tell whether the simplified polynomial is a monomial, *a* binomial, *a* trinomial, *or* none of these. *See Example 3.*

▶ 33. $6x^4 - 9x$ **34.** $7t^3 - 3t$

35. $5m^4 - 3m^2 + 6m^4 - 7m^3$ **36.** $6p^5 + 4p^3 - 8p^5 + 10p^2$

37. $\dfrac{5}{3}x^4 - \dfrac{2}{3}x^4$ **38.** $\dfrac{4}{5}r^6 + \dfrac{1}{5}r^6$

39. $0.8x^4 - 0.3x^4 - 0.5x^4 + 7$ **40.** $1.2t^3 - 0.9t^3 - 0.3t^3 + 9$

41. $-11ab + 2ab - 4ab$ **42.** $5xy + 13xy - 12xy$

Find the value of each polynomial for **(a)** $x = 2$ *and* **(b)** $x = -1$. *See Example 4.*

▶ 43. $2x^2 - 3x - 5$ **44.** $x^2 + 5x - 10$ **45.** $-3x^2 + 14x - 2$

46. $-2x^2 + 5x - 1$ **47.** $2x^5 - 4x^4 + 5x^3 - x^2$ **48.** $x^4 - 6x^3 + x^2 - x$

Add. See Examples 5 and 6.

49. $\begin{aligned} 2x^2 - 4x \\ \underline{3x^2 + 2x} \end{aligned}$ **50.** $\begin{aligned} -5y^3 + 3y \\ \underline{8y^3 - 4y} \end{aligned}$ **▶ 51.** $\begin{aligned} 3m^2 + 5m + 6 \\ \underline{2m^2 - 2m - 4} \end{aligned}$

52. $\begin{aligned} 4a^3 - 4a^2 - 4 \\ \underline{6a^3 + 5a^2 - 8} \end{aligned}$ **53.** $\begin{aligned} \frac{2}{3}x^2 + \frac{1}{5}x + \frac{1}{6} \\ \underline{\frac{1}{2}x^2 - \frac{1}{3}x + \frac{2}{3}} \end{aligned}$ **54.** $\begin{aligned} \frac{4}{7}y^2 - \frac{1}{5}y + \frac{7}{9} \\ \underline{\frac{1}{3}y^2 - \frac{1}{3}y + \frac{2}{5}} \end{aligned}$

55. $9m^3 - 5m^2 + 4m - 8$ and $-3m^3 + 6m^2 - 6$

56. $12r^5 + 11r^4 - 7r^3 - 2r^2$ and $-8r^5 + 3r^3 + 2r^2$

Subtract. See Example 8.

57. $\begin{aligned} 5y^3 - 3y^2 \\ \underline{2y^3 + 8y^2} \end{aligned}$ **58.** $\begin{aligned} -6t^3 + 4t^2 \\ \underline{8t^3 - 6t^2} \end{aligned}$

59. $\begin{aligned} 12x^4 - x^2 + x \\ \underline{8x^4 + 3x^2 - 3x} \end{aligned}$ **60.** $\begin{aligned} 13y^5 - y^3 - 8y^2 \\ \underline{7y^5 + 5y^3 + y^2} \end{aligned}$

▶ 61. $\begin{aligned} 12m^3 - 8m^2 + 6m + 7 \\ \underline{-3m^3 + 5m^2 - 2m - 4} \end{aligned}$ **62.** $\begin{aligned} 5a^4 - 3a^3 + 2a^2 - a + 6 \\ \underline{-6a^4 + a^3 - a^2 + a - 1} \end{aligned}$

Perform each indicated operation. See Examples 6, 7, and 9.

▶ 63. $(8m^2 - 7m) - (3m^2 + 7m - 6)$ **64.** $(x^2 + x) - (3x^2 + 2x - 1)$

▶ **65.** $(16x^3 - x^2 + 3x) + (-12x^3 + 3x^2 + 2x)$

66. $(-2b^6 + 3b^4 - b^2) + (b^6 + 2b^4 + 2b^2)$

67. Subtract $18y^4 - 5y^2 + y$ from $7y^4 + 3y^2 + 2y$.

68. Subtract $19t^5 - 6t^3 + t$ from $8t^5 + 3t^3 + 5t$.

69. $(9a^4 - 3a^2 + 2) + (4a^4 - 4a^2 + 2) + (-12a^4 + 6a^2 - 3)$

70. $(4m^2 - 3m + 2) + (5m^2 + 13m - 4) + (-16m^2 - 4m + 3)$

71. $[(8m^2 + 4m - 7) - (2m^2 - 5m + 2)] - (m^2 + m + 1)$

72. $[(9b^3 - 4b^2 + 3b + 2) - (-2b^3 - 3b^2 + b)] - (8b^3 + 6b + 4)$

73. $[(3x^2 - 2x + 7) - (4x^2 + 2x - 3)] - [(9x^2 + 4x - 6) + (-4x^2 + 4x + 4)]$

74. $[(6t^2 - 3t + 1) - (12t^2 + 2t - 6)] - [(4t^2 - 3t - 8) + (-6t^2 + 10t - 12)]$

75. *Concept Check*   Without actually performing the operations, determine mentally the coefficient of the x^2-term in the simplified form of
$$(-4x^2 + 2x - 3) - (-2x^2 + x - 1) + (-8x^2 + 3x - 4).$$

76. *Concept Check*   Without actually performing the operations, determine mentally the coefficient of the x-term in the simplified form of
$$(-8x^2 - 3x + 2) - (4x^2 - 3x + 8) - (-2x^2 - x + 7).$$

Add or subtract as indicated. *See Example 10.*

▶ **77.** $(6b + 3c) + (-2b - 8c)$ **78.** $(-5t + 13s) + (8t - 3s)$

79. $(4x + 2xy - 3) - (-2x + 3xy + 4)$ **80.** $(8ab + 2a - 3b) - (6ab - 2a + 3b)$

81. $(5x^2y - 2xy + 9xy^2) - (8x^2y + 13xy + 12xy^2)$

82. $(16t^3s^2 + 8t^2s^3 + 9ts^4) - (-24t^3s^2 + 3t^2s^3 - 18ts^4)$

Find a polynomial that represents the perimeter of each rectangle, square, or triangle.

83.
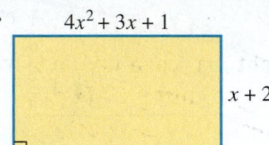
$4x^2 + 3x + 1$
$x + 2$

84.
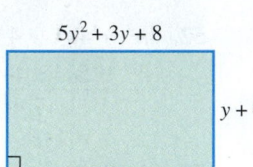
$5y^2 + 3y + 8$
$y + 4$

85.
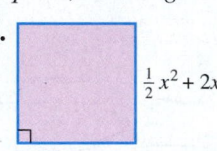
$\frac{1}{2}x^2 + 2x$

86.
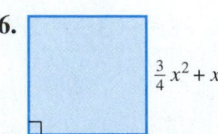
$\frac{3}{4}x^2 + x$

87.
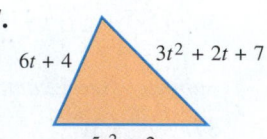
$6t + 4$ $3t^2 + 2t + 7$
$5t^2 + 2$

88.
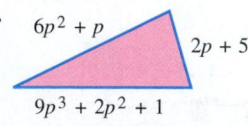
$6p^2 + p$ $2p + 5$
$9p^3 + 2p^2 + 1$

*Find **(a)** a polynomial that represents the perimeter of each triangle and **(b)** the degree measures of the angles of the triangle. (Hint: The sum of the measures of the angles of any triangle is 180°.)*

89.
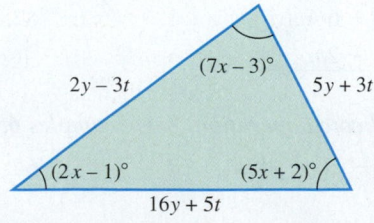
$2y - 3t$ $(7x - 3)°$ $5y + 3t$
$(2x - 1)°$ $(5x + 2)°$
$16y + 5t$

90.
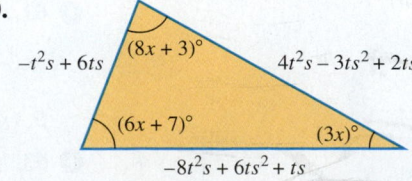
$-t^2s + 6ts$ $(8x + 3)°$ $4t^2s - 3ts^2 + 2ts$
$(6x + 7)°$ $(3x)°$
$-8t^2s + 6ts^2 + ts$

Extending Skills   *Perform each indicated operation.*

91. Find the difference of the sum of $5x^2 + 2x - 3$ and $x^2 - 8x + 2$ and the sum of $7x^2 - 3x + 6$ and $-x^2 + 4x - 6$.

92. Subtract the sum of $9t^3 - 3t + 8$ and $t^2 - 8t + 4$ from the sum of $12t + 8$ and $t^2 - 10t + 3$.

Graph each equation by completing the table of values. See Example 11.

▶ **93.** $y = x^2 - 4$

x	y
-2	
-1	
0	
1	
2	

94. $y = x^2 - 6$

x	y
-2	
-1	
0	
1	
2	

95. $y = 2x^2 - 1$

x	y
-2	
-1	
0	
1	
2	

96. $y = 2x^2 + 2$

x	y
-2	
-1	
0	
1	
2	

97. $y = -x^2 + 4$

x	y
-2	
-1	
0	
1	
2	

98. $y = -x^2 + 2$

x	y
-2	
-1	
0	
1	
2	

99. $y = (x + 3)^2$

x	-5	-4	-3	-2	-1
y					

100. $y = (x - 4)^2$

x	2	3	4	5	6
y					

RELATING CONCEPTS   For Individual or Group Work   (Exercises 101–104)

The age of a dog in human years y is given by the polynomial equation

$$y = -0.1133x^2 + 6.966x + 4.915,$$

where x represents age in dog years (based on data from www.dogyears.com). For example if a dog is 4 in dog years, then we let x = 4 to find that y ≈ 31. (Verify this.) The dog is about 31 yr old in human years. This illustrates the important mathematical concept of a **function**—*for each input value x, we obtain one and only one output value y.*

   Exercises 101–104 further illustrate the function concept with polynomials. **Work them in order.**

101. It used to be thought that each dog year was about 7 human years, so that $y = 7x$ gave the number of human years for x dog years. Evaluate y for $x = 9$, and interpret the result.

102. Use the polynomial equation in the directions above to find the number of human years equivalent to each given number of dog years. Round to the nearest whole number.

   (a) 5   **(b)** 11   **(c)** 14

103. If an object is projected upward under certain conditions, its height y in feet is given by

$$y = -16x^2 + 60x + 80,$$

where x is in seconds. Evaluate y for $x = 2.5$. Use the result to fill in the blanks: If _____ seconds have elapsed, the height of the object is _____ feet.

104. If it costs \$15 to rent a chain saw, plus \$2 per day, the equation

$$y = 2x + 15$$

gives the cost y in dollars to rent the chain saw for x days. Evaluate y for $x = 6$. Use the result to fill in the blanks: If the saw is rented for _____ days, the cost is _____.

4.5 Multiplying Polynomials

OBJECTIVES

1 Multiply a monomial and a polynomial.

2 Multiply two polynomials.

3 Multiply binomials by the FOIL method.

VOCABULARY
☐ FOIL method
☐ outer product
☐ inner product

NOW TRY
EXERCISE 1
Find the product.

$-3x^5(2x^3 - 5x^2 + 10)$

OBJECTIVE 1   Multiply a monomial and a polynomial.

In **Section 4.1,** we found the product of two monomials as follows.

$$(-8m^6)(-9n^6)$$

$$= (-8)(-9)(m^6)(n^6) \quad \text{Commutative and associative properties}$$

$$= 72m^6n^6 \quad \text{Multiply.}$$

⚠ **CAUTION**   *Do not confuse addition of terms with multiplication of terms.*

$7q^5 + 2q^5$	$(7q^5)(2q^5)$	Commutative property; product rule
$= (7+2)q^5$ Distributive property	$= 7 \cdot 2q^{5+5}$	
$= 9q^5$ Add.	$= 14q^{10}$ Multiply. Add.	

To find the product of a monomial and a polynomial with more than one term, we use the distributive property and multiplication of monomials.

EXAMPLE 1   Multiplying Monomials and Polynomials

Find each product.

(a) $4x^2(3x + 5)$ $\qquad a(b+c) = ab + ac$

$$= 4x^2(3x) + 4x^2(5) \quad \text{Distributive property}$$

$$= 12x^3 + 20x^2 \quad \text{Multiply monomials.}$$

(b) $-8m^3(4m^3 + 3m^2 + 2m - 1)$

$$= -8m^3(4m^3) + (-8m^3)(3m^2)$$

$$+ (-8m^3)(2m) + (-8m^3)(-1) \quad \text{Distributive property}$$

$$= -32m^6 - 24m^5 - 16m^4 + 8m^3 \quad \text{Multiply monomials.} \qquad \textbf{NOW TRY}$$

OBJECTIVE 2   Multiply two polynomials.

To find the product of the polynomials $x^2 + 3x + 5$ and $x - 4$, we can think of $x - 4$ as a single quantity and use the distributive property as follows.

$$(x^2 + 3x + 5)(x - 4)$$

$$= x^2(x - 4) + 3x(x - 4) + 5(x - 4) \quad \text{Distributive property}$$

$$= x^2(x) + x^2(-4) + 3x(x) + 3x(-4) + 5(x) + 5(-4)$$
$$\qquad\qquad\qquad\qquad\qquad\qquad \text{Distributive property again}$$

$$= x^3 - 4x^2 + 3x^2 - 12x + 5x - 20 \quad \text{Multiply monomials.}$$

$$= x^3 - x^2 - 7x - 20 \quad \text{Combine like terms.}$$

Multiplying Polynomials

To multiply two polynomials, multiply each term of the second polynomial by each term of the first polynomial and add the products.

NOW TRY ANSWER
1. $-6x^8 + 15x^7 - 30x^5$

NOW TRY EXERCISE 2

Multiply.

$(x^2 - 4)(2x^2 - 5x + 3)$

EXAMPLE 2 Multiplying Two Polynomials

Multiply $(m^2 + 5)(4m^3 - 2m^2 + 4m)$.

$(m^2 + 5)(4m^3 - 2m^2 + 4m)$　　Multiply each term of the second polynomial by each term of the first.

$= m^2(4m^3) + m^2(-2m^2) + m^2(4m) + 5(4m^3) + 5(-2m^2) + 5(4m)$　　Distributive properly

$= 4m^5 - 2m^4 + 4m^3 + 20m^3 - 10m^2 + 20m$　　Multiply monomials.

$= 4m^5 - 2m^4 + 24m^3 - 10m^2 + 20m$　　Combine like terms.

NOW TRY

NOW TRY EXERCISE 3

Multiply.

$$5t^2 - 7t + 4$$
$$\underline{2t - 6}$$

EXAMPLE 3 Multiplying Polynomials Vertically

Multiply $(x^3 + 2x^2 + 4x + 1)(3x + 5)$ vertically.

$$x^3 + 2x^2 + 4x + 1$$
$$\underline{3x + 5}$$

Write the polynomials vertically.

Begin by multiplying each of the terms in the top row by 5.

$$x^3 + 2x^2 + 4x + 1$$
$$\underline{3x + 5}$$
$$5x^3 + 10x^2 + 20x + 5$$　　$5(x^3 + 2x^2 + 4x + 1)$

Now multiply each term in the top row by $3x$. Then add like terms.

$$x^3 + 2x^2 + 4x + 1$$
$$\underline{3x + 5}$$
$$5x^3 + 10x^2 + 20x + 5$$
$$\underline{3x^4 + 6x^3 + 12x^2 + 3x}$$
$$3x^4 + 11x^3 + 22x^2 + 23x + 5$$

Place *like* terms in columns so they can be added.

This process is similar to multiplication of whole numbers.

$3x(x^3 + 2x^2 + 4x + 1)$

Add in columns.　　**NOW TRY**

NOW TRY EXERCISE 4

Find the product of
$9x^3 - 12x^2 + 3$ and $\frac{1}{3}x^2 - \frac{2}{3}$.

EXAMPLE 4 Multiplying Polynomials Vertically (Fractional Coefficients)

Find the product of $4m^3 - 2m^2 + 4m$ and $\frac{1}{2}m^2 + \frac{5}{2}$.

$$4m^3 - 2m^2 + 4m$$
$$\frac{1}{2}m^2 + \frac{5}{2}$$
$$\underline{}$$
$$10m^3 - 5m^2 + 10m$$　　Terms of top row are multiplied by $\frac{5}{2}$.
$$\underline{2m^5 - m^4 + 2m^3}$$　　Terms of top row are multiplied by $\frac{1}{2}m^2$.
$$2m^5 - m^4 + 12m^3 - 5m^2 + 10m$$　　Add in columns.　　**NOW TRY**

We can use a rectangle to model polynomial multiplication. For example, to find

$$(2x + 1)(3x + 2),$$

we label a rectangle with each term, as shown below on the left. Then put the product of each pair of monomials in the appropriate box, as shown on the right.

NOW TRY ANSWERS
2. $2x^4 - 5x^3 - 5x^2 + 20x - 12$
3. $10t^3 - 44t^2 + 50t - 24$
4. $3x^5 - 4x^4 - 6x^3 + 9x^2 - 2$

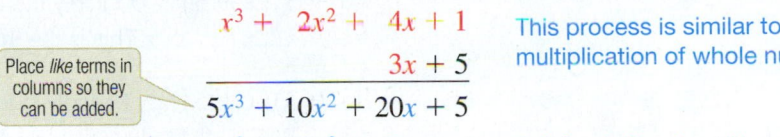

	3x	2
2x	6x²	4x
1	3x	2

The product of the binomials is the sum of the four monomial products.

$$(2x + 1)(3x + 2)$$

$$= 6x^2 + 4x + 3x + 2$$

$$= 6x^2 + 7x + 2 \qquad \text{Combine like terms.}$$

This approach can be extended to polynomials with any number of terms.

OBJECTIVE 3 Multiply binomials by the FOIL method.

When multiplying binomials, the **FOIL method** reduces the rectangle method to a systematic approach without the rectangle. Consider this example.

$$(x + 3)(x + 5)$$

$$= (x + 3)x + (x + 3)5 \qquad \text{Distributive property}$$

$$= x(x) + 3(x) + x(5) + 3(5) \qquad \text{Distributive property again}$$

$$= x^2 + 3x + 5x + 15 \qquad \text{Multiply.}$$

$$= x^2 + 8x + 15 \qquad \text{Combine like terms.}$$

The letters of the word FOIL refer to the positions of the terms.

$(x + 3)(x + 5)$ Multiply the **First terms:** $x(x)$.   **F**

$(x + 3)(x + 5)$ Multiply the **Outer terms:** $x(5)$.   **O**
This is the **outer product.**

$(x + 3)(x + 5)$ Multiply the **Inner terms:** $3(x)$.   **I**
This is the **inner product.**

$(x + 3)(x + 5)$ Multiply the **Last terms:** $3(5)$.   **L**

We add the outer product, $5x$, and the inner product, $3x$, to obtain $8x$ so that the three terms of the answer can be written without extra steps.

$$(x + 3)(x + 5)$$

$$= x^2 + 8x + 15$$

Multiplying Binomials by the FOIL Method

Step 1 Multiply the two **First** terms of the binomials to obtain the first term of the product.

Step 2 Find the **Outer** product and the **Inner** product and combine them (when possible) to obtain the middle term of the product.

Step 3 Multiply the two **Last** terms of the binomials to obtain the last term of the product.

$$\mathbf{F} = x^2 \qquad \mathbf{L} = 15$$

$$(x + 3)(x + 5) \qquad (x + 3)(x + 5)$$

$$\qquad\qquad\qquad\qquad\qquad = x^2 + 8x + 15$$

$$\mathbf{I} \qquad 3x$$
$$\mathbf{O} \qquad \underline{5x}$$
$$\qquad 8x \qquad \text{Combine like terms.}$$

Add the terms found in Steps 1–3.

NOW TRY
EXERCISE 5

Use the FOIL method to find the product.

$$(t - 6)(t + 5)$$

EXAMPLE 5 Using the FOIL Method

Use the FOIL method to find the product $(x + 8)(x - 6)$.

Step 1 **F** Multiply the First terms: $x(x) = x^2$.

Step 2 **O** Find the Outer product: $x(-6) = -6x$.

I Find the Inner product: $8(x) = 8x$.

Combine the outer and inner products mentally: $-6x + 8x = 2x$.

Step 3 **L** Multiply the Last terms: $8(-6) = -48$.

The product $(x + 8)(x - 6)$ is $x^2 + 2x - 48$. Add the terms found in Steps 1–3.

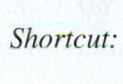 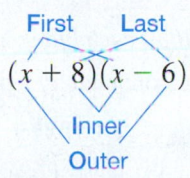 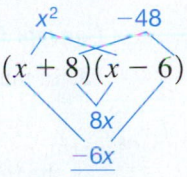

Shortcut:

$$(x + 8)(x - 6) \qquad (x + 8)(x - 6) \qquad (x + 8)(x - 6)$$
$$= x^2 + 2x - 48$$

Combine like terms. **NOW TRY**

NOW TRY
EXERCISE 6

Multiply.

$$(7y - 3)(2x + 5)$$

EXAMPLE 6 Using the FOIL Method

Multiply $(9x - 2)(3y + 1)$.

First	$(9x - 2)(3y + 1)$	$27xy$
Outer	$(9x - 2)(3y + 1)$	$9x$
Inner	$(9x - 2)(3y + 1)$	$-6y$
Last	$(9x - 2)(3y + 1)$	-2

These unlike terms *cannot* be combined.

$$\begin{array}{cccc} \text{F} & \text{O} & \text{I} & \text{L} \end{array}$$

The product $(9x - 2)(3y + 1)$ is $27xy + 9x - 6y - 2$. **NOW TRY**

NOW TRY
EXERCISE 7

Find each product.

(a) $(3p - 5q)(4p - q)$

(b) $5x^2(3x + 1)(x - 5)$

EXAMPLE 7 Using the FOIL Method

Find each product.

(a) $(2k + 5y)(k + 3y)$

$$\begin{array}{cccc} \text{F} & \text{O} & \text{I} & \text{L} \end{array}$$
$$= 2k(k) + 2k(3y) + 5y(k) + 5y(3y)$$
$$= 2k^2 + 6ky + 5ky + 15y^2 \qquad \text{Multiply.}$$
$$= 2k^2 + 11ky + 15y^2 \qquad \text{Combine like terms.}$$

(b) $(7p + 2q)(3p - q)$

$$= 21p^2 - pq - 2q^2 \qquad \begin{array}{l}\text{FOIL}\\ \text{method}\end{array}$$

(c) $2x^2(x - 3)(3x + 4)$

$$= 2x^2(3x^2 - 5x - 12) \qquad \text{FOIL method}$$
$$= 6x^4 - 10x^3 - 24x^2 \qquad \begin{array}{l}\text{Distributive}\\ \text{property}\end{array}$$

NOW TRY

NOTE Alternatively, the factors in **Example 7(c)** can be multiplied as follows.

$$2x^2(x - 3)(3x + 4) \qquad \text{Multiply } 2x^2 \text{ and } x - 3 \text{ first.}$$
$$= (2x^3 - 6x^2)(3x + 4) \qquad \text{Multiply that product and } 3x + 4.$$
$$= 6x^4 - 10x^3 - 24x^2 \qquad \text{The same answer results.}$$

4.5 Exercises

    ▶ MyMathLab®

▶ *Complete solution available in MyMathLab*

Concept Check   *In Exercises 1 and 2, match each product in Column I with the correct polynomial in Column II.*

I	II	I	II
1. (a) $5x^3(6x^7)$	**A.** $125x^{21}$	**2. (a)** $(x-5)(x+4)$	**A.** $x^2 + 9x + 20$
(b) $-5x^7(6x^3)$	**B.** $30x^{10}$	**(b)** $(x+5)(x+4)$	**B.** $x^2 - 9x + 20$
(c) $(5x^7)^3$	**C.** $-216x^9$	**(c)** $(x-5)(x-4)$	**C.** $x^2 - x - 20$
(d) $(-6x^3)^3$	**D.** $-30x^{10}$	**(d)** $(x+5)(x-4)$	**D.** $x^2 + x - 20$

Concept Check   *Fill in each blank with the correct response.*

3. In multiplying a monomial by a polynomial, such as in

$$4x(3x^2 + 7x^3) = 4x(3x^2) + 4x(7x^3),$$

the first property that is used is the _____ property.

4. The FOIL method can only be used to multiply two polynomials when both polynomials are _____ .

5. The product $2x^2(-3x^5)$ has exactly _____ term(s) after the multiplication is performed.

6. The product $(a + b)(c + d)$ has exactly _____ term(s) after the multiplication is performed.

Find each product. ***See Objective 1.***

7. $5y^4(3y^7)$ **8.** $10p^2(5p^3)$ **9.** $-15a^4(-2a^5)$

10. $-3m^6(-5m^4)$ **11.** $5p(3q^2)$ **12.** $4a^3(3b^2)$

13. $-6m^3(3n^2)$ **14.** $9r^3(-2s^2)$ **15.** $y^5 \cdot 9y \cdot y^4$

16. $x^2 \cdot 3x^3 \cdot 2x$ **17.** $(4x^3)(2x^2)(-x^5)$ **18.** $(7t^5)(3t^4)(-t^8)$

Find each product. ***See Example 1.***

▶ **19.** $2m(3m + 2)$ **20.** $4x(5x + 3)$ **21.** $3p(-2p^3 + 4p^2)$

22. $4x(3 + 2x + 5x^3)$ **23.** $-8z(2z + 3z^2 + 3z^3)$ **24.** $-7y(3 + 5y^2 - 2y^3)$

25. $2y^3(3 + 2y + 5y^4)$ **26.** $2m^4(6 + 5m + 3m^2)$

27. $-4r^3(-7r^2 + 8r - 9)$ **28.** $-9a^5(-3a^6 - 2a^4 + 8a^2)$

29. $3a^2(2a^2 - 4ab + 5b^2)$ **30.** $4z^3(8z^2 + 5zy - 3y^2)$

Concept Check   *Multiply.*

31. $7m^3n^2(3m^2 + 2mn - n^3)$

$= 7m^3n^2(\underline{\quad}) + 7m^3n^2(\underline{\quad})$
$+ 7m^3n^2(\underline{\quad})$
$= \underline{\hspace{3cm}}$

32. $2p^2q(3p^2q^2 - 5p + 2q^2)$

$= \underline{\quad}(3p^2q^2) + \underline{\quad}(-5p)$
$+ \underline{\quad}(2q^2)$
$= \underline{\hspace{3cm}}$

Find each product. ***See Examples 2–4.***

▶ **33.** $(6x + 1)(2x^2 + 4x + 1)$ **34.** $(9a + 2)(9a^2 + a + 1)$

35. $(9y - 2)(8y^2 - 6y + 1)$ **36.** $(2r - 1)(3r^2 + 4r - 4)$

▶ **37.** $(4m + 3)(5m^3 - 4m^2 + m - 5)$ **38.** $(2y + 8)(3y^4 - 2y^2 + 1)$

39. $(2x - 1)(3x^5 - 2x^3 + x^2 - 2x + 3)$ **40.** $(2a + 3)(a^4 - a^3 + a^2 - a + 1)$

41. $(5x^2 + 2x + 1)(x^2 - 3x + 5)$ **42.** $(2m^2 + m - 3)(m^2 - 4m + 5)$

▶ **43.** $(6x^4 - 4x^2 + 8x)\left(\dfrac{1}{2}x + 3\right)$ **44.** $(8y^6 + 4y^4 - 12y^2)\left(\dfrac{3}{4}y^2 + 2\right)$

Find each product using the rectangle method shown in the text. Determine the individual terms that should appear on the blanks or in the rectangles, and then give the final product.

45. $(x + 3)(x + 4)$

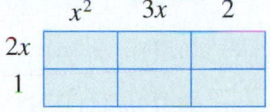

Product: _____

46. $(x + 5)(x + 2)$

Product: _____

47. $(2x + 1)(x^2 + 3x + 2)$

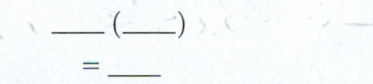

Product: _____

48. $(x + 4)(3x^2 + 2x + 1)$

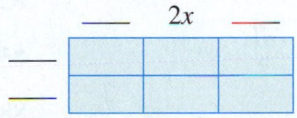

Product: _____

49. *Concept Check*   For the product

$$(2p - 5)(3p + 7),$$

find and simplify the following.

(a) Product of first terms

_____ (_____)

= _____

(b) Outer product

_____ (_____)

= _____

(c) Inner product

_____ (_____)

= _____

(d) Product of last terms

_____ (_____)

= _____

(e) Complete product in simplified form

_____

50. *Concept Check*   Repeat the process of **Exercise 49** for $(2p - 5)(2p + 5)$, and compare the results in parts (b) and (c). What do you notice? What is the complete product in simplified form?

Find each product. See Examples 5–7.

▶ **51.** $(m + 7)(m + 5)$ **52.** $(n + 9)(n + 3)$ **53.** $(n - 1)(n + 4)$

54. $(t - 3)(t + 8)$ **55.** $(2x + 3)(6x - 4)$ **56.** $(3y + 5)(8y - 6)$

57. $(9 + t)(9 - t)$ **58.** $(10 + r)(10 - r)$ **59.** $(3x - 2)(3x - 2)$

60. $(4m + 3)(4m + 3)$ **61.** $(5a + 1)(2a + 7)$ **62.** $(b + 8)(6b - 2)$

63. $(6 - 5m)(2 + 3m)$ **64.** $(8 - 3a)(2 + a)$ **65.** $(5 - 3x)(4 + x)$

66. $(6 - 5x)(2 + x)$ **67.** $(3t - 4s)(t + 3s)$ **68.** $(2m - 3n)(m + 5n)$

▶ **69.** $(4x + 3)(2y - 1)$ **70.** $(5x + 7)(3y - 8)$ ▶ **71.** $(3x + 2y)(5x - 3y)$

72. $(5a + 3b)(5a - 4b)$ **73.** $3y^3(2y + 3)(y - 5)$ **74.** $2x^2(2x - 5)(x + 3)$

75. $-8r^3(5r^2 + 2)(5r^2 - 2)$ **76.** $-5t^4(2t^4 + 1)(2t^4 - 1)$

Find polynomials that represent (a) the area and (b) the perimeter of each square or rectangle. (If necessary, refer to the formulas found inside the back cover of this book.)

77.

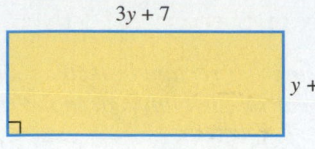

78.

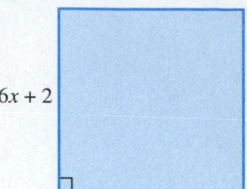

Extending Skills   Find each product. In Exercises 87–90, 95, and 96, apply the meaning of exponents.

79. $(x + 7)^2$ **80.** $(m + 6)^2$ **81.** $(a - 4)(a + 4)$

82. $(b - 10)(b + 10)$ **83.** $(2p - 5)^2$ **84.** $(3m - 1)^2$

85. $(5k + 3q)^2$ **86.** $(8m + 3n)^2$ **87.** $(m - 5)^3$

88. $(p - 3)^3$ **89.** $(2a + 1)^3$ **90.** $(3m + 1)^3$

91. $-3a(3a + 1)(a - 4)$ **92.** $-4r(3r + 2)(2r - 5)$

93. $7(4m - 3)(2m + 1)$ **94.** $5(3k - 7)(5k + 2)$

95. $(3r - 2s)^4$ **96.** $(2z - 5y)^4$

97. $3p^3(2p^2 + 5p)(p^3 + 2p + 1)$ **98.** $5k^2(k^3 - 3)(k^2 - k + 4)$

99. $-2x^5(3x^2 + 2x - 5)(4x + 2)$ **100.** $-4x^3(3x^4 + 2x^2 - x)(-2x + 1)$

101. $\left(3p^2 + \dfrac{5}{4}q\right)\left(2p^2 - \dfrac{5}{3}q\right)$ **102.** $\left(2x^2 + \dfrac{2}{3}y\right)\left(3x^2 - \dfrac{3}{4}y\right)$

The figures in Exercises 103–106 are composed of triangles, squares, rectangles, and circles. Find a polynomial that represents the area of each shaded region. In Exercises 105 and 106, leave π in the answers. (If necessary, refer to the formulas found inside the back cover of this book.)

103.

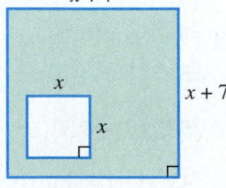

104.

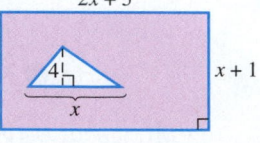

105.

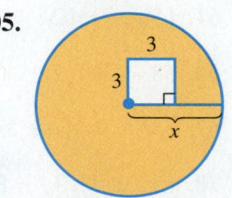

106.

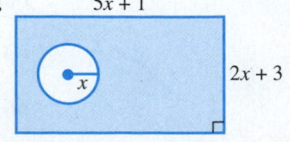

RELATING CONCEPTS For Individual or Group Work (Exercises 107–112)

Work Exercises 107–112 in order. (All units are in feet.)

107. Find a polynomial that represents the area, in square feet, of the rectangle.

108. Suppose we know that the area of the rectangle is 600 ft^2. Use this information and the polynomial from **Exercise 107** to write an equation in x, and solve it.

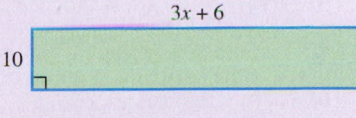

$3x + 6$

10

109. Refer to **Exercise 108**. What are the dimensions of the rectangle?

110. Use the result of **Exercise 109** to find the perimeter of the rectangle.

111. Suppose the rectangle represents a strip of lawn and it costs $0.75 per square foot to lay sod on the lawn. How much will it cost to sod the entire lawn?

112. Suppose it costs $20.50 per linear foot for fencing. How much will it cost to fence the entire lawn?

4.6 Special Products

OBJECTIVES

1. Square binomials.
2. Find the product of the sum and difference of two terms.
3. Find greater powers of binomials.

VOCABULARY

☐ conjugates

NOW TRY EXERCISE 1

Find $(x + 5)^2$.

OBJECTIVE 1 Square binomials.

EXAMPLE 1 Squaring a Binomial

Find $(m + 3)^2$.

$$(m + 3)(m + 3)$$

> $(m + 3)^2$ means $(m + 3)(m + 3)$.

$$= m^2 + 3m + 3m + 9 \qquad \text{FOIL method}$$

$$= m^2 + 6m + 9 \qquad \text{Combine like terms.}$$

This is the answer.

This result has the squares of the first and the last terms of the binomial.

$$m^2 = m^2 \quad \text{and} \quad 3^2 = 9$$

The middle term, 6m, is twice the product of the two terms of the binomial, because the outer and inner products are $m(3)$ and $3(m)$. Then we find their sum.

$$m(3) + 3(m)$$

$$= 2(m)(3)$$

$$= 6m$$

NOW TRY

Example 1 suggests the following rules.

Square of a Binomial

The square of a binomial is a trinomial consisting of

the square of the first term $+$ twice the product of the two terms $+$ the square of the last term.

For x and y, the following hold.

$$(x + y)^2 = x^2 + 2xy + y^2$$

$$(x - y)^2 = x^2 - 2xy + y^2$$

NOW TRY ANSWER

1. $x^2 + 10x + 25$

NOW TRY
EXERCISE 2

Square each binomial.

(a) $(3x - 1)^2$

(b) $(4p - 5q)^2$

(c) $\left(6t - \frac{1}{3}\right)^2$

(d) $-(3y + 2)^2$

(e) $m(2m + 3)^2$

EXAMPLE 2 **Squaring Binomials**

Square each binomial.

$$(x - y)^2 = x^2 - 2 \cdot x \cdot y + y^2$$
$$\downarrow \quad \downarrow \quad\quad \downarrow \quad\quad \downarrow \; \downarrow \quad \downarrow \quad\quad \downarrow$$

(a) $(5z - 1)^2 = (5z)^2 - 2(5z)(1) + (1)^2$

$ = 25z^2 - 10z + 1$ $(5z)^2 = 5^2z^2 = 25z^2$

(b) $(3b + 5r)^2$

> Be careful to square $3b$ and $5r$ correctly.

$ = (3b)^2 + 2(3b)(5r) + (5r)^2$

$ = 9b^2 + 30br + 25r^2$

(c) $(2a - 9x)^2$

$ = (2a)^2 - 2(2a)(9x) + (9x)^2$

$ = 4a^2 - 36ax + 81x^2$

(d) $\left(4m + \frac{1}{2}\right)^2$

$ = (4m)^2 + 2(4m)\left(\frac{1}{2}\right) + \left(\frac{1}{2}\right)^2$

$ = 16m^2 + 4m + \frac{1}{4}$

(e) $-(2x - 3)^2$

$ = -\left[(2x)^2 - 2(2x)(3) + 3^2\right]$

$ = -(4x^2 - 12x + 9)$

$ = -4x^2 + 12x - 9$

(f) $x(4x - 3)^2$ Remember the middle term.

$ = x(16x^2 - 24x + 9)$ Square the binomial.

$ = 16x^3 - 24x^2 + 9x$ Distributive property   **NOW TRY**

> *In the square of a sum, all of the terms are positive,* as in Examples 2(b) and (d). *In the square of a difference, the middle term is negative,* as in Examples 2(a), (c), and (f).

! CAUTION   A common error when squaring a binomial is to forget the middle term of the product. In general, remember the following.

$$(x + y)^2 = x^2 + 2xy + y^2, \quad \textit{not} \quad x^2 + y^2,$$

and

$$(x - y)^2 = x^2 - 2xy + y^2, \quad \textit{not} \quad x^2 - y^2.$$

OBJECTIVE 2 **Find the product of the sum and difference of two terms.**

In binomial products of the form $(x + y)(x - y)$, one binomial is a sum of two terms. The other is a difference of the *same* two terms. Consider the following.

$$(x + 2)(x - 2)$$

$ = x^2 - 2x + 2x - 4$ FOIL method

$ = x^2 - 4$ Combine like terms.

Thus, the product of $x + y$ and $x - y$ is a difference of two squares.

NOW TRY ANSWERS

2. **(a)** $9x^2 - 6x + 1$

 (b) $16p^2 - 40pq + 25q^2$

 (c) $36t^2 - 4t + \frac{1}{9}$

 (d) $-9y^2 - 12y - 4$

 (e) $4m^3 + 12m^2 + 9m$

Product of a Sum and Difference of Two Terms

$$(x + y)(x - y) = x^2 - y^2$$

NOTE The expressions $x + y$ and $x - y$, a sum and difference of the *same* two terms, are **conjugates**. In the preceding example, $x + 2$ and $x - 2$ are conjugates.

 **NOW TRY EXERCISE 3**

Find the product.

$(t + 10)(t - 10)$

EXAMPLE 3 Finding the Product of a Sum and Difference of Two Terms

Find each product.

(a)
$$(x + 4)(x - 4) \longleftarrow \text{This is a sum and difference of two terms.}$$
$$= x^2 - 4^2 \qquad (x + y)(x - y) = x^2 - y^2$$
$$= x^2 - 16 \qquad \text{Square 4.}$$

(b) $\left(\dfrac{2}{3} - w\right)\left(\dfrac{2}{3} + w\right)$

$$= \left(\frac{2}{3} + w\right)\left(\frac{2}{3} - w\right) \qquad \text{Commutative property}$$

$$= \left(\frac{2}{3}\right)^2 - w^2 \qquad (x + y)(x - y) = x^2 - y^2$$

$$= \frac{4}{9} - w^2 \qquad \text{Square } \frac{2}{3}.$$

(c) $x(x + 2)(x - 2)$

$$= x(x^2 - 4) \qquad \text{Find the product of the sum and difference of two terms.}$$
$$= x^3 - 4x \qquad \text{Distributive property} \qquad \text{NOW TRY} \ \text{}$$

 **NOW TRY EXERCISE 4**

Find each product.

(a) $(4x - 6)(4x + 6)$

(b) $\left(5r - \dfrac{4}{5}\right)\left(5r + \dfrac{4}{5}\right)$

(c) $y(3y + 1)(3y - 1)$

(d) $-5(p + q^2)(p - q^2)$

EXAMPLE 4 Finding the Product of a Sum and Difference of Two Terms

Find each product.

(a)
$$(x \ + \ y) \ (x \ - \ y)$$
$$\downarrow \quad \downarrow \quad \downarrow \quad \downarrow$$
$$(5m + 3)(5m - 3) \longleftarrow \text{This indicates the product of a sum and difference of two terms.}$$
$$= (5m)^2 - 3^2 \qquad (x + y)(x - y) = x^2 - y^2$$
$$= 25m^2 - 9 \qquad \text{Apply the exponents.}$$

> Be careful to square 5m correctly.

(b) $(4x + y)(4x - y)$
$$= (4x)^2 - y^2$$
$$= 16x^2 - y^2$$

(c) $\left(z - \dfrac{1}{4}\right)\left(z + \dfrac{1}{4}\right)$
$$= z^2 - \frac{1}{16}$$

(d) $p(2p + 1)(2p - 1)$
$$= p(4p^2 - 1)$$
$$= 4p^3 - p \qquad \begin{array}{l}\text{Distributive}\\\text{property}\end{array}$$

(e) $-3(x + y^2)(x - y^2)$
$$= -3(x^2 - y^4)$$
$$= -3x^2 + 3y^4$$

NOW TRY

NOW TRY ANSWERS

3. $t^2 - 100$

4. (a) $16x^2 - 36$

  (b) $25r^2 - \dfrac{16}{25}$

  (c) $9y^3 - y$

  (d) $-5p^2 + 5q^4$

OBJECTIVE 3 Find greater powers of binomials.

The methods used in the previous section and this section can be combined to find greater powers of binomials.

**NOW TRY
EXERCISE 5**

Find the product.

$(2m - 1)^3$

EXAMPLE 5   Finding Greater Powers of Binomials

Find each product.

(a) $(x + 5)^3$

$$= (x + 5)(x + 5)^2 \qquad\qquad a^3 = a \cdot a^2$$

$$= (x + 5)(x^2 + 10x + 25) \qquad \text{Square the binomial.}$$

$$= x^3 + 10x^2 + 25x + 5x^2 + 50x + 125 \qquad \text{Multiply polynomials.}$$

$$= x^3 + 15x^2 + 75x + 125 \qquad \text{Combine like terms.}$$

(b) $(2y - 3)^4$

$$= (2y - 3)^2(2y - 3)^2 \qquad\qquad a^4 = a^2 \cdot a^2$$

$$= (4y^2 - 12y + 9)(4y^2 - 12y + 9) \qquad \text{Square each binomial.}$$

$$= 16y^4 - 48y^3 + 36y^2 - 48y^3 + 144y^2 \qquad \text{Multiply polynomials.}$$

$$\quad - 108y + 36y^2 - 108y + 81$$

$$= 16y^4 - 96y^3 + 216y^2 - 216y + 81 \qquad \text{Combine like terms.}$$

(c) $-2r(r + 2)^3$

$$= -2r(r + 2)(r + 2)^2 \qquad\qquad a^3 = a \cdot a^2$$

$$= -2r(r + 2)(r^2 + 4r + 4) \qquad \text{Square the binomial.}$$

$$= -2r(r^3 + 4r^2 + 4r + 2r^2 + 8r + 8) \qquad \text{Multiply polynomials.}$$

$$= -2r(r^3 + 6r^2 + 12r + 8) \qquad \text{Combine like terms.}$$

$$= -2r^4 - 12r^3 - 24r^2 - 16r \qquad \text{Multiply.}$$

NOW TRY

NOW TRY ANSWER

5. $8m^3 - 12m^2 + 6m - 1$

4.6 Exercises

FOR EXTRA HELP

 MyMathLab®

● *Complete solution available in MyMathLab*

1. *Concept Check*   Consider the square of the binomial $4x + 3$:

$$(4x + 3)^2.$$

(a) What is the first term of the binomial? Square it.

(b) Find twice the product of the two terms of the binomial: $2(\underline{\quad})(\underline{\quad}) = \underline{\quad}$.

(c) What is the last term of the binomial? Square it.

(d) Use the results of parts (a)–(c) to find $(4x + 3)^2$.

2. *Concept Check*   Consider the product of $(7x + 3y)$ and $(7x - 3y)$:

$$(7x + 3y)(7x - 3y).$$

(a) What is the first term of each binomial factor? Square it.

(b) What is the product of the outer terms? The inner terms? Add them.

(c) What are the last terms of the binomial factors? Multiply them.

(d) Use the results of parts (a)–(c) to find $(7x + 3y)(7x - 3y)$.

Find each product. **See Examples 1 and 2.**

3. $(m + 2)^2$ **4.** $(x + 8)^2$ **5.** $(r - 3)^2$ **6.** $(z - 5)^2$

● **7.** $(x + 2y)^2$ **8.** $(p - 3m)^2$ **9.** $(5p + 2q)^2$ **10.** $(8a + 3b)^2$

11. $(4x - 3)^2$ **12.** $(9x - 4)^2$ **13.** $(4a + 5b)^2$ **14.** $(9y + 4z)^2$

15. $\left(6m - \dfrac{4}{5}n\right)^2$ **16.** $\left(5x + \dfrac{2}{5}y\right)^2$ **17.** $\left(\dfrac{1}{2}x + \dfrac{1}{3}\right)^2$ **18.** $\left(\dfrac{1}{4}x + \dfrac{1}{5}\right)^2$

19. $2(x + 6)^2$ **20.** $4(x + 3)^2$ **21.** $t(3t - 1)^2$ **22.** $x(2x + 5)^2$

23. $3t(4t + 1)^2$ **24.** $2x(7x - 2)^2$ **25.** $-(4r - 2)^2$ **26.** $-(3y - 8)^2$

Find each product. See Examples 3 and 4.

27. $(k + 5)(k - 5)$ **28.** $(a + 8)(a - 8)$ **29.** $(4 - 3t)(4 + 3t)$

30. $(7 - 2x)(7 + 2x)$ **31.** $(5x + 2)(5x - 2)$ **32.** $(2m + 5)(2m - 5)$

33. $(5y + 3x)(5y - 3x)$ **34.** $(3x + 4y)(3x - 4y)$ **35.** $(10x + 3y)(10x - 3y)$

36. $(13r + 2z)(13r - 2z)$ **37.** $(2x^2 - 5)(2x^2 + 5)$ **38.** $(9y^2 - 2)(9y^2 + 2)$

39. $\left(\dfrac{3}{4} - x\right)\left(\dfrac{3}{4} + x\right)$ **40.** $\left(\dfrac{2}{3} + r\right)\left(\dfrac{2}{3} - r\right)$ **41.** $\left(9y + \dfrac{2}{3}\right)\left(9y - \dfrac{2}{3}\right)$

42. $\left(7x + \dfrac{3}{7}\right)\left(7x - \dfrac{3}{7}\right)$ **43.** $q(5q - 1)(5q + 1)$ **44.** $p(3p + 7)(3p - 7)$

45. $-5(a - b^3)(a + b^3)$ **46.** $-6(r - s^4)(r + s^4)$

47. $\dfrac{1}{2}(2k - 1)(2k + 1)$ **48.** $\dfrac{1}{3}(3m - 5)(3m + 5)$

49. $-\dfrac{1}{100}(10x + 10)(10x - 10)$ **50.** $-\dfrac{1}{200}(20y + 20)(20y - 20)$

Find each product. See Example 5.

51. $(x + 1)^3$ **52.** $(y + 2)^3$ **53.** $(t - 3)^3$ **54.** $(m - 5)^3$

55. $(r + 5)^3$ **56.** $(p + 3)^3$ **57.** $(2a + 1)^3$ **58.** $(3m + 1)^3$

59. $(4x - 1)^4$ **60.** $(2x - 1)^4$ **61.** $(3r - 2t)^4$ **62.** $(2z + 5y)^4$

63. $2x(x + 1)^3$ **64.** $3y(y + 2)^3$ **65.** $-4t(t + 3)^3$

66. $-5r(r + 1)^3$ **67.** $(x + y)^2(x - y)^2$ **68.** $(s + 2)^2(s - 2)^2$

69. *Concept Check*   Does $(a + b)^n = a^n + b^n$ hold true in general?

70. *Concept Check*   Give values for a, b, and n for which $(a + b)^n = a^n + b^n$ is true.

The special product $(x + y)(x - y) = x^2 - y^2$ *can be used to perform some multiplication problems. Here are two examples.*

$$
\begin{array}{c|c}
51 \times 49 = (50 + 1)(50 - 1) & 102 \times 98 = (100 + 2)(100 - 2) \\
= 50^2 - 1^2 & = 100^2 - 2^2 \\
= 2500 - 1 & = 10{,}000 - 4 \\
= 2499 & = 9996
\end{array}
$$

Once these patterns are recognized, multiplications of this type can be done mentally. Use this method to calculate each product mentally.

71. 101×99 **72.** 103×97 **73.** 201×199

74. 301×299 **75.** $20\dfrac{1}{2} \times 19\dfrac{1}{2}$ **76.** $30\dfrac{1}{3} \times 29\dfrac{2}{3}$

Find a polynomial that represents the area of each figure. (If necessary, refer to the formulas found inside the back cover of this book.)

77.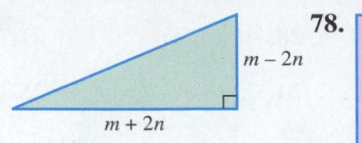

78.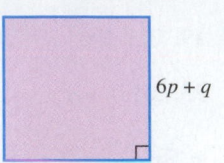

79.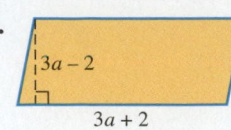

80.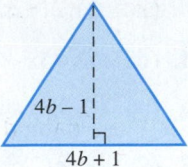

81.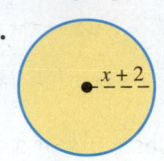

82.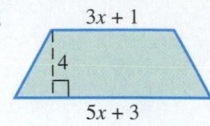

In Exercises 83 and 84, refer to the figure shown here.

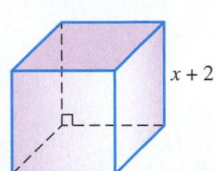

83. Find a polynomial that represents the volume of the cube (in cubic units).

84. If the value of x is 6, what is the volume of the cube (in cubic units)?

RELATING CONCEPTS For Individual or Group Work (Exercises 85–94)

Special products can be illustrated by using areas of rectangles. Use the figure, and ***work Exercises 85–90 in order*** *to justify*

$$(a + b)^2 = a^2 + 2ab + b^2.$$

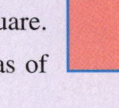

85. Express the area of the large square as the square of a binomial.

86. Give the monomial that represents the area of the red square.

87. Give the monomial that represents the sum of the areas of the blue rectangles.

88. Give the monomial that represents the area of the yellow square.

89. What is the sum of the monomials obtained in **Exercises 86–88?**

90. Explain why the binomial square found in **Exercise 85** must equal the polynomial found in **Exercise 89.**

To understand how the special product $(a + b)^2 = a^2 + 2ab + b^2$ can be applied to a purely numerical problem, ***work Exercises 91–94 in order.***

91. Evaluate 35^2, using either traditional paper-and-pencil methods or a calculator.

92. The number 35 can be written as $30 + 5$. Therefore, $35^2 = (30 + 5)^2$. Use the special product for squaring a binomial with $a = 30$ and $b = 5$ to write an expression for $(30 + 5)^2$. Do not simplify at this time.

93. Use the order of operations to simplify the expression found in **Exercise 92.**

94. How do the answers in **Exercises 91 and 94** compare?

4.7 Dividing Polynomials

OBJECTIVES

1. Divide a polynomial by a monomial.
2. Divide a polynomial by a polynomial.
3. Use division in a geometry application.

OBJECTIVE 1 Divide a polynomial by a monomial.

We add two fractions with a common denominator as follows.

$$\frac{a}{c} + \frac{b}{c} = \frac{a + b}{c}$$

In reverse, this statement gives a rule for dividing a polynomial by a monomial.

Dividing a Polynomial by a Monomial

To divide a polynomial by a monomial, divide each term of the polynomial by the monomial.

$$\frac{a + b}{c} = \frac{a}{c} + \frac{b}{c} \quad (\text{where } c \neq 0)$$

Examples: $\quad \dfrac{2 + 5}{3} = \dfrac{2}{3} + \dfrac{5}{3} \quad$ and $\quad \dfrac{x + 3z}{2y} = \dfrac{x}{2y} + \dfrac{3z}{2y}$

The parts of a division problem are named as follows.

Dividend $\longrightarrow$ $\dfrac{12x^2 + 6x}{6x} = 2x + 1 \longleftarrow$ Quotient

Divisor $\longrightarrow$

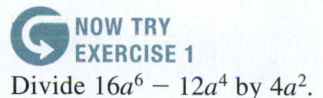
**NOW TRY
EXERCISE 1**

Divide $16a^6 - 12a^4$ by $4a^2$.

EXAMPLE 1   Dividing a Polynomial by a Monomial

Divide $5m^5 - 10m^3$ by $5m^2$.

$$\frac{5m^5 - 10m^3}{5m^2} \quad \text{A fraction bar means division.}$$

$$= \frac{5m^5}{5m^2} - \frac{10m^3}{5m^2} \quad \text{Use the preceding rule, with } + \text{ replaced by } -.$$

$$= m^3 - 2m \quad \text{Quotient rule}$$

CHECK   Multiply $\quad 5m^2 \cdot (m^3 - 2m) = 5m^5 - 10m^3 \quad$ ✓

         Divisor   Quotient   Original polynomial (Dividend)

Because division by 0 is undefined, the quotient $\frac{5m^5 - 10m^3}{5m^2}$ is undefined if $5m^2 = 0$, or $m = 0$. From now on, we assume that no denominators are 0.   **NOW TRY**

**NOW TRY
EXERCISE 2**

Divide.

$$\frac{36x^5 + 24x^4 - 12x^3}{6x^4}$$

EXAMPLE 2   Dividing a Polynomial by a Monomial

Divide.

$$\frac{16a^5 - 12a^4 + 8a^2}{4a^3} \quad \text{This becomes } \frac{2}{a}, \text{ not } 2a.$$

$$= \frac{16a^5}{4a^3} - \frac{12a^4}{4a^3} + \frac{8a^2}{4a^3} \quad \text{Divide each term by } 4a^3.$$

$$= 4a^2 - 3a + \frac{2}{a} \quad \text{Quotient rule}$$

The quotient $4a^2 - 3a + \frac{2}{a}$ is *not* a polynomial because $\frac{2}{a}$ has a variable in the denominator. While the sum, difference, and product of two polynomials are always polynomials, the quotient of two polynomials may not be a polynomial.

CHECK $4a^3 \left(4a^2 - 3a + \dfrac{2}{a} \right) \quad$ Divisor $\times$ Quotient should equal Dividend.

NOW TRY ANSWERS

1. $4a^4 - 3a^2$

2. $6x + 4 - \dfrac{2}{x}$

$$= 4a^3(4a^2) + 4a^3(-3a) + 4a^3\left(\frac{2}{a}\right) \quad \text{Distributive property}$$

$$= 16a^5 - 12a^4 + 8a^2 \text{ ✓} \quad \text{Dividend} \quad\quad \text{NOW TRY}$$

⚠ **CAUTION**   The most frequent error in a problem like that in **Example 2** is with the last term of the quotient.

$$\frac{8a^2}{4a^3} = \frac{8}{4}a^{2-3} = 2a^{-1} = 2\left(\frac{1}{a}\right) = \frac{2}{a}$$

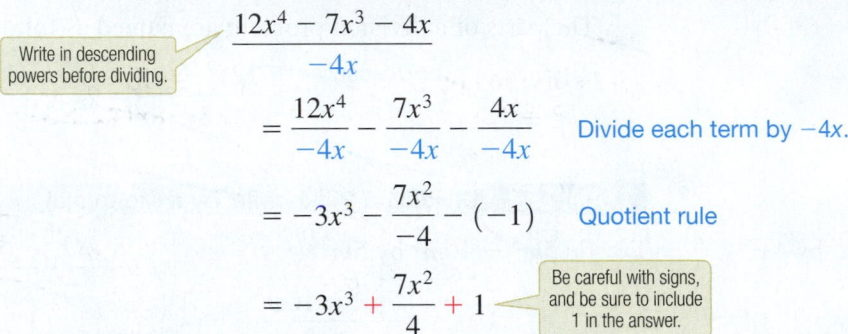

NOW TRY EXERCISE 3

Divide $7y^4 - 40y^5 + 100y^2$ by $-5y^2$.

EXAMPLE 3   Dividing a Polynomial by a Monomial (Negative Coefficient)

Divide $-7x^3 + 12x^4 - 4x$ by $-4x$.

Write the dividend polynomial in descending powers as $12x^4 - 7x^3 - 4x$.

$$\frac{12x^4 - 7x^3 - 4x}{-4x}$$

Write in descending powers before dividing.

$$= \frac{12x^4}{-4x} - \frac{7x^3}{-4x} - \frac{4x}{-4x} \qquad \text{Divide each term by } -4x.$$

$$= -3x^3 - \frac{7x^2}{-4} - (-1) \qquad \text{Quotient rule}$$

$$= -3x^3 + \frac{7x^2}{4} + 1 \qquad \text{Be careful with signs, and be sure to include 1 in the answer.}$$

Check by multiplying this quotient by $-4x$ to obtain the dividend.   **NOW TRY** ↻

NOW TRY EXERCISE 4

Divide $35m^5n^4 - 49m^2n^3 + 12mn$ by $7m^2n$.

EXAMPLE 4   Dividing a Polynomial by a Monomial

Divide $180x^4y^{10} - 150x^3y^8 + 120x^2y^6 - 90xy^4 + 100y$ by $30xy^2$.

$$\frac{180x^4y^{10} - 150x^3y^8 + 120x^2y^6 - 90xy^4 + 100y}{30xy^2}$$

$$= \frac{180x^4y^{10}}{30xy^2} - \frac{150x^3y^8}{30xy^2} + \frac{120x^2y^6}{30xy^2} - \frac{90xy^4}{30xy^2} + \frac{100y}{30xy^2}$$

$$= 6x^3y^8 - 5x^2y^6 + 4xy^4 - 3y^2 + \frac{10}{3xy} \qquad \textbf{NOW TRY} ↻$$

OBJECTIVE 2   Divide a polynomial by a polynomial.

We use a method of "long division" to divide a polynomial by a polynomial (other than a monomial). ***Both polynomials must first be written in descending powers.***

Dividing Whole Numbers	Dividing Polynomials
Step 1 Divide 6696 by 27. $27\overline{)6696}$	Divide $8x^3 - 4x^2 - 14x + 15$ by $2x + 3$. $2x + 3\overline{)8x^3 - 4x^2 - 14x + 15}$
Step 2 66 divided by 27 = 2. $2 \cdot 27 = 54$ 2 $27\overline{)6696}$ 54	$8x^3$ divided by $2x = 4x^2$. $4x^2(2x + 3) = 8x^3 + 12x^2$ $4x^2$ $2x + 3\overline{)8x^3 - 4x^2 - 14x + 15}$ $8x^3 + 12x^2$

NOW TRY ANSWERS

3. $8y^3 - \dfrac{7y^2}{5} - 20$

4. $5m^3n^3 - 7n^2 + \dfrac{12}{7m}$

Step 3

Subtract. Then bring down the next digit.

$$\begin{array}{r} 2 \\ 27\overline{)6696} \\ 54\downarrow \\ \hline 129 \end{array}$$

Subtract. Then bring down the next term.

$$\begin{array}{r} 4x^2 \\ 2x+3\overline{)8x^3 - 4x^2 - 14x + 15} \\ 8x^3 + 12x^2\downarrow \\ \hline -16x^2 - 14x \end{array}$$

(To subtract two polynomials, change the signs of the second and then add.)

Step 4

129 divided by 27 = 4.
4 · 27 = 108

$$\begin{array}{r} 24 \\ 27\overline{)6696} \\ 54 \\ \hline 129 \\ 108 \\ \hline \end{array}$$

$-16x^2$ divided by $2x = -8x$.
$-8x(2x + 3) = -16x^2 - 24x$

$$\begin{array}{r} 4x^2 - 8x \\ 2x+3\overline{)8x^3 - 4x^2 - 14x + 15} \\ 8x^3 + 12x^2 \\ \hline -16x^2 - 14x \\ -16x^2 - 24x \\ \hline \end{array}$$

Step 5

Subtract. Then bring down the next digit.

$$\begin{array}{r} 24 \\ 27\overline{)6696} \\ 54 \\ \hline 129 \\ 108\downarrow \\ \hline 216 \end{array}$$

Subtract. Then bring down the next term.

$$\begin{array}{r} 4x^2 - 8x \\ 2x+3\overline{)8x^3 - 4x^2 - 14x + 15} \\ 8x^3 + 12x^2 \\ \hline -16x^2 - 14x \\ -16x^2 - 24x\downarrow \\ \hline 10x + 15 \end{array}$$

Step 6

216 divided by 27 = 8.
8 · 27 = 216

$$\begin{array}{r} 248 \\ 27\overline{)6696} \\ 54 \\ \hline 129 \\ 108 \\ \hline 216 \\ 216 \\ \hline \end{array}$$

Remainder $\longrightarrow 0$

6696 divided by 27 is 248. The remainder is 0.

$10x$ divided by $2x = 5$.
$5(2x + 3) = 10x + 15$

$$\begin{array}{r} 4x^2 - 8x + 5 \\ 2x+3\overline{)8x^3 - 4x^2 - 14x + 15} \\ 8x^3 + 12x^2 \\ \hline -16x^2 - 14x \\ -16x^2 - 24x \\ \hline 10x + 15 \\ 10x + 15 \\ \hline \end{array}$$

Remainder $\longrightarrow 0$

$8x^3 - 4x^2 - 14x + 15$ divided by $2x + 3$ is $4x^2 - 8x + 5$. The remainder is 0.

Step 7

CHECK Multiply.

$$27 \cdot 248 = 6696 \quad \checkmark$$

CHECK Multiply.

$$(2x + 3)(4x^2 - 8x + 5)$$
$$= 8x^3 - 4x^2 - 14x + 15 \quad \checkmark$$

**NOW TRY
EXERCISE 5**

Divide.

$$\frac{4x^2 + x - 18}{x - 2}$$

EXAMPLE 5 Dividing a Polynomial by a Polynomial

Divide. $\dfrac{3x^2 - 5x - 28}{x - 4}$

Step 1 $3x^2$ divided by x is $3x$.
$3x(x - 4) = 3x^2 - 12x$

Step 2 Subtract $3x^2 - 12x$ from $3x^2 - 5x$. Bring down -28.

Step 3 $7x$ divided by x is 7.
$7(x - 4) = 7x - 28$

Divisor

$$
\begin{array}{r}
3x + 7 \quad \leftarrow \text{Quotient}\\
x - 4 \overline{)3x^2 - 5x - 28} \quad \leftarrow \text{Dividend}\\
\underline{3x^2 - 12x}\\
7x - 28\\
\underline{7x - 28}\\
0
\end{array}
$$

Step 4 Subtract $7x - 28$ from $7x - 28$. The remainder is 0.

CHECK Multiply the divisor, $x - 4$, by the quotient, $3x + 7$. The product must be the original dividend, $3x^2 - 5x - 28$.

$$\underset{\text{Divisor} \quad \text{Quotient}}{(x - 4)(3x + 7)} = 3x^2 + 7x - 12x - 28$$
$$= \underset{\text{Dividend}}{3x^2 - 5x - 28} \checkmark$$

NOW TRY

EXAMPLE 6 Dividing a Polynomial by a Polynomial

Divide. $\dfrac{5x + 4x^3 - 8 - 4x^2}{2x - 1}$

The first polynomial must be written in descending powers as $4x^3 - 4x^2 + 5x - 8$. Then divide by $2x - 1$.

$$
\begin{array}{r}
2x^2 - x + 2\\
2x - 1 \overline{)4x^3 - 4x^2 + 5x - 8}\\
\underline{4x^3 - 2x^2}\\
-2x^2 + 5x\\
\underline{-2x^2 + x}\\
4x - 8\\
\underline{4x - 2}\\
-6 \quad \leftarrow \text{Remainder}
\end{array}
$$

Write in descending powers.

In each subtraction, add the opposite.

Step 1 $4x^3$ divided by $2x$ is $2x^2$. $2x^2(2x - 1) = 4x^3 - 2x^2$

Step 2 Subtract. Bring down the next term.

Step 3 $-2x^2$ divided by $2x$ is $-x$. $-x(2x - 1) = -2x^2 + x$

Step 4 Subtract. Bring down the next term.

Step 5 $4x$ divided by $2x$ is 2. $2(2x - 1) = 4x - 2$

Step 6 Subtract. The remainder is -6. Write the remainder as the numerator of a fraction that has $2x - 1$ as its denominator. Because there is a nonzero remainder, the answer is not a polynomial.

Remember to add $\frac{\text{remainder}}{\text{divisor}}$. Don't forget the + sign.

$$\underset{\text{Divisor} \longrightarrow}{\overset{\text{Dividend} \longrightarrow}{\frac{4x^3 - 4x^2 + 5x - 8}{2x - 1}}} = \underset{\substack{\text{Quotient} \\ \text{polynomial}}}{2x^2 - x + 2} + \underset{\substack{\text{Fractional part} \\ \text{of quotient}}}{\frac{\overset{\text{Remainder}}{-6}}{2x - 1}} \underset{\longleftarrow \text{Divisor}}{}$$

NOW TRY ANSWER
5. $4x + 9$

NOW TRY
EXERCISE 6

Divide.

$$\frac{6k^3 - 20k - k^2 + 1}{2k - 3}$$

Step 7 CHECK

$$(2x - 1)\left(2x^2 - x + 2 + \frac{-6}{2x - 1}\right)$$ Multiply Divisor × (Quotient including the Remainder).

$$= (2x - 1)(2x^2) + (2x - 1)(-x) + (2x - 1)(2) + (2x - 1)\left(\frac{-6}{2x - 1}\right)$$

$$= 4x^3 - 2x^2 - 2x^2 + x + 4x - 2 - 6$$

$$= 4x^3 - 4x^2 + 5x - 8 \checkmark$$

NOW TRY

NOW TRY
EXERCISE 7

Divide $m^3 - 1000$ by $m - 10$.

EXAMPLE 7 Dividing into a Polynomial with Missing Terms

Divide $x^3 - 1$ by $x - 1$.

Here, the dividend, $x^3 - 1$, is missing the x^2-term and the x-term. We use 0 as the coefficient for each missing term. Thus, $x^3 - 1 = x^3 + 0x^2 + 0x - 1$.

$$\require{enclose}
\begin{array}{r}
x^2 + x + 1 \\
x - 1 \enclose{longdiv}{x^3 + 0x^2 + 0x - 1} \\
\underline{x^3 - x^2} \\
x^2 + 0x \\
\underline{x^2 - x} \\
x - 1 \\
\underline{x - 1} \\
0
\end{array}$$

Insert placeholders for the missing terms.

The remainder is 0. The quotient is $x^2 + x + 1$.

CHECK $(x - 1)(x^2 + x + 1)$

$$= x^3 + x^2 + x - x^2 - x - 1$$

$$= x^3 - 1 \checkmark \quad \text{Divisor × Quotient = Dividend}$$

NOW TRY

NOW TRY
EXERCISE 8

Divide

$$y^4 - 5y^3 + 6y^2 + y - 4$$

by $y^2 + 2$.

EXAMPLE 8 Dividing by a Polynomial with Missing Terms

Divide $x^4 + 2x^3 + 2x^2 - x - 1$ by $x^2 + 1$.

Because the divisor, $x^2 + 1$, has a missing x-term, write it as $x^2 + 0x + 1$.

$$\require{enclose}
\begin{array}{r}
x^2 + 2x + 1 \\
x^2 + 0x + 1 \enclose{longdiv}{x^4 + 2x^3 + 2x^2 - x - 1} \\
\underline{x^4 + 0x^3 + x^2} \\
2x^3 + x^2 - x \\
\underline{2x^3 + 0x^2 + 2x} \\
x^2 - 3x - 1 \\
\underline{x^2 + 0x + 1} \\
-3x - 2 \longleftarrow \text{Remainder}
\end{array}$$

Insert a placeholder for the missing term.

When the result of subtracting ($-3x - 2$ here) is a constant or a polynomial of degree less than the divisor ($x^2 + 0x + 1$), that constant or polynomial is the remainder. We write the answer as follows.

NOW TRY ANSWERS

6. $3k^2 + 4k - 4 + \dfrac{-11}{2k - 3}$

7. $m^2 + 10m + 100$

8. $y^2 - 5y + 4 + \dfrac{11y - 12}{y^2 + 2}$

$$x^2 + 2x + 1 + \frac{-3x - 2}{x^2 + 1}$$ Remember to write "$+ \frac{\text{remainder}}{\text{divisor}}$"

Multiply to check that this is correct.

NOW TRY

**NOW TRY
EXERCISE 9**

Divide $10x^3 + 21x^2 + 5x - 8$
by $2x + 4$.

EXAMPLE 9 Dividing a Polynomial When the Quotient Has Fractional Coefficients

Divide $4x^3 + 2x^2 + 3x + 2$ by $4x - 4$.

$$
\frac{6x^2}{4x} = \frac{3}{2}x
$$
$$
\frac{9x}{4x} = \frac{9}{4}
$$

$$
\begin{array}{r}
x^2 + \frac{3}{2}x + \frac{9}{4} \\
4x - 4 \overline{)4x^3 + 2x^2 + 3x + 2} \\
\underline{4x^3 - 4x^2} \\
6x^2 + 3x \\
\underline{6x^2 - 6x} \\
9x + 2 \\
\underline{9x - 9} \\
11
\end{array}
$$

The answer is $x^2 + \frac{3}{2}x + \frac{9}{4} + \frac{11}{4x - 4}$.

NOW TRY

OBJECTIVE 3 Use division in a geometry application.

**NOW TRY
EXERCISE 10**

The area of a rectangle is given by $(x^3 + 7x^2 + 17x + 20)$ sq. units. The width is given by $(x + 4)$ units. What is its length?

EXAMPLE 10 Using an Area Formula

The area of the rectangle in **FIGURE 5** is given by $(x^3 + 4x^2 + 8x + 8)$ sq. units. The width is given by $(x + 2)$ units. What is its length?

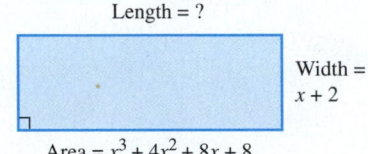

Length = ?

Width = $x + 2$

Area = $x^3 + 4x^2 + 8x + 8$

FIGURE 5

For a rectangle, $\mathcal{A} = LW$. Solving for L gives $L = \frac{\mathcal{A}}{W}$. Divide the area, $x^3 + 4x^2 + 8x + 8$, by the width, $x + 2$, to find the length.

$$
\begin{array}{r}
x^2 + 2x + 4 \\
x + 2 \overline{)x^3 + 4x^2 + 8x + 8} \\
\underline{x^3 + 2x^2} \\
2x^2 + 8x \\
\underline{2x^2 + 4x} \\
4x + 8 \\
\underline{4x + 8} \\
0
\end{array}
$$

NOW TRY ANSWERS

9. $5x^2 + \frac{1}{2}x + \frac{3}{2} + \frac{-14}{2x + 4}$

10. $(x^2 + 3x + 5)$ units

The quotient $(x^2 + 2x + 4)$ units represents the length.

NOW TRY

4.7 Exercises

 MyMathLab®

 Complete solution available in MyMathLab

Concept Check Complete each statement.

1. In the statement $\frac{10x^2 + 8}{2} = 5x^2 + 4$, _____ is the dividend, _____ is the divisor, and _____ is the quotient.

2. The expression $\frac{3x + 13}{x}$ is undefined for $x = $ _____ .

3. To check the division shown in **Exercise 1,** multiply _____ by _____ and show that the product is _____ .

4. The expression $5x^2 - 4x + 6 + \frac{2}{x}$ *(is / is not)* a polynomial.

Concept Check *Divide.*

5. $\dfrac{6p^4 + 18p^7}{3p^2}$

$= \dfrac{\overline{}}{3p^2} + \dfrac{\overline{}}{3p^2}$

$= \underline{}$

6. $\dfrac{20x^4 - 25x^3 + 5x}{5x^2}$

$= \dfrac{\overline{}}{\underline{}} - \dfrac{\overline{}}{\underline{}} + \dfrac{\overline{}}{\underline{}}$

$= \underline{}$

Perform each division. See Examples 1–3.

7. $\dfrac{60x^4 - 20x^2 + 10x}{2x}$

8. $\dfrac{120x^6 - 60x^3 + 80x^2}{2x}$

9. $\dfrac{20m^5 - 10m^4 + 5m^2}{5m^2}$

10. $\dfrac{12t^5 - 6t^3 + 6t^2}{6t^2}$

11. $\dfrac{8t^5 - 4t^3 + 4t^2}{2t}$

12. $\dfrac{8r^4 - 4r^3 + 6r^2}{2r}$

▶ **13.** $\dfrac{4a^5 - 4a^2 + 8}{4a}$

14. $\dfrac{5t^8 + 5t^7 + 15}{5t}$

15. $\dfrac{18p^5 + 12p^3 - 6p^2}{-6p^3}$

16. $\dfrac{32x^8 + 24x^5 - 8x}{-8x^2}$

17. $\dfrac{-7r^7 + 6r^5 - r^4}{-r^5}$

18. $\dfrac{-13t^9 + 8t^6 - t^5}{-t^6}$

Divide each polynomial by $3x^2$. See Examples 1–3.

▶ **19.** $12x^5 - 9x^4 + 6x^3$

20. $24x^6 - 12x^5 + 30x^4$

21. $3x^2 + 15x^3 - 27x^4$

22. $3x^2 - 18x^4 + 30x^5$

23. $36x + 24x^2 + 6x^3$

24. $9x - 12x^2 + 9x^3$

25. $4x^4 + 3x^3 + 2x$

26. $5x^4 - 6x^3 + 8x$

27. $-81x^5 + 30x^4 + 12x^2$

28. *Concept Check* If $-60x^5 - 30x^4 + 20x^3$ is divided by $3x^2$, what is the sum of the coefficients of the third- and second-degree terms in the quotient?

Perform each division. See Examples 1–4.

▶ **29.** $\dfrac{-27r^4 + 36r^3 - 6r^2 - 26r + 2}{-3r}$

30. $\dfrac{-8k^4 + 12k^3 + 2k^2 - 7k + 3}{-2k}$

31. $\dfrac{2m^5 - 6m^4 + 8m^2}{-2m^3}$

32. $\dfrac{6r^5 - 8r^4 + 10r^2}{-2r^3}$

33. $(20a^4 - 15a^5 + 25a^3) \div (5a^4)$

34. $(36y^2 - 12y^3 + 20y) \div (4y^2)$

35. $(120x^{11} - 60x^{10} + 140x^9 - 100x^8) \div (10x^{12})$

36. $(45y^7 + 9y^6 - 6y^5 + 12y^4) \div (3y^8)$

▶ **37.** $(120x^5y^4 - 80x^2y^3 + 40x^2y^4 - 20x^5y^3) \div (20xy^2)$

38. $(200a^5b^6 - 160a^4b^7 - 120a^3b^9 + 40a^2b^2) \div (40a^2b)$

Perform each division using the "long division" process. **See Examples 5 and 6.**

39. $\dfrac{x^2 - x - 6}{x - 3}$

40. $\dfrac{m^2 - 2m - 24}{m - 6}$

41. $\dfrac{2y^2 + 9y - 35}{y + 7}$

42. $\dfrac{2y^2 + 9y + 7}{y + 1}$

43. $\dfrac{p^2 + 2p + 20}{p + 6}$

44. $\dfrac{x^2 + 11x + 16}{x + 8}$

45. $\dfrac{12m^2 - 20m + 3}{2m - 3}$

46. $\dfrac{12y^2 + 20y + 7}{2y + 1}$

47. $\dfrac{4a^2 - 22a + 32}{2a + 3}$

48. $\dfrac{9w^2 + 6w + 10}{3w - 2}$

▶ 49. $\dfrac{8x^3 - 10x^2 - x + 3}{2x + 1}$

50. $\dfrac{12t^3 - 11t^2 + 9t + 18}{4t + 3}$

51. $\dfrac{8k^4 - 12k^3 - 2k^2 + 7k - 6}{2k - 3}$

52. $\dfrac{27r^4 - 36r^3 - 6r^2 + 26r - 24}{3r - 4}$

53. $\dfrac{5y^4 + 5y^3 + 2y^2 - y - 8}{y + 1}$

54. $\dfrac{2r^3 - 5r^2 - 6r + 15}{r - 3}$

▶ 55. $\dfrac{3k^3 - 4k^2 - 6k + 10}{k - 2}$

56. $\dfrac{5z^3 - z^2 + 10z + 2}{z + 2}$

57. $\dfrac{6p^4 - 16p^3 + 15p^2 - 5p + 10}{3p + 1}$

58. $\dfrac{6r^4 - 11r^3 - r^2 + 16r - 8}{2r - 3}$

Perform each division. **See Examples 6–9.**

▶ 59. $(x^3 + 2x^2 - 3) \div (x - 1)$

60. $(x^3 - 2x^2 - 9) \div (x - 3)$

61. $(2x^3 + x + 2) \div (x + 3)$

62. $(3x^3 + x + 5) \div (x + 1)$

63. $\dfrac{5 - 2r^2 + r^4}{r^2 - 4}$

64. $\dfrac{4t^2 + t^4 + 7}{t^2 - 4}$

65. $\dfrac{-4x + 3x^3 + 2}{x - 1}$

66. $\dfrac{-5x + 6x^3 + 5}{x - 1}$

67. $\dfrac{y^3 + 27}{y + 3}$

68. $\dfrac{y^3 - 64}{y - 4}$

69. $\dfrac{a^4 - 25}{a^2 - 5}$

70. $\dfrac{a^4 - 36}{a^2 + 6}$

▶ 71. $\dfrac{x^4 - 4x^3 + 5x^2 - 3x + 2}{x^2 + 3}$

72. $\dfrac{3t^4 + 5t^3 - 8t^2 - 13t + 2}{t^2 - 5}$

73. $\dfrac{2x^5 + 9x^4 + 8x^3 + 10x^2 + 14x + 5}{2x^2 + 3x + 1}$

74. $\dfrac{4t^5 - 11t^4 - 6t^3 + 5t^2 - t + 3}{4t^2 + t - 3}$

75. $(3a^2 - 11a + 17) \div (2a + 6)$

76. $(4x^2 + 11x - 8) \div (3x + 6)$

77. $\dfrac{3x^3 + 5x^2 - 9x + 5}{3x - 3}$

78. $\dfrac{5x^3 + 4x^2 + 10x + 20}{5x + 5}$

In Exercises 79–84, if necessary, refer to the formulas found inside the back cover of this book. **See Example 10.**

79. The area of the rectangle is given by the polynomial

$$5x^3 + 7x^2 - 13x - 6.$$

What polynomial expresses the length (in appropriate units)?

80. The area of the rectangle is given by the polynomial

$$15x^3 + 12x^2 - 9x + 3.$$

What polynomial expresses the length (in appropriate units)?

81. The area of the triangle is given by the polynomial

$$24m^3 + 48m^2 + 12m.$$

What polynomial expresses the length of the base (in appropriate units)?

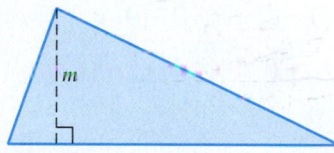

82. The area of the parallelogram is given by the polynomial

$$2x^3 + 2x^2 - 3x - 1.$$

What polynomial expresses the length of the base (in appropriate units)?

83. If the distance traveled is $(5x^3 - 6x^2 + 3x + 14)$ miles and the rate is $(x + 1)$ mph, write an expression, in hours, for the time traveled.

84. If it costs $(4x^5 + 3x^4 + 2x^3 + 9x^2 - 29x + 2)$ dollars to fertilize a garden, and fertilizer costs $(x + 2)$ dollars per square yard, write an expression, in square yards, for the area of the garden.

Chapter 4   Summary

Key Terms

4.1
base
exponent (power)
exponential expression

4.3
scientific notation
standard notation

4.4
term
leading term
numerical coefficient
  (coefficient)
like terms
unlike terms
polynomial

descending powers
degree of a term
degree of a polynomial
monomial
binomial
trinomial
parabola
vertex
axis of symmetry (axis)

4.5
FOIL method
outer product
inner product

4.6
conjugates

New Symbols

x^{-n} x to the negative n
       power

Test Your Word Power

See how well you have learned the vocabulary in this chapter.

1. A **polynomial** is an algebraic expression made up of
   A. a term or a finite product of terms with positive coefficients and exponents
   B. a term or a finite sum of terms with real coefficients and whole number exponents
   C. the product of two or more terms with positive exponents
   D. the sum of two or more terms with whole number coefficients and exponents.

2. The **degree of a term** is
   A. the number of variables in the term

   B. the product of the exponents on the variables
   C. the least exponent on the variables
   D. the sum of the exponents on the variables.

3. The **FOIL** method is used when
   A. adding two binomials
   B. adding two trinomials
   C. multiplying two binomials
   D. multiplying two trinomials.

4. A **binomial** is a polynomial with
   A. only one term
   B. exactly two terms
   C. exactly three terms
   D. more than three terms.

5. A **monomial** is a polynomial with
   A. only one term
   B. exactly two terms
   C. exactly three terms
   D. more than three terms.

6. A **trinomial** is a polynomial with
   A. only one term
   B. exactly two terms
   C. exactly three terms
   D. more than three terms.

ANSWERS

1. B; *Example:* $5x^3 + 2x^2 - 7$ 2. D; *Examples:* The term 6 has degree 0, $3x$ has degree 1, $-2x^8$ has degree 8, and $5x^2y^4$ has degree 6.

                                                F   O   I   L

3. C; *Example:* $(m + 4)(m - 3) = m(m) - 3m + 4m + 4(-3) = m^2 + m - 12$ 4. B; *Example:* $3t^3 + 5t$ 5. A; *Examples:* -5 and $4xy^5$
6. C; *Example:* $2a^2 - 3ab + b^2$

Quick Review

CONCEPTS	EXAMPLES

4.1 The Product Rule and Power Rules for Exponents

For any integers m and n, the following hold.

Product Rule $a^m \cdot a^n = a^{m+n}$

Power Rules   (a) $(a^m)^n = a^{mn}$

(b) $(ab)^m = a^m b^m$

(c) $\left(\dfrac{a}{b}\right)^m = \dfrac{a^m}{b^m}$ (where $b \neq 0$)

Simplify by using the rules for exponents.

$$2^4 \cdot 2^5 = 2^{4+5} = 2^9$$

$$(3^4)^2 = 3^{4 \cdot 2} = 3^8$$

$$(6a)^5 = 6^5 a^5$$

$$\left(\frac{2}{3}\right)^4 = \frac{2^4}{3^4}$$

4.2 Integer Exponents and the Quotient Rule

If $a \neq 0$, then for integers m and n, the following hold.

Zero Exponent $a^0 = 1$

Negative Exponent $a^{-n} = \dfrac{1}{a^n}$

Quotient Rule $\dfrac{a^m}{a^n} = a^{m-n}$

Negative-to-Positive Rules $\dfrac{a^{-m}}{b^{-n}} = \dfrac{b^n}{a^m}$ (where $b \neq 0$)

$\left(\dfrac{a}{b}\right)^{-m} = \left(\dfrac{b}{a}\right)^m$ (where $b \neq 0$)

Simplify by using the rules for exponents.

$$15^0 = 1$$

$$5^{-2} = \frac{1}{5^2} = \frac{1}{25}$$

$$\frac{4^8}{4^3} = 4^{8-3} = 4^5$$

$$\frac{4^{-2}}{3^{-5}} = \frac{3^5}{4^2}$$

$$\left(\frac{6}{5}\right)^{-3} = \left(\frac{5}{6}\right)^3$$

CONCEPTS	**EXAMPLES**

4.3 Scientific Notation

To write a positive number in scientific notation

$$a \times 10^n, \quad \text{where} \quad 1 \le |a| < 10,$$

move the decimal point to follow the first nonzero digit.

1. If moving the decimal point makes the number less, then n is positive.
2. If it makes the number greater, n is negative.
3. If the decimal point is not moved, then n is 0.

For a negative number, follow these steps using the absolute value of the number. Then make the result negative.

Write in scientific notation.

$$247 = 2.47 \times 10^2$$
$$0.0051 = 5.1 \times 10^{-3}$$
$$-4.8 = -4.8 \times 10^0$$

Write in standard notation.

$$3.25 \times 10^5 = 325,000$$
$$8.44 \times 10^{-6} = 0.00000844$$

4.4 Adding, Subtracting, and Graphing Polynomials

Adding Polynomials

Add like terms.

Add.
$$2x^2 + 5x - 3$$
$$5x^2 - 2x + 7$$
$$\overline{7x^2 + 3x + 4}$$

Subtracting Polynomials

Change the signs of the terms in the subtrahend (second polynomial) and add the result to the minuend (first polynomial).

Subtract.
$$(2x^2 + 5x - 3) - (5x^2 - 2x + 7)$$
$$= (2x^2 + 5x - 3) + (-5x^2 + 2x - 7)$$
$$= -3x^2 + 7x - 10$$

Graphing Polynomials

To graph a simple polynomial equation such as

$$y = x^2 - 2,$$

plot points near the vertex. (In this chapter, all parabolas have a vertex on the x-axis or the y-axis.)

Graph $y = x^2 - 2$.

x	y
-2	2
-1	-1
0	-2
1	-1
2	2

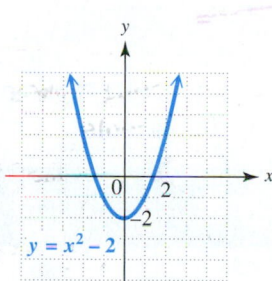

4.5 Multiplying Polynomials

General Method for Multiplying Polynomials

Multiply each term of the first polynomial by each term of the second polynomial. Then add like terms.

Multiply.
$$3x^3 - 4x^2 + 2x - 7$$
$$4x + 3$$
$$\overline{9x^3 - 12x^2 + 6x - 21}$$
$$12x^4 - 16x^3 + 8x^2 - 28x$$
$$\overline{12x^4 - 7x^3 - 4x^2 - 22x - 21}$$

FOIL Method for Multiplying Binomials

Step 1 Multiply the two **F**irst terms to obtain the first term of the product.

Step 2 Find the **O**uter product and the **I**nner product and combine them (when possible) to obtain the middle term of the product.

Step 3 Multiply the two **L**ast terms to obtain the last term of the product.

Add the terms found in Steps 1–3.

Multiply. $(2x + 3)(5x - 4)$

$$2x(5x) = 10x^2 \quad \textbf{F}$$

$$2x(-4) + 3(5x) = 7x \quad \textbf{O, I}$$

$$3(-4) = -12 \quad \textbf{L}$$

The product is $10x^2 + 7x - 12$.

CONCEPTS	EXAMPLES

4.6 Special Products

Square of a Binomial

$$(x + y)^2 = x^2 + 2xy + y^2$$
$$(x - y)^2 = x^2 - 2xy + y^2$$

Product of a Sum and Difference of Two Terms

$$(x + y)(x - y) = x^2 - y^2$$

Multiply.

$(3x + 1)^2$

$= (3x)^2 + 2(3x)(1) + 1^2$

$= 9x^2 + 6x + 1$

$(2m - 5n)^2$

$= (2m)^2 - 2(2m)(5n) + (5n)^2$

$= 4m^2 - 20mn + 25n^2$

$(4a + 3)(4a - 3)$

$= (4a)^2 - 3^2$

$= 16a^2 - 9$

4.7 Dividing Polynomials

Dividing a Polynomial by a Monomial
Divide each term of the polynomial by the monomial.

$$\frac{a + b}{c} = \frac{a}{c} + \frac{b}{c} \quad \text{(where } c \neq 0\text{)}$$

Dividing a Polynomial by a Polynomial
Use "long division."

Divide.

$$\frac{4x^3 - 2x^2 + 6x - 9}{2x} = 2x^2 - x + 3 - \frac{9}{2x}$$

Divide each term in the numerator by $2x$.

$$\begin{array}{r} 2x - 5 \\ 3x + 4 \overline{)6x^2 - 7x - 21} \\ \underline{6x^2 + 8x} \\ -15x - 21 \\ \underline{-15x - 20} \\ -1 \leftarrow \text{Remainder} \end{array}$$

The answer is $2x - 5 + \dfrac{-1}{3x + 4}$.

Chapter 4   Review Exercises

4.1 *Use the product rule, power rules, or both to simplify each expression. Write each answer in exponential form.*

1. $4^3 \cdot 4^8$ **2.** $(-5)^6(-5)^5$ **3.** $(-8x^4)(9x^3)$

4. $(2x^2)(5x^3)(x^9)$ **5.** $(19x)^5$ **6.** $(-4y)^7$

7. $5(pt)^4$ **8.** $\left(\dfrac{7}{5}\right)^6$ **9.** $(3x^2y^3)^3$

10. $(t^4)^8(t^2)^5$ **11.** $(6x^2z^4)^2(x^3yz^2)^4$ **12.** $\left(\dfrac{2m^3n}{p^2}\right)^3$

4.2 *Evaluate each expression.*

13. -10^0 **14.** $-(-23)^0$ **15.** $6^0 + (-6)^0$ **16.** $-3^0 - 2^0$

Simplify each expression. Assume that all variables represent nonzero real numbers.

17. -7^{-2} **18.** $\left(\dfrac{5}{8}\right)^{-2}$ **19.** $(2^{-2})^{-3}$

Answers (margin):

1. 4^{11} **2.** $(-5)^{11}$
3. $-72x^7$ **4.** $10x^{14}$
5. 19^5x^5 **6.** $(-4)^7y^7$
7. $5p^4t^4$ **8.** $\dfrac{7^6}{5^6}$
9. $3^3x^6y^9$ **10.** t^{42}
11. $6^2x^{16}y^4z^{16}$ **12.** $\dfrac{2^3m^9n^3}{p^6}$
13. -1 **14.** -1
15. 2 **16.** -2
17. $-\dfrac{1}{49}$ **18.** $\dfrac{64}{25}$
19. 64

20. $\dfrac{1}{81}$

21. $\dfrac{3}{4}$ **22.** $\dfrac{1}{36}$

23. r^2 **24.** y^7

25. $\dfrac{r^8}{81}$ **26.** $\dfrac{3^5}{p^3}$

27. $\dfrac{1}{a^3b^5}$ **28.** $72r^5$

29. 4.8×10^7 **30.** 2.8988×10^{10}
31. 8.24×10^{-8} **32.** -4.82×10^6
33. $24{,}000$ **34.** $78{,}300{,}000$
35. 0.000000897 **36.** -0.00076
37. 800 **38.** 5
39. $4{,}000{,}000$ **40.** 0.025
41. $81{,}887{,}000{,}000{,}000{,}000$
42. $37{,}217{,}400$
43. 1.0086×10^{15}
44. 6.78×10^{13}

45. $20m^2$; degree 2; monomial
46. $p^3 - p^2 - 4p$; degree 3; trinomial
47. $-8y^5 - 7y^4$; degree 5; binomial
48. $-r^3 - 2r + 7$
49. $13x^3y^2 - 5xy^5 + 21x^2$
50. $a^3 + 4a^2$
51. $y^2 - 10y + 9$
52. $-13k^4 - 15k^2 + 18k$

---

20. $9^3 \cdot 9^{-5}$ **21.** $2^{-1} + 4^{-1}$ **22.** $\dfrac{6^{-5}}{6^{-3}}$

23. $\dfrac{x^{-7}}{x^{-9}}$ **24.** $\dfrac{y^4 \cdot y^{-2}}{y^{-5}}$ **25.** $(3r^{-2})^{-4}$

26. $(3p)^4(3p^{-7})$ **27.** $\dfrac{ab^{-3}}{a^4b^2}$ **28.** $\dfrac{(6r^{-1})^2(2r^{-4})}{r^{-5}(r^2)^{-3}}$

4.3 *Write each number in scientific notation.*

29. $48{,}000{,}000$ **30.** $28{,}988{,}000{,}000$

31. 0.0000000824 **32.** $-4{,}820{,}000$

Write each number in standard notation.

33. 2.4×10^4 **34.** 7.83×10^7 **35.** 8.97×10^{-7} **36.** -7.6×10^{-4}

Perform the indicated operations. Write each answer in standard notation.

37. $(2 \times 10^{-3})(4 \times 10^5)$ **38.** $(2.5 \times 10^{-51})(2.0 \times 10^{51})$

39. $\dfrac{8 \times 10^4}{2 \times 10^{-2}}$ **40.** $\dfrac{60 \times 10^{-1}}{24 \times 10}$

Write each boldface italic number in standard form.

41. In a recent year, China was the world's largest energy producer. China accounted for ***8.1887 × 10^{16}*** Btu. (*Source: World Almanac and Book of Facts.*)

42. The 2011 population of Tokyo, Japan, was ***3.72174 × 10^7***. (*Source: World Almanac and Book of Facts.*)

Write each boldface italic number in scientific notation.

43. As of July 2013, Japan's outstanding public debt was ***1,008,600,000,000,000*** yen. (*Source:* www.bloomberg.com)

44. In 2011, the budget of the U.S. Department of Defense was ***67,800,000,000,000*** dollars. (*Source:* www.defense.gov)

4.4 *In Exercises 45–47, simplify by combining like terms whenever possible. Write the result in descending powers of the variable. Then give the degree and tell whether the simplified polynomial is a* monomial, *a* binomial, *a* trinomial, *or* none of these.

45. $9m^2 + 11m^2$ **46.** $-4p + p^3 - p^2$ **47.** $-7y^5 - 8y^4 - y^5 + y^4$

Add or subtract as indicated.

48. $(12r^4 - 7r^3 + 2r^2) - (5r^4 - 3r^3 + 2r^2 - 1) - (7r^4 - 3r^3 + 2r - 6)$

49. $(5x^3y^2 - 3xy^5 + 12x^2) - (-9x^2 - 8x^3y^2 + 2xy^5)$

50. Add.   **51.** Subtract.   **52.** Subtract.

$$\begin{array}{r} -2a^3 + 5a^2 \\ 3a^3 - a^2 \\ \hline \end{array} \qquad \begin{array}{r} 6y^2 - 8y + 2 \\ 5y^2 + 2y - 7 \\ \hline \end{array} \qquad \begin{array}{r} -12k^4 - 8k^2 + 7k \\ k^4 + 7k^2 - 11k \\ \hline \end{array}$$

53. 1, 4, 5, 4, 1

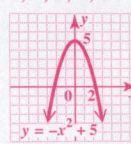

54. 10, 1, −2, 1, 10

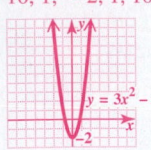

55. $a^3 - 2a^2 - 7a + 2$
56. $6r^3 + 8r^2 - 17r + 6$
57. $5p^5 - 2p^4 - 3p^3 + 25p^2 + 15p$
58. $m^2 - 7m - 18$
59. $6k^2 - 9k - 6$
60. $2a^2 + 5ab - 3b^2$
61. $12k^2 - 32kq - 35q^2$
62. $s^3 - 3s^2 + 3s - 1$

63. $a^2 + 8a + 16$
64. $4r^2 + 20rt + 25t^2$
65. $36m^2 - 25$
66. $25a^2 - 36b^2$
67. $r^3 + 6r^2 + 12r + 8$
68. $25t^3 - 30t^2 + 9t$
69. (a) Answers will vary. For example, let $x = 1$ and $y = 2$.
$(1 + 2)^2 \neq 1^2 + 2^2$,
because $9 \neq 5$.
  (b) Answers will vary. For example, let $x = 1$ and $y = 2$.
$(1 + 2)^3 \neq 1^3 + 2^3$,
because $27 \neq 9$.
70. Find the third power of a binomial, such as $(a + b)^3$, as follows.
$(a + b)^3$
$\quad = (a + b)(a + b)^2$
$\quad = (a + b)(a^2 + 2ab + b^2)$
$\quad = a^3 + 2a^2b + ab^2 + a^2b +$
$\qquad 2ab^2 + b^3$
$\quad = a^3 + 3a^2b + 3ab^2 + b^3$
71. $x^6 + 6x^4 + 12x^2 + 8$
72. $\dfrac{4}{3}\pi x^3 + 4\pi x^2 + 4\pi x + \dfrac{4}{3}\pi$
73. $\dfrac{-5y^2}{3}$
74. $-2m^2n + mn + \dfrac{6n^3}{5}$
75. $y^3 - 2y + 3$
76. $-6r^5s - 3r^4 + \dfrac{2}{r^2s^5}$
77. $2mn + 3m^4n^2 - 4n$
78. The friend wrote the second term of the quotient as $-12x$ rather than $-2x$.
$\dfrac{6x^2 - 12x}{6} = \dfrac{6x^2}{6} - \dfrac{12x}{6}$
$\qquad\qquad = x^2 - 2x$

Graph each equation by completing the table of values.

53. $y = -x^2 + 5$

x	−2	−1	0	1	2
y					

54. $y = 3x^2 - 2$

x	−2	−1	0	1	2
y					

4.5 *Find each product.*

55. $(a + 2)(a^2 - 4a + 1)$

56. $(3r - 2)(2r^2 + 4r - 3)$

57. $(5p^2 + 3p)(p^3 - p^2 + 5)$

58. $(m - 9)(m + 2)$

59. $(3k - 6)(2k + 1)$

60. $(a + 3b)(2a - b)$

61. $(6k + 5q)(2k - 7q)$

62. $(s - 1)^3$

4.6 *Find each product.*

63. $(a + 4)^2$

64. $(2r + 5t)^2$

65. $(6m - 5)(6m + 5)$

66. $(5a + 6b)(5a - 6b)$

67. $(r + 2)^3$

68. $t(5t - 3)^2$

69. Choose values for x and y to show that, in general, the following hold true.

(a) $(x + y)^2 \neq x^2 + y^2$

(b) $(x + y)^3 \neq x^3 + y^3$

70. Write an explanation on how to raise a binomial to the third power. Give an example.

In Exercises 71 and 72, refer to the formulas found inside the back cover of this book, if necessary.

71. Find a polynomial that represents, in cubic centimeters, the volume of a cube with one side having length $(x^2 + 2)$ centimeters.

72. Find a polynomial that represents, in cubic inches, the volume of a sphere with radius $(x + 1)$ inches.

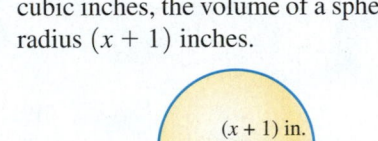

A cube

$(x^2 + 2)$ cm

$(x + 1)$ in.

A sphere

4.7 *Perform each division.*

73. $\dfrac{-15y^4}{9y^2}$

74. $(-10m^4n^2 + 5m^3n^2 + 6m^2n^4) \div (5m^2n)$

75. $\dfrac{6y^4 - 12y^2 + 18y}{6y}$

76. $\dfrac{24r^8s^6 + 12r^7s^5 - 8r}{-4r^3s^5}$

77. What polynomial, when multiplied by $6m^2n$, gives the product

$$12m^3n^2 + 18m^6n^3 - 24m^2n^2?$$

78. One of your friends in class simplified

$$\dfrac{6x^2 - 12x}{6} \quad \text{as} \quad x^2 - 12x.$$

WHAT WENT WRONG? Give the correct answer.

79. $2r + 7$

80. $2a^2 + 3a - 1 + \dfrac{6}{5a - 3}$

81. $x^2 + 3x - 4$

82. $m^2 + 4m - 2$

83. $4x - 5$

84. $5y - 10$

85. $y^2 + 2y + 4$

86. $100x^4 - 10x^2 + 1$

87. $2y^2 - 5y + 4 + \dfrac{-5}{3y^2 + 1}$

88. $x^3 - 2x^2 + 4 + \dfrac{-3}{4x^2 - 3}$

Perform each division.

79. $\dfrac{2r^2 + 3r - 14}{r - 2}$

80. $\dfrac{10a^3 + 9a^2 - 14a + 9}{5a - 3}$

81. $\dfrac{x^4 - 5x^2 + 3x^3 - 3x + 4}{x^2 - 1}$

82. $\dfrac{m^4 + 4m^3 - 12m - 5m^2 + 6}{m^2 - 3}$

83. $\dfrac{16x^2 - 25}{4x + 5}$

84. $\dfrac{25y^2 - 100}{5y + 10}$

85. $\dfrac{y^3 - 8}{y - 2}$

86. $\dfrac{1000x^6 + 1}{10x^2 + 1}$

87. $\dfrac{6y^4 - 15y^3 + 14y^2 - 5y - 1}{3y^2 + 1}$

88. $\dfrac{4x^5 - 8x^4 - 3x^3 + 22x^2 - 15}{4x^2 - 3}$

Chapter 4   Mixed Review Exercises

Perform each indicated operation, or simplify each expression. Assume that all variables represent nonzero real numbers.

1. 2

2. $\dfrac{216r^6p^3}{5^3}$

3. $144a^2 - 1$

4. $\dfrac{1}{16}$

5. $\dfrac{1}{256}$

6. $p - 3 + \dfrac{5}{2p}$

7. $\dfrac{2}{3m^3}$

8. $6k^3 - 21k - 6$

9. r^{13}

10. $4r^2 + 20rs + 25s^2$

11. $y^2 + 5y + 1$

12. $10r^2 + 21r - 10$

13. $-y^2 - 4y + 4$

14. $\dfrac{5}{2} - \dfrac{4}{5xy} + \dfrac{3x}{2y^2}$

15. $10p^2 - 3p - 5$

16. $3x^2 + 9x + 25 + \dfrac{80}{x - 3}$

17. $49 - 28k + 4k^2$

18. $\dfrac{1}{x^4y^{12}}$

19. (a) $6x - 2$

    (b) $2x^2 + x - 6$

20. (a) $20x^4 + 8x^2$

    (b) $25x^8 + 20x^6 + 4x^4$

1. $5^0 + 7^0$

2. $\left(\dfrac{6r^2p}{5}\right)^3$

3. $(12a + 1)(12a - 1)$

4. 2^{-4}

5. $(4^{-2})^2$

6. $\dfrac{2p^3 - 6p^2 + 5p}{2p^2}$

7. $\dfrac{(2m^{-5})(3m^2)^{-1}}{m^{-2}(m^{-1})^2}$

8. $(3k - 6)(2k^2 + 4k + 1)$

9. $\dfrac{r^9 \cdot r^{-5}}{r^{-2} \cdot r^{-7}}$

10. $(2r + 5s)^2$

11. $\dfrac{2y^3 + 17y^2 + 37y + 7}{2y + 7}$

12. $(2r + 5)(5r - 2)$

13. $(-5y^2 + 3y - 11) + (4y^2 - 7y + 15)$

14. $(25x^2y^3 - 8xy^2 + 15x^3y) \div (10x^2y^3)$

15. $(6p^2 - p - 8) - (-4p^2 + 2p - 3)$

16. $\dfrac{3x^3 - 2x + 5}{x - 3}$

17. $(-7 + 2k)^2$

18. $\left(\dfrac{x}{y^{-3}}\right)^{-4}$

Find polynomials that represent **(a)** *the perimeter and* **(b)** *the area of each square or rectangle.*

19.

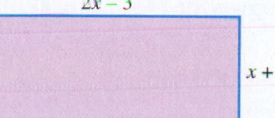

$2x - 3$

$x + 2$

20.

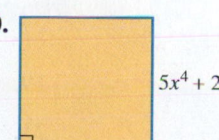

$5x^4 + 2x^2$

| Chapter 4 | Test | FOR EXTRA HELP | *Step-by-step test solutions are found on the Chapter Test Prep Videos available in* MyMathLab®, *or on* YouTube™. |

▶ View the complete solutions to all Chapter Test exercises in MyMathLab.

[4.1, 4.2]

1. $\dfrac{1}{625}$ 2. 2

3. $\dfrac{7}{12}$ 4. $9x^3y^5$

5. 8^5 6. x^2y^6

7. (a) positive   (b) positive
   (c) negative   (d) positive
   (e) zero   (f) negative

[4.3]

8. (a) 4.5×10^{10}
   (b) 0.0000036
   (c) 0.00019

9. (a) 1×10^3; 5.89×10^{12}
   (b) 5.89×10^{15} mi

[4.4]

10. $-7x^2 + 8x$; 2; binomial

11. $4n^4 + 13n^3 - 10n^2$; 4; trinomial

12. $4, -2, -4, -2, 4$

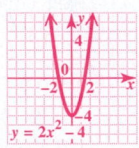

13. $-2y^2 - 9y + 17$

14. $-21a^3b^2 + 7ab^5 - 5a^2b^2$

15. $16r^2 - 19$

16. $-12t^2 + 5t + 8$

[4.5]

17. $-27x^5 + 18x^4 - 6x^3 + 3x^2$

18. $t^2 - 5t - 24$

19. $8x^2 + 2xy - 3y^2$

[4.6]

20. $25x^2 - 20xy + 4y^2$

21. $100v^2 - 9w^2$

[4.5]

22. $2r^3 + r^2 - 16r + 15$

Evaluate each expression.

1. 5^{-4} 2. $(-3)^0 + 4^0$ 3. $4^{-1} + 3^{-1}$

4. Simplify $\dfrac{(3x^2y)^2(xy^3)^2}{(xy)^3}$. Assume that x and y represent nonzero numbers.

Simplify each expression. Assume that all variables represent nonzero real numbers.

5. $\dfrac{8^{-1} \cdot 8^4}{8^{-2}}$ 6. $\dfrac{(x^{-3})^{-2}(x^{-1}y)^2}{(xy^{-2})^2}$

7. Determine whether each expression represents a number that is *positive, negative,* or *zero.*

   (a) 3^{-4} (b) $(-3)^4$ (c) -3^4 (d) 3^0 (e) $(-3)^0 - 3^0$ (f) $(-3)^{-3}$

8. (a) Write 45,000,000,000 using scientific notation.

   (b) Write 3.6×10^{-6} using standard notation.

   (c) Write the quotient $\dfrac{9.5 \times 10^{-1}}{5 \times 10^3}$ using standard notation.

9. A satellite galaxy of the Milky Way, known as the Large Magellanic Cloud, is *1000* light-years across. A *light-year* is equal to *5,890,000,000,000* mi. (*Source:* "Images of Brightest Nebula Unveiled," *USA Today.*)

   (a) Write the two boldface italic numbers in scientific notation.

   (b) How many miles across is the Large Magellanic Cloud?

For each polynomial, simplify by combining like terms whenever possible. Write the result in descending powers of the variable. Then give the degree and tell whether the simplified polynomial is a monomial, *a* binomial, *a* trinomial, *or* none of these.

10. $5x^2 + 8x - 12x^2$ 11. $13n^3 - n^2 + n^4 + 3n^4 - 9n^2$

12. Graph the equation $y = 2x^2 - 4$ by completing the table of values.

x	-2	-1	0	1	2
y					

Perform each indicated operation.

13. $(2y^2 - 8y + 8) + (-3y^2 + 2y + 3) - (y^2 + 3y - 6)$

14. $(-9a^3b^2 + 13ab^5 + 5a^2b^2) - (6ab^5 + 12a^3b^2 + 10a^2b^2)$

15. Add.

   $-6r^5 + 4r^2 - 3$
   $\underline{6r^5 + 12r^2 - 16}$

16. Subtract.

   $9t^3 - 4t^2 + 2t + 2$
   $\underline{9t^3 + 8t^2 - 3t - 6}$

17. $3x^2(-9x^3 + 6x^2 - 2x + 1)$ 18. $(t - 8)(t + 3)$

19. $(4x + 3y)(2x - y)$ 20. $(5x - 2y)^2$

21. $(10v + 3w)(10v - 3w)$ 22. $(2r - 3)(r^2 + 2r - 5)$

[4.6]
23. $12x + 36$
24. $9x^2 + 54x + 81$

[4.7]
25. $4y^2 - 3y + 2 + \dfrac{5}{y}$
26. $3xy^2 + 2x^3y^2 + 4y^2$
27. $x - 2$
28. $3x^2 + 6x + 11 + \dfrac{26}{x - 2}$

Refer to the square below. Find polynomials that represent the following.

23. The perimeter

24. The area

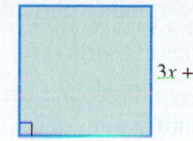

$3x + 9$

Perform each division.

25. $\dfrac{8y^3 - 6y^2 + 4y + 10}{2y}$

26. $(-9x^2y^3 + 6x^4y^3 + 12xy^3) \div (3xy)$

27. $\dfrac{5x^2 - x - 18}{5x + 9}$

28. $(3x^3 - x + 4) \div (x - 2)$

Chapters R–4   Cumulative Review Exercises

[R.1]
1. $\dfrac{7}{4}$ **2.** 5

3. $31\dfrac{1}{4}$ yd^3

[R.2]
4. $1836

[1.4, 1.5]
5. 1, 3, 5, 9, 15, 45
6. -8

7. $\dfrac{1}{2}$ **8.** -4

[1.6]
9. associative property
10. distributive property

[1.7]
11. $-10x^2 + 21x - 29$

[2.1–2.3]
12. $\left\{\dfrac{13}{4}\right\}$

13. $\varnothing$

[2.5]
14. $r = \dfrac{d}{t}$

[2.6]
15. $\{-5\}$

[2.1–2.3]
16. $\{0\}$
17. $\{20\}$ **18.** $\{-12\}$
19. {all real numbers}

Write each fraction in lowest terms.

1. $\dfrac{28}{16}$

2. $\dfrac{55}{11}$

3. A contractor installs sheds. Each requires $1\dfrac{1}{4}$ yd^3 of concrete. How much concrete would be needed for 25 sheds?

4. A retailer has $34,000 invested in her business. She finds that last year she earned 5.4% on this investment. How much did she earn?

5. List all positive integer factors of 45.

6. Find the value of $\dfrac{4x - 2y}{x + y}$ for $x = -2$ and $y = 4$.

Perform each indicated operation.

7. $\dfrac{(-13 + 15) - (3 + 2)}{6 - 12}$

8. $-7 - 3[2 + (5 - 8)]$

Name the property illustrated.

9. $(9 + 2) + 3 = 9 + (2 + 3)$

10. $6(4 + 2) = 6(4) + 6(2)$

11. Simplify the expression $-3(2x^2 - 8x + 9) - (4x^2 + 3x + 2)$.

Solve each equation.

12. $2 - 3(t - 5) = 4 + t$

13. $2(5x + 1) = 10x + 4$

14. $d = rt$ for r

15. $\dfrac{x}{5} = \dfrac{x - 2}{7}$

16. $3x - (4 + 2x) = -4$

17. $0.05x + 0.15(50 - x) = 5.50$

18. $\dfrac{1}{3}p - \dfrac{1}{6}p = -2$

19. $4 - (3x + 12) = -7 - (3x + 1)$

[2.4]
20. exertion: 9443 calories;
    regulating body temperature:
    1757 calories

[2.8]
21. 11 ft and 22 ft

22. $\left(-\infty, -\dfrac{14}{5}\right)$

23. $[-4, 2)$

[3.2]
24.

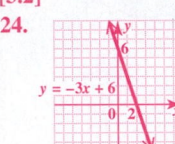

[3.3–3.5]
25. (a) 1   **(b)** $y = x + 6$

[4.1, 4.2]
26. $\dfrac{5}{4}$

27. 1   **28.** $\dfrac{2b}{a^{10}}$

[4.3]
29. 10,800,000 km

[4.4]
30. $11x^3 - 14x^2 - x + 14$

[4.5]
31. $63x^2 + 57x + 12$

[4.7]
32. $y^2 - 2y + 6$

Solve each problem.

20. A husky running the Iditarod in Alaska burns $5\frac{3}{8}$ calories in exertion for every 1 calorie burned in thermoregulation in extreme cold. According to one scientific study, a husky in top condition burns an amazing total of 11,200 calories per day. How many calories are burned for exertion, and how many are burned for regulation of body temperature? Round answers to the nearest whole number.

21. One side of a triangle is twice as long as a second side. The third side of the triangle is 17 ft long. The perimeter of the triangle cannot be more than 50 ft. Find the longest possible values for the other two sides of the triangle.

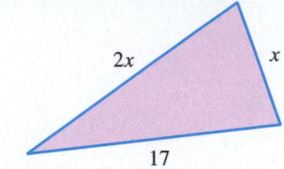

Solve each inequality.

22. $-2(x + 4) > 3x + 6$ **23.** $-3 \le 2x + 5 < 9$

24. Graph $y = -3x + 6$.

25. Consider the two points $(-1, 5)$ and $(2, 8)$.

  (a) Find the slope of the line passing through them.

  (b) Find the equation of the line passing through them.

Evaluate each expression.

26. $4^{-1} + 3^0$ **27.** $\dfrac{8^{-5} \cdot 8^7}{8^2}$

28. Write $\dfrac{(a^{-3}b^2)^2}{(2a^{-4}b^{-3})^{-1}}$ with positive exponents only.

29. It takes about 3.6×10^1 sec at a speed of 3.0×10^5 km per sec for light from the sun to reach Venus. How far is Venus from the sun? (*Source: World Almanac and Book of Facts.*)

Perform each indicated operation.

30. $(7x^3 - 12x^2 - 3x + 8) + (6x^2 + 4) - (-4x^3 + 8x^2 - 2x - 2)$

31. $(7x + 4)(9x + 3)$ **32.** $\dfrac{y^3 - 3y^2 + 8y - 6}{y - 1}$

5

Factoring and Applications

Formulas associated with the mathematicians Pythagoras (c. 380–300 B.C.) and Galileo (1564–1642) are used in applications that involve *factoring* polynomials, the subject of this chapter.

5.1 The Greatest Common Factor; Factoring by Grouping

OBJECTIVES

1   Find the greatest common factor of a list of terms.

2   Factor out the greatest common factor.

3   Factor by grouping.

VOCABULARY

☐ factor
☐ factored form
☐ common factor
☐ greatest common factor (GCF)

To **factor** a number means to write it as a product of two or more numbers. The product is a **factored form** of the number. Consider an example.

$$\underset{\text{Factored form}}{\underbrace{12 = 6 \cdot 2}} \qquad \text{Factors}$$

Factoring is a process that "undoes" multiplying. We multiply $6 \cdot 2$ to obtain 12, but we factor 12 by writing it as $6 \cdot 2$. Other factored forms of 12 are

$$-6(-2), \quad 3 \cdot 4, \quad -3(-4), \quad 12 \cdot 1, \quad -12(-1), \quad \text{and} \quad 2 \cdot 2 \cdot 3.$$

OBJECTIVE 1   Find the greatest common factor of a list of terms.

An integer that is a factor of two or more integers is a **common factor** of those integers. For example, 6 is a common factor of 18 and 24 because 6 is a factor of both 18 and 24. Other common factors of 18 and 24 are 1, 2, and 3.

The **greatest common factor (GCF)** of a list of integers is the largest common factor of those integers. Thus, 6 is the greatest common factor of 18 and 24 because it is the largest of their common factors.

Finding the Greatest Common Factor (GCF)

Step 1   **Factor.** Write each number in prime factored form.

Step 2   **List common factors.** List each prime number or each variable that is a factor of every term in the list. (If a prime does not appear in one of the prime factored forms, it *cannot* appear in the greatest common factor.)

Step 3   **Choose least exponents.** Use as exponents on the common prime factors the *least* exponents from the prime factored forms.

Step 4   **Multiply** the primes from Step 3. If there are no primes left after Step 3, the greatest common factor is 1.

NOTE   *Factors* of a number are also ***divisors*** of the number. The ***greatest common factor*** is the same as the ***greatest common divisor***. Divisibility tests are useful for deciding what numbers divide into a given number.

▼ Divisibility Tests

A Whole Number Divisible by	Must Have the Following Property:
2	Ends in 0, 2, 4, 6, or 8
3	Sum of digits divisible by 3
4	Last two digits form a number divisible by 4
5	Ends in 0 or 5
6	Divisible by both 2 and 3
8	Last three digits form a number divisible by 8
9	Sum of digits divisible by 9
10	Ends in 0

**NOW TRY
EXERCISE 1**

Find the greatest common factor for each list of numbers.

(a) 24, 36

(b) 54, 90, 108

(c) 15, 19, 25

EXAMPLE 1 Finding the Greatest Common Factor (Numbers)

Find the greatest common factor for each list of numbers.

(a) 30, 45

$$30 = 2 \cdot 3 \cdot 5$$
$$45 = 3 \cdot 3 \cdot 5$$

Write the prime factored form of each number.

*Use each prime the **least** number of times it appears in **all** the factored forms.* There is no 2 in the prime factored form of 45, so there will be no 2 in the greatest common factor. The least number of times 3 appears in all the factored forms is 1. The least number of times 5 appears is also 1.

$$\text{GCF} = 3^1 \cdot 5^1 = 15 \qquad 3^1 = 3 \text{ and } 5^1 = 5.$$

(b) 72, 120, 432

$$72 = 2 \cdot 2 \cdot 2 \cdot 3 \cdot 3$$
$$120 = 2 \cdot 2 \cdot 2 \cdot 3 \cdot 5$$
$$432 = 2 \cdot 2 \cdot 2 \cdot 2 \cdot 3 \cdot 3 \cdot 3$$

Write the prime factored form of each number.

The least number of times 2 appears in all the factored forms is 3, and the least number of times 3 appears is 1. There is no 5 in the prime factored form of either 72 or 432.

$$\text{GCF} = 2^3 \cdot 3^1 = 24 \qquad 2^3 = 8 \text{ and } 3^1 = 3.$$

(c) 10, 11, 14

$$10 = 2 \cdot 5$$
$$11 = 11$$
$$14 = 2 \cdot 7$$

Write the prime factored form of each number.

There are no primes common to all three numbers, so the GCF is 1.   **NOW TRY**

The greatest common factor can also be found for a list of variable terms. For example, the terms x^4, x^5, x^6, and x^7 have x^4 as the greatest common factor because the least exponent on the variable x in the factored forms is 4.

$$x^4 = 1 \cdot x^4, \quad x^5 = x \cdot x^4, \quad x^6 = x^2 \cdot x^4, \quad x^7 = x^3 \cdot x^4$$

$$\text{GCF} = x^4$$

NOTE The exponent on a variable in the GCF is the **least** exponent that appears on that variable in **all** the terms.

EXAMPLE 2 Finding the Greatest Common Factor (Variable Terms)

Find the greatest common factor for each list of terms.

(a) $21m^7$, $18m^6$, $45m^8$, $24m^5$

$$21m^7 = 3 \cdot 7 \cdot m^7$$
$$18m^6 = 2 \cdot 3 \cdot 3 \cdot m^6$$
$$45m^8 = 3 \cdot 3 \cdot 5 \cdot m^8$$
$$24m^5 = 2 \cdot 2 \cdot 2 \cdot 3 \cdot m^5$$

Here, 3 is the greatest common factor of the coefficients 21, 18, 45, and 24. The least exponent on m is 5.

$$\text{GCF} = 3m^5$$

NOW TRY ANSWERS
1. (a) 12 (b) 18 (c) 1

NOW TRY
EXERCISE 2
Find the greatest common
factor for each list of terms.

(a) $25k^3, 15k^2, 35k^5$

(b) m^3n^5, m^4n^4, m^5n^2

(b) $x^4y^2, \quad x^7y^5, \quad x^3y^7, \quad y^{15}$

$x^4y^2 = x^4 \cdot y^2$

$x^7y^5 = x^7 \cdot y^5$

$x^3y^7 = x^3 \cdot y^7$

$y^{15} = y^{15}$

There is no x in the last term, y^{15}, so x will not appear in the greatest common factor. There is a y in each term, however, and 2 is the least exponent on y.

$$\text{GCF} = y^2$$

NOW TRY

OBJECTIVE 2 Factor out the greatest common factor.

Factoring a polynomial is the process of writing a polynomial sum in factored form as a product. For example, the polynomial

$$3m + 12$$

has two terms, $3m$ and 12. The greatest common factor of these two terms is 3. We can write $3m + 12$ so that each term is a product with 3 as one factor.

$$3m + 12$$
$$= 3 \cdot m + 3 \cdot 4 \qquad \text{GCF} = 3$$
$$= 3(m + 4) \qquad \begin{array}{l}\text{Distributive property,}\\ a \cdot b + a \cdot c = a(b + c)\end{array}$$

The factored form of $3m + 12$ is $3(m + 4)$. This process is called **factoring out the greatest common factor.**

> **! CAUTION** The polynomial $3m + 12$ is *not* in factored form when written as
>
> $$3 \cdot m + 3 \cdot 4. \quad \text{Not in factored form}$$
>
> **The terms are factored, but the polynomial is not.** The factored form of $3m + 12$ is the *product*
>
> The factors here are 3 and $m + 4$. $\quad 3(m + 4). \quad$ In factored form

---

EXAMPLE 3 Factoring Out the Greatest Common Factor

Write in factored form by factoring out the greatest common factor.

(a) $5y^2 + 10y$

$$= 5y(y) + 5y(2) \qquad \text{GCF} = 5y$$
$$= 5y(y + 2) \qquad \text{Distributive property}$$

CHECK $\quad 5y(y + 2) \qquad$ Multiply the factored form.

$$= 5y(y) + 5y(2) \qquad \text{Distributive property}$$
$$= 5y^2 + 10y \checkmark \qquad \text{Original polynomial}$$

(b) $20m^5 + 10m^4 - 15m^3$

$$= 5m^3(4m^2) + 5m^3(2m) - 5m^3(3) \qquad \text{GCF} = 5m^3$$
$$= 5m^3(4m^2 + 2m - 3) \qquad \text{Factor out } 5m^3.$$

NOW TRY ANSWERS
2. (a) $5k^2$ **(b)** m^3n^2

NOW TRY
EXERCISE 3
Write in factored form by factoring out the greatest common factor.

(a) $7t^4 - 14t^3$

(b) $8x^6 - 20x^5 + 28x^4$

(c) $30m^4n^3 - 42m^2n^2$

CHECK
$$5m^3(4m^2 + 2m - 3)$$
$$= 5m^3(4m^2) + 5m^3(2m) + 5m^3(-3) \quad \text{Distributive property}$$
$$= 20m^5 + 10m^4 - 15m^3 \; \checkmark \quad \text{Original polynomial}$$

(c) $x^5 + x^3$
$$= x^3(x^2) + x^3(1) \quad \text{GCF} = x^3$$
$$= x^3(x^2 + 1) \quad \boxed{\text{Don't forget the 1.}}$$

Check mentally by distributing x^3 over each term inside the parentheses.

(d) $20m^7p^2 - 36m^3p^4$
$$= 4m^3p^2(5m^4) - 4m^3p^2(9p^2) \quad \text{GCF} = 4m^3p^2$$
$$= 4m^3p^2(5m^4 - 9p^2) \quad \text{Factor out } 4m^3p^2. \qquad \text{NOW TRY} \; \circlearrowleft$$

> **⚠ CAUTION** Be sure to include the 1 in a problem like **Example 3(c)**. *Check that the factored form can be multiplied out to give the original polynomial.*

NOW TRY
EXERCISE 4
Write
$$-14b^2 - 21b^3 + 7b$$
in factored form by factoring out a negative common factor.

EXAMPLE 4   Factoring Out a Negative Common Factor

Write $-8x^4 + 16x^3 - 4x^2$ in factored form.

We can factor out either $4x^2$ or $-4x^2$ here. So that the coefficient of the leading (first) term in the trinomial factor will be positive, we factor out $-4x^2$.

$$-8x^4 + 16x^3 - 4x^2 \quad \boxed{\text{Be careful with signs.}}$$
$$= -4x^2(2x^2) - 4x^2(-4x) - 4x^2(1) \quad -4x^2 \text{ is a common factor.}$$
$$= -4x^2(\underset{\uparrow \text{— Positive coefficient}}{2x^2} - 4x + 1) \quad \text{Factor out } -4x^2.$$

CHECK
$$-4x^2(2x^2 - 4x + 1)$$
$$= -4x^2(2x^2) - 4x^2(-4x) - 4x^2(1) \quad \text{Distributive property}$$
$$= -8x^4 + 16x^3 - 4x^2 \; \checkmark \quad \text{Original polynomial}$$

NOW TRY $\circlearrowleft$

> **NOTE** Whenever we factor a polynomial in which the coefficient of the leading term is negative, we will factor out the negative common factor, even if it is just -1. However, it would also be correct to factor out $4x^2$ in **Example 4** to obtain
> $$4x^2(\underset{\uparrow \text{— Negative coefficient}}{-2x^2} + 4x - 1).$$

EXAMPLE 5   Factoring Out the Greatest Common Factor

Write in factored form by factoring out the greatest common factor.

(a) $a\underset{\text{Same}}{(a + 3)} + 4\underset{}{(a + 3)} \quad \text{The binomial } a + 3 \text{ is the greatest common factor.}$
$$= (a + 3)(a + 4) \quad \text{Factor out } a + 3.$$

NOW TRY
EXERCISE 5

Write in factored form by factoring out the greatest common factor.

(a) $x(x + 2) + 5(x + 2)$

(b) $a(t + 10) - b(t + 10)$

(b) $x^2(x + 1) - 5(x + 1)$

$\qquad = (x + 1)(x^2 - 5)$ Factor out $x + 1$.

NOW TRY

NOTE In factored forms like those in **Example 5,** the order of the factors does not matter because of the commutative property of multiplication, $ab = ba$.

$\qquad (a + 3)(a + 4)$ can also be written $(a + 4)(a + 3)$.

OBJECTIVE 3 Factor by grouping.

When a polynomial has four terms, common factors can sometimes be used to factor by grouping.

EXAMPLE 6   Factoring by Grouping

Factor by grouping.

(a) $2x + 6 + ax + 3a$

Group the first two terms and the last two terms because the first two terms have a common factor of 2 and the last two terms have a common factor of a.

$$2x + 6 + ax + 3a$$
$$= (2x + 6) + (ax + 3a) \qquad \text{Group the terms.}$$
$$= 2(x + 3) + a(x + 3) \qquad \text{Factor each group.}$$

The expression is still not in factored form because it is the *sum* of two terms. Now, however, $x + 3$ is a common factor and can be factored out.

$$= 2(x + 3) + a(x + 3) \qquad x + 3 \text{ is a common factor.}$$

$(2 + a)(x + 3)$ is also correct.

$$= (x + 3)(2 + a) \qquad \text{Factor out } x + 3.$$

The final result $(x + 3)(2 + a)$ is in factored form because it is a ***product.***

CHECK $(x + 3)(2 + a)$

$$= x(2) + x(a) + 3(2) + 3(a) \qquad \text{Multiply using the FOIL method.}$$
$$\qquad \qquad \qquad \qquad \qquad \qquad \text{(Section 4.5)}$$
$$= 2x + ax + 6 + 3a \qquad \text{Simplify.}$$
$$= 2x + 6 + ax + 3a \;\checkmark \qquad \text{Rearrange terms to obtain the original polynomial.}$$

(b) $6ax + 24x + a + 4$

$$= (6ax + 24x) + (a + 4) \qquad \text{Group the terms.}$$
$$= 6x(a + 4) + 1(a + 4) \qquad \text{Factor each group.}$$

Remember the 1.

$$= (a + 4)(6x + 1) \qquad \text{Factor out } a + 4.$$

CHECK $(a + 4)(6x + 1)$

$$= 6ax + a + 24x + 4 \qquad \text{FOIL method}$$
$$= 6ax + 24x + a + 4 \;\checkmark \qquad \text{Rearrange terms to obtain the original polynomial.}$$

NOW TRY ANSWERS

5. (a) $(x + 2)(x + 5)$

   (b) $(t + 10)(a - b)$

**NOW TRY
EXERCISE 6**

Factor by grouping.

(a) $ab + 3a + 5b + 15$

(b) $12xy + 3x + 4y + 1$

(c) $x^3 + 5x^2 - 8x - 40$

(c) $2x^2 - 10x + 3xy - 15y$

$\qquad = (2x^2 - 10x) + (3xy - 15y)$ Group the terms.

$\qquad = 2x(x - 5) + 3y(x - 5)$ Factor each group.

$\qquad = (x - 5)(2x + 3y)$ Factor out $x - 5$.

CHECK $(x - 5)(2x + 3y)$

$\qquad\qquad = 2x^2 + 3xy - 10x - 15y$ FOIL method

$\qquad\qquad = 2x^2 - 10x + 3xy - 15y$ ✓ Original polynomial

(d) $t^3 + 2t^2 - 3t - 6$ 〔Write a + sign between the groups.〕

$\qquad = (t^3 + 2t^2) + (-3t - 6)$ Group the terms.

$\qquad = t^2(t + 2) - 3(t + 2)$ Factor out -3 so there is a common factor, $t + 2$. Check: $-3(t + 2) = -3t - 6$

〔Be careful with signs.〕

$\qquad = (t + 2)(t^2 - 3)$ Factor out $t + 2$.

Check by multiplying using the FOIL method.   **NOW TRY**

⚠ **CAUTION** *Be careful with signs when grouping* in a problem like **Example 6(d).** It is wise to check the factoring in the second step, as shown in the example side comment, before continuing.

Factoring a Polynomial with Four Terms by Grouping

Step 1 Group the terms. Collect the terms into two groups so that each group has a common factor.

Step 2 Factor within the groups. Factor out the greatest common factor from each group.

Step 3 If possible, factor the entire polynomial. Factor out a common binomial factor from the results of Step 2.

Step 4 If necessary, rearrange terms. If Step 2 does not result in a common binomial factor, try a different grouping.

Always check the factored form by multiplying.

EXAMPLE 7 **Rearranging Terms before Factoring by Grouping**

Factor by grouping.

(a) $10x^2 - 12y + 15x - 8xy$

Factoring out a common factor of 2 from the first two terms and a common factor of x from the last two terms gives the following.

$$(10x^2 - 12y) + (15x - 8xy)\qquad \text{Group the terms.}$$

$$= 2(5x^2 - 6y) + x(15 - 8y)\qquad \text{Factor each group.}$$

This does not lead to a common factor, so we try rearranging the terms. There is usually more than one way to do this.

 **NOW TRY EXERCISE 7**

Factor by grouping.

(a) $12p^2 - 28q - 16pq + 21p$

(b) $5xy - 6 - 15x + 2y$

We try the following.

$$10x^2 - 12y + 15x - 8xy \qquad \text{Original polynomial}$$
$$= 10x^2 - 8xy - 12y + 15x \qquad \text{Commutative property}$$
$$= (10x^2 - 8xy) + (-12y + 15x) \qquad \text{Group the terms.}$$
$$= 2x(5x - 4y) + 3(-4y + 5x) \qquad \text{Factor each group.}$$
$$= 2x(5x - 4y) + 3(5x - 4y) \qquad \text{Rewrite } -4y + 5x.$$
$$= (5x - 4y)(2x + 3) \qquad \text{Factor out } 5x - 4y.$$

CHECK $(5x - 4y)(2x + 3)$

$$= 10x^2 + 15x - 8xy - 12y \qquad \text{FOIL method}$$
$$= 10x^2 - 12y + 15x - 8xy \checkmark \qquad \text{Original polynomial}$$

(b) $2xy + 12 - 3y - 8x$

We must rearrange these terms to obtain two groups that each have a common factor. Trial and error suggests the following grouping.

$$2xy + 12 - 3y - 8x \quad \boxed{\text{Write a } + \text{ sign between the groups.}}$$

$$= (2xy - 3y) + (-8x + 12) \qquad \text{Rearrange and group the terms.}$$

$$= y(2x - 3) - 4(2x - 3) \qquad \begin{array}{l}\text{Factor each group.}\\ \textit{Check: } -4(2x - 3) = -8x + 12\end{array}$$

$$\boxed{\text{Be careful with signs.}}$$

$$= (2x - 3)(y - 4) \qquad \text{Factor out } 2x - 3.$$

Because the quantities in parentheses in the second step must be the same, we factored out -4 rather than 4.

CHECK $(2x - 3)(y - 4)$

$$= 2xy - 8x - 3y + 12 \qquad \text{FOIL method}$$
$$= 2xy + 12 - 3y - 8x \checkmark \qquad \text{Original polynomial} \qquad \text{NOW TRY} \; \circlearrowleft$$

NOW TRY ANSWERS

7. **(a)** $(3p - 4q)(4p + 7)$

     (b) $(5x + 2)(y - 3)$

5.1 Exercises

FOR EXTRA HELP   MyMathLab®

 Complete solution available in MyMathLab

Concept Check *Complete each statement.*

1. To factor a number or quantity means to write it as a(n) _____. Factoring is the opposite, or inverse, process of _____.

2. An integer or variable expression that is a factor of two or more terms is a(n) _____. For example, 12 (*is/is not*) a common factor of both 36 and 72 because it _____ evenly into both integers.

Find the greatest common factor for each list of numbers. ***See Example 1.***

3. 12, 16   **4.** 18, 24   **5.** 40, 20, 4   **6.** 50, 30, 5

7. 18, 24, 36, 48   **8.** 15, 30, 45, 75   **9.** 6, 8, 9   **10.** 20, 22, 23

Find the greatest common factor for each list of terms. **See Examples 1 and 2.**

11. $16y$, 24

12. $18w$, 27

13. $30x^3$, $40x^6$, $50x^7$

14. $60z^4$, $70z^8$, $90z^9$

15. x^4y^3, xy^2

16. a^4b^5, a^3b

17. $42ab^3$, $36a$, $90b$, $48ab$

18. $45c^3d$, $75c$, $90d$, $105cd$

19. $12m^3n^2$, $18m^5n^4$, $36m^8n^3$

20. $25p^5r^7$, $30p^7r^8$, $50p^5r^3$

Concept Check   *An expression is factored when it is written as a product, not a sum. Decide whether each expression is* factored *or* not factored.

21. $2k^2(5k)$

22. $2k^2(5k + 1)$

23. $2k^2 + (5k + 1)$

24. $(2k^2 + 5k) + 1$

25. *Concept Check*   A student factored as follows.

$$18x^3y^2 + 9xy$$

$$= 9xy(2x^2y)$$

WHAT WENT WRONG? Factor correctly.

26. How can we check an answer when we factor a polynomial?

Complete each factoring by writing each polynomial as the product of two factors. **See Example 3.**

27. $9m^4$

$= 3m^2(\underline{\hspace{1cm}})$

28. $12p^5$

$= 6p^3(\underline{\hspace{1cm}})$

29. $-8z^9$

$= -4z^5(\underline{\hspace{1cm}})$

30. $-15k^{11}$

$= -5k^8(\underline{\hspace{1cm}})$

31. $6m^4n^5$

$= 3m^3n(\underline{\hspace{1cm}})$

32. $27a^3b^2$

$= 9a^2b(\underline{\hspace{1cm}})$

33. $12y + 24$

$= 12(\underline{\hspace{1cm}})$

34. $18p + 36$

$= 18(\underline{\hspace{1cm}})$

35. $10a^2 - 20a$

$= 10a(\underline{\hspace{1cm}})$

36. $15x^2 - 30x$

$= 15x(\underline{\hspace{1cm}})$

37. $8x^2y + 12x^3y^2$

$= 4x^2y(\underline{\hspace{1cm}})$

38. $18s^3t^2 + 10st$

$= 2st(\underline{\hspace{1cm}})$

Write in factored form by factoring out the greatest common factor. **See Examples 3–5.**

39. $x^2 - 4x$

40. $m^2 - 7m$

41. $6t^2 + 15t$

42. $8x^2 + 6x$

43. $27m^3 - 9m$

44. $12p^3 - 4p$

45. $m^3 - m^2$

46. $p^3 - p^2$

▶ **47.** $16z^4 + 24z^2$

48. $25k^4 + 15k^2$

49. $-12x^3 - 6x^2$

50. $-21b^3 - 7b^2$

51. $65y^{10} + 35y^6$

52. $100a^5 + 16a^3$

53. $11w^3 - 100$

54. $13z^5 - 80$

55. $8mn^3 + 24m^2n^3$

56. $19p^2y + 38p^2y^3$

57. $13y^8 + 26y^4 - 39y^2$

58. $5x^5 + 25x^4 - 20x^3$

59. $-4x^3 + 10x^2 - 6x$

60. $-9z^3 + 6z^2 - 12z$

61. $36p^6q + 45p^5q^4 + 81p^3q^2$

62. $125a^3z^5 + 60a^4z^4 + 85a^5z^2$

63. $a^5 + 2a^3b^2 - 3a^5b^2 + 4a^4b^3$

64. $x^6 + 5x^4y^3 - 6xy^4 + 10xy$

▶ **65.** $c(x + 2) - d(x + 2)$

66. $r(x + 5) - t(x + 5)$

67. $m(m + 2n) + n(m + 2n)$

68. $q(q + 4p) + p(q + 4p)$

69. $q^2(p - 4) + 1(p - 4)$

70. $y^2(x - 9) + 1(x - 9)$

Students often have difficulty when factoring by grouping because they are not able to tell when a polynomial is completely factored. For example,

$$5y(2x - 3) + 8t(2x - 3) \qquad \text{Not in factored form}$$

is not in factored form, because it is the *sum* of two terms: $5y(2x - 3)$ and $8t(2x - 3)$. However, because $2x - 3$ is a common factor of these two terms, the expression can now be factored.

$$(2x - 3)(5y + 8t) \qquad \text{In factored form}$$

The factored form is a *product* of the two factors $2x - 3$ and $5y + 8t$.

Concept Check   Determine whether each expression is in factored form *or is* not in factored form. *If it is not in factored form, factor it if possible.*

71. $8(7t + 4) + x(7t + 4)$

72. $3r(5x - 1) + 7(5x - 1)$

73. $(8 + x)(7t + 4)$

74. $(3r + 7)(5x - 1)$

75. $18x^2(y + 4) + 7(y - 4)$

76. $12k^3(s - 3) + 7(s + 3)$

77. *Concept Check*   A student factored as follows.

$$x^3 + 4x^2 - 2x - 8$$
$$= (x^3 + 4x^2) + (-2x - 8)$$
$$= x^2(x + 4) + 2(-x - 4)$$

The student could not find a common factor of the two terms. **WHAT WENT WRONG?** Complete the factoring.

78. *Concept Check*   A student factored as follows.

$$10xy + 18 + 12x + 15y$$
$$= (10xy + 18) + (12x + 15y)$$
$$= 2(5xy + 9) + 3(4x + 5y)$$

The student could not find a common factor of the two terms. **WHAT WENT WRONG?** Complete the factoring.

Factor by grouping. ***See Examples 6 and 7.***

79. $p^2 + 4p + pq + 4q$

80. $m^2 + 2m + mn + 2n$

▶ **81.** $a^2 - 2a + ab - 2b$

82. $y^2 - 6y + yw - 6w$

83. $7z^2 + 14z - az - 2a$

84. $5m^2 + 15mp - 2mr - 6pr$

85. $18r^2 + 12ry - 3xr - 2xy$

86. $8s^2 + 6sy - 4st - 3yt$

87. $3a^3 + 3ab^2 + 2a^2b + 2b^3$

88. $4x^3 + 4xy^2 + 3x^2y + 3y^3$

89. $12 - 4a - 3b + ab$

90. $6 - 3x - 2y + xy$

91. $16m^3 - 4m^2p^2 - 4mp + p^3$

92. $10t^3 - 2t^2s^2 - 5ts + s^3$

93. $y^2 + 3x + 3y + xy$

94. $m^2 + 14p + 7m + 2mp$

▶ **95.** $5m - 6p - 2mp + 15$

96. $7y - 9x - 3xy + 21$

97. $18r^2 - 2ty + 12ry - 3rt$

98. $12a^2 - 4bc + 16ac - 3ab$

99. $a^5 - 3 + 2a^5b - 6b$

100. $b^3 - 2 + 5ab^3 - 10a$

Extending Skills   Factor each polynomial. (Hint: As the first step, factor out the greatest common factor.)

101. $16a^2 + 40ab^2 + 16ab + 40b^3$

102. $18x^2 + 12xy^2 + 18xy + 12y^3$

103. $2p^2q^2 - 2p^2q + 2p^3 - 2pq^3$

104. $4m^2n^2 - 4mn^2 - 4m^3n + 4n^3$

5.2   Factoring Trinomials

OBJECTIVES

1   Factor trinomials with coefficient 1 for the second-degree term.

2   Factor such trinomials after factoring out the greatest common factor.

VOCABULARY

☐ prime polynomial

Using the FOIL method, we can find the product of the binomials $k - 3$ and $k + 1$.

$$(k - 3)(k + 1) = k^2 - 2k - 3 \qquad \text{Multiplying}$$

Suppose instead that we are given the polynomial $k^2 - 2k - 3$ and want to write it as the product $(k - 3)(k + 1)$.

$$k^2 - 2k - 3 = (k - 3)(k + 1) \qquad \text{Factoring}$$

Recall that *factoring* is a process that reverses, or "undoes," multiplying.

OBJECTIVE 1   Factor trinomials with coefficient 1 for the second-degree term.

When factoring polynomials with integer coefficients, we use only integers in the factors. For example, we can factor $x^2 + 5x + 6$ by finding integers m and n such that

$$x^2 + 5x + 6 \quad \text{is written as} \quad (x + m)(x + n).$$

To find these integers m and n, we multiply the two binomials on the right.

$$(x + m)(x + n)$$
$$= x^2 + nx + mx + mn \qquad \text{FOIL method}$$
$$= x^2 + (n + m)x + mn \qquad \text{Distributive property}$$

Comparing this result with $x^2 + 5x + 6$ shows that we must find integers m and n having a sum of 5 and a product of 6.

Product of m and n is 6.

$$x^2 + 5x + 6 = x^2 + (n + m)x + mn$$

Sum of m and n is 5.

Because many pairs of integers have a sum of 5, it is best to begin by listing those pairs of integers whose product is 6. Both 5 and 6 are positive, so we consider only pairs in which both integers are positive.

Factors of 6	Sums of Factors
6, 1	$6 + 1 = 7$
3, 2	$3 + 2 = 5$

Sum is 5.

Both pairs have a product of 6, but only the pair 3 and 2 has a sum of 5. So 3 and 2 are the required integers.

$$x^2 + 5x + 6 \quad \text{factors as} \quad (x + 3)(x + 2).$$

Check by using the FOIL method to multiply the binomials. *Make sure that the sum of the outer and inner products produces the correct middle term.*

CHECK $\qquad (x + 3)(x + 2) = x^2 + 5x + 6$ ✓ Correct

$3x$

$2x$

$5x$ Add.

This method can be used only to factor trinomials that have 1 as the coefficient of the second-degree (squared variable) term.

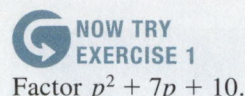
**NOW TRY
EXERCISE 1**

Factor $p^2 + 7p + 10$.

EXAMPLE 1   Factoring a Trinomial (All Positive Terms)

Factor $m^2 + 9m + 14$.

Look for two integers whose product is 14 and whose sum is 9. List pairs of integers whose product is 14, and examine the sums. Only positive integers are needed because all signs in $m^2 + 9m + 14$ are positive.

Factors of 14	Sums of Factors
14, 1	$14 + 1 = 15$
7, 2	$7 + 2 = 9$

Sum is 9.

The required integers are 7 and 2 because $7 \cdot 2 = 14$ and $7 + 2 = 9$.

$$m^2 + 9m + 14 \quad \text{factors as} \quad (m + 7)(m + 2).$$

$(m + 2)(m + 7)$ is also correct.

CHECK $(m + 7)(m + 2)$

$= m^2 + 2m + 7m + 14$ FOIL method

$= m^2 + 9m + 14 \checkmark$ Original polynomial   NOW TRY

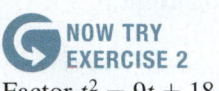
**NOW TRY
EXERCISE 2**

Factor $t^2 - 9t + 18$.

EXAMPLE 2   Factoring a Trinomial (Negative Middle Term)

Factor $x^2 - 9x + 20$.

We must find two integers whose product is 20 and whose sum is -9. Because the numbers we are looking for have a *positive product* and a *negative sum,* we consider only pairs of negative integers.

Factors of 20	Sums of Factors
$-20, -1$	$-20 + (-1) = -21$
$-10, -2$	$-10 + (-2) = -12$
$-5, -4$	$-5 + (-4) = -9$

Sum is -9.

The required integers are -5 and -4.

$$x^2 - 9x + 20 \quad \text{factors as} \quad (x - 5)(x - 4).$$

The order of the factors does not matter.

CHECK $(x - 5)(x - 4)$

$= x^2 - 4x - 5x + 20$ FOIL method

$= x^2 - 9x + 20 \checkmark$ Original polynomial   NOW TRY

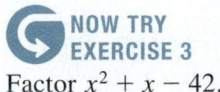
**NOW TRY
EXERCISE 3**

Factor $x^2 + x - 42$.

EXAMPLE 3   Factoring a Trinomial (Negative Last (Constant) Term)

Factor $x^2 + x - 6$.

We must find two integers whose product is -6 and whose sum is 1 (because the coefficient of x, or $1x$, is 1). To obtain a *negative product,* the pairs of integers must have different signs.

Factors of -6	Sums of Factors
6, -1	$6 + (-1) = 5$
$-6, 1$	$-6 + 1 = -5$
3, -2	$3 + (-2) = 1$

Sum is 1.

Once we find the required pair, we can stop listing factors.

NOW TRY ANSWERS
1. $(p + 5)(p + 2)$
2. $(t - 6)(t - 3)$
3. $(x + 7)(x - 6)$

The required integers are 3 and -2.

To check, multiply the factored form.

$$x^2 + x - 6 \quad \text{factors as} \quad (x + 3)(x - 2).$$ NOW TRY

**NOW TRY
EXERCISE 4**

Factor $x^2 - 4x - 21$.

EXAMPLE 4 Factoring a Trinomial (Two Negative Terms)

Factor $p^2 - 2p - 15$.

Find two integers whose product is -15 and whose sum is -2. Because the constant term, -15, is negative, list pairs of integers with different signs.

Factors of −15	Sums of Factors
15, −1	$15 + (-1) = 14$
−15, 1	$-15 + 1 = -14$
5, −3	$5 + (-3) = 2$
−5, 3	$-5 + 3 = -2$

Sum is -2.

The required integers are -5 and 3.

To check, multiply the factored form.

$$p^2 - 2p - 15 \quad \text{factors as} \quad (p - 5)(p + 3).$$

NOW TRY

NOTE In **Examples 1–4,** we listed factors in descending order (disregarding their signs) when we were looking for the required pair of integers. This helps avoid skipping the correct combination.

Trinomials that cannot be factored using only integers are **prime polynomials.**

**NOW TRY
EXERCISE 5**

Factor each trinomial if possible.

(a) $m^2 + 5m + 8$

(b) $t^2 + 11t - 24$

EXAMPLE 5 Deciding Whether Polynomials Are Prime

Factor each trinomial if possible.

(a) $x^2 - 5x + 12$

As in **Example 2,** both factors must be negative to give a positive product and a negative sum. List pairs of negative integers whose product is 12, and examine the sums.

Factors of 12	Sums of Factors
−12, −1	$-12 + (-1) = -13$
−6, −2	$-6 + (-2) = -8$
−4, −3	$-4 + (-3) = -7$

No sum is -5.

None of the pairs of integers has a sum of -5. Therefore, the trinomial $x^2 - 5x + 12$ *cannot be factored using only integers.* It is a prime polynomial.

(b) $k^2 - 8k + 11$

There is no pair of integers whose product is 11 and whose sum is -8, so $k^2 - 8k + 11$ is a prime polynomial.

NOW TRY

Guidelines for Factoring $x^2 + bx + c$

Find two integers whose product is c and whose sum is b.

1. Both integers must be positive if b and c are positive. (See **Example 1.**)

2. Both integers must be negative if c is positive and b is negative. (See **Example 2.**)

3. One integer must be positive and one must be negative if c is negative. (See **Examples 3 and 4.**)

NOW TRY ANSWERS
4. $(x - 7)(x + 3)$
5. **(a)** prime **(b)** prime

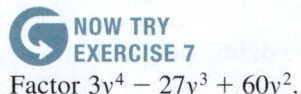

**NOW TRY
EXERCISE 6**

Factor $a^2 + 2ab - 15b^2$.

EXAMPLE 6   Factoring a Multivariable Trinomial

Factor $z^2 - 2bz - 3b^2$.

Here, the coefficient of z in the middle term is $-2b$, so we need to find two expressions whose product is $-3b^2$ and whose sum is $-2b$.

Factors of $-3b^2$	Sums of Factors
$3b, -b$	$3b + (-b) = 2b$
$-3b, b$	$-3b + b = -2b$

Sum is $-2b$.

$$z^2 - 2bz - 3b^2 \quad \text{factors as} \quad (z - 3b)(z + b).$$

CHECK $(z - 3b)(z + b)$

$$= z^2 + zb - 3bz - 3b^2 \qquad \text{FOIL method}$$

$$= z^2 + 1bz - 3bz - 3b^2 \qquad \text{Identity and commutative properties}$$

$$= z^2 - 2bz - 3b^2 \checkmark \qquad \text{Combine like terms.} \qquad \textbf{NOW TRY}$$

OBJECTIVE 2   Factor such trinomials after factoring out the greatest common factor.

**NOW TRY
EXERCISE 7**

Factor $3y^4 - 27y^3 + 60y^2$.

EXAMPLE 7   Factoring a Trinomial with a Common Factor

Factor $4x^5 - 28x^4 + 40x^3$.

The terms have a common factor.

$$4x^5 - 28x^4 + 40x^3$$

$$= 4x^3(x^2 - 7x + 10) \qquad \text{Factor out the greatest common factor, } 4x^3.$$

Factor $x^2 - 7x + 10$. The integers -5 and -2 have a product of 10 and a sum of -7.

Include $4x^3$. $= 4x^3(x - 5)(x - 2)$ Completely factored form

CHECK $4x^3(x - 5)(x - 2)$

$$= 4x^3(x^2 - 2x - 5x + 10) \qquad \text{FOIL method}$$

$$= 4x^3(x^2 - 7x + 10) \qquad \text{Combine like terms.}$$

$$= 4x^5 - 28x^4 + 40x^3 \checkmark \qquad \text{Distributive property} \qquad \textbf{NOW TRY}$$

NOW TRY ANSWERS

6. $(a + 5b)(a - 3b)$

7. $3y^2(y - 5)(y - 4)$

! CAUTION   *When factoring, always look for a common factor first.* Remember to include the common factor as part of the answer. Check by multiplying out the completely factored form.

5.2 Exercises

FOR EXTRA HELP

  MyMathLab®

▶ *Complete solution available in MyMathLab*

Concept Check   *Answer each question.*

1. When factoring a trinomial in x as $(x + a)(x + b)$, what must be true of a and b if the coefficient of the constant term of the trinomial is negative?

2. In **Exercise 1,** what must be true of a and b if the coefficient of the constant term is positive?

3. Which is the correct factored form of $x^2 - 12x + 32$?

    A. $(x - 8)(x + 4)$ **B.** $(x + 8)(x - 4)$

    C. $(x - 8)(x - 4)$ **D.** $(x + 8)(x + 4)$

4. What is the suggested first step in factoring

$$2x^3 + 8x^2 - 10x? \quad \text{(See Example 7.)}$$

5. What polynomial can be factored as $(a + 9)(a + 4)$?

6. What polynomial can be factored as $(y - 7)(y + 3)$?

List all pairs of integers with the given product. Then find the pair whose sum is given. **See the tables in Examples 1–4.**

7. Product: 48;   Sum: -19 8. Product: 18;   Sum: 9

9. Product: -24;   Sum: -5 10. Product: -36;   Sum: -16

Concept Check   *Complete each factoring.*

11. To factor $y^2 + 12y + 20$, find two integers whose product is _____ and whose sum is _____. Complete the table.

Factors of 20	Sums of Factors
20, 1	$20 + 1 = 21$
10, ___	$10 + $ ___ $ = $ ___
5, ___	$5 + $ ___ $ = $ ___

Which pair of factors has the required sum? _____
Now factor the trinomial.

12. To factor $t^2 - 12t + 32$, find two integers whose product is _____ and whose sum is _____. Complete the table.

Factors of 32	Sums of Factors
$-32, -1$	$-32 + (-1) = -33$
$-16, $ ___	$-16 + ($ ___ $) = $ ___
$-8, $ ___	$-8 + ($ ___ $) = $ ___

Which pair of factors has the required sum? _____
Now factor the trinomial.

Complete each factoring. **See Examples 1–4.**

13. $p^2 + 11p + 30$
    $= (p + 5)(\underline{\hspace{1cm}})$

14. $x^2 + 10x + 21$
    $= (x + 7)(\underline{\hspace{1cm}})$

15. $x^2 + 15x + 44$
    $= (x + 4)(\underline{\hspace{1cm}})$

16. $r^2 + 15r + 56$
    $= (r + 7)(\underline{\hspace{1cm}})$

17. $x^2 - 9x + 8$
    $= (x - 1)(\underline{\hspace{1cm}})$

18. $t^2 - 14t + 24$
    $= (t - 2)(\underline{\hspace{1cm}})$

19. $y^2 - 2y - 15$
    $= (y + 3)(\underline{\hspace{1cm}})$

20. $t^2 - t - 42$
    $= (t + 6)(\underline{\hspace{1cm}})$

21. $x^2 + 9x - 22$
    $= (x - 2)(\underline{\hspace{1cm}})$

22. $x^2 + 6x - 27$
    $= (x - 3)(\underline{\hspace{1cm}})$

23. $y^2 - 7y - 18$
    $= (y + 2)(\underline{\hspace{1cm}})$

24. $y^2 - 2y - 24$
    $= (y + 4)(\underline{\hspace{1cm}})$

Factor completely. If a polynomial cannot be factored, write prime. **See Examples 1–5.**
(Hint: In Exercises 43 and 44, first write the trinomial in descending powers and then factor.)

25. $y^2 + 9y + 8$ 26. $a^2 + 9a + 20$

▶ 27. $b^2 + 8b + 15$ 28. $x^2 + 6x + 8$

29. $m^2 + m - 20$

30. $p^2 + 4p - 5$

▶ **31.** $y^2 - 8y + 15$

32. $y^2 - 6y + 8$

▶ **33.** $x^2 + 4x + 5$

34. $t^2 + 11t + 12$

35. $z^2 - 15z + 56$

36. $x^2 - 13x + 36$

▶ **37.** $r^2 - r - 30$

38. $q^2 - q - 42$

39. $a^2 - 8a - 48$

40. $d^2 - 4d - 45$

41. $x^2 + 3x - 39$

42. $m^2 + 10m - 30$

43. $-32 + 14x + x^2$

44. $-39 + 10x + x^2$

Factor completely. **See Example 6.**

45. $r^2 + 3ra + 2a^2$

46. $x^2 + 5xa + 4a^2$

47. $x^2 + 4xy + 3y^2$

48. $p^2 + 9pq + 8q^2$

▶ **49.** $t^2 - tz - 6z^2$

50. $a^2 - ab - 12b^2$

51. $v^2 - 11vw + 30w^2$

52. $v^2 - 11vx + 24x^2$

53. $m^2 + 4mn - 12n^2$

54. $x^2 + 6xy - 16y^2$

55. $a^2 - 9ab + 18b^2$

56. $h^2 - 11hk + 28k^2$

Factor completely. **See Example 7.**

57. $4x^2 + 12x - 40$

58. $5y^2 + 5y - 30$

▶ **59.** $2t^3 + 8t^2 + 6t$

60. $3t^3 + 27t^2 + 24t$

61. $2x^6 + 8x^5 - 42x^4$

62. $4y^5 + 12y^4 - 40y^3$

63. $6z^4 - 24z^3 + 18z^2$

64. $5x^4 - 35x^3 + 30x^2$

65. $5m^5 - 25m^4 + 40m^2$

66. $12k^5 - 6k^3 + 10k^2$

67. $x^3 - 7x^2y + 12xy^2$

68. $p^3 - 8p^2q + 15pq^2$

69. $a^5 + 3a^4b - 4a^3b^2$

70. $k^7 - 2k^6m - 15k^5m^2$

71. $z^{10} - 4z^9y - 21z^8y^2$

72. $x^9 + 5x^8w - 24x^7w^2$

73. $m^3n - 10m^2n^2 + 24mn^3$

74. $y^3z + 3y^2z^2 - 54yz^3$

75. $y^3z + y^2z^2 - 6yz^3$

76. $m^3n - 2m^2n^2 - 3mn^3$

Extending Skills   *Factor each polynomial.*

77. $(a + b)x^2 + (a + b)x - 12(a + b)$

78. $(x + y)n^2 + (x + y)n - 20(x + y)$

79. $(2p + q)r^2 - 12(2p + q)r + 27(2p + q)$

80. $(3m - n)k^2 - 13(3m - n)k + 40(3m - n)$

5.3   More on Factoring Trinomials

OBJECTIVES

1 Factor trinomials by grouping when the coefficient of the second-degree term is not 1.

2 Factor trinomials using the FOIL method.

OBJECTIVES

1 Factor trinomials by grouping when the coefficient of the second-degree term is not 1.

2 Factor trinomials using the FOIL method.

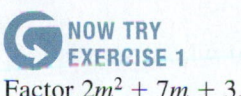
**NOW TRY
EXERCISE 1**
Factor $2m^2 + 7m + 3$.

OBJECTIVE 1 Factor trinomials by grouping when the coefficient of the second-degree term is not 1.

We factor a trinomial in which the coefficient of the second-degree term is *not* 1, such as

$$2x^2 + 7x + 6,$$

by extending our work from the previous sections.

EXAMPLE 1 Factoring by Grouping (Coefficient of the Second-Degree Term Not 1)

Factor $2x^2 + 7x + 6$.

To factor this trinomial, we look for two positive integers whose product is $2 \cdot 6 = 12$ and whose sum is 7.

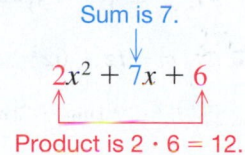

Sum is 7.

$$2x^2 + 7x + 6$$

Product is $2 \cdot 6 = 12$.

The required integers are 3 and 4. We use these integers to write the middle term $7x$ as $3x + 4x$.

$$2x^2 + 7x + 6$$
$$= 2x^2 + \underbrace{3x + 4x}_{7x} + 6$$
$$= (2x^2 + 3x) + (4x + 6) \qquad \text{Group the terms.}$$
$$= x(2x + 3) + 2(2x + 3) \qquad \text{Factor each group.}$$

Must be the same factor

$$= (2x + 3)(x + 2) \qquad \text{Factor out } 2x + 3.$$

CHECK   Multiply $(2x + 3)(x + 2)$ to obtain $2x^2 + 7x + 6.$ ✓   **NOW TRY**

NOTE   In **Example 1,** we could have written $7x$ as $4x + 3x$, rather than as $3x + 4x$. Factoring by grouping would give the same answer. Try this.

EXAMPLE 2 Factoring Trinomials by Grouping

Factor each trinomial.

(a) $6r^2 + r - 1$

We must find two integers with a product of $6(-1) = -6$ and a sum of 1.

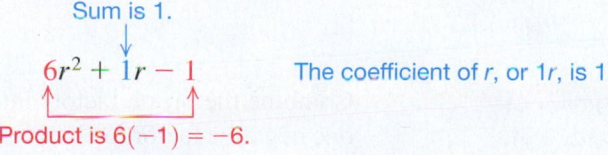

Sum is 1.

$$6r^2 + 1r - 1 \qquad \text{The coefficient of } r, \text{ or } 1r, \text{ is } 1.$$

Product is $6(-1) = -6$.

NOW TRY ANSWER
1. $(2m + 1)(m + 3)$

NOW TRY
EXERCISE 2

Factor.

(a) $2z^2 + 5z + 3$

(b) $15m^2 + m - 2$

(c) $8x^2 - 2xy - 3y^2$

The integers -2 and 3 have a product of -6 and a sum of 1. We write the middle term r as $-2r + 3r$.

$$6r^2 + r - 1$$

$$= 6r^2 - 2r + 3r - 1 \qquad r = -2r + 3r$$

$$= (6r^2 - 2r) + (3r - 1) \qquad \text{Group the terms.}$$

$$= 2r(3r - 1) + 1(3r - 1) \qquad \text{The binomials must be the same.}$$

Remember the 1.

$$= (3r - 1)(2r + 1) \qquad \text{Factor out } 3r - 1.$$

CHECK   Multiply $(3r - 1)(2r + 1)$ to obtain $6r^2 + r - 1$. ✓

(b) $12z^2 - 5z - 2$

Look for two integers whose product is $12(-2) = -24$ and whose sum is -5. The required integers are 3 and -8.

$$12z^2 - 5z - 2$$

$$= 12z^2 + 3z - 8z - 2 \qquad -5z = 3z - 8z$$

$$= (12z^2 + 3z) + (-8z - 2) \qquad \text{Group the terms.}$$

$$= 3z(4z + 1) - 2(4z + 1) \qquad \text{Factor each group.}$$

Be careful with signs.

$$= (4z + 1)(3z - 2) \qquad \text{Factor out } 4z + 1.$$

CHECK   Multiply $(4z + 1)(3z - 2)$ to obtain $12z^2 - 5z - 2$. ✓

(c) $10m^2 + mn - 3n^2$

Two integers whose product is $10(-3) = -30$ and whose sum is 1 are -5 and 6.

$$10m^2 + mn - 3n^2$$

$$= 10m^2 - 5mn + 6mn - 3n^2 \qquad mn = -5mn + 6mn$$

$$= (10m^2 - 5mn) + (6mn - 3n^2) \qquad \text{Group the terms.}$$

$$= 5m(2m - n) + 3n(2m - n) \qquad \text{Factor each group.}$$

$$= (2m - n)(5m + 3n) \qquad \text{Factor out } 2m - n.$$

CHECK   Multiply $(2m - n)(5m + 3n)$ to obtain $10m^2 + mn - 3n^2$. ✓

NOW TRY

EXAMPLE 3   **Factoring a Trinomial with a Common Factor by Grouping**

Factor $28x^5 - 58x^4 - 30x^3$.

$$28x^5 - 58x^4 - 30x^3$$

$$= 2x^3(14x^2 - 29x - 15) \qquad \text{Factor out the greatest common factor, } 2x^3.$$

To factor $14x^2 - 29x - 15$, find two integers whose product is $14(-15) = -210$ and whose sum is -29. Factoring 210 into prime factors helps find these integers.

$$210 = 2 \cdot 3 \cdot 5 \cdot 7$$

Combine the prime factors into pairs in different ways, using one positive factor and one negative factor to obtain -210. The factors 6 and -35 have the correct sum, -29.

NOW TRY ANSWERS

2. (a) $(2z + 3)(z + 1)$

 (b) $(3m - 1)(5m + 2)$

 (c) $(4x - 3y)(2x + y)$

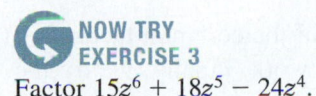

**NOW TRY
EXERCISE 3**

Factor $15z^6 + 18z^5 - 24z^4$.

$$28x^5 - 58x^4 - 30x^3$$

$= 2x^3(14x^2 - 29x - 15)$ Factor out the GCF.

Remember the common factor. $= 2x^3(14x^2 + 6x - 35x - 15)$ $-29x = 6x - 35x$

$= 2x^3[(14x^2 + 6x) + (-35x - 15)]$ Group the terms.

$= 2x^3[2x(7x + 3) - 5(7x + 3)]$ Factor each group.

$= 2x^3[(7x + 3)(2x - 5)]$ Factor out $7x + 3$.

$= 2x^3(7x + 3)(2x - 5)$

CHECK   One way to check is to first multiply $2x^3(7x + 3)$ to obtain $(14x^4 + 6x^3)$. Then multiply

$(14x^4 + 6x^3)(2x - 5)$ to obtain $28x^5 - 58x^4 - 30x^3$.   ✓   **NOW TRY**

OBJECTIVE 2   Factor trinomials using the FOIL method.

There is an alternative method of factoring trinomials that uses trial and error.

EXAMPLE 4   **Factoring Using FOIL (Coefficient of the Second-Degree Term Not 1)**

Factor $2x^2 + 7x + 6$. (We factored this trinomial by grouping in **Example 1**.)

We want to write $2x^2 + 7x + 6$ as the product of two binomials.

$$2x^2 + 7x + 6$$

$= (\underline{})(\underline{})$ We use the FOIL method in reverse.

The product of the two first terms of the binomials must be $2x^2$. The possible factors of $2x^2$ are $2x$ and x, or $-2x$ and $-x$. Because all terms of the trinomial are positive, we consider only positive factors. Thus, we have the following.

$$2x^2 + 7x + 6$$

$= (2x\underline{})(x\underline{})$

The product of the two last terms of the binomials must be 6. It can be factored as $1 \cdot 6$, $6 \cdot 1$, $2 \cdot 3$, or $3 \cdot 2$. Beginning with 1 and 6, we try each pair of factors in $(2x\underline{})(x\underline{})$ to find the pair that gives the correct middle term, $7x$.

$(2x + 1)(x + 6)$ Incorrect
x
$12x$
$13x$ Add. (Wrong middle term)

Now try the pair 6 and 1 in $(2x\underline{})(x\underline{})$.

$(2x + 6)(x + 1)$ Incorrect
$6x$
$2x$
$8x$ Add. (Wrong middle term)

Because $2x + 6 = 2(x + 3)$, the terms of the binomial $2x + 6$ have a common factor of 2, while the terms of $2x^2 + 7x + 6$ have no common factor other than 1. The product $(2x + 6)(x + 1)$ cannot be correct.

NOW TRY ANSWER

3. $3z^4(5z - 4)(z + 2)$

If the terms of the original polynomial have greatest common factor 1, then each of its factors will also have terms with GCF 1.

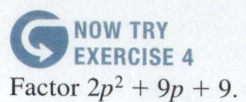
NOW TRY
EXERCISE 4
Factor $2p^2 + 9p + 9$.

We try the pair 2 and 3 in $(2x\underline{\hspace{1cm}})(x\underline{\hspace{1cm}})$. Because of the common factor 2 in the terms of $2x + 2$, the product $(2x + 2)(x + 3)$ will not work. Finally, we try the pair 3 and 2 in $(2x\underline{\hspace{1cm}})(x\underline{\hspace{1cm}})$.

$$(2x + 3)(x + 2) = 2x^2 + 7x + 6 \qquad \text{Correct}$$

$3x$

$4x$

$7x$ Add. (Correct middle term)

Thus, $2x^2 + 7x + 6$ factors as $(2x + 3)(x + 2)$.

CHECK   Multiply $(2x + 3)(x + 2)$ to obtain $2x^2 + 7x + 6$. ✓   NOW TRY

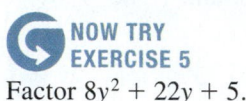
NOW TRY
EXERCISE 5
Factor $8y^2 + 22y + 5$.

EXAMPLE 5   Factoring a Trinomial Using FOIL (All Positive Terms)

Factor $8p^2 + 14p + 5$.

The number 8 has several possible pairs of factors, but 5 has only 1 and 5 or -1 and -5, so we begin by considering the factors of 5. We ignore the negative factors because all coefficients in the trinomial are positive. If $8p^2 + 14p + 5$ can be factored, the factors will have this form.

$$(\underline{\hspace{1cm}} + 5)(\underline{\hspace{1cm}} + 1)$$

The possible pairs of factors of $8p^2$ are $8p$ and p, or $4p$ and $2p$. We try various combinations, checking to see if the middle term is $14p$.

$(8p + 5)(p + 1)$ Incorrect | $(p + 5)(8p + 1)$ Incorrect | $(4p + 5)(2p + 1)$ Correct

$5p$ | $40p$ | $10p$

$8p$ | p | $4p$

$13p$ Add. | $41p$ Add. | $14p$ Add.

The combination on the right produces $14p$, the correct middle term.

$$8p^2 + 14p + 5 \quad \text{factors as} \quad (4p + 5)(2p + 1).$$

CHECK   Multiply $(4p + 5)(2p + 1)$ to obtain $8p^2 + 14p + 5$. ✓   NOW TRY

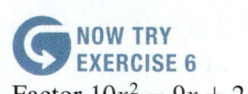
NOW TRY
EXERCISE 6
Factor $10x^2 - 9x + 2$.

EXAMPLE 6   Factoring a Trinomial Using FOIL (Negative Middle Term)

Factor $6x^2 - 11x + 3$.

Because 3 has only 1 and 3 or -1 and -3 as factors, it is better here to begin by factoring 3. The last (constant) term of the trinomial $6x^2 - 11x + 3$ is positive and the middle term has a negative coefficient, so we consider only negative factors. We need two negative factors, because the *product* of two negative factors is positive and their *sum* is negative, as required.

We try -3 and -1 as factors of 3.

$$(\underline{\hspace{1cm}} - 3)(\underline{\hspace{1cm}} - 1)$$

The factors of $6x^2$ may be either $6x$ and x, or $2x$ and $3x$.

$(6x - 3)(x - 1)$ Incorrect | $(2x - 3)(3x - 1)$ Correct

$-3x$ | $-9x$

$-6x$ | $-2x$

$-9x$ Add. | $-11x$ Add.

NOW TRY ANSWERS
4. $(2p + 3)(p + 3)$
5. $(4y + 1)(2y + 5)$
6. $(5x - 2)(2x - 1)$

The factors $2x$ and $3x$ produce $-11x$, the correct middle term.    *Check* by multiplying.

$$6x^2 - 11x + 3 \quad \text{factors as} \quad (2x - 3)(3x - 1).$$ NOW TRY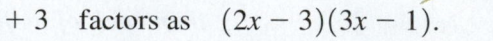

NOTE In **Example 6,** our initial attempt to factor $6x^2 - 11x + 3$ as $(6x - 3)(x - 1)$ *cannot* be correct because the terms of $6x - 3$ have a common factor of 3, while those of the original polynomial do not.

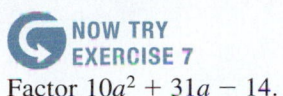

**NOW TRY
EXERCISE 7**

Factor $10a^2 + 31a - 14$.

EXAMPLE 7 Factoring a Trinomial Using FOIL (Negative Constant Term)

Factor $8x^2 + 6x - 9$.

The integer 8 has several possible pairs of factors, as does -9. Because the constant term is negative, one positive factor and one negative factor of -9 are needed. The coefficient of the middle term is relatively small, so we avoid large factors such as 8 or 9. We try $4x$ and $2x$ as factors of $8x^2$, and 3 and -3 as factors of -9.

The combination on the right produces $6x$, the correct middle term.

$$8x^2 + 6x - 9 \quad \text{factors as} \quad (4x - 3)(2x + 3).$$

Check by multiplying.

NOW TRY

**NOW TRY
EXERCISE 8**

Factor $8z^2 + 2wz - 15w^2$.

EXAMPLE 8 Factoring a Multivariable Trinomial

Factor $12a^2 - ab - 20b^2$.

There are several pairs of factors of $12a^2$, including

$$12a \text{ and } a, \quad 6a \text{ and } 2a, \quad \text{and} \quad 3a \text{ and } 4a.$$

There are also many pairs of factors of $-20b^2$, including

$$20b \text{ and } -b, \quad -20b \text{ and } b, \quad 10b \text{ and } -2b, \quad -10b \text{ and } 2b,$$

$$4b \text{ and } -5b, \quad \text{and} \quad -4b \text{ and } 5b.$$

Once again, because the coefficient of the desired middle term is relatively small, avoid the larger factors. Try the factors $6a$ and $2a$, and $4b$ and $-5b$.

$$(6a + 4b)(2a - 5b)$$

This cannot be correct because the terms of $6a + 4b$ have 2 as a common factor, while the terms of the given trinomial do not. Try $3a$ and $4a$ with $4b$ and $-5b$.

$$(3a + 4b)(4a - 5b)$$

$$= 12a^2 + ab - 20b^2 \qquad \text{Incorrect}$$

Here the middle term is ab rather than $-ab$, so we interchange the signs of the last two terms in the factors.

Check by multiplying.

$$12a^2 - ab - 20b^2 \quad \text{factors as} \quad (3a - 4b)(4a + 5b). \qquad \text{NOW TRY} \quad \text{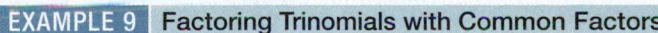}$$

EXAMPLE 9 Factoring Trinomials with Common Factors

Factor each trinomial.

(a) $15y^3 + 55y^2 + 30y$

$$= 5y(3y^2 + 11y + 6) \qquad \text{Factor out the greatest common factor, } 5y.$$

To factor $3y^2 + 11y + 6$, try $3y$ and y as factors of $3y^2$, and 2 and 3 as factors of 6.

$$(3y + 2)(y + 3)$$

$$= 3y^2 + 11y + 6 \qquad \text{Correct}$$

NOW TRY ANSWERS

7. $(5a - 2)(2a + 7)$

8. $(4z - 5w)(2z + 3w)$

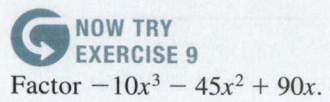

NOW TRY
EXERCISE 9

Factor $-10x^3 - 45x^2 + 90x$.

This leads to the completely factored form.

$$15y^3 + 55y^2 + 30y$$

> Remember the common factor.

$$= 5y(3y + 2)(y + 3)$$

CHECK $5y(3y + 2)(y + 3)$

$\qquad = 5y(3y^2 + 9y + 2y + 6)$ FOIL method

$\qquad = 5y(3y^2 + 11y + 6)$ Combine like terms.

$\qquad = 15y^3 + 55y^2 + 30y$ ✓   Distributive property

(b) $-24a^3 - 42a^2 + 45a$

The common factor could be $3a$ or $-3a$. If we factor out $-3a$, the leading term of the trinomial will be positive, which makes it easier to factor the remaining trinomial.

$$-24a^3 - 42a^2 + 45a$$

$$= -3a(8a^2 + 14a - 15)$$ Factor out $-3a$.

$$= -3a(4a - 3)(2a + 5)$$ Factor the trinomial.

CHECK $-3a(4a - 3)(2a + 5)$

NOW TRY ANSWER

9. $-5x(2x - 3)(x + 6)$

> We can multiply $-3a(4a - 3)$ first.

$\qquad = (-12a^2 + 9a)(2a + 5)$ Distributive property

$\qquad = -24a^3 - 42a^2 + 45a$ ✓   FOIL method

NOW TRY

5.3 Exercises

> FOR EXTRA HELP   ▶ MyMathLab®

▶ *Complete solution available in MyMathLab*

Concept Check   *The middle term of each trinomial has been rewritten. Now factor by grouping.*

1. $10t^2 + 9t + 2$
$\quad = 10t^2 + 5t + 4t + 2$

2. $6x^2 + 13x + 6$
$\quad = 6x^2 + 9x + 4x + 6$

3. $15z^2 - 19z + 6$
$\quad = 15z^2 - 10z - 9z + 6$

4. $12p^2 - 17p + 6$
$\quad = 12p^2 - 9p - 8p + 6$

5. $8s^2 + 2st - 3t^2$
$\quad = 8s^2 - 4st + 6st - 3t^2$

6. $3x^2 - xy - 14y^2$
$\quad = 3x^2 - 7xy + 6xy - 14y^2$

Concept Check   *Complete the steps to factor each trinomial by grouping.*

7. $2m^2 + 11m + 12$

(a) Find two integers whose product is _____ · _____ = _____ and whose sum is _____.

(b) The required integers are _____ and _____.

(c) Now write the middle term, $11m$, as _____ + _____.

(d) Rewrite the given trinomial using four terms as _____.

(e) Factor the polynomial in part (d) by grouping.

(f) Check by multiplying.

8. $6y^2 - 19y + 10$

(a) Find two integers whose product is _____ · _____ = _____ and whose sum is _____.

(b) The required integers are _____ and _____.

(c) Now write the middle term, $-19y$, as _____ + _____.

(d) Rewrite the given trinomial using four terms as _____.

(e) Factor the polynomial in part (d) by grouping.

(f) Check by multiplying.

9. *Concept Check* Which pair of integers would be used to rewrite the middle term when factoring $12y^2 + 5y - 2$ by grouping?

    A. $-8, 3$ **B.** $8, -3$ **C.** $-6, 4$ **D.** $6, -4$

10. *Concept Check* Which pair of integers would be used to rewrite the middle term when factoring $20b^2 - 13b + 2$ by grouping?

    A. $10, 3$ **B.** $-10, -3$ **C.** $8, 5$ **D.** $-8, -5$

Concept Check *Which is the correct factored form of the given polynomial?*

11. $2x^2 - x - 1$

    A. $(2x - 1)(x + 1)$

    B. $(2x + 1)(x - 1)$

12. $3a^2 - 5a - 2$

    A. $(3a + 1)(a - 2)$

    B. $(3a - 1)(a + 2)$

13. $4y^2 + 17y - 15$

    A. $(y + 5)(4y - 3)$

    B. $(2y - 5)(2y + 3)$

14. $12c^2 - 7c - 12$

    A. $(6c - 2)(2c + 6)$

    B. $(4c + 3)(3c - 4)$

15. *Concept Check* A student factoring the trinomial

$$12x^2 + 7x - 12$$

wrote $(4x + 4)$ as one binomial factor. **WHAT WENT WRONG?** Factor correctly.

16. *Concept Check* Another student factored $3k^3 - 12k^2 - 15k$ by first factoring out the common factor $3k$ to obtain $3k(k^2 - 4k - 5)$. Then she wrote the following.

$$k^2 - 4k - 5$$
$$= k^2 - 5k + k - 5$$
$$= k(k - 5) + 1(k - 5)$$
$$= (k - 5)(k + 1) \qquad \text{Her answer}$$

WHAT WENT WRONG? What is the correct factored form?

Complete each factoring. **See Examples 1–9.**

17. $6a^2 + 7ab - 20b^2$

    $= (3a - 4b)(\underline{\hspace{1.5cm}})$

18. $9m^2 + 6mn - 8n^2$

    $= (3m - 2n)(\underline{\hspace{1.5cm}})$

19. $2x^2 + 6x - 8$

    $= 2(\underline{\hspace{2cm}})$

    $= 2(\underline{\hspace{1cm}})(\underline{\hspace{1cm}})$

20. $3x^2 + 9x - 30$

    $= 3(\underline{\hspace{2cm}})$

    $= 3(\underline{\hspace{1cm}})(\underline{\hspace{1cm}})$

21. $4z^3 - 10z^2 - 6z$

    $= 2z(\underline{\hspace{2cm}})$

    $= 2z(\underline{\hspace{1cm}})(\underline{\hspace{1cm}})$

22. $15r^3 - 39r^2 - 18r$

    $= 3r(\underline{\hspace{2cm}})$

    $= 3r(\underline{\hspace{1cm}})(\underline{\hspace{1cm}})$

Factor each trinomial completely. **See Examples 1–9.** *(Hint: In Exercises 57 and 58, first write the trinomial in descending powers and then factor.)*

▶ 23. $3a^2 + 10a + 7$ 24. $7r^2 + 8r + 1$ ▶ 25. $2y^2 + 7y + 6$

26. $5z^2 + 12z + 4$ 27. $15m^2 + m - 2$ 28. $6x^2 + x - 1$

29. $12s^2 + 11s - 5$ 30. $20x^2 + 11x - 3$ ▶ 31. $10m^2 - 23m + 12$

32. $6x^2 - 17x + 12$ 33. $8w^2 - 14w + 3$ 34. $9p^2 - 18p + 8$

▶ 35. $20y^2 - 39y - 11$ 36. $10x^2 - 11x - 6$ 37. $3x^2 - 15x + 16$

38. $2t^2 - 14t + 15$

39. $20x^2 + 22x + 6$

40. $36y^2 + 81y + 45$

41. $24x^2 - 42x + 9$

42. $48b^2 - 74b - 10$

43. $-40m^2q - mq + 6q$

44. $-15a^2b - 22ab - 8b$

▶ **45.** $15n^4 - 39n^3 + 18n^2$

46. $24a^4 + 10a^3 - 4a^2$

▶ **47.** $15x^2y^2 - 7xy^2 - 4y^2$

48. $14a^2b^3 + 15ab^3 - 9b^3$

49. $5a^2 - 7ab - 6b^2$

50. $6x^2 - 5xy - y^2$

▶ **51.** $12s^2 + 11st - 5t^2$

52. $25a^2 + 25ab + 6b^2$

53. $6m^6n + 7m^5n^2 + 2m^4n^3$

54. $12k^3q^4 - 4k^2q^5 - kq^6$

55. $x^2 - 6x - 5$

56. $x^2 - 8x - 7$

57. $16 + 16x + 3x^2$

58. $18 + 65x + 7x^2$

59. $-10x^3 + 5x^2 + 140x$

60. $-18k^3 - 48k^2 + 66k$

61. $12x^2 - 7x - 4$

62. $12x^2 - 9x - 10$

63. $24y^2 - 41xy - 14x^2$

64. $24x^2 + 19xy - 5y^2$

65. $36x^4 - 64x^2y + 15y^2$

66. $36x^4 + 59x^2y + 24y^2$

67. $48a^2 - 94ab - 4b^2$

68. $48t^2 - 147ts + 9s^2$

69. $10x^4y^5 + 39x^3y^5 - 4x^2y^5$

70. $14x^7y^4 - 31x^6y^4 + 6x^5y^4$

71. $36a^3b^2 - 104a^2b^2 - 12ab^2$

72. $36p^4q + 129p^3q - 60p^2q$

73. $24x^2 - 46x + 15$

74. $24x^2 - 94x + 35$

75. $24x^4 + 55x^2 - 24$

76. $24x^4 + 17x^2 - 20$

77. $24x^2 + 38xy + 15y^2$

78. $24x^2 + 62xy + 33y^2$

If a trinomial has a negative coefficient for the second-degree term, such as $-2x^2 + 11x - 12$, it is usually easier to factor by first factoring out the common factor -1.

$$-2x^2 + 11x - 12$$

$$= -1(2x^2 - 11x + 12) \qquad \text{Factor out } -1.$$

$$= -1(2x - 3)(x - 4) \qquad \text{Factor the trinomial.}$$

Use this method to factor each trinomial. **See Example 9(b).**

79. $-x^2 - 4x + 21$

80. $-x^2 + x + 72$

81. $-3x^2 - x + 4$

82. $-5x^2 + 2x + 16$

83. $-2a^2 - 5ab - 2b^2$

84. $-3p^2 + 13pq - 4q^2$

Extending Skills   Factor each polynomial. (Hint: As the first step, factor out the greatest common factor.)

85. $25q^2(m + 1)^3 - 5q(m + 1)^3 - 2(m + 1)^3$

86. $18x^2(y - 3)^2 - 21x(y - 3)^2 - 4(y - 3)^2$

87. $9x^2(r + 3)^3 + 12xy(r + 3)^3 + 4y^2(r + 3)^3$

88. $4t^2(k + 9)^7 + 20ts(k + 9)^7 + 25s^2(k + 9)^7$

Extending Skills   Find all integers k so that the trinomial can be factored by the methods of this section.

89. $5x^2 + kx - 1$

90. $2x^2 + kx - 3$

91. $2m^2 + km + 5$

92. $3y^2 + ky + 4$

5.4 Special Factoring Techniques

OBJECTIVES

1. Factor a difference of squares.
2. Factor a perfect square trinomial.
3. Factor a difference of cubes.
4. Factor a sum of cubes.

VOCABULARY

☐ perfect square
☐ perfect square trinomial
☐ perfect cube

By reversing the rules for multiplication of binomials from **Section 4.6,** we obtain rules for factoring polynomials in certain forms.

OBJECTIVE 1 Factor a difference of squares.

The rule for finding the product of a sum and difference of the same two terms is

$$(x + y)(x - y) = x^2 - y^2.$$

Reversing this rule leads to the following special factoring rule.

Factoring a Difference of Squares

$$x^2 - y^2 = (x + y)(x - y)$$

For example,

$$m^2 - 4$$
$$= m^2 - 2^2$$
$$= (m + 2)(m - 2).$$

Two conditions must be true for a binomial to be a difference of squares.

1. Both terms of the binomial must be **perfect squares,** such as

$$x^2, \quad 9y^2 = (3y)^2, \quad m^4 = (m^2)^2, \quad 1 = 1^2, \quad 25 = 5^2, \quad 144 = 12^2.$$

2. The terms of the binomial must have different signs (one positive and one negative).

NOW TRY EXERCISE 1

Factor each binomial if possible.

(a) $x^2 - 100$

(b) $x^2 + 100$

(c) $x^2 - 32$

EXAMPLE 1 Factoring Binomials

Factor each binomial if possible.

$$x^2 - y^2 = (x + y)(x - y)$$
$$\downarrow \quad \downarrow \quad \quad \downarrow \quad \downarrow \quad \downarrow \quad \downarrow$$

(a) $p^2 - 16 = p^2 - 4^2 = (p + 4)(p - 4)$

(b) $x^2 - 8$

Because 8 is not the square of an integer, this binomial does not satisfy Condition 1 above. It cannot be factored, so it is a prime polynomial.

(c) $p^2 + 16$

The binomial $p^2 + 16$ does not satisfy Condition 2 above. It is a *sum* of squares— it is *not* equal to $(p + 4)(p - 4)$. (See part (a).) We can use the FOIL method and try the following.

$$(p - 4)(p - 4) \qquad \qquad (p + 4)(p + 4)$$
$$= p^2 - 8p + 16, \quad \text{not} \quad p^2 + 16. \qquad = p^2 + 8p + 16, \quad \text{not} \quad p^2 + 16.$$

Thus, $p^2 + 16$ is a prime polynomial.

NOW TRY ANSWERS

1. (a) $(x + 10)(x - 10)$
   (b) prime   (c) prime

> **Sum of Squares**
>
> **If x and y have no common factors,** then the following holds.
>
> **A sum of squares $x^2 + y^2$ cannot be factored using real numbers.**
>
> That is, $x^2 + y^2$ is prime. (See **Example 1(c).**)

NOW TRY
EXERCISE 2

Factor each binomial.

(a) $9t^2 - 100$

(b) $36a^2 - 49b^2$

EXAMPLE 2 Factoring Differences of Squares

Factor each binomial.

$$x^2 \;\; - \;\; y^2 \;\; = \;\; (x \;\; + \;\; y) \;\; (x \;\; - \;\; y)$$

(a) $25m^2 - 4 = (5m)^2 - 2^2 = (5m + 2)(5m - 2)$

(b) $49z^2 - 64t^2$

$\qquad = (7z)^2 - (8t)^2$ Write each term as a square.

$\qquad = (7z + 8t)(7z - 8t)$ Factor the difference of squares.

CHECK $(7z + 8t)(7z - 8t)$

$\qquad\qquad = 49z^2 - 56zt + 56tz - 64t^2$ FOIL method

$\qquad\qquad = 49z^2 - 64t^2$ ✓   Commutative property; Combine like terms.

NOW TRY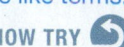

NOTE   *Always check a factored form by multiplying.*

NOW TRY
EXERCISE 3

Factor each binomial completely.

(a) $16k^2 - 64$

(b) $m^4 - 144$

(c) $v^4 - 625$

EXAMPLE 3 Factoring More Complex Differences of Squares

Factor each binomial completely.

(a) $81y^2 - 36$ *Always check for a common factor first.*

$\qquad = 9(9y^2 - 4)$ Factor out the GCF, 9.

$\qquad = 9[(3y)^2 - 2^2]$ Write each term as a square.

$\qquad = 9(3y + 2)(3y - 2)$ Factor the difference of squares.

(b) $\qquad\qquad p^4 - 36$

$\qquad\qquad = (p^2)^2 - 6^2$ Write each term as a square.

Neither binomial can be factored further. $\;= (p^2 + 6)(p^2 - 6)$ Factor the difference of squares.

(c) $\qquad\quad m^4 - 16$

$\qquad\qquad = (m^2)^2 - 4^2$ Write each term as a square.

$\qquad\qquad = (m^2 + 4)(m^2 - 4)$ Factor the difference of squares.

Don't stop here. $\;= (m^2 + 4)(m + 2)(m - 2)$ Factor the difference of squares again.

NOW TRY

NOW TRY ANSWERS

2. (a) $(3t + 10)(3t - 10)$

   (b) $(6a + 7b)(6a - 7b)$

3. (a) $16(k + 2)(k - 2)$

   (b) $(m^2 + 12)(m^2 - 12)$

   (c) $(v^2 + 25)(v + 5)(v - 5)$

⚠ **CAUTION** *Factor again when any of the factors is a difference of squares,* as in **Example 3(c).** Check by multiplying.

OBJECTIVE 2 Factor a perfect square trinomial.

Recall the rules for squaring binomials from **Section 4.6.**

$(x + y)^2$	Squared binomial	$(x - y)^2$	Squared binomial
$= (x + y)(x + y)$		$= (x - y)(x - y)$	
$= x^2 + 2xy + y^2$	Perfect square trinomial	$= x^2 - 2xy + y^2$	Perfect square trinomial

A **perfect square trinomial** is a trinomial that is the square of a binomial. For example, $x^2 + 8x + 16$ is a perfect square trinomial because it is the square of the binomial $x + 4$.

$$(x + 4)^2 \quad \text{Squared binomial}$$
$$= (x + 4)(x + 4)$$
$$= x^2 + 8x + 16 \quad \text{Perfect square trinomial}$$

Two conditions must be true for a trinomial to be a perfect square trinomial.

1. Two of its terms must be perfect squares. In the perfect square trinomial $x^2 + 8x + 16$, the terms x^2 and $16 = 4^2$ are perfect squares.

2. *The remaining (middle) term of a perfect square trinomial is always twice the product of the two terms in the squared binomial.* For example,

$$x^2 + 8x + 16$$
$$= x^2 + 2(x)(4) + 4^2 \quad 8x = 2(x)(4)$$
$$= (x + 4)^2. \quad \text{Factor.}$$

The following are *not* perfect square trinomials.

$16x^2 + 4x + 15$ — Violates Condition 1 (Only $16x^2 = (4x)^2$ is a perfect square; 15 is not.)

$x^2 + 6x + 36$ — Violates Condition 2 (x^2 and $36 = 6^2$ are perfect squares, but $2(x)(6) = 12x$, *not* $6x$.)

Reversing the rules for squaring binomials leads to the following special factoring rules.

> **Factoring Perfect Square Trinomials**
>
> $$x^2 + 2xy + y^2 = (x + y)^2$$
> $$x^2 - 2xy + y^2 = (x - y)^2$$

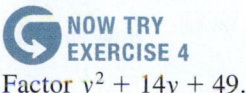
NOW TRY EXERCISE 4
Factor $y^2 + 14y + 49$.

EXAMPLE 4 Factoring a Perfect Square Trinomial

Factor $x^2 + 10x + 25$.

The x^2-term is a perfect square, and so is 25, which equals 5^2.

Try to factor $\quad x^2 + 10x + 25 \quad$ as the squared binomial $\quad (x + 5)^2$.

To check, take twice the product of the two terms in the squared binomial.

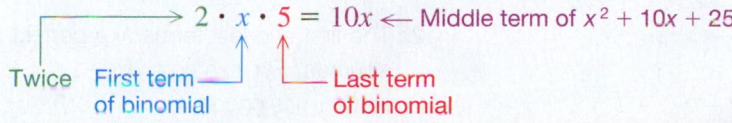

$$2 \cdot x \cdot 5 = 10x \leftarrow \text{Middle term of } x^2 + 10x + 25$$

Twice   First term of binomial   Last term of binomial

Because $10x$ is the middle term of the trinomial, the trinomial is a perfect square.

$$x^2 + 10x + 25 \quad \text{factors as} \quad (x + 5)^2.$$

NOW TRY ANSWER
4. $(y + 7)^2$

NOW TRY

**NOW TRY
EXERCISE 5**
Factor each trinomial.

(a) $t^2 - 18t + 81$

(b) $4p^2 - 28p + 49$

(c) $9x^2 + 6x + 4$

(d) $80x^3 + 120x^2 + 45x$

EXAMPLE 5   Factoring Perfect Square Trinomials

Factor each trinomial.

(a) $x^2 - 22x + 121$

The first and last terms are perfect squares ($121 = 11^2$ or $(-11)^2$). Check to see whether the middle term of $x^2 - 22x + 121$ is twice the product of the first and last terms of the binomial $x - 11$.

$$2 \cdot x \cdot (-11) = -22x \leftarrow \text{Middle term of } x^2 - 22x + 121$$

Twice ⎯⎯⎯ First term ⎯⎯⎯ Last term

Thus, $x^2 - 22x + 121$ is a perfect square trinomial.

> Check by squaring the binomial.

$$x^2 - 22x + 121 \quad \text{factors as} \quad (x - 11)^2.$$

Same sign

Notice that the sign of the second term in the squared binomial is the same as the sign of the middle term in the trinomial.

(b) $9m^2 - 24m + 16 = (3m)^2 + 2(3m)(-4) + (-4)^2$

Perfect squares ⎯⎯⎯ Twice ⎯⎯⎯ First term ⎯⎯⎯ Last term

$$= (3m - 4)^2$$

(c) $25y^2 + 20y + 16$

The first and last terms are perfect squares.

$$25y^2 = (5y)^2 \quad \text{and} \quad 16 = 4^2$$

Twice the product of the first and last terms of the binomial $5y + 4$ is

$$2 \cdot 5y \cdot 4 = 40y,$$

which is *not* the middle term of

$$25y^2 + 20y + 16.$$

This trinomial is not a perfect square. In fact, the trinomial cannot be factored even with the methods of the previous sections. It is a prime polynomial.

(d) $12z^3 + 60z^2 + 75z$

$$= 3z(4z^2 + 20z + 25)$$ Factor out the common factor, $3z$.

$$= 3z[(2z)^2 + 2(2z)(5) + 5^2]$$ $4z^2 + 20z + 25$ is a perfect square trinomial.

$$= 3z(2z + 5)^2$$ Factor.   **NOW TRY**

NOTE   Keep the following in mind when factoring perfect square trinomials.

1. The sign of the second term in the squared binomial is always the same as the sign of the middle term in the trinomial.

2. The first and last terms of a perfect square trinomial must be *positive* because they are squares. For example, $x^2 - 2x - 1$ cannot be a perfect square trinomial because the last term is negative.

3. Perfect square trinomials can also be factored by grouping or the FOIL method. Using the method of this section is often easier.

NOW TRY ANSWERS

5. (a) $(t - 9)^2$

(b) $(2p - 7)^2$

(c) prime

(d) $5x(4x + 3)^2$

OBJECTIVE 3   Factor a difference of cubes.

In a difference of cubes $x^3 - y^3$, both terms of the binomial must be **perfect cubes,** such as

$$x^3, \quad 8p^3 = (2p)^3, \quad s^6 = (s^2)^3, \quad 1 = 1^3, \quad 27 = 3^3, \quad 216 = 6^3.$$

We can factor a **difference of cubes** using the following rule.

Factoring a Difference of Cubes

$$x^3 - y^3 = (x - y)(x^2 + xy + y^2)$$

This rule for factoring a difference of cubes should be memorized. To see that the rule is correct, multiply $(x - y)(x^2 + xy + y^2)$.

$$
\begin{array}{llr}
\quad x^2 + xy \;+ y^2 & \text{Multiply vertically.} \\
\underline{\qquad\qquad x \;- y} & \text{(Section 4.5)} \\
-x^2y - xy^2 - y^3 & -y(x^2 + xy + y^2) \\
\underline{x^3 + x^2y + xy^2 \qquad} & x(x^2 + xy + y^2) \\
x^3 \qquad\qquad - y^3 & \text{Add.}
\end{array}
$$

Notice the pattern of the terms in the factored form of $x^3 - y^3$.

- $x^3 - y^3$ factors as (a binomial factor) · (a trinomial factor).

- The binomial factor has the difference of the cube roots of the given terms. (*Note:* A cube root of 1 is 1 because $1^3 = 1$, a cube root of 8 is 2 because $2^3 = 8$, and so on.)

- The terms in the trinomial factor are all positive.

- The terms in the binomial factor help to determine the trinomial factor.

$$
\begin{array}{ccccccc}
& & \text{First term} & & \substack{\text{positive} \\ \text{product of}} & & \text{second term} \\
& & \text{squared} & + & \text{the terms} & + & \text{squared} \\
x^3 - y^3 = (x - y)(& & x^2 & + & xy & + & y^2 \;)
\end{array}
$$

⚠ **CAUTION**   The polynomial $x^3 - y^3$ is **not** equivalent to $(x - y)^3$.

$x^3 - y^3$	$(x - y)^3$
$= (x - y)(x^2 + xy + y^2)$	$= (x - y)(x - y)(x - y)$
	$= (x - y)(x^2 - 2xy + y^2)$

EXAMPLE 6   Factoring Differences of Cubes

Factor each binomial.

$$x^3 - y^3 = (x - y)(x^2 + xy + y^2)$$

(a) $m^3 - 125 = m^3 - 5^3 = (m - 5)(m^2 + 5m + 5^2)$ Let $x = m$ and $y = 5$.

$$= (m - 5)(m^2 + 5m + 25) \qquad\qquad 5^2 = 25$$

NOW TRY
EXERCISE 6

Factor each binomial.

(a) $a^3 - 27$

(b) $8t^3 - 125$

(c) $3k^3 - 192$

(d) $125x^3 - 343y^6$

(b) $8p^3 - 27$

$= (2p)^3 - 3^3$ $8p^3 = (2p)^3$ and $27 = 3^3$.

$= (2p - 3)[(2p)^2 + (2p)3 + 3^2]$ Let $x = 2p$ and $y = 3$.

$= (2p - 3)(4p^2 + 6p + 9)$ Apply the exponents. Multiply.

> $(2p)^2 = 2^2p^2 = 4p^2$,
> **not** $2p^2$.

(c) $4m^3 - 32$

$= 4(m^3 - 8)$ Factor out the common factor, 4.

$= 4(m^3 - 2^3)$ $8 = 2^3$

$= 4(m - 2)(m^2 + 2m + 4)$ Factor the difference of cubes.

(d) $125t^3 - 216s^6$

$= (5t)^3 - (6s^2)^3$ Write each term as a cube.

$= (5t - 6s^2)[(5t)^2 + 5t(6s^2) + (6s^2)^2]$ Factor the difference of cubes.

$= (5t - 6s^2)(25t^2 + 30ts^2 + 36s^4)$ Apply the exponents. Multiply.

> Square carefully.
> $(6s^2)^2 = 6^2(s^2)^2 = 36s^4$

NOW TRY

! CAUTION A common error when factoring $x^3 - y^3 = (x - y)(x^2 + xy + y^2)$ is to try to factor $x^2 + xy + y^2$. This is usually not possible.

OBJECTIVE 4 Factor a sum of cubes.

A *sum of squares*, such as $m^2 + 25$, *cannot* be factored using real numbers, but a **sum of cubes** can.

Factoring a Sum of Cubes

$$x^3 + y^3 = (x + y)(x^2 - xy + y^2)$$

Compare the rule for the *sum* of cubes with that for the *difference* of cubes.

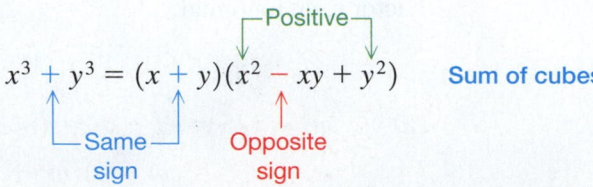

$$x^3 - y^3 = (x - y)(x^2 + xy + y^2)$$ **Difference of cubes**

Same sign   Opposite sign

The only difference between the rules is the positive and negative signs.

$$x^3 + y^3 = (x + y)(x^2 - xy + y^2)$$ **Sum of cubes**

Same sign   Opposite sign

NOW TRY ANSWERS

6. (a) $(a - 3)(a^2 + 3a + 9)$
(b) $(2t - 5)(4t^2 + 10t + 25)$
(c) $3(k - 4)(k^2 + 4k + 16)$
(d) $(5x - 7y^2) \cdot$
$\quad (25x^2 + 35xy^2 + 49y^4)$

NOW TRY
EXERCISE 7

Factor each binomial.

(a) $x^3 + 125$

(b) $27a^3 + 8b^3$

EXAMPLE 7 Factoring Sums of Cubes

Factor each binomial.

(a) $k^3 + 27$

$$= k^3 + 3^3 \qquad\qquad 27 = 3^3$$

$$= (k + 3)(k^2 - 3k + 3^2) \qquad \text{Factor the sum of cubes.}$$

$$= (k + 3)(k^2 - 3k + 9) \qquad \text{Apply the exponent.}$$

(b) $8m^3 + 125n^3$

$$= (2m)^3 + (5n)^3 \qquad\qquad 8m^3 = (2m)^3 \text{ and } 125n^3 = (5n)^3.$$

$$= (2m + 5n)[(2m)^2 - 2m(5n) + (5n)^2] \qquad \text{Factor the sum of cubes.}$$

$$= (2m + 5n)(4m^2 - 10mn + 25n^2) \quad \boxed{\text{Be careful:} \\ (2m)^2 = 2^2 m^2 \\ \text{and } (5n)^2 = 5^2 n^2.}$$

(c) $1000a^6 + 27b^3$

$$= (10a^2)^3 + (3b)^3$$

$$= (10a^2 + 3b)[(10a^2)^2 - (10a^2)(3b) + (3b)^2] \qquad \text{Factor the sum of cubes.}$$

$$= (10a^2 + 3b)(100a^4 - 30a^2b + 9b^2) \qquad (10a^2)^2 = 10^2(a^2)^2 = 100a^4$$

NOW TRY

The methods of factoring discussed in this section are summarized here.

Special Factoring Rules

Difference of squares	$x^2 - y^2 = (x + y)(x - y)$
Perfect square trinomials	$x^2 + 2xy + y^2 = (x + y)^2$
	$x^2 - 2xy + y^2 = (x - y)^2$
Difference of cubes	$x^3 - y^3 = (x - y)(x^2 + xy + y^2)$
Sum of cubes	$x^3 + y^3 = (x + y)(x^2 - xy + y^2)$

A sum of squares can be factored only if the terms have a common factor.

NOW TRY ANSWERS

7. (a) $(x + 5)(x^2 - 5x + 25)$

(b) $(3a + 2b)(9a^2 - 6ab + 4b^2)$

5.4 Exercises

FOR EXTRA HELP

  MyMathLab®

▶ *Complete solution available in MyMathLab*

Concept Check Work each problem.

1. To help factor differences of squares, complete the following list of perfect squares.

$1^2 =$ _____ $2^2 =$ _____ $3^2 =$ _____ $4^2 =$ _____ $5^2 =$ _____

$6^2 =$ _____ $7^2 =$ _____ $8^2 =$ _____ $9^2 =$ _____ $10^2 =$ _____

$11^2 =$ _____ $12^2 =$ _____ $13^2 =$ _____ $14^2 =$ _____ $15^2 =$ _____

$16^2 =$ _____ $17^2 =$ _____ $18^2 =$ _____ $19^2 =$ _____ $20^2 =$ _____

2. The following powers of x are all perfect squares:

$$x^2, \quad x^4, \quad x^6, \quad x^8, \quad x^{10}.$$

On the basis of this observation, we may make a conjecture (an educated guess) that if the power of a variable is divisible by _____ (with 0 remainder), then we have a perfect square.

3. Which of the following are differences of squares?

   A. $x^2 - 4$ **B.** $y^2 + 9$ **C.** $2a^2 - 25$ **D.** $9m^2 - 1$

4. Which of the following binomial sums can be factored?

   A. $x^2 + 36$ **B.** $x^3 + x$ **C.** $3x^2 + 12$ **D.** $25x^2 + 49$

5. On a quiz, a student indicated *prime* when asked to factor $4x^2 + 16$ because she said that a sum of a squares cannot be factored. **WHAT WENT WRONG?**

6. When a student was directed to factor $k^4 - 81$ completely, his teacher did not give him full credit.

$$(k^2 + 9)(k^2 - 9) \qquad \text{His answer}$$

The student argued that since his answer does indeed give $k^4 - 81$ when multiplied out, he should be given full credit. **WHAT WENT WRONG?** Give the correct factored form.

Factor each binomial completely. If the binomial is prime, say so. Use the answers from Exercises 1 and 2 as necessary. See Examples 1–3.

▶ **7.** $y^2 - 25$ **8.** $t^2 - 36$ **9.** $x^2 - 144$ **10.** $x^2 - 400$

11. $m^2 - 12$ **12.** $k^2 - 18$ **13.** $m^2 + 64$ **14.** $k^2 + 49$

15. $4m^2 + 16$ **16.** $9x^2 + 81$ ▶ **17.** $9r^2 - 4$

18. $4x^2 - 9$ ▶ **19.** $36x^2 - 16$ **20.** $32a^2 - 8$

21. $196p^2 - 225$ **22.** $361q^2 - 400$ **23.** $16r^2 - 25a^2$

24. $49m^2 - 100p^2$ **25.** $81x^2 - 49y^2$ **26.** $36y^2 - 121z^2$

27. $54x^2 - 6y^2$ **28.** $48m^2 - 75n^2$ **29.** $100x^2 + 49$

30. $81w^2 + 16$ **31.** $4 - x^2$ **32.** $25 - x^2$

33. $36 - 25t^2$ **34.** $16 - 49p^2$ **35.** $x^3 + 4x$

36. $z^3 + 25z$ **37.** $x^4 - x^2$ **38.** $y^4 - 9y^2$

39. $p^4 - 49$ **40.** $r^4 - 25$ **41.** $x^4 - 1$

42. $y^4 - 10,000$ **43.** $p^4 - 256$ **44.** $k^4 - 81$

Concept Check *Work each problem.*

45. Which of the following are perfect square trinomials?

   A. $y^2 - 13y + 36$ **B.** $x^2 + 6x + 9$ **C.** $4z^2 - 4z + 1$ **D.** $16m^2 + 10m + 1$

46. In the polynomial $9y^2 + 14y + 25$, the first and last terms are perfect squares. Can the polynomial be factored? If it can, factor it. If it cannot, explain why it is not a perfect square trinomial.

Concept Check *Find the value of the indicated variable.*

47. Find b so that $x^2 + bx + 25$ factors as $(x + 5)^2$.

48. Find c so that $4m^2 - 12m + c$ factors as $(2m - 3)^2$.

49. Find a so that $ay^2 - 12y + 4$ factors as $(3y - 2)^2$.

50. Find b so that $100a^2 + ba + 9$ factors as $(10a + 3)^2$.

Factor each trinomial completely. See Examples 4 and 5.

▶ **51.** $w^2 + 2w + 1$ **52.** $p^2 + 4p + 4$ ▶ **53.** $x^2 - 8x + 16$

54. $x^2 - 10x + 25$ **55.** $x^2 - 10x + 100$ **56.** $x^2 - 18x + 36$

57. $2x^2 + 24x + 72$

58. $3y^2 + 48y + 192$

59. $4x^2 + 12x + 9$

60. $25x^2 + 10x + 1$

61. $16x^2 - 40x + 25$

62. $36y^2 - 60y + 25$

63. $49x^2 - 28xy + 4y^2$

64. $4z^2 - 12zw + 9w^2$

65. $64x^2 + 48xy + 9y^2$

66. $9t^2 + 24tr + 16r^2$

67. $50h^2 - 40hy + 8y^2$

68. $18x^2 - 48xy + 32y^2$

69. $4k^3 - 4k^2 + 9k$

70. $9r^3 - 6r^2 + 16r$

71. $25z^4 + 5z^3 + z^2$

72. $4x^4 + 2x^3 + x^2$

Concept Check *Work each problem.*

73. To help factor sums or differences of cubes, complete the following list of perfect cubes.

$1^3 = $ ____ $2^3 = $ ____ $3^3 = $ ____ $4^3 = $ ____ $5^3 = $ ____

$6^3 = $ ____ $7^3 = $ ____ $8^3 = $ ____ $9^3 = $ ____ $10^3 = $ ____

74. The following powers of x are all perfect cubes:

$$x^3, \quad x^6, \quad x^9, \quad x^{12}, \quad x^{15}.$$

On the basis of this observation, we may make a conjecture that if the power of a variable is divisible by ____ (with 0 remainder), then we have a perfect cube.

75. Which of the following are differences of cubes?

   A. $9x^3 - 125$ **B.** $x^3 - 16$ **C.** $x^3 - 1$ **D.** $8x^3 - 27y^3$

76. Which of the following are sums of cubes?

   A. $x^3 + 1$ **B.** $x^3 + 36$ **C.** $12x^3 + 27$ **D.** $64x^3 + 216y^3$

77. Identify each monomial as a *perfect square*, a *perfect cube*, *both of these*, or *neither of these*.

   (a) $4x^3$ **(b)** $8y^6$ **(c)** $49x^{12}$ **(d)** $81r^{10}$ **(e)** $64x^6y^{12}$ **(f)** $125t^6$

78. What must be true for x^n to be both a perfect square and a perfect cube?

*Factor each binomial completely. Use the answers from **Exercises 73 and 74** as necessary. See Examples 6 and 7.*

▶ **79.** $a^3 - 1$

80. $m^3 - 8$

▶ **81.** $m^3 + 8$

82. $b^3 + 1$

83. $y^3 - 216$

84. $x^3 - 343$

85. $k^3 + 1000$

86. $p^3 + 512$

87. $27x^3 - 64$

88. $64y^3 - 27$

89. $6p^3 + 6$

90. $81x^3 + 3$

91. $5x^3 + 40$

92. $128y^3 + 54$

93. $y^3 - 8x^3$

94. $w^3 - 216z^3$

95. $2x^3 - 16y^3$

96. $27w^3 - 216z^3$

97. $8p^3 + 729q^3$

98. $64x^3 + 125y^3$

99. $27a^3 + 64b^3$

100. $125m^3 + 8p^3$

101. $125t^3 + 8s^3$

102. $27r^3 + 1000s^3$

103. $8x^3 - 125y^6$

104. $27t^3 - 64s^6$

105. $27m^6 + 8n^3$

106. $1000r^6 + 27s^3$

107. $x^9 + y^9$

108. $x^9 - y^9$

Extending Skills   *Although we usually factor polynomials using integers, we can apply the same concepts to factoring using fractions and decimals.*

$$z^2 - \frac{9}{16}$$

$$= z^2 - \left(\frac{3}{4}\right)^2 \qquad \frac{9}{16} = \left(\frac{3}{4}\right)^2$$

$$= \left(z + \frac{3}{4}\right)\left(z - \frac{3}{4}\right) \qquad \text{Factor the difference of squares.}$$

Apply the special factoring rules of this section to factor each polynomial.

109. $p^2 - \dfrac{1}{9}$
110. $q^2 - \dfrac{1}{4}$
111. $36m^2 - \dfrac{16}{25}$

112. $100b^2 - \dfrac{4}{49}$
113. $x^2 - 0.64$
114. $y^2 - 0.36$

115. $t^2 + t + \dfrac{1}{4}$
116. $m^2 + \dfrac{2}{3}m + \dfrac{1}{9}$
117. $x^2 - 1.8x + 0.81$

118. $y^2 - 1.4y + 0.49$
119. $x^3 + \dfrac{1}{8}$
120. $x^3 + \dfrac{1}{64}$

Extending Skills   *Factor each polynomial completely.*

121. $(m + n)^2 - (m - n)^2$
122. $(a - b)^3 - (a + b)^3$

123. $m^2 - p^2 + 2m + 2p$
124. $3r - 3k + 3r^2 - 3k^2$

SUMMARY EXERCISES   Recognizing and Applying Factoring Strategies

When factoring a polynomial, ask these questions to decide on a suitable factoring technique.

Factoring a Polynomial

Question 1   **Is there a common factor other than 1?** If so, factor it out.

Question 2   **How many terms are in the polynomial?**

> *Two terms:* Is it a difference of squares or a sum or difference of cubes? If so, factor as in **Section 5.4.**

> *Three terms:* Is it a perfect square trinomial? In this case, factor as in **Section 5.4.**

> If the trinomial is not a perfect square trinomial, what is the coefficient of the second-degree term?

> • If it is 1, use the factoring method of **Section 5.2.**

> • If it is not 1, use the general factoring methods of **Section 5.3.**

> *Four terms:* Try to factor by grouping, as in **Section 5.1.**

Question 3   **Can any factors be factored further?** If so, factor them.

⚠️ **CAUTION** Be careful when checking the answer to a factoring problem.

1. Check that the product of all the factors does indeed yield the original polynomial.

2. Check that the original polynomial has been factored **completely.**

🔄 **NOW TRY EXERCISE**

Factor completely.

$24m^2 - 42my + 9y^2$

NOW TRY ANSWER

$3(4m - y)(2m - 3y)$

EXAMPLE Applying Factoring Strategies

Factor $12x^2 + 26xy + 12y^2$ completely.

Question 1 **Is there a common factor other than 1?**

Yes, 2 is a common factor, so factor it out.

$$12x^2 + 26xy + 12y^2$$
$$= 2(6x^2 + 13xy + 6y^2)$$

Question 2 **How many terms are in the polynomial?**

The polynomial $6x^2 + 13xy + 6y^2$ has three terms. It is not a perfect square trinomial. To factor by grouping, we find two integers with a product of $6 \cdot 6$, or 36, and a sum of 13. These integers are 4 and 9.

$$12x^2 + 26xy + 12y^2$$

$= 2(6x^2 + 13xy + 6y^2)$	Factor out the GCF, 2.
$= 2(6x^2 + 4xy + 9xy + 6y^2)$	$4 \cdot 9 = 36; 4 + 9 = 13$
$= 2[(6x^2 + 4xy) + (9xy + 6y^2)]$	Group the terms.
$= 2[2x(3x + 2y) + 3y(3x + 2y)]$	Factor each group.
$= 2(3x + 2y)(2x + 3y)$	Factor out the common factor, $3x + 2y$.

We could also have factored the trinomial $6x^2 + 13xy + 6y^2$ by trial and error, using the FOIL method in reverse, as in **Section 5.3.**

Question 3 **Can any factors be factored further?**

No. The original polynomial has been factored completely.

NOW TRY 🔄

Match each polynomial in Column I with the best choice for factoring it in Column II. The choices in Column II may be used once, more than once, or not at all.

I

1. G	2. H
3. A	4. B
5. E	6. I
7. C	8. F
9. I	10. E

1. $12x^2 + 20x + 8$

2. $x^2 - 17x + 72$

3. $16m^2n + 24mn - 40mn^2$

4. $64a^2 - 121b^2$

5. $36p^2 - 60pq + 25q^2$

6. $z^2 - 4z + 6$

7. $8r^3 - 125$

8. $x^6 + 4x^4 - 3x^2 - 12$

9. $4w^2 + 49$

10. $z^2 - 24z + 144$

II

A. Factor out the GCF. No further factoring is possible.

B. Factor a difference of squares.

C. Factor a difference of cubes.

D. Factor a sum of cubes.

E. Factor a perfect square trinomial.

F. Factor by grouping.

G. Factor out the GCF. Then factor a trinomial by grouping or trial and error.

H. Factor into two binomials by finding two integers whose product is the constant in the trinomial and whose sum is the coefficient of the middle term.

I. The polynomial is prime.

11. $(a - 6)(a + 2)$
12. $(a + 8)(a + 9)$
13. $6(y - 2)(y + 1)$
14. $7y^4(y + 6)(y - 4)$
15. $6(a + 2b + 3c)$
16. $(m - 4n)(m + n)$
17. $(p - 11)(p - 6)$
18. $(z + 7)(z - 6)$
19. $(5z - 6)(2z + 1)$
20. $2(m - 8)(m + 3)$
21. $17xy(x^2y + 3)$
22. $5(3y + 1)$
23. $8a^3(a - 3)(a + 2)$
24. $(4k + 1)(2k - 3)$
25. $(z - 5a)(z + 2a)$
26. $50(z^2 - 2)$
27. $(x - 5)(x - 4)$
28. prime
29. $(3n - 2)(2n - 5)$
30. $(3y - 1)(3y + 5)$
31. $4(4x + 5)$
32. $(m + 5)(m - 3)$
33. $(3y - 4)(2y + 1)$
34. $(m + 9)(m - 9)$
35. $(6z + 1)(z + 5)$
36. $(12x - 1)(x + 4)$
37. $(2k - 3)^2$
38. $(8p - 1)(p + 3)$
39. $6(3m + 2z)(3m - 2z)$
40. $(4m - 3)(2m + 1)$
41. $(3k - 2)(k + 2)$
42. $(2a - 3)(4a^2 + 6a + 9)$
43. $7k(2k + 5)(k - 2)$
44. $(5 + r)(1 - s)$
45. $(y^2 + 4)(y + 2)(y - 2)$
46. prime
47. $8m(1 - 2m)$
48. $(k + 4)(k - 4)$
49. $(z - 2)(z^2 + 2z + 4)$
50. $(y - 8)(y + 7)$
51. prime
52. $9p^8(3p + 7)(p - 4)$
53. $8m^3(4m^6 + 2m^2 + 3)$
54. $(2m + 5)(4m^2 - 10m + 25)$
55. $(4r + 3m)^2$ **56.** $(z - 6)^2$
57. $(5h + 7g)(3h - 2g)$
58. $5z(z - 7)(z - 2)$
59. $(k - 5)(k - 6)$
60. $4(4p - 5m)(4p + 5m)$
61. $3k(k - 5)(k + 1)$
62. $(y - 6k)(y + 2k)$
63. $(10p + 3)(100p^2 - 30p + 9)$
64. $(4r - 7)(16r^2 + 28r + 49)$
65. $(2 + m)(3 + p)$
66. $(2m - 3n)(m + 5n)$
67. $(4z - 1)^2$
68. $(a^2 + 25)(a + 5)(a - 5)$
69. $3(6m - 1)^2$
70. $(10a + 9y)(10a - 9y)$
71. prime
72. $(2y + 5)(2y - 5)$
73. $8z(4z - 1)(z + 2)$
74. $5(2m - 3)(m + 4)$

Factor each polynomial completely.

11. $a^2 - 4a - 12$

12. $a^2 + 17a + 72$

13. $6y^2 - 6y - 12$

14. $7y^6 + 14y^5 - 168y^4$

15. $6a + 12b + 18c$

16. $m^2 - 3mn - 4n^2$

17. $p^2 - 17p + 66$

18. $z^2 - 6z + 7z - 42$

19. $10z^2 - 7z - 6$

20. $2m^2 - 10m - 48$

21. $17x^3y^2 + 51xy$

22. $15y + 5$

23. $8a^5 - 8a^4 - 48a^3$

24. $8k^2 - 10k - 3$

25. $z^2 - 3za - 10a^2$

26. $50z^2 - 100$

27. $x^2 - 4x - 5x + 20$

28. $x^2 + 2x + 16$

29. $6n^2 - 19n + 10$

30. $9y^2 + 12y - 5$

31. $16x + 20$

32. $m^2 + 2m - 15$

33. $6y^2 - 5y - 4$

34. $m^2 - 81$

35. $6z^2 + 31z + 5$

36. $12x^2 + 47x - 4$

37. $4k^2 - 12k + 9$

38. $8p^2 + 23p - 3$

39. $54m^2 - 24z^2$

40. $8m^2 - 2m - 3$

41. $3k^2 + 4k - 4$

42. $8a^3 - 27$

43. $14k^3 + 7k^2 - 70k$

44. $5 + r - 5s - rs$

45. $y^4 - 16$

46. $9z^2 + 64$

47. $8m - 16m^2$

48. $k^2 - 16$

49. $z^3 - 8$

50. $y^2 - y - 56$

51. $k^2 + 9$

52. $27p^{10} - 45p^9 - 252p^8$

53. $32m^9 + 16m^5 + 24m^3$

54. $8m^3 + 125$

55. $16r^2 + 24rm + 9m^2$

56. $z^2 - 12z + 36$

57. $15h^2 + 11hg - 14g^2$

58. $5z^3 - 45z^2 + 70z$

59. $k^2 - 11k + 30$

60. $64p^2 - 100m^2$

61. $3k^3 - 12k^2 - 15k$

62. $y^2 - 4yk - 12k^2$

63. $1000p^3 + 27$

64. $64r^3 - 343$

65. $6 + 3m + 2p + mp$

66. $2m^2 + 7mn - 15n^2$

67. $16z^2 - 8z + 1$

68. $a^4 - 625$

69. $108m^2 - 36m + 3$

70. $100a^2 - 81y^2$

71. $x^2 - xy + y^2$

72. $4y^2 - 25$

73. $32z^3 + 56z^2 - 16z$

74. $10m^2 + 25m - 60$

75. $(8m - 5n)^2$
76. $(2 - q)(2 - 3p)$
77. $2(3a - 1)(a + 2)$
78. $6y^4(3y + 4)(2y - 5)$
79. prime
80. $4(2k - 3)^2$
81. $(4 + m)(5 + 3n)$
82. $12y^2(6yz^2 + 1 - 2y^2z^2)$
83. $(4k - 3h)(2k + h)$
84. $(2a + 5)(a - 6)$
85. $2(x + 4)(x^2 - 4x + 16)$
86. $15a^3b^2(3b^3 - 4a + 5a^3b^2)$
87. $(5y - 6z)(2y + z)$
88. $(m - 2)^2$
89. $(8a - b)(a + 3b)$
90. $5m^2(5m - 3n)(5m - 13n)$

75. $64m^2 - 80mn + 25n^2$

76. $4 - 2q - 6p + 3pq$

77. $6a^2 + 10a - 4$

78. $36y^6 - 42y^5 - 120y^4$

79. $36x^2 + 32x + 9$

80. $16k^2 - 48k + 36$

81. $20 + 5m + 12n + 3mn$

82. $72y^3z^2 + 12y^2 - 24y^4z^2$

83. $8k^2 - 2kh - 3h^2$

84. $2a^2 - 7a - 30$

85. $2x^3 + 128$

86. $45a^3b^5 - 60a^4b^2 + 75a^6b^4$

87. $10y^2 - 7yz - 6z^2$

88. $m^2 - 4m + 4$

89. $8a^2 + 23ab - 3b^2$

90. $125m^4 - 400m^3n + 195m^2n^2$

5.5 Solving Quadratic Equations Using the Zero-Factor Property

OBJECTIVES

1 Solve quadratic equations using the zero-factor property.

2 Solve other equations using the zero-factor property.

VOCABULARY

☐ quadratic equation
☐ double solution

Galileo Galilei (1564–1642)

Galileo Galilei developed theories to explain physical phenomena. According to legend, Galileo dropped objects of different weights from the Leaning Tower of Pisa to disprove the belief that heavier objects fall faster than lighter objects. He developed a formula that describes the motion of freely falling objects,

$$d = 16t^2,$$

where d is the distance in feet that an object falls (disregarding air resistance) in t seconds, regardless of weight. The equation $d = 16t^2$ is a *quadratic equation*.

Quadratic Equation

A **quadratic equation** (in x here) can be written in the form

$$ax^2 + bx + c = 0,$$

where a, b, and c are real numbers and $a \neq 0$. The given form is called **standard form.**

Examples: $x^2 + 5x + 6 = 0$, $2x^2 - 5x = 3$, $x^2 = 4$ Quadratic equations

A quadratic equation has a second-degree term and no terms of greater degree. Of the above examples, only $x^2 + 5x + 6 = 0$ is in standard form.

We have factored many quadratic *expressions* of the form $ax^2 + bx + c$. In this section, we use factored quadratic expressions to solve quadratic *equations*.

OBJECTIVE 1   Solve quadratic equations using the zero-factor property.

We use the following property to solve some quadratic equations.

Zero-Factor Property

If a and b are real numbers and if $ab = 0$, then $a = 0$ or $b = 0$.

That is, if the product of two numbers is 0, then at least one of the numbers must be 0. One number *must* be 0, but both *may* be 0.

NOW TRY
EXERCISE 1

Solve each equation.

(a) $(x - 4)(3x + 1) = 0$

(b) $y(4y - 5) = 0$

EXAMPLE 1 Using the Zero-Factor Property

Solve each equation.

(a) $(x + 3)(2x - 1) = 0$

The product $(x + 3)(2x - 1)$ is equal to 0. By the zero-factor property, the product of these two factors will equal 0 only if at least one of the factors equals 0. Therefore, either $x + 3 = 0$ or $2x - 1 = 0$.

$$x + 3 = 0 \quad \text{or} \quad 2x - 1 = 0 \qquad \text{Zero-factor property}$$

$$x = -3 \qquad\qquad 2x = 1 \qquad \text{Solve each equation.}$$

$$x = \frac{1}{2} \qquad \text{Divide each side by 2.}$$

Check these values by substituting -3 for x in the original equation. ***Then start over*** and substitute $\frac{1}{2}$ for x.

CHECK   Let $x = -3$.   Let $x = \frac{1}{2}$.

$$(x + 3)(2x - 1) = 0 \qquad\qquad (x + 3)(2x - 1) = 0$$

$$(-3 + 3)[2(-3) - 1] \stackrel{?}{=} 0 \qquad \left(\frac{1}{2} + 3\right)\left(2 \cdot \frac{1}{2} - 1\right) \stackrel{?}{=} 0$$

$$0(-7) \stackrel{?}{=} 0 \qquad\qquad\qquad \frac{7}{2}(0) \stackrel{?}{=} 0$$

$$0 = 0 \ \checkmark \ \text{True} \qquad\qquad\qquad 0 = 0 \ \checkmark \ \text{True}$$

Because true statements result, the solution set is $\left\{-3, \frac{1}{2}\right\}$.   *Include **both** solutions in the solution set.*

(b) $$y(3y - 4) = 0$$

$$y = 0 \quad \text{or} \quad 3y - 4 = 0 \qquad \text{Zero-factor property}$$

Don't forget that 0 is a solution.

$$3y = 4 \qquad \text{Add 4.}$$

$$y = \frac{4}{3} \qquad \text{Divide by 3.}$$

CHECK   Let $y = 0$.   Let $y = \frac{4}{3}$.

$$y(3y - 4) = 0 \qquad\qquad\qquad y(3y - 4) = 0$$

$$0(3 \cdot 0 - 4) \stackrel{?}{=} 0 \qquad\qquad \frac{4}{3}\left(3 \cdot \frac{4}{3} - 4\right) \stackrel{?}{=} 0$$

$$0(-4) \stackrel{?}{=} 0 \qquad\qquad\qquad \frac{4}{3}(0) \stackrel{?}{=} 0$$

$$0 = 0 \ \checkmark \ \text{True} \qquad\qquad\qquad 0 = 0 \ \checkmark \ \text{True}$$

True statements result. The solution set is $\left\{0, \frac{4}{3}\right\}$.   **NOW TRY**

NOTE   The word *or* as used in **Example 1** means "one or the other or both."

NOW TRY ANSWERS

1. (a) $\left\{-\frac{1}{3}, 4\right\}$ **(b)** $\left\{0, \frac{5}{4}\right\}$

If the polynomial in an equation is not already factored, first make sure that the equation is in standard form. Then factor and solve.

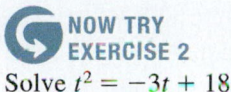

NOW TRY EXERCISE 2

Solve $t^2 = -3t + 18$.

EXAMPLE 2 Solving Quadratic Equations

Solve each equation.

(a) $x^2 - 5x = -6$

First, write the equation in standard form $ax^2 + bx + c = 0$.

> *Don't factor x out at this step.*

$$x^2 - 5x = -6$$

$$x^2 - 5x + 6 = 0 \qquad \text{Add 6 to each side.}$$

Now factor $x^2 - 5x + 6$. Find two numbers whose product is 6 and whose sum is -5. These two numbers are -2 and -3, so we factor as follows.

$$(x - 2)(x - 3) = 0 \qquad \text{Factor.}$$

$$x - 2 = 0 \quad \text{or} \quad x - 3 = 0 \qquad \text{Zero-factor property}$$

$$x = 2 \quad \text{or} \qquad x = 3 \qquad \text{Solve each equation.}$$

CHECK   Let $x = 2$.

$$x^2 - 5x = -6$$
$$2^2 - 5(2) \overset{?}{=} -6$$
$$4 - 10 \overset{?}{=} -6$$
$$-6 = -6 \quad \checkmark \quad \text{True}$$

Let $x = 3$.

$$x^2 - 5x = -6$$
$$3^2 - 5(3) \overset{?}{=} -6$$
$$9 - 15 \overset{?}{=} -6$$
$$-6 = -6 \quad \checkmark \quad \text{True}$$

Both values check, so the solution set is $\{2, 3\}$.

(b) $\qquad\qquad y^2 = y + 20$

> *Write this equation in standard form.*

Standard form $\longrightarrow y^2 - y - 20 = 0 \qquad \text{Subtract } y \text{ and 20.}$

$$(y - 5)(y + 4) = 0 \qquad \text{Factor.}$$

$$y - 5 = 0 \quad \text{or} \quad y + 4 = 0 \qquad \text{Zero-factor property}$$

$$y = 5 \quad \text{or} \qquad y = -4 \qquad \text{Solve each equation.}$$

Check each result to verify that the solution set is $\{-4, 5\}$.   **NOW TRY**

Solving a Quadratic Equation Using the Zero-Factor Property

Step 1 **Write the equation in standard form**—that is, with all terms on one side of the equality symbol in descending powers of the variable and 0 on the other side.

Step 2 **Factor** completely.

Step 3 **Apply the zero-factor property.** Set each factor with a variable equal to 0.

Step 4 **Solve** the resulting equations.

Step 5 **Check** each result in the original equation. Write the solution set.

NOW TRY ANSWER

2. $\{-6, 3\}$

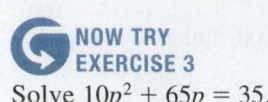

NOW TRY
EXERCISE 3

Solve $10p^2 + 65p = 35$.

EXAMPLE 3   Solving a Quadratic Equation (Common Factor)

Solve $4x^2 + 40 = 26x$.

$$4x^2 + 40 = 26x$$ *Write this equation in the form $ax^2 + bx + c = 0$.*

Step 1 $4x^2 - 26x + 40 = 0$ Standard form

This 2 is not a solution of the equation. $2(2x^2 - 13x + 20) = 0$ Factor out 2.

$$2x^2 - 13x + 20 = 0$$ Divide each side by 2.

Step 2 $(2x - 5)(x - 4) = 0$ Factor.

Step 3 $2x - 5 = 0$ or $x - 4 = 0$ Zero-factor property

Step 4 $2x = 5$ $x = 4$ Solve each equation.

$$x = \frac{5}{2}$$

Step 5   *Check* each result to verify that the solution set is $\left\{\frac{5}{2}, 4\right\}$.   **NOW TRY**

⚠ **CAUTION**   A common error is to include the common factor 2 as a solution in **Example 3**. *Only factors containing variables lead to solutions,* such as the factor y in the equation $y(3y - 4) = 0$ in **Example 1(b)**.

EXAMPLE 4   Solving Quadratic Equations

Solve each equation.

(a) $16m^2 - 25 = 0$ *This equation is in standard form $ax^2 + bx + c = 0$. There is no first-degree term because $b = 0$.*

$$(4m + 5)(4m - 5) = 0$$ Factor the difference of squares. (Section 5.4)

$4m + 5 = 0$ or $4m - 5 = 0$ Zero-factor property

$4m = -5$ or $4m = 5$ Solve each equation.

$$m = -\frac{5}{4} \quad \text{or} \quad m = \frac{5}{4}$$

Check $-\frac{5}{4}$ and $\frac{5}{4}$ in the original equation. The solution set is $\left\{-\frac{5}{4}, \frac{5}{4}\right\}$.

(b) $y^2 = 2y$ *This equation is in the form $ax^2 + bx + c = 0$. Here, $c = 0$.*

$$y^2 - 2y = 0$$

$$y(y - 2) = 0$$ Factor.

Don't forget to set the variable factor y equal to 0. $y = 0$ or $y - 2 = 0$ Zero-factor property

$$y = 2$$ Solve.

A check confirms that the solution set is $\{0, 2\}$.

(c) $k(2k + 1) = 3$ *To be in standard form, 0 must be on the right side.*

$$2k^2 + k = 3$$ Distributive property

Standard form ⟶ $2k^2 + k - 3 = 0$ Subtract 3.

$$(2k + 3)(k - 1) = 0$$ Factor.

NOW TRY ANSWER

3. $\left\{-7, \frac{1}{2}\right\}$

NOW TRY
EXERCISE 4

Solve each equation.

(a) $9x^2 - 64 = 0$

(b) $m^2 = 5m$

(c) $p(6p - 1) = 2$

$$2k + 3 = 0 \quad \text{or} \quad k - 1 = 0 \quad \text{Zero-factor property}$$
$$2k = -3 \qquad\qquad k = 1 \qquad \text{Solve each equation.}$$
$$k = -\frac{3}{2}$$

A check confirms that the solution set is $\left\{-\frac{3}{2}, 1\right\}$.

NOW TRY

⚠️ **CAUTION**   In **Example 4(b)**, it is tempting to begin by dividing both sides of

$$y^2 = 2y$$

by y to obtain $y = 2$. Note, however, that we do not find the other solution, 0, if we divide by a variable. (We *may* divide each side of an equation by a *nonzero* real number, however. In **Example 3** we divided each side by 2.)

In **Example 4(c)**, we cannot directly apply the zero-factor property to solve

$$k(2k + 1) = 3$$

in its given form because of the 3 on the right side of the equation. *We can apply the zero-factor property only to a product that equals 0.*

EXAMPLE 5   Solving Quadratic Equations (Double Solutions)

Solve each equation.

(a)

$$z^2 - 22z + 121 = 0 \quad \text{This is a perfect square trinomial.}$$
$$(z - 11)^2 = 0 \qquad \text{Factor.}$$
$$(z - 11)(z - 11) = 0 \qquad a^2 = a \cdot a$$
$$z - 11 = 0 \quad \text{or} \quad z - 11 = 0 \qquad \text{Zero-factor property}$$

Because the two factors are identical, they both lead to the same solution, called a **double solution.**

$$z = 11 \qquad \text{Add 11.}$$

CHECK
$$z^2 - 22z + 121 = 0 \qquad \text{Original equation}$$
$$11^2 - 22(11) + 121 \stackrel{?}{=} 0 \qquad \text{Let } z = 11.$$
$$121 - 242 + 121 \stackrel{?}{=} 0 \qquad \text{Apply the exponent. Multiply.}$$
$$0 = 0 \ \checkmark \quad \text{True}$$

The solution set is $\{11\}$.

(b)
$$9t^2 - 30t = -25$$
$$9t^2 - 30t + 25 = 0 \qquad \text{Standard form}$$
$$(3t - 5)^2 = 0 \qquad \text{Factor the perfect square trinomial.}$$
$$3t - 5 = 0 \quad \text{or} \quad 3t - 5 = 0 \qquad \text{Zero-factor property}$$
$$3t = 5 \qquad \text{Solve the equation.}$$
$$t = \frac{5}{3} \qquad \tfrac{5}{3} \text{ is a double solution.}$$

NOW TRY ANSWERS

4. **(a)** $\left\{-\frac{8}{3}, \frac{8}{3}\right\}$ **(b)** $\{0, 5\}$

   (c) $\left\{-\frac{1}{2}, \frac{2}{3}\right\}$

**NOW TRY
EXERCISE 5**

Solve.

$$4x^2 - 4x + 1 = 0$$

CHECK

$$9t^2 - 30t = -25 \qquad \text{Original equation}$$

$$9\left(\frac{5}{3}\right)^2 - 30\left(\frac{5}{3}\right) \stackrel{?}{=} -25 \qquad \text{Let } t = \frac{5}{3}.$$

$$9\left(\frac{25}{9}\right) - 30\left(\frac{5}{3}\right) \stackrel{?}{=} -25 \qquad \text{Apply the exponent.}$$

$$25 - 50 \stackrel{?}{=} -25 \qquad \text{Multiply.}$$

$$-25 = -25 \quad \checkmark \quad \text{True}$$

The solution set is $\left\{\frac{5}{3}\right\}$.   **NOW TRY**

⚠ CAUTION   Each of the equations in **Example 5** has only *one* distinct solution. *There is no need to write the same number more than once in a solution set.*

OBJECTIVE 2   Solve other equations using the zero-factor property.

We can also use the zero-factor property to solve equations that involve more than two factors with variables. (These equations will have at least one term greater than second degree. They are *not* quadratic equations.)

**NOW TRY
EXERCISE 6**

Solve each equation.

(a) $3x^3 - 27x = 0$

(b) $(3a - 1) \cdot$
$(2a^2 - 5a - 12) = 0$

EXAMPLE 6   Solving Equations with More Than Two Variable Factors

Solve each equation.

(a)
$$6z^3 - 6z = 0$$

$$6z(z^2 - 1) = 0 \qquad \text{Factor out } 6z.$$

$$6z(z + 1)(z - 1) = 0 \qquad \text{Factor } z^2 - 1.$$

By an extension of the zero-factor property, this product can equal 0 only if at least one of the factors equals 0. Write and solve three equations, one for each factor with a variable.

$$6z = 0 \quad \text{or} \quad z + 1 = 0 \quad \text{or} \quad z - 1 = 0$$

$$z = 0 \quad \text{or} \qquad z = -1 \quad \text{or} \qquad z = 1$$

Check by substituting, in turn, 0, -1, and 1 into the original equation. The solution set is $\{-1, 0, 1\}$.

(b)
$$(3x - 1)(x^2 - 9x + 20) = 0$$

> The product of the factors is 0, as required. Do *not* multiply.

$$(3x - 1)(x - 5)(x - 4) = 0 \qquad \text{Factor the trinomial.}$$

$$3x - 1 = 0 \quad \text{or} \quad x - 5 = 0 \quad \text{or} \quad x - 4 = 0 \qquad \text{Zero-factor property}$$

$$x = \frac{1}{3} \quad \text{or} \qquad x = 5 \quad \text{or} \qquad x = 4 \qquad \text{Solve each equation.}$$

Check to verify that the solution set is $\left\{\frac{1}{3}, 4, 5\right\}$.   **NOW TRY**

NOW TRY ANSWERS

5. $\left\{\frac{1}{2}\right\}$

6. (a) $\{-3, 0, 3\}$

  (b) $\left\{-\frac{3}{2}, \frac{1}{3}, 4\right\}$

⚠ CAUTION   In **Example 6(b)**, it would be unproductive to begin by multiplying the two factors together. The zero-factor property requires the *product* of two or more factors to equal 0. *Always consider first whether an equation is given in an appropriate form to apply the zero-factor property.*

NOW TRY
EXERCISE 7

Solve.

$x(4x - 9) = (x - 2)^2 + 24$

EXAMPLE 7 Solving an Equation Requiring Multiplication before Factoring

Solve $(3x + 1)x = (x + 1)^2 + 5$.

The zero-factor property requires the *product* of two or more factors to equal 0.

$$(3x + 1)x = (x + 1)^2 + 5 \quad \boxed{\text{This equation is } not \text{ in the correct form.}}$$

$$3x^2 + x = x^2 + 2x + 1 + 5 \qquad \text{Multiply on the left.}$$
$$\text{Square } x + 1 \text{ on the right.}$$

$$3x^2 + x = x^2 + 2x + 6 \qquad \text{Combine like terms.}$$

$$2x^2 - x - 6 = 0 \qquad \text{Standard form}$$

$$\boxed{\text{The product of the factors is now 0.}} \quad (2x + 3)(x - 2) = 0 \qquad \text{Factor.}$$

$$2x + 3 = 0 \quad \text{or} \quad x - 2 = 0 \qquad \text{Zero-factor property}$$

$$x = -\frac{3}{2} \quad \text{or} \quad x = 2 \qquad \text{Solve each equation.}$$

NOW TRY ANSWER
7. $\left\{-\frac{7}{3}, 4\right\}$

Check to verify that the solution set is $\left\{-\frac{3}{2}, 2\right\}$.

NOW TRY

NOTE Not all quadratic equations can be solved using the zero-factor property. A more general method for solving such equations is given in **Chapter 11**.

5.5 Exercises

FOR EXTRA HELP

▶ MyMathLab®

▶ *Complete solution available in MyMathLab*

Concept Check *Fill in each blank with the correct response.*

1. A quadratic equation in x can be written in the form _____ $= 0$.

2. The form $ax^2 + bx + c = 0$ is called _____ form.

3. If the product of two numbers is 0, then at least one of the numbers is _____. This is the _____ property.

4. If a quadratic equation is in standard form, to solve the equation we should begin by attempting to _____ the polynomial.

5. The equation $x^3 + x^2 + x = 0$ is not a quadratic equation because _____.

6. If a quadratic equation $ax^2 + bx + c = 0$ has $c = 0$, then _____ *must* be a solution because _____ is a factor of the polynomial.

Concept Check *Work each problem.*

7. Identify each equation as *linear* or *quadratic*.

   (a) $2x - 5 = 6$ (b) $x^2 - 5 = -4$

   (c) $x^2 + 2x - 3 = 2x^2 - 2$ (d) $5^2x + 2 = 0$

8. The number 9 is a *double solution* of the following equation. Why is this so?

   $$(x - 9)^2 = 0$$

9. Look at this "solution."
   WHAT WENT WRONG?

$$x(7x - 1) = 0$$

$$7x - 1 = 0 \qquad \text{Zero-factor property}$$

$$x = \frac{1}{7}$$

The solution set is $\left\{\frac{1}{7}\right\}$.

10. Look at this "solution."
   WHAT WENT WRONG?

$$3x(5x - 4) = 0$$

$$x = 3 \quad \text{or} \quad x = 0 \quad \text{or} \quad 5x - 4 = 0$$

$$x = \frac{4}{5}$$

The solution set is $\left\{3, 0, \frac{4}{5}\right\}$.

Solve each equation, and check the solutions. **See Example 1.**

11. $(x + 5)(x - 2) = 0$

12. $(x - 1)(x + 8) = 0$

▶ **13.** $(2m - 7)(m - 3) = 0$

14. $(6x + 5)(x + 4) = 0$

15. $(2x + 1)(6x - 1) = 0$

16. $(3x + 2)(10x - 1) = 0$

17. $t(6t + 5) = 0$

18. $w(4w + 1) = 0$

19. $2x(3x - 4) = 0$

20. $6y(4y + 9) = 0$

21. $(x - 6)(x - 6) = 0$

22. $(y + 1)(y + 1) = 0$

Solve each equation, and check the solutions. **See Examples 2–7.**

23. $y^2 + 3y + 2 = 0$

24. $p^2 + 8p + 7 = 0$

25. $y^2 - 3y + 2 = 0$

26. $r^2 - 4r + 3 = 0$

▶ **27.** $x^2 = 24 - 5x$

28. $t^2 = 2t + 15$

29. $x^2 = 3 + 2x$

30. $x^2 = 4 + 3x$

31. $z^2 + 3z = -2$

32. $p^2 - 2p = 3$

33. $m^2 + 8m + 16 = 0$

34. $x^2 - 6x + 9 = 0$

35. $3x^2 + 5x - 2 = 0$

36. $6r^2 - r - 2 = 0$

▶ **37.** $12p^2 = 8 - 10p$

38. $18x^2 = 12 + 15x$

39. $9s^2 + 12s = -4$

40. $36x^2 + 60x = -25$

41. $y^2 - 9 = 0$

42. $m^2 - 100 = 0$

43. $16x^2 - 49 = 0$

44. $4w^2 - 9 = 0$

45. $n^2 = 121$

46. $x^2 = 400$

47. $x^2 + 6x = 0$

48. $x^2 + 4x = 0$

49. $x^2 = 7x$

50. $t^2 = 9t$

51. $6r^2 = 3r$

52. $10y^2 = -5y$

▶ **53.** $x(x - 7) = -10$

54. $r(r - 5) = -6$

55. $3z(2z + 7) = 12$

56. $4x(2x + 3) = 36$

57. $2y(y + 13) = 136$

58. $t(3t - 20) = -12$

59. $(x - 8)(x + 6) = 6x$

60. $(x - 2)(x + 9) = 4x$

61. $(x + 4)(x + 7) = 10$

62. $(x + 2)(x + 5) = 4$

63. $9y^3 - 49y = 0$

64. $16r^3 - 9r = 0$

65. $r^3 - 2r^2 - 8r = 0$

66. $x^3 - x^2 - 6x = 0$

67. $x^3 + x^2 - 20x = 0$

68. $y^3 - 6y^2 + 8y = 0$

69. $4x^3 - 18x^2 + 8x = 0$

70. $9x^3 - 24x^2 + 12x = 0$

71. $r^4 = 2r^3 + 15r^2$

72. $x^4 = 3x^2 + 2x^3$

▶ **73.** $(2r + 5)(3r^2 - 16r + 5) = 0$

74. $(3m + 4)(6m^2 + m - 2) = 0$

75. $(2x + 7)(x^2 + 2x - 3) = 0$

76. $(x + 1)(6x^2 + x - 12) = 0$

77. $3x(x + 1) = (2x + 3)(x + 1)$

78. $2x(x + 3) = (3x + 1)(x + 3)$

79. $x^2 + (x + 1)^2 = (x + 2)^2$

80. $(x - 7)^2 + x^2 = (x + 1)^2$

Extending Skills   *Solve each equation, and check the solutions.*

81. $(2x)^2 = (2x + 4)^2 - (x + 5)^2$

82. $5 - (x - 1)^2 = (x - 2)^2$

83. $(x + 3)^2 - (2x - 1)^2 = 0$

84. $(4y - 3)^3 - 9(4y - 3) = 0$

85. $6p^2(p + 1) = 4(p + 1) - 5p(p + 1)$

86. $6x^2(2x + 3) = 4(2x + 3) + 5x(2x + 3)$

Galileo's formula describing the motion of freely falling objects is

$$d = 16t^2.$$

The distance d in feet an object falls depends on the time t elapsed, in seconds. (This is an example of an important mathematical concept, the **function.***)*

87. **(a)** Use Galileo's formula and complete the following table. (*Hint:* Substitute each given value into the formula and solve for the unknown value.)

t in seconds	0	1	2	3		
d in feet	0	16			256	576

   (b) When $t = 0$, we find that $d = 0$. Explain this in the context of the problem.

88. Refer to **Exercise 87.** When 256 was substituted for d and the formula was solved for t, there should have been two solutions, 4 and -4. Why doesn't -4 make sense as an answer?

5.6   Applications of Quadratic Equations

VOCABULARY

☐ consecutive integers
☐ consecutive even (odd) integers
☐ legs
☐ hypotenuse

We use the zero-factor property to solve quadratic equations that arise in application problems. We follow the same six problem-solving steps given in **Section 2.4.**

Solving an Applied Problem

Step 1 **Read** the problem carefully. *What information is given? What is to be found?*

Step 2 **Assign a variable** to represent the unknown value. Make a sketch, diagram, or table, as needed. If necessary, express any other unknown values in terms of the variable.

Step 3 **Write an equation** using the variable expression(s).

Step 4 **Solve** the equation.

Step 5 **State the answer.** Label it appropriately. *Does it seem reasonable?*

Step 6 **Check** the answer in the words of the *original* problem.

PROBLEM-SOLVING HINT   Refer to the formulas inside the back cover of the text as needed when solving application problems.

**NOW TRY
EXERCISE 1**

A right triangle has one leg that is 4 ft shorter than the other leg. The area of the triangle is 6 ft². Determine the lengths of the legs.

OBJECTIVE 1   Solve problems involving geometric figures.

EXAMPLE 1   Solving an Area Problem

Abe wants to plant a triangular flower bed in a corner of his garden. One leg of the right-triangular flower bed will be 2 m shorter than the other leg. He wants the bed to have an area of 24 m². Find the lengths of the legs.

Step 1   **Read** the problem. We need to find the lengths of the legs of a right triangle with area 24 m².

Step 2   **Assign a variable.**

Let x = the length of one leg.

Then $x - 2$ = the length of the other leg.

See **FIGURE 1**.

$x - 2$

x

FIGURE 1

Step 3   **Write an equation.** In a right triangle, the legs are the base and height, so we substitute 24 for the area, x for the base, and $x - 2$ for the height in the formula for the area of a triangle.

$$\mathcal{A} = \frac{1}{2}bh \qquad \text{Formula for the area of a triangle}$$

$$24 = \frac{1}{2}x(x - 2) \qquad \text{Let } \mathcal{A} = 24, b = x, h = x - 2.$$

Step 4   **Solve.**

$$24 = \frac{1}{2}x^2 - x \qquad \text{Distributive property}$$

$$48 = x^2 - 2x \qquad \text{Multiply each term by 2.}$$

$$x^2 - 2x - 48 = 0 \qquad \text{Standard form}$$

$$(x + 6)(x - 8) = 0 \qquad \text{Factor.}$$

$$x + 6 = 0 \quad \text{or} \quad x - 8 = 0 \qquad \text{Zero-factor property}$$

$$x = -6 \quad \text{or} \qquad x = 8 \qquad \text{Solve each equation.}$$

Step 5   **State the answer.** The solutions are -6 and 8. Because a triangle cannot have a side of negative length, we discard the solution -6. Then the lengths of the legs will be 8 m and $8 - 2 = 6$ m.

Step 6   **Check.** The length of one leg is 2 m less than the length of the other leg, and the area is

$$\frac{1}{2}(8)(6) = 24 \text{ m}^2, \quad \text{as required.} \qquad \textbf{NOW TRY} \; \text{}$$

! CAUTION   *In solving applied problems, always check solutions against physical facts and discard any answers that are not appropriate.*

OBJECTIVE 2   Solve problems involving consecutive integers.

Recall from **Section 2.4** that **consecutive integers** are integers that are next to each other on a number line, such as

3 and 4,   or -11 and -10.

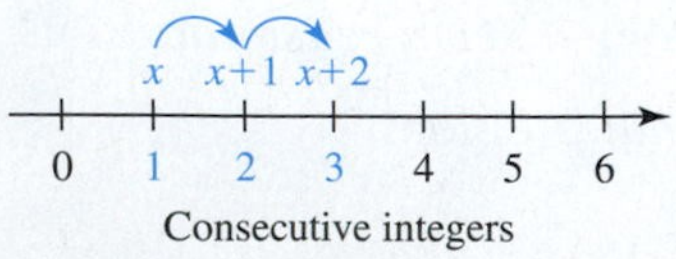

$$x \quad x+1 \; x+2$$

0 1 2 3 4 5 6

Consecutive integers

FIGURE 2

See **FIGURE 2**.

NOW TRY ANSWER

1. 2 ft, 6 ft

Consecutive even integers

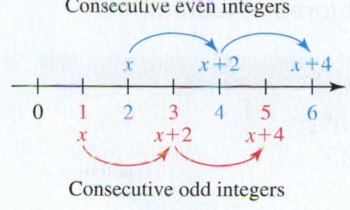

Consecutive odd integers

FIGURE 3

Consecutive even integers are *even* integers that are next to each other on a number line, such as

$$4 \text{ and } 6, \quad \text{or} \quad -10 \text{ and } -8.$$

Consecutive odd integers are defined similarly—for example, 3 and 5 are consecutive *odd* integers, as are -13 and -11. See **FIGURE 3**.

PROBLEM-SOLVING HINT   If $x =$ the lesser (least) integer in a consecutive integer problem, then the following apply.

- For two consecutive integers, use $x, \ x + 1.$
- For three consecutive integers, use $x, \ x + 1, \ x + 2.$
- For two consecutive even or odd integers, use $x, \ x + 2.$
- For three consecutive even or odd integers, use $x, \ x + 2, \ x + 4.$

In this book, we list consecutive integers in increasing order.

NOW TRY EXERCISE 2

The product of the first and second of three consecutive integers is 2 more than 8 times the third integer. Find the integers.

EXAMPLE 2 Solving a Consecutive Integer Problem

The product of the second and third of three consecutive integers is 2 more than 7 times the first integer. Find the integers.

Step 1 **Read** the problem. Note that the integers are consecutive.

Step 2 **Assign a variable.**

Let $x =$ the first integer.

Then $x + 1 =$ the second integer,

and $x + 2 =$ the third integer.

Step 3 **Write an equation.**

The product of the second and third   is   2 more than 7 times the first.

$$(x + 1)(x + 2) \qquad\qquad = \qquad\qquad 7x + 2$$

Step 4 **Solve.**

$$x^2 + 3x + 2 = 7x + 2 \qquad \text{Multiply.}$$

$$x^2 - 4x = 0 \qquad \text{Standard form}$$

$$x(x - 4) = 0 \qquad \text{Factor.}$$

$$x = 0 \quad \text{or} \quad x - 4 = 0 \qquad \text{Zero-factor property}$$

$$x = 4 \qquad \text{Add 4.}$$

Step 5 **State the answer.** The values 0 and 4 each lead to a distinct answer.

If $x = 0$, then $x + 1 = 1$ and $x + 2 = 2$. The integers are 0, 1, 2.

If $x = 4$, then $x + 1 = 5$ and $x + 2 = 6$. The integers are 4, 5, 6.

Step 6 **Check.** The product of the second and third integers must equal 2 more than 7 times the first. Because

$$1 \cdot 2 = 7 \cdot 0 + 2 \quad \text{and} \quad 5 \cdot 6 = 7 \cdot 4 + 2 \quad \text{are both true,}$$

both sets of consecutive integers satisfy the statement of the problem.

NOW TRY

OBJECTIVE 3   Solve problems by applying the Pythagorean theorem.

Pythagorean Theorem

If a and b are the lengths of the shorter sides of a right triangle (a triangle with a 90° angle) and c is the length of the longest side, then

$$a^2 + b^2 = c^2.$$

The two shorter sides are the **legs** of the triangle, and the longest side, the **hypotenuse,** is opposite the right angle.

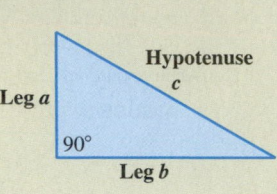

PROBLEM-SOLVING HINT   In solving a problem involving the Pythagorean theorem, be sure that the expressions for the sides are properly placed.

$$(\text{one leg})^2 + (\text{other leg})^2 = \text{hypotenuse}^2$$

EXAMPLE 3   **Applying the Pythagorean Theorem**

Patricia and Ali leave their office, with Patricia traveling north and Ali traveling east. When Ali is 1 mi farther than Patricia from the office, the distance between them is 2 mi more than Patricia's distance from the office. Find their distances from the office and the distance between them.

Step 1   **Read** the problem again. We must find three distances.

Step 2   **Assign a variable.**

Let $x =$ Patricia's distance from the office.

Then $x + 1 =$ Ali's distance from the office,

and $x + 2 =$ the distance between them.

Place these expressions on a right triangle, as in **FIGURE 4.**

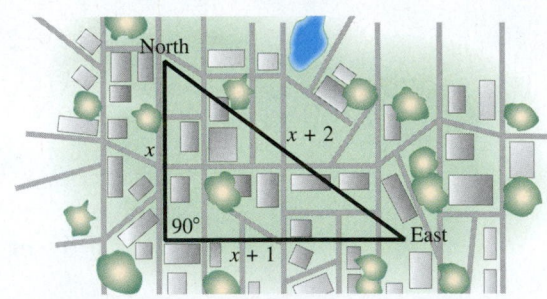

FIGURE 4

Step 3   **Write an equation.** Substitute into the Pythagorean theorem.

$$a^2 + b^2 = c^2$$

$$x^2 + (x + 1)^2 = (x + 2)^2 \qquad \text{Be careful to substitute properly.}$$

Step 4   **Solve.** $x^2 + x^2 + 2x + 1 = x^2 + 4x + 4$ Square each binomial.

> Remember the middle terms when squaring the binomials.

$$x^2 - 2x - 3 = 0 \qquad \text{Standard form}$$

$$(x - 3)(x + 1) = 0 \qquad \text{Factor.}$$

$$x - 3 = 0 \quad \text{or} \quad x + 1 = 0 \qquad \text{Zero-factor property}$$

$$x = 3 \quad \text{or} \qquad x = -1 \qquad \text{Solve each equation.}$$

NOW TRY EXERCISE 3

The longer leg of a right triangle is 7 ft longer than the shorter leg and the hypotenuse is 8 ft longer than the shorter leg. Find the lengths of the sides of the triangle.

Step 5 **State the answer.** Because -1 cannot represent a distance, 3 is the only possible answer. Patricia's distance is 3 mi, Ali's distance is $3 + 1 = 4$ mi, and the distance between them is $3 + 2 = 5$ mi.

Step 6 **Check.** Because $3^2 + 4^2 = 5^2$ is true, the answers are correct. **NOW TRY**

OBJECTIVE 4 Solve problems using given quadratic models.

In **Examples 1–3**, we wrote quadratic equations to model, or mathematically describe, various situations and then solved the equations. In the remaining examples, we are given quadratic models and must use them to determine data.

NOW TRY EXERCISE 4

Refer to **Example 4.** How long will it take for the ball to reach a height of 50 ft?

EXAMPLE 4 Finding the Height of a Ball

A tennis player's serve travels 180 ft per sec (123 mph). If she hits a ball directly upward, the height h of the ball in feet at time t in seconds is modeled by the quadratic equation

$$h = -16t^2 + 180t + 6.$$

How long will it take for the ball to reach a height of 206 ft?

A height of 206 ft means that $h = 206$, so we substitute 206 for h in the equation and solve for t.

$$h = -16t^2 + 180t + 6$$

$$206 = -16t^2 + 180t + 6 \qquad \text{Let } h = 206.$$

$$-16t^2 + 180t + 6 = 206 \qquad \text{Interchange sides.}$$

$$-16t^2 + 180t - 200 = 0 \qquad \text{Standard form}$$

$$4t^2 - 45t + 50 = 0 \qquad \text{Divide by } -4.$$

$$(4t - 5)(t - 10) = 0 \qquad \text{Factor.}$$

$$4t - 5 = 0 \quad \text{or} \quad t - 10 = 0 \qquad \text{Zero-factor property}$$

$$4t = 5 \quad \text{or} \qquad t = 10 \qquad \text{Solve each equation.}$$

$$t = \frac{5}{4}$$

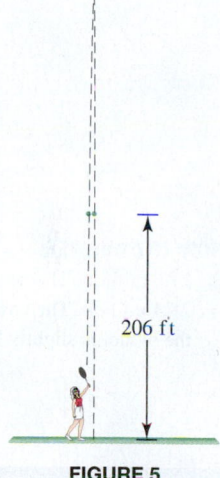

206 ft

FIGURE 5

Because we found two acceptable answers, the ball will be 206 ft above the ground twice—once on its way up and once on its way down—at $\frac{5}{4}$ sec and at 10 sec after it is hit. See **FIGURE 5**.

NOW TRY

EXAMPLE 5 Modeling the Foreign-Born Population of the United States

The foreign-born population of the United States over the years 1930–2010 can be modeled by the quadratic equation

$$y = 0.009665x^2 - 0.4942x + 15.12,$$

where $x = 0$ represents 1930, $x = 10$ represents 1940, and so on, and y is the number of people in millions. (*Source:* U.S. Census Bureau.)

(a) Use the model to find the foreign-born population in 1980 to the nearest tenth of a million.

NOW TRY ANSWERS

3. 5 ft, 12 ft, 13 ft

4. $\frac{1}{4}$ sec and 11 sec

NOW TRY EXERCISE 5

Use the model in **Example 5** to find the foreign-born population of the United States in the year 2000. Give the answer to the nearest tenth of a million. How does it compare to the actual value from the table?

Because $x = 0$ represents 1930, $x = 50$ represents 1980. Substitute 50 for x in the given equation.

$y = 0.009665x^2 - 0.4942x + 15.12$ Given quadratic model

$y = 0.009665(50)^2 - 0.4942(50) + 15.12$ Let $x = 50$.

$y = 14.6$ Round to the nearest tenth.

In 1980, the foreign-born population of the United States was about 14.6 million.

(b) Repeat part (a) for 2010.

$y = 0.009665(80)^2 - 0.4942(80) + 15.12$ For 2010, let $x = 80$.

$y = 37.4$ Round to the nearest tenth.

In 2010, the U.S. foreign-born population was about 37.4 million.

(c) The model used above was developed from the data in the table. How do the results in parts (a) and (b) compare to the actual data from the table?

Year	Foreign-Born Population (in millions)
1930	14.2
1940	11.6
1950	10.3
1960	9.7
1970	9.6
1980	14.1
1990	19.8
2000	28.4
2010	37.6

NOW TRY ANSWER

5. 27.9 million; The actual value is 28.4 million. The answer using the model is slightly low.

From the table, the actual value for 1980 is 14.1 million. Our answer in part (a), 14.6 million, is slightly high. For 2010, the actual value is 37.6 million, so our answer of 37.4 million in part (b) is slightly low, but a good estimate.   **NOW TRY**

5.6 Exercises

FOR EXTRA HELP   ▶ MyMathLab®

▶ *Complete solution available in MyMathLab*

1. *Concept Check* To review the six problem-solving steps first introduced in **Section 2.4,** complete each statement.

Step 1: _____ the problem carefully.

Step 2: Assign a _____ to represent the unknown value.

Step 3: Write a(n) _____ using the variable expression(s).

Step 4: _____ the equation.

Step 5: State the _____ .

Step 6: _____ the answer in the words of the _____ problem.

2. *Concept Check* A student solves an applied problem and gets 6 or -3 for the length of the side of a square. Which of these answers is reasonable? Why?

*In Exercises 3–6, a geometric figure is given. Write the indicated formula, and using x as the variable, complete Steps 3–6 for each problem. (Refer to the steps in **Exercise 1** as needed.)*

3.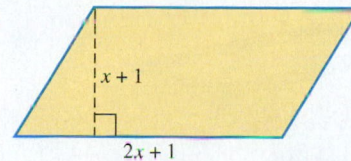

The area of this parallelogram is 45 sq. units. Find its base and height.

Formula for the area of a parallelogram:

_____

Step 3: $45 = $ _____

Step 4: $x = $ ____ or $x = $ ____

Step 5: base: ____ units;
height: ____ units

Step 6: _____ $= 45$

4.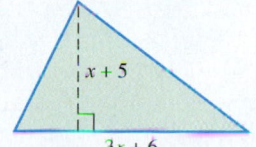

The area of this triangle is 60 sq. units. Find its base and height.

Formula for the area of a triangle:

_____

Step 3: $60 = $ _____

Step 4: $x = $ ____ or $x = $ ____

Step 5: base: ____ units;
height: ____ units

Step 6: _____ $= 60$

5.

The area of this rug is 80 sq. units. Find its length and width.

Formula for the area of a rectangle:

_____

Step 3: ____ $= (x + 8)$ _____

Step 4: $x = $ ____ or $x = $ ____

Step 5: length: ____ units;
width: ____ units

Step 6: _____ $= 80$

6.

The volume of this box is 192 cu. units. Find its length and width.

Formula for the volume of a rectangular solid: _____

Step 3: ____ $= $ ____ $(x + 2)$

Step 4: $x = $ ____ or $x = $ ____

Step 5: length: ____ units;
width: ____ units

Step 6: _____ $\cdot 4 = $ ____

Solve each problem. Check answers to be sure that they are reasonable. Refer to the formulas inside the back cover of this book as needed. **See Example 1.**

▶ 7. The length of a standard jewel case is 2 cm more than its width. The area of the rectangular top of the case is 168 cm². Find the length and width of the jewel case.

8. A standard DVD case is 6 cm longer than it is wide. The area of the rectangular top of the case is 247 cm². Find the length and width of the case.

9. The area of a triangle is 30 in.². The base of the triangle measures 2 in. more than twice the height of the triangle. Find the measures of the base and the height.

10. A certain triangle has its base equal in measure to its height. The area of the triangle is 72 m². Find the equal base and height measure.

11. A 10-gal aquarium is 3 in. higher than it is wide. Its length is 21 in., and its volume is 2730 in.^{3}. What are the height and width of the aquarium?

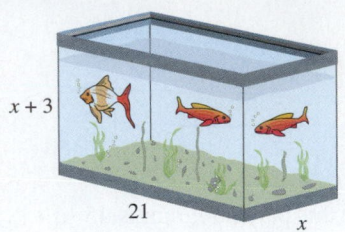

12. A toolbox is 2 ft high, and its width is 3 ft less than its length. If its volume is 80 ft^3, find the length and width of the box.

13. The dimensions of a rectangular monitor screen are such that its length is 3 in. more than its width. If the length were doubled and if the width were decreased by 1 in., the area would be increased by 150 in.^{2}. What are the length and width of the screen?

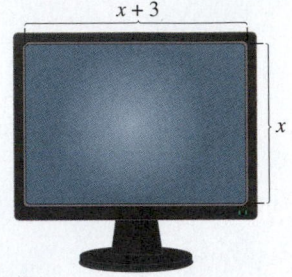

14. A computer keyboard is 11 in. longer than it is wide. If the length were doubled and if 2 in. were added to the width, the area would be increased by 198 in.^{2}. What are the length and width of the keyboard?

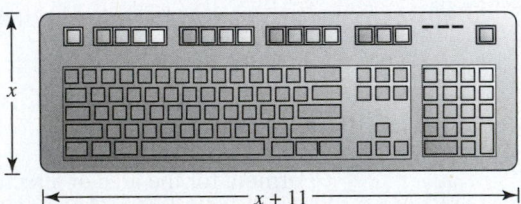

15. A square mirror has sides measuring 2 ft less than the sides of a square painting. If the difference between their areas is 32 ft^2, find the lengths of the sides of the mirror and the painting.

16. The sides of one square have length 3 m more than the sides of a second square. If the area of the larger square is subtracted from 4 times the area of the smaller square, the result is 36 m^2. What are the lengths of the sides of each square?

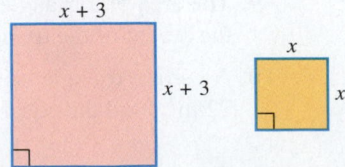

Solve each problem. See Example 2.

17. The product of the numbers on two consecutive volumes of research data is 420. Find the volume numbers.

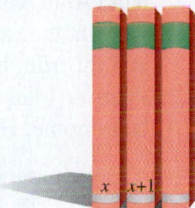

18. The product of the page numbers on two facing pages of a book is 600. Find the page numbers.

▶ **19.** The product of the second and third of three consecutive integers is 2 more than 10 times the first integer. Find the integers.

20. The product of the first and third of three consecutive integers is 3 more than 3 times the second integer. Find the integers.

21. Find two consecutive odd integers such that their product is 15 more than three times their sum.

22. Find two consecutive odd integers such that five times their sum is 23 less than their product.

23. Find three consecutive odd integers such that 3 times the sum of all three is 18 more than the product of the first and second integers.

24. Find three consecutive odd integers such that the sum of all three is 42 less than the product of the second and third integers.

25. Find three consecutive even integers such that the sum of the squares of the first and second integers is equal to the square of the third integer.

26. Find three consecutive even integers such that the square of the sum of the first and second integers is equal to twice the third integer.

Solve each problem. See Example 3.

▶ **27.** The hypotenuse of a right triangle is 1 cm longer than the longer leg. The shorter leg is 7 cm shorter than the longer leg. Find the length of the longer leg of the triangle.

28. The longer leg of a right triangle is 1 m longer than the shorter leg. The hypotenuse is 1 m shorter than twice the shorter leg. Find the length of the shorter leg of the triangle.

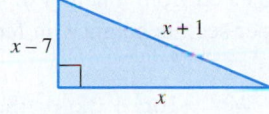

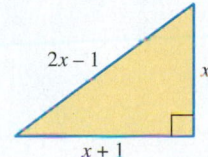

29. The length of a rectangle is 5 in. longer than its width. The diagonal is 5 in. shorter than twice the width. Find the length, width, and diagonal measures of the rectangle.

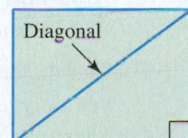

30. The length of a rectangle is 4 in. longer than its width. The diagonal is 8 in. longer than the width. Find the length, width, and diagonal measures of the rectangle.

31. Tram works due north of home. Her husband Alan works due east. They leave for work at the same time. By the time Tram is 5 mi from home, the distance between them is 1 mi more than Alan's distance from home. How far from home is Alan?

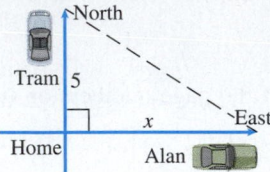

32. Two cars left an intersection at the same time. One traveled north. The other traveled 14 mi farther, but to the east. How far apart were they at that time if the distance between them was 4 mi more than the distance traveled east?

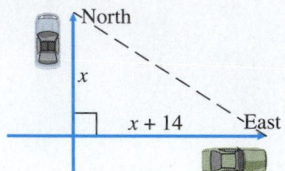

33. A ladder is leaning against a building. The distance from the bottom of the ladder to the building is 4 ft less than the length of the ladder. How high up the side of the building is the top of the ladder if that distance is 2 ft less than the length of the ladder?

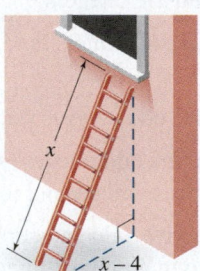

34. A lot has the shape of a right triangle with one leg 2 m longer than the other. The hypotenuse is 2 m less than twice the length of the shorter leg. Find the length of the shorter leg.

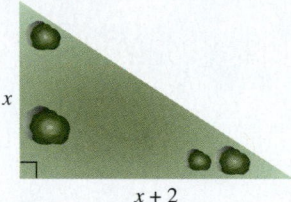

If an object is projected upward with an initial velocity of 128 *ft per sec, its height h in feet after t seconds is given by the quadratic equation*

$$h = -16t^2 + 128t.$$

Find the height of the object after each time listed. ***See Example 4.***

35. 1 sec   **36.** 2 sec   **37.** 4 sec

38. How long does it take the object just described to return to the ground? (*Hint:* When the object hits the ground, $h = 0$.)

Solve each problem. ***See Example 4.***

▶ **39.** If an object is projected from a height of 48 ft with an initial velocity of 32 ft per sec, its height h in feet after t seconds is given by

$$h = -16t^2 + 32t + 48.$$

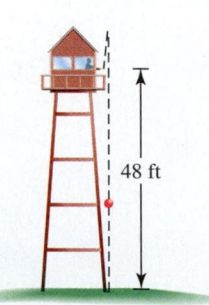

(a) After how many seconds is the height 64 ft? (*Hint:* Let $h = 64$ and solve.)

(b) After how many seconds is the height 60 ft?

(c) After how many seconds does the object hit the ground? (*Hint:* When the object hits the ground, $h = 0$.)

(d) The quadratic equation from part (c) has two solutions, yet only one of them is appropriate for answering the question. Why is this so?

40. If an object is projected upward from ground level with an initial velocity of 64 ft per sec, its height h in feet t seconds later is given by

$$h = -16t^2 + 64t.$$

(a) After how many seconds is the height 48 ft?

(b) The object reaches its maximum height 2 sec after it is projected. What is this maximum height?

(c) After how many seconds does the object hit the ground? (*Hint:* When the object hits the ground, $h = 0$.)

(d) The quadratic equation from part (c) has two solutions, yet only one of them is appropriate for answering the question. Why is this so?

(e) Find the number of seconds after which the height is 60 ft.

(f) What is the physical interpretation of why part (e) has two answers?

Solve each problem. See Example 5.

41. The table shows the number of cellular phone subscribers (in millions) in the United States.

Year	Subscribers (in millions)
1990	5
1992	11
1994	24
1996	44
1998	69
2000	109
2002	141
2004	182
2006	233
2008	270
2010	296
2012	326

Source: CTIA.

We used the data to develop the quadratic equation

$$y = 0.339x^2 + 8.50x - 8.26,$$

which models the number of cellular phone subscribers y (in millions) in the year x, where $x = 0$ represents 1990, $x = 2$ represents 1992, and so on.

(a) What value of x corresponds to 2000?

(b) Use the model to find the number of subscribers in 2000, to the nearest million. How does the result compare with the actual data in the table?

(c) What value of x corresponds to 2012?

(d) Use the model to find the number of cellular phone subscribers in 2012, to the nearest million. How does the result compare with the actual data in the table?

(e) Assuming that the trend in the data continues, what value of x would correspond to 2014?

(f) Use the model to find the number of cellular phone subscribers in 2014, to the nearest million.

42. World population (in billions) is shown in the table. Using the data, we developed the quadratic equation

$$y = 0.0002x^2 + 0.0593x + 2.501,$$

which models population y (in billions) in the year x, where $x = 0$ represents 1950, $x = 10$ represents 1960, and so on.

Year	Population (in billions)
1950	2.6
1960	3.0
1970	3.7
1980	4.5
1990	5.3
2000	6.1
2010	6.8
2013	7.1

Source: www.worldpopulationstatistics.com

(a) What value of x corresponds to the year 2000? To the year 2013?

(b) Use the model to find world population in 2000 and 2013, to the nearest tenth. How do the results compare with the actual data in the table?

(c) World population is projected to reach 8.0 billion in 2025. What value of x corresponds to the year 2025?

(d) Use the model to project world population in 2025, to the nearest tenth. How does the result compare to the projection given in part (c)?

Chapter 5   Summary

Key Terms

5.1
factor
factored form
common factor
greatest common factor
    (GCF)

5.2
prime polynomial

5.4
perfect square
perfect square trinomial
perfect cube

5.5
quadratic equation
double solution

5.6
consecutive integers
consecutive even (odd)
    integers

legs
hypotenuse

Test Your Word Power

See how well you have learned the vocabulary in this chapter.

1. Factoring is
   A. a method of multiplying polynomials
   B. the process of writing a polynomial as a product
   C. the answer in a multiplication problem
   D. a way to add the terms of a polynomial.

2. A polynomial is in **factored form** when
   A. it is prime
   B. it is written as a sum

   C. the second-degree term has a coefficient of 1
   D. it is written as a product.

3. A **perfect square trinomial** is a trinomial
   A. that can be factored as the square of a binomial
   B. that cannot be factored
   C. that is multiplied by a binomial
   D. where all terms are perfect squares.

4. A **quadratic equation** is an equation that can be written in the form
   A. $y = mx + b$
   B. $ax^2 + bx + c = 0$ $(a \neq 0)$
   C. $Ax + By = C$
   D. $x = k$.

5. A **hypotenuse** is
   A. either of the two shorter sides of a triangle
   B. the shortest side of a triangle
   C. the side opposite the right angle in a triangle
   D. the longest side in any triangle.

ANSWERS

1. B; *Example:* $x^2 - 5x - 14$ factors as $(x - 7)(x + 2)$.   **2.** D; *Example:* The factored form of $x^2 - 5x - 14$ is $(x - 7)(x + 2)$.   **3.** A; *Example:* $a^2 + 2a + 1$ is a perfect square trinomial. Its factored form is $(a + 1)^2$.   **4.** B; *Examples:* $y^2 - 3y + 2 = 0$, $x^2 - 9 = 0$, $2m^2 = 6m + 8$ **5.** C; *Example:* In **FIGURE 4** of **Section 5.6**, the hypotenuse is the side labeled $x + 2$.

Quick Review

CONCEPTS

EXAMPLES

5.1 The Greatest Common Factor; Factoring by Grouping

Finding the Greatest Common Factor (GCF)

Step 1 Write each number in prime factored form.

Step 2 List each prime number or each variable that is a factor of every term in the list.

Step 3 Use as exponents on the common prime factors the *least* exponents from the prime factored forms.

Step 4 Multiply the primes from Step 3.

Find the greatest common factor of $4x^2y$, $6x^2y^3$, and $2xy^2$.

$$4x^2y = 2 \cdot 2 \cdot x^2 \cdot y$$

$$6x^2y^3 = 2 \cdot 3 \cdot x^2 \cdot y^3$$

$$2xy^2 = 2 \cdot x \cdot y^2$$

The greatest common factor is $2xy$.

Factoring by Grouping

Step 1 Group the terms.

Step 2 Factor out the greatest common factor from each group.

Step 3 Factor out a common binomial factor from the results of Step 2.

Step 4 If necessary, rearrange terms and try a different grouping.

Factor by grouping.

$$3x^2 + 5x - 24xy - 40y$$

$$= (3x^2 + 5x) + (-24xy - 40y) \qquad \text{Group the terms.}$$

$$= x(3x + 5) - 8y(3x + 5) \qquad \text{Factor each group.}$$

$$= (3x + 5)(x - 8y) \qquad \text{Factor out } 3x + 5.$$

5.2 Factoring Trinomials

To factor $x^2 + bx + c$, find two integers m and n such that $mn = c$ and $m + n = b$.

$$mn = c$$
$$x^2 + bx + c$$
$$m + n = b$$

Then $x^2 + bx + c$ factors as $(x + m)(x + n)$.

Check by multiplying.

Factor $x^2 + 6x + 8$.

$$mn = 8$$
$$x^2 + 6x + 8 \qquad \text{Here, } m = 2 \text{ and } n = 4.$$
$$m + n = 6$$

$x^2 + 6x + 8$ factors as $(x + 2)(x + 4)$.

CHECK $(x + 2)(x + 4)$

$$= x^2 + 4x + 2x + 8 \qquad \text{FOIL method}$$

$$= x^2 + 6x + 8 \checkmark \qquad \text{Combine like terms.}$$

5.3 More on Factoring Trinomials

To factor $ax^2 + bx + c$, use one of the following methods.

Factoring by Grouping

Find m and n such that $mn = ac$ and $m + n = b$.

$$mn = ac$$
$$ax^2 + bx + c$$
$$m + n = b$$

Factor $3x^2 + 14x - 5$ by grouping.

$$3x^2 + 14x - 5 \qquad \text{Here, } mn = -15 \text{ and } m + n = 14.$$
$$\underline{\qquad -15 \qquad}$$

The required integers are $m = -1$ and $n = 15$.

$$3x^2 + 14x - 5$$

$$= 3x^2 - x + 15x - 5 \qquad 14x = -x + 15x$$

$$= (3x^2 - x) + (15x - 5) \qquad \text{Group the terms.}$$

$$= x(3x - 1) + 5(3x - 1) \qquad \text{Factor each group.}$$

$$= (3x - 1)(x + 5) \qquad \text{Factor out } 3x - 1.$$

CONCEPTS	EXAMPLES

Factoring by Trial and Error
Use the FOIL method in reverse.

Factor $3x^2 + 14x - 5$ by trial and error.

Because the only positive factors of 3 are 3 and 1, and -5 has possible factors of 1 and -5, or -1 and 5, the possible factored forms for this trinomial follow.

$(3x - 5)(x + 1)$	Incorrect	$(3x + 5)(x - 1)$	Incorrect
$(3x + 1)(x - 5)$	Incorrect	$(3x - 1)(x + 5)$	Correct

Using grouping or trial and error,

$$3x^2 + 14x - 5 \quad \text{factors as} \quad (3x - 1)(x + 5).$$

5.4   Special Factoring Techniques

Difference of Squares
$$x^2 - y^2 = (x + y)(x - y)$$

Perfect Square Trinomials
$$x^2 + 2xy + y^2 = (x + y)^2$$
$$x^2 - 2xy + y^2 = (x - y)^2$$

Difference of Cubes
$$x^3 - y^3 = (x - y)(x^2 + xy + y^2)$$

Sum of Cubes
$$x^3 + y^3 = (x + y)(x^2 - xy + y^2)$$

Factor.

$$4x^2 - 9$$
$$= (2x + 3)(2x - 3)$$

$9x^2 + 6x + 1$	$4x^2 - 20x + 25$
$= (3x + 1)^2$	$= (2x - 5)^2$

$m^3 - 8$	$z^3 + 27$
$= m^3 - 2^3$	$= z^3 + 3^3$
$= (m - 2)(m^2 + 2m + 4)$	$= (z + 3)(z^2 - 3z + 9)$

5.5   Solving Quadratic Equations Using the Zero-Factor Property

Zero-Factor Property
If a and b are real numbers and if $ab = 0$, then $a = 0$ or $b = 0$.

If $(x - 2)(x + 3) = 0$, then $x - 2 = 0$ or $x + 3 = 0$.

Solving a Quadratic Equation Using the Zero-Factor Property

Step 1   Write the equation in standard form.

Step 2   Factor.

Step 3   Apply the zero-factor property.

Step 4   Solve the resulting equations.

Solve $2x^2 = 7x + 15$.

$$2x^2 - 7x - 15 = 0 \qquad \text{Standard form}$$

$$(2x + 3)(x - 5) = 0 \qquad \text{Factor.}$$

$$2x + 3 = 0 \quad \text{or} \quad x - 5 = 0 \qquad \text{Zero-factor property}$$

$$2x = -3 \qquad\qquad x = 5 \qquad \text{Solve each equation.}$$

$$x = -\frac{3}{2}$$

Step 5   Check. Write the solution set.

CHECK $2x^2 = 7x + 15$

$$2(5)^2 \stackrel{?}{=} 7(5) + 15 \qquad \text{Let } x = 5.$$

$$50 \stackrel{?}{=} 35 + 15$$

$$50 = 50 \quad \checkmark \qquad \text{True}$$

The other value also checks. The solution set is $\left\{ -\frac{3}{2}, 5 \right\}$.

CONCEPTS	**EXAMPLES**

5.6 Applications of Quadratic Equations

Pythagorean Theorem

In a right triangle, the sum of the squares of the legs equals the square of the hypotenuse.

$$a^2 + b^2 = c^2$$

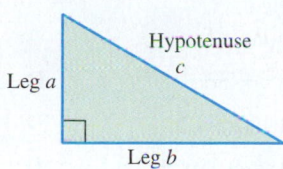

The longer leg of a right triangle is 2 ft longer than the shorter leg. The hypotenuse is 4 ft longer than the shorter leg. Find the lengths of the sides of the triangle.

Let x = the length of the shorter leg.
Then $x + 2$ = the length of the longer leg,
and $x + 4$ = the length of the hypotenuse.

$$x^2 + (x + 2)^2 = (x + 4)^2$$

$x^2 + x^2 + 4x + 4 = x^2 + 8x + 16$ Square each binomial.

$x^2 - 4x - 12 = 0$ Standard form

$(x - 6)(x + 2) = 0$ Factor.

$x - 6 = 0$ or $x + 2 = 0$ Zero-factor property

$x = 6$ or $x = -2$ Solve each equation.

Discard -2 as a solution. Check that the sides have lengths

6 ft, $6 + 2 = 8$ ft,   and $6 + 4 = 10$ ft.

Chapter 5   Review Exercises

Answers (left margin):

1. $7(t + 2)$
2. $30z(2z^2 + 1)$
3. $-3x(x^2 - 2x - 1)$
4. $50m^2n^2(2n - mn^2 + 3)$
5. $(2y + 3)(x - 4)$
6. $(3y + 2x)(2y + 3)$
7. $(x + 3)(x + 2)$
8. $(y - 5)(y - 8)$
9. $(q + 9)(q - 3)$
10. $(r - 8)(r + 7)$
11. prime
12. $8p(p + 2)(p - 5)$
13. $3x^2(x + 2)(x + 8)$
14. $(r + 8s)(r - 12s)$
15. $(p + 12q)(p - 10q)$
16. $p^5(p - 2q)(p + q)$
17. $3r^3(r + 3s)(r - 5s)$
18. $2x^5(x - 2y)(x + 3y)$
19. r and $6r$, $2r$ and $3r$
20. Factor out z.
21. $(2k - 1)(k - 2)$
22. $(3r - 1)(r + 4)$
23. $(3r + 2)(2r - 3)$
24. $(5z + 1)(2z - 1)$

5.1 *Factor out the greatest common factor, or factor by grouping.*

1. $7t + 14$
2. $60z^3 + 30z$
3. $-3x^3 + 6x^2 + 3x$
4. $100m^2n^3 - 50m^3n^4 + 150m^2n^2$
5. $2xy - 8y + 3x - 12$
6. $6y^2 + 9y + 4xy + 6x$

5.2 *Factor completely.*

7. $x^2 + 5x + 6$
8. $y^2 - 13y + 40$
9. $q^2 + 6q - 27$
10. $r^2 - r - 56$
11. $x^2 + x + 1$
12. $8p^3 - 24p^2 - 80p$
13. $3x^4 + 30x^3 + 48x^2$
14. $r^2 - 4rs - 96s^2$
15. $p^2 + 2pq - 120q^2$
16. $p^7 - p^6q - 2p^5q^2$
17. $3r^5 - 6r^4s - 45r^3s^2$
18. $2x^7 + 2x^6y - 12x^5y^2$

5.3 *Answer each question.*

19. To begin factoring $6r^2 - 5r - 6$, what are the possible first terms of the two binomial factors if we consider only positive integer coefficients?

20. What is the first step to factor $2z^3 + 9z^2 - 5z$?

Factor completely.

21. $2k^2 - 5k + 2$
22. $3r^2 + 11r - 4$
23. $6r^2 - 5r - 6$
24. $10z^2 - 3z - 1$
25. $5t^2 - 11t + 12$
26. $24x^5 - 20x^4 + 4x^3$
27. $-6x^2 + 3x + 30$
28. $10r^3s + 17r^2s^2 + 6rs^3$
29. $48x^4y + 4x^3y^2 - 4x^2y^3$

25. prime
26. $4x^3(3x - 1)(2x - 1)$
27. $-3(2x - 5)(x + 2)$
28. $rs(5r + 6s)(2r + s)$
29. $4x^2y(3x + y)(4x - y)$
30. The student stopped too soon. He needs to factor out the common factor $4x - 1$ to obtain

$$(4x - 1)(4x - 5)$$

as the correct answer.

31. B   **32.** D
33. $(n + 7)(n - 7)$
34. $(5b + 11)(5b - 11)$
35. $(7y + 5w)(7y - 5w)$
36. $36(2p + q)(2p - q)$
37. prime   **38.** $(r - 6)^2$
39. $(3t - 7)^2$
40. $(m + 10)(m^2 - 10m + 100)$
41. $(5k + 4x)(25k^2 - 20kx + 16x^2)$
42. $(7x - 4)(49x^2 + 28x + 16)$
43. $(10 - 3x^2)(100 + 30x^2 + 9x^4)$
44. $(x - y)(x + y)(x^2 + xy + y^2) \cdot$
     $(x^2 - xy + y^2)$

45. $\left\{-\frac{3}{4}, 1\right\}$ **46.** $\left\{0, \frac{5}{2}\right\}$

47. $\{-3, -1\}$ **48.** $\{1, 4\}$

49. $\{3, 5\}$ **50.** $\left\{-\frac{4}{3}, 5\right\}$

51. $\left\{-\frac{8}{9}, \frac{8}{9}\right\}$ **52.** $\{0, 8\}$

53. $\{-1, 6\}$ **54.** $\{7\}$

55. $\{6\}$ **56.** $\{-3, 3\}$

57. $\left\{-2, -1, -\frac{2}{5}\right\}$

58. $\left\{-\frac{3}{8}, 0, \frac{3}{8}\right\}$

59. length: 10 ft; width: 4 ft
60. 5 ft
61. 26 mi
62. length: 6 m; width: 4 m

---

30. On a quiz, a student factored $16x^2 - 24x + 5$ by grouping as follows.

$$16x^2 - 24x + 5$$
$$= 16x^2 - 4x - 20x + 5$$
$$= 4x(4x - 1) - 5(4x - 1) \qquad \text{His answer}$$

He thought his answer was correct because it checked by multiplication. **WHAT WENT WRONG?** Give the correct factored form.

5.4 *Answer each question.*

31. Which one of the following is a difference of squares?

   A. $32x^2 - 1$ **B.** $4x^2y^2 - 25z^2$ **C.** $x^2 + 36$ **D.** $25y^3 - 1$

32. Which one of the following is a perfect square trinomial?

   A. $x^2 + x + 1$ **B.** $y^2 - 4y + 9$ **C.** $4x^2 + 10x + 25$ **D.** $x^2 - 20x + 100$

Factor completely.

33. $n^2 - 49$ **34.** $25b^2 - 121$ **35.** $49y^2 - 25w^2$

36. $144p^2 - 36q^2$ **37.** $x^2 + 100$ **38.** $r^2 - 12r + 36$

39. $9t^2 - 42t + 49$ **40.** $m^3 + 1000$ **41.** $125k^3 + 64x^3$

42. $343x^3 - 64$ **43.** $1000 - 27x^6$ **44.** $x^6 - y^6$

5.5 *Solve each equation, and check the solutions.*

45. $(4t + 3)(t - 1) = 0$ **46.** $x(2x - 5) = 0$ **47.** $z^2 + 4z + 3 = 0$

48. $m^2 - 5m + 4 = 0$ **49.** $x^2 = -15 + 8x$ **50.** $3z^2 - 11z - 20 = 0$

51. $81t^2 - 64 = 0$ **52.** $y^2 = 8y$ **53.** $n(n - 5) = 6$

54. $t^2 - 14t + 49 = 0$ **55.** $t^2 = 12(t - 3)$ **56.** $x^2 = 9$

57. $(5z + 2)(z^2 + 3z + 2) = 0$ **58.** $64x^3 - 9x = 0$

5.6 *Solve each problem.*

59. The length of a rug is 6 ft more than the width. The area is 40 ft². Find the length and width of the rug.

60. A treasure chest from a sunken galleon has the dimensions shown in the figure. Its surface area is 650 ft². Find its width.

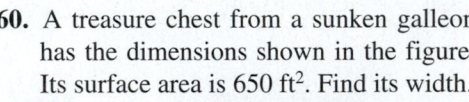
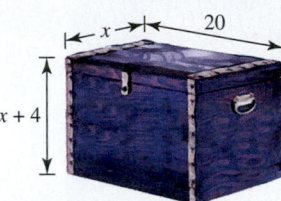

61. Two cars left an intersection at the same time. One traveled west, and the other traveled 14 mi less, but to the south. How far apart were they at that time, if the distance between them was 16 mi more than the distance traveled south?

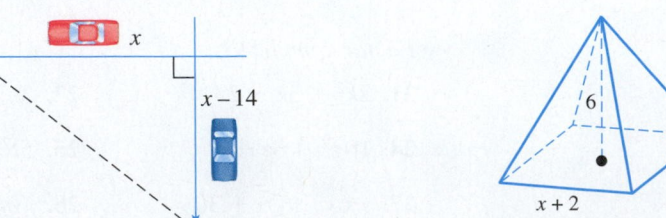

62. A pyramid has a rectangular base with a length that is 2 m more than its width. The height of the pyramid is 6 m, and its volume is 48 m³. Find the length and width of the base.

63. $-5, -4, -3$ or $5, 6, 7$

63. The product of the first and second of three consecutive integers is equal to 23 plus the third. Find the integers.

64. (a) 256 ft (b) 1024 ft

64. If an object is dropped, the distance d in feet it falls in t seconds (disregarding air resistance) is given by the quadratic equation

$$d = 16t^2.$$

Find the distance an object would fall in (a) 4 sec and (b) 8 sec.

65. (a) 2007: 704 thousand;
2011: 1148 thousand

(b) 2007: The result is slightly higher than the actual number;
2011: The result is lower than the actual number.

65. The numbers of alternative-fueled vehicles in use in the United States are given in the table.

Year	Alternative-Fueled Vehicles (in thousands)
2001	425
2003	534
2005	592
2007	696
2009	826
2011	1192

Source: Energy Information Administration.

Using the data, we developed the quadratic equation

$$y = 7.02x^2 - 15.5x + 469,$$

which models the number of vehicles y (in thousands) in the year x, where $x = 1$ represents 2001, $x = 3$ represents 2003, and so on.

(a) Use the model to find the number of alternative-fueled vehicles in 2007 and 2011, to the nearest thousand.

(b) How do the results in part (a) compare with the actual data in the table?

Chapter 5	Mixed Review Exercises

1. D

1. Which of the following is *not* factored completely?

A. $3(7t)$ **B.** $3x(7t + 4)$ **C.** $(3 + x)(7t + 4)$ **D.** $3(7t + 4) + x(7t + 4)$

2. The factor $(2x + 8)$ has a factor of 2. The completely factored form is

$$2(x + 4)(3x - 4).$$

2. A student did not receive full credit for factoring

$$6x^2 + 16x - 32 \quad \text{as} \quad (2x + 8)(3x - 4).$$

WHAT WENT WRONG? Give the completely factored form.

3. $(3k + 5)(k + 2)$

4. $(z - x)(z - 10x)$

5. $(y^2 + 25)(y + 5)(y - 5)$

6. $3m(2m + 3)(m - 5)$

7. prime

8. $2a^3(a + 2)(a - 6)$

9. $(3m + 4)(5m - 4p)$

10. $8abc(3b^2c - 7ac^2 + 9ab)$

11. $6xyz(2xz^2 + 2y - 5x^2yz^3)$

12. $(2r + 3q)(6r - 5)$

13. $(7t + 4)^2$

14. $(10a + 3)(100a^2 - 30a + 9)$

Factor completely.

3. $3k^2 + 11k + 10$

4. $z^2 - 11zx + 10x^2$

5. $y^4 - 625$

6. $6m^3 - 21m^2 - 45m$

7. $25a^2 + 15ab + 9b^2$

8. $2a^5 - 8a^4 - 24a^3$

9. $15m^2 + 20m - 12mp - 16p$

10. $24ab^3c^2 - 56a^2bc^3 + 72a^2b^2c$

11. $12x^2yz^3 + 12xy^2z - 30x^3y^2z^4$

12. $12r^2 + 18rq - 10r - 15q$

13. $49t^2 + 56t + 16$

14. $1000a^3 + 27$

15. $\{0, 7\}$ **16.** $\{-5, 2\}$

17. $\left\{-\dfrac{2}{5}\right\}$

Solve each equation.

15. $t(t - 7) = 0$ **16.** $x^2 + 3x = 10$ **17.** $25x^2 + 20x + 4 = 0$

18. 15 m, 36 m, 39 m

19. 6 m

20. width: 10 m; length: 17 m

Solve each problem.

18. A lot is in the shape of a right triangle. The hypotenuse is 3 m longer than the longer leg. The longer leg is 6 m longer than twice the length of the shorter leg. Find the lengths of the sides of the lot.

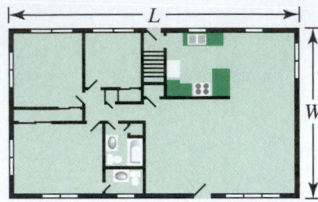

19. The triangular sail of a schooner has an area of 30 m². The height of the sail is 4 m more than the base. Find the base of the sail.

20. The floor plan for a house is a rectangle with length 7 m more than its width. The area is 170 m². Find the width and length of the house.

| Chapter 5 | Test | FOR EXTRA HELP | *Step-by-step test solutions are found on the Chapter Test Prep Videos available in* MyMathLab®, *or on* YouTube. |

▶ *View the complete solutions to all Chapter Test exercises in MyMathLab.*

[5.1–5.4]

1. D   **2.** $6x(2x - 5)$

3. $m^2n(2mn + 3m - 5n)$

4. $(2x + y)(a - b)$

5. $(x + 3)(x - 8)$

6. $(2x + 3)(x - 1)$

7. $(5z - 1)(2z - 3)$

8. $3(x + 1)(x - 5)$

9. prime   **10.** prime

11. $(2 - a)(6 + b)$

12. $(3y + 8)(3y - 8)$

13. $(9a + 11b)(9a - 11b)$

14. $(x + 8)^2$

15. $(2x - 7y)^2$

16. $3t^2(2t + 9)(t - 4)$

17. $(r - 5)(r^2 + 5r + 25)$

18. $8(k + 2)(k^2 - 2k + 4)$

19. $(x^2 + 9)(x + 3)(x - 3)$

20. $(3x + 2y)(3x - 2y)(9x^2 + 4y^2)$

[5.5]

21. $\{-3, 9\}$ **22.** $\left\{\frac{1}{2}, 6\right\}$

23. $\left\{-\frac{2}{5}, \frac{2}{5}\right\}$ **24.** $\{0, 9\}$

25. $\{10\}$

26. $\left\{-8, -\frac{5}{2}, \frac{1}{3}\right\}$

1. Which one of the following is the correct, completely factored form of $2x^2 - 2x - 24$?

   A. $(2x + 6)(x - 4)$ **B.** $(x + 3)(2x - 8)$

   C. $2(x + 4)(x - 3)$ **D.** $2(x + 3)(x - 4)$

Factor completely. If the polynomial is prime, say so.

2. $12x^2 - 30x$ **3.** $2m^3n^2 + 3m^3n - 5m^2n^2$ **4.** $2ax - 2bx + ay - by$

5. $x^2 - 5x - 24$ **6.** $2x^2 + x - 3$ **7.** $10z^2 - 17z + 3$

8. $3x^2 - 12x - 15$ **9.** $t^2 + 2t + 3$ **10.** $x^2 + 36$

11. $12 - 6a + 2b - ab$ **12.** $9y^2 - 64$ **13.** $81a^2 - 121b^2$

14. $x^2 + 16x + 64$ **15.** $4x^2 - 28xy + 49y^2$ **16.** $6t^4 + 3t^3 - 108t^2$

17. $r^3 - 125$ **18.** $8k^3 + 64$ **19.** $x^4 - 81$ **20.** $81x^4 - 16y^4$

Solve each equation.

21. $(x + 3)(x - 9) = 0$ **22.** $2r^2 - 13r + 6 = 0$

23. $25x^2 - 4 = 0$ **24.** $t^2 = 9t$

25. $x(x - 20) = -100$ **26.** $(s + 8)(6s^2 + 13s - 5) = 0$

Solve each problem.

27. The length of a rectangular flower bed is 3 ft less than twice its width. The area of the bed is 54 ft². Find the dimensions of the flower bed.

28. Find two consecutive integers such that the square of the sum of the two integers is 11 more than the first integer.

29. A carpenter needs to cut a brace to support a wall stud, as shown in the figure. The brace should be 7 ft less than three times the length of the stud. If the brace will be anchored on the floor 15 ft away from the stud, how long should the brace be?

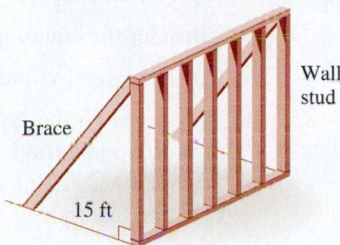

30. The public debt y (in billions of dollars) of the United States from 2000 through 2014 can be approximated by the quadratic equation

$$y = 57.53x^2 - 72.93x + 3417,$$

where $x = 0$ represents 2000, $x = 1$ represents 2001, and so on. Use the model to estimate the public debt, to the nearest billion dollars, in the year 2014. (*Source:* Bureau of Public Debt.)

Chapters R–5   Cumulative Review Exercises

Solve each equation.

1. $3x + 2(x - 4) = 4(x - 2)$

2. $0.3x + 0.9x = 0.06$

3. $\dfrac{2}{3}m - \dfrac{1}{2}(m - 4) = 3$

4. Solve for P: $A = P + Prt$.

5. Find the measures of the marked angles.

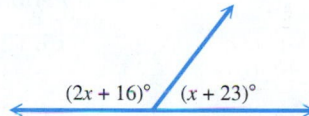

Solve each problem.

6. At the 2014 Winter Olympics in Sochi, Russia, the United States won a total of 28 medals. The United States won 2 more gold medals than silver and 5 fewer silver medals than bronze. Find the number of each type of medal won. (*Source: The Gazette.*)

7. From a list of "technology-related items," 500 adults were surveyed as to those items they couldn't live without. Complete the results shown in the table.

Item	Percent That Couldn't Live Without	Number That Couldn't Live Without
Personal computer	46%	
Cell phone	41%	
High-speed Internet		190
MP3 player		60

(Other items included digital cable, HDTV, and electronic gaming console.)
Source: Ipsos for AP.

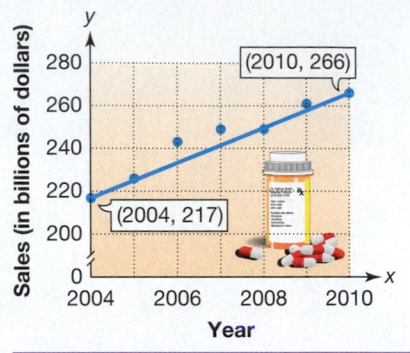

8. Fill in each blank with *positive* or *negative*. The point with coordinates (a, b) is in

  (a) quadrant II if a is _____ and b is _____.

  (b) quadrant III if a is _____ and b is _____.

9. Consider the equation $y = -2x - 4$. Find the following.

  (a) The x- and y-intercepts   **(b)** The slope   **(c)** The graph

10. The points on the graph show total retail sales of prescription drugs in the United States in the years 2004–2010, along with a graph of a linear equation that models the data.

  (a) Use the ordered pairs shown on the graph to find the slope of the line to the nearest whole number. Interpret the slope.

  (b) Use the graph to estimate sales in the year 2008. Write your answer as an ordered pair of the form (year, sales in billions of dollars).

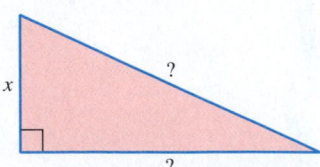

Retail Prescription Drug Sales

(2010, 266)

(2004, 217)

Sales (in billions of dollars)

Year

Source: National Association of Chain Drug Stores.

Evaluate each expression.

11. $\left(\dfrac{3}{4}\right)^{-2}$

12. $\left(\dfrac{4^{-3} \cdot 4^4}{4^5}\right)^{-1}$

Simplify each expression, and write the answer using only positive exponents. Assume that no denominators are 0.

13. $\dfrac{(p^2)^3 p^{-4}}{(p^{-3})^{-1} p}$

14. $\dfrac{(m^{-2})^3 m}{m^5 m^{-4}}$

Perform each indicated operation.

15. $(2k^2 + 4k) - (5k^2 - 2) - (k^2 + 8k - 6)$ **16.** $(9x + 6)(5x - 3)$

17. $(3p + 2)^2$

18. $\dfrac{8x^4 + 12x^3 - 6x^2 + 20x}{2x}$

19. To make a pound of honey, bees may travel 55,000 mi and visit more than 2,000,000 flowers. Write the two given numbers in scientific notation. (*Source: Home & Garden.*)

Factor completely.

20. $2a^2 + 7a - 4$ **21.** $10m^2 + 19m + 6$ **22.** $8t^2 + 10tv + 3v^2$

23. $4p^2 - 12p + 9$ **24.** $25r^2 - 81t^2$ **25.** $2pq + 6p^3q + 8p^2q$

Solve.

26. $6m^2 + m - 2 = 0$ **27.** $8x^2 = 64x$

28. The length of the hypotenuse of a right triangle is twice the length of the shorter leg, plus 3 m. The longer leg is 7 m longer than the shorter leg. Find the lengths of the sides.

6

Rational Expressions and Applications

The formula that gives the rate of a speeding car in terms of its distance and its time traveled involves a *rational expression* (or *fraction*), the subject of this chapter.

6.1 The Fundamental Property of Rational Expressions

OBJECTIVES

1 Find the numerical value of a rational expression.

2 Find the values of the variable for which a rational expression is undefined.

3 Write rational expressions in lowest terms.

4 Recognize equivalent forms of rational expressions.

VOCABULARY

☐ rational expression
☐ lowest terms

The quotient of two integers (with denominator not 0), such as $\frac{2}{3}$ or $-\frac{3}{4}$, is a *rational number*. In the same way, the quotient of two polynomials with denominator not equal to 0 is a *rational expression*.

> **Rational Expression**
>
> A **rational expression** is an expression of the form $\frac{P}{Q}$, where P and Q are polynomials and $Q \neq 0$.
>
> *Examples:* $\dfrac{-6x}{x^3 + 8}$, $\dfrac{9x}{y + 3}$, and $\dfrac{2m^3}{8}$ Rational expressions

Our work with rational expressions requires much of what we learned in **Chapters 4 and 5** on polynomials and factoring, as well as the rules for fractions from **Section R.1.**

OBJECTIVE 1 Find the numerical value of a rational expression.

Remember that to *evaluate* an expression means to find its *value*. We use substitution to evaluate a rational expression for a given value of the variable.

NOW TRY EXERCISE 1

Find the numerical value of each expression for $x = -3$.

(a) $\dfrac{2x - 1}{x + 4}$ (b) $\dfrac{x + 3}{4}$

(c) $\dfrac{4}{x + 3}$

EXAMPLE 1 Evaluating Rational Expressions

Find the numerical value of $\frac{3x + 6}{2x - 4}$ for each value of x.

(a) $x = 1$

$$\frac{3x + 6}{2x - 4}$$

$$= \frac{3(1) + 6}{2(1) - 4} \quad \text{Let } x = 1.$$

$$= \frac{9}{-2}$$

$$= -\frac{9}{2} \quad \frac{a}{-b} = -\frac{a}{b}$$

(b) $x = 0$

$$\frac{3x + 6}{2x - 4}$$

$$= \frac{3(0) + 6}{2(0) - 4} \quad \text{Let } x = 0.$$

$$= \frac{6}{-4}$$

$$= -\frac{3}{2} \quad \text{Lowest terms}$$

(c) $x = 2$

$$\frac{3x + 6}{2x - 4}$$

$$= \frac{3(2) + 6}{2(2) - 4} \quad \text{Let } x = 2.$$

$$= \frac{12}{0} \quad \boxed{\text{The expression is undefined for } x = 2.}$$

(d) $x = -2$

$$\frac{3x + 6}{2x - 4}$$

$$= \frac{3(-2) + 6}{2(-2) - 4} \quad \text{Let } x = -2.$$

$$= \frac{0}{-8}$$

$$= 0 \quad \frac{0}{b} = 0$$

NOW TRY

NOW TRY ANSWERS

1. (a) -7 (b) 0
   (c) The expression is undefined for $x = -3$.

NOTE *The numerator of a rational expression may be any real number.* If the numerator equals 0 and the denominator does not equal 0, then the rational expression equals 0. See Example 1(d).

OBJECTIVE 2 Find the values of the variable for which a rational expression is undefined.

In the definition of a rational expression $\dfrac{P}{Q}$, Q cannot equal 0. *The denominator of a rational expression cannot equal 0 because division by 0 is undefined.*

For instance, in the rational expression

$$\frac{8x^2}{x-3}, \quad \leftarrow \text{Denominator cannot equal 0.}$$

the variable x can take on any real number value except 3. If x is 3, then the denominator becomes $3-3=0$, making the expression undefined. Thus, x cannot equal 3. We indicate this restriction by writing $x \neq 3$.

Determining When a Rational Expression Is Undefined

Step 1 Set the denominator of the rational expression equal to 0.

Step 2 Solve this equation.

Step 3 The solutions of the equation are the values that make the rational expression undefined. The variable *cannot* equal these values.

EXAMPLE 2 Finding Values That Make Rational Expressions Undefined

Find any values of the variable for which each rational expression is undefined.

(a) $\dfrac{x+5}{3x+2}$ We must find any value of x that makes the *denominator* equal to 0 because division by 0 is undefined.

Step 1 Set the denominator equal to 0.

$$3x+2=0$$

Step 2 Solve. $\qquad 3x=-2 \qquad$ Subtract 2.

$$x=-\frac{2}{3} \qquad \text{Divide by 3.}$$

Step 3 The given expression is undefined for $-\frac{2}{3}$, so $x \neq -\frac{2}{3}$.

(b) $\dfrac{8x^2+1}{x-3}$ The denominator $x-3=0$ when x is 3. The given expression is undefined for 3, so $x \neq 3$.

(c) $\dfrac{9m^2}{m^2-5m+6}$

$$m^2-5m+6=0 \qquad \text{Set the denominator equal to 0.}$$

$$(m-2)(m-3)=0 \qquad \text{Factor.}$$

$$m-2=0 \quad \text{or} \quad m-3=0 \qquad \text{Zero-factor property}$$

$$m=2 \quad \text{or} \qquad m=3 \qquad \text{Solve for } m.$$

The given expression is undefined for 2 and 3, so $m \neq 2$, $m \neq 3$.

**NOW TRY
EXERCISE 2**

Find any values of the variable for which each rational expression is undefined.

(a) $\dfrac{k-4}{2k-1}$

(b) $\dfrac{2x}{x^2+5x-14}$

(c) $\dfrac{y+10}{y^2+10}$

(d) $\dfrac{2r}{r^2+1}$ This denominator will not equal 0 for any value of r, because r^2 is always greater than or equal to 0, and adding 1 makes the sum greater than or equal to 1. There are no values for which this expression is undefined.

NOW TRY

OBJECTIVE 3   Write rational expressions in lowest terms.

A fraction such as $\frac{2}{3}$ is said to be in *lowest terms*.

> **Lowest Terms**
>
> A rational expression $\dfrac{P}{Q}$ (where $Q \neq 0$) is in **lowest terms** if the greatest common factor of its numerator and denominator is 1.

We use the **fundamental property of rational expressions** to write a rational expression in lowest terms.

> **Fundamental Property of Rational Expressions**
>
> If $\dfrac{P}{Q}$ (where $Q \neq 0$) is a rational expression and if K represents any polynomial (where $K \neq 0$), then the following holds.
>
> $$\frac{PK}{QK} = \frac{P}{Q}$$

This property is based on the identity property of multiplication.

$$\frac{PK}{QK} = \frac{P}{Q} \cdot \frac{K}{K} = \frac{P}{Q} \cdot 1 = \frac{P}{Q}$$

**NOW TRY
EXERCISE 3**

Write each rational expression in lowest terms.

(a) $\dfrac{20}{48}$ (b) $\dfrac{21y^5}{7y^2}$

EXAMPLE 3   Writing in Lowest Terms

Write each rational expression in lowest terms.

(a) $\dfrac{30}{72}$

Begin by factoring.

$$\frac{30}{72} = \frac{2 \cdot 3 \cdot 5}{2 \cdot 2 \cdot 2 \cdot 3 \cdot 3}$$

Group any factors common to the numerator and denominator.

$$= \frac{5 \cdot (2 \cdot 3)}{2 \cdot 2 \cdot 3 \cdot (2 \cdot 3)}$$

Use the fundamental property.

$$= \frac{5}{2 \cdot 2 \cdot 3}$$

$$= \frac{5}{12}$$

(b) $\dfrac{14k^2}{2k^3}$

Write k^2 as $k \cdot k$ and k^3 as $k \cdot k \cdot k$.

$$\frac{14k^2}{2k^3} = \frac{2 \cdot 7 \cdot k \cdot k}{2 \cdot k \cdot k \cdot k}$$

$$= \frac{7(2 \cdot k \cdot k)}{k(2 \cdot k \cdot k)}$$

$$= \frac{7}{k}$$

NOW TRY

NOW TRY ANSWERS

2. (a) $k \neq \frac{1}{2}$ (b) $x \neq -7, x \neq 2$

   (c) It is never undefined.

3. (a) $\frac{5}{12}$ (b) $3y^3$

> **Writing a Rational Expression in Lowest Terms**
>
> **Step 1** **Factor** the numerator and denominator completely.
>
> **Step 2** Use **the fundamental property** to divide out any common factors.

NOW TRY EXERCISE 4

Write each rational expression in lowest terms.

(a) $\dfrac{3x + 15}{5x + 25}$

(b) $\dfrac{k^2 - 36}{k^2 + 8k + 12}$

EXAMPLE 4 Writing in Lowest Terms

Write each rational expression in lowest terms.

(a) $\dfrac{3x - 12}{5x - 20}$ *x ≠ 4 because the denominator is 0 for this value.*

$= \dfrac{3(x - 4)}{5(x - 4)}$ Factor. (Step 1)

$= \dfrac{3}{5}$ Fundamental property (Step 2)

The given expression is equal to $\frac{3}{5}$ for all values of x, where $x \neq 4$ (because the denominator of the original rational expression is 0 when x is 4).

(b) $\dfrac{2y^2 - 8}{2y + 4}$ *y ≠ −2 because the denominator is 0 for this value.*

$= \dfrac{2(y^2 - 4)}{2(y + 2)}$ Factor. (Step 1)

$= \dfrac{2(y + 2)(y - 2)}{2(y + 2)}$ Factor the numerator completely.

$= y - 2$ Fundamental property (Step 2)

(c) $\dfrac{m^2 + 2m - 8}{2m^2 - m - 6}$ *$m \neq -\frac{3}{2}, m \neq 2$*

$= \dfrac{(m + 4)(m - 2)}{(2m + 3)(m - 2)}$ Factor. (Step 1)

$= \dfrac{m + 4}{2m + 3}$ Fundamental property (Step 2)   **NOW TRY**

From now on, we write statements of equality of rational expressions with the understanding that they apply only to real numbers that make neither denominator equal to 0.

⚠ **CAUTION** *Rational expressions cannot be written in lowest terms until after the numerator and denominator have been factored. Only common* factors *(not* terms*) can be divided out.*

$$\frac{6x + 9}{4x + 6} = \frac{3(2x + 3)}{2(2x + 3)} = \frac{3}{2} \qquad\qquad \frac{6 + x}{4x} \;\leftarrow\; \text{Numerator cannot be factored.}$$

↑ Divide out the common factor.   Already in lowest terms

NOW TRY ANSWERS
4. (a) $\frac{3}{5}$ (b) $\frac{k - 6}{k + 2}$

NOW TRY
EXERCISE 5

Write in lowest terms.

$$\frac{10 - a^2}{a^2 - 10}$$

EXAMPLE 5 **Writing in Lowest Terms (Factors Are Opposites)**

Write $\dfrac{x - y}{y - x}$ in lowest terms.

To find a common factor, the denominator $y - x$ can be factored as follows.

$$y - x \qquad \boxed{\text{We are factoring out } -1, \text{ NOT multiplying by it.}}$$

$$= -1(-y + x) \qquad \text{Factor out } -1.$$

$$= -1(x - y) \qquad \text{Commutative property, } a + b = b + a.$$

With this result in mind, we simplify as follows.

$$\frac{x - y}{y - x}$$

$$= \frac{1(x - y)}{-1(x - y)} \qquad y - x = -1(x - y) \text{ from above.}$$

$$= \frac{1}{-1} \qquad \text{Fundamental property}$$

$$= -1 \qquad \text{Lowest terms} \qquad \text{NOW TRY}$$

NOTE The numerator *or* the denominator could have been factored in the first step in **Example 5.** Factor -1 from the numerator, and confirm that the result is the same.

In **Example 5,** notice that $y - x$ is the **opposite** (or **additive inverse**) of $x - y$. A general rule for this situation follows.

> **Quotient of Opposites**
>
> If the numerator and the denominator of a rational expression are opposites, such as in $\dfrac{x - y}{y - x}$, then the rational expression is equal to -1.

Based on this result, the following are true.

Numerator and denominator are opposites. $\longrightarrow \dfrac{q - 7}{7 - q} = -1$ and $\dfrac{-5a + 2b}{5a - 2b} = -1$

However, the following expression cannot be simplified further.

$$\frac{x - 2}{x + 2} \longleftarrow \text{Numerator and denominator are } \textit{not} \text{ opposites.}$$

EXAMPLE 6 **Writing in Lowest Terms (Factors Are Opposites)**

Write each rational expression in lowest terms.

(a) $\dfrac{2 - m}{m - 2}$

NOW TRY ANSWER
5. -1

Because $2 - m$ and $m - 2$ are opposites, this expression equals -1.

**NOW TRY
EXERCISE 6**

Write each rational expression
in lowest terms.

(a) $\dfrac{p-4}{4-p}$ **(b)** $\dfrac{4m^2-n^2}{2n-4m}$

(c) $\dfrac{x+y}{x-y}$

(b) $\dfrac{4x^2-9}{6-4x}$

$= \dfrac{(2x+3)(2x-3)}{2(3-2x)}$ Factor the numerator and denominator.

$= \dfrac{(2x+3)(2x-3)}{2(-1)(2x-3)}$ Write $3-2x$ in the denominator as $-1(2x-3)$.

$= \dfrac{2x+3}{2(-1)}$ Fundamental property

$= \dfrac{2x+3}{-2}$ Multiply in the denominator.

$= -\dfrac{2x+3}{2}$ $\dfrac{a}{-b} = -\dfrac{a}{b}$

(c) $\dfrac{3+r}{3-r}$ $3-r$ is *not* the opposite of $3+r$.

This rational expression is already in lowest terms.   **NOW TRY**

OBJECTIVE 4 Recognize equivalent forms of rational expressions.

It is important in algebra to recognize equivalent forms of expressions. For example,

$$0.5, \quad \frac{1}{2}, \quad 50\%, \quad \text{and} \quad \frac{50}{100}$$

all represent the *same* real number. On a number line, the exact same point would apply to all four of them.

A similar situation exists with negative common fractions. The common fraction $-\dfrac{5}{6}$ can also be written $\dfrac{-5}{6}$ and $\dfrac{5}{-6}$, with the negative sign appearing in any of three different positions. All represent the *same* rational number.

Consider the following rational expression.

$$-\frac{2x+3}{2} \qquad \text{Final result from Example 6(b)}$$

The $-$ sign representing the factor -1 is in front of the expression, aligned with the fraction bar. To obtain other equivalent forms of this rational expression, the factor -1 may instead be placed in the numerator or in the denominator.

Use parentheses.

$$\frac{-(2x+3)}{2} \quad \text{and} \quad \frac{2x+3}{-2}$$

In the first of these two expressions, the distributive property can be applied. Thus,

$$\frac{-(2x+3)}{2} \quad \text{can also be written} \quad \frac{-2x-3}{2}.$$

Multiply *each* term in the binomial by -1.

NOW TRY ANSWERS

6. (a) -1

(b) $\dfrac{2m+n}{-2}$, or $-\dfrac{2m+n}{2}$

(c) It is already in lowest terms.

⚠ **CAUTION** $\dfrac{-2x+3}{2}$ is *not* an equivalent form of $\dfrac{-(2x+3)}{2}$. **Be careful to apply the distributive property correctly.**

**NOW TRY
EXERCISE 7**

Write four equivalent forms
of the rational expression.

$$-\frac{4k - 9}{k + 3}$$

EXAMPLE 7　Writing Equivalent Forms of a Rational Expression

Write four equivalent forms of the rational expression.

$$-\frac{3x + 2}{x - 6}$$

If we apply the negative sign to the numerator, we obtain these equivalent forms.

　$\dfrac{-(3x + 2)}{x - 6}$　and, by the distributive property,　$\dfrac{-3x - 2}{x - 6}$

If we apply the negative sign to the denominator, we obtain two more forms.

③→ $\dfrac{3x + 2}{-(x - 6)}$　and, by distributing once again,　$\dfrac{3x + 2}{-x + 6}$ ←④

NOW TRY

NOW TRY ANSWER

7. $\dfrac{-(4k - 9)}{k + 3},\ \dfrac{-4k + 9}{k + 3},\ \dfrac{4k - 9}{-(k + 3)},$
$\dfrac{4k - 9}{-k - 3}$

> ⚠ **CAUTION**　Recall that $-\frac{5}{6} \neq \frac{-5}{-6}$. Thus, in **Example 7,** it would be incorrect to distribute the negative sign in $-\frac{3x + 2}{x - 6}$ to *both* the numerator *and* the denominator. (Doing this would actually lead to the *opposite* of the original expression.)

6.1 Exercises

FOR EXTRA HELP　▶　MyMathLab®

▶ *Complete solution available
in MyMathLab*

Concept Check　Work each problem.

1. Fill in each blank with the correct response:　The rational expression $\frac{x + 5}{x - 3}$ is undefined when x is _____, so $x \neq$ _____. This rational expression is equal to 0 when $x =$ _____.

2. Which one of these rational expressions can be simplified?

   A. $\dfrac{x^2 + 2}{x^2}$　　**B.** $\dfrac{x^2 + 2}{2}$　　**C.** $\dfrac{x^2 + y^2}{y^2}$　　**D.** $\dfrac{x^2 - 5x}{x}$

3. Which two of the following rational expressions equal -1?

   A. $\dfrac{2x + 3}{2x - 3}$　　**B.** $\dfrac{2x - 3}{3 - 2x}$　　**C.** $\dfrac{2x + 3}{3 + 2x}$　　**D.** $\dfrac{2x + 3}{-2x - 3}$

4. Make the correct choice: $\frac{4 - r^2}{4 + r^2}$ *(is / is not)* equal to -1.

5. Which one of these rational expressions is *not* equivalent to $\frac{x - 3}{4 - x}$?

   A. $\dfrac{3 - x}{x - 4}$　　**B.** $\dfrac{x + 3}{4 + x}$　　**C.** $-\dfrac{3 - x}{4 - x}$　　**D.** $-\dfrac{x - 3}{x - 4}$

6. Make the correct choice: $\frac{5 + 2x}{3 - x}$ and $\frac{-5 - 2x}{x - 3}$ *(are / are not)* equivalent rational expressions.

7. Find the numerical value of the rational expression for $x = -3$.

   $$\frac{x}{2x + 1}$$

   $$= \frac{\underline{\quad}}{2(\underline{\quad}) + 1} \qquad \text{Let } x = -3.$$

   $$= \frac{\underline{\quad}}{\underline{\quad} + 1}$$

   $$= \underline{\quad}$$

8. Find any values of the variable for which the rational expression is undefined.

   $$\frac{x + 2}{x - 5}$$

   Step 1　$\underline{\quad} = 0$

   Step 2　$x = \underline{\quad}$

   Step 3　The given expression is undefined for $\underline{\quad}$. Thus, $x\ (=/\neq)\ 5$.

Find the numerical value of each rational expression for (a) $x = 2$ and (b) $x = -3$. See Example 1.

▶ 9. $\dfrac{3x + 1}{5x}$ **10.** $\dfrac{5x - 2}{4x}$ **11.** $\dfrac{x^2 - 4}{2x + 1}$ **12.** $\dfrac{2x^2 - 4x}{3x - 1}$

13. $\dfrac{(-2x)^3}{3x + 9}$ **14.** $\dfrac{(-3x)^2}{4x + 12}$ **15.** $\dfrac{7 - 3x}{3x^2 - 7x + 2}$ **16.** $\dfrac{5x + 2}{4x^2 - 5x - 6}$

17. $\dfrac{(x + 3)(x - 2)}{500x}$ **18.** $\dfrac{(x - 2)(x + 3)}{1000x}$ **19.** $\dfrac{x^2 - 4}{x^2 - 9}$ **20.** $\dfrac{x^2 - 9}{x^2 - 4}$

Find any values of the variable for which each rational expression is undefined. Write answers with the symbol $\neq$. See Example 2.

21. $-\dfrac{5}{x}$ **22.** $-\dfrac{2}{y}$ **23.** $\dfrac{12}{5y}$ **24.** $\dfrac{-7}{3z}$ **25.** $\dfrac{x + 1}{x - 6}$ **26.** $\dfrac{m - 2}{m - 5}$

▶ 27. $\dfrac{4x^2}{3x + 5}$ **28.** $\dfrac{2x^3}{3x + 4}$ **29.** $\dfrac{5m + 2}{m^2 + m - 6}$ **30.** $\dfrac{2r - 5}{r^2 - 5r + 4}$

31. $\dfrac{x^2 + 3x}{4}$ **32.** $\dfrac{x^2 - 4x}{6}$ **33.** $\dfrac{3x - 1}{x^2 + 2}$ **34.** $\dfrac{4q + 2}{q^2 + 9}$

Concept Check   *Work each problem.*

35. Identify the two *terms* in the numerator and the two *terms* in the denominator of the rational expression $\dfrac{x^2 + 4x}{x + 4}$.

36. Describe the steps you would use to write the rational expression in **Exercise 35** in lowest terms. (*Hint:* It simplifies to x.)

Write each rational expression in lowest terms. See Examples 3 and 4.

▶ 37. $\dfrac{18r^3}{6r}$ **38.** $\dfrac{27p^4}{3p}$ **39.** $\dfrac{4(y - 2)}{10(y - 2)}$ **40.** $\dfrac{15(m - 1)}{9(m - 1)}$

41. $\dfrac{(x + 1)(x - 1)}{(x + 1)^2}$ **42.** $\dfrac{(t + 5)(t - 3)}{(t + 5)^2}$ **▶ 43.** $\dfrac{7m + 14}{5m + 10}$ **44.** $\dfrac{16x + 8}{14x + 7}$

45. $\dfrac{6m - 18}{7m - 21}$ **46.** $\dfrac{5r + 20}{3r + 12}$ **47.** $\dfrac{m^2 - n^2}{m + n}$ **48.** $\dfrac{a^2 - b^2}{a - b}$

49. $\dfrac{2t + 6}{t^2 - 9}$ **50.** $\dfrac{5s - 25}{s^2 - 25}$ **51.** $\dfrac{12m^2 - 3}{8m - 4}$ **52.** $\dfrac{20p^2 - 45}{6p - 9}$

53. $\dfrac{3m^2 - 3m}{5m - 5}$ **54.** $\dfrac{6t^2 - 6t}{5t - 5}$ **55.** $\dfrac{9r^2 - 4s^2}{9r + 6s}$ **56.** $\dfrac{16x^2 - 9y^2}{12x - 9y}$

57. $\dfrac{x - 6}{x^2 - 36}$ **58.** $\dfrac{x - 8}{x^2 - 64}$ **59.** $\dfrac{x^2 - 9}{x^2 - 6x + 9}$ **60.** $\dfrac{x^2 - 16}{x^2 - 8x + 16}$

61. $\dfrac{13x^2 - 39x^3}{7x - 21x^2}$ **62.** $\dfrac{30x^3 - 15x^5}{22x^2 - 11x^4}$ **63.** $\dfrac{5k^2 - 13k - 6}{5k + 2}$

64. $\dfrac{7t^2 - 31t - 20}{7t + 4}$ **65.** $\dfrac{x^2 + 2x - 15}{x^2 + 6x + 5}$ **66.** $\dfrac{y^2 - 5y - 14}{y^2 + y - 2}$

67. $\dfrac{2x^2 - 3x - 5}{2x^2 - 7x + 5}$ **68.** $\dfrac{3x^2 + 8x + 4}{3x^2 - 4x - 4}$ **69.** $\dfrac{3x^3 + 13x^2 + 14x}{3x^3 - 5x^2 - 28x}$

70. $\dfrac{2x^3 + 7x^2 - 30x}{2x^3 - 11x^2 + 15x}$ **71.** $\dfrac{-3t + 6t^2 - 3t^3}{7t^2 - 14t^3 + 7t^4}$ **72.** $\dfrac{-20r - 20r^2 - 5r^3}{24r^2 + 24r^3 + 6r^4}$

Extending Skills   *Exercises 73–94 involve factoring by grouping (**Section 5.1**) and factoring sums and differences of cubes (**Section 5.4**). Write each rational expression in lowest terms.*

73. $\dfrac{zw + 4z - 3w - 12}{zw + 4z + 5w + 20}$ **74.** $\dfrac{km + 4k - 4m - 16}{km + 4k + 5m + 20}$ **75.** $\dfrac{pr + qr + ps + qs}{pr + qr - ps - qs}$

76. $\dfrac{wt + ws + xt + xs}{wt - xs - xt + ws}$ **77.** $\dfrac{ac - ad + bc - bd}{ac - ad - bc + bd}$ **78.** $\dfrac{ac - bc - ad + bd}{ac - ad - bd + bc}$

79. $\dfrac{m^2 - n^2 - 4m - 4n}{2m - 2n - 8}$ **80.** $\dfrac{x^2 - y^2 - 7y - 7x}{3x - 3y - 21}$ **81.** $\dfrac{x^2y + y + x^2z + z}{xy + xz}$

82. $\dfrac{y^2k + pk - y^2z - pz}{yk - yz}$ **83.** $\dfrac{1 + p^3}{1 + p}$ **84.** $\dfrac{8 + x^3}{2 + x}$

85. $\dfrac{x^3 - 27}{x - 3}$ **86.** $\dfrac{r^3 - 1000}{r - 10}$ **87.** $\dfrac{b^3 - a^3}{a^2 - b^2}$

88. $\dfrac{8y^3 - 27z^3}{9z^2 - 4y^2}$ **89.** $\dfrac{k^3 + 8}{k^2 - 4}$ **90.** $\dfrac{r^3 + 27}{r^2 - 9}$

91. $\dfrac{z^3 + 27}{z^3 - 3z^2 + 9z}$ **92.** $\dfrac{t^3 + 64}{t^3 - 4t^2 + 16t}$ **93.** $\dfrac{1 - 8r^3}{8r^2 + 4r + 2}$ **94.** $\dfrac{8 - 27x^3}{27x^2 + 18x + 12}$

*Write each rational expression in lowest terms. **See Examples 5 and 6.***

▶ **95.** $\dfrac{6 - t}{t - 6}$ **96.** $\dfrac{2 - k}{k - 2}$ ▶ **97.** $\dfrac{m^2 - 1}{1 - m}$ **98.** $\dfrac{a^2 - b^2}{b - a}$

99. $\dfrac{q^2 - 4q}{4q - q^2}$ **100.** $\dfrac{z^2 - 5z}{5z - z^2}$ **101.** $\dfrac{p + 6}{p - 6}$

102. $\dfrac{5 - x}{5 + x}$ **103.** $\dfrac{-2m + 2n}{m - n}$ **104.** $\dfrac{-5p + 5q}{p - q}$

*Write four equivalent forms for each rational expression. **See Example 7.***

▶ **105.** $-\dfrac{x + 4}{x - 3}$ **106.** $-\dfrac{x + 6}{x - 1}$ **107.** $-\dfrac{2x - 3}{x + 3}$

108. $-\dfrac{5x - 6}{x + 4}$ **109.** $-\dfrac{3x - 1}{5x - 6}$ **110.** $-\dfrac{2x - 9}{7x - 1}$

Solve each problem.

111. The area of the rectangle is represented by

$$x^4 + 10x^2 + 21.$$

What is the width? $\left(\textit{Hint: Use } W = \dfrac{A}{L}.\right)$

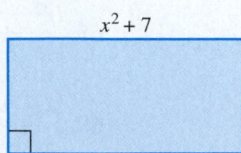

$x^2 + 7$

112. The volume of the box is represented by

$$(x^2 + 8x + 15)(x + 4).$$

Find the polynomial that represents the area of the bottom of the box.

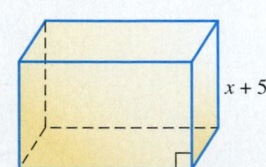

$x + 5$

Solve each problem.

113. The average number of vehicles waiting in line to enter a sports arena parking area is approximated by the rational expression

$$\frac{x^2}{2(1-x)},$$

where x is a quantity between 0 and 1 known as the **traffic intensity.** (*Source:* Mannering, F., and W. Kilareski, *Principles of Highway Engineering and Traffic Control,* John Wiley and Sons.)

To the nearest tenth, find the average number of vehicles waiting if the traffic intensity is the given number.

(a) 0.1   **(b)** 0.8   **(c)** 0.9

(d) What happens to the number of vehicles waiting as traffic intensity increases?

114. The percent of deaths caused by smoking is modeled by the rational expression

$$\frac{x-1}{x},$$

where x is the number of times a smoker is more likely than a nonsmoker to die of lung cancer. This is called the **incidence rate.** (*Source:* Walker, A., *Observation and Inference: An Introduction to the Methods of Epidemiology,* Epidemiology Resources Inc.) For example, $x = 10$ means that a smoker is 10 times more likely than a nonsmoker to die of lung cancer.

Find the percent of deaths if the incidence rate is the given number.

(a) 5   **(b)** 10   **(c)** 20

(d) Can the incidence rate equal 0? Explain.

RELATING CONCEPTS For Individual or Group Work (Exercises 115–118)

In Section 4.7, we used long division to find a quotient of two polynomials. We obtain the same quotient by expressing a division problem as a rational expression (fraction) and writing this rational expression in lowest terms, as shown below.

$$
\begin{array}{r}
x + 4 \\
2x - 3\overline{)2x^2 + 5x - 12} \\
\underline{2x^2 - 3x} \\
8x - 12 \\
\underline{8x - 12} \\
0
\end{array}
$$

$$\frac{2x^2 + 5x - 12}{2x - 3}$$

$$= \frac{(2x-3)(x+4)}{2x-3} \quad \text{Factor.}$$

$$= x + 4 \quad \text{Fundamental property}$$

Show that performing the long division and simplifying the rational expression yield the same result.

115. $4x + 7\overline{)8x^2 + 26x + 21}$

and $\dfrac{8x^2 + 26x + 21}{4x + 7}$

116. $6x + 5\overline{)12x^2 + 16x + 5}$

and $\dfrac{12x^2 + 16x + 5}{6x + 5}$

117. $x + 1\overline{)x^3 + x^2 + x + 1}$

and $\dfrac{x^3 + x^2 + x + 1}{x + 1}$

118. $x + 1\overline{)x^3 + x^2 + 2x + 2}$

and $\dfrac{x^3 + x^2 + 2x + 2}{x + 1}$

6.2 Multiplying and Dividing Rational Expressions

OBJECTIVES

1 Multiply rational expressions.

2 Divide rational expressions.

OBJECTIVE 1 Multiply rational expressions.

The product of two fractions is found by multiplying the numerators and multiplying the denominators. Rational expressions are multiplied in the same way.

Multiplying Rational Expressions

The product of the rational expressions $\frac{P}{Q}$ and $\frac{R}{S}$ is defined as follows.

$$\frac{P}{Q} \cdot \frac{R}{S} = \frac{PR}{QS}$$

That is, to multiply rational expressions, multiply the numerators and multiply the denominators.

NOW TRY EXERCISE 1

Multiply. Write each answer in lowest terms.

(a) $\frac{7}{18} \cdot \frac{9}{14}$ (b) $\frac{4k^2}{7} \cdot \frac{14}{11k}$

EXAMPLE 1 Multiplying Rational Expressions

Multiply. Write each answer in lowest terms.

(a) $\dfrac{3}{10} \cdot \dfrac{5}{9}$ (b) $\dfrac{6}{x} \cdot \dfrac{x^2}{12}$

Indicate the product of the numerators and the product of the denominators.

$$= \frac{3 \cdot 5}{10 \cdot 9} \qquad\qquad = \frac{6 \cdot x^2}{x \cdot 12}$$

Leave the products in factored form. Factor the numerator and denominator to further identify any common factors. Then use the fundamental property to divide out any common factors and write each product in lowest terms.

$$= \frac{3 \cdot 5}{2 \cdot 5 \cdot 3 \cdot 3} \qquad\qquad = \frac{6 \cdot x \cdot x}{x \cdot 2 \cdot 6}$$

$$= \frac{1}{6} \quad \text{Remember to write 1 in the numerator.} \qquad = \frac{x}{2}$$

NOW TRY

NOTE It is also possible to divide out common factors in the numerator and denominator *before* multiplying the rational expressions. Consider the following.

$$\frac{3}{10} \cdot \frac{5}{9} \qquad\qquad \text{Example 1(a)}$$

$$= \frac{3}{5 \cdot 2} \cdot \frac{5}{3 \cdot 3} \qquad \text{Identify the common factors.}$$

$$= \frac{1}{2 \cdot 3} \qquad\qquad \begin{array}{l}\text{Divide out the common factors.}\\\text{Insert a factor of 1 in the numerator.}\end{array}$$

$$= \frac{1}{6} \qquad\qquad \text{Multiply.}$$

NOW TRY ANSWERS

1. (a) $\frac{1}{4}$ (b) $\frac{8k}{11}$

 **NOW TRY**
EXERCISE 2
Multiply. Write the answer in lowest terms.

$$\frac{m-3}{3m} \cdot \frac{9m^2}{8(m-3)^2}$$

EXAMPLE 2 Multiplying Rational Expressions

Multiply. Write the answer in lowest terms.

$$\frac{x+y}{2x} \cdot \frac{x^2}{(x+y)^2}$$

> Use parentheses here around $x + y$.

$$= \frac{(x+y)x^2}{2x(x+y)^2}$$ Multiply numerators.
Multiply denominators.

$$= \frac{(x+y)x \cdot x}{2x(x+y)(x+y)}$$ Factor. Identify the common factors.

$$= \frac{x}{2(x+y)}$$ $\frac{(x+y)x}{x(x+y)} = 1$; Lowest terms   **NOW TRY**

 **NOW TRY**
EXERCISE 3
Multiply. Write the answer in lowest terms.

$$\frac{y^2-3y-28}{y^2-9y+14} \cdot \frac{y^2-7y+10}{y^2+4y}$$

EXAMPLE 3 Multiplying Rational Expressions

Multiply. Write the answer in lowest terms.

$$\frac{x^2+3x}{x^2-3x-4} \cdot \frac{x^2-5x+4}{x^2+2x-3}$$

$$= \frac{(x^2+3x)(x^2-5x+4)}{(x^2-3x-4)(x^2+2x-3)}$$ Definition of multiplication

$$= \frac{x(x+3)(x-4)(x-1)}{(x-4)(x+1)(x+3)(x-1)}$$ Factor.

$$= \frac{x}{x+1}$$ Divide out the common factors; lowest terms

The quotients $\frac{x+3}{x+3}$, $\frac{x-4}{x-4}$, and $\frac{x-1}{x-1}$ all equal 1, justifying the final product $\frac{x}{x+1}$.

NOW TRY

OBJECTIVE 2 Divide rational expressions.

Suppose we have $\frac{7}{8}$ gal of milk and want to find how many quarts we have. Because 1 qt is $\frac{1}{4}$ gal, we ask, "How many $\frac{1}{4}$s are there in $\frac{7}{8}$?" This would be interpreted as follows.

$$\frac{7}{8} \div \frac{1}{4}, \quad \text{or} \quad \frac{\frac{7}{8}}{\frac{1}{4}} \leftarrow \text{The fraction bar means division.}$$

The fundamental property of rational expressions discussed earlier can be applied to rational number values of P, Q, and K.

$$\frac{P}{Q} = \frac{P \cdot K}{Q \cdot K} = \frac{\frac{7}{8} \cdot 4}{\frac{1}{4} \cdot 4} = \frac{\frac{7}{8} \cdot 4}{1} = \frac{7}{8} \cdot \frac{4}{1}$$ Let $P = \frac{7}{8}$, $Q = \frac{1}{4}$, and $K = 4$.
$\left(K \text{ is the reciprocal of } Q = \frac{1}{4}.\right)$

NOW TRY ANSWERS
2. $\frac{3m}{8(m-3)}$
3. $\frac{y-5}{y}$

Therefore, to divide $\frac{7}{8}$ by $\frac{1}{4}$, we multiply $\frac{7}{8}$ by the reciprocal of $\frac{1}{4}$, namely 4. Because $\frac{7}{8}(4) = \frac{7}{2}$, there are $\frac{7}{2}$ qt, or $3\frac{1}{2}$ qt, in $\frac{7}{8}$ gal.

The preceding discussion illustrates dividing common fractions. Division of rational expressions is defined in the same way.

> **Dividing Rational Expressions**
>
> If $\frac{P}{Q}$ and $\frac{R}{S}$ are any two rational expressions where $\frac{R}{S} \neq 0$, then their quotient is defined as follows.
>
> $$\frac{P}{Q} \div \frac{R}{S} = \frac{P}{Q} \cdot \frac{S}{R} = \frac{PS}{QR}$$
>
> That is, to divide one rational expression by another rational expression, multiply the first rational expression (dividend) by the reciprocal of the second rational expression (divisor).

NOW TRY
EXERCISE 4

Divide. Write each answer in lowest terms.

(a) $\dfrac{3}{10} \div \dfrac{11}{20}$

(b) $\dfrac{2x-5}{3x^2} \div \dfrac{2x-5}{12x}$

EXAMPLE 4   Dividing Rational Expressions

Divide. Write each answer in lowest terms.

(a) $\dfrac{5}{8} \div \dfrac{7}{16}$

(b) $\dfrac{y}{y+3} \div \dfrac{4y}{y+5}$

Multiply the dividend by the reciprocal of the divisor.

$= \dfrac{5}{8} \cdot \dfrac{16}{7}$ ← Reciprocal of $\frac{7}{16}$

$= \dfrac{5 \cdot 16}{8 \cdot 7}$ Multiply.

$= \dfrac{5 \cdot 8 \cdot 2}{8 \cdot 7}$ Factor 16.

$= \dfrac{10}{7}$ Lowest terms

$= \dfrac{y}{y+3} \cdot \dfrac{y+5}{4y}$ ← Reciprocal of $\frac{4y}{y+5}$

$= \dfrac{y(y+5)}{(y+3)(4y)}$ Multiply.

$= \dfrac{y+5}{4(y+3)}$ Lowest terms

NOW TRY

NOW TRY
EXERCISE 5

Divide. Write the answer in lowest terms.

$\dfrac{(3k)^3}{2j^4} \div \dfrac{9k^2}{6j}$

EXAMPLE 5   Dividing Rational Expressions

Divide. Write the answer in lowest terms.

$$\dfrac{(3m)^2}{(2p)^3} \div \dfrac{6m^3}{16p^2}$$

$= \dfrac{(3m)^2}{(2p)^3} \cdot \dfrac{16p^2}{6m^3}$ Multiply by the reciprocal of the divisor.

$(3m)^2 = 3^2m^2;$
$(2p)^3 = 2^3p^3$

$= \dfrac{9m^2}{8p^3} \cdot \dfrac{16p^2}{6m^3}$ Power rule for exponents

$= \dfrac{9 \cdot 16m^2p^2}{8 \cdot 6p^3m^3}$ Multiply numerators.
Multiply denominators.

$= \dfrac{3}{mp}$ Lowest terms

NOW TRY

NOW TRY ANSWERS

4. (a) $\dfrac{6}{11}$ (b) $\dfrac{4}{x}$ 5. $\dfrac{9k}{j^3}$

NOW TRY EXERCISE 6

Divide. Write the answer in lowest terms.

$$\frac{(t+2)(t-5)}{-4t} \div \frac{t^2-25}{(t+5)(t+2)}$$

EXAMPLE 6   Dividing Rational Expressions

Divide. Write the answer in lowest terms.

$$\frac{x^2-4}{(x+3)(x-2)} \div \frac{(x+2)(x+3)}{-2x}$$

$$= \frac{x^2-4}{(x+3)(x-2)} \cdot \frac{-2x}{(x+2)(x+3)} \qquad \text{Multiply by the reciprocal of the divisor.}$$

$$= \frac{-2x(x^2-4)}{(x+3)(x-2)(x+2)(x+3)} \qquad \begin{array}{l}\text{Multiply numerators.}\\ \text{Multiply denominators.}\end{array}$$

$$= \frac{-2x(x+2)(x-2)}{(x+3)(x-2)(x+2)(x+3)} \qquad \text{Factor the numerator.}$$

$$= \frac{-2x}{(x+3)^2} \qquad \begin{array}{l}\text{Divide out the common factors;}\\ a \cdot a = a^2\end{array}$$

$$= -\frac{2x}{(x+3)^2} \qquad \frac{-a}{b} = -\frac{a}{b}; \text{ Lowest terms}$$

NOW TRY

NOW TRY EXERCISE 7

Divide. Write the answer in lowest terms.

$$\frac{7-x}{2x+6} \div \frac{x^2-49}{x^2+6x+9}$$

EXAMPLE 7   Dividing Rational Expressions (Factors Are Opposites)

Divide. Write the answer in lowest terms.

$$\frac{m^2-4}{m^2-1} \div \frac{2m^2+4m}{1-m}$$

$$= \frac{m^2-4}{m^2-1} \cdot \frac{1-m}{2m^2+4m} \qquad \text{Multiply by the reciprocal of the divisor.}$$

$$= \frac{(m^2-4)(1-m)}{(m^2-1)(2m^2+4m)} \qquad \begin{array}{l}\text{Multiply numerators.}\\ \text{Multiply denominators.}\end{array}$$

$$= \frac{(m+2)(m-2)(1-m)}{(m+1)(m-1)(2m)(m+2)} \qquad \text{Factor. } 1-m \text{ and } m-1 \text{ are opposites.}$$

$$= \frac{-1(m-2)}{2m(m+1)} \qquad \begin{array}{l}\text{Divide out the common factors.}\\ \text{From Section 6.1, } \frac{1-m}{m-1} = -1.\end{array}$$

$$= \frac{-m+2}{2m(m+1)} \qquad \text{Distribute } -1 \text{ in the numerator.}$$

$$= \frac{2-m}{2m(m+1)} \qquad \text{Rewrite } -m+2 \text{ as } 2-m. \qquad \text{NOW TRY}$$

NOW TRY ANSWERS

6. $-\dfrac{(t+2)^2}{4t}$

7. $-\dfrac{x+3}{2(x+7)}$

Multiplying or Dividing Rational Expressions

Step 1   **Note the operation.** If the operation is division, use the definition of division to rewrite it as multiplication.

Step 2   **Multiply** numerators and multiply denominators.

Step 3   **Factor** all numerators and denominators completely.

Step 4   **Write in lowest terms** using the fundamental property.

Note: Steps 2 and 3 may be interchanged based on personal preference.

6.2 Exercises

 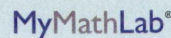

▶ *Complete solution available in MyMathLab*

1. *Concept Check* Match each multiplication problem in Column I with the correct product in Column II.

I		II
(a) $\dfrac{5x^3}{10x^4} \cdot \dfrac{10x^7}{4x}$		**A.** $\dfrac{4}{5x^5}$
(b) $\dfrac{10x^4}{5x^3} \cdot \dfrac{10x^7}{4x}$		**B.** $\dfrac{5x^5}{4}$
(c) $\dfrac{5x^3}{10x^4} \cdot \dfrac{4x}{10x^7}$		**C.** $\dfrac{1}{5x^7}$
(d) $\dfrac{10x^4}{5x^3} \cdot \dfrac{4x}{10x^7}$		**D.** $5x^7$

2. *Concept Check* Match each division problem in Column I with the correct quotient in Column II.

I		II
(a) $\dfrac{5x^3}{10x^4} \div \dfrac{10x^7}{4x}$		**A.** $\dfrac{5x^5}{4}$
(b) $\dfrac{10x^4}{5x^3} \div \dfrac{10x^7}{4x}$		**B.** $5x^7$
(c) $\dfrac{5x^3}{10x^4} \div \dfrac{4x}{10x^7}$		**C.** $\dfrac{4}{5x^5}$
(d) $\dfrac{10x^4}{5x^3} \div \dfrac{4x}{10x^7}$		**D.** $\dfrac{1}{5x^7}$

Multiply. Write each answer in lowest terms. **See Examples 1 and 2.**

▶ **3.** $\dfrac{15a^2}{14} \cdot \dfrac{7}{5a}$

4. $\dfrac{21b^6}{18} \cdot \dfrac{9}{7b^4}$

5. $\dfrac{12x^4}{18x^3} \cdot \dfrac{-8x^5}{4x^2}$

6. $\dfrac{12m^5}{-2m^2} \cdot \dfrac{6m^6}{28m^3}$

7. $\dfrac{2(c+d)}{3} \cdot \dfrac{18}{6(c+d)^2}$

8. $\dfrac{4(y-2)}{x} \cdot \dfrac{3x}{6(y-2)^2}$

▶ **9.** $\dfrac{(x-y)^2}{2} \cdot \dfrac{24}{3(x-y)}$

10. $\dfrac{(a+b)^2}{5} \cdot \dfrac{30}{2(a+b)}$

11. $\dfrac{t-4}{8} \cdot \dfrac{4t^2}{t-4}$

12. $\dfrac{z+9}{12} \cdot \dfrac{3z^2}{z+9}$

13. $\dfrac{3x}{x+3} \cdot \dfrac{(x+3)^2}{6x^2}$

14. $\dfrac{(t-2)^2}{4t^2} \cdot \dfrac{2t}{t-2}$

Concept Check *Multiply or divide. Write each answer in lowest terms.*

15. $\dfrac{5x-10}{6} \cdot \dfrac{9}{10x-20}$

$= \dfrac{5(\underline{\quad})}{6} \cdot \dfrac{3 \cdot \underline{\quad}}{10(\underline{\quad})}$

$= \dfrac{5(x-2) \cdot 3 \cdot 3}{2 \cdot 3 \cdot 2 \cdot \underline{\quad} \cdot (x-2)}$

$= \underline{\quad}$

16. $\dfrac{6x-4}{3} \div \dfrac{15x-10}{9}$

$= \dfrac{6x-4}{3} \cdot \dfrac{\underline{\quad}}{\underline{\quad}}$

$= \dfrac{2(\underline{\quad})}{3} \cdot \dfrac{9}{5(\underline{\quad})}$

$= \dfrac{2(3x-2) \cdot 3 \cdot 3}{\underline{\quad} \cdot 5(3x-2)}$

$= \underline{\quad}$

Divide. Write each answer in lowest terms. **See Examples 4 and 5.**

17. $\dfrac{9z^4}{3z^5} \div \dfrac{3z^2}{5z^3}$

18. $\dfrac{35x^8}{7x^9} \div \dfrac{5x^5}{9x^6}$

▶ **19.** $\dfrac{4t^4}{2t^5} \div \dfrac{(2t)^3}{-6}$

20. $\dfrac{-12a^6}{3a^2} \div \dfrac{(2a)^3}{27a}$

▶ **21.** $\dfrac{3}{2y-6} \div \dfrac{6}{y-3}$

22. $\dfrac{4m+16}{10} \div \dfrac{3m+12}{18}$

23. $\dfrac{7t+7}{-6} \div \dfrac{4t+4}{15}$

24. $\dfrac{8z-16}{-20} \div \dfrac{3z-6}{40}$

25. $\dfrac{2x}{x-1} \div \dfrac{x^2}{x+2}$

26. $\dfrac{y^2}{y+1} \div \dfrac{3y}{y-3}$

27. $\dfrac{(x-3)^2}{6x} \div \dfrac{x-3}{x^2}$

28. $\dfrac{2a}{a+4} \div \dfrac{a^2}{(a+4)^2}$

29. $\dfrac{5x^3}{x^2 - 16} \div \dfrac{x^5}{(x-4)^2}$ **30.** $\dfrac{8x^4}{x^2 - 25} \div \dfrac{x^7}{(x-5)^2}$ **31.** $\dfrac{-4t^3}{t^2 - 1} \div \dfrac{t^2}{(t+1)^2}$

32. *Concept Check*   After factoring numerators and denominators in a multiplication problem, a student obtained the following.

$$\dfrac{(x+3)^2}{(x+3)} \cdot \dfrac{(x+5)}{(x+5)^3}$$

Is it permissible for the student to divide out common factors within the same fractions here?

Multiply or divide. Write each answer in lowest terms. **See Examples 3, 6, and 7.**

33. $\dfrac{5x - 15}{3x + 9} \cdot \dfrac{4x + 12}{6x - 18}$ **34.** $\dfrac{8r + 16}{24r - 24} \cdot \dfrac{6r - 6}{3r + 6}$ **35.** $\dfrac{2 - t}{8} \div \dfrac{t - 2}{6}$

36. $\dfrac{m - 2}{4} \div \dfrac{2 - m}{6}$ **37.** $\dfrac{27 - 3z}{4} \cdot \dfrac{12}{2z - 18}$ **38.** $\dfrac{35 - 5x}{6} \cdot \dfrac{12}{3x - 21}$

▶ **39.** $\dfrac{p^2 + 4p - 5}{p^2 + 7p + 10} \div \dfrac{p - 1}{p + 4}$ **40.** $\dfrac{z^2 - 3z + 2}{z^2 + 4z + 3} \div \dfrac{z - 1}{z + 1}$ ▶ **41.** $\dfrac{m^2 - 4}{16 - 8m} \div \dfrac{m + 2}{8}$

42. $\dfrac{r^2 - 36}{54 - 9r} \div \dfrac{r + 6}{9}$ **43.** $\dfrac{m^2 - 4}{16 - 8m} \div \dfrac{m^2 + 3m + 2}{8m + 16}$ **44.** $\dfrac{t^2 - 49}{42 - 6t} \div \dfrac{t^2 + 10t + 21}{6t + 42}$

45. $\dfrac{2x^2 - 7x + 3}{x - 3} \cdot \dfrac{x + 2}{x - 1}$ **46.** $\dfrac{3x^2 - 5x - 2}{x - 2} \cdot \dfrac{x - 3}{x + 1}$

47. $\dfrac{2k^2 - k - 1}{2k^2 + 5k + 3} \div \dfrac{4k^2 - 1}{2k^2 + k - 3}$ **48.** $\dfrac{3t^2 - 4t - 4}{3t^2 + 10t + 8} \div \dfrac{9t^2 + 21t + 10}{3t^2 - t - 10}$

▶ **49.** $\dfrac{2k^2 + 3k - 2}{6k^2 - 7k + 2} \cdot \dfrac{4k^2 - 5k + 1}{k^2 + k - 2}$ **50.** $\dfrac{2m^2 - 5m - 12}{m^2 - 10m + 24} \cdot \dfrac{m^2 - 9m + 18}{4m^2 - 9}$

51. $\dfrac{m^2 + 2mp - 3p^2}{m^2 - 3mp + 2p^2} \div \dfrac{m^2 + 4mp + 3p^2}{m^2 + 2mp - 8p^2}$ **52.** $\dfrac{x^2 - 2xy - 3y^2}{x^2 + xy - 30y^2} \div \dfrac{x^2 + xy - 12y^2}{x^2 - xy - 20y^2}$

53. $\dfrac{m^2 + 3m + 2}{m^2 + 5m + 4} \cdot \dfrac{m^2 + 10m + 24}{m^2 + 5m + 6}$ **54.** $\dfrac{z^2 - z - 6}{z^2 - 2z - 8} \cdot \dfrac{z^2 + 7z + 12}{z^2 - 9}$

55. $\dfrac{y^2 + y - 2}{y^2 + 3y - 4} \div \dfrac{y + 2}{y + 3}$ **56.** $\dfrac{r^2 + r - 6}{r^2 + 4r - 12} \div \dfrac{r + 3}{r - 1}$

57. $\dfrac{2m^2 + 7m + 3}{m^2 - 9} \cdot \dfrac{m^2 - 3m}{2m^2 + 11m + 5}$ **58.** $\dfrac{6s^2 + 17s + 10}{s^2 - 4} \cdot \dfrac{s^2 - 2s}{6s^2 + 29s + 20}$

59. $\dfrac{r^2 + rs - 12s^2}{r^2 - rs - 20s^2} \div \dfrac{r^2 - 2rs - 3s^2}{r^2 + rs - 30s^2}$ **60.** $\dfrac{m^2 + 8mn + 7n^2}{m^2 + mn - 42n^2} \div \dfrac{m^2 - 3mn - 4n^2}{m^2 - mn - 30n^2}$

61. $\dfrac{(q-3)^4(q+2)}{q^2 + 3q + 2} \div \dfrac{q^2 - 6q + 9}{q^2 + 4q + 4}$ **62.** $\dfrac{(x+4)^3(x-3)}{x^2 - 9} \div \dfrac{x^2 + 8x + 16}{x^2 + 6x + 9}$

Extending Skills   *Exercises 63–68 involve grouping symbols (**Section 1.1**), factoring by grouping (**Section 5.1**), and factoring sums and differences of cubes (**Section 5.4**). Multiply or divide as indicated. Write each answer in lowest terms.*

63. $\dfrac{3a - 3b - a^2 + b^2}{4a^2 - 4ab + b^2} \cdot \dfrac{4a^2 - b^2}{2a^2 - ab - b^2}$ **64.** $\dfrac{4r^2 - t^2 + 10r - 5t}{2r^2 + rt + 5r} \cdot \dfrac{4r^3 + 4r^2t + rt^2}{2r + t}$

65. $\dfrac{-x^3 - y^3}{x^2 - 2xy + y^2} \div \dfrac{3y^2 - 3xy}{x^2 - y^2}$

66. $\dfrac{b^3 - 8a^3}{4a^3 + 4a^2b + ab^2} \div \dfrac{4a^2 + 2ab + b^2}{-a^3 - ab^3}$

67. $\dfrac{x + 5}{x + 10} \div \left(\dfrac{x^2 + 10x + 25}{x^2 + 10x} \cdot \dfrac{10x}{x^2 + 15x + 50} \right)$

68. $\dfrac{m - 8}{m - 4} \div \left(\dfrac{m^2 - 12m + 32}{8m} \cdot \dfrac{m^2 - 8m}{m^2 - 8m + 16} \right)$

Answer each question.

69. If the rational expression $\dfrac{5x^2y^3}{2pq}$ represents the area of a rectangle and $\dfrac{2xy}{p}$ represents the length, what rational expression represents the width?

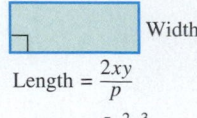
Width
Length $= \dfrac{2xy}{p}$
The area is $\dfrac{5x^2y^3}{2pq}$.

70. Given the following problem, what polynomial is represented by the red question mark?

$$\dfrac{4y + 12}{2y - 10} \div \dfrac{?}{y^2 - y - 20} = \dfrac{2(y + 4)}{y - 3}$$

6.3 Least Common Denominators

OBJECTIVES

1 Find the least common denominator for a group of fractions.

2 Write equivalent rational expressions.

VOCABULARY

☐ least common denominator (LCD)

OBJECTIVE 1 Find the least common denominator for a group of fractions.

Adding or subtracting rational expressions often requires finding the **least common denominator (LCD)**. The LCD is the simplest expression that is divisible by all of the denominators in all of the expressions. For example, the fractions

$$\frac{2}{9} \quad \text{and} \quad \frac{5}{12} \quad \text{have LCD} \quad 36,$$

because 36 is the least positive number divisible by both 9 and 12.

We can often find least common denominators by inspection. In other cases, we find the LCD using the following procedure.

> **Finding the Least Common Denominator (LCD)**
>
> *Step 1* **Factor** each denominator into prime factors.
>
> *Step 2* **List each different denominator factor** the *greatest* number of times it appears in any of the denominators.
>
> *Step 3* **Multiply** the denominator factors from Step 2 to find the LCD.

When each denominator is factored into prime factors, every prime factor must be a factor of the LCD.

> **⚠ CAUTION** When finding the LCD, use each factor the *greatest* number of times it appears in any *single* denominator, not the *total* number of times it appears. For instance, the greatest number of times 2 appears as a factor in one denominator in **Example 1(b)** on the next page is 3, *not* 4.

NOW TRY
EXERCISE 1

Find the LCD for each pair of fractions.

(a) $\dfrac{5}{48}, \dfrac{1}{30}$ (b) $\dfrac{3}{10y}, \dfrac{1}{6y}$

EXAMPLE 1 Finding Least Common Denominators

Find the LCD for each pair of fractions.

(a) $\dfrac{1}{24}, \dfrac{7}{15}$

(b) $\dfrac{1}{8x}, \dfrac{3}{10x}$

Step 1 Write each denominator in factored form with numerical coefficients in prime factored form.

$$24 = 2 \cdot 2 \cdot 2 \cdot 3 = 2^3 \cdot 3$$
$$15 = 3 \cdot 5$$

$$8x = 2 \cdot 2 \cdot 2 \cdot x = 2^3 \cdot x$$
$$10x = 2 \cdot 5 \cdot x$$

Step 2 Find the LCD by taking each different factor the *greatest* number of times it appears as a factor in any of the denominators.

The factor 2 appears three times in one product and not at all in the other, so the greatest number of times 2 appears is three. The greatest number of times both 3 and 5 appear is one.

Here, 2 appears three times in one product and once in the other, so the greatest number of times 2 appears is three. The greatest number of times 5 appears is one. The greatest number of times x appears in either product is one.

Step 3 $\text{LCD} = 2 \cdot 2 \cdot 2 \cdot 3 \cdot 5$
$\phantom{\text{LCD}} = 2^3 \cdot 3 \cdot 5$
$\phantom{\text{LCD}} = 120$

$\text{LCD} = 2 \cdot 2 \cdot 2 \cdot 5 \cdot x$
$\phantom{\text{LCD}} = 2^3 \cdot 5 \cdot x$
$\phantom{\text{LCD}} = 40x$ **NOW TRY**

NOW TRY
EXERCISE 2

Find the LCD for the pair of fractions.

$\dfrac{5}{6x^4}$ and $\dfrac{7}{8x^3}$

EXAMPLE 2 Finding the LCD

Find the LCD for $\dfrac{5}{6r^2}$ and $\dfrac{3}{4r^3}$.

Step 1 Factor each denominator.

$$6r^2 = 2 \cdot 3 \cdot r^2$$
$$4r^3 = 2 \cdot 2 \cdot r^3 = 2^2 \cdot r^3$$

Step 2 The greatest number of times 2 appears is two, the greatest number of times 3 appears is one, and the greatest number of times r appears is three.

Step 3 $\text{LCD} = 2^2 \cdot 3 \cdot r^3 = 12r^3$ **NOW TRY**

EXAMPLE 3 Finding LCDs

Find the LCD for the fractions in each list.

(a) $\dfrac{6}{5m}, \dfrac{4}{m^2 - 3m}$

$\left. \begin{array}{l} 5m = 5 \cdot m \\ m^2 - 3m = m(m - 3) \end{array} \right\}$ Factor each denominator.

Use each different factor the greatest number of times it appears.

$\text{LCD} = 5 \cdot m \cdot (m - 3) = 5m(m - 3)$  Be sure to include m as a factor in the LCD.

NOW TRY ANSWERS
1. (a) 240 **(b)** $30y$
2. $24x^4$

Because m is not a *factor* of $m - 3$, *both* m and $m - 3$ must appear in the LCD.

NOW TRY EXERCISE 3

Find the LCD for the fractions in each list.

(a) $\dfrac{3t}{2t^2 - 10t}, \dfrac{t + 4}{t^2 - 25}$

(b) $\dfrac{1}{x^2 + 7x + 12},$
$\dfrac{2}{x^2 + 6x + 9}, \dfrac{5}{x^2 + 2x - 8}$

(c) $\dfrac{2}{a - 4}, \dfrac{1}{4 - a}$

(b) $\dfrac{1}{r^2 - 4r - 5}, \dfrac{3}{r^2 - r - 20}, \dfrac{1}{r^2 - 10r + 25}$

$$\left.\begin{aligned} r^2 - 4r - 5 &= (r - 5)(r + 1) \\ r^2 - r - 20 &= (r - 5)(r + 4) \\ r^2 - 10r + 25 &= (r - 5)^2 \end{aligned}\right\} \quad \text{Factor each denominator.}$$

Use each different factor the greatest number of times it appears as a factor.

$$\text{LCD} = (r - 5)^2(r + 1)(r + 4) \quad \longleftarrow \begin{array}{l}\text{Be sure to include the exponent}\\ \text{2 on the factor } (r - 5).\end{array}$$

(c) $\dfrac{1}{q - 5}, \dfrac{3}{5 - q}$

The expressions $q - 5$ and $5 - q$ are opposites of each other. This means that if we multiply $q - 5$ by -1, we will obtain $5 - q$.

$$-(q - 5) = -q + 5 = 5 - q$$

Therefore, either $q - 5$ or $5 - q$ can be used as the LCD.

NOW TRY

OBJECTIVE 2 Write equivalent rational expressions.

Once we have the LCD, the next step in preparing to add or subtract two rational expressions is to use the fundamental property to write equivalent rational expressions.

> **Writing a Rational Expression with a Specified Denominator**
>
> **Step 1** **Factor** both denominators.
>
> **Step 2** **Decide what factor(s) the denominator must be multiplied by** in order to equal the specified denominator.
>
> **Step 3** **Multiply** the rational expression by that factor divided by itself. (That is, multiply by 1.)

NOW TRY EXERCISE 4

Write each rational expression with the indicated denominator.

(a) $\dfrac{2}{9} = \dfrac{?}{27}$ **(b)** $\dfrac{4t}{11} = \dfrac{?}{33t}$

EXAMPLE 4 Writing Equivalent Rational Expressions

Write each rational expression with the indicated denominator.

(a) $\dfrac{3}{8} = \dfrac{?}{40}$ **(b)** $\dfrac{9k}{25} = \dfrac{?}{50k}$

Step 1 For each example, first factor the denominator on the right. Then compare the denominator on the left with the one on the right to decide what factors are missing. (It may sometimes be necessary to factor both denominators.)

$$\dfrac{3}{8} = \dfrac{?}{5 \cdot 8} \qquad\qquad \dfrac{9k}{25} = \dfrac{?}{25 \cdot 2k}$$

Step 2 A factor of 5 is missing. Factors of 2 and k are missing.

Step 3 Multiply $\dfrac{3}{8}$ by $\dfrac{5}{5}$. Multiply $\dfrac{9k}{25}$ by $\dfrac{2k}{2k}$.

$$\dfrac{3}{8} = \dfrac{3}{8} \cdot \dfrac{5}{5} = \dfrac{15}{40} \qquad\qquad \dfrac{9k}{25} = \dfrac{9k}{25} \cdot \dfrac{2k}{2k} = \dfrac{18k^2}{50k}$$

$$\dfrac{5}{5} = 1 \uparrow\!\!\!\!\!\!\!\longrightarrow \qquad\qquad \dfrac{2k}{2k} = 1 \uparrow\!\!\!\!\!\!\!\longrightarrow$$

NOW TRY

NOW TRY ANSWERS
3. **(a)** $2t(t - 5)(t + 5)$
   (b) $(x + 3)^2(x + 4)(x - 2)$
   (c) either $a - 4$ or $4 - a$
4. **(a)** $\dfrac{6}{27}$ **(b)** $\dfrac{12t^2}{33t}$

 **NOW TRY
EXERCISE 5**

Write each rational expression with the indicated denominator.

(a) $\dfrac{8k}{5k - 2} = \dfrac{?}{25k - 10}$

(b) $\dfrac{2t - 1}{t^2 + 4t} = \dfrac{?}{t^3 + 12t^2 + 32t}$

EXAMPLE 5 **Writing Equivalent Rational Expressions**

Write each rational expression with the indicated denominator.

(a) $\dfrac{8}{3x + 1} = \dfrac{?}{12x + 4}$

$\dfrac{8}{3x + 1} = \dfrac{?}{4(3x + 1)}$ ← Factor the denominator on the right.

The missing factor is 4, so multiply the fraction on the left by $\dfrac{4}{4}$.

$$\dfrac{8}{3x + 1} \cdot \dfrac{4}{4} = \dfrac{32}{12x + 4} \qquad \text{Fundamental property}$$

(b) $\dfrac{12p}{p^2 + 8p} = \dfrac{?}{p^3 + 4p^2 - 32p}$

Factor the denominator in each rational expression.

$$\dfrac{12p}{p(p + 8)} = \dfrac{?}{p(p + 8)(p - 4)}$$

$\begin{aligned} p^3 + 4p^2 - 32p \\ = p(p^2 + 4p - 32) \\ = p(p + 8)(p - 4) \end{aligned}$

The factor $p - 4$ is missing, so multiply $\dfrac{12p}{p(p + 8)}$ by $\dfrac{p - 4}{p - 4}$.

$$\dfrac{12p}{p^2 + 8p} = \dfrac{12p}{p(p + 8)} \cdot \dfrac{p - 4}{p - 4} \qquad \text{Fundamental property}$$

$$= \dfrac{12p(p - 4)}{p(p + 8)(p - 4)} \qquad \begin{aligned}&\text{Multiply numerators.}\\&\text{Multiply denominators.}\end{aligned}$$

$$= \dfrac{12p^2 - 48p}{p^3 + 4p^2 - 32p} \qquad \text{Multiply the factors.}$$

NOW TRY

NOW TRY ANSWERS

5. (a) $\dfrac{40k}{25k - 10}$

(b) $\dfrac{2t^2 + 15t - 8}{t^3 + 12t^2 + 32t}$

NOTE While it is beneficial to leave the denominator in factored form, we multiplied the factors in the denominator in **Example 5** to give the answer in the same form as the original problem.

6.3 Exercises

FOR EXTRA HELP

 MyMathLab®

▶ *Complete solution available in MyMathLab*

Concept Check *Choose the correct response.*

1. Suppose that the greatest common factor of x and y is 1. What is the least common denominator for $\dfrac{1}{x}$ and $\dfrac{1}{y}$?

    A. x **B.** y **C.** xy **D.** 1

2. If x is a factor of y, what is the least common denominator for $\dfrac{1}{x}$ and $\dfrac{1}{y}$?

    A. x **B.** y **C.** xy **D.** 1

3. What is the least common denominator for $\dfrac{9}{20}$ and $\dfrac{1}{2}$?

    A. 40   **B.** 2   **C.** 20   **D.** None of these

4. Suppose that we wish to write the fraction $\frac{1}{(x-4)^2(y-3)}$ with denominator $(x-4)^3(y-3)^2$. By what must we multiply both the numerator and the denominator?

A. $(x-4)(y-3)$ **B.** $(x-4)^2$ **C.** $x-4$ **D.** $(x-4)^2(y-3)$

5. *Concept Check*   Find the LCD for the pair of fractions.

$$\frac{7}{10}, \frac{1}{25}$$

Step 1 $10 = 2 \cdot$ _____

$25 =$ _____ $\cdot\, 5$

Step 2   The greatest number of times 2 appears is _____. The greatest number of times _____ appears is two.

Step 3   LCD $=$ _____ $\cdot$ _____ $\cdot\, 5$

LCD $=$ _____

6. *Concept Check*   Write the rational expression as an equivalent expression with the indicated denominator.

$$\frac{7k}{5} = \frac{?}{30p}$$

Step 1   Factor the denominator on the right.

$$\frac{7k}{5} = \frac{?}{5 \cdot \text{_____}}$$

Step 2   Factors of _____ and _____ are missing.

Step 3 $\dfrac{7k}{5} \cdot \dfrac{\overline{}}{\underline{}} = \dfrac{\overline{}}{30p}$

*Find the LCD for the fractions in each list. **See Examples 1 and 2.***

▶ 7. $\dfrac{7}{15}, \dfrac{21}{20}$

8. $\dfrac{9}{10}, \dfrac{13}{25}$

9. $\dfrac{17}{100}, \dfrac{23}{120}, \dfrac{43}{180}$

10. $\dfrac{17}{250}, \dfrac{21}{300}, \dfrac{1}{360}$

11. $\dfrac{9}{x^2}, \dfrac{8}{x^5}$

12. $\dfrac{12}{m^7}, \dfrac{14}{m^8}$

13. $\dfrac{-2}{5p}, \dfrac{13}{6p}$

14. $\dfrac{-14}{15k}, \dfrac{11}{4k}$

▶ 15. $\dfrac{17}{15y^2}, \dfrac{55}{36y^4}$

16. $\dfrac{4}{25m^3}, \dfrac{7}{10m^4}$

17. $\dfrac{5}{21r^3}, \dfrac{7}{12r^5}$

18. $\dfrac{6}{35t^2}, \dfrac{5}{49t^6}$

19. $\dfrac{13}{5a^2b^3}, \dfrac{29}{15a^5b}$

20. $\dfrac{7}{3r^4s^5}, \dfrac{23}{9r^6s^8}$

21. $\dfrac{1}{r^2t^3}, \dfrac{1}{r^5t}, \dfrac{1}{r^9t^2}$

22. $\dfrac{5}{x^8y^4}, \dfrac{5}{x^9y^3}, \dfrac{5}{xy^2}$

23. $\dfrac{7}{x+1}, \dfrac{9}{x-1}$

24. $\dfrac{3}{y+3}, \dfrac{2}{y-3}$

*Find the LCD for the fractions in each list. **See Example 3.***

▶ 25. $\dfrac{7}{6p}, \dfrac{15}{4p-8}$

26. $\dfrac{7}{8k}, \dfrac{28}{12k-24}$

27. $\dfrac{9}{28m^2}, \dfrac{3}{12m-20}$

28. $\dfrac{14}{27a^3}, \dfrac{7}{9a-45}$

29. $\dfrac{7}{5b-10}, \dfrac{11}{6b-12}$

30. $\dfrac{3}{7x^2+21x}, \dfrac{2}{5x^2+15x}$

31. $\dfrac{37}{6r-12}, \dfrac{25}{9r-18}$

32. $\dfrac{14}{5p-30}, \dfrac{11}{6p-36}$

33. $\dfrac{5}{c-d}, \dfrac{8}{d-c}$

34. $\dfrac{4}{y-x}, \dfrac{8}{x-y}$

35. $\dfrac{12}{m-3}, \dfrac{-4}{3-m}$

36. $\dfrac{3}{a-8}, \dfrac{-17}{8-a}$

37. $\dfrac{29}{p-q}, \dfrac{18}{q-p}$

38. $\dfrac{16}{z-x}, \dfrac{9}{x-z}$

39. $\dfrac{13}{x^2-1}, \dfrac{-5}{2x+2}$

40. $\dfrac{9}{y^2-9}, \dfrac{-2}{2y+6}$

41. $\dfrac{4x^2}{(x-4)^2}, \dfrac{17x}{3x-12}$

42. $\dfrac{3y^2}{(y+6)^2}, \dfrac{5y}{2y+12}$

43. $\dfrac{5}{12p + 60}, \dfrac{-17}{p^2 + 5p}, \dfrac{-16}{p^2 + 10p + 25}$

44. $\dfrac{13}{r^2 + 7r}, \dfrac{-3}{5r + 35}, \dfrac{-4}{r^2 + 14r + 49}$

45. $\dfrac{-3}{8y + 16}, \dfrac{-22}{y^2 + 3y + 2}$

46. $\dfrac{-2}{9m - 18}, \dfrac{-6}{m^2 - 7m + 10}$

47. $\dfrac{3}{k^2 + 5k}, \dfrac{2}{k^2 + 3k - 10}$

48. $\dfrac{1}{z^2 - 4z}, \dfrac{9}{z^2 - 3z - 4}$

49. $\dfrac{6}{a^2 + 6a}, \dfrac{-5}{a^2 + 3a - 18}$

50. $\dfrac{8}{y^2 - 5y}, \dfrac{-5}{y^2 - 2y - 15}$

51. $\dfrac{5}{p^2 + 8p + 15}, \dfrac{3}{p^2 - 3p - 18}, \dfrac{12}{p^2 - p - 30}$

52. $\dfrac{10}{y^2 - 10y + 21}, \dfrac{2}{y^2 - 2y - 3}, \dfrac{15}{y^2 - 6y - 7}$

53. $\dfrac{-5}{k^2 + 2k - 35}, \dfrac{-8}{k^2 + 3k - 40}, \dfrac{19}{k^2 - 2k - 15}$

54. $\dfrac{-19}{z^2 + 4z - 12}, \dfrac{-16}{z^2 + z - 30}, \dfrac{16}{z^2 + 2z - 24}$

Write each rational expression with the indicated denominator. **See Examples 4 and 5.**

▶ **55.** $\dfrac{4}{11} = \dfrac{?}{55}$

56. $\dfrac{8}{7} = \dfrac{?}{42}$

57. $\dfrac{-5}{k} = \dfrac{?}{9k}$

58. $\dfrac{-4}{q} = \dfrac{?}{6q}$

59. $\dfrac{15m^2}{8k} = \dfrac{?}{32k^4}$

60. $\dfrac{7t^2}{3y} = \dfrac{?}{9y^2}$

▶ **61.** $\dfrac{19z}{2z - 6} = \dfrac{?}{6z - 18}$

62. $\dfrac{3r}{5r - 5} = \dfrac{?}{15r - 15}$

63. $\dfrac{-2a}{9a - 18} = \dfrac{?}{18a - 36}$

64. $\dfrac{-7y}{6y + 18} = \dfrac{?}{24y + 72}$

65. $\dfrac{6}{k^2 - 4k} = \dfrac{?}{k(k - 4)(k + 1)}$

66. $\dfrac{25}{m^2 - 9m} = \dfrac{?}{m(m - 9)(m + 8)}$

67. $\dfrac{4r - t}{r^2 + rt + t^2} = \dfrac{?}{t^3 - r^3}$

68. $\dfrac{3x - 1}{x^2 + 2x + 4} = \dfrac{?}{x^3 - 8}$

69. $\dfrac{2(z - y)}{y^2 + yz + z^2} = \dfrac{?}{y^4 - z^3 y}$

70. $\dfrac{2p + 3q}{p^2 + 2pq + q^2} = \dfrac{?}{(p + q)(p^3 + q^3)}$

Extending Skills *Write each rational expression with the indicated denominator.* **See Examples 4 and 5.**

71. $\dfrac{36r}{r^2 - r - 6} = \dfrac{?}{(r - 3)(r + 2)(r + 1)}$

72. $\dfrac{4m}{m^2 + m - 2} = \dfrac{?}{(m - 1)(m - 3)(m + 2)}$

73. $\dfrac{a + 2b}{2a^2 + ab - b^2} = \dfrac{?}{2a^3 b + a^2 b^2 - ab^3}$

74. $\dfrac{m - 4}{6m^2 + 7m - 3} = \dfrac{?}{12m^3 + 14m^2 - 6m}$

Work Exercises 75–80 in order.

75. Suppose that we want to write $\frac{3}{4}$ as an equivalent fraction with denominator 28. By what number must we multiply both the numerator and the denominator?

76. If we write $\frac{3}{4}$ as an equivalent fraction with denominator 28, by what number are we actually multiplying the fraction?

77. What property of multiplication is being used when we write a common fraction as an equivalent one with a larger denominator? (See **Section 1.6.**)

78. Suppose that we want to write $\frac{2x+5}{x-4}$ as an equivalent fraction with denominator $7x - 28$. By what number must we multiply both the numerator and the denominator?

79. If we write $\frac{2x+5}{x-4}$ as an equivalent fraction with denominator $7x - 28$, by what number are we actually multiplying the fraction?

80. Repeat **Exercise 77,** changing "a common" to "an algebraic."

6.4 Adding and Subtracting Rational Expressions

OBJECTIVES

1. Add rational expressions having the same denominator.
2. Add rational expressions having different denominators.
3. Subtract rational expressions.

OBJECTIVE 1 Add rational expressions having the same denominator.

We find the sum of two rational expressions with the same denominator using the same procedure that we used in **Section R.1** for adding two common fractions.

Adding Rational Expressions (Same Denominator)

The rational expressions $\frac{P}{Q}$ and $\frac{R}{Q}$ (where $Q \neq 0$) are added as follows.

$$\frac{P}{Q} + \frac{R}{Q} = \frac{P + R}{Q}$$

That is, to add rational expressions with the same denominator, add the numerators and keep the same denominator.

NOW TRY EXERCISE 1

Add. Write each answer in lowest terms.

(a) $\dfrac{2}{7k} + \dfrac{4}{7k}$

(b) $\dfrac{4y}{y+3} + \dfrac{12}{y+3}$

EXAMPLE 1 Adding Rational Expressions (Same Denominator)

Add. Write each answer in lowest terms.

(a) $\dfrac{4}{9} + \dfrac{2}{9}$ 　　　　　　　**(b)** $\dfrac{3x}{x+1} + \dfrac{3}{x+1}$

The denominators are the same, so the sum is found by adding the two numerators and keeping the same (common) denominator.

$$= \frac{4+2}{9} \quad \text{Add.}$$

$$= \frac{6}{9}$$

$$= \frac{2 \cdot 3}{3 \cdot 3} \quad \text{Factor.}$$

$$= \frac{2}{3} \quad \text{Lowest terms}$$

$$= \frac{3x+3}{x+1} \quad \text{Add.}$$

$$= \frac{3(x+1)}{x+1} \quad \text{Factor.}$$

$$= 3 \quad \text{Lowest terms}$$

NOW TRY ANSWERS

1. **(a)** $\frac{6}{7k}$ **(b)** 4

NOW TRY ↻

OBJECTIVE 2 Add rational expressions having different denominators.

As in **Section R.1,** we use the following steps to add fractions having different denominators.

> **Adding Rational Expressions (Different Denominators)**
>
> **Step 1** **Find the least common denominator (LCD).**
>
> **Step 2** **Write each rational expression** as an equivalent rational expression with the LCD as the denominator.
>
> **Step 3** **Add** the numerators to obtain the numerator of the sum. The LCD is the denominator of the sum.
>
> **Step 4** **Write in lowest terms** using the fundamental property.

NOW TRY
EXERCISE 2

Add. Write each answer in lowest terms.

(a) $\dfrac{5}{12} + \dfrac{3}{20}$ **(b)** $\dfrac{3}{5x} + \dfrac{2}{7x}$

EXAMPLE 2 Adding Rational Expressions (Different Denominators)

Add. Write each answer in lowest terms.

(a) $\dfrac{1}{12} + \dfrac{7}{15}$ **(b)** $\dfrac{2}{3y} + \dfrac{1}{4y}$

Step 1 Find the LCD, using the methods of the previous section.

$$12 = 2 \cdot 2 \cdot 3 = 2^2 \cdot 3 \qquad\qquad 3y = 3 \cdot y$$
$$15 = 3 \cdot 5 \qquad\qquad\qquad 4y = 2 \cdot 2 \cdot y = 2^2 \cdot y$$
$$\text{LCD} = 2^2 \cdot 3 \cdot 5 = 60 \qquad\qquad \text{LCD} = 2^2 \cdot 3 \cdot y = 12y$$

Step 2 Write each rational expression as a fraction with the LCD (60 and $12y$, respectively) as the denominator.

$$\dfrac{1}{12} + \dfrac{7}{15} \quad \text{The LCD is 60.} \qquad\qquad \dfrac{2}{3y} + \dfrac{1}{4y} \quad \text{The LCD is } 12y.$$

$$= \dfrac{1(5)}{12(5)} + \dfrac{7(4)}{15(4)} \qquad\qquad = \dfrac{2(4)}{3y(4)} + \dfrac{1(3)}{4y(3)}$$

$$= \dfrac{5}{60} + \dfrac{28}{60} \qquad\qquad\qquad = \dfrac{8}{12y} + \dfrac{3}{12y}$$

Step 3 Add the numerators. The LCD is the denominator.

Step 4 Write in lowest terms if necessary.

$$= \dfrac{5 + 28}{60} \qquad\qquad\qquad = \dfrac{8 + 3}{12y}$$

$$= \dfrac{33}{60} \qquad\qquad\qquad\qquad = \dfrac{11}{12y}$$

$$= \dfrac{11}{20}$$

NOW TRY

**NOW TRY
EXERCISE 3**

Add. Write the answer in lowest terms.

$$\frac{6t}{t^2 - 9} + \frac{-3}{t + 3}$$

EXAMPLE 3 Adding Rational Expressions

Add. Write the answer in lowest terms.

$$\frac{2x}{x^2 - 1} + \frac{-1}{x + 1}$$

Step 1 The denominators are different, so find the LCD.

$$\left. \begin{array}{l} x^2 - 1 = (x + 1)(x - 1) \\ x + 1 \text{ is prime.} \end{array} \right\} \quad \text{The LCD is } (x + 1)(x - 1).$$

Step 2 Write each rational expression with the LCD as the denominator.

$$\frac{2x}{x^2 - 1} + \frac{-1}{x + 1} \qquad \text{LCD} = (x + 1)(x - 1)$$

$$= \frac{2x}{(x + 1)(x - 1)} + \frac{-1(x - 1)}{(x + 1)(x - 1)} \qquad \text{Multiply the second fraction by } \frac{x - 1}{x - 1}.$$

$$= \frac{2x}{(x + 1)(x - 1)} + \frac{-x + 1}{(x + 1)(x - 1)} \qquad \text{Distributive property}$$

Step 3 $$= \frac{2x - x + 1}{(x + 1)(x - 1)} \qquad \begin{array}{l} \text{Add numerators.} \\ \text{Keep the same denominator.} \end{array}$$

$$= \frac{x + 1}{(x + 1)(x - 1)} \qquad \text{Combine like terms.}$$

Step 4 $$= \frac{1(x + 1)}{(x + 1)(x - 1)} \qquad \begin{array}{l} \text{Identity property of} \\ \text{multiplication} \end{array}$$

Remember to write 1 in the numerator. $$= \frac{1}{x - 1} \qquad \text{Lowest terms} \qquad \text{NOW TRY} \; \circlearrowleft$$

**NOW TRY
EXERCISE 4**

Add. Write the answer in lowest terms.

$$\frac{x - 1}{x^2 + 6x + 8} + \frac{4x}{x^2 + x - 12}$$

EXAMPLE 4 Adding Rational Expressions

Add. Write the answer in lowest terms.

$$\frac{2x}{x^2 + 5x + 6} + \frac{x + 1}{x^2 + 2x - 3}$$

$$= \frac{2x}{(x + 2)(x + 3)} + \frac{x + 1}{(x + 3)(x - 1)} \qquad \text{Factor the denominators.}$$

$$= \frac{2x(x - 1)}{(x + 2)(x + 3)(x - 1)} + \frac{(x + 1)(x + 2)}{(x + 2)(x + 3)(x - 1)} \qquad \begin{array}{l} \text{The LCD is} \\ (x + 2)(x + 3)(x - 1). \end{array}$$

$$= \frac{2x(x - 1) + (x + 1)(x + 2)}{(x + 2)(x + 3)(x - 1)} \qquad \begin{array}{l} \text{Add numerators.} \\ \text{Keep the same denominator.} \end{array}$$

$$= \frac{2x^2 - 2x + x^2 + 3x + 2}{(x + 2)(x + 3)(x - 1)} \qquad \text{Multiply.}$$

$$= \frac{3x^2 + x + 2}{(x + 2)(x + 3)(x - 1)} \qquad \text{Combine like terms.}$$

The numerator cannot be factored here, so the expression is in lowest terms.

NOW TRY $\circlearrowleft$

NOW TRY ANSWERS

3. $\dfrac{3}{t - 3}$

4. $\dfrac{5x^2 + 4x + 3}{(x + 4)(x + 2)(x - 3)}$

▶ 17. $\dfrac{5m}{m+1} - \dfrac{1+4m}{m+1}$

18. $\dfrac{4x}{x+2} - \dfrac{2+3x}{x+2}$

19. $\dfrac{a+b}{2} - \dfrac{a-b}{2}$

20. $\dfrac{x-y}{2} - \dfrac{x+y}{2}$

21. $\dfrac{x^2}{x+5} + \dfrac{5x}{x+5}$

22. $\dfrac{t^2}{t-3} + \dfrac{-3t}{t-3}$

23. $\dfrac{y^2-3y}{y+3} + \dfrac{-18}{y+3}$

24. $\dfrac{r^2-8r}{r-5} + \dfrac{15}{r-5}$

25. $\dfrac{x}{x^2-9} - \dfrac{-3}{x^2-9}$

26. $\dfrac{-4}{y^2-16} - \dfrac{-y}{y^2-16}$

27. $\dfrac{y^2+x^2}{x^2-y^2} - \dfrac{2x^2}{x^2-y^2}$

28. $\dfrac{3a^2+b^2}{a^2-b^2} - \dfrac{4a^2}{a^2-b^2}$

Add or subtract. Write each answer in lowest terms. **See Examples 2, 3, 4, and 7.**

▶ 29. $\dfrac{z}{5} + \dfrac{1}{3}$

30. $\dfrac{p}{8} + \dfrac{4}{5}$

31. $\dfrac{5}{7} - \dfrac{r}{2}$

32. $\dfrac{20}{9} - \dfrac{z}{3}$

33. $-\dfrac{3}{4} - \dfrac{1}{2x}$

34. $-\dfrac{7}{8} - \dfrac{3}{2a}$

35. $\dfrac{7}{4t} + \dfrac{3}{7t}$

36. $\dfrac{8}{3r} + \dfrac{2}{5r}$

37. $\dfrac{x+1}{6} + \dfrac{3x+3}{9}$

38. $\dfrac{2x-6}{4} + \dfrac{x+5}{6}$

39. $\dfrac{x+3}{3x} + \dfrac{2x+2}{4x}$

40. $\dfrac{x+2}{5x} + \dfrac{6x+3}{3x}$

41. $\dfrac{7}{3p^2} - \dfrac{2}{p}$

42. $\dfrac{12}{5m^2} - \dfrac{5}{m}$

▶ 43. $\dfrac{1}{k+4} - \dfrac{2}{k}$

44. $\dfrac{3}{m+1} - \dfrac{4}{m}$

▶ 45. $\dfrac{x}{x-2} + \dfrac{-8}{x^2-4}$

46. $\dfrac{2x}{x-1} + \dfrac{-4}{x^2-1}$

▶ 47. $\dfrac{4m}{m^2+3m+2} + \dfrac{2m-1}{m^2+6m+5}$

48. $\dfrac{a}{a^2+3a-4} + \dfrac{4a}{a^2+7a+12}$

49. $\dfrac{4y}{y^2-1} - \dfrac{5}{y^2+2y+1}$

50. $\dfrac{2x}{x^2-16} - \dfrac{3}{x^2+8x+16}$

51. $\dfrac{t}{t+2} + \dfrac{5-t}{t} - \dfrac{4}{t^2+2t}$

52. $\dfrac{2p}{p-3} + \dfrac{2+p}{p} - \dfrac{-6}{p^2-3p}$

Concept Check *Answer each question.*

53. What are the *two* possible LCDs that could be used for the sum $\dfrac{10}{m-2} + \dfrac{5}{2-m}$?

54. If one form of the correct answer to a sum or difference of rational expressions is $\dfrac{4}{k-3}$, what would an alternative form of the answer be if the denominator is $3-k$?

Add or subtract. Write each answer in lowest terms. **See Examples 5 and 8.**

▶ 55. $\dfrac{4}{x-5} + \dfrac{6}{5-x}$

56. $\dfrac{10}{m-2} + \dfrac{5}{2-m}$

▶ 57. $\dfrac{-1}{1-y} - \dfrac{4y-3}{y-1}$

58. $\dfrac{-4}{p-3} - \dfrac{p+1}{3-p}$

59. $\dfrac{2}{x-y^2} + \dfrac{7}{y^2-x}$

60. $\dfrac{-8}{p-q^2} + \dfrac{3}{q^2-p}$

61. $\dfrac{x}{5x-3y} - \dfrac{y}{3y-5x}$

62. $\dfrac{t}{8t-9s} - \dfrac{s}{9s-8t}$

63. $\dfrac{3}{4p-5} + \dfrac{9}{5-4p}$

64. $\dfrac{8}{3 - 7y} - \dfrac{2}{7y - 3}$

65. $\dfrac{15x}{5x - 7} - \dfrac{-21}{7 - 5x}$

66. $\dfrac{24y}{6y - 5} - \dfrac{-20}{5 - 6y}$

In these subtraction problems, the rational expression that follows the subtraction sign has a numerator with more than one term. **Be careful with signs** *and find each difference. See* **Example 9.**

67. $\dfrac{2m}{m - n} - \dfrac{5m + n}{2m - 2n}$

68. $\dfrac{5p}{p - q} - \dfrac{3p + 1}{4p - 4q}$

▶ **69.** $\dfrac{5}{x^2 - 9} - \dfrac{x + 2}{x^2 + 4x + 3}$

70. $\dfrac{1}{a^2 - 1} - \dfrac{a - 1}{a^2 + 3a - 4}$

71. $\dfrac{2q + 1}{3q^2 + 10q - 8} - \dfrac{3q + 5}{2q^2 + 5q - 12}$

72. $\dfrac{4y - 1}{2y^2 + 5y - 3} - \dfrac{y + 3}{6y^2 + y - 2}$

Perform each indicated operation. See Examples 1–10.

73. $\dfrac{y^2}{y - 2} - \dfrac{9y - 14}{y - 2}$

74. $\dfrac{y^2}{y - 4} - \dfrac{y + 12}{y - 4}$

75. $\dfrac{3}{x + 4} + 7$

76. $\dfrac{9}{x + 7} + 2$

77. $\dfrac{-x + 2}{x} - \dfrac{x - 5}{4x}$

78. $\dfrac{-y + 3}{y} - \dfrac{y + 4}{3y}$

79. $\dfrac{5x}{x - 7} - \dfrac{3x}{x - 3}$

80. $\dfrac{6t}{t + 4} - \dfrac{2t}{t + 1}$

81. $\dfrac{5a}{3a - 6} - \dfrac{a - 7}{a - 2}$

82. $\dfrac{4a}{5a - 15} - \dfrac{a - 1}{a - 3}$

83. $\dfrac{4}{3 - x} + \dfrac{x}{2x - 6}$

84. $\dfrac{5}{4 - x} + \dfrac{x}{2x - 8}$

85. $\dfrac{5x + 11}{x^2 - 11x + 18} - \dfrac{4x + 20}{x^2 - 11x + 18}$

86. $\dfrac{4x + 7}{x^2 + 2x - 3} - \dfrac{3x + 4}{x^2 + 2x - 3}$

87. $\dfrac{4}{r^2 - r} + \dfrac{6}{r^2 + 2r} - \dfrac{1}{r^2 + r - 2}$

88. $\dfrac{6}{k^2 + 3k} - \dfrac{1}{k^2 - k} + \dfrac{2}{k^2 + 2k - 3}$

89. $\dfrac{x + 3y}{x^2 + 2xy + y^2} + \dfrac{x - y}{x^2 + 4xy + 3y^2}$

90. $\dfrac{m}{m^2 - 1} + \dfrac{m - 1}{m^2 + 2m + 1}$

91. $\dfrac{r + y}{18r^2 + 9ry - 2y^2} + \dfrac{3r - y}{36r^2 - y^2}$

92. $\dfrac{2x - z}{2x^2 + xz - 10z^2} - \dfrac{x + z}{x^2 - 4z^2}$

Work each problem.

93. Refer to the rectangle in the figure.

   (a) Find an expression that represents its perimeter. Give the simplified form.

   (b) Find an expression that represents its area. Give the simplified form.

94. Refer to the triangle in the figure. Find an expression that represents its perimeter. Give the simplified form.

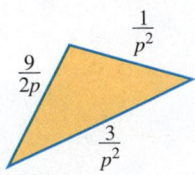

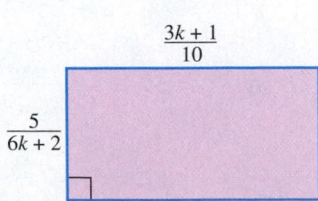

A concours d'elegance is a competition in which a maximum of 100 *points is awarded to a car based on its general attractiveness. The rational expression*

$$\frac{1010}{49(101 - x)} - \frac{10}{49}$$

approximates the cost, in thousands of dollars, of restoring a car so that it will win x points.
Use this information to work Exercises 95 and 96.

95. Simplify the given expression by performing the indicated subtraction.

96. Use the simplified expression from **Exercise 95** to determine how much it would cost to win 95 points.

6.5 Complex Fractions

OBJECTIVES

1 Define and recognize a complex fraction.

2 Simplify a complex fraction by writing it as a division problem (Method 1).

3 Simplify a complex fraction by multiplying numerator and denominator by the LCD (Method 2).

4 Simplify rational expressions with negative exponents.

VOCABULARY

☐ complex fraction

OBJECTIVE 1 Define and recognize a complex fraction.

The quotient of two mixed numbers in arithmetic, such as $2\frac{1}{2} \div 3\frac{1}{4}$, can be written as a fraction.

$$2\frac{1}{2} \div 3\frac{1}{4}$$

$$= \frac{2\frac{1}{2}}{3\frac{1}{4}}$$

$$= \frac{2 + \frac{1}{2}}{3 + \frac{1}{4}}$$ We do this to illustrate a *complex fraction.*

Some rational expressions in algebra have fractions in the numerator, or denominator, or both.

Complex Fraction

A quotient with one or more fractions in the numerator, or denominator, or both, is a **complex fraction.**

Examples: $\dfrac{2 + \dfrac{1}{2}}{3 + \dfrac{1}{4}}, \quad \dfrac{\dfrac{3x^2 - 5x}{6x^2}}{2x - \dfrac{1}{x}}, \quad$ and $\quad \dfrac{3 + x}{5 - \dfrac{2}{x}}$ Complex fractions

---

The parts of a complex fraction are named as follows.

$$\left.\dfrac{\dfrac{2}{p} - \dfrac{1}{q}}{\dfrac{3}{p} + \dfrac{5}{q}}\right.$$
← Numerator of complex fraction
← Main fraction bar
← Denominator of complex fraction

OBJECTIVE 2 Simplify a complex fraction by writing it as a division problem (Method 1).

Because the main fraction bar represents division in a complex fraction, one method of simplifying a complex fraction involves division.

> **Simplifying a Complex Fraction (Method 1)**
>
> **Step 1** Write both the numerator and denominator as single fractions.
>
> **Step 2** Change the complex fraction to a division problem.
>
> **Step 3** Perform the indicated division.

EXAMPLE 1 Simplifying Complex Fractions (Method 1)

Simplify each complex fraction.

(a) $\dfrac{\dfrac{2}{3}+\dfrac{5}{9}}{\dfrac{1}{4}+\dfrac{1}{12}}$
(b) $\dfrac{6+\dfrac{3}{x}}{\dfrac{x}{4}+\dfrac{1}{8}}$

Step 1 First, write each numerator as a single fraction.

$$\frac{2}{3}+\frac{5}{9}$$

$$=\frac{2(3)}{3(3)}+\frac{5}{9}$$

$$=\frac{6}{9}+\frac{5}{9}$$

$$=\frac{11}{9}$$

$$6+\frac{3}{x}$$

$$=\frac{6}{1}+\frac{3}{x}$$

$$=\frac{6x}{x}+\frac{3}{x}$$

$$=\frac{6x+3}{x}$$

Repeat the process for each denominator.

$$\frac{1}{4}+\frac{1}{12}$$

$$=\frac{1(3)}{4(3)}+\frac{1}{12}$$

$$=\frac{3}{12}+\frac{1}{12}$$

$$=\frac{4}{12}$$

$$\frac{x}{4}+\frac{1}{8}$$

$$=\frac{x(2)}{4(2)}+\frac{1}{8}$$

$$=\frac{2x}{8}+\frac{1}{8}$$

$$=\frac{2x+1}{8}$$

Step 2 Write the equivalent complex fraction as a division problem.

$$\frac{\dfrac{11}{9}}{\dfrac{4}{12}}$$

$$=\frac{11}{9}\div\frac{4}{12}$$

$$\frac{\dfrac{6x+3}{x}}{\dfrac{2x+1}{8}}$$

$$=\frac{6x+3}{x}\div\frac{2x+1}{8}$$

**NOW TRY
EXERCISE 1**

Simplify each complex fraction.

(a) $\dfrac{\dfrac{2}{5} + \dfrac{1}{4}}{\dfrac{1}{6} + \dfrac{3}{8}}$ **(b)** $\dfrac{2 + \dfrac{4}{x}}{\dfrac{5}{6} + \dfrac{5x}{12}}$

Step 3 Now use the definition of division and multiply by the reciprocal. Then write in lowest terms using the fundamental property.

$$= \frac{11}{9} \cdot \frac{12}{4} \qquad\qquad = \frac{6x+3}{x} \cdot \frac{8}{2x+1}$$

$$= \frac{11 \cdot 3 \cdot 4}{3 \cdot 3 \cdot 4} \qquad\qquad = \frac{3(2x+1)}{x} \cdot \frac{8}{2x+1}$$

$$= \frac{11}{3} \qquad\qquad = \frac{24}{x}$$

 **NOW TRY**

**NOW TRY
EXERCISE 2**

Simplify the complex fraction.

$$\frac{\dfrac{a^2 b}{c}}{\dfrac{ab^2}{c^3}}$$

EXAMPLE 2 Simplifying a Complex Fraction (Method 1)

Simplify the complex fraction.

$\dfrac{\dfrac{xp}{q^3}}{\dfrac{p^2}{qx^2}}$ The numerator and denominator are single fractions, so use the definition of division and then the fundamental property. $\dfrac{xp}{q^3} \div \dfrac{p^2}{qx^2}$

$$= \frac{xp}{q^3} \cdot \frac{qx^2}{p^2}$$

$$= \frac{x^3}{q^2 p} \qquad \text{NOW TRY}$$

**NOW TRY
EXERCISE 3**

Simplify the complex fraction.

$$\frac{5 + \dfrac{2}{a-3}}{\dfrac{1}{a-3} - 2}$$

EXAMPLE 3 Simplifying a Complex Fraction (Method 1)

Simplify the complex fraction.

$$\frac{\dfrac{3}{x+2} - 4}{\dfrac{2}{x+2} + 1}$$ *Find a common denominator before subtracting in the numerator and denominator.*

$$= \frac{\dfrac{3}{x+2} - \dfrac{4(x+2)}{x+2}}{\dfrac{2}{x+2} + \dfrac{1(x+2)}{x+2}} \qquad \text{Write both second terms with a denominator of } x+2.$$

$$= \frac{\dfrac{3 - 4(x+2)}{x+2}}{\dfrac{2 + 1(x+2)}{x+2}} \qquad \begin{array}{l}\text{Subtract in the numerator.}\\[1.5em]\text{Add in the denominator.}\end{array}$$

Be careful with signs.

$$= \frac{\dfrac{3 - 4x - 8}{x+2}}{\dfrac{2 + x + 2}{x+2}} \qquad \text{Distributive property}$$

$$= \frac{\dfrac{-5 - 4x}{x+2}}{\dfrac{4 + x}{x+2}} \qquad \text{Combine like terms.}$$

$$= \frac{-5 - 4x}{x+2} \cdot \frac{x+2}{4+x} \qquad \begin{array}{l}\text{Multiply by the reciprocal of the}\\\text{denominator (divisor).}\end{array}$$

$$= \frac{-5 - 4x}{4+x} \qquad \begin{array}{l}\text{Divide out the}\\\text{common factor.}\end{array} \qquad \text{NOW TRY}$$

NOW TRY ANSWERS

1. (a) $\dfrac{6}{5}$ (b) $\dfrac{24}{5x}$

2. $\dfrac{ac^2}{b}$

3. $\dfrac{5a - 13}{7 - 2a}$

OBJECTIVE 3 Simplify a complex fraction by multiplying numerator and denominator by the LCD (Method 2).

If we multiply both the numerator and the denominator of a complex fraction by the LCD of all the fractions within the complex fraction, the result will no longer be complex. This is Method 2.

Simplifying a Complex Fraction (Method 2)

Step 1   Find the LCD of all fractions within the complex fraction.

Step 2   Multiply both the numerator and the denominator of the complex fraction by this LCD using the distributive property as necessary. Write in lowest terms.

NOW TRY
EXERCISE 4

Simplify each complex fraction.

(a) $\dfrac{\dfrac{3}{5} - \dfrac{1}{4}}{\dfrac{1}{8} + \dfrac{3}{20}}$ **(b)** $\dfrac{\dfrac{2}{x} - 3}{7 + \dfrac{x}{5}}$

EXAMPLE 4   Simplifying Complex Fractions (Method 2)

Simplify each complex fraction.

(a) $\dfrac{\dfrac{2}{3} + \dfrac{5}{9}}{\dfrac{1}{4} + \dfrac{1}{12}}$ **(b)** $\dfrac{6 + \dfrac{3}{x}}{\dfrac{x}{4} + \dfrac{1}{8}}$ (In **Example 1**, we simplified these same fractions using Method 1.)

Step 1   Find the LCD for all denominators in the complex fraction.

The LCD for 3, 9, 4, and 12 is 36. | The LCD for x, 4, and 8 is $8x$.

Step 2
$$\dfrac{\dfrac{2}{3} + \dfrac{5}{9}}{\dfrac{1}{4} + \dfrac{1}{12}}$$
$$\dfrac{6 + \dfrac{3}{x}}{\dfrac{x}{4} + \dfrac{1}{8}}$$ Multiply numerator and denominator of the complex fraction by the LCD.

$$= \dfrac{36\left(\dfrac{2}{3} + \dfrac{5}{9}\right)}{36\left(\dfrac{1}{4} + \dfrac{1}{12}\right)}$$
$$= \dfrac{8x\left(6 + \dfrac{3}{x}\right)}{8x\left(\dfrac{x}{4} + \dfrac{1}{8}\right)}$$

Multiply each term by 36.
$$= \dfrac{36\left(\dfrac{2}{3}\right) + 36\left(\dfrac{5}{9}\right)}{36\left(\dfrac{1}{4}\right) + 36\left(\dfrac{1}{12}\right)}$$
Multiply each term by $8x$.
$$= \dfrac{8x(6) + 8x\left(\dfrac{3}{x}\right)}{8x\left(\dfrac{x}{4}\right) + 8x\left(\dfrac{1}{8}\right)}$$ Distributive property

$$= \dfrac{24 + 20}{9 + 3}$$ Multiply.
$$= \dfrac{48x + 24}{2x^2 + x}$$ Multiply.

$$= \dfrac{44}{12}$$ Add.
$$= \dfrac{24(2x + 1)}{x(2x + 1)}$$ Factor.

$$= \dfrac{4 \cdot 11}{4 \cdot 3}$$ Factor.
$$= \dfrac{24}{x}$$ Lowest terms

$$= \dfrac{11}{3}$$ Lowest terms

NOW TRY ANSWERS

4. (a) $\dfrac{14}{11}$ (b) $\dfrac{10 - 15x}{x^2 + 35x}$

NOW TRY

**NOW TRY
EXERCISE 5**

Simplify the complex fraction.

$$\dfrac{\dfrac{1}{y} + \dfrac{2}{3y^2}}{\dfrac{5}{4y^2} - \dfrac{3}{2y^3}}$$

EXAMPLE 5 Simplifying a Complex Fraction (Method 2)

Simplify the complex fraction.

$$\dfrac{\dfrac{3}{5m} - \dfrac{2}{m^2}}{\dfrac{9}{2m} + \dfrac{3}{4m^2}}$$ The LCD for $5m$, m^2, $2m$, and $4m^2$ is $20m^2$.

$$= \dfrac{20m^2\left(\dfrac{3}{5m} - \dfrac{2}{m^2}\right)}{20m^2\left(\dfrac{9}{2m} + \dfrac{3}{4m^2}\right)}$$ Multiply numerator and denominator by $20m^2$.

$$= \dfrac{20m^2\left(\dfrac{3}{5m}\right) - 20m^2\left(\dfrac{2}{m^2}\right)}{20m^2\left(\dfrac{9}{2m}\right) + 20m^2\left(\dfrac{3}{4m^2}\right)}$$ Distributive property

$$= \dfrac{12m - 40}{90m + 15}$$ Multiply. Divide out the common factors.

$$= \dfrac{4(3m - 10)}{5(18m + 3)}$$ Factor.

NOW TRY

Some students prefer Method 1 for problems like **Example 2,** which is the quotient of two fractions, and Method 2 for problems like **Examples 1, 3, 4, and 5,** which have sums or differences in the numerators, or denominators, or both. (However, either method can be used for *any* complex fraction.)

EXAMPLE 6 Simplifying Complex Fractions

Simplify each complex fraction. Use either method.

(a) $\dfrac{\dfrac{1}{y} + \dfrac{2}{y + 2}}{\dfrac{4}{y} - \dfrac{3}{y + 2}}$ There are sums and differences in the numerator and denominator. Use Method 2.

$$= \dfrac{\left(\dfrac{1}{y} + \dfrac{2}{y + 2}\right) \cdot y(y + 2)}{\left(\dfrac{4}{y} - \dfrac{3}{y + 2}\right) \cdot y(y + 2)}$$ Multiply numerator and denominator by the LCD, $y(y + 2)$. Because y appears in two denominators, it must be a factor in the LCD.

$$= \dfrac{\left(\dfrac{1}{y}\right)y(y + 2) + \left(\dfrac{2}{y + 2}\right)y(y + 2)}{\left(\dfrac{4}{y}\right)y(y + 2) - \left(\dfrac{3}{y + 2}\right)y(y + 2)}$$ Distributive property

$$= \dfrac{1(y + 2) + 2y}{4(y + 2) - 3y}$$ Multiply. Divide out the common factors.

$$= \dfrac{y + 2 + 2y}{4y + 8 - 3y}$$ Distributive property

$$= \dfrac{3y + 2}{y + 8}$$ Combine like terms.

NOW TRY ANSWER

5. $\dfrac{12y^2 + 8y}{15y - 18}$,   or $\dfrac{4y(3y + 2)}{3(5y - 6)}$

**NOW TRY
EXERCISE 6**
Simplify each complex fraction.

(a) $\dfrac{1 - \dfrac{2}{x} - \dfrac{15}{x^2}}{1 + \dfrac{5}{x} + \dfrac{6}{x^2}}$

(b) $\dfrac{\dfrac{9y^2 - 16}{y^2 - 100}}{\dfrac{3y - 4}{y + 10}}$

(b) $\dfrac{1 - \dfrac{2}{x} - \dfrac{3}{x^2}}{1 - \dfrac{5}{x} + \dfrac{6}{x^2}}$ There are sums and differences in the numerator and denominator. Use Method 2.

$= \dfrac{\left(1 - \dfrac{2}{x} - \dfrac{3}{x^2}\right)x^2}{\left(1 - \dfrac{5}{x} + \dfrac{6}{x^2}\right)x^2}$ Multiply numerator and denominator by the LCD, x^2.

$= \dfrac{(1)x^2 - \left(\dfrac{2}{x}\right)x^2 - \left(\dfrac{3}{x^2}\right)x^2}{(1)x^2 - \left(\dfrac{5}{x}\right)x^2 + \left(\dfrac{6}{x^2}\right)x^2}$ Distributive property

$= \dfrac{x^2 - 2x - 3}{x^2 - 5x + 6}$ Multiply.

$= \dfrac{(x - 3)(x + 1)}{(x - 3)(x - 2)}$ Factor.

$= \dfrac{x + 1}{x - 2}$ Lowest terms

(c) $\dfrac{\dfrac{x + 2}{x - 3}}{\dfrac{x^2 - 4}{x^2 - 9}}$ This is a quotient of two rational expressions. Use Method 1.

$= \dfrac{x + 2}{x - 3} \div \dfrac{x^2 - 4}{x^2 - 9}$ Write as a division problem.

$= \dfrac{x + 2}{x - 3} \cdot \dfrac{x^2 - 9}{x^2 - 4}$ Multiply by the reciprocal of the divisor.

$= \dfrac{(x + 2)(x + 3)(x - 3)}{(x - 3)(x + 2)(x - 2)}$ Multiply, and then factor.

$= \dfrac{x + 3}{x - 2}$ Lowest terms   **NOW TRY**

OBJECTIVE 4 Simplify rational expressions with negative exponents.

We begin by rewriting the expressions with only positive exponents. Recall from **Section 4.2** that for any nonzero real number a and any integer n,

$$a^{-n} = \dfrac{1}{a^n}. \quad \text{Definition of negative exponent}$$

! CAUTION $a^{-1} + b^{-1} = \dfrac{1}{a} + \dfrac{1}{b}$, **not** $\dfrac{1}{a + b}$.   Avoid this common error.

NOW TRY ANSWERS
6. **(a)** $\frac{x - 5}{x + 2}$ **(b)** $\frac{3y + 4}{y - 10}$

NOW TRY
EXERCISE 7

Simplify each expression, using only positive exponents in the answer.

(a) $\dfrac{r^{-2} - s^{-1}}{4r^{-1} + s^{-2}}$

(b) $\dfrac{2y^{-1} - 3y^{-2}}{y^{-2} + 3x^{-1}}$

EXAMPLE 7   Simplifying Rational Expressions with Negative Exponents

Simplify each expression, using only positive exponents in the answer.

(a) $\dfrac{m^{-1} + p^{-2}}{2m^{-2} - p^{-1}}$ $a^{-n} = \dfrac{1}{a^n}$ **(Section 4.2)**

$$= \dfrac{\dfrac{1}{m} + \dfrac{1}{p^2}}{\dfrac{2}{m^2} - \dfrac{1}{p}} \qquad$$

Write with positive exponents.

The base of $2m^{-2}$ is m, not $2m$: $2m^{-2} = \dfrac{2}{m^2}$. $2m^{-2} = 2 \cdot m^{-2} = \dfrac{2}{1} \cdot \dfrac{1}{m^2} = \dfrac{2}{m^2}$

$$= \dfrac{m^2 p^2 \left(\dfrac{1}{m} + \dfrac{1}{p^2}\right)}{m^2 p^2 \left(\dfrac{2}{m^2} - \dfrac{1}{p}\right)}$$

Simplify by Method 2. Multiply the numerator and denominator by the LCD, $m^2 p^2$.

$$= \dfrac{m^2 p^2 \cdot \dfrac{1}{m} + m^2 p^2 \cdot \dfrac{1}{p^2}}{m^2 p^2 \cdot \dfrac{2}{m^2} - m^2 p^2 \cdot \dfrac{1}{p}}$$

Distributive property

$$= \dfrac{mp^2 + m^2}{2p^2 - m^2 p}$$

Write in lowest terms.

(b) $\dfrac{x^{-2} - 2y^{-1}}{y - 2x^2}$

The 2 does *not* go in the denominator of this fraction.

$$= \dfrac{\dfrac{1}{x^2} - \dfrac{2}{y}}{y - 2x^2}$$

Write with positive exponents.

$$= \dfrac{\left(\dfrac{1}{x^2} - \dfrac{2}{y}\right) x^2 y}{(y - 2x^2)\, x^2 y}$$

Use Method 2. Multiply by the LCD, $x^2 y$.

$$= \dfrac{\left(\dfrac{1}{x^2}\right) x^2 y - \left(\dfrac{2}{y}\right) x^2 y}{(y - 2x^2) x^2 y}$$

Use the distributive property in the numerator.

$$= \dfrac{y - 2x^2}{(y - 2x^2) x^2 y}$$

Multiply in the numerator.

$$= \dfrac{1}{x^2 y} \qquad$$

Remember to write 1 in the numerator.

Write in lowest terms.   **NOW TRY**

NOW TRY ANSWERS

7. (a) $\dfrac{s^2 - r^2 s}{4rs^2 + r^2}$ **(b)** $\dfrac{2xy - 3x}{x + 3y^2}$

6.5 Exercises

        MyMathLab®

▶ *Complete solution available in MyMathLab*

Concept Check   In Exercises 1 and 2, consider the following complex fraction.

$$\dfrac{\dfrac{1}{2} - \dfrac{1}{3}}{\dfrac{5}{6} - \dfrac{1}{12}}$$

1. Answer each part, outlining Method 1 for simplifying the complex fraction.

   (a) To combine the terms in the numerator, we must find the LCD of $\frac{1}{2}$ and $\frac{1}{3}$.

   What is this LCD?

   Determine the simplified form of the numerator of the complex fraction.

   (b) To combine the terms in the denominator, we must find the LCD of $\frac{5}{6}$ and $\frac{1}{12}$.

   What is this LCD?

   Determine the simplified form of the denominator of the complex fraction.

   (c) Now use the results from parts (a) and (b) to write the complex fraction as a division problem using the symbol $\div$.

   (d) Perform the operation from part (c) to obtain the final simplification.

2. Answer each part, outlining Method 2 for simplifying the complex fraction.

   (a) We must determine the LCD of all the fractions within the complex fraction.

   What is this LCD?

   (b) Multiply every term in the complex fraction by the LCD found in part (a), but at this time do not combine the terms in the numerator and the denominator.

   (c) Now combine the terms from part (b) to obtain the simplified form of the complex fraction.

Concept Check   Work each problem.

3. Which complex fraction is equivalent to $\dfrac{2 - \frac{1}{4}}{3 - \frac{1}{2}}$? Answer this question without showing any work, and explain your reasoning.

   A. $\dfrac{2 + \frac{1}{4}}{3 + \frac{1}{2}}$ **B.** $\dfrac{2 - \frac{1}{4}}{-3 + \frac{1}{2}}$ **C.** $\dfrac{-2 - \frac{1}{4}}{-3 - \frac{1}{2}}$ **D.** $\dfrac{-2 + \frac{1}{4}}{-3 + \frac{1}{2}}$

4. Only one of these choices is equal to $\dfrac{\frac{1}{3} + \frac{1}{12}}{\frac{1}{2} + \frac{1}{4}}$. Which one is it? Answer this question without showing any work, and explain your reasoning.

   A. $\dfrac{5}{9}$ **B.** $-\dfrac{5}{9}$ **C.** $-\dfrac{9}{5}$ **D.** $-\dfrac{1}{12}$

Concept Check   Find the slope of the line that passes through each pair of points. (Hint: This will involve simplifying complex fractions. Recall that slope $m = \frac{y_2 - y_1}{x_2 - x_1}$.)

5. $\left(-\dfrac{5}{2}, \dfrac{1}{6}\right)$ and $\left(\dfrac{5}{3}, \dfrac{3}{8}\right)$

6. $\left(-\dfrac{5}{6}, -\dfrac{1}{2}\right)$ and $\left(-\dfrac{1}{3}, -\dfrac{3}{2}\right)$

Simplify each complex fraction. Use either method. See Examples 1–6.

7. $\dfrac{-\dfrac{4}{3}}{\dfrac{2}{9}}$

8. $\dfrac{-\dfrac{5}{6}}{\dfrac{5}{4}}$

▶ 9. $\dfrac{\dfrac{x}{y^2}}{\dfrac{x^2}{y}}$

10. $\dfrac{\dfrac{p^4}{r}}{\dfrac{p^2}{r^2}}$

11. $\dfrac{\dfrac{4a^4b^3}{3a}}{\dfrac{2ab^4}{b^2}}$

12. $\dfrac{\dfrac{2r^4t^2}{3t}}{\dfrac{5r^2t^5}{3r}}$

13. $\dfrac{\dfrac{m+2}{3}}{\dfrac{m-4}{m}}$

14. $\dfrac{\dfrac{q-5}{q}}{\dfrac{q+5}{3}}$

15. $\dfrac{\dfrac{2}{x}-3}{\dfrac{2-3x}{2}}$

16. $\dfrac{6+\dfrac{2}{r}}{\dfrac{3r+1}{4}}$

17. $\dfrac{\dfrac{1}{x}+x}{\dfrac{x^2+1}{8}}$

18. $\dfrac{\dfrac{3}{m}-m}{\dfrac{3-m^2}{4}}$

19. $\dfrac{a-\dfrac{5}{a}}{a+\dfrac{1}{a}}$

20. $\dfrac{q+\dfrac{1}{q}}{q+\dfrac{4}{q}}$

21. $\dfrac{\dfrac{5}{8}+\dfrac{2}{3}}{\dfrac{7}{3}-\dfrac{1}{4}}$

22. $\dfrac{\dfrac{6}{5}-\dfrac{1}{9}}{\dfrac{2}{5}+\dfrac{5}{3}}$

23. $\dfrac{\dfrac{1}{x^2}+\dfrac{1}{y^2}}{\dfrac{1}{x}-\dfrac{1}{y}}$

24. $\dfrac{\dfrac{1}{a^2}-\dfrac{1}{b^2}}{\dfrac{1}{a}-\dfrac{1}{b}}$

▶ 25. $\dfrac{\dfrac{2}{p^2}-\dfrac{3}{5p}}{\dfrac{4}{p}+\dfrac{1}{4p}}$

26. $\dfrac{\dfrac{2}{m^2}-\dfrac{3}{m}}{\dfrac{2}{5m^2}+\dfrac{1}{3m}}$

27. $\dfrac{\dfrac{5}{x^2y}-\dfrac{2}{xy^2}}{\dfrac{3}{x^2y^2}+\dfrac{4}{xy}}$

28. $\dfrac{\dfrac{1}{m^3p}+\dfrac{2}{mp^2}}{\dfrac{4}{mp}+\dfrac{1}{m^2p}}$

▶ 29. $\dfrac{\dfrac{1}{4}-\dfrac{1}{a^2}}{\dfrac{1}{2}+\dfrac{1}{a}}$

30. $\dfrac{\dfrac{1}{9}-\dfrac{1}{m^2}}{\dfrac{1}{3}+\dfrac{1}{m}}$

31. $\dfrac{\dfrac{1}{z+5}}{\dfrac{4}{z^2-25}}$

32. $\dfrac{\dfrac{1}{a+1}}{\dfrac{2}{a^2-1}}$

▶ 33. $\dfrac{\dfrac{1}{m+1}-1}{\dfrac{1}{m+1}+1}$

34. $\dfrac{\dfrac{2}{x-1}+2}{\dfrac{2}{x-1}-2}$

35. $\dfrac{\dfrac{12}{x+2}+2}{\dfrac{18}{x+2}-2}$

36. $\dfrac{\dfrac{6}{x+3}+3}{\dfrac{9}{x+3}-3}$

37. $\dfrac{\dfrac{x}{y}+\dfrac{y}{x}}{\dfrac{x}{y}-\dfrac{y}{x}}$

38. $\dfrac{\dfrac{x}{y} - \dfrac{y}{x}}{\dfrac{x}{y} + \dfrac{y}{x}}$

39. $\dfrac{1}{\dfrac{1}{a} + \dfrac{1}{b}}$

40. $\dfrac{-1}{\dfrac{1}{a} - \dfrac{1}{b}}$

41. $\dfrac{\dfrac{1}{m-1} + \dfrac{2}{m+2}}{\dfrac{2}{m+2} - \dfrac{1}{m-3}}$

42. $\dfrac{\dfrac{5}{r+3} - \dfrac{1}{r-1}}{\dfrac{2}{r+2} + \dfrac{3}{r+3}}$

43. $\dfrac{2 + \dfrac{1}{x} - \dfrac{28}{x^2}}{3 + \dfrac{13}{x} + \dfrac{4}{x^2}}$

44. $\dfrac{4 - \dfrac{11}{x} - \dfrac{3}{x^2}}{2 - \dfrac{1}{x} - \dfrac{15}{x^2}}$

45. $\dfrac{\dfrac{y+8}{y-4}}{\dfrac{y^2-64}{y^2-16}}$

46. $\dfrac{\dfrac{t+5}{t-8}}{\dfrac{t^2-25}{t^2-64}}$

47. $\dfrac{\dfrac{15a^2 + 15b^2}{5}}{\dfrac{a^4 - b^4}{10}}$

48. $\dfrac{\dfrac{14x^2 + 14y^2}{21}}{\dfrac{x^4 - y^4}{27}}$

49. $\dfrac{\dfrac{1}{x^3 - y^3}}{\dfrac{1}{x^2 - y^2}}$

50. *Concept Check*   What property of real numbers justifies Method 2 of simplifying complex fractions?

Simplify each expression, using only positive exponents in the answer. **See Example 7.**

51. $\dfrac{1}{x^{-2} + y^{-2}}$

52. $\dfrac{1}{p^{-2} - q^{-2}}$

▶ **53.** $\dfrac{x^{-2} + y^{-2}}{x^{-1} + y^{-1}}$

54. $\dfrac{x^{-1} - y^{-1}}{x^{-2} - y^{-2}}$

55. $\dfrac{k^{-1} + p^{-2}}{k^{-1} - 3p^{-2}}$

56. $\dfrac{x^{-2} - y^{-1}}{x^{-2} + 4y^{-1}}$

57. $\dfrac{x^{-1} + 2y^{-1}}{2y + 4x}$

58. $\dfrac{a^{-2} - 4b^{-2}}{3b - 6a}$

Extending Skills   The fractions in Exercises 59–64 are **continued fractions**. Simplify by starting at "the bottom" and working upward.

59. $1 + \dfrac{1}{1 + \dfrac{1}{1+1}}$

60. $5 + \dfrac{5}{5 + \dfrac{5}{5+5}}$

61. $7 - \dfrac{3}{5 + \dfrac{2}{4-2}}$

62. $3 - \dfrac{2}{4 + \dfrac{2}{4-2}}$

63. $r + \dfrac{r}{4 - \dfrac{2}{6+2}}$

64. $\dfrac{2q}{7} - \dfrac{q}{6 + \dfrac{8}{4+4}}$

RELATING CONCEPTS   For Individual or Group Work (Exercises 65–68)

To find the average of two numbers, we add them and divide by 2. Suppose that we wish to find the average of $\frac{3}{8}$ and $\frac{5}{6}$. **Work Exercises 65–68 in order,** to see how a complex fraction occurs in a problem like this.

65. Write in symbols: The sum of $\frac{3}{8}$ and $\frac{5}{6}$, divided by 2. The result should be a complex fraction.

66. Use Method 1 to simplify the complex fraction from **Exercise 65.**

67. Use Method 2 to simplify the complex fraction from **Exercise 65.**

68. The answers in **Exercises 66 and 67** should be the same. Which method did you prefer? Why?

6.6 Solving Equations with Rational Expressions

VOCABULARY

☐ proposed solution
☐ extraneous solution (extraneous value)

NOW TRY EXERCISE 1

Identify each of the following as an *expression* or an *equation*. Then simplify the expression or solve the equation.

(a) $\dfrac{3}{2}t - \dfrac{5}{7}t = \dfrac{11}{7}$

(b) $\dfrac{3}{2}t - \dfrac{5}{7}t$

OBJECTIVE 1 Distinguish between operations with rational expressions and equations with terms that are rational expressions.

Before solving equations with rational expressions, we emphasize the distinction between sums and differences of terms with rational coefficients, or rational *expressions*, and *equations* with terms that are rational expressions.

Sums and differences are expressions to simplify. Equations are solved.

EXAMPLE 1 Distinguishing between Expressions and Equations

Identify each of the following as an *expression* or an *equation*. Then simplify the expression or solve the equation.

(a) $\dfrac{3}{4}x - \dfrac{2}{3}x$ This is a difference of two terms. It represents an *expression* to simplify because there is no equality symbol.

$$= \dfrac{3 \cdot 3}{3 \cdot 4}x - \dfrac{4 \cdot 2}{4 \cdot 3}x$$ The LCD is 12. Write each coefficient with this LCD.

$$= \dfrac{9}{12}x - \dfrac{8}{12}x$$ Multiply.

$$= \dfrac{1}{12}x$$ Combine like terms, using the distributive property: $\dfrac{9}{12}x - \dfrac{8}{12}x = \left(\dfrac{9}{12} - \dfrac{8}{12}\right)x$.

(b) $\qquad\qquad \dfrac{3}{4}x - \dfrac{2}{3}x = \dfrac{1}{2}$ Because there is an equality symbol, this is an *equation* to be solved.

$$12\left(\dfrac{3}{4}x - \dfrac{2}{3}x\right) = 12\left(\dfrac{1}{2}\right)$$ Use the multiplication property of equality to clear fractions. Multiply by 12, the LCD.

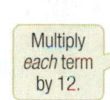
Multiply *each* term by 12.

$$12\left(\dfrac{3}{4}x\right) - 12\left(\dfrac{2}{3}x\right) = 12\left(\dfrac{1}{2}\right)$$ Distributive property

$$9x - 8x = 6$$ Multiply.

$$x = 6$$ Combine like terms.

CHECK $\qquad\qquad \dfrac{3}{4}x - \dfrac{2}{3}x = \dfrac{1}{2}$ Original equation

$$\dfrac{3}{4}(6) - \dfrac{2}{3}(6) \stackrel{?}{=} \dfrac{1}{2}$$ Let $x = 6$.

$$\dfrac{9}{2} - 4 \stackrel{?}{=} \dfrac{1}{2}$$ Multiply.

$$\dfrac{1}{2} = \dfrac{1}{2} \checkmark$$ True

Because a true statement results, {6} is the solution set of the equation.

NOW TRY ANSWERS
1. (a) equation; {2}
   (b) expression; $\dfrac{11}{14}t$

NOW TRY

The ideas of **Example 1** can be summarized as follows.

> ## Uses of the LCD
>
> When adding or subtracting rational expressions, keep the LCD throughout the simplification. (See **Example 1(a).**)
>
> When solving an equation with terms that are rational expressions, multiply each side by the LCD so that denominators are eliminated. (See **Example 1(b).**)

OBJECTIVE 2 Solve equations with rational expressions.

When an equation involves fractions, as in **Example 1(b)**, we use the multiplication property of equality to clear the fractions. When we choose the LCD of all denominators as the multiplier, the resulting equation contains no fractions.

NOW TRY EXERCISE 2

Solve, and check the solution.

$$\frac{x}{6} + \frac{x}{3} = 6 + x$$

EXAMPLE 2 Solving an Equation with Rational Expressions

Solve, and check the solution.

$$\frac{x}{3} + \frac{x}{4} = 10 + x$$

$$12\left(\frac{x}{3} + \frac{x}{4}\right) = 12(10 + x) \qquad \text{Multiply by the LCD, 12, to clear fractions.}$$

$$12\left(\frac{x}{3}\right) + 12\left(\frac{x}{4}\right) = 12(10) + 12x \qquad \text{Distributive property}$$

$$4x + 3x = 120 + 12x \qquad \text{Multiply.}$$

$$7x = 120 + 12x \qquad \text{Combine like terms.}$$

$$-5x = 120 \qquad \text{Subtract } 12x.$$

$$x = -24 \qquad \text{Divide by } -5.$$

CHECK

$$\frac{x}{3} + \frac{x}{4} = 10 + x \qquad \text{Original equation}$$

$$\frac{-24}{3} + \frac{-24}{4} \stackrel{?}{=} 10 - 24 \qquad \text{Let } x = -24.$$

$$-8 + (-6) \stackrel{?}{=} -14 \qquad \text{Divide. Subtract.}$$

$$-14 = -14 \ \checkmark \qquad \text{True}$$

The solution set is $\{-24\}$.   **NOW TRY**

⚠️ **CAUTION**   *Be careful not to confuse the following procedures.*

- In **Examples 2 and 3,** we use the multiplication property of equality to multiply each side of an *equation* by the LCD.

- In **Section 6.5,** we used the fundamental property to multiply a *fraction* (an expression) by another fraction that had the LCD as both its numerator and denominator.

**NOW TRY
EXERCISE 3**

Solve, and check the solution.

$$\frac{x}{7} - \frac{x+5}{5} = -\frac{3}{7}$$

EXAMPLE 3 Solving an Equation with Rational Expressions

Solve, and check the solution.

$$\frac{p}{2} - \frac{p-1}{3} = 1$$

$$6\left(\frac{p}{2} - \frac{p-1}{3}\right) = 6(1) \qquad \text{Multiply each side by the LCD, 6.}$$

$$6\left(\frac{p}{2}\right) - 6\left(\frac{p-1}{3}\right) = 6(1) \qquad \text{Distributive property}$$

$$3p - 2(p-1) = 6 \qquad \text{Use parentheses around } p - 1 \text{ to avoid errors.}$$

$$3p - 2(p) - 2(-1) = 6 \qquad \text{Distributive property}$$

Be careful with signs.

$$3p - 2p + 2 = 6 \qquad \text{Multiply.}$$

$$p + 2 = 6 \qquad \text{Combine like terms.}$$

$$p = 4 \qquad \text{Subtract 2.}$$

CHECK

$$\frac{p}{2} - \frac{p-1}{3} = 1 \qquad \text{Original equation}$$

$$\frac{4}{2} - \frac{4-1}{3} \overset{?}{=} 1 \qquad \text{Let } p = 4.$$

$$2 - 1 \overset{?}{=} 1 \qquad \text{Simplify.}$$

$$1 = 1 \ \checkmark \qquad \text{True}$$

The solution set is $\{4\}$.

NOW TRY

Recall that division by 0 is undefined. *When solving an equation with rational expressions that have variables in the denominator, the solution cannot be a number that makes the denominator equal 0.*

A value of the variable that appears to be a solution after both sides of a rational equation are multiplied by a variable expression is a **proposed solution.** *All proposed solutions must be checked in the original equation.*

EXAMPLE 4 Solving an Equation with Rational Expressions

Solve, and check the proposed solution.

$$\frac{x}{x-2} = \frac{2}{x-2} + 2 \qquad \begin{array}{l} x \text{ cannot equal 2, because 2} \\ \text{causes both denominators} \\ \text{to equal 0.} \end{array}$$

$$(x-2)\left(\frac{x}{x-2}\right) = (x-2)\left(\frac{2}{x-2} + 2\right) \qquad \begin{array}{l} \text{Multiply each side by the LCD,} \\ x - 2. \end{array}$$

$$(x-2)\left(\frac{x}{x-2}\right) = (x-2)\left(\frac{2}{x-2}\right) + (x-2)(2) \qquad \text{Distributive property}$$

$$x = 2 + 2x - 4 \qquad \text{Simplify.}$$

$$x = -2 + 2x \qquad \text{Combine like terms.}$$

$$-x = -2 \qquad \text{Subtract } 2x.$$

Proposed solution $\rightarrow x = 2 \qquad \text{Multiply by } -1.$

NOW TRY ANSWER
3. $\{-10\}$

 NOW TRY
EXERCISE 4

Solve, and check the proposed solution.

$$4 + \frac{6}{x - 3} = \frac{2x}{x - 3}$$

As noted, x cannot equal 2 because replacing x with 2 in the original equation causes the denominators to equal 0. We see this in the following check.

CHECK

$$\frac{x}{x - 2} = \frac{2}{x - 2} + 2 \qquad \text{Original equation}$$

$$\frac{2}{2 - 2} \stackrel{?}{=} \frac{2}{2 - 2} + 2 \qquad \text{Let } x = 2.$$

Division by 0 is undefined.

$$\frac{2}{0} \stackrel{?}{=} \frac{2}{0} + 2 \qquad \text{Subtract in the denominators.}$$

Thus, the proposed solution 2 must be rejected. The solution set is $\varnothing$.　**NOW TRY**

A proposed solution that is not an actual solution of the original equation, such as 2 in **Example 4,** is an **extraneous solution,** or **extraneous value.** Some students like to determine which numbers cannot be solutions *before* solving the equation, as we did in **Example 4.**

Solving an Equation with Rational Expressions

Step 1　**Multiply each side of the equation by the LCD** to clear the equation of fractions. Be sure to distribute to *every* term on *both* sides.

Step 2　**Solve** the resulting equation for proposed solutions.

Step 3　**Check** each proposed solution by substituting it into the original equation. Reject any that cause a denominator to equal 0.

EXAMPLE 5　Solving an Equation with Rational Expressions

Solve, and check the proposed solution.

$$\frac{2}{x^2 - x} = \frac{1}{x^2 - 1}$$

Step 1　Factor the denominators to find the LCD.

$$\frac{2}{x(x - 1)} = \frac{1}{(x + 1)(x - 1)}$$

The LCD is $x(x + 1)(x - 1)$. Notice that 0, 1, and -1 cannot be solutions. Otherwise a denominator will equal 0. Multiply both sides of the equation by the LCD to clear the fractions.

$$x(x + 1)(x - 1)\frac{2}{x(x - 1)} = x(x + 1)(x - 1)\frac{1}{(x + 1)(x - 1)} \qquad \text{Multiply by the LCD.}$$

Step 2　$\qquad 2(x + 1) = x \qquad$ Divide out the common factors.

$$2x + 2 = x \qquad \text{Distributive property}$$

$$x + 2 = 0 \qquad \text{Subtract } x.$$

$$x = -2 \qquad \text{Subtract 2.}$$

NOW TRY ANSWER
4. $\varnothing$

Step 3　The proposed solution is -2, which does not make any denominator equal 0.

NOW TRY EXERCISE 5

Solve, and check the proposed solution.

$$\frac{3}{2x^2 - 8x} = \frac{1}{x^2 - 16}$$

CHECK

$$\frac{2}{x^2 - x} = \frac{1}{x^2 - 1} \qquad \text{Original equation}$$

$$\frac{2}{(-2)^2 - (-2)} \stackrel{?}{=} \frac{1}{(-2)^2 - 1} \qquad \text{Let } x = -2.$$

$$\frac{2}{4 + 2} \stackrel{?}{=} \frac{1}{4 - 1} \qquad \begin{array}{l}\text{Apply the exponents;}\\ \text{definition of subtraction}\end{array}$$

$$\frac{1}{3} = \frac{1}{3} \checkmark \qquad \text{True}$$

The solution set is $\{-2\}$.

NOW TRY

NOW TRY EXERCISE 6

Solve, and check the proposed solution.

$$\frac{2y}{y^2 - 25} = \frac{8}{y + 5} - \frac{1}{y - 5}$$

EXAMPLE 6 Solving an Equation with Rational Expressions

Solve, and check the proposed solution.

$$\frac{2m}{m^2 - 4} + \frac{1}{m - 2} = \frac{2}{m + 2}$$

$$\frac{2m}{(m + 2)(m - 2)} + \frac{1}{m - 2} = \frac{2}{m + 2} \qquad \begin{array}{l}\text{Factor the first denominator}\\ \text{on the left to find the LCD,}\\ (m + 2)(m - 2).\end{array}$$

Notice that -2 and 2 cannot be solutions of this equation.

$$(m + 2)(m - 2)\left(\frac{2m}{(m + 2)(m - 2)} + \frac{1}{m - 2}\right)$$

$$\qquad\qquad\qquad\qquad\qquad\qquad \text{Multiply by the LCD.}$$

$$= (m + 2)(m - 2)\frac{2}{m + 2}$$

$$(m + 2)(m - 2)\frac{2m}{(m + 2)(m - 2)} + (m + 2)(m - 2)\frac{1}{m - 2}$$

$$= (m + 2)(m - 2)\frac{2}{m + 2} \qquad \text{Distributive property}$$

$$2m + m + 2 = 2(m - 2) \qquad \text{Divide out the common factors.}$$

$$3m + 2 = 2m - 4 \qquad \text{Combine like terms; distributive property}$$

$$m + 2 = -4 \qquad \text{Subtract } 2m.$$

$$m = -6 \qquad \text{Subtract 2.}$$

CHECK

$$\frac{2m}{m^2 - 4} + \frac{1}{m - 2} = \frac{2}{m + 2} \qquad \text{Original equation}$$

$$\frac{2(-6)}{(-6)^2 - 4} + \frac{1}{-6 - 2} \stackrel{?}{=} \frac{2}{-6 + 2} \qquad \text{Let } m = -6.$$

$$\frac{-12}{32} + \frac{1}{-8} \stackrel{?}{=} \frac{2}{-4} \qquad \begin{array}{l}\text{Apply the exponent.}\\ \text{Subtract and add.}\end{array}$$

$$-\frac{1}{2} = -\frac{1}{2} \checkmark \qquad \text{True}$$

The solution set is $\{-6\}$.

NOW TRY

**NOW TRY
EXERCISE 7**

Solve, and check the proposed solution(s).

$$\frac{3}{m^2 - 9} = \frac{1}{2(m - 3)} - \frac{1}{4}$$

EXAMPLE 7 Solving an Equation with Rational Expressions

Solve, and check the proposed solution(s).

$$\frac{1}{x - 1} + \frac{1}{2} = \frac{2}{x^2 - 1}$$

> $x \neq 1, -1.$ Otherwise, a denominator is 0.

$$\frac{1}{x - 1} + \frac{1}{2} = \frac{2}{(x + 1)(x - 1)}$$

Factor the denominator on the right. The LCD is $2(x + 1)(x - 1)$.

$$2(x + 1)(x - 1)\left(\frac{1}{x - 1} + \frac{1}{2}\right) = 2(x + 1)(x - 1)\frac{2}{(x + 1)(x - 1)}$$

Multiply by the LCD.

$$2(x + 1)(x - 1)\frac{1}{x - 1} + 2(x + 1)(x - 1)\frac{1}{2} = 2(x + 1)(x - 1)\frac{2}{(x + 1)(x - 1)}$$

Distributive property

$$2(x + 1) + (x + 1)(x - 1) = 2(2)$$ Divide out the common factors.

$$2x + 2 + x^2 - 1 = 4$$ Multiply.

> Write in standard form.

$$x^2 + 2x - 3 = 0$$ Subtract 4. Combine like terms.

$$(x + 3)(x - 1) = 0$$ Factor.

$$x + 3 = 0 \quad \text{or} \quad x - 1 = 0$$ Zero-factor property

$$x = -3 \quad \text{or} \quad x = 1 \leftarrow \text{Proposed solutions}$$

Because 1 makes a denominator equal 0, 1 is an extraneous value. Check that -3 is a solution.

CHECK

$$\frac{1}{x - 1} + \frac{1}{2} = \frac{2}{x^2 - 1}$$ Original equation

$$\frac{1}{-3 - 1} + \frac{1}{2} \stackrel{?}{=} \frac{2}{(-3)^2 - 1}$$ Let $x = -3$.

$$\frac{1}{-4} + \frac{1}{2} \stackrel{?}{=} \frac{2}{9 - 1}$$ Subtract. Apply the exponent.

$$\frac{1}{4} = \frac{1}{4} \quad \checkmark$$ True

The solution set is $\{-3\}$.

NOW TRY

EXAMPLE 8 Solving an Equation with Rational Expressions

Solve, and check the proposed solution.

$$\frac{1}{k^2 + 4k + 3} + \frac{1}{2k + 2} = \frac{3}{4k + 12}$$

$$\frac{1}{(k + 1)(k + 3)} + \frac{1}{2(k + 1)} = \frac{3}{4(k + 3)}$$ Factor each denominator. The LCD is $4(k + 1)(k + 3)$.

> $k \neq -1, -3$

$$4(k + 1)(k + 3)\left(\frac{1}{(k + 1)(k + 3)} + \frac{1}{2(k + 1)}\right)$$

NOW TRY ANSWER
7. $\{-1\}$

$$= 4(k + 1)(k + 3)\frac{3}{4(k + 3)}$$ Multiply by the LCD.

**NOW TRY
EXERCISE 8**

Solve, and check the proposed solution.

$$\frac{5}{k^2 + k - 2} = \frac{1}{3k - 3} - \frac{1}{k + 2}$$

$$4(k+1)(k+3)\frac{1}{(k+1)(k+3)} + 2 \cdot 2(k+1)(k+3)\frac{1}{2(k+1)}$$

$$= 4(k+1)(k+3)\frac{3}{4(k+3)} \qquad \text{Distributive property}$$

> Do *not* add
> $4 + 2$ here.

$$4 + 2(k+3) = 3(k+1) \qquad \text{Divide out the common factors.}$$

$$4 + 2k + 6 = 3k + 3 \qquad \text{Distributive property}$$

$$2k + 10 = 3k + 3 \qquad \text{Combine like terms.}$$

$$10 = k + 3 \qquad \text{Subtract } 2k.$$

$$7 = k \qquad \text{Subtract 3.}$$

The proposed solution, 7, does not make an original denominator equal 0. A check shows that the algebra is correct. (See **Exercise 82.**) The solution set is $\{7\}$.

NOW TRY

OBJECTIVE 3 Solve a formula for a specified variable.

When solving a formula for a specified variable, *remember to treat the variable for which you are solving as if it were the only variable, and all others as if they were constants.*

EXAMPLE 9 Solving for a Specified Variable

Solve each formula for the specified variable.

(a) $a = \dfrac{v - w}{t}$ for v

$$a = \frac{v - w}{t} \qquad \boxed{\text{Our goal is to isolate } v.}$$

$$at = \left(\frac{v - w}{t}\right)t \qquad \text{Multiply by } t \text{ to clear the fraction.}$$

$$at = v - w \qquad \text{Divide out the common factor.}$$

$$at + w = v, \quad \text{or} \quad v = at + w \qquad \text{Add } w. \text{ Rewrite.}$$

(b) $F = \dfrac{k}{d - D}$ for d

$$F = \frac{k}{d - D} \qquad \boxed{\text{We must isolate } d.} \qquad \text{Given equation}$$

$$F(d - D) = \frac{k}{d - D}(d - D) \qquad \text{Multiply by } d - D \text{ to clear the fraction.}$$

$$F(d - D) = k \qquad \text{Divide out the common factor.}$$

$$Fd - FD = k \qquad \text{Distributive property}$$

$$Fd = k + FD \qquad \text{Add } FD.$$

$$d = \frac{k + FD}{F} \qquad \text{Divide by } F.$$

NOW TRY ANSWER
8. $\{-5\}$

**NOW TRY
EXERCISE 9**

Solve each formula for the specified variable.

(a) $p = \dfrac{x - y}{z}$ for x

(b) $a = \dfrac{b}{c + d}$ for d

We can write an equivalent form of this answer as follows.

$$d = \frac{k + FD}{F} \qquad \text{Answer on the preceding page}$$

$$d = \frac{k}{F} + \frac{FD}{F} \qquad \text{Definition of addition of fractions, } \frac{a + b}{c} = \frac{a}{c} + \frac{b}{c}$$

$$d = \frac{k}{F} + D \qquad \text{Divide out the common factor from } \frac{FD}{F}.$$

Either answer is correct.   NOW TRY

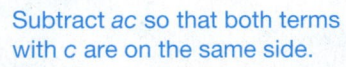 **EXAMPLE 10**   Solving for a Specified Variable

**NOW TRY
EXERCISE 10**

Solve the following formula for x.

$$\frac{2}{w} = \frac{1}{x} - \frac{3}{y}$$

Solve the following formula for c.

$$\frac{1}{a} = \frac{1}{b} + \frac{1}{c} \qquad \boxed{\text{Goal: Isolate } c, \text{ the specified variable.}}$$

$$abc\left(\frac{1}{a}\right) = abc\left(\frac{1}{b} + \frac{1}{c}\right) \qquad \text{Multiply by the LCD, } abc, \text{ to clear the fractions.}$$

$$abc\left(\frac{1}{a}\right) = abc\left(\frac{1}{b}\right) + abc\left(\frac{1}{c}\right) \qquad \text{Distributive property}$$

$$bc = ac + ab \qquad \text{Divide out the common factors.}$$

$$bc - ac = ab \qquad \begin{array}{l}\text{Subtract } ac \text{ so that both terms} \\ \text{with } c \text{ are on the same side.}\end{array}$$

$\boxed{\text{Pay careful attention here.}}$ $\quad c(b - a) = ab \qquad \text{Factor out } c.$

$$c = \frac{ab}{b - a} \qquad \text{Divide by } b - a. \qquad \text{NOW TRY}$$

NOW TRY ANSWERS

9. (a) $x = pz + y$

(b) $d = \dfrac{b - ac}{a}$

10. $x = \dfrac{wy}{2y + 3w}$

⚠ **CAUTION**   Students often have trouble in the step that involves factoring out the variable for which they are solving. In **Example 10**, we needed to transform so that both terms with c are on the same side of the equation. This allowed us to factor out c on the left, and then isolate it by dividing each side by $b - a$.

When solving an equation for a specified variable, be sure that the specified variable appears alone on only one side of the equality symbol in the final equation.

6.6 Exercises

FOR EXTRA HELP

▶ MyMathLab®

▶ *Complete solution available in MyMathLab*

Concept Check   *Provide a short answer to each of the following.*

1. What is the least positive whole number by which we can multiply both sides of the equation

$$\frac{2}{3}x + \frac{1}{4}x = 6$$

to obtain an equation with only integer coefficients?

2. Before even beginning to solve the equation

$$\frac{1}{x - 3} + \frac{2}{3 - x} = 4,$$

what number do we know cannot be a solution? Why?

3. What is the simplest monomial by which we can multiply both sides of the equation

$$\frac{1}{x} - \frac{1}{y} = \frac{1}{z}$$

so that there are no variables in the denominators?

4. If we are solving an equation for the variable k, and our steps lead to the equation

$$kr - mr = km,$$

what would be the next step?

5. Suppose an equation includes the rational expression $\frac{1}{3-x}$. Is it acceptable to replace this expression by $\frac{-1}{x-3}$? Why or why not?

6. To combine the terms in $\frac{2}{3x} + \frac{7}{5y}$, is it acceptable to simply multiply through by $15xy$? Why or why not?

Identify each of the following as an expression or an equation. Then simplify each expression or solve each equation. See Examples 1 and 2.

▶ 7. $\frac{7}{8}x + \frac{1}{5}x$

8. $\frac{4}{7}x + \frac{4}{5}x$

9. $\frac{7x}{8} + \frac{x}{5} = 1$

10. $\frac{4x}{7} + \frac{4x}{5} = 1$

11. $\frac{3}{5}x - \frac{7}{10}x$

12. $\frac{2}{3}x - \frac{9}{4}x$

13. $\frac{3}{5}x - \frac{7}{10}x = 1$

14. $\frac{2}{3}x - \frac{9}{4}x = -19$

15. $\frac{3}{4}x - \frac{1}{2}x = 0$

16. *Concept Check*   Why is the equation in **Exercise 15** easy to check?

When solving an equation with variables in denominators, we must determine the values that cause these denominators to equal 0, so that we can reject these extraneous values if they appear as potential solutions. Find all values for which at least one denominator is equal to 0. Write answers using the symbol $\neq$. Do not solve. See Examples 4–8.

17. $\frac{3}{x+2} - \frac{5}{x} = 1$

18. $\frac{7}{x} + \frac{9}{x-4} = 5$

19. $\frac{-1}{(x+3)(x-4)} = \frac{1}{2x+1}$

20. $\frac{8}{(x-7)(x+3)} = \frac{7}{3x-10}$

21. $\frac{4}{x^2+8x-9} + \frac{1}{x^2-4} = 0$

22. $\frac{-3}{x^2+9x-10} - \frac{12}{x^2-49} = 0$

Solve each equation, and check the solutions. See Examples 1–4.

23. $\frac{5}{m} - \frac{3}{m} = 8$

24. $\frac{4}{y} + \frac{1}{y} = 2$

25. $\frac{5}{y} + 4 = \frac{2}{y}$

26. $\frac{11}{q} - 3 = \frac{1}{q}$

27. $\frac{3x}{5} - 6 = x$

28. $\frac{5t}{4} + t = 9$

29. $\frac{4m}{7} + m = 11$

30. $x - \frac{3x}{2} = 1$

31. $\frac{z-1}{4} = \frac{z+3}{3}$

32. $\frac{r-5}{2} = \frac{r+2}{3}$

33. $\frac{3p+6}{8} = \frac{3p-3}{16}$

34. $\frac{2z+1}{5} = \frac{7z+5}{15}$

35. $\frac{2x+3}{x} = \frac{3}{2}$

36. $\frac{7-2x}{x} = \frac{-17}{5}$

37. $\dfrac{k}{k-4} - 5 = \dfrac{4}{k-4}$

38. $\dfrac{-5}{a+5} - 2 = \dfrac{a}{a+5}$

39. $\dfrac{q+2}{3} + \dfrac{q-5}{5} = \dfrac{7}{3}$

40. $\dfrac{x-6}{6} + \dfrac{x+2}{8} = \dfrac{11}{4}$

41. $\dfrac{x}{2} = \dfrac{5}{4} + \dfrac{x-1}{4}$

42. $\dfrac{8p}{5} = \dfrac{3p-4}{2} + \dfrac{5}{2}$

43. $x + \dfrac{17}{2} = \dfrac{x}{2} + x + 6$

44. $t + \dfrac{8}{3} = \dfrac{t}{3} + t + \dfrac{14}{3}$

45. $\dfrac{9}{3x+4} = \dfrac{36-27x}{16-9x^2}$

46. $\dfrac{25}{5x-6} = \dfrac{-150-125x}{36-25x^2}$

Solve each equation, and check the solutions. **Be careful with signs. See Example 3.**

47. $\dfrac{a+7}{8} - \dfrac{a-2}{3} = \dfrac{4}{3}$

48. $\dfrac{x+3}{7} - \dfrac{x+2}{6} = \dfrac{1}{6}$

49. $\dfrac{p}{2} - \dfrac{p-1}{4} = \dfrac{5}{4}$

50. $\dfrac{r}{6} - \dfrac{r-2}{3} = -\dfrac{4}{3}$

51. $\dfrac{3x}{5} - \dfrac{x-5}{7} = 3$

52. $\dfrac{8k}{5} - \dfrac{3k-4}{2} = \dfrac{5}{2}$

Solve each equation, and check the solutions. **See Examples 4–8.**

53. $\dfrac{4}{x^2-3x} = \dfrac{1}{x^2-9}$

54. $\dfrac{2}{t^2-4} = \dfrac{3}{t^2-2t}$

55. $\dfrac{2}{m} = \dfrac{m}{5m+12}$

56. $\dfrac{x}{4-x} = \dfrac{2}{x}$

57. $\dfrac{-2}{z+5} + \dfrac{3}{z-5} = \dfrac{20}{z^2-25}$

58. $\dfrac{3}{r+3} - \dfrac{2}{r-3} = \dfrac{-12}{r^2-9}$

59. $\dfrac{3}{x-1} + \dfrac{2}{4x-4} = \dfrac{7}{4}$

60. $\dfrac{2}{p+3} + \dfrac{3}{8} = \dfrac{5}{4p+12}$

61. $\dfrac{x}{3x+3} = \dfrac{2x-3}{x+1} - \dfrac{2x}{3x+3}$

62. $\dfrac{2k+3}{k+1} - \dfrac{3k}{2k+2} = \dfrac{-2k}{2k+2}$

63. $\dfrac{2p}{p^2-1} = \dfrac{2}{p+1} - \dfrac{1}{p-1}$

64. $\dfrac{2x}{x^2-16} - \dfrac{2}{x-4} = \dfrac{4}{x+4}$

65. $\dfrac{5x}{14x+3} = \dfrac{1}{x}$

66. $\dfrac{m}{8m+3} = \dfrac{1}{3m}$

67. $\dfrac{2}{x-1} - \dfrac{2}{3} = \dfrac{-1}{x+1}$

68. $\dfrac{5}{p-2} = 7 - \dfrac{10}{p+2}$

69. $\dfrac{x}{2x+2} = \dfrac{-2x}{4x+4} + \dfrac{2x-3}{x+1}$

70. $\dfrac{5t+1}{3t+3} = \dfrac{5t-5}{5t+5} + \dfrac{3t-1}{t+1}$

71. $\dfrac{8x+3}{x} = 3x$

72. $\dfrac{10x-24}{x} = x$

73. $\dfrac{1}{x+4} + \dfrac{x}{x-4} = \dfrac{-8}{x^2-16}$

74. $\dfrac{x}{x-3} + \dfrac{4}{x+3} = \dfrac{18}{x^2-9}$

75. $\dfrac{4}{3x+6} - \dfrac{3}{x+3} = \dfrac{8}{x^2+5x+6}$

76. $\dfrac{-13}{t^2+6t+8} + \dfrac{4}{t+2} = \dfrac{3}{2t+8}$

77. $\dfrac{3x}{x^2 + 5x + 6} = \dfrac{5x}{x^2 + 2x - 3} - \dfrac{2}{x^2 + x - 2}$

78. $\dfrac{m}{m^2 + m - 2} + \dfrac{m}{m^2 - 1} = \dfrac{m}{m^2 + 3m + 2}$

79. $\dfrac{x + 4}{x^2 - 3x + 2} - \dfrac{5}{x^2 - 4x + 3} = \dfrac{x - 4}{x^2 - 5x + 6}$

80. $\dfrac{3}{r^2 + r - 2} - \dfrac{1}{r^2 - 1} = \dfrac{7}{2(r^2 + 3r + 2)}$

81. $\dfrac{1}{x^2 - 1} = \dfrac{2}{x - 1} - \dfrac{1}{x - 1}$

82. Refer to **Example 8,** and show that 7 is a solution.

Solve each formula for the specified variable. See Examples 9 and 10.

83. $m = \dfrac{kF}{a}$ for F

84. $I = \dfrac{kE}{R}$ for E

85. $m = \dfrac{kF}{a}$ for a

86. $I = \dfrac{kE}{R}$ for R

87. $I = \dfrac{E}{R + r}$ for R

88. $I = \dfrac{E}{R + r}$ for r

89. $h = \dfrac{2\mathcal{A}}{B + b}$ for $\mathcal{A}$

90. $d = \dfrac{2S}{n(a + L)}$ for S

91. $d = \dfrac{2S}{n(a + L)}$ for a

92. $h = \dfrac{2\mathcal{A}}{B + b}$ for B

93. $\dfrac{1}{x} = \dfrac{1}{y} - \dfrac{1}{z}$ for y

94. $\dfrac{3}{k} = \dfrac{1}{p} + \dfrac{1}{q}$ for q

95. $\dfrac{2}{r} + \dfrac{3}{s} + \dfrac{1}{t} = 1$ for t

96. $\dfrac{5}{p} + \dfrac{2}{q} + \dfrac{3}{r} = 1$ for r

97. $9x + \dfrac{3}{z} = \dfrac{5}{y}$ for z

98. $-3t - \dfrac{4}{p} = \dfrac{6}{s}$ for p

99. $\dfrac{t}{x - 1} - \dfrac{2}{x + 1} = \dfrac{1}{x^2 - 1}$ for t

100. $\dfrac{5}{y + 2} - \dfrac{r}{y - 2} = \dfrac{3}{y^2 - 4}$ for r

RELATING CONCEPTS   For Individual or Group Work   (Exercises 101–108)

In these exercises, we summarize various concepts involving rational expressions. **Work Exercises 101–108 in order.**

Let P, Q, and R be rational expressions defined as follows.

$$P = \dfrac{6}{x + 3}, \qquad Q = \dfrac{5}{x + 1}, \qquad R = \dfrac{4x}{x^2 + 4x + 3}$$

101. Find the values for which each expression is undefined.

    (a) P **(b)** Q **(c)** R

102. Find and express $(P \cdot Q) \div R$ in lowest terms.

103. Why is $(P \cdot Q) \div R$ not defined if $x = 0$?

104. Find the LCD for P, Q, and R.

105. Perform the operations and express $P + Q - R$ in lowest terms.

106. Simplify the complex fraction $\dfrac{P + Q}{R}$.

107. Solve the equation $P + Q = R$.

108. How does the answer to **Exercise 101** help when working **Exercise 107**?

Students often confuse *simplifying expressions* with *solving equations*. We review the four operations applied to the rational expressions $\frac{1}{x}$ and $\frac{1}{x-2}$ as follows.

Add: $\dfrac{1}{x}+\dfrac{1}{x-2}$

$=\dfrac{1(x-2)}{x(x-2)}+\dfrac{x(1)}{x(x-2)}$　Write with a common denominator.

$=\dfrac{x-2+x}{x(x-2)}$　Add numerators. Keep the same denominator.

$=\dfrac{2x-2}{x(x-2)}$　Combine like terms.

Subtract: $\dfrac{1}{x}-\dfrac{1}{x-2}$

$=\dfrac{1(x-2)}{x(x-2)}-\dfrac{x(1)}{x(x-2)}$　Write with a common denominator.

$=\dfrac{x-2-x}{x(x-2)}$　Subtract numerators. Keep the same denominator.

$=\dfrac{-2}{x(x-2)}$　Combine like terms.

Multiply: $\dfrac{1}{x}\cdot\dfrac{1}{x-2}$

$=\dfrac{1}{x(x-2)}$　Multiply numerators. Multiply denominators.

Divide: $\dfrac{1}{x}\div\dfrac{1}{x-2}$

$=\dfrac{1}{x}\cdot\dfrac{x-2}{1}$　Multiply by the reciprocal of the divisor.

$=\dfrac{x-2}{x}$　Multiply numerators. Multiply denominators.

By contrast, consider the following *equation*.

$$\dfrac{1}{x}+\dfrac{1}{x-2}=\dfrac{3}{4}$$　$x\neq 0,2$ because a denominator is 0 for these values.

$$4x(x-2)\left(\dfrac{1}{x}+\dfrac{1}{x-2}\right)=4x(x-2)\dfrac{3}{4}$$　Multiply each side by the LCD, $4x(x-2)$, to clear fractions.

$$4x(x-2)\dfrac{1}{x}+4x(x-2)\dfrac{1}{x-2}=4x(x-2)\dfrac{3}{4}$$　Distributive property

$$4(x-2)+4x=3x(x-2)$$　Divide out the common factors.

$$4x-8+4x=3x^2-6x$$　Distributive property

$$3x^2 - 14x + 8 = 0 \qquad \text{Standard form}$$

$$(3x - 2)(x - 4) = 0 \qquad \text{Factor.}$$

$$3x - 2 = 0 \quad \text{or} \quad x - 4 = 0 \qquad \text{Zero-factor property}$$

Proposed solutions $\rightarrow$ $x = \dfrac{2}{3}$ or $x = 4$ $\qquad$ Solve for x.

Neither $\dfrac{2}{3}$ nor 4 makes a denominator equal 0. Check to confirm that the solution set is $\left\{\dfrac{2}{3}, 4\right\}$.

Points to Remember

1. When simplifying rational expressions, the fundamental property is applied only after numerators and denominators have been *factored*.

2. When adding and subtracting rational expressions, the common denominator must be kept throughout the problem and in the final result.

3. When simplifying rational expressions, always check to see if the answer is in lowest terms. If it is not, use the fundamental property.

4. When solving equations with rational expressions, the LCD is used to clear the equation of fractions. Multiply each side by the LCD. (Notice how this use differs from that of the LCD in Point 2.)

5. When solving equations with rational expressions, reject any proposed solution that causes an original denominator to equal 0.

For each exercise, indicate "expression" if an expression is to be simplified or "equation" if an equation is to be solved. Then simplify the expression or solve the equation.

1. expression; $\dfrac{10}{p}$

2. expression; $\dfrac{y^3}{x^3}$

3. expression; $\dfrac{1}{2x^2(x + 2)}$

4. equation; $\{9\}$

5. equation; $\{39\}$

6. expression; $\dfrac{5k + 8}{k(k - 4)(k + 4)}$

7. expression; $\dfrac{y + 2}{y - 1}$

8. expression; $\dfrac{t - 5}{3(2t + 1)}$

9. expression; $\dfrac{13}{3(p + 2)}$

10. equation; $\left\{-1, \dfrac{12}{5}\right\}$

11. equation; $\left\{\dfrac{1}{7}, 2\right\}$

12. expression; $\dfrac{16}{3k}$

13. expression; $\dfrac{7}{12z}$

14. equation; $\{13\}$

15. expression; $\dfrac{3m + 5}{(m + 3)(m + 2)(m + 1)}$

16. expression; $\dfrac{k + 3}{5(k - 1)}$

17. equation; $\varnothing$

18. equation; $\varnothing$

19. expression; $\dfrac{t + 2}{2(2t + 1)}$

20. equation; $\{-7\}$

1. $\dfrac{4}{p} + \dfrac{6}{p}$

2. $\dfrac{x^3 y^2}{x^2 y^4} \cdot \dfrac{y^5}{x^4}$

3. $\dfrac{1}{x^2 + x - 2} \div \dfrac{4x^2}{2x - 2}$

4. $\dfrac{8}{t - 5} = 2$

5. $\dfrac{x - 4}{5} = \dfrac{x + 3}{6}$

6. $\dfrac{2}{k^2 - 4k} + \dfrac{3}{k^2 - 16}$

7. $\dfrac{2y^2 + y - 6}{2y^2 - 9y + 9} \cdot \dfrac{y^2 - 2y - 3}{y^2 - 1}$

8. $\dfrac{3t^2 - t}{6t^2 + 15t} \div \dfrac{6t^2 + t - 1}{2t^2 - 5t - 25}$

9. $\dfrac{4}{p + 2} + \dfrac{1}{3p + 6}$

10. $\dfrac{1}{x} + \dfrac{1}{x - 3} = -\dfrac{5}{4}$

11. $\dfrac{3}{t - 1} + \dfrac{1}{t} = \dfrac{7}{2}$

12. $\dfrac{6}{k} - \dfrac{2}{3k}$

13. $\dfrac{5}{4z} - \dfrac{2}{3z}$

14. $\dfrac{x + 2}{3} = \dfrac{2x - 1}{5}$

15. $\dfrac{1}{m^2 + 5m + 6} + \dfrac{2}{m^2 + 4m + 3}$

16. $\dfrac{2k^2 - 3k}{20k^2 - 5k} \div \dfrac{2k^2 - 5k + 3}{4k^2 + 11k - 3}$

17. $\dfrac{2}{x + 1} + \dfrac{5}{x - 1} = \dfrac{10}{x^2 - 1}$

18. $\dfrac{3}{x + 3} + \dfrac{4}{x + 6} = \dfrac{9}{x^2 + 9x + 18}$

19. $\dfrac{4t^2 - t}{6t^2 + 10t} \div \dfrac{8t^2 + 2t - 1}{3t^2 + 11t + 10}$

20. $\dfrac{x}{x - 2} + \dfrac{3}{x + 2} = \dfrac{8}{x^2 - 4}$

6.7 Applications of Rational Expressions

OBJECTIVES

1 Solve problems about numbers.

2 Solve problems about distance, rate, and time.

3 Solve problems about work.

NOW TRY EXERCISE 1

In a certain fraction, the numerator is 4 less than the denominator. If 7 is added to both the numerator and denominator, the resulting fraction is equivalent to $\frac{7}{8}$. What is the original fraction?

For applications that lead to rational equations, the six-step problem-solving method of **Section 2.4** still applies.

 **Solve problems about numbers.**

EXAMPLE 1 Solving a Problem about an Unknown Number

If the same number is added to both the numerator and the denominator of the fraction $\frac{2}{5}$, the result is equivalent to $\frac{2}{3}$. Find the number.

Step 1 **Read** the problem carefully. We are trying to find a number.

Step 2 **Assign a variable.**

Let x = the number added to the numerator and the denominator.

Step 3 **Write an equation.** The fraction

$$\frac{2 + x}{5 + x}$$

represents the result of adding the same number to both the numerator and the denominator. This result is equivalent to $\frac{2}{3}$, so the equation is written as follows.

$$\frac{2 + x}{5 + x} = \frac{2}{3}$$

Step 4 **Solve.** $3(5 + x)\dfrac{2 + x}{5 + x} = 3(5 + x)\dfrac{2}{3}$ Multiply by the LCD, $3(5 + x)$.

$3(2 + x) = 2(5 + x)$ Divide out the common factors.

$6 + 3x = 10 + 2x$ Distributive property

$x = 4$ Subtract $2x$. Subtract 6.

Step 5 **State the answer.** The number is 4.

Step 6 **Check** the solution in the words of the original problem. If 4 is added to both the numerator and the denominator of $\frac{2}{5}$, the result is $\frac{2 + 4}{5 + 4} = \frac{6}{9} = \frac{2}{3}$, as required.

NOW TRY

OBJECTIVE 2 **Solve problems about distance, rate, and time.**

Recall the following formulas relating distance, rate, and time. Refer to **Example 6** in **Section 2.7** to review the basic use of these formulas.

Forms of the Distance Formula

$$d = rt \qquad r = \frac{d}{t} \qquad t = \frac{d}{r}$$

NOW TRY ANSWER

1. $\frac{21}{25}$

EXAMPLE 2   Solving a Problem about Distance, Rate, and Time

The Tickfaw River has a current of 3 mph. A motorboat takes as long to travel 12 mi downstream as to travel 8 mi upstream. What is the rate of the boat in still water?

Step 1   **Read** the problem again. We must find the rate (speed) of the boat in still water.

Step 2   **Assign a variable.**

Let x = the rate of the boat in still water.

Because the current pushes the boat when the boat is going downstream, the rate of the boat downstream will be the *sum* of the rate of the boat and the rate of the current, $(x + 3)$ mph.

Because the current slows the boat down when the boat is going upstream, the boat's rate going upstream will be the *difference* between the rate of the boat in still water and the rate of the current, $(x - 3)$ mph. See **FIGURE 1**.

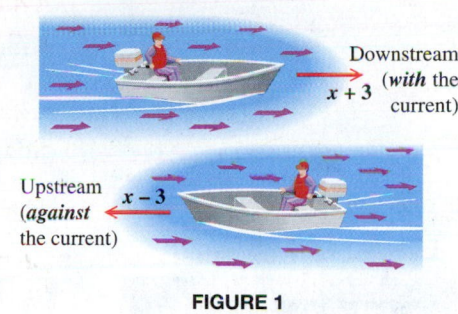

FIGURE 1

This information is summarized in the following table.

	d	r	t
Downstream	12	$x + 3$	
Upstream	8	$x - 3$	

Fill in the times using the formula $t = \dfrac{d}{r}$.

The time downstream is the distance divided by the rate.

$$t = \frac{d}{r} = \frac{12}{x + 3} \qquad \text{Time downstream}$$

The time upstream is that distance divided by that rate.

$$t = \frac{d}{r} = \frac{8}{x - 3} \qquad \text{Time upstream}$$

	d	r	t
Downstream	12	$x + 3$	$\dfrac{12}{x + 3}$
Upstream	8	$x - 3$	$\dfrac{8}{x - 3}$

Times are equal.

Step 3   **Write an equation.**

$$\frac{12}{x + 3} = \frac{8}{x - 3} \qquad \text{The time downstream equals the time upstream, so the two times from the table must be equal.}$$

Step 4   **Solve.**

$$(x + 3)(x - 3)\frac{12}{x + 3} = (x + 3)(x - 3)\frac{8}{x - 3} \qquad \text{Multiply by the LCD, } (x + 3)(x - 3).$$

$$12(x - 3) = 8(x + 3) \qquad \text{Divide out the common factors.}$$

$$12x - 36 = 8x + 24 \qquad \text{Distributive property}$$

$$4x = 60 \qquad \text{Subtract } 8x. \text{ Add 36.}$$

$$x = 15 \qquad \text{Divide by 4.}$$

NOW TRY
EXERCISE 2

In her small boat, Jennifer can travel 12 mi downstream in the same amount of time that she can travel 4 mi upstream. The rate of the current is 2 mph. Find the rate of Jennifer's boat in still water.

Step 5 **State the answer.** The rate of the boat in still water is 15 mph.

Step 6 **Check.** The rate of the boat downstream is $15 + 3 = 18$ mph. Divide 12 mi by 18 mph to find the time.

$$t = \frac{d}{r} = \frac{12}{18} = \frac{2}{3} \text{ hr}$$

The rate of the boat upstream is $15 - 3 = 12$ mph. Divide 8 mi by 12 mph to find the time.

$$t = \frac{d}{r} = \frac{8}{12} = \frac{2}{3} \text{ hr}$$

The time upstream equals the time downstream, as required.

NOW TRY

OBJECTIVE 3 **Solve problems about work.**

Suppose that we can mow a lawn in 4 hr. Then after 1 hr, we will have mowed $\frac{1}{4}$ of the lawn. After 2 hr, we will have mowed $\frac{2}{4}$, or $\frac{1}{2}$, of the lawn, and so on. This idea is generalized as follows.

> **Rate of Work**
>
> If a job can be completed in t units of time, then the rate of work is
>
> $$\frac{1}{t} \text{ job per unit of time.}$$

PROBLEM-SOLVING HINT Recall that the formula $d = rt$ says that distance traveled is equal to rate of travel multiplied by time traveled. Similarly, the fractional part of a job accomplished is equal to the rate of work multiplied by the time worked.

In the lawn-mowing example, after 3 hr, the fractional part of the job done is as follows.

$$\underbrace{\frac{1}{4}}_{\substack{\text{Rate of} \\ \text{work}}} \cdot \underbrace{3}_{\substack{\text{Time} \\ \text{worked}}} = \underbrace{\frac{3}{4}}_{\substack{\text{Fractional part} \\ \text{of job done}}}$$

After 4 hr, $\frac{1}{4}(4) = 1$ whole job has been done.

EXAMPLE 3 **Solving a Problem about Work Rates**

"If Joe can paint a house in 3 hr and Sam can paint the same house in 5 hr, how long does it take for them to do it together?" (*Source:* The movie *Little Big League.*)

Step 1 **Read** the problem again. We are looking for time working together.

Step 2 **Assign a variable.**

Let x = the number of hours it takes Joe and Sam to paint the house, working together.

NOW TRY ANSWER
2. 4 mph

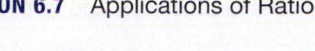

**NOW TRY
EXERCISE 3**

Sarah can proofread a manuscript in 10 hr, while Joyce can proofread the same manuscript in 12 hr. How long will it take them to proofread the manuscript if they work together?

Certainly, x will be less than 3 because Joe alone can complete the job in 3 hr. We begin by making a table. Based on the preceding discussion, Joe's rate alone is $\frac{1}{3}$ job per hour, and Sam's rate is $\frac{1}{5}$ job per hour.

	Rate	Time Working Together	Fractional Part of the Job Done When Working Together
Joe	$\frac{1}{3}$	x	$\frac{1}{3}x$
Sam	$\frac{1}{5}$	x	$\frac{1}{5}x$

Sum is 1 whole job.

Step 3 **Write an equation.**

$$\underbrace{\frac{1}{3}x}_{\text{Fractional part done by Joe}} + \underbrace{\frac{1}{5}x}_{\text{Fractional part done by Sam}} = \underbrace{1}_{\text{1 whole job.}}$$

Together, Joe and Sam complete 1 whole job. Add their individual fractional parts and set the sum equal to 1.

Step 4 **Solve.**

$$15\left(\frac{1}{3}x + \frac{1}{5}x\right) = 15(1) \qquad \text{Multiply by the LCD, 15.}$$

$$15\left(\frac{1}{3}x\right) + 15\left(\frac{1}{5}x\right) = 15(1) \qquad \text{Distributive property}$$

$$5x + 3x = 15 \qquad \text{Simplify.}$$

$$8x = 15 \qquad \text{Combine like terms.}$$

$$x = \frac{15}{8} \qquad \text{Divide by 8.}$$

Step 5 **State the answer.** Working together, Joe and Sam can paint the house in $\frac{15}{8}$ hr, or $1\frac{7}{8}$ hr.

Step 6 **Check.** Substitute $\frac{15}{8}$ for x in the equation from Step 3.

$$\frac{1}{3}x + \frac{1}{5}x = 1 \qquad \text{Equation from Step 3}$$

$$\frac{1}{3}\left(\frac{15}{8}\right) + \frac{1}{5}\left(\frac{15}{8}\right) \stackrel{?}{=} 1 \qquad \text{Let } x = \frac{15}{8}.$$

$$\frac{5}{8} + \frac{3}{8} \stackrel{?}{=} 1 \qquad \text{Multiply.}$$

$$1 = 1 \quad \checkmark \quad \text{True}$$

The answer $\frac{15}{8}$ hr, or $1\frac{7}{8}$ hr, is correct.

NOW TRY

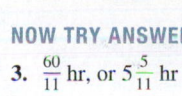

NOW TRY ANSWER

3. $\frac{60}{11}$ hr, or $5\frac{5}{11}$ hr

NOTE An alternative approach in work problems is to consider the part of the job that can be done in 1 hr. For instance, in **Example 3** Joe can do the entire job in 3 hr and Sam can do it in 5 hr. Thus, their work rates, as we saw in **Example 3,** are $\frac{1}{3}$ and $\frac{1}{5}$, respectively. Because it takes them x hours to complete the job working together, in 1 hr they can paint $\frac{1}{x}$ of the house.

The amount painted by Joe in 1 hr plus the amount painted by Sam in 1 hr must equal the amount they can paint together in 1 hr. This relationship leads to the equation

$$\text{Amount by Joe} \rightarrow \frac{1}{3} + \underset{\uparrow}{\frac{1}{5}} = \frac{1}{x}. \leftarrow \text{Amount together}$$
$$\text{Amount by Sam}$$

Compare this equation with the one in **Example 3.** Multiplying each side by $15x$ leads to the same equation found in the third line of Step 4 in the example,

$$5x + 3x = 15.$$

The same solution results.

6.7 Exercises

FOR EXTRA HELP   ▶ MyMathLab®

▶ *Complete solution available in MyMathLab*

Concept Check *Answer each question.*

1. If a migrating hawk travels m mph in still air, what is its rate when it flies into a steady headwind of 5 mph? What is its rate with a tailwind of 5 mph?

2. Suppose Stephanie walks D miles at R mph in the same time that Wally walks d miles at r mph. What is an equation relating D, R, d, and r?

3. If it takes Katherine 10 hr to do a job, what is her rate?

4. If it takes Clayton 12 hr to do a job, how much of the job does he do in 8 hr?

Use Steps 2 and 3 of the six-step problem-solving method to set up an equation to use in solving each problem. (Remember that Step 1 is to read the problem carefully.) Do not actually solve the equation. ***See Example 1.***

5. The numerator of the fraction $\frac{5}{6}$ is increased by an amount so that the value of the resulting fraction is equivalent to $\frac{13}{3}$. By what amount was the numerator increased?

    (a) Let $x = $ _____ . (*Step 2*)

    (b) Write an expression for "the numerator of the fraction $\frac{5}{6}$ is increased by an amount."

    (c) Set up an equation to solve the problem. (*Step 3*)

6. If the same number is added to the numerator and subtracted from the denominator of the fraction $\frac{23}{12}$, the resulting fraction is equivalent to $\frac{3}{2}$. What is the number?

    (a) Let $x = $ _____ . (*Step 2*)

    (b) Write an expression for "a number is added to the numerator of $\frac{23}{12}$." Then write an expression for "the same number is subtracted from the denominator of $\frac{23}{12}$."

    (c) Set up an equation to solve the problem. (*Step 3*)

In each problem, state what x represents, write an equation, and answer the question. See Example 1.

7. In a certain fraction, the denominator is 4 less than the numerator. If 3 is added to both the numerator and the denominator, the resulting fraction is equivalent to $\frac{3}{2}$. What was the original fraction?

8. In a certain fraction, the denominator is 6 more than the numerator. If 3 is added to both the numerator and the denominator, the resulting fraction is equivalent to $\frac{5}{7}$. What was the original fraction (*not* written in lowest terms)?

9. The denominator of a certain fraction is three times the numerator. If 2 is added to the numerator and subtracted from the denominator, the resulting fraction is equivalent to 1. What was the original fraction (*not* written in lowest terms)?

10. The numerator of a certain fraction is four times the denominator. If 6 is added to both the numerator and the denominator, the resulting fraction is equivalent to 2. What was the original fraction (*not* written in lowest terms)?

11. One-sixth of a number is 5 more than the same number. What is the number?

12. One-third of a number is 2 more than one-sixth of the same number. What is the number?

13. A quantity, its $\frac{3}{4}$, its $\frac{1}{2}$, and its $\frac{1}{3}$, added together, becomes 93. What is the quantity? (*Source: Rhind Mathematical Papyrus.*)

14. A quantity, its $\frac{2}{3}$, its $\frac{1}{2}$, and its $\frac{1}{7}$, added together, becomes 33. What is the quantity? (*Source: Rhind Mathematical Papyrus.*)

*Solve each problem. **See Example 6 in Section 2.7.***

15. British explorer and endurance swimmer Lewis Gordon Pugh was the first person to swim at the North Pole. He swam 0.6 mi at 0.0319 mi per min in waters created by melted sea ice. What was his time (to three decimal places)? (*Source: The Gazette.*)

16. In the 2012 Summer Olympics, Ranomi Kromowidjojo of the Netherlands won the women's 100-m freestyle swimming event. Her rate was 1.8868 m per sec. What was her time (to two decimal places)? (*Source:* www.olympic.org)

17. Meseret Defar of Ethiopia won the women's 5000-m run in the 2012 Olympics with a time of 15.071 min. What was her rate (to three decimal places)? (*Source:* www.olympic.org)

18. Asli Cakir Alptekin of Turkey won the women's 1500-m run in the 2012 Olympics with a time of 4.1705 min. What was her rate (to three decimal places)? (*Source:* www.olympic.org)

19. In 2012, Matt Kenseth drove his Ford to victory in the Daytona 500 (mile) race with a rate of 140.256 mph. What was his time (to the nearest thousandth of an hour)? (*Source:* www.cbssports.com)

20. In 2011, Trevor Bayne drove his Ford to victory in the Daytona 500 (mile) race. His rate was 130.326 mph. What was his time (to the nearest thousandth of an hour)? (*Source:* www.cbssports.com)

Set up an equation to solve each problem. Do not actually solve the equation. See Example 2.

21. Mitch flew his airplane 500 mi against the wind in the same time it took him to fly 600 mi with the wind. If the speed of the wind was 10 mph, what was the rate of his plane in still air? (Let x = rate of the plane in still air.)

	d	r	t
Against the Wind	500	$x - 10$	
With the Wind	600	$x + 10$	

22. Janet can row her boat 4 mph in still water. She takes as long to row 8 mi upstream as 24 mi downstream. What is the rate of the current? (Let x = rate of the current.)

	d	r	t
Upstream	8	$4 - x$	
Downstream	24	$4 + x$	

Solve each problem. See Example 2.

▶ **23.** A boat can travel 20 mi against a current in the same time that it can travel 60 mi with the current. The rate of the current is 4 mph. Find the rate of the boat in still water.

24. Vince can fly his plane 200 mi against the wind in the same time it takes him to fly 300 mi with the wind. The wind blows at 30 mph. Find the rate of his plane in still air.

25. The sanderling is a small shorebird about 6.5 in. long, with a thin, dark bill and a wide, white wing stripe. If a sanderling can fly 30 mi with the wind in the same time it can fly 18 mi against the wind when the wind speed is 8 mph, what is the rate of the bird in still air?

26. Airplanes usually fly faster from west to east than from east to west because the prevailing winds go from west to east. The air distance between Chicago and London is about 4000 mi, while the air distance between New York and London is about 3500 mi. If a jet can fly eastbound from Chicago to London in the same time it can fly westbound from London to New York in a 35-mph wind, what is the rate of the plane in still air?

27. An airplane maintaining a constant airspeed takes as long to travel 450 mi with the wind as it does to travel 375 mi against the wind. If the wind is blowing at 15 mph, what is the rate of the plane in still air?

	d	r	t
Against the Wind			
With the Wind			

28. A river has a current of 4 km per hr. Find the rate of Jai's boat in still water if it travels 40 km downstream in the same time that it takes to travel 24 km upstream.

	d	r	t
Upstream			
Downstream			

29. Connie's boat travels at 12 mph. Find the rate of the current of the river if she can travel 6 mi upstream in the same amount of time she can travel 10 mi downstream.

30. Mohammed can travel 8 mi upstream in the same time it takes him to travel 12 mi downstream. His boat travels 15 mph in still water. What is the rate of the current?

31. The distance from Seattle, Washington, to Victoria, British Columbia, is about 148 mi by ferry. It takes about 4 hr less to travel by the same ferry from Victoria to Vancouver, British Columbia, a distance of about 74 mi. What is the average rate of the ferry?

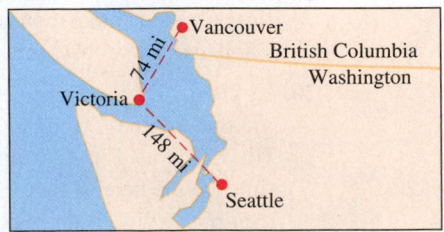

32. Driving from Tulsa to Detroit, Dean averaged 50 mph. He figured that if he had averaged 60 mph, his driving time would have decreased 3 hr. How far is it from Tulsa to Detroit?

Set up an equation to solve each problem. Do not actually solve the equation. **See Example 3.**

33. Working alone, Edward can paint a room in 8 hr. Abdalla can paint the same room working alone in 6 hr. How long will it take them if they work together? (Let *t* represent the time they work together.)

	r	t	w
Edward		t	
Abdalla		t	

34. Donald can tune up his Chevy in 2 hr working alone. Jeff can do the same job in 3 hr working alone. How long would it take them if they worked together? (Let *t* represent the time they work together.)

	r	t	w
Donald		t	
Jeff		t	

Solve each problem. **See Example 3.**

▶ 35. Heather, a high school mathematics teacher, gave a test on perimeter, area, and volume to her geometry classes. Working alone, it would take her 4 hr to grade the tests. Her student teacher, Courtney, would take 6 hr to grade the same tests. How long would it take them to grade these tests if they work together?

36. Zachary and Samuel are brothers who share a bedroom. By himself, Zachary can completely mess up their room in 20 min, while it would take Samuel only 12 min to do the same thing. How long would it take them to mess up the room together?

37. A pump can pump the water out of a flooded basement in 10 hr. A smaller pump takes 12 hr. How long would it take to pump the water from the basement with both pumps?

38. Lou's copier can do a printing job in 7 hr. Nora's copier can do the same job in 12 hr. How long would it take to do the job with both copiers?

39. An experienced employee can enter tax data into a computer twice as fast as a new employee. Working together, it takes the employees 2 hr. How long would it take the experienced employee working alone?

40. One roofer can put a roof on a house three times faster than another. Working together, they can roof a house in 4 days. How long would it take the faster roofer working alone?

41. One pipe can fill a swimming pool in 6 hr, and another pipe can do it in 9 hr. How long will it take the two pipes working together to fill the pool $\frac{3}{4}$ full?

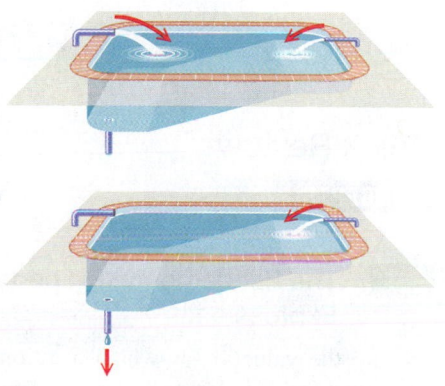

42. An inlet pipe can fill a swimming pool in 9 hr, and an outlet pipe can empty the pool in 12 hr. Through an error, both pipes are left open. How long will it take to fill the pool?

Extending Skills   *Extend the concepts of* **Example 3** *to solve each problem.*

43. A cold-water faucet can fill a sink in 12 min, and a hot-water faucet can fill it in 15 min. The drain can empty the sink in 25 min. If both faucets are on and the drain is open, how long will it take to fill the sink?

44. Refer to **Exercise 42.** Assume that the error was discovered after both pipes had been running for 3 hr and the outlet pipe was then closed. How much more time would then be required to fill the pool? (*Hint:* Consider how much of the job had been done when the error was discovered.)

Chapter 6	Summary

Key Terms

6.1
rational expression
lowest terms

6.3
least common
    denominator (LCD)

6.5
complex fraction

6.6
proposed solution
extraneous solution
    (extraneous value)

Test Your Word Power

See how well you have learned the vocabulary in this chapter.

1. A **rational expression** is
   A. an algebraic expression made up of a term or the sum of a finite number of terms with real coefficients and whole number exponents
   B. a polynomial equation of degree 2
   C. an expression with one or more fractions in the numerator, or denominator, or both
   D. the quotient of two polynomials with denominator not 0.

2. In a given set of fractions, the **least common denominator** is
   A. the smallest denominator of all the denominators
   B. the smallest expression that is divisible by all the denominators
   C. the largest integer that evenly divides the numerator and denominator of all the fractions
   D. the largest denominator of all the denominators.

3. A **complex fraction** is
   A. an algebraic expression made up of a term or the sum of a finite number of terms with real coefficients and whole number exponents
   B. a polynomial equation of degree 2
   C. a quotient with one or more fractions in the numerator, or denominator, or both
   D. the quotient of two polynomials with denominator not 0.

ANSWERS

1. D; *Examples:* $-\dfrac{3}{4y}, \dfrac{5x^3}{x+2}, \dfrac{a+3}{a^2-4a-5}$ 2. B; *Example:* The LCD of $\dfrac{1}{x}, \dfrac{2}{3}$, and $\dfrac{5}{x+1}$ is $3x(x+1)$.   3. C; *Examples:* $\dfrac{\frac{2}{3}}{\frac{4}{7}}, \dfrac{x-\frac{1}{y}}{x+\frac{1}{y}}, \dfrac{\frac{2}{a+1}}{a^2-1}$

Quick Review

CONCEPTS	EXAMPLES

6.1 **The Fundamental Property of Rational Expressions**

To find the value(s) for which a rational expression is undefined, set the denominator equal to 0 and solve the equation.

Find the values for which the expression $\dfrac{x-4}{x^2-16}$ is undefined.

$$x^2 - 16 = 0$$

$$(x-4)(x+4) = 0 \qquad \text{Factor.}$$

$$x - 4 = 0 \quad \text{or} \quad x + 4 = 0 \qquad \text{Zero-factor property}$$

$$x = 4 \quad \text{or} \qquad x = -4 \qquad \text{Solve for } x.$$

The rational expression is undefined for 4 and -4, so $x \neq 4$ and $x \neq -4$.

CONCEPTS	EXAMPLES

Writing a Rational Expression in Lowest Terms

Step 1 Factor the numerator and denominator.

Step 2 Use the fundamental property to divide out common factors.

Write in lowest terms. $\dfrac{x^2 - 1}{(x - 1)^2}$

$$= \frac{(x - 1)(x + 1)}{(x - 1)(x - 1)} \qquad \text{Factor.}$$

$$= \frac{x + 1}{x - 1} \qquad \text{Lowest terms}$$

There are often several different equivalent forms of a rational expression.

Give four equivalent forms of $-\dfrac{x - 1}{x + 2}$.

$$① \to \frac{-(x - 1)}{x + 2}, \quad \text{or} \quad \frac{-x + 1}{x + 2} \gets ②$$

Distribute the negative sign in the numerator.

$$③ \to \frac{x - 1}{-(x + 2)}, \quad \text{or} \quad \frac{x - 1}{-x - 2} \gets ④$$

Distribute the negative sign in the denominator.

6.2 Multiplying and Dividing Rational Expressions

Multiplying or Dividing Rational Expressions

Step 1 Note the operation. If the operation is division, use the definition of division to rewrite as multiplication.

Step 2 Multiply numerators and multiply denominators.

Step 3 Factor numerators and denominators completely.

Step 4 Write in lowest terms, using the fundamental property.

Note: Steps 2 and 3 may be interchanged based on personal preference.

Multiply. $\dfrac{3x + 9}{x - 5} \cdot \dfrac{x^2 - 3x - 10}{x^2 - 9}$

$$= \frac{(3x + 9)(x^2 - 3x - 10)}{(x - 5)(x^2 - 9)} \qquad \begin{array}{l}\text{Multiply numerators}\\\text{and denominators.}\end{array}$$

$$= \frac{3(x + 3)(x - 5)(x + 2)}{(x - 5)(x + 3)(x - 3)} \qquad \text{Factor.}$$

$$= \frac{3(x + 2)}{x - 3} \qquad \text{Lowest terms}$$

Divide. $\dfrac{2x + 1}{x + 5} \div \dfrac{6x^2 - x - 2}{x^2 - 25}$

$$= \frac{2x + 1}{x + 5} \cdot \frac{x^2 - 25}{6x^2 - x - 2} \qquad \begin{array}{l}\text{Multiply by the}\\\text{reciprocal of}\\\text{the divisor.}\end{array}$$

$$= \frac{(2x + 1)(x^2 - 25)}{(x + 5)(6x^2 - x - 2)} \qquad \begin{array}{l}\text{Multiply numerators}\\\text{and denominators.}\end{array}$$

$$= \frac{(2x + 1)(x + 5)(x - 5)}{(x + 5)(2x + 1)(3x - 2)} \qquad \text{Factor.}$$

$$= \frac{x - 5}{3x - 2} \qquad \text{Lowest terms}$$

6.3 Least Common Denominators

Finding the LCD

Step 1 Factor each denominator into prime factors.

Step 2 List each different factor the greatest number of times it appears.

Step 3 Multiply the factors from Step 2 to find the LCD.

Find the LCD for $\dfrac{3}{k^2 - 8k + 16}$ and $\dfrac{1}{4k^2 - 16k}$.

$$\left.\begin{array}{l}k^2 - 8k + 16 = (k - 4)^2 \\ 4k^2 - 16k = 4k(k - 4)\end{array}\right\} \quad \begin{array}{l}\text{Factor each}\\\text{denominator.}\end{array}$$

$$\text{LCD} = (k - 4)^2 \cdot 4 \cdot k$$

$$= 4k(k - 4)^2$$

CONCEPTS	EXAMPLES

Writing a Rational Expression with a Specified Denominator

Step 1   Factor both denominators.

Step 2   Decide what factor(s) the denominator must be multiplied by in order to equal the specified denominator.

Step 3   Multiply the rational expression by that factor divided by itself. (That is, multiply by 1.)

Find the numerator. $\dfrac{5}{2z^2 - 6z} = \dfrac{?}{4z^3 - 12z^2}$

$$\dfrac{5}{2z(z - 3)} = \dfrac{?}{4z^2(z - 3)}$$

$2z(z - 3)$ must be multiplied by $2z$ in order to obtain $4z^2(z - 3)$.

$$\dfrac{5}{2z(z - 3)} \cdot \dfrac{2z}{2z}$$

$$= \dfrac{10z}{4z^2(z - 3)}, \quad \text{or} \quad \dfrac{10z}{4z^3 - 12z^2}$$

6.4   Adding and Subtracting Rational Expressions

Adding Rational Expressions

Step 1   Find the LCD.

Step 2   Write each rational expression with the LCD as denominator.

Step 3   Add the numerators to obtain the numerator of the sum. The LCD is the denominator of the sum.

Step 4   Write in lowest terms.

Add. $\dfrac{2}{3m + 6} + \dfrac{m}{m^2 - 4}$

$$\begin{aligned} 3m + 6 &= 3(m + 2) \\ m^2 - 4 &= (m + 2)(m - 2) \end{aligned} \Bigg\} \quad \begin{array}{l} \text{The LCD is} \\ 3(m + 2)(m - 2). \end{array}$$

$$= \dfrac{2(m - 2)}{3(m + 2)(m - 2)} + \dfrac{3m}{3(m + 2)(m - 2)} \quad \begin{array}{l}\text{Write with}\\ \text{the LCD.}\end{array}$$

$$= \dfrac{2(m - 2) + 3m}{3(m + 2)(m - 2)} \quad \begin{array}{l}\text{Add numerators.}\\ \text{Keep the same denominator.}\end{array}$$

$$= \dfrac{2m - 4 + 3m}{3(m + 2)(m - 2)} \quad \text{Distributive property}$$

$$= \dfrac{5m - 4}{3(m + 2)(m - 2)} \quad \text{Combine like terms.}$$

This last expression is in lowest terms.

Subtracting Rational Expressions

Follow the same steps as for addition, but subtract in Step 3.

Subtract. $\dfrac{6}{k + 4} - \dfrac{2}{k}$ The LCD is $k(k + 4)$.

$$= \dfrac{6k}{(k + 4)k} - \dfrac{2(k + 4)}{k(k + 4)} \quad \text{Write with the LCD.}$$

$$= \dfrac{6k - 2(k + 4)}{k(k + 4)} \quad \begin{array}{l}\text{Subtract numerators.}\\ \text{Keep the same denominator.}\end{array}$$

$$= \dfrac{6k - 2k - 8}{k(k + 4)} \quad \text{Distributive property}$$

$$= \dfrac{4k - 8}{k(k + 4)} \quad \text{Combine like terms.}$$

$$= \dfrac{4(k - 2)}{k(k + 4)} \quad \text{Factor.}$$

This last expression is in lowest terms.

CONCEPTS	EXAMPLES

6.5 Complex Fractions

Simplifying Complex Fractions

Method 1 Simplify the numerator and denominator separately. Then divide the simplified numerator by the simplified denominator.

Method 2 Multiply the numerator and denominator of the complex fraction by the LCD of all the fractions in the numerator and denominator of the complex fraction. Write in lowest terms.

Simplify.   **Method 1**   **Method 2**

$$\dfrac{\dfrac{1}{a} - a}{1 - a} \qquad\qquad \dfrac{\dfrac{1}{a} - a}{1 - a}$$

Method 1

$$= \dfrac{\dfrac{1}{a} - \dfrac{a^2}{a}}{1 - a}$$

$$= \dfrac{\dfrac{1 - a^2}{a}}{1 - a}$$

$$= \dfrac{1 - a^2}{a} \div (1 - a)$$

$$= \dfrac{1 - a^2}{a} \cdot \dfrac{1}{1 - a}$$

$$= \dfrac{(1 - a)(1 + a)}{a(1 - a)}$$

$$= \dfrac{1 + a}{a}$$

Method 2

$$= \dfrac{\left(\dfrac{1}{a} - a\right)a}{(1 - a)a}$$

$$= \dfrac{\dfrac{a}{a} - a^2}{(1 - a)a}$$

$$= \dfrac{1 - a^2}{(1 - a)a}$$

$$= \dfrac{(1 + a)(1 - a)}{(1 - a)a}$$

$$= \dfrac{1 + a}{a}$$

6.6 Solving Equations with Rational Expressions

Solving Equations with Rational Expressions

Step 1 Multiply each side of the equation by the LCD to clear the equation of fractions. Be sure to distribute to *every* term on *both* sides.

Solve. $\dfrac{x}{x - 3} + \dfrac{4}{x + 3} = \dfrac{18}{x^2 - 9}$

$$\dfrac{x}{x - 3} + \dfrac{4}{x + 3} = \dfrac{18}{(x - 3)(x + 3)} \qquad \text{Factor.}$$

The LCD is $(x - 3)(x + 3)$. Note that 3 and -3 cannot be solutions, as they cause a denominator to equal 0.

$$(x - 3)(x + 3)\left(\dfrac{x}{x - 3} + \dfrac{4}{x + 3}\right)$$

Multiply by the LCD, $(x - 3)(x + 3)$.

$$= (x - 3)(x + 3)\dfrac{18}{(x - 3)(x + 3)}$$

Step 2 Solve the resulting equation.

$x(x + 3) + 4(x - 3) = 18$	Distributive property
$x^2 + 3x + 4x - 12 = 18$	Distributive property
$x^2 + 7x - 30 = 0$	Standard form
$(x - 3)(x + 10) = 0$	Factor.
$x - 3 = 0 \quad \text{or} \quad x + 10 = 0$	Zero-factor property
Reject $\longrightarrow$ $x = 3 \quad \text{or} \qquad x = -10$	Solve.

Step 3 Check each proposed solution by substituting it in the original equation. Reject any value that causes an original denominator to equal 0.

Because 3 causes denominators to equal 0, it is an extraneous value. Check that the only solution is -10. Thus, $\{-10\}$ is the solution set.

CONCEPTS	**EXAMPLES**

6.7 Applications of Rational Expressions

Solving Problems about Distance, Rate, and Time

Use the formulas relating d, r, and t.

$$d = rt, \quad r = \frac{d}{t}, \quad t = \frac{d}{r}$$

A small plane flew from Chicago to Kansas City averaging 145 mph. The trip took 3.5 hr. What is the distance between Chicago and Kansas City?

$$145 \cdot 3.5 = 507.5 \text{ mi}$$

$$\uparrow \qquad \uparrow \qquad \uparrow$$

Rate   Time   Distance

Solving Problems about Work

Step 1 Read the problem carefully.

Step 2 Assign a variable. State what the variable represents. Organize the information from the problem in a table. If a job is done in t units of time, then the rate is $\frac{1}{t}$.

It takes the regular mail carrier 6 hr to cover her route. A substitute takes 8 hr to cover the same route. How long would it take them to cover the route together?

Let $x =$ the number of hours required to cover the route together.

	Rate	Time	Part of the Job Done
Regular	$\frac{1}{6}$	x	$\frac{1}{6}x$
Substitute	$\frac{1}{8}$	x	$\frac{1}{8}x$

Multiply rate by time to find the fractional part done.

Step 3 Write an equation. The sum of the fractional parts should equal 1 (whole job).

Step 4 Solve the equation.

$$\frac{1}{6}x + \frac{1}{8}x = 1 \qquad \text{The parts add to 1 whole job.}$$

$$24\left(\frac{1}{6}x + \frac{1}{8}x\right) = 24(1) \qquad \text{The LCD is 24.}$$

$$24\left(\frac{1}{6}x\right) + 24\left(\frac{1}{8}x\right) = 24 \qquad \text{Distributive property; Multiply.}$$

$$4x + 3x = 24 \qquad \text{Multiply.}$$

$$7x = 24 \qquad \text{Combine like terms.}$$

$$x = \frac{24}{7} \qquad \text{Divide by 7.}$$

Step 5 State the answer.

To cover the route together, it would take them

$$\frac{24}{7} \text{ hr}, \quad \text{or} \quad 3\frac{3}{7} \text{ hr.}$$

Step 6 Check the solution.

The solution checks because

$$\frac{1}{6}\left(\frac{24}{7}\right) + \frac{1}{8}\left(\frac{24}{7}\right) = 1 \quad \text{is true.}$$

Chapter 6   Review Exercises

6.1 *Find the numerical value of each rational expression for* **(a)** $x = -2$ *and* **(b)** $x = 4$.

1. (a) $\dfrac{11}{8}$ (b) $\dfrac{13}{22}$

1. $\dfrac{4x - 3}{5x + 2}$

2. $\dfrac{3x}{x^2 - 4}$

2. (a) undefined   (b) 1

3. Set the denominator equal to 0 and solve the equation. Any solutions are values for which the rational expression is undefined.

3. Explain the process used to determine the values of the variable for which a rational expression is undefined.

4. $x \neq 3$

5. $y \neq 0$

Find any values of the variable for which each rational expression is undefined. Write answers with the symbol $\neq$.

6. $k \neq -5, -\dfrac{2}{3}$

4. $\dfrac{4}{x - 3}$

5. $\dfrac{y + 3}{2y}$

6. $\dfrac{2k + 1}{3k^2 + 17k + 10}$

7. $\dfrac{b}{3a}$ **8.** -1

Write each rational expression in lowest terms.

9. $\dfrac{-(2x + 3)}{2}$ **10.** $\dfrac{2p + 5q}{5p + q}$

7. $\dfrac{5a^3b^3}{15a^4b^2}$

8. $\dfrac{m - 4}{4 - m}$

9. $\dfrac{4x^2 - 9}{6 - 4x}$

10. $\dfrac{4p^2 + 8pq - 5q^2}{10p^2 - 3pq - q^2}$

Answers may vary in Exercises 11 and 12.

11. $\dfrac{-(4x - 9)}{2x + 3}$, $\dfrac{-4x + 9}{2x + 3}$, $\dfrac{4x - 9}{-(2x + 3)}$, $\dfrac{4x - 9}{-2x - 3}$

Write four equivalent forms for each rational expression.

11. $-\dfrac{4x - 9}{2x + 3}$

12. $-\dfrac{8 - 3x}{3 - 6x}$

12. $\dfrac{-(8 - 3x)}{3 - 6x}$, $\dfrac{-8 + 3x}{3 - 6x}$, $\dfrac{8 - 3x}{-(3 - 6x)}$, $\dfrac{8 - 3x}{-3 + 6x}$

6.2 *Multiply or divide. Write each answer in lowest terms.*

13. $\dfrac{72}{p}$ **14.** 2

13. $\dfrac{18p^3}{6} \cdot \dfrac{24}{p^4}$

14. $\dfrac{8x^2}{12x^5} \cdot \dfrac{6x^4}{2x}$

15. $\dfrac{5}{8}$ **16.** $\dfrac{r + 4}{3}$

15. $\dfrac{x - 3}{4} \cdot \dfrac{5}{2x - 6}$

16. $\dfrac{2r + 3}{r - 4} \cdot \dfrac{r^2 - 16}{6r + 9}$

17. $\dfrac{3a - 1}{a + 5}$ **18.** $\dfrac{y - 2}{y - 3}$

17. $\dfrac{6a^2 + 7a - 3}{2a^2 - a - 6} \div \dfrac{a + 5}{a - 2}$

18. $\dfrac{y^2 - 6y + 8}{y^2 + 3y - 18} \div \dfrac{y - 4}{y + 6}$

19. $\dfrac{p + 5}{p + 1}$ **20.** $\dfrac{3z + 1}{z + 3}$

19. $\dfrac{2p^2 + 13p + 20}{p^2 + p - 12} \cdot \dfrac{p^2 + 2p - 15}{2p^2 + 7p + 5}$

20. $\dfrac{3z^2 + 5z - 2}{9z^2 - 1} \cdot \dfrac{9z^2 + 6z + 1}{z^2 + 5z + 6}$

21. $108y^4$

22. $(x + 3)(x + 1)(x + 4)$

6.3 *Find the LCD for the fractions in each list.*

23. $\dfrac{15a}{10a^4}$ **24.** $\dfrac{-54}{18 - 6x}$

21. $\dfrac{4}{9y}$, $\dfrac{7}{12y^2}$, $\dfrac{5}{27y^4}$

22. $\dfrac{3}{x^2 + 4x + 3}$, $\dfrac{5}{x^2 + 5x + 4}$

25. $\dfrac{15y}{50 - 10y}$

Write each rational expression with the given denominator.

26. $\dfrac{4b(b + 2)}{(b + 3)(b - 1)(b + 2)}$

23. $\dfrac{3}{2a^3} = \dfrac{?}{10a^4}$

24. $\dfrac{9}{x - 3} = \dfrac{?}{18 - 6x}$

27. $\dfrac{15}{x}$ **28.** $-\dfrac{2}{p}$

25. $\dfrac{-3y}{2y - 10} = \dfrac{?}{50 - 10y}$

26. $\dfrac{4b}{b^2 + 2b - 3} = \dfrac{?}{(b + 3)(b - 1)(b + 2)}$

29. $\dfrac{4k - 45}{k(k - 5)}$

6.4 *Add or subtract. Write each answer in lowest terms.*

27. $\dfrac{10}{x} + \dfrac{5}{x}$

28. $\dfrac{6}{3p} - \dfrac{12}{3p}$

29. $\dfrac{9}{k} - \dfrac{5}{k - 5}$

30. $\dfrac{28 + 11y}{y(7 + y)}$

31. $\dfrac{-2 - 3m}{6}$ **32.** $\dfrac{3(16 - x)}{4x^2}$

33. $\dfrac{7a + 6b}{(a - 2b)(a + 2b)}$

34. $\dfrac{-k^2 - 6k + 3}{3(k + 3)(k - 3)}$

35. $\dfrac{5z - 16}{z(z + 6)(z - 2)}$

36. $\dfrac{-13p + 33}{p(p - 2)(p - 3)}$

37. $\dfrac{4(y - 3)}{y + 3}$ **38.** $\dfrac{10}{13}$

39. $\dfrac{xw + 1}{xw - 1}$ **40.** $\dfrac{(q - p)^2}{pq}$

41. $(x - 5)(x - 3)$, or $x^2 - 8x + 15$

42. $\dfrac{y + x}{xy}$

43. $\varnothing$ **44.** $\{-16\}$

45. $\{0\}$ **46.** $\{3\}$

47. $t = \dfrac{Ry}{m}$ **48.** $y = \dfrac{4x + 5}{3}$

49. $m = \dfrac{4 + p^2 q}{3p^2}$

50. $\dfrac{20}{15}$

51. $\dfrac{3}{18}$ **52.** 10 mph

53. $3\dfrac{1}{13}$ hr   **54.** 2 hr

30. $\dfrac{4}{y} + \dfrac{7}{7 + y}$

31. $\dfrac{m}{3} - \dfrac{2 + 5m}{6}$

32. $\dfrac{12}{x^2} - \dfrac{3}{4x}$

33. $\dfrac{5}{a - 2b} + \dfrac{2}{a + 2b}$

34. $\dfrac{4}{k^2 - 9} - \dfrac{k + 3}{3k - 9}$

35. $\dfrac{8}{z^2 + 6z} - \dfrac{3}{z^2 + 4z - 12}$

36. $\dfrac{11}{2p - p^2} - \dfrac{2}{p^2 - 5p + 6}$

6.5 *Simplify each complex fraction.*

37. $\dfrac{\dfrac{y - 3}{y}}{\dfrac{y + 3}{4y}}$

38. $\dfrac{\dfrac{2}{3} - \dfrac{1}{6}}{\dfrac{1}{4} + \dfrac{2}{5}}$

39. $\dfrac{x + \dfrac{1}{w}}{x - \dfrac{1}{w}}$

40. $\dfrac{\dfrac{1}{p} - \dfrac{1}{q}}{\dfrac{1}{q - p}}$

41. $\dfrac{\dfrac{x^2 - 25}{x + 3}}{\dfrac{x + 5}{x^2 - 9}}$

42. $\dfrac{x^{-2} - y^{-2}}{x^{-1} - y^{-1}}$

6.6 *Solve each equation, and check the solutions.*

43. $\dfrac{3x - 1}{x - 2} = \dfrac{5}{x - 2} + 1$

44. $\dfrac{4 - z}{z} + \dfrac{3}{2} = \dfrac{-4}{z}$

45. $\dfrac{3}{x + 4} - \dfrac{2x}{5} = \dfrac{3}{x + 4}$

46. $\dfrac{3}{m - 2} + \dfrac{1}{m - 1} = \dfrac{7}{m^2 - 3m + 2}$

Solve each formula for the specified variable.

47. $m = \dfrac{Ry}{t}$ for t

48. $x = \dfrac{3y - 5}{4}$ for y

49. $p^2 = \dfrac{4}{3m - q}$ for m

6.7 *Solve each problem.*

50. In a certain fraction, the denominator is 5 less than the numerator. If 5 is added to both the numerator and the denominator, the resulting fraction is equivalent to $\dfrac{5}{4}$. Find the original fraction (*not* written in lowest terms).

51. The denominator of a certain fraction is six times the numerator. If 3 is added to the numerator and subtracted from the denominator, the resulting fraction is equivalent to $\dfrac{2}{5}$. Find the original fraction (*not* written in lowest terms).

52. A plane flies 350 mi with the wind in the same time that it can fly 310 mi against the wind. The plane has a speed of 165 mph in still air. Find the speed of the wind.

53. Sarita can plant her garden in 5 hr working alone. A friend can do the same job in 8 hr. How long would it take them if they worked together?

54. The head gardener can mow the lawns in the city park twice as fast as his assistant. Working together, they can complete the job in $1\dfrac{1}{3}$ hr. How long would it take the head gardener working alone?

Chapter 6   Mixed Review Exercises

Perform each indicated operation.

Answers (left column):

1. $\dfrac{m+7}{(m-1)(m+1)}$

2. $8p^2$

3. $\dfrac{1}{6}$ 4. 3

5. $\dfrac{z+7}{(z+1)(z-1)^2}$

6. $\dfrac{-t-1}{(t+2)(t-2)}$,

   or $\dfrac{t+1}{(2+t)(2-t)}$

7. (a) equation; $\{-6\}$

   (b) expression; $\dfrac{7}{6}x$

8. $\{-2, 3\}$

9. $\{2\}$ 10. $v - at + w$

11. 150 km per hr

12. $5\dfrac{1}{11}$ hr

---

1. $\dfrac{4}{m-1} - \dfrac{3}{m+1}$

2. $\dfrac{8p^5}{5} \div \dfrac{2p^3}{10}$

3. $\dfrac{r-3}{8} \div \dfrac{3r-9}{4}$

4. $\dfrac{\dfrac{5}{x} - 1}{\dfrac{5-x}{3x}}$

5. $\dfrac{4}{z^2 - 2z + 1} - \dfrac{3}{z^2 - 1}$

6. $\dfrac{1}{t^2 - 4} + \dfrac{1}{2 - t}$

7. Identify each of the following as an *expression* or an *equation*. Then simplify the expression and solve the equation.

(a) $\dfrac{2}{3}x + \dfrac{1}{2}x = -7$ (b) $\dfrac{2}{3}x + \dfrac{1}{2}x$

Solve each equation.

8. $\dfrac{2}{z} - \dfrac{z}{z+3} = \dfrac{1}{z+3}$

9. $\dfrac{1}{x^2 - 1} = \dfrac{1}{x+1}$

10. $a = \dfrac{v-w}{t}$ for v

Solve each problem.

11. Rob flew his plane 400 km with the wind in the same time it took him to travel 200 km against the wind. The speed of the wind is 50 km per hr. Find the rate of the plane in still air.

12. With spraying equipment, Lizette can paint the woodwork in a small house in 8 hr. Seyed needs 14 hr to complete the same job painting by hand. If Lizette and Seyed work together, how long will it take them to paint the woodwork?

Chapter 6   Test

FOR EXTRA HELP

Step-by-step test solutions are found on the Chapter Test Prep Videos available in MyMathLab® *or on* YouTube™.

▶ *View the complete solutions to all Chapter Test exercises in MyMathLab.*

[6.1]

1. (a) $\dfrac{11}{6}$ (b) undefined

2. $x \neq -2, 4$

3. (Answers may vary.)

   $\dfrac{-(6x-5)}{2x+3}$, $\dfrac{-6x+5}{2x+3}$,

   $\dfrac{6x-5}{-(2x+3)}$, $\dfrac{6x-5}{-2x-3}$

4. $-3x^2y^3$ 5. $\dfrac{3a+2}{a-1}$

[6.2]

6. $\dfrac{25}{27}$ 7. $\dfrac{3k-2}{3k+2}$

8. $\dfrac{a-1}{a+4}$ 9. $\dfrac{x-5}{3-x}$

[6.3]

10. $150p^5$

11. $(2r+3)(r+2)(r-5)$

---

1. Find the numerical value of $\dfrac{6r+1}{2r^2 - 3r - 20}$ for (a) $r = -2$ and (b) $r = 4$.

2. Find any values for which $\dfrac{3x-1}{x^2 - 2x - 8}$ is undefined. Write the answer with the symbol $\neq$.

3. Write four equivalent forms of the rational expression $-\dfrac{6x-5}{2x+3}$.

Write each rational expression in lowest terms.

4. $\dfrac{-15x^6y^4}{5x^4y}$

5. $\dfrac{6a^2 + a - 2}{2a^2 - 3a + 1}$

Multiply or divide. Write each answer in lowest terms.

6. $\dfrac{5(d-2)}{9} \div \dfrac{3(d-2)}{5}$

7. $\dfrac{6k^2 - k - 2}{8k^2 + 10k + 3} \cdot \dfrac{4k^2 + 7k + 3}{3k^2 + 5k + 2}$

8. $\dfrac{4a^2 + 9a + 2}{3a^2 + 11a + 10} \div \dfrac{4a^2 + 17a + 4}{3a^2 + 2a - 5}$

9. $\dfrac{x^2 - 10x + 25}{9 - 6x + x^2} \cdot \dfrac{x-3}{5-x}$

Find the least common denominator for the fractions in each list.

10. $\dfrac{-3}{10p^2}, \dfrac{21}{25p^3}, \dfrac{-7}{30p^5}$

11. $\dfrac{r+1}{2r^2 + 7r + 6}, \dfrac{-2r+1}{2r^2 - 7r - 15}$

12. $\dfrac{240p^2}{64p^3}$

13. $\dfrac{21}{42m - 84}$

[6.4]

14. 2

15. $\dfrac{-14}{5(y+2)}$

16. $\dfrac{-x^2 + x + 1}{3 - x}$, or $\dfrac{x^2 - x - 1}{x - 3}$

17. $\dfrac{-m^2 + 7m + 2}{(2m + 1)(m - 5)(m - 1)}$

[6.5]

18. $\dfrac{2k}{3p}$

19. $\dfrac{-2 - x}{4 + x}$

[6.6]

20. $\left\{-\dfrac{1}{2}, 1\right\}$

21. $\left\{-\dfrac{1}{2}\right\}$

22. $D = \dfrac{dF - k}{F}$, or $D = d - \dfrac{k}{F}$

[6.7]

23. 3 mph   **24.** $2\dfrac{2}{9}$ hr

Write each rational expression with the given denominator.

12. $\dfrac{15}{4p} = \dfrac{?}{64p^3}$

13. $\dfrac{3}{6m - 12} = \dfrac{?}{42m - 84}$

Add or subtract. Write each answer in lowest terms.

14. $\dfrac{4x + 2}{x + 5} + \dfrac{-2x + 8}{x + 5}$

15. $\dfrac{-4}{y + 2} + \dfrac{6}{5y + 10}$

16. $\dfrac{x + 1}{3 - x} + \dfrac{x^2}{x - 3}$

17. $\dfrac{3}{2m^2 - 9m - 5} - \dfrac{m + 1}{2m^2 - m - 1}$

Simplify each complex fraction.

18. $\dfrac{\dfrac{2p}{k^2}}{\dfrac{3p^2}{k^3}}$

19. $\dfrac{\dfrac{1}{x + 3} - 1}{1 + \dfrac{1}{x + 3}}$

Solve each equation.

20. $\dfrac{3x}{x + 1} = \dfrac{3}{2x}$

21. $\dfrac{2x}{x - 3} + \dfrac{1}{x + 3} = \dfrac{-6}{x^2 - 9}$

22. $F = \dfrac{k}{d - D}$ for D

Solve each problem.

23. A boat travels 7 mph in still water. It takes as long to travel 20 mi upstream as 50 mi downstream. Find the rate of the current.

24. Sanford can paint a room in his house, working alone, in 5 hr. His neighbor can do the job in 4 hr. How long will it take them to paint the room if they work together?

Chapters R–6   Cumulative Review Exercises

[1.3]

1. (a) 9   **(b)** 0, 9
(c) $-8, 0, 9$
(d) $-8, -\dfrac{2}{3}, 0, \dfrac{4}{5}, 9, 10.\overline{6}$
(e) $-\sqrt{6}$
(f) All are real numbers.

[R.1, 1.7]   **[1.7]**
2. 2   **3.** $13k + 42$

[1.6]
4. commutative property
5. distributive property

[2.3]
6. $\{17\}$ **7.** $\varnothing$
8. $\{15\}$

[2.5]   **[2.6]**
9. $b = \dfrac{2\mathcal{A}}{h}$ **10.** $\left\{-\dfrac{2}{7}\right\}$

[2.8]
11. $[-8, \infty)$

1. Let $A = \left\{-8, -\dfrac{2}{3}, -\sqrt{6}, 0, \dfrac{4}{5}, 9, 10.\overline{6}\right\}$. Simplify the elements of A as necessary and then list the elements that belong to each set.

(a) Natural numbers   **(b)** Whole numbers   **(c)** Integers

(d) Rational numbers   **(e)** Irrational numbers   **(f)** Real numbers

2. Evaluate $3 + 4\left(\dfrac{1}{2} - \dfrac{3}{4}\right)$.

3. Simplify $-(3k + 8) - 2(4k - 7) + 3(8k + 12)$.

Identify the property of real numbers illustrated by each equation.

4. $(a + b) + 4 = 4 + (a + b)$

5. $4x + 12x = (4 + 12)x$

Solve.

6. $3(2y - 5) = 2 + 5y$

7. $4(2x - 6) + 3(x - 2) = 11x + 1$

8. $0.06x + 0.03(100 + x) = 4.35$

9. $\mathcal{A} = \dfrac{1}{2}bh$ for b

10. $\dfrac{2 + m}{3} = \dfrac{2 - m}{4}$

11. $5y \le 6y + 8$

[2.7]
12. $5000 at 5%; $7000 at 6%

[2.5]
13. 44 mg

[3.2, 3.3]
14. (a) $(-3, 0)$ **(b)** $(0, -4)$

   (c) $-\dfrac{4}{3}$

[3.2]   **[4.4]**
15.   **16.**

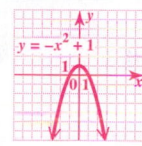
$y = -3x + 2$

$y = -x^2 + 1$

[4.1, 4.2]
17. $\dfrac{1}{2^4 x^7}$ **18.** $\dfrac{1}{m^6}$

[4.4]
19. $k^2 + 2k + 1$

[4.6]
20. $4a^2 - 4ab + b^2$

[4.5]
21. $3y^3 + 8y^2 + 12y - 5$

[4.7]
22. $6p^2 + 7p + 1 + \dfrac{3}{p - 1}$

[5.3]
23. $(4t + 3v)(2t + v)$
24. prime

[5.4]
25. $(4x^2 + 1)(2x + 1)(2x - 1)$

[5.5]
26. $\{-3, 5\}$ **27.** $\left\{5, -\dfrac{1}{2}, \dfrac{2}{3}\right\}$

[5.6]
28. -2 or -1 **29.** 6 m

[6.1]
30. A   **31.** D

Solve each problem.

12. Shamil invested some money at 5% simple interest and $2000 more than that amount at 6%. The interest for the year totaled $670. How much was invested at each rate?

13. Clark's rule, a formula used in reducing drug dosage according to weight from the recommended adult dosage to a child dosage, is

$$\frac{\text{weight of child in pounds}}{150} \times \text{adult dose} = \text{child's dose}.$$

Find a child's dosage if the child weighs 55 lb and the recommended adult dosage is 120 mg.

14. Consider the graph of $4x + 3y = -12$.

   (a) What is the x-intercept?   **(b)** What is the y-intercept?

   (c) What is the slope?

Graph each equation.

15. $y = -3x + 2$ **16.** $y = -x^2 + 1$

Simplify each expression. Write answers with only positive exponents.

17. $\dfrac{(2x^3)^{-1} \cdot x}{2^3 x^5}$ **18.** $\dfrac{(m^{-2})^3 m}{m^5 m^{-4}}$

Perform each indicated operation.

19. $(2k^2 + 3k) - (k^2 + k - 1)$ **20.** $(2a - b)^2$

21. $(y^2 + 3y + 5)(3y - 1)$ **22.** $\dfrac{12p^3 + 2p^2 - 12p + 4}{2p - 2}$

Factor completely.

23. $8t^2 + 10tv + 3v^2$ **24.** $8r^2 - 9rs + 12s^2$ **25.** $16x^4 - 1$

Solve each equation.

26. $r^2 = 2r + 15$ **27.** $(r - 5)(2r + 1)(3r - 2) = 0$

Solve each problem.

28. One number is 4 greater than another. The product of the numbers is 2 less than the lesser number. Find the lesser number.

29. The length of a rectangle is 2 m less than twice the width. The area is 60 m². Find the width of the rectangle.

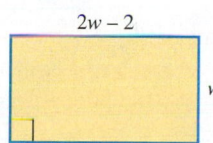
$2w - 2$
w

30. Which one of the following is equal to 1 for *all* real numbers?

   A. $\dfrac{k^2 + 2}{k^2 + 2}$ **B.** $\dfrac{4 - m}{4 - m}$ **C.** $\dfrac{2x + 9}{2x + 9}$ **D.** $\dfrac{x^2 - 1}{x^2 - 1}$

31. Which one of the following rational expressions is *not* equivalent to $\dfrac{4 - 3x}{7}$?

   A. $-\dfrac{-4 + 3x}{7}$ **B.** $-\dfrac{4 - 3x}{-7}$ **C.** $\dfrac{-4 + 3x}{-7}$ **D.** $\dfrac{-(3x + 4)}{7}$

Perform each operation, and write the answer in lowest terms.

32. $\dfrac{5}{q} - \dfrac{1}{q}$

33. $\dfrac{3}{7} + \dfrac{4}{r}$

34. $\dfrac{4}{5q - 20} - \dfrac{1}{3q - 12}$

35. $\dfrac{2}{k^2 + k} - \dfrac{3}{k^2 - k}$

36. $\dfrac{7z^2 + 49z + 70}{16z^2 + 72z - 40} \div \dfrac{3z + 6}{4z^2 - 1}$

37. $\dfrac{\dfrac{4}{a} + \dfrac{5}{2a}}{\dfrac{7}{6a} - \dfrac{1}{5a}}$

Solve each equation. Check the solutions.

38. $\dfrac{r + 2}{5} = \dfrac{r - 3}{3}$

39. $\dfrac{1}{x} = \dfrac{1}{x + 1} + \dfrac{1}{2}$

40. Jody can weed the yard in 3 hr. Pat can weed the same yard in 2 hr. How long will it take them if they work together?

	Rate	Time Working Together	Fractional Part of the Job Done
Jody			
Pat			

7

Graphs, Linear Equations, and Systems

Just as the *intersection* of two streets consists of the region common to both, a solution of a *system* of two linear equations (represented graphically by lines) is an ordered pair found in the solution sets of *both* of the individual equations.

7.1 Review of Graphs and Slopes of Lines

This section and the next review and extend some of the main topics of linear equations in two variables, first introduced in **Chapter 3.**

OBJECTIVE 1 Plot ordered pairs.

Each of the pairs of numbers $(3, 2)$, $(-5, 6)$, and $(4, -1)$ is an example of an **ordered pair**—that is, a pair of numbers written within parentheses in which the order of the numbers is important. We graph an ordered pair by using two perpendicular number lines that intersect at their 0 points, as shown in **FIGURE 1**. The common 0 point is the **origin.**

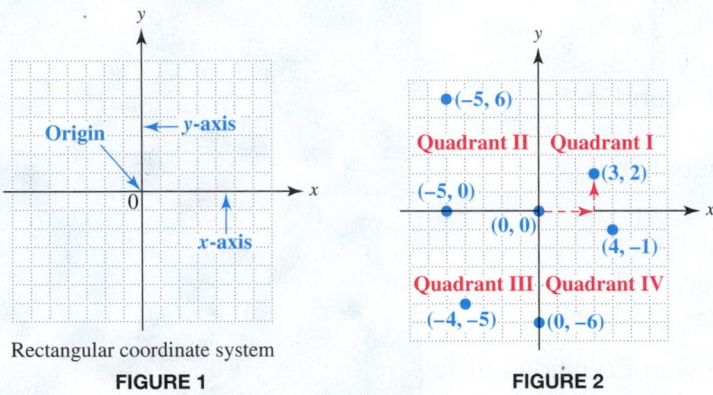

Rectangular coordinate system
FIGURE 1   **FIGURE 2**

The position of any point in this coordinate plane is determined by referring to the horizontal number line, or **x-axis,** and the vertical number line, or **y-axis.** The *x*-axis and the *y*-axis make up a **rectangular** (or **Cartesian,** for René Descartes) **coordinate system.**

The numbers in an ordered pair (x, y) are its **components.** The first component indicates position relative to the *x*-axis, and the second component indicates position relative to the *y*-axis. For example, to locate, or **plot,** the point on the graph that corresponds to the ordered pair $(3, 2)$, we move three units from 0 to the right along the *x*-axis and then two units up parallel to the *y*-axis. See **FIGURE 2**. The numbers in an ordered pair are the **coordinates** of the corresponding point.

The four regions of the graph, shown in **FIGURE 2**, are **quadrants I, II, III,** and **IV,** reading counterclockwise from the upper right quadrant. *The points on the x-axis and y-axis do not belong to any quadrant.*

OBJECTIVE 2 Graph lines and find intercepts.

Each solution of an equation with two variables, such as

$$2x + 3y = 6, \quad \text{Equation with two variables } x \text{ and } y$$

includes two numbers, one for each variable. We write the solutions as ordered pairs. *(If x and y are used as the variables, the x-value is given first.)* For example, we can show that $(6, -2)$ is a solution of $2x + 3y = 6$ by substitution.

$$2x + 3y = 6$$
$$2(6) + 3(-2) \overset{?}{=} 6 \qquad \text{Let } x = 6, y = -2.$$
$$12 - 6 \overset{?}{=} 6 \qquad \text{Multiply.}$$
$$6 = 6 \;\checkmark \qquad \text{True}$$

Use parentheses to avoid errors.

Because the ordered pair $(6, -2)$ makes the equation true, it is a solution.

On the other hand, $(5, 1)$ is *not* a solution of the equation $2x + 3y = 6$.

$$2x + 3y = 6$$

$$2(5) + 3(1) \stackrel{?}{=} 6 \qquad \text{Let } x = 5, y = 1.$$

$$10 + 3 \stackrel{?}{=} 6 \qquad \text{Multiply.}$$

$$13 = 6 \qquad \text{False}$$

To find ordered pairs that satisfy an equation, select a number for one of the variables, substitute it into the equation for that variable, and solve for the other variable.

Because any real number could be selected for one variable and would lead to a real number for the other variable, an equation with two variables such as $2x + 3y = 6$ has an infinite number of solutions.

The **graph of an equation** is the set of points corresponding to *all* ordered pairs that satisfy the equation. It gives a "picture" of the equation. The graph of $2x + 3y = 6$ is shown in **FIGURE 3** along with a **table of ordered pairs** (or **table of values**).

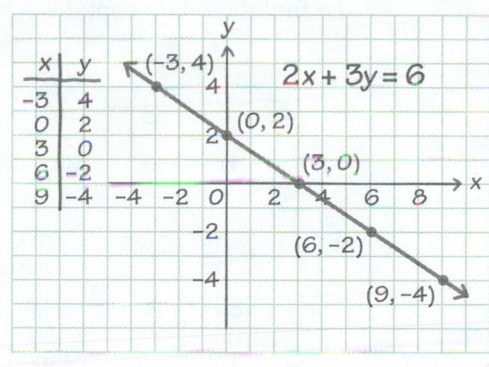

FIGURE 3

The equation $2x + 3y = 6$ is a **first-degree equation** because it has no term with a variable to a power greater than 1.

The graph of any first-degree equation in two variables is a straight line.

Because first-degree equations with two variables have straight-line graphs, they are called *linear equations in two variables.*

> ### Linear Equation in Two Variables
>
> A **linear equation in two variables** (here x and y) can be written in the form
>
> $$Ax + By = C,$$
>
> where A, B, and C are real numbers and A and B are not both 0. This form is called **standard form.**

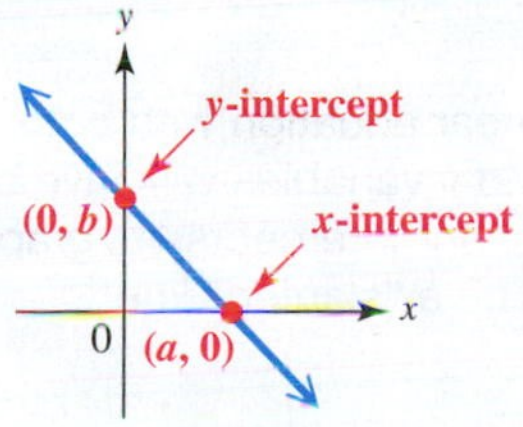

A straight line is determined if any two different points on the line are known. Therefore, finding two different points is sufficient to graph the line.

   Two useful points for graphing are the x- and y-intercepts. The **x-intercept** is the point (if any) where a line intersects the x-axis. The **y-intercept** is the point (if any) where a line intersects the y-axis.* See **FIGURE 4**.

FIGURE 4

---

*Some texts define an intercept as a number, not a point. For example, "y-intercept $(0, 4)$" would be given as "y-intercept 4."

Remember the following.

- The y-value of the point where a line intersects the x-axis is always 0.
- The x-value of the point where a line intersects the y-axis is always 0.

This suggests a method for finding the x- and y-intercepts.

Finding Intercepts

When graphing the equation of a line, find the intercepts as follows.

Let $y = 0$ to find the x-intercept.

Let $x = 0$ to find the y-intercept.

NOW TRY
EXERCISE 1

Find the x- and y-intercepts, and graph the equation.

$$x - 2y = 4$$

EXAMPLE 1   Finding Intercepts

Find the x- and y-intercepts of $4x - y = -3$, and graph the equation.

To find the x-intercept, let $y = 0$.

$$4x - y = -3$$
$$4x - 0 = -3 \qquad \text{Let } y = 0.$$
$$4x = -3 \qquad \text{Subtract.}$$
$$x = -\frac{3}{4} \qquad \text{Divide by 4.}$$

The x-intercept is $\left(-\frac{3}{4}, 0\right)$.

To find the y-intercept, let $x = 0$.

$$4x - y = -3$$
$$4(0) - y = -3 \qquad \text{Let } x = 0.$$
$$-y = -3 \qquad \text{Multiply. Subtract.}$$
$$y = 3 \qquad \text{Multiply by } -1.$$

The y-intercept is $(0, 3)$.

To guard against errors when graphing the equation, it is a good idea to find a third point. We arbitrarily choose $x = -2$, and substitute this value in the equation to find the corresponding value of y.

$$4x - y = -3$$
$$4(-2) - y = -3 \qquad \text{Let } x = -2.$$
$$-8 - y = -3 \qquad \text{Multiply.}$$
$$-y = 5 \qquad \text{Add 8.}$$
$$y = -5 \qquad \text{Multiply by } -1.$$

The ordered pair $(-2, -5)$ lies on the graph. We plot the three ordered pairs and draw a line through them. See **FIGURE 5.**

NOW TRY ANSWER

1. x-intercept: $(4, 0)$;
   y-intercept: $(0, -2)$

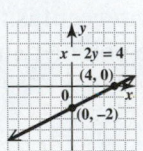

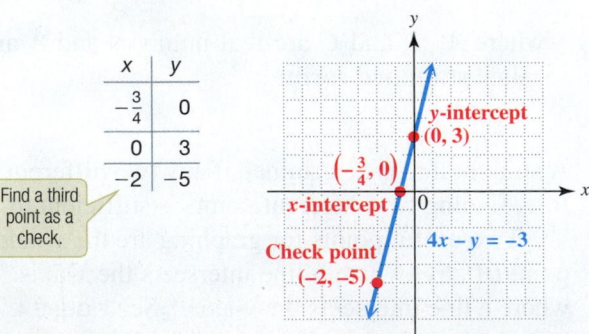

FIGURE 5

A linear equation with both x and y variables will have both x- and y-intercepts. Its graph will be a "slanted" line.

NOW TRY

**NOW TRY
EXERCISE 2**

Graph $2x + 3y = 0$.

EXAMPLE 2 Graphing a Line That Passes through the Origin

Graph $x + 2y = 0$.

Find the *x*-intercept.	Find the *y*-intercept.
$x + 2y = 0$	$x + 2y = 0$
$x + 2(0) = 0$ Let $y = 0$.	$0 + 2y = 0$ Let $x = 0$.
$x + 0 = 0$ Multiply.	$2y = 0$ Add.
$x = 0$ *x*-intercept is $(0, 0)$.	$y = 0$ *y*-intercept is $(0, 0)$.

Both intercepts are the *same* point, $(0, 0)$, which means that the graph passes through the origin. To find a second point, we choose any nonzero number for x or y and solve for the other variable. We arbitrarily choose $x = 4$.

$$x + 2y = 0$$
$$4 + 2y = 0 \quad \text{Let } x = 4.$$
$$2y = -4 \quad \text{Subtract 4.}$$
$$y = -2 \quad \text{Divide by 2.}$$

x	y
-2	1
0	0
4	-2

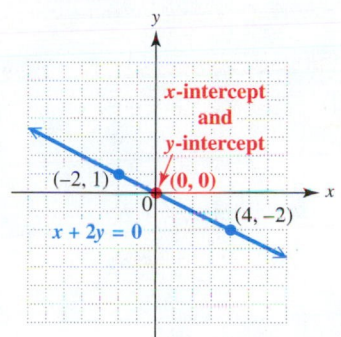

FIGURE 6   NOW TRY

This gives the ordered pair $(4, -2)$. As a final check, substitute 1 for y in the equation and verify that $(-2, 1)$ also lies on the line. The graph is shown in **FIGURE 6**.

OBJECTIVE 3 Recognize equations of horizontal and vertical lines.

**NOW TRY
EXERCISE 3**

Graph each equation.

(a) $y = -2$ **(b)** $x + 3 = 0$

EXAMPLE 3 Graphing Horizontal and Vertical Lines

Graph each equation.

(a) $y = 2$ (This equation can be written as $0x + y = 2$.)

Because y *always* equals 2, there is no value of x corresponding to $y = 0$, and the graph has no *x*-intercept. One value where $y = 2$ is on the *y*-axis, so the *y*-intercept is $(0, 2)$. Plot any two other points with *y*-coordinate 2, such as $(-1, 2)$ and $(3, 2)$.

The graph is shown in **FIGURE 7**. It is a horizontal line.

x	y
-1	2
0	2
3	2

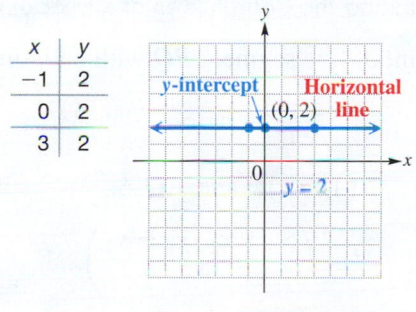

x	y
-1	-4
-1	0
-1	5

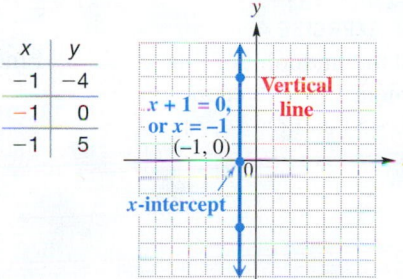

FIGURE 7   **FIGURE 8**

(b) $x + 1 = 0$ (This equation can be written as $x = -1$ or $x + 0y = -1$.)

Because x *always* equals -1, there is no value of y that makes $x = 0$, and the graph has no *y*-intercept. One value where $x = -1$ is on the *x*-axis, so the *x*-intercept is $(-1, 0)$. Plot any two other points with *x*-coordinate -1, such as $(-1, -4)$ and $(-1, 5)$.

The graph is shown in **FIGURE 8**. It is a vertical line.

NOW TRY

NOW TRY ANSWERS

2.

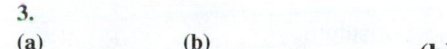

3.

(a)   **(b)**

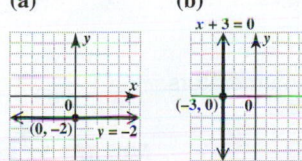

OBJECTIVE 4 Use the midpoint formula.

If the coordinates of the endpoints of a line segment are known, then the coordinates of the *midpoint* of the segment can be found.

FIGURE 9 shows a line segment PQ with endpoints $P(-8, 4)$ and $Q(3, -2)$. R is the point with the same x-coordinate as P and the same y-coordinate as Q. So the coordinates of R are $(-8, -2)$.

The x-coordinate of the midpoint M of PQ is the same as the x-coordinate of the midpoint of RQ. Because RQ is horizontal, the x-coordinate of its midpoint is the *average* (or *mean*) of the x-coordinates of its endpoints.

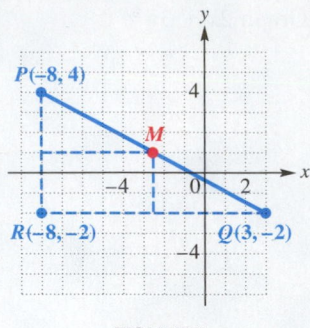

FIGURE 9

$$\frac{1}{2}(-8 + 3) = -2.5$$

The y-coordinate of M is the average (or *mean*) of the y-coordinates of the midpoint of PR.

$$\frac{1}{2}(4 + (-2)) = 1$$

The midpoint of PQ is $M(-2.5, 1)$.

This discussion leads to the *midpoint formula*.

Midpoint Formula

The midpoint M of a line segment PQ with endpoints (x_1, y_1) and (x_2, y_2) is found as follows.

$$M = \left(\frac{x_1 + x_2}{2}, \frac{y_1 + y_2}{2}\right)$$

Recall that the two nonspecific points (x_1, y_1) and (x_2, y_2) use **subscript notation.** Read (x_1, y_1) as **"x-sub-one, y-sub-one."**

NOW TRY
EXERCISE 4

Find the midpoint of line segment PQ with endpoints $P(2, -5)$ and $Q(-4, 7)$.

EXAMPLE 4 Finding the Coordinates of a Midpoint

Find the midpoint of line segment PQ with endpoints $P(4, -3)$ and $Q(6, -1)$.

$$\underset{\downarrow \quad \downarrow}{(x_1, \quad y_1)} \qquad \underset{\downarrow \quad \downarrow}{(x_2, \quad y_2)}$$

$$P(4, -3) \quad \text{and} \quad Q(6, -1) \qquad \text{Label the points.}$$

$$M = \left(\frac{x_1 + x_2}{2}, \frac{y_1 + y_2}{2}\right) \qquad \text{Midpoint formula}$$

> We are finding the average of the x-coordinates and the average of the y-coordinates.

$$= \left(\frac{4 + 6}{2}, \frac{-3 + (-1)}{2}\right) \qquad \text{Substitute.}$$

$$= \left(\frac{10}{2}, \frac{-4}{2}\right) \qquad \text{Add in the numerators.}$$

NOW TRY ANSWER
4. $(-1, 1)$

$$= (5, -2) \leftarrow \text{Midpoint of segment } PQ \qquad \text{NOW TRY}$$

NOTE When graphing with a graphing calculator, we "set up" a rectangular coordinate system. In the screen in **FIGURE 10**, which shows the **standard viewing window**, minimum *x*- and *y*-values are −10 and maximum *x*- and *y*-values are 10. The **scale** on each axis, here 1, determines the distance between the tick marks.

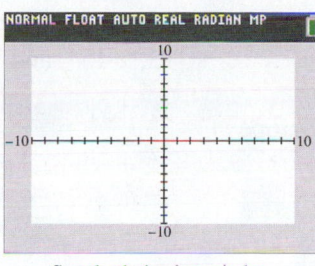

Standard viewing window
FIGURE 10

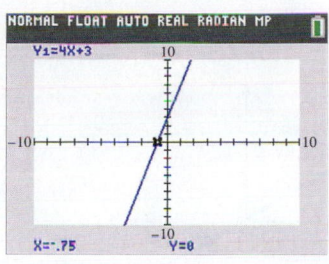

FIGURE 11

To graph an equation such as $4x - y = -3$, we use the intercepts $(-0.75, 0)$ and $(0, 3)$ to determine an appropriate window. Here, we choose the standard viewing window. We solve the equation for *y* to obtain $y = 4x + 3$ and enter it into the calculator. The graph in **FIGURE 11** also gives the *x*-intercept at the bottom of the screen.

OBJECTIVE 5 Find the slope of a line.

Slope (steepness) is used in many practical ways, as shown in **FIGURE 12**.

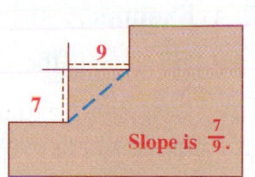
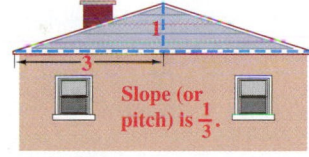

FIGURE 12

Slope is the ratio of vertical change, or **rise,** to horizontal change, or **run.** A simple way to remember this is to think, *"Slope is rise over run."*

To obtain a formal definition of the slope of a line, we designate two different points on the line as (x_1, y_1) and (x_2, y_2). See **FIGURE 13**. As we move along the line in **FIGURE 13** from (x_1, y_1) to (x_2, y_2), the *y*-value changes (vertically) from y_1 to y_2, an amount equal to $y_2 - y_1$. As *y* changes from y_1 to y_2, the value of *x* changes (horizontally) from x_1 to x_2 by the amount $x_2 - x_1$.

The ratio of the change in *y* to the change in *x* ("rise over run," or $\frac{\text{rise}}{\text{run}}$) is the *slope* of the line, with the letter *m* traditionally used for slope.

The Greek letter **delta** Δ denotes "change in," so Δy and Δx represent the change in *y* and the change in *x*, respectively.

FIGURE 13

Slope Formula

The **slope** *m* of the line passing through the distinct points (x_1, y_1) and (x_2, y_2) is defined as follows.

$$m = \frac{\text{rise}}{\text{run}} = \frac{\text{change in } y}{\text{change in } x} = \frac{\Delta y}{\Delta x} = \frac{y_2 - y_1}{x_2 - x_1} \quad (\text{where } x_1 \neq x_2)$$

NOW TRY
EXERCISE 5

Find the slope of the line passing through the points $(2, -6)$ and $(-3, 5)$.

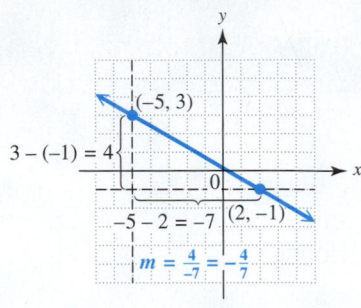

FIGURE 14

EXAMPLE 5   Finding the Slope of a Line

Find the slope of the line passing through the points $(2, -1)$ and $(-5, 3)$.

Label the points, and then apply the slope formula.

$$(x_1, \ y_1) \qquad\qquad (x_2, \ y_2)$$
$$\downarrow \ \downarrow \qquad\qquad\quad \downarrow \ \downarrow$$
$$(2, -1) \quad \text{and} \quad (-5, 3)$$

$$\text{slope } m = \frac{y_2 - y_1}{x_2 - x_1} = \frac{3 - (-1)}{-5 - 2} \qquad \text{Substitute.}$$

$$= \frac{4}{-7}, \quad \text{or} \quad -\frac{4}{7} \qquad \text{Subtract; } \frac{a}{-b} = -\frac{a}{b}$$

The slope is $-\dfrac{4}{7}$. See **FIGURE 14**.

The same slope is obtained if we label the points in reverse order. ***It makes no difference which point is identified as (x_1, y_1) or (x_2, y_2).***

$$(x_2, \ y_2) \qquad\qquad (x_1, \ y_1)$$
$$\downarrow \ \downarrow \qquad\qquad\quad \downarrow \ \downarrow$$
$$(2, -1) \quad \text{and} \quad (-5, 3)$$

> *y*-values are in the numerator, *x*-values in the denominator.

$$\text{slope } m = \frac{y_2 - y_1}{x_2 - x_1} = \frac{-1 - 3}{2 - (-5)} \qquad \text{Substitute.}$$

$$= \frac{-4}{7}, \quad \text{or} \quad -\frac{4}{7} \qquad \text{Subtract; } \frac{-a}{b} = -\frac{a}{b} \qquad \textbf{NOW TRY}$$

Example 5 suggests the following important ideas regarding slope.

1. The slope is the same no matter which point we consider first.

2. Using similar triangles from geometry, we can show that the slope is the same no matter which two different points on the line we choose.

⚠ **CAUTION**   *When calculating slope, remember that the change in y (rise) is the numerator and the change in x (run) is the denominator. Be careful to subtract the y-values and the x-values in the same order.*

Correct	Incorrect
$\dfrac{y_2 - y_1}{x_2 - x_1}$	$\dfrac{x_2 - x_1}{y_2 - y_1}$ or $\dfrac{y_2 - y_1}{x_1 - x_2}$ or $\dfrac{y_1 - y_2}{x_2 - x_1}$

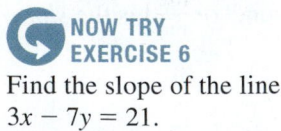
NOW TRY
EXERCISE 6

Find the slope of the line $3x - 7y = 21$.

EXAMPLE 6   Finding the Slope of a Line

Find the slope of the line $4x - y = -8$.

The intercepts can be used as the two points needed to find the slope. Let $y = 0$ to find that the *x*-intercept is $(-2, 0)$. Let $x = 0$ to find that the *y*-intercept is $(0, 8)$.

$$\text{slope } m = \frac{y_2 - y_1}{x_2 - x_1} = \frac{8 - 0}{0 - (-2)} \qquad \begin{array}{l} (x_1, y_1) = (-2, 0) \\ (x_2, y_2) = (0, 8) \end{array}$$

$$= \frac{8}{2} \qquad \text{Subtract.}$$

$$= 4 \qquad \text{Divide.} \qquad \textbf{NOW TRY}$$

The slope of a line can also be found directly from its equation. Consider the equation $4x - y = -8$ from **Example 6.** Solve this equation for y.

$$4x - y = -8 \qquad \text{Equation from Example 6}$$

$$-y = -4x - 8 \qquad \text{Subtract } 4x.$$

$$y = 4x + 8 \qquad \text{Multiply by } -1.$$

The slope, 4, found with the slope formula in **Example 6** is the same number as the coefficient of x in the equation $y = 4x + 8$. We will see in the next section that this is true in general, *as long as the equation is solved for y.*

NOW TRY
EXERCISE 7

Find the slope of the graph of $5x - 4y = 7$.

EXAMPLE 7 Finding the Slope from an Equation

Find the slope of the graph of $3x - 5y = 8$.

Solve the equation for y.

$$3x - 5y = 8 \quad \triangleleft \boxed{\text{Solve for } y.}$$

$$-5y = -3x + 8 \qquad \text{Subtract } 3x.$$

$$\frac{-5y}{-5} = \frac{-3x + 8}{-5} \qquad \text{Divide each side by } -5.$$

$$\boxed{\frac{-3x}{-5} = \frac{-3}{-5} \cdot \frac{x}{1} = \frac{3}{5}x} \quad y = \frac{3}{5}x - \frac{8}{5} \qquad \frac{a+b}{c} = \frac{a}{c} + \frac{b}{c}$$

The slope is given by the coefficient of x, so the slope is $\frac{3}{5}$. **NOW TRY**

NOTE We can solve the standard form of a linear equation $Ax + By = C$ (where $B \neq 0$) for y to show that, in general, **the slope of a line in this form is $-\frac{A}{B}$.**

$$Ax + By = C \qquad \text{Standard form}$$

$$By = -Ax + C \qquad \text{Subtract } Ax.$$

$$y = -\frac{A}{B}x + \frac{C}{B} \qquad \text{Divide each term by } B.$$

The slope is given by the coefficient of x, $-\frac{A}{B}$. In the equation $3x - 5y = 8$ from **Example 7,** $A = 3$ and $B = -5$, so the slope is

$$-\frac{A}{B} = -\frac{3}{-5} = \frac{3}{5}. \qquad \text{The same slope results.}$$

We review the following special cases of slope.

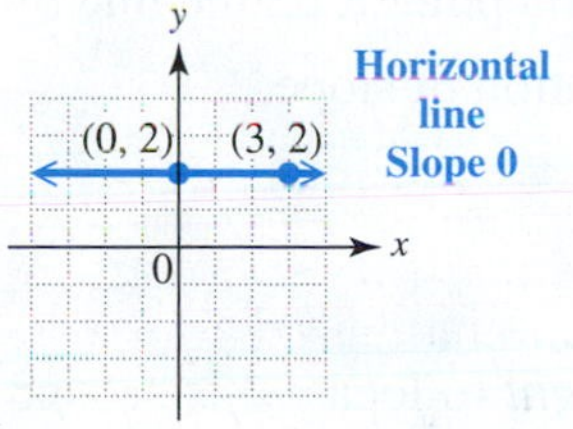

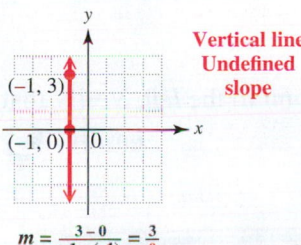

$$m = \frac{2-2}{3-0} = \frac{0}{3} = 0$$

FIGURE 15

$$m = \frac{3-0}{-1-(-1)} = \frac{3}{0}$$

FIGURE 16

Horizontal and Vertical Lines

- An equation of the form $y = b$ always intersects the y-axis at the point $(0, b)$. A line with this equation is **horizontal** and has **slope 0.** See **FIGURE 15.**

- An equation of the form $x = a$ always intersects the x-axis at the point $(a, 0)$. A line with this equation is **vertical** and has **undefined slope.** See **FIGURE 16.**

NOW TRY ANSWER

7. $\frac{5}{4}$

OBJECTIVE 6 Graph a line given its slope and a point on the line.

EXAMPLE 8   Using the Slope and a Point to Graph Lines

Graph each line described.

(a) With slope $\frac{2}{3}$ and y-intercept $(0, -4)$

Begin by plotting the point $P(0, -4)$, as shown in **FIGURE 17**. Then use the geometric interpretation of slope to find a second point.

$$m = \frac{\text{change in } y}{\text{change in } x} = \frac{2}{3} \begin{matrix} \leftarrow \text{rise} \\ \leftarrow \text{run} \end{matrix}$$

We move 2 units *up* from $(0, -4)$ and then 3 units to the *right* to locate another point on the graph, $R(3, -2)$. The line through $P(0, -4)$ and R is the graph.

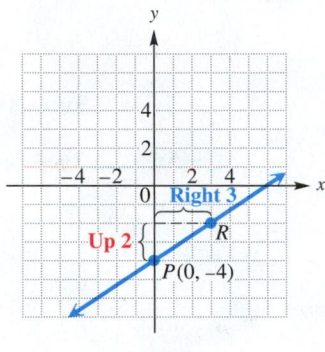
FIGURE 17

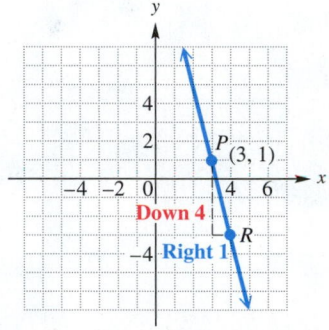
FIGURE 18

(b) Through $(3, 1)$ with slope -4

Plot the point $P(3, 1)$, as shown in **FIGURE 18**. Find a second point R on the line by writing the slope -4 as $\frac{-4}{1}$ and using the geometric interpretation of slope.

$$m = \frac{\text{change in } y}{\text{change in } x} = \frac{-4}{1} \begin{matrix} \leftarrow \text{rise} \\ \leftarrow \text{run} \end{matrix}$$

We move 4 units *down* from $(3, 1)$ and then 1 unit to the *right* to locate this second point $R(4, -3)$. The line through $P(3, 1)$ and R is the graph.

The slope -4 also could be written as

$$m = \frac{\text{change in } y}{\text{change in } x} = \frac{4}{-1}.$$

In this case, the second point R is located 4 units *up* and 1 unit to the *left*. Verify that this approach also produces the line in **FIGURE 18**.   **NOW TRY**

Orientation of a Line in the Plane

A positive slope indicates that the line slants *up* (rises) from left to right. See **FIGURE 17**.

A negative slope indicates that the line slants *down* (falls) from left to right. See **FIGURE 18**.

FIGURE 19 summarizes the four cases for slopes of lines.

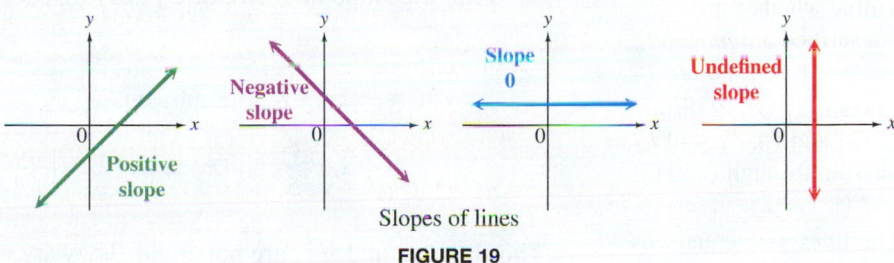

Slopes of lines

FIGURE 19

OBJECTIVE 7 Use slopes to determine whether two lines are parallel, perpendicular, or neither.

Recall that the slopes of a pair of parallel or perpendicular lines are related.

> ## Slopes of Parallel Lines and Perpendicular Lines
>
> - Two nonvertical lines with the same slope are parallel. Two nonvertical parallel lines have the same slope.
>
> - Two perpendicular lines, neither of which is vertical, have slopes that are negative reciprocals—that is, their product is -1. Also, lines with slopes that are negative reciprocals are perpendicular.
>
> - A line with slope 0 is perpendicular to a line with undefined slope.

EXAMPLE 9 Determining Whether Two Lines Are Parallel, Perpendicular, or Neither

Determine whether the two lines described are *parallel, perpendicular*, or *neither*.

(a) Lines L_1, passing through $(-2, 1)$ and $(4, 5)$, and L_2, passing through $(3, 0)$ and $(0, -2)$

$$\text{Slope of } L_1: \quad m_1 = \frac{5 - 1}{4 - (-2)} = \frac{4}{6} = \frac{2}{3} \qquad \begin{array}{l}(x_1, y_1) = (-2, 1) \\ (x_2, y_2) = (4, 5)\end{array}$$

$$\text{Slope of } L_2: \quad m_2 = \frac{-2 - 0}{0 - 3} = \frac{-2}{-3} = \frac{2}{3} \qquad \begin{array}{l}(x_1, y_1) = (3, 0) \\ (x_2, y_2) = (0, -2)\end{array}$$

Because the slopes are equal, the two lines are parallel.

(b) The lines with equations $2y = 3x - 6$ and $2x + 3y = -6$

Find the slope of each line by solving each equation for y.

$2y = 3x - 6$	$2x + 3y = -6$
$y = \dfrac{3}{2}x - 3$ Divide by 2.	$3y = -2x - 6$ Subtract $2x$.
$\uparrow$ Slope	$y = -\dfrac{2}{3}x - 2$ Divide by 3.
	$\uparrow$ Slope

The slopes are negative reciprocals because their product is $\frac{3}{2}\left(-\frac{2}{3}\right) = -1$. The lines are perpendicular.

**NOW TRY
EXERCISE 9**

Determine whether the two lines described are *parallel*, *perpendicular*, or *neither*.

(a) Lines L_1, passing through $(2, 5)$ and $(4, 8)$, and L_2, passing through $(2, 0)$ and $(-1, -2)$

(b) The lines with equations
$$x + 2y = 7$$
and $$2x = y - 4$$

(c) The lines with equations
$$2x - y = 4$$
and $$-2x + y = 6$$

(c) The lines with equations $2x - 5y = 8$ and $2x + 5y = 8$

Find the slope of each line by solving each equation for y.

$2x - 5y = 8$	$2x + 5y = 8$
$-5y = -2x + 8$ Subtract $2x$.	$5y = -2x + 8$ Subtract $2x$.
$y = \dfrac{2}{5}x - \dfrac{8}{5}$ Divide by -5.	$y = -\dfrac{2}{5}x + \dfrac{8}{5}$ Divide by 5.

The slopes, $\frac{2}{5}$ and $-\frac{2}{5}$, are not equal. They are not negative reciprocals—their product is $-\frac{4}{25}$, *not* -1. The two lines are *neither* parallel nor perpendicular.   **NOW TRY**

OBJECTIVE 8   **Solve problems involving average rate of change.**

The slope formula applied to any two points on a line gives the **average rate of change** in y per unit change in x, where the value of y depends on the value of x.

For example, suppose the height of a boy increased from 60 to 68 in. between the ages of 12 and 16, as shown in **FIGURE 20**.

Change in height $y \longrightarrow$ $\dfrac{68 - 60}{16 - 12} = \dfrac{8}{4} = 2$ in.
Change in age $x \longrightarrow$

Boy's average growth rate (or average change in height) *per year*

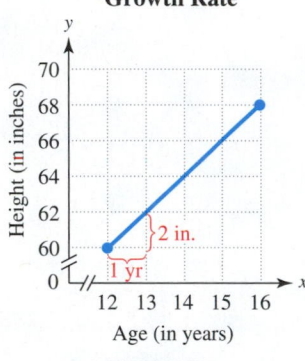

Growth Rate

FIGURE 20

The boy may actually have grown more than 2 in. during some years and less than 2 in. during other years. If we plotted ordered pairs (age, height) for those years and drew a line connecting any two of the points, the average rate of change would likely be slightly different than that found above. However, using the data for ages 12 and 16, the boy's *average* change in height was 2 in. per year over these years.

EXAMPLE 10   Interpreting Slope as Average Rate of Change

The graph in **FIGURE 21** shows the number of digital cable TV customers in the United States from 2007 to 2012. Find the average rate of change in number of customers per year.

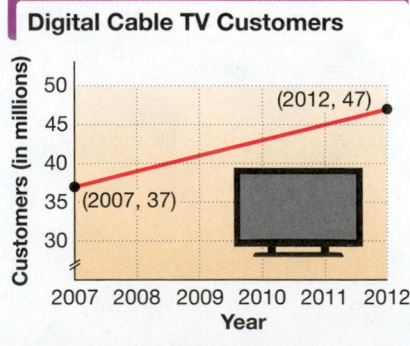

Digital Cable TV Customers

Source: SNL Kagan.

FIGURE 21

**NOW TRY
EXERCISE 10**

There were approximately 40 million digital cable TV customers in 2008. Using this number for 2008 and the number for 2012 from the graph in **FIGURE 21**, find the average rate of change from 2008 to 2012. How does it compare with the average rate of change found in **Example 10?**

**NOW TRY
EXERCISE 11**

In 2010, sales of digital camcorders in the United States totaled $1150 million. In 2013, sales totaled $137 million. Find the average rate of change in sales of digital camcorders per year, to the nearest million dollars. (*Source:* Consumer Electronics Association.)

NOW TRY ANSWERS

10. 1.75 million customers per year; It is less than the average rate of change from 2007 to 2012.

11. −$338 million per year

To find the average rate of change, we need two pairs of data. From the graph, we have the ordered pairs (2007, 37) and (2012, 47). We use the slope formula.

$$\text{average rate of change} = \frac{47 - 37}{2012 - 2007} = \frac{10}{5} = 2$$

A positive slope indicates an increase.

This means that the number of digital TV customers *increased* by an average of 2 million customers per year from 2007 to 2012.

NOW TRY

EXAMPLE 11 Interpreting Slope as Average Rate of Change

In 2006, there were 65 million basic cable TV customers in the United States. There were 56 million such customers in 2012. Find the average rate of change in the number of customers per year. (*Source:* SNL Kagan.)

To use the slope formula, we let one ordered pair be (2006, 65) and the other be (2012, 56).

$$\text{average rate of change} = \frac{56 - 65}{2012 - 2006} = \frac{-9}{6} = -1.5$$

A negative slope indicates a decrease.

The graph in **FIGURE 22** confirms that the line through the ordered pairs falls from left to right and therefore has negative slope. Thus, the number of basic cable TV customers *decreased* by an average of 1.5 million customers per year from 2006 to 2012.

The negative sign in −1.5 denotes the *decrease*. (We say "The number of customers decreased by 1.5 million per year." It is *incorrect* to say "The number of customers decreased by −1.5 million per year.")

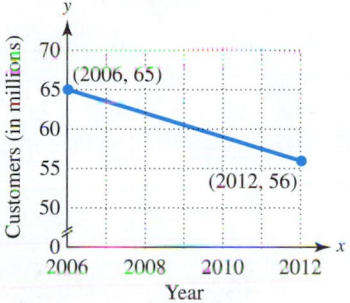

Basic Cable TV Customers

FIGURE 22

NOW TRY

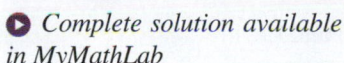

7.1 Exercises

FOR EXTRA HELP

 MyMathLab®

▶ *Complete solution available in MyMathLab*

1. *Concept Check* Name the quadrant, if any, in which each point is located.
   (a) (1, 6) (b) (−4, −2) (c) (−3, 6) (d) (7, −5) (e) (−3, 0) (f) (0, −8)

2. *Concept Check* Plot each point in a rectangular coordinate system.
   (a) (2, 3) (b) (−3, −2) (c) (0, 5) (d) (−2, 4) (e) (−2, 0) (f) (3, −3)

In the following exercises, (a) complete the given table for each equation and then (b) graph the equation. See Objective 2 and FIGURE 3.

▶ **3.** $x - y = 3$

x	y
0	
	0
5	
2	

4. $x + 2y = 5$

x	y
0	
	0
2	
	2

5. $4x - 5y = 20$

x	y
0	
	0
2	
	−3

6. $y = -2x + 3$

x	y
0	
1	
2	
3	

Find the x- and y-intercepts. Then graph each equation. See Examples 1–3.

▶ **7.** $2x + 3y = 12$ **8.** $5x + 2y = 10$ **9.** $x - 3y = 6$

10. $x - 2y = -4$ **11.** $5x + 6y = -10$ **12.** $3x - 7y = 9$

▶ **13.** $y = 5$ **14.** $y = -3$ ▶ **15.** $x + 4 = 0$ **16.** $x - 4 = 0$

▶ **17.** $x + 5y = 0$ **18.** $x - 3y = 0$ **19.** $2x = 3y$ **20.** $4y = 3x$

Each table of values gives several points that lie on a line.

(a) What is the x-intercept of the line?   The y-intercept?

(b) Which equation in choices A–D corresponds to the given table of values?

(c) Graph the equation.

21.

x	y
-4	-3
-2	0
0	3
2	6

  A. $3x + 2y = 6$
  B. $3x - 2y = -6$
  C. $3x + 2y = -6$
  D. $3x - 2y = 6$

22.

x	y
-1	6
0	4
1	2
2	0

  A. $2x - y = 4$
  B. $2x + y = -4$
  C. $2x + y = 4$
  D. $2x - y = -4$

23. *Concept Check*   Match each equation in parts (a)–(d) with its graph in choices A–D. (Coordinates of the points shown are integers.)

  (a) $x + 3y = 3$ **(b)** $x - 3y = -3$ **(c)** $x - 3y = 3$ **(d)** $x + 3y = -3$

A. 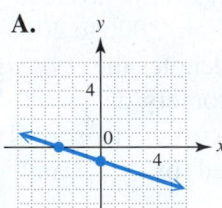   **B.** 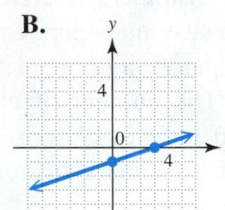   **C.** 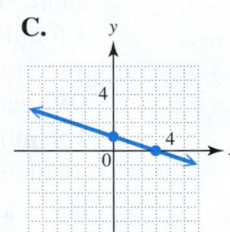   **D.**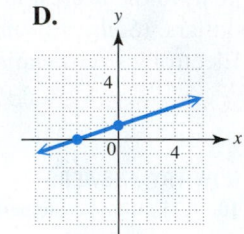

24. *Concept Check*   Which of the following equations have a graph that is a horizontal line?   A vertical line?

  A. $x - 6 = 0$ **B.** $x + y = 0$ **C.** $y + 3 = 0$ **D.** $y = -10$ **E.** $x + 1 = 5$

Find the midpoint of each segment with the given endpoints. See Example 4.

▶ **25.** $(-8, 4)$ and $(-2, -6)$ **26.** $(5, 2)$ and $(-1, 8)$

27. $(3, -6)$ and $(6, 3)$ **28.** $(-10, 4)$ and $(7, 1)$

29. $(-9, 3)$ and $(9, 8)$ **30.** $(4, -3)$ and $(-1, 3)$

31. $(2.5, 3.1)$ and $(1.7, -1.3)$ **32.** $(6.2, 5.8)$ and $(1.4, -0.6)$

Concept Check   *Answer each question.*

33. A hill rises 30 ft for every horizontal 100 ft. Which of the following express its slope (or grade)? (There are several correct choices.)

  A. 0.3 **B.** $\dfrac{3}{10}$ **C.** $3\dfrac{1}{3}$

  D. $\dfrac{30}{100}$ **E.** $\dfrac{10}{3}$ **F.** 30%

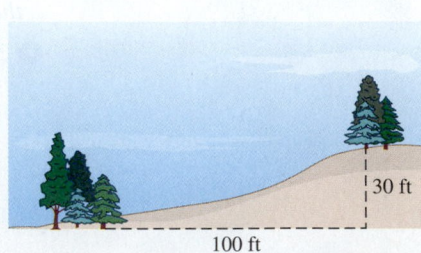

34. If a walkway rises 2 ft for every 24 ft on the horizontal, which of the following express its slope (or grade)? (There are several correct choices.)

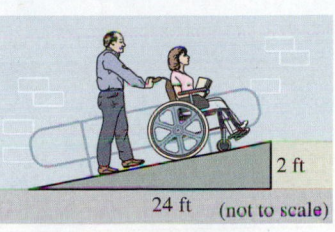

A. 12%   **B.** $\dfrac{2}{24}$ **C.** $\dfrac{1}{12}$

D. 12   **E.** $8.\overline{3}\%$ **F.** $\dfrac{24}{2}$

2 ft

24 ft   (not to scale)

35. *Concept Check*   Determine the slope of each line segment in the given figure.

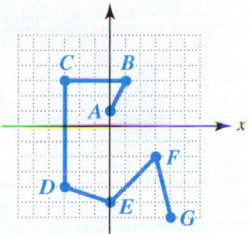

   (a) *AB*   **(b)** *BC*   **(c)** *CD*

   (d) *DE*   **(e)** *EF*   **(f)** *FG*

   (g) If *A* and *F* were joined by a line segment in the figure, what would be its slope?

   (h) If *B* and *D* were joined by a line segment in the figure, what would be its slope?

36. *Concept Check*   Which forms of the slope formula are correct? Explain.

A. $m = \dfrac{y_1 - y_2}{x_2 - x_1}$ **B.** $m = \dfrac{y_1 - y_2}{x_1 - x_2}$ **C.** $m = \dfrac{x_2 - x_1}{y_2 - y_1}$ **D.** $m = \dfrac{y_2 - y_1}{x_2 - x_1}$

*In the following exercises, **(a)** find the slope of the line passing through each pair of points, if possible, and **(b)** based on the slope, indicate whether the line* rises *from left to right,* falls *from left to right, is* horizontal, *or is* vertical. *See Example 5 and* **FIGURE 19**.

37. $(-2, -3)$ and $(-1, 5)$ **38.** $(-4, 1)$ and $(-3, 4)$ ▶ **39.** $(-4, 1)$ and $(2, 6)$

40. $(-3, -3)$ and $(5, 6)$ **41.** $(2, 4)$ and $(-4, 4)$ **42.** $(-6, 3)$ and $(2, 3)$

43. $(-2, 2)$ and $(4, -1)$ **44.** $(-3, 1)$ and $(6, -2)$ **45.** $(5, -3)$ and $(5, 2)$

46. $(4, -1)$ and $(4, 3)$ **47.** $(1.5, 2.6)$ and $(0.5, 3.6)$ **48.** $(3.4, 4.2)$ and $(1.4, 10.2)$

Each table of values gives several points that lie on a line. Find the slope of the line.

49.

x	y
−1	8
0	6
2	2
3	0

50.

x	y
−3	6
−1	0
0	−3
2	−9

51.

x	y
−6	−4
−3	0
0	4
3	8

52.

x	y
−5	−4
0	−2
5	0
10	2

Use the geometric interpretation of slope ("rise over run") to find the slope of each line. (Coordinates of the points shown are integers.)

53.

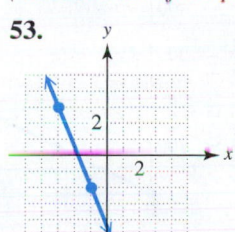

54.

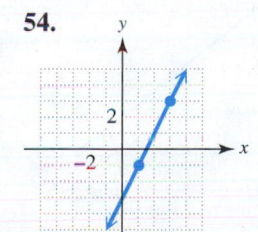

55.

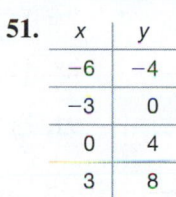

56.

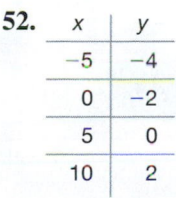
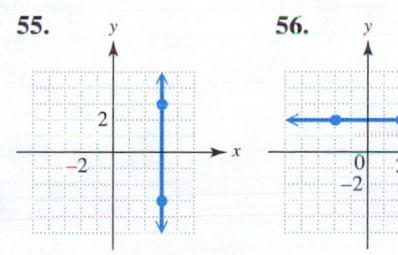

Find the slope of each line in three ways by doing the following.

*(a) Give any two points that lie on the line, and use them to determine the slope. **See Example 6.***

*(b) Solve the equation for y, and identify the slope from the equation. **See Example 7.***

*(c) For the form $Ax + By = C$, calculate $-\frac{A}{B}$. **See the Note following Example 7.***

57. $2x - y = 8$ **58.** $3x + 4y = 12$ **59.** $x + y = -3$ **60.** $x - y = 4$

*Find the slope of each line, and sketch its graph. **See Examples 5–7.***

▶ **61.** $x + 2y = 4$ **62.** $x + 3y = -6$ ▶ **63.** $5x - 2y = 10$

64. $4x - y = 4$ **65.** $y = 4x$ **66.** $y = -3x$

▶ **67.** $x - 3 = 0$ **68.** $x + 2 = 0$ ▶ **69.** $y = -5$ **70.** $y = -4$

*Graph each line described. **See Example 8.***

71. Through $(-4, 2)$; $m = \frac{1}{2}$ **72.** Through $(-2, -3)$; $m = \frac{5}{4}$

▶ **73.** *y*-intercept $(0, -2)$; $m = -\frac{2}{3}$ **74.** *y*-intercept $(0, -4)$; $m = -\frac{3}{2}$

75. Through $(-1, -2)$; $m = 3$ **76.** Through $(-2, -4)$; $m = 4$

77. $m = 0$; through $(2, -5)$ **78.** $m = 0$; through $(5, 3)$

79. Undefined slope; through $(-3, 1)$ **80.** Undefined slope; through $(-4, 1)$

Determine whether each pair of lines is parallel, perpendicular, *or* neither. ***See Example 9.***

▶ **81.** The line passing through $(15, 9)$ and $(12, -7)$ and the line passing through $(8, -4)$ and $(5, -20)$

82. The line passing through $(4, 6)$ and $(-8, 7)$ and the line passing through $(-5, 5)$ and $(7, 4)$

▶ **83.** $x + 4y = 7$ and $4x - y = 3$ **84.** $2x + 5y = -7$ and $5x - 2y = 1$

85. $4x - 3y = 6$ and $3x - 4y = 2$ **86.** $2x + y = 6$ and $x - y = 4$

87. $x = 6$ and $6 - x = 8$ **88.** $3x = y$ and $2y - 6x = 5$

89. $4x + y = 0$ and $5x - 8 = 2y$ **90.** $2x + 5y = -8$ and $6 + 2x = 5y$

91. $2x = y + 3$ and $2y + x = 3$ **92.** $4x - 3y = 8$ and $4y + 3x = 12$

*Find and interpret the average rate of change illustrated in each graph. **See Objective 8 and*** **FIGURE 20**.

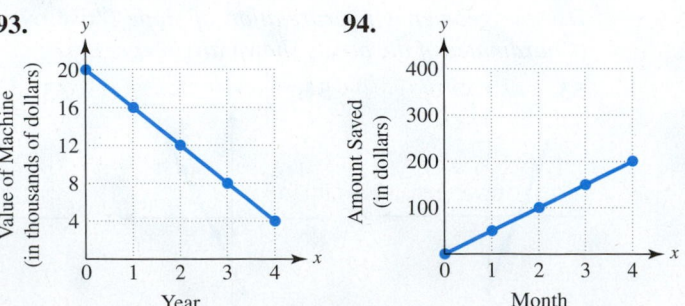

93.   **94.**   **95.**

96. *Concept Check* If the graph of a linear equation rises from left to right, then the average rate of change is (*positive* / *negative*). If the graph of a linear equation falls from left to right, then the average rate of change is (*positive* / *negative*).

*Solve each problem. **See Examples 10 and 11.***

97. The graph shows the number of wireless subscriber connections (that is, active devices, including smartphones, feature phones, tablets, etc.) in millions in the United States for the years 2007 to 2012.

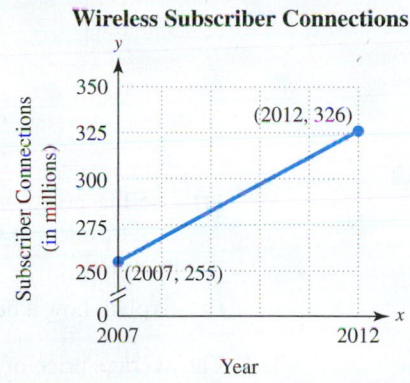

Wireless Subscriber Connections

Source: CTIA.

(a) In the context of this graph, what does the ordered pair (2012, 326) mean?

(b) Use the given ordered pairs to find the slope of the line.

(c) Interpret the slope in the context of this problem.

98. The graph shows the percent of households in the United States that were wireless-only households for the years 2007 to 2012.

(a) In the context of this graph, what does the ordered pair (2012, 38) mean?

(b) Use the given ordered pairs to find the slope of the line.

(c) Interpret the slope in the context of this problem.

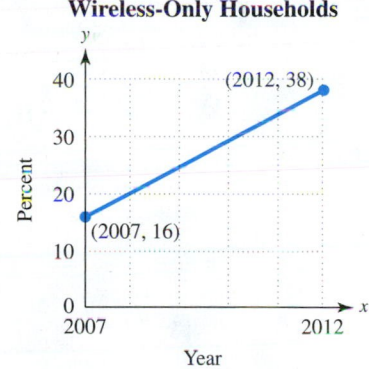

Wireless-Only Households

Source: CTIA.

99. The graph shows the number of drive-in movie theaters in the United States from 2005 through 2012.

(a) Use the given ordered pairs to find the average rate of change in the number of drive-in theaters per year during this period. Round the answer to the nearest whole number.

(b) Explain how a negative slope is interpreted in this situation.

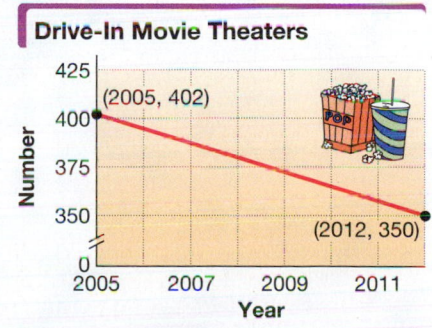

Drive-In Movie Theaters

Source: www.drive-ins.com

100. The graph shows the number of U.S. travelers to Canada (in thousands) from 2000 through 2011.

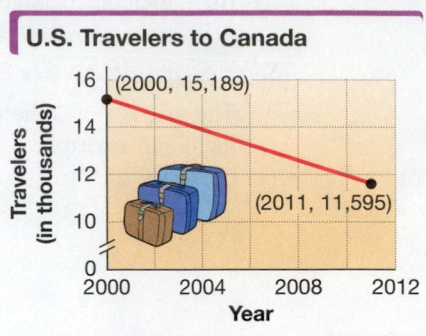

U.S. Travelers to Canada

Source: U.S. Department of Commerce.

(a) Use the given ordered pairs to find the average rate of change in the number of U.S. travelers to Canada per year during this period. Round the answer to the nearest thousand.

(b) Explain how a negative slope is interpreted in this situation.

101. The average price of a gallon of gasoline in 1980 was $1.22. In 2012, the average price was $3.70. Find and interpret the average rate of change in the price of a gallon of gasoline per year to the nearest cent. (*Source:* Energy Information Administration.)

102. The average price of a movie ticket in 1990 was $4.23. In 2012, the average price was $7.96. Find and interpret the average rate of change in the price of a movie ticket per year to the nearest cent. (*Source:* Motion Picture Association of America.)

103. In 2010, the number of digital cameras sold in the United States totaled 7246 thousand. There were 1670 thousand sold in 2013. Find and interpret the average rate of change in the number of digital cameras sold per year to the nearest thousand. (*Source:* Consumer Electronics Association.)

104. In 2010, sales of desktop computers in the United States totaled $7390 million. In 2013, sales were $6876 million. Find and interpret the average rate of change in sales of desktop computers per year to the nearest million dollars. (*Source:* Consumer Electronics Association.)

Extending Skills   *Use your knowledge of the slopes of parallel and perpendicular lines.*

105. Show that $(-13, -9)$, $(-11, -1)$, $(2, -2)$, and $(4, 6)$ are the vertices of a parallelogram. (*Hint:* A parallelogram is a four-sided figure with opposite sides parallel.)

106. Is the figure with vertices at $(-11, -5)$, $(-2, -19)$, $(12, -10)$, and $(3, 4)$ a parallelogram? Is it a rectangle? (*Hint:* A rectangle is a parallelogram with a right angle.)

RELATING CONCEPTS   For Individual or Group Work   (Exercises 107–112)

Three points that lie on the same straight line are said to be **collinear.** *Consider the points* $A(3, 1)$, $B(6, 2)$, *and* $C(9, 3)$. ***Work Exercises 107–112 in order.***

107. Find the slope of segment AB.   **108.** Find the slope of segment BC.

109. Find the slope of segment AC.

110. If slope of segment AB = slope of segment BC = slope of segment AC, then A, B, and C are collinear. Use the results of **Exercises 107–109** to show that this statement is satisfied.

111. Use the slope formula to determine whether the points $(1, -2)$, $(3, -1)$, and $(5, 0)$ are collinear.

112. Repeat **Exercise 111** for the points $(0, 6)$, $(4, -5)$, and $(-2, 12)$.

7.2 Review of Equations of Lines; Linear Models

OBJECTIVE 1 Write an equation of a line given its slope and *y*-intercept.

Recall that we can find the slope of a line from its equation by solving the equation for *y*. For example, we found that the slope of the line with equation

$$y = 4x + 8$$

is 4, the coefficient of *x*. *What does the number 8 represent?*

To answer this question, suppose a line has slope *m* and *y*-intercept $(0, b)$. We can find an equation of this line by choosing another point (x, y) on the line, as shown in **FIGURE 23**, and applying the slope formula.

$$m = \frac{y - b}{x - 0} \quad \leftarrow \text{Change in } y \atop \leftarrow \text{Change in } x$$

$$m = \frac{y - b}{x} \qquad \text{Subtract.}$$

$$mx = y - b \qquad \text{Multiply by } x.$$

$$mx + b = y \qquad \text{Add } b.$$

$$y = mx + b \qquad \text{Interchange sides.}$$

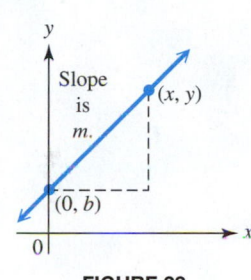

FIGURE 23

This last equation is the *slope-intercept form* of the equation of a line, because we can identify the slope *m* and *y*-intercept $(0, b)$ at a glance. In the line with equation $y = 4x + 8$, the number 8 indicates that the *y*-intercept is $(0, 8)$.

Slope-Intercept Form

The **slope-intercept form** of the equation of a line with slope *m* and *y*-intercept $(0, b)$ is

$$y = mx + b.$$

Slope —↑   ↑— $(0, b)$ is the *y*-intercept.

NOW TRY EXERCISE 1

Write an equation of the line with slope $\frac{2}{3}$ and *y*-intercept $(0, 1)$.

EXAMPLE 1 Writing an Equation of a Line

Write an equation of the line with slope $-\frac{4}{5}$ and *y*-intercept $(0, -2)$.

Here, $m = -\frac{4}{5}$ and $b = -2$. Substitute these values into the slope-intercept form.

$$y = mx + b \qquad \text{Slope-intercept form}$$

$$y = -\frac{4}{5}x + (-2) \qquad \text{Let } m = -\frac{4}{5} \text{ and } b = -2.$$

$$y = -\frac{4}{5}x - 2 \qquad \text{Definition of subtraction} \qquad \text{NOW TRY} \ \text{}$$

NOW TRY ANSWER

1. $y = \frac{2}{3}x + 1$

NOTE Every linear equation (of a nonvertical line) has a *unique* (one and only one) slope-intercept form. *Linear functions* are defined later using this form.

**NOW TRY
EXERCISE 2**

Graph the line using the slope and y-intercept.

$$4x + 3y = 6$$

We used this approach in **Example 8(a)** of **Section 7.1.**

EXAMPLE 2   Graphing Lines Using Slope and y-Intercept

Graph each line using the slope and y-intercept.

(a) $y = 3x - 6$ (In slope-intercept form)

Here, $m = 3$ and $b = -6$. Plot the y-intercept $(0, -6)$. The slope 3 can be interpreted geometrically.

$$m = \frac{\text{rise}}{\text{run}} = \frac{\text{change in } y}{\text{change in } x} = \frac{3}{1}$$

From $(0, -6)$, move 3 units *up* and 1 unit to the *right,* and plot a second point at $(1, -3)$. Join the two points with a straight line. See **FIGURE 24.**

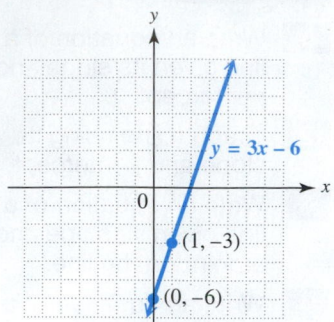

FIGURE 24

(b) $3y + 2x = 9$ (*Not* in slope-intercept form)

Write the equation in slope-intercept form by solving for y.

$$3y + 2x = 9$$

$$3y = -2x + 9 \qquad \text{\textcolor{blue}{Subtract } } 2x.$$

$$y = -\frac{2}{3}x + 3 \qquad \text{\textcolor{blue}{Divide by 3.}}$$

Slope $\uparrow$ $\uparrow$ y-intercept is $(0, 3)$.

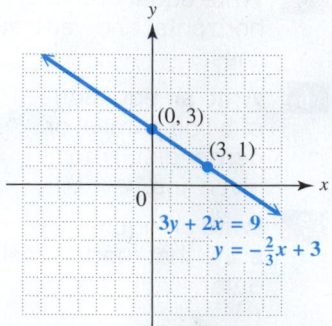

FIGURE 25

Plot the y-intercept $(0, 3)$. The slope can be interpreted as either $\frac{-2}{3}$ or $\frac{2}{-3}$. Using $\frac{-2}{3}$, begin at $(0, 3)$ and move 2 units *down* and 3 units to the *right* to locate the point $(3, 1)$. The line through these two points is the required graph. See **FIGURE 25.** (Verify that the point obtained using $\frac{2}{-3}$ as the slope is also on this line.)   **NOW TRY**

OBJECTIVE 3 Write an equation of a line given its slope and a point on the line.

Let m represent the slope of a line and (x_1, y_1) represent a given point on the line. Let (x, y) represent any other point on the line. See **FIGURE 26.**

$$m = \frac{y - y_1}{x - x_1} \qquad \text{\textcolor{blue}{Slope formula}}$$

$$m(x - x_1) = y - y_1 \qquad \text{\textcolor{blue}{Multiply each side by } } x - x_1.$$

$$y - y_1 = m(x - x_1) \qquad \text{\textcolor{blue}{Interchange sides.}}$$

This last equation is the *point-slope form* of the equation of a line.

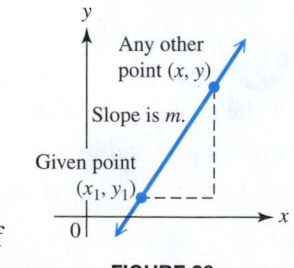

FIGURE 26

NOW TRY ANSWER

2.

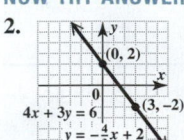

Point-Slope Form

The **point-slope form** of the equation of a line with slope m passing through the point (x_1, y_1) is

Slope
$\downarrow$

$$y - y_1 = m(x - x_1).$$

$\llcorner$ Given point $\lrcorner$

NOW TRY EXERCISE 3

Write an equation of the line with slope $-\frac{1}{5}$ passing through the point $(5, -3)$.

EXAMPLE 3   **Writing an Equation of a Line Given Its Slope and a Point**

Write an equation of the line with slope $\frac{1}{3}$ passing through the point $(-2, 5)$.

Method 1   Use point-slope form with $(x_1, y_1) = (-2, 5)$ and $m = \frac{1}{3}$.

$$y - y_1 = m(x - x_1) \qquad \text{Point-slope form}$$

$$y - 5 = \frac{1}{3}[x - (-2)] \qquad \text{Let } y_1 = 5, \ m = \tfrac{1}{3}, \text{ and } x_1 = -2.$$

$$y - 5 = \frac{1}{3}(x + 2) \qquad \text{Definition of subtraction}$$

$$y - 5 = \frac{1}{3}x + \frac{2}{3} \qquad \text{(*)}\quad \text{Distributive property}$$

Slope-intercept form $\rightarrow$ $\quad y = \frac{1}{3}x + \frac{17}{3} \qquad 5 = \tfrac{15}{3}; \text{ Add } \tfrac{15}{3}.$

Method 2   Use slope-intercept form with $(x, y) = (-2, 5)$ and $m = \frac{1}{3}$.

$$y = mx + b \qquad \text{Slope-intercept form}$$

$$5 = \frac{1}{3}(-2) + b \qquad \text{Let } y = 5, \ m = \tfrac{1}{3}, \text{ and } x = -2.$$

Solve for b. $\quad 5 = -\dfrac{2}{3} + b \qquad \text{Multiply.}$

$$\frac{17}{3} = b, \quad \text{or} \quad b = \frac{17}{3} \qquad 5 = \tfrac{15}{3}; \text{ Add } \tfrac{2}{3}.$$

Substitute to obtain the equation $y = \frac{1}{3}x + \frac{17}{3}$, as above.   **NOW TRY**

Earlier we defined *standard form* for a linear equation.

$$Ax + By = C \qquad \text{Standard form}$$

Here A, B, and C are real numbers and A and B are not both 0. (In most cases in this book, A, B, and C are rational numbers.) For consistency, we give answers so that A, B, and C are integers with greatest common factor 1, and $A \geq 0$. (If $A = 0$, then we give $B > 0$.) For example, the equation in **Example 3** is written in standard form as follows.

$$y - 5 = \frac{1}{3}x + \frac{2}{3} \qquad \text{Equation (*) from } \textbf{Example 3}$$

$$3y - 15 = x + 2 \qquad \text{Multiply each term by 3.}$$

$$-x + 3y = 17 \qquad \text{Subtract } x. \text{ Add 15.}$$

Standard form $\rightarrow$ $\quad x - 3y = -17 \qquad \text{Multiply by } -1.$

NOTE   "Standard form" is not standard among texts. A linear equation can be written in many different, equally correct ways. For example, $2x + 3y = 8$ can be written as

$$2x = 8 - 3y, \quad 3y = 8 - 2x, \quad x + \frac{3}{2}y = 4, \quad \text{and} \quad 4x + 6y = 16.$$

NOW TRY ANSWER

3. $y = -\frac{1}{5}x - 2$

We prefer the standard form $2x + 3y = 8$ over any multiples of each side, such as $4x + 6y = 16$. (To write $4x + 6y = 16$ in this preferred form, divide each side by 2.)

NOW TRY
EXERCISE 4

Write an equation of the line passing through the points $(3, -4)$ and $(-2, -1)$. Give the final answer in standard form.

OBJECTIVE 4 Write an equation of a line given two points on the line.

EXAMPLE 4   Writing an Equation of a Line Given Two Points

Write an equation of the line passing through the points $(-4, 3)$ and $(5, -7)$. Give the final answer in standard form.

First find the slope using the slope formula.

$$m = \frac{-7 - 3}{5 - (-4)} = -\frac{10}{9}$$

Use either $(-4, 3)$ or $(5, -7)$ as (x_1, y_1) in the point-slope form of the equation of a line. We choose $(-4, 3)$, so $-4 = x_1$ and $3 = y_1$.

$y - y_1 = m(x - x_1)$	Point-slope form
$y - 3 = -\dfrac{10}{9}[x - (-4)]$	Let $y_1 = 3$, $m = -\frac{10}{9}$, and $x_1 = -4$.
$y - 3 = -\dfrac{10}{9}(x + 4)$	Definition of subtraction
$y - 3 = -\dfrac{10}{9}x - \dfrac{40}{9}$	Distributive property
$9y - 27 = -10x - 40$	Multiply each term by 9.
Standard form $\longrightarrow$ $10x + 9y = -13$	Add $10x$. Add 27.

Verify that if $(5, -7)$ were used, the same equation would result.   **NOW TRY**

NOTE   Once the slope is found in **Example 4,** the equation of the line could also be determined using Method 2 from **Example 3.**

OBJECTIVE 5 Write equations of horizontal and vertical lines.

A horizontal line has slope 0. Using point-slope form, we can find the equation of a horizontal line through the point (a, b).

$y - y_1 = m(x - x_1)$	Point-slope form
$y - b = 0(x - a)$	$y_1 = b$, $m = 0$, $x_1 = a$
$y - b = 0$	Multiplication property of 0
Horizontal line $\longrightarrow$ $y = b$	Add b.

Point-slope form does not apply to a vertical line because the slope of a vertical line is undefined. A vertical line through the point (a, b) has equation $x = a$.

Equations of Horizontal and Vertical Lines

A **horizontal line** through the point (a, b) has equation $y = b$.

A **vertical line** through the point (a, b) has equation $x = a$.

NOW TRY ANSWER
4. $3x + 5y = -11$

NOW TRY
EXERCISE 5

Write an equation of the line passing through the point $(4, -4)$ that satisfies the given condition.

(a) The line has undefined slope.

(b) The line has slope 0.

EXAMPLE 5   Writing Equations of Horizontal and Vertical Lines

Write an equation of the line passing through the point $(-3, 3)$ that satisfies the given condition.

(a) The line has slope 0.

Because the slope is 0, this is a horizontal line. A horizontal line through the point (a, b) has equation $y = b$. In $(-3, 3)$, the y-coordinate is 3, so the equation is $y = 3$.

(b) The line has undefined slope.

This is a vertical line because the slope is undefined. A vertical line through the point (a, b) has equation $x = a$. In $(-3, 3)$, the x-coordinate is -3, so the equation is $x = -3$.

Both lines are graphed in **FIGURE 27**.

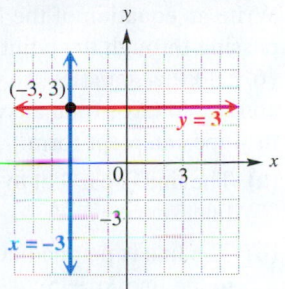

FIGURE 27

NOW TRY

OBJECTIVE 6   Write an equation of a line parallel or perpendicular to a given line.

Recall that parallel lines have the same slope and perpendicular lines have slopes that are negative reciprocals.

EXAMPLE 6   Writing Equations of Parallel or Perpendicular Lines

Write an equation of the line passing through the point $(-3, 6)$ that satisfies the given condition. Give final answers in slope-intercept form.

(a) The line is parallel to the line $2x + 3y = 6$.

We can find the slope of the given line by solving for y.

$$2x + 3y = 6$$
$$3y = -2x + 6 \qquad \text{Subtract } 2x.$$
$$y = -\frac{2}{3}x + 2 \qquad \text{Divide by 3.}$$
$$\underset{\text{Slope}}{\uparrow}$$

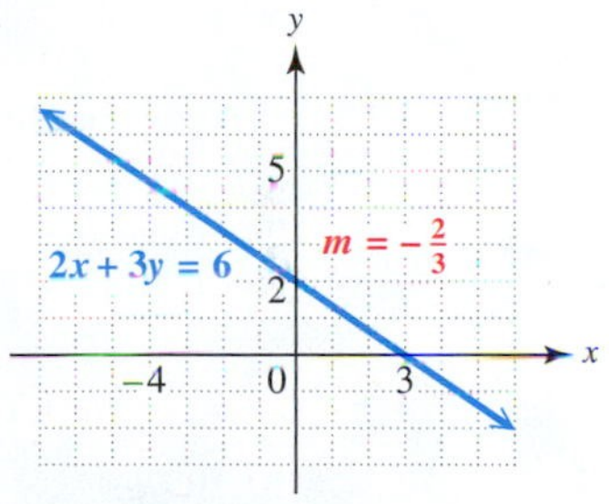

The slope of the line is given by the coefficient of x, so $m = -\frac{2}{3}$. See **FIGURE 28**.

FIGURE 28

The required equation of the line through $(-3, 6)$ and parallel to $2x + 3y = 6$ must also have slope $-\frac{2}{3}$. To find this equation, we use the point-slope form with $(x_1, y_1) = (-3, 6)$ and $m = -\frac{2}{3}$.

$$y - 6 = -\frac{2}{3}[x - (-3)] \qquad y_1 = 6, \ m = -\frac{2}{3}, \ x_1 = -3$$
$$y - 6 = -\frac{2}{3}(x + 3) \qquad \text{Definition of subtraction}$$
$$y - 6 = -\frac{2}{3}x - 2 \qquad \text{Distributive property}$$
$$y = -\frac{2}{3}x + 4 \qquad \text{Add 6.}$$

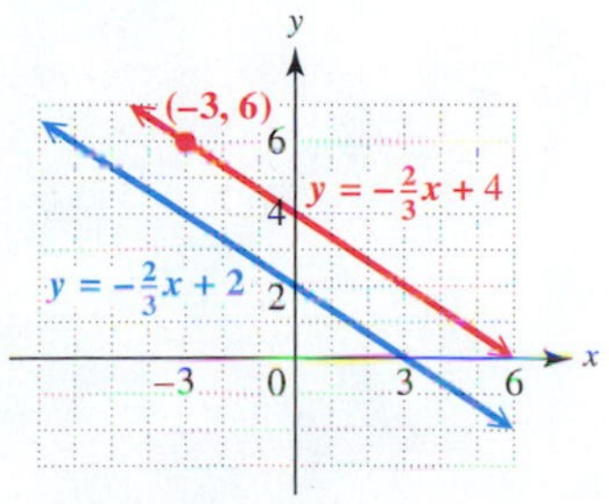

FIGURE 29

We did not clear the fraction because we want the final equation in slope-intercept form—that is, solved for y. Both lines are shown in **FIGURE 29**.

NOW TRY ANSWERS
5. (a) $x = 4$ (b) $y = -4$

NOW TRY
EXERCISE 6

Write an equation of the line passing through the point $(6, -1)$ that satisfies the given condition. Give final answers in slope-intercept form.

(a) The line is parallel to the line $3x - 5y = 7$.

(b) The line is perpendicular to the line $3x - 5y = 7$.

(b) The line is perpendicular to the line $2x + 3y = 6$.

In part (a), we wrote the equation $2x + 3y = 6$ in slope-intercept form.

$$y = -\frac{2}{3}x + 2$$

$\longrightarrow$ Slope

To be perpendicular to the line $2x + 3y = 6$, a line must have slope $\frac{3}{2}$, the negative reciprocal of $-\frac{2}{3}$.

We use $(-3, 6)$ and slope $\frac{3}{2}$ in the point-slope form to find the equation of the perpendicular line shown in **FIGURE 30**.

$$y - 6 = \frac{3}{2}[x - (-3)] \qquad y_1 = 6,\ m = \frac{3}{2},\ x_1 = -3$$

$$y - 6 = \frac{3}{2}(x + 3) \qquad \text{Definition of subtraction}$$

$$y - 6 = \frac{3}{2}x + \frac{9}{2} \qquad \text{Distributive property}$$

$$y = \frac{3}{2}x + \frac{21}{2} \qquad \text{Add } 6 = \frac{12}{2}.$$

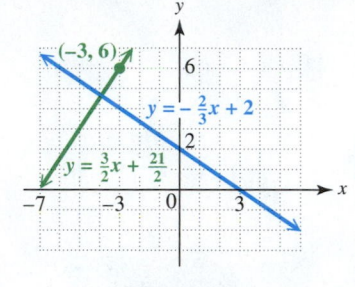

FIGURE 30

Again, we did not clear the fractions because we want the final equation in slope-intercept form.

NOW TRY

▼ **Summary of Forms of Linear Equations**

Equation	Description	When to Use
$y = mx + b$	**Slope-Intercept Form** Slope is m. y-intercept is $(0, b)$.	The slope and y-intercept can be easily identified and used to quickly graph the equation.
$y - y_1 = m(x - x_1)$	**Point-Slope Form** Slope is m. Line passes through (x_1, y_1).	This form is ideal for finding the equation of a line if the slope and a point on the line or two points on the line are known.
$Ax + By = C$	**Standard Form** (A, B, and C integers, $A \geq 0$) Slope is $-\frac{A}{B}$ ($B \neq 0$). x-intercept is $\left(\frac{C}{A}, 0\right)$ ($A \neq 0$). y-intercept is $\left(0, \frac{C}{B}\right)$ ($B \neq 0$).	The x- and y-intercepts can be found quickly and used to graph the equation. The slope must be calculated.
$y = b$	**Horizontal Line** Slope is 0. y-intercept is $(0, b)$.	If the graph intersects only the y-axis, then y is the only variable in the equation.
$x = a$	**Vertical Line** Slope is undefined. x-intercept is $(a, 0)$.	If the graph intersects only the x-axis, then x is the only variable in the equation.

NOW TRY ANSWERS

6. **(a)** $y = \frac{3}{5}x - \frac{23}{5}$

    (b) $y = -\frac{5}{3}x + 9$

OBJECTIVE 7 Write an equation of a line that models real data.

If a given set of data changes at a fairly constant rate, the data may fit a linear pattern, where the rate of change is the slope of the line.

**NOW TRY
EXERCISE 7**

A cell phone plan costs $100 for the telephone plus $85 per month for service. Write an equation that gives the cost y in dollars for x months of cell phone service using this plan.

EXAMPLE 7 Writing a Linear Equation to Describe Real Data

A local gasoline station is selling 89-octane gas for $3.50 per gal.

(a) Write an equation that describes the cost y to buy x gallons of gas.

The total cost is determined by the number of gallons we buy multiplied by the price per gallon (in this case, $3.50). As the gas is pumped, two sets of numbers spin by: the number of gallons pumped and the cost of that number of gallons.

The table illustrates this situation.

Number of Gallons Pumped	Cost of This Number of Gallons
0	0($3.50) = $ 0.00
1	1($3.50) = $ 3.50
2	2($3.50) = $ 7.00
3	3($3.50) = $10.50
4	4($3.50) = $14.00

If we let x denote the number of gallons pumped, then the total cost y in dollars can be found using the following linear equation.

Total cost ⟶   ⟵ Number of gallons
$$y = 3.50x$$

Theoretically, there are infinitely many ordered pairs (x, y) that satisfy this equation, but here we are limited to nonnegative values for x because we cannot have a negative number of gallons. In this situation, there is also a practical maximum value for x, which varies from one car to another—the size of the gas tank.

(b) A car wash at this gas station costs an additional $3.00. Write an equation that defines the cost of gas and a car wash.

The cost will be $3.50x + 3.00$ dollars for x gallons of gas and a car wash.

$$y = 3.5x + 3 \qquad \text{Final 0's need not be included.}$$

(c) Interpret the ordered pairs $(5, 20.5)$ and $(10, 38)$ in relation to the equation from part (b).

$(5, 20.5)$ indicates that 5 gal of gas and a car wash cost $20.50.

$(10, 38)$ indicates that 10 gal of gas and a car wash cost $38.00.

NOW TRY

NOTE In **Example 7(a),** the ordered pair $(0, 0)$ satisfied the equation $y = 3.50x$, so the linear equation has the form

$$y = mx, \quad \text{where } b = 0.$$

If a realistic situation involves an initial charge b plus a charge per unit, m, as in the equation $y = 3.5x + 3$ in **Example 7(b),** then the equation has the form

$$y = mx + b, \quad \text{where } b \neq 0.$$

NOW TRY ANSWER
7. $y = 85x + 100$

**NOW TRY
EXERCISE 8**

Refer to **Example 8.**

(a) Use the ordered pairs
(8, 243) and (12, 263)
to write an equation that
models the data.

(b) Use the equation from
part (a) to estimate retail
spending on prescription
drugs in 2015.

EXAMPLE 8   **Writing an Equation of a Line That Models Data**

Retail spending (in billions of dollars) on prescription drugs in the United States is shown in the graph in **FIGURE 31**.

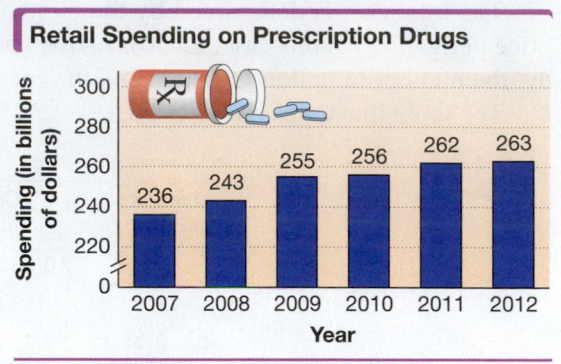

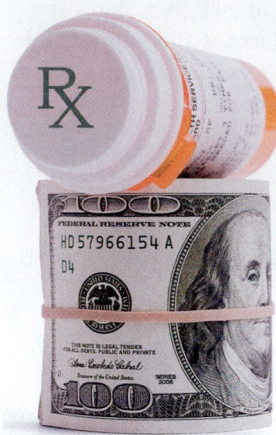

Source: Centers for Medicare & Medicaid Services.

FIGURE 31

(a) Write an equation that models the data.

The data increase linearly—that is, a straight line through the tops of any two bars in the graph would be close to the top of each bar. To model the relationship between year x and spending on prescription drugs y, we let $x = 7$ represent 2007, $x = 8$ represent 2008, and so on. The given data for 2007 and 2012 can be written as the ordered pairs $(7, 236)$ and $(12, 263)$.

$$m = \frac{263 - 236}{12 - 7} = \frac{27}{5} = 5.4 \qquad \text{Find the slope of the line through } (7, 236) \text{ and } (12, 263).$$

Thus, spending increased by about $5.4 billion per year. To write an equation, we substitute this slope and the point $(7, 236)$ into the point-slope form.

$$y - y_1 = m(x - x_1) \qquad \text{Point-slope form}$$

$$y - 236 = 5.4(x - 7) \qquad (x_1, y_1) = (7, 236); m = 5.4$$

> Either point can be used here. (12, 263) provides the same answer.

$$y - 236 = 5.4x - 37.8 \qquad \text{Distributive property}$$

$$y = 5.4x + 198.2 \qquad \text{Add 236.}$$

Retail spending y (in billions of dollars) on prescription drugs in the United States in year x can be approximated by the equation $y = 5.4x + 198.2$.

(b) Use the equation from part (a) to estimate retail spending on prescription drugs in the United States in 2015. (Assume a constant rate of change.)

Because we let $x = 7$ represent 2007, $x = 15$ represents 2015.

$$y = 5.4x + 198.2 \qquad \text{Equation from part (a)}$$

$$y = 5.4(15) + 198.2 \qquad \text{Substitute 15 for } x.$$

$$y = 279.2 \qquad \text{Multiply, and then add.}$$

About $279 billion was spent on prescription drugs in 2015.   **NOW TRY**

NOW TRY ANSWERS
8. **(a)** $y = 5x + 203$
    (b) $278 billion

7.2 Exercises

 FOR EXTRA HELP

 MyMathLab®

Complete solution available in MyMathLab

Concept Check *Provide the appropriate response.*

1. The following equations all represent the same line. Which one is in standard form as specified in this section?

   A. $3x - 2y = 5$ **B.** $2y = 3x - 5$ **C.** $\frac{3}{5}x - \frac{2}{5}y = 1$ **D.** $3x = 2y + 5$

2. Which equation is in point-slope form?

   A. $y = 6x + 2$ **B.** $4x + y = 9$ **C.** $y - 3 = 2(x - 1)$ **D.** $2y = 3x - 7$

3. Which equation in **Exercise 2** is in slope-intercept form?

4. Write the equation $y + 2 = -3(x - 4)$ in slope-intercept form.

5. Write the equation from **Exercise 4** in standard form.

6. Write the equation $10x - 7y = 70$ in slope-intercept form.

Concept Check *Match each equation with the graph that it most closely resembles. (Hint: Determining the signs of m and b will help in each case.)*

7. $y = 2x + 3$

8. $y = -2x + 3$

9. $y = -2x - 3$

10. $y = 2x - 3$

11. $y = 2x$

12. $y = -2x$

13. $y = 3$

14. $y = -3$

A. 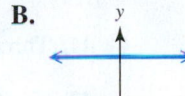   **B.** 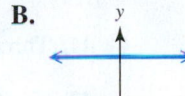   **C.**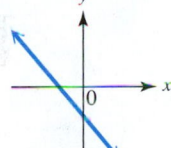

D. 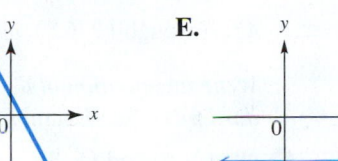   **E.** 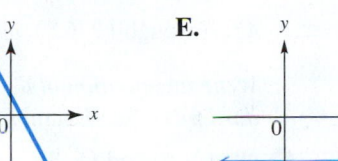   **F.**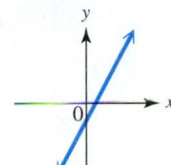

G. 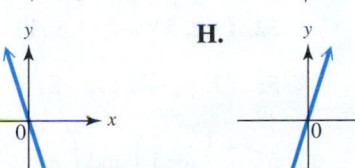   **H.** 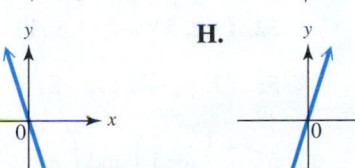   **I.**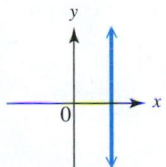

Write an equation in slope-intercept form of the line that satisfies the given conditions. **See Example 1.**

15. $m = 5;\ b = 15$ 16. $m = 2;\ b = 12$

17. $m = -\frac{2}{3};\ b = \frac{4}{5}$ 18. $m = -\frac{5}{8};\ b = -\frac{1}{3}$

19. Slope 1; y-intercept $(0, -1)$ 20. Slope -1; y-intercept $(0, -3)$

21. Slope $\frac{2}{5}$; y-intercept $(0, 5)$ 22. Slope $-\frac{3}{4}$; y-intercept $(0, 7)$

Write an equation in slope-intercept form of the line shown in each graph. (Coordinates of the points shown are integers.)

23. 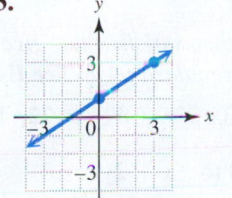   24. 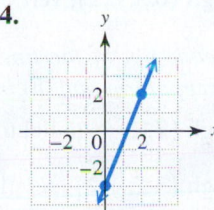   25. 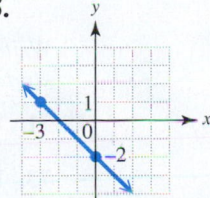   26.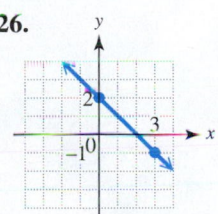

Each table of values gives several points that lie on a line. Write an equation in slope-intercept form of the line.

27.

x	y
−2	−8
0	−4
1	−2
3	2

28.

x	y
−2	−3
0	3
2	9
3	12

29.

x	y
−5	6
0	3
5	0
10	−3

30.

x	y
−4	5
−2	0
0	−5
2	−10

For each equation, (a) write it in slope-intercept form, (b) give the slope of the line, (c) give the y-intercept, and (d) graph the line. See Example 2.

31. $-x + y = 4$

32. $-x + y = 6$

▶ **33.** $6x + 5y = 30$

34. $3x + 4y = 12$

35. $4x - 5y = 20$

36. $7x - 3y = 3$

37. $x + 2y = -4$

38. $x + 3y = -9$

Write an equation of the line that satisfies the given conditions. Give the equation (a) in slope-intercept form and (b) in standard form. See Example 3 and the discussion on standard form.

39. Through $(5, 8)$; slope -2

40. Through $(12, 10)$; slope 1

▶ **41.** Through $(-2, 4)$; slope $-\dfrac{3}{4}$

42. Through $(-1, 6)$; slope $-\dfrac{5}{6}$

43. Through $(-5, 4)$; slope $\dfrac{1}{2}$

44. Through $(7, -2)$; slope $\dfrac{1}{4}$

45. x-intercept $(3, 0)$; slope 4

46. x-intercept $(-2, 0)$; slope -5

47. Through $(2, 6.8)$; slope 1.4

48. Through $(6, -1.2)$; slope 0.8

Write an equation of the line passing through the given points. Give the final answer in standard form. See Example 4.

▶ **49.** $(3, 4)$ and $(5, 8)$

50. $(5, -2)$ and $(-3, 14)$

51. $(6, 1)$ and $(-2, 5)$

52. $(-2, 5)$ and $(-8, 1)$

53. $(2, 5)$ and $(1, 5)$

54. $(-2, 2)$ and $(4, 2)$

55. $(7, 6)$ and $(7, -8)$

56. $(13, 5)$ and $(13, -1)$

57. $\left(\dfrac{1}{2}, -3\right)$ and $\left(-\dfrac{2}{3}, -3\right)$

58. $\left(-\dfrac{4}{9}, -6\right)$ and $\left(\dfrac{12}{7}, -6\right)$

59. $\left(-\dfrac{2}{5}, \dfrac{2}{5}\right)$ and $\left(\dfrac{4}{3}, \dfrac{2}{3}\right)$

60. $\left(\dfrac{3}{4}, \dfrac{8}{3}\right)$ and $\left(\dfrac{2}{5}, \dfrac{2}{3}\right)$

Write an equation of the line that satisfies the given conditions. See Example 5.

▶ **61.** Through $(9, 5)$; slope 0

62. Through $(-4, -2)$; slope 0

63. Through $(9, 10)$; undefined slope

64. Through $(-2, 8)$; undefined slope

65. Through $\left(-\dfrac{3}{4}, -\dfrac{3}{2}\right)$; slope 0

66. Through $\left(-\dfrac{5}{8}, -\dfrac{9}{2}\right)$; slope 0

67. Through $(-7, 8)$; horizontal

68. Through $(2, -7)$; horizontal

69. Through $(0.5, 0.2)$; vertical

70. Through $(0.1, 0.4)$; vertical

Write an equation of the line that satisfies the given conditions. Give the equation (a) in slope-intercept form and (b) in standard form. See Example 6.

▶ **71.** Through $(7, 2)$; parallel to $3x - y = 8$

72. Through $(4, 1)$; parallel to $2x + 5y = 10$

73. Through $(-2, -2)$; parallel to $-x + 2y = 10$

74. Through $(-1, 3)$; parallel to $-x + 3y = 12$

▶ **75.** Through $(8, 5)$; perpendicular to $2x - y = 7$

76. Through $(2, -7)$; perpendicular to $5x + 2y = 18$

77. Through $(-2, 7)$; perpendicular to $x = 9$

78. Through $(8, 4)$; perpendicular to $x = -3$

Write an equation in the form $y = mx$ for each situation. Then give the three ordered pairs associated with the equation for x-values 0, 5, and 10. **See Example 7(a).**

79. x represents the number of hours traveling at 45 mph, and y represents the distance traveled (in miles).

80. x represents the number of t-shirts sold at $26 each, and y represents the total cost of the t-shirts (in dollars).

81. x represents the number of gallons of gas sold at $3.75 per gal, and y represents the total cost of the gasoline (in dollars).

82. x represents the number of days a DVD movie is rented at $4.50 per day, and y represents the total charge for the rental (in dollars).

83. x represents the number of credit hours taken at Kirkwood Community College at $140 per credit hour, and y represents the total tuition paid for the credit hours (in dollars). (*Source:* www.kirkwood.edu)

84. x represents the number of tickets to a performance of *Jersey Boys* at the Des Moines Civic Center purchased at $125 per ticket, and y represents the total paid for the tickets (in dollars). (*Source:* Ticketmaster.)

For each situation, do the following.

(a) Write an equation in the form $y = mx + b$.

(b) Find and interpret the ordered pair associated with the equation for $x = 5$.

(c) Answer the question posed in the problem.

See Examples 7(b) and 7(c).

85. A ticket for the Diving Board Tour, featuring Elton John, costs $149. A parking pass costs $15. Let x represent the number of tickets and y represent the cost in dollars. How much does it cost for 2 tickets and a parking pass? (*Source:* Ticketmaster.)

86. Resident tuition at Broward College is $105.90 per credit hour. There is also a $20 health science application fee. Let x represent the number of credit hours and y represent the cost in dollars. How much does it cost for a student in health science to take 15 credit hours? (*Source:* www.broward.edu)

87. A health club membership costs $99, plus $41 per month. Let x represent the number of months and y represent the cost in dollars. How much does the first year's membership cost? (*Source:* Midwest Athletic Club.)

88. An Executive VIP/Gold membership to a health club costs $159, plus $57 per month. Let x represent the number of months and y represent the cost in dollars. How much does a one-year membership cost? (*Source:* Midwest Athletic Club.)

89. A wireless plan includes unlimited talk and text plus 2 GB of data for $95 per month. There is a $36 activation fee. Let x represent the number of months and y represent the cost in dollars. Over a two-year contract, how much will this plan cost? (*Source:* AT&T.)

90. Another wireless plan includes unlimited talk and text plus 4 GB of data for $110 per month. There is a $36 activation fee and a $99 charge for an Apple iPhone 5c. Let *x* represent the number of months and *y* represent the cost in dollars. Over a two-year contract, how much will this plan cost? (*Source*: AT&T.)

91. There is a $30 fee to rent a chain saw, plus $6 per day. Let *x* represent the number of days the saw is rented and *y* represent the charge to the user in dollars. If the total charge is $138, for how many days is the saw rented?

92. A rental car costs $50, plus $0.45 per mile. Let *x* represent the number of miles driven and *y* represent the total charge to the renter in dollars. How many miles was the car driven if the renter paid $127.85?

Solve each problem. In part (a), give equations in slope-intercept form. **See Example 8.**

93. Total sales of portable media/MP3 players in the United States (in millions of dollars) are shown in the graph, where the year 2010 corresponds to *x* = 0.

(a) Use the ordered pairs from the graph to write an equation that models the data. Interpret the slope in the context of this problem.

(b) Use the equation from part (a) to approximate sales of portable media/MP3 players in the United States in 2011, the year data was unavailable.

Media/MP3 Player Sales

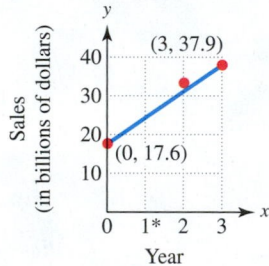

*Data for this year is unavailable.

Source: Consumer Electronics Association.

94. Total sales of smartphones in the United States (in billions of dollars) are shown in the graph, where the year 2010 corresponds to *x* = 0.

(a) Use the ordered pairs from the graph to write an equation that models the data. (Round the slope to the nearest tenth.) Interpret the slope in the context of this problem.

(b) Use the equation from part (a) to approximate smartphone sales in the United States in 2011, the year data was unavailable.

Smartphone Sales

*Data for this year is unavailable.

Source: Consumer Electronics Association.

95. Expenditures for home health care in the United States are shown in the graph.

(a) Use the information given for the years 2008 and 2012, letting *x* = 8 represent 2008 and *x* = 12 represent 2012, and letting *y* represent spending (in billions of dollars), to write an equation that models the data.

(b) Use the equation from part (a) to approximate the amount spent on home health care in 2011 to the nearest tenth. How does the result compare with the actual value, $74.0 billion?

Spending on Home Health Care

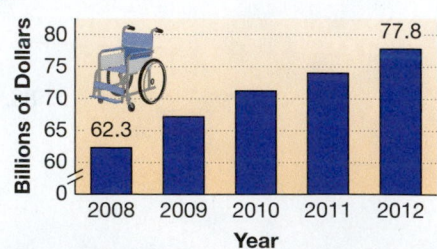

Source: Centers for Medicare & Medicaid Services.

96. The number of pieces of first class mail delivered in the United States is shown in the graph.

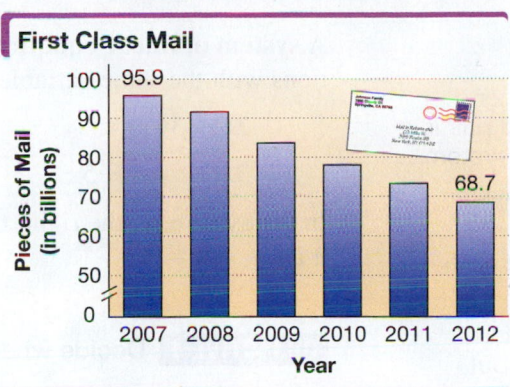

First Class Mail

Source: U.S. Postal Service.

(a) Use the information given for the years 2007 and 2012, letting $x = 7$ represent 2007 and $x = 12$ represent 2012, and letting y represent the number of pieces of mail (in billions), to write an equation that models the data.

(b) Use the equation from part (a) to approximate the number of pieces of first class mail delivered in 2010 to the nearest tenth. How does this result compare to the actual value, 78.2 billion?

RELATING CONCEPTS For Individual or Group Work (Exercises 97–104)

In Section 2.5, we worked with formulas. **Work Exercises 97–104 in order,** *to see how the formula that relates Celsius and Fahrenheit temperatures is derived.*

97. There is a linear relationship between Celsius and Fahrenheit temperatures.

When $C = 0°$, $F = $ _____ °.

When $C = 100°$, $F = $ _____ °.

98. Think of ordered pairs of temperatures (C, F), where C and F represent corresponding Celsius and Fahrenheit temperatures. The equation that relates the two scales has a straight-line graph that contains the two points determined in **Exercise 97.** What are these two points?

99. Find the slope of the line described in **Exercise 98.**

100. Use the slope found in **Exercise 99** and one of the two points determined earlier, and write an equation that gives F in terms of C. (*Hint:* Use the point-slope form, where C replaces x and F replaces y.)

101. To obtain another form of the formula, use the equation found in **Exercise 100** and solve for C in terms of F.

102. Use the equation from **Exercise 100** to find the Fahrenheit temperature when $C = 30$.

103. Use the equation from **Exercise 101** to find the Celsius temperature when $F = 50$.

104. For what temperature does $F = C$? (Use the photo to confirm this temperature.)

(7.3) Solving Systems of Linear Equations by Graphing

VOCABULARY

☐ system of linear equations (linear system)
☐ solution of a system
☐ solution set of a system
☐ consistent system
☐ inconsistent system
☐ independent equations
☐ dependent equations

NOW TRY EXERCISE 1

Decide whether the ordered pair $(5, 2)$ is a solution of each system.

(a) $2x + 5y = 20$
$\quad\; x - \; y = 7$

(b) $3x - y = 13$
$\quad\; 2x + y = 12$

A **system of linear equations,** or **linear system,** consists of two or more linear equations with the same variables.

$$2x + 3y = 4 \qquad x + 3y = 1 \qquad x - y = 1$$
$$3x - \; y = -5 \qquad -y = 4 - 2x \qquad y = 3$$

Linear systems

In the system on the right, think of $y = 3$ as an equation in two variables by writing it as $0x + y = 3$.

OBJECTIVE 1   Decide whether a given ordered pair is a solution of a system.

A **solution of a system** of linear equations is an ordered pair that makes both equations true at the same time. A solution of an equation is said to *satisfy* the equation.

EXAMPLE 1   Determining Whether an Ordered Pair Is a Solution

Decide whether the ordered pair $(4, -3)$ is a solution of each system.

(a) $x + 4y = -8$

$3x + 2y = 6$

To decide whether $(4, -3)$ is a solution of the system, substitute 4 for x and -3 for y in each equation.

$x + 4y = -8$		$3x + 2y = 6$	
$4 + 4(-3) \stackrel{?}{=} -8$	Substitute.	$3(4) + 2(-3) \stackrel{?}{=} 6$	Substitute.
$4 + (-12) \stackrel{?}{=} -8$	Multiply.	$12 + (-6) \stackrel{?}{=} 6$	Multiply.
$-8 = -8$ ✓	True	$6 = 6$ ✓	True

Because $(4, -3)$ satisfies both equations, it is a solution of the system.

(b) $2x + 5y = -7$

$3x + 4y = 2$

Again, substitute 4 for x and -3 for y in each equation.

$2x + 5y = -7$		$3x + 4y = 2$	
$2(4) + 5(-3) \stackrel{?}{=} -7$	Substitute.	$3(4) + 4(-3) \stackrel{?}{=} 2$	Substitute.
$8 + (-15) \stackrel{?}{=} -7$	Multiply.	$12 + (-12) \stackrel{?}{=} 2$	Multiply.
$-7 = -7$ ✓	True	$0 = 2$	False

The ordered pair $(4, -3)$ is not a solution of this system because it does not satisfy the second equation.

NOW TRY

OBJECTIVE 2   Solve linear systems by graphing.

The set of all ordered pairs that are solutions of a system is its **solution set.** One way to find the solution set of a system of two linear equations is to graph both equations on the same axes. Any intersection point would be on both lines and would therefore be a solution of *both* equations. ***Thus, the coordinates of any point at which the lines intersect give a solution of the system.***

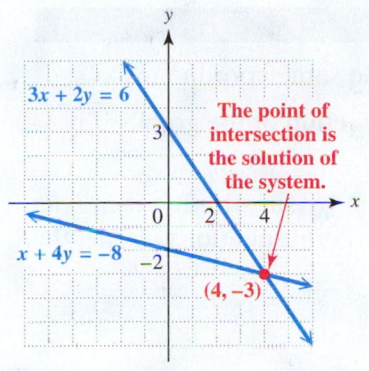

FIGURE 32

The graph in **FIGURE 32** shows that the solution of the system in **Example 1(a)** is the intersection point $(4, -3)$. Because two *different* straight lines can intersect at no more than one point, there can never be more than one solution for such a system.

EXAMPLE 2 Solving a System by Graphing

Solve the system by graphing.

$$2x + 3y = 4$$
$$3x - y = -5$$

We graph these two lines by plotting several points for each line. The intercepts are often convenient choices. We show finding the intercepts for $2x + 3y = 4$.

To find the y-intercept, let $x = 0$.

$$2x + 3y = 4$$
$$2(0) + 3y = 4 \quad \text{Let } x = 0.$$
$$3y = 4$$
$$y = \frac{4}{3} \quad \begin{array}{l} y\text{-intercept} \\ \left(0, \frac{4}{3}\right) \end{array}$$

To find the x-intercept, let $y = 0$.

$$2x + 3y = 4$$
$$2x + 3(0) = 4 \quad \text{Let } y = 0.$$
$$2x = 4$$
$$x = 2 \quad \begin{array}{l} x\text{-intercept} \\ (2, 0) \end{array}$$

The tables show the intercepts and a check point for each graph.

$2x + 3y = 4$

x	y
0	$\frac{4}{3}$ ← y-intercept
2	0 ← x-intercept
-2	$\frac{8}{3}$

Find a third ordered pair as a check.

$3x - y = -5$

x	y
0	5 ← y-intercept
$-\frac{5}{3}$	0 ← x-intercept
-2	-1

The lines in **FIGURE 33** suggest that the graphs intersect at the point $(-1, 2)$. We check by substituting -1 for x and 2 for y in *both* equations.

CHECK

$$2x + 3y = 4 \qquad \text{First equation}$$
$$2(-1) + 3(2) \stackrel{?}{=} 4 \qquad \text{Substitute.}$$
$$4 = 4 \checkmark \qquad \text{True}$$
$$3x - y = -5 \qquad \text{Second equation}$$
$$3(-1) - 2 \stackrel{?}{=} -5 \qquad \text{Substitute.}$$
$$-5 = -5 \checkmark \qquad \text{True}$$

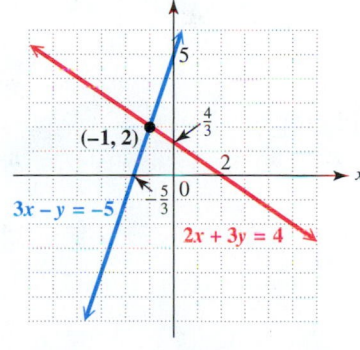

FIGURE 33

Because $(-1, 2)$ satisfies both equations, the solution set of the system is $\{(-1, 2)\}$.

NOW TRY

NOTE We can also write each equation in a system in slope-intercept form and use the slope and y-intercept to graph each line. See **Example 2.**

$2x + 3y = 4 \qquad$ becomes $\qquad y = -\frac{2}{3}x + \frac{4}{3}. \qquad$ y-intercept $\left(0, \frac{4}{3}\right)$; slope $-\frac{2}{3}$

$3x - y = -5 \qquad$ becomes $\qquad y = 3x + 5. \qquad$ y-intercept $(0, 5)$; slope 3, or $\frac{3}{1}$

Confirm that graphing these equations gives the same results shown in **FIGURE 33.**

NOW TRY EXERCISE 2

Solve the system by graphing.

$$x - 2y = 4$$
$$2x + y = 3$$

NOW TRY ANSWER

2. $\{(2, -1)\}$

> ### Solving a Linear System by Graphing
>
> *Step 1* **Graph each equation** of the system on the same coordinate axes.
>
> *Step 2* **Find the coordinates of the point of intersection** of the graphs if possible, and write it as an ordered pair.
>
> *Step 3* **Check** that the ordered pair is the solution by substituting it in *both* of the *original* equations. If it satisfies *both* equations, write the solution set.

⚠ **CAUTION** We recommend using graph paper and a straightedge when solving systems of equations graphically. It may not be possible to determine from the graph the exact coordinates of the point that represents the solution, particularly if those coordinates are not integers. The graphing method does, however, show geometrically how solutions are found and is useful when approximate answers will suffice.

OBJECTIVE 3 Solve special systems by graphing.

NOW TRY EXERCISE 3

Solve each system by graphing.

(a) $5x - 3y = 2$
    $10x - 6y = 4$

(b) $4x + y = 7$
    $12x + 3y = 10$

EXAMPLE 3 Solving Special Systems by Graphing

Solve each system by graphing.

(a) $2x + y = 2$

$2x + y = 8$

See the graphs in **FIGURE 34**. The two lines are parallel and have no points in common. For such a system, there is no solution. The solution set is $\varnothing$.

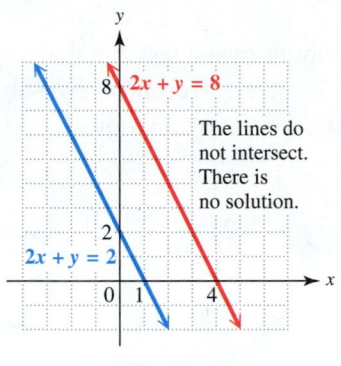

FIGURE 34

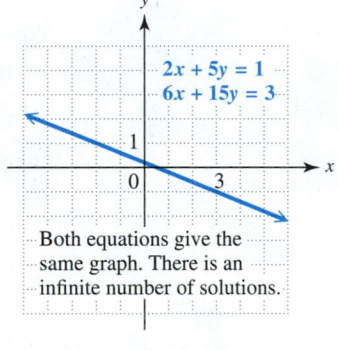

FIGURE 35

(b) $2x + 5y = 1$

$6x + 15y = 3$

The graphs of these two equations are the same line. See **FIGURE 35**. We can obtain the second equation by multiplying each side of the first equation by 3. In this case, every point on the line is a solution of the system, and the solution set contains an infinite number of ordered pairs, each of which satisfies both equations of the system. We write the solution set as

$$\{(x, y) \mid 2x + 5y = 1\},$$

This is the first equation in the system. See the Note on the next page.

read "the set of ordered pairs (x, y) such that $2x + 5y = 1$." Recall from **Section 1.3** that this notation is called **set-builder notation**.

NOW TRY ↻

> **NOTE**   When a system has an infinite number of solutions, as in **Example 3(b),** either equation of the system could be used to write the solution set. *We prefer to use the equation in standard form with integer coefficients having greatest common factor 1.* If neither of the given equations is in this form, we use an *equivalent* equation that is in standard form with integer coefficients having greatest common factor 1.

The system in **Example 2** has exactly one solution. A system with at least one solution is a **consistent system.** A system with no solution, such as the one in **Example 3(a),** is an **inconsistent system.**

The equations in **Example 2** are **independent equations** with different graphs. The equations of the system in **Example 3(b)** have the same graph and are equivalent. Because they are different forms of the same equation, these equations are **dependent equations.**

Examples 2 and 3 illustrate the three cases that may occur when solving a system of equations with two variables.

Three Cases for Solutions of Linear Systems with Two Variables

Case 1   The graphs intersect at exactly one point, which gives the (single) ordered-pair solution of the system. The **system is consistent** and the **equations are independent.** See **FIGURE 36(a).**

Case 2   The graphs are parallel lines, so there is no solution and the solution set is $\varnothing$. The **system is inconsistent** and the **equations are independent.** See **FIGURE 36(b).**

Case 3   The graphs are the same line. There is an infinite number of solutions, and the solution set is written in set-builder notation as

$$\{(x, y) \mid \underline{\qquad\qquad}\},$$

where a form of one of the equations follows the $\mid$ symbol. The **system is consistent** and the **equations are dependent.** See **FIGURE 36(c).**

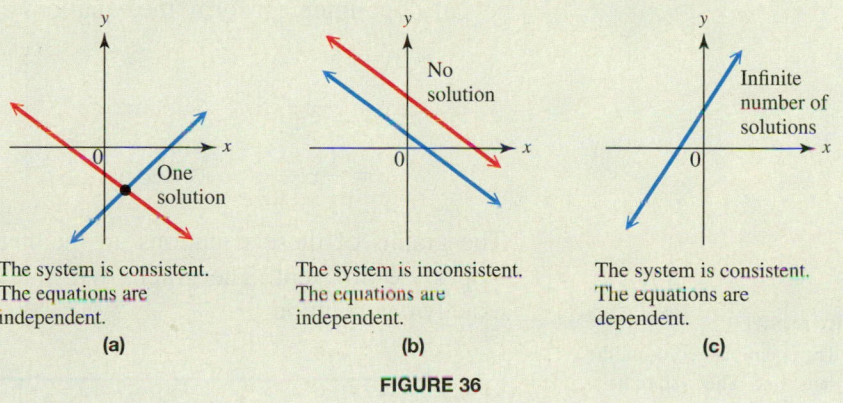

The system is consistent.
The equations are
independent.
(a)

The system is inconsistent.
The equations are
independent.
(b)

The system is consistent.
The equations are
dependent.
(c)

FIGURE 36

OBJECTIVE 4   Identify special systems without graphing.

We can recognize special systems without graphing by comparing their slopes and y-intercepts. We do this by writing each equation in slope-intercept form, solving for y.

**NOW TRY
EXERCISE 4**

Describe each system without graphing. State the number of solutions.

(a) $5x - 8y = 4$

$x - \frac{8}{5}y = \frac{4}{5}$

(b) $2x + y = 7$

$3y = -6x - 12$

(c) $y - 3x = 7$

$3y - \ x = 0$

EXAMPLE 4 Identifying the Three Cases Using Slopes

Describe each system without graphing. State the number of solutions.

(a) $\quad 3x + 2y = 6$

$-2y = 3x - 5$

Write each equation in slope-intercept form $y = mx + b$.

$3x + 2y = 6$ ◁ Solve for y.

$\quad 2y = -3x + 6$ Subtract $3x$.

$\quad y = -\frac{3}{2}x + 3$ Divide *each* term by 2.

$-2y = 3x - 5$ ◁ Solve for y.

$\quad y = -\frac{3}{2}x + \frac{5}{2}$ Divide *each* term by -2.

Both equations have slope $-\frac{3}{2}$, but they have different y-intercepts, $(0, 3)$ and $\left(0, \frac{5}{2}\right)$. Such lines are parallel (**Section 7.2**), so these equations have graphs that are parallel lines, which do not intersect. Thus, the system has no solution.

(b) $2x - y = 4$

$x = \frac{y}{2} + 2$

Again, write the equations in slope-intercept form.

$2x - y = 4$

$\quad -y = -2x + 4$ Subtract $2x$.

$\quad y = 2x - 4$ Multiply by -1.

$\frac{y}{2} + 2 = x$ Interchange sides.

$\frac{y}{2} = x - 2$ Subtract 2.

$\quad y = 2x - 4$ Multiply by 2.

The equations are exactly the same—their graphs are the same line. Any ordered-pair solution of one equation is also a solution of the other equation. Thus, the system has an infinite number of solutions.

(c) $\quad x - 3y = 5$

$2x + \ y = 8$

In slope-intercept form, the equations are as follows.

$x - 3y = 5$

$\quad -3y = -x + 5$ Subtract x.

$\quad y = \frac{1}{3}x - \frac{5}{3}$ Divide by -3.

$2x + y = 8$

$\quad y = -2x + 8$ Subtract $2x$.

The graphs of these equations are neither parallel nor the same line because the slopes are different. The graphs will intersect in one point—thus, the system has exactly one solution.

NOW TRY

NOW TRY ANSWERS

4. (a) The equations represent the same line. The system has an infinite number of solutions.
 (b) The equations represent parallel lines. The system has no solution.
 (c) The equations represent lines that are neither parallel nor the same line. The system has exactly one solution.

NOTE The solution set of the system in **Example 4(a)** is $\varnothing$ because the graphs of the equations of the system are parallel lines. The solution set of the system in **Example 4(b)**, written using set-builder notation and the first equation, is

$$\{(x, y) \mid 2x - y = 4\}.$$

If we try to solve the system in **Example 4(c)** by graphing, we will have difficulty identifying the point of intersection of the graphs. We introduce an algebraic method for solving systems like this in **Section 7.4**.

7.3 Exercises

FOR EXTRA HELP

 MyMathLab®

▶ *Complete solution available in MyMathLab*

Concept Check *Complete each statement. The following terms may be used once, more than once, or not at all.*

| consistent | system of linear equations | inconsistent | solution |
| ordered pair | independent | linear equation | dependent |

1. A(n) _____ consists of two or more linear equations with the (*same/different*) variables.

2. A solution of a system of linear equations is a(n) _____ that makes all equations of the system (*true/false*) at the same time.

3. The equations of two parallel lines form a(n) _____ system that has (*one/no/infinitely many*) solution(s). The equations are _____ because their graphs are different.

4. If the graphs of a linear system intersect in one point, the point of intersection is the _____ of the system. The system is _____ and the equations are independent.

5. If two equations of a linear system have the same graph, the equations are _____. The system is _____ and has (*one/no/infinitely many*) solution(s).

6. If a linear system is inconsistent, the graphs of the two equations are (*intersecting/parallel/the same*) line(s). The system has no _____.

Concept Check   *Work each problem.*

7. A student determined that the ordered pair $(1, -2)$ is a solution of the following system. His reasoning was that the ordered pair satisfies the equation $x + y = -1$ because $1 + (-2) = -1$. **WHAT WENT WRONG?**

$$x + y = -1$$
$$2x + y = 4$$

8. The following system has infinitely many solutions. Write its solution set using set-builder notation as described in **Example 3(b).**

$$6x - 4y = 8$$
$$3x - 2y = 4$$

9. Which ordered pair could not be a solution of the system graphed? Why is it the only valid choice?

   A. $(-4, -4)$ **B.** $(-2, 2)$

   C. $(-4, 4)$ **D.** $(-3, 3)$

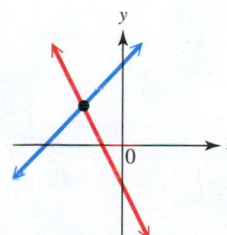

10. Which ordered pair could be a solution of the system graphed? Why is it the only valid choice?

    A. $(2, 0)$ **B.** $(0, 2)$

    C. $(-2, 0)$ **D.** $(0, -2)$

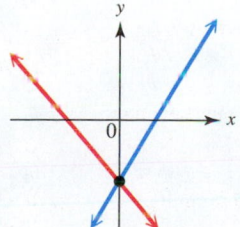

11. Each ordered pair in parts (a)–(d) is a solution of one of the systems graphed in choices A–D. Use the location of the point of intersection to determine the correct system for each solution. Match each system from A–D with its solution from (a)–(d).

(a) $(3, 4)$

(b) $(-2, 3)$

(c) $(-3, 2)$

(d) $(5, -2)$

A.

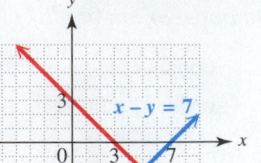

$$x - y = 7$$
$$x + y = 3$$

B.

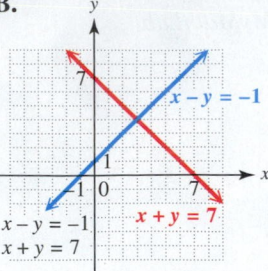

$$x - y = -1$$
$$x + y = 7$$

C.

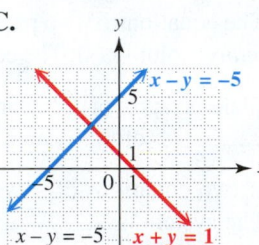

$$x - y = -5$$
$$x + y = 1$$

D.

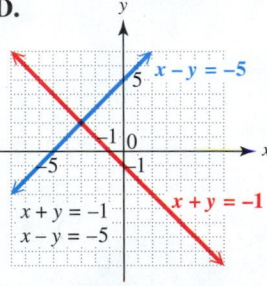

$$x + y = -1$$
$$x - y = -5$$

12. Solve the system by graphing. Can the solution be checked? Why or why not?

$$2x + 3y = 6$$
$$x - 3y = 5$$

Decide whether the given ordered pair is a solution of the given system. **See Example 1.**

13. $(2, -3)$
$$x + \ y = -1$$
$$2x + 5y = 19$$

14. $(4, 3)$
$$x + 2y = 10$$
$$3x + 5y = 3$$

15. $(-1, -3)$
$$3x + 5y = -18$$
$$4x + 2y = -10$$

16. $(-9, -2)$
$$2x - 5y = -8$$
$$3x + 6y = -39$$

17. $(7, -2)$
$$4x = 26 - \ y$$
$$3x = 29 + 4y$$

18. $(9, 1)$
$$2x = 23 - 5y$$
$$3x = 24 + 3y$$

19. $(6, -8)$
$$-2y = \ x + 10$$
$$3y = 2x + 30$$

20. $(-5, 2)$
$$5y = \ 3x + 20$$
$$3y = -2x - 4$$

21. $(0, 0)$
$$4x + 2y = 0$$
$$x + \ y = 0$$

22. $(-1, -1)$
$$-4x + 4y = 0$$
$$x - \ y = 0$$

Solve each system by graphing. If the system is inconsistent or the equations are dependent, say so. **See Examples 2 and 3.**

23. $x - y = 2$
$$x + y = 6$$

24. $x - y = 3$
$$x + y = -1$$

25. $x + y = 4$
$$y - x = 4$$

26. $x + y = -5$
$$y - x = -5$$

27. $x - 2y = 6$
$$x + 2y = 2$$

28. $2x - y = 4$
$$4x + y = 2$$

29. $3x - 2y = -3$
$$-3x - \ y = -6$$

30. $2x - \ y = 4$
$$2x + 3y = 12$$

31. $3x + \ y = 5$
$$6x + 2y = 10$$

32. $2x - \ y = 4$
$$4x - 2y = 8$$

33. $2x - 3y = -6$
$$y = -3x + 2$$

34. $-3x + y = -3$
$$y = x - 3$$

35. $2x - \ y = 6$
$$4x - 2y = 8$$

36. $x + 2y = 4$
$$2x + 4y = 12$$

37. $2y - 6x = 12$
$3x - y = -6$

38. $-8y - 2x = -8$
$x + 4y = 4$

39. $3x - 4y = 24$
$y = -\dfrac{3}{2}x + 3$

40. $4x + y = 5$
$y = \dfrac{3}{2}x - 6$

41. $2x = y - 4$
$4x + 4 = 2y$

42. $3x = y + 5$
$6x - 5 = 2y$

Without graphing, answer the following questions for each linear system. **See Example 4.**

(a) *Is the system inconsistent, are the equations dependent, or neither?*
(b) *Is the graph a pair of intersecting lines, a pair of parallel lines, or one line?*
(c) *Does the system have one solution, no solution, or an infinite number of solutions?*

▶ **43.** $y - x = -5$
$x + y = 1$

44. $y + 2x = 6$
$x - 3y = -4$

45. $x + 2y = 0$
$4y = -2x$

46. $2x - y = 4$
$y + 4 = 2x$

47. $x - 3y = 5$
$2x + y = 8$

48. $2x + 3y = 12$
$2x - y = 4$

49. $5x + 4y = 7$
$10x + 8y = 4$

50. $3x + 2y = 5$
$6x + 4y = 3$

Work each problem using the graph provided.

51. The numbers of daily morning and evening newspapers in the United States in selected years over the period 1980–2012 are shown in the graph.

(a) For which years were there more evening dailies than morning dailies?

(b) Estimate the year in which the number of evening and morning dailies was closest to the same. About how many newspapers of each type were there in that year?

(c) Express the point of intersection of the two graphs as an ordered pair written in the form (year, number of newspapers).

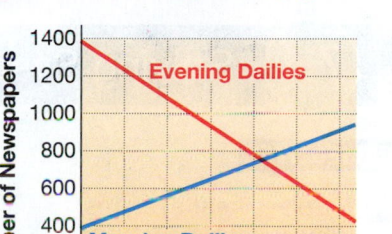

Number of Daily Newspapers

Source: Editor & Publisher International Year Book.

52. The graph shows how sales of music CDs and digital downloads of single songs (in millions) in the United States have changed over the years 2004 through 2012.

(a) In what year did Americans purchase about the same number of CDs as single digital downloads? How many units was this?

(b) Express the point of intersection of the two graphs as an ordered pair written in the form (year, units sold in millions).

(c) Describe the trend in sales of music CDs over the years 2004 to 2012. If a straight line were used to approximate its graph, would the slope of the line be positive, negative, or zero? Explain.

(d) If a straight line were used to approximate the graph of sales of digital downloads over the years 2004 to 2012, would the slope of the line be positive, negative, or zero? Explain.

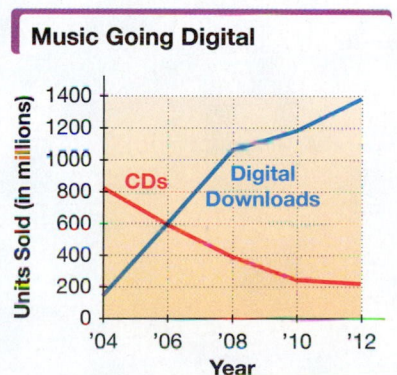

Music Going Digital

Source: Recording Industry Association of America.

*Economics deals with **supply** and **demand**. Typically, as the price of an item increases, the demand for the item decreases and the supply increases. If supply and demand can be described by straight-line equations, the point at which the lines intersect determines the **equilibrium supply** and **equilibrium demand**.*

The price per unit, p, and the demand, x, for a particular aluminum siding are related by the linear equation

$$p = 60 - \frac{3}{4}x,$$

while the price and the supply are related by

$$p = \frac{3}{4}x.$$

Supply and Demand

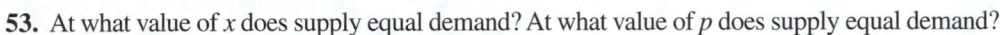

Use the graph to work Exercises 53–56.

53. At what value of x does supply equal demand? At what value of p does supply equal demand?

54. Express the equilibrium supply and equilibrium demand as an ordered pair of the form (quantity, price).

55. When $x > 40$, does demand exceed supply or does supply exceed demand?

56. When $x < 40$, does demand exceed supply or does supply exceed demand?

7.4 Solving Systems of Linear Equations by Substitution

OBJECTIVES

1. Solve linear systems by substitution.
2. Solve special systems by substitution.
3. Solve linear systems with fractions and decimals.

OBJECTIVE 1 Solve linear systems by substitution.

Graphing to solve a system of equations has a serious drawback. For example, consider the system graphed in **FIGURE 37**. It is difficult to determine an accurate solution of the system from the graph.

As a result, there are algebraic methods for solving systems of equations. The **substitution method,** which gets its name from the fact that an expression in one variable is *substituted* for the other variable, is one such method.

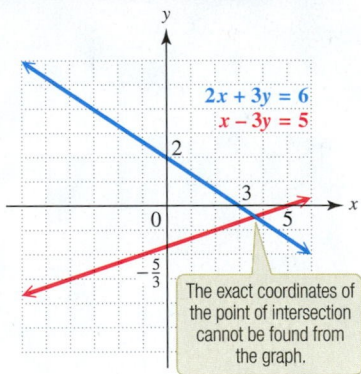

> The exact coordinates of the point of intersection cannot be found from the graph.

FIGURE 37

EXAMPLE 1 Using the Substitution Method

Solve the system by the substitution method.

$$3x + 5y = 26 \qquad (1)$$
$$y = 2x \qquad (2)$$

> We number the equations for reference in our discussion.

Equation (2) is already solved for y. This equation says that y is equal to $2x$, so we substitute $2x$ for y in equation (1).

$$3x + 5y = 26 \qquad (1)$$
$$3x + 5(2x) = 26 \qquad \text{Let } y = 2x.$$
$$3x + 10x = 26 \qquad \text{Multiply.}$$
$$13x = 26 \qquad \text{Combine like terms.}$$
$$x = 2 \qquad \text{Divide by 13.}$$

> Don't stop here.

**NOW TRY
EXERCISE 1**

Solve the system by the substitution method.

$$2x - 4y = 28$$
$$y = -3x$$

Now we can find the value of y by substituting 2 for x in either equation. We choose equation (2) because the substitution is easier.

$$y = 2x \qquad (2)$$
$$y = 2(2) \qquad \text{Let } x = 2.$$
$$y = 4 \qquad \text{Multiply.}$$

We check that the ordered pair $(2, 4)$ is the solution by substituting 2 for x and 4 for y in *both* equations.

CHECK
$$3x + 5y = 26 \qquad (1) \qquad\qquad y = 2x \qquad (2)$$
$$3(2) + 5(4) \overset{?}{=} 26 \quad \text{Substitute.} \qquad 4 \overset{?}{=} 2(2) \quad \text{Substitute.}$$
$$6 + 20 \overset{?}{=} 26 \quad \text{Multiply.} \qquad 4 = 4 \checkmark \quad \text{True}$$
$$26 = 26 \checkmark \quad \text{True}$$

Because $(2, 4)$ satisfies both equations, the solution set of the system is $\{(2, 4)\}$.

NOW TRY

⚠ **CAUTION** *A system is not completely solved until values for both x and y are found.* Write the solution set as a set containing an ordered pair.

**NOW TRY
EXERCISE 2**

Solve the system by the substitution method.

$$4x + 9y = 1$$
$$x = y - 3$$

EXAMPLE 2 Using the Substitution Method

Solve the system by the substitution method.

$$2x + 5y = 7 \qquad (1)$$
$$x = -1 - y \qquad (2)$$

Equation (2) gives x in terms of y. Substitute $-1 - y$ for x in equation (1).

$$2x + 5y = 7 \qquad (1) \quad \boxed{\text{Be sure to substitute in the } \textit{other} \text{ equation.}}$$
$$2(-1 - y) + 5y = 7 \qquad \text{Let } x = -1 - y.$$
$$\boxed{\text{Distribute 2 to } \textit{both} -1 \text{ and } -y.} \quad -2 - 2y + 5y = 7 \qquad \text{Distributive property}$$
$$-2 + 3y = 7 \qquad \text{Combine like terms.}$$
$$3y = 9 \qquad \text{Add 2.}$$
$$y = 3 \qquad \text{Divide by 3.}$$

To find x, substitute 3 for y in equation (2).

$$x = -1 - y \qquad (2)$$
$$x = -1 - 3 \qquad \text{Let } y = 3.$$
$$x = -4 \qquad \text{Subtract.}$$

Check that $(-4, 3)$ is the solution.

CHECK
$$2x + 5y = 7 \qquad (1) \qquad\qquad x = -1 - y \qquad (2)$$
$$2(-4) + 5(3) \overset{?}{=} 7 \quad \text{Substitute.} \qquad -4 \overset{?}{=} -1 - 3 \quad \text{Substitute.}$$
$$-8 + 15 \overset{?}{=} 7 \quad \text{Multiply.} \qquad -4 = -4 \checkmark \quad \text{True}$$
$$7 = 7 \checkmark \quad \text{True} \qquad\qquad \boxed{\text{Write the } x\text{-coordinate first.}}$$

Both results are true. The solution set of the system is $\{(-4, 3)\}$. **NOW TRY**

> ⚠️ **CAUTION** Even though we found y first in **Example 2,** *the x-coordinate is always written first in the ordered-pair solution of a system.* The ordered pair $(-4, 3)$ is **not** the same as $(3, -4)$.

Solving a Linear System by Substitution

Step 1   **Solve one equation for either variable.** If one of the equations has a variable term with coefficient 1 or -1, choose it because the substitution method is usually easier.

Step 2   **Substitute** for that variable in the other equation. The result should be an equation with just one variable.

Step 3   **Solve** the equation from Step 2.

Step 4   **Find the other value.** Substitute the result from Step 3 into the equation from Step 1 and solve for the other variable.

Step 5   **Check** the values in *both* of the *original* equations. Then write the solution set as a set containing an ordered pair.

EXAMPLE 3   Using the Substitution Method

Solve the system by the substitution method.

$$2x = 4 - y \qquad (1)$$
$$5x + 3y = 10 \qquad (2)$$

Step 1   We must solve one of the equations for either x or y. Because the coefficient of y in equation (1) is -1, we avoid fractions by solving this equation for y.

$$2x = 4 - y \qquad (1)$$
$$y + 2x = 4 \qquad \text{Add } y.$$
$$y = -2x + 4 \qquad \text{Subtract } 2x.$$

Step 2   Now substitute $-2x + 4$ for y in equation (2).

$$5x + 3y = 10 \qquad (2)$$
$$5x + 3(-2x + 4) = 10 \qquad \text{Let } y = -2x + 4.$$

Step 3   Solve the equation from Step 2.

$$5x - 6x + 12 = 10 \qquad \text{Distributive property}$$

> Distribute 3 to *both* $-2x$ and 4.

$$-x + 12 = 10 \qquad \text{Combine like terms.}$$
$$-x = -2 \qquad \text{Subtract 12.}$$
$$x = 2 \qquad \text{Multiply by } -1.$$

Step 4   We solved equation (1) for y in Step 1. Substitute 2 for x in this equation to find y.

$$y = -2x + 4 \qquad \text{Equation (1) solved for } y$$
$$y = -2(2) + 4 \qquad \text{Let } x = 2.$$
$$y = 0 \qquad \text{Multiply, and then add.}$$

NOW TRY
EXERCISE 3
Solve the system by the substitution method.

$$2y = x - 2$$
$$4x - 5y = -4$$

Step 5   Check that $(2, 0)$ is the solution.

CHECK $2x = 4 - y$ (1) $5x + 3y = 10$ (2)

$2(2) \overset{?}{=} 4 - 0$ Substitute. $5(2) + 3(0) \overset{?}{=} 10$ Substituto.

$\qquad 4 = 4$ ✓   True $\qquad 10 = 10$ ✓   True

Because both results are true, the solution set of the system is $\{(2, 0)\}$.

NOW TRY

OBJECTIVE 2 Solve special systems by substitution.

NOW TRY
EXERCISE 4
Solve the system by the substitution method.

$$8x - 2y = 1$$
$$y = 4x - 8$$

EXAMPLE 4   Solving an Inconsistent System Using Substitution

Solve the system by the substitution method.

$$x = 5 - 2y \qquad (1)$$
$$2x + 4y = 6 \qquad (2)$$

Because equation (1) is solved for x, we substitute $5 - 2y$ for x in equation (2).

$2x + 4y = 6$ (2)

$2(5 - 2y) + 4y = 6$ Let $x = 5 - 2y$.

$10 - 4y + 4y = 6$ Distributive property

$10 = 6$ False

The false result means that the equations in the system have graphs that are parallel lines. The system is inconsistent and has no solution, so the solution set is $\varnothing$.

CHECK   We can confirm the solution set by writing each equation in slope-intercept form—that is, solved for y. (See **Section 7.3, Example 4.**)

$x = 5 - 2y$ (1) $2x + 4y = 6$ (2)

$2y = -x + 5$ $4y = -2x + 6$

$y = -\dfrac{1}{2}x + \dfrac{5}{2}$ $y = -\dfrac{1}{2}x + \dfrac{3}{2}$

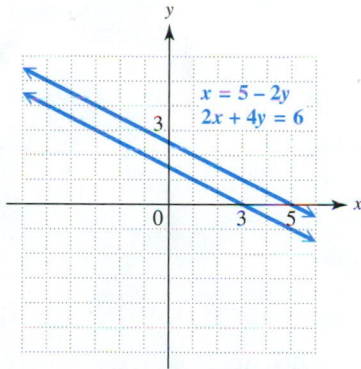

The two lines have the same slope, but different y-intercepts. Therefore, they are parallel and do not intersect, confirming that the solution set is $\varnothing$. See **FIGURE 38**. ✓

FIGURE 38

NOW TRY

EXAMPLE 5   Solving a System with Dependent Equations Using Substitution

Solve the system by the substitution method.

$$3x - y = 4 \qquad (1)$$
$$-9x + 3y = -12 \qquad (2)$$

Begin by solving equation (1) for y to obtain

$$y = 3x - 4. \qquad \text{Equation (1) solved for } y$$

We substitute $3x - 4$ for y in equation (2) and solve the resulting equation.

NOW TRY ANSWERS
3. $\{(-6, -4)\}$ **4.** $\varnothing$

**NOW TRY
EXERCISE 5**

Solve the system by the substitution method.

$$5x - y = 6$$
$$-10x + 2y = -12$$

$$-9x + 3y = -12 \quad (2)$$
$$-9x + 3(3x - 4) = -12 \quad \text{Let } y = 3x - 4.$$
$$-9x + 9x - 12 = -12 \quad \text{Distributive property}$$
$$0 = 0 \quad \text{Add 12. Combine like terms.}$$

This true result means that every solution of one equation is also a solution of the other, so the system has an infinite number of solutions. The solution set, written in set-builder notation using equation (1), is

$$\{(x, y) \mid 3x - y = 4\}.$$

CHECK   If we multiply equation (1) by -3, we obtain equation (2). Therefore,

$$3x - y = 4 \quad \text{and} \quad -9x + 3y = -12$$

are equivalent equations. They represent the same line. All of the ordered pairs corresponding to points that lie on the common graph are solutions. See **FIGURE 39**. ✓

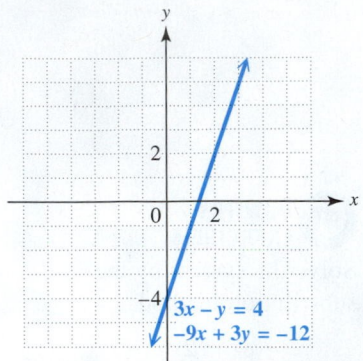

FIGURE 39

NOW TRY

⚠ **CAUTION**   Avoid these common mistakes.

1. Do not give "false" as the solution of an inconsistent system. The correct response is ∅. (See **Example 4**.)

2. Do not give "true" as the solution of a system of dependent equations. In this book, we write the solution set in set-builder notation using the equation in the system (or an equivalent equation) that is in standard form with integer coefficients having greatest common factor 1. (See **Example 5**.)

OBJECTIVE 3   Solve linear systems with fractions and decimals.

EXAMPLE 6   Using the Substitution Method (Fractional Coefficients)

Solve the system by the substitution method.

$$3x + \frac{1}{4}y = 2 \quad (1)$$

$$\frac{1}{2}x + \frac{3}{4}y = -\frac{5}{2} \quad (2)$$

Clear equation (1) of fractions by multiplying each side by 4.

$$3x + \frac{1}{4}y = 2 \quad (1)$$

$$4\left(3x + \frac{1}{4}y\right) = 4(2) \quad \text{Multiply by 4.}$$

$$4(3x) + 4\left(\frac{1}{4}y\right) = 4(2) \quad \text{Distributive property}$$

$$12x + y = 8 \quad (3)$$

NOW TRY ANSWER
5. $\{(x, y) \mid 5x - y = 6\}$

NOW TRY
EXERCISE 6
Solve the system by the substitution method.

$$x + \frac{1}{2}y = \frac{1}{2}$$

$$\frac{1}{6}x - \frac{1}{3}y = \frac{4}{3}$$

Now clear equation (2) of fractions by multiplying each side by 4.

$$\frac{1}{2}x + \frac{3}{4}y = -\frac{5}{2} \qquad (2)$$

$$4\left(\frac{1}{2}x + \frac{3}{4}y\right) = 4\left(-\frac{5}{2}\right) \qquad \text{Multiply by 4, the common denominator.}$$

$$4\left(\frac{1}{2}x\right) + 4\left(\frac{3}{4}y\right) = 4\left(-\frac{5}{2}\right) \qquad \text{Distributive property}$$

$$2x + 3y = -10 \qquad (4)$$

The given system of equations has been simplified to an equivalent system.

$$12x + y = 8 \qquad (3)$$

$$2x + 3y = -10 \qquad (4)$$

To solve this system by substitution, solve equation (3) for y.

$$12x + y = 8 \qquad (3)$$

$$y = -12x + 8 \qquad \text{Subtract } 12x.$$

Now substitute this result for y in equation (4).

$$2x + 3y = -10 \qquad (4)$$

$$2x + 3(-12x + 8) = -10 \qquad \text{Let } y = -12x + 8.$$

$$2x - 36x + 24 = -10 \qquad \text{Distributive property}$$

$$-34x = -34 \qquad \text{Combine like terms. Subtract 24.}$$

$$x = 1 \qquad \text{Divide by } -34.$$

Substitute 1 for x in $y = -12x + 8$ (equation (3) solved for y).

$$y = -12(1) + 8 \qquad \text{Let } x = 1.$$

$$y = -4 \qquad \text{Multiply, and then add.}$$

Check $(1, -4)$ in both of the original equations. The solution set is $\{(1, -4)\}$.

NOW TRY

EXAMPLE 7 Using the Substitution Method (Decimal Coefficients)

Solve the system by the substitution method.

$$0.5x + 2.4y = 4.2 \qquad (1)$$

$$-0.1x + 1.5y = 5.1 \qquad (2)$$

Clear each equation of decimals by multiplying by 10.

$$10(0.5x + 2.4y) = 10(4.2) \qquad \text{Multiply equation (1) by 10.}$$

$$10(0.5x) + 10(2.4y) = 10(4.2) \qquad \text{Distributive property}$$

$$5x + 24y = 42 \qquad (3)$$

$$10(-0.1x + 1.5y) = 10(5.1) \qquad \text{Multiply equation (2) by 10.}$$

$$10(-0.1x) + 10(1.5y) = 10(5.1) \qquad \text{Distributive property}$$

 $\boxed{10(-0.1x) = -1x = -x}$ $-x + 15y = 51 \qquad (4)$

NOW TRY ANSWER
6. $\{(2, -3)\}$

**NOW TRY
EXERCISE 7**

Solve the system by the substitution method.

$$0.2x + 0.3y = 0.5$$
$$0.3x - 0.1y = 1.3$$

Now solve the equivalent system of equations by substitution.

$$5x + 24y = 42 \quad \text{(3)}$$
$$-x + 15y = 51 \quad \text{(4)}$$

Equation (4) can be solved for x.

$$x = 15y - 51 \quad \text{Equation (4) solved for } x$$

Substitute this result for x in equation (3).

$$5x + 24y = 42 \quad \text{(3)}$$
$$5(15y - 51) + 24y = 42 \quad \text{Let } x = 15y - 51.$$
$$75y - 255 + 24y = 42 \quad \text{Distributive property}$$
$$99y = 297 \quad \text{Combine like terms. Add 255.}$$
$$y = 3 \quad \text{Divide by 99.}$$

Equation (4) solved for x is $x = 15y - 51$. Substitute 3 for y.

$$x = 15(3) - 51 \quad \text{Let } y = 3.$$
$$x = -6 \quad \text{Multiply, and then subtract.}$$

Check $(-6, 3)$ in both of the original equations. The solution set is $\{(-6, 3)\}$.

NOW TRY ANSWER
7. $\{(4, -1)\}$

NOW TRY

7.4 Exercises

▶ *Complete solution available in MyMathLab*

Concept Check   Work each problem.

1. A student solves the following system and finds that $x = 3$, which is correct. The student gives the solution set as $\{3\}$. **WHAT WENT WRONG?**

$$5x - y = 15$$
$$7x + y = 21$$

2. A student solves the following system and obtains the equation $0 = 0$. The student gives the solution set as $\{(0, 0)\}$. **WHAT WENT WRONG?**

$$x + y = 4$$
$$2x + 2y = 8$$

3. When we use the substitution method, how can we tell that a system has no solution?

4. When we use the substitution method, how can we tell that a system has an infinite number of solutions?

Solve each system by the substitution method. Check each solution. **See Examples 1–5.**

5. $x + y = 12$
$y = 3x$

6. $x + 3y = -28$
$y = -5x$

7. $3x + 2y = 27$
$x = y + 4$

8. $4x + 3y = -5$
$x = y - 3$

9. $3x + 4 = -y$
$2x + y = 0$

10. $2x - 5 = -y$
$x + 3y = 0$

11. $7x + 4y = 13$
$x + y = 1$

12. $3x - 2y = 19$
$x + y = 8$

13. $3x + 5y = 25$
$x - 2y = -10$

14. $5x + 2y = -15$
$2x - y = -6$

15. $3x - y = 5$
$y = 3x - 5$

16. $4x - y = -3$
$y = 4x + 3$

17. $2x + y = 0$
$4x - 2y = 2$

18. $x + y = 0$
$4x + 2y = 3$

19. $2x + 8y = 3$
$x = 8 - 4y$

20. $2x + 10y = 3$
$x = 1 - 5y$

21. $2y = 4x + 24$
$2x - y = -12$

22. $2y = 14 - 6x$
$3x + y = 7$

23. $y = 6 - x$
$y = 2x + 3$

24. $y = 4x - 4$
$y = -3x - 11$

25. $x = y - 4$
$x - y = 1$

26. $x = 2 - y$
$x + y = -5$

27. $x + y = 0$
$3x - 3y = 0$

28. $5x + y = 0$
$x - y = 0$

Solve each system by the substitution method. Check each solution. **See Examples 6 and 7.**

29. $\dfrac{1}{2}x + \dfrac{1}{3}y = 3$
$y = 3x$

30. $\dfrac{1}{4}x - \dfrac{1}{5}y = 9$
$y = 5x$

31. $\dfrac{1}{2}x + \dfrac{1}{3}y = -\dfrac{1}{3}$
$\dfrac{1}{2}x + 2y = -7$

32. $\dfrac{1}{6}x + \dfrac{1}{6}y = 1$
$-\dfrac{1}{2}x - \dfrac{1}{3}y = -5$

33. $\dfrac{x}{5} + 2y = \dfrac{8}{5}$
$\dfrac{3x}{5} + \dfrac{y}{2} = -\dfrac{7}{10}$

34. $\dfrac{x}{2} + \dfrac{y}{3} = \dfrac{7}{6}$
$\dfrac{x}{4} - \dfrac{3y}{2} = \dfrac{9}{4}$

35. $\dfrac{1}{6}x + \dfrac{1}{3}y = 8$
$\dfrac{1}{4}x + \dfrac{1}{2}y = 12$

36. $\dfrac{1}{2}x - \dfrac{1}{8}y = -\dfrac{1}{4}$
$\dfrac{1}{3}x - \dfrac{1}{12}y = -\dfrac{1}{6}$

37. $0.2x - 1.3y = -3.2$
$-0.1x + 2.7y = 9.8$

38. $0.1x + 0.9y = -2$
$0.5x - 0.2y = 4.1$

39. $0.3x - 0.1y = 2.1$
$0.6x + 0.3y = -0.3$

40. $0.8x - 0.1y = 1.3$
$2.2x + 1.5y = 8.9$

RELATING CONCEPTS For Individual or Group Work (Exercises 41–44)

A system of linear equations can be used to model the cost and the revenue of a business.
Work Exercises 41–44 in order.

41. Suppose that it costs $5000 to start a business manufacturing and selling bicycles. Each bicycle will cost $400 to manufacture. Explain why the linear equation

$$y_1 = 400x + 5000 \quad (y_1 \text{ in dollars})$$

gives the *total* cost to manufacture x bicycles.

42. We decide to sell each bicycle for $600. Write an equation using y_2 (in dollars) to express the revenue when we sell x bicycles.

43. Form a system from the two equations in **Exercises 41 and 42.** Solve the system.

44. The value of x from **Exercise 43** is the number of bicycles it takes to *break even.* Fill in the blanks: When _____ bicycles are sold, the break-even point is reached. At that point, we have spent _____ dollars and taken in _____ dollars.

7.5 Solving Systems of Linear Equations by Elimination

NOW TRY EXERCISE 1

Solve the system by the elimination method.

$$x - y = 4$$
$$3x + y = 8$$

OBJECTIVE 1 Solve linear systems by elimination.

Adding the same quantity to each side of an equation results in equal sums.

$$\text{If } \quad A = B, \quad \text{then} \quad A + C = B + C.$$

We can take this addition a step further. Adding *equal* quantities, rather than the *same* quantity, to each side of an equation also results in equal sums.

$$\text{If } \quad A = B \quad \text{and} \quad C = D, \quad \text{then} \quad A + C = B + D.$$

The **elimination method** uses the addition property of equality to solve systems of equations.

EXAMPLE 1 Using the Elimination Method

Solve the system by the elimination method.

$$x + y = 5 \quad (1)$$
$$x - y = 3 \quad (2)$$

Each equation in this system is a statement of equality, so the sum of the left sides equals the sum of the right sides. Adding vertically in this way gives the following.

$$
\begin{array}{rl}
x + y = 5 & (1) \\
\underline{x - y = 3} & (2) \\
2x \quad\;\; = 8 & \text{Add left sides and add right sides.} \\
x = 4 & \text{Divide by 2.}
\end{array}
$$

Notice that y has been eliminated. The result, $x = 4$, gives the x-value of an ordered pair. To find the corresponding y-value, substitute 4 for x in either of the two equations of the system. We choose equation (1).

$$
\begin{array}{rl}
x + y = 5 & (1) \\
4 + y = 5 & \text{Let } x = 4. \\
y = 1 & \text{Subtract 4.}
\end{array}
$$

Check the ordered pair $(4, 1)$ in both equations of the given system.

CHECK

$x + y = 5$	(1)		$x - y = 3$	(2)
$4 + 1 \stackrel{?}{=} 5$	Substitute.		$4 - 1 \stackrel{?}{=} 3$	Substitute.
$5 = 5$ ✓	True		$3 = 3$ ✓	True

Both results are true, so the solution set of the system is $\{(4, 1)\}$.   **NOW TRY**

With the elimination method, the idea is to *eliminate* one of the two variables in a system.

To do this, one pair of variable terms in the two equations must have coefficients that are opposites (additive inverses).

> ### Solving a Linear System by Elimination
>
> **Step 1** Write both equations in the form $Ax + By = C$.
>
> **Step 2** **Transform the equations as needed so that the coefficients of one pair of variable terms are opposites.** Multiply one or both equations by appropriate numbers so that the sum of the coefficients of either the x- or y-terms is 0.
>
> **Step 3** **Add** the new equations to eliminate a variable. The sum should be an equation with just one variable.
>
> **Step 4** **Solve** the equation from Step 3 for the remaining variable.
>
> **Step 5** **Find the other value.** Substitute the result from Step 4 into either of the original equations, and solve for the other variable.
>
> **Step 6** **Check** the values in *both* of the *original* equations. Then write the solution set as a set containing an ordered pair.

It does not matter which variable is eliminated first. Usually, we choose the one that is more convenient to work with.

 **NOW TRY EXERCISE 2**

Solve the system.

$$2x - 6 = -3y$$
$$5x - 3y = -27$$

EXAMPLE 2 Using the Elimination Method

Solve the system.

$$y + 11 = 2x \quad (1)$$
$$5x = y + 26 \quad (2)$$

Step 1 Write both equations in the form $Ax + By = C$.

$$-2x + y = -11 \qquad \text{Subtract } 2x \text{ and } 11 \text{ in equation (1).}$$
$$5x - y = 26 \qquad \text{Subtract } y \text{ in equation (2).}$$

Step 2 Because the coefficients of y are 1 and -1, adding will eliminate y. It is not necessary to multiply either equation by a number.

Step 3 Add the two equations.

$$
\begin{array}{rcl}
-2x + y &=& -11 \\
5x - y &=& \ \ 26 \\
\hline
3x &=& \ \ 15 \qquad \text{Add in columns.}
\end{array}
$$

Step 4 Solve. $\qquad\qquad\qquad x = 5 \qquad$ Divide by 3.

Step 5 Find the value of y by substituting 5 for x in either of the original equations.

$$y + 11 = 2x \qquad (1)$$
$$y + 11 = 2(5) \qquad \text{Let } x = 5.$$
$$y + 11 = 10 \qquad \text{Multiply.}$$
$$y = -1 \qquad \text{Subtract 11.}$$

Step 6 Check the ordered pair $(5, -1)$ in both of the original equations.

CHECK

$y + 11 = 2x \qquad (1)$	$5x = y + 26 \qquad (2)$
$(-1) + 11 \overset{?}{=} 2(5) \quad$ Substitute.	$5(5) = -1 + 26 \quad$ Substitute.
$10 = 10 \ \checkmark \quad$ True	$25 = 25 \ \checkmark \quad$ True

Because $(5, -1)$ is a solution of *both* equations, the solution set is $\{(5, -1)\}$.

NOW TRY

OBJECTIVE 2   Multiply when using the elimination method.

Sometimes we need to multiply each side of one or both equations in a system by a number before adding will eliminate a variable.

NOW TRY
EXERCISE 3
Solve the system.

$$3x - 5y = 25$$
$$2x + 8y = -6$$

EXAMPLE 3   Using the Elimination Method

Solve the system.

$$2x + 3y = -15 \qquad (1)$$
$$5x + 2y = 1 \qquad (2)$$

Adding the two equations gives $7x + 5y = -14$, which does not eliminate either variable. However, we can multiply each equation by a suitable number so that the coefficients of one of the two variables are opposites. For example, to eliminate x, we multiply each side of $2x + 3y = -15$ (equation (1)) by 5 and each side of $5x + 2y = 1$ (equation (2)) by -2.

$$10x + 15y = -75 \qquad \text{Multiply } both\ sides \text{ of equation (1) by 5.}$$
$$\underline{-10x - 4y = -2} \qquad \text{Multiply } both\ sides \text{ of equation (2) by } -2.$$

The coefficients of x are opposites.

$$11y = -77 \qquad \text{Add.}$$
$$y = -7 \qquad \text{Divide by 11.}$$

Find the value of x by substituting -7 for y in either equation (1) or (2).

$$5x + 2y = 1 \qquad (2)$$
$$5x + 2(-7) = 1 \qquad \text{Let } y = -7.$$
$$5x - 14 = 1 \qquad \text{Multiply.}$$
$$5x = 15 \qquad \text{Add 14.}$$
$$x = 3 \qquad \text{Divide by 5.}$$

Check that the solution set of the system is $\{(3, -7)\}$.   Write the x-value first.   **NOW TRY**

NOTE   In **Example 3,** we eliminated the variable x. Alternatively, we could multiply each equation of the system by a suitable number so that the variable y is eliminated.

$$2x + 3y = -15 \quad (1) \xrightarrow{\text{Multiply by 2.}} \quad 4x + 6y = -30$$
$$5x + 2y = 1 \quad (2) \xrightarrow{\text{Multiply by } -3.} \quad -15x - 6y = -3$$

Complete this approach and confirm that the same solution results.

⊘ CAUTION   When using the elimination method, remember to *multiply both sides* of an equation by the same nonzero number.

OBJECTIVE 3   Use an alternative method to find the second value in a solution.

Sometimes it is easier to find the value of the second variable in a solution using the elimination method twice.

NOW TRY ANSWER
3. $\{(5, -2)\}$

NOW TRY EXERCISE 4

Solve the system.

$$4x + 9y = 3$$
$$5y = 6 - 3x$$

EXAMPLE 4 Finding the Second Value Using an Alternative Method

Solve the system.

$$4x = 9 - 3y \quad (1)$$
$$5x - 2y = 8 \quad (2)$$

Write equation (1) in the form $Ax + By = C$ by adding $3y$ to each side.

$$4x + 3y = 9 \quad (3)$$
$$5x - 2y = 8 \quad (2)$$

One way to proceed is to eliminate y by multiplying each side of equation (3) by 2 and each side of equation (2) by 3 and then adding.

$$8x + 6y = 18 \qquad \text{Multiply equation (3) by 2.}$$
$$15x - 6y = 24 \qquad \text{Multiply equation (2) by 3.}$$
$$\overline{23x \qquad \quad = 42} \qquad \text{Add.}$$

The coefficients of y are opposites.

$$x = \frac{42}{23} \qquad \text{Divide by 23.}$$

Substituting $\frac{42}{23}$ for x in one of the given equations would give y, but the arithmetic would be complicated. Instead, solve for y by starting over again with the original equations written in $Ax + By = C$ form (equations (3) and (2)) and eliminating x.

$$20x + 15y = 45 \qquad \text{Multiply equation (3) by 5.}$$
$$-20x + 8y = -32 \qquad \text{Multiply equation (2) by } -4.$$
$$\overline{23y = 13} \qquad \text{Add.}$$

The coefficients of x are opposites.

$$y = \frac{13}{23} \qquad \text{Divide by 23.}$$

Check that the solution set is $\left\{\left(\frac{42}{23}, \frac{13}{23}\right)\right\}$.

NOW TRY

NOTE When the value of the first variable is a fraction, the method used in **Example 4** helps avoid arithmetic errors. This method could be used to solve any system.

OBJECTIVE 4 Solve special systems by elimination.

EXAMPLE 5 Solving Special Systems Using the Elimination Method

Solve each system by the elimination method.

(a)
$$2x + 4y = 5 \quad (1)$$
$$4x + 8y = -9 \quad (2)$$

Multiply each side of equation (1) by -2. Then add the two equations.

$$-4x - 8y = -10 \qquad \text{Multiply equation (1) by } -2.$$
$$4x + 8y = -9 \qquad (2)$$
$$\overline{0 = -19} \qquad \text{False}$$

The false statement $0 = -19$ indicates that the given system has solution set $\varnothing$.

NOW TRY ANSWER
4. $\left\{\left(\frac{39}{7}, -\frac{15}{7}\right)\right\}$

**NOW TRY
EXERCISE 5**

Solve each system by the elimination method.

(a) $x - y = 2$
   $5x - 5y = 10$

(b) $4x + 3y = 0$
   $-4x - 3y = -1$

NOW TRY ANSWERS

5. **(a)** $\{(x, y) \mid x - y = 2\}$
   (b) $\varnothing$

(b)
$$3x - y = 4 \quad\quad (1)$$
$$-9x + 3y = -12 \quad\quad (2)$$

Multiply each side of equation (1) by 3. Then add the two equations.

$$9x - 3y = 12 \quad\quad \text{Multiply equation (1) by 3.}$$
$$\underline{-9x + 3y = -12} \quad\quad (2)$$
$$0 = 0 \quad\quad \text{True}$$

A true statement occurs when the equations are equivalent. This indicates that every solution of one equation is also a solution of the other. The solution set is

$$\{(x, y) \mid 3x - y = 4\}.$$

(See **Section 7.4, Example 5,** where this same system was solved by substitution.)

NOW TRY

7.5 Exercises

FOR EXTRA HELP   ▶ **MyMathLab**

▶ *Complete solution available in MyMathLab*

Concept Check   *Answer* true *or* false *for each statement. If false, tell why.*

1. If the elimination method leads to $0 = -1$, the solution set of the system is $\{(0, -1)\}$.

2. A system that includes the equation $5x - 4y = 0$ cannot have $(4, -5)$ as a solution.

Solve each system by the elimination method. Check each solution. **See Examples 1 and 2.**

▶ 3. $x - y = -2$
   $x + y = 10$

4. $x + y = 10$
   $x - y = -6$

5. $2x + y = -5$
   $x - y = 2$

6. $2x + y = -15$
   $-x - y = 10$

▶ 7. $2y = -3x$
   $-3x - y = 3$

8. $5x = y + 5$
   $-5x + 2y = 0$

9. $6x - y = -1$
   $5y = 17 + 6x$

10. $y = 9 - 6x$
   $-6x + 3y = 15$

Solve each system by the elimination method. Check each solution. **See Examples 3–5.**

11. $2x - y = 12$
   $3x + 2y = -3$

12. $x + y = 3$
   $-3x + 2y = -19$

13. $x + 4y = 16$
   $3x + 5y = 20$

14. $2x + y = 8$
   $5x - 2y = -16$

15. $2x - 8y = 0$
   $4x + 5y = 0$

16. $3x - 15y = 0$
   $6x + 10y = 0$

▶ 17. $3x + 3y = 33$
   $5x - 2y = 27$

18. $4x - 3y = -19$
   $3x + 2y = 24$

19. $5x + 4y = 12$
   $3x + 5y = 15$

20. $2x + 3y = 21$
   $5x - 2y = -14$

21. $5x - 4y = 15$
   $-3x + 6y = -9$

22. $4x + 5y = -16$
   $5x - 6y = -20$

▶ 23. $-x + 3y = 4$
   $-2x + 6y = 8$

24. $6x - 2y = 24$
   $-3x + y = -12$

25. $5x - 2y = 3$
   $10x - 4y = 5$

26. $3x - 5y = 1$
$6x - 10y = 4$

27. $6x - 2y = -22$
$-3x + 4y = 17$

28. $5x - 4y = -1$
$x + 8y = -9$

29. $3x = 3 + 2y$
$-\dfrac{4}{3}x + y = \dfrac{1}{3}$

30. $3x = 27 + 2y$
$x - \dfrac{7}{2}y = -25$

31. $\dfrac{1}{5}x + y = \dfrac{6}{5}$
$\dfrac{1}{10}x + \dfrac{1}{3}y = \dfrac{5}{6}$

32. $\dfrac{1}{3}x + \dfrac{1}{2}y = \dfrac{13}{6}$
$\dfrac{1}{2}x - \dfrac{1}{4}y = -\dfrac{3}{4}$

33. $2.4x + 1.7y = 7.6$
$1.2x - 0.5y = 9.2$

34. $0.5x + 3.4y = 13$
$1.5x - 2.6y = -25$

35. $x + 3y = 6$
$-2x + 12 = 6y$

36. $7x + 2y = 0$
$4y = -14x$

37. $4x - 3y = 1$
$8x = 3 + 6y$

38. $5x + 8y = 10$
$24y = -15x - 10$

▶ **39.** $4x = 3y - 2$
$5x + 3 = 2y$

40. $2x + 3y = 0$
$4x + 12 = 9y$

41. $24x + 12y = -7$
$16x - 18y = 17$

42. $9x + 4y = -3$
$6x + 6y = -7$

SUMMARY EXERCISES   Applying Techniques for Solving Systems of Linear Equations

Guidelines for Choosing a Method to Solve a System of Linear Equations

1. If one of the equations of the system is already solved for one of the variables, as indicated by the arrows in the following systems, the substitution method is the better choice.

$$3x + 4y = 9$$
$$\rightarrow y = 2x - 6$$
and
$$\rightarrow x = 3y - 7$$
$$-5x + 3y = 9$$

2. If both equations are in the form $Ax + By = C$ and none of the variables has coefficient -1 or 1, as in the following system, the elimination method is the better choice.

$$4x - 11y = 3$$
$$-2x + 3y = 4$$

3. If one or both of the equations are in the form $Ax + By = C$ and the coefficient of one of the variables is -1 or 1, as indicated by the arrows in the following systems, either method is appropriate.

$$\rightarrow 3x + y = -2$$
$$-5x + 2y = 4$$
and
$$3x - 2y = 8$$
$$\rightarrow -x + 3y = -4$$

1. (a) Use substitution because the second equation is solved for y.
(b) Use elimination because the coefficients of the y-terms are opposites.
(c) Use elimination because the equations are in $Ax + By = C$ form with no coefficients of 1 or -1. Solving by substitution would involve fractions.

2. System B is easier to solve by substitution because the second equation is already solved for y.

3. (a) $\{(1, 4)\}$ **(b)** $\{(1, 4)\}$
(c) Answers will vary.
4. (a) $\{(-5, 2)\}$
(b) $\{(-5, 2)\}$
(c) Answers will vary.

5. $\{(3, 12)\}$ **6.** $\{(-3, 2)\}$

7. $\left\{\left(\dfrac{1}{3}, \dfrac{1}{2}\right)\right\}$ **8.** $\varnothing$

9. $\{(3, -2)\}$ **10.** $\{(-1, -11)\}$
11. $\{(x, y) \mid 2x - 3y = 5\}$
12. $\{(9, 4)\}$

13. $\left\{\left(\dfrac{45}{31}, \dfrac{4}{31}\right)\right\}$

14. $\{(4, -5)\}$
15. $\varnothing$ **16.** $\{(-4, 6)\}$

17. $\{(0, 0)\}$ **18.** $\left\{\left(\dfrac{22}{13}, -\dfrac{23}{13}\right)\right\}$

19. $\{(2, -3)\}$ **20.** $\{(24, -12)\}$
21. $\{(3, 2)\}$ **22.** $\{(10, -12)\}$
23. $\{(-4, 2)\}$ **24.** $\{(5, 3)\}$

Concept Check   *Use the preceding guidelines to solve each problem.*

1. To minimize the amount of work required, tell whether you would use the substitution or elimination method to solve each system, and why. *Do not actually solve.*

(a) $3x + 5y = 69$
$\quad\; y = 4x$

(b) $3x + y = -7$
$\quad\; x - y = -5$

(c) $3x - 2y = 0$
$\quad\; 9x + 8y = 7$

2. Which system would be easier to solve with the substitution method? Why?

$\quad$ *System A:* $5x - 3y = 7$
$\qquad\qquad\qquad 2x + 8y = 3$

$\quad$ *System B:* $7x + 2y = 4$
$\qquad\qquad\qquad y = -3x + 1$

In Exercises 3 and 4, (a) solve the system by the elimination method, (b) solve the system by the substitution method, and (c) tell which method you prefer for that particular system and why.

3. $4x - 3y = -8$
$\quad\; x + 3y = 13$

4. $2x + 5y = 0$
$\quad\; x = -3y + 1$

*Solve each system by any method. (For Exercises 5–7, see the answers to **Exercise 1**.)*

5. $3x + 5y = 69$
$\quad\; y = 4x$

6. $3x + y = -7$
$\quad\; x - y = -5$

7. $3x - 2y = 0$
$\quad\; 9x + 8y = 7$

8. $x + y = 7$
$\quad\; x = -3 - y$

9. $6x + 7y = 4$
$\quad\; 5x + 8y = -1$

10. $6x - y = 5$
$\qquad y = 11x$

11. $\quad 4x - 6y = 10$
$\quad -10x + 15y = -25$

12. $3x - 5y = 7$
$\quad\; 2x + 3y = 30$

13. $5x = 7 + 2y$
$\quad\; 5y = 5 - 3x$

14. $4x + 3y = 1$
$\quad\; 3x + 2y = 2$

15. $\quad 2x - 3y = 7$
$\quad -4x + 6y = 14$

16. $\quad 2x + 3y = 10$
$\quad -3x + \; y = 18$

17. $7x - 4y = 0$
$\quad\; 3x = 2y$

18. $\quad x - 3y = 7$
$\qquad 4x + \; y = 5$

Solve each system by any method. First clear all fractions or decimals.

19. $\dfrac{1}{5}x + \dfrac{2}{3}y = -\dfrac{8}{5}$
$\qquad 3x - \quad y = 9$

20. $\dfrac{1}{6}x + \dfrac{1}{6}y = 2$
$\qquad -\dfrac{1}{2}x - \dfrac{1}{3}y = -8$

21. $\dfrac{x}{3} - \dfrac{3y}{4} = -\dfrac{1}{2}$
$\qquad \dfrac{x}{6} + \dfrac{y}{8} = \dfrac{3}{4}$

22. $\dfrac{x}{2} - \dfrac{y}{3} = 9$
$\qquad \dfrac{x}{5} - \dfrac{y}{4} = 5$

23. $0.1x + \quad y = 1.6$
$\qquad 0.6x + 0.5y = -1.4$

24. $0.2x - 0.3y = 0.1$
$\qquad 0.3x - 0.2y = 0.9$

7.6   Systems of Linear Equations in Three Variables

VOCABULARY
☐ ordered triple
☐ focus variable
☐ working equation

A solution of an equation in three variables, such as

$$2x + 3y - z = 4, \qquad \text{Linear equation in three variables}$$

is an **ordered triple** and is written (x, y, z). For example, the ordered triple $(0, 1, -1)$ is a solution of the preceding equation, because

$$2(0) + 3(1) - (-1) = 4 \qquad \text{is a true statement.}$$

Verify that another solution of this equation is $(10, -3, 7)$.

We now extend the term *linear equation* to equations of the form

$$Ax + By + Cz + \cdots + Dw = K,$$

where not all the coefficients $A, B, C, \ldots, D$ equal 0. For example,

$$2x + 3y - 5z = 7 \quad \text{and} \quad x - 2y - z + 3w = 8$$

are linear equations, the first with three variables and the second with four.

OBJECTIVE 1   Understand the geometry of systems of three equations in three variables.

Consider the solution of a system such as the following.

$$4x + 8y + \ z = 2$$
$$x + 7y - 3z = -14 \qquad \text{System of linear equations in three variables}$$
$$2x - 3y + 2z = 3$$

Theoretically, a system of this type can be solved by graphing. However, the graph of a linear equation with three variables is a *plane*, not a line. Because visualizing a plane requires three-dimensional graphing, the method of graphing is not practical with these systems. However, it does illustrate the number of solutions possible for such systems, as shown in **FIGURE 40**.

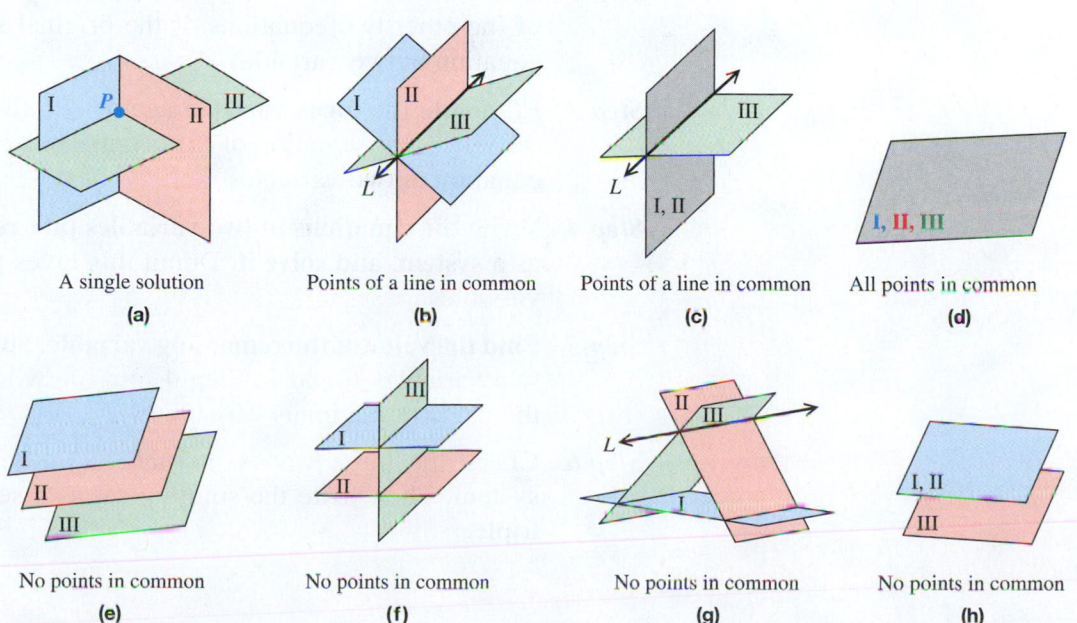

| A single solution | Points of a line in common | Points of a line in common | All points in common |
| (a) | (b) | (c) | (d) |

| No points in common | No points in common | No points in common | No points in common |
| (e) | (f) | (g) | (h) |

FIGURE 40

FIGURE 40 illustrates the following cases.

Graphs of Linear Systems in Three Variables

Case 1   **The three planes may meet at a single, common point.** This point is the solution of the system. See **FIGURE 40(a)**.

Case 2   **The three planes may have the points of a line in common.** The infinite set of points that satisfy the equation of the line is the solution of the system. See **FIGURES 40(b) AND (c)**.

Case 3   **The three planes may coincide.** The solution of the system is the set of all points on a plane. See **FIGURE 40(d)**.

Case 4   **The planes may have no points common to all three.** There is no solution of the system. See **FIGURES 40(e)–(h)**.

OBJECTIVE 2   Solve linear systems (with three equations and three variables) by elimination.

Because graphing to find the solution set of a system of three equations in three variables is impractical, these systems are solved with an extension of the elimination method from **Section 7.5.**

In the steps that follow, we use the term **focus variable** to identify the first variable to be eliminated in the process. The focus variable will always be present in the **working equation,** which will be used twice to eliminate this variable.

Solving a Linear System in Three Variables*

Step 1   **Select a variable and an equation.** A good choice for the variable, which we call the *focus variable,* is one that has coefficient 1 or −1. Then select an equation, one that contains the focus variable, as the *working equation.*

Step 2   **Eliminate the focus variable.** Use the working equation and one of the other two equations of the original system. The result is an equation in two variables.

Step 3   **Eliminate the focus variable again.** Use the working equation and the remaining equation of the original system. The result is another equation in two variables.

Step 4   **Write the equations in two variables that result from Steps 2 and 3 as a system, and solve it.** Doing this gives the values of two of the variables.

Step 5   **Find the value of the remaining variable.** Substitute the values of the two variables found in Step 4 into the working equation to obtain the value of the focus variable.

Step 6   **Check** the three values in *each* of the *original* equations of the system. Then write the solution set as a set containing an ordered triple.

---

*The authors wish to thank Christine Heinecke Lehmann of Purdue University North Central for her suggestions here.

EXAMPLE 1 Solving a System in Three Variables

Solve the system.

$$4x + 8y + z = 2 \quad (1)$$
$$x + 7y - 3z = -14 \quad (2)$$
$$2x - 3y + 2z = 3 \quad (3)$$

Step 1 Because z in equation (1) has coefficient 1, we choose z as the focus variable and (1) as the working equation. (Another option would be to choose x as the focus variable—it also has coefficient 1—and use (2) as the working equation.)

Focus variable

$$4x + 8y + z = 2 \quad (1) \leftarrow \text{Working equation}$$

Step 2 Multiply working equation (1) by 3 and add the result to equation (2).

$$12x + 24y + 3z = 6 \qquad \text{Multiply each side of (1) by 3.}$$
$$\underline{x + 7y - 3z = -14 \quad (2)}$$
$$13x + 31y \qquad = -8 \qquad \text{Add.} \quad (4)$$

Focus variable z was eliminated.

Step 3 Multiply working equation (1) by -2 and add the result to remaining equation (3) to again eliminate focus variable z.

$$-8x - 16y - 2z = -4 \qquad \text{Multiply each side of (1) by } -2.$$
$$\underline{2x - 3y + 2z = 3 \qquad (3)}$$
$$-6x - 19y \qquad = -1 \qquad \text{Add.} \quad (5)$$

Focus variable z was eliminated.

Step 4 Write the equations in two variables that result in Steps 2 and 3 as a system.

Make sure these equations have the same two variables.

$$13x + 31y = -8 \quad (4) \quad \text{The result from Step 2}$$
$$-6x - 19y = -1 \quad (5) \quad \text{The result from Step 3}$$

Now solve this system. We choose to eliminate x.

$$78x + 186y = -48 \qquad \text{Multiply each side of (4) by 6.}$$
$$\underline{-78x - 247y = -13 \qquad \text{Multiply each side of (5) by 13.}}$$
$$-61y = -61 \qquad \text{Add.}$$
$$y = 1 \qquad \text{Divide by } -61.$$

Substitute 1 for y in either equation (4) or (5) to find x.

$$-6x - 19y = -1 \qquad (5)$$
$$-6x - 19(1) = -1 \qquad \text{Let } y = 1.$$
$$-6x - 19 = -1 \qquad \text{Multiply.}$$
$$-6x = 18 \qquad \text{Add 19.}$$
$$x = -3 \qquad \text{Divide by } -6.$$

Step 5 Now substitute the two values we found in Step 4 in working equation (1) to find the value of the remaining variable, focus variable z.

$$4x + 8y + z = 2 \qquad (1)$$
$$4(-3) + 8(1) + z = 2 \qquad \text{Let } x = -3 \text{ and } y = 1.$$
$$-4 + z = 2 \qquad \text{Multiply, and then add.}$$
$$z = 6 \qquad \text{Add 4.}$$

**NOW TRY
EXERCISE 1**

Solve the system.

$$x - y + 2z = 1$$
$$3x + 2y + 7z = 8$$
$$-3x - 4y + 9z = -10$$

> Write the values of x, y, and z in the correct order.

Step 6 It appears that $(-3, 1, 6)$ is the only solution of the system. We must check that this ordered triple satisfies all three original equations of the system. We begin with equation (1).

CHECK

$$4x + 8y + z = 2 \qquad (1)$$
$$4(-3) + 8(1) + 6 \overset{?}{=} 2 \qquad \text{Substitute.}$$
$$-12 + 8 + 6 \overset{?}{=} 2 \qquad \text{Multiply.}$$
$$2 = 2 \ \checkmark \quad \text{True}$$

Because $(-3, 1, 6)$ also satisfies equations (2) and (3), $\{(-3, 1, 6)\}$ is the solution set. This is Case 1 as shown in **FIGURE 40(a)** at the beginning of this section.

NOW TRY ↩

OBJECTIVE 3 Solve linear systems (with three equations and three variables) in which some of the equations have missing terms.

If a linear system includes an equation that is missing a term or terms, one elimination step can be omitted.

EXAMPLE 2 Solving a System of Equations with Missing Terms

Solve the system.

$$6x - 12y = -5 \qquad (1) \quad \text{Missing } z$$
$$8y + z = 0 \qquad (2) \quad \text{Missing } x$$
$$9x - z = 12 \qquad (3) \quad \text{Missing } y$$

Equation (3) is missing the variable y, so one way to begin is to eliminate y again, using equations (1) and (2).

> Leave space for the missing terms.

$$12x - 24y \qquad = -10 \qquad \text{Multiply each side of (1) by 2.}$$
$$\underline{\qquad 24y + 3z = \quad 0} \qquad \text{Multiply each side of (2) by 3.}$$
$$12x \qquad + 3z = -10 \qquad \text{Add. \quad (4)}$$

Use resulting equation (4) in x and z, together with equation (3), $9x - z = 12$, to eliminate z.

$$27x - 3z = \quad 36 \qquad \text{Multiply each side of (3) by 3.}$$
$$\underline{12x + 3z = -10} \qquad (4)$$
$$39x \qquad = \quad 26 \qquad \text{Add.}$$
$$x = \frac{26}{39} \qquad \text{Divide by 39.}$$
$$x = \frac{2}{3} \qquad \text{Lowest terms}$$

We can find z by substituting this value for x in equation (3).

$$9x - z = 12 \qquad (3)$$
$$9\left(\frac{2}{3}\right) - z = 12 \qquad \text{Let } x = \tfrac{2}{3}.$$
$$6 - z = 12 \qquad \text{Multiply.}$$
$$z = -6 \qquad \text{Subtract 6. Multiply by } -1.$$

NOW TRY ANSWER
1. $\{(2, 1, 0)\}$

**NOW TRY
EXERCISE 2**

Solve the system.

$$3x - z = -10$$
$$4y + 5z = 24$$
$$x - 6y = -8$$

We can find y by substituting -6 for z in equation (2).

$$8y + z = 0 \qquad (2)$$
$$8y - 6 = 0 \qquad \text{Let } z = -6.$$
$$8y = 6 \qquad \text{Add 6.}$$
$$y = \frac{6}{8} \qquad \text{Divide by 8.}$$
$$y = \frac{3}{4} \qquad \text{Lowest terms}$$

Check to verify that the solution set is $\left\{\left(\frac{2}{3}, \frac{3}{4}, -6\right)\right\}$. This is also an example of Case 1.

NOW TRY

NOTE   Another way to solve the system in **Example 2** is to begin by eliminating the variable z from equations (2) and (3). The resulting equation together with equation (1) forms a system of two equations in the variables x and y. Try working **Example 2** this way to see that the same solution results.

There are often multiple ways to solve a system of equations. Some ways may involve more work than others.

OBJECTIVE 4 Solve special systems.

**NOW TRY
EXERCISE 3**

Solve the system.

$$x - 3y + 2z = 10$$
$$-2x + 6y - 4z = -20$$
$$\frac{1}{2}x - \frac{3}{2}y + z = 5$$

EXAMPLE 3   Solving a System of Dependent Equations with Three Variables

Solve the system.

$$2x - 3y + 4z = 8 \qquad (1)$$
$$-x + \frac{3}{2}y - 2z = -4 \qquad (2)$$
$$6x - 9y + 12z = 24 \qquad (3)$$

Multiplying each side of equation (1) by 3 gives equation (3). Multiplying each side of equation (2) by -6 also gives equation (3). Because of this, the equations are dependent.

When solving a system such as this, attempting to eliminate one variable results in elimination of *all* variables, leading to the true statement $0 = 0$. This indicates that all three equations have the same graph, as illustrated in **FIGURE 40(d)**. This is Case 3. The solution set is written as follows.

$$\{(x, y, z) \mid 2x - 3y + 4z = 8\} \qquad \text{Set-builder notation}$$

Although any one of the three equations could be used to write the solution set, we use the equation in standard form with coefficients that are integers with greatest common factor 1, as explained in **Section 7.3**.

NOW TRY

EXAMPLE 4   Solving an Inconsistent System with Three Variables

Solve the system.

$$2x - 4y + 6z = 5 \qquad (1)$$
$$-x + 3y - 2z = -1 \qquad (2)$$
$$x - 2y + 3z = 1 \qquad (3)$$

Use as the working equation, with focus variable x.

NOW TRY ANSWERS
2. $\{(-2, 1, 4)\}$
3. $\{(x, y, z) \mid x - 3y + 2z = 10\}$

Eliminate the focus variable, x, using equations (1) and (3).

**NOW TRY
EXERCISE 4**

Solve the system.

$$x - 5y + 2z = 4$$
$$3x + y - z = 6$$
$$-2x + 10y - 4z = 7$$

$$-2x + 4y - 6z = -2 \quad \text{Multiply each side of (3) by } -2.$$
$$\underline{2x - 4y + 6z = 5} \quad (1)$$
$$0 = 3 \quad \text{False}$$

The resulting false statement indicates that equations (1) and (3) have no common solution. Thus, the system is inconsistent and the solution set is $\varnothing$. The graph of this system would show two planes parallel to one another and a third plane which intersects both, as in **FIGURE 40(f)**. This is Case 4.   **NOW TRY**

NOTE   If a false statement results when adding, as in **Example 4**, it is not necessary to go any further with the solution. Because two of the three planes are parallel, it is not possible for the three planes to have any points in common.

**NOW TRY
EXERCISE 5**

Solve the system.

$$x - 3y + 2z = 4$$
$$\frac{1}{3}x - y + \frac{2}{3}z = 7$$
$$\frac{1}{2}x - \frac{3}{2}y + z = 2$$

EXAMPLE 5   Solving Another Special System

Solve the system.

$$2x - y + 3z = 6 \quad (1)$$
$$x - \frac{1}{2}y + \frac{3}{2}z = 3 \quad (2)$$
$$4x - 2y + 6z = 1 \quad (3)$$

Multiplying each side of equation (2) by 2 gives equation (1), so these two equations are dependent. Equations (1) and (3) are not equivalent, however. Multiplying equation (3) by $\frac{1}{2}$ does *not* give equation (1). Instead, we obtain two equations with the same coefficients, but with different constant terms.

The graphs of equations (1) and (3) have no points in common (that is, the planes are parallel). Thus, the system is inconsistent and the solution set is $\varnothing$, as illustrated in **FIGURE 40(h)**. This is another example of Case 4.   **NOW TRY**

NOW TRY ANSWERS
4. $\varnothing$ 5. $\varnothing$

7.6 Exercises

FOR EXTRA HELP   ▶ **MyMathLab®**

▶ *Complete solution available in MyMathLab*

Concept Check   *Answer each of the following.*

1. Using your immediate surroundings, give an example of three planes that satisfy the condition.

   (a) They intersect in a single point.

   (b) They do not intersect.

   (c) They intersect in infinitely many points.

2. Suppose that a system has infinitely many ordered triple solutions of the form (x, y, z) such that

   $$x + y + 2z = 1.$$

   Give three specific ordered triples that are solutions of the system.

3. Explain what the following statement means: "The solution set of the following system is $\{(-1, 2, 3)\}$."

   $$2x + y + z = 3$$
   $$3x - y + z = -2$$
   $$4x - y + 2z = 0$$

4. The following two equations have a common solution of $(1, 2, 3)$.

$$x + y + z = 6$$
$$2x - y + z = 3$$

Which equation would complete a system of three linear equations in three variables having solution set $\{(1, 2, 3)\}$?

A. $3x + 2y - z = 1$ **B.** $3x + 2y - z = 4$

C. $3x + 2y - z = 5$ **D.** $3x + 2y - z = 6$

5. What constant should replace the question mark in this system so that the solution set is $\{(1, 1, 1)\}$?

$$2x - 3y + z = 0$$
$$-5x + 2y - z = -4$$
$$x + y + 2z = ?$$

6. Complete the work of **Example 1** and show that the ordered triple $(-3, 1, 6)$ is also a solution of equations (2) and (3).

$$x + 7y - 3z = -14 \qquad \text{Equation (2)}$$
$$2x - 3y + 2z = 3 \qquad \text{Equation (3)}$$

Solve each system. See Example 1.

▶ 7. $2x - 5y + 3z = -1$
$\ x + 4y - 2z = 9$
$\ x - 2y - 4z = -5$

8. $x + 3y - 6z = 1$
$\ 2x - y + z = 7$
$\ x + 2y + 2z = 14$

9. $3x + 2y + z = 8$
$\ 2x - 3y + 2z = -16$
$\ x + 4y - z = 20$

10. $-3x + y - z = -10$
$\ -4x + 2y + 3z = -1$
$\ 2x + 3y - 2z = -5$

11. $2x + 5y + 2z = 0$
$\ 4x - 7y - 3z = 1$
$\ 3x - 8y - 2z = -6$

12. $5x - 2y + 3z = -9$
$\ 4x + 3y + 5z = 4$
$\ 2x + 4y - 2z = 14$

13. $x + 2y + z = 4$
$\ 2x + y - z = -1$
$\ x - y - z = -2$

14. $x - 2y + 5z = -7$
$\ -2x - 3y + 4z = -14$
$\ -3x + 5y - z = -7$

15. $-x + 2y + 6z = 2$
$\ 3x + 2y + 6z = 6$
$\ x + 4y - 3z = 1$

16. $2x + y + 2z = 1$
$\ x + 2y + z = 2$
$\ x - y - z = 0$

17. $x + y - z = -2$
$\ 2x - y + z = -5$
$\ -x + 2y - 3z = -4$

18. $x + 2y + 3z = 1$
$\ -x - y + 3z = 2$
$\ -6x + y + z = -2$

19. $\dfrac{1}{3}x + \dfrac{1}{6}y - \dfrac{2}{3}z = -1$

$\ -\dfrac{3}{4}x - \dfrac{1}{3}y - \dfrac{1}{4}z = 3$

$\ \dfrac{1}{2}x + \dfrac{3}{2}y + \dfrac{3}{4}z = 21$

20. $\dfrac{2}{3}x - \dfrac{1}{4}y + \dfrac{5}{8}z = 0$

$\ \dfrac{1}{5}x + \dfrac{2}{3}y - \dfrac{1}{4}z = -7$

$\ -\dfrac{3}{5}x + \dfrac{4}{3}y - \dfrac{7}{8}z = -5$

21. $5.5x - 2.5y + 1.6z = 11.83$
$\ 2.2x + 5.0y - 0.1z = -5.97$
$\ 3.3x - 7.5y + 3.2z = 21.25$

22. $6.2x - 1.4y + 2.4z = -1.80$
$\ 3.1x + 2.8y - 0.2z = 5.68$
$\ 9.3x - 8.4y - 4.8z = -34.20$

Solve each system. See Example 2.

23. $2x - 3y + 2z = -1$
$\ x + 2y + z = 17$
$\ 2y - z = 7$

24. $2x - y + 3z = 6$
$\ x + 2y - z = 8$
$\ 2y + z = 1$

25. $4x + 2y - 3z = 6$
$\ x - 4y + z = -4$
$\ -x + 2z = 2$

26. $2x + 3y - 4z = 4$
$x - 6y + z = -16$
$-x + 3z = 8$

▶ 27. $2x + y = 6$
$3y - 2z = -4$
$3x - 5z = -7$

28. $4x - 8y = -7$
$4y + z = 7$
$-8x + z = -4$

29. $-5x + 2y + z = 5$
$-3x - 2y - z = 3$
$-x + 6y = 1$

30. $-4x + 3y - z = 4$
$-5x - 3y + z = -4$
$-2x - 3z = 12$

31. $7x - 3z = -34$
$2y + 4z = 20$
$\dfrac{3}{4}x + \dfrac{1}{6}y = -2$

32. $5x - 2z = 8$
$4y + 3z = -9$
$\dfrac{1}{2}x + \dfrac{2}{3}y = -1$

33. $4x - z = -6$
$\dfrac{3}{5}y + \dfrac{1}{2}z = 0$
$\dfrac{1}{3}x + \dfrac{2}{3}z = -5$

34. $5x - z = 38$
$\dfrac{2}{3}y + \dfrac{1}{4}z = -17$
$\dfrac{1}{5}y + \dfrac{5}{6}z = 4$

Solve each system. If the system is inconsistent or has dependent equations, say so. **See Examples 1, 3, 4, and 5.**

▶ 35. $2x + 2y - 6z = 5$
$-3x + y - z = -2$
$-x - y + 3z = 4$

36. $-2x + 5y + z = -3$
$5x + 14y - z = -11$
$7x + 9y - 2z = -5$

37. $-5x + 5y - 20z = -40$
$x - y + 4z = 8$
$3x - 3y + 12z = 24$

38. $x + 4y - z = 3$
$-2x - 8y + 2z = -6$
$3x + 12y - 3z = 9$

39. $x + 5y - 2z = -1$
$-2x + 8y + z = -4$
$3x - y + 5z = 19$

40. $x + 3y + z = 2$
$4x + y + 2z = -4$
$5x + 2y + 3z = -2$

▶ 41. $2x + y - z = 6$
$4x + 2y - 2z = 12$
$-x - \dfrac{1}{2}y + \dfrac{1}{2}z = -3$

42. $2x - 8y + 2z = -10$
$-x + 4y - z = 5$
$\dfrac{1}{8}x - \dfrac{1}{2}y + \dfrac{1}{8}z = -\dfrac{5}{8}$

43. $x + y - 2z = 0$
$3x - y + z = 0$
$4x + 2y - z = 0$

44. $2x + 3y - z = 0$
$x - 4y + 2z = 0$
$3x - 5y - z = 0$

▶ 45. $x - 2y + \dfrac{1}{3}z = 4$
$3x - 6y + z = 12$
$-6x + 12y - 2z = -3$

46. $4x + y - 2z = 3$
$x + \dfrac{1}{4}y - \dfrac{1}{2}z = \dfrac{3}{4}$
$2x + \dfrac{1}{2}y - z = 1$

Extending Skills   *Extend the method of this section to solve each system. Express the solution in the form (x, y, z, w).*

47. $x + y + z - w = 5$
$2x + y - z + w = 3$
$x - 2y + 3z + w = 18$
$-x - y + z + 2w = 8$

48. $3x + y - z + 2w = 9$
$x + y + 2z - w = 10$
$x - y - z + 3w = -2$
$-x + y - z + w = -6$

49. $3x + y - z + w = -3$
$2x + 4y + z - w = -7$
$-2x + 3y - 5z + w = 3$
$5x + 4y - 5z + 2w = -7$

50. $x - 3y + 7z + w = 11$
$2x + 4y + 6z - 3w = -3$
$3x + 2y + z + 2w = 19$
$4x + y - 3z + w = 22$

7.7 Applications of Systems of Linear Equations

Although some problems with two unknowns can be solved using just one variable, it is often easier to use two variables and a system of equations. The following problem, which can be solved with a system, appeared in a Hindu work that dates back to about A.D. 850. (See **Exercise 39.**)

The mixed price of 9 citrons (a lemonlike fruit) and 7 fragrant wood apples is 107; again, the mixed price of 7 citrons and 9 fragrant wood apples is 101. O you arithmetician, tell me quickly the price of a citron and the price of a wood apple here, having distinctly separated those prices well.

PROBLEM-SOLVING HINT When solving an applied problem using two variables, it is a good idea to pick letters that correspond to the descriptions of the unknown quantities. In the example above, we could choose c to represent the number of citrons, and w to represent the number of wood apples.

The following steps are based on the problem-solving method of **Section 2.4.**

Solving an Applied Problem Using a System of Equations

Step 1 **Read** the problem carefully. *What information is given? What is to be found?*

Step 2 **Assign variables** to represent the unknown values. Write down what each variable represents. Make a sketch, diagram, or table, as needed.

Step 3 **Write a system of equations** using both variables.

Step 4 **Solve** the system of equations.

Step 5 **State the answer.** Label it appropriately. *Does it seem reasonable?*

Step 6 **Check** the answer in the words of the *original* problem.

OBJECTIVE 1 Solve geometry problems using two variables.

EXAMPLE 1 Finding the Dimensions of a Soccer Field

A rectangular soccer field may have a width between 50 and 100 yd and a length between 100 and 130 yd. One particular soccer field has a perimeter of 320 yd. Its length measures 40 yd more than its width. What are the dimensions of this field? (*Source:* www.soccer-training-guide.com)

Step 1 **Read** the problem again. We must find the dimensions of the field.

Step 2 **Assign variables.** A sketch may be helpful. See **FIGURE 41** on the next page.

Let L = the length and W = the width.

**NOW TRY
EXERCISE 1**

A rectangular parking lot has a length that is 10 ft more than twice its width. The perimeter of the parking lot is 620 ft. What are the dimensions of the parking lot?

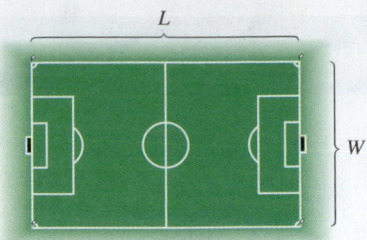

L

W

FIGURE 41

Step 3 **Write a system of equations.** Because the perimeter is 320 yd, we find one equation by using the perimeter formula.

$$2L + 2W = 320 \qquad 2L + 2W = P$$

For a second equation, use the information given about the width.

$$L = W + 40 \qquad \text{The length is 40 yd more than the width.}$$

These two equations form a system of equations.

$$2L + 2W = 320 \qquad (1)$$
$$L = W + 40 \qquad (2)$$

Step 4 **Solve** the system of equations. Because equation (2), $L = W + 40$, is solved for L, we can substitute $W + 40$ for L in equation (1) and solve for W.

$$2L + 2W = 320 \qquad (1)$$
$$2(W + 40) + 2W = 320 \qquad \text{Let } L = W + 40.$$

> Be sure to use parentheses around $W + 40$.

$$2W + 80 + 2W = 320 \qquad \text{Distributive property}$$
$$4W + 80 = 320 \qquad \text{Combine like terms.}$$
$$4W = 240 \qquad \text{Subtract 80.}$$

> Don't stop here. → $W = 60 \qquad \text{Divide by 4.}$

Let $W = 60$ in the equation $L = W + 40$ to find L.

$$L = 60 + 40 = 100$$

Step 5 **State the answer.** The length is 100 yd, and the width is 60 yd. Both dimensions are within the ranges given in the problem.

Step 6 **Check** using the words of the problem. The answer is correct.

$$2(100) + 2(60) = 320 \qquad \text{The perimeter is 320 yd.}$$
$$100 - 60 = 40 \qquad \text{Length is 40 yd more than width.} \qquad \textbf{NOW TRY}$$

PROBLEM-SOLVING HINT There is often more than one way to write the equations in a system used to solve an application. In **Example 1,** we might write the second equation as

$$W = L - 40. \qquad (2)$$

In this case, we would substitute $L - 40$ for W in equation (1) to obtain

$$2L + 2(L - 40) = 320 \qquad \text{Let } W = L - 40.$$

and solve for L first (instead of W, as in Step 4 above). The *same* answers result.

NOW TRY ANSWER
1. length: 210 ft; width: 100 ft

OBJECTIVE 2 Solve money problems using two variables.

NOW TRY EXERCISE 2

For the 2013 season at Six Flags St. Louis, two general admission tickets and three tickets for children cost $239.95. One general admission ticket and four tickets for children cost $224.95. Determine the ticket prices for general admission and for children. (*Source:* www.sixflags.com)

EXAMPLE 2 Solving a Problem about Ticket Prices

For the 2012–2013 National Hockey League and National Basketball Association seasons, two hockey tickets and one basketball ticket purchased at their average prices cost $173.01. One hockey ticket and two basketball tickets cost $162.99. What were the average ticket prices for the two sports? (*Source:* Team Marketing Report.)

Step 1 **Read** the problem again. There are two unknowns.

Step 2 **Assign variables.**

$$\text{Let } h = \text{the average price for a hockey ticket}$$
$$\text{and } b = \text{the average price for a basketball ticket.}$$

Step 3 **Write a system of equations.** Because two hockey tickets and one basketball ticket cost a total of $173.01, one equation for the system is

$$2h + b = 173.01.$$

By similar reasoning, the second equation is

$$h + 2b = 162.99.$$

These two equations form a system of equations.

$$2h + b = 173.01 \quad (1)$$
$$h + 2b = 162.99 \quad (2)$$

Step 4 **Solve** the system. To eliminate h, multiply equation (2) by -2 and add.

$$\begin{array}{ll} 2h + b = 173.01 & (1) \\ \underline{-2h - 4b = -325.98} & \text{Multiply each side of (2) by } -2. \\ {-3b} = -152.97 & \text{Add.} \\ b = 50.99 & \text{Divide by } -3. \end{array}$$

To find the value of h, let $b = 50.99$ in equation (2).

$$h + 2b = 162.99 \quad (2)$$
$$h + 2(50.99) = 162.99 \quad \text{Let } b = 50.99.$$
$$h + 101.98 = 162.99 \quad \text{Multiply.}$$
$$h = 61.01 \quad \text{Subtract } 101.98.$$

Step 5 **State the answer.** The average price for one basketball ticket was $50.99. For one hockey ticket, the average price was $61.01.

Step 6 **Check** that these values satisfy the problem conditions.

$$2(\$61.01) + \$50.99 = \$173.01, \quad \text{as required.}$$
$$\$61.01 + 2(\$50.99) = \$162.99, \quad \text{as required.} \qquad \text{NOW TRY}$$

NOW TRY ANSWER
2. general admission: $56.99; children: $41.99

NOTE In **Example 2**, we could have solved the system using the substitution method.

OBJECTIVE 3   Solve mixture problems using two variables.

We solved mixture problems in **Section 2.7** using one variable. For many mixture problems we can use more than one variable and a system of equations.

EXAMPLE 3   Solving a Mixture Problem

How many ounces each of 5% hydrochloric acid and 20% hydrochloric acid must be combined to obtain 10 oz of solution that is 12.5% hydrochloric acid?

Step 1   **Read** the problem. Two solutions of different strengths are being mixed to obtain a specific amount of a solution with an "in-between" strength.

Step 2   **Assign variables.**

Let $x =$ the number of ounces of 5% solution

and $y =$ the number of ounces of 20% solution.

Use a table to summarize the information from the problem.

Ounces of Solution	Percent (as a decimal)	Ounces of Pure Acid
x	5% = 0.05	0.05x
y	20% = 0.20	0.20y
10	12.5% = 0.125	(0.125)10

Multiply the amount of each solution (given in the first column) by its concentration of acid (given in the second column) to find the amount of acid in that solution (given in the third column).

Gives equation (1)   Gives equation (2)

FIGURE 42 illustrates what is happening in the problem.

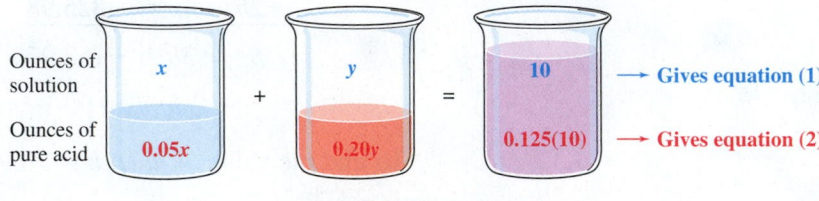

Ounces of solution   Ounces of pure acid

FIGURE 42

Step 3   **Write a system of equations.** When x ounces of 5% solution and y ounces of 20% solution are combined, the total number of ounces is 10.

$$x + y = 10$$

The number of ounces of acid in the 5% solution $(0.05x)$ added to the number of ounces of acid in the 20% solution $(0.20y)$ must equal the total number of ounces of acid in the mixture, which is $(0.125)10$, or 1.25.

$$0.05x + 0.20y = 1.25$$

Notice that these equations can be quickly determined by reading down the table or using the labels in **FIGURE 42**.

Multiply the equation in red by 100 to clear the decimals and obtain an equivalent system of equations.

$$x + y = 10 \quad \text{(1)}$$
$$5x + 20y = 125 \quad \text{(2)}$$

This is the system to solve.

**NOW TRY
EXERCISE 3**

How many liters each of a 15% acid solution and a 25% acid solution should be mixed to obtain 30 L of an 18% acid solution?

Step 4 **Solve** the system. To eliminate x, multiply equation (1), $x + y = 10$, by -5.

$$-5x - 5y = -50 \qquad \text{Multiply each side of (1) by } -5.$$
$$\underline{5x + 20y = 125} \qquad (2)$$
$$15y = 75 \qquad \text{Add.}$$

Ounces of 20% solution $\rightarrow y = 5$ \qquad Divide by 15.

Substitute 5 for y in equation (1) to find the value of x.

$$x + y = 10 \qquad (1)$$
$$x + 5 = 10 \qquad \text{Let } y = 5.$$

Ounces of 5% solution $\rightarrow x = 5$ \qquad Subtract 5.

Step 5 **State the answer.** The desired mixture will require 5 oz of the 5% solution and 5 oz of the 20% solution.

Step 6 **Check.**

Total amount of solution: $\quad x + y = 5 \text{ oz} + 5 \text{ oz}$
$$= 10 \text{ oz}, \quad \text{as required.}$$

Total amount of acid: $\quad$ 5% of 5 oz + 20% of 5 oz
$$= 0.05(5) + 0.20(5)$$
$$= 1.25 \text{ oz}$$

Percent of acid in solution:

Total acid $\longrightarrow$
Total solution $\longrightarrow$ $\dfrac{1.25}{10} = 0.125, \quad \text{or} \quad 12.5\%, \quad \text{as required.}$ **NOW TRY**

OBJECTIVE 4 Solve distance-rate-time problems using two variables.

Motion problems require the distance formula $d = rt$, where d is distance, r is rate (or speed), and t is time.

EXAMPLE 4 Solving a Motion Problem

A car travels 250 km in the same time that a truck travels 225 km. If the rate of the car is 8 km per hr faster than the rate of the truck, find both rates.

Step 1 **Read** the problem again. Given the distances traveled, we need to find the rate of each vehicle.

Step 2 **Assign variables.**

Let $x = $ the rate of the car,

and $y = $ the rate of the truck.

As in **Example 3,** a table helps organize the information. Fill in the distance for each vehicle, and the variables for the unknown rates.

	d	r	t
Car	250	x	$\frac{250}{x}$
Truck	225	y	$\frac{225}{y}$

To find the expressions for time, we solved the distance formula $d = rt$ for t. Thus, $\frac{d}{r} = t$.

NOW TRY
EXERCISE 4
On a bicycle ride, Vann can travel 50 mi in the same amount of time that Ivy can travel 40 mi. Determine both bicyclists' rates, if Vann's rate is 2 mph faster than Ivy's.

Step 3 **Write a system of equations.** The car travels 8 km per hr faster than the truck. Because the two rates are x and y,

$$x = y + 8. \quad \text{(1)}$$

Both vehicles travel for the *same* time, so the times must be equal.

$$\text{Time for car} \longrightarrow \frac{250}{x} = \frac{225}{y} \longleftarrow \text{Time for truck}$$

Multiply both sides by xy to obtain an equivalent equation with no variable denominators.

$$xy \cdot \frac{250}{x} = \frac{225}{y} \cdot xy \qquad \text{Multiply by the LCD, } xy.$$

$$\frac{250xy}{x} = \frac{225xy}{y}$$

$$250y = 225x \qquad \text{Divide out the common factors.} \quad \text{(2)}$$

We now have a system of linear equations.

$$x = y + 8 \qquad \text{(1)}$$

$$250y = 225x \qquad \text{(2)}$$

Step 4 **Solve** the system by substitution. Replace x with $y + 8$ in equation (2).

$$250y = 225x \qquad \qquad \text{(2)}$$

$$250y = 225(y + 8) \qquad \text{Let } x = y + 8.$$

Be sure to use parentheses around $y + 8$.

$$250y = 225y + 1800 \qquad \text{Distributive property}$$

$$25y = 1800 \qquad \text{Subtract } 225y.$$

$$\text{Truck's rate} \longrightarrow y = 72 \qquad \text{Divide by 25.}$$

Let $y = 72$ in the equation $x = y + 8$ to find x.

$$\text{Car's rate} \longrightarrow x = 72 + 8 = 80$$

Step 5 **State the answer.** The rate of the car is 80 km per hr, and the rate of the truck is 72 km per hr.

Step 6 **Check.**

$$Car: \quad t = \frac{d}{r} = \frac{250}{80} = 3.125 \longleftarrow$$

Times are equal, as required.

$$Truck: \quad t = \frac{d}{r} = \frac{225}{72} = 3.125 \longleftarrow$$

The rate of the car, 80 km per hr, is 8 km per hour greater than that of the truck, 72 km per hr, as required.

NOW TRY

NOW TRY ANSWER
4. Vann: 10 mph; Ivy: 8 mph

PROBLEM-SOLVING HINT   When solving a problem as in **Example 4,** where one quantity is compared to another (e.g., the car travels 8 km per hr faster than the truck), be sure to translate correctly in terms of the two variables.

NOW TRY EXERCISE 5

In his motorboat, Ed travels 42 mi upstream at top speed in 2.1 hr. Still at top speed, the return trip to the same spot takes only 1.5 hr. Find the rate of Ed's boat in still water and the rate of the current.

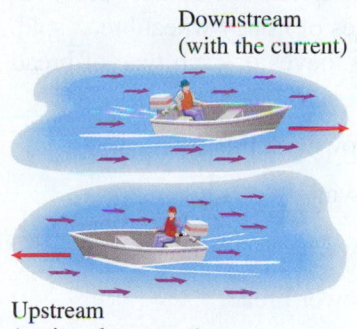

Downstream
(with the current)

Upstream
(against the current)

FIGURE 43

EXAMPLE 5 Solving a Motion Problem

While kayaking on the Blackledge River, Rebecca traveled 9 mi upstream (against the current) in 2.25 hr. It only took her 1 hr paddling downstream (with the current) back to the spot where she started. Find Rebecca's kayaking rate in still water and the rate of the current.

Step 1 **Read** the problem. We must find two rates—Rebecca's kayaking rate in still water and the rate of the current.

Step 2 **Assign variables.**

$$\text{Let } x = \text{Rebecca's kayaking rate in still water}$$

$$\text{and } y = \text{the rate of the current.}$$

When the kayak is traveling *against* the current, the current slows it down. The rate of the kayak is the *difference* between its rate in still water and the rate of the current, which is $(x - y)$ mph.

When the kayak is traveling *with* the current, the current speeds it up. The rate of the kayak is the *sum* of its rate in still water and the rate of the current, which is $(x + y)$ mph.

$$\text{Thus, } \quad x - y = \text{ the rate of the kayak } \textit{against} \text{ the current,}$$

$$\text{and} \quad x + y = \text{ the rate of the kayak } \textit{with} \text{ the current.}$$

See **FIGURE 43**. Make a table. Use the formula $d = rt$, or $rt = d$.

	r	t	d	
Upstream	$x - y$	2.25	$2.25(x - y)$	← The distance is the same
Downstream	$x + y$	1	$1(x + y)$	← in each direction, 9 mi.

Step 3 **Write a system of equations.**

$$2.25(x - y) = 9 \quad \text{Upstream}$$

$$1(x + y) = 9 \quad \text{Downstream}$$

Clear parentheses in each equation, and then divide each term in the first equation by 2.25 to obtain an equivalent system.

$$\begin{array}{lr} x - y = 4 & (1) \\ \underline{x + y = 9} & (2) \end{array}$$

Step 4 **Solve.**

$$2x = 13 \quad \text{Add.}$$

$$\text{Rebecca's rate} \rightarrow x = 6.5 \quad \text{Divide by 2.}$$

Substitute 6.5 for x in equation (2) and solve for y.

$$x + y = 9 \quad (2)$$

$$6.5 + y = 9 \quad \text{Let } x = 6.5.$$

$$\text{Rate of current} \rightarrow y = 2.5 \quad \text{Subtract 6.5.}$$

Step 5 **State the answer.** Rebecca's rate in still water was 6.5 mph, and the rate of the current was 2.5 mph.

Step 6 **Check.**

Distance upstream: $\quad 2.25(6.5 - 2.5) = 9$ ← True statements

Distance downstream: $\quad 1(6.5 + 2.5) = 9$ ← result.

NOW TRY

NOW TRY ANSWER

5. boat: 24 mph; current: 4 mph

OBJECTIVE 5 Solve problems with three variables using a system of three equations.

PROBLEM-SOLVING HINT   If an application has *three* unknown quantities, we can use a system of *three* equations to solve it. We extend the method used for two unknowns.

EXAMPLE 6   Solving a Problem Involving Prices

At Panera Bread, a loaf of honey wheat bread costs $2.95, a loaf of sunflower bread costs $2.99, and a loaf of French bread costs $5.79. On a recent day, three times as many loaves of honey wheat bread were sold as sunflower bread. The number of loaves of French bread sold was 5 less than the number of loaves of honey wheat bread sold. Total receipts for these breads were $87.89. How many loaves of each type of bread were sold? (*Source:* Panera Bread menu.)

Step 1   **Read** the problem again. There are three unknowns in this problem.

Step 2   **Assign variables** to represent the three unknowns.

> Let x = the number of loaves of honey wheat bread,
>
> y = the number of loaves of sunflower bread,
>
> and z = the number of loaves of French bread.

Step 3   **Write a system of three equations.** Three times as many loaves of honey wheat bread were sold as sunflower bread.

$$x = 3y, \quad \text{or} \quad x - 3y = 0 \qquad \text{Subtract } 3y. \quad (1)$$

Also, we have the information needed for another equation.

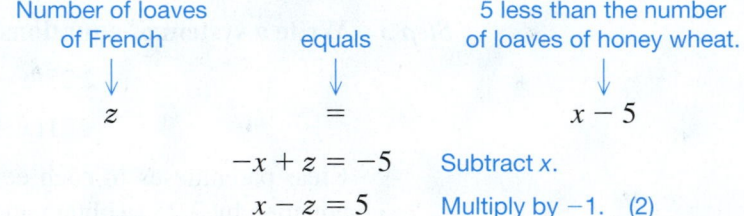

$$-x + z = -5 \qquad \text{Subtract } x.$$
$$x - z = 5 \qquad \text{Multiply by } -1. \quad (2)$$

Multiplying the cost of a loaf of each kind of bread by the number of loaves of that kind sold and adding gives an equation for the total receipts.

$$2.95x + 2.99y + 5.79z = 87.89$$
$$295x + 299y + 579z = 8789 \qquad \begin{array}{l}\text{Multiply each term by}\\ \text{100 to clear decimals.} \quad (3)\end{array}$$

Step 4   **Solve** the system of three equations.

$$x - 3y = 0 \qquad (1)$$
$$x - z = 5 \qquad (2)$$
$$295x + 299y + 579z = 8789 \qquad (3)$$

Equation (1) is missing the variable z, so one way to begin is to eliminate z again, using equations (2) and (3).

$$\begin{array}{lll} 579x \qquad\quad - 579z = \quad 2895 & \text{Multiply (2) by 579.} \\ 295x + 299y + 579z = \quad 8789 & (3) \\ \hline 874x + 299y \qquad\quad\; = 11{,}684 & \text{Add.} \quad (4) \end{array}$$

NOW TRY EXERCISE 6

At Panera Bread, a loaf of white bread costs $3.69, a loaf of cheese bread costs $4.29, and a loaf of whole grain bread costs $7.69. On a recent day, twice as many loaves of white bread were sold as cheese bread. The number of loaves of whole grain bread sold was 3 less than the number of loaves of white bread sold. Total receipts for these breads were $139.23. How many loaves of each type of bread were sold? (*Source:* Panera Bread menu.)

Use the resulting equation (4) in x and y, together with equation (1), $x - 3y = 0$, to eliminate x.

$$-874x + 2622y = 0 \qquad \text{Multiply (1) by } -874.$$
$$\underline{874x + 299y = 11{,}684} \qquad \text{(4)}$$
$$2921y = 11{,}684 \qquad \text{Add.}$$
$$y = 4 \qquad \text{Divide by 2921.}$$

We can find x by substituting this value for y in equation (1).

$$x - 3y = 0 \qquad \text{(1)}$$
$$x - 3(4) = 0 \qquad \text{Let } y = 4,$$
$$x - 12 = 0 \qquad \text{Multiply.}$$
$$x = 12 \qquad \text{Add 12.}$$

We can find z by substituting this value for x in equation (2).

$$x - z = 5 \qquad \text{(2)}$$
$$12 - z = 5 \qquad \text{Let } x = 12.$$
$$z = 7 \qquad \text{Subtract 12. Multiply by } -1.$$

Thus, $x = 12$, $y = 4$, and $z = 7$.

Step 5 **State the answer.** There were 12 loaves of honey wheat bread, 4 loaves of sunflower bread, and 7 loaves of French bread sold.

Step 6 **Check.** Because $12 = 3 \cdot 4$, the number of loaves of honey wheat bread is three times the number of loaves of sunflower bread. Also, $12 - 7 = 5$, so the number of loaves of French bread is 5 less than the number of loaves of honey wheat bread. Multiply the appropriate cost per loaf by the number of loaves sold and add the results to check that total receipts were $87.89.

NOW TRY

EXAMPLE 7 Solving a Business Production Problem

A company produces three flat screen television sets: models X, Y, and Z.

- Each model X set requires 2 hr of electronics work, 2 hr of assembly time, and 1 hr of finishing time.

- Each model Y requires 1 hr of electronics work, 3 hr of assembly time, and 1 hr of finishing time.

- Each model Z requires 3 hr of electronics work, 2 hr of assembly time, and 2 hr of finishing time.

There are 100 hr available for electronics, 100 hr available for assembly, and 65 hr available for finishing per week. How many of each model should be produced each week if all available time must be used?

Step 1 **Read** the problem again. There are three unknowns.

Step 2 **Assign variables.** Then organize the information in a table.

Let $x =$ the number of model X produced per week,

$y =$ the number of model Y produced per week,

and $z =$ the number of model Z produced per week.

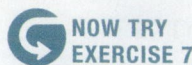

NOW TRY
EXERCISE 7

Katherine has a quilting shop and makes three kinds of quilts: the lone star quilt, the bandana quilt, and the log cabin quilt.

- Each lone star quilt requires 8 hr of piecework, 4 hr of machine quilting, and 2 hr of finishing.

- Each bandana quilt requires 2 hr of piecework, 2 hr of machine quilting, and 2 hr of finishing.

- Each log cabin quilt requires 10 hr of piecework, 5 hr of machine quilting, and 2 hr of finishing.

Katherine allocates 74 hr for piecework, 42 hr for machine quilting, and 24 hr for finishing quilts each month. How many of each type of quilt should be made each month if all available time must be used?

	Each Model X	Each Model Y	Each Model Z	Totals	
Hours of Electronics Work	2	1	3	100	→ Gives equation (1)
Hours of Assembly Time	2	3	2	100	→ Gives equation (2)
Hours of Finishing Time	1	1	2	65	→ Gives equation (3)

Step 3 **Write a system of three equations.** The x model X sets require $2x$ hours of electronics, the y model Y sets require $1y$ (or y) hours of electronics, and the z model Z sets require $3z$ hours of electronics. There are 100 hr available for electronics. This gives one equation.

$$2x + y + 3z = 100$$

By similar reasoning, we write two more equations using the fact that there are 100 hr available for assembly and 65 hr available for finishing.

$$2x + \ y + 3z = 100 \quad \text{Electronics} \quad (1)$$
$$2x + 3y + 2z = 100 \quad \text{Assembly} \quad (2)$$
$$x + \ y + 2z = \ 65 \quad \text{Finishing} \quad (3)$$

Notice that by reading *across* the table, we can easily determine the coefficients and constants in the equations of the system.

Step 4 **Solve** the system of equations (1), (2), and (3). Because x in equation (3) has coefficient 1, we choose x as the focus variable and (3) as the working equation.

$$
\begin{aligned}
2x + \ y + 3z &= \ \ 100 \quad (1) \\
-2x - 2y - 4z &= -130 \quad \text{Multiply (3) by } -2. \\
\hline
-y - \ z &= \ -30 \quad \text{Add. \ (4)}
\end{aligned}
$$

Eliminate x again, using equations (2) and (3).

$$
\begin{aligned}
2x + 3y + 2z &= \ \ 100 \quad (2) \\
-2x - 2y - 4z &= -130 \quad \text{Multiply (3) by } -2. \\
\hline
y - 2z &= \ -30 \quad \text{Add. \ (5)}
\end{aligned}
$$

Solve the system of two equations (4) and (5).

$$
\begin{aligned}
-y - \ z &= -30 \quad (4) \\
y - 2z &= -30 \quad (5) \\
\hline
-3z &= -60 \quad \text{Add.} \\
z &= 20 \quad \text{Divide by } -3.
\end{aligned}
$$

We can find y by substituting this value for z in equation (5).

$$
\begin{aligned}
y - 2z &= -30 \quad (5) \\
y - 2(20) &= -30 \quad \text{Let } z = 20. \\
y - 40 &= -30 \quad \text{Multiply.} \\
y &= 10 \quad \text{Add 40.}
\end{aligned}
$$

NOW TRY ANSWER
7. lone star quilts: 3;
   bandana quilts: 5;
   log cabin quilts: 4

NOW TRY EXERCISE 7
is on the preceding page.

We can find x by substituting the values for y and z in equation (3).

$$x + y + 2z = 65 \quad \text{(3)}$$

$$x + 10 + 2(20) = 65 \quad \text{Let } y = 10 \text{ and } z = 20.$$

$$x + 50 = 65 \quad \text{Multiply. Add.}$$

$$x = 15 \quad \text{Subtract 50.}$$

Thus, $x = 15$, $y = 10$, and $z = 20$.

Step 5 **State the answer.** The company should produce 15 model X, 10 model Y, and 20 model Z sets per week.

Step 6 **Check** that these values satisfy the conditions of the problem. **NOW TRY**

7.7 Exercises

FOR EXTRA HELP ▶ MyMathLab®

▶ *Complete solution available in MyMathLab*

Concept Check *Answer each question.*

1. If a container of liquid contains 60 oz of solution, what is the number of ounces of pure acid if the given solution contains the following acid concentrations?

   (a) 10%   **(b)** 25%   **(c)** 40%   **(d)** 50%

2. If $5000 is invested in an account paying simple annual interest, how much interest will be earned during the first year at the following rates?

   (a) 2%   **(b)** 3%   **(c)** 4%   **(d)** 3.5%

3. If one pound of turkey costs $1.89, how much will x pounds cost?

4. If one ticket to the movie *12 Years a Slave* costs $13.50 and y tickets are sold, how much is collected from the sale?

5. If the rate of a boat in still water is 10 mph, and the rate of the current of a river is x mph, what is the rate of the boat in each case?

   (a) The boat is going upstream (that is, against the current, which slows the boat down).

   (b) The boat is going downstream (that is, with the current, which speeds the boat up).

6. The swimming rate of a whale is 25 mph.

   (a) If the whale swims for y hours, what is its distance?

   (b) If the whale travels 10 mi, what is its time?

Solve each problem. See Example 1.

7. During the 2013 Major League Baseball season, the Los Angeles Dodgers played 162 games. They won 22 more games than they lost. What was their win-loss record that year?

8. Refer to **Exercise 7.** During the same 162-game season, the Colorado Rockies lost 14 more games than they won. What was the team's win-loss record?

Team	W	L
L.A. Dodgers	——	——
Arizona	81	70
San Diego	76	86
San Francisco	76	86
Colorado	——	——

Source: Major League Baseball.

▶ 9. Venus and Serena measured a tennis court and found that it was 42 ft longer than it was wide and had a perimeter of 228 ft. What were the length and the width of the tennis court?

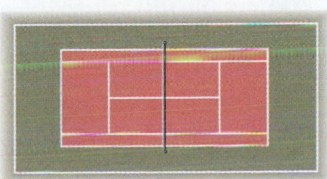

10. LeBron and Shaq measured a basketball court and found that the width of the court was 44 ft less than the length. If the perimeter was 288 ft, what were the length and the width of the basketball court?

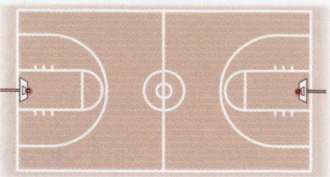

11. In 2012, the two American telecommunication companies with the greatest revenues were AT&T and Verizon. The two companies had combined revenues of $242.2 billion. AT&T's revenue was $10.6 billion more than that of Verizon. What was the revenue for each company? (*Source:* Verizon and AT&T Annual Reports.)

12. In 2012, U.S. exports to Canada were $76.6 billion more than exports to Mexico. Together, exports to these two countries totaled $508.4 billion. How much were exports to each country? (*Source:* U.S. Census Bureau.)

Find the measures of the angles marked x and y. Remember that **(1)** *the sum of the measures of the angles of a triangle is* 180°, **(2)** *supplementary angles have a sum of* 180°, *and* **(3)** *vertical angles have equal measures.*

13.

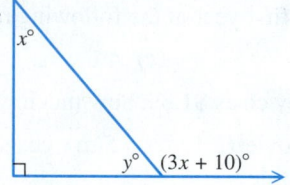

14.

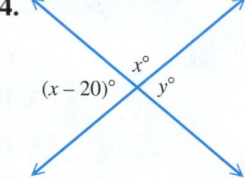

The Fan Cost Index (FCI) represents the cost of four average-price tickets (two adult, two child), four small soft drinks, two small beers, four hot dogs, parking for one car, two game programs, and two souvenir caps to a sporting event. (Source: Team Marketing Report.)
   Use the concept of FCI in Exercises 15 and 16. **See Example 2.**

15. For the 2013 season, the FCI prices for the National Hockey League and the National Basketball Association totaled $670.50. The hockey FCI was $39.18 more than that of basketball. What were the FCIs for these sports?

16. In 2013, the FCI prices for Major League Baseball and the National Football League totaled $667.45. The football FCI was $251.85 more than that of baseball. What were the FCIs for these sports?

Solve each problem. **See Example 2.**

17. Leanna is a waitress at Bonefish Grill. During one particular day she sold 15 ribeye steak dinners and 20 grilled salmon dinners, totaling $559.15. Another day she sold 25 ribeye steak dinners and 10 grilled salmon dinners, totaling $582.15. How much did each type of dinner cost? (*Source:* Bonefish Grill Menu.)

18. Two days at Busch Gardens Williamsburg (Virginia) and 3 days at Universal Studios Florida (Orlando) cost $420, while 4 days at Busch Gardens and 2 days at Universal Studios cost $472. (Prices are based on single-day admissions.) What was the cost per day for each park? (*Sources:* Busch Gardens and Universal Studios.)

19. The movie *Saving Mr. Banks* was available in both DVD and Blu-ray formats. The price for 3 DVDs and 2 Blu-ray discs was $77.86, while the price for 2 DVDs and 3 Blu-ray discs was $84.39. How much did each format cost?

20. On the basis of average total costs per day for business travel to New York City and Washington, DC (which include a hotel room, car rental, and three meals), 2 days in New York and 3 days in Washington cost $2936, while 4 days in New York and 2 days in Washington cost $3616. What was the average cost per day in each city? (*Source: Business Travel News.*)

Solve each problem. **See Example 3.**

▶ 21. How many gallons each of 25% alcohol and 35% alcohol should be mixed to obtain 20 gal of 32% alcohol?

Gallons of Solution	Percent (as a decimal)	Gallons of Pure Alcohol
x	25% = 0.25	
y	35% = 0.35	
20	32% =	

22. How many liters each of 15% acid and 33% acid should be mixed to obtain 120 L of 21% acid?

Liters of Solution	Percent (as a decimal)	Liters of Pure Acid
x	15% = 0.15	
y	33% =	
120	21% =	

23. Pure acid is to be added to a 10% acid solution to obtain 54 L of a 20% acid solution. What amounts of each should be used?

24. A truck radiator holds 36 L of fluid. How much pure antifreeze must be added to a mixture that is 4% antifreeze to fill the radiator with a mixture that is 20% antifreeze?

25. A party mix is made by adding nuts that sell for $2.50 per kg to a cereal mixture that sells for $1 per kg. How much of each should be added to obtain 30 kg of a mix that will sell for $1.70 per kg?

	Number of Kilograms	Price per Kilogram (in dollars)	Value (in dollars)
Nuts	*x*	2.50	
Cereal	*y*	1.00	
Mixture		1.70	

26. A fruit drink is made by mixing fruit juices. Such a drink with 50% juice is to be mixed with another drink that is 30% juice to obtain 200 L of a drink that is 45% juice. How much of each should be used?

	Liters of Drink	Percent (as a decimal)	Liters of Pure Juice
50% Juice	*x*	0.50	
30% Juice	*y*	0.30	
Mixture		0.45	

27. A total of $3000 is invested, part at 2% simple interest and part at 4%. If the total annual return from the two investments is $100, how much is invested at each rate?

Principal (in dollars)	Rate (as a decimal)	Interest (in dollars)
x	0.02	0.02x
y	0.04	0.04y
3000		100

28. An investor will invest a total of $15,000 in two accounts, one paying 4% annual simple interest and the other 3%. If he wants to earn $550 annual interest, how much should he invest at each rate?

Principal (in dollars)	Rate (as a decimal)	Interest (in dollars)
x	0.04	
y	0.03	
15,000		

Solve each problem. See Examples 4 and 5.

▶ **29.** A train travels 150 km in the same time that a plane travels 400 km. If the rate of the plane is 20 km per hr less than three times the rate of the train, find both rates.

	r	t	d
Train	x		150
Plane	y		400

30. A freight train and an express train leave towns 390 km apart, traveling toward one another. The freight train travels 30 km per hr slower than the express train. They pass one another 3 hr later. What are their rates?

	r	t	d
Freight Train	x	3	
Express Train	y	3	

31. A motor scooter travels 20 mi in the same time that a bicycle travels 8 mi. If the rate of the scooter is 5 mph more than twice the rate of the bicycle, find both rates.

32. A plane travels 1000 mi in the same time that a car travels 300 mi. If the rate of the plane is 20 mph greater than three times the rate of the car, find both rates.

33. In his motorboat, Bill travels upstream at top speed to his favorite fishing spot, a distance of 36 mi, in 2 hr. Returning, he finds that the trip downstream, still at top speed, takes only 1.5 hr. Find the rate of Bill's boat and the rate of the current. Let x = the rate of the boat and y = the rate of the current.

	r	t	d
Upstream	$x - y$	2	
Downstream	$x + y$		

34. Traveling for 3 hr into a steady head wind, a plane flies 1650 mi. The pilot determines that flying *with* the same wind for 2 hr, he could make a trip of 1300 mi. Find the rate of the plane and the wind speed.

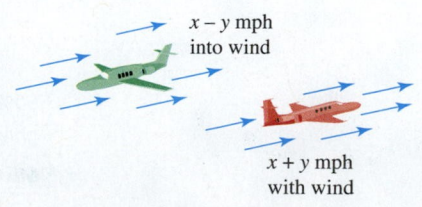

$x - y$ mph
into wind

$x + y$ mph
with wind

Solve each problem. See Examples 1–5.

35. How many pounds of candy that sells for $0.75 per lb must be mixed with candy that sells for $1.25 per lb to obtain 9 lb of a mixture that should sell for $0.96 per lb? (*Source: The Bill Cosby Show.*)

36. The top-grossing tour on the North American concert circuit for 2012 was Madonna, followed by Bruce Springsteen and the E Street Band. Together, they took in $238.4 million from ticket sales. If Springsteen took in $29 million less than Madonna, how much did each band generate? (*Source:* Pollstar.)

37. Tickets to a production of *A Midsummer Night's Dream* at Broward College cost $5 for general admission or $4 with a student ID. If 184 people paid to see a performance and $812 was collected, how many of each type of ticket were sold?

38. At a business meeting at Panera Bread, the bill for two cappuccinos and three house lattes was $17.55. At another table, the bill for one cappuccino and two house lattes was $10.57. How much did each type of beverage cost? (*Source:* Panera Bread menu.)

39. The mixed price of 9 citrons and 7 fragrant wood apples is 107; again, the mixed price of 7 citrons and 9 fragrant wood apples is 101. O you arithmetician, tell me quickly the price of a citron and the price of a wood apple here, having distinctly separated those prices well. (*Source:* Hindu work, A.D. 850.)

40. Braving blizzard conditions on the planet Hoth, Luke Skywalker sets out in his snow speeder for a rebel base 4800 mi away. He travels into a steady head wind and makes the trip in 3 hr. Returning, he finds that the trip back, now with a tailwind, takes only 2 hr. Find the rate of Luke's snow speeder and the speed of the wind.

	r	t	d
Into Head Wind			
With Tailwind			

Solve each problem. See Examples 6 and 7. (In Exercises 41–44, remember that the sum of the measures of the angles of a triangle is 180°.)

41. In the triangle below, $z = x + 10$ and $x + y = 100$. Determine a third equation involving x, y, and z, and then find the measures of the three angles.

42. In the triangle below, x is 10 less than y and 20 less than z. Write a system of equations and find the measures of the three angles.

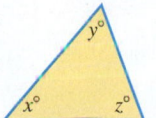

43. In a certain triangle, the measure of the second angle is 10° greater than three times the first. The third angle measure is equal to the sum of the measures of the other two. Find the measures of the three angles.

44. The measure of the largest angle of a triangle is 12° less than the sum of the measures of the other two. The smallest angle measures 58° less than the largest. Find the measures of the angles.

45. The perimeter of a triangle is 70 cm. The longest side is 4 cm less than the sum of the other two sides. Twice the shortest side is 9 cm less than the longest side. Find the length of each side of the triangle.

46. The perimeter of a triangle is 56 in. The longest side measures 4 in. less than the sum of the other two sides. Three times the shortest side is 4 in. more than the longest side. Find the lengths of the three sides.

47. In the 2014 Winter Olympics in Sochi, Russia, host Russia earned 4 more gold medals than bronze. The number of silver medals earned was 7 less than twice the number of bronze medals. Russia earned a total of 33 medals. How many of each kind of medal did Russia earn? (*Source:* www.sochi.com)

48. In a random sample of Americans of voting age conducted recently, 17% more people identified themselves as Independents than as Republicans, while 6% fewer people identified themselves as Republicans than as Democrats. Of those sampled, 2% did not identify with any of the three categories. What percent of the people in the sample identified themselves with each of the three political affiliations? (*Source:* Gallup, Inc.)

49. Tickets for a Harlem Globetrotters show cost $28 general admission, $43 courtside, or $173 bench seats. Nine times as many general admission tickets were sold as bench tickets, and the number of general admission tickets sold was 55 more than the sum of the number of courtside tickets and bench tickets. Sales of all three kinds of tickets totaled $97,605. How many of each kind of ticket were sold? (*Source:* www.harlemglobetrotters.com)

50. Three kinds of tickets are available for a rock concert: "up close," "in the middle," and "far out." "Up close" tickets cost $10 more than "in the middle" tickets. "In the middle" tickets cost $10 more than "far out" tickets. Twice the cost of an "up close" ticket is $20 more than three times the cost of a "far out" ticket. Find the price of each kind of ticket.

▶ **51.** A wholesaler supplies college t-shirts to three college bookstores: A, B, and C. The wholesaler recently shipped a total of 800 t-shirts to the three bookstores. Twice as many t-shirts were shipped to bookstore B as to bookstore A, and the number shipped to bookstore C was 40 less than the sum of the numbers shipped to the other two bookstores. How many t-shirts were shipped to each bookstore?

52. An office supply store sells three models of computer desks: A, B, and C. In one month, the store sold a total of 85 computer desks. The number of model B desks was five more than the number of model C desks. The number of model A desks was four more than twice the number of model C desks. How many of each model did the store sell that month?

53. A plant food is to be made from three chemicals. The mix must include 60% of the first and second chemicals. The second and third chemicals must be in the ratio of 4 to 3 by weight. How much of each chemical is needed to make 750 kg of the plant food?

54. How many ounces of 5% hydrochloric acid, 20% hydrochloric acid, and water must be combined to obtain 10 oz of solution that is 8.5% hydrochloric acid if the amount of water used must equal the total amount of the other two solutions?

The National Hockey League uses a point system to determine team standings. A team is awarded 2 points for a win (W), 0 points for a loss in regulation play (L), and 1 point for an overtime loss (OTL). Use this information to solve each problem.

55. During the 2012–2013 NHL regular season, the Anaheim Ducks played 48 games. Their wins and overtime losses resulted in a total of 66 points. They had 6 more losses in regulation play than overtime losses. How many wins, losses, and overtime losses did they have that year?

Team	GP	W	L	OTL	Points
Anaheim	48	___	___	___	66
Los Angeles	48	27	16	5	59
San Jose	48	25	16	7	57
Phoenix	48	21	18	9	51
Dallas	48	___	___	___	48

Source: World Almanac and Book of Facts.

56. During the same NHL regular season, the Dallas Stars also played 48 games. Their wins and overtime losses resulted in a total of 48 points. They had 4 more total losses (in regulation play and overtime) than wins. How many wins, losses, and overtime losses did they have that year?

The following exercises are based on the "Bait Box Puzzle," featured in a recent "Ask Marilyn" column in Parade Magazine.

57. There are three kinds of bait boxes—those containing fish, those containing bugs, and those containing worms. Although the bait boxes look alike, each kind has a different weight. Use the information in **FIGURE A** to write a system of three equations, and solve it to determine the weight of each kind of bait box.

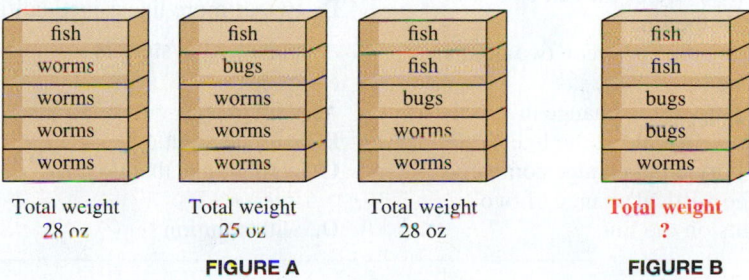

Total weight 28 oz	Total weight 25 oz	Total weight 28 oz	Total weight ?

FIGURE A　　　　　　　　**FIGURE B**

58. Write an equation that describes the situation depicted in **FIGURE B**. Use the same three variables used in **Exercise 57.** Then determine the total weight of the bait boxes in **FIGURE B**.

Chapter 7 | Summary

Key Terms

7.1
ordered pair
origin
x-axis
y-axis
rectangular (Cartesian)
  coordinate system
components
plot
coordinate
quadrant

graph of an equation
table of ordered pairs
  (table of values)
first-degree equation
linear equation in two
  variables
x-intercept
y-intercept
rise
run
slope

7.3
system of linear equations
  (linear system)
solution of a system
solution set of a system
consistent system
inconsistent system
independent equations
dependent equations

7.6
ordered triple
focus variable
working equation

New Symbols

(x, y) ordered pair

(x_1, y_1) subscript notation
    (read "x-sub-one,
    y-sub-one")

Δ Greek letter delta
m slope

(x, y, z) ordered triple

Test Your Word Power

See how well you have learned the vocabulary in this chapter.

1. A **linear equation in two variables** is an equation that can be written in the form
   A. $Ax + By < C$
   B. $ax = b$
   C. $y = x^2$
   D. $Ax + By = C$.

2. The **slope** of a line is
   A. the measure of the run over the rise of the line
   B. the distance between two points on the line
   C. the ratio of the change in y to the change in x along the line
   D. the horizontal change compared to the vertical change of two points on the line.

3. A **system of linear equations** consists of
   A. at least two linear equations with different variables
   B. two or more linear equations that have an infinite number of solutions
   C. two or more linear equations with the same variables
   D. two or more linear inequalities.

4. A **consistent system** is a system of equations
   A. with at least one solution
   B. with no solution
   C. with graphs that do not intersect
   D. with solution set $\varnothing$.

5. An **inconsistent system** is a system of equations
   A. with one solution
   B. with no solution
   C. with an infinite number of solutions
   D. that have the same graph.

6. **Dependent equations**
   A. have different graphs
   B. have no solution
   C. have one solution
   D. are different forms of the same equation.

ANSWERS

1. D; *Examples:* $3x + 2y = 6$, $x = y - 7$ 2. C; *Example:* The line through $(3, 6)$ and $(5, 4)$ has slope $\frac{4-6}{5-3} = \frac{-2}{2} = -1$.
3. C; *Example:* $2x + y = 7$, $3x - y = 3$ 4. A; *Example:* The system in **Answer 3** is consistent. The graphs of the equations intersect at exactly one point—in this case, the solution $(2, 3)$.   5. B; *Example:* The equations of two parallel lines make up an inconsistent system. Their graphs never intersect, so there is no solution to the system.   6. D; *Example:* The equations $4x - y = 8$ and $8x - 2y = 16$ are dependent because their graphs are the same line.

Quick Review

CONCEPTS

EXAMPLES

7.1 Review of Graphs and Slopes of Lines

Finding Intercepts

To find the *x*-intercept, let $y = 0$ and solve for *x*.

To find the *y*-intercept, let $x = 0$ and solve for *y*.

Find the intercepts of the graph of $2x + 3y = 12$.

$2x + 3(0) = 12$ Let $y = 0$. | $2(0) + 3y = 12$ Let $x = 0$.

$2x = 12$ | $3y = 12$

$x = 6$ | $y = 4$

The *x*-intercept is $(6, 0)$. | The *y*-intercept is $(0, 4)$.

Midpoint Formula

The midpoint *M* of a line segment *PQ* with endpoints (x_1, y_1) and (x_2, y_2) is found as follows.

$$M = \left(\frac{x_1 + x_2}{2}, \frac{y_1 + y_2}{2} \right)$$

Find the midpoint of the segment with endpoints $(4, -7)$ and $(-10, -13)$.

$$M = \left(\frac{4 + (-10)}{2}, \frac{-7 + (-13)}{2} \right)$$

$$= (-3, -10)$$

Slope Formula

The slope *m* of the line passing through the points (x_1, y_1) and (x_2, y_2) is defined as follows.

$$\text{slope } m = \frac{\text{rise}}{\text{run}} = \frac{\text{change in } y}{\text{change in } x} = \frac{\Delta y}{\Delta x} = \frac{y_2 - y_1}{x_2 - x_1}$$

(where $x_1 \neq x_2$)

Find the slope of the graph of $2x + 3y = 12$.

Use the intercepts $(6, 0)$ and $(0, 4)$ and the slope formula.

$$m = \frac{4 - 0}{0 - 6} = \frac{4}{-6} = -\frac{2}{3} \qquad \begin{array}{l} (x_1, y_1) = (6, 0) \\ (x_2, y_2) = (0, 4) \end{array}$$

A horizontal line has slope 0.

A vertical line has undefined slope.

Parallel lines have equal slopes.

The graph of the horizontal line $y = -5$ has slope $m = 0$.

The graph of the vertical line $x = 3$ has undefined slope.

The lines $y = 2x + 3$ and $4x - 2y = 6$ are parallel.
Both have $m = 2$.

$y = 2x + 3$ | $4x - 2y = 6$ Solve for *y*.

                       | $-2y = -4x + 6$

                       | $y = 2x - 3$

Perpendicular lines, neither of which is vertical, **have slopes that are negative reciprocals** (that is, have a product of -1).

The lines $y = 3x - 1$ and $x + 3y = 4$ are perpendicular.
Their slopes are negative reciprocals.

$y = 3x - 1$ | $x + 3y = 4$ Solve for *y*.

                       | $3y = -x + 4$

                       | $y = -\frac{1}{3}x + \frac{4}{3}$

Slope gives the **average rate of change** in *y* per unit change in *x*, where the value of *y* depends on the value of *x*.

The weight of a young child increased from 30 lb to 60 lb between the ages of 3 and 8. What was the child's average change in weight per year over these years?

$$\frac{60 - 30}{8 - 3} = \frac{30}{5} = 6 \text{ lb per yr}$$

CONCEPTS	EXAMPLES

7.2 Review of Equations of Lines; Linear Models

Slope-Intercept Form

$y = mx + b$

$y = 2x + 3$ $m = 2$, y-intercept is $(0, 3)$.

Point-Slope Form

$y - y_1 = m(x - x_1)$

$y - 3 = 4(x - 5)$ $(5, 3)$ is on the line, $m = 4$.

Standard Form

$Ax + By = C$, where A, B, and C are real numbers and A and B are not both 0. (We give A, B, and C integers, with greatest common factor 1 and $A \geq 0$.)

$2x - 5y = 8$ Standard form

Horizontal Line

$y = b$

$y = 4$ Horizontal line

Vertical Line

$x = a$

$x = -1$ Vertical line

7.3 Solving Systems of Linear Equations by Graphing

An ordered pair is a solution of a system if it makes all equations of the system true at the same time.

Is $(4, -1)$ a solution of the following system?

$$x + y = 3$$
$$2x - y = 9$$

Yes, because $4 + (-1) = 3$ and $2(4) - (-1) = 9$ are both true, $(4, -1)$ is a solution.

To solve a linear system by graphing, follow these steps.

Solve the system by graphing.

Step 1 Graph each equation of the system on the same axes.

$$x + y = 5$$
$$2x - y = 4$$

Step 2 Find the coordinates of the point of intersection.

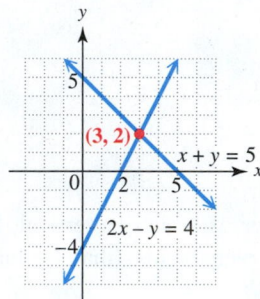

Step 3 Check. Write the solution set.

The ordered pair $(3, 2)$ satisfies *both* equations, so $\{(3, 2)\}$ is the solution set.

7.4 Solving Systems of Linear Equations by Substitution

Solve the system by substitution.

Step 1 Solve one equation for either variable.

$$x + 2y = -5 \quad (1)$$
$$y = -2x - 1 \quad (2)$$

Equation (2) is already solved for y.

CONCEPTS	EXAMPLES

Step 2   Substitute for that variable in the other equation to obtain an equation in one variable.

Substitute $-2x - 1$ for y in equation (1).

$$x + 2y = -5 \quad (1)$$
$$x + 2(-2x - 1) = -5 \quad \text{Let } y = -2x - 1 \text{ in (1).}$$
$$x - 4x - 2 = -5 \quad \text{Distributive property}$$
$$-3x - 2 = -5 \quad \text{Combine like terms.}$$
$$-3x = -3 \quad \text{Add 2.}$$
$$x = 1 \quad \text{Divide by } -3.$$

Step 3   Solve the equation from Step 2.

Step 4   Find the other value by substituting the result from Step 3 into the equation from Step 1 and solving for the remaining variable.

To find y, let $x = 1$ in equation (2).

$$y = -2x - 1 \quad (2)$$
$$y = -2(1) - 1 \quad \text{Let } x = 1.$$
$$y = -3 \quad \text{Multiply, and then subtract.}$$

Step 5   Check. Write the solution set.

A check confirms that $\{(1, -3)\}$ is the solution set.

7.5   **Solving Systems of Linear Equations by Elimination**

Step 1   Write both equations in the form $Ax + By = C$.

Solve the system by elimination.

$$x + 3y = 7 \quad (1)$$
$$3x - y = 1 \quad (2)$$

Step 2   Multiply to transform the equations so that the coefficients of one pair of variable terms are opposites.

Multiply equation (1) by -3 to eliminate the x-terms.

$$\begin{array}{ll} -3x - 9y = -21 & \text{Multiply equation (1) by } -3. \\ \underline{3x - y = 1} & (2) \\ -10y = -20 & \text{Add.} \\ y = 2 & \text{Divide by } -10. \end{array}$$

Step 3   Add the equations to get an equation with only one variable.

Step 4   Solve the equation from Step 3.

Step 5   Find the other value by substituting the result from Step 4 into either of the original equations and solving for the remaining variable.

Substitute to find the value of x.

$$x + 3y = 7 \quad (1)$$
$$x + 3(2) = 7 \quad \text{Let } y = 2.$$
$$x + 6 = 7 \quad \text{Multiply.}$$
$$x = 1 \quad \text{Subtract 6.}$$

Step 6   Check. Write the solution set.

A check confirms that $\{(1, 2)\}$ is the solution set.

If the result of the addition step (Step 3) is a false statement, such as $0 = 4$, the graphs are parallel lines and *there is no solution.*

         The solution set is $\varnothing$.

If the result is a true statement, such as $0 = 0$, the graphs are the same line, and an *infinite number of ordered pairs are solutions. The solution set is written in set-builder notation as*

$$\{(x, y) \,|\, \underline{}\},$$

where a form of one of the equations is written in the blank.

$$\begin{array}{l} x - 2y = 6 \\ \underline{-x + 2y = -2} \\ 0 = 4 \end{array}$$
Solution set: $\varnothing$

$$\begin{array}{l} x - 2y = 6 \\ \underline{-x + 2y = -6} \\ 0 = 0 \end{array}$$
Solution set:
$$\{(x, y) \,|\, x - 2y = 6\}$$

CONCEPTS	EXAMPLES

7.6 Systems of Linear Equations in Three Variables

Solving a Linear System in Three Variables

Step 1 Select a focus variable, preferably one with coefficient 1 or −1, and a working equation.

Step 2 Eliminate the focus variable, using the working equation and one of the equations of the system.

Step 3 Eliminate the focus variable again, using the working equation and the remaining equation of the system.

Step 4 Solve the system of two equations in two variables formed by the equations from Steps 2 and 3.

Step 5 Find the value of the remaining variable.

Step 6 Check the values in *each* of the *original* equations of the system. Then write the solution set as a set containing an ordered triple.

Solve the system.

$$x + 2y - z = 6 \quad (1)$$
$$x + y + z = 6 \quad (2)$$
$$2x + y - z = 7 \quad (3)$$

We choose z as the focus variable and (2) as the working equation.

Add equations (1) and (2).

$$2x + 3y = 12 \quad (4)$$

Add equations (2) and (3).

$$3x + 2y = 13 \quad (5)$$

Use equations (4) and (5) to eliminate x.

$$-6x - 9y = -36 \quad \text{Multiply (4) by } -3.$$
$$\underline{6x + 4y = 26} \quad \text{Multiply (5) by 2.}$$
$$-5y = -10 \quad \text{Add.}$$
$$y = 2 \quad \text{Divide by } -5.$$

To find x, substitute 2 for y in equation (4).

$$2x + 3(2) = 12 \quad \text{Let } y = 2 \text{ in (4).}$$
$$2x + 6 = 12 \quad \text{Multiply.}$$
$$2x = 6 \quad \text{Subtract 6.}$$
$$x = 3 \quad \text{Divide by 2.}$$

Substitute 3 for x and 2 for y in working equation (2).

$$x + y + z = 6 \quad (2)$$
$$3 + 2 + z = 6 \quad \text{Substitute.}$$
$$z = 1 \quad \text{Subtract 5.}$$

A check of the ordered triple $(3, 2, 1)$ confirms that the solution set is $\{(3, 2, 1)\}$.

7.7 Applications of Systems of Linear Equations

Use the six-step problem-solving method.

Step 1 Read the problem carefully.

Step 2 Assign variables.

Step 3 Write a system of equations.

Step 4 Solve the system.

Step 5 State the answer.

Step 6 Check.

The perimeter of a rectangle is 18 ft. The length is 3 ft more than twice the width. What are the dimensions of the rectangle?

Let $x =$ the length and $y =$ the width. Write a system of equations.

$$2x + 2y = 18 \quad \text{From the perimeter formula}$$
$$x = 2y + 3 \quad \text{Length is 3 ft more than twice the width.}$$

Solve to find that

$$x = 7 \quad \text{and} \quad y = 2.$$

The length is 7 ft, and the width is 2 ft. The answer checks because

$$2(7) + 2(2) = 18 \quad \text{and} \quad 2(2) + 3 = 7, \quad \text{as required.}$$

Chapter 7 | Review Exercises

7.1 *Complete the table of ordered pairs for each equation. Then graph the equation.*

1. $3x + 2y = 10$

x	y
0	
	0
2	
	-2

2. $x - y = 8$

x	y
2	
	-3
3	
	-2

Find the x- and y-intercepts. Then graph each equation.

3. $4x - 3y = 12$ **4.** $5x + 7y = 28$ **5.** $2x + 5y = 20$ **6.** $x - 4y = 8$

Find the midpoint of each segment with the given endpoints.

7. $(-8, -12)$ and $(8, 16)$

8. $(0, -5)$ and $(-9, 8)$

Find the slope of each line. (In Exercises 17 and 18, coordinates of the points shown are integers.)

9. Through $(-1, 2)$ and $(4, -5)$ **10.** Through $(0, 3)$ and $(-2, 4)$

11. $y = 2x + 3$ **12.** $3x - 4y = 5$

13. $x = 5$ **14.** Parallel to $3y = 2x + 5$

15. Perpendicular to $3x - y = 4$ **16.** Through $(-1, 5)$ and $(-1, -4)$

17.
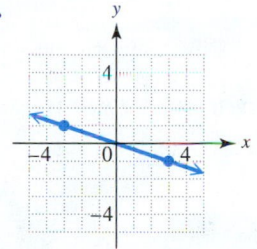

18.
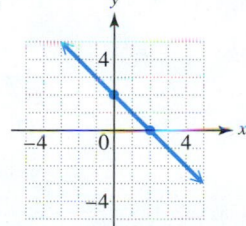

Tell whether the slope of the line is positive, negative, 0, or undefined.

19.
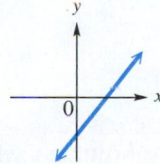

20.
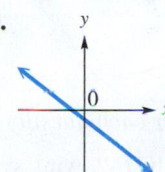

21.
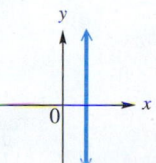

22.
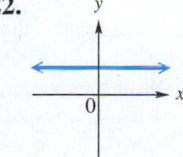

Solve each problem.

23. If the pitch of a roof is $\frac{1}{4}$, how many feet in the horizontal direction correspond to a rise of 3 ft?

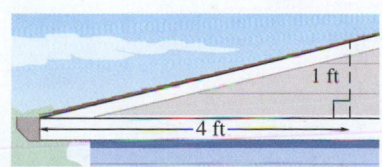

24. In 1980, the median family income in the United States was about $21,000 per year. In 2012, it was about $51,017 per year. Find the average rate of change in median family income per year to the nearest dollar during this period. (*Source:* U.S. Census Bureau.)

Answers (left margin):

1.

x	y
0	5
$\frac{10}{3}$	0
2	2
$\frac{14}{3}$	-2

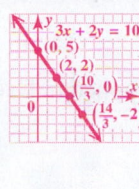

2.

x	y
2	-6
5	-3
3	-5
6	-2

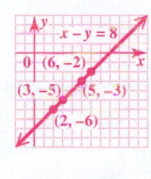

3. $(3, 0); (0, -4)$ **4.** $\left(\frac{28}{5}, 0\right); (0, 4)$

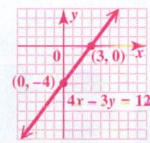 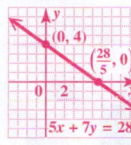

5. $(10, 0); (0, 4)$ **6.** $(8, 0); (0, -2)$

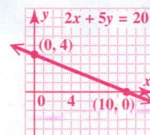 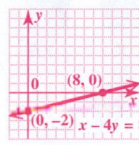

7. $(0, 2)$ **8.** $\left(-\frac{9}{2}, \frac{3}{2}\right)$

9. $-\frac{7}{5}$ **10.** $-\frac{1}{2}$

11. 2   **12.** $\frac{3}{4}$

13. undefined   **14.** $\frac{2}{3}$

15. $-\frac{1}{3}$ **16.** undefined

17. $-\frac{1}{3}$ **18.** -1

19. positive   **20.** negative

21. undefined   **22.** 0

23. 12 ft   **24.** $938 per year

25. (a) $y = -\dfrac{1}{3}x - 1$

   (b) $x + 3y = -3$

26. (a) $y = -2$

   (b) $y = -2$

27. (a) $y = -\dfrac{4}{3}x + \dfrac{29}{3}$

   (b) $4x + 3y = 29$

28. (a) $y = 3x + 7$

   (b) $3x - y = -7$

29. (a) not possible

   (b) $x = 2$

30. (a) $y = -9x + 13$

   (b) $9x + y = 13$

31. (a) $y = \dfrac{7}{5}x + \dfrac{16}{5}$

   (b) $7x - 5y = -16$

32. (a) $y = -x + 2$

   (b) $x + y = 2$

33. (a) $y = 4x - 29$

   (b) $4x - y = 29$

34. (a) $y = -\dfrac{5}{2}x + 13$

   (b) $5x + 2y = 26$

35. $y = 47x + 159$; $723

36. (a) $y = 38x + 2172$;

   The revenue from skiing facilities increased by an average of $38 million per year from 2008 to 2012.

   (b) $2552 million

37. yes   **38.** no

39. $\{(3, 1)\}$ **40.** $\{(0, -2)\}$

41. $\varnothing$

42. $\{(x, y) \mid x - 2y = 2\}$

43. It would be easiest to solve for x in the second equation because its coefficient is -1. No fractions would be involved.

44. The true statement $0 = 0$ is an indication that the system has an infinite number of solutions. Write the solution set using set-builder notation and the equation of the system that is in standard form with integer coefficients having greatest common factor 1.

45. $\{(2, 1)\}$ **46.** $\{(3, 5)\}$

47. $\{(6, 4)\}$

48. $\{(x, y) \mid x + 3y = 6\}$

7.2   *Write an equation of the line that satisfies the given conditions. Give the equation (a) in slope-intercept form if possible and (b) in standard form.*

25. Slope $-\dfrac{1}{3}$; y-intercept $(0, -1)$

26. Slope 0; y-intercept $(0, -2)$

27. Slope $-\dfrac{4}{3}$; through $(2, 7)$

28. Slope 3; through $(-1, 4)$

29. Vertical; through $(2, 5)$

30. Through $(2, -5)$ and $(1, 4)$

31. Through $(-3, -1)$ and $(2, 6)$

32. The line graphed in **Exercise 18**

33. Through $(7, -1)$; parallel to $4x - y = 3$

34. Through $(4, 3)$; perpendicular to $2x - 5y = 7$

Solve each problem.

35. An Executive Regular/Silver membership to a health club costs $159, plus $47 per month. Let x represent the number of months and y represent the cost. How much will a one-year membership cost? (*Source:* Midwest Athletic Club.)

36. Revenue for skiing facilities in the United States is shown in the graph.

   (a) Use the information given for the years 2008 and 2012, letting $x = 8$ represent 2008 and $x = 12$ represent 2012, and letting y represent revenue (in millions of dollars), to write an equation that models the data. Write the equation in slope-intercept form. Interpret the slope.

   (b) Use the equation from part (a) to approximate revenue for skiing facilities in 2010.

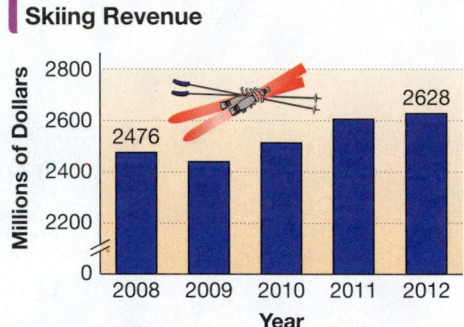

Skiing Revenue

Source: U.S. Department of Commerce.

7.3   *Decide whether the given ordered pair is a solution of the given system.*

37. $(3, 4)$

   $4x - 2y = 4$

   $5x + y = 19$

38. $(-5, 2)$

   $x - 4y = -13$

   $2x + 3y = 4$

Solve each system by graphing.

39. $x + y = 4$

   $2x - y = 5$

40. $x - 2y = 4$

   $2x + y = -2$

41. $2x + 4 = 2y$

   $y - x = -3$

42. $x - 2 = 2y$

   $2x - 4y = 4$

7.4   *Answer each question.*

43. To solve the following system by substitution, which variable in which equation would be easiest to solve for in the first step?

$$5x - 3y = 7$$
$$-x + 2y = 4$$

44. After solving a system of linear equations by the substitution method, a student obtained the equation "$0 = 0$." He gave the solution set of the system as $\{(0, 0)\}$. **WHAT WENT WRONG?**

Solve each system by the substitution method.

45. $3x + y = 7$

   $x = 2y$

46. $2x - 5y = -19$

   $y = x + 2$

47. $4x + 5y = 44$

   $x + 2 = 2y$

48. $5x + 15y = 30$

   $x + 3y = 6$

49. C

50. (a) 2 (b) 9

51. $\{(7,1)\}$ **52.** $\{(-4,3)\}$

53. $\{(x,y)\,|\,3x-4y=9\}$

54. $\varnothing$

55. $\{(-4,1)\}$

56. $\{(x,y)\,|\,2x-3y=0\}$; dependent equations

57. $\{(9,2)\}$ **58.** $\{(8,9)\}$

59. $\{(2,1)\}$ **60.** $\{(-3,2)\}$

61. $\{(1,2,3)\}$

62. $\{(1,-5,3)\}$

63. $\varnothing$; inconsistent system

64. length: 200 ft; width: 85 ft

65. New York Yankees: $51.55; Boston Red Sox: $53.38

66. plane: 300 mph; wind: 20 mph

7.5 *Answer each question.*

49. Which system does not require that we multiply one or both equations by a constant to solve the system by the elimination method?

A. $-4x+3y=7$
$3x-4y=4$

B. $5x+8y=13$
$12x+24y=36$

C. $2x+3y=5$
$x-3y=12$

D. $x+2y=9$
$3x-\;\;y=6$

50. If we were to multiply equation (1) by -3 in the system below, by what number would we have to multiply equation (2) in order to do the following?

$$2x+12y=7 \qquad (1)$$
$$3x+4y=1 \qquad (2)$$

(a) Eliminate the x-terms when solving by the elimination method.

(b) Eliminate the y-terms when solving by the elimination method.

Solve each system by the elimination method.

51. $2x-y=13$
$\;\;\;x+y=8$

52. $-4x+3y=25$
$\;\;\;6x-5y=-39$

53. $3x-4y=9$
$6x-8y=18$

54. $2x+\;\;y=3$
$-4x-2y=6$

7.3–7.5 *Solve each system by any method. If a system is inconsistent or has dependent equations, say so.*

55. $2x+3y=-5$
$3x+4y=-8$

56. $6x-9y=0$
$2x-3y=0$

57. $x-2y=5$
$y=x-7$

58. $\dfrac{x}{2}+\dfrac{y}{3}=7$
$\dfrac{x}{4}+\dfrac{2y}{3}=8$

59. $\dfrac{3}{4}x-\dfrac{1}{3}y=\dfrac{7}{6}$
$\dfrac{1}{2}x+\dfrac{2}{3}y=\dfrac{5}{3}$

60. $0.4x-0.5y=-2.2$
$0.3x+0.2y=-0.5$

7.6 *Solve each system. If a system is inconsistent or has dependent equations, say so.*

61. $4x-\;\;y=2$
$3y+\;\;z=9$
$\;\;x+2z=7$

62. $2x+3y-\;\;z=-16$
$\;\;x+2y+2z=-3$
$-3x+\;\;y+\;\;z=-5$

63. $3x-\;\;y-\;\;z=-8$
$4x+2y+3z=15$
$-6x+2y+2z=10$

7.7 *Solve each problem.*

64. A regulation National Hockey League ice rink has perimeter 570 ft. The length of the rink is 30 ft longer than twice the width. What are the dimensions of an NHL ice rink? (*Source:* www.nhl.com)

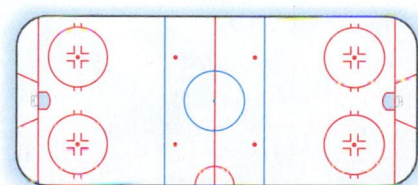

65. During a recent Major League Baseball season, two New York Yankees tickets and three Boston Red Sox tickets purchased at their average prices cost $263.24. Three Yankees tickets and two Red Sox tickets cost $261.41. Find the average ticket price for a Yankees ticket and a Red Sox ticket. (*Source:* Team Marketing Report.)

66. A plane flies 560 mi in 1.75 hr traveling with the wind. The return trip later against the same wind takes the plane 2 hr. Find the rate of the plane and the wind speed. Let x = the rate of the plane and y = the wind speed.

	r	t	d
With Wind	$x+y$	1.75	
Against Wind		2	

67. $2-per-lb nuts: 30 lb;
$1-per-lb candy: 70 lb

68. 85°, 60°, 35°

69. $40,000 at 10%;
$100,000 at 6%;
$140,000 at 5%

70. Mantle: 54;
Maris: 61;
Berra: 22

67. For Valentine's Day, Ms. Sweet will mix some $2-per-lb nuts with some $1-per-lb chocolate candy to obtain 100 lb of mix, which she will sell at $1.30 per lb. How many pounds of each should she use?

	Number of Pounds	Price per Pound (in dollars)	Value (in dollars)
Nuts	x		
Chocolate	y		
Mixture	100		

68. The sum of the measures of the angles of a triangle is 180°. The largest angle measures 10° less than the sum of the other two. The measure of the middle-sized angle is the average of the other two. Find the measures of the three angles.

69. Noemi sells real estate. On three recent sales, she made 10% commission, 6% commission, and 5% commission. Her total commissions on these sales were $17,000, and she sold property worth $280,000. If the 5% sale amounted to the sum of the other two, what were the three sales prices?

70. In the great baseball year of 1961, Yankee teammates Mickey Mantle, Roger Maris, and Yogi Berra combined for 137 home runs. Mantle hit 7 fewer than Maris, and Maris hit 39 more than Berra. What were the home run totals for each player? (*Source:* www.mlb.com)

Chapter 7 | Mixed Review Exercises

1. perpendicular

2. parallel

3. −1.8 lb per year;
Per capita consumption of potatoes decreased by an average of 1.8 lb per year from 2003 to 2011.

4. $y = -1.8x + 46.8$

5. $x + 2y = 6$

6. $y = -3$

7. $\{(0, 4)\}$

8. $\left\{\left(\dfrac{82}{23}, -\dfrac{4}{23}\right)\right\}$

9. $\{(5, 3)\}$ **10.** $\{(3, -1)\}$

11. $\{(12, 9)\}$ **12.** $\varnothing$

Determine whether each pair of lines is parallel, perpendicular, *or* neither.

1. $3x + y = 4$ and $3y = x - 6$

2. $4x + 3y = 8$ and $6y = 7 - 8x$

The graph shows per capita consumption of potatoes (in pounds) in the United States from 2003 to 2011.

3. Use the given ordered pairs to find and interpret the average rate of change in per capita potato consumption per year to the nearest tenth during this period.

4. Write an equation in slope-intercept form that models per capita consumption of potatoes y (in pounds) in year x, where $x = 0$ represents 2003.

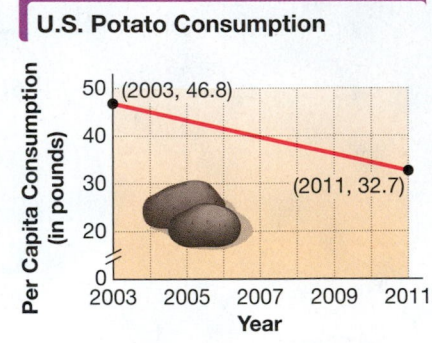

U.S. Potato Consumption

Source: U.S. Department of Agriculture.

Write an equation of a line (in the form specified, if given) that satisfies the given conditions.

5. Through $(0, 3)$ and $(-2, 4)$ (standard form)

6. Through $(2, -3)$; perpendicular to $x = 2$

Solve by any method.

7. $-7x + 3y = 12$
$5x + 2y = 8$

8. $2x - 5y = 8$
$3x + 4y = 10$

9. $x + 4y = 17$
$-3x + 2y = -9$

10. $x = 7y + 10$
$2x + 3y = 3$

11. $\dfrac{2}{3}x + \dfrac{1}{6}y = \dfrac{19}{2}$
$\dfrac{1}{3}x - \dfrac{2}{9}y = 2$

12. $2x + 5y - z = 12$
$-x + y - 4z = -10$
$-8x - 20y + 4z = 31$

13. *B*; The second equation is already solved for *y*.

14. 20 L

15. United States: 104;
China: 88;
Russia: 82

Solve each problem.

13. Which system would be easier to solve using the substitution method? Why?

$$\text{System A: } 5x - 3y = 7 \qquad \text{System B: } 7x + 2y = 4$$
$$2x + 8y = 3 \qquad\qquad y = -3x + 1$$

14. To make a 10% acid solution, Jeffrey wants to mix some 5% solution with 10 L of 20% solution. How many liters of 5% solution should he use?

15. In the 2012 Summer Olympics, China, the United States, and Russia won a combined total of 274 medals. China won 16 fewer medals than the United States, while Russia won 22 fewer medals than the United States. How many medals did each country win? (*Source: World Almanac and Book of Facts.*)

| Chapter 7 | Test | FOR EXTRA HELP | *Step-by-step test solutions are found on the Chapter Test Prep Videos available in* MyMathLab® *or on* YouTube. |

▶ *View the complete solutions to all Chapter Test exercises in MyMathLab.*

[7.1]

1. $\left(\dfrac{20}{3}, 0\right)$; $(0, -10)$

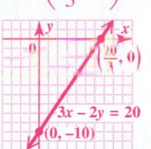

2. none; $(0, 5)$ **3.** $(2, 0)$; none

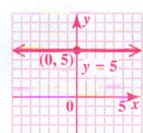

4. $\dfrac{1}{2}$ **5.** perpendicular

6. -838 farms per yr;
The number of farms decreased an average of 838 farms per year from 1980 to 2012.

[7.2]

7. (a) $y = -5x + 19$
(b) $5x + y = 19$

8. (a) $y = 14$ **(b)** $y = 14$

9. (a) not possible **(b)** $x = 5$

10. (a) $y = -\dfrac{3}{5}x - \dfrac{11}{5}$
(b) $3x + 5y = -11$

11. B

12. (a) $y = 142.75x + 45$
(b) $901.50

[7.3]

13. $\{(6, 1)\}$

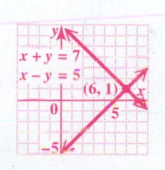

Find the x- and y-intercepts. Then graph each equation.

1. $3x - 2y = 20$ **2.** $y = 5$ **3.** $x = 2$

Solve each problem.

4. Find the slope of the line passing through the points $(6, 4)$ and $(-4, -1)$.

5. Determine whether the pair of lines $5x - y = 8$ and $5y = -x + 3$ is *parallel*, *perpendicular*, or *neither*.

6. In 1980, there were 119,000 farms in Iowa. As of 2012, there were 92,200. Find and interpret the average rate of change in the number of farms per year, to the nearest whole number. (*Source:* U.S. Department of Agriculture.)

Write an equation of the line that satisfies the given conditions. Give the equation (a) in slope-intercept form if possible and (b) in standard form.

7. Through $(4, -1)$; $m = -5$ **8.** Through $(-3, 14)$; horizontal

9. Through $(5, -6)$; vertical   **10.** Through $(-7, 2)$; parallel to $3x + 5y = 6$

Solve each problem.

11. Which line has positive slope and negative *y*-coordinate for its *y*-intercept?

A. 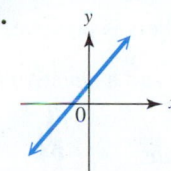 **B.** 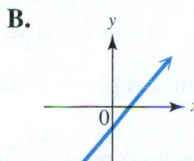 **C.** 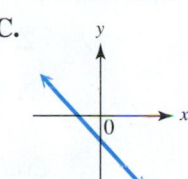 **D.**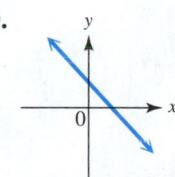

12. A ticket to a rock concert costs $142.75. An advance parking pass costs $45. Let *x* represent the number of tickets purchased and *y* represent the total cost in dollars.

(a) Write an equation in slope-intercept form that represents this situation.

(b) How much does it cost for 6 tickets and a parking pass?

13. Use a graph to solve the system. $\begin{aligned} x + y &= 7 \\ x - y &= 5 \end{aligned}$

Solve each system. If a system is inconsistent or has dependent equations, say so.

14. $\begin{aligned} 2x - 3y &= 24 \\ y &= -\dfrac{2}{3}x \end{aligned}$ **15.** $\begin{aligned} 3x - y &= -8 \\ 2x + 6y &= 3 \end{aligned}$ **16.** $\begin{aligned} 12x - 5y &= 8 \\ 3x &= \dfrac{5}{4}y + 2 \end{aligned}$

[7.4, 7.5]
14. $\{(6, -4)\}$

15. $\left\{\left(-\dfrac{9}{4}, \dfrac{5}{4}\right)\right\}$

16. $\{(x, y) \mid 12x - 5y = 8\}$; dependent equations

17. $\{(3, 3)\}$ **18.** $\{(0, -2)\}$

19. $\varnothing$; inconsistent system

[7.6]

20. $\left\{\left(-\dfrac{2}{3}, \dfrac{4}{5}, 0\right)\right\}$

21. $\{(3, -2, 1)\}$

[7.7]
22. *Iron Man 3:* \$409 million; *Man of Steel:* \$291 million

23. 45 mph, 75 mph

24. 20% solution: 4 L; 50% solution: 8 L

25. Orange Pekoe: 60 oz; Irish Breakfast: 30 oz; Earl Grey: 10 oz

17. $3x + y = 12$
$2x - y = 3$

18. $-5x + 2y = -4$
$6x + 3y = -6$

19. $3x + 4y = 8$
$8y = 7 - 6x$

20. $3x + 5y + 3z = 2$
$6x + 5y + z = 0$
$3x + 10y - 2z = 6$

21. $4x + y + z = 11$
$x - y - z = 4$
$y + 2z = 0$

Solve each problem.

22. Two top-grossing super hero films, *Iron Man 3* and *Man of Steel,* earned \$700 million together. If *Man of Steel* grossed \$118 million less than *Iron Man 3,* how much did each film gross? (*Source:* www.the-numbers.com)

23. Two cars start from points 420 mi apart and travel toward each other. They meet after 3.5 hr. Find the average rate of each car if one travels 30 mph slower than the other.

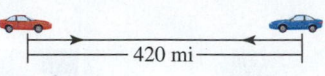
420 mi

24. A chemist needs 12 L of a 40% alcohol solution. She must mix a 20% solution and a 50% solution. How many liters of each will be required to obtain what she needs?

Liters of Solution	Percent (as a decimal)	Liters of Pure Alcohol

25. The owner of a tea shop wants to mix three kinds of tea to make 100 oz of a mixture that will sell for \$0.83 per oz. He uses Orange Pekoe, which sells for \$0.80 per oz, Irish Breakfast, for \$0.85 per oz, and Earl Grey, for \$0.95 per oz. If he wants to use twice as much Orange Pekoe as Irish Breakfast, how much of each kind of tea should he use?

Chapters R–7   Cumulative Review Exercises

[1.3, 1.4]
1. always true   **2.** never true
3. sometimes true;
For example, $3 + (-3) = 0$, but $3 + (-1) = 2$ and $2 \neq 0$.

[1.1, 4.1]
4. 81   **5.** -81
6. -81

[1.5]
7. -199 **8.** 455

[1.6]
9. commutative property

[2.3]
10. $\left\{-\dfrac{15}{4}\right\}$ **11.** $\{11\}$

[2.5]
12. $x = \dfrac{d - by}{a}$

[2.8]
13. $\left(-\infty, \dfrac{240}{13}\right]$

Decide whether each statement is always true, *sometimes true,* or never true. *If the statement is* sometimes true, *give examples for which it is true and for which it is false.*

1. The absolute value of a negative number equals the additive inverse of the number.

2. The sum of two negative numbers is positive.

3. The sum of a positive number and a negative number is 0.

Evaluate each expression if possible.

4. $(-3)^4$ **5.** -3^4 **6.** $-(-3)^4$

Evaluate for $x = -4$, $y = 3$, *and* $z = 6$.

7. $|2x| + 3y - z^3$ **8.** $-5(x^3 - y^3)$

9. Which property of real numbers justifies the statement $5 + (3 \cdot 6) = 5 + (6 \cdot 3)$?

Solve each equation or inequality.

10. $7(2x + 3) - 4(2x + 1) = 2(x + 1)$ **11.** $0.04x + 0.06(x - 1) = 1.04$

12. $ax + by = d$ for x **13.** $\dfrac{2}{3}x + \dfrac{5}{12}x \leq 20$

[R.2, 2.6]
14. 2010; 1813; 62.8%; 57.2%

[2.4, 2.7]
15. pennies: 35;
nickels: 29;
dimes: 30
16. 46°, 46°, 88°

[3.4, 7.1]
17. $y = 6$ **18.** $x = 4$

[3.3, 7.1]
19. $-\dfrac{4}{3}$ **20.** $\dfrac{3}{4}$

[3.5, 7.2]
21. $4x + 3y = 10$

[3.4, 7.1]
22.

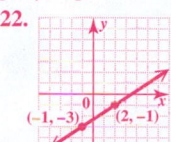

[4.1, 4.2]
23. $\dfrac{y}{18x}$

[4.6]
24. $49x^2 + 42xy + 9y^2$

[4.4]
25. $x^3 + 12x^2 - 3x - 7$

[4.7]
26. $m^2 - 2m + 3$

[5.1–5.4]
27. $(2w + 7z)(8w - 3z)$
28. $(10x^2 + 9)(10x^2 - 9)$
29. $(2p + 3)(4p^2 - 6p + 9)$

[5.5]
30. $\left\{\dfrac{1}{3}\right\}$

Solve each problem.

14. A survey measured public recognition of some popular contemporary advertising slogans. Complete the results shown in the table if 2500 people were surveyed.

Slogan (product or company)	Percent Recognition (nearest tenth of a percent)	Actual Number Who Recognized Slogan (nearest whole number)
Please Don't Squeeze the . . . (Charmin)	80.4%	
The Breakfast of Champions (Wheaties)	72.5%	
The King of Beers (Budweiser)		1570
Like a Good Neighbor (State Farm)		1430

Source: Department of Integrated Marketing Communications, Northwestern University.

15. A jar contains only pennies, nickels, and dimes. The number of dimes is one more than the number of nickels, and the number of pennies is six more than the number of nickels. How many of each denomination are in the jar, if the total value is $4.80?

16. Two angles of a triangle have the same measure. The measure of the third angle is 4° less than twice the measure of each of the equal angles. Find the measures of the three angles.

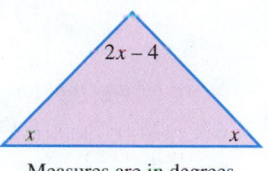

Measures are in degrees.

In Exercises 17–21, point A has coordinates $(-2, 6)$ and point B has coordinates $(4, -2)$.

17. What is the equation of the horizontal line through A?

18. What is the equation of the vertical line through B?

19. What is the slope of line AB?

20. What is the slope of a line perpendicular to line AB?

21. What is the standard form of the equation of line AB?

22. Graph the line having slope $\dfrac{2}{3}$ and passing through the point $(-1, -3)$.

Perform the indicated operations. In Exercise 23, assume that variables represent nonzero real numbers.

23. $(3x^2y^{-1})^{-2}(2x^{-3}y)^{-1}$ **24.** $(7x + 3y)^2$

25. $(3x^3 + 4x^2 - 7) - (2x^3 - 8x^2 + 3x)$ **26.** $\dfrac{m^3 - 3m^2 + 5m - 3}{m - 1}$

Factor.

27. $16w^2 + 50wz - 21z^2$ **28.** $100x^4 - 81$ **29.** $8p^3 + 27$

30. Solve $9x^2 = 6x - 1$.

Solve each problem.

31. A sign is to have the shape of a triangle with a height 3 ft greater than the length of the base. How long should the base be if the area is to be 14 ft²?

32. A game board has the shape of a rectangle. The longer sides are each 2 in. longer than the distance between them. The area of the board is 288 in.². Find the length of the longer sides and the distance between them.

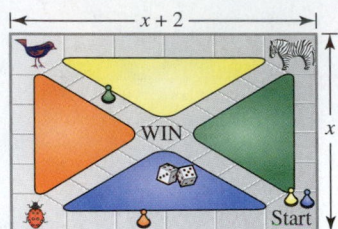

Perform each indicated operation. Write answers in lowest terms.

33. $\dfrac{8}{x + 1} - \dfrac{2}{x + 3}$

34. $\dfrac{x^2 + 5x + 6}{3x} \div \dfrac{x^2 - 4}{x^2 + x - 6}$

35. Simplify $\dfrac{\dfrac{12}{x + 6}}{\dfrac{4}{2x + 12}}$.

36. Solve $\dfrac{2}{x - 1} = \dfrac{5}{x - 1} - \dfrac{3}{4}$.

Solve by any method.

37. $-2x + 3y = -15$
$4x - y = 15$

38. $x - 3y = 7$
$2x - 6y = 14$

39. $x + y + z = 10$
$x - y - z = 0$
$-x + y - z = -4$

40. The graph shows a company's costs to produce computer parts and the revenue from the sale of computer parts.

(a) At what production level does the cost equal the revenue? What is the revenue at that point?

(b) Profit is revenue less cost. Estimate the profit on the sale of 1100 parts.

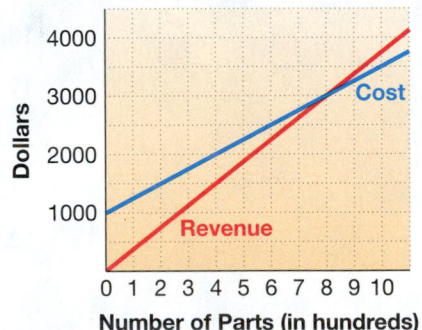

8

Inequalities and Absolute Value

In manufacturing and quality control situations, relative error, or tolerance, in measurements uses *inequality* and *absolute value*, two topics covered in this chapter.

8.1 Review of Linear Inequalities in One Variable

OBJECTIVES

1 Review inequalities and interval notation.

2 Solve linear inequalities using the addition property.

3 Solve linear inequalities using the multiplication property.

4 Solve linear inequalities with three parts.

VOCABULARY

☐ inequality
☐ interval
☐ linear inequality in one variable
☐ three-part inequality

OBJECTIVE 1 Review inequalities and interval notation.

An **inequality** consists of algebraic expressions related by one of the symbols $<$ ("*is less than*"), $\leq$ ("*is less than or equal to*"), $>$ ("*is greater than*"), or $\geq$ ("*is greater than or equal to*"). We **solve an inequality** by finding all real number solutions for it.

For example, the solution set of $x \leq 2$ includes *all* real numbers that are less than or equal to 2. This set of numbers is an example of an **interval** on a number line. We write this interval using **interval notation** as

$$(-\infty, 2]. \quad \text{Interval notation}$$

The negative infinity symbol $-\infty$ shows that the interval includes *all* real numbers less than 2. The square bracket indicates that 2 is included in the solution set, which is graphed in **FIGURE 1**.

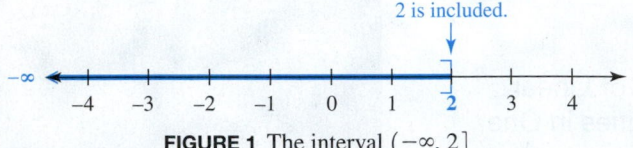

FIGURE 1 The interval $(-\infty, 2]$

Remember the following important concepts regarding interval notation.

- A parenthesis indicates that an endpoint is *not* included.

- A square bracket indicates that an endpoint is included.

- A parenthesis is always used next to an infinity symbol, $-\infty$ or ∞.

- The set of real numbers is written in interval notation as $(-\infty, \infty)$.

▼ Summary of Types of Intervals

Type of Interval	Set-Builder Notation	Interval Notation	Graph
Open interval	$\{x \mid a < x < b\}$	(a, b)	
Closed interval	$\{x \mid a \leq x \leq b\}$	$[a, b]$	
Half-open (or half-closed) interval	$\{x \mid a \leq x < b\}$	$[a, b)$	
	$\{x \mid a < x \leq b\}$	$(a, b]$	
Disjoint interval	$\{x \mid x < a \text{ or } x > b\}$	$(-\infty, a) \cup (b, \infty)$	
Infinite interval	$\{x \mid x > a\}$	(a, ∞)	
	$\{x \mid x \geq a\}$	$[a, \infty)$	
	$\{x \mid x < a\}$	$(-\infty, a)$	
	$\{x \mid x \leq a\}$	$(-\infty, a]$	
	$\{x \mid x \text{ is a real number}\}$	$(-\infty, \infty)$	

Linear Inequality in One Variable

A **linear inequality in one variable** (here x) can be written in the form

$$Ax + B < C, \quad Ax + B \leq C, \quad Ax + B > C, \quad \text{or} \quad Ax + B \geq C,$$

where A, B, and C are real numbers and $A \neq 0$.

Examples: $x + 5 < 2$, $x - 3 \geq 5$, and $2x + 5 \leq 10$ Linear inequalities

OBJECTIVE 2 Solve linear inequalities using the addition property.

We solve an inequality by finding all numbers that make the inequality true. Usually, an inequality has an infinite number of solutions.

We use two important properties to solve inequalities.

Addition Property of Inequality

If A, B, and C represent real numbers, then the inequalities

$$A < B \quad \text{and} \quad A + C < B + C \quad \text{are equivalent.*}$$

That is, the same number may be added to (or subtracted from) each side of an inequality without changing the solution set.

*This also applies to $A \leq B$, $A > B$, and $A \geq B$.

NOW TRY EXERCISE 1

Solve $x - 10 > -7$, and graph the solution set.

EXAMPLE 1 Using the Addition Property of Inequality

Solve $x - 7 < -12$, and graph the solution set.

$$x - 7 < -12$$
$$x - 7 + 7 < -12 + 7 \qquad \text{Add 7.}$$
$$x < -5 \qquad \text{Combine like terms.}$$

CHECK Substitute -5 for x in the *equation* $x - 7 = -12$.

$$x - 7 = -12$$
$$-5 - 7 \overset{?}{=} -12 \qquad \text{Let } x = -5.$$
$$-12 = -12 \checkmark \quad \text{True}$$

The result, a true statement, shows that -5 is the boundary point. Now test a number on each side of -5 to verify that numbers *less than* -5 make the inequality true. We choose -4 and -6.

$$x - 7 < -12$$

$-4 - 7 \overset{?}{<} -12$ Let $x = -4$.	$-6 - 7 \overset{?}{<} -12$ Let $x = -6$.
$-11 < -12$ False	$-13 < -12 \checkmark$ True
-4 *is not* in the solution set.	-6 *is* in the solution set.

The check confirms that the interval $(-\infty, -5)$ is the solution set. See **FIGURE 2**.

NOW TRY ANSWER
1. $(3, \infty)$

FIGURE 2

NOW TRY

OBJECTIVE 3 Solve linear inequalities using the multiplication property.

Consider the following true statement.

$$-2 < 5$$

Multiply each side by some positive number—for example, 8.

$$-2(8) < 5(8) \qquad \text{Multiply by 8.}$$

$$-16 < 40 \qquad \text{True}$$

The result is true. Start again with $-2 < 5$, and multiply each side by some negative number—for example, -8.

$$-2(-8) < 5(-8) \qquad \text{Multiply by } -8.$$

$$16 < -40 \qquad \text{False}$$

The result, $16 < -40$, is false. To make it true, we must change the direction of the inequality symbol.

$$16 > -40 \qquad \text{True}$$

As these examples suggest, multiplying each side of an inequality by a *negative* number requires reversing the direction of the inequality symbol. The same is true for dividing by a negative number because division is defined in terms of multiplication.

Multiplication Property of Inequality

Let A, B, and C represent real numbers, where $C \neq 0$.

(a) If C is *positive,* then the inequalities

$$A < B \quad \text{and} \quad AC < BC \quad \text{are equivalent.*}$$

(b) If C is *negative,* then the inequalities

$$A < B \quad \text{and} \quad AC > BC \quad \text{are equivalent.*}$$

That is, each side of an inequality may be multiplied (or divided) by the same *positive* number without changing the direction of the inequality symbol. *If the multiplier is negative, we must reverse the direction of the inequality symbol.*

*This also applies to $A \leq B$, $A > B$, and $A \geq B$.

EXAMPLE 2 Using the Multiplication Property of Inequality

Solve each inequality, and graph the solution set.

(a) $5x \leq -30$

Divide each side by 5. *Because $5 > 0$, do not reverse the direction of the inequality symbol.*

$$5x \leq -30$$

$$\frac{5x}{5} \leq \frac{-30}{5} \qquad \text{Divide by 5.}$$

$$x \leq -6$$

Check that the solution set is the interval $(-\infty, -6]$, graphed in **FIGURE 3.**

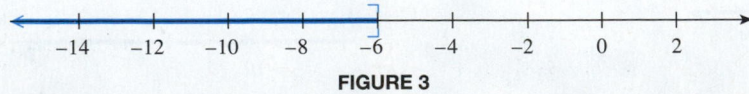

FIGURE 3

NOW TRY
EXERCISE 2

Solve each inequality, and graph the solution set.

(a) $8x \geq -40$

(b) $-20x > -60$

(b) $-4x \leq 32$

Divide each side by -4. *Because $-4 < 0$, reverse the direction of the inequality symbol.*

$$-4x \leq 32$$

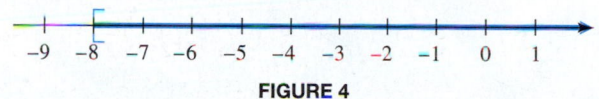

$$\frac{-4x}{-4} \geq \frac{32}{-4} \qquad \text{Divide by } -4.$$
Reverse the direction of the symbol.

Reverse the inequality symbol when dividing by a *negative* number.

$$x \geq -8$$

Check that the solution set is $[-8, \infty)$, graphed in **FIGURE 4**.

FIGURE 4   NOW TRY

To solve a linear inequality in one variable, use the following steps.

Solving a Linear Inequality in One Variable

Step 1 Simplify each side separately. Use the distributive property as needed.
- Clear any parentheses.
- Clear any fractions or decimals.
- Combine like terms.

Step 2 Isolate the variable terms on one side. Use the addition property of inequality so that all terms with variables are on one side of the inequality and all constants (numbers) are on the other side.

Step 3 Isolate the variable. Use the multiplication property of inequality to obtain an inequality in one of the following forms, where k is a constant (number).

$$\text{variable} < k, \quad \text{variable} \leq k, \quad \text{variable} > k, \quad \text{or} \quad \text{variable} \geq k$$

Remember: Reverse the direction of the inequality symbol only when multiplying or dividing each side of an inequality by a negative number.

EXAMPLE 3 Solving a Linear Inequality

Solve $-3(x + 4) + 2 \geq 7 - x$, and graph the solution set.

Step 1 $-3(x + 4) + 2 \geq 7 - x$

$-3x - 3(4) + 2 \geq 7 - x$ Distributive property

$-3x - 12 + 2 \geq 7 - x$ Multiply.

$-3x - 10 \geq 7 - x$ Combine like terms.

Step 2 $-3x - 10 + x \geq 7 - x + x$ Add x.

$-2x - 10 \geq 7$ Combine like terms.

NOW TRY ANSWERS

2. (a) $[-5, \infty)$

(b) $(-\infty, 3)$

NOW TRY
EXERCISE 3

Solve and graph the solution set.

$5 - 2(x - 4) \le 11 - 4x$

$$-2x - 10 + 10 \ge 7 + 10 \qquad \text{Add 10.}$$

$$-2x \ge 17 \qquad \text{Combine like terms.}$$

Step 3 $\qquad \dfrac{-2x}{-2} \le \dfrac{17}{-2} \qquad$ Divide by -2.
Change $\ge$ to $\le$.

Be sure to reverse the direction of the inequality symbol.

$$x \le -\dfrac{17}{2} \qquad \dfrac{a}{-b} = -\dfrac{a}{b}$$

FIGURE 5 shows the graph of the solution set, $\left(-\infty, -\dfrac{17}{2}\right]$.

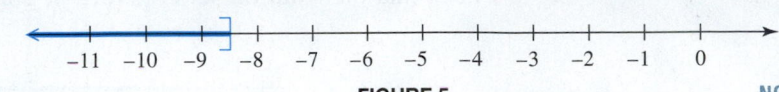

FIGURE 5

NOW TRY

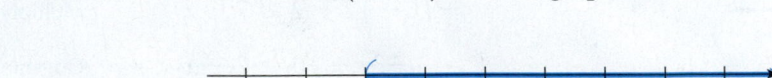

EXAMPLE 4 **Solving a Linear Inequality with Fractions**

NOW TRY
EXERCISE 4

Solve and graph the solution set.

$\dfrac{3}{4}(x - 2) + \dfrac{1}{2} > \dfrac{1}{5}(x - 8)$

Solve $-\dfrac{2}{3}(x - 3) - \dfrac{1}{2} < \dfrac{1}{2}(5 - x)$, and graph the solution set.

$$-\dfrac{2}{3}(x - 3) - \dfrac{1}{2} < \dfrac{1}{2}(5 - x)$$

Step 1 $\qquad -\dfrac{2}{3}x + 2 - \dfrac{1}{2} < \dfrac{5}{2} - \dfrac{1}{2}x \qquad$ Clear parentheses.

$$6\left(-\dfrac{2}{3}x + 2 - \dfrac{1}{2}\right) < 6\left(\dfrac{5}{2} - \dfrac{1}{2}x\right) \qquad \begin{array}{l}\text{To clear the fractions,}\\ \text{multiply by 6, the LCD.}\end{array}$$

$$6\left(-\dfrac{2}{3}x\right) + 6(2) + 6\left(-\dfrac{1}{2}\right) < 6\left(\dfrac{5}{2}\right) + 6\left(-\dfrac{1}{2}x\right) \qquad \text{Distributive property}$$

$$-4x + 12 - 3 < 15 - 3x \qquad \text{Multiply.}$$

$$-4x + 9 < 15 - 3x \qquad \text{Combine like terms.}$$

Step 2 $\qquad -4x + 9 + 3x < 15 - 3x + 3x \qquad$ Add $3x$.

$$-x + 9 < 15 \qquad \text{Combine like terms.}$$

$$-x + 9 - 9 < 15 - 9 \qquad \text{Subtract 9.}$$

$$-x < 6 \qquad \text{Combine like terms.}$$

Step 3 $\qquad -1(-x) > -1(6) \qquad \begin{array}{l}\text{Multiply by } -1.\\ \text{Change } < \text{ to } >.\end{array}$

Reverse the inequality symbol when multiplying by a *negative* number.

$$x > -6$$

Check that the solution set is $(-6, \infty)$. See the graph in **FIGURE 6**.

NOW TRY ANSWERS

3. $(-\infty, -1]$

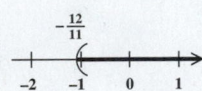

4. $\left(-\dfrac{12}{11}, \infty\right)$

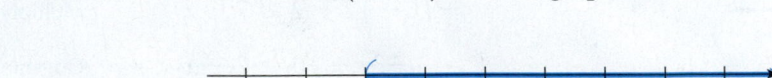

FIGURE 6

NOW TRY

OBJECTIVE 4 Solve linear inequalities with three parts.

Some applications involve a **three-part inequality** such as

$$3 < x + 2 < 8,$$

where $x + 2$ is *between* 3 and 8.

NOW TRY
EXERCISE 5
Solve and graph the solution set.

$$-1 < x - 2 < 3$$

EXAMPLE 5   Solving a Three-Part Inequality

Solve $3 < x + 2 < 8$, and graph the solution set.

$$3 < \quad x + 2 \quad < 8$$

$$3 - 2 < \ x + 2 - 2 < 8 - 2 \quad \text{Subtract 2 from all three parts.}$$

$$1 < \quad x \quad < 6$$

Thus, x must be between 1 and 6 so that $x + 2$ will be between 3 and 8. The solution set, $(1, 6)$, is graphed in **FIGURE 7**.

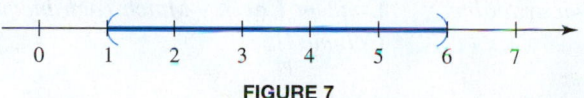

FIGURE 7   NOW TRY

⚠ **CAUTION**   *In three-part inequalities, the order of the parts is important.* For example, do not write $8 < x + 2 < 3$ because this would imply that $8 < 3$, a false statement. *Write three-part inequalities so that the symbols point in the same direction, and both point toward the lesser number.*

NOW TRY
EXERCISE 6
Solve and graph the solution set.

$$-2 < -4x - 5 \le 7$$

EXAMPLE 6   Solving a Three-Part Inequality

Solve $-2 \le -3x - 1 \le 5$ and graph the solution set.

$$-2 \le \quad -3x - 1 \quad \le 5$$

$$-2 + 1 \le \ -3x - 1 + 1 \le 5 + 1 \quad \text{Add 1 to each part.}$$

$$-1 \le \quad -3x \quad \le 6$$

$$\frac{-1}{-3} \ge \frac{-3x}{-3} \ge \frac{6}{-3} \quad \begin{array}{l}\text{Divide each part by } -3.\\ \text{Reverse the direction of the}\\ \text{inequality symbols.}\end{array}$$

$$\frac{1}{3} \ge \quad x \quad \ge -2$$

$$-2 \le \quad x \quad \le \frac{1}{3} \quad \begin{array}{l}\text{Rewrite in the}\\ \text{order on the}\\ \text{number line.}\end{array}$$

NOW TRY ANSWERS
5. $(1, 5)$

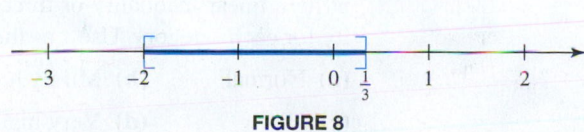

6. $\left[-3, -\frac{3}{4}\right)$

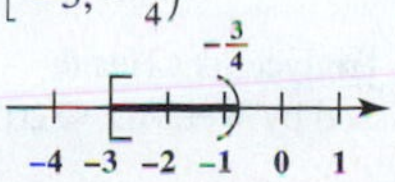

Check that the solution set is $\left[-2, \frac{1}{3}\right]$, as shown in **FIGURE 8**.

FIGURE 8   NOW TRY

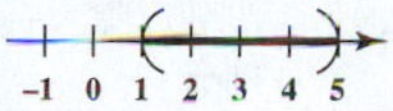

▼ **Summary of Solution Sets of Equations and Inequalities**

Equation or Inequality	Typical Solution Set	Graph of Solution Set
Linear equation $5x + 4 = 14$	$\{2\}$	
Linear inequality $5x + 4 < 14$	$(-\infty, 2)$	
$5x + 4 > 14$	$(2, \infty)$	
Three-part inequality $-1 \le 5x + 4 \le 14$	$[-1, 2]$	

8.1 Exercises

FOR EXTRA HELP

▶ MyMathLab®

▶ *Complete solution available in MyMathLab*

Concept Check   *Match each inequality in Column I with the correct graph or interval in Column II.*

I

1. $x \le 3$

2. $x > 3$

3. $x < 3$

4. $x \ge 3$

5. $-3 \le x \le 3$

6. $-3 < x < 3$

II

A.

B.

C. $(3, \infty)$

D. $(-\infty, 3]$

E. $(-3, 3)$

F. $[-3, 3]$

Concept Check   *Work each problem involving inequalities.*

7. A high level of LDL cholesterol ("bad cholesterol") in the blood increases a person's risk of heart disease. The table shows how LDL levels affect risk.

   If x represents the LDL cholesterol number, write a linear inequality or three-part inequality for each category. Use x as the variable.

   (a) Optimal

   (b) Near optimal/above optimal

   (c) Borderline high

   (d) High   **(e)** Very high

LDL Cholesterol	Risk Category
Less than 100	Optimal
100–129	Near optimal/ above optimal
130–159	Borderline high
160–189	High
190 and above	Very high

Source: Cholesterol & Triglycerides Health Center, WebMD, published by WebMD, © 2014.

8. A high level of triglycerides in the blood also increases a person's risk of heart disease. The table shows how triglyceride levels affect risk.

   If x represents the triglycerides number, write a linear inequality or three-part inequality for each category. Use x as the variable.

   (a) Normal   **(b)** Mildly high

   (c) High   **(d)** Very high

Triglycerides	Risk Category
Less than 100	Normal
100–199	Mildly high
200–499	High
500 or higher	Very high

Source: Cholesterol & Triglycerides Health Center, WebMD, published by WebMD, © 2014.

*Solve each inequality. Give the solution set in both interval and graph forms. **See Examples 1–4.***

▶ 9. $x - 4 \geq 12$

10. $x - 3 \geq 7$

11. $3k + 1 > 22$

12. $5x + 6 < 76$

▶ 13. $4x < -16$

14. $2x > -10$

15. $-\dfrac{3}{4}x \geq 30$

16. $-\dfrac{2}{3}x \leq 12$

17. $-1.3x \geq -5.2$

18. $-2.5x \leq -1.25$

19. $5x + 2 \leq -48$

20. $4x + 1 \leq -31$

21. $\dfrac{5x - 6}{8} < 8$

22. $\dfrac{3x - 1}{4} > 5$

23. $\dfrac{2x - 5}{-4} > 5$

24. $\dfrac{3x - 2}{-5} < 6$

▶ 25. $6x - 4 \geq -2x$

26. $2x - 8 \geq -2x$

27. $x - 2(x - 4) \leq 3x$

28. $x - 3(x + 1) \leq 4x$

▶ 29. $-(4 + r) + 2 - 3r < -14$

30. $-(9 + x) - 5 + 4x \geq 4$

31. $-3(x - 6) > 2x - 2$

32. $-2(x + 4) \leq 6x + 16$

33. $\dfrac{2}{3}(3x - 1) \geq \dfrac{3}{2}(2x - 3)$

34. $\dfrac{7}{5}(10x - 1) < \dfrac{2}{3}(6x + 5)$

▶ 35. $-\dfrac{1}{4}(p + 6) + \dfrac{3}{2}(2p - 5) < 10$

36. $\dfrac{3}{5}(t - 2) - \dfrac{1}{4}(2t - 7) \leq 3$

37. $3(2x - 4) - 4x < 2x + 3$

38. $7(4 - x) + 5x < 2(16 - x)$

39. $8\left(\dfrac{1}{2}x + 3\right) < 8\left(\dfrac{1}{2}x - 1\right)$

40. $10\left(\dfrac{1}{5}x + 2\right) < 10\left(\dfrac{1}{5}x + 1\right)$

41. *Concept Check*   Which one is the graph of $-2 < x$?   Of $-x > 2$?

A. 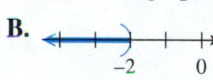   **B.** 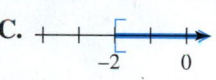   **C.**    **D.**

42. *Concept Check*   A student solved the following inequality as shown.

$$4x \geq -64$$

$$\dfrac{4x}{4} \leq \dfrac{-64}{4}$$

$$x \leq -16 \qquad \text{Solution set: } (-\infty, -16]$$

WHAT WENT WRONG?   Give the correct solution set.

*Solve each inequality. Give the solution set in both interval and graph forms. **See Examples 5 and 6.***

43. $-4 < x - 5 < 6$

44. $-1 < x + 1 < 8$

45. $-9 \leq x + 5 \leq 15$

46. $-4 \leq x + 3 \leq 10$

47. $-6 \leq 2x + 4 \leq 16$

48. $-15 < 3x + 6 < -12$

49. $-19 \leq 3x - 5 \leq 1$

50. $-16 < 3x + 2 < -10$

▶ 51. $4 \leq -9x + 5 < 8$

52. $4 \leq -2x + 3 < 8$

53. $-8 \leq -4x + 2 \leq 6$

54. $-12 \leq -6x + 3 \leq 15$

55. $-1 \leq \dfrac{2x - 5}{6} \leq 5$

56. $-3 \leq \dfrac{3x + 1}{4} \leq 3$

8.2 Set Operations and Compound Inequalities

VOCABULARY

☐ intersection
☐ compound inequality
☐ union

OBJECTIVE 1 Recognize set intersection and union.

Consider the two sets A and B defined as follows.

$$A = \{1, 2, 3\}, \qquad B = \{2, 3, 4\}$$

The set of all elements that belong to both A **and** B, called their *intersection* and symbolized $A \cap B$, is given by

$$A \cap B = \{2, 3\}. \qquad \text{Intersection}$$

The set of all elements that belong to either A **or** B, or both, called their *union* and symbolized $A \cup B$, is given by

$$A \cup B = \{1, 2, 3, 4\}. \qquad \text{Union}$$

OBJECTIVE 2 Find the intersection of two sets.

The intersection of two sets is defined with the word *and*.

Intersection of Sets

For any two sets A and B, the **intersection** of A and B, symbolized $A \cap B$, is defined as follows.

$$A \cap B = \{x \mid x \text{ is an element of } A \textbf{ and } x \text{ is an element of } B\}$$

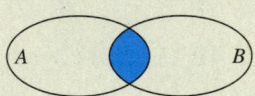

NOW TRY EXERCISE 1

Let $A = \{2, 4, 6, 8\}$ and $B = \{0, 2, 6, 8\}$.
Find $A \cap B$.

EXAMPLE 1 Finding the Intersection of Two Sets

Let $A = \{1, 2, 3, 4\}$ and $B = \{2, 4, 6\}$. Find $A \cap B$.

The set $A \cap B$ contains those elements that belong to both A *and* B.

$$A \cap B = \{1, 2, 3, 4\} \cap \{2, 4, 6\}$$
$$= \{2, 4\}$$

NOW TRY

OBJECTIVE 3 Solve compound inequalities with the word *and*.

A **compound inequality** consists of two inequalities linked by a connective word.

$$x + 1 \leq 9 \quad \text{and} \quad x - 2 \geq 3$$
$$2x > 4 \quad \text{or} \quad 3x - 6 < 5$$

Compound inequalities

Solving a Compound Inequality with *and*

Step 1 Solve each inequality individually.

Step 2 Because the inequalities are joined with *and*, the solution set of the compound inequality will include all numbers that satisfy both inequalities in Step 1 (the *intersection* of the solution sets).

NOW TRY ANSWER
1. $\{2, 6, 8\}$

**NOW TRY
EXERCISE 2**

Solve the compound inequality, and graph the solution set.

$$x - 2 \le 5 \quad \text{and} \quad x + 5 \ge 9$$

EXAMPLE 2 Solving a Compound Inequality with *and*

Solve the compound inequality, and graph the solution set.

$$x + 1 \le 9 \quad \text{and} \quad x - 2 \ge 3$$

Step 1 Solve each inequality individually.

$$x + 1 \le 9 \qquad \text{and} \qquad x - 2 \ge 3$$
$$x + 1 - 1 \le 9 - 1 \quad \text{and} \quad x - 2 + 2 \ge 3 + 2$$
$$x \le 8 \qquad \text{and} \qquad x \ge 5$$

Step 2 The solution set will include all numbers that satisfy *both* inequalities in Step 1 at the same time. The compound inequality is true whenever $x \le 8$ and $x \ge 5$ are both true. See the graphs in **FIGURE 9**.

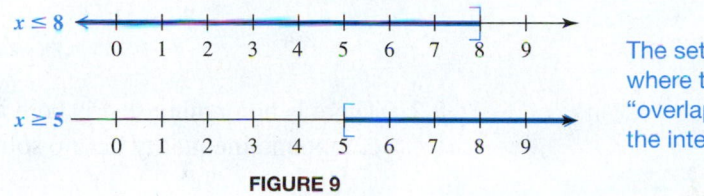

The set of points where the graphs "overlap" represents the intersection.

FIGURE 9

The intersection of the two graphs in **FIGURE 9** is the solution set. **FIGURE 10** shows this solution set, $[5, 8]$.

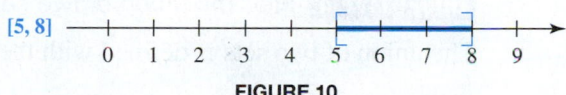

FIGURE 10

NOW TRY

**NOW TRY
EXERCISE 3**

Solve and graph.

$$-4x - 1 < 7 \quad \text{and}$$
$$3x + 4 \ge -5$$

EXAMPLE 3 Solving a Compound Inequality with *and*

Solve the compound inequality, and graph the solution set.

$$-3x - 2 > 5 \quad \text{and} \quad 5x - 1 \le -21$$

Step 1 Solve each inequality individually.

$$-3x - 2 > 5 \qquad \text{and} \qquad 5x - 1 \le -21$$

Remember to reverse the direction of the inequality symbol.

$$-3x > 7 \qquad \text{and} \qquad 5x \le -20$$
$$x < -\frac{7}{3} \qquad \text{and} \qquad x \le -4$$

The graphs of $x < -\frac{7}{3}$ and $x \le -4$ are shown in **FIGURE 11**.

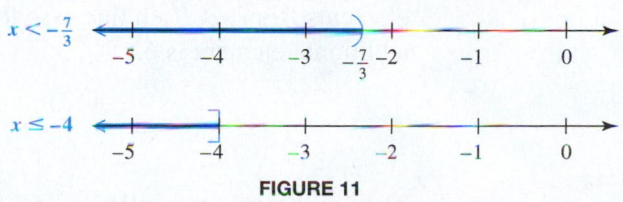

FIGURE 11

NOW TRY ANSWERS

2. $[4, 7]$

3. $(-2, \infty)$

Step 2 Now find all values of x that are less than $-\frac{7}{3}$ and also less than or equal to -4. As shown in **FIGURE 12**, the solution set is $(-\infty, -4]$.

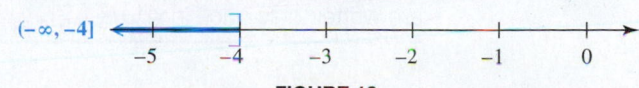

FIGURE 12

NOW TRY

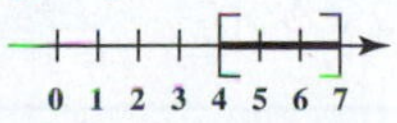

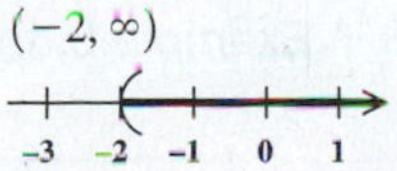

NOW TRY
EXERCISE 4

Solve and graph.

$$x - 7 < -12 \quad \text{and}$$
$$2x + 1 > 5$$

EXAMPLE 4 Solving a Compound Inequality with *and*

Solve the compound inequality, and graph the solution set.

$$x + 2 < 5 \quad \text{and} \quad x - 10 > 2$$

Step 1 Solve each inequality individually.

$$x + 2 < 5 \quad \text{and} \quad x - 10 > 2$$
$$x < 3 \quad \text{and} \quad x > 12$$

The graphs of $x < 3$ and $x > 12$ are shown in **FIGURE 13**.

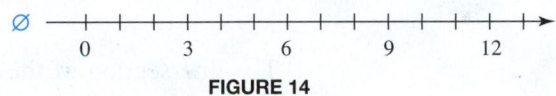

$x < 3$

$x > 12$

FIGURE 13

Step 2 There is no number that is both less than 3 *and* greater than 12, so the given compound inequality has no solution. The solution set is $\varnothing$. See **FIGURE 14**.

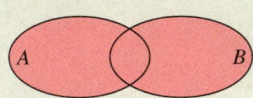

$\varnothing$

FIGURE 14

 NOW TRY

OBJECTIVE 4 Find the union of two sets.

The union of two sets is defined with the word *or*.

Union of Sets

For any two sets A and B, the **union** of A and B, symbolized $A \cup B$, is defined as follows.

$$A \cup B = \{x \mid x \text{ is an element of } A \text{ } \mathbf{or} \text{ } x \text{ is an element of } B\}$$

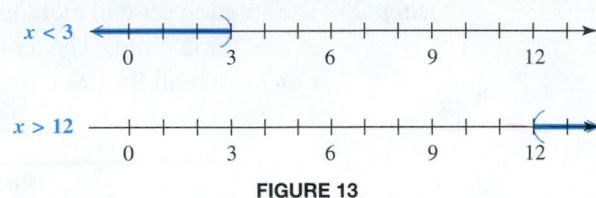

NOW TRY
EXERCISE 5

Let $A = \{5, 10, 15, 20\}$ and $B = \{5, 15, 25\}$.
Find $A \cup B$.

EXAMPLE 5 Finding the Union of Two Sets

Let $A = \{1, 2, 3, 4\}$ and $B = \{2, 4, 6\}$. Find $A \cup B$.

Begin by listing all the elements of set A: 1, 2, 3, 4. Then list any additional elements from set B. In this case the elements 2 and 4 are already listed, so the only additional element is 6.

$$A \cup B = \{1, 2, 3, 4\} \cup \{2, 4, 6\}$$
$$= \{1, 2, 3, 4, 6\}$$

The union consists of all elements in either A or B (or both).

 NOW TRY

NOW TRY ANSWERS
4. $\varnothing$
5. $\{5, 10, 15, 20, 25\}$

NOTE Although the elements 2 and 4 appeared in both sets A and B in **Example 5,** they are written only once in $A \cup B$.

OBJECTIVE 5 Solve compound inequalities with the word *or*.

> **Solving a Compound Inequality with *or***
>
> **Step 1** Solve each inequality individually.
>
> **Step 2** Because the inequalities are joined with *or*, the solution set of the compound inequality includes all numbers that satisfy either one of the two inequalities in Step 1 (the *union* of the solution sets).

NOW TRY EXERCISE 6

Solve and graph.

$-12x \le -24$ or $x + 9 < 8$

EXAMPLE 6 Solving a Compound Inequality with *or*

Solve the compound inequality, and graph the solution set.

$$6x - 4 < 2x \quad \text{or} \quad -3x \le -9$$

Step 1 Solve each inequality individually.

$$6x - 4 < 2x \quad \text{or} \quad -3x \le -9$$

$$4x < 4$$

Remember to reverse the inequality symbol.

$$x < 1 \quad \text{or} \quad x \ge 3$$

The graphs of these two inequalities are shown in **FIGURE 15**.

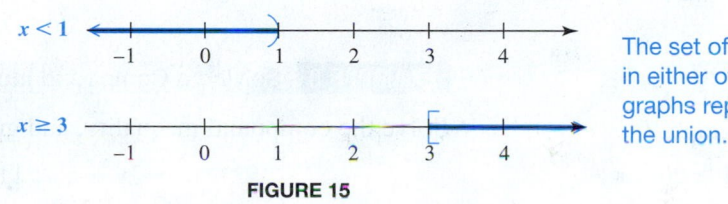

FIGURE 15

The set of points in either of the graphs represents the union.

Step 2 Because the inequalities are joined with *or*, find the union of the two solution sets. The union is the disjoint interval in **FIGURE 16**.

$$(-\infty, 1) \cup [3, \infty)$$

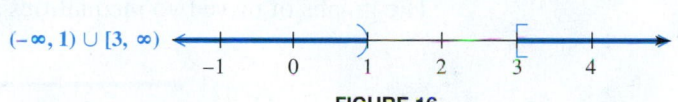

FIGURE 16

Always pay particular attention to the end points of the solution sets and whether parentheses, brackets, or one of each should be used. **NOW TRY**

⚠️ **CAUTION** When inequalities are used to write the solution set in **Example 6**, it *must* be written using two separate inequalities.

$$x < 1 \quad \text{or} \quad x \ge 3,$$

Writing $3 \le x < 1$, which translates using *and*, would imply that

$$3 \le 1, \quad \text{which is } \textit{FALSE}.$$

NOW TRY ANSWER

6. $(-\infty, -1) \cup [2, \infty)$

NOW TRY EXERCISE 7

Solve and graph.

$-x + 2 < 6$ or $6x - 8 \geq 10$

EXAMPLE 7 Solving a Compound Inequality with *or*

Solve the compound inequality, and graph the solution set.

$$-4x + 1 \geq 9 \quad \text{or} \quad 5x + 3 \leq -12$$

Step 1 Solve each inequality individually.

$$-4x + 1 \geq 9 \qquad \text{or} \qquad 5x + 3 \leq -12$$
$$-4x \geq 8 \qquad \text{or} \qquad 5x \leq -15$$
$$x \leq -2 \qquad \text{or} \qquad x \leq -3$$

The graphs of these two inequalities are shown in **FIGURE 17**.

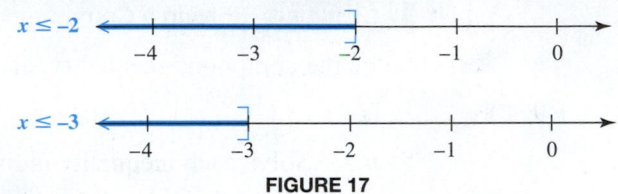

FIGURE 17

Step 2 We take the union to obtain $(-\infty, -2]$. See **FIGURE 18**.

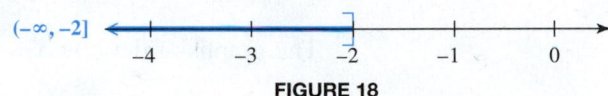

FIGURE 18

NOW TRY

NOW TRY EXERCISE 8

Solve and graph.

$8x - 4 \geq 20$ or

$\quad -2x + 1 > -9$

EXAMPLE 8 Solving a Compound Inequality with *or*

Solve the compound inequality, and graph the solution set.

$$-2x + 5 \geq 11 \quad \text{or} \quad 4x - 7 \geq -27$$

Step 1 Solve each inequality individually.

$$-2x + 5 \geq 11 \quad \text{or} \quad 4x - 7 \geq -27$$
$$-2x \geq 6 \quad \text{or} \quad 4x \geq -20$$
$$x \leq -3 \quad \text{or} \quad x \geq -5$$

The graphs of these two inequalities are shown in **FIGURE 19**.

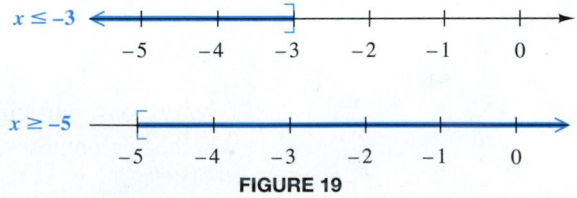

FIGURE 19

Step 2 By taking the union, we obtain every real number as a solution because every real number satisfies at least one of the two inequalities. The set of all real numbers is written in interval notation as $(-\infty, \infty)$ and graphed as in **FIGURE 20**.

NOW TRY ANSWERS

7. $(-4, \infty)$

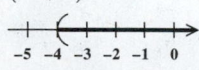

8. $(-\infty, \infty)$

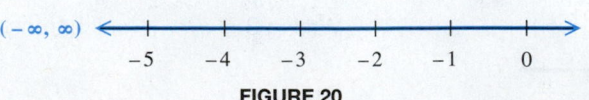

FIGURE 20

NOW TRY

**NOW TRY
EXERCISE 9**

In **Example 9**, list the elements of each set.

(a) The set of countries to which exports were greater than $100,000 million and from which imports were less than $400,000 million

(b) The set of countries to which exports were less than $200,000 million or from which imports were greater than $250,000 million

EXAMPLE 9 Applying Intersection and Union

The five top U.S. trading partners for 2012 are listed in the table. Amounts are in millions of dollars. (*Source:* U.S. Census Bureau.)

Country	U.S. Exports to Country	U.S. Imports from Country
Canada	292,540	323,937
China	110,484	425,579
Mexico	215,931	277,570
Japan	69,955	146,392
Germany	48,797	108,708

List the elements of the following sets.

(a) The set of countries to which exports were greater than $200,000 million *and* from which imports were less than $300,000 million

The only country that satisfies both conditions is Mexico, so the set is

$$\{\text{Mexico}\}.$$

(b) The set of countries to which exports were less than $100,000 million *or* from which imports were greater than $200,000 million

Here, any country that satisfies at least one of the conditions is in the set. This set includes all five countries:

$$\{\text{Canada, China, Mexico, Japan, Germany}\}. \qquad \textbf{NOW TRY} \; \text{}$$

NOW TRY ANSWERS
9. (a) {Canada, Mexico}
  (b) {Canada, China, Mexico, Japan, Germany}

8.2 Exercises

FOR EXTRA HELP ▶ MyMathLab®

▶ *Complete solution available in MyMathLab*

Concept Check *Decide whether each statement is* true *or* false. *If it is false, explain why.*

1. The union of the solution sets of $x + 1 = 6$, $x + 1 < 6$, and $x + 1 > 6$ is $(-\infty, \infty)$.

2. The intersection of the sets $\{x \mid x \geq 9\}$ and $\{x \mid x \leq 9\}$ is $\varnothing$.

3. The union of the sets $(-\infty, 7)$ and $(7, \infty)$ is $\{7\}$.

4. The intersection of the sets $(-\infty, 7]$ and $[7, \infty)$ is $\{7\}$.

5. The intersection of the set of rational numbers and the set of irrational numbers is $\{0\}$.

6. The union of the set of rational numbers and the set of irrational numbers is the set of real numbers.

Let $A = \{1, 2, 3, 4, 5, 6\}$, $B = \{1, 3, 5\}$, $C = \{1, 6\}$, *and* $D = \{4\}$. *Find each set.* **See Examples 1 and 5.**

▶ **7.** $B \cap A$ **8.** $A \cap B$ **9.** $A \cap D$ **10.** $B \cap C$

11. $B \cap \varnothing$ **12.** $A \cap \varnothing$ ▶ **13.** $A \cup B$ **14.** $B \cup D$

Concept Check *Two sets are specified by graphs. Graph the intersection of the two sets.*

15.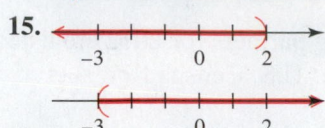

16.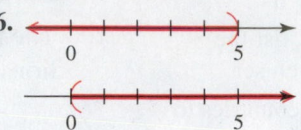

17.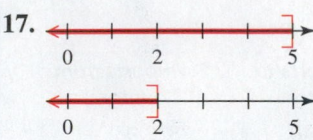

18.

Solve each compound inequality. Give the solution set in both interval and graph forms. **See Examples 2–4.**

19. $x < 2$ and $x > -3$

20. $x < 5$ and $x > 0$

21. $x \le 2$ and $x \le 5$

22. $x \ge 3$ and $x \ge 6$

▶ **23.** $x \le 3$ and $x \ge 6$

24. $x \le -1$ and $x \ge 3$

▶ **25.** $x - 3 \le 6$ and $x + 2 \ge 7$

26. $x + 5 \le 11$ and $x - 3 \ge -1$

27. $-3x > 3$ and $x + 3 > 0$

28. $-3x < 3$ and $x + 2 < 6$

▶ **29.** $3x - 4 \le 8$ and $-4x + 1 \ge -15$

30. $7x + 6 \le 48$ and $-4x \ge -24$

Concept Check *Two sets are specified by graphs. Graph the union of the two sets.*

31.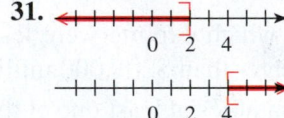

32.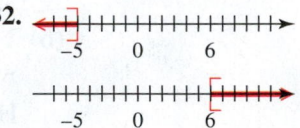

33.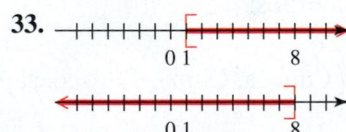

34.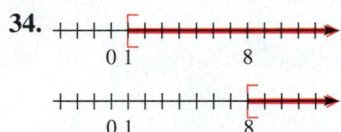

Solve each compound inequality. Give the solution set in both interval and graph forms. **See Examples 6–8.**

▶ **35.** $x \le 1$ or $x \le 8$

36. $x \ge 1$ or $x \ge 8$

37. $x \ge -2$ or $x \ge 5$

38. $x \le -2$ or $x \le 6$

39. $x \ge -2$ or $x \le 4$

40. $x \ge 5$ or $x \le 7$

▶ **41.** $x + 2 > 7$ or $1 - x > 6$

42. $x + 1 > 3$ or $x + 4 < 2$

43. $x + 1 > 3$ or $-4x + 1 > 5$

44. $3x < x + 12$ or $x + 1 > 10$

▶ **45.** $4x + 1 \ge -7$ or $-2x + 3 \ge 5$

46. $3x + 2 \le -7$ or $-2x + 1 \le 9$

Concept Check *Express each set in simplest interval form. (Hint: Graph each set and look for the intersection or union.)*

47. $(-\infty, -1] \cap [-4, \infty)$

48. $[-1, \infty) \cap (-\infty, 9]$

49. $(-\infty, -6] \cap [-9, \infty)$

50. $(5, 11] \cap [6, \infty)$

51. $(-\infty, 3) \cup (-\infty, -2)$

52. $[-9, 1] \cup (-\infty, -3)$

53. $[3, 6] \cup (4, 9)$

54. $[-1, 2] \cup (0, 5)$

Solve each compound inequality. Give the solution set in both interval and graph forms. **See Examples 2–4 and 6–8.**

55. $x < -1$ and $x > -5$

56. $x > -1$ and $x < 7$

57. $x < 4$ or $x < -2$

58. $x < 5$ or $x < -3$

59. $-3x \le -6$ or $-3x \ge 0$

60. $2x - 6 \le -18$ and $2x \ge -18$

61. $x + 1 \ge 5$ and $x - 2 \le 10$

62. $-8x \le -24$ or $-5x \ge 15$

Average expenses for full-time resident college students at 4-year institutions during the 2011–2012 academic year are shown in the table.

▼ **College Expenses (in Dollars)**

Type of Expense	Public Schools (in-state)	Private Schools
Tuition and fees	7701	23,479
Board rates	4052	4591
Dormitory charges	5036	5646

Source: National Center for Education Statistics.

Refer to the table, and list the elements of each set. **See Example 9.**

63. The set of expenses that are less than $8000 for public schools *and* are greater than $15,000 for private schools

64. The set of expenses that are greater than $4000 for public schools *and* are less than $5000 for private schools

65. The set of expenses that are less than $8000 for public schools *or* are greater than $15,000 for private schools

66. The set of expenses that are greater than $15,000 *or* are between $7000 and $8000

RELATING CONCEPTS For Individual or Group Work (Exercises 67–72)

The figures represent the backyards of neighbors Luigi, Maria, Than, and Joe. Find the area and the perimeter of each yard. Suppose that each resident has 150 ft of fencing and enough sod to cover 1400 ft^2 of lawn. Give the name or names of the residents whose yards satisfy each description. **Work Exercises 67–72 in order.**

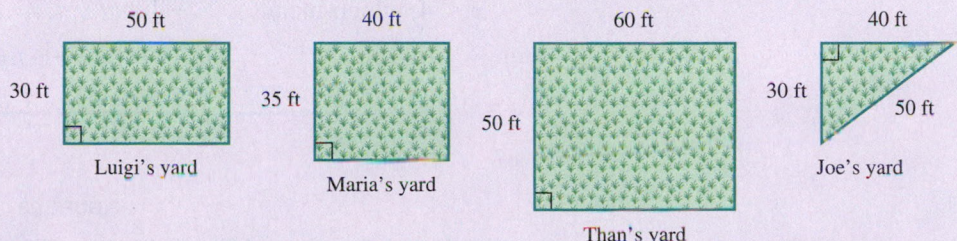

67. The yard can be fenced *and* the yard can be sodded.

68. The yard can be fenced *and* the yard cannot be sodded.

69. The yard cannot be fenced *and* the yard can be sodded.

70. The yard cannot be fenced *and* the yard cannot be sodded.

71. The yard can be fenced *or* the yard can be sodded.

72. The yard cannot be fenced *or* the yard can be sodded.

8.3 Absolute Value Equations and Inequalities

VOCABULARY

☐ absolute value equation
☐ absolute value inequality

Suppose the government of a country decides that it will comply with a restriction on greenhouse gas emissions *within* 3 years of 2020. This means that the *difference* between the year it will comply and 2020 is less than 3, *without regard to sign*. We state this mathematically as follows, where x represents the year in which it complies.

$$|x - 2020| < 3 \qquad \text{Absolute value inequality}$$

We can intuitively reason that the year must be between 2017 and 2023, and thus $2017 < x < 2023$ makes this inequality true.

OBJECTIVE 1 Use the distance definition of absolute value.

In **Section 1.3,** we saw that the absolute value of a number x, written $|x|$, represents the undirected distance from x to 0 on a number line. For example, the solutions of $|x| = 4$ are 4 and -4, as shown in **FIGURE 21.**

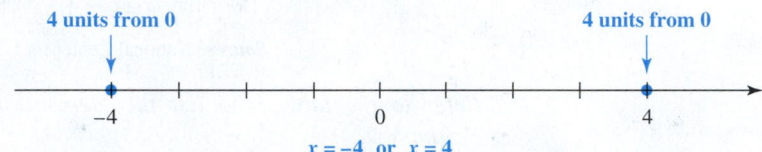

FIGURE 21

Because absolute value represents distance from 0, we interpret the solutions of $|x| > 4$ to be all numbers that are *more* than four units from 0 on a number line. The set $(-\infty, -4) \cup (4, \infty)$ fits this description. **FIGURE 22** shows the graph of the solution set of $|x| > 4$. The graph consists of two separate intervals, which means $x < -4$ *or* $x > 4$.

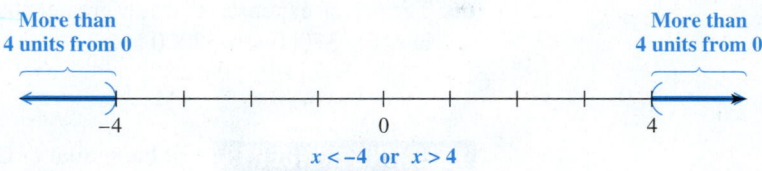

FIGURE 22

The solution set of $|x| < 4$ consists of all numbers that are *less* than 4 units from 0 on a number line. This is represented by all numbers *between* -4 and 4. This set of numbers is given by $(-4, 4)$, as shown in **FIGURE 23.** Here, the graph shows that $-4 < x < 4$, which means $x > -4$ *and* $x < 4$.

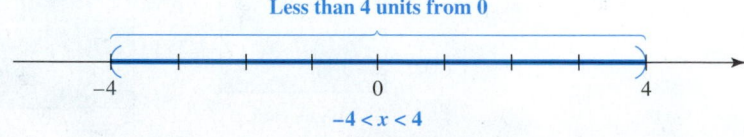

FIGURE 23

Absolute value equations and inequalities generally take the form

$$|ax + b| = k, \qquad |ax + b| > k, \qquad \text{or} \qquad |ax + b| < k,$$

where k is a positive number. From **FIGURES 21–23,** we see that

$|x| = 4$ has the same solution set as $x = -4$ **or** $x = 4$,

$|x| > 4$ has the same solution set as $x < -4$ **or** $x > 4$,

$|x| < 4$ has the same solution set as $x > -4$ **and** $x < 4$.

> This is equivalent to $-4 < x < 4$.

Solving Absolute Value Equations and Inequalities

Let k be a positive real number, and p and q be real numbers.

Case 1 To solve $|ax + b| = k$, solve the following compound equation.

$$ax + b = k \quad \text{or} \quad ax + b = -k$$

The solution set is usually of the form $\{p, q\}$, which includes two numbers.

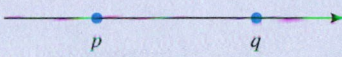

Case 2 To solve $|ax + b| > k$*, solve the following compound inequality.

$$ax + b > k \quad \text{or} \quad ax + b < -k$$

The solution set is of the form $(-\infty, p) \cup (q, \infty)$, which is a disjoint interval.

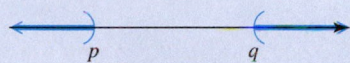

Case 3 To solve $|ax + b| < k$**, solve the following three-part inequality.

$$-k < ax + b < k$$

The solution set is of the form (p, q), which is a single interval.

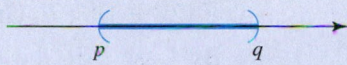

*This also applies to $|ax + b| \geq k$. The solution set *includes* the endpoints, using brackets rather than parentheses.

**This also applies to $|ax + b| \leq k$. The solution set *includes* the endpoints, using brackets rather than parentheses.

NOTE It is acceptable to write the compound statements in Cases 1 and 2 of the preceding box as follows. These forms produce the same results.

$ax + b = k \quad$ or $\quad -(ax + b) = k \qquad$ Alternative for Case 1

$ax + b > k \quad$ or $\quad -(ax + b) > k \qquad$ Alternative for Case 2

OBJECTIVE 2 Solve equations of the form $|ax + b| = k$, for $k > 0$.

Remember that because absolute value refers to distance from the origin, an absolute value equation will have two parts.

EXAMPLE 1 Solving an Absolute Value Equation (Case 1)

Solve $|2x + 1| = 7$. Graph the solution set.

For $|2x + 1|$ to equal 7, $2x + 1$ must be 7 units from 0 on a number line. This can happen only when

$$2x + 1 = 7 \quad \text{or} \quad 2x + 1 = -7.$$

This is Case 1 in the box above. Solve this compound equation as follows.

$2x + 1 = 7 \quad$ or $\quad 2x + 1 = -7$

$2x = 6 \quad$ or $\qquad 2x = -8 \qquad$ Subtract 1.

$x = 3 \quad$ or $\qquad x = -4 \qquad$ Divide by 2.

NOW TRY
EXERCISE 1

Solve $|4x - 1| = 11$.

CHECK $\qquad\qquad\qquad |2x + 1| = 7$

$$|2(3) + 1| \overset{?}{=} 7 \qquad \text{Let } x = 3. \qquad\qquad |2(-4) + 1| \overset{?}{=} 7 \qquad \text{Let } x = -4.$$

$$|6 + 1| \overset{?}{=} 7 \qquad\qquad\qquad\qquad |-8 + 1| \overset{?}{=} 7$$

$$|7| \overset{?}{=} 7 \qquad\qquad\qquad\qquad\qquad |-7| \overset{?}{=} 7$$

$$7 = 7 \ \checkmark \quad \text{True} \qquad\qquad\qquad\qquad 7 = 7 \ \checkmark \quad \text{True}$$

The solution set is $\{-4, 3\}$. The graph is shown in **FIGURE 24**.

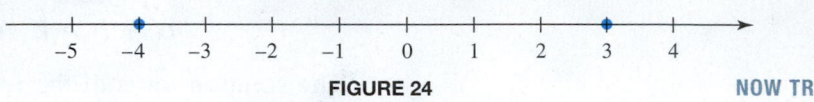

FIGURE 24

NOW TRY

OBJECTIVE 3 Solve inequalities of the form $|ax + b| < k$ and of the form $|ax + b| > k$, for $k > 0$.

NOW TRY
EXERCISE 2

Solve $|4x - 1| > 11$.

EXAMPLE 2 Solving an Absolute Value Inequality (Case 2)

Solve $|2x + 1| > 7$. Graph the solution set.

By Case 2 in the box on the preceding page, this absolute value inequality is rewritten as

$$2x + 1 > 7 \quad \text{or} \quad 2x + 1 < -7$$

because $2x + 1$ must represent a number that is *more* than 7 units from 0 on either side of a number line. Now solve the compound inequality.

$$2x + 1 > 7 \quad \text{or} \quad 2x + 1 < -7$$

$$2x > 6 \quad \text{or} \qquad 2x < -8 \qquad \text{Subtract 1.}$$

$$x > 3 \quad \text{or} \qquad x < -4 \qquad \text{Divide by 2.}$$

CHECK The excluded endpoints -4 and 3 are correct because from **Example 1** we know that the solutions of the related equation are -4 and 3. We now choose a test point in each of the three intervals $(-\infty, -4)$, $(-4, 3)$, and $(3, \infty)$.

For $(-\infty, -4)$, let $x = -5$. $\quad$ For $(-4, 3)$, let $x = 0$. $\quad$ For $(3, \infty)$, let $x = 4$.

$$|2x + 1| > 7 \qquad\qquad |2x + 1| > 7 \qquad\qquad |2x + 1| > 7$$

$$|2(-5) + 1| \overset{?}{>} 7 \qquad |2(0) + 1| \overset{?}{>} 7 \qquad |2(4) + 1| \overset{?}{>} 7$$

$$|-9| \overset{?}{>} 7 \qquad\qquad |1| \overset{?}{>} 7 \qquad\qquad |9| \overset{?}{>} 7$$

$$9 > 7 \ \checkmark \ \text{True} \qquad\quad 1 > 7 \qquad \text{False} \qquad 9 > 7 \ \checkmark \ \text{True}$$

The solution set, $(-\infty, -4) \cup (3, \infty)$, is a disjoint interval. See **FIGURE 25**.

NOW TRY ANSWERS

1. $\left\{-\frac{5}{2}, 3\right\}$

2. $\left(-\infty, -\frac{5}{2}\right) \cup (3, \infty)$

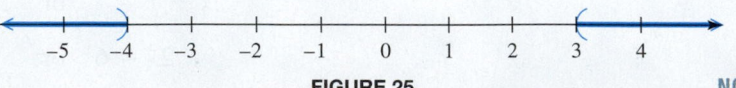

FIGURE 25

NOW TRY

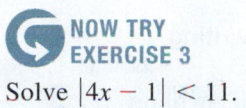
NOW TRY EXERCISE 3

Solve $|4x - 1| < 11$.

EXAMPLE 3 Solving an Absolute Value Inequality (Case 3)

Solve $|2x + 1| < 7$. Graph the solution set.

The expression $2x + 1$ must represent a number that is less than 7 units from 0 on either side of a number line. That is, $2x + 1$ must be between -7 and 7. As Case 3 in the earlier box shows, this is written as a three-part inequality.

$$-7 < 2x + 1 < 7$$

$$-8 < \quad 2x \quad < 6 \qquad \text{Subtract 1 from each part.}$$

$$-4 < \quad x \quad < 3 \qquad \text{Divide each part by 2.}$$

Check that the solution set is $(-4, 3)$. The graph is the open interval in **FIGURE 26**.

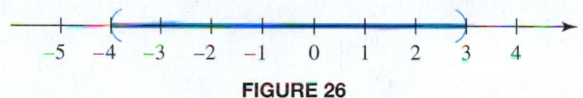

FIGURE 26   NOW TRY

Look back at **FIGURES 24, 25, AND 26,** with the graphs of the solution sets of

$$|2x + 1| = 7, \quad |2x + 1| > 7, \quad \text{and} \quad |2x + 1| < 7.$$

If we find the union of the three sets, we obtain the set of all real numbers. For any value of x, $|2x + 1|$ will satisfy *one and only one* of the following: It is equal to 7, greater than 7, or less than 7.

⚠ **CAUTION**   Remember the following when solving absolute value equations and inequalities.

1. The methods described apply when the constant is alone on one side of the equation or inequality and is *positive*.

2. Absolute value equations $|ax + b| = k$ and inequalities of the form $|ax + b| > k$ translate into "or" compound statements.

3. Absolute value inequalities of the form $|ax + b| < k$ translate into "and" compound statements, which may be written as three-part inequalities.

4. An "or" statement *cannot* be written in three parts. It would be **incorrect** to write $-7 > 2x + 1 > 7$ in **Example 2,** because this would imply that $-7 > 7$, which is *false*.

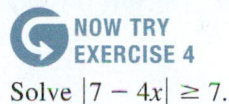
NOW TRY EXERCISE 4

Solve $|7 - 4x| \geq 7$.

EXAMPLE 4 Solving an Absolute Value Inequality (Case 2, for $\geq$)

Solve $|5 - 2x| \geq 5$.

Case 2 is applied. Notice that the endpoints are included because equality is part of the symbol $\geq$.

$$5 - 2x \geq 5 \quad \text{or} \quad 5 - 2x \leq -5$$

$$-2x \geq 0 \quad \text{or} \quad -2x \leq -10 \qquad \text{Subtract 5.}$$

$$x \leq 0 \quad \text{or} \quad x \geq 5 \qquad \begin{array}{l}\text{Divide by } -2. \text{ Reverse the} \\ \text{direction of the inequality symbols.}\end{array}$$

Check that the solution set is $(-\infty, 0] \cup [5, \infty)$. See **FIGURE 27**.

NOW TRY ANSWERS

3. $\left(-\frac{5}{2}, 3\right)$

4. $(-\infty, 0] \cup \left[\frac{7}{2}, \infty\right)$

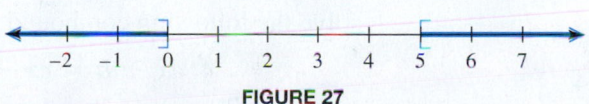

FIGURE 27   NOW TRY

OBJECTIVE 4 Solve absolute value equations that involve rewriting.

NOW TRY
EXERCISE 5

Solve $|10x - 2| - 2 = 12$.

EXAMPLE 5 Solving an Absolute Value Equation That Requires Rewriting

Solve $|x + 3| + 5 = 12$.

Isolate the absolute value expression on one side of the equality symbol.

$$|x + 3| + 5 = 12$$

$$|x + 3| + 5 - 5 = 12 - 5 \qquad \text{Subtract 5.}$$

$$|x + 3| = 7 \qquad \text{Combine like terms.}$$

Now use the method shown in **Example 1** for Case 1 to solve $|x + 3| = 7$.

$$x + 3 = 7 \quad \text{or} \quad x + 3 = -7$$

$$x = 4 \quad \text{or} \qquad x = -10 \qquad \text{Subtract 3.}$$

CHECK $\qquad\qquad\qquad |x + 3| + 5 = 12$

$\|4 + 3\| + 5 \stackrel{?}{=} 12$ Let $x = 4$.	$\|-10 + 3\| + 5 \stackrel{?}{=} 12$ Let $x = -10$.
$\|7\| + 5 \stackrel{?}{=} 12$	$\|-7\| + 5 \stackrel{?}{=} 12$
$12 = 12$ ✓ True	$12 = 12$ ✓ True

The check confirms that the solution set is $\{-10, 4\}$.   **NOW TRY**

NOW TRY
EXERCISE 6

Solve each inequality.

(a) $|x - 1| - 4 \leq 2$

(b) $|x - 1| - 4 \geq 2$

EXAMPLE 6 Solving Absolute Value Inequalities That Require Rewriting

Solve each inequality.

(a) $\qquad |x + 3| + 5 \geq 12$

$\qquad\qquad |x + 3| \geq 7 \qquad$ (Case 2)

$x + 3 \geq 7 \quad \text{or} \quad x + 3 \leq -7$

$x \geq 4 \quad \text{or} \qquad x \leq -10$

The solution set is $(-\infty, -10] \cup [4, \infty)$.

(b) $\qquad |x + 3| + 5 \leq 12$

$\qquad\qquad |x + 3| \leq 7 \qquad$ (Case 3)

$-7 \leq x + 3 \leq 7$

$-10 \leq \quad x \quad \leq 4$

The solution set is $[-10, 4]$.

NOW TRY

OBJECTIVE 5 Solve equations of the form $|ax + b| = |cx + d|$.

If two expressions have the same absolute value, they must either be equal or be negatives of each other.

Solving $|ax + b| = |cx + d|$

To solve an absolute value equation of the form

$$|ax + b| = |cx + d|,$$

solve the following compound equation.

$$ax + b = cx + d \quad \text{or} \quad ax + b = -(cx + d)$$

NOW TRY ANSWERS

5. $\left\{-\frac{6}{5}, \frac{8}{5}\right\}$

6. (a) $[-5, 7]$

   (b) $(-\infty, -5] \cup [7, \infty)$

**NOW TRY
EXERCISE 7**

Solve

$|3x - 4| = |5x + 12|.$

EXAMPLE 7 Solving an Equation Involving Two Absolute Values

Solve $|x + 6| = |2x - 3|$.

This equation is satisfied either if $x + 6$ and $2x - 3$ are equal to each other, or if $x + 6$ and $2x - 3$ are negatives of each other.

$$x + 6 = 2x - 3 \quad \text{or} \quad x + 6 = -(2x - 3)$$
$$x + 9 = 2x \quad \text{or} \quad x + 6 = -2x + 3$$
$$9 = x \quad \text{or} \quad 3x = -3$$
$$x = 9 \quad \text{or} \quad x = -1$$

CHECK $\qquad |x + 6| = |2x - 3|$

$|9 + 6| \overset{?}{=} |2(9) - 3|$ Let $x = 9$.	$|-1 + 6| \overset{?}{=} |2(-1) - 3|$ Let $x = -1$.
$|15| \overset{?}{=} |18 - 3|$	$|5| \overset{?}{=} |-2 - 3|$
$|15| \overset{?}{=} |15|$	$|5| \overset{?}{=} |-5|$
$15 = 15$ ✓ True	$5 = 5$ ✓ True

The check confirms that the solution set is $\{-1, 9\}$.   **NOW TRY**

OBJECTIVE 6 Solve special cases of absolute value equations and inequalities.

When an absolute value equation or inequality involves a *negative constant or 0* alone on one side, use the properties of absolute value to solve the equation or inequality.

Special Cases of Absolute Value

Case 1 The absolute value of an expression can never be negative—that is, $|a| \geq 0$ for all real numbers a.

Case 2 The absolute value of an expression equals 0 only when the expression is equal to 0.

**NOW TRY
EXERCISE 8**

Solve each equation.
(a) $|3x - 8| = -2$
(b) $|7x + 12| = 0$

EXAMPLE 8 Solving Special Cases of Absolute Value Equations

Solve each equation.

(a) $|5x - 3| = -4$

See Case 1 in the preceding box. ***The absolute value of an expression can never be negative,*** so there are no solutions for this equation. The solution set is $\varnothing$.

(b) $|7x - 3| = 0$

See Case 2 in the preceding box. The expression $|7x - 3|$ will equal 0 *only* if $7x - 3 = 0$.

$$7x - 3 = 0$$
$$7x = 3 \qquad \text{Add 3.}$$
$$x = \frac{3}{7} \qquad \text{Divide by 7.}$$

Check by substituting in the original equation.

NOW TRY ANSWERS
7. $\{-8, -1\}$
8. **(a)** $\varnothing$ **(b)** $\left\{-\frac{12}{7}\right\}$

The solution set $\left\{\frac{3}{7}\right\}$ consists of only one element.   **NOW TRY**

NOW TRY
EXERCISE 9

Solve each inequality.

(a) $|x| > -10$

(b) $|4x + 1| + 5 < 4$

(c) $|x - 2| - 3 \leq -3$

EXAMPLE 9   Solving Special Cases of Absolute Value Inequalities

Solve each inequality.

(a) $|x| \geq -4$

The absolute value of a number is always greater than or equal to 0. Thus, $|x| \geq -4$ is true for *all* real numbers. The solution set is $(-\infty, \infty)$.

(b)
$$|x + 6| - 3 < -5$$
$$|x + 6| < -2 \qquad \text{Add 3 to each side.}$$

There is no number whose absolute value is less than -2, so this inequality has no solution. The solution set is $\varnothing$.

(c)
$$|x - 7| + 4 \leq 4$$
$$|x - 7| \leq 0 \qquad \text{Subtract 4 from each side.}$$

The value of $|x - 7|$ will never be less than 0. However, $|x - 7|$ will equal 0 when $x = 7$. Therefore, the solution set is $\{7\}$.   **NOW TRY**

OBJECTIVE 7   Solve an application involving relative error.

Absolute value is used to find the **relative error,** or **tolerance,** in a measurement. If x_t represents the expected measurement and x represents the actual measurement, then the relative error in x equals the absolute value of the difference of x_t and x, divided by x_t.

$$\textbf{relative error in } x \; = \; \left| \frac{x_t - x}{x_t} \right|$$

In quality control situations, the relative error must often be less than some predetermined amount.

NOW TRY
EXERCISE 10

Suppose a machine filling *quart* milk cartons is set for a relative error that *is no greater than* 0.032 oz. How many ounces may a filled carton contain?

EXAMPLE 10   Solving an Application Involving Relative Error

Suppose a machine filling *quart* milk cartons is set for a relative error that *is no greater than* 0.05 oz. How many ounces may a filled carton contain?

Here $x_t = 32$ oz (because 1 qt = 32 oz), the relative error = 0.05 oz, and we must find x, given the following condition.

$$\left| \frac{32 - x}{32} \right| \leq 0.05 \qquad \textit{Is no greater than } \text{translates as } \leq.$$

$$-0.05 \leq \frac{32 - x}{32} \leq 0.05 \qquad \text{(Case 3)}$$

$$-1.6 \leq 32 - x \leq 1.6 \qquad \text{Multiply by 32.}$$

$$-33.6 \leq \quad -x \quad \leq -30.4 \qquad \text{Subtract 32.}$$

$$33.6 \geq \quad x \quad \geq 30.4 \qquad \begin{array}{l}\text{Multiply by } -1, \text{ and reverse the}\\ \text{direction of the inequality symbols.}\end{array}$$

$$30.4 \leq \quad x \quad \leq 33.6 \qquad \text{Rewrite.}$$

The filled carton may contain between 30.4 and 33.6 oz, inclusive.   **NOW TRY**

NOW TRY ANSWERS

9. (a) $(-\infty, \infty)$ **(b)** $\varnothing$ **(c)** $\{2\}$

10. between 30.976 and 33.024 oz, inclusive

8.3 Exercises

 FOR EXTRA HELP MyMathLab®

● *Complete solution available in MyMathLab*

Concept Check Match each absolute value equation or inequality in Column I with the graph of its solution set in Column II.

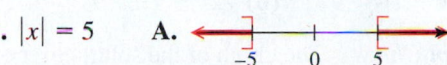

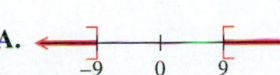

3. *Concept Check* How many solutions will $|ax + b| = k$ have for each situation?

(a) $k = 0$ **(b)** $k > 0$ **(c)** $k < 0$

4. *Concept Check* Explain when to use *and* and when to use *or* for solving an absolute value equation or inequality of the form $|ax + b| = k$, $|ax + b| < k$, or $|ax + b| > k$, where k is a positive number.

Solve each equation. See Example 1.

5. $|x| = 12$ **6.** $|x| = 14$ **7.** $|4x| = 20$ **8.** $|5x| = 30$

9. $|x - 3| = 9$ **10.** $|x - 5| = 13$ ● **11.** $|2x - 1| = 11$ **12.** $|2x + 3| = 19$

13. $|4x - 5| = 17$ **14.** $|5x - 1| = 21$ **15.** $|2x + 5| = 14$ **16.** $|2x - 9| = 18$

17. $|-3x + 8| = 1$ **18.** $|-6x + 5| = 4$ **19.** $\left|12 - \frac{1}{2}x\right| = 6$

20. $\left|14 - \frac{1}{3}x\right| = 8$ **21.** $|0.5x| = 6$ **22.** $|0.3x| = 9$

23. $\left|\frac{1}{2}x + 3\right| = 2$ **24.** $\left|\frac{2}{3}x - 1\right| = 5$ **25.** $\left|1 + \frac{3}{4}x\right| = 7$

26. $\left|2 - \frac{5}{2}x\right| = 14$ **27.** $|0.02x - 1| = 2.50$ **28.** $|0.04x - 3| = 5.96$

Solve each inequality, and graph the solution set. See Example 2.

29. $|x| > 3$ **30.** $|x| > 5$ **31.** $|x| \geq 4$

32. $|x| \geq 6$ ● **33.** $|r + 5| \geq 20$ **34.** $|x + 4| \geq 8$

35. $|5x + 2| > 10$ **36.** $|4x + 1| \geq 21$ **37.** $|3 - x| > 5$

38. $|5 - x| > 3$ **39.** $|-5x + 3| \geq 12$ **40.** $|-2x - 4| \geq 5$

41. *Concept Check* The graph of the solution set of $|2x + 1| = 9$ is given here.

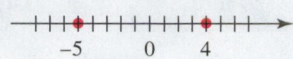

Without actually doing the algebraic work, graph the solution set of each inequality, referring to the graph shown.

(a) $|2x + 1| < 9$ **(b)** $|2x + 1| > 9$

42. *Concept Check* The graph of the solution set of $|3x - 4| < 5$ is given here.

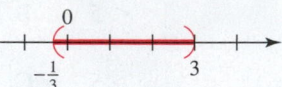

Without actually doing the algebraic work, graph the solution set of the following, referring to the graph shown.

(a) $|3x - 4| = 5$ **(b)** $|3x - 4| > 5$

*Solve each inequality, and graph the solution set. **See Example 3.** (Hint: Compare the answers with those in **Exercises 29–40.**)*

43. $|x| \leq 3$ **44.** $|x| \leq 5$ **45.** $|x| < 4$

46. $|x| < 6$ **47.** $|r + 5| < 20$ **48.** $|x + 4| < 8$

49. $|5x + 2| \leq 10$ **50.** $|4x + 1| < 21$ **51.** $|3 - x| \leq 5$

52. $|5 - x| \leq 3$ **53.** $|-5x + 3| < 12$ **54.** $|-2x - 4| < 5$

*In Exercises 55–78, decide which method of solution applies, and find the solution set. In Exercises 55–66, graph the solution set. **See Examples 1–4.***

55. $|-4 + x| > 9$ **56.** $|-3 + x| > 8$ **57.** $|3x + 2| < 11$

58. $|2x - 1| < 7$ **59.** $|7 + 2x| = 5$ **60.** $|9 - 3x| = 3$

61. $|3x - 1| \leq 11$ **62.** $|2x - 6| \leq 6$ **63.** $|-6x - 6| \leq 1$

64. $|-2x - 6| \leq 5$ **65.** $|-8 + x| \leq 5$ **66.** $|-4 + x| \leq 9$

67. $|10 - 12x| \geq 4$ **68.** $|8 - 10x| \geq 2$ **69.** $|3(x - 1)| = 8$

70. $|7(x - 2)| = 4$ **71.** $|0.1x - 1| > 3$ **72.** $|0.1x + 1| > 2$

73. $|x + 2| = 5 - 2$ **74.** $|x + 3| = 12 - 2$ **75.** $3|x - 6| = 9$

76. $5|x - 4| = 5$ **77.** $|2 - 0.2x| = 2$ **78.** $|5 - 0.5x| = 4$

*Solve each equation or inequality. **See Examples 5 and 6.***

79. $|x| - 1 = 4$ **80.** $|x| + 3 = 10$ ▶ **81.** $|x + 4| + 1 = 2$

82. $|x + 5| - 2 = 12$ **83.** $|2x + 1| + 3 > 8$ **84.** $|6x - 1| - 2 > 6$

85. $|x + 5| - 6 \leq -1$ **86.** $|x - 2| - 3 \leq 4$

87. $|0.1x - 2.5| + 0.3 \geq 0.8$ **88.** $|0.5x - 3.5| + 0.2 \geq 0.6$

89. $\left| \dfrac{1}{2}x + \dfrac{1}{3} \right| + \dfrac{1}{4} = \dfrac{3}{4}$ **90.** $\left| \dfrac{2}{3}x + \dfrac{1}{6} \right| + \dfrac{1}{2} = \dfrac{5}{2}$

Solve each equation. See Example 7.

91. $|3x + 1| = |2x + 4|$

92. $|7x + 12| = |x - 8|$

93. $\left| x - \dfrac{1}{2} \right| = \left| \dfrac{1}{2}x - 2 \right|$

94. $\left| \dfrac{2}{3}x - 2 \right| = \left| \dfrac{1}{3}x + 3 \right|$

95. $|6x| = |9x + 1|$

96. $|13x| = |2x + 1|$

97. $|2x - 6| = |2x + 11|$

98. $|3x - 1| = |3x + 9|$

Solve each equation or inequality. See Examples 8 and 9.

99. $|x| \geq -10$

100. $|x| \geq -15$

101. $|12t - 3| = -8$

102. $|13x + 1| = -3$

103. $|4x + 1| = 0$

104. $|6x - 2| = 0$

105. $|2x - 1| = -6$

106. $|8x + 4| = -4$

107. $|x + 5| > -9$

108. $|x + 9| > -3$

109. $|7x + 3| \leq 0$

110. $|4x - 1| \leq 0$

111. $|5x - 2| = 0$

112. $|7x + 4| = 0$

113. $|x - 2| + 3 \geq 2$

114. $|x - 4| + 5 \geq 4$

115. $|10x + 7| + 3 < 1$

116. $|4x + 1| - 2 < -5$

In Exercises 117–120, determine the number of ounces a filled 32 oz carton may contain for the given relative error. See Example 10.

117. no greater than 0.04 oz

118. no greater than 0.03 oz

119. no greater than 0.025 oz

120. no greater than 0.015 oz

In later courses in mathematics, it is sometimes necessary to find an interval in which x must lie in order to keep y within a given difference of some number. For example, suppose

$$y = 2x + 1$$

and we want y to be within 0.01 unit of 4. This criterion can be written as

$$|y - 4| < 0.01.$$

Solving this inequality shows that x must lie in the interval (1.495, 1.505) *to satisfy the requirement.*

  In Exercises 121–124, find the open interval in which x must lie in order for the given condition to hold.

121. $y = 2x + 1$, and the difference of y and 1 is less than 0.1.

122. $y = 4x - 6$, and the difference of y and 2 is less than 0.02.

123. $y = 4x - 8$, and the difference of y and 3 is less than 0.001.

124. $y = 5x + 12$, and the difference of y and 4 is less than 0.0001.

Work each problem.

125. Dr. Mosely has determined that 99% of the babies he has delivered have weighed x pounds, where

$$|x - 8.3| < 1.5.$$

What range of weights corresponds to this inequality?

126. The Celsius temperatures x on Mars approximately satisfy the inequality

$$|x + 85| \leq 55.$$

What range of temperatures corresponds to this inequality?

127. The recommended daily intake (RDI) of calcium for females aged 19–50 is 1000 mg. Actual needs vary from person to person. Write this statement as an absolute value inequality, with x representing the RDI, to express the RDI plus or minus 100 mg, and solve the inequality. (*Source:* National Academy of Sciences—Institute of Medicine.)

128. The average clotting time of blood is 7.45 sec, with a variation of plus or minus 3.6 sec. Write this statement as an absolute value inequality, with x representing the time, and solve the inequality.

RELATING CONCEPTS   For Individual or Group Work   (Exercises 129–132)

The 10 tallest buildings in Houston, Texas, are listed along with their heights.

Building	Height (in feet)
JPMorgan Chase Tower	1002
Wells Fargo Plaza	992
Williams Tower	901
Bank of America Center	780
Texaco Heritage Plaza	762
Enterprise Plaza	756
Centerpoint Energy Plaza	741
Continental Center I	732
Fulbright Tower	725
One Shell Plaza	714

Source: World Almanac and Book of Facts.

*Use this information to **work Exercises 129–132 in order.***

129. To find the average of a group of numbers, we add the numbers and then divide by the number of numbers added. Use a calculator to find the average of the heights.

130. Let k represent the average height of these buildings. If a height x satisfies the inequality

$$|x - k| < t,$$

then the height is said to be within t feet of the average. Using the result from **Exercise 129,** list the buildings that are within 50 ft of the average.

131. Repeat **Exercise 130,** but list the buildings that are within 95 ft of the average.

132. **(a)** Write an absolute value inequality that describes the height of a building that is *not* within 95 ft of the average.

(b) Solve the inequality from part (a).

(c) Use the result of part (b) to list the buildings that are not within 95 ft of the average.

(d) Confirm that the answer to part (c) makes sense by comparing it with the answer to **Exercise 131.**

Solve each equation or inequality. Give the solution set in set notation for equations and in interval notation for inequalities.

1. $4x + 1 = 49$

2. $|x - 1| = 6$

3. $6x - 9 = 12 + 3x$

4. $3x + 7 = 9 + 8x$

5. $|x + 3| = -4$

6. $2x + 1 \le x$

7. $8x + 2 \ge 5x$

8. $4(x - 11) + 3x = 20x - 31$

9. $2x - 1 = -7$

10. $|3x - 7| - 4 = 0$

11. $6x - 5 \le 3x + 10$

12. $|5x - 8| + 9 \ge 7$

13. $9x - 3(x + 1) = 8x - 7$

14. $|x| \ge 8$

15. $9x - 5 \ge 9x + 3$

16. $13x - 5 > 13x - 8$

17. $|x| < 5.5$

18. $4x - 1 = 12 + x$

19. $\dfrac{2}{3}x + 8 = \dfrac{1}{4}x$

20. $-\dfrac{5}{8}x \ge -20$

21. $\dfrac{1}{4}x < -6$

22. $\dfrac{1}{2} \le \dfrac{2}{3}x \le \dfrac{5}{4}$

23. $\dfrac{3}{5}x - \dfrac{1}{10} = 2$

24. $\dfrac{x}{6} - \dfrac{3x}{5} = x - 86$

25. $x + 9 + 7x = 4(3 + 2x) - 3$

26. $6 - 3(2 - x) < 2(1 + x) + 3$

27. $-6 \le \dfrac{3}{2} - x \le 6$

28. $\dfrac{x}{4} - \dfrac{2x}{3} = -10$

29. $|5x + 1| \le 0$

30. $5x - (3 + x) \ge 2(3x + 1)$

31. $-2 \le 3x - 1 \le 8$

32. $-1 \le 6 - x \le 5$

33. $|7x - 1| = |5x + 3|$

34. $|x + 2| = |x + 4|$

35. $|1 - 3x| \ge 4$

36. $7x - 3 + 2x = 9x - 8x$

37. $-(x + 4) + 2 = 3x + 8$

38. $|x - 1| < 7$

39. $|2x - 3| > 11$

40. $|5 - x| < 4$

41. $|x - 1| \ge -6$

42. $|2x - 5| = |x + 4|$

43. $8x - (1 - x) = 3(1 + 3x) - 4$

44. $8x - (x + 3) = -(2x + 1) - 12$

45. $|x - 5| = |x + 9|$

46. $|x + 2| < -3$

47. $2x + 1 > 5 \quad \text{or} \quad 3x + 4 < 1$

48. $1 - 2x \ge 5 \quad \text{and} \quad 7 + 3x \ge -2$

1. $\{12\}$ **2.** $\{-5, 7\}$

3. $\{7\}$ **4.** $\left\{-\dfrac{2}{5}\right\}$

5. $\varnothing$ **6.** $(-\infty, -1]$

7. $\left[-\dfrac{2}{3}, \infty\right)$ **8.** $\{-1\}$

9. $\{-3\}$ **10.** $\left\{1, \dfrac{11}{3}\right\}$

11. $(-\infty, 5]$ **12.** $(-\infty, \infty)$

13. $\{2\}$

14. $(-\infty, -8] \cup [8, \infty)$

15. $\varnothing$ **16.** $(-\infty, \infty)$

17. $(-5.5, 5.5)$ **18.** $\left\{\dfrac{13}{3}\right\}$

19. $\left\{-\dfrac{96}{5}\right\}$ **20.** $(-\infty, 32]$

21. $(-\infty, -24)$ **22.** $\left[\dfrac{3}{4}, \dfrac{15}{8}\right]$

23. $\left\{\dfrac{7}{2}\right\}$ **24.** $\{60\}$

25. $\{\text{all real numbers}\}$

26. $(-\infty, 5)$

27. $\left[-\dfrac{9}{2}, \dfrac{15}{2}\right]$ **28.** $\{24\}$

29. $\left\{-\dfrac{1}{5}\right\}$ **30.** $\left(-\infty, -\dfrac{5}{2}\right]$

31. $\left[-\dfrac{1}{3}, 3\right]$ **32.** $[1, 7]$

33. $\left\{-\dfrac{1}{6}, 2\right\}$ **34.** $\{-3\}$

35. $(-\infty, -1] \cup \left[\dfrac{5}{3}, \infty\right)$

36. $\left\{\dfrac{3}{8}\right\}$

37. $\left\{-\dfrac{5}{2}\right\}$ **38.** $(-6, 8)$

39. $(-\infty, -4) \cup (7, \infty)$

40. $(1, 9)$

41. $(-\infty, \infty)$ **42.** $\left\{\dfrac{1}{3}, 9\right\}$

43. $\{\text{all real numbers}\}$

44. $\left\{-\dfrac{10}{9}\right\}$

45. $\{-2\}$ **46.** $\varnothing$

47. $(-\infty, -1) \cup (2, \infty)$

48. $[-3, -2]$

8.4   Linear Inequalities and Systems in Two Variables

VOCABULARY

☐ linear inequality in two variables
☐ boundary line
☐ system of linear inequalities
☐ solution set of a system of linear inequalities

OBJECTIVE 1   Graph linear inequalities in two variables.

In our earlier work in this chapter, we graphed linear inequalities in *one* variable on a *number line*. In this section, we graph linear inequalities in *two* variables on a *rectangular coordinate system.*

> **Linear Inequality in Two Variables**
>
> A **linear inequality in two variables** (here x and y) can be written in the form
>
> $$Ax + By < C, \quad Ax + By \le C, \quad Ax + By > C, \quad \text{or} \quad Ax + By \ge C,$$
>
> where A, B, and C are real numbers and A and B are not both 0.

Consider the graph in **FIGURE 28**. The graph of the line $x + y = 5$ divides the points in the rectangular coordinate system into three sets of points.

1. Those points that lie *on* the line itself and satisfy the equation $x + y = 5$, such as $(0, 5)$, $(2, 3)$, and $(5, 0)$;

2. Those points that lie in the region *above* the line and satisfy the inequality $x + y > 5$, such as $(5, 3)$ and $(2, 4)$;

3. Those points that lie in the region *below* the line and satisfy the inequality $x + y < 5$, such as $(0, 0)$ and $(-3, -1)$.

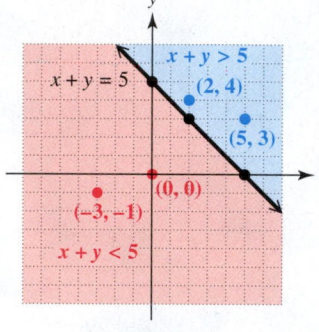

FIGURE 28

The graph of the line $x + y = 5$ is the **boundary line** for the two inequalities

$$x + y > 5 \quad \text{and} \quad x + y < 5.$$

A graph of a linear inequality in two variables is a region in the real number plane that may or may not include the boundary line.

To graph a linear inequality in two variables, follow these steps.

> **Graphing a Linear Inequality in Two Variables**
>
> *Step 1*   **Draw the graph of the straight line that is the boundary.**
>
> - Make the line solid if the inequality involves $\le$ or $\ge$.
> - Make the line dashed if the inequality involves $<$ or $>$.
>
> *Step 2*   **Choose a test point.** Choose any point not on the line, and substitute the coordinates of that point in the inequality.
>
> *Step 3*   **Shade the appropriate region.** Shade the region that includes the test point if it satisfies the original inequality. Otherwise, shade the region on the other side of the boundary line.

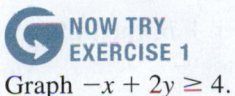

NOW TRY EXERCISE 1

Graph $-x + 2y \geq 4$.

EXAMPLE 1   Graphing a Linear Inequality

Graph $3x + 2y \geq 6$.

Step 1   First graph the boundary line $3x + 2y = 6$, which has intercepts $(2, 0)$ and $(0, 3)$, as shown in **FIGURE 29**.

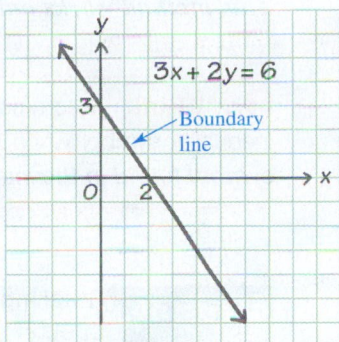

FIGURE 29

Step 2   The graph of the inequality $3x + 2y \geq 6$ includes the points of the boundary line $3x + 2y = 6$ (because the inequality symbol $\geq$ includes equality) and either the points *above* that line or the points *below* it. To decide which, select any point *not* on the boundary line to use as a test point. Substitute the values from the test point for x and y in the inequality.

$$3x + 2y > 6 \quad \text{← We are testing the region.}$$

$(0, 0)$ is a convenient test point.

$$3(0) + 2(0) \overset{?}{>} 6 \quad \text{Let } x = 0 \text{ and } y = 0.$$

$$0 > 6 \quad \text{False}$$

Step 3   Because the result is false, $(0, 0)$ does *not* satisfy the inequality. The solution set includes all points in the region on the *other* side of the line. See **FIGURE 30**.

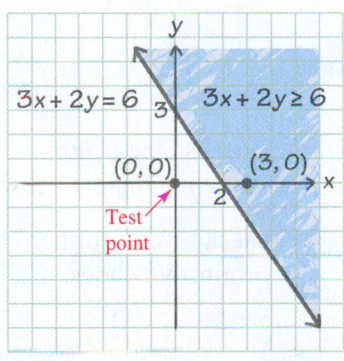

FIGURE 30

As a further check, select a test point in the shaded region, such as $(3, 0)$, and confirm that it does indeed satisfy the inequality. See **FIGURE 30**.

NOW TRY

NOW TRY ANSWER

1.

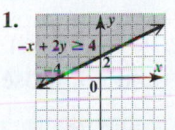

⚠ **CAUTION**   When drawing the boundary line in Step 1, be careful to draw a solid line if the inequality includes equality ($\leq$, $\geq$) or a dashed line if equality is not included ($<$, $>$).

If an inequality is written in the form $y > mx + b$ or $y < mx + b$, then the inequality symbol indicates which region to shade.

If $y > mx + b$, then shade above the boundary line.

If $y < mx + b$, then shade below the boundary line.

This method works only if the inequality is solved for y.

⚠ **CAUTION**   A common error in using the method just described is to use the original inequality symbol when deciding which region to shade. ***Be sure to use the inequality symbol found in the inequality after it is solved for y.***

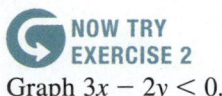

NOW TRY EXERCISE 2
Graph $3x - 2y < 0$.

EXAMPLE 2   Graphing a Linear Inequality with Boundary Passing through the Origin

Graph $3x - 4y > 0$.

First graph the boundary line. The x- and y-intercepts are the same point, $(0, 0)$. Thus, this line passes through the origin. Two other points on the line are $(4, 3)$ and $(-4, -3)$. The points of the boundary line do *not* belong to the inequality

$$3x - 4y > 0$$

(because the inequality symbol is $>$, *not* $\geq$). For this reason, the line is dashed. See **FIGURE 31**.

To use the method explained above, we solve the inequality for y.

$$3x - 4y > 0 \qquad \text{Original inequality}$$

$$-4y > -3x \qquad \text{Subtract } 3x.$$

> Use this equivalent inequality to decide which region to shade.

$$y < \frac{3}{4}x \qquad \begin{array}{l}\text{Divide by } -4.\\ \text{Change } > \text{ to } <.\end{array}$$

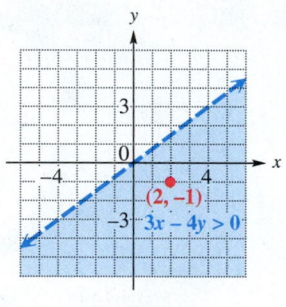

FIGURE 31

Because the *is less than* symbol occurs ***when the original inequality is solved for y***, shade the region *below* the boundary line.

CHECK   As a further check, choose a test point not on the line, which rules out the origin. We choose $(2, -1)$.

$$3x - 4y > 0 \qquad \text{Original inequality}$$

$$3(2) - 4(-1) \overset{?}{>} 0 \qquad \text{Let } x = 2 \text{ and } y = -1.$$

$$6 + 4 \overset{?}{>} 0 \qquad \text{Multiply.}$$

$$10 > 0 \; \checkmark \qquad \text{True}$$

This result agrees with the decision to shade below the line. The solution set, graphed in **FIGURE 31**, includes only those points in the shaded region (and *not* those on the line).

NOW TRY ANSWER
2.
(graph showing $3x - 2y < 0$)

NOW TRY ↻

NOW TRY EXERCISE 3

Graph $x + 2 > 0$.

EXAMPLE 3 Graphing a Linear Inequality

Graph $x - 3 < 1$.

We graph $x - 3 = 1$, which is equivalent to $x = 4$, as a dashed vertical line passing through the point $(4, 0)$. To determine which region to shade, we choose $(0, 0)$ as a test point.

$$x - 3 < 1$$
$$0 - 3 \overset{?}{<} 1 \quad \text{Let } x = 0.$$
$$-3 < 1 \quad \text{True}$$

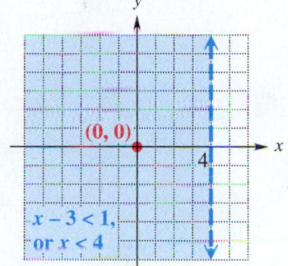

Because a true statement results, we shade the region containing $(0, 0)$. See **FIGURE 32**.

FIGURE 32

NOW TRY

OBJECTIVE 2 Solve systems of linear inequalities by graphing.

A **system of linear inequalities** consists of two or more linear inequalities. The **solution set of a system of linear inequalities** includes all ordered pairs that make all inequalities of the system true at the same time.

> **Solving a System of Linear Inequalities**
>
> **Step 1** **Graph each inequality.** Use the method of **Objective 1.**
>
> **Step 2** **Choose the intersection.** Indicate the solution set by shading the intersection of the graphs—that is, the region where the graphs overlap.

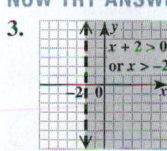

(a)

EXAMPLE 4 Solving a System of Linear Inequalities

Graph the solution set of the system.

$$4x + 3y \leq 12$$
$$2x - 5y \geq 10$$

Step 1 To graph $4x + 3y \leq 12$, graph the solid boundary line $4x + 3y = 12$ using the intercepts $(0, 4)$ and $(3, 0)$. Determine the region to shade.

$$4x + 3y < 12 \quad \boxed{\text{We are testing the region.}}$$
$$4(0) + 3(0) \overset{?}{<} 12 \quad \text{Use } (0, 0) \text{ as a test point.}$$
$$0 < 12 \quad \text{True}$$

Shade the region that includes $(0, 0)$. See **FIGURE 33(a)**.

Now graph $2x - 5y \geq 10$ with solid boundary line $2x - 5y = 10$ using the intercepts $(0, -2)$ and $(5, 0)$. Determine the region to shade.

$$2x - 5y > 10$$
$$2(0) - 5(0) \overset{?}{>} 10 \quad \text{Use } (0, 0) \text{ as a test point.}$$
$$0 > 10 \quad \text{False}$$

Shade the region that does *not* include $(0, 0)$. See **FIGURE 33(b)**.

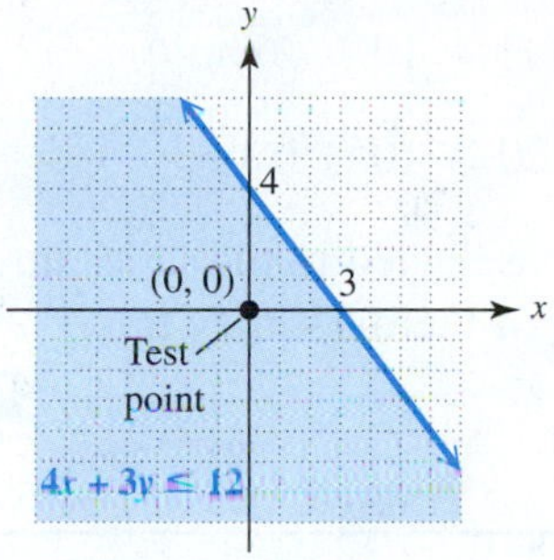

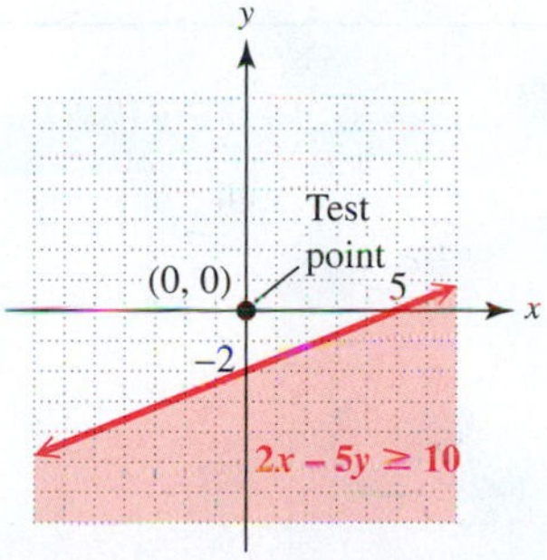

(b)

FIGURE 33

NOW TRY ANSWER

3.

**NOW TRY
EXERCISE 4**

Graph the solution set of
the system.

$$4x - 2y \le 8$$
$$x + 3y \ge 3$$

Step 2 The solution set of this system includes all points in the intersection—that is, the overlap—of the graphs of the two inequalities. As shown in **FIGURE 34**, this intersection is the gray shaded region and portions of the two boundary lines that surround it.

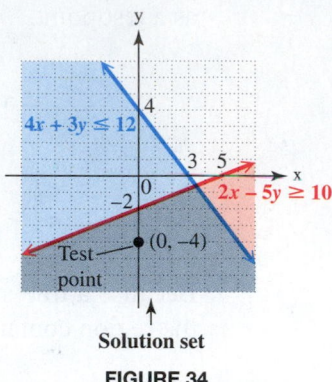

Solution set

FIGURE 34

CHECK To confirm the solution set in **FIGURE 34**, select a test point in the gray shaded region, such as $(0, -4)$, and substitute it into *both* inequalities to make sure that true statements result. (Using an ordered pair that has one coordinate 0 makes the substitution easier.)

$$4x + 3y < 12 \qquad\qquad 2x - 5y > 10$$
$$4(0) + 3(-4) \overset{?}{<} 12 \quad \text{Test } (0, -4). \qquad 2(0) - 5(-4) \overset{?}{>} 10 \quad \text{Test } (0, -4).$$
$$-12 < 12 \quad \text{True} \qquad\qquad 20 > 10 \quad \text{True}$$

The gray shaded region in **FIGURE 34** is correct. Test points selected in the other three regions will satisfy only one of the inequalities or neither of them. (Verify this.) ✔

NOW TRY

> **NOTE** We usually do all the work on one set of axes. Be sure that the region of the final solution set is clearly indicated.

**NOW TRY
EXERCISE 5**

Graph the solution set of the
system.

$$2x + 5y > 10$$
$$x - 2y < 0$$

NOW TRY ANSWERS

4.
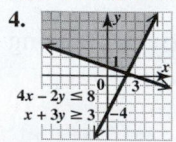

$4x - 2y \le 8$
$x + 3y \ge 3$

5.
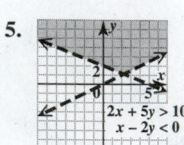

$2x + 5y > 10$
$x - 2y < 0$

EXAMPLE 5 Solving a System of Linear Inequalities

Graph the solution set of the system.

$$x - y > 5$$
$$2x + y < 2$$

FIGURE 35 shows the graphs of both $x - y > 5$ and $2x + y < 2$. Dashed lines indicate that the graphs of the inequalities do not include their boundary lines. Use $(0, 0)$ as a test point to determine the region to shade for each inequality.

The solution set of the system is the region with gray shading. Neither boundary line is included. (Use $(0, -6)$ in the gray shaded region as a test point to confirm the solution set.)

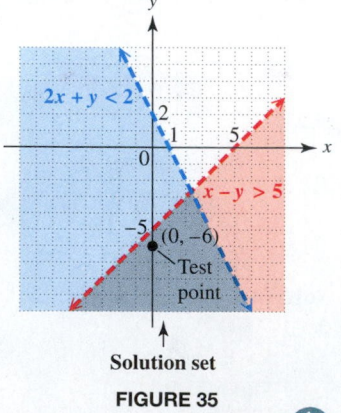

Solution set

FIGURE 35

NOW TRY

8.4 Exercises

  MyMathLab®

● *Complete solution available in MyMathLab*

Concept Check Decide whether each ordered pair is a solution of the given inequality.

1. $x - 2y \leq 4$

   (a) $(0, 0)$ (b) $(2, -1)$

   (c) $(7, 1)$ (d) $(0, 2)$

2. $x + y > 0$

   (a) $(0, 0)$ (b) $(-2, 1)$

   (c) $(2, -1)$ (d) $(-4, 6)$

3. $x - 5 > 0$

   (a) $(0, 0)$ (b) $(5, 0)$

   (c) $(-1, 3)$ (d) $(6, 2)$

4. $y \leq 1$

   (a) $(0, 0)$ (b) $(3, 1)$

   (c) $(2, -1)$ (d) $(-3, 3)$

Concept Check In each statement, fill in the first blank with either **solid** or **dashed**. Fill in the second blank with either **above** or **below**.

5. The boundary of the graph of $y \leq -x + 2$ will be a _____ line, and the shading will be _____ the line.

6. The boundary of the graph of $y < -x + 2$ will be a _____ line, and the shading will be _____ the line.

7. The boundary of the graph of $y > -x + 2$ will be a _____ line, and the shading will be _____ the line.

8. The boundary of the graph of $y \geq -x + 2$ will be a _____ line, and the shading will be _____ the line.

Concept Check Refer to the given graph, and complete each statement with the correct inequality symbol $<$, $\leq$, $>$, or $\geq$.

9. x _____ 4

10. y _____ -3

11. y _____ $3x - 2$

12. y _____ $-x + 3$

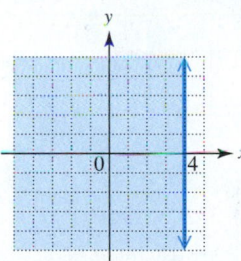 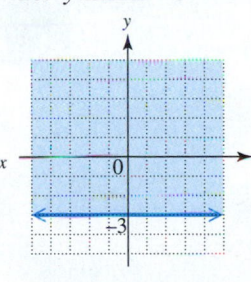 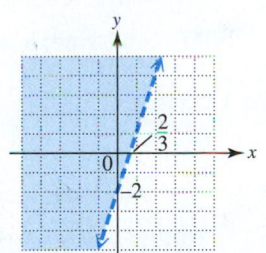 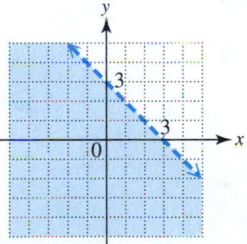

Graph each linear inequality in two variables. See Examples 1–3.

● 13. $x + y \leq 2$

14. $x + y \leq -3$

● 15. $4x - y < 4$

16. $3x - y < 3$

17. $x + 3y \geq -2$

18. $x + 4y \geq -3$

19. $y < \dfrac{1}{2}x + 3$

20. $y < \dfrac{1}{3}x - 2$

21. $y \geq -\dfrac{2}{5}x + 2$

22. $y \geq -\dfrac{3}{2}x + 3$

23. $2x + 3y \geq 6$

24. $3x + 4y \geq 12$

25. $5x - 3y > 15$

26. $4x - 5y > 20$

27. $x + y > 0$

28. $x + 2y > 0$

29. $x - 3y \leq 0$

30. $x - 5y \leq 0$

31. $y < x$

32. $y \leq 5x$

33. $x + 3 \geq 0$

34. $x - 1 \leq 0$

35. $y + 5 < 2$

36. $y - 1 > 3$

Extending Skills Complete each of the following to write an inequality for the graph shown.

37. Determine the following for the boundary line.

Slope: _____

y-intercept: _____

Equation: y = _____

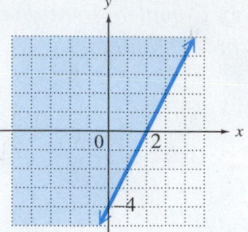

The boundary line here is (*solid* / *dashed*), and the region (*above* / *below*) it is shaded.

The inequality symbol to indicate this is ($<$ / $\leq$ / $>$ / $\geq$).

Inequality for the graph: y _____

38. Determine the following for the boundary line.

Slope: _____

y-intercept: _____

Equation: y = _____

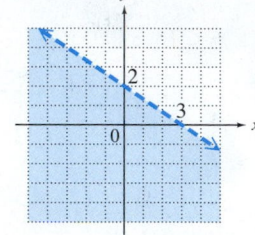

The boundary line here is (*solid* / *dashed*), and the region (*above* / *below*) it is shaded.

The inequality symbol to indicate this is ($<$ / $\leq$ / $>$ / $\geq$).

Inequality for the graph: y _____

Concept Check Match each system of inequalities with the correct graph from choices A–D.

39. $x \geq 5$

$y \leq -3$

A.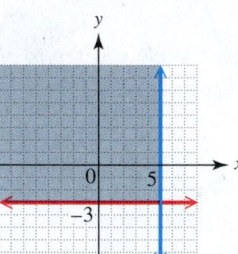

B.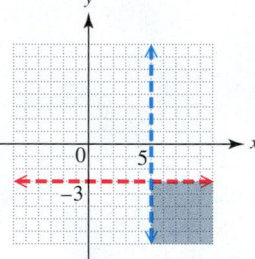

40. $x \leq 5$

$y \geq -3$

41. $x > 5$

$y < -3$

C.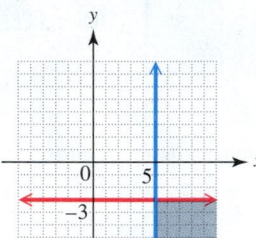

D.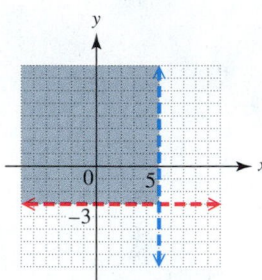

42. $x < 5$

$y > -3$

Concept Check Decide whether each ordered pair is a solution of the given system of inequalities. Then shade the solution set of each system. Boundary lines are already graphed.

43. $x + y > 4$

$5x - 3y < 15$

(a) $(0, 0)$

(b) $(3, 3)$

(c) $(5, 0)$

44. $x - 3y \leq 6$

$x \geq -4$

(a) $(-5, -4)$

(b) $(0, -4)$

(c) $(0, 0)$

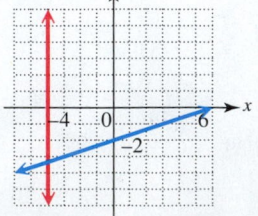

Graph the solution set of each system of linear inequalities. See Examples 4 and 5.

45. $x + y \le 6$
$x - y \ge 1$

46. $x + y \le 2$
$x - y \ge 3$

47. $4x + 5y \ge 20$
$x - 2y \le 5$

48. $x + 4y \le 8$
$2x - y \ge 4$

49. $2x + 3y < 6$
$x - y < 5$

50. $x + 2y < 4$
$x - y < -1$

51. $y \le 2x - 5$
$x < 3y + 2$

52. $x \ge 2y + 6$
$y > -2x + 4$

53. $4x + 3y < 6$
$x - 2y > 4$

54. $3x + y > 4$
$x + 2y < 2$

55. $x \le 2y + 3$
$x + y < 0$

56. $x \le 4y + 3$
$x + y > 0$

57. $x - 3y \le 6$
$x \ge -5$

58. $x - 2y \ge 2$
$x \le -3$

59. $-3x + y \ge 1$
$6x - 2y \ge -10$

60. $x + y < 4$
$-2x - 2y < 4$

61. $2x + 3y < 6$
$4x + 6y > 18$

62. $2x - y < -3$
$6x - 3y > 9$

Extending Skills *Graph the solution set of each system of linear inequalities.*

63. $4x + 5y < 8$
$y > -2$
$x > -4$

64. $x - 2y \ge -2$
$y \ge -2$
$x \le 3$

65. $x + y \ge -3$
$x - y \le 3$
$y \le 3$

66. $x + y < 4$
$x - y > -4$
$y > -1$

RELATING CONCEPTS For Individual or Group Work (Exercises 67–72)

Linear programming *is a method for finding the optimal (best possible) solution that meets all the conditions for a problem such as the following.*

A factory can have no more than 200 workers on a shift, but must have at least 100 and must manufacture at least 3000 units at minimum cost. How many workers should be on a shift in order to produce the required units at minimal cost?

Let x represent the number of workers and y represent the number of units manufactured. **Work Exercises 67–72 in order.**

67. Write three inequalities expressing the problem conditions.

68. Graph the inequalities from **Exercise 67** using the axes at the right, and shade the intersection.

69. The cost per worker is $50 per day and the cost to manufacture 1 unit is $100. Write an equation in x, y, and C representing the total daily cost C.

70. Find values of x and y for several points in or on the boundary of the shaded region. Include any "corner points," where C is maximized or minimized.

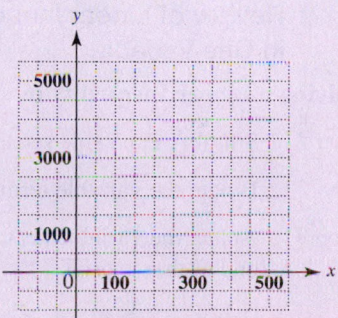

71. Of the values of x and y found in **Exercise 70,** which ones give the least value when substituted in the cost equation from **Exercise 69?**

72. What does the answer in **Exercise 71** mean in terms of the given problem?

Chapter 8	Summary

Key Terms

8.1

inequality
interval
linear inequality in one
    variable
three-part inequality

8.2

intersection
compound inequality
union

8.3

absolute value equation
absolute value inequality

8.4

linear inequality in two
    variables
boundary line
system of linear inequalities
solution set of a system of
    linear inequalities

New Symbols

∞ infinity
$-\infty$ negative infinity

$(-\infty, \infty)$ the set of real
           numbers
(a, b) interval notation
           for $a < x < b$

$[a, b]$ interval notation
          for $a \le x \le b$

$\cap$ set intersection
$\cup$ set union

Test Your Word Power

See how well you have learned the vocabulary in this chapter.

1. The **intersection** of two sets A and B
is the set of elements that belong
  A. to both A and B
  B. to either A or B, or both
  C. to either A or B, but not both
  D. to just A.

2. The **union** of two sets A and B is the
set of elements that belong
  A. to both A and B
  B. to either A or B, or both
  C. to either A or B, but not both
  D. to just B.

3. A **linear inequality in two variables**
is an inequality that can be written in
the form
  A. $Ax + By < C$ or $Ax + By > C$
     ($\le$ or $\ge$ can be used)
  B. $ax < b$
  C. $y \ge x^2$
  D. $Ax + By = C$.

ANSWERS

1. A; *Example:* If $A = \{2, 4, 6, 8\}$ and $B = \{1, 2, 3\}$, then $A \cap B = \{2\}$. **2.** B; *Example:* Using the sets A and B
from Answer 1, $A \cup B = \{1, 2, 3, 4, 6, 8\}$. **3.** A; *Examples:* $4x + 3y < 12$, $x > 6y$, $2x \ge 4y + 5$

Quick Review

CONCEPTS

EXAMPLES

8.1 **Review of Linear Inequalities
in One Variable**

Solving a Linear Inequality in One Variable

Step 1   Simplify each side separately.

Step 2   Isolate the variable terms on one side.

Step 3   Isolate the variable (x) to write the inequality in
one of these forms.

$$x < k, \quad x \le k, \quad x > k, \quad \text{or} \quad x \ge k$$

*If an inequality is multiplied or divided by a negative
number, the direction of the inequality symbol must be
reversed.*

Solve each inequality.

$$3(x + 2) - 5x \le 12$$

$\quad 3x + 6 - 5x \le 12$ Distributive property

$\quad\quad\quad -2x + 6 \le 12$ Combine like terms.

$\quad -2x + 6 - 6 \le 12 - 6$ Subtract 6.

$\quad\quad\quad\quad\quad -2x \le 6$ Combine like terms.

$\quad\quad\quad\quad \dfrac{-2x}{-2} \ge \dfrac{6}{-2}$ Divide by -2.
                         Change $\le$ to $\ge$.

$\quad\quad\quad\quad\quad\quad x \ge -3$

The solution set $[-3, \infty)$ is graphed here.

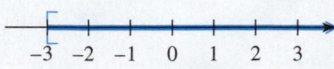

CONCEPTS	EXAMPLES
Solving a Three-Part Inequality Work with all three parts at the same time.	$-4 < \quad 2x + 3 \quad \le 7$ $-4 - 3 < 2x + 3 - 3 \le 7 - 3$ Subtract 3. $-7 < \quad 2x \quad \le 4$ $\dfrac{-7}{2} < \quad \dfrac{2x}{2} \quad \le \dfrac{4}{2}$ Divide by 2. $-\dfrac{7}{2} < \quad x \quad \le 2$ The solution set $\left(-\dfrac{7}{2}, 2\right]$ is graphed here. 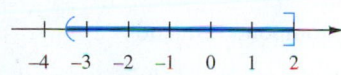

8.2   Set Operations and Compound Inequalities

Solving a Compound Inequality

Step 1 Solve each inequality in the compound inequality individually.

Step 2 If the inequalities are joined with *and,* then the solution set is the intersection of the two individual solution sets.

If the inequalities are joined with *or,* then the solution set is the union of the two individual solution sets.

Solve each compound inequality.

$$x + 1 > 2 \quad \text{and} \quad 2x < 6$$

$$x > 1 \quad \text{and} \quad x < 3$$

The solution set is $(1, 3)$.

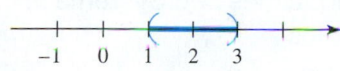

$$2(x + 3) - 2 \le 4 \quad \text{or} \quad -4x \le -16$$

$$2x + 6 - 2 \le 4 \quad \text{or} \quad \frac{-4x}{-4} \ge \frac{-16}{-4}$$

$$2x + 4 \le 4 \quad \text{or} \quad x \ge 4$$

$$x \le 0$$

The solution set is $(-\infty, 0] \cup [4, \infty)$.

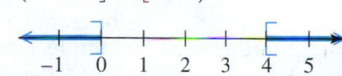

8.3   Absolute Value Equations and Inequalities

Solving Absolute Value Equations and Inequalities
Let k be a positive number.

To solve $|ax + b| = k,$ solve the following compound equation.

$$ax + b = k \quad \text{or} \quad ax + b = -k$$

To solve $|ax + b| > k,$ solve the following compound inequality.

$$ax + b > k \quad \text{or} \quad ax + b < -k$$

Solve each equation or inequality.

$$|x - 7| = 3$$

$$x - 7 = 3 \quad \text{or} \quad x - 7 = -3$$

$$x = 10 \quad \text{or} \quad x = 4 \qquad \text{Add 7.}$$

The solution set is $\{4, 10\}$.

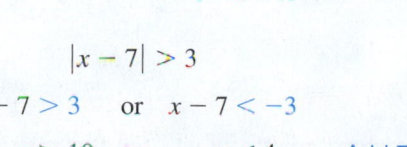

$$|x - 7| > 3$$

$$x - 7 > 3 \quad \text{or} \quad x - 7 < -3$$

$$x > 10 \quad \text{or} \quad x < 4 \qquad \text{Add 7.}$$

The solution set is $(-\infty, 4) \cup (10, \infty)$.

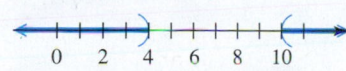

CONCEPTS	EXAMPLES

To solve $|ax + b| < k,$ solve the following compound inequality.

$$-k < ax + b < k$$

$$|x - 7| < 3$$
$$-3 < x - 7 < 3$$
$$4 < \quad x \quad < 10 \qquad \text{Add 7.}$$

The solution set is $(4, 10)$.

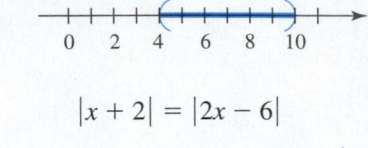

To solve an absolute value equation of the form

$$|ax + b| = |cx + d|,$$

solve the following compound equation.

$$ax + b = cx + d \quad \text{or} \quad ax + b = -(cx + d)$$

$$|x + 2| = |2x - 6|$$
$$x + 2 = 2x - 6 \quad \text{or} \quad x + 2 = -(2x - 6)$$
$$x = 8 \qquad \text{or} \quad x + 2 = -2x + 6$$
$$3x = 4$$
$$x = \frac{4}{3}$$

The solution set is $\left\{\frac{4}{3}, 8\right\}$.

8.4   Linear Inequalities and Systems in Two Variables

Graphing a Linear Inequality

Step 1   Draw the graph of the straight line that is the boundary. Make the line solid if the inequality involves $\leq$ or $\geq$. Make the line dashed if the inequality involves $<$ or $>$.

Step 2   Choose any point not on the line as a test point. Substitute the coordinates of that point in the inequality.

Step 3   Shade the region that includes the test point if it satisfies the original inequality. Otherwise, shade the region on the other side of the boundary line.

Graph $2x - 3y \leq 6$.

Draw the graph of $2x - 3y = 6$. Use a solid line because of the inclusion of equality in the symbol $\leq$.

Choose $(0, 0)$ as a test point.

$$2(0) - 3(0) \overset{?}{<} 6$$
$$0 < 6 \qquad \text{True}$$

Shade the region that includes $(0, 0)$.

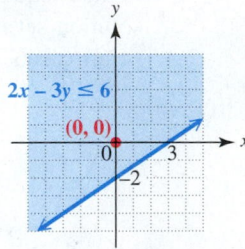

Solving a System of Linear Inequalities

Step 1   Graph each inequality on the same axes.

Step 2   Choose the intersection. The solution set of the system is formed by the overlap of the regions of the graphs.

Graph the solution set of the system.

$$2x + 4y \geq 5$$
$$x \geq 1$$

First graph the solid boundary lines

$$2x + 4y = 5 \quad \text{and} \quad x = 1.$$

Then use a test point, such as $(0, 0)$, to determine the region to shade for each inequality. The intersection, the gray shaded region, is the solution set of the system.

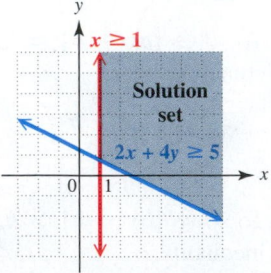

Chapter 8   Review Exercises

1. $(-9, \infty)$ **2.** $(-\infty, -3]$

3. $\left(\dfrac{3}{2}, \infty\right)$ **4.** $[-3, \infty)$

5. $[3, 5)$ **6.** $\left(\dfrac{59}{31}, \infty\right)$

7. $\{a, c\}$ **8.** $\{a\}$

9. $\{a, c, e, f, g\}$

10. $\{a, b, c, d, e, f, g\}$

11. $(6, 9)$

12. $(8, 14)$

13. $(-\infty, -3] \cup (5, \infty)$

14. $(-\infty, \infty)$

15. $\varnothing$

16. $(-\infty, -2] \cup [7, \infty)$

17. $(-3, 4)$ **18.** $(-\infty, 2)$

19. $(4, \infty)$ **20.** $(1, \infty)$

21. $\{-7, 7\}$ **22.** $\{-11, 7\}$

23. $\left\{-\dfrac{1}{3}, 5\right\}$ **24.** $\varnothing$

25. $\{0, 7\}$ **26.** $\left\{-\dfrac{3}{2}, \dfrac{1}{2}\right\}$

27. $\left\{-\dfrac{3}{4}, \dfrac{1}{2}\right\}$ **28.** $\left\{-\dfrac{1}{2}\right\}$

29. $(-14, 14)$ **30.** $[-1, 13]$

31. $[-3, -2]$ **32.** $(-\infty, \infty)$

33. $\varnothing$ **34.** $(-\infty, \infty)$

35. 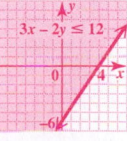   **36.**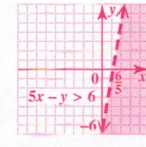

37. 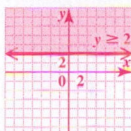   **38.**

39.    **40.**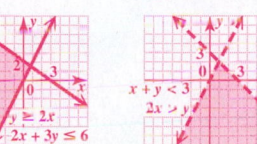

8.1 *Solve each inequality. Give the solution set in interval form.*

1. $-\dfrac{2}{3}x < 6$ **2.** $-5x - 4 \geq 11$

3. $\dfrac{6x + 3}{-4} < -3$ **4.** $5 - (6 - 4x) \geq 2x - 7$

5. $8 \leq 3x - 1 < 14$ **6.** $\dfrac{5}{3}(x - 2) + \dfrac{2}{5}(x + 1) > 1$

8.2 *Let $A = \{a, b, c, d\}$, $B = \{a, c, e, f\}$, and $C = \{a, e, f, g\}$. Find each set.*

7. $A \cap B$ **8.** $A \cap C$ **9.** $B \cup C$ **10.** $A \cup C$

Solve each compound inequality. Give the solution set in both interval and graph forms.

11. $x > 6$ and $x < 9$ **12.** $x + 4 > 12$ and $x - 2 < 12$

13. $x > 5$ or $x \leq -3$ **14.** $x \geq -2$ or $x < 2$

15. $x - 4 > 6$ and $x + 3 \leq 10$ **16.** $-5x + 1 \geq 11$ or $3x + 5 \geq 26$

Express each set in simplest interval form.

17. $(-3, \infty) \cap (-\infty, 4)$ **18.** $(-\infty, 6) \cap (-\infty, 2)$

19. $(4, \infty) \cup (9, \infty)$ **20.** $(1, 2) \cup (1, \infty)$

8.3 *Solve each equation.*

21. $|x| = 7$ **22.** $|x + 2| = 9$

23. $|3x - 7| = 8$ **24.** $|x - 4| = -12$

25. $|2x - 7| + 4 = 11$ **26.** $|4x + 2| - 7 = -3$

27. $|3x + 1| = |x + 2|$ **28.** $|2x - 1| = |2x + 3|$

Solve each inequality.

29. $|x| < 14$ **30.** $|-x + 6| \leq 7$

31. $|2x + 5| \leq 1$ **32.** $|x + 1| \geq -3$

33. $|3 - 4x| + 7 < -4$ **34.** $|-8 - 3x| - 7 > -8$

8.4 *Graph each linear inequality in two variables.*

35. $3x - 2y \leq 12$ **36.** $5x - y > 6$ **37.** $y \geq 2$

Graph the solution set of each system of linear inequalities.

38. $x + y \geq 2$ **39.** $y \geq 2x$ **40.** $x + y < 3$

       $x - y \leq 4$ $2x + 3y \leq 6$ $2x > y$

Chapter 8   Mixed Review Exercises

Solve.

1. $(-2, \infty)$ **2.** $[-2, 3)$

3. $(-\infty, \infty)$ **4.** $(-\infty, 2]$

5. $\left\{-\dfrac{7}{3}, 1\right\}$ **6.** $[-16, 10]$

7. $\left(-\infty, -\dfrac{13}{5}\right) \cup (3, \infty)$

8. $(-\infty, \infty)$

9. $\left\{-4, -\dfrac{2}{3}\right\}$ **10.** $\left\{1, \dfrac{11}{3}\right\}$

11. $(6, 8)$

12. $(-\infty, -2] \cup [7, \infty)$

13. D   **14.** B

15.

1. $5 - (6 - 4x) > 2x - 5$ **2.** $x + 4 < 7$ and $x + 5 \geq 3$

3. $|3x + 6| \geq 0$ **4.** $-5x \geq -10$

5. $|3x + 2| + 4 = 9$ **6.** $|x + 3| \leq 13$

7. $|5x - 1| > 14$ **8.** $x \geq -2$ or $x < 4$

9. $|x - 1| = |2x + 3|$ **10.** $|3x - 7| = 4$

11. $-5x < -30$ and $-7x > -56$ **12.** $-5x + 1 \geq 11$ or $3x + 5 \geq 26$

Solve each problem.

13. Which inequality has as its graph a dashed boundary line and shading below the line?

     A. $y \geq 4x + 3$ **B.** $y > 4x + 3$ **C.** $y \leq 4x + 3$ **D.** $y < 4x + 3$

14. Which system of linear inequalities is graphed in the figure?

     A. $x \leq 3$ **B.** $x \leq 3$ **C.** $x \geq 3$ **D.** $x \geq 3$

           $y \leq 1$ $y \geq 1$ $y \leq 1$ $y \geq 1$

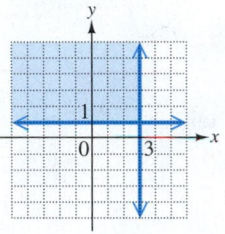

15. Graph the solution set of the system of linear inequalities.

$$x + y < 5$$
$$x - y \geq 2$$

Chapter 8   Test

FOR EXTRA HELP   *Step-by-step test solutions are found on the Chapter Test Prep Videos available in* MyMathLab® *or on* YouTube.

▶ *View the complete solutions to all Chapter Test exercises in MyMathLab.*

[8.1]

1. $[1, \infty)$

2. $(-\infty, 28)$

3. $(1, 2)$

4. C

[8.2]

5. (a) $\{1, 5\}$

     (b) $\{1, 2, 5, 7, 9, 12\}$

6. $[2, 9)$

7. $(-\infty, 3) \cup [6, \infty)$

[8.3]

8. $\left[-\dfrac{5}{2}, 1\right]$

9. $\left(-\infty, -\dfrac{7}{6}\right) \cup \left(\dfrac{17}{6}, \infty\right)$

10. $\{1, 5\}$

11. $\left(\dfrac{1}{3}, \dfrac{7}{3}\right)$ **12.** $\varnothing$

Solve each inequality. Give the solution set in both interval and graph forms.

1. $4 - 6(x + 3) \leq -2 - 3(x + 6) + 3x$ **2.** $-\dfrac{4}{7}x > -16$

3. $-1 < 3x - 4 < 2$

Solve each problem.

4. Which one of the following inequalities is equivalent to $x < -3$?

     A. $-3x < 9$ **B.** $-3x > -9$ **C.** $-3x > 9$ **D.** $-3x < -9$

5. Let $A = \{1, 2, 5, 7\}$ and $B = \{1, 5, 9, 12\}$. Find each set.

     (a) $A \cap B$ **(b)** $A \cup B$

Solve each compound or absolute value equation or inequality.

6. $3x \geq 6$ and $x < 9$ **7.** $-4x \leq -24$ or $4x < 12$

8. $|4x + 3| \leq 7$ **9.** $|5 - 6x| > 12$

10. $|3x - 9| = 6$ **11.** $|-3x + 4| - 4 < -1$

12. $|7 - x| \leq -1$ **13.** $|3x - 2| + 1 = 8$ **14.** $|3 - 5x| = |2x + 8|$

15. If $k < 0$, what is the solution set of each of the following?

     (a) $|8x - 5| < k$ **(b)** $|8x - 5| > k$ **(c)** $|8x - 5| = k$

13. $\left\{-\dfrac{5}{3}, 3\right\}$ **14.** $\left\{-\dfrac{5}{7}, \dfrac{11}{3}\right\}$

15. (a) $\varnothing$ (b) $(-\infty, \infty)$ (c) $\varnothing$

[8.4]

16. 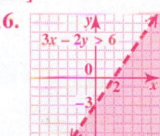 **17.**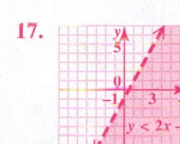

18. 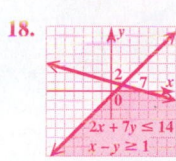 **19.**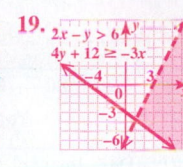

20. B

Graph each linear inequality in two variables.

16. $3x - 2y > 6$ **17.** $y < 2x - 1$

Graph the solution set of each system of linear inequalities.

18. $2x + 7y \le 14$
   $x - y \ge 1$

19. $2x - y > 6$
   $4y + 12 \ge -3x$

20. Without actually graphing, determine which one of the following systems of inequalities has no solution.

A. $x \ge 4$
   $y \le 3$

B. $x + y > 4$
   $x + y < 3$

C. $x > 2$
   $y < 1$

D. $x + y > 4$
   $x - y < 3$

Chapters R–8   Cumulative Review Exercises

[1.3]

1. (a) A, B, C, D, F
   (h) B, C, D, F
   (c) D, F (d) C, D, F
   (e) E, F (f) D, F

[1.1]   **[1.3, 1.4]**
2. 32   **3.** 0

[2.3]
4. $\{-65\}$
5. {all real numbers}

[2.5]   **[2.8, 8.1]**
6. $t = \dfrac{A - p}{pr}$ **7.** $(-\infty, 6)$

[R.2, 2.6]
8. 1650; 32%; 21%; 700

[3.3, 7.1]
9. $-\dfrac{4}{3}$ **10.** 0

[3.4, 3.5, 7.2]
11. (a) $y = -4x + 15$
   (b) $4x + y = 15$
12. (a) $y = 4x$ (b) $4x - y = 0$

[3.2, 7.1]   **[8.4]**
13.   **14.**

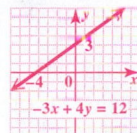 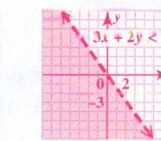

---

1. Match each number in Column I with the choice or choices of sets of numbers in Column II to which the number belongs.

I		II	
(a) 34	(b) 0	**A.** Natural numbers	**B.** Whole numbers
(c) 2.16	(d) −6	**C.** Integers	**D.** Rational numbers
(e) $\sqrt{13}$	(f) $-\dfrac{4}{5}$	**E.** Irrational numbers	**F.** Real numbers

Evaluate.

2. $9 \cdot 4 - 16 \div 4$

3. $-|8 - 13| - |-4| + |-9|$

Solve.

4. $-5(8 - 2z) + 4(7 - z) = 7(8 + z) - 3$ **5.** $3(x + 2) - 5(x + 2) = -2x - 4$

6. $A = p + prt$ for t

7. $2(m + 5) - 3m + 1 > 5$

8. A survey polled Internet users age 12 and older about their most frequent daily Internet activities. Complete the results shown in the table if 5000 such users were surveyed.

Internet Activity	Percent	Number
Look for news	33%	
Check email		1600
Find or check a fact		1050
Play games	14%	

Source: The 2013 Digital Future Report.

Find the slope of each line described.

9. Through $(-4, 5)$ and $(2, -3)$ **10.** Through $(4, 5)$; horizontal

Write an equation of each line that satisfies the given conditions. Give the equation (a) in slope-intercept form and (b) in standard form.

11. Through $(4, -1)$; $m = -4$ **12.** Through $(0, 0)$ and $(1, 4)$

Graph each equation or inequality.

13. $-3x + 4y = 12$ **14.** $3x + 2y < 0$

15. Per capita consumption of whole milk in the United States (in gallons) is shown in the graph, where $x = 0$ represents 1970.

(a) Use the given ordered pairs to find the average rate of change in per capita consumption of whole milk (in gallons, to the nearest hundredth) per year during this period. Interpret the answer.

(b) Use the answer from part (a) to write an equation of the line in slope-intercept form that models per capita consumption of whole milk y (in gallons).

(c) Use the equation from part (b) to approximate per capita consumption of whole milk in 2000.

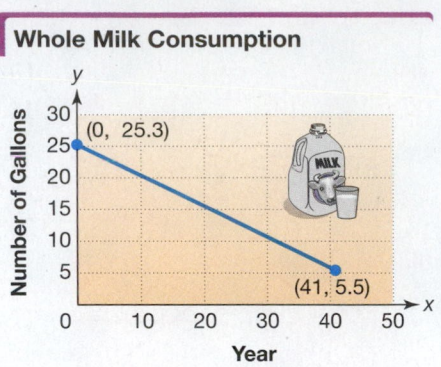

Whole Milk Consumption

Source: U.S. Department of Agriculture.

Simplify. Write answers with only positive exponents. Assume that all variables represent positive real numbers.

16. $\left(\dfrac{2m^3n}{p^2}\right)^3$

17. $\dfrac{x^{-6}y^3z^{-1}}{x^7y^{-4}z}$

Perform the indicated operations.

18. $(3x^2 - 8x + 1) - (x^2 - 3x - 9)$

19. $(x + 2y)(x^2 - 2xy + 4y^2)$

20. $(3x + 2y)(5x - y)$

21. $\dfrac{16x^3y^5 - 8x^2y^2 + 4}{4x^2y}$

Factor each polyomial completely.

22. $m^2 + 12m + 32$

23. $25t^4 - 36$

24. $81z^2 + 72z + 16$

Perform each indicated operation. Express answers in lowest terms.

25. $\dfrac{x^2 - 3x - 4}{x^2 + 3x} \cdot \dfrac{x^2 + 2x - 3}{x^2 - 5x + 4}$

26. $\dfrac{t^2 + 4t - 5}{t + 5} \div \dfrac{t - 1}{t^2 + 8t + 15}$

27. $\dfrac{2}{x + 3} - \dfrac{4}{x - 1}$

28. $\dfrac{\dfrac{2}{3} + \dfrac{1}{2}}{\dfrac{1}{9} - \dfrac{1}{6}}$

Solve each equation.

29. $(x + 4)(x - 1) = -6$

30. $\dfrac{x}{x + 8} - \dfrac{3}{x - 8} = \dfrac{128}{x^2 - 64}$

Solve each system.

31. $3x - 4y = 1$
    $2x + 3y = 12$

32. $3x - 2y = 4$
    $-6x + 4y = 7$

33. $x + 3y - 6z = 7$
    $2x - y + z = 1$
    $x + 2y + 2z = -1$

34. The Star-Spangled Banner that flew over Fort McHenry during the War of 1812 had a perimeter of 144 ft. Its length measured 12 ft more than its width. Find the dimensions of this flag, which is displayed in the Smithsonian Institution's Museum of American History in Washington, DC. (*Source:* National Park Service brochure.)

Solve each equation or inequality.

35. $x > -4$ and $x < 4$

36. $2x + 1 > 5$ or $2 - x \geq 2$

37. $|3x - 1| = 2$

38. $|3z + 1| \geq 7$

Relations and Functions

Linear equations whose graphs are straight (nonvertical) lines define *linear functions,* one of the topics of this chapter. We use the concept of slope, or steepness, to graph such functions.

9.1 Introduction to Relations and Functions

VOCABULARY

☐ relation
☐ function
☐ dependent variable
☐ independent variable
☐ domain
☐ range

OBJECTIVE 1 Define and identify relations and functions.

Consider the relationship illustrated in the following table between number of hours worked and paycheck amount for an hourly worker.

Number of Hours Worked	Paycheck Amount (in dollars)	Ordered Pairs
5	40	⟶ (5, 40)
10	80	⟶ (10, 80)
20	160	⟶ (20, 160)
40	320	⟶ (40, 320)

The data from the table can be represented by a set of ordered pairs.

$$\{(5, 40), (10, 80), (20, 160), (40, 320)\}$$

Number of hours worked ↑ ↑ Paycheck amount in dollars

Each first component of the ordered pairs represents a number of hours worked, and each second component represents the corresponding paycheck amount. Such a set of ordered pairs is a *relation*.

Relation

A **relation** is any set of ordered pairs.

**NOW TRY
EXERCISE 1**

Write the relation as a set of ordered pairs.

Year	Average Gas Price per Gallon (in dollars)
2000	1.56
2005	2.34
2010	2.84
2015	3.39

Source: Energy Information Administration.

EXAMPLE 1 Writing Ordered Pairs for a Relation

Write the relation as a set of ordered pairs.

Number of Gallons of Gas	Cost (in dollars)
0	0
1	3.50
2	7.00
3	10.50
4	14.00

This table is from **Section 7.2, Example 7.**

The data in the table defines a relation between number of gallons of gas and cost and can be written as the following set of ordered pairs.

$$\{(0, 0), (1, 3.50), (2, 7.00), (3, 10.50), (4, 14.00)\}$$

Number of gallons of gas ↑ ↑ Cost in dollars

NOW TRY

A *function* is a special kind of relation.

NOW TRY ANSWER
1. {(2000, 1.56), (2005, 2.34), (2010, 2.84), (2015, 3.39)}

> **Function**
>
> A **function** is a relation in which, for each distinct value of the first component of the ordered pairs, there is *exactly one value* of the second component.

**NOW TRY
EXERCISE 2**

Determine whether each relation defines a function.

(a) $\{(1, 5), (3, 5), (5, 5)\}$

(b) $\{(-1, -3), (0, 2), (-1, 6)\}$

EXAMPLE 2 Determining Whether Relations Are Functions

Determine whether each relation defines a function.

(a) $F = \{(1, 2), (-2, 4), (3, -1)\}$

Look at the ordered pairs that define this relation.

> For $x = 1$, there is only one value of y, 2.
>
> For $x = -2$, there is only one value of y, 4.
>
> For $x = 3$, there is only one value of y, -1.

Relation F is a function—for each distinct x-value, there is *exactly one y*-value.

(b) $G = \{(-2, -1), (-1, 0), (0, 1), (1, 2), (2, 2)\}$

Relation G is also a function. Although the last two ordered pairs have the same y-value (1 is paired with 2, and 2 is paired with 2), this does not violate the definition of a function. The first components (x-values) are distinct, and each is paired with only one second component (y-value).

(c) $H = \{(-4, 1), (-2, 1), (-2, 0)\}$

In relation H, the last two ordered pairs have the **same** x-value paired with **two different** y-values (-2 is paired with both 1 and 0). H is a relation, but *not* a function. *In a function, no two ordered pairs have the same first component and different second components.*

Different y-values

Relation $H = \{(-4, 1), (-2, 1), (-2, 0)\}$ Not a function

Same x-value   **NOW TRY**

Relations may be defined in several different ways.

- **A relation may be defined as a set of ordered pairs. (See Example 2.)**

Relation $F = \{(1, 2), (-2, 4), (3, -1)\}$ Function

Relation $H = \{(-4, 1), (-2, 1), (-2, 0)\}$ Not a function

- **A relation may be defined as a correspondence or *mapping*.**

See **FIGURE 1**. In the mapping for relation F from **Example 2(a),** 1 is mapped to 2, -2 is mapped to 4, and 3 is mapped to -1. Thus, F is a function—each first component of an ordered pair is paired with exactly one second component.

In the mapping for relation H from **Example 2(c),** which is *not* a function, the first component -2 is paired with two different second components.

NOW TRY ANSWERS
2. (a) function
 (b) not a function

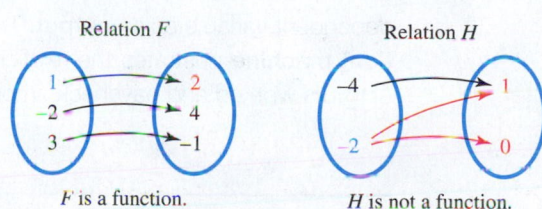

Relation F Relation H

F is a function. H is not a function.

FIGURE 1

- **A relation may be defined as a table.**
- **A relation may be defined as a graph.**

  FIGURE 2 includes a table and graph for relation F, which is a function, from **Example 2(a).**

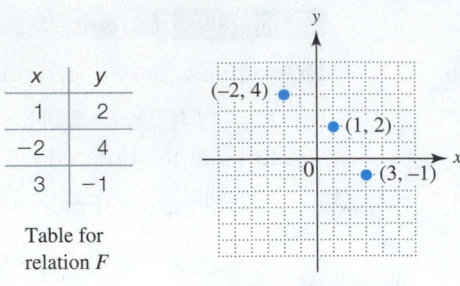

x	y
1	2
-2	4
3	-1

Table for
relation F

Graph of relation F

FIGURE 2

- **A relation may be defined as an equation (or rule).**

  The solutions of an equation give an infinite set of ordered pairs. For example, if the value of y is twice the value of x, the equation is

  $$y = 2x.$$

  The infinite number of ordered-pair solutions (x, y) can be represented by the graph in **FIGURE 3**.

  In the equation $y = 2x$, the value of y *depends* on the value of x. Thus, the variable y is the **dependent variable.** The variable x is the **independent variable.**

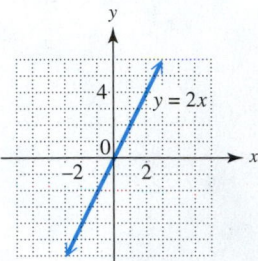

Graph of the relation $y = 2x$

FIGURE 3

Dependent variable $\longrightarrow y = 2x \longleftarrow$ Independent variable

An equation tells how to determine the value of the dependent variable for a specific value of the independent variable.

NOTE   An equation that describes the relationship given at the beginning of this section between number of hours worked and paycheck amount is

$$y = 8x.   \text{8 represents the hourly rate, \$8.}$$

Paycheck amount y *depends* on number of hours worked x. Thus, paycheck amount is the dependent variable and number of hours worked is the independent variable.

In a function, there is exactly one value of the dependent variable, the second component, for each value of the independent variable, the first component.

NOTE   Another way to think of a function relationship is to think of the independent variable as an **input** and the dependent variable as an **output.** This **input-output (function) machine** illustrates the relationship between number of hours worked and paycheck amount.

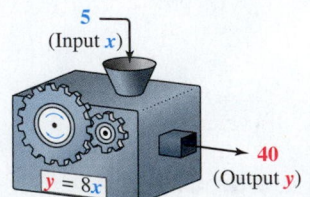

Function machine

OBJECTIVE 2 Find the domain and range.

> **Domain and Range**
>
> For every relation defined by a set of ordered pairs (x, y), there are two important sets of elements.
>
> - The set of all values of the independent variable (x) is the **domain.**
>
> - The set of all values of the dependent variable (y) is the **range.**

**NOW TRY
EXERCISE 3**

Give the domain and range of each relation. Decide whether the relation defines a function.

(a) $\{(2, 2), (2, 5), (4, 8)\}$

(b) The table from **Objective 1**

Number of Hours Worked	Paycheck Amount (in dollars)
5	40
10	80
20	160
40	320

EXAMPLE 3 Finding Domains and Ranges of Relations

Give the domain and range of each relation. Decide whether the relation defines a function.

(a) $\{(3, -1), (4, 2), (4, 5), (6, 8)\}$ *Only list 4 once.*

Domain: $\{3, 4, 6\}$ *Set of x-values*

Range: $\{-1, 2, 5, 8\}$ *Set of y-values*

This relation is not a function because the same x-value 4 is paired with two different y-values, 2 and 5.

(b)

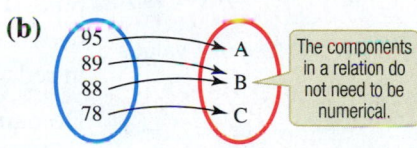

The components in a relation do not need to be numerical.

This mapping represents the following set of ordered pairs.

$$\{(95, A), (89, B), (88, B), (78, C)\}$$

Domain: $\{95, 89, 88, 78\}$ *Set of first components*

Range: $\{A, B, C\}$ *Set of second components*

The mapping defines a function—each domain value corresponds to exactly one range value.

(c)

x	y
-5	2
0	2
5	2

This table represents the following set of ordered pairs.

$$\{(-5, 2), (0, 2), (5, 2)\}$$

Domain: $\{-5, 0, 5\}$ *Set of x-values*

Range: $\{2\}$ *Set of y-values*

The table defines a function—each distinct x-value corresponds to exactly one y-value (even though it is the same y-value).

NOW TRY ANSWERS
3. **(a)** domain: $\{2, 4\}$;
   range: $\{2, 5, 8\}$;
   not a function
   (b) domain: $\{5, 10, 20, 40\}$;
   range: $\{40, 80, 160, 320\}$;
   function

NOW TRY

A graph gives a "picture" of a relation and can be used to determine its domain and range.

NOTE   Pay particular attention to the use of color to interpret domain and range in **Example 4**—blue for domain and red for range.

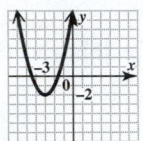

**NOW TRY
EXERCISE 4**

Give the domain and range of the relation.

EXAMPLE 4   Finding Domains and Ranges from Graphs

Give the domain and range of each relation.

(a)

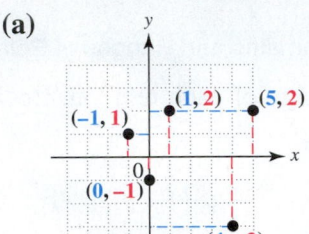

This relation includes the five ordered pairs that are graphed.

$$\{(-1, 1), (0, -1), (1, 2), (4, -3), (5, 2)\}$$

Domain: $\{-1, 0, 1, 4, 5\}$ Set of *x*-values

Range: $\{1, -1, 2, -3\}$ Set of *y*-values

> Only list 2 once.

(b)

The *x*-values of the ordered pairs that form the graph include all numbers between -4 and 4, inclusive, as shown in blue. The *y*-values include all numbers between -6 and 6, inclusive, as shown in red.

Domain: $[-4, 4]$ Use interval notation.

Range: $[-6, 6]$ (Section 8.1)

(c)

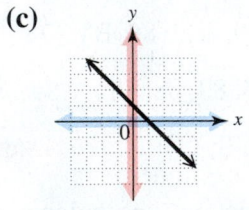

The arrowheads on the graphed line indicate that the line extends indefinitely left and right, as well as up and down. Therefore, both the domain, shown in blue, (the set of *x*-values) and the range, shown in red, (the set of *y*-values) include all real numbers.

Domain: $(-\infty, \infty)$ Range: $(-\infty, \infty)$

(d)

The graphed curve extends indefinitely left and right, as well as upward. The domain, shown in blue, includes all real numbers. Because there is a least *y*-value, -3, the range, shown in red, includes all numbers greater than or equal to -3.

Domain: $(-\infty, \infty)$ Range: $[-3, \infty)$

NOW TRY

OBJECTIVE 3   Identify functions defined by graphs and equations.

Because each value of *x* in a function corresponds to only one value of *y*, any vertical line drawn through the graph of a function must intersect the graph in at most one point. This is the *vertical line test* for a function.

FIGURE 4 on the next page illustrates this test with the graphs of two relations.

NOW TRY ANSWER
4. domain: $(-\infty, \infty)$;
   range: $[-2, \infty)$

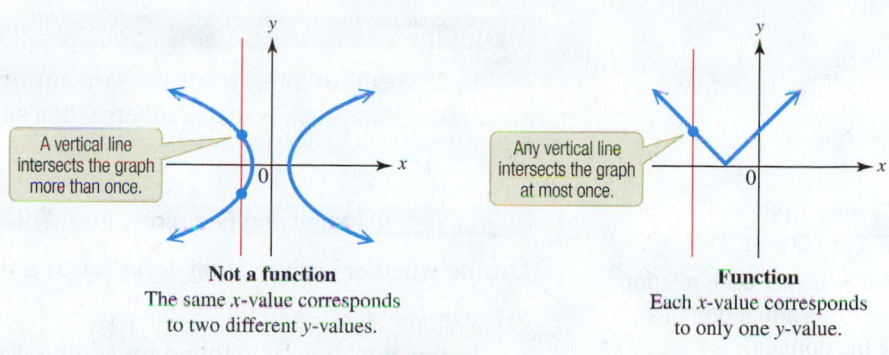

FIGURE 4

Vertical Line Test

If every vertical line intersects the graph of a relation in no more than one point, then the relation represents a function.

NOW TRY EXERCISE 5

Use the vertical line test to determine whether the relation is a function.

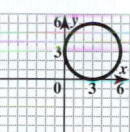

EXAMPLE 5 Using the Vertical Line Test

Use the vertical line test to determine whether each relation graphed in **Example 4** is a function. (We repeat the graphs here.)

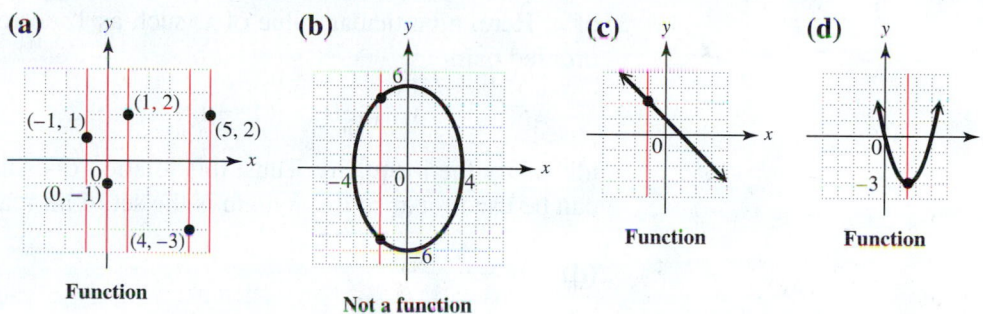

The graphs in (a), (c), and (d) satisfy the vertical line test and represent functions. The graph in (b) fails the vertical line test because a vertical line intersects the graph more than once—that is, the same x-value corresponds to two different y-values. This is not the graph of a function.

NOW TRY

NOTE Graphs that do not represent functions are still relations. *All equations and graphs represent relations, and all relations have a domain and range.*

If a relation is defined by an equation involving a fraction, keep the following in mind when finding its domain.

Exclude from the domain any values that make the denominator equal to 0.

Example: The function $y = \dfrac{1}{x}$ has all real numbers *except* 0 as its domain because division by 0 is undefined.

NOTE As we will see in Section 10.1, we must also exclude from the domain any values that result in an even root of a negative number.

Example: The function $y = \sqrt{x}$ has all *nonnegative* real numbers as its domain because the square root of a negative number is not real.

Agreement on Domain

Unless specified otherwise, the domain of a relation is assumed to be all real numbers that produce real numbers when substituted for the independent variable.

**NOW TRY
EXERCISE 6**

Decide whether each relation defines y as a function of x. Give the domain.

(a) $y = 4x - 3$

(b) $y = \dfrac{1}{x - 2}$

(c) $y < 3x + 1$

EXAMPLE 6   Identifying Functions from Their Equations

Decide whether each relation defines y as a function of x. Give the domain.

(a) $y = x + 4$

In this equation, y is found by adding 4 to x. Thus, each value of x corresponds to just one value of y, and the relation defines a function. Because x can be any real number, the domain is $(-\infty, \infty)$.

(b) $y^2 = x$

The ordered pairs $(16, 4)$ and $(16, -4)$ both satisfy this equation. One value of x, 16, corresponds to two values of y, 4 and -4, so this equation does not define a function. Because x is equal to the square of y, the values of x must always be nonnegative. The domain of the relation is $[0, \infty)$.

(c) $y \leq x - 1$

By definition, y is a function of x if every value of x leads to exactly one value of y. Here, a particular value of x, such as 1, corresponds to many values of y. The ordered pairs

$$(1, 0), \quad (1, -1), \quad (1, -2), \quad (1, -3), \quad \text{and so on}$$

all satisfy the inequality. Thus, this relation does not define a function. Any number can be used for x, so the domain is the set of all real numbers, $(-\infty, \infty)$.

(d) $y = \dfrac{5}{x - 1}$

Given any value of x in the domain, we find y by subtracting 1 and then dividing the result into 5. This process produces exactly one value of y for each value in the domain, so the given equation defines a function.

The domain includes all real numbers except those which make the denominator 0.

$$x - 1 = 0 \qquad \text{Set the denominator equal to 0.}$$

$$x = 1 \qquad \text{Add 1.}$$

The domain includes all real numbers *except* 1, written $(-\infty, 1) \cup (1, \infty)$.

NOW TRY

In summary, we give three variations of the definition of a function.

Variations of the Definition of a Function

1. A **function** is a relation in which, for each distinct value of the first component of the ordered pairs, there is exactly one value of the second component.

2. A **function** is a set of distinct ordered pairs in which no first component is repeated.

3. A **function** is a correspondence (mapping) or an equation (rule) that assigns exactly one range value to each distinct domain value.

NOW TRY ANSWERS

6. (a) yes; $(-\infty, \infty)$
   (b) yes; $(-\infty, 2) \cup (2, \infty)$
   (c) no; $(-\infty, \infty)$

9.1 Exercises

  MyMathLab®

▶ *Complete solution available in MyMathLab*

Concept Check *Complete each statement. Choices may be used more than once.*

> function   independent variable   vertical line test   relation
>
> domain   ordered pairs   dependent variable   range

1. A(n) _____ is any set of _____ $\{(x, y)\}$.

2. A(n) _____ is a relation in which, for each distinct value of the first component of the _____, there is exactly one value of the second component.

3. In a relation $\{(x, y)\}$, the _____ is the set of x-values, and the _____ is the set of y-values.

4. The relation $\{(0, -2), (2, -1), (2, -4), (5, 3)\}$ (*does / does not*) define a function. The set $\{0, 2, 5\}$ is its _____, and the set $\{-2, -1, -4, 3\}$ is its _____.

5. Consider the function $d = 50t$, where d represents distance and t represents time. The value of d depends on the value of t, so the variable t is the _____, and the variable d is the _____.

6. The _____ is used to determine whether a graph is that of a function. It says that any vertical line can intersect the graph of a(n) _____ in no more than (*zero / one / two*) point(s).

Write each relation as a set of ordered pairs. **See Example 1 and Objective 1.**

7.

x	y
2	-2
2	0
2	1

8.

x	y
-1	-1
0	-1
1	-1

9.

Year	Average Movie Ticket Price (in dollars)
1960	0.76
1980	2.69
2000	5.39
2013	8.38

Source: Motion Picture Association of America.

10.

Year	Average ACT Composite Score
2010	21.0
2011	21.1
2012	21.1
2013	20.9

Source: ACT.

11.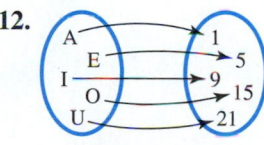

12.

Concept Check *Express each relation using a different form. (For example, if the given form is a set of ordered pairs, use a graph.) There is more than one correct way to do this.* **See Objective 1.**

13. $\{(0, 2), (2, 4), (4, 6)\}$

14.

x	y
-1	-3
0	-1
1	1
3	3

15.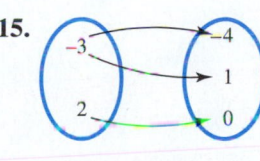

16. Concept Check Does the relation given in **Exercise 15** define a function? Why or why not?

Decide whether each relation defines a function, and give the domain and range. ***See Examples 2–5.***

▶ **17.** $\{(5, 1), (3, 2), (4, 9), (7, 6)\}$ **18.** $\{(8, 0), (5, 4), (9, 3), (3, 8)\}$

▶ **19.** $\{(2, 4), (0, 2), (2, 5)\}$ **20.** $\{(9, -2), (-3, 5), (9, 2)\}$

21. $\{(-3, 1), (4, 1), (-2, 7)\}$ **22.** $\{(-12, 5), (-10, 3), (8, 3)\}$

▶ **23.** $\{(1, 1), (1, -1), (0, 0), (2, 4), (2, -4)\}$ **24.** $\{(2, 5), (3, 7), (4, 9), (5, 11)\}$

25. 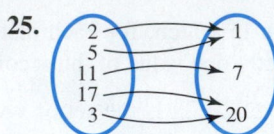   **26.**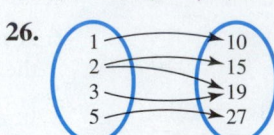

27.	x	y
	1	5
	1	2
	1	-1
	1	-4

28.	x	y
	-4	-4
	-4	0
	-4	4
	-4	8

29.	x	y
	4	-3
	2	-3
	0	-3
	-2	-3

30.	x	y
	-3	-6
	-1	-6
	1	-6
	3	-6

31. 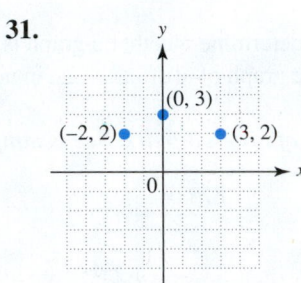   **32.** 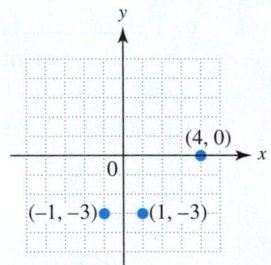   ▶ **33.**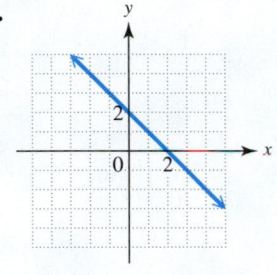

34. 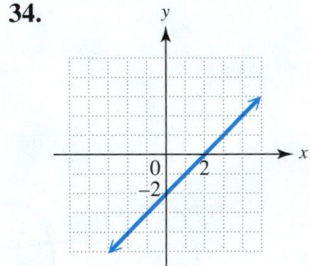   **35.** 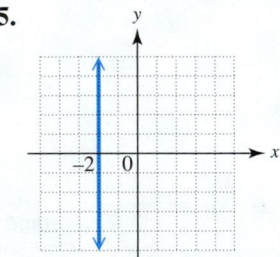   **36.**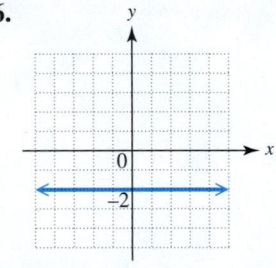

37. 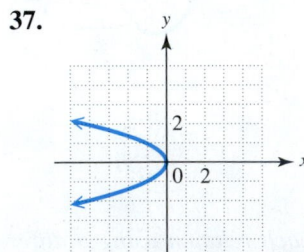   **38.** 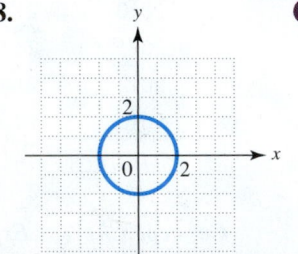   ▶ **39.**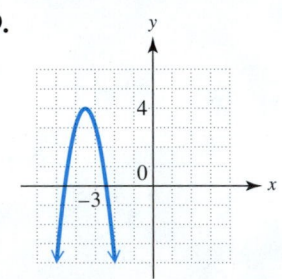

40. 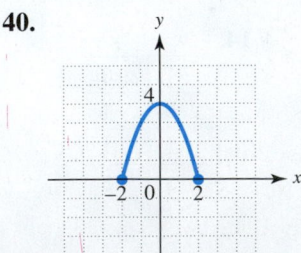   ▶ **41.** 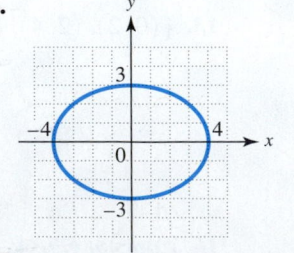   **42.**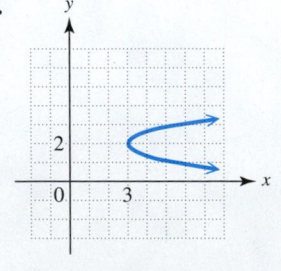

43.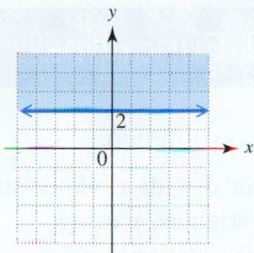

44.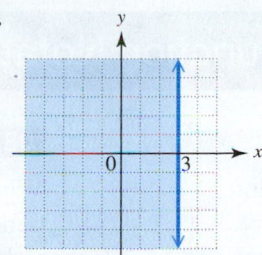

Decide whether each relation defines y as a function of x. Give the domain. **See Example 6.**

45. $y = -6x$

46. $y = -9x$

▶ **47.** $y = 2x - 6$

48. $y = 6x + 8$

49. $y = x^2$

50. $y = x^3$

51. $x = y^6$

52. $x = y^4$

53. $x + y < 4$

54. $x - y < 3$

55. $x = y$

56. $-x = y$

57. $y = \dfrac{x + 4}{5}$

58. $y = \dfrac{x - 3}{2}$

59. $y = -\dfrac{2}{x}$

60. $y = -\dfrac{6}{x}$

61. $y = \dfrac{2}{x - 4}$

62. $y = \dfrac{7}{x - 2}$

63. $y = \dfrac{1}{4x + 2}$

64. $y = \dfrac{1}{2x + 9}$

65. $x = y^2 + 1$

66. $x = y^2 - 3$

67. $xy = 1$

68. $xy = 3$

Solve each problem.

69. The table shows the percentage of students at 4-year public colleges who graduated within 5 years.

Year	Percentage
2009	44.0
2010	43.4
2011	43.1
2012	42.9
2013	43.1

Source: ACT.

(a) Does the table define a function?

(b) What are the domain and range?

(c) What is the range element that corresponds to 2012? The domain element that corresponds to 43.4?

(d) Call this function f. Give two ordered pairs that belong to f.

70. The table shows the percentage of full-time college freshmen who said that they frequently smoked cigarettes in the last year.

Year	Percentage
2008	4.4
2009	4.2
2010	3.7
2011	2.8
2012	2.6

Source: Higher Education Research Institute.

(a) Does the table define a function?

(b) What are the domain and range?

(c) What is the range element that corresponds to 2012? The domain element that corresponds to 2.8?

(d) Call this function g. Give two ordered pairs that belong to g.

9.2 Function Notation and Linear Functions

VOCABULARY

☐ linear function
☐ constant function

OBJECTIVE 1 Use function notation.

When a function f is defined with a rule or an equation using x and y for the independent and dependent variables, we say, "*y is a function of x*" to emphasize that *y depends on x*. We use the notation

$$y = f(x),$$

> The parentheses here do *not* indicate multiplication.

called **function notation,** to express this and read $f(x)$ as "**f of x,**" or "**f at x.**" The letter f is a name for this particular function. For example, if $y = 3x - 5$, we can name this function f and write the following.

$$f(x) = 3x - 5$$

> f is the name of the function.
> x is a value from the domain.
> $f(x)$ is the function value (or y-value) that corresponds to x.

$f(x)$ is just another name for the dependent variable y.

We evaluate a function at different values of x by substituting x-values from the domain into the function.

NOW TRY EXERCISE 1

Let $f(x) = 4x + 3$. Find the value of function f for each value of x.

(a) $x = -2$ (b) $x = 0$

EXAMPLE 1 Evaluating a Function

Let $f(x) = 3x - 5$. Find the value of function f for each value of x.

(a) $x = 2$

$$f(x) = 3x - 5$$

> Read $f(2)$ as "f of 2" or "f at 2."

$$f(2) = 3 \cdot 2 - 5 \qquad \text{Replace } x \text{ with 2.}$$
$$f(2) = 6 - 5 \qquad \text{Multiply.}$$
$$f(2) = 1 \qquad \text{Subtract.}$$

Thus, for $x = 2$, the corresponding function value (or y-value) is 1. $f(2) = 1$ is an abbreviation for the statement "If $x = 2$ in the function f, then $y = 1$" and is represented by the ordered pair $(2, 1)$.

(b) $x = -1$

$$f(x) = 3x - 5$$

> Use parentheses to avoid errors.

$$f(-1) = 3(-1) - 5 \qquad \text{Replace } x \text{ with } -1.$$
$$f(-1) = -3 - 5 \qquad \text{Multiply.}$$
$$f(-1) = -8 \qquad \text{Subtract.}$$

Thus, $f(-1) = -8$ and the ordered pair $(-1, -8)$ belongs to f.   **NOW TRY**

> **⚠ CAUTION** The symbol $f(x)$ *does not* indicate "f times x," but represents the y-value associated with the indicated x-value. As shown in **Example 1(a),** $f(2)$ is the y-value that corresponds to the x-value 2 in f.

NOW TRY ANSWERS
1. (a) -5 (b) 3

These ideas can be illustrated as follows.

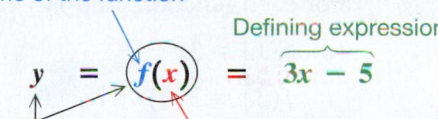

Name of the function

Defining expression

$$y = f(x) = 3x - 5$$

Value of the function   Name of the independent variable

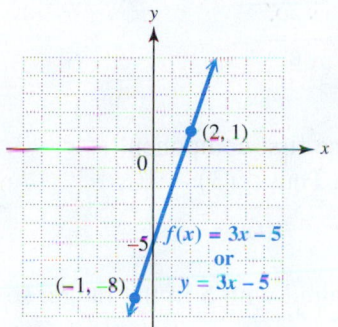

FIGURE 5

> **NOTE** In the function $f(x) = 3x - 5$ in **Example 1**, $f(2) = 1$ and $f(-1) = -8$ correspond to the ordered pairs $(2, 1)$ and $(-1, -8)$. Because the domain of f is $(-\infty, \infty)$—that is, x can be any real number—this function defines an infinite set of ordered pairs whose graph is a line with slope 3 and y-intercept $(0, -5)$. See **FIGURE 5**. This makes sense because $f(x)$ is another name for y, and
>
> $$f(x) = 3x - 5 \quad \text{is equivalent to} \quad y = 3x - 5.$$

As we will see in **Objective 2**, f is a *linear function*.

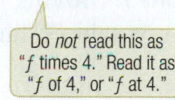
**NOW TRY
EXERCISE 2**

Let $f(x) = 2x^2 - 4x + 1$. Find the following.

(a) $f(-2)$ **(b)** $f(a)$

EXAMPLE 2   Evaluating a Function

Let $f(x) = -x^2 + 5x - 3$. Find the following.

(a) $f(4)$

> Do *not* read this as "f times 4." Read it as "f of 4," or "f at 4."

$$f(x) = -x^2 + 5x - 3 \qquad \text{The base in } -x^2 \text{ is } x, \text{ not } (-x).$$

$$f(4) = -4^2 + 5 \cdot 4 - 3 \qquad \text{Replace } x \text{ with } 4.$$

$$f(4) = -16 + 20 - 3 \qquad \text{Apply the exponent. Multiply.}$$

$$f(4) = 1 \qquad \text{Add and subtract.}$$

Thus, $f(4) = 1$, and the ordered pair $(4, 1)$ belongs to f.

(b) $f(q)$

$$f(x) = -x^2 + 5x - 3$$

$$f(q) = -q^2 + 5q - 3 \qquad \text{Replace } x \text{ with } q.$$

The replacement of one variable with another is important in later courses.

NOW TRY

Sometimes letters other than f, such as g, h, or capital letters F, G, and H are used to name functions.

**NOW TRY
EXERCISE 3**

Let $g(x) = 8x - 5$. Find and simplify $g(a - 2)$.

EXAMPLE 3   Evaluating a Function

Let $g(x) = 2x + 3$. Find and simplify $g(a + 1)$.

$$g(x) = 2x + 3$$

$$g(a + 1) = 2(a + 1) + 3 \qquad \text{Replace } x \text{ with } a + 1.$$

$$g(a + 1) = 2a + 2 + 3 \qquad \text{Distributive property}$$

$$g(a + 1) = 2a + 5 \qquad \text{Add.}$$

NOW TRY

NOW TRY ANSWERS
2. (a) 17 **(b)** $2a^2 - 4a + 1$
3. $8a - 21$

NOW TRY
EXERCISE 4

Find $f(-1)$ for each function.

(a) $f = \{(-5, -1), (-3, 2),$
    $(-1, 4)\}$

(b) $f(x) = x^2 - 12$

EXAMPLE 4   Evaluating Functions

For each function, find $f(3)$.

(a) $f(x) = 3x - 7$

   $f(3) = 3(3) - 7$ Replace *x* with 3.

   $f(3) = 9 - 7$ Multiply.

   $f(3) = 2$ Subtract.

(b)

x	$y = f(x)$
6	−12
3	−6
0	0
−3	6

← Here, $f(3) = -6$.

(c) $f = \{(-3, 5), (0, 3), (3, 1), (6, -1)\}$

   We want $f(3)$, the *y*-value of the ordered pair whose first component is $x = 3$. As indicated by the ordered pair $(3, 1)$, for $x = 3$, $y = 1$. Thus, $f(3) = 1$.

(d)

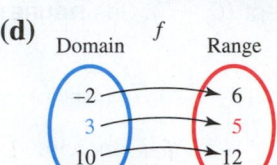

The domain element 3 is paired with 5 in the range, so

$$f(3) = 5.$$

NOW TRY

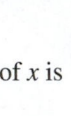

NOW TRY
EXERCISE 5

Refer to the function graphed in **FIGURE 6**.

(a) Find $f(-1)$.

(b) For what value of *x* is $f(x) = 2$?

EXAMPLE 5   Finding Function Values from a Graph

Refer to the function graphed in **FIGURE 6**.

(a) Find $f(3)$.

   Locate 3 on the *x*-axis. See **FIGURE 7**. Moving up to the graph of *f* and over to the *y*-axis gives 4 for the corresponding *y*-value. Thus, $f(3) = 4$, which corresponds to the ordered pair $(3, 4)$.

(b) Find $f(0)$.

   Refer to **FIGURE 7** to see that $f(0) = 1$.

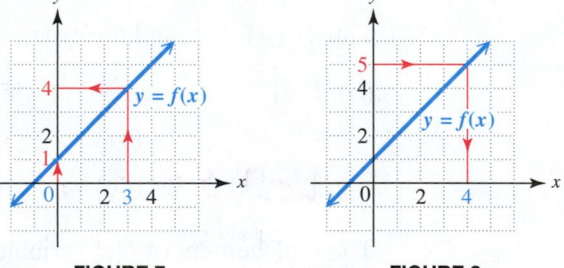

FIGURE 7   FIGURE 8

(c) For what value of *x* is $f(x) = 5$?

   Because $f(x) = y$, we want the value of *x* that corresponds to $y = 5$. See **FIGURE 8**, and locate 5 on the *y*-axis. Moving across to the graph of *f* and down to the *x*-axis gives $x = 4$. Thus, $f(4) = 5$, which corresponds to the ordered pair $(4, 5)$.

NOW TRY

   If a function *f* is defined by an equation in *x* and *y* instead of function notation, use the following steps to find $f(x)$.

Writing an Equation Using Function Notation

Step 1   Solve the equation for *y*.

Step 2   Replace *y* with $f(x)$.

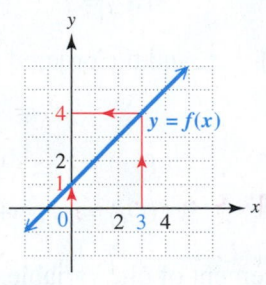

FIGURE 6

NOW TRY ANSWERS

4. (a) 4   (b) −11

5. (a) 0   (b) 1

**NOW TRY
EXERCISE 6**

Write the equation using function notation $f(x)$. Then find $f(-3)$ and $f(h)$.

$$-4x^2 + y = 5$$

EXAMPLE 6 Writing Equations Using Function Notation

Write each equation using function notation $f(x)$. Then find $f(-2)$ and $f(a)$.

(a) $y = x^2 + 1$ ⟵ This equation is already solved for y.   (Step 1)

$$f(x) = x^2 + 1 \qquad \text{Replace } y \text{ with } f(x). \quad \text{(Step 2)}$$

To find $f(-2)$, let $x = -2$.

$$f(x) = x^2 + 1$$
$$f(-2) = (-2)^2 + 1 \qquad \text{Let } x = -2.$$
$$f(-2) = 4 + 1 \qquad (-2)^2 = -2(-2)$$
$$f(-2) = 5 \qquad \text{Add.}$$

To find $f(a)$, let $x = a$.

$$f(a) = a^2 + 1$$

(b) $x - 4y = 5$ ⟵——— Solve this equation for y.

Step 1 $-4y = -x + 5$ Subtract x.

$$y = \frac{1}{4}x - \frac{5}{4} \qquad \text{Divide by } -4.$$

Step 2 $f(x) = \frac{1}{4}x - \frac{5}{4}$ Replace y with $f(x)$.

Now find $f(-2)$ and $f(a)$.

$$f(-2) = \frac{1}{4}(-2) - \frac{5}{4} = -\frac{7}{4} \qquad \text{Let } x = -2.$$

$$f(a) = \frac{1}{4}a - \frac{5}{4} \qquad \text{Let } x = a. \qquad \text{NOW TRY} \;\; \text{⟳}$$

OBJECTIVE 2 Graph linear and constant functions.

Linear equations (except for vertical lines with equations of the form $x = a$) define *linear functions.*

Linear Function

A function f that can be written in the form

$$f(x) = ax + b,$$

where a and b are real numbers, is a **linear function.** The value of a is the slope m of the graph of the function. The domain of a linear function is $(-\infty, \infty)$, unless specified otherwise.

A linear function whose graph is a horizontal line has the form

$$f(x) = b \qquad \text{Constant function}$$

and is a **constant function.** While the range of any nonconstant linear function is $(-\infty, \infty)$, the range of a constant function $f(x) = b$ is $\{b\}$.

NOW TRY ANSWER
6. $f(x) = 4x^2 + 5$;
  $f(-3) = 41$;
  $f(h) = 4h^2 + 5$

**NOW TRY
EXERCISE 7**

Graph the function. Give the domain and range.

$$f(x) = \frac{1}{3}x - 2$$

NOW TRY ANSWER

7.

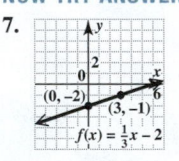

domain: $(-\infty, \infty)$;
range: $(-\infty, \infty)$

EXAMPLE 7   Graphing Linear and Constant Functions

Graph each function. Give the domain and range.

(a) $f(x) = \dfrac{1}{4}x - \dfrac{5}{4}$ (from **Example 6(b)**)

Slope ⟶   ⟵ *y*-intercept is $\left(0, -\frac{5}{4}\right)$.

To graph this function, plot the *y*-intercept $\left(0, -\frac{5}{4}\right)$. Use the geometric definition of slope as $\dfrac{\text{rise}}{\text{run}}$ to find a second point on the line. The slope is $\frac{1}{4}$, so we move 1 unit up from $\left(0, -\frac{5}{4}\right)$ and 4 units to the right to the point $\left(4, -\frac{1}{4}\right)$. Draw the straight line through these points. See **FIGURE 9**. The domain and range are both $(-\infty, \infty)$.

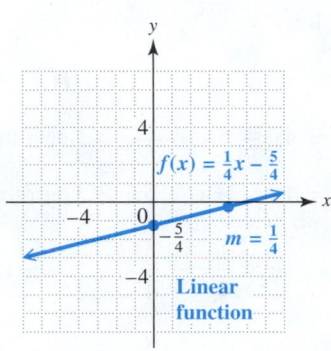

FIGURE 9

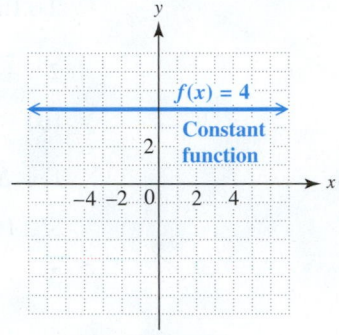

FIGURE 10

(b) $f(x) = 4$

The graph of this constant function is the horizontal line containing all points with *y*-coordinate 4. See **FIGURE 10**. The domain is $(-\infty, \infty)$ and the range is $\{4\}$.

NOW TRY

9.2 Exercises

FOR
EXTRA
HELP

▶ MyMathLab®

▶ *Complete solution available
in MyMathLab*

Concept Check   *Work each problem.*

1. To emphasize that "*y* is a function of *x*" for a given function *f*, we use function notation and write *y* = _____. Here, *f* is the name of the _____, *x* is a value from the _____, and *f*(*x*) is the function value (or *y*-value) that corresponds to _____. We read *f*(*x*) as "_____."

2. Choose the correct response.

For a function *f*, the notation *f*(3) means _____.

A. the variable *f* times 3, or 3*f*.

B. the value of the dependent variable when the independent variable is 3.

C. the value of the independent variable when the dependent variable is 3.

D. *f* equals 3.

3. Fill in each blank with the correct response.

The equation $2x + y = 4$ has a straight _____ as its graph. One point that lies on the graph is $(3, \_\_\_)$. If we solve the equation for *y* and use function notation, we have a(n) _____ function $f(x) = \_\_\_$. For this function, $f(3) = \_\_\_$, meaning that the point $(\_\_\_, \_\_\_)$ lies on the graph of the function.

4. Which of the following defines y as a linear function of x?

    A. $y = \dfrac{1}{4}x - \dfrac{5}{4}$ **B.** $y = \dfrac{1}{x}$ **C.** $y = x^2$ **D.** $y = x^3$

*Let $f(x) = -3x + 4$ and $g(x) = -x^2 + 4x + 1$. Find the following. **See Examples 1–3.***

▶ **5.** $f(0)$ **6.** $g(0)$ **7.** $f(-3)$ **8.** $f(-5)$

  9. $g(-2)$ **10.** $g(-1)$ **11.** $g(3)$ **12.** $g(10)$

13. $f(100)$ **14.** $f(-100)$ **15.** $f\left(\dfrac{1}{3}\right)$ **16.** $f\left(\dfrac{7}{3}\right)$

17. $g(0.5)$ **18.** $g(1.5)$ **19.** $f(p)$ **20.** $g(k)$

21. $f(-x)$ **22.** $g(-x)$ ▶ **23.** $f(x + 2)$ **24.** $f(x - 2)$

25. $f(2t + 1)$ **26.** $f(3t - 2)$ **27.** $g(\pi)$ **28.** $g(t)$

29. $f(x + h)$ **30.** $f(a + b)$ **31.** $g\left(\dfrac{p}{3}\right)$ **32.** $g\left(\dfrac{1}{x}\right)$

*For each function, find **(a)** $f(2)$ and **(b)** $f(-1)$. **See Examples 4, 5(a), and 5(b).***

33. $f = \{(-2, 2), (-1, -1), (2, -1)\}$ **34.** $f = \{(-1, -5), (0, 5), (2, -5)\}$

▶ **35.** $f = \{(-1, 3), (4, 7), (0, 6), (2, 2)\}$ **36.** $f = \{(2, 5), (3, 9), (-1, 11), (5, 3)\}$

37. f **38.** f

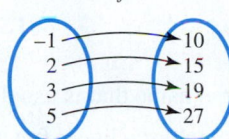 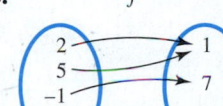

39.

x	$y = f(x)$
2	4
1	1
0	0
-1	1
-2	4

40.

x	$y = f(x)$
8	6
5	3
2	0
-1	-3
-4	-6

41.

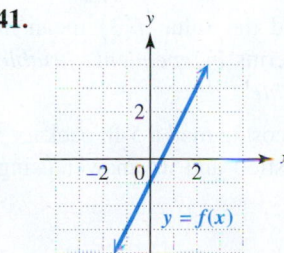

42.

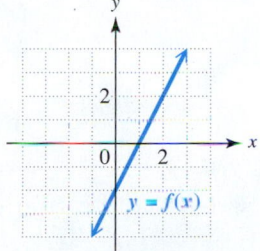

43.

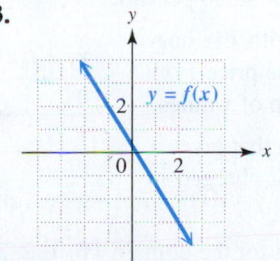

44.

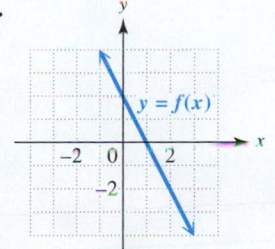

Refer to the given graph. Find the value of x for each value of f(x). See Example 5(c).

45. (a) $f(x) = 3$

(b) $f(x) = -1$

(c) $f(x) = -3$

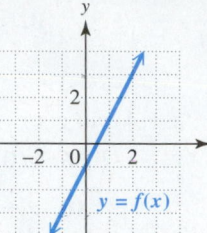

46. (a) $f(x) = 4$

(b) $f(x) = -2$

(c) $f(x) = 0$

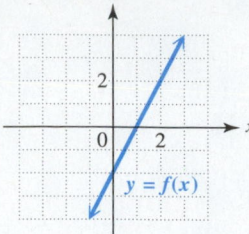

An equation that defines y as a function f of x is given. (a) Solve for y in terms of x, and replace y with function notation f(x). (b) Find f(3). See Example 6.

▶ **47.** $x + 3y = 12$ **48.** $x - 4y = 8$ **49.** $y + 2x^2 = 3$

50. $y - 3x^2 = 2$ **51.** $4x - 3y = 8$ **52.** $-2x + 5y = 9$

Graph each linear function. Give the domain and range. See Example 7.

▶ **53.** $f(x) = -2x + 5$ **54.** $g(x) = 4x - 1$ **55.** $h(x) = \frac{1}{2}x + 2$

56. $F(x) = -\frac{1}{4}x + 1$ **57.** $G(x) = 2x$ **58.** $H(x) = -3x$

▶ **59.** $g(x) = -4$ **60.** $f(x) = 5$ **61.** $f(x) = 0$ **62.** $f(x) = -2.5$

63. *Concept Check*   What is the name that is usually given to the graph in **Exercise 61?**

64. *Concept Check*   Can the graph of a linear function have an undefined slope? Explain.

Solve each problem.

65. A package weighing x pounds costs $f(x)$ dollars to mail to a given location, where

$$f(x) = 3.75x.$$

(a) Evaluate $f(3)$.

(b) Describe what 3 and the value $f(3)$ mean in part (a), using the terms *independent variable* and *dependent variable*.

(c) How much would it cost to mail a 5-lb package? Then write this question and its answer using function notation.

66. A taxicab driver charges \$2.50 per mile.

(a) Fill in the table with the correct response for the price $f(x)$ he charges for a trip of x miles.

(b) The linear function that gives a rule for the amount charged is $f(x) = $ _____ .

x	f(x)
0	
1	
2	
3	

(c) Graph this function for the domain $\{0, 1, 2, 3\}$.

67. To print t-shirts, there is a $100 set-up fee, plus a $12 charge per t-shirt. Let x represent the number of t-shirts printed and $f(x)$ represent the total charge.

   (a) Write a linear function that models this situation.

   (b) Find $f(125)$. Interpret your answer in the context of this problem.

   (c) Find the value of x if $f(x) = 1000$. Express this situation using function notation, and interpret it in the context of this problem.

68. Rental on a car is $150, plus $0.50 per mile. Let x represent the number of miles driven and $f(x)$ represent the total cost to rent the car.

   (a) Write a linear function that models this situation.

   (b) How much would it cost to drive 250 mi? Interpret this question and answer, using function notation.

   (c) Find the value of x if $f(x) = 400$. Interpret your answer in the context of this problem.

69. The table represents a linear function.

   (a) What is $f(2)$?

   (b) If $f(x) = -2.5$, what is the value of x?

   (c) What is the slope of the line?

   (d) What is the y-intercept of the line?

   (e) Using your answers from parts (c) and (d), write an equation for $f(x)$.

x	$y = f(x)$
0	3.5
1	2.3
2	1.1
3	−0.1
4	−1.3
5	−2.5

70. The table represents a linear function.

   (a) What is $f(2)$?

   (b) If $f(x) = 2.1$, what is the value of x?

   (c) What is the slope of the line?

   (d) What is the y-intercept of the line?

   (e) Using your answers from parts (c) and (d), write an equation for $f(x)$.

x	$y = f(x)$
−1	−3.9
0	−2.4
1	−0.9
2	0.6
3	2.1

71. The graph shows water in a swimming pool over time.

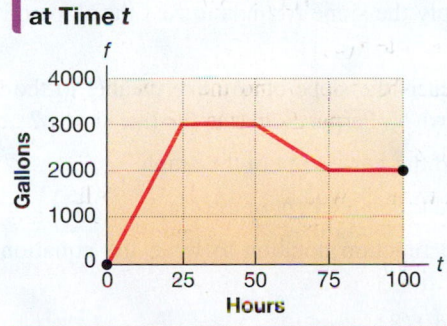

Gallons of Water in a Pool at Time t

   (a) What numbers are possible values of the independent variable? The dependent variable?

   (b) For how long is the water level increasing? Decreasing?

   (c) How many gallons of water are in the pool after 90 hr?

   (d) Call this function f. What is $f(0)$? What does it mean?

   (e) What is $f(25)$? What does it mean?

72. The graph shows electricity use on a summer day.

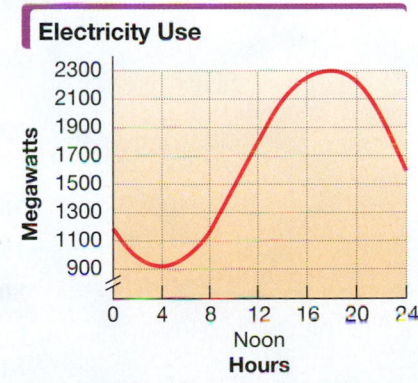

Electricity Use

   (a) Why is this the graph of a function?

   (b) What is the domain?

   (c) Estimate the number of megawatts used at 8 A.M.

   (d) At what time was the most electricity used? The least electricity?

   (e) Call this function f. What is $f(12)$? What does it mean?

73. Forensic scientists use the lengths of certain bones to calculate the height of a person. Two such bones are the tibia (t), the bone from the ankle to the knee, and the femur (r), the bone from the knee to the hip socket. A person's height (h) in centimeters is determined from the lengths of these bones by using the following functions.

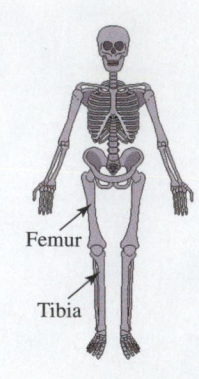

Femur

Tibia

For men: $h(r) = 69.09 + 2.24r$ or $h(t) = 81.69 + 2.39t$

For women: $h(r) = 61.41 + 2.32r$ or $h(t) = 72.57 + 2.53t$

(a) Find the height of a man with a femur measuring 56 cm.

(b) Find the height of a man with a tibia measuring 40 cm.

(c) Find the height of a woman with a femur measuring 50 cm.

(d) Find the height of a woman with a tibia measuring 36 cm.

74. Based on federal regulations, a pool to house sea otters must have a volume that is "the square of the sea otter's average adult length (in meters) multiplied by 3.14 and by 0.91 meter." If x represents the sea otter's average adult length and $f(x)$ represents the volume (in cubic meters) of the corresponding pool size, this formula can be written as the function

$$f(x) = 0.91(3.14)x^2.$$

Find the volume of the pool for each adult sea otter length (in meters). Round answers to the nearest hundredth.

(a) 0.8   **(b)** 1.0   **(c)** 1.2   **(d)** 1.5

RELATING CONCEPTS   For Individual or Group Work   (Exercises 75–82)

Refer to the straight-line graph and ***work Exercises 75–82 in order.***

75. By just looking at the graph, how can we tell whether the slope is positive, negative, 0, or undefined?

76. Apply the slope formula to find the slope of the line.

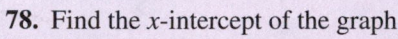

77. What is the slope of any line parallel to the line shown?   Perpendicular to the line shown?

78. Find the x-intercept of the graph.

79. Find the y-intercept of the graph.

80. Use function notation to write the equation of the line. Use f to designate the function.

81. Find $f(8)$.

82. If $f(x) = -8$, what is the value of x?

9.3 Polynomial Functions, Operations, and Composition

OBJECTIVES

1 Recognize and evaluate polynomial functions.
2 Perform operations on polynomial functions.
3 Find the composition of functions.

VOCABULARY

☐ polynomial function
☐ composite function (composition of functions)

OBJECTIVE 1 Recognize and evaluate polynomial functions.

In **Section 9.2**, we studied linear (first-degree polynomial) functions $f(x) = ax + b$. Now we consider more general polynomial functions.

Polynomial Function

A **polynomial function of degree n** is defined by

$$f(x) = a_n x^n + a_{n-1} x^{n-1} + \cdots + a_1 x + a_0,$$

for real numbers $a_n, a_{n-1}, \ldots, a_1$, and a_0, where $a_n \neq 0$ and n is a whole number.

Another way of describing a polynomial function is to say that it is a function defined by a polynomial in one variable, consisting of one or more terms. It is usually written in descending powers of the variable, and its degree is the degree of the polynomial that defines it.

We can evaluate a polynomial function $f(x)$ at different values of the variable x.

NOW TRY EXERCISE 1

Let $f(x) = x^3 - 2x^2 + 7$.
Find $f(-3)$.

EXAMPLE 1 Evaluating Polynomial Functions

Let $f(x) = 4x^3 - x^2 + 5$. Find each value.

(a) $f(3)$

Read this as "f of 3," not "f times 3."

$f(x) = 4x^3 - x^2 + 5$	Given function
$f(3) = 4(3)^3 - 3^2 + 5$	Substitute 3 for x.
$f(3) = 4(27) - 9 + 5$	Apply the exponents.
$f(3) = 108 - 9 + 5$	Multiply.
$f(3) = 104$	Subtract, and then add.

Thus, $f(3) = 104$ and the ordered pair $(3, 104)$ belongs to f.

(b) $f(-4)$

$f(x) = 4x^3 - x^2 + 5$	Use parentheses.
$f(-4) = 4(-4)^3 - (-4)^2 + 5$	Let $x = -4$.
$f(-4) = 4(-64) - 16 + 5$	Be careful with signs.
$f(-4) = -256 - 16 + 5$	Multiply.
$f(-4) = -267$	Subtract, and then add.

So, $f(-4) = -267$. The ordered pair $(-4, -267)$ belongs to f.   NOW TRY

The capital letter P is sometimes used for polynomial functions. The function

$$P(x) = 4x^3 - x^2 + 5$$

yields the same ordered pairs as the function f in **Example 1**.

NOW TRY ANSWER
1. -38

OBJECTIVE 2 Perform operations on polynomial functions.

The operations of addition, subtraction, multiplication, and division are also defined for functions. For example, the graph in **FIGURE 11** shows dollars (in billions) spent for general science and for space/other technologies over a 20-year period.

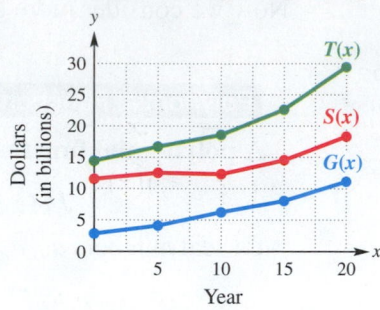

Science and Space Spending

Source: U.S. Office of Management and Budget.

FIGURE 11

$G(x)$ represents dollars spent for general science.

$S(x)$ represents dollars spent for space/other technologies.

$T(x)$ represents total expenditures for these two categories.

The total expenditures function can be found by *adding* the spending functions for the two individual categories.

$$T(x) = G(x) + S(x)$$

As another example, businesses use the equation "profit equals revenue minus cost," which can be written using function notation.

$$P(x) = R(x) - C(x) \qquad \text{x is the number of items produced and sold.}$$

Profit   Revenue   Cost
function   function   function

The profit function is found by *subtracting* the cost function from the revenue function.

We define the following **operations on functions.**

Operations on Functions

If $f(x)$ and $g(x)$ define functions, then

$$(f + g)(x) = f(x) + g(x), \qquad \text{Sum function}$$
$$(f - g)(x) = f(x) - g(x), \qquad \text{Difference function}$$
$$(fg)(x) = f(x) \cdot g(x), \qquad \text{Product function}$$

and
$$\left(\frac{f}{g}\right)x = \frac{f(x)}{g(x)}, \quad g(x) \neq 0. \qquad \text{Quotient function}$$

In each case, the domain of the new function is the intersection of the domains of $f(x)$ and $g(x)$. Additionally, the domain of the quotient function must exclude any values of x for which $g(x) = 0$.

NOW TRY
EXERCISE 2

For $f(x) = x^3 - 3x^2 + 4$
and $g(x) = -2x^3 + x^2 - 12$,
find each of the following.

(a) $(f + g)(x)$

(b) $(f - g)(x)$

EXAMPLE 2 Adding and Subtracting Functions

Find each of the following for the polynomial functions f and g as defined.

$$f(x) = x^2 - 3x + 7 \quad \text{and} \quad g(x) = -3x^2 - 7x + 7$$

(a) $(f + g)(x)$ 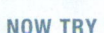 *This notation does* not *indicate the distributive property.*

$= f(x) + g(x)$ Use the definition.

$= (x^2 - 3x + 7) + (-3x^2 - 7x + 7)$ Substitute.

$= -2x^2 - 10x + 14$ Add the polynomials.

(b) $(f - g)(x)$

$= f(x) - g(x)$ Use the definition.

$= (x^2 - 3x + 7) - (-3x^2 - 7x + 7)$ Substitute.

$= (x^2 - 3x + 7) + (3x^2 + 7x - 7)$ Change subtraction to addition.

$= 4x^2 + 4x$ Add.   **NOW TRY**

NOW TRY
EXERCISE 3

For $f(x) = x^2 - 4$
and $g(x) = -6x^2$,
find each of the following.

(a) $(f + g)(x)$

(b) $(f - g)(-4)$

EXAMPLE 3 Adding and Subtracting Functions

Find each of the following for the polynomial functions f and g as defined.

$$f(x) = 10x^2 - 2x \quad \text{and} \quad g(x) = 2x$$

(a) $(f + g)(2)$

$= f(2) + g(2)$ Use the definition.

$f(x) = 10x^2 - 2x \qquad g(x) = 2x$

$= [10(2)^2 - 2(2)] + 2(2)$ Substitute.

This is a key step.

$= [40 - 4] + 4$ Order of operations

$= 40$ Subtract, and then add.

Alternative method: $(f + g)(x)$ Find $(f + g)(x)$.

$= f(x) + g(x)$ Use the definition.

$= (10x^2 - 2x) + 2x$ Substitute.

$= 10x^2$ Combine like terms.

$(f + g)(2)$ Now find $(f + g)(2)$.

$= 10(2)^2$ $(f + g)(x) = 10x^2$; Substitute.

$= 40$ The result is the same.

(b) $(f - g)(x)$

$= f(x) - g(x)$ Use the definition.

$= (10x^2 - 2x) - 2x$ Substitute.

$= 10x^2 - 4x$ Combine like terms.

NOW TRY ANSWERS

2. (a) $-x^3 - 2x^2 - 8$
   (b) $3x^3 - 4x^2 + 16$
3. (a) $-5x^2 - 4$
   (b) 108

(c) $(f - g)(1)$ Now find $(f - g)(1)$.

Confirm that $f(1) - g(1)$ gives the same result.

$= 10(1)^2 - 4(1)$ $(f - g)(x) = 10x^2 - 4x$ from part (b).

$= 6$ Perform the operations.   **NOW TRY**

**NOW TRY
EXERCISE 4**
For $f(x) = 3x^2 - 1$
and $g(x) = 8x + 7$,
find $(fg)(x)$ and $(fg)(-2)$.

EXAMPLE 4   Multiplying Polynomial Functions

For $f(x) = 3x + 4$ and $g(x) = 2x^2 + x$, find $(fg)(x)$ and $(fg)(-1)$.

$(fg)(x)$

> This notation indicates function multiplication.

$$= f(x) \cdot g(x) \qquad\qquad \text{Use the definition.}$$

$$= (3x + 4)(2x^2 + x) \qquad \text{Substitute.}$$

$$= 6x^3 + 3x^2 + 8x^2 + 4x \qquad \text{FOIL method}$$

$$= 6x^3 + 11x^2 + 4x \qquad \text{Combine like terms.}$$

$(fg)(-1)$ $(fg)(x) = 6x^3 + 11x^2 + 4x$

$$= 6(-1)^3 + 11(-1)^2 + 4(-1) \qquad \text{Let } x = -1 \text{ in } (fg)(x).$$

> Be careful with signs.

$$= -6 + 11 - 4 \qquad\qquad \text{Apply the exponents. Multiply.}$$

$$= 1 \qquad\qquad\qquad \text{Add and subtract.}$$

Another way to find $(fg)(-1)$ is to find $f(-1)$ and $g(-1)$ and multiply the results. Verify that $f(-1) \cdot g(-1) = 1$. This follows from the definition.   **NOW TRY**

**NOW TRY
EXERCISE 5**
For $f(x) = 8x^2 + 2x - 3$
and $g(x) = 2x - 1$,
find $\left(\frac{f}{g}\right)(x)$ and $\left(\frac{f}{g}\right)(8)$.

EXAMPLE 5   Dividing Polynomial Functions

For $f(x) = 2x^2 + x - 10$ and $g(x) = x - 2$, find $\left(\frac{f}{g}\right)(x)$ and $\left(\frac{f}{g}\right)(-3)$. What value of x is not in the domain of the quotient function?

$$\left(\frac{f}{g}\right)(x) = \frac{f(x)}{g(x)} = \frac{2x^2 + x - 10}{x - 2}$$

To find the quotient, divide as follows.

$$\begin{array}{r} 2x + 5 \\ x - 2 \overline{\smash{)}\, 2x^2 + x - 10} \end{array}$$

> To subtract, add the opposite.

$$\underline{2x^2 - 4x} \qquad 2x(x - 2)$$
$$5x - 10 \qquad \text{Subtract.}$$
$$\underline{5x - 10} \qquad 5(x - 2)$$
$$0$$

This quotient is $2x + 5$.

Thus, $\left(\frac{f}{g}\right)(x) = 2x + 5, \quad x \neq 2.$ ← 2 is not in the domain. It causes denominator $g(x) = x - 2$ to equal 0.

Then $\left(\frac{f}{g}\right)(-3) = 2(-3) + 5 = -1.$ Let $x = -3$ in $\left(\frac{f}{g}\right)(x) = 2x + 5.$

Verify that the same value is found by evaluating $\frac{f(-3)}{g(-3)}$.   **NOW TRY**

OBJECTIVE 3   Find the composition of functions.

NOW TRY ANSWERS

4. $24x^3 + 21x^2 - 8x - 7$; -99

5. $4x + 3, \quad x \neq \frac{1}{2}$; 35

The diagram in **FIGURE 12** on the next page shows a function g that assigns, to each element x of set X, some element y of set Y. Suppose that a function f takes each element of set Y and assigns a value z of set Z. Then f and g together assign an element x in X to an element z in Z.

The result of this process is a new function h that takes an element x in X and assigns it an element z in Z.

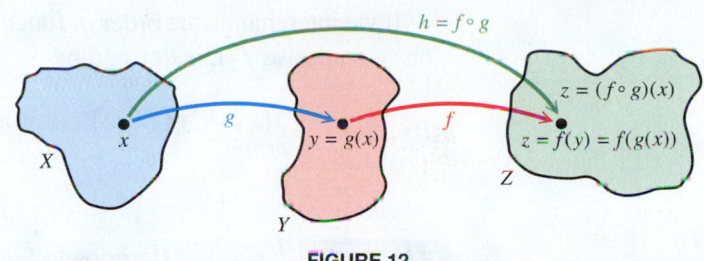

FIGURE 12

Function h is the *composition* of functions f and g, written $f \circ g$.

Composition of Functions

The **composite function,** or **composition,** of functions f and g is defined by

$$(f \circ g)(x) = f(g(x)),$$

for all x in the domain of g such that $g(x)$ is in the domain of f.

Read $f \circ g$ **as** "f **of** g" (**or** "f **compose** g").

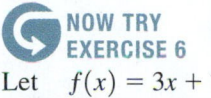

As a real-life example of how composite functions occur, consider the following.

A $40 pair of blue jeans is on sale for 25% off. If we purchase the jeans before noon, the retailer offers an additional 10% off. What is the final sale price of the blue jeans?

We might be tempted to say that the blue jeans are $25\% + 10\% = 35\%$ off and calculate $\$40(0.35) = \$14$, giving a final sale price of

$$\$40 - \$14 = \$26. \qquad \text{This is not correct.}$$

To find the correct final sale price, we must first find the price after taking 25% off, and then take an additional 10% off *that* price.

$\$40(0.25) = \$10$, giving a sale price of $\$40 - \$10 = \$30$.   Take 25% off original price.

$\$30(0.10) = \$3$, giving a ***final sale price*** of $\$30 - \$3 = \$27$.   Take additional 10% off.

This is the idea behind composition of functions.

NOW TRY EXERCISE 6

Let $f(x) = 3x + 7$ and $g(x) = x - 2$. Find $(f \circ g)(7)$.

EXAMPLE 6   Finding a Composite Function

Let $f(x) = x^2$ and $g(x) = x + 3$. Find $(f \circ g)(4)$.

$(f \circ g)(4)$ Evaluate the "inside" function value first.

$= f(g(4))$ Definition of composition

$= f(4 + 3)$ Use the rule for $g(x)$; $g(4) = 4 + 3$.

$= f(7)$ Add.

Now evaluate the "outside" function. $= 7^2$ Use the rule for $f(x)$; $f(7) = 7^2$.

$= 49$ Square 7.

NOW TRY ANSWER
6. 22

In this composition, g is the innermost "operation" and acts on x (here 4) first. Then the output value of g (here 7) becomes the input (domain) value of f.   **NOW TRY**

If we interchange the order of functions f and g, the composition $g \circ f$, read "g of f" (or "g compose f"), is defined by

$$(g \circ f)(x) = g(f(x)),$$ for all x in the domain of f such that $f(x)$ is in the domain of g.

NOW TRY EXERCISE 7

As in **Now Try Exercise 6,**
let $f(x) = 3x + 7$
and $g(x) = x - 2$.
Find $(g \circ f)(7)$.

EXAMPLE 7   Finding a Composite Function

Find $(g \circ f)(4)$ for the functions $f(x) = x^2$ and $g(x) = x + 3$ from **Example 5.**

$(g \circ f)(4)$ *(Evaluate the "inside" function value first.)*

$= g(f(4))$ Definition of composition

$= g(4^2)$ Use the rule for $f(x)$; $f(4) = 4^2$.

$= g(16)$ Square 4.

(Now evaluate the "outside" function.) $= 16 + 3$ Use the rule for $g(x)$; $g(16) = 16 + 3$.

$= 19$ Add.

In this composition, f is the innermost "operation" and acts on x (again 4) first. Then the output value of f (here 16) becomes the input (domain) value of g.   **NOW TRY**

We see in **Examples 6 and 7** that

$$(f \circ g)(4) \neq (g \circ f)(4)   \text{because}   49 \neq 19.$$

In general,

$$(f \circ g)(x) \neq (g \circ f)(x).$$

EXAMPLE 8   Finding Composite Functions

Let $f(x) = 4x - 1$ and $g(x) = x^2 + 5$. Find each of the following.

(a) $(f \circ g)(2)$

$= f(g(2))$ Definition of composition

$= f(2^2 + 5)$ $g(x) = x^2 + 5$

$= f(9)$ Work inside the parentheses.

$= 4(9) - 1$ $f(x) = 4x - 1$

$= 35$ Multiply, and then subtract.

(b) $(f \circ g)(x)$

$= f(g(x))$ Use $g(x)$ as the input for function f.

$= 4(g(x)) - 1$ Use the rule for $f(x)$; $f(x) = 4x - 1$.

$= 4(x^2 + 5) - 1$ $g(x) = x^2 + 5$

$= 4x^2 + 20 - 1$ Distributive property

$= 4x^2 + 19$ Combine like terms.

NOW TRY ANSWER
7. 26

NOW TRY
EXERCISE 8

Let $f(x) = x - 5$
and $g(x) = -x^2 + 2$.
Find each of the following.
(a) $(g \circ f)(-1)$
(b) $(f \circ g)(x)$

NOW TRY ANSWERS
8. (a) -34 **(b)** $-x^2 - 3$

(c) Find $(f \circ g)(2)$ again, this time using the rule obtained in part (b).

$$(f \circ g)(x) = 4x^2 + 19 \qquad \text{From part (b)}$$

$$(f \circ g)(2) = 4(2)^2 + 19 \qquad \text{Let } x = 2.$$

$$= 4(4) + 19 \qquad \text{Square 2.}$$

$$= 16 + 19 \qquad \text{Multiply.}$$

Same result as in part (a) → $= 35 \qquad$ Add.   **NOW TRY**

9.3 Exercises

FOR EXTRA HELP ▶ MyMathLab®

▶ *Complete solution available in MyMathLab*

1. *Concept Check*   A polynomial function is a function defined by a _____ in (*one / two / three*) variable(s), consisting of one or more (*factors / terms*) and usually written in descending _____ of the variable.

2. *Concept Check*   Which of the following are *not* polynomial functions?

A. $P(x) = x^{-2} - 2x$ **B.** $f(x) = \dfrac{1}{2}x^2 + x - 1$

C. $g(x) = -4x + 1.5$ **D.** $p(x) = x^3 - x^2 - \dfrac{5}{x}$

For each polynomial function, find (a) $f(-1)$, (b) $f(2)$, and (c) $f(0)$. See Example 1.

3. $f(x) = 6x - 4$ **4.** $f(x) = -2x + 5$ **5.** $f(x) = x^2 - 7x$

6. $f(x) = x^2 + 5x$ **7.** $f(x) = x^2 - 3x + 4$ **8.** $f(x) = x^2 - 5x - 4$

9. $f(x) = 2x^2 - 4x + 1$ **10.** $f(x) = 3x^2 + x - 5$ **11.** $f(x) = 5x^4 - 3x^2 + 6$

12. $f(x) = 4x^4 + 2x^2 - 1$ ▶ **13.** $f(x) = -x^2 + 2x^3 - 8$ **14.** $f(x) = -x^2 - x^3 + 11$

For each pair of functions, find (a) $(f + g)(x)$ and (b) $(f - g)(x)$. See Example 2.

15. $f(x) = 5x - 10, \quad g(x) = 3x + 7$

16. $f(x) = -4x + 1, \quad g(x) = 6x + 2$

▶ **17.** $f(x) = 4x^2 + 8x - 3, \quad g(x) = -5x^2 + 4x - 9$

18. $f(x) = 3x^2 - 9x + 10, \quad g(x) = -4x^2 + 2x + 12$

Concept Check   Find two polynomial functions defined by $f(x)$ and $g(x)$ such that each statement is true.

19. $(f + g)(x) = 3x^3 - x + 3$ **20.** $(f - g)(x) = -x^2 + x - 5$

Let $f(x) = x^2 - 9$, $g(x) = 2x$, and $h(x) = x - 3$. Find each of the following. See Example 3.

▶ **21.** $(f + g)(x)$ **22.** $(f - g)(x)$ **23.** $(f + g)(3)$ **24.** $(f - g)(-3)$

25. $(f - h)(x)$ **26.** $(f + h)(x)$ **27.** $(f - h)(-3)$ **28.** $(f + h)(-2)$

29. $(g + h)(-10)$ **30.** $(g - h)(10)$ **31.** $(g - h)(-3)$ **32.** $(g + h)(1)$

33. $(g + h)\left(\dfrac{1}{4}\right)$ **34.** $(g + h)\left(\dfrac{1}{3}\right)$ **35.** $(g + h)\left(-\dfrac{1}{2}\right)$ **36.** $(g + h)\left(-\dfrac{1}{4}\right)$

*Solve each problem. **See Objective 2.***

37. The cost in dollars to produce x t-shirts is $C(x) = 2.5x + 50$. The revenue in dollars from sales of x t-shirts is $R(x) = 10.99x$.

 (a) Write and simplify a function P that gives profit in terms of x.

 (b) Find the profit if 100 t-shirts are produced and sold.

38. The cost in dollars to produce x baseball caps is $C(x) = 4.3x + 75$. The revenue in dollars from sales of x caps is $R(x) = 25x$.

 (a) Write and simplify a function P that gives profit in terms of x.

 (b) Find the profit if 50 caps are produced and sold.

*For each pair of functions, find $(fg)(x)$. **See Example 4.***

39. $f(x) = 2x, \quad g(x) = 5x - 1$ **40.** $f(x) = 3x, \quad g(x) = 6x - 8$

41. $f(x) = x + 1, \quad g(x) = 2x - 3$ **42.** $f(x) = x - 7, \quad g(x) = 4x + 5$

43. $f(x) = 2x - 3, \quad g(x) = 4x^2 + 6x + 9$

44. $f(x) = 3x + 4, \quad g(x) = 9x^2 - 12x + 16$

*Let $f(x) = x^2 - 9$, $g(x) = 2x$, and $h(x) = x - 3$. Find each of the following. **See Example 4.***

▶ **45.** $(fg)(x)$ **46.** $(fh)(x)$ **47.** $(fg)(2)$

48. $(fh)(1)$ **49.** $(gh)(x)$ **50.** $(fh)(-1)$

51. $(gh)(-3)$ **52.** $(fg)(-2)$ **53.** $(fg)\left(-\dfrac{1}{2}\right)$

54. $(fg)\left(-\dfrac{1}{3}\right)$ **55.** $(fh)\left(-\dfrac{1}{4}\right)$ **56.** $(fh)\left(-\dfrac{1}{5}\right)$

*For each pair of functions, find $\left(\dfrac{f}{g}\right)(x)$ and give any x-values that are not in the domain of the quotient function. **See Example 5.***

57. $f(x) = 10x^2 - 2x, \quad g(x) = 2x$ **58.** $f(x) = 18x^2 - 24x, \quad g(x) = 3x$

59. $f(x) = 2x^2 - x - 3, \quad g(x) = x + 1$ **60.** $f(x) = 4x^2 - 23x - 35, \quad g(x) = x - 7$

61. $f(x) = 8x^3 - 27, \quad g(x) = 2x - 3$ **62.** $f(x) = 27x^3 + 64, \quad g(x) = 3x + 4$

*Let $f(x) = x^2 - 9$, $g(x) = 2x$, and $h(x) = x - 3$. Find each of the following. **See Example 5.***

▶ **63.** $\left(\dfrac{f}{g}\right)(x)$ **64.** $\left(\dfrac{f}{h}\right)(x)$ **65.** $\left(\dfrac{f}{g}\right)(2)$

66. $\left(\dfrac{f}{h}\right)(1)$ **67.** $\left(\dfrac{h}{g}\right)(x)$ **68.** $\left(\dfrac{g}{h}\right)(x)$

69. $\left(\dfrac{h}{g}\right)(3)$ **70.** $\left(\dfrac{g}{h}\right)(-1)$ **71.** $\left(\dfrac{f}{g}\right)\left(\dfrac{1}{2}\right)$

72. $\left(\dfrac{f}{g}\right)\left(\dfrac{3}{2}\right)$ **73.** $\left(\dfrac{h}{g}\right)\left(-\dfrac{1}{2}\right)$ **74.** $\left(\dfrac{h}{g}\right)\left(-\dfrac{3}{2}\right)$

Concept Check   Let $f(x) = x^2$ and $g(x) = 2x - 1$. *Match each expression in Column I with the description of how to evaluate it in Column II.*

I

II

75. $(f \circ g)(5)$ **A.** Square 5. Take the result and square it.

76. $(g \circ f)(5)$ **B.** Double 5 and subtract 1. Take the result and square it.

77. $(f \circ f)(5)$ **C.** Double 5 and subtract 1. Take the result, double it, and subtract 1.

78. $(g \circ g)(5)$ **D.** Square 5. Take the result, double it, and subtract 1.

Let $f(x) = x^2 + 4$, $g(x) = 2x + 3$, and $h(x) = x - 5$. Find each value or expression. **See Examples 6–8.**

79. $(h \circ g)(4)$ **80.** $(f \circ g)(4)$ **81.** $(g \circ f)(6)$ **82.** $(h \circ f)(6)$

83. $(f \circ h)(-2)$ **84.** $(h \circ g)(-2)$ **85.** $(f \circ g)(0)$ **86.** $(f \circ h)(0)$

87. $(g \circ f)(x)$ **88.** $(g \circ h)(x)$ **89.** $(h \circ g)(x)$ **90.** $(h \circ f)(x)$

91. $(f \circ h)\left(\dfrac{1}{2}\right)$ **92.** $(h \circ f)\left(\dfrac{1}{2}\right)$ **93.** $(f \circ g)\left(-\dfrac{1}{2}\right)$ **94.** $(g \circ f)\left(-\dfrac{1}{2}\right)$

Extending Skills   *The tables give some selected ordered pairs for functions f and g.*

x	3	4	6	8
$f(x)$	1	3	9	2

x	2	7	1	9
$g(x)$	3	6	9	12

Tables like these can be used to evaluate composite functions. For example, to evaluate $(g \circ f)(6)$, use the first table to find $f(6) = 9$. Then use the second table to find

$$(g \circ f)(6) = g(f(6)) = g(9) = 12.$$

Find each of the following.

95. $(f \circ g)(2)$ **96.** $(f \circ g)(7)$ **97.** $(g \circ f)(3)$

98. $(g \circ f)(8)$ **99.** $(f \circ f)(4)$ **100.** $(g \circ g)(1)$

Solve each problem. **See Objective 3.**

101. The function $f(x) = 12x$ computes the number of inches in x feet, and the function $g(x) = 5280x$ computes the number of feet in x miles. Find and simplify $(f \circ g)(x)$. What does it compute?

102. The function $f(x) = 60x$ computes the number of minutes in x hours, and the function $g(x) = 24x$ computes the number of hours in x days. Find and simplify $(f \circ g)(x)$. What does it compute?

103. The perimeter x of a square with sides of length s is given by the formula $x = 4s$.

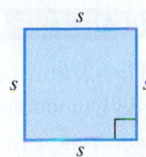

   (a) Solve for s in terms of x.

   (b) If y represents the area of this square, write y as a function of the perimeter x.

   (c) Use the composite function of part (b) to find the area of a square with perimeter 6.

104. The perimeter x of an equilateral triangle with sides of length s is given by the formula $x = 3s$.

(a) Solve for s in terms of x.

(b) The area y of an equilateral triangle with sides of length s is given by the formula $y = \dfrac{s^2\sqrt{3}}{4}$. Write y as a function of the perimeter x.

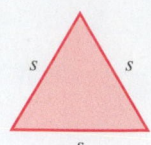

(c) Use the composite function of part (b) to find the area of an equilateral triangle with perimeter 12.

105. When a thermal inversion layer is over a city (as happens often in Los Angeles), pollutants cannot rise vertically, but are trapped below the layer and must disperse horizontally.

Assume that a factory smokestack begins emitting a pollutant at 8 A.M and that the pollutant disperses horizontally over a circular area. Suppose that t represents the time, in hours, since the factory began emitting pollutants ($t = 0$ represents 8 A.M.), and assume that the radius of the circle of pollution is $r(t) = 2t$ miles. Let $\mathcal{A}(r) = \pi r^2$ represent the area of a circle of radius r. Find and interpret $(\mathcal{A} \circ r)(t)$.

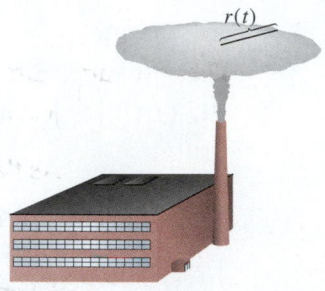

106. An oil well is leaking, with the leak spreading oil over the surface as a circle. At any time t, in minutes, after the beginning of the leak, the radius of the circular oil slick on the surface is $r(t) = 4t$ feet. Let $\mathcal{A}(r) = \pi r^2$ represent the area of a circle of radius r. Find and interpret $(\mathcal{A} \circ r)(t)$.

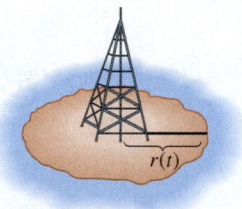

9.4 Variation

OBJECTIVES

1. Write an equation expressing direct variation.

2. Find the constant of variation, and solve direct variation problems.

3. Solve inverse variation problems.

4. Solve joint variation problems.

5. Solve combined variation problems.

Functions in which y *depends on a multiple of x* or y *depends on a number divided by x* occur in business, mathematics, and the physical sciences and are explored in this section.

OBJECTIVE 1 Write an equation expressing direct variation.

The circumference of a circle is given by the formula $C = 2\pi r$, where r is the radius of the circle. See **FIGURE 13**. The circumference is always a constant multiple of the radius—that is, C is always found by multiplying r by the constant 2π.

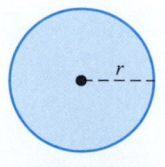

$C = 2\pi r$

FIGURE 13

As the *radius increases,* the *circumference increases.*

As the *radius decreases,* the *circumference decreases.*

Because of these relationships, the circumference is said to *vary directly* as the radius.

VOCABULARY

☐ direct variation
☐ constant of variation
☐ inverse variation
☐ joint variation
☐ combined variation

Direct Variation

y varies directly as *x* if there exists a real number *k* such that

$$y = kx.$$

y is said to be *proportional to x*. The number *k* is the **constant of variation.**

In direct variation, for k > 0, as the value of x increases, the value of y increases. Similarly, as x decreases, y decreases.

OBJECTIVE 2 Find the constant of variation, and solve direct variation problems.

The direct variation equation y = kx defines a linear function, where the constant of variation k is the slope of the line. For example, the following equation describes the cost *y* to buy *x* gallons of gasoline.

$$y = 3.50x$$

The cost varies directly as, or is proportional to, the number of gallons purchased.

> As the *number of gallons increases,* the *cost increases.*
>
> As the *number of gallons decreases,* the *cost decreases.*

The constant of variation *k* is 3.50, the cost of 1 gal of gasoline.

NOW TRY EXERCISE 1

One week Morgan sold 8 dozen eggs for $20. How much does she charge for one dozen eggs?

EXAMPLE 1 Solving a Direct Variation Problem

Eva is paid an hourly wage. One week she worked 43 hr and was paid $795.50. How much does she earn per hour?

$$\text{Let } h = \text{the number of hours she works}$$
$$\text{and } P = \text{her corresponding pay.}$$

Write a variation equation.

k represents Eva's hourly wage. → $P = kh$	*P* varies directly as *h*.
$795.50 = 43k$	Substitute 795.50 for *P* and 43 for *h*.
This is the constant of variation. → $k = 18.50$	Use a calculator.

Her hourly wage is $18.50, and *P* and *h* are related by the equation

$$P = 18.50h.$$

We can use this equation to find her pay for any number of hours worked.

NOW TRY

EXAMPLE 2 Solving a Direct Variation Problem

Hooke's law for an elastic spring states that the distance a spring stretches is directly proportional to the force applied. If a force of 150 newtons* stretches a certain spring 8 cm, how much will a force of 400 newtons stretch the spring? See **FIGURE 14**.

$$\text{Let } d = \text{the distance the spring stretches}$$
$$\text{and } f = \text{the force applied.}$$

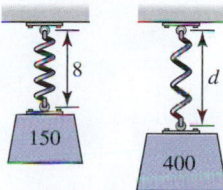

FIGURE 14

---

*A newton is a unit of measure of force used in physics.

**NOW TRY
EXERCISE 2**

For a constant height, the area of a parallelogram is directly proportional to its base. If the area is 20 cm² when the base is 4 cm, find the area when the base is 7 cm.

Then $d = kf$ for some constant k. A force of 150 newtons stretches the spring 8 cm, so we use these values to find k.

$$d = kf \qquad \text{Variation equation}$$

Solve for k. $\qquad 8 = k \cdot 150 \qquad$ Let $d = 8$ and $f = 150$.

$$k = \frac{8}{150} \qquad \text{Solve for } k.$$

$$k = \frac{4}{75} \qquad \text{Write in lowest terms.}$$

Now we rewrite the variation equation $d = kf$ using $\frac{4}{75}$ for k.

$$d = \frac{4}{75}f \qquad \text{Here, } k = \frac{4}{75}.$$

For a force of 400 newtons, substitute 400 for f.

$$d = \frac{4}{75}(400) = \frac{64}{3} \qquad \text{Let } f = 400.$$

The spring will stretch $\frac{64}{3}$ cm, or $21\frac{1}{3}$ cm, if a force of 400 newtons is applied.

NOW TRY

Solving a Variation Problem

Step 1 Write a variation equation.

Step 2 Substitute the initial values and solve for k.

Step 3 Rewrite the variation equation with the value of k from Step 2.

Step 4 Substitute the remaining values, solve for the unknown, and find the required answer.

One variable can be proportional to a power of another variable.

Direct Variation as a Power

y varies directly as the nth power of x if there exists a real number k such that

$$y = kx^n.$$

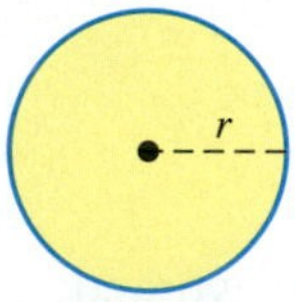

$\mathscr{A} = \pi r^2$

FIGURE 15

An example of direct variation as a power is the formula for the area of a circle, $\mathscr{A} = \pi r^2$. See **FIGURE 15**. Here, π is the constant of variation, and the area varies directly as the *square* of the radius.

EXAMPLE 3   Solving a Direct Variation Problem

The distance a body falls from rest varies directly as the square of the time it falls (disregarding air resistance). If a skydiver falls 64 ft in 2 sec, how far will she fall in 8 sec?

Step 1   Let $d =$ the distance the skydiver falls

                      and $t =$ the time it takes to fall.

Then d is a function of t for some constant k.

$$d = kt^2 \qquad d \text{ varies directly as the square of } t.$$

NOW TRY ANSWER
2. 35 cm²

 **NOW TRY EXERCISE 3**

Suppose y varies directly as the square of x, and $y = 200$ when $x = 5$. Find y when $x = 7$.

Step 2 To find the value of k, use the fact that the skydiver falls 64 ft in 2 sec.

$$d = kt^2 \qquad \text{Variation equation}$$
$$64 = k(2)^2 \qquad \text{Let } d = 64 \text{ and } t = 2.$$
$$k = 16 \qquad \text{Find } k.$$

Step 3 Now we rewrite the variation equation $d = kt^2$ using 16 for k.

$$d = 16t^2 \qquad \text{Here, } k = 16.$$

Step 4 Let $t = 8$ to find the number of feet the skydiver will fall in 8 sec.

$$d = 16(8)^2 = 1024 \qquad \text{Let } t = 8.$$

The skydiver will fall 1024 ft in 8 sec.

NOW TRY

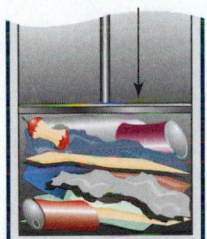

As pressure on trash increases, volume of trash decreases.

FIGURE 16

OBJECTIVE 3 Solve inverse variation problems.

Another type of variation is *inverse variation*. **With inverse variation, for $k > 0$, as one variable increases, the other variable decreases.**

For example, in a closed space, volume decreases as pressure increases, which can be illustrated by a trash compactor. See **FIGURE 16**. As the compactor presses down, the pressure on the trash increases, and in turn, the trash occupies a smaller space.

Inverse Variation

y **varies inversely as** x if there exists a real number k such that

$$y = \frac{k}{x}.$$

y **varies inversely as the nth power of** x if there exists a real number k such that

$$y = \frac{k}{x^n}.$$

The inverse variation equation defines a rational function. Another example of inverse variation comes from the distance formula.

$$d = rt \qquad \text{Distance formula}$$

$$t = \frac{d}{r} \qquad \text{Divide each side by } r.$$

Here, t (time) varies inversely as r (rate or speed), with d (distance) serving as the constant of variation. For example, if the distance between Chicago and Des Moines is 300 mi, then

$$t = \frac{300}{r}.$$

The values of r and t might be any of the following.

$$\left.\begin{array}{l} r = 50, t = 6 \\ r = 60, t = 5 \\ r = 75, t = 4 \end{array}\right\} \begin{array}{l} \text{As } r \text{ increases,} \\ t \text{ decreases.} \end{array} \qquad \left.\begin{array}{l} r = 30, t = 10 \\ r = 25, t = 12 \\ r = 20, t = 15 \end{array}\right\} \begin{array}{l} \text{As } r \text{ decreases,} \\ t \text{ increases.} \end{array}$$

NOW TRY ANSWER
3. 392

If we *increase* the rate (speed) at which we drive, time *decreases*. If we *decrease* the rate (speed) at which we drive, time *increases*.

NOW TRY
EXERCISE 4

For a constant area, the height of a triangle varies inversely as the base. If the height is 7 cm when the base is 8 cm, find the height when the base is 14 cm.

EXAMPLE 4   Solving an Inverse Variation Problem

In the manufacture of a phone-charging device, the cost of producing the device varies inversely as the number produced. If 10,000 units are produced, the cost is $2 per unit. Find the cost per unit to produce 25,000 units.

$$\text{Let} \quad x = \text{the number of units produced}$$

$$\text{and} \quad c = \text{the cost per unit.}$$

Here, as production increases, cost decreases, and as production decreases, cost increases. We write a variation equation using the variables c and x and the constant k.

$$c = \frac{k}{x} \qquad \textit{c varies inversely as x.}$$

To find k, we replace c with 2 and x with 10,000.

$$2 = \frac{k}{10,000} \qquad \textit{Substitute in the variation equation.}$$

$$20,000 = k \qquad \textit{Multiply by 10,000.}$$

Thus, the variation equation $c = \frac{k}{x}$ becomes $c = \frac{20,000}{x}$. When $x = 25,000$,

$$c = \frac{20,000}{25,000} = 0.80. \qquad \textit{Let x = 25,000.}$$

The cost per unit to make 25,000 units is $0.80.   **NOW TRY**

NOW TRY
EXERCISE 5

The weight of an object above Earth varies inversely as the square of its distance from the center of Earth. If an object weighs 150 lb on the surface of Earth, and the radius of Earth is about 3960 mi, how much does it weigh when it is 1000 mi above Earth's surface? Round to the nearest pound.

EXAMPLE 5   Solving an Inverse Variation Problem

The weight of an object above Earth varies inversely as the square of its distance from the center of Earth. A space shuttle in an elliptical orbit has a maximum distance from the center of Earth (**apogee**) of 6700 mi. Its minimum distance from the center of Earth (**perigee**) is 4090 mi. See **FIGURE 17**. If an astronaut in the shuttle weighs 57 lb at its apogee, what does the astronaut weigh at its perigee?

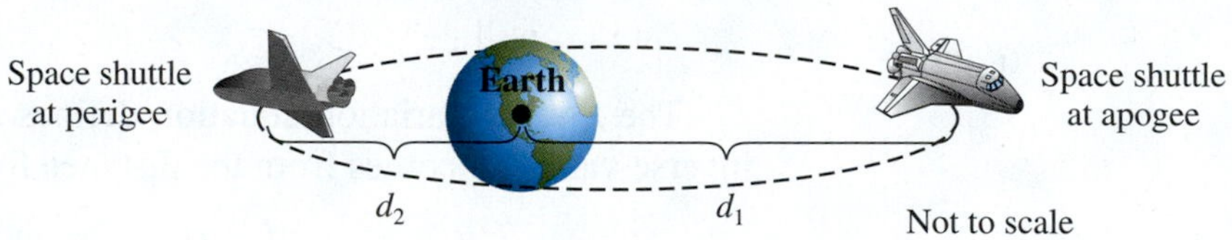

Space shuttle at perigee   **Earth**   Space shuttle at apogee

d_2 d_1 Not to scale

FIGURE 17

Let $w = $ the weight and $d = $ the distance from the center of Earth, for some constant k. We write a variation equation using these variables.

$$w = \frac{k}{d^2} \qquad \textit{w varies inversely as the square of d.}$$

At the apogee, the astronaut weighs 57 lb, and the distance from the center of Earth is 6700 mi. Use these values to find k.

$$57 = \frac{k}{(6700)^2} \qquad \textit{Let w = 57 and d = 6700.}$$

$$k = 57(6700)^2 \qquad \textit{Solve for k.}$$

Substitute $k = 57(6700)^2$ and $d = 4090$ to find the weight at the perigee.

NOW TRY ANSWERS
4. 4 cm   **5.** 96 lb

$$w = \frac{57(6700)^2}{(4090)^2} = 153 \text{ lb} \qquad \textit{Use a calculator. Round to the nearest pound.} \qquad \textbf{NOW TRY}$$

OBJECTIVE 4 Solve joint variation problems.

If one variable varies directly as the *product* of several other variables (perhaps raised to powers), the first variable is said to *vary jointly* as the others.

> **Joint Variation**
>
> **y varies jointly as x and z** if there exists a real number k such that
> $$y = kxz.$$

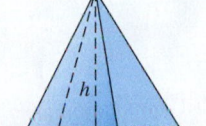

B = area of the base

FIGURE 18

An example of joint variation is the formula for the volume of a right pyramid, $V = \frac{1}{3}Bh$. See **FIGURE 18**. Here, $\frac{1}{3}$ is the constant of variation, and the volume varies jointly as the area of the base and the height.

EXAMPLE 6 Solving a Joint Variation Problem

The interest on a loan or an investment is given by the formula $I = prt$. Here, for a given principal p, the interest earned, I, varies jointly as the interest rate r and the time t the principal is left earning interest. If an investment earns $100 interest at 5% for 2 yr, how much interest will the same principal earn at 4.5% for 3 yr?

We use the formula $I = prt$, where p is the constant of variation because it is the same for both investments.

$$I = prt \qquad \text{Here, } p \text{ is the constant of variation.}$$

Solve for p. ▸ $100 = p(0.05)(2) \qquad$ Let $I = 100$, $r = 0.05$, and $t = 2$.

$$100 = 0.1p \qquad \text{Multiply.}$$

$$p = 1000 \qquad \text{Divide by 0.1. Rewrite.}$$

Now we find I when $p = 1000$, $r = 0.045$, and $t = 3$.

$$I = 1000(0.045)(3) = 135 \qquad \text{Let } p = 1000, r = 0.045, \text{ and } t = 3.$$

The interest will be $135. **NOW TRY**

> ⛔ **CAUTION** Note that *and* in the expression "y varies directly as *x and z*" translates as a product in the variation equation $y = kxz$. The word *and* does **not** indicate addition here.

OBJECTIVE 5 Solve combined variation problems.

There are combinations of direct and inverse variation, called **combined variation.**

EXAMPLE 7 Solving a Combined Variation Problem

Body mass index (BMI) is used to assess whether a person's weight is healthy. A BMI from 19 through 25 is considered desirable. BMI varies directly as an individual's weight in pounds and inversely as the square of their height in inches.

A person who weighs 118 lb and is 64 in. tall has a BMI of 20. (BMI is rounded to the nearest whole number.) Find the BMI of a man who weighs 165 lb and is 70 in. tall. (*Source: Washington Post.*)

Let B = BMI, w = weight, and h = height. Write a variation equation.

$$B = \frac{kw}{h^2} \qquad \begin{array}{l} \longleftarrow \text{ BMI varies directly as the weight.} \\ \longleftarrow \text{ BMI varies inversely as the square of the height.} \end{array}$$

NOW TRY EXERCISE 6

The volume of a right pyramid varies jointly as the height and the area of the base. If the volume is 100 ft³ when the area of the base is 30 ft² and the height is 10 ft, find the volume when the area of the base is 90 ft² and the height is 20 ft.

NOW TRY ANSWER

6. 600 ft³

**NOW TRY
EXERCISE 7**

In statistics, the sample size used to estimate a population mean varies directly as the variance and inversely as the square of the maximum error of the estimate. If the sample size is 200 when the variance is 25 m^2 and the maximum error of the estimate is 0.5 m, find the sample size when the variance is 25 m^2 and the maximum error of the estimate is 0.1 m.

NOW TRY ANSWER
7. 5000

To find k, let $B = 20$, $w = 118$, and $h = 64$.

$$20 = \frac{k(118)}{64^2} \qquad B = \frac{kw}{h^2}$$

$$k = \frac{20(64^2)}{118} \qquad \text{Multiply by } 64^2.\\ \text{Divide by 118.}$$

$$k = 694 \qquad \text{Use a calculator. Round to the nearest whole number.}$$

Now find B when $k = 694$, $w = 165$, and $h = 70$.

$$B = \frac{694(165)}{70^2} = 23 \qquad \text{Round to the nearest whole number.}$$

The man's BMI is 23.

NOW TRY

9.4 Exercises

FOR EXTRA HELP

 MyMathLab®

▶ *Complete solution available in MyMathLab*

Concept Check   *Fill in each blank with the correct response.*

1. For $k > 0$, if y varies directly as x, then when x increases, y _____, and when x decreases, y _____.

2. For $k > 0$, if y varies inversely as x, then when x increases, y _____, and when x decreases, y _____.

Concept Check   *Use personal experience or intuition to determine whether the situation suggests* direct *or* inverse *variation.*

3. The number of movie tickets purchased and the total price for the tickets

4. The rate and the distance traveled by a pickup truck in 3 hr

5. The amount of pressure put on the accelerator of a car and the speed of the car

6. The percentage off an item that is on sale and the price of the item

7. Your age and the probability that you believe in the tooth fairy

8. The surface area of a balloon and its diameter

9. The demand for an item and the price of the item

10. The number of hours worked by an hourly worker and the amount of money earned

Concept Check   *Determine whether each equation represents* direct, inverse, joint, *or* combined *variation.*

11. $y = \dfrac{3}{x}$ 12. $y = \dfrac{8}{x}$ 13. $y = 10x^2$ 14. $y = 2x^3$

15. $y = 3xz^4$ 16. $y = 6x^3z^2$ 17. $y = \dfrac{4x}{wz}$ 18. $y = \dfrac{6x}{st}$

Concept Check   *Write each formula using the "language" of variation. For example, the formula for the circumference of a circle, $C = 2\pi r$, can be written as*

"The circumference of a circle varies directly as the length of its radius."

19. $P = 4s$, where P is the perimeter of a square with side of length s

20. $d = 2r$, where d is the diameter of a circle with radius r

21. $S = 4\pi r^2$, where S is the surface area of a sphere with radius r

22. $V = \frac{4}{3}\pi r^3$, where V is the volume of a sphere with radius r

23. $\mathcal{A} = \frac{1}{2}bh$, where $\mathcal{A}$ is the area of a triangle with base b and height h

24. $V = \frac{1}{3}\pi r^2 h$, where V is the volume of a cone with radius r and height h

25. *Concept Check*   What is the constant of variation in each of the variation equations in **Exercises 19–24?**

26. *Concept Check*   What is meant by the constant of variation in a direct variation problem? If we were to graph the linear equation $y = kx$ for some nonnegative constant k, what role would k play in the graph?

Write a variation equation for each situation. Use k as the constant of variation. **See Examples 1–6.**

27. A varies directly as b.

28. W varies directly as f.

29. h varies inversely as t.

30. p varies inversely as s.

31. M varies directly as the square of d.

32. P varies inversely as the cube of x.

33. I varies jointly as g and h.

34. C varies jointly as a and the square of b.

Solve each problem. **See Examples 1–7.**

35. If x varies directly as y, and $x = 9$ when $y = 3$, find x when $y = 12$.

36. If x varies directly as y, and $x = 10$ when $y = 7$, find y when $x = 50$.

37. If a varies directly as the square of b, and $a = 4$ when $b = 3$, find a when $b = 2$.

38. If h varies directly as the square of m, and $h = 15$ when $m = 5$, find h when $m = 7$.

39. If z varies inversely as w, and $z = 10$ when $w = 0.5$, find z when $w = 8$.

40. If t varies inversely as s, and $t = 3$ when $s = 5$, find s when $t = 5$.

41. If m varies inversely as p^2, and $m = 20$ when $p = 2$, find m when $p = 5$.

42. If a varies inversely as b^2, and $a = 48$ when $b = 4$, find a when $b = 7$.

43. p varies jointly as q and r^2, and $p = 200$ when $q = 2$ and $r = 3$. Find p when $q = 5$ and $r = 2$.

44. f varies jointly as g^2 and h, and $f = 50$ when $g = 4$ and $h = 2$. Find f when $g = 3$ and $h = 6$.

Solve each problem. **See Examples 1–7.**

▶ **45.** Ben bought 8.5 gal of gasoline and paid \$33.32. What is the price of gasoline per gallon?

46. Sara gives horseback rides at Shadow Mountain Ranch. A 2.5-hr ride costs \$50.00. What is the price per hour?

▶ **47.** The weight of an object on Earth is directly proportional to the weight of that same object on the moon. A 200-lb astronaut would weigh 32 lb on the moon. How much would a 50-lb dog weigh on the moon?

48. The pressure exerted by a certain liquid at a given point is directly proportional to the depth of the point beneath the surface of the liquid. The pressure at 30 m is 80 newtons. What pressure is exerted at 50 m?

49. The volume of a can of tomatoes is directly proportional to the height of the can. If the volume of the can is 300 cm^3 when its height is 10.62 cm, find the volume to the nearest whole number of a can with height 15.92 cm.

50. The force required to compress a spring is directly proportional to the change in length of the spring. If a force of 20 newtons is required to compress a certain spring 2 cm, how much force is required to compress the spring from 20 cm to 8 cm?

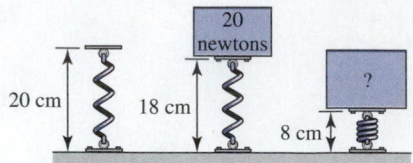

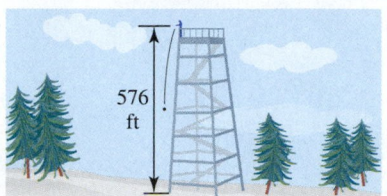

▶ 51. For a body falling freely from rest (disregarding air resistance), the distance the body falls varies directly as the square of the time. If an object is dropped from the top of a tower 576 ft high and hits the ground in 6 sec, how far did it fall in the first 4 sec?

52. The amount of water emptied by a pipe varies directly as the square of the diameter of the pipe. For a certain constant water flow, a pipe emptying into a canal will allow 200 gal of water to escape in an hour. The diameter of the pipe is 6 in. How much water would a 12-in. pipe empty into the canal in an hour, assuming the same water flow?

▶ 53. Over a specified distance, rate varies inversely with time. If a Dodge Viper on a test track goes a certain distance in one-half minute at 160 mph, what rate is needed to go the same distance in three-fourths minute?

54. For a constant area, the length of a rectangle varies inversely as the width. The length of a rectangle is 27 ft when the width is 10 ft. Find the width of a rectangle with the same area if the length is 18 ft.

55. The frequency of a vibrating string varies inversely as its length. That is, a longer string vibrates fewer times in a second than a shorter string. Suppose a piano string 2 ft long vibrates 250 cycles per sec. What frequency would a string 5 ft long have?

56. The current in a simple electrical circuit varies inversely as the resistance. If the current is 20 amps when the resistance is 5 ohms, find the current when the resistance is 7.5 ohms.

▶ 57. The amount of light (measured in foot-candles) produced by a light source varies inversely as the square of the distance from the source. If the illumination produced 1 m from a light source is 768 foot-candles, find the illumination produced 6 m from the same source.

58. The force with which Earth attracts an object above Earth's surface varies inversely as the square of the distance of the object from the center of Earth. If an object 4000 mi from the center of Earth is attracted with a force of 160 lb, find the force of attraction if the object were 6000 mi from the center of Earth.

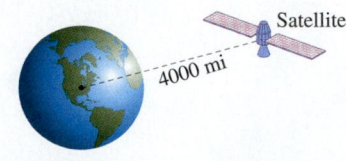

▶ 59. For a given interest rate, simple interest varies jointly as principal and time. If $2000 left in an account for 4 yr earned interest of $280, how much interest would be earned in 6 yr?

60. The collision impact of an automobile varies jointly as its mass and the square of its speed. Suppose a 2000-lb car traveling at 55 mph has a collision impact of 6.1. What is the collision impact (to the nearest tenth) of the same car at 65 mph?

61. The weight of a bass varies jointly as its girth and the square of its length. (**Girth** is the distance around the body of the fish.) A prize-winning bass weighed in at 22.7 lb and measured 36 in. long with a 21-in. girth. How much (to the nearest tenth of a pound) would a bass 28 in. long with an 18-in. girth weigh?

62. The weight of a trout varies jointly as its length and the square of its girth. One angler caught a trout that weighed 10.5 lb and measured 26 in. long with an 18-in. girth. Find the weight (to the nearest tenth of a pound) of a trout that is 22 in. long with a 15-in. girth.

63. The force needed to keep a car from skidding on a curve varies inversely as the radius of the curve and jointly as the weight of the car and the square of the speed. If 242 lb of force keeps a 2000-lb car from skidding on a curve of radius 500 ft at 30 mph, what force (to the nearest tenth of a pound) would keep the same car from skidding on a curve of radius 750 ft at 50 mph?

64. The maximum load that a cylindrical column with a circular cross section can hold varies directly as the fourth power of the diameter of the cross section and inversely as the square of the height. A 9-m column 1 m in diameter will support 8 metric tons. How many metric tons can be supported by a column 12 m high and $\frac{2}{3}$ m in diameter?

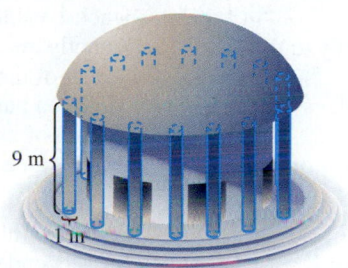

9 m

1 m

Load = 8 metric tons

65. The number of long-distance phone calls between two cities during a certain period varies jointly as the populations of the cities, p_1 and p_2, and inversely as the distance between them. If 80,000 calls are made between two cities 400 mi apart, with populations of 70,000 and 100,000, how many calls (to the nearest hundred) are made between cities with populations of 50,000 and 75,000 that are 250 mi apart?

66. The volume of gas varies inversely as the pressure and directly as the temperature. (Temperature must be measured in *Kelvin* (K), a unit of measurement used in physics.) If a certain gas occupies a volume of 1.3 L at 300 K and a pressure of 18 newtons, find the volume at 340 K and a pressure of 24 newtons.

▶ **67.** A body mass index from 27 through 29 carries a slight risk of weight-related health problems, while one of 30 or more indicates a great increase in risk. Use your own height and weight and the information in **Example 7** to determine your BMI and whether you are at risk.

68. The maximum load of a horizontal beam that is supported at both ends varies directly as the width and the square of the height and inversely as the length between the supports. A beam 6 m long, 0.1 m wide, and 0.06 m high supports a load of 360 kg. What is the maximum load supported by a beam 16 m long, 0.2 m wide, and 0.08 m high?

Chapter 9   Summary

Key Terms

9.1
relation
function
dependent variable
independent variable
domain
range

9.2
linear function
constant function

9.3
polynomial function
composite function
(composition of functions)

9.4
direct variation
constant of variation
inverse variation
joint variation
combined variation

New Symbols

$f(x)$ function notation;
function of x (read
"f of x" or "f at x")

$(f \circ g)(x) = f(g(x))$ composite function

Test Your Word Power

See how well you have learned the vocabulary in this chapter.

1. A **relation** is
  A. a set of ordered pairs
  B. the ratio of the change in y to the change in x along a line
  C. the set of all possible values of the independent variable
  D. all the second components of a set of ordered pairs.

2. A **function** is
  A. a pair of numbers in an ordered pair

  B. a set of ordered pairs in which each x-value corresponds to exactly one y-value
  C. a pair of numbers written between parentheses
  D. the set of all ordered pairs that satisfy an equation.

3. The **domain** of a relation is
  A. the set of all possible values of the dependent variable y
  B. a set of ordered pairs

C. the difference between the x-values
  D. the set of all possible values of the independent variable x.

4. The **range** of a relation is
  A. the set of all possible values of the dependent variable y
  B. a set of ordered pairs
  C. the difference between the y-values
  D. the set of all possible values of the independent variable x.

ANSWERS

1. A; *Example:* The set $\{(2, 0), (4, 3), (6, 6)\}$ defines a relation.   **2.** B; *Example:* The relation given in Answer 1 is a function.
3. D; *Example:* In the relation in Answer 1, the domain is the set of x-values, $\{2, 4, 6\}$.   **4.** A; *Example:* In the relation in
Answer 1, the range is the set of y-values, $\{0, 3, 6\}$.

Quick Review

CONCEPTS

EXAMPLES

9.1   Introduction to Relations and Functions

A **relation** is any set of ordered pairs. A **function** is a set of ordered pairs such that, for each distinct first component, there is one and only one second component.

The set of first components is the **domain.**

The set of second components is the **range.**

The set of ordered pairs $\{(-1, 4), (0, 6), (1, 4)\}$ defines a function.

    Domain: $\{-1, 0, 1\}$ Set of x-values

    Range: $\{4, 6\}$ Set of y-values

The equation $y = x^2$ defines a function.

    Domain: $(-\infty, \infty)$ Range: $[0, \infty)$

9.2 Function Notation and Linear Functions

To evaluate a function f, where $f(x)$ defines the range value for a given value of x in the domain, substitute the value wherever x appears.

$$f(x) = x^2 - 7x + 12$$
$$f(1) = 1^2 - 7(1) + 12 \qquad \text{Let } x = 1.$$
$$f(1) = 6$$

To write an equation that defines a function f in function notation, follow these steps.

Write $2x + 3y = 12$ using function notation for a function f.

Step 1 Solve the equation for y.

$$3y = -2x + 12 \qquad \text{Subtract } 2x.$$
$$y = -\frac{2}{3}x + 4 \qquad \text{Divide by 3.}$$

Step 2 Replace y with $f(x)$.

$$f(x) = -\frac{2}{3}x + 4 \qquad y = f(x)$$

9.3 Polynomial Functions, Operations, and Composition

Operations on Functions

If $f(x)$ and $g(x)$ define functions, then

$$(f + g)(x) = f(x) + g(x),$$
$$(f - g)(x) = f(x) - g(x),$$
$$(fg)(x) = f(x) \cdot g(x),$$

and

$$\left(\frac{f}{g}\right)(x) = \frac{f(x)}{g(x)}, \quad g(x) \neq 0.$$

Let $f(x) = x^2$ and $g(x) = 2x + 1$. Find the following.

$$(f + g)(x) \qquad (f - g)(x)$$
$$= f(x) + g(x) \qquad = f(x) - g(x)$$
$$= x^2 + 2x + 1 \qquad = x^2 - (2x + 1)$$
$$\qquad\qquad = x^2 - 2x - 1$$

$$(fg)(x) \qquad \left(\frac{f}{g}\right)(x)$$
$$= f(x) \cdot g(x) \qquad = \frac{f(x)}{g(x)}$$
$$= x^2(2x + 1) \qquad = \frac{x^2}{2x + 1}, \quad x \neq -\frac{1}{2}$$
$$= 2x^3 + x^2$$

Composition of f and g

$$(f \circ g)(x) = f(g(x))$$
$$(g \circ f)(x) = g(f(x))$$

Let $f(x) = x^2$ and $g(x) = 2x + 1$. Find $(f \circ g)(x)$ and $(g \circ f)(x)$.

$$(f \circ g)(x) = f(g(x)) \qquad (g \circ f)(x) = g(f(x))$$
$$= f(2x + 1) \qquad = g(x^2)$$
$$= (2x + 1)^2 \qquad = 2x^2 + 1$$

9.4 Variation

Let k be a real number.

If $y = kx^n$, then y varies directly as x^n.

The area of a circle varies directly as the square of the radius.

$$\mathcal{A} = kr^2 \qquad \text{Here, } k = \pi.$$

If $y = \dfrac{k}{x^n}$, then y varies inversely as x^n.

Pressure varies inversely as volume.

$$p = \frac{k}{V}$$

If $y = kxz$, then y varies jointly as x and z.

For a given principal, interest varies jointly as interest rate and time.

$$I = krt \qquad k \text{ is the given principal.}$$

Chapter 9   Review Exercises

1. not a function;
domain: $\{-4, 1\}$;
range: $\{2, -2, 5, -5\}$

2. function;
domain: $\{9, 11, 4, 17, 25\}$;
range: $\{32, 47, 69, 14\}$

3. function;
domain: $[-4, 4]$;
range: $[0, 2]$

4. not a function;
domain: $(-\infty, 0]$;
range: $(-\infty, \infty)$

5. function;
domain: $(-\infty, \infty)$;
linear function

6. not a function;
domain: $(-\infty, \infty)$

7. not a function;
domain: $[0, \infty)$

8. function;
domain: $(-\infty, 6) \cup (6, \infty)$

9. -6 **10.** -8.52

11. -8 **12.** $-2k^2 + 3k - 6$

13. $f(x) = 2x^2$; 18

14. C

15. It is a horizontal line.

16. (a) yes
(b) domain: $\{1960, 1970, 1980,$
$1990, 2000, 2010\}$;
range: $\{69.7, 70.8, 73.7, 75.4,$
$76.8, 78.7\}$
(c) Answers will vary.
Two possible answers are
$(1960, 69.7)$ and $(2010, 78.7)$.
(d) 73.7;
In 1980, life expectancy at
birth was 73.7 yr.
(e) 2000

17. (a) -11 **(b)** 4 **(c)** 7

18. $f(x)$ and $g(x)$ can be any
two polynomials that have a
sum of $5x^2 - 3x - 1$, such
as $f(x) = 4x^2 - x + 2$ and
$g(x) = x^2 - 2x - 3$.

9.1 *Decide whether each relation defines a function, and give the domain and range.*

1. $\{(-4, 2), (-4, -2), (1, 5), (1, -5)\}$

2. 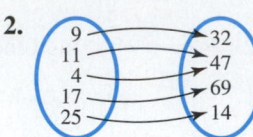   **3.** 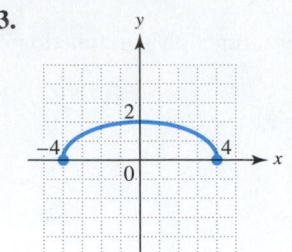   **4.**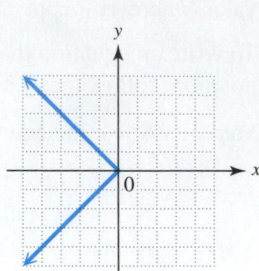

9.1, 9.2 *Decide whether each relation defines y as a function of x. Give the domain. Identify any linear functions.*

5. $y = 3x - 3$ **6.** $y < x + 2$ **7.** $x = y^2$ **8.** $y = \dfrac{7}{x - 6}$

9.2 *Let $f(x) = -2x^2 + 3x - 6$. Find the following.*

9. $f(0)$ **10.** $f(2.1)$ **11.** $f\left(-\dfrac{1}{2}\right)$ **12.** $f(k)$

Solve each problem.

13. The equation $2x^2 - y = 0$ defines y as a function f of x. Write it using function notation, and find $f(3)$.

14. Suppose that $2x - 5y = 7$ defines y as a function f of x. If $y = f(x)$, which one of the following defines the same function?

A. $f(x) = -\dfrac{2}{5}x + \dfrac{7}{5}$ **B.** $f(x) = -\dfrac{2}{5}x - \dfrac{7}{5}$

C. $f(x) = \dfrac{2}{5}x - \dfrac{7}{5}$ **D.** $f(x) = \dfrac{2}{5}x + \dfrac{7}{5}$

15. Describe the graph of a constant function.

16. The table shows life expectancy at birth in the United States for selected years.

(a) Does the table define a function?

(b) What are the domain and range?

(c) Call this function f. Give two ordered pairs that belong to f.

(d) Find $f(1980)$. What does this mean?

(e) If $f(x) = 76.8$, what does x equal?

Year	Life Expectancy at Birth (years)
1960	69.7
1970	70.8
1980	73.7
1990	75.4
2000	76.8
2010	78.7

Source: National Center for Health Statistics.

9.3 *Work each problem.*

17. For the polynomial function $f(x) = -2x^2 + 5x + 7$, find each value.

(a) $f(-2)$ **(b)** $f(3)$ **(c)** $f(0)$

18. Find two polynomial functions defined by $f(x)$ and $g(x)$ such that

$$(f + g)(x) = 5x^2 - 3x - 1.$$

19. (a) $5x^2 - x + 5$ **(b)** -9
20. (a) $-5x^2 + 5x + 1$ **(b)** 11
21. (a) $36x^3 - 9x^2$ **(b)** -45
22. (a) $4x - 1,\ \ x \neq 0$ **(b)** 7
23. (a) $75x^2 + 220x + 160$
    (b) 1495 **(c)** 20
24. (a) $13x^2 + 10x + 2$
    (b) 167 **(c)** 42

25. C   **26.** 430 mm
27. 5.59 vibrations per sec
28. 22.5 ft³

Find each of the following for the polynomial functions

$$f(x) = 2x + 3 \quad and \quad g(x) = 5x^2 - 3x + 2.$$

19. (a) $(f + g)(x)$ **(b)** $(f - g)(-1)$ **20. (a)** $(f - g)(x)$ **(b)** $(f + g)(-1)$

Find each of the following for the polynomial functions

$$f(x) = 12x^2 - 3x \quad and \quad g(x) = 3x.$$

21. (a) $(fg)(x)$ **(b)** $(fg)(-1)$ **22. (a)** $\left(\dfrac{f}{g}\right)(x)$ **(b)** $\left(\dfrac{f}{g}\right)(2)$

Find each of the following for the polynomial functions

$$f(x) = 3x^2 + 2x - 1 \quad and \quad g(x) = 5x + 7.$$

23. (a) $(f \circ g)(x)$ **(b)** $(f \circ g)(3)$ **(c)** $(f \circ g)(-2)$
24. (a) $(g \circ f)(x)$ **(b)** $(g \circ f)(3)$ **(c)** $(g \circ f)(-2)$

9.4

25. In which one of the following does y vary inversely as x?

  A. $y = 2x$ **B.** $y = \dfrac{x}{3}$ **C.** $y = \dfrac{3}{x}$ **D.** $y = x^2$

Solve each problem.

26. For a particular camera, the viewing distance varies directly as the amount of enlargement. A picture that is taken with this camera and enlarged 5 times should be viewed from a distance of 250 mm. Suppose a print 8.6 times the size of the negative is made. From what distance should it be viewed?

27. The frequency (number of vibrations per second) of a vibrating guitar string varies inversely as its length. That is, a longer string vibrates fewer times in a second than a shorter string. Suppose a guitar string 0.65 m long vibrates 4.3 times per sec. What frequency would a string 0.5 m long have?

28. The volume of a rectangular box of a given height is proportional to its width and length. A box with width 2 ft and length 4 ft has volume 12 ft³. Find the volume of a box with the same height that is 3 ft wide and 5 ft long.

Chapter 9 | Mixed Review Exercises

1. domain: $\{14, 91, 75, 23\}$;
range: $\{9, 70, 56, 5\}$;
not a function;
75 in the domain is paired with
two different values, 70 and 56,
in the range.

2. (a) -1 **(b)** -2
    (c) 2
    (d) $(-\infty, \infty)$; $(-\infty, \infty)$

Work each problem.

1. Give the domain and range of the relation. Does it define a function? Why or why not?

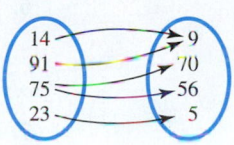

2. Refer to the graph of function f.

  (a) Find $f(-2)$.

  (b) Find $f(0)$.

  (c) For what value of x is $f(x) = -3$?

  (d) Give the domain and range.

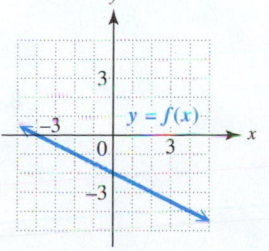

3. −2   **4.** 24
5. −112   **6.** 1
7. 94   **8.** 13

9. 12 ft²   **10.** 32.97 in.

Let $f(x) = 2x^2 - 4$ and $g(x) = 3x + 1$. Find each of the following.

3. $f(-1)$ **4.** $(f + g)(3)$ **5.** $(gf)(-3)$

6. $\left(\dfrac{g}{f}\right)(-1)$ **7.** $(f \circ g)(2)$ **8.** $(g \circ f)(2)$

Solve each problem.

9. The area of a triangle varies jointly as the lengths of the base and height. A triangle with base 10 ft and height 4 ft has area 20 ft². Find the area of a triangle with base 3 ft and height 8 ft.

10. The circumference of a circle varies directly as its radius. A circle with circumference 9.42 in. has radius approximately 1.5 in. Find the circumference of a circle with radius 5.25 in. Give the answer to the nearest hundredth.

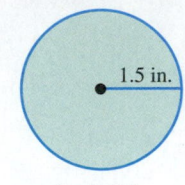

$C = 9.42$ in.

Chapter 9   Test

FOR EXTRA HELP *Step-by-step test solutions are found on the Chapter Test Prep Videos available in* **MyMathLab®** *or on* **YouTube**.

▶ *View the complete solutions to all Chapter Test exercises in MyMathLab.*

[9.1]
1. D   **2.** C
3. domain: $[0, \infty)$;
  range: $(-\infty, \infty)$
4. domain: $\{0, -2, 4\}$;
  range: $\{1, 3, 8\}$

[9.2]
5. 0; $-a^2 + 2a - 1$
6.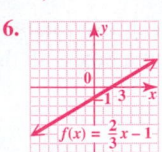

domain: $(-\infty, \infty)$;
range: $(-\infty, \infty)$

[9.3]
7. (a) −18
  (b) $-2x^2 + 12x - 9$
  (c) $-2x^2 - 2x - 3$
  (d) −7
8. (a) $x^3 + 4x^2 + 5x + 2$
  (b) 0

1. Which one of the following is the graph of a function?

A. 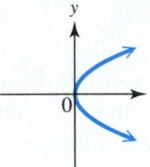   **B.** 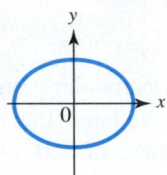   **C.** 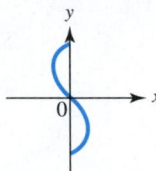   **D.**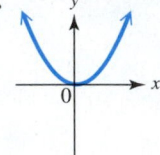

2. Which of the following does not define y as a function of x?

  A. $\{(0, 1), (-2, 3), (4, 8)\}$

  B. $y = 2x - 6$

  C.

x	y
0	1
3	2
0	2
6	3

Give the domain and range of the relation shown in each of the following.

3. Choice A of **Exercise 1**   **4.** Choice A of **Exercise 2**

5. For $f(x) = -x^2 + 2x - 1$, find $f(1)$ and $f(a)$.

6. Graph the linear function $f(x) = \dfrac{2}{3}x - 1$. Give the domain and range.

Solve each problem.

7. Find each of the following for the polynomial functions

$$f(x) = -2x^2 + 5x - 6 \quad \text{and} \quad g(x) = 7x - 3.$$

  (a) $f(4)$ **(b)** $(f + g)(x)$ **(c)** $(f - g)(x)$ **(d)** $(f - g)(-2)$

8. Find each of the following for the polynomial functions

$$f(x) = x^2 + 3x + 2 \quad \text{and} \quad g(x) = x + 1.$$

  (a) $(fg)(x)$ **(b)** $(fg)(-2)$

9. (a) $x + 2$, $x \neq -1$
 (b) 0
10. (a) 23 (b) $3x^2 + 11$
 (c) $9x^2 + 30x + 27$

[9.4]
11. 200 amps 12. 0.8 lb

9. Use $f(x)$ and $g(x)$ from **Exercise 8** to find each of the following.

(a) $\left(\dfrac{f}{g}\right)(x)$ (b) $\left(\dfrac{f}{g}\right)(-2)$

10. Find each of the following for the polynomial functions

$$f(x) = 3x + 5 \quad \text{and} \quad g(x) = x^2 + 2.$$

(a) $(f \circ g)(-2)$ (b) $(f \circ g)(x)$ (c) $(g \circ f)(x)$

11. The current in a simple electrical circuit is inversely proportional to the resistance. If the current is 80 amps when the resistance is 30 ohms, find the current when the resistance is 12 ohms.

12. The force of the wind blowing on a vertical surface varies jointly as the area of the surface and the square of the velocity. If a wind blowing at 40 mph exerts a force of 50 lb on a surface of 500 ft², how much force will a wind of 80 mph place on a surface of 2 ft²?

Chapters R–9 Cumulative Review Exercises

[1.5] **[2.3]**
1. -199 2. $\left\{-\dfrac{15}{4}\right\}$

1. Evaluate $|2x| + 3y - z^3$ for $x = -4$, $y = 3$, and $z = 6$.

Solve each equation or inequality.

[2.8, 8.1] **[2.5]**
3. $\left(-\infty, \dfrac{240}{13}\right]$ 4. 6 m

2. $7(2x + 3) - 4(2x + 1) = 2(x + 1)$ 3. $\dfrac{2}{3}y + \dfrac{5}{12}y \leq 20$

[2.7]
5. $4000 at 4\%$; $8000 at 3\%$

Solve each problem.

4. A triangle has area 42 m². The base is 14 m long. Find the height of the triangle.

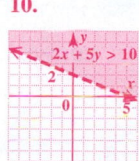

14 m

[3.3, 7.1]
6. $-\dfrac{3}{2}$ 7. $-\dfrac{3}{4}$

5. Abushieba invested some money at 4% interest and twice as much at 3% interest. His interest for the first year was $400. How much did he invest at each rate?

[3.5, 7.2]
8. $y = -\dfrac{3}{2}x + \dfrac{1}{2}$

[3.2, 7.1] **[8.4]**
9. 10.

Find the slope of each line described.

6. Through $(-5, 8)$ and $(-1, 2)$ 7. Perpendicular to $4x - 3y = 12$

8. Write an equation of the line in **Exercise 6**. Give the equation in the form $y = mx + b$.

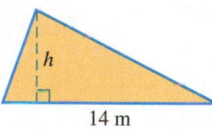

x-intercept: $(-2, 0)$;
y-intercept: $(0, 4)$

Graph each equation or inequality.

9. $-4x + 2y = 8$ (Give the intercepts.) 10. $2x + 5y > 10$

[4.1, 4.2] **[4.4]**
11. $\dfrac{m}{n}$ 12. $4y^2 - 7y - 6$

11. Simplify $\left(\dfrac{m^{-4}n^2}{m^2n^{-3}}\right) \cdot \left(\dfrac{m^5n^{-1}}{m^{-2}n^5}\right)$. Write the answer with only positive exponents. Assume that all variables represent nonzero real numbers.

[4.5]
13. $12f^2 + 5f - 3$

[4.6]
14. $\dfrac{1}{16}x^2 + \dfrac{5}{2}x + 25$

Perform the indicated operations.

12. $(3y^2 - 2y + 6) - (-y^2 + 5y + 12)$ 13. $(4f + 3)(3f - 1)$

14. $\left(\dfrac{1}{4}x + 5\right)^2$

[4.7]
15. $x^2 + 4x - 7$

15. $(3x^3 + 13x^2 - 17x - 7) \div (3x + 1)$

[5.3]
16. $(2x + 5)(x - 9)$

[5.4]
17. $(2p + 5)(4p^2 - 10p + 25)$

[6.1]   **[6.2]**

18. $\dfrac{y + 4}{y - 4}$ **19.** $\dfrac{a(a - b)}{2(a + b)}$

[6.2]

20. $\dfrac{2(x + 3)}{(x + 2)(x^2 + 3x + 9)}$

[6.4]   **[5.5]**

21. 3   **22.** $\left\{-\dfrac{7}{3}, 1\right\}$

[6.6]   **[8.3]**

23. $\{-4\}$ **24.** $\left\{\dfrac{2}{3}, 2\right\}$

[8.3]

25. $(-\infty, -2] \cup \left[\dfrac{2}{3}, \infty\right)$

[7.4, 7.5]
26. $\{(-1, 3)\}$

[7.6]
27. $\{(-2, 3, 1)\}$

[9.1]
28. function;
   domain: $(-\infty, \infty)$
   range: $(-\infty, \infty)$

[9.2]
29. (a) $\dfrac{5}{3}x - \dfrac{8}{3}$ (b) -1

[9.3]
30. (a) $2x^3 - 2x^2 + 6x - 4$
   (b) $2x^3 - 4x^2 + 2x + 2$
   (c) -14
   (d) $x^4 + 2x^2 - 3$

[6.7]   **[9.4]**
31. $\dfrac{6}{5}$ hr   **32.** \$9.92

Factor each polynomial completely.

16. $2x^2 - 13x - 45$ **17.** $8p^3 + 125$

18. Write $\dfrac{y^2 - 16}{y^2 - 8y + 16}$ in lowest terms.

Perform the indicated operations. Express answers in lowest terms.

19. $\dfrac{2a^2}{a + b} \cdot \dfrac{a - b}{4a}$ **20.** $\dfrac{x^2 - 9}{2x + 4} \div \dfrac{x^3 - 27}{4}$ **21.** $\dfrac{x + 4}{x - 2} + \dfrac{2x - 10}{x - 2}$

Solve each equation or inequality.

22. $3x^2 + 4x = 7$ **23.** $\dfrac{-3x}{x + 1} + \dfrac{4x + 1}{x} = \dfrac{-3}{x^2 + x}$

24. $|6x - 8| - 4 = 0$ **25.** $|3x + 2| \geq 4$

Solve each system.

26. $4x - y = -7$ **27.** $x + y - 2z = -1$
$\ 5x + 2y = 1$ $\ 2x - y + z = -6$
    $\ 3x + 2y - 3z = -3$

Solve each problem.

28. Decide whether the relation $y = -x + 2$ is a function, and give its domain and range.

29. Suppose that $y = f(x)$ and $5x - 3y = 8$.
   (a) Find an equation that defines $f(x)$. That is, $f(x) = $ _____.
   (b) Find $f(1)$.

30. Find each of the following for the polynomial functions

$$f(x) = x^2 + 2x - 3, \quad g(x) = 2x^3 - 3x^2 + 4x - 1, \quad \text{and} \quad h(x) = x^2.$$

   (a) $(f + g)(x)$ **(b)** $(g - f)(x)$ **(c)** $(f + g)(-1)$ **(d)** $(f \circ h)(x)$

31. Machine A can complete a certain job in 2 hr. To speed up the work, Machine B, which can complete the job alone in 3 hr, is brought in to help. How long will it take the two machines to complete the job working together?

32. The cost of a pizza varies directly as the square of its radius. If a pizza with a 7-in. radius costs \$6.00, how much should a pizza with a 9-in. radius cost?

10

Roots, Radicals, and Root Functions

The formula for calculating the distance one can see to the horizon from the top of a tall building involves a *square root radical,* one of the topics covered in this chapter.

10.1 Radical Expressions and Graphs

OBJECTIVE 1　Find square roots.

Recall that *squaring* a number means multiplying the number by itself.

7^2　means　$7 \cdot 7$,　which equals　49.　*The square of 7 is 49.*

The opposite (inverse) of squaring a number is taking its *square root*. This is equivalent to asking

"What number when multiplied by itself equals 49?"

From the example above, one answer is 7 because $7 \cdot 7 = 49$.

> **Square Root**
>
> A number b is a **square root** of a if $b^2 = a$ (that is, $b \cdot b = a$).

EXAMPLE 1　Finding All Square Roots of a Number

Find all square roots of 49.

We ask, "What number when multiplied by itself equals 49?" As mentioned above, one square root is 7. Another square root of 49 is -7, because

$$(-7)(-7) = 49.$$

Thus, the number 49 has *two* square roots, 7 and -7. One square root is positive, and one is negative.

NOW TRY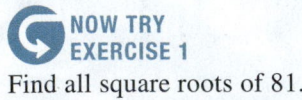

The **positive** or **principal square root** of a number is written with the symbol $\sqrt{}$. For example, the positive square root of 121 is 11.

$$\sqrt{121} = 11 \qquad 11^2 = 121$$

The symbol $-\sqrt{}$ is used for the **negative square root** of a number. For example, the negative square root of 121 is -11.

$$-\sqrt{121} = -11 \qquad (-11)^2 = 121$$

The **radical symbol** $\sqrt{}$ always represents the positive square root $\left(\text{except that } \sqrt{0} = 0\right)$. The number inside the radical symbol is the **radicand,** and the entire expression—radical symbol and radicand—is a **radical.**

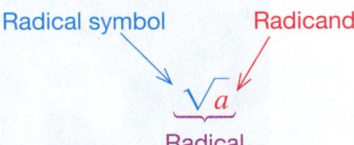

Radical symbol　Radicand
$\sqrt{a}$
Radical

An algebraic expression containing a radical is a **radical expression.**

The radical symbol $\sqrt{}$ has been used since 16th-century Germany and was likely derived from the letter *R*. The radical symbol at the right comes from the Latin word *radix,* for *root.* It was first used by Leonardo of Pisa (Fibonacci) in 1220.

Early radical symbol

NOW TRY ANSWER
1. 9, −9

We summarize our discussion of square roots as follows.

Square Roots of *a*

Let *a* be a positive real number.

$$\sqrt{a} \text{ is the positive or principal square root of } a.$$

$$-\sqrt{a} \text{ is the negative square root of } a.$$

For nonnegative *a*, the following hold.

$$\sqrt{a} \cdot \sqrt{a} = \left(\sqrt{a}\right)^2 = a \quad \text{and} \quad -\sqrt{a} \cdot \left(-\sqrt{a}\right) = \left(-\sqrt{a}\right)^2 = a$$

Also, $\sqrt{0} = 0$.

NOW TRY
EXERCISE 2

Find each square root.

(a) $\sqrt{400}$ **(b)** $-\sqrt{169}$

(c) $\sqrt{\dfrac{100}{121}}$

EXAMPLE 2 Finding Square Roots

Find each square root.

(a) $\sqrt{144}$

The radical $\sqrt{144}$ represents the positive or principal square root of 144. Think of a positive number whose square is 144.

$$12^2 = 144, \quad \text{so} \quad \sqrt{144} = 12.$$

(b) $-\sqrt{1024}$

This symbol represents the negative square root of 1024. A calculator with a square root key can be used to find $\sqrt{1024} = 32$. Therefore,

$$-\sqrt{1024} = -32.$$

(c) $\sqrt{\dfrac{4}{9}} = \dfrac{2}{3}$ **(d)** $-\sqrt{\dfrac{16}{49}} = -\dfrac{4}{7}$ **(e)** $\sqrt{0.81} = 0.9$ **NOW TRY**

$(0.9)^2 = 0.9 \cdot 0.9$
$= 0.81$

⊘ CAUTION By definition, $\sqrt{4} = 2$ because $2^2 = 4$. *In general, however, the square root of a number is not half the number.*

As shown in the preceding definition, when the square root of a positive real number is squared, the result is that positive real number. $\left(\text{Also, } \left(\sqrt{0}\right)^2 = 0.\right)$

NOW TRY
EXERCISE 3

Find the square of each radical expression.

(a) $\sqrt{15}$ **(b)** $-\sqrt{23}$

(c) $\sqrt{2k^2 + 5}$

EXAMPLE 3 Squaring Radical Expressions

Find the square of each radical expression.

(a) $\sqrt{13}$

The square of $\sqrt{13}$ is $\left(\sqrt{13}\right)^2 = 13$.   Definition of square root

(b) $-\sqrt{29}$

$$\left(-\sqrt{29}\right)^2 = 29 \quad \text{The square of a \textit{negative} number is positive.}$$

(c) $\sqrt{p^2 + 1}$

$$\left(\sqrt{p^2 + 1}\right)^2 = p^2 + 1$$

NOW TRY

NOW TRY ANSWERS

2. (a) 20 **(b)** -13 **(c)** $\dfrac{10}{11}$

3. (a) 15 **(b)** 23 **(c)** $2k^2 + 5$

OBJECTIVE 2 Decide whether a given root is rational, irrational, or not a real number.

Numbers with square roots that are rational are **perfect squares.**

Perfect squares		Rational square roots
25		$\sqrt{25} = 5$
144	are perfect squares because	$\sqrt{144} = 12$
$\dfrac{4}{9}$		$\sqrt{\dfrac{4}{9}} = \dfrac{2}{3}$

A number that is not a perfect square has a square root that is not a rational number. For example, $\sqrt{5}$ is not a rational number because it cannot be written as the ratio of two integers. Its decimal equivalent neither terminates nor repeats. However, $\sqrt{5}$ is a real number and corresponds to a point on the number line.

A real number that is not rational is an **irrational number.** The number $\sqrt{5}$ is irrational. *Many square roots of integers are irrational.*

If a is a *positive* real number that is *not* a perfect square, then $\sqrt{a}$ is irrational.

Not every number has a real number square root. For example, there is no real number that can be squared to obtain -36. (The square of a real number can never be negative.) Because of this, $\sqrt{-36}$ *is not a real number.*

If a is a *negative* real number, then $\sqrt{a}$ is *not* a real number.

⚠ **CAUTION** Do not confuse $\sqrt{-36}$ and $-\sqrt{36}$. $\sqrt{-36}$ is not a real number because there is no real number that can be squared to obtain -36. However, $-\sqrt{36}$ is the negative square root of 36, which is -6.

⟳ **NOW TRY EXERCISE 4**

Tell whether each number is *rational, irrational,* or *not a real number.*

(a) $\sqrt{31}$ **(b)** $\sqrt{900}$

(c) $\sqrt{-16}$

EXAMPLE 4 Identifying Types of Square Roots

Tell whether each number is *rational, irrational,* or *not a real number.*

(a) $\sqrt{17}$ Because 17 is not a perfect square, $\sqrt{17}$ is irrational.

(b) $\sqrt{64}$ The number 64 is a perfect square, 8^2, so $\sqrt{64} = 8$, a rational number.

(c) $\sqrt{-25}$ There is no real number whose square is -25. Therefore, $\sqrt{-25}$ is not a real number.

NOW TRY ⟳

NOW TRY ANSWERS
4. (a) irrational   **(b)** rational
   (c) not a real number

NOTE Not all irrational numbers are square roots of integers. For example, the number π (approximately 3.14159) is an irrational number that is not a square root of any integer.

OBJECTIVE 3 Find cube, fourth, and other roots.

Finding the square root of a number is the inverse (opposite) of squaring a number. In a similar way, there are inverses to finding the cube of a number and to finding the fourth or greater power of a number. These inverses are, respectively, the **cube root**, $\sqrt[3]{a}$, and the **fourth root**, $\sqrt[4]{a}$. Similar symbols are used for other roots.

$$\sqrt[n]{a}$$

The *n*th root of *a*, written $\sqrt[n]{a}$, is a number whose *n*th power equals *a*. That is,

$$\sqrt[n]{a} = b \quad \text{means} \quad b^n = a.$$

In $\sqrt[n]{a}$, the number *n* is the **index**, or **order**, of the radical.

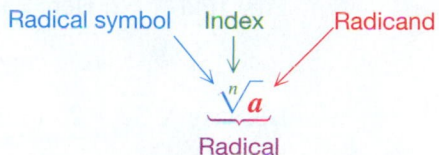

Radical symbol   Index   Radicand

$$\sqrt[n]{a}$$

Radical

We could write $\sqrt[2]{a}$ instead of $\sqrt{a}$, but the simpler symbol $\sqrt{a}$ is customary because the square root is the most commonly used root.

NOTE When working with cube roots or fourth roots, it is helpful to learn the first few **perfect cubes** ($1^3 = 1$, $2^3 = 8$, $3^3 = 27$, and so on) and the first few **perfect fourth powers** ($1^4 = 1$, $2^4 = 16$, $3^4 = 81$, and so on). See **Exercises 67 and 68.**

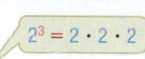
**NOW TRY
EXERCISE 5**
Find each cube root.

(a) $\sqrt[3]{343}$ **(b)** $\sqrt[3]{-1000}$

(c) $\sqrt[3]{27}$

EXAMPLE 5 Finding Cube Roots

Find each cube root.

$2^3 = 2 \cdot 2 \cdot 2$

(a) $\sqrt[3]{8}$ What number can be cubed to give 8? Because $2^3 = 8$, $\sqrt[3]{8} = 2$.

(b) $\sqrt[3]{-8}$ Because $(-2)^3 = -8$, $\sqrt[3]{-8} = -2$.

(c) $\sqrt[3]{216}$ Because $6^3 = 216$, $\sqrt[3]{216} = 6$.   **NOW TRY**

In **Example 5(b)**, $\sqrt[3]{-8} = -2$—that is, we can find the *cube root* of a negative number. (Contrast this with the *square root* of a negative number, which is not real.) In fact, the cube root of a positive number is positive, and the cube root of a negative number is negative. ***There is only one real number cube root for each real number.***

When a radical has an ***even index*** (square root, fourth root, and so on), ***the radicand must be nonnegative*** to yield a real number root. Also, for $a > 0$,

$$\sqrt{a}, \ \sqrt[4]{a}, \ \sqrt[6]{a}, \text{ and so on are positive (principal) roots.}$$

NOW TRY ANSWERS
5. **(a)** 7   **(b)** -10 **(c)** 3

$$-\sqrt{a}, \ -\sqrt[4]{a}, \ -\sqrt[6]{a}, \text{ and so on are negative roots.}$$

NOW TRY
EXERCISE 6

Find each root.

(a) $\sqrt[4]{625}$ **(b)** $\sqrt[4]{-625}$

(c) $-\sqrt[4]{625}$ **(d)** $\sqrt[5]{3125}$

(e) $\sqrt[5]{-3125}$

EXAMPLE 6   **Finding Other Roots**

Find each root.

 $2^4 = 2 \cdot 2 \cdot 2 \cdot 2$

(a) $\sqrt[4]{16}$ Because 2 is positive and $2^4 = 16$, $\sqrt[4]{16} = 2$.

(b) $-\sqrt[4]{16}$

From part (a), $\sqrt[4]{16} = 2$, so the negative root is

$$-\sqrt[4]{16} = -2.$$

(c) $\sqrt[4]{-16}$

For a fourth root to be a real number, the radicand must be nonnegative. There is no real number that equals $\sqrt[4]{-16}$.

(d) $-\sqrt[5]{32}$

First find $\sqrt[5]{32}$. Here 2 is the number whose fifth power is 32, so

$$\sqrt[5]{32} = 2.$$

Because $\sqrt[5]{32} = 2$, it follows that

$$-\sqrt[5]{32} = -2.$$

(e) $\sqrt[5]{-32}$ Because $(-2)^5 = -32$, $\sqrt[5]{-32} = -2$.   **NOW TRY** ⟳

OBJECTIVE 4 Graph functions defined by radical expressions.

A radical expression is an algebraic expression that contains radicals.

$$3 - \sqrt{x}, \quad \sqrt[3]{x}, \quad \text{and} \quad \sqrt{2x - 1} \quad \text{Radical expressions}$$

In earlier chapters, we graphed functions defined by polynomial and rational expressions. Now we examine the graphs of functions defined by the basic radical expressions, such as $f(x) = \sqrt{x}$ and $f(x) = \sqrt[3]{x}$.

FIGURE 1 shows the graph of the **square root function,**

$$f(x) = \sqrt{x},$$

together with a table of selected points. Only nonnegative values can be used for x, so the domain is $[0, \infty)$. Because $\sqrt{x}$ is the principal square root of x, it always has a nonnegative value, so the range is also $[0, \infty)$.

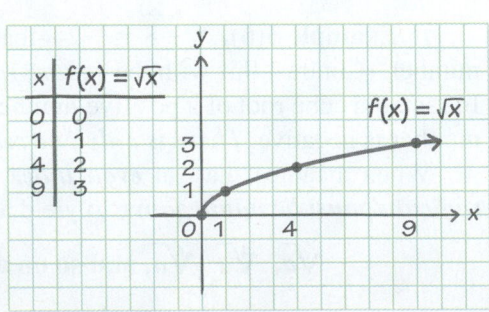

Square root function

$$f(x) = \sqrt{x}$$

Domain: $[0, \infty)$

Range: $[0, \infty)$

FIGURE 1

NOW TRY ANSWERS
6. **(a)** 5   **(b)** not a real number
(c) -5 **(d)** 5   **(e)** -5

FIGURE 2 shows the graph of the **cube root function**

$$f(x) = \sqrt[3]{x}.$$

Any real number (positive, negative, or 0) can be used for x in the cube root function, so $\sqrt[3]{x}$ can be positive, negative, or 0. Thus, both the domain and the range of the cube root function are $(-\infty, \infty)$.

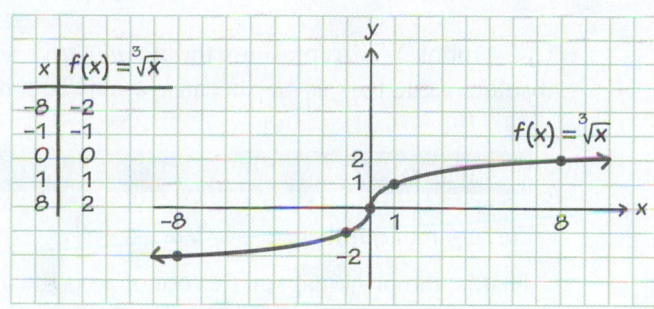

FIGURE 2

Cube root function

$$f(x) = \sqrt[3]{x}$$

Domain: $(-\infty, \infty)$

Range: $(-\infty, \infty)$

**NOW TRY
EXERCISE 7**

Graph each function, and give its domain and range.

(a) $f(x) = \sqrt{x + 1}$

(b) $f(x) = \sqrt[3]{x} - 1$

EXAMPLE 7 Graphing Functions Defined with Radicals

Graph each function, and give its domain and range.

(a) $f(x) = \sqrt{x - 3}$

Create a table of values as given with the graph in **FIGURE 3**. The x-values were chosen in such a way that the function values are all integers. For the radicand to be nonnegative, we must have

$$x - 3 \geq 0, \quad \text{or} \quad x \geq 3.$$

Therefore, the domain of this function is $[3, \infty)$. Function values are positive or 0, so the range is $[0, \infty)$.

x	$f(x) = \sqrt{x - 3}$
3	$\sqrt{3 - 3} = 0$
4	$\sqrt{4 - 3} = 1$
7	$\sqrt{7 - 3} = 2$

This graph is shifted 3 units to the right compared to the graph of $y = \sqrt{x}$.

FIGURE 3

NOW TRY ANSWERS

7. (a)

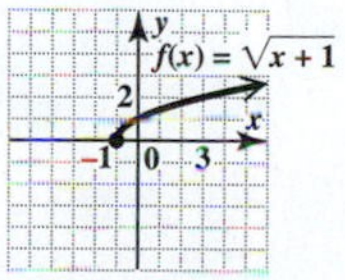

domain: $[-1, \infty)$;
range: $[0, \infty)$

(b)

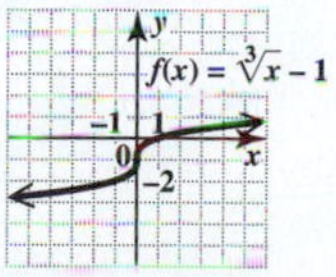

domain: $(-\infty, \infty)$;
range: $(-\infty, \infty)$

(b) $f(x) = \sqrt[3]{x} + 2$

See **FIGURE 4**. Both the domain and the range are $(-\infty, \infty)$.

x	$f(x) = \sqrt[3]{x} + 2$
-8	$\sqrt[3]{-8} + 2 = 0$
-1	$\sqrt[3]{-1} + 2 = 1$
0	$\sqrt[3]{0} + 2 = 2$
1	$\sqrt[3]{1} + 2 = 3$
8	$\sqrt[3]{8} + 2 = 4$

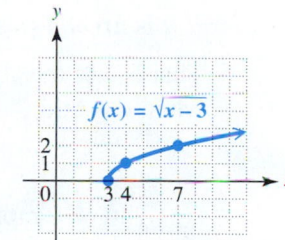

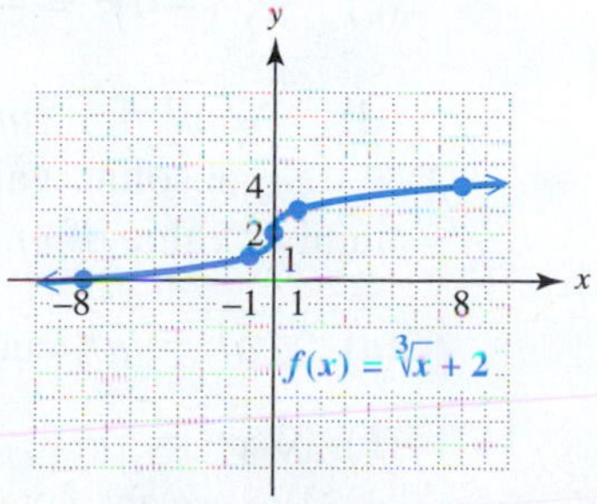

This graph is shifted 2 units up compared to the graph of $y = \sqrt[3]{x}$.

FIGURE 4

NOW TRY

OBJECTIVE 5 Find *n*th roots of *n*th powers.

Consider the expression $\sqrt{a^2}$. At first glance, we might think that it is equivalent to *a*. However, this is not necessarily true. For example, consider the following.

If $a = 6$,   then $\sqrt{a^2} = \sqrt{6^2} = \sqrt{36} = 6$.

If $a = -6$,   then $\sqrt{a^2} = \sqrt{(-6)^2} = \sqrt{36} = 6$. ← Instead of −6, we get 6, the *absolute value* of −6.

The symbol $\sqrt{a^2}$ represents the *nonnegative* square root, so we express $\sqrt{a^2}$ with absolute value bars as $|a|$ because *a* may be a negative number.

> **Meaning of $\sqrt{a^2}$**
>
> For any real number *a*, $\sqrt{a^2} = |a|$.
>
> That is, the principal square root of a^2 is the absolute value of *a*.

NOW TRY EXERCISE 8

Find each square root.

(a) $\sqrt{11^2}$ **(b)** $\sqrt{(-11)^2}$

(c) $\sqrt{z^2}$ **(d)** $\sqrt{(-z)^2}$

EXAMPLE 8 Simplifying Square Roots Using Absolute Value

Find each square root.

(a) $\sqrt{7^2} = |7| = 7$ **(b)** $\sqrt{(-7)^2} = |-7| = 7$

(c) $\sqrt{k^2} = |k|$ **(d)** $\sqrt{(-k)^2} = |-k| = |k|$ **NOW TRY**

We can generalize this idea to any *n*th root.

> **Meaning of $\sqrt[n]{a^n}$**
>
> If *n* is an *even* positive integer, then $\sqrt[n]{a^n} = |a|$.
>
> If *n* is an *odd* positive integer, then $\sqrt[n]{a^n} = a$.
>
> That is, use the absolute value symbol when *n* is even. Absolute value is not used when *n* is odd.

NOW TRY EXERCISE 9

Simplify each root.

(a) $\sqrt[8]{(-2)^8}$ **(b)** $\sqrt[3]{(-9)^3}$

(c) $-\sqrt[4]{(-10)^4}$ **(d)** $-\sqrt{m^8}$

(e) $\sqrt[3]{x^{18}}$ **(f)** $\sqrt[4]{t^{20}}$

EXAMPLE 9 Simplifying Higher Roots Using Absolute Value

Simplify each root.

(a) $\sqrt[6]{(-3)^6} = |-3| = 3$ *n* is even. Use absolute value.

(b) $\sqrt[5]{(-4)^5} = -4$ *n* is odd.

(c) $-\sqrt[4]{(-9)^4} = -|-9| = -9$ *n* is even. Use absolute value.

(d) $-\sqrt{m^4} = -|m^2| = -m^2$ For all *m*, $|m^2| = m^2$.

No absolute value bars are needed here, because m^2 is nonnegative for any real number value of *m*.

(e) $\sqrt[3]{a^{12}} = a^4$, because $a^{12} = (a^4)^3$.

(f) $\sqrt[4]{x^{12}} = |x^3|$

We use absolute value to guarantee that the result is not negative (because x^3 is negative when *x* is negative). If desired, $|x^3|$ can be written as $x^2 \cdot |x|$.   **NOW TRY**

NOW TRY ANSWERS

8. **(a)** 11 **(b)** 11 **(c)** $|z|$
   (d) $|z|$
9. **(a)** 2 **(b)** −9 **(c)** −10
   (d) $-m^4$ **(e)** x^6 **(f)** $|t^5|$

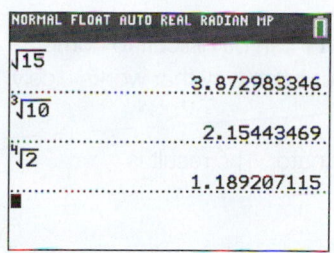

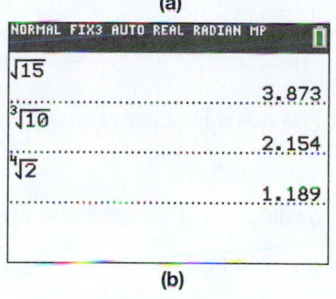

(b)

FIGURE 5

OBJECTIVE 6 Use a calculator to find roots.

While numbers such as $\sqrt{9}$ and $\sqrt[3]{-8}$ are rational, radicals are often irrational numbers. To find approximations of such radicals, we usually use a scientific or graphing calculator. For example,

$$\sqrt{15} \approx 3.872983346, \quad \sqrt[3]{10} \approx 2.15443469, \quad \text{and} \quad \sqrt[4]{2} \approx 1.189207115,$$

where the symbol $\approx$ means "is approximately equal to." In this book, we often show approximations rounded to three decimal places. Thus,

$$\sqrt{15} \approx 3.873, \quad \sqrt[3]{10} \approx 2.154, \quad \text{and} \quad \sqrt[4]{2} \approx 1.189.$$

FIGURE 5 shows how the preceding approximations are displayed on a TI-83/84 Plus graphing calculator. In **FIGURE 5(a)**, eight or nine decimal places are shown, while in **FIGURE 5(b)**, the number of decimal places is fixed at three.

There is a simple way to check that a calculator approximation is "in the ballpark." For example, because 16 is a little larger than 15, $\sqrt{16} = 4$ should be a little larger than $\sqrt{15}$. Thus, 3.873 is reasonable as an approximation for $\sqrt{15}$.

NOTE The methods for finding approximations differ among makes and models of calculators. *Always consult your owner's manual for keystroke instructions.* Be aware that graphing calculators often differ from scientific calculators in the order in which keystrokes are made.

 **NOW TRY**
EXERCISE 10

Use a calculator to approximate each radical to three decimal places.

(a) $-\sqrt{92}$ **(b)** $\sqrt[4]{39}$

(c) $\sqrt[5]{33}$

EXAMPLE 10 Finding Approximations for Roots

Use a calculator to verify that each approximation is correct.

(a) $\sqrt{39} \approx 6.245$ **(b)** $-\sqrt{72} \approx -8.485$

(c) $\sqrt[3]{93} \approx 4.531$ **(d)** $\sqrt[4]{39} \approx 2.499$ **NOW TRY**

 **NOW TRY**
EXERCISE 11

Use the formula in **Example 11** to approximate f to the nearest thousand if

$$L = 7 \times 10^{-5}$$

and $\qquad C = 3 \times 10^{-9}$.

EXAMPLE 11 Using Roots to Calculate Resonant Frequency

In electronics, the resonant frequency f of a circuit may be found using the formula

$$f = \frac{1}{2\pi\sqrt{LC}},$$

where f is in cycles per second, L is in henrys, and C is in farads. (Henrys and farads are units of measure in electronics.) Find the resonant frequency f if $L = 5 \times 10^{-4}$ henry and $C = 3 \times 10^{-10}$ farad. Give the answer to the nearest thousand.

Find the value of f when $L = 5 \times 10^{-4}$ and $C = 3 \times 10^{-10}$.

$$f = \frac{1}{2\pi\sqrt{LC}} \qquad \text{Given formula}$$

$$f = \frac{1}{2\pi\sqrt{(5 \times 10^{-4})(3 \times 10^{-10})}} \qquad \text{Substitute for } L \text{ and } C.$$

$$f \approx 411{,}000 \qquad \text{Use a calculator.}$$

The resonant frequency f is approximately 411,000 cycles per sec. **NOW TRY**

NOW TRY ANSWERS
10. (a) -9.592 (b) 2.499
(c) 2.012
11. $347{,}000$ cycles per sec

NOTE   The expression in the second-to-last line of **Example 11** can be difficult to compute on a calculator. There are several ways to approach it, but here is one way that works nicely. Use a calculator to verify the following steps.

Step 1   Calculate the product under the radical in the denominator. The result is

$$1.5 \times 10^{-13}.$$

Step 2   Use the square root function ($\sqrt{x}$) to find the square root of this number. The result is approximately

$$3.872983346 \times 10^{-7}.$$

Step 3   Multiply by 2π, using the π function of the calculator. The result is approximately

$$2.433467206 \times 10^{-6}.$$

Step 4   Now use the reciprocal function (labeled $\frac{1}{x}$, or x^{-1}) of the calculator. The result should be 410936.296, which, rounded to the nearest thousand, is 411,000.

If the numerator had not been 1, and perhaps an expression to evaluate, one approach would be to save the result found in Step 3 in the calculator memory, evaluate the numerator, and then divide by the number saved in the memory.

10.1 Exercises

FOR EXTRA HELP

▶ MyMathLab®

● *Complete solution available in MyMathLab*

Concept Check   *Decide whether each statement is* true *or* false. *If false, tell why.*

1. Every positive number has two real square roots.

2. A negative number has negative real square roots.

3. Every nonnegative number has two real square roots.

4. The positive square root of a positive number is its principal square root.

5. The cube root of every nonzero real number has the same sign as the number itself.

6. Every positive number has three real cube roots.

Concept Check   *What must be true about the variable a for each statement to be true?*

7. $\sqrt{a}$ represents a positive number.

8. $-\sqrt{a}$ represents a negative number.

9. $\sqrt{a}$ is not a real number.

10. $-\sqrt{a}$ is not a real number.

Find all square roots of each number. **See Example 1.**

● **11.** 9	**12.** 16	**13.** 64	**14.** 100	**15.** 169
16. 225	**17.** $\dfrac{25}{196}$	**18.** $\dfrac{81}{400}$	**19.** 900	**20.** 1600

Find each square root. **See Examples 2 and 4(c).**

21. $\sqrt{1}$	**22.** $\sqrt{4}$	● **23.** $\sqrt{49}$	**24.** $\sqrt{81}$	**25.** $\sqrt{100}$
26. $\sqrt{400}$	**27.** $-\sqrt{16}$	**28.** $-\sqrt{64}$	**29.** $-\sqrt{256}$	**30.** $-\sqrt{196}$
● **31.** $-\sqrt{\dfrac{144}{121}}$	**32.** $-\sqrt{\dfrac{49}{36}}$	**33.** $\sqrt{0.64}$	**34.** $\sqrt{0.16}$	
35. $\sqrt{-121}$	**36.** $\sqrt{-64}$	**37.** $-\sqrt{-49}$	**38.** $-\sqrt{-100}$	

Find the square of each radical expression. See Example 3.

39. $\sqrt{19}$ **40.** $\sqrt{59}$ **41.** $-\sqrt{19}$ **42.** $-\sqrt{59}$

43. $\sqrt{\dfrac{2}{3}}$ **44.** $\sqrt{\dfrac{5}{7}}$ **45.** $\sqrt{3x^2 + 4}$ **46.** $\sqrt{9y^2 + 3}$

Determine whether each number is rational, irrational, *or* not a real number. *If a number is rational, give its exact value. If a number is irrational, give a decimal approximation to the nearest thousandth. Use a calculator as necessary. See Examples 4 and 10.*

47. $\sqrt{25}$ **48.** $\sqrt{169}$ **49.** $\sqrt{29}$ **50.** $\sqrt{33}$

51. $-\sqrt{64}$ **52.** $-\sqrt{81}$ **53.** $-\sqrt{300}$ **54.** $-\sqrt{500}$

55. $\sqrt{-29}$ **56.** $\sqrt{-47}$ **57.** $\sqrt{1200}$ **58.** $\sqrt{1500}$

Concept Check *Without using a calculator, determine between which two consecutive integers each square root lies. For example,*

$\sqrt{75}$ *is between* 8 *and* 9, *because* $\sqrt{64} = 8$, $\sqrt{81} = 9$, *and* $64 < 75 < 81$.

59. $\sqrt{94}$ **60.** $\sqrt{43}$ **61.** $\sqrt{51}$ **62.** $\sqrt{30}$

63. $-\sqrt{40}$ **64.** $-\sqrt{63}$ **65.** $\sqrt{23.2}$ **66.** $\sqrt{10.3}$

67. *Concept Check* To help find cube roots, complete this list of perfect cubes.

$1^3 =$ _____ $2^3 =$ _____ $3^3 =$ _____ $4^3 =$ _____ $5^3 =$ _____

$6^3 =$ _____ $7^3 =$ _____ $8^3 =$ _____ $9^3 =$ _____ $10^3 =$ _____

68. *Concept Check* To help find fourth roots, complete this list of perfect fourth powers.

$1^4 =$ _____ $2^4 =$ _____ $3^4 =$ _____ $4^4 =$ _____ $5^4 =$ _____

$6^4 =$ _____ $7^4 =$ _____ $8^4 =$ _____ $9^4 =$ _____ $10^4 =$ _____

69. Match each expression from Column I with the equivalent choice from Column II. Answers may be used once, more than once, or not at all. **See Examples 5 and 6.**

I		II	
(a) $-\sqrt{16}$	**(b)** $\sqrt{-16}$	**A.** 3	**B.** -2
(c) $\sqrt[3]{-27}$	**(d)** $\sqrt[5]{-32}$	**C.** 2	**D.** -3
(e) $\sqrt[4]{16}$	**(f)** $-\sqrt[3]{64}$	**E.** -4	**F.** Not a real number

70. *Concept Check* If n is odd, under what conditions is $\sqrt[n]{a}$ the following?

(a) positive **(b)** negative **(c)** 0

Find each root. See Examples 5 and 6.

71. $-\sqrt{81}$ **72.** $-\sqrt{121}$ **73.** $\sqrt[3]{216}$ **74.** $\sqrt[3]{343}$

75. $\sqrt[3]{-64}$ **76.** $\sqrt[3]{-125}$ **77.** $-\sqrt[3]{512}$ **78.** $-\sqrt[3]{1000}$

79. $\sqrt[4]{1296}$ **80.** $\sqrt[4]{625}$ **81.** $-\sqrt[4]{16}$ **82.** $-\sqrt[4]{256}$

83. $\sqrt[4]{-625}$ **84.** $\sqrt[4]{-256}$ **85.** $\sqrt[6]{64}$ **86.** $\sqrt[6]{729}$

87. $\sqrt[6]{-32}$ **88.** $\sqrt[8]{-1}$ **89.** $\sqrt{\dfrac{64}{81}}$ **90.** $\sqrt{\dfrac{100}{9}}$

91. $\sqrt[3]{\dfrac{64}{27}}$ **92.** $\sqrt[4]{\dfrac{81}{16}}$ **93.** $-\sqrt[6]{\dfrac{1}{64}}$ **94.** $-\sqrt[5]{\dfrac{1}{32}}$

95. $-\sqrt[3]{-27}$ **96.** $-\sqrt[3]{-64}$ **97.** $\sqrt{0.25}$ **98.** $\sqrt{0.36}$

99. $-\sqrt{0.49}$ **100.** $-\sqrt{0.81}$ **101.** $\sqrt[3]{0.001}$ **102.** $\sqrt[3]{0.125}$

Graph each function, and give its domain and range. ***See Example 7.***

▶ **103.** $f(x) = \sqrt{x + 3}$ **104.** $f(x) = \sqrt{x - 5}$ **105.** $f(x) = \sqrt{x - 2}$

106. $f(x) = \sqrt{x + 4}$ **107.** $f(x) = \sqrt[3]{x - 3}$ **108.** $f(x) = \sqrt[3]{x + 1}$

109. $f(x) = \sqrt[3]{x - 3}$ **110.** $f(x) = \sqrt[3]{x + 1}$

Simplify each root. ***See Examples 8 and 9.***

▶ **111.** $\sqrt{12^2}$ **112.** $\sqrt{19^2}$ **113.** $\sqrt{(-10)^2}$ **114.** $\sqrt{(-13)^2}$

▶ **115.** $\sqrt[6]{(-2)^6}$ **116.** $\sqrt[6]{(-4)^6}$ **117.** $\sqrt[5]{(-9)^5}$ **118.** $\sqrt[5]{(-8)^5}$

119. $-\sqrt[6]{(-5)^6}$ **120.** $-\sqrt[6]{(-7)^6}$ **121.** $\sqrt{x^2}$ **122.** $-\sqrt{x^2}$

123. $\sqrt{(-z)^2}$ **124.** $\sqrt{(-q)^2}$ **125.** $\sqrt[3]{x^3}$ **126.** $-\sqrt[3]{x^3}$

127. $\sqrt[3]{x^{15}}$ **128.** $\sqrt[3]{m^9}$ **129.** $\sqrt[6]{x^{30}}$ **130.** $\sqrt[4]{k^{20}}$

Concept Check   *Refer to the figure to answer each question.*

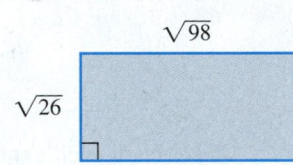

$\sqrt{98}$

$\sqrt{26}$

131. Which one of the following is the best estimate of its area?

    A. 2500   **B.** 250   **C.** 50   **D.** 100

132. Which one of the following is the best estimate of its perimeter?

    A. 15   **B.** 250   **C.** 100   **D.** 30

Find a decimal approximation for each radical. Round answers to three decimal places. ***See Example 10.***

▶ **133.** $\sqrt{9483}$ **134.** $\sqrt{6825}$ **135.** $\sqrt{284.361}$ **136.** $\sqrt{846.104}$

137. $-\sqrt{82}$ **138.** $-\sqrt{91}$ **139.** $\sqrt[3]{423}$ **140.** $\sqrt[3]{555}$

141. $\sqrt[4]{100}$ **142.** $\sqrt[4]{250}$ **143.** $\sqrt[5]{23.8}$ **144.** $\sqrt[5]{98.4}$

Solve each problem. ***See Example 11.***

▶ **145.** Use the formula from **Example 11** to calculate the resonant frequency of a circuit to the nearest thousand if $L = 7.237 \times 10^{-5}$ henry and $C = 2.5 \times 10^{-10}$ farad.

$$f = \dfrac{1}{2\pi\sqrt{LC}}$$

146. The threshold weight T for a person is the weight above which the risk of death increases greatly. The threshold weight in pounds for men aged 40–49 is related to height h in inches by the formula

$$h = 12.3\sqrt[3]{T}.$$

What height corresponds to a threshold weight of 216 lb for a 43-year-old man? Round the answer to the nearest inch and then to the nearest tenth of a foot.

147. According to an article in *The World Scanner Report*, the distance D, in miles, to the horizon from an observer's point of view over water or "flat" earth is given by

$$D = \sqrt{2H},$$

where H is the height of the point of view, in feet. If a person whose eyes are 6 ft above ground level is standing at the top of a hill 44 ft above "flat" earth, approximately how far to the horizon will she be able to see?

148. The time t in seconds for one complete swing of a simple pendulum, where L is the length of the pendulum in feet, and g, the acceleration due to gravity, is about 32 ft per sec², is

$$t = 2\pi\sqrt{\frac{L}{g}}.$$

Find the time of a complete swing of a 2-ft pendulum to the nearest tenth of a second.

149. Heron's formula gives a method of finding the area of a triangle if the lengths of its sides are known. Suppose that a, b, and c are the lengths of the sides. Let s denote one-half of the perimeter of the triangle (called the **semiperimeter**)—that is, $s = \frac{1}{2}(a + b + c)$. Then the area of the triangle is

$$\mathcal{A} = \sqrt{s(s - a)(s - b)(s - c)}.$$

Find the area of the Bermuda Triangle, to the nearest thousand square miles, if the "sides" of this triangle measure approximately 850 mi, 925 mi, and 1300 mi.

150. Use Heron's formula from **Exercise 149** to find the area of a triangle with sides of lengths $a = 11$ m, $b = 60$ m, and $c = 61$ m.

151. The coefficient of self-induction L (in henrys), the energy P stored in an electronic circuit (in joules), and the current I (in amps) are related by this formula.

$$I = \sqrt{\frac{2P}{L}}$$

(a) Find I if $P = 120$ and $L = 80$.   **(b)** Find I if $P = 100$ and $L = 40$.

152. The Vietnam Veterans Memorial in Washington, DC, is in the shape of an unenclosed isosceles triangle with equal sides of length 246.75 ft. If the triangle were enclosed, the third side would have length 438.14 ft. Use Heron's formula from **Exercise 149** to find the area of this enclosure to the nearest hundred square feet. (*Source:* Information pamphlet obtained at the Vietnam Veterans Memorial.)

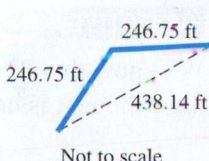

Not to scale

10.2 Rational Exponents

OBJECTIVES

1. Use exponential notation for *n*th roots.
2. Define and use expressions of the form $a^{m/n}$.
3. Convert between radicals and rational exponents.
4. Use the rules for exponents with rational exponents.

OBJECTIVE 1 Use exponential notation for *n*th roots.

Consider the product $(3^{1/2})^2 = 3^{1/2} \cdot 3^{1/2}$. We can simplify this product as follows.

$$(3^{1/2})^2 = 3^{1/2} \cdot 3^{1/2}$$
$$= 3^{1/2+1/2} \qquad \text{Product rule: } a^m \cdot a^n = a^{m+n}$$
$$= 3^1 \qquad \text{Add exponents.}$$
$$= 3 \qquad a^1 = a$$

Also, by definition,

$$(\sqrt{3})^2 = \sqrt{3} \cdot \sqrt{3} = 3.$$

Because both $(3^{1/2})^2$ and $(\sqrt{3})^2$ are equal to 3, it seems reasonable to define

$$3^{1/2} = \sqrt{3}.$$

This suggests the following generalization.

> **Meaning of $a^{1/n}$**
>
> If $\sqrt[n]{a}$ is a real number, then $\qquad a^{1/n} = \sqrt[n]{a}.$
>
> *Examples:* $\quad 4^{1/2} = \sqrt{4}, \quad 8^{1/3} = \sqrt[3]{8}, \quad \text{and} \quad 16^{1/4} = \sqrt[4]{16}$
>
> *Notice that the denominator of the rational exponent is the index of the radical.*

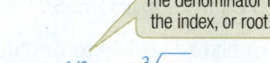 **NOW TRY EXERCISE 1**

Evaluate each exponential.

(a) $81^{1/2}$ **(b)** $125^{1/3}$

(c) $-625^{1/4}$ **(d)** $(-625)^{1/4}$

(e) $(-125)^{1/3}$ **(f)** $\left(\dfrac{1}{16}\right)^{1/4}$

EXAMPLE 1 Evaluating Exponentials of the Form $a^{1/n}$

Evaluate each exponential.

The denominator is the index, or root.

(a) $64^{1/3} = \sqrt[3]{64} = 4$

The denominator is the index, or root. $\sqrt{}$ means $\sqrt[2]{}$.

(b) $100^{1/2} = \sqrt{100} = 10$

(c) $-256^{1/4} = -\sqrt[4]{256} = -4$

(d) $(-256)^{1/4} = \sqrt[4]{-256}$ is not a real number because the radicand, -256, is negative and the index is even.

(e) $(-32)^{1/5} = \sqrt[5]{-32} = -2$

(f) $\left(\dfrac{1}{8}\right)^{1/3} = \sqrt[3]{\dfrac{1}{8}} = \dfrac{1}{2}$ **NOW TRY**

> ⚠ **CAUTION** Notice the distinction between **Examples 1(c) and (d).** The radical in part (c) is the *negative fourth root of a positive number,* while the radical in part (d) is the *principal fourth root of a negative number, which is not a real number.*

OBJECTIVE 2 Define and use expressions of the form $a^{m/n}$.

We know that $8^{1/3} = \sqrt[3]{8}$. We can define a number like $8^{2/3}$, where the numerator of the exponent is not 1. For past rules of exponents to be valid,

$$8^{2/3} = 8^{(1/3)2} = (8^{1/3})^2.$$

Because $8^{1/3} = \sqrt[3]{8}$,

$$8^{2/3} = \left(\sqrt[3]{8}\right)^2 = 2^2 = 4.$$

Generalizing from this example, we define $a^{m/n}$ as follows.

Meaning of $a^{m/n}$

If m and n are positive integers with m/n in lowest terms, then

$$a^{m/n} = \left(a^{1/n}\right)^m,$$

provided that $a^{1/n}$ is a real number. If $a^{1/n}$ is not a real number, then $a^{m/n}$ is not a real number.

NOW TRY EXERCISE 2

Evaluate each exponential.

(a) $32^{2/5}$ (b) $8^{5/3}$

(c) $-100^{3/2}$ (d) $(-121)^{3/2}$

(e) $(-125)^{4/3}$

EXAMPLE 2 Evaluating Exponentials of the Form $a^{m/n}$

Evaluate each exponential.

Think:
$36^{1/2} = \sqrt{36} = 6$

Think:
$125^{1/3} = \sqrt[3]{125} = 5$

(a) $36^{3/2} = (36^{1/2})^3 = 6^3 = 216$

(b) $125^{2/3} = (125^{1/3})^2 = 5^2 = 25$

Be careful. The base is 4.

(c) $-4^{5/2} = -(4^{5/2}) = -(4^{1/2})^5 = -(2)^5 = -32$

Because the base here is 4, the negative sign is *not* affected by the exponent.

(d) $(-27)^{2/3} = [(-27)^{1/3}]^2 = (-3)^2 = 9$

Notice in part (c) that we first evaluate the exponential and then find its negative. In part (d), the $-$ sign is part of the base, -27.

(e) $(-100)^{3/2} = [(-100)^{1/2}]^3$, which is not a real number, because

$$(-100)^{1/2}, \quad \text{or} \quad \sqrt{-100}, \quad \text{is not a real number.} \qquad \textbf{NOW TRY} \, \text{}$$

Recall from **Section 4.2** that for any natural number n,

$$a^{-n} = \frac{1}{a^n} \quad (\text{where } a \neq 0).$$

When a rational exponent is negative, this earlier interpretation of negative exponents is applied.

NOW TRY ANSWERS

2. (a) 4 (b) 32 (c) -1000
   (d) It is not a real number.
   (e) 625

Meaning of $a^{-m/n}$

If $a^{m/n}$ is a real number, then

$$a^{-m/n} = \frac{1}{a^{m/n}} \quad (\text{where } a \neq 0).$$

NOW TRY EXERCISE 3

Evaluate each exponential.

(a) $243^{-3/5}$ **(b)** $4^{-5/2}$

(c) $\left(\dfrac{216}{125}\right)^{-2/3}$

EXAMPLE 3   Evaluating Exponentials of the Form $a^{-m/n}$

Evaluate each exponential.

(a) $16^{-3/4} = \dfrac{1}{16^{3/4}} = \dfrac{1}{\left(16^{1/4}\right)^3} = \dfrac{1}{\left(\sqrt[4]{16}\right)^3} = \dfrac{1}{2^3} = \dfrac{1}{8}$

> The denominator of 3/4 is the index and the numerator is the exponent.

(b) $25^{-3/2} = \dfrac{1}{25^{3/2}} = \dfrac{1}{\left(25^{1/2}\right)^3} = \dfrac{1}{\left(\sqrt{25}\right)^3} = \dfrac{1}{5^3} = \dfrac{1}{125}$

(c) $\left(\dfrac{8}{27}\right)^{-2/3} = \dfrac{1}{\left(\dfrac{8}{27}\right)^{2/3}} = \dfrac{1}{\left(\sqrt[3]{\dfrac{8}{27}}\right)^2} = \dfrac{1}{\left(\dfrac{2}{3}\right)^2} = \dfrac{1}{\dfrac{4}{9}} = \dfrac{9}{4}$

> $\dfrac{1}{\frac{4}{9}} = 1 \div \dfrac{4}{9} = 1 \cdot \dfrac{9}{4}$

We can also use the rule $\left(\dfrac{b}{a}\right)^{-m} = \left(\dfrac{a}{b}\right)^{m}$ here, as follows.

$$\left(\dfrac{8}{27}\right)^{-2/3} = \left(\dfrac{27}{8}\right)^{2/3} = \left(\sqrt[3]{\dfrac{27}{8}}\right)^2 = \left(\dfrac{3}{2}\right)^2 = \dfrac{9}{4}$$ The result is the same.

> Take the reciprocal only of the base, *not* the exponent.

NOW TRY

⚠ **CAUTION**   Be careful to distinguish between exponential expressions like the following.

$16^{-1/4}$, which equals $\dfrac{1}{2}$, $-16^{1/4}$, which equals -2,   and $-16^{-1/4}$, which equals $-\dfrac{1}{2}$

A negative exponent does not necessarily lead to a negative result. Negative exponents lead to reciprocals, which may be positive.

We obtain an alternative definition of $a^{m/n}$ using the power rule for exponents differently than in the earlier definition. If all indicated roots are real numbers, then

$$a^{m/n} = a^{m(1/n)} = (a^m)^{1/n}, \quad \text{so} \quad a^{m/n} = (a^m)^{1/n}.$$

Alternative Meaning of $a^{m/n}$

If all indicated roots are real numbers, then

$$a^{m/n} = (a^{1/n})^m = (a^m)^{1/n}.$$

We can now evaluate an expression such as $27^{2/3}$ in two ways.

$$27^{2/3} = (27^{1/3})^2 = 3^2 = 9$$

or $27^{2/3} = (27^2)^{1/3} = 729^{1/3} = 9$ The result is the same.

In most cases, it is easier to use $(a^{1/n})^m$.

NOW TRY ANSWERS

3. **(a)** $\dfrac{1}{27}$ **(b)** $\dfrac{1}{32}$ **(c)** $\dfrac{25}{36}$

Radical Form of $a^{m/n}$

If all indicated roots are real numbers, then
$$a^{m/n} = \sqrt[n]{a^m} = \left(\sqrt[n]{a}\right)^m.$$

That is, raise a to the mth power and then take the nth root, or take the nth root of a and then raise to the mth power.

For example, $\quad 8^{2/3} = \sqrt[3]{8^2} = \sqrt[3]{64} = 4, \quad$ and $\quad 8^{2/3} = \left(\sqrt[3]{8}\right)^2 = 2^2 = 4,$

so
$$8^{2/3} = \sqrt[3]{8^2} = \left(\sqrt[3]{8}\right)^2.$$

OBJECTIVE 3 Convert between radicals and rational exponents.

Using the definition of rational exponents, we can simplify many problems involving radicals by converting the radicals to numbers with rational exponents. After simplifying, we can convert the answer back to radical form if required.

> **NOTE** The ability to convert between radicals and rational exponents is important in the study of exponential and logarithmic functions later in this book.

NOW TRY EXERCISE 4

Write each exponential as a radical. Assume that all variables represent positive real numbers.

(a) $21^{1/2}$ **(b)** $17^{5/4}$

(c) $4t^{3/5} + (4t)^{2/3}$

(d) $w^{-2/5}$ **(e)** $(a^2 - b^2)^{1/4}$

In parts (f)–(h), write each radical as an exponential. Simplify. Assume that all variables represent positive real numbers.

(f) $\sqrt[3]{15}$ **(g)** $\sqrt[4]{4^2}$

(h) $\sqrt[4]{x^4}$

EXAMPLE 4 Converting between Rational Exponents and Radicals

Write each exponential as a radical. Assume that all variables represent positive real numbers. Use the definition that takes the root first.

(a) $13^{1/2} = \sqrt{13}$ **(b)** $6^{3/4} = \left(\sqrt[4]{6}\right)^3$ **(c)** $9m^{5/8} = 9\left(\sqrt[8]{m}\right)^5$

(d) $6x^{2/3} - (4x)^{3/5} = 6\left(\sqrt[3]{x}\right)^2 - \left(\sqrt[5]{4x}\right)^3$

(e) $r^{-2/3} = \dfrac{1}{r^{2/3}} = \dfrac{1}{\left(\sqrt[3]{r}\right)^2}$

(f) $(a^2 + b^2)^{1/2} = \sqrt{a^2 + b^2}$ ◁ $\sqrt{a^2 + b^2} \neq a + b$

In parts (g)–(i), write each radical as an exponential. Simplify. Assume that all variables represent positive real numbers.

(g) $\sqrt{10} = 10^{1/2}$ **(h)** $\sqrt[4]{3^8} = 3^{8/4} = 3^2 = 9$

(i) $\sqrt[6]{z^6} = z$ because z is positive.

NOW TRY

> **NOTE** In **Example 4(i)**, it is not necessary to use absolute value bars because the directions specifically state that the variable represents a positive real number. The absolute value of the positive real number z is z itself, so the result is simply z.

OBJECTIVE 4 Use the rules for exponents with rational exponents.

The definition of rational exponents allows us to apply the rules for exponents from **Sections 4.1 and 4.2.**

Rules for Rational Exponents

Let r and s be rational numbers. For all real numbers a and b for which the indicated expressions exist, the following hold true.

$$a^r \cdot a^s = a^{r+s} \qquad a^{-r} = \frac{1}{a^r} \qquad \frac{a^r}{a^s} = a^{r-s} \qquad \left(\frac{a}{b}\right)^{-r} = \frac{b^r}{a^r}$$

$$(a^r)^s = a^{rs} \qquad (ab)^r = a^r b^r \qquad \left(\frac{a}{b}\right)^r = \frac{a^r}{b^r} \qquad a^{-r} = \left(\frac{1}{a}\right)^r$$

EXAMPLE 5   **Applying Rules for Rational Exponents**

Simplify each expression. Write answers with only positive exponents. Assume that all variables represent positive real numbers.

(a) $2^{1/2} \cdot 2^{1/4}$

$= 2^{1/2+1/4}$ Product rule

$= 2^{3/4}$ Add exponents.

(b) $\dfrac{5^{2/3}}{5^{7/3}}$

$= 5^{2/3-7/3}$ Quotient rule

$= 5^{-5/3}$ Subtract exponents.

$= \dfrac{1}{5^{5/3}}$ Definition of negative exponent

(c) $\dfrac{\left(x^{1/2}y^{2/3}\right)^4}{y}$

$= \dfrac{\left(x^{1/2}\right)^4\left(y^{2/3}\right)^4}{y}$ Power rule

$= \dfrac{x^2 y^{8/3}}{y^1}$ Power rule; $y = y^1$

$= x^2 y^{8/3-1}$ Quotient rule

$= x^2 y^{5/3}$ $\frac{8}{3} - 1 = \frac{8}{3} - \frac{3}{3} = \frac{5}{3}$

(d) $\left(\dfrac{x^4 y^{-6}}{x^{-2}y^{1/3}}\right)^{-2/3}$

$= \dfrac{\left(x^4\right)^{-2/3}\left(y^{-6}\right)^{-2/3}}{\left(x^{-2}\right)^{-2/3}\left(y^{1/3}\right)^{-2/3}}$ Power rule

$= \dfrac{x^{-8/3}y^4}{x^{4/3}y^{-2/9}}$ Power rule

$= x^{-8/3-4/3}y^{4-(-2/9)}$ Quotient rule

$= x^{-4}y^{38/9}$ Use parentheses to avoid errors. $4 - \left(-\frac{2}{9}\right) = \frac{36}{9} + \frac{2}{9} = \frac{38}{9}$

$= \dfrac{y^{38/9}}{x^4}$ Definition of negative exponent

The same result is obtained if we simplify within the parentheses first, as follows.

NOW TRY EXERCISE 5

Simplify each expression. Write answers with only positive exponents. Assume that all variables represent positive real numbers.

(a) $5^{1/4} \cdot 5^{2/3}$ **(b)** $\dfrac{9^{3/5}}{9^{7/5}}$

(c) $\dfrac{(r^{2/3}t^{1/4})^8}{t}$

(d) $\left(\dfrac{2x^{1/2}y^{-2/3}}{x^{-3/5}y^{-1/5}}\right)^{-3}$

(e) $y^{2/3}(y^{1/3} + y^{5/3})$

$$\left(\frac{x^4y^{-6}}{x^{-2}y^{1/3}}\right)^{-2/3}$$

$$= (x^{4-(-2)}y^{-6-1/3})^{-2/3} \qquad \text{Quotient rule}$$

$$= (x^6 y^{-19/3})^{-2/3} \qquad \qquad -6 - \frac{1}{3} = -\frac{18}{3} - \frac{1}{3} = -\frac{19}{3}$$

$$= (x^6)^{-2/3}(y^{-19/3})^{-2/3} \qquad \text{Power rule}$$

$$= x^{-4}y^{38/9} \qquad \qquad \text{Power rule}$$

$$= \frac{y^{38/9}}{x^4} \qquad \qquad \text{Definition of negative exponent;}$$
$$\text{The result is the same.}$$

(e) $\qquad\qquad m^{3/4}(m^{5/4} - m^{1/4})$

> Do not make the common mistake of multiplying exponents in the first step.

$$= m^{3/4}(m^{5/4}) - m^{3/4}(m^{1/4}) \qquad \text{Distributive property}$$

$$= m^{3/4+5/4} - m^{3/4+1/4} \qquad \text{Product rule}$$

$$= m^{8/4} - m^{4/4} \qquad \qquad \text{Add exponents.}$$

$$= m^2 - m \qquad \qquad \text{Write the exponents in lowest terms.}$$

NOW TRY

NOW TRY EXERCISE 6

Write each radical as an exponential, and then simplify. Leave answers in exponential form. Assume that all variables represent positive real numbers.

(a) $\sqrt[5]{y^3} \cdot \sqrt[3]{y}$ **(b)** $\dfrac{\sqrt[4]{y^3}}{\sqrt{y^5}}$

(c) $\sqrt{\sqrt[3]{y}}$

EXAMPLE 6 Applying Rules for Rational Exponents

Write each radical as an exponential, and then simplify. Leave answers in exponential form. Assume that all variables represent positive real numbers.

(a) $\sqrt[3]{x^2} \cdot \sqrt[4]{x}$

$$= x^{2/3} \cdot x^{1/4} \qquad \text{Convert to rational exponents.}$$

$$= x^{2/3+1/4} \qquad \text{Product rule}$$

$$= x^{8/12+3/12} \qquad \text{Write exponents with a common denominator.}$$

$$= x^{11/12} \qquad \text{Add exponents.}$$

(b) $\dfrac{\sqrt{x^3}}{\sqrt[3]{x^2}}$

$$= \frac{x^{3/2}}{x^{2/3}} \qquad \text{Convert to rational exponents.}$$

$$= x^{3/2-2/3} \qquad \text{Quotient rule}$$

$$= x^{5/6} \qquad \frac{3}{2} - \frac{2}{3} = \frac{9}{6} - \frac{4}{6} = \frac{5}{6}$$

(c) $\sqrt{\sqrt[4]{z}}$

$$= \sqrt{z^{1/4}} \qquad \text{Convert the inside radical to a rational exponent.}$$

$$= (z^{1/4})^{1/2} \qquad \text{Convert the square root to a rational exponent.}$$

$$= z^{1/8} \qquad \text{Power rule}$$

NOW TRY ANSWERS

5. (a) $5^{11/12}$ **(b)** $\dfrac{1}{9^{4/5}}$

 (c) $r^{16/3}t$ **(d)** $\dfrac{y^{7/5}}{8x^{33/10}}$

 (e) $y + y^{7/3}$

6. (a) $y^{14/15}$ **(b)** $\dfrac{1}{y^{7/4}}$ **(c)** $y^{1/6}$

NOW TRY

10.2 Exercises

    MyMathLab®

▶ *Complete solution available in MyMathLab*

Concept Check *Match each expression from Column I with the equivalent choice from Column II.*

I

1. $3^{1/2}$	**2.** $(-27)^{1/3}$
3. $-16^{1/2}$	**4.** $(-25)^{1/2}$
5. $(-32)^{1/5}$	**6.** $(-32)^{2/5}$
7. $4^{3/2}$	**8.** $6^{2/4}$
9. $-6^{2/4}$	**10.** $36^{0.5}$

II

A. -4	**B.** 8
C. $\sqrt{3}$	**D.** $-\sqrt{6}$
E. -3	**F.** $\sqrt{6}$
G. 4	**H.** -2
I. 6	**J.** Not a real number

Evaluate each exponential. **See Examples 1–3.**

▶ **11.** $169^{1/2}$ **12.** $121^{1/2}$ **13.** $729^{1/3}$ **14.** $512^{1/3}$

15. $16^{1/4}$ **16.** $625^{1/4}$ **17.** $\left(\dfrac{64}{81}\right)^{1/2}$ **18.** $\left(\dfrac{8}{27}\right)^{1/3}$

▶ **19.** $(-27)^{1/3}$ **20.** $(-32)^{1/5}$ ▶ **21.** $(-144)^{1/2}$ **22.** $(-36)^{1/2}$

▶ **23.** $100^{3/2}$ **24.** $64^{3/2}$ **25.** $81^{3/4}$ **26.** $216^{2/3}$

27. $-16^{5/2}$ **28.** $-32^{3/5}$ **29.** $(-8)^{4/3}$ **30.** $(-243)^{2/5}$

▶ **31.** $32^{-3/5}$ **32.** $27^{-4/3}$ **33.** $64^{-3/2}$ **34.** $81^{-3/2}$

35. $\left(\dfrac{125}{27}\right)^{-2/3}$ **36.** $\left(\dfrac{64}{125}\right)^{-2/3}$ **37.** $\left(\dfrac{16}{81}\right)^{-3/4}$ **38.** $\left(\dfrac{729}{64}\right)^{-5/6}$

Write each exponential as a radical. Assume that all variables represent positive real numbers. **See Example 4.**

▶ **39.** $10^{1/2}$ **40.** $3^{1/2}$ **41.** $8^{3/4}$

42. $7^{2/3}$ ▶ **43.** $(9q)^{5/8} - (2x)^{2/3}$ **44.** $(3p)^{3/4} + (4x)^{1/3}$

45. $(2m)^{-3/2}$ **46.** $(5y)^{-3/5}$ **47.** $(2y + x)^{2/3}$

48. $(r + 2z)^{3/2}$ **49.** $(3m^4 + 2k^2)^{-2/3}$ **50.** $(5x^2 + 3z^3)^{-5/6}$

Write each radical as an exponential. Simplify. Assume that all variables represent positive real numbers. **See Example 4.**

51. $\sqrt{2^{12}}$ **52.** $\sqrt{5^{10}}$ ▶ **53.** $\sqrt[3]{4^9}$ **54.** $\sqrt[4]{6^8}$ **55.** $\sqrt{x^{20}}$

56. $\sqrt{r^{50}}$ **57.** $\sqrt[3]{x} \cdot \sqrt{x}$ **58.** $\sqrt[4]{y} \cdot \sqrt[5]{y^2}$ **59.** $\dfrac{\sqrt[3]{t^4}}{\sqrt[5]{t^4}}$ **60.** $\dfrac{\sqrt[4]{w^3}}{\sqrt[6]{w}}$

Simplify each expression. Write answers with only positive exponents. Assume that all variables represent positive real numbers. **See Example 5.**

▶ **61.** $3^{1/2} \cdot 3^{3/2}$ **62.** $6^{4/3} \cdot 6^{2/3}$ **63.** $\dfrac{64^{5/3}}{64^{4/3}}$

64. $\dfrac{125^{7/3}}{125^{5/3}}$ **65.** $y^{7/3} \cdot y^{-4/3}$ **66.** $r^{-8/9} \cdot r^{17/9}$

67. $x^{2/3} \cdot x^{-1/4}$

68. $x^{2/5} \cdot x^{-1/3}$

69. $\dfrac{k^{1/3}}{k^{2/3} \cdot k^{-1}}$

70. $\dfrac{z^{3/4}}{z^{5/4} \cdot z^{-2}}$

71. $\dfrac{(x^{1/4}y^{2/5})^{20}}{x^2}$

72. $\dfrac{(r^{1/5}s^{2/3})^{15}}{r^2}$

73. $\dfrac{(x^{2/3})^2}{(x^2)^{7/3}}$

74. $\dfrac{(p^3)^{1/4}}{(p^{5/4})^2}$

75. $\dfrac{m^{3/4}n^{-1/4}}{(m^2n)^{1/2}}$

76. $\dfrac{(a^2b^5)^{-1/4}}{(a^{-3}b^2)^{1/6}}$

77. $\dfrac{p^{1/5}p^{7/10}p^{1/2}}{(p^3)^{-1/5}}$

78. $\dfrac{z^{1/3}z^{-2/3}z^{1/6}}{(z^{-1/6})^3}$

79. $\left(\dfrac{b^{-3/2}}{c^{-5/3}}\right)^2 (b^{-1/4}c^{-1/3})^{-1}$

80. $\left(\dfrac{m^{-2/3}}{a^{-3/4}}\right)^4 (m^{-3/8}a^{1/4})^{-2}$

81. $\left(\dfrac{p^{-1/4}q^{-3/2}}{3^{-1}p^{-2}q^{-2/3}}\right)^{-2}$

82. $\left(\dfrac{2^{-2}w^{-3/4}x^{-5/8}}{w^{3/4}x^{-1/2}}\right)^{-3}$

83. $p^{2/3}(p^{1/3} + 2p^{4/3})$

84. $z^{5/8}(3z^{5/8} + 5z^{11/8})$

85. $k^{1/4}(k^{3/2} - k^{1/2})$

86. $r^{3/5}(r^{1/2} + r^{3/4})$

87. $6a^{7/4}(a^{-7/4} + 3a^{-3/4})$

88. $4m^{5/3}(m^{-2/3} - 4m^{-5/3})$

89. $-5x^{7/6}(x^{5/6} - x^{-1/6})$

90. $-8y^{11/7}(y^{3/7} - y^{-4/7})$

Write each radical as an exponential, and then simplify. Leave answers in exponential form. Assume that all variables represent positive numbers. **See Example 6.**

▶ **91.** $\sqrt[5]{x^3} \cdot \sqrt[4]{x}$

92. $\sqrt[6]{y^5} \cdot \sqrt[3]{y^2}$

93. $\dfrac{\sqrt{x^5}}{\sqrt{x^8}}$

94. $\dfrac{\sqrt[3]{k^5}}{\sqrt[3]{k^7}}$

95. $\sqrt{y} \cdot \sqrt[3]{yz}$

96. $\sqrt[3]{xz} \cdot \sqrt{z}$

97. $\sqrt[4]{\sqrt[3]{m}}$

98. $\sqrt[3]{\sqrt{k}}$

99. $\sqrt{\sqrt{\sqrt{x}}}$

100. $\sqrt{\sqrt{\sqrt{\sqrt{x}}}}$

101. $\sqrt{\sqrt[3]{\sqrt[4]{x}}}$

102. $\sqrt[3]{\sqrt[5]{\sqrt{y}}}$

Concept Check Work each problem.

103. Replace a with 3 and b with 4 to show that, in general,

$$\sqrt{a^2 + b^2} \neq a + b.$$

104. Suppose someone claims that $\sqrt[n]{a^n + b^n}$ must equal $a + b$, because when $a = 1$ and $b = 0$, a true statement results:

$$\sqrt[n]{a^n + b^n} = \sqrt[n]{1^n + 0^n} = \sqrt[n]{1^n} = 1 = 1 + 0 = a + b.$$

Explain why this is faulty reasoning.

Solve each problem.

105. Meteorologists can determine the duration of a storm using the function

$$T(d) = 0.07d^{3/2},$$

where d is the diameter of the storm in miles and T is the time in hours. Find the duration of a storm with a diameter of 16 mi. Round the answer to the nearest tenth of an hour.

106. The threshold weight t, in pounds, for a person is the weight above which the risk of death increases greatly. The threshold weight in pounds for men aged 40–49 is related to height H in inches by the function

$$H(t) = (1860.867t)^{1/3}.$$

What height corresponds to a threshold weight of 200 lb for a 46-yr-old man? Round the answer to the nearest inch and then to the nearest tenth of a foot.

*The **windchill factor** is a measure of the cooling effect that the wind has on a person's skin. It calculates the equivalent cooling temperature if there were no wind. The National Weather Service uses the formula*

$$\text{Windchill temperature} = 35.74 + 0.6215T - 35.75V^{4/25} + 0.4275TV^{4/25},$$

where T is the temperature in °F and V is the wind speed in miles per hour, to calculate windchill. The chart gives the windchill factor for various wind speeds and temperatures at which frostbite is a risk, and how quickly it may occur.

Temperature (°F)

Calm	40	30	20	10	0	–10	–20	–30	–40
5	36	25	13	1	–11	–22	–34	–46	–57
10	34	21	9	–4	–16	–28	–41	–53	–66
15	32	19	6	–7	–19	–32	–45	–58	–71
20	30	17	4	–9	–22	–35	–48	–61	–74
25	29	16	3	–11	–24	–37	–51	–64	–78
30	28	15	1	–12	–26	–39	–53	–67	–80
35	28	14	0	–14	–27	–41	–55	–69	–82
40	27	13	–1	–15	–29	–43	–57	–71	–84

Wind speed (mph)

Frostbites times: ▦ 30 minutes   ▦ 10 minutes   ▦ 5 minutes

Source: National Oceanic and Atmospheric Administration, National Weather Service.

Use the formula and a calculator to determine the windchill to the nearest tenth of a degree, given the following conditions. Compare answers with the appropriate entries in the table.

107. 30°F, 15-mph wind
108. 10°F, 30-mph wind
109. 20°F, 20-mph wind
110. 40°F, 10-mph wind

10.3   Simplifying Radicals, the Distance Formula, and Circles

OBJECTIVES

1   Use the product rule for radicals.

2   Use the quotient rule for radicals.

3   Simplify radicals.

4   Simplify products and quotients of radicals with different indexes.

5   Use the Pythagorean theorem.

6   Use the distance formula.

7   Find an equation of a circle given its center and radius.

VOCABULARY

☐ hypotenuse
☐ legs (of a right triangle)
☐ circle
☐ center
☐ radius

OBJECTIVE 1   Use the product rule for radicals.

Consider the expressions $\sqrt{36 \cdot 4}$ and $\sqrt{36} \cdot \sqrt{4}$. Are they equal?

$$\sqrt{36 \cdot 4} = \sqrt{144} = 12$$
$$\sqrt{36} \cdot \sqrt{4} = 6 \cdot 2 = 12$$

The result is the same.

This is an example of the **product rule for radicals.**

Product Rule for Radicals

If $\sqrt[n]{a}$ and $\sqrt[n]{b}$ are real numbers and n is a natural number, then the following holds.

$$\sqrt[n]{a} \cdot \sqrt[n]{b} = \sqrt[n]{ab}$$

That is, the product of two *n*th roots is the *n*th root of the product.

We justify the product rule using the rules for rational exponents. Because $\sqrt[n]{a} = a^{1/n}$ and $\sqrt[n]{b} = b^{1/n}$,

$$\sqrt[n]{a} \cdot \sqrt[n]{b} = a^{1/n} \cdot b^{1/n} = (ab)^{1/n} = \sqrt[n]{ab}.$$

⊟ **CAUTION** *Use the product rule only when the radicals have the same index.*

NOW TRY
EXERCISE 1

Multiply. Assume that all variables represent positive real numbers.

(a) $\sqrt{7} \cdot \sqrt{11}$

(b) $\sqrt{2mn} \cdot \sqrt{15}$

EXAMPLE 1 **Using the Product Rule**

Multiply. Assume that all variables represent positive real numbers.

(a) $\sqrt{5} \cdot \sqrt{7}$

$= \sqrt{5 \cdot 7}$

$= \sqrt{35}$

(b) $\sqrt{11} \cdot \sqrt{p}$

$= \sqrt{11p}$

(c) $\sqrt{7} \cdot \sqrt{11xyz}$

$= \sqrt{77xyz}$

NOW TRY

NOW TRY
EXERCISE 2

Multiply. Assume that all variables represent positive real numbers.

(a) $\sqrt[3]{4} \cdot \sqrt[3]{5}$

(b) $\sqrt[4]{5t} \cdot \sqrt[4]{6r^3}$

(c) $\sqrt[7]{20x} \cdot \sqrt[7]{3xy^3}$

(d) $\sqrt[3]{5} \cdot \sqrt[4]{9}$

EXAMPLE 2 **Using the Product Rule**

Multiply. Assume that all variables represent positive real numbers.

(a) $\sqrt[3]{3} \cdot \sqrt[3]{12}$

$= \sqrt[3]{3 \cdot 12}$

$= \sqrt[3]{36}$ ⟵ Remember to write the index.

(b) $\sqrt[4]{8y} \cdot \sqrt[4]{3r^2}$

$= \sqrt[4]{24yr^2}$

(c) $\sqrt[6]{10m^4} \cdot \sqrt[6]{5m}$

$= \sqrt[6]{50m^5}$

(d) $\sqrt[4]{2} \cdot \sqrt[5]{2}$ This product cannot be simplified using the product rule for radicals because the indexes (4 and 5) are different.

NOW TRY

OBJECTIVE 2 **Use the quotient rule for radicals.**

The **quotient rule for radicals** is similar to the product rule.

> **Quotient Rule for Radicals**
>
> If $\sqrt[n]{a}$ and $\sqrt[n]{b}$ are real numbers, and n is a natural number, then the following holds.
>
> $$\sqrt[n]{\frac{a}{b}} = \frac{\sqrt[n]{a}}{\sqrt[n]{b}} \quad \text{(where } b \neq 0\text{)}$$
>
> That is, the *n*th root of a quotient is the quotient of the *n*th roots.

EXAMPLE 3 **Using the Quotient Rule**

Simplify. Assume that all variables represent positive real numbers.

(a) $\sqrt{\frac{16}{25}} = \frac{\sqrt{16}}{\sqrt{25}} = \frac{4}{5}$

(b) $\sqrt{\frac{7}{36}} = \frac{\sqrt{7}}{\sqrt{36}} = \frac{\sqrt{7}}{6}$

(c) $\sqrt[3]{-\frac{8}{125}} = \sqrt[3]{\frac{-8}{125}} = \frac{\sqrt[3]{-8}}{\sqrt[3]{125}} = \frac{-2}{5} = -\frac{2}{5}$ $\frac{-a}{b} = -\frac{a}{b}$

(d) $\sqrt[3]{\frac{7}{216}} = \frac{\sqrt[3]{7}}{\sqrt[3]{216}} = \frac{\sqrt[3]{7}}{6}$

(e) $\sqrt[5]{\frac{x}{32}} = \frac{\sqrt[5]{x}}{\sqrt[5]{32}} = \frac{\sqrt[5]{x}}{2}$

NOW TRY ANSWERS

1. **(a)** $\sqrt{77}$ **(b)** $\sqrt{30mn}$
2. **(a)** $\sqrt[3]{20}$ **(b)** $\sqrt[4]{30tr^3}$
   (c) $\sqrt[7]{60x^2y^3}$
   (d) This expression cannot be simplified using the product rule.

**NOW TRY
EXERCISE 3**

Simplify. Assume that all variables represent positive real numbers.

(a) $\sqrt{\dfrac{49}{36}}$ **(b)** $\sqrt{\dfrac{5}{144}}$

(c) $\sqrt[3]{-\dfrac{27}{1000}}$ **(d)** $\sqrt[4]{\dfrac{t}{16}}$

(e) $-\sqrt[5]{\dfrac{m^{15}}{243}}$

(f) $-\sqrt[3]{\dfrac{m^6}{125}} = -\dfrac{\sqrt[3]{m^6}}{\sqrt[3]{125}} = -\dfrac{m^2}{5}$ Think: $\sqrt[3]{m^6} = m^{6/3} = m^2$ **NOW TRY**

OBJECTIVE 3 Simplify radicals.

We use the product and quotient rules to simplify radicals. A radical is **simplified** if the following four conditions are met.

> **Conditions for a Simplified Radical**
>
> 1. The radicand has no factor raised to a power greater than or equal to the index.
>
> 2. The radicand has no fractions.
>
> 3. No denominator contains a radical.
>
> 4. Exponents in the radicand and the index of the radical have greatest common factor 1.
>
> *Examples:* $\sqrt{22}, \quad \sqrt{15xy}, \quad \sqrt[3]{18}, \quad \dfrac{\sqrt[4]{m^3}}{m}$ These radicals are simplified.
>
> $\sqrt{28}, \quad \sqrt[3]{\dfrac{3}{5}}, \quad \dfrac{7}{\sqrt{7}}, \quad \sqrt[3]{r^{12}}$ These radicals are not simplified. Each violates one of the above conditions.

EXAMPLE 4 Simplifying Roots of Numbers

Simplify.

(a) $\sqrt{24}$

Check to see whether 24 is divisible by a perfect square (the square of a natural number) such as 4, 9, 16, … . The greatest perfect square that divides into 24 is 4.

$$\sqrt{24}$$
$$= \sqrt{4 \cdot 6} \qquad \text{Factor; 4 is a perfect square.}$$
$$= \sqrt{4} \cdot \sqrt{6} \qquad \text{Product rule}$$
$$= 2\sqrt{6} \qquad \sqrt{4} = 2$$

(b) $\sqrt{108}$

As shown on the left, the number 108 is divisible by the perfect square 36. If this perfect square is not immediately clear, try factoring 108 into its prime factors, as shown on the right.

$\sqrt{108}$		$\sqrt{108}$	
$= \sqrt{36 \cdot 3}$	Factor.	$= \sqrt{2^2 \cdot 3^3}$	
$= \sqrt{36} \cdot \sqrt{3}$	Product rule	$= \sqrt{2^2 \cdot 3^2 \cdot 3}$	$a^3 = a^2 \cdot a$
$= 6\sqrt{3}$	$\sqrt{36} = 6$	$= \sqrt{2^2} \cdot \sqrt{3^2} \cdot \sqrt{3}$	Product rule
		$= 2 \cdot 3 \cdot \sqrt{3}$	$\sqrt{2^2} = 2, \sqrt{3^2} = 3$
		$= 6\sqrt{3}$	Multiply.

(c) $\sqrt{10}$ No perfect square (other than 1) divides into 10, so $\sqrt{10}$ cannot be simplified further.

NOW TRY ANSWERS

3. **(a)** $\dfrac{7}{6}$ **(b)** $\dfrac{\sqrt{5}}{12}$ **(c)** $-\dfrac{3}{10}$

(d) $\dfrac{\sqrt[4]{t}}{2}$ **(e)** $-\dfrac{m^3}{3}$

**NOW TRY
EXERCISE 4**

Simplify.

(a) $\sqrt{50}$ **(b)** $\sqrt{192}$

(c) $\sqrt{42}$ **(d)** $\sqrt[3]{108}$

(e) $-\sqrt[4]{80}$

(d) $\sqrt[3]{16}$

The greatest perfect *cube* that divides into 16 is 8, so factor 16 as $8 \cdot 2$.

$$\sqrt[3]{16}$$

> Remember to write the index.

$$= \sqrt[3]{8 \cdot 2} \qquad \text{8 is a perfect cube.}$$

$$= \sqrt[3]{8} \cdot \sqrt[3]{2} \qquad \text{Product rule}$$

$$= 2\sqrt[3]{2} \qquad \sqrt[3]{8} = 2$$

(e) $\qquad -\sqrt[4]{162}$

> Remember the negative sign in each line.

$$= -\sqrt[4]{81 \cdot 2} \qquad \text{81 is a perfect 4th power.}$$

$$= -\sqrt[4]{81} \cdot \sqrt[4]{2} \qquad \text{Product rule}$$

$$= -3\sqrt[4]{2} \qquad \sqrt[4]{81} = 3 \qquad \text{NOW TRY}$$

⚠ **CAUTION**   *Be careful with which factors belong outside the radical symbol and which belong inside.* Note on the right in **Example 4(b)** how $2 \cdot 3$ is written outside because $\sqrt{2^2} = 2$ and $\sqrt{3^2} = 3$, while the remaining 3 is left inside the radical.

**NOW TRY
EXERCISE 5**

Simplify. Assume that all variables represent positive real numbers.

(a) $\sqrt{36x^5}$ **(b)** $\sqrt{32m^5n^4}$

(c) $\sqrt[3]{-125k^3p^7}$

(d) $-\sqrt[4]{162x^7y^8}$

| **EXAMPLE 5** Simplifying Radicals Involving Variables |

Simplify. Assume that all variables represent positive real numbers.

(a) $\sqrt{16m^3}$

$$= \sqrt{16m^2 \cdot m} \qquad \text{Factor.}$$

$$= \sqrt{16m^2} \cdot \sqrt{m} \qquad \text{Product rule}$$

$$= 4m\sqrt{m} \qquad \text{Take the square root.}$$

Absolute value bars are not needed around the m in color because all the variables represent *positive* real numbers.

(b) $\sqrt{200k^7q^8}$

$$= \sqrt{10^2 \cdot 2 \cdot (k^3)^2 \cdot k \cdot (q^4)^2} \qquad \text{Factor into perfect squares.}$$

$$= 10k^3q^4\sqrt{2k} \qquad \text{Take the square root.}$$

(c) $\sqrt[3]{-8x^4y^5}$

$$= \sqrt[3]{(-8x^3y^3)(xy^2)} \qquad \begin{array}{l}\text{Choose } -8x^3y^3 \text{ as the perfect cube that} \\ \text{divides into } -8x^4y^5.\end{array}$$

$$= \sqrt[3]{-8x^3y^3} \cdot \sqrt[3]{xy^2} \qquad \text{Product rule}$$

$$= -2xy\sqrt[3]{xy^2} \qquad \text{Take the cube root.}$$

NOW TRY ANSWERS

4. **(a)** $5\sqrt{2}$ **(b)** $8\sqrt{3}$
    (c) $\sqrt{42}$ cannot be simplified
      further.
    (d) $3\sqrt[3]{4}$ **(e)** $-2\sqrt[4]{5}$
5. **(a)** $6x^2\sqrt{x}$ **(b)** $4m^2n^2\sqrt{2m}$
    (c) $-5kp^2\sqrt[3]{p}$ **(d)** $-3xy^2\sqrt[4]{2x^3}$

(d) $-\sqrt[4]{32y^9}$

$$= -\sqrt[4]{(16y^8)(2y)} \qquad \begin{array}{l}16y^8 \text{ is the greatest 4th power that} \\ \text{divides into } 32y^9.\end{array}$$

$$= -\sqrt[4]{16y^8} \cdot \sqrt[4]{2y} \qquad \text{Product rule}$$

$$= -2y^2\sqrt[4]{2y} \qquad \text{Take the fourth root.} \qquad \text{NOW TRY}$$

NOTE   From **Example 5,** we see that if a variable is raised to a power with an exponent divisible by 2, it is a perfect square. If it is raised to a power with an exponent divisible by 3, it is a perfect cube. *In general, if it is raised to a power with an exponent divisible by n, it is a perfect nth power.*

The conditions for a simplified radical given earlier state that an exponent in the radicand and the index of the radical should have greatest common factor 1.

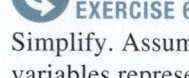
NOW TRY
EXERCISE 6
Simplify. Assume that all variables represent positive real numbers.

(a) $\sqrt[6]{7^2}$ **(b)** $\sqrt[6]{y^4}$

EXAMPLE 6   Simplifying Radicals Using Lesser Indexes

Simplify. Assume that all variables represent positive real numbers.

(a) $\sqrt[9]{5^6}$

We write this radical using rational exponents and then write the exponent in lowest terms. We then express the answer as a radical.

$$\sqrt[9]{5^6} = (5^6)^{1/9} = 5^{6/9} = 5^{2/3} = \sqrt[3]{5^2}, \quad \text{or} \quad \sqrt[3]{25}$$

(b) $\sqrt[4]{p^2} = (p^2)^{1/4} = p^{2/4} = p^{1/2} = \sqrt{p}$ (Recall the assumption that $p > 0$.)

NOW TRY

These examples suggest the following rule.

Meaning of $\sqrt[kn]{a^{km}}$

If m is an integer, n and k are natural numbers, and all indicated roots exist, then the following holds.

$$\sqrt[kn]{a^{km}} = \sqrt[n]{a^m}$$

OBJECTIVE 4   Simplify products and quotients of radicals with different indexes.

We multiply and divide radicals with different indexes using rational exponents.

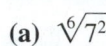
NOW TRY
EXERCISE 7
Simplify $\sqrt[3]{3} \cdot \sqrt{6}$.

EXAMPLE 7   Multiplying Radicals with Different Indexes

Simplify $\sqrt{7} \cdot \sqrt[3]{2}$.

Because the different indexes, 2 and 3, have a least common multiple of 6, use rational exponents to write each radical as a *sixth* root.

$$\sqrt{7} = 7^{1/2} = 7^{3/6} = \sqrt[6]{7^3} = \sqrt[6]{343}$$
$$\sqrt[3]{2} = 2^{1/3} = 2^{2/6} = \sqrt[6]{2^2} = \sqrt[6]{4}$$

Now we can multiply.

$$\sqrt{7} \cdot \sqrt[3]{2}$$
$$= \sqrt[6]{343} \cdot \sqrt[6]{4} \qquad \text{Substitute; } \sqrt{7} = \sqrt[6]{343}, \; \sqrt[3]{2} = \sqrt[6]{4}$$
$$= \sqrt[6]{1372} \qquad \text{Product rule} \qquad \textbf{NOW TRY}$$

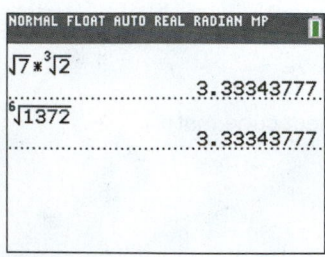

FIGURE 6

Results such as the one in **Example 7** can be supported with a calculator, as shown in **FIGURE 6**. Notice that the calculator gives the same *approximation* for the initial product and the final radical that we obtained.

NOW TRY ANSWERS
6. **(a)** $\sqrt[3]{7}$ **(b)** $\sqrt[3]{y^2}$
7. $\sqrt[6]{1944}$

⚠ **CAUTION** The computation in **FIGURE 6** is not *proof* that the two expressions are equal. The algebra in **Example 7,** however, is valid proof of their equality.

OBJECTIVE 5 Use the Pythagorean theorem.

The **Pythagorean theorem** provides an equation that relates the lengths of the three sides of a right triangle.

Pythagorean Theorem

If a and b are the lengths of the shorter sides of a right triangle and c is the length of the longest side, then the following holds.

$$a^2 + b^2 = c^2$$

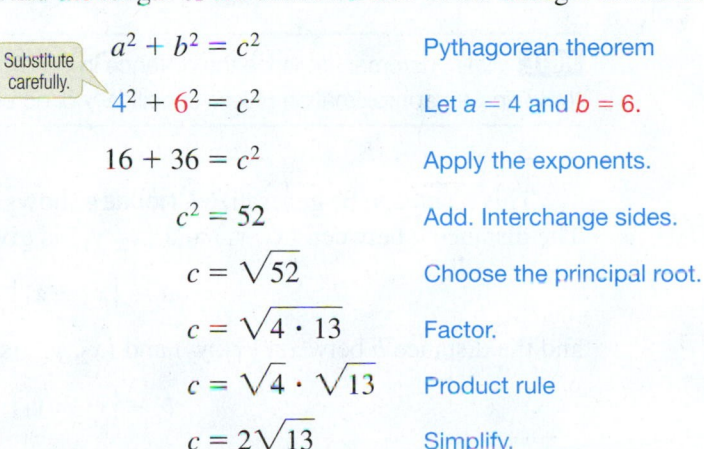

The two shorter sides are the **legs** of the triangle, and the longest side is the **hypotenuse.** The hypotenuse is the side opposite the right angle.

$$\textbf{leg}^2 + \textbf{leg}^2 = \textbf{hypotenuse}^2$$

In **Section 11.1** we will see that an equation such as $x^2 = 7$ has two solutions: $\sqrt{7}$ (the principal, or positive, square root of 7) and $-\sqrt{7}$. Similarly, $c^2 = 52$ has two solutions, $\pm\sqrt{52} = \pm 2\sqrt{13}$. In applications we often choose only the principal (positive) square root.

🌀 **NOW TRY**
EXERCISE 8
Find the length of the unknown side in each triangle.

(a)

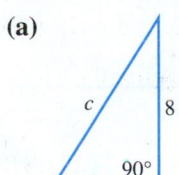

(b)

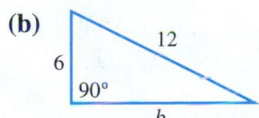

EXAMPLE 8 Using the Pythagorean Theorem

Find the length of the unknown side of the triangle in **FIGURE 7.**

$a^2 + b^2 = c^2$	Pythagorean theorem
$4^2 + 6^2 = c^2$	Let $a = 4$ and $b = 6$.
$16 + 36 = c^2$	Apply the exponents.
$c^2 = 52$	Add. Interchange sides.
$c = \sqrt{52}$	Choose the principal root.
$c = \sqrt{4 \cdot 13}$	Factor.
$c = \sqrt{4} \cdot \sqrt{13}$	Product rule
$c = 2\sqrt{13}$	Simplify.

Substitute carefully.

FIGURE 7

The length of the hypotenuse is $2\sqrt{13}$.

NOW TRY

⚠ **CAUTION** In the equation $a^2 + b^2 = c^2$, be sure that the length of the hypotenuse is substituted for c and the lengths of the legs are substituted for a and b.

NOW TRY ANSWERS
8. (a) $\sqrt{89}$ (b) $6\sqrt{3}$

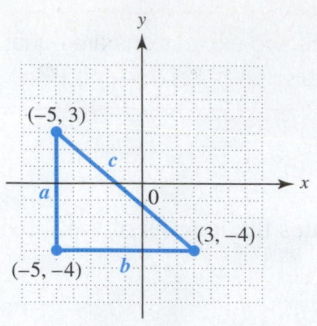

FIGURE 8

OBJECTIVE 6 Use the distance formula.

The *distance formula* allows us to find the distance between two points in the coordinate plane, or the length of the line segment joining those two points.

FIGURE 8 shows the points $(3, -4)$ and $(-5, 3)$. The vertical line through $(-5, 3)$ and the horizontal line through $(3, -4)$ intersect at the point $(-5, -4)$. Thus, the point $(-5, -4)$ becomes the vertex of the right angle in a right triangle.

By the Pythagorean theorem, the square of the length of the hypotenuse c of the right triangle in **FIGURE 8** is equal to the sum of the squares of the lengths of the two legs a and b.

$$a^2 + b^2 = c^2$$

The length a is the difference between the y-coordinates of the endpoints. The x-coordinate of both points in **FIGURE 8** is -5, so the side is vertical, and we can find a by finding the difference of the y-coordinates. We subtract -4 from 3 to obtain a positive value for a.

$$a = 3 - (-4) = 7$$

Similarly, we find b by subtracting -5 from 3.

$$b = 3 - (-5) = 8$$

Now substitute these values into the equation.

$$c^2 = a^2 + b^2$$
$$c^2 = 7^2 + 8^2 \qquad \text{Let } a = 7 \text{ and } b = 8.$$
$$c^2 = 49 + 64 \qquad \text{Apply the exponents.}$$
$$c^2 = 113 \qquad \text{Add.}$$
$$c = \sqrt{113} \qquad \text{Choose the principal root.}$$

We choose the principal root because here the distance cannot be negative. Therefore, the distance between $(-5, 3)$ and $(3, -4)$ is $\sqrt{113}$.

NOTE It is customary to leave the distance in simplified radical form. Do not use a calculator to find an approximation unless specifically directed to do so.

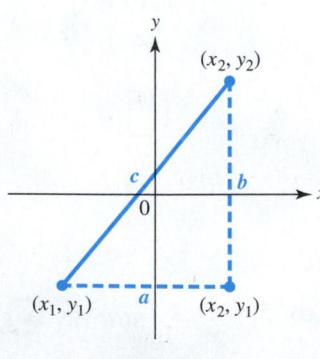

FIGURE 9

This result can be generalized. **FIGURE 9** shows the two points (x_1, y_1) and (x_2, y_2). The distance a between (x_1, y_1) and (x_2, y_1) is given by

$$a = |x_2 - x_1|,$$

and the distance b between (x_2, y_2) and (x_2, y_1) is given by

$$b = |y_2 - y_1|.$$

From the Pythagorean theorem, we obtain the following.

$$c^2 = a^2 + b^2$$
$$c^2 = (x_2 - x_1)^2 + (y_2 - y_1)^2 \qquad \text{For all real numbers } a, \ |a|^2 = a^2.$$

Choosing the principal square root gives the **distance formula.** In this formula, we use d (to denote distance) rather than c.

> **Distance Formula**
>
> The distance d between the points (x_1, y_1) and (x_2, y_2) is given by the following.
>
> $$d = \sqrt{(x_2 - x_1)^2 + (y_2 - y_1)^2}$$

NOW TRY EXERCISE 9

Find the distance between the points $(-4, -3)$ and $(-8, 6)$.

EXAMPLE 9 Using the Distance Formula

Find the distance between the points $(-3, 5)$ and $(6, 4)$.

We arbitrarily choose to let $(x_1, y_1) = (-3, 5)$ and $(x_2, y_2) = (6, 4)$.

$$d = \sqrt{(x_2 - x_1)^2 + (y_2 - y_1)^2} \qquad \text{Distance formula}$$

$$d = \sqrt{[6 - (-3)]^2 + (4 - 5)^2} \qquad \text{Let } x_2 = 6,\, y_2 = 4,\, x_1 = -3,\, y_1 = 5.$$

Substitute carefully.

$$d = \sqrt{9^2 + (-1)^2}$$

$$d = \sqrt{82} \qquad \text{Leave in radical form.}$$

The distance is $\sqrt{82}$.

NOW TRY

OBJECTIVE 7 Find an equation of a circle given its center and radius.

A **circle** is the set of all points in a plane that lie a fixed distance from a fixed point. The fixed point is the **center,** and the fixed distance is the **radius.** We use the distance formula to find an equation of a circle.

NOW TRY EXERCISE 10

Find an equation of the circle with radius 6 and center $(0, 0)$, and graph it.

EXAMPLE 10 Finding an Equation of a Circle and Graphing It

Find an equation of the circle with radius 3 and center $(0, 0)$, and graph it.

If the point (x, y) is on the circle, then the distance from (x, y) to the center $(0, 0)$ is the radius 3.

$$\sqrt{(x_2 - x_1)^2 + (y_2 - y_1)^2} = d \qquad \text{Distance formula}$$

$$\sqrt{(x - 0)^2 + (y - 0)^2} = 3 \qquad \text{Let } x_1 = 0,\, y_1 = 0, \text{ and } d = 3.$$

$$x^2 + y^2 = 9 \qquad \text{Square each side.}$$

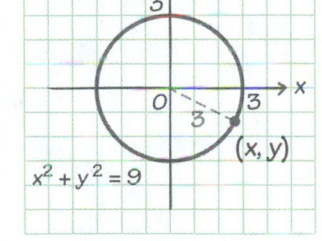

FIGURE 10

An equation of this circle is $x^2 + y^2 = 9$. The graph is shown in **FIGURE 10**.

NOW TRY

A circle may not be centered at the origin, as seen in the next example.

NOW TRY ANSWERS

9. $\sqrt{97}$
10. $x^2 + y^2 = 36$

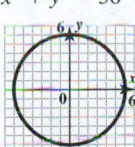

EXAMPLE 11 Finding an Equation of a Circle and Graphing It

Find an equation of the circle with center $(4, -3)$ and radius 5, and graph it.

$$\sqrt{(x_2 - x_1)^2 + (y_2 - y_1)^2} = d \qquad \text{Distance formula}$$

$$\sqrt{(x - 4)^2 + [y - (-3)]^2} = 5 \qquad \text{Let } x_1 = 4,\, y_1 = -3, \text{ and } d = 5.$$

$$(x - 4)^2 + (y + 3)^2 = 25 \qquad \text{Square each side.}$$

NOW TRY
EXERCISE 11
Find an equation of the circle with center at $(-2, 2)$ and radius 3, and graph it.

To graph the circle with equation

$$(x - 4)^2 + (y + 3)^2 = 25,$$

plot the center $(4, -3)$, and then, because the radius is 5, move 5 units right, left, up, and down from the center, plotting the points

$$(9, -3), \quad (-1, -3), \quad (4, 2), \quad \text{and} \quad (4, -8).$$

Draw a smooth curve through these four points, sketching one quarter of the circle at a time. See **FIGURE 11**.

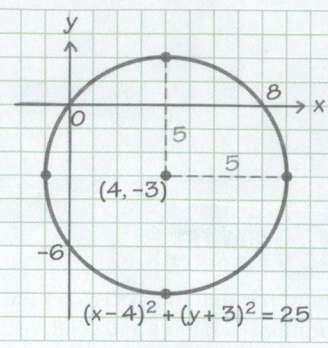

FIGURE 11

NOW TRY

Examples 10 and 11 suggest the form of an equation of a circle with radius r and center (h, k). If (x, y) is a point on the circle, then the distance from the center (h, k) to the point (x, y) is r. By the distance formula,

$$\sqrt{(x - h)^2 + (y - k)^2} = r.$$

Squaring both sides gives the **center-radius form** of the equation of a circle.

Equation of a Circle (Center-Radius Form)

An equation of a circle with radius r and center (h, k) is given by the following.

$$(x - h)^2 + (y - k)^2 = r^2$$

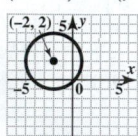
NOW TRY
EXERCISE 12
Find an equation of the circle with center $(-5, 4)$ and radius $\sqrt{6}$.

EXAMPLE 12 Using the Center-Radius Form of the Equation of a Circle

Find an equation of the circle with center $(-1, 2)$ and radius $\sqrt{7}$.

$(x - h)^2 + (y - k)^2 = r^2$	Center-radius form
$[x - (-1)]^2 + (y - 2)^2 = (\sqrt{7})^2$	Let $h = -1$, $k = 2$, and $r = \sqrt{7}$.
$(x + 1)^2 + (y - 2)^2 = 7$	Simplify; $(\sqrt{a})^2 = a$ **NOW TRY**

Pay attention to signs here.

NOW TRY ANSWERS
11. $(x + 2)^2 + (y - 2)^2 = 9$

12. $(x + 5)^2 + (y - 4)^2 = 6$

NOTE If a circle has its center at the origin $(0, 0)$ and radius r, then its equation is as follows.

$$(x - 0)^2 + (y - 0)^2 = r^2 \qquad \text{Let } h = 0, k = 0 \text{ in the center-radius form.}$$

$$x^2 + y^2 = r^2 \qquad \text{See Example 10.}$$

10.3 Exercises

FOR EXTRA HELP

MyMathLab®

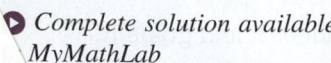
Complete solution available MyMathLab

Concept Check   *Choose the correct response.*

1. Which is the greatest perfect square factor of 128?

    A. 12   **B.** 16   **C.** 32   **D.** 64

2. Which is the greatest perfect cube factor of $81a^7$?

    A. $8a^3$ **B.** $27a^3$ **C.** $81a^6$ **D.** $27a^6$

3. Which radical can be simplified?

    A. $\sqrt{21}$ **B.** $\sqrt{48}$ **C.** $\sqrt[3]{12}$ **D.** $\sqrt[4]{10}$

4. Which radical *cannot* be simplified?

    A. $\sqrt[3]{30}$ **B.** $\sqrt[3]{27a^2b}$ **C.** $\sqrt{\dfrac{25}{81}}$ **D.** $\dfrac{2}{\sqrt{7}}$

5. Which one of the following is *not* equal to $\sqrt{\dfrac{1}{2}}$? (Do not use calculator approximations.)

    A. $\sqrt{0.5}$ **B.** $\sqrt{\dfrac{2}{4}}$ **C.** $\sqrt{\dfrac{3}{6}}$ **D.** $\dfrac{\sqrt{4}}{\sqrt{16}}$

6. Which one of the following is *not* equal to $\sqrt[3]{\dfrac{2}{5}}$? (Do not use calculator approximations.)

    A. $\sqrt[3]{\dfrac{6}{15}}$ **B.** $\dfrac{\sqrt[3]{50}}{5}$ **C.** $\dfrac{\sqrt[3]{10}}{\sqrt[3]{25}}$ **D.** $\dfrac{\sqrt[3]{10}}{5}$

7. Why is $\sqrt[3]{x} \cdot \sqrt[3]{x}$ not equal to x? What is it equal to?

8. Why is $\sqrt[4]{x} \cdot \sqrt[4]{x}$ not equal to x? Explain why it *is* equal to $\sqrt{x}$, for $x \geq 0$.

Multiply, if possible, using the product rule. Assume that all variables represent positive real numbers. See Examples 1 and 2.

  9. $\sqrt{3} \cdot \sqrt{3}$ **10.** $\sqrt{5} \cdot \sqrt{5}$ **11.** $\sqrt{18} \cdot \sqrt{2}$ **12.** $\sqrt{12} \cdot \sqrt{3}$

▶ **13.** $\sqrt{5} \cdot \sqrt{6}$ **14.** $\sqrt{10} \cdot \sqrt{3}$ **15.** $\sqrt{14} \cdot \sqrt{x}$ **16.** $\sqrt{23} \cdot \sqrt{t}$

  17. $\sqrt{14} \cdot \sqrt{3pqr}$ **18.** $\sqrt{7} \cdot \sqrt{5xt}$ **19.** $\sqrt[3]{2} \cdot \sqrt[3]{5}$ **20.** $\sqrt[3]{3} \cdot \sqrt[3]{6}$

▶ **21.** $\sqrt[3]{7x} \cdot \sqrt[3]{2y}$ **22.** $\sqrt[3]{9x} \cdot \sqrt[3]{4y}$ **23.** $\sqrt[4]{11} \cdot \sqrt[4]{3}$ **24.** $\sqrt[4]{6} \cdot \sqrt[4]{9}$

  25. $\sqrt[4]{2x} \cdot \sqrt[4]{3x^2}$ **26.** $\sqrt[4]{3y^2} \cdot \sqrt[4]{6y}$ **27.** $\sqrt[3]{7} \cdot \sqrt[4]{3}$ **28.** $\sqrt[5]{8} \cdot \sqrt[6]{12}$

Simplify. Assume that all variables represent positive real numbers. See Example 3.

▶ **29.** $\sqrt{\dfrac{64}{121}}$ **30.** $\sqrt{\dfrac{16}{49}}$ **31.** $\sqrt{\dfrac{3}{25}}$ **32.** $\sqrt{\dfrac{13}{49}}$

  33. $\sqrt{\dfrac{x}{25}}$ **34.** $\sqrt{\dfrac{k}{100}}$ **35.** $\sqrt{\dfrac{p^6}{81}}$ **36.** $\sqrt{\dfrac{w^{10}}{36}}$

  37. $\sqrt[3]{-\dfrac{27}{64}}$ **38.** $\sqrt[3]{-\dfrac{216}{125}}$ **39.** $\sqrt[3]{\dfrac{r^2}{8}}$ **40.** $\sqrt[3]{\dfrac{t}{125}}$

  41. $-\sqrt[4]{\dfrac{81}{x^4}}$ **42.** $-\sqrt[4]{\dfrac{625}{y^4}}$ **43.** $\sqrt[5]{\dfrac{1}{x^{15}}}$ **44.** $\sqrt[5]{\dfrac{32}{y^{20}}}$

Simplify. See Example 4.

▶ **45.** $\sqrt{12}$ **46.** $\sqrt{18}$ **47.** $\sqrt{288}$ **48.** $\sqrt{72}$ **49.** $-\sqrt{32}$

  50. $-\sqrt{48}$ **51.** $-\sqrt{28}$ **52.** $-\sqrt{24}$ **53.** $\sqrt{30}$ **54.** $\sqrt{46}$

  55. $\sqrt[3]{128}$ **56.** $\sqrt[3]{24}$ **57.** $\sqrt[3]{-16}$ **58.** $\sqrt[3]{-250}$ **59.** $\sqrt[3]{40}$

  60. $\sqrt[3]{375}$ **61.** $-\sqrt[4]{512}$ **62.** $-\sqrt[4]{1250}$ **63.** $\sqrt[5]{64}$ **64.** $\sqrt[5]{128}$

  65. $-\sqrt[5]{486}$ **66.** $-\sqrt[5]{2048}$ **67.** $\sqrt[6]{128}$ **68.** $\sqrt[6]{1458}$

Simplify. Assume that all variables represent positive real numbers. **See Example 5.**

69. $\sqrt{72k^2}$

70. $\sqrt{18m^2}$

71. $\sqrt{144x^3y^9}$

72. $\sqrt{169s^5t^{10}}$

73. $\sqrt{121x^6}$

74. $\sqrt{256z^{12}}$

75. $-\sqrt[3]{27t^{12}}$

76. $-\sqrt[3]{64y^{18}}$

77. $-\sqrt{100m^8z^4}$

78. $-\sqrt{25t^6s^{20}}$

79. $-\sqrt[3]{-125a^6b^9c^{12}}$

80. $-\sqrt[3]{-216y^{15}x^6z^3}$

81. $\sqrt[4]{\dfrac{1}{16}r^8t^{20}}$

82. $\sqrt[4]{\dfrac{81}{256}t^{12}u^8}$

▶ **83.** $\sqrt{50x^3}$

84. $\sqrt{300z^3}$

85. $-\sqrt{500r^{11}}$

86. $-\sqrt{200p^{13}}$

87. $\sqrt{13x^7y^8}$

88. $\sqrt{23k^9p^{14}}$

89. $\sqrt[3]{8z^6w^9}$

90. $\sqrt[3]{64a^{15}b^{12}}$

91. $\sqrt[3]{-16z^5t^7}$

92. $\sqrt[3]{-81m^4n^{10}}$

93. $\sqrt[4]{81x^{12}y^{16}}$

94. $\sqrt[4]{81t^8u^{28}}$

95. $-\sqrt[4]{162r^{15}s^{10}}$

96. $-\sqrt[4]{32k^5m^{10}}$

97. $\sqrt{\dfrac{y^{11}}{36}}$

98. $\sqrt{\dfrac{v^{13}}{49}}$

99. $\sqrt[3]{\dfrac{x^{16}}{27}}$

100. $\sqrt[3]{\dfrac{y^{17}}{125}}$

Simplify. Assume that $x \geq 0$. **See Example 6.**

▶ **101.** $\sqrt[4]{48^2}$

102. $\sqrt[4]{50^2}$

103. $\sqrt[4]{25}$

104. $\sqrt[6]{8}$

105. $\sqrt[10]{x^{25}}$

106. $\sqrt[12]{x^{44}}$

Simplify by first writing the radicals as radicals with the same index. Then multiply. Assume that all variables represent positive real numbers. **See Example 7.**

▶ **107.** $\sqrt[3]{4} \cdot \sqrt{3}$

108. $\sqrt[3]{5} \cdot \sqrt{6}$

109. $\sqrt[4]{3} \cdot \sqrt[3]{4}$

110. $\sqrt[5]{7} \cdot \sqrt[4]{5}$

111. $\sqrt{x} \cdot \sqrt[3]{x}$

112. $\sqrt[3]{y} \cdot \sqrt[4]{y}$

Find the length of the unknown side in each right triangle. Simplify answers if possible. **See Example 8.**

▶ **113.**

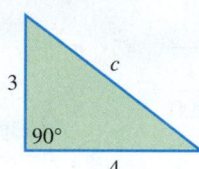

114.

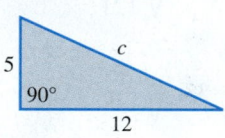

115.

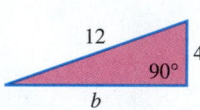

116.

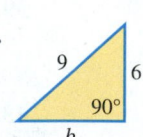

117.

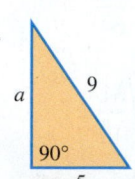

118.

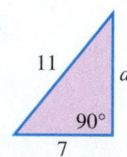

Find the distance between each pair of points. **See Example 9.**

119. $(6, 13)$ and $(1, 1)$

120. $(8, 13)$ and $(2, 5)$

▶ **121.** $(-6, 5)$ and $(3, -4)$

122. $(-1, 5)$ and $(-7, 7)$

123. $(-8, 2)$ and $(-4, 1)$

124. $(-1, 2)$ and $(5, 3)$

125. $(4.7, 2.3)$ and $(1.7, -1.7)$

126. $(-2.9, 18.2)$ and $(2.1, 6.2)$

127. $\left(\sqrt{2}, \sqrt{6}\right)$ and $\left(-2\sqrt{2}, 4\sqrt{6}\right)$ **128.** $\left(\sqrt{7}, 9\sqrt{3}\right)$ and $\left(-\sqrt{7}, 4\sqrt{3}\right)$

129. $(x + y, y)$ and $(x - y, x)$ **130.** $(c, c - d)$ and $(d, c + d)$

Concept Check   *Work each problem.*

131. Match each equation with the correct graph.

(a) $(x - 3)^2 + (y - 2)^2 = 25$ **A.** 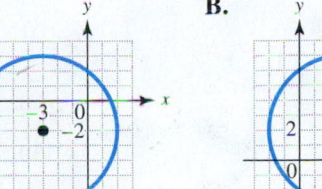   **B.**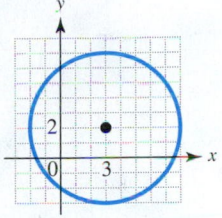

(b) $(x - 3)^2 + (y + 2)^2 = 25$

(c) $(x + 3)^2 + (y - 2)^2 = 25$ **C.** 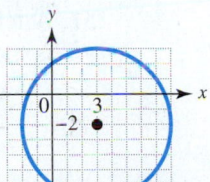   **D.**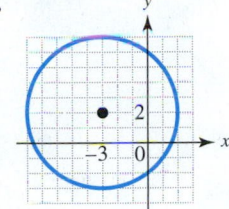

(d) $(x + 3)^2 + (y + 2)^2 = 25$

132. A circle can be drawn on a piece of posterboard by fastening one end of a string with a thumbtack, pulling the string taut with a pencil, and tracing a curve, as shown in the figure. Explain why this method works.

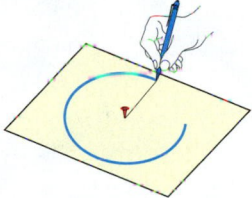

*Find the equation of a circle satisfying the given conditions. **See Examples 10–12.***

133. Center: $(0, 0)$; radius: 12   **134.** Center: $(0, 0)$; radius: 9

135. Center: $(-4, 3)$; radius: 2   **136.** Center: $(5, -2)$; radius: 4

137. Center: $(-8, -5)$; radius: $\sqrt{5}$ **138.** Center: $(-12, 13)$; radius: $\sqrt{7}$

*Graph each circle. Identify the center. **See Examples 10–12.***

139. $x^2 + y^2 = 9$ **140.** $x^2 + y^2 = 4$

141. $x^2 + y^2 = 16$ **142.** $x^2 + y^2 = 25$

143. $(x + 3)^2 + (y - 2)^2 = 9$ **144.** $(x - 1)^2 + (y + 3)^2 = 16$

145. $(x - 2)^2 + (y - 3)^2 = 4$ **146.** $(x + 4)^2 + (y + 1)^2 = 25$

Extending Skills   *Find the perimeter of each triangle.* $\left(\text{Hint: For Exercise 147, use } \sqrt{k} + \sqrt{k} = 2\sqrt{k}.\right)$

147. 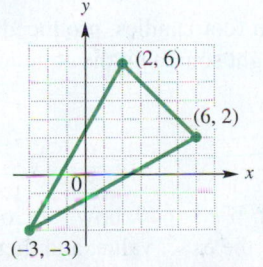   **148.**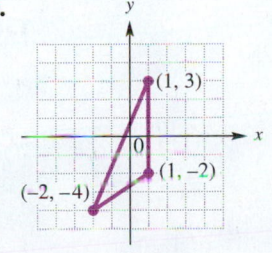

Solve each problem.

149. Ted is lucky enough to own a condominium in a high rise building on the shore of Lake Michigan, and has a beautiful view of the lake from his window. He lives on the 14th floor, which is 150 ft above the ground. He discovered that he can find the number of miles to the horizon by multiplying 1.224 by the square root of his eye level in feet from the ground. Ted's eyes are 6 ft above his floor.

(a) Write a formula that could be used to calculate the distance d in miles to the horizon from a height h in feet from the ground.

(b) Use the formula from part (a) to calculate the distance, to the nearest tenth of a mile, that Ted can see to the horizon from his condominium window.

150. Refer to **Exercise 149.** Ted's neighbor Sheri lives on a floor that is 100 ft above the ground. Assuming that her eyes are 5 ft above the ground, to the nearest tenth of a mile, how far can she see to the horizon?

151. A television has a rectangular screen with a 21.7-in. width. Its height is 16 in. What is the measure of the diagonal of the screen, to the nearest tenth of an inch? (*Source: Actual measurements of the author's television.*)

21.7 in.

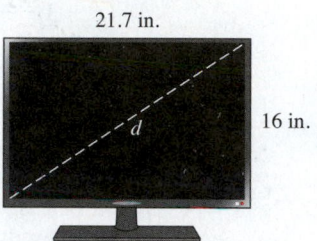

16 in.

152. The length of the diagonal of a box is given by

$$D = \sqrt{L^2 + W^2 + H^2},$$

where L, W, and H are, respectively, the length, width, and height of the box. Find the length of the diagonal D of a box that is 4 ft long, 2 ft wide, and 3 ft high. Give the exact value, and then round to the nearest tenth of a foot.

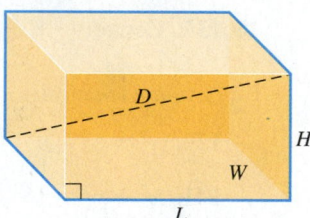

153. In the study of sound, one version of the law of tensions is

$$f_1 = f_2\sqrt{\frac{F_1}{F_2}}.$$

If $F_1 = 300$, $F_2 = 60$, and $f_2 = 260$, find f_1 to the nearest unit.

154. The illumination I, in foot-candles, produced by a light source is related to the distance d, in feet, from the light source by the equation

$$d = \sqrt{\frac{k}{I}},$$

where k is a constant. If $k = 640$, how far from the light source will the illumination be 2 foot-candles? Give the exact value, and then round to the nearest tenth of a foot.

10.4 Adding and Subtracting Radical Expressions

OBJECTIVE

1 Simplify radical expressions involving addition and subtraction.

OBJECTIVE 1 Simplify radical expressions involving addition and subtraction.

Expressions such as $4\sqrt{2} + 3\sqrt{2}$ and $2\sqrt{3} - 5\sqrt{3}$ can be simplified using the distributive property.

$$4\sqrt{2} + 3\sqrt{2}$$
$$= (4 + 3)\sqrt{2} = 7\sqrt{2}$$

This is similar to simplifying $4x + 3x$ to $7x$.

$$2\sqrt{3} - 5\sqrt{3}$$
$$= (2 - 5)\sqrt{3} = -3\sqrt{3}$$

This is similar to simplifying $2x - 5x$ to $-3x$.

> **! CAUTION** *Only radical expressions with the same index and the same radicand may be combined.*

NOW TRY EXERCISE 1

Add or subtract to simplify each radical expression.

(a) $\sqrt{12} + \sqrt{75}$

(b) $-\sqrt{63t} + 3\sqrt{28t}$, $t \geq 0$

(c) $6\sqrt{7} - 2\sqrt{3}$

EXAMPLE 1 Adding and Subtracting Radicals

Add or subtract to simplify each radical expression.

(a) $3\sqrt{24} + \sqrt{54}$ *(Simplify each individual radical.)*

$$= 3\sqrt{4} \cdot \sqrt{6} + \sqrt{9} \cdot \sqrt{6}$$ Product rule

$$= 3 \cdot 2\sqrt{6} + 3\sqrt{6}$$ $\sqrt{4} = 2$; $\sqrt{9} = 3$

$$= 6\sqrt{6} + 3\sqrt{6}$$ Multiply.

$$= (6 + 3)\sqrt{6}$$ Distributive property

$$= 9\sqrt{6}$$ Add.

(b) $2\sqrt{20x} - \sqrt{45x}$, $x \geq 0$

$$= 2\sqrt{4} \cdot \sqrt{5x} - \sqrt{9} \cdot \sqrt{5x}$$ Product rule

$$= 2 \cdot 2\sqrt{5x} - 3\sqrt{5x}$$ $\sqrt{4} = 2$; $\sqrt{9} = 3$

$$= 4\sqrt{5x} - 3\sqrt{5x}$$ Multiply.

$$= \sqrt{5x}$$ Combine like terms.

(4 − 3)$\sqrt{5x}$ = 1$\sqrt{5x}$, or $\sqrt{5x}$

(c) $2\sqrt{3} - 4\sqrt{5}$ The radicands differ and are already simplified, so $2\sqrt{3} - 4\sqrt{5}$ cannot be simplified further.

NOW TRY

> **! CAUTION** *The root of a sum does not equal the sum of the roots.* For example,
> $$\sqrt{9 + 16} \neq \sqrt{9} + \sqrt{16}$$
> because $\sqrt{9 + 16} = \sqrt{25} = 5$, but $\sqrt{9} + \sqrt{16} = 3 + 4 = 7$.

NOW TRY ANSWERS

1. (a) $7\sqrt{3}$ (b) $3\sqrt{7t}$
   (c) The expression cannot be simplified further.

NOW TRY
EXERCISE 2

Add or subtract to simplify each radical expression. Assume that all variables represent positive real numbers.

(a) $3\sqrt[3]{2000} - 4\sqrt[3]{128}$

(b) $5\sqrt[4]{a^5b^3} + \sqrt[4]{81ab^7}$

(c) $\sqrt[3]{128t^4} - 2\sqrt{72t^3}$

EXAMPLE 2 Adding and Subtracting Radicals with Higher Indexes

Add or subtract to simplify each radical expression. Assume that all variables represent positive real numbers.

(a) $2\sqrt[3]{16} - 5\sqrt[3]{54}$ *Remember to write the index with each radical.*

$$= 2\sqrt[3]{8 \cdot 2} - 5\sqrt[3]{27 \cdot 2} \qquad \text{Factor.}$$

$$= 2\sqrt[3]{8} \cdot \sqrt[3]{2} - 5\sqrt[3]{27} \cdot \sqrt[3]{2} \qquad \text{Product rule}$$

$$= 2 \cdot 2 \cdot \sqrt[3]{2} - 5 \cdot 3 \cdot \sqrt[3]{2} \qquad \text{Find the cube roots.}$$

$$= 4\sqrt[3]{2} - 15\sqrt[3]{2} \qquad \text{Multiply.}$$

$$= (4 - 15)\sqrt[3]{2} \qquad \text{Distributive property}$$

$$= -11\sqrt[3]{2} \qquad \text{Combine like terms.}$$

(b) $\qquad 2\sqrt[3]{x^2y} + \sqrt[3]{8x^5y^4}$

$$= 2\sqrt[3]{x^2y} + \sqrt[3]{(8x^3y^3)x^2y} \qquad \text{Factor.}$$

$$= 2\sqrt[3]{x^2y} + \sqrt[3]{8x^3y^3} \cdot \sqrt[3]{x^2y} \qquad \text{Product rule}$$

$$= 2\sqrt[3]{x^2y} + 2xy\sqrt[3]{x^2y} \qquad \text{Find the cube root.}$$

This result cannot be simplified further.

$$= (2 + 2xy)\sqrt[3]{x^2y} \qquad \text{Distributive property}$$

(c) $5\sqrt{4x^3} + 3\sqrt[3]{64x^4}$ *Be careful. The indexes are different.*

$$= 5\sqrt{4x^2 \cdot x} + 3\sqrt[3]{64x^3 \cdot x} \qquad \text{Factor.}$$

$$= 5\sqrt{4x^2} \cdot \sqrt{x} + 3\sqrt[3]{64x^3} \cdot \sqrt[3]{x} \qquad \text{Product rule}$$

$$= 5 \cdot 2x\sqrt{x} + 3 \cdot 4x\sqrt[3]{x} \qquad \textit{Keep track of the indexes.}$$

$$= 10x\sqrt{x} + 12x\sqrt[3]{x} \qquad \text{These cannot be combined.} \qquad \text{NOW TRY}$$

EXAMPLE 3 Adding and Subtracting Radicals with Fractions

Perform the indicated operations. Assume that all variables represent positive real numbers.

(a) $2\sqrt{\dfrac{75}{16}} + 4\dfrac{\sqrt{8}}{\sqrt{32}}$

$$= 2\dfrac{\sqrt{25 \cdot 3}}{\sqrt{16}} + 4\dfrac{\sqrt{4 \cdot 2}}{\sqrt{16 \cdot 2}} \qquad \text{Quotient rule; factor.}$$

$$= 2\left(\dfrac{5\sqrt{3}}{4}\right) + 4\left(\dfrac{2\sqrt{2}}{4\sqrt{2}}\right) \qquad \text{Product rule; find the square roots.}$$

$$= \dfrac{5\sqrt{3}}{2} + 2 \qquad \text{Multiply; } \dfrac{\sqrt{2}}{\sqrt{2}} = 1.$$

NOW TRY ANSWERS

2. (a) $14\sqrt[3]{2}$

  (b) $(5a + 3b)\sqrt[4]{ab^3}$

  (c) $4t\sqrt[3]{2t} - 12t\sqrt{2t}$

 **NOW TRY EXERCISE 3**

Perform the indicated operations. Assume that all variables represent positive real numbers.

(a) $5\dfrac{\sqrt{5}}{\sqrt{45}} - 4\sqrt{\dfrac{28}{9}}$

(b) $6\sqrt[3]{\dfrac{16}{x^{12}}} + 7\sqrt[3]{\dfrac{9}{x^9}}$

$= \dfrac{5\sqrt{3}}{2} + \dfrac{4}{2}$ Write with a common denominator; $2 = \dfrac{4}{2}$

$= \dfrac{5\sqrt{3}+4}{2}$ $\dfrac{a}{c} + \dfrac{b}{c} = \dfrac{a+b}{c}$

(b) $10\sqrt[3]{\dfrac{5}{x^6}} - 3\sqrt[3]{\dfrac{4}{x^9}}$

$= 10\dfrac{\sqrt[3]{5}}{\sqrt[3]{x^6}} - 3\dfrac{\sqrt[3]{4}}{\sqrt[3]{x^9}}$ Quotient rule

$= \dfrac{10\sqrt[3]{5}}{x^2} - \dfrac{3\sqrt[3]{4}}{x^3}$ Simplify denominators.

This equals x^3 so there is a common denominator.

$= \dfrac{10\sqrt[3]{5} \cdot x}{x^2 \cdot x} - \dfrac{3\sqrt[3]{4}}{x^3}$ Write with a common denominator.

$= \dfrac{10x\sqrt[3]{5} - 3\sqrt[3]{4}}{x^3}$ Subtract fractions.   **NOW TRY**

NOW TRY ANSWERS

3. (a) $\dfrac{5 - 8\sqrt{7}}{3}$

(b) $\dfrac{12\sqrt[3]{2} + 7x\sqrt[3]{9}}{x^4}$

10.4 Exercises

FOR EXTRA HELP   ▶ MyMathLab®

● *Complete solution available in MyMathLab*

Concept Check   *Choose the correct response.*

1. Which sum can be simplified without first simplifying the individual radical expressions?

A. $\sqrt{50} + \sqrt{32}$ **B.** $3\sqrt{6} + 9\sqrt{6}$

C. $\sqrt[3]{32} - \sqrt[3]{108}$ **D.** $\sqrt[5]{6} - \sqrt[5]{192}$

2. Which difference can be simplified without first simplifying the individual radical expressions?

A. $\sqrt{81} - \sqrt{18}$ **B.** $\sqrt[3]{8} - \sqrt[3]{16}$

C. $4\sqrt[3]{7} - 9\sqrt[3]{7}$ **D.** $\sqrt{75} - \sqrt{12}$

3. *Concept Check*   Even though the indexes of the terms are not equal, the sum

$$\sqrt{64} + \sqrt[3]{125} + \sqrt[4]{16}$$

can be simplified easily. What is this sum? Why can these terms be easily combined?

4. *Concept Check*   On an algebra quiz, Erin gave the difference $28 - 4\sqrt{2}$ as $24\sqrt{2}$. Her teacher did not give her any credit for this answer. **WHAT WENT WRONG?**

Add or subtract to simplify each radical expression. Assume that all variables represent positive real numbers. **See Examples 1 and 2.**

5. $\sqrt{36} - \sqrt{100}$ **6.** $\sqrt{25} - \sqrt{81}$ ● **7.** $-2\sqrt{48} + 3\sqrt{75}$

8. $4\sqrt{32} - 2\sqrt{8}$ **9.** $\sqrt[3]{16} + 4\sqrt[3]{54}$ **10.** $3\sqrt[3]{24} - 2\sqrt[3]{192}$

11. $\sqrt[4]{32} + 3\sqrt[4]{2}$

12. $\sqrt[4]{405} - 2\sqrt[4]{5}$

13. $6\sqrt{18} - \sqrt{32} + 2\sqrt{50}$

14. $5\sqrt{8} + 3\sqrt{72} - 3\sqrt{50}$

15. $5\sqrt{6} + 2\sqrt{10}$

16. $3\sqrt{11} - 5\sqrt{13}$

17. $2\sqrt{5} + 3\sqrt{20} + 4\sqrt{45}$

18. $5\sqrt{54} - 2\sqrt{24} - 2\sqrt{96}$

19. $\sqrt{72x} - \sqrt{8x}$

20. $\sqrt{18k} - \sqrt{72k}$

21. $3\sqrt{72m^2} - 5\sqrt{32m^2} - 3\sqrt{18m^2}$

22. $9\sqrt{27p^2} - 14\sqrt{108p^2} + 2\sqrt{48p^2}$

23. $2\sqrt[3]{16} + \sqrt[3]{54}$

24. $15\sqrt[3]{81} + 4\sqrt[3]{24}$

▶ **25.** $2\sqrt[3]{27x} - 2\sqrt[3]{8x}$

26. $6\sqrt[3]{128m} - 3\sqrt[3]{16m}$

27. $3\sqrt[3]{x^2y} - 5\sqrt[3]{8x^2y}$

28. $3\sqrt[3]{x^2y^2} - 2\sqrt[3]{64x^2y^2}$

29. $3x\sqrt[3]{xy^2} - 2\sqrt[3]{8x^4y^2}$

30. $6q^2\sqrt[3]{5q} - 2q\sqrt[3]{40q^4}$

31. $5\sqrt[4]{32} + 3\sqrt[4]{162}$

32. $2\sqrt[4]{512} + 4\sqrt[4]{32}$

33. $3\sqrt[4]{x^5y} - 2x\sqrt[4]{xy}$

34. $2\sqrt[4]{m^9p^6} - 3m^2p\sqrt[4]{mp^2}$

35. $2\sqrt[4]{32a^3} + 5\sqrt[4]{2a^3}$

36. $5\sqrt[4]{243x^3} + 2\sqrt[4]{3x^3}$

37. $\sqrt[3]{64xy^2} + \sqrt[3]{27x^4y^5}$

38. $\sqrt[4]{625s^3t} + \sqrt[4]{81s^7t^5}$

39. $\sqrt[3]{192st^4} - \sqrt{27s^3t}$

40. $\sqrt{125a^5b^5} + \sqrt[3]{125a^4b^4}$

41. $2\sqrt[3]{8x^4} + 3\sqrt[4]{16x^5}$

42. $3\sqrt[3]{64m^4} + 5\sqrt[4]{81m^5}$

Perform the indicated operations. Assume that all variables represent positive real numbers. See Example 3.

43. $\sqrt{8} - \dfrac{\sqrt{64}}{\sqrt{16}}$

44. $\sqrt{48} - \dfrac{\sqrt{81}}{\sqrt{9}}$

45. $\dfrac{2\sqrt{5}}{3} + \dfrac{\sqrt{5}}{6}$

46. $\dfrac{4\sqrt{3}}{3} + \dfrac{2\sqrt{3}}{9}$

47. $\sqrt{\dfrac{8}{9}} + \sqrt{\dfrac{18}{36}}$

48. $\sqrt{\dfrac{12}{16}} + \sqrt{\dfrac{48}{64}}$

49. $\dfrac{\sqrt{32}}{3} + \dfrac{2\sqrt{2}}{3} - \dfrac{\sqrt{2}}{\sqrt{9}}$

50. $\dfrac{\sqrt{27}}{2} - \dfrac{3\sqrt{3}}{2} + \dfrac{\sqrt{3}}{\sqrt{4}}$

▶ **51.** $3\sqrt{\dfrac{50}{9}} + 8\dfrac{\sqrt{2}}{\sqrt{8}}$

52. $5\sqrt{\dfrac{288}{25}} + 21\dfrac{\sqrt{2}}{\sqrt{18}}$

53. $\sqrt{\dfrac{25}{x^8}} + \sqrt{\dfrac{9}{x^6}}$

54. $\sqrt{\dfrac{100}{y^4}} + \sqrt{\dfrac{81}{y^{10}}}$

55. $3\sqrt[3]{\dfrac{m^5}{27}} - 2m\sqrt[3]{\dfrac{m^2}{64}}$

56. $2a\sqrt[4]{\dfrac{a}{16}} - 5a\sqrt[4]{\dfrac{a}{81}}$

57. $3\sqrt[3]{\dfrac{2}{x^6}} - 4\sqrt[3]{\dfrac{5}{x^9}}$

58. $-4\sqrt[3]{\dfrac{4}{t^9}} + 3\sqrt[3]{\dfrac{9}{t^{12}}}$

59. *Concept Check* Consider the expression

$$\sqrt{63} + \sqrt{112} - \sqrt{252}.$$

(a) Simplify this expression using the methods of this section.

(b) Use a calculator to approximate the given expression.

(c) Use a calculator to approximate the simplified expression in part (a).

(d) Complete the following: Assuming the work in part (a) is correct, the approximations in parts (b) and (c) should be (*equal/unequal*).

60. *Concept Check* Let $a = 1$ and let $b = 64$.

(a) Evaluate $\sqrt{a} + \sqrt{b}$. Then find $\sqrt{a + b}$. Are they equal?

(b) Evaluate $\sqrt[3]{a} + \sqrt[3]{b}$. Then find $\sqrt[3]{a + b}$. Are they equal?

(c) Complete the following: In general,

$$\sqrt[n]{a} + \sqrt[n]{b} \neq \underline{\hspace{3cm}},$$

based on the observations in parts (a) and (b) of this exercise.

Solve each problem.

61. A rectangular yard has a length of $\sqrt{192}$ m and a width of $\sqrt{48}$ m. Choose the best estimate of its dimensions. Then estimate the perimeter.

A. 14 m by 7 m   **B.** 5 m by 7 m   **C.** 14 m by 8 m   **D.** 15 m by 8 m

62. If the sides of a triangle are $\sqrt{65}$ in., $\sqrt{35}$ in., and $\sqrt{26}$ in., which one of the following is the best estimate of its perimeter?

A. 20 in.   **B.** 26 in.   **C.** 19 in.   **D.** 24 in.

Solve each problem. Give answers as simplified radical expressions.

63. Find the perimeter of the triangle.   **64.** Find the perimeter of the rectangle.

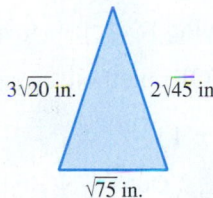

3√20 in.   2√45 in.

√75 in.

√192 m

√48 m

65. What is the perimeter of the computer graphic?   **66.** Find the area of the trapezoid.

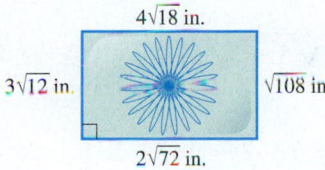

4√18 in.

3√12 in.   √108 in.

2√72 in.

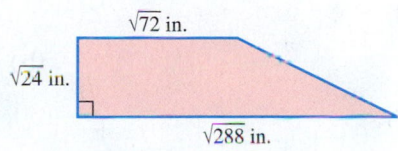

√72 in.

√24 in.

√288 in.

10.5 Multiplying and Dividing Radical Expressions

VOCABULARY

☐ conjugates

NOW TRY EXERCISE 1

Multiply.

(a) $\sqrt{10}(4 + \sqrt{7})$

(b) $3(\sqrt{20} - \sqrt{45})$

OBJECTIVE 1 Multiply radical expressions.

The distributive property may be used when multiplying radical expressions.

EXAMPLE 1 Using the Distributive Property with Radicals

Multiply.

(a) $\sqrt{5}(2 + \sqrt{6})$

$= \sqrt{5} \cdot 2 + \sqrt{5} \cdot \sqrt{6}$ Distributive property: $a(b + c) = ab + ac$

$= 2\sqrt{5} + \sqrt{30}$ Commutative property; product rule

(b) $4(\sqrt{12} - \sqrt{27})$

$= 4\sqrt{12} - 4\sqrt{27}$ Distributive property

$= 4\sqrt{4 \cdot 3} - 4\sqrt{9 \cdot 3}$ Factor the radicands so that one factor is a perfect square.

$= 4 \cdot 2\sqrt{3} - 4 \cdot 3\sqrt{3}$ $\sqrt{4} = 2; \sqrt{9} = 3$

$= 8\sqrt{3} - 12\sqrt{3}$ Multiply.

$= -4\sqrt{3}$ $8\sqrt{3} - 12\sqrt{3} = (8 - 12)\sqrt{3}$ **NOW TRY**

We multiply binomial expressions involving radicals using the FOIL method from **Section 4.5.** Recall that the acronym **FOIL** refers to the positions of the terms. We multiply the **F**irst terms, **O**uter terms, **I**nner terms, and **L**ast terms of the binomials.

EXAMPLE 2 Multiplying Binomials Involving Radical Expressions

Multiply, using the FOIL method.

(a) $(\sqrt{5} + 3)(\sqrt{6} + 1)$

$\quad\quad\quad$ First $\quad$ Outer $\quad$ Inner $\quad$ Last

$= \sqrt{5} \cdot \sqrt{6} + \sqrt{5} \cdot 1 + 3 \cdot \sqrt{6} + 3 \cdot 1$

$= \sqrt{30} + \sqrt{5} + 3\sqrt{6} + 3$ ← This result cannot be simplified further.

(b) $(7 - \sqrt{3})(\sqrt{5} + \sqrt{2})$

$\quad\quad\quad$ F $\quad$ O $\quad$ I $\quad$ L

$= 7\sqrt{5} + 7\sqrt{2} - \sqrt{3} \cdot \sqrt{5} - \sqrt{3} \cdot \sqrt{2}$

$= 7\sqrt{5} + 7\sqrt{2} - \sqrt{15} - \sqrt{6}$ Product rule

(c) $(\sqrt{10} + \sqrt{3})(\sqrt{10} - \sqrt{3})$

$= \sqrt{10} \cdot \sqrt{10} - \sqrt{10} \cdot \sqrt{3} + \sqrt{10} \cdot \sqrt{3} - \sqrt{3} \cdot \sqrt{3}$ FOIL method

$= 10 - 3$ Product rule; $-\sqrt{30} + \sqrt{30} = 0$

$= 7$ Subtract.

NOW TRY ANSWERS

1. (a) $4\sqrt{10} + \sqrt{70}$
   (b) $-3\sqrt{5}$

**NOW TRY
EXERCISE 2**

Multiply, using the FOIL method.

(a) $(8 - \sqrt{5})(9 - \sqrt{2})$

(b) $(\sqrt{7} + \sqrt{5})(\sqrt{7} - \sqrt{5})$

(c) $(\sqrt{15} - 4)^2$

(d) $(8 + \sqrt[3]{5})(8 - \sqrt[3]{5})$

(e) $(\sqrt{m} - \sqrt{n})(\sqrt{m} + \sqrt{n})$
$(m \geq 0$ and $n \geq 0)$

The product $(\sqrt{10} + \sqrt{3})(\sqrt{10} - \sqrt{3}) = (\sqrt{10})^2 - (\sqrt{3})^2$ in part (c) is a difference of squares.

$$(x + y)(x - y) = x^2 - y^2 \qquad \text{Here, } x = \sqrt{10} \text{ and } y = \sqrt{3}.$$

(d) $(\sqrt{7} - 3)^2$

$= (\sqrt{7} - 3)(\sqrt{7} - 3)$ $a^2 = a \cdot a$

$= \sqrt{7} \cdot \sqrt{7} - 3\sqrt{7} - 3\sqrt{7} + 3 \cdot 3$ FOIL method

$= 7 - 6\sqrt{7} + 9$ Multiply. Combine like terms.

$= 16 - 6\sqrt{7}$ ◁ Be careful. These terms cannot be combined.   Add.

(e) $(5 - \sqrt[3]{3})(5 + \sqrt[3]{3})$ ◁ Remember to write the index 3 in *each* radical.

$= 5 \cdot 5 + 5\sqrt[3]{3} - 5\sqrt[3]{3} - \sqrt[3]{3} \cdot \sqrt[3]{3}$ FOIL method

$= 25 - \sqrt[3]{3^2}$ Multiply. Combine like terms.

$= 25 - \sqrt[3]{9}$ Apply the exponent.

(f) $(\sqrt{k} + \sqrt{y})(\sqrt{k} - \sqrt{y})$

$= (\sqrt{k})^2 - (\sqrt{y})^2$ Difference of squares

$= k - y$ $(k \geq 0$ and $y \geq 0)$ NOW TRY

NOTE In **Example 2(d),** we could have used the formula for the square of a binomial to obtain the same result.

$$(\sqrt{7} - 3)^2$$

$= (\sqrt{7})^2 - 2(\sqrt{7})(3) + 3^2$ $(x - y)^2 = x^2 - 2xy + y^2$

$= 7 - 6\sqrt{7} + 9$ Apply the exponents. Multiply.

$= 16 - 6\sqrt{7}$ Add.

OBJECTIVE 2 Rationalize denominators with one radical term.

A simplified radical expression has no radical in the denominator. The origin of this agreement no doubt occurred before the days of high-speed calculation, when computation was a tedious process performed by hand.

Consider the expression $\dfrac{1}{\sqrt{2}}$. To find a decimal approximation by hand, it is necessary to divide 1 by a decimal approximation for $\sqrt{2}$, such as 1.414. It is much easier if the divisor is a whole number. This can be accomplished by multiplying $\dfrac{1}{\sqrt{2}}$ by 1 in the form $\dfrac{\sqrt{2}}{\sqrt{2}}$. *Multiplying by 1 in any form does not change the value of the original expression.*

NOW TRY ANSWERS

2. (a) $72 - 8\sqrt{2} - 9\sqrt{5} + \sqrt{10}$
(b) 2   (c) $31 - 8\sqrt{15}$
(d) $64 - \sqrt[3]{25}$ (e) $m - n$

$$\frac{1}{\sqrt{2}} \cdot \frac{\sqrt{2}}{\sqrt{2}} = \frac{\sqrt{2}}{2} \qquad \text{Multiply by 1; } \tfrac{\sqrt{2}}{\sqrt{2}} = 1$$

Now the computation requires dividing 1.414 by 2 to obtain 0.707, which is easier.

With current technology, either form $\frac{1}{\sqrt{2}}$ or $\frac{\sqrt{2}}{2}$ can be approximated with the same number of keystrokes. See **FIGURE 12**, which shows how a calculator gives the same approximation for both forms of the expression.

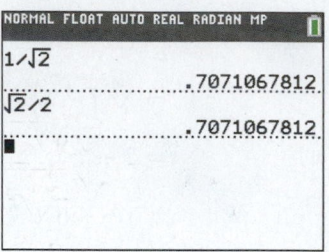

FIGURE 12

Rationalizing the Denominator

A common way of "standardizing" the form of a radical expression is to have the denominator contain no radicals. The process of removing radicals from a denominator so that the denominator contains only rational numbers is called **rationalizing the denominator.** This is done by multiplying by a form of 1.

**NOW TRY
EXERCISE 3**

Rationalize each denominator.

(a) $\dfrac{8}{\sqrt{13}}$　**(b)** $\dfrac{9\sqrt{7}}{\sqrt{3}}$

(c) $\dfrac{-10}{\sqrt{20}}$

EXAMPLE 3　Rationalizing Denominators with Square Roots

Rationalize each denominator.

(a) $\dfrac{3}{\sqrt{7}}$

Multiply the numerator and denominator by $\sqrt{7}$. This is, in effect, multiplying by 1.

$$\frac{3}{\sqrt{7}} = \frac{3 \cdot \sqrt{7}}{\sqrt{7} \cdot \sqrt{7}} = \frac{3\sqrt{7}}{7}$$

In the denominator,
$\sqrt{7} \cdot \sqrt{7} = \sqrt{7 \cdot 7} = \sqrt{49} = 7.$
The final denominator is now a rational number.

(b) $\dfrac{5\sqrt{2}}{\sqrt{5}} = \dfrac{5\sqrt{2} \cdot \sqrt{5}}{\sqrt{5} \cdot \sqrt{5}} = \dfrac{5\sqrt{10}}{5} = \sqrt{10}$

(c) $\dfrac{-6}{\sqrt{12}}$

Less work is involved if the radical in the denominator is simplified first.

$$\frac{-6}{\sqrt{12}} = \frac{-6}{\sqrt{4 \cdot 3}} = \frac{-6}{2\sqrt{3}} = \frac{-3}{\sqrt{3}}$$

NOW TRY ANSWERS

3. **(a)** $\dfrac{8\sqrt{13}}{13}$　**(b)** $3\sqrt{21}$

　(c) $-\sqrt{5}$

Now we rationalize the denominator.

$$\frac{-3}{\sqrt{3}} = \frac{-3 \cdot \sqrt{3}}{\sqrt{3} \cdot \sqrt{3}} = \frac{-3\sqrt{3}}{3} = -\sqrt{3}$$

 **NOW TRY**

NOW TRY
EXERCISE 4

Simplify each radical. In part (b), $y > 0$.

(a) $-\sqrt{\dfrac{27}{80}}$ **(b)** $\sqrt{\dfrac{48x^8}{y^3}}$

EXAMPLE 4 Rationalizing Denominators in Roots of Fractions

Simplify each radical. In part (b), $p > 0$.

(a) $-\sqrt{\dfrac{18}{125}}$

$$= -\frac{\sqrt{18}}{\sqrt{125}} \qquad \text{Quotient rule}$$

$$= -\frac{\sqrt{9 \cdot 2}}{\sqrt{25 \cdot 5}} \qquad \text{Factor.}$$

$$= -\frac{3\sqrt{2}}{5\sqrt{5}} \qquad \text{Product rule}$$

$$= -\frac{3\sqrt{2} \cdot \sqrt{5}}{5\sqrt{5} \cdot \sqrt{5}} \qquad \text{Multiply by } \frac{\sqrt{5}}{\sqrt{5}}.$$

$$= -\frac{3\sqrt{10}}{5 \cdot 5} \qquad \text{Product rule}$$

$$= -\frac{3\sqrt{10}}{25} \qquad \text{Multiply.}$$

(b) $\sqrt{\dfrac{50m^4}{p^5}}$

$$= \frac{\sqrt{50m^4}}{\sqrt{p^5}} \qquad \text{Quotient rule}$$

$$= \frac{\sqrt{25m^4 \cdot 2}}{\sqrt{p^4 \cdot p}} \qquad \text{Factor.}$$

$$= \frac{5m^2\sqrt{2}}{p^2\sqrt{p}} \qquad \text{Product rule}$$

$$= \frac{5m^2\sqrt{2} \cdot \sqrt{p}}{p^2\sqrt{p} \cdot \sqrt{p}} \qquad \text{Multiply by } \frac{\sqrt{p}}{\sqrt{p}}.$$

$$= \frac{5m^2\sqrt{2p}}{p^2 \cdot p} \qquad \text{Product rule}$$

$$= \frac{5m^2\sqrt{2p}}{p^3} \qquad \text{Multiply.}$$

NOW TRY

EXAMPLE 5 Rationalizing Denominators with Higher Roots

Simplify.

(a) $\sqrt[3]{\dfrac{27}{16}}$

Use the quotient rule, and simplify the numerator and denominator.

$$\sqrt[3]{\frac{27}{16}} = \frac{\sqrt[3]{27}}{\sqrt[3]{16}} = \frac{3}{\sqrt[3]{8} \cdot \sqrt[3]{2}} = \frac{3}{2\sqrt[3]{2}}$$

Because $2 \cdot 4 = 8$ is a perfect cube, multiply the numerator and denominator by $\sqrt[3]{4}$.

$$\frac{3}{2\sqrt[3]{2}} \qquad \sqrt[3]{\frac{27}{16}} = \frac{3}{2\sqrt[3]{2}} \text{ from above}$$

$$= \frac{3 \cdot \sqrt[3]{4}}{2\sqrt[3]{2} \cdot \sqrt[3]{4}} \qquad \begin{array}{l}\text{Multiply by } \sqrt[3]{4} \text{ in numerator} \\ \text{and denominator, This will give} \\ \sqrt[3]{8} = 2 \text{ in the denominator.}\end{array}$$

$$= \frac{3\sqrt[3]{4}}{2\sqrt[3]{8}} \qquad \text{Multiply.}$$

$$= \frac{3\sqrt[3]{4}}{2 \cdot 2} \qquad \sqrt[3]{8} = 2$$

$$= \frac{3\sqrt[3]{4}}{4} \qquad \text{Multiply.}$$

NOW TRY ANSWERS

4. **(a)** $-\dfrac{3\sqrt{15}}{20}$ **(b)** $\dfrac{4x^4\sqrt{3y}}{y^2}$

**NOW TRY
EXERCISE 5**

Simplify.

(a) $\sqrt[3]{\dfrac{8}{81}}$

(b) $\sqrt[4]{\dfrac{7x}{y}}$ $(x \geq 0, y > 0)$

(b) $\sqrt[4]{\dfrac{5x}{z}}$

$= \dfrac{\sqrt[4]{5x}}{\sqrt[4]{z}}$ Quotient rule

$= \dfrac{\sqrt[4]{5x}}{\sqrt[4]{z}} \cdot \dfrac{\sqrt[4]{z^3}}{\sqrt[4]{z^3}}$ Multiply by 1.

> $\sqrt[4]{z} \cdot \sqrt[4]{z^3}$
> will give $\sqrt[4]{z^4}$.

$= \dfrac{\sqrt[4]{5xz^3}}{\sqrt[4]{z^4}}$ Product rule

$= \dfrac{\sqrt[4]{5xz^3}}{z}$ $(x \geq 0, z > 0)$ **NOW TRY**

⚠ CAUTION   In **Example 5(a),** a typical error is to multiply the numerator and denominator by $\sqrt[3]{2}$, forgetting that $\sqrt[3]{2} \cdot \sqrt[3]{2} = \sqrt[3]{2^2}$, which does **not** equal 2. We need **three** factors of 2 to obtain 2^3 under the radical.

$$\sqrt[3]{2} \cdot \sqrt[3]{2} \cdot \sqrt[3]{2} = \sqrt[3]{2^3} \quad \text{which does equal} \quad 2.$$

OBJECTIVE 3   Rationalize denominators with binomials involving radicals.

Recall the special product

$$(x + y)(x - y) = x^2 - y^2.$$

To rationalize a denominator that contains a binomial expression (one that contains exactly two terms) involving radicals, such as

$$\frac{3}{1 + \sqrt{2}},$$

we must use *conjugates*. The conjugate of $1 + \sqrt{2}$ is $1 - \sqrt{2}$. In general,

$$x + y \text{ and } x - y \text{ are } \textbf{conjugates.}$$

Specifically, if a and b represent nonnegative rational numbers, the product

$$\left(\sqrt{a} + \sqrt{b}\right)\left(\sqrt{a} - \sqrt{b}\right) \quad \text{produces the rational number} \quad a - b.$$

NOW TRY ANSWERS

5. **(a)** $\dfrac{2\sqrt[3]{9}}{9}$ **(b)** $\dfrac{\sqrt[4]{7xy^3}}{y}$

Rationalizing a Binomial Denominator

Whenever a radical expression has a sum or difference with square root radicals in the denominator, rationalize the denominator by multiplying both the numerator and denominator by the conjugate of the denominator.

**NOW TRY
EXERCISE 6**

Rationalize each denominator.

(a) $\dfrac{4}{1 + \sqrt{3}}$ **(b)** $\dfrac{4}{5 + \sqrt{7}}$

(c) $\dfrac{\sqrt{3} + \sqrt{7}}{\sqrt{5} - \sqrt{2}}$

(d) $\dfrac{8}{\sqrt{3x} - \sqrt{y}}$

     $(3x \neq y, x > 0, y > 0)$

EXAMPLE 6   **Rationalizing Binomial Denominators**

Rationalize each denominator.

(a) $\dfrac{3}{1 + \sqrt{2}}$

> Again, we are multiplying by a form of 1.

$= \dfrac{3(1 - \sqrt{2})}{(1 + \sqrt{2})(1 - \sqrt{2})}$ Multiply the numerator and denominator by $1 - \sqrt{2}$, the conjugate of the denominator.

$\begin{aligned} (1 + \sqrt{2})(1 - \sqrt{2}) \\ = 1^2 - (\sqrt{2})^2 \\ = 1 - 2, \text{ or } -1 \end{aligned}$

> The denominator is now a rational number.

$= \dfrac{3(1 - \sqrt{2})}{-1}$

$= \dfrac{3}{-1}(1 - \sqrt{2})$ $\dfrac{a \cdot b}{c} = \dfrac{a}{c} \cdot b$

$= -3(1 - \sqrt{2})$ Simplify.

> Either of these forms is correct.

$= -3 + 3\sqrt{2}$ Distributive property

(b) $\dfrac{5}{4 - \sqrt{3}}$

$= \dfrac{5(4 + \sqrt{3})}{(4 - \sqrt{3})(4 + \sqrt{3})}$ Multiply the numerator and denominator by $4 + \sqrt{3}$.

$= \dfrac{5(4 + \sqrt{3})}{16 - 3}$ Multiply in the denominator.

$= \dfrac{5(4 + \sqrt{3})}{13}$ Subtract in the denominator.

Notice that we leave the numerator in factored form. This makes it easier to determine whether the expression is written in lowest terms.

(c) $\dfrac{\sqrt{2} - \sqrt{3}}{\sqrt{5} + \sqrt{3}}$

$= \dfrac{(\sqrt{2} - \sqrt{3})(\sqrt{5} - \sqrt{3})}{(\sqrt{5} + \sqrt{3})(\sqrt{5} - \sqrt{3})}$ Multiply the numerator and denominator by $\sqrt{5} - \sqrt{3}$.

$= \dfrac{\sqrt{10} - \sqrt{6} - \sqrt{15} + 3}{5 - 3}$ Multiply.

$= \dfrac{\sqrt{10} - \sqrt{6} - \sqrt{15} + 3}{2}$ Subtract in the denominator.

NOW TRY ANSWERS

6. (a) $-2 + 2\sqrt{3}$

(b) $\dfrac{2(5 - \sqrt{7})}{9}$

(c) $\dfrac{\sqrt{15} + \sqrt{6} + \sqrt{35} + \sqrt{14}}{3}$

(d) $\dfrac{8(\sqrt{3x} + \sqrt{y})}{3x - y}$

(d) $\dfrac{3}{\sqrt{5m} - \sqrt{p}}$ $(5m \neq p, m > 0, p > 0)$

$= \dfrac{3(\sqrt{5m} + \sqrt{p})}{(\sqrt{5m} - \sqrt{p})(\sqrt{5m} + \sqrt{p})}$ Multiply the numerator and denominator by $\sqrt{5m} + \sqrt{p}$.

$= \dfrac{3(\sqrt{5m} + \sqrt{p})}{5m - p}$ Multiply in the denominator.   **NOW TRY**

OBJECTIVE 4   Write radical quotients in lowest terms.

**NOW TRY
EXERCISE 7**

Write each quotient in lowest terms.

(a) $\dfrac{15 - 6\sqrt{2}}{18}$

(b) $\dfrac{15k + \sqrt{50k^2}}{20k}$ $(k > 0)$

EXAMPLE 7   Writing Radical Quotients in Lowest Terms

Write each quotient in lowest terms.

(a) $\dfrac{6 + 2\sqrt{5}}{4}$

> This is a key step.

$= \dfrac{2(3 + \sqrt{5})}{2 \cdot 2}$ Factor the numerator and denominator.

$= \dfrac{3 + \sqrt{5}}{2}$ Divide out the common factor.

Here is an alternative method for writing this expression in lowest terms.

$$\frac{6 + 2\sqrt{5}}{4} = \frac{6}{4} + \frac{2\sqrt{5}}{4} = \frac{3}{2} + \frac{\sqrt{5}}{2} = \frac{3 + \sqrt{5}}{2}$$

(b) $\dfrac{5y - \sqrt{8y^2}}{6y}$ $(y > 0)$

$= \dfrac{5y - 2y\sqrt{2}}{6y}$ $\sqrt{8y^2} = \sqrt{4y^2 \cdot 2} = 2y\sqrt{2}$

$= \dfrac{y(5 - 2\sqrt{2})}{6y}$ Factor the numerator.

$= \dfrac{5 - 2\sqrt{2}}{6}$ Divide out the common factor.   **NOW TRY**

> ! **CAUTION**   *Be careful to factor before writing a quotient in lowest terms.*

NOW TRY ANSWERS

7. **(a)** $\dfrac{5 - 2\sqrt{2}}{6}$ **(b)** $\dfrac{3 + \sqrt{2}}{4}$

10.5 Exercises

FOR EXTRA HELP   ▶ MyMathLab®

▶ *Complete solution available in MyMathLab*

Concept Check   *Match each part of a rule for a special product in Column I with the other part in Column II. Assume that A and B represent positive real numbers.*

	I		II
1.	$(A + \sqrt{B})(A - \sqrt{B})$	**A.**	$A - B$
2.	$(\sqrt{A} + B)(\sqrt{A} - B)$	**B.**	$A + 2B\sqrt{A} + B^2$
3.	$(\sqrt{A} + \sqrt{B})(\sqrt{A} - \sqrt{B})$	**C.**	$A - B^2$
4.	$(\sqrt{A} + \sqrt{B})^2$	**D.**	$A - 2\sqrt{AB} + B$
5.	$(\sqrt{A} - \sqrt{B})^2$	**E.**	$A^2 - B$
6.	$(\sqrt{A} + B)^2$	**F.**	$A + 2\sqrt{AB} + B$

*Multiply, and then simplify. Assume that all variables represent positive real numbers. **See Examples 1 and 2.***

7. $\sqrt{6}(3 + \sqrt{2})$

8. $\sqrt{10}(5 - \sqrt{3})$

9. $5(\sqrt{72} - \sqrt{8})$

10. $7(\sqrt{50} - \sqrt{18})$

11. $(\sqrt{7} + 3)(\sqrt{7} - 3)$

12. $(\sqrt{3} - 5)(\sqrt{3} + 5)$

⦿ **13.** $(\sqrt{2} - \sqrt{3})(\sqrt{2} + \sqrt{3})$

14. $(\sqrt{7} + \sqrt{14})(\sqrt{7} - \sqrt{14})$

15. $(\sqrt{8} - \sqrt{2})(\sqrt{8} + \sqrt{2})$

16. $(\sqrt{20} - \sqrt{5})(\sqrt{20} + \sqrt{5})$

17. $(\sqrt{2} + 1)(\sqrt{3} - 1)$

18. $(\sqrt{3} + 3)(\sqrt{5} - 2)$

19. $(\sqrt{11} - \sqrt{7})(\sqrt{2} + \sqrt{5})$

20. $(\sqrt{13} - \sqrt{7})(\sqrt{3} + \sqrt{11})$

21. $(2\sqrt{3} + \sqrt{5})(3\sqrt{3} - 2\sqrt{5})$

22. $(\sqrt{7} - \sqrt{11})(2\sqrt{7} + 3\sqrt{11})$

23. $(\sqrt{5} + 2)^2$

24. $(\sqrt{11} - 1)^2$

25. $(\sqrt{21} - \sqrt{5})^2$

26. $(\sqrt{6} - \sqrt{2})^2$

27. $(2 + \sqrt[3]{6})(2 - \sqrt[3]{6})$

28. $(\sqrt[3]{3} + 6)(\sqrt[3]{3} - 6)$

29. $(2 + \sqrt[3]{2})(4 - 2\sqrt[3]{2} + \sqrt[3]{4})$

30. $(\sqrt[3]{3} - 1)(\sqrt[3]{9} + \sqrt[3]{3} + 1)$

31. $(3\sqrt{x} - \sqrt{5})(2\sqrt{x} + 1)$

32. $(4\sqrt{p} + \sqrt{7})(\sqrt{p} - 9)$

33. $(3\sqrt{r} - \sqrt{s})(3\sqrt{r} + \sqrt{s})$

34. $(\sqrt{k} + 4\sqrt{m})(\sqrt{k} - 4\sqrt{m})$

35. $(\sqrt[3]{2y} - 5)(4\sqrt[3]{2y} + 1)$

36. $(\sqrt[3]{9z} - 2)(5\sqrt[3]{9z} + 7)$

37. $(\sqrt{3x} + 2)(\sqrt{3x} - 2)$

38. $(\sqrt{6y} - 4)(\sqrt{6y} + 4)$

39. $(2\sqrt{x} + \sqrt{y})(2\sqrt{x} - \sqrt{y})$

40. $(\sqrt{p} + 5\sqrt{s})(\sqrt{p} - 5\sqrt{s})$

41. $\left[(\sqrt{2} + \sqrt{3}) - \sqrt{6}\right]\left[(\sqrt{2} + \sqrt{3}) + \sqrt{6}\right]$

42. $\left[(\sqrt{5} - \sqrt{2}) - \sqrt{3}\right]\left[(\sqrt{5} - \sqrt{2}) + \sqrt{3}\right]$

*Rationalize each denominator. Assume that all variables represent positive real numbers. **See Examples 3 and 4.***

⦿ **43.** $\dfrac{7}{\sqrt{7}}$

44. $\dfrac{11}{\sqrt{11}}$

45. $\dfrac{15}{\sqrt{3}}$

46. $\dfrac{12}{\sqrt{6}}$

47. $\dfrac{\sqrt{3}}{\sqrt{2}}$

48. $\dfrac{\sqrt{7}}{\sqrt{6}}$

49. $\dfrac{9\sqrt{3}}{\sqrt{5}}$

50. $\dfrac{3\sqrt{2}}{\sqrt{11}}$

51. $\dfrac{-7}{\sqrt{48}}$

52. $\dfrac{-5}{\sqrt{24}}$

53. $\sqrt{\dfrac{7}{2}}$

54. $\sqrt{\dfrac{10}{3}}$

⦿ **55.** $-\sqrt{\dfrac{7}{50}}$

56. $-\sqrt{\dfrac{13}{75}}$

57. $\sqrt{\dfrac{24}{x}}$

58. $\sqrt{\dfrac{52}{y}}$

59. $\dfrac{-8\sqrt{3}}{\sqrt{k}}$

60. $\dfrac{-4\sqrt{13}}{\sqrt{m}}$

61. $-\sqrt{\dfrac{150m^5}{n^3}}$

62. $-\sqrt{\dfrac{98r^3}{s^5}}$

63. $\sqrt{\dfrac{288x^7}{y^9}}$

64. $\sqrt{\dfrac{242t^9}{u^{11}}}$

65. $\dfrac{5\sqrt{2m}}{\sqrt{y^3}}$

66. $\dfrac{2\sqrt{5r}}{\sqrt{m^3}}$

67. $-\sqrt{\dfrac{48k^2}{z}}$

68. $-\sqrt{\dfrac{75m^3}{p}}$

Simplify. Assume that all variables represent positive real numbers. **See Example 5.**

69. $\sqrt[3]{\dfrac{2}{3}}$

70. $\sqrt[3]{\dfrac{4}{5}}$

▶ **71.** $\sqrt[3]{\dfrac{4}{9}}$

72. $\sqrt[3]{\dfrac{5}{16}}$

73. $\sqrt[3]{\dfrac{9}{32}}$

74. $\sqrt[3]{\dfrac{10}{9}}$

75. $-\sqrt[3]{\dfrac{2p}{r^2}}$

76. $-\sqrt[3]{\dfrac{6x}{y^2}}$

77. $\sqrt[3]{\dfrac{x^6}{y}}$

78. $\sqrt[3]{\dfrac{m^9}{q}}$

79. $\sqrt[4]{\dfrac{16}{x}}$

80. $\sqrt[4]{\dfrac{81}{y}}$

81. $\sqrt[4]{\dfrac{2y}{z}}$

82. $\sqrt[4]{\dfrac{7t}{s^2}}$

Rationalize each denominator. Assume that all variables represent positive real numbers and no denominators are 0. **See Example 6.**

▶ **83.** $\dfrac{3}{4+\sqrt{5}}$

84. $\dfrac{4}{5+\sqrt{6}}$

85. $\dfrac{\sqrt{8}}{3-\sqrt{2}}$

86. $\dfrac{\sqrt{27}}{3-\sqrt{3}}$

87. $\dfrac{2}{3\sqrt{5}+2\sqrt{3}}$

88. $\dfrac{-1}{3\sqrt{2}-2\sqrt{7}}$

89. $\dfrac{\sqrt{2}-\sqrt{3}}{\sqrt{6}-\sqrt{5}}$

90. $\dfrac{\sqrt{5}+\sqrt{6}}{\sqrt{3}-\sqrt{2}}$

91. $\dfrac{m-4}{\sqrt{m}+2}$

92. $\dfrac{r-9}{\sqrt{r}-3}$

93. $\dfrac{4}{\sqrt{x}-2\sqrt{y}}$

94. $\dfrac{5}{3\sqrt{r}+\sqrt{s}}$

95. $\dfrac{\sqrt{x}-\sqrt{y}}{\sqrt{x}+\sqrt{y}}$

96. $\dfrac{\sqrt{a}+\sqrt{b}}{\sqrt{a}-\sqrt{b}}$

97. $\dfrac{5\sqrt{k}}{2\sqrt{k}+\sqrt{q}}$

98. $\dfrac{3\sqrt{x}}{\sqrt{x}-2\sqrt{y}}$

Write each quotient in lowest terms. Assume that all variables represent positive real numbers. **See Example 7.**

99. $\dfrac{30-20\sqrt{6}}{10}$

100. $\dfrac{24+12\sqrt{5}}{12}$

101. $\dfrac{3-3\sqrt{5}}{3}$

102. $\dfrac{-5+5\sqrt{2}}{5}$

▶ **103.** $\dfrac{16-4\sqrt{8}}{12}$

104. $\dfrac{12-9\sqrt{72}}{18}$

105. $\dfrac{6p+\sqrt{24p^3}}{3p}$

106. $\dfrac{11y-\sqrt{242y^5}}{22y}$

Extending Skills   *Rationalize each denominator. Assume that all radicals represent real numbers and no denominators are 0.*

107. $\dfrac{3}{\sqrt{x+y}}$

108. $\dfrac{5}{\sqrt{m-n}}$

109. $\dfrac{p}{\sqrt{p+2}}$

110. $\dfrac{q}{\sqrt{5+q}}$

Solve each problem.

111. The following expression occurs in a standard problem in trigonometry.

$$\frac{1}{\sqrt{2}} \cdot \frac{\sqrt{3}}{2} - \frac{1}{\sqrt{2}} \cdot \frac{1}{2}$$

Show that it simplifies to $\dfrac{\sqrt{6} - \sqrt{2}}{4}$. Then verify, using a calculator approximation.

112. The following expression occurs in a standard problem in trigonometry.

$$\frac{\sqrt{3} + 1}{1 - \sqrt{3}}$$

Show that it simplifies to $-2 - \sqrt{3}$. Then verify, using a calculator approximation.

*Extending Skills   In calculus, it is sometimes desirable to **rationalize the numerator.** For example, to rationalize the numerator of*

$$\frac{6 - \sqrt{2}}{4},$$

we multiply the numerator and the denominator by the conjugate of the numerator.

$$\frac{6 - \sqrt{2}}{4} = \frac{\left(6 - \sqrt{2}\right)\left(6 + \sqrt{2}\right)}{4\left(6 + \sqrt{2}\right)} = \frac{36 - 2}{4\left(6 + \sqrt{2}\right)} = \frac{34}{4\left(6 + \sqrt{2}\right)} = \frac{17}{2\left(6 + \sqrt{2}\right)}$$

Rationalize each numerator. Assume that all variables represent positive real numbers.

113. $\dfrac{6 - \sqrt{3}}{8}$ **114.** $\dfrac{2\sqrt{5} - 3}{2}$ **115.** $\dfrac{2\sqrt{x} - \sqrt{y}}{3x}$ **116.** $\dfrac{\sqrt{p} - 3\sqrt{q}}{4q}$

SUMMARY EXERCISES   Performing Operations with Radicals and Rational Exponents

Conditions for a Simplified Radical

1. The radicand has no factor raised to a power greater than or equal to the index.

2. The radicand has no fractions.

3. No denominator contains a radical.

4. Exponents in the radicand and the index of the radical have greatest common factor 1.

Concept Check   Give the reason why each radical is not simplified.

1. $\sqrt{\dfrac{2}{5}}$ **2.** $\sqrt[15]{x^5}$ **3.** $\dfrac{5}{\sqrt[3]{10}}$ **4.** $\sqrt[3]{x^5 y^6}$

Perform all indicated operations, and express each answer in simplest form with positive exponents. Assume that all variables represent positive real numbers.

5. $6\sqrt{10} - 12\sqrt{10}$ **6.** $\sqrt{7}\left(\sqrt{7} - \sqrt{2}\right)$ **7.** $\left(1 - \sqrt{3}\right)\left(2 + \sqrt{6}\right)$

8. $\sqrt{50} - \sqrt{98} + \sqrt{72}$ **9.** $\left(3\sqrt{5} + 2\sqrt{7}\right)^2$ **10.** $\dfrac{-3}{\sqrt{6}}$

1. The radicand is a fraction, $\frac{2}{5}$.

2. The exponent in the radicand and the index of the radical have greatest common factor 5.

3. The denominator contains a radical, $\sqrt[3]{10}$.

4. The radicand has two factors, x and y, that are raised to powers greater than the index, 3.

5. $-6\sqrt{10}$ **6.** $7 - \sqrt{14}$

7. $2 + \sqrt{6} - 2\sqrt{3} - 3\sqrt{2}$

8. $4\sqrt{2}$

9. $73 + 12\sqrt{35}$

10. $\dfrac{-\sqrt{6}}{2}$

11. $4\left(\sqrt{7}-\sqrt{5}\right)$

12. $-3+2\sqrt{2}$

13. -44

14. $\dfrac{\sqrt{x}+\sqrt{5}}{x-5}$

15. $2abc^3\sqrt[3]{b^2}$

16. $5\sqrt[3]{3}$

17. $3\left(\sqrt{5}-2\right)$

18. $\dfrac{\sqrt{15x}}{5x}$

19. $\dfrac{8}{5}$

20. $\dfrac{\sqrt{2}}{8}$

21. $-\sqrt[3]{100}$

22. $11+2\sqrt{30}$

23. $-3\sqrt{3x}$

24. $52-30\sqrt{3}$

25. $\dfrac{\sqrt[3]{117}}{9}$

26. $3\sqrt{2}+\sqrt{15}+\sqrt{42}+\sqrt{35}$

27. $2\sqrt[4]{27}$

28. $\dfrac{1+\sqrt[3]{3}+\sqrt[3]{9}}{-2}$

29. $\dfrac{x\sqrt[3]{x^2}}{y}$

30. $-4\sqrt{3}-3$

31. $xy^{6/5}$

32. $x^{10}y$

33. $\dfrac{1}{25x^2}$

34. $7+4\cdot3^{1/2}$, or $7+4\sqrt{3}$

35. $3\sqrt[3]{2x^2}$

36. -2

37. (a) 8   (b) $\{-8,8\}$

38. (a) 10   (b) $\{-10,10\}$

39. (a) $\{-4,4\}$
   (b) -4

40. (a) $\{-5,5\}$
   (b) -5

---

11. $\dfrac{8}{\sqrt{7}+\sqrt{5}}$

12. $\dfrac{1-\sqrt{2}}{1+\sqrt{2}}$

13. $\left(\sqrt{5}+7\right)\left(\sqrt{5}-7\right)$

14. $\dfrac{1}{\sqrt{x}-\sqrt{5}}, \quad x\neq5$

15. $\sqrt[3]{8a^3b^5c^9}$

16. $\dfrac{15}{\sqrt[3]{9}}$

17. $\dfrac{3}{\sqrt{5}+2}$

18. $\sqrt{\dfrac{3}{5x}}$

19. $\dfrac{16\sqrt{3}}{5\sqrt{12}}$

20. $\dfrac{2\sqrt{25}}{8\sqrt{50}}$

21. $\dfrac{-10}{\sqrt[3]{10}}$

22. $\dfrac{\sqrt{6}+\sqrt{5}}{\sqrt{6}-\sqrt{5}}$

23. $\sqrt{12x}-\sqrt{75x}$

24. $\left(5-3\sqrt{3}\right)^2$

25. $\sqrt[3]{\dfrac{13}{81}}$

26. $\dfrac{\sqrt{3}+\sqrt{7}}{\sqrt{6}-\sqrt{5}}$

27. $\dfrac{6}{\sqrt[4]{3}}$

28. $\dfrac{1}{1-\sqrt[3]{3}}$

29. $\sqrt[3]{\dfrac{x^2y}{x^{-3}y^4}}$

30. $\sqrt{12}-\sqrt{108}-\sqrt[3]{27}$

31. $\dfrac{x^{-2/3}y^{4/5}}{x^{-5/3}y^{-2/5}}$

32. $\left(\dfrac{x^{3/4}y^{2/3}}{x^{1/3}y^{5/8}}\right)^{24}$

33. $\left(125x^3\right)^{-2/3}$

34. $\dfrac{4^{1/2}+3^{1/2}}{4^{1/2}-3^{1/2}}$

35. $\sqrt[3]{16x^2}-\sqrt[3]{54x^2}+\sqrt[3]{128x^2}$

36. $\left(1-\sqrt[3]{3}\right)\left(1+\sqrt[3]{3}+\sqrt[3]{9}\right)$

Students often have trouble distinguishing between the following two types of problems.

Simplifying a Radical Involving a Square Root

Exercise: Simplify $\sqrt{25}$.

Answer: 5

In this situation, $\sqrt{25}$ represents the positive square root of 25, namely 5.

Solving an Equation Using Square Roots

Exercise: Solve $x^2=25$.

Answer: $\{-5,5\}$

In this situation, $x^2=25$ has two solutions, the negative square root of 25 or the positive square root of 25: $-5, 5$.

In Exercises 37–40, provide the appropriate responses.

37. (a) Simplify $\sqrt{64}$.

   (b) Solve $x^2=64$.

38. (a) Simplify $\sqrt{100}$.

   (b) Solve $x^2=100$.

39. (a) Solve $x^2=16$.

   (b) Simplify $-\sqrt{16}$.

40. (a) Solve $x^2=25$.

   (b) Simplify $-\sqrt{25}$.

10.6 Solving Equations with Radicals

OBJECTIVES

1. Solve radical equations using the power rule.
2. Solve radical equations with indexes greater than 2.
3. Use the power rule to solve a formula for a specified variable.

VOCABULARY

☐ radical equation
☐ proposed solution
☐ extraneous solution

OBJECTIVE 1 Solve radical equations using the power rule.

An equation that includes one or more radical expressions with a variable is called a **radical equation.**

$$\sqrt{x-4} = 8, \quad \sqrt{5x+12} = 3\sqrt{2x-1}, \quad \text{and} \quad \sqrt[3]{6+x} = 27 \qquad \text{Radical equations}$$

Solving radical equations involves a process that we have not yet seen, and it requires careful application. Notice that the equation $x = 1$ has only one solution. Its solution set is $\{1\}$. If we square both sides of this equation, we get $x^2 = 1$. This new equation has *two* solutions: -1 and 1. The solution of the original equation is also a solution of the equation following squaring. However, that equation has another solution, -1, that is *not* a solution of the original equation.

When solving equations with radicals, we use this idea of raising both sides to a power, which is an application of the **power rule.**

Power Rule for Solving an Equation with Radicals

If both sides of an equation are raised to the same power, all solutions of the original equation are also solutions of the new equation.

The power rule does not say that all solutions of the new equation are solutions of the original equation. They may or may not be. A value of the variable that appears to be a solution is a **proposed solution.** Such solutions that do not satisfy the original equation are **extraneous solutions.** They must be rejected.

⚠ **CAUTION** *When the power rule is used to solve an equation, every solution of the new equation must be checked in the original equation.*

NOW TRY EXERCISE 1

Solve $\sqrt{9x+7} = 5$.

EXAMPLE 1 Using the Power Rule

Solve $\sqrt{3x+4} = 8$.

$$\left(\sqrt{3x+4}\right)^2 = 8^2 \qquad \text{Use the power rule and square each side.}$$

$(\sqrt{a})^2 = \sqrt{a} \cdot \sqrt{a} = a$

$$3x + 4 = 64 \qquad \text{Apply the exponents.}$$

$$3x = 60 \qquad \text{Subtract 4.}$$

$$x = 20 \qquad \text{Divide by 3.}$$

CHECK

$$\sqrt{3x+4} = 8 \qquad \text{Original equation}$$

$$\sqrt{3 \cdot 20 + 4} \stackrel{?}{=} 8 \qquad \text{Let } x = 20.$$

$$\sqrt{64} \stackrel{?}{=} 8 \qquad \text{Simplify.}$$

$$8 = 8 \quad \checkmark \quad \text{True}$$

NOW TRY ANSWER
1. $\{2\}$

Because 20 satisfies the *original* equation, the solution set is $\{20\}$. **NOW TRY** ↻

Solving an Equation with Radicals

Step 1 **Isolate the radical.** Make sure that one radical term is alone on one side of the equation.

Step 2 **Apply the power rule.** Raise each side of the equation to a power that is the same as the index of the radical.

Step 3 **Solve** the resulting equation. If it still contains a radical, repeat Steps 1 and 2.

Step 4 **Check** all proposed solutions in the original equation. Discard any values that are not solutions of the original equation.

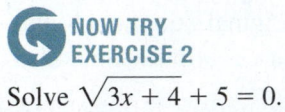

**NOW TRY
EXERCISE 2**

Solve $\sqrt{3x + 4} + 5 = 0$.

EXAMPLE 2   Using the Power Rule

Solve $\sqrt{5x - 1} + 3 = 0$.

Step 1	$\sqrt{5x - 1} = -3$	To isolate the radical on one side, subtract 3 from each side.
Step 2	$(\sqrt{5x - 1})^2 = (-3)^2$	Square each side.
Step 3	$5x - 1 = 9$	Apply the exponents.
	$5x = 10$	Add 1.
	$x = 2$	Divide by 5.

Step 4   **CHECK** $\sqrt{5x - 1} + 3 = 0$ Original equation

Be sure to check the proposed solution.

$$\sqrt{5 \cdot 2 - 1} + 3 \overset{?}{=} 0 \qquad \text{Let } x = 2.$$

$$3 + 3 = 0 \qquad \text{False}$$

This false result shows that the *proposed* solution 2 is *not* a solution of the original equation. It is extraneous. The solution set is $\varnothing$.   **NOW TRY**

NOTE   We could have determined after Step 1 that the equation in **Example 2** has no solution because the expression on the left cannot be negative.

The next examples involve finding the square of a binomial. Recall the rule from **Section 4.6.**

$$(x + y)^2 = x^2 + 2xy + y^2$$

EXAMPLE 3   Using the Power Rule (Squaring a Binomial)

Solve $\sqrt{4 - x} = x + 2$.

Step 1   The radical is isolated on the left side of the equation.

Step 2   Square each side. The square of $x + 2$ is $(x + 2)^2 = x^2 + 2(x)(2) + 4$.

$$(\sqrt{4 - x})^2 = (x + 2)^2 \qquad \text{Remember the middle term.}$$

$$4 - x = x^2 + 4x + 4$$

Twice the product of 2 and x

**NOW TRY
EXERCISE 3**

Solve $\sqrt{16 - x} = x + 4$.

Step 3 The new equation is quadratic, so write it in standard form.

$$x^2 + 5x = 0 \qquad \text{Standard form of } 4 - x = x^2 + 4x + 4$$

$$x(x + 5) = 0 \qquad \text{Factor.}$$

Set *each* factor equal to 0. $\longrightarrow$ $x = 0 \quad$ or $\quad x + 5 = 0 \qquad$ Zero-factor property

$$x = -5 \qquad \text{Solve for } x.$$

Step 4 Check each proposed solution in the original equation.

CHECK $\quad \sqrt{4 - x} = x + 2$ $\qquad\qquad\qquad\quad \sqrt{4 - x} = x + 2$

$\sqrt{4 - 0} \stackrel{?}{=} 0 + 2 \qquad$ Let $x = 0$. $\qquad \sqrt{4 - (-5)} \stackrel{?}{=} -5 + 2 \qquad$ Let $x = -5$.

$\sqrt{4} \stackrel{?}{=} 2$ $\qquad\qquad\qquad\qquad\qquad \sqrt{9} \stackrel{?}{=} -3$

$2 = 2 \checkmark \qquad$ True $\qquad\qquad\qquad\qquad 3 = -3 \qquad$ False

The number 0 is a solution, but the proposed solution -5 is extraneous. The solution set is $\{0\}$.

NOW TRY

**NOW TRY
EXERCISE 4**

Solve

$\sqrt{x^2 - 3x + 18} = x + 3$.

EXAMPLE 4 Using the Power Rule (Squaring a Binomial)

Solve $\sqrt{x^2 - 4x + 9} = x - 1$.

Squaring gives $(x - 1)^2 = x^2 - 2(x)(1) + 1^2$ on the right.

$$\left(\sqrt{x^2 - 4x + 9}\right)^2 = (x - 1)^2 \qquad \text{Remember the middle term.}$$

$$x^2 - 4x + 9 = x^2 - 2x + 1$$

$\qquad\qquad\qquad\qquad\qquad$ Twice the product of x and -1

$$-2x = -8 \qquad \text{Subtract } x^2 \text{ and } 9. \text{ Add } 2x.$$

$$x = 4 \qquad \text{Divide by } -2.$$

CHECK $\qquad\qquad \sqrt{x^2 - 4x + 9} = x - 1 \qquad$ Original equation

$\sqrt{4^2 - 4 \cdot 4 + 9} \stackrel{?}{=} 4 - 1 \qquad$ Let $x = 4$.

$3 = 3 \checkmark \qquad$ True

The solution set is $\{4\}$.

NOW TRY

EXAMPLE 5 Using the Power Rule (Squaring Twice)

Solve $\sqrt{5x + 6} + \sqrt{3x + 4} = 2$.

Isolate one radical on one side of the equation by subtracting $\sqrt{3x + 4}$ from each side.

$$\sqrt{5x + 6} = 2 - \sqrt{3x + 4} \qquad \text{Subtract } \sqrt{3x + 4}.$$

$$\left(\sqrt{5x + 6}\right)^2 = \left(2 - \sqrt{3x + 4}\right)^2 \qquad \text{Square each side.}$$

$$5x + 6 = 4 - 4\sqrt{3x + 4} + (3x + 4) \qquad \text{Be careful here.}$$

Remember the middle term. $\qquad$ Twice the product of 2 and $-\sqrt{3x + 4}$

The equation still contains a radical, so isolate the radical term on the right and square both sides again.

NOW TRY
EXERCISE 5
Solve
$\sqrt{3x + 1} - \sqrt{x + 4} = 1$.

$5x + 6 = 4 - 4\sqrt{3x + 4} + 3x + 4$	Result after squaring
$5x + 6 = 8 - 4\sqrt{3x + 4} + 3x$	Combine like terms.
$2x - 2 = -4\sqrt{3x + 4}$	Subtract 8 and $3x$.
$x - 1 = -2\sqrt{3x + 4}$	Divide by 2.
$(x - 1)^2 = \left(-2\sqrt{3x + 4}\right)^2$	Square each side again.
$x^2 - 2x + 1 = (-2)^2\left(\sqrt{3x + 4}\right)^2$	On the right, $(ab)^2 = a^2b^2$.
$x^2 - 2x + 1 = 4(3x + 4)$	Apply the exponents.
$x^2 - 2x + 1 = 12x + 16$	Distributive property
$x^2 - 14x - 15 = 0$	Standard form
$(x - 15)(x + 1) = 0$	Factor.
$x - 15 = 0$ or $x + 1 = 0$	Zero-factor property
$x = 15$ or $x = -1$	Solve each equation.

> Divide *each* term by 2.

CHECK   First check the proposed solution $x = 15$.

$\sqrt{5x + 6} + \sqrt{3x + 4} = 2$	Original equation
$\sqrt{5(15) + 6} + \sqrt{3(15) + 4} \stackrel{?}{=} 2$	Let $x = 15$.
$\sqrt{81} + \sqrt{49} \stackrel{?}{=} 2$	Simplify.
$9 + 7 \stackrel{?}{=} 2$	Take square roots.
$16 = 2$	**False**

Now check the proposed solution -1.

$\sqrt{5x + 6} + \sqrt{3x + 4} = 2$	Original equation
$\sqrt{5(-1) + 6} + \sqrt{3(-1) + 4} \stackrel{?}{=} 2$	Let $x = -1$.
$\sqrt{1} + \sqrt{1} \stackrel{?}{=} 2$	Evaluate the radicands.
$1 + 1 \stackrel{?}{=} 2$	Take square roots.
$2 = 2$ ✓	True

The proposed solution -1 is valid, but 15 is extraneous and must be rejected. Thus, the solution set is $\{-1\}$.   **NOW TRY** ↩

OBJECTIVE 2 Solve radical equations with indexes greater than 2.

EXAMPLE 6 Using the Power Rule for a Power Greater Than 2

Solve $\sqrt[3]{z + 5} = \sqrt[3]{2z - 6}$.

$\left(\sqrt[3]{z + 5}\right)^3 = \left(\sqrt[3]{2z - 6}\right)^3$	Cube each side.
$z + 5 = 2z - 6$	Apply the exponents.
$11 = z$	Subtract z. Add 6.

NOW TRY ANSWER
5. $\{5\}$

NOW TRY EXERCISE 6

Solve $\sqrt[3]{4x - 5} = \sqrt[3]{3x + 2}$.

NOW TRY EXERCISE 7

Solve the formula for a.

$$x = \sqrt{\frac{y + 2}{a}}$$

NOW TRY ANSWERS

6. $\{7\}$ 7. $a = \dfrac{y + 2}{x^2}$

CHECK

$$\sqrt[3]{z + 5} = \sqrt[3]{2z - 6} \qquad \text{Original equation}$$

$$\sqrt[3]{11 + 5} \overset{?}{=} \sqrt[3]{2 \cdot 11 - 6} \qquad \text{Let } z = 11.$$

$$\sqrt[3]{16} = \sqrt[3]{16} \ \checkmark \qquad \text{True}$$

The solution set is $\{11\}$.

NOW TRY

OBJECTIVE 3 Use the power rule to solve a formula for a specified variable.

EXAMPLE 7 Solving a Formula from Electronics for a Variable

An important property of a radio-frequency transmission line is its **characteristic impedance,** represented by Z and measured in ohms. If L and C are the inductance and capacitance, respectively, per unit of length of the line, then these quantities are related by the formula $Z = \sqrt{\dfrac{L}{C}}$. Solve this formula for C.

$$Z = \sqrt{\frac{L}{C}} \qquad \boxed{\text{Our goal is to isolate } C \text{ on one side of the equality symbol.}}$$

$$Z^2 = \left(\sqrt{\frac{L}{C}}\right)^2 \qquad \text{Square each side.}$$

$$Z^2 = \frac{L}{C} \qquad (\sqrt{a})^2 = a$$

$$CZ^2 = L \qquad \text{Multiply by } C.$$

$$C = \frac{L}{Z^2} \qquad \text{Divide by } Z^2.$$

NOW TRY

10.6 Exercises

FOR EXTRA HELP   ▶ MyMathLab®

▶ *Complete solution available in MyMathLab*

Concept Check Check each equation to see if the given value for x is a solution.

1. $\sqrt{3x + 18} - x = 0$
   (a) 6   (b) -3

2. $\sqrt{3x - 3} - x + 1 = 0$
   (a) 1   (b) 4

3. $\sqrt{x + 2} - \sqrt{9x - 2} = -2\sqrt{x - 1}$
   (a) 2   (b) 7

4. $\sqrt{8x - 3} - 2x = 0$
   (a) $\dfrac{3}{2}$ (b) $\dfrac{1}{2}$

5. *Concept Check* Is 9 a solution of the following equation?

   $$\sqrt{x} = -3$$

   If not, can there be a solution of this equation?

6. *Concept Check* Before even attempting to solve

   $$\sqrt{3x + 18} = x,$$

   how can we be sure that the equation cannot have a negative solution?

Solve each equation. See Examples 1–4.

7. $\sqrt{x - 2} = 3$

8. $\sqrt{x + 1} = 7$

▶ 9. $\sqrt{6k - 1} = 1$

10. $\sqrt{7x - 3} = 6$

▶ 11. $\sqrt{4r + 3} + 1 = 0$

12. $\sqrt{5k - 3} + 2 = 0$

13. $\sqrt{3x + 1} - 4 = 0$

14. $\sqrt{5x + 1} - 11 = 0$

15. $4 - \sqrt{x - 2} = 0$

16. $9 - \sqrt{4x + 1} = 0$

17. $\sqrt{9x - 4} = \sqrt{8x + 1}$

18. $\sqrt{4x - 2} = \sqrt{3x + 5}$

19. $2\sqrt{x} = \sqrt{3x + 4}$

20. $2\sqrt{x} = \sqrt{5x - 16}$

21. $3\sqrt{x - 1} = 2\sqrt{2x + 2}$

22. $5\sqrt{4x + 1} = 3\sqrt{10x + 25}$

23. $x = \sqrt{x^2 + 4x - 20}$

24. $x = \sqrt{x^2 - 3x + 18}$

25. $x = \sqrt{x^2 + 3x + 9}$

26. $x = \sqrt{x^2 - 4x - 8}$

▶ 27. $\sqrt{9 - x} = x + 3$

28. $\sqrt{5 - x} = x + 1$

▶ 29. $\sqrt{k^2 + 2k + 9} = k + 3$

30. $\sqrt{x^2 - 3x + 3} = x - 1$

31. $\sqrt{x^2 + 12x - 4} = x - 4$

32. $\sqrt{x^2 - 15x + 15} = x - 5$

33. $\sqrt{r^2 + 9r + 15} - r - 4 = 0$

34. $\sqrt{m^2 + 3m + 12} - m - 2 = 0$

35. *Concept Check*   In solving the equation $\sqrt{3x + 4} = 8 - x$, a student wrote the following for her first step. **WHAT WENT WRONG?** Solve the given equation correctly.

$$3x + 4 = 64 + x^2$$

36. *Concept Check*   In solving the equation $\sqrt{5x + 6} - \sqrt{x + 3} = 3$, a student wrote the following for his first step. **WHAT WENT WRONG?** Solve the given equation correctly.

$$(5x + 6) + (x + 3) = 9$$

*Solve each equation. **See Examples 5 and 6.***

▶ 37. $\sqrt[3]{2x + 5} = \sqrt[3]{6x + 1}$

38. $\sqrt[3]{p + 5} = \sqrt[3]{2p - 4}$

39. $\sqrt[3]{x^2 + 5x + 1} = \sqrt[3]{x^2 + 4x}$

40. $\sqrt[3]{r^2 + 2r + 8} = \sqrt[3]{r^2 + 3r + 12}$

41. $\sqrt[3]{2m - 1} = \sqrt[3]{m + 13}$

42. $\sqrt[3]{2k - 11} = \sqrt[3]{5k + 1}$

43. $\sqrt[4]{x + 12} = \sqrt[4]{3x - 4}$

44. $\sqrt[4]{z + 11} = \sqrt[4]{2z + 6}$

45. $\sqrt[3]{x - 8} + 2 = 0$

46. $\sqrt[3]{r + 1} + 1 = 0$

47. $\sqrt[4]{2k - 5} + 4 = 0$

48. $\sqrt[4]{8z - 3} + 2 = 0$

49. $\sqrt{k + 2} - \sqrt{k - 3} = 1$

50. $\sqrt{r + 6} - \sqrt{r - 2} = 2$

▶ 51. $\sqrt{2r + 11} - \sqrt{5r + 1} = -1$

52. $\sqrt{3x - 2} - \sqrt{x + 3} = 1$

53. $\sqrt{3p + 4} - \sqrt{2p - 4} = 2$

54. $\sqrt{4x + 5} - \sqrt{2x + 2} = 1$

55. $\sqrt{3 - 3p} - 3 = \sqrt{3p + 2}$

56. $\sqrt{4x + 7} - 4 = \sqrt{4x - 1}$

57. $\sqrt{2\sqrt{x + 11}} = \sqrt{4x + 2}$

58. $\sqrt{1 + \sqrt{24 - 10x}} = \sqrt{3x + 5}$

Extending Skills   *For each equation, write the expressions with rational exponents as radical expressions, and then solve, using the procedures explained in this section.*

59. $(2x - 9)^{1/2} = 2 + (x - 8)^{1/2}$

60. $(3w + 7)^{1/2} = 1 + (w + 2)^{1/2}$

61. $(2w - 1)^{2/3} - w^{1/3} = 0$

62. $(x^2 - 2x)^{1/3} - x^{1/3} = 0$

*Solve each formula for the indicated variable. **See Example 7.** (Source: Cooke, N., and J. Orleans, Mathematics Essential to Electricity and Radio, McGraw-Hill.)*

63. $Z = \sqrt{\dfrac{L}{C}}$ for L

64. $r = \sqrt{\dfrac{\mathscr{A}}{\pi}}$ for $\mathscr{A}$

65. $V = \sqrt{\dfrac{2K}{m}}$ for K

66. $V = \sqrt{\dfrac{2K}{m}}$ for m

67. $r = \sqrt{\dfrac{Mm}{F}}$ for M

68. $r = \sqrt{\dfrac{Mm}{F}}$ for F

The formula

$$N = \frac{1}{2\pi}\sqrt{\frac{a}{r}}$$

is used to find the rotational rate N of a space station. Here, a is the acceleration and r represents the radius of the space station, in meters. To find the value of r that will make N simulate the effect of gravity on Earth, the equation must be solved for r, using the required value of N. (Source: Kastner, B., Space Mathematics, NASA.)

69. Solve the equation for r.

70. (a) Approximate the value of r so that $N = 0.063$ rotation per sec if $a = 9.8$ m per sec^2.

   (b) Approximate the value of r so that $N = 0.04$ rotation per sec if $a = 9.8$ m per sec^2.

10.7 Complex Numbers

OBJECTIVES

1. Simplify numbers of the form $\sqrt{-b}$, where $b > 0$.
2. Recognize subsets of the complex numbers.
3. Add and subtract complex numbers.
4. Multiply complex numbers.
5. Divide complex numbers.
6. Simplify powers of i.

VOCABULARY

☐ complex number
☐ real part
☐ imaginary part
☐ pure imaginary number
☐ nonreal complex number
☐ complex conjugates

OBJECTIVE 1 Simplify numbers of the form $\sqrt{-b}$, where $b > 0$.

The equation $x^2 + 1 = 0$ has no real number solution because any solution must be a number whose square is -1. In the set of real numbers, all squares are nonnegative numbers because the product of two positive numbers or two negative numbers is positive and $0^2 = 0$. To provide a solution of the equation

$$x^2 + 1 = 0,$$

we introduce a new number i.

Imaginary Unit i

The **imaginary unit i** is defined as follows.

$$i = \sqrt{-1}, \quad \text{and thus} \quad i^2 = -1$$

That is, i is the principal square root of -1.

We can use this definition to define any square root of a negative real number.

Meaning of $\sqrt{-b}$

For any positive real number b, $\qquad \sqrt{-b} = i\sqrt{b}.$

NOW TRY
EXERCISE 1
Write each number as a product of a real number and i.

(a) $\sqrt{-49}$ **(b)** $-\sqrt{-121}$

(c) $\sqrt{-3}$ **(d)** $\sqrt{-32}$

EXAMPLE 1 Simplifying Square Roots of Negative Numbers

Write each number as a product of a real number and i.

(a) $\sqrt{-100} = i\sqrt{100} = 10i$ **(b)** $-\sqrt{-36} = -i\sqrt{36} = -6i$

(c) $\sqrt{-2} = i\sqrt{2}$ **(d)** $\sqrt{-54} = i\sqrt{54} = i\sqrt{9 \cdot 6} = 3i\sqrt{6}$

 **NOW TRY**

> **⊘ CAUTION** It is easy to mistake $\sqrt{2}i$ for $\sqrt{2i}$, with the i under the radical. For this reason, we usually write $\sqrt{2}i$ as $i\sqrt{2}$, as in the definition of $\sqrt{-b}$.

When finding a product such as $\sqrt{-4} \cdot \sqrt{-9}$, we cannot use the product rule for radicals because it applies only to *nonnegative* radicands.

For this reason, we change $\sqrt{-b}$ to the form $i\sqrt{b}$ before performing any multiplications or divisions.

NOW TRY
EXERCISE 2
Multiply.

(a) $\sqrt{-4} \cdot \sqrt{-16}$

(b) $\sqrt{-5} \cdot \sqrt{-11}$

(c) $\sqrt{-3} \cdot \sqrt{-12}$

(d) $\sqrt{13} \cdot \sqrt{-2}$

EXAMPLE 2 Multiplying Square Roots of Negative Numbers

Multiply.

(a) $\sqrt{-4} \cdot \sqrt{-9}$

> First write all square roots in terms of i.

$= i\sqrt{4} \cdot i\sqrt{9}$ $\sqrt{-b} = i\sqrt{b}$

$= i \cdot 2 \cdot i \cdot 3$ Take square roots.

$= 6i^2$ Multiply.

$= 6(-1)$ Substitute -1 for i^2.

$= -6$

(b) $\sqrt{-3} \cdot \sqrt{-7}$

$= i\sqrt{3} \cdot i\sqrt{7}$ $\sqrt{-b} = i\sqrt{b}$

$= i^2\sqrt{3 \cdot 7}$ Product rule

$= (-1)\sqrt{21}$ Substitute -1 for i^2.

$= -\sqrt{21}$ $(-1)a = -a$

(c) $\sqrt{-2} \cdot \sqrt{-8}$ **(d)** $\sqrt{-5} \cdot \sqrt{6}$

$= i\sqrt{2} \cdot i\sqrt{8}$ $\sqrt{-b} = i\sqrt{b}$ $= i\sqrt{5} \cdot \sqrt{6}$

$= i^2\sqrt{2 \cdot 8}$ Product rule $= i\sqrt{30}$

$= (-1)\sqrt{16}$ $i^2 = -1$

$= -4$ Take the square root.  **NOW TRY**

> **⊘ CAUTION** Using the product rule for radicals *before* using the definition of $\sqrt{-b}$ gives an *incorrect* answer. **Example 2(a)** shows that
>
> $\sqrt{-4} \cdot \sqrt{-9} = -6,$ Correct (Example 2(a))
>
> but $\sqrt{-4(-9)} = \sqrt{36} = 6.$ Incorrect
>
> Thus $\sqrt{-4} \cdot \sqrt{-9} \neq \sqrt{-4(-9)}.$

NOW TRY ANSWERS
1. **(a)** $7i$ **(b)** $-11i$
   (c) $i\sqrt{3}$ **(d)** $4i\sqrt{2}$
2. **(a)** -8 **(b)** $-\sqrt{55}$
   (c) -6 **(d)** $i\sqrt{26}$

NOW TRY EXERCISE 3

Divide.

(a) $\dfrac{\sqrt{-72}}{\sqrt{-8}}$ (b) $\dfrac{\sqrt{-48}}{\sqrt{3}}$

EXAMPLE 3 Dividing Square Roots of Negative Numbers

Divide.

(a) $\dfrac{\sqrt{-75}}{\sqrt{-3}}$

$= \dfrac{i\sqrt{75}}{i\sqrt{3}}$ ⟵ First write all square roots in terms of i.

$= \sqrt{\dfrac{75}{3}}$ $\dfrac{i}{i} = 1$; Quotient rule

$= \sqrt{25}$ Divide.

$= 5$

(b) $\dfrac{\sqrt{-32}}{\sqrt{8}}$

$= \dfrac{i\sqrt{32}}{\sqrt{8}}$ $\sqrt{-32} = i\sqrt{32}$

$= i\sqrt{\dfrac{32}{8}}$ Quotient rule

$= i\sqrt{4}$ Divide.

$= 2i$ **NOW TRY**

OBJECTIVE 2 Recognize subsets of the complex numbers.

A new set of numbers, the *complex numbers,* are defined as follows.

> ### Complex Number
>
> If a and b are real numbers, then any number of the form $a + bi$ is a **complex number.** In the complex number $a + bi$, the number a is the **real part** and b is the **imaginary part.***

For a complex number $a + bi$, if $b = 0$, then $a + bi = a$, which is a real number.

Thus, the set of real numbers is a subset of the set of complex numbers.

If $a = 0$ and $b \neq 0$, the complex number is a **pure imaginary number.** For example, $3i$ is a pure imaginary number. A number such as $7 + 2i$ is a **nonreal complex number.**

A complex number written in the form $a + bi$ is in **standard form.** In this section, most answers will be given in standard form, but if a or b is 0, we consider answers such as a or bi to be in standard form.

The relationships among the various sets of numbers are shown in **FIGURE 13.**

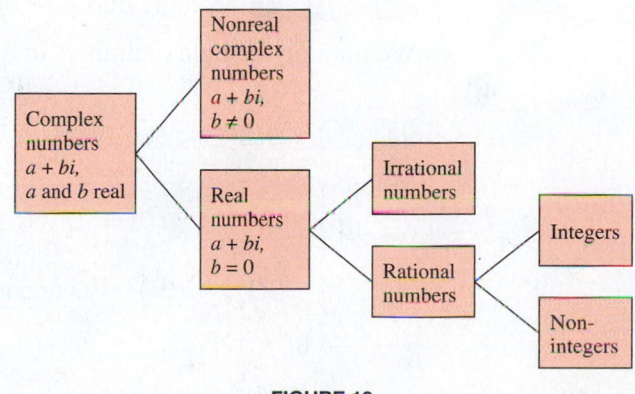

FIGURE 13

*Some texts define bi as the imaginary part of the complex number $a + bi$.

OBJECTIVE 3 Add and subtract complex numbers.

The commutative, associative, and distributive properties for real numbers are also valid for complex numbers. ***Thus, to add complex numbers, we add their real parts and add their imaginary parts.***

NOW TRY EXERCISE 4

Add.

(a) $(-3 + 2i) + (4 + 7i)$

(b) $(5 - i) + (-3 + 3i)$
$\quad + (6 - 4i)$

EXAMPLE 4 Adding Complex Numbers

Add.

(a) $(2 + 3i) + (6 + 4i)$

$\quad = (2 + 6) + (3 + 4)i$ Commutative, associative, and distributive properties

$\quad = 8 + 7i$ Add real parts. Add imaginary parts.

(b) $(4 + 2i) + (3 - i) + (-6 + 3i)$

$\quad = [4 + 3 + (-6)] + [2 + (-1) + 3]i$ Associative property

$\quad = 1 + 4i$ Add real parts.
Add imaginary parts.   **NOW TRY**

To subtract complex numbers, we subtract their real parts and subtract their imaginary parts.

NOW TRY EXERCISE 5

Subtract.

(a) $(7 + 10i) - (3 + 5i)$

(b) $(5 - 2i) - (9 - 7i)$

(c) $(-1 + 12i) - (-1 - i)$

EXAMPLE 5 Subtracting Complex Numbers

Subtract.

(a) $(6 + 5i) - (3 + 2i)$

$\quad = (6 - 3) + (5 - 2)i$ Properties of real numbers

$\quad = 3 + 3i$ Subtract real parts. Subtract imaginary parts.

(b) $(7 - 3i) - (8 - 6i)$

$\quad = (7 - 8) + [-3 - (-6)]i$

$\quad = -1 + 3i$

(c) $(-9 + 4i) - (-9 + 8i)$

$\quad = (-9 + 9) + (4 - 8)i$ Be careful.

$\quad = 0 - 4i$

$\quad = -4i$ **NOW TRY**

OBJECTIVE 4 Multiply complex numbers.

We multiply complex numbers in the same way that we multiply polynomials.

EXAMPLE 6 Multiplying Complex Numbers

Multiply.

(a) $4i(2 + 3i)$

$\quad = 4i(2) + 4i(3i)$ Distributive property

$\quad = 8i + 12i^2$ Multiply.

$\quad = 8i + 12(-1)$ Substitute -1 for i^2.

$\quad = -12 + 8i$ Standard form

NOW TRY ANSWERS

4. (a) $1 + 9i$ (b) $8 - 2i$
5. (a) $4 + 5i$ (b) $-4 + 5i$
  (c) $13i$

**NOW TRY
EXERCISE 6**

Multiply.

(a) $8i(3 - 5i)$

(b) $(7 - 2i)(4 + 3i)$

(b) $(3 + 5i)(4 - 2i)$

$$= \underbrace{3(4)}_{\text{First}} + \underbrace{3(-2i)}_{\text{Outer}} + \underbrace{5i(4)}_{\text{Inner}} + \underbrace{5i(-2i)}_{\text{Last}} \qquad \text{Use the FOIL method.}$$

$= 12 - 6i + 20i - 10i^2 \qquad$ Multiply.

$= 12 + 14i - 10(-1) \qquad$ Add imaginary parts. Substitute -1 for i^2.

$= 12 + 14i + 10 \qquad$ Multiply.

$= 22 + 14i \qquad$ Add real parts.

(c) $(2 + 3i)(1 - 5i)$

$= 2(1) + 2(-5i) + 3i(1) + 3i(-5i) \qquad$ FOIL method

$= 2 - 10i + 3i - 15i^2 \qquad$ Multiply.

$= 2 - 7i - 15(-1) \qquad$ Use parentheses around -1 to avoid errors. $\qquad i^2 = -1$

$= 2 - 7i + 15 \qquad$ Multiply.

$= 17 - 7i \qquad$ Add real parts. **NOW TRY**

The two complex numbers $a + bi$ and $a - bi$ are **complex conjugates,** or simply *conjugates*, of each other. **The product of a complex number and its conjugate is always a real number,** as shown here.

$$(a + bi)(a - bi)$$

$= a^2 - abi + abi - b^2i^2 \qquad$ FOIL method

$= a^2 - b^2(-1) \qquad$ Combine like terms; $i^2 = -1$

$= a^2 + b^2 \qquad$ The product eliminates i.

For example, $(3 + 7i)(3 - 7i) = 3^2 + 7^2 = 9 + 49 = 58.$

OBJECTIVE 5 Divide complex numbers.

EXAMPLE 7 Dividing Complex Numbers

Find each quotient.

(a) $\dfrac{8 + 9i}{5 + 2i}$

$= \dfrac{(8 + 9i)(5 - 2i)}{(5 + 2i)(5 - 2i)} \qquad$ Multiply numerator and denominator by $5 - 2i$, the conjugate of the denominator.

$= \dfrac{40 - 16i + 45i - 18i^2}{5^2 + 2^2} \qquad$ In the denominator, $(a + bi)(a - bi) = a^2 + b^2$.

$= \dfrac{40 + 29i - 18(-1)}{25 + 4} \qquad$ In the numerator, add imaginary parts; $i^2 = -1$

$= \dfrac{58 + 29i}{29} \qquad$ Multiply. Add real parts. Add in the denominator.

$= \dfrac{29(2 + i)}{29} \qquad$ Factor the numerator.

Factor first. Then divide out the common factor.

$= 2 + i \qquad$ Lowest terms

NOW TRY ANSWERS

6. (a) $40 + 24i$ **(b)** $34 + 13i$

**NOW TRY
EXERCISE 7**

Find each quotient.

(a) $\dfrac{4 + 2i}{1 + 3i}$ (b) $\dfrac{5 - 4i}{i}$

(b) $\dfrac{1 + i}{i}$

$\quad = \dfrac{(1 + i)(-i)}{i(-i)}$ Multiply numerator and denominator by $-i$, the conjugate of i.

$\quad = \dfrac{-i - i^2}{-i^2}$ Use the distributive property in the numerator. Multiply in the denominator.

$\quad = \dfrac{-i - (-1)}{-(-1)}$ Substitute -1 for i^2.

> Use parentheses to avoid errors.

$\quad = \dfrac{-i + 1}{1}$

$\quad = 1 - i$ $\dfrac{a}{1} = a$

 **NOW TRY**

OBJECTIVE 6   Simplify powers of i.

Because i^2 is defined to be -1, we can find greater powers of i as shown in the following examples.

$$i^3 = i \cdot i^2 = i(-1) = -i \qquad i^6 = i^2 \cdot i^4 = (-1) \cdot 1 = -1$$
$$i^4 = i^2 \cdot i^2 = (-1)(-1) = 1 \qquad i^7 = i^3 \cdot i^4 = (-i) \cdot 1 = -i$$
$$i^5 = i \cdot i^4 = i \cdot 1 = i \qquad i^8 = i^4 \cdot i^4 = 1 \cdot 1 = 1$$

Notice that the powers of i rotate through the four numbers $i, -1, -i,$ and 1. Greater powers of i can be simplified using the fact that $i^4 = 1$.

**NOW TRY
EXERCISE 8**

Find each power of i.

(a) i^{16} (b) i^{21}
(c) i^{-6} (d) i^{-13}

NOW TRY ANSWERS
7. (a) $1 - i$ (b) $-4 - 5i$
8. (a) 1 (b) i (c) -1 (d) $-i$

EXAMPLE 8   Simplifying Powers of i

Find each power of i.

(a) $i^{12} = (i^4)^3 = 1^3 = 1$ $\quad i^4 = 1$

(b) $i^{39} = i^{36} \cdot i^3 = (i^4)^9 \cdot i^3 = 1^9 \cdot (-i) = -i$

(c) $i^{-2} = \dfrac{1}{i^2} = \dfrac{1}{-1} = -1$

(d) $i^{-1} = \dfrac{1}{i} = \dfrac{1(-i)}{i(-i)} = \dfrac{-i}{-i^2} = \dfrac{-i}{-(-1)} = \dfrac{-i}{1} = -i$

 **NOW TRY**

10.7 Exercises

FOR EXTRA HELP

 MyMathLab®

 *Complete solution available in MyMathLab*

Concept Check   *List all of the following sets to which each number belongs. A number may belong to more than one set.*

real numbers   pure imaginary numbers   nonreal complex numbers   complex numbers

1. $3 + 5i$ **2.** $-7i$ **3.** $\sqrt{2}$

4. $\dfrac{13}{3}$ **5.** $\sqrt{-49}$ **6.** $-\sqrt{-8}$

Concept Check   *Decide whether each expression is equal to $1, -1, i,$ or $-i$.*

7. $\sqrt{-1}$ **8.** $-\sqrt{-1}$ **9.** i^2 **10.** $-i^2$ **11.** $\dfrac{1}{i}$ **12.** $(-i)^2$

*Write each number as a product of a real number and i. Simplify all radical expressions. **See Example 1.***

▶ 13. $\sqrt{-169}$ **14.** $\sqrt{-225}$ **15.** $-\sqrt{-144}$ **16.** $-\sqrt{-196}$

17. $\sqrt{-5}$ **18.** $\sqrt{-21}$ **19.** $\sqrt{-48}$ **20.** $\sqrt{-96}$

*Multiply or divide as indicated. **See Examples 2 and 3.***

▶ 21. $\sqrt{-7} \cdot \sqrt{-15}$ **22.** $\sqrt{-3} \cdot \sqrt{-19}$ **23.** $\sqrt{-4} \cdot \sqrt{-25}$ **24.** $\sqrt{-9} \cdot \sqrt{-81}$

25. $\sqrt{-3} \cdot \sqrt{11}$ **26.** $\sqrt{-10} \cdot \sqrt{2}$ **▶ 27.** $\dfrac{\sqrt{-300}}{\sqrt{-100}}$ **28.** $\dfrac{\sqrt{-40}}{\sqrt{-10}}$

29. $\dfrac{\sqrt{-75}}{\sqrt{3}}$ **30.** $\dfrac{\sqrt{-160}}{\sqrt{10}}$ **31.** $\dfrac{-\sqrt{-64}}{\sqrt{-16}}$ **32.** $\dfrac{-\sqrt{-100}}{\sqrt{-25}}$

*Add or subtract as indicated. Give answers in standard form. **See Examples 4 and 5.***

▶ 33. $(3 + 2i) + (-4 + 5i)$ **34.** $(7 + 15i) + (-11 + 14i)$

35. $(5 - i) + (-5 + i)$ **36.** $(-2 + 6i) + (2 - 6i)$

▶ 37. $(4 + i) - (-3 - 2i)$ **38.** $(9 + i) - (3 + 2i)$

39. $(-3 - 4i) - (-1 - 4i)$ **40.** $(-2 - 3i) - (-5 - 3i)$

41. $(-4 + 11i) + (-2 - 4i) + (7 + 6i)$ **42.** $(-1 + i) + (2 + 5i) + (3 + 2i)$

43. $[(7 + 3i) - (4 - 2i)] + (3 + i)$ **44.** $[(7 + 2i) + (-4 - i)] - (2 + 5i)$

Concept Check Fill in the blank with the correct response.

45. Because $(4 + 2i) - (3 + i) = 1 + i$, using the definition of subtraction we can check this to find that

$$(1 + i) + (3 + i) = \underline{\hspace{2cm}}.$$

46. Because $\dfrac{-5}{2 - i} = -2 - i$, using the definition of division we can check this to find that

$$(-2 - i)(2 - i) = \underline{\hspace{2cm}}.$$

*Multiply. **See Example 6.***

47. $(3i)(27i)$ **48.** $(5i)(125i)$ **49.** $(-8i)(-2i)$

50. $(-32i)(-2i)$ **▶ 51.** $5i(-6 + 2i)$ **52.** $3i(4 + 9i)$

53. $(4 + 3i)(1 - 2i)$ **54.** $(7 - 2i)(3 + i)$ **55.** $(4 + 5i)^2$

56. $(3 + 2i)^2$ **57.** $2i(-4 - i)^2$ **58.** $3i(-3 - i)^2$

59. $(12 + 3i)(12 - 3i)$ **60.** $(6 + 7i)(6 - 7i)$ **61.** $(4 + 9i)(4 - 9i)$

62. $(7 + 2i)(7 - 2i)$ **63.** $(1 + i)^2(1 - i)^2$ **64.** $(2 - i)^2(2 + i)^2$

Concept Check Answer each of the following.

65. Let a and b represent real numbers.

  (a) What is the conjugate of $a + bi$?

  (b) If we multiply $a + bi$ by its conjugate, we obtain $\underline{\hspace{1cm}} + \underline{\hspace{1cm}}$, which is always a real number.

66. By what complex number should we multiply the numerator and denominator of $\frac{2 + i\sqrt{2}}{2 - i\sqrt{2}}$ to write the quotient in standard form?

    A. $\sqrt{2}$ **B.** $i\sqrt{2}$ **C.** $2 + i\sqrt{2}$ **D.** $2 - i\sqrt{2}$

Find each quotient. See Example 7.

67. $\dfrac{2}{1 - i}$ **68.** $\dfrac{2}{1 + i}$ **69.** $\dfrac{8i}{2 + 2i}$ **70.** $\dfrac{-8i}{1 + i}$

▶ **71.** $\dfrac{-7 + 4i}{3 + 2i}$ **72.** $\dfrac{-38 - 8i}{7 + 3i}$ **73.** $\dfrac{2 - 3i}{2 + 3i}$ **74.** $\dfrac{-1 + 5i}{3 + 2i}$

75. $\dfrac{3 + i}{i}$ **76.** $\dfrac{5 - i}{i}$ **77.** $\dfrac{3 - i}{-i}$ **78.** $\dfrac{5 + i}{-i}$

Find each power of i. See Example 8.

▶ **79.** i^{18} **80.** i^{26} **81.** i^{89} **82.** i^{48} **83.** i^{38}

84. i^{102} **85.** i^{43} **86.** i^{83} **87.** i^{-5} **88.** i^{-17}

Ohm's law for the current I in a circuit with voltage E, resistance R, capacitive reactance X_c, and inductive reactance X_L is

$$I = \frac{E}{R + (X_L - X_c)i}.$$

Use this law to work each exercise.

89. Find I if $E = 2 + 3i$, $R = 5$, $X_L = 4$, and $X_c = 3$.

90. Find E if $I = 1 - i$, $R = 2$, $X_L = 3$, and $X_c = 1$.

Chapter 10	Summary

Key Terms

10.1	fourth root	circle	**10.7**
square root	index (order)	center	complex number
positive (principal)	perfect cube	radius	real part
square root	perfect fourth		imaginary part
negative square root	power	**10.5**	pure imaginary
radicand	square root function	conjugates	number
radical	cube root function		nonreal complex
radical expression		**10.6**	number
perfect square	**10.3**	radical equation	complex conjugates
irrational number	hypotenuse	proposed solution	
cube root	legs (of a right triangle)	extraneous solution	

New Symbols

$\sqrt{}$ radical symbol

$\sqrt[n]{a}$ radical; principal nth root of a

$\pm$ "positive or negative" or "plus or minus"

$\approx$ is approximately equal to

$a^{1/n}$ a to the power $\frac{1}{n}$

$a^{m/n}$ a to the power $\frac{m}{n}$

i imaginary unit

Test Your Word Power

See how well you have learned the vocabulary in this chapter.

1. A **radicand** is
   A. the index of a radical
   B. the number or expression under the radical symbol
   C. the positive root of a number
   D. the radical symbol.

2. The **Pythagorean theorem** states that, in a right triangle,
   A. the sum of the measures of the angles is 180°
   B. the sum of the lengths of the two shorter sides equals the length of the longest side
   C. the longest side is opposite the right angle
   D. the square of the length of the longest side equals the sum of the squares of the lengths of the two shorter sides.

3. A **hypotenuse** is
   A. either of the two shorter sides of a triangle
   B. the shortest side of a triangle
   C. the side opposite the right angle in a triangle
   D. the longest side in any triangle.

4. **Rationalizing the denominator** is the process of
   A. eliminating fractions from a radical expression
   B. changing the denominator of a fraction from a radical to a rational number
   C. clearing a radical expression of radicals
   D. multiplying radical expressions.

5. An **extraneous solution** is a value
   A. that does not satisfy the original equation
   B. that makes an equation true
   C. that makes an expression equal 0
   D. that checks in the original equation.

6. A **complex number** is
   A. a real number that includes a complex fraction
   B. a zero multiple of i
   C. a number of the form $a + bi$, where a and b are real numbers
   D. the square root of -1.

ANSWERS

1. B; *Example:* In $\sqrt{3xy}$, $3xy$ is the radicand.   2. D; *Example:* In a right triangle where $a = 6$, $b = 8$, and $c = 10$, $6^2 + 8^2 = 10^2$.
3. C; *Example:* In a right triangle where the sides measure 9, 12, and 15 units, the hypotenuse is the side opposite the right angle, with measure 15 units.   4. B; *Example:* To rationalize the denominator of $\frac{5}{\sqrt{3}+1}$, multiply both the numerator and denominator by $\sqrt{3} - 1$ to obtain $\frac{5(\sqrt{3}-1)}{2}$.
5. A; *Example:* The proposed solution 2 is extraneous in $\sqrt{5x - 1} + 3 = 0$.   6. C; *Examples:* -5 (or $-5 + 0i$), $7i$ (or $0 + 7i$), $\sqrt{2} - 4i$

Quick Review

CONCEPTS

EXAMPLES

10.1 Radical Expressions and Graphs

$\sqrt[n]{a} = b$ means $b^n = a$.

$\sqrt[n]{a}$ is the **principal nth root** of a.

$\sqrt[n]{a^n} = |a|$ if n is even. $\sqrt[n]{a^n} = a$ if n is odd.

The two square roots of 64 are $\sqrt{64} = 8$ (the principal square root) and $-\sqrt{64} = -8$.

$$\sqrt[4]{(-2)^4} = |-2| = 2 \qquad \sqrt[3]{-27} = -3$$

Functions Defined by Radical Expressions
The square root function

$$f(x) = \sqrt{x}$$

and the cube root function

$$f(x) = \sqrt[3]{x}$$

are two important functions defined by radical expressions.

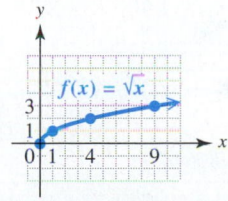
Square root function

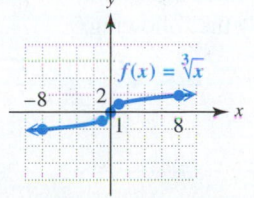
Cube root function

CONCEPTS	EXAMPLES

10.2 Rational Exponents

$a^{1/n} = \sqrt[n]{a}$ whenever $\sqrt[n]{a}$ exists.

If m and n are positive integers with $\frac{m}{n}$ in lowest terms, then $a^{m/n} = (a^{1/n})^m$, provided that $a^{1/n}$ is a real number.

All of the usual definitions and rules for exponents are valid for rational exponents.

Apply the rules for rational exponents.

$$81^{1/2} = \sqrt{81} = 9 \qquad -64^{1/3} = -\sqrt[3]{64} = -4$$

$$8^{5/3} = (8^{1/3})^5 = 2^5 = 32 \qquad (y^{2/5})^{10} = y^4$$

Write with positive exponents.

$$5^{-1/2} \cdot 5^{1/4} = 5^{-1/2+1/4} \qquad\qquad \frac{x^{-1/3}}{x^{-1/2}} = x^{-1/3-(-1/2)}$$

$$= 5^{-1/4} \qquad\qquad\qquad = x^{-1/3+1/2}$$

$$= \frac{1}{5^{1/4}} \qquad\qquad\qquad = x^{1/6}, \quad x > 0$$

10.3 Simplifying Radicals, the Distance Formula, and Circles

Product and Quotient Rules for Radicals

If $\sqrt[n]{a}$ and $\sqrt[n]{b}$ are real numbers and n is a natural number, then the following hold.

$$\sqrt[n]{a} \cdot \sqrt[n]{b} = \sqrt[n]{ab} \quad \text{and} \quad \sqrt[n]{\frac{a}{b}} = \frac{\sqrt[n]{a}}{\sqrt[n]{b}} \quad (\text{where } b \neq 0)$$

Conditions for a Simplified Radical

1. The radicand has no factor raised to a power greater than or equal to the index.

2. The radicand has no fractions.

3. No denominator contains a radical.

4. Exponents in the radicand and the index of the radical have greatest common factor 1.

Simplify.

$$\sqrt{3} \cdot \sqrt{7} = \sqrt{21} \qquad \sqrt[5]{x^3y} \cdot \sqrt[5]{xy^2} = \sqrt[5]{x^4y^3}$$

$$\frac{\sqrt{x^5}}{\sqrt{x^4}} = \sqrt{\frac{x^5}{x^4}} = \sqrt{x}, \quad x > 0$$

$$\sqrt{18} = \sqrt{9 \cdot 2} = 3\sqrt{2}$$

$$\sqrt[3]{54x^5y^3} = \sqrt[3]{27x^3y^3 \cdot 2x^2} = 3xy\sqrt[3]{2x^2}$$

$$\sqrt{\frac{7}{4}} = \frac{\sqrt{7}}{\sqrt{4}} = \frac{\sqrt{7}}{2}$$

$$\sqrt[9]{x^3} = x^{3/9} = x^{1/3}, \quad \text{or} \quad \sqrt[3]{x}$$

Pythagorean Theorem

If a and b are the lengths of the shorter sides of a right triangle and c is the length of the longest side, then the following holds.

$$a^2 + b^2 = c^2$$

Find b for the triangle in the figure.

$$10^2 + b^2 = \left(2\sqrt{61}\right)^2 \qquad \text{Let } a = 10, c = 2\sqrt{61}.$$

$$b^2 = 4(61) - 100 \qquad \text{Square and then subtract 100.}$$

$$b^2 = 144 \qquad\qquad \text{Simplify.}$$

$$b = 12 \qquad\qquad 12^2 = 144, \text{ and } 12 > 0.$$

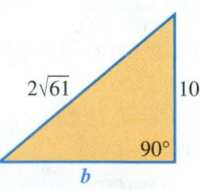

Distance Formula

The distance d between the points (x_1, y_1) and (x_2, y_2) is given by the following.

$$d = \sqrt{(x_2 - x_1)^2 + (y_2 - y_1)^2}$$

Find the distance between $(3, -2)$ and $(-1, 1)$.

$$\sqrt{(-1-3)^2 + [1-(-2)]^2} \qquad \text{Substitute.}$$

$$= \sqrt{(-4)^2 + 3^2} \qquad\qquad \text{Subtract.}$$

$$= \sqrt{16 + 9} \qquad\qquad\quad \text{Square.}$$

$$= \sqrt{25} \qquad\qquad\qquad \text{Add.}$$

$$= 5 \qquad\qquad\qquad\quad \text{Take the square root.}$$

CONCEPTS	EXAMPLES

Circles

The circle with radius r and center at (h, k) has an equation of the form

$$(x - h)^2 + (y - k)^2 = r^2.$$

The circle with equation $(x + 2)^2 + (y - 3)^2 = 25$, which can be written

$$[x - (-2)]^2 + (y - 3)^2 = 5^2,$$

has center $(-2, 3)$ and radius 5.

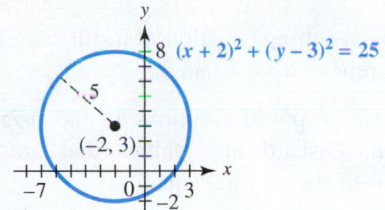

10.4 Adding and Subtracting Radical Expressions

Only radical expressions with the same index and the same radicand may be combined.

Simplify.

$$2\sqrt{28} - 3\sqrt{63} + 8\sqrt{112}$$

$$= 2\sqrt{4 \cdot 7} - 3\sqrt{9 \cdot 7} + 8\sqrt{16 \cdot 7}$$

$$= 2 \cdot 2\sqrt{7} - 3 \cdot 3\sqrt{7} + 8 \cdot 4\sqrt{7}$$

$$= 4\sqrt{7} - 9\sqrt{7} + 32\sqrt{7}$$

$$= (4 - 9 + 32)\sqrt{7}$$

$$= 27\sqrt{7}$$

$$\left.\begin{array}{l}\sqrt{15} + \sqrt{30} \\ \sqrt{3} + \sqrt[3]{9}\end{array}\right\} \text{ Cannot be simplified further}$$

10.5 Multiplying and Dividing Radical Expressions

Multiply binomial radical expressions by using the FOIL method. Special products from **Section 4.6** may apply.

Perform the operations and simplify.

$$(\sqrt{2} + \sqrt{7})(\sqrt{3} - \sqrt{6})$$
$$= \sqrt{6} - 2\sqrt{3} + \sqrt{21} - \sqrt{42} \qquad \sqrt{12} = 2\sqrt{3}$$

$$\begin{array}{l|l}(\sqrt{5} - \sqrt{10})(\sqrt{5} + \sqrt{10}) & (\sqrt{3} - \sqrt{2})^2 \\ = 5 - 10 & = 3 - 2\sqrt{3} \cdot \sqrt{2} + 2 \\ = -5 & = 5 - 2\sqrt{6}\end{array}$$

Rationalize the denominator by multiplying both the numerator and the denominator by the same expression, one that will yield a rational number in the final denominator.

$$\frac{\sqrt{7}}{\sqrt{5}} = \frac{\sqrt{7} \cdot \sqrt{5}}{\sqrt{5} \cdot \sqrt{5}} = \frac{\sqrt{35}}{5}$$

$$\frac{4}{\sqrt{5} - \sqrt{2}} = \frac{4(\sqrt{5} + \sqrt{2})}{(\sqrt{5} - \sqrt{2})(\sqrt{5} + \sqrt{2})}$$

$$= \frac{4(\sqrt{5} + \sqrt{2})}{5 - 2} = \frac{4(\sqrt{5} + \sqrt{2})}{3}$$

To write a radical quotient in lowest terms, factor the numerator and denominator and then divide out any common factor(s).

$$\frac{5 + 15\sqrt{6}}{10} = \frac{5(1 + 3\sqrt{6})}{5 \cdot 2} = \frac{1 + 3\sqrt{6}}{2}$$

CONCEPTS	EXAMPLES

10.6 Solving Equations with Radicals

Solving an Equation with Radicals

Step 1 Isolate one radical on one side of the equation.

Step 2 Raise both sides of the equation to a power that is the same as the index of the radical.

Step 3 Solve the resulting equation. If it still contains a radical, repeat Steps 1 and 2.

Step 4 Check all proposed solutions in the *original* equation. Discard any values that are not solutions of the original equation.

Solve $\sqrt{2x + 3} - x = 0$.

$$\sqrt{2x + 3} = x \qquad \text{Add } x.$$
$$\left(\sqrt{2x + 3}\right)^2 = x^2 \qquad \text{Square each side.}$$
$$2x + 3 = x^2 \qquad \text{Apply the exponents.}$$
$$x^2 - 2x - 3 = 0 \qquad \text{Standard form}$$
$$(x - 3)(x + 1) = 0 \qquad \text{Factor.}$$
$$x - 3 = 0 \quad \text{or} \quad x + 1 = 0 \qquad \text{Zero-factor property}$$
$$x = 3 \quad \text{or} \qquad x = -1 \qquad \text{Solve each equation.}$$

A check shows that 3 is a solution, but -1 is extraneous. The solution set is $\{3\}$.

10.7 Complex Numbers

$i = \sqrt{-1}$, where $i^2 = -1$.

For any positive number b, $\sqrt{-b} = i\sqrt{b}$.

To multiply radicals with negative radicands, first change each factor to the form $i\sqrt{b}$ and then multiply. The same procedure applies to quotients.

Express each in $a + bi$ form.

$$\sqrt{-25} = i\sqrt{25} = 5i$$

$$\sqrt{-3} \cdot \sqrt{-27}$$
$$= i\sqrt{3} \cdot i\sqrt{27} \qquad \sqrt{-b} = i\sqrt{b}$$
$$= i^2\sqrt{81}$$
$$= -1 \cdot 9 \qquad i^2 = -1$$
$$= -9$$

$$\frac{\sqrt{-18}}{\sqrt{-2}} = \frac{i\sqrt{18}}{i\sqrt{2}} = \sqrt{\frac{18}{2}} = \sqrt{9} = 3$$

Adding and Subtracting Complex Numbers

Add (or subtract) the real parts and add (or subtract) the imaginary parts.

Perform the operations.

$$\begin{array}{c|c}(5 + 3i) + (8 - 7i) & (5 + 3i) - (8 - 7i) \\ = 13 - 4i & = -3 + 10i\end{array}$$

Multiplying Complex Numbers

Multiply complex numbers using the FOIL method.

Multiply. $(2 + i)(5 - 3i)$

$$= 10 - 6i + 5i - 3i^2 \qquad \text{Use the FOIL method.}$$
$$= 10 - i - 3(-1) \qquad i^2 = -1$$
$$= 10 - i + 3 \qquad \text{Multiply.}$$
$$= 13 - i \qquad \text{Add real parts.}$$

Dividing Complex Numbers

Divide complex numbers by multiplying the numerator and the denominator by the conjugate of the denominator.

Divide. $\dfrac{20}{3 + i}$

$$= \frac{20(3 - i)}{(3 + i)(3 - i)} \qquad \begin{array}{l}\text{Multiply both numerator} \\ \text{and denominator by the} \\ \text{conjugate of the denominator.}\end{array}$$

$$= \frac{20(3 - i)}{9 - i^2} \qquad (a + b)(a - b) = a^2 - b^2$$

$$= \frac{20(3 - i)}{10} \qquad i^2 = -1 \text{ and } 9 - (-1) = 10.$$

$$= 2(3 - i) \qquad \text{Divide out the common factor, 10.}$$

$$= 6 - 2i \qquad \text{Distributive property}$$

Chapter 10 Review Exercises

10.1 *Find each root.*

1. $\sqrt{1764}$

2. $-\sqrt{289}$

3. $\sqrt[3]{216}$

4. $\sqrt[3]{-125}$

5. $-\sqrt[3]{27}$

6. $\sqrt[5]{-32}$

7. Under what conditions is $\sqrt[n]{a}$ not a real number?

Simplify each root so that no radicals appear. Assume that x represents any real number.

8. $\sqrt{x^2}$

9. $-\sqrt{x^2}$

10. $\sqrt[3]{x^3}$

Find a decimal approximation for each radical. Round answers to three decimal places.

11. $-\sqrt{47}$

12. $\sqrt[3]{-129}$

13. $\sqrt[4]{605}$

14. $\sqrt[4]{500}$

15. $-\sqrt[3]{500}$

16. $-\sqrt{28}$

Graph each function and give its domain and range.

17. $f(x) = \sqrt{x - 1}$

18. $f(x) = \sqrt[3]{x} + 4$

10.2

19. Fill in the blanks with the correct responses: One way to evaluate $8^{2/3}$ is to first find the _____ root of _____, which is _____. Then raise that result to the _____ power, to obtain an answer of _____. Therefore, $8^{2/3} =$ _____.

20. Which one of the following is a positive number?

  A. $(-27)^{2/3}$ **B.** $(-64)^{5/3}$ **C.** $(-100)^{1/2}$ **D.** $(-32)^{1/5}$

21. If a is a negative number and n is odd, then what must be true about m for $a^{m/n}$ to be
  (a) positive **(b)** negative?

22. If a is negative and n is even, is $a^{1/n}$ a real number?

Simplify. If the expression does not represent a real number, say so.

23. $49^{1/2}$

24. $-121^{1/2}$

25. $16^{5/4}$

26. $-8^{2/3}$

27. $-\left(\dfrac{36}{25}\right)^{3/2}$

28. $\left(-\dfrac{1}{8}\right)^{-5/3}$

29. $\left(\dfrac{81}{10,000}\right)^{-3/4}$

30. $(-16)^{3/4}$

Write each exponential as a radical.

31. $(m + 3n)^{1/2}$

32. $(3a + b)^{-5/3}$

Write each radical as an exponential.

33. $\sqrt{7^9}$

34. $\sqrt[5]{p^4}$

Simplify each expression. Write answers with only positive exponents. Assume that all variables represent positive real numbers.

35. $5^{1/4} \cdot 5^{7/4}$

36. $\dfrac{96^{2/3}}{96^{-1/3}}$

37. $\dfrac{(a^{1/3})^4}{a^{2/3}}$

38. $\dfrac{y^{-1/3} \cdot y^{5/6}}{y}$

39. $\left(\dfrac{z^{-1}x^{-3/5}}{2^{-2}z^{-1/2}x}\right)^{-1}$

40. $r^{-1/2}(r + r^{3/2})$

---

1. 42 **2.** -17

3. 6 **4.** -5

5. -3 **6.** -2

7. $\sqrt[n]{a}$ is not a real number if n is even and a is negative.

8. $|x|$

9. $-|x|$ **10.** x

11. -6.856 **12.** -5.053

13. 4.960 **14.** 4.729

15. -7.937 **16.** -5.292

17. 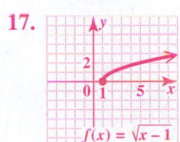 domain: $[1, \infty)$; range: $[0, \infty)$

$f(x) = \sqrt{x - 1}$

18. domain: $(-\infty, \infty)$; range: $(-\infty, \infty)$

$f(x) = \sqrt[3]{x} + 4$

19. cube (third); 8; 2; second; 4; 4

20. A

21. (a) m must be even.
  (b) m must be odd.

22. no

23. 7 **24.** -11

25. 32 **26.** -4

27. $-\dfrac{216}{125}$ **28.** -32

29. $\dfrac{1000}{27}$

30. It is not a real number.

31. $\sqrt{m + 3n}$

32. $\dfrac{1}{(\sqrt[3]{3a + b})^5}$, or $\dfrac{1}{\sqrt[3]{(3a + b)^5}}$

33. $7^{9/2}$ **34.** $p^{4/5}$

35. 5^2, or 25 **36.** 96

37. $a^{2/3}$ **38.** $\dfrac{1}{y^{1/2}}$

39. $\dfrac{z^{1/2}x^{8/5}}{4}$ **40.** $r^{1/2} + r$

41. $s^{1/2}$ **42.** $r^{3/2}$

43. $p^{1/2}$ **44.** $k^{9/4}$

45. $m^{13/3}$ **46.** $z^{1/12}$

47. $x^{1/8}$ **48.** $x^{1/15}$

49. $x^{1/36}$

50. The product rule for exponents applies only if the bases are the same.

51. $\sqrt{66}$ **52.** $\sqrt{5r}$

53. $\sqrt[3]{30}$ **54.** $\sqrt[4]{21}$

55. $2\sqrt{5}$ **56.** $5\sqrt{3}$

57. $-5\sqrt{5}$ **58.** $-3\sqrt[3]{4}$

59. $10y^3\sqrt{y}$ **60.** $4pq^2\sqrt[3]{p}$

61. $3a^2b\sqrt[3]{4a^2b^2}$

62. $2r^2t\sqrt[3]{79r^2t}$

63. $\dfrac{y\sqrt{y}}{12}$ **64.** $\dfrac{m^5}{3}$

65. $\dfrac{\sqrt[3]{r^2}}{2}$ **66.** $\dfrac{a^2\sqrt[4]{a}}{3}$

67. $\sqrt{15}$ **68.** $p\sqrt{p}$

69. $\sqrt[12]{2000}$ **70.** $\sqrt[10]{x^7}$

71. 10   **72.** $\sqrt{197}$

73. $x^2 + y^2 = 121$

74. $(x + 2)^2 + (y - 4)^2 = 9$

75. $(x + 1)^2 + (y + 3)^2 = 25$

76. $(x - 4)^2 + (y - 2)^2 = 36$

77.   **78.**

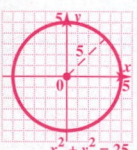   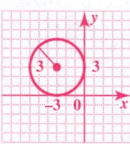

$x^2 + y^2 = 25$ $(x + 3)^2 + (y - 3)^2 = 9$

center: (0, 0)   center: (−3, 3)

79.

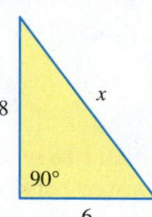

$(x - 2)^2 + (y + 5)^2 = 9$

center: (2, −5)

80. It is impossible for the sum of the squares of two real numbers to be negative.

81. $-11\sqrt{2}$ **82.** $23\sqrt{5}$

83. $7\sqrt{3y}$ **84.** $26m\sqrt{6m}$

85. $19\sqrt[3]{2}$ **86.** $-8\sqrt[4]{2}$

87. $1 - \sqrt{3}$ **88.** 2

89. $9 - 7\sqrt{2}$ **90.** $15 - 2\sqrt{26}$

91. 29

92. $2\sqrt[3]{2y^2} + 2\sqrt[3]{4y} - 3$

Write each radical as an exponential and then simplify. Leave answers in exponential form. Assume that all variables represent positive real numbers.

41. $\sqrt[8]{s^4}$

42. $\sqrt[6]{r^9}$

43. $\dfrac{\sqrt{p^5}}{p^2}$

44. $\sqrt[4]{k^3} \cdot \sqrt{k^3}$

45. $\sqrt[3]{m^5} \cdot \sqrt[3]{m^8}$

46. $\sqrt[4]{\sqrt[3]{z}}$

47. $\sqrt{\sqrt{\sqrt{x}}}$

48. $\sqrt[3]{\sqrt[5]{x}}$

49. $\sqrt{\sqrt[6]{\sqrt[3]{x}}}$

50. The product rule does not apply to $3^{1/4} \cdot 2^{1/5}$. Why?

10.3 *Simplify. Assume that all variables represent positive real numbers.*

51. $\sqrt{6} \cdot \sqrt{11}$ **52.** $\sqrt{5} \cdot \sqrt{r}$ **53.** $\sqrt[3]{6} \cdot \sqrt[3]{5}$ **54.** $\sqrt[4]{7} \cdot \sqrt[4]{3}$

55. $\sqrt{20}$ **56.** $\sqrt{75}$ **57.** $-\sqrt{125}$ **58.** $\sqrt[3]{-108}$

59. $\sqrt{100y^7}$ **60.** $\sqrt[3]{64p^4q^6}$ **61.** $\sqrt[3]{108a^8b^5}$ **62.** $\sqrt[3]{632r^8t^4}$

63. $\sqrt{\dfrac{y^3}{144}}$ **64.** $\sqrt[3]{\dfrac{m^{15}}{27}}$ **65.** $\sqrt[3]{\dfrac{r^2}{8}}$ **66.** $\sqrt[4]{\dfrac{a^9}{81}}$

Simplify each radical expression.

67. $\sqrt[6]{15^3}$ **68.** $\sqrt[4]{p^6}$ **69.** $\sqrt[3]{2} \cdot \sqrt[4]{5}$ **70.** $\sqrt{x} \cdot \sqrt[5]{x}$

71. Find the length x of the hypotenuse of the right triangle shown.

72. Find the distance between the points $(-4, 7)$ and $(10, 6)$.

Find the equation of a circle satisfying the given conditions.

73. Center $(0, 0)$, $r = 11$ **74.** Center $(-2, 4)$, $r = 3$

75. Center $(-1, -3)$, $r = 5$ **76.** Center $(4, 2)$, $r = 6$

Graph each circle. Identify the center.

77. $x^2 + y^2 = 25$ **78.** $(x + 3)^2 + (y - 3)^2 = 9$ **79.** $(x - 2)^2 + (y + 5)^2 = 9$

80. Why does the equation $x^2 + y^2 = -1$ have no points on its graph?

10.4 *Perform the indicated operations. Assume that all variables represent positive real numbers.*

81. $2\sqrt{8} - 3\sqrt{50}$ **82.** $8\sqrt{80} - 3\sqrt{45}$ **83.** $-\sqrt{27y} + 2\sqrt{75y}$

84. $2\sqrt{54m^3} + 5\sqrt{96m^3}$ **85.** $3\sqrt[3]{54} + 5\sqrt[3]{16}$ **86.** $-6\sqrt[4]{32} + \sqrt[4]{512}$

10.5 *Multiply, and then simplify.*

87. $\left(\sqrt{3} + 1\right)\left(\sqrt{3} - 2\right)$ **88.** $\left(\sqrt{7} + \sqrt{5}\right)\left(\sqrt{7} - \sqrt{5}\right)$

89. $\left(3\sqrt{2} + 1\right)\left(2\sqrt{2} - 3\right)$ **90.** $\left(\sqrt{13} - \sqrt{2}\right)^2$

91. $\left(\sqrt[3]{2} + 3\right)\left(\sqrt[3]{4} - 3\sqrt[3]{2} + 9\right)$ **92.** $\left(\sqrt[3]{4y} - 1\right)\left(\sqrt[3]{4y} + 3\right)$

93. $\dfrac{\sqrt{30}}{5}$ **94.** $-3\sqrt{6}$

95. $\dfrac{3\sqrt{7py}}{y}$ **96.** $\dfrac{\sqrt{22}}{4}$

97. $-\dfrac{\sqrt[3]{45}}{5}$ **98.** $\dfrac{3m\sqrt[3]{4n}}{n^2}$

99. $\dfrac{\sqrt{2}-\sqrt{7}}{-5}$ **100.** $\dfrac{5(\sqrt{6}+3)}{3}$

101. $\dfrac{1-\sqrt{5}}{4}$ **102.** $\dfrac{-6+\sqrt{3}}{2}$

103. $\{2\}$ **104.** $\{6\}$

105. $\{0,5\}$ **106.** $\{9\}$

107. $\{3\}$ **108.** $\{7\}$

109. $\left\{-\dfrac{1}{2}\right\}$ **110.** $\{14\}$

111. $\varnothing$ **112.** $\{7\}$

113. $H=\sqrt{L^2-W^2}$

114. 7.9 ft

115. $4i$ **116.** $10i\sqrt{2}$

117. $-10-2i$ **118.** $14+7i$

119. $-\sqrt{35}$ **120.** -45

121. 3 **122.** $5+i$

123. $32-24i$ **124.** $1-i$

125. $-i$ **126.** 1

127. -1 **128.** 1

Rationalize each denominator. Assume that all variables represent positive real numbers.

93. $\dfrac{\sqrt{6}}{\sqrt{5}}$ **94.** $\dfrac{-6\sqrt{3}}{\sqrt{2}}$ **95.** $\dfrac{3\sqrt{7p}}{\sqrt{y}}$ **96.** $\sqrt{\dfrac{11}{8}}$

97. $-\sqrt[3]{\dfrac{9}{25}}$ **98.** $\sqrt[3]{\dfrac{108m^3}{n^5}}$ **99.** $\dfrac{1}{\sqrt{2}+\sqrt{7}}$ **100.** $\dfrac{-5}{\sqrt{6}-3}$

Write each quotient in lowest terms.

101. $\dfrac{2-2\sqrt{5}}{8}$ **102.** $\dfrac{-18+\sqrt{27}}{6}$

10.6 *Solve each equation.*

103. $\sqrt{8x+9}=5$ **104.** $\sqrt{2x-3}-3=0$

105. $\sqrt{7x+1}=x+1$ **106.** $3\sqrt{x}=\sqrt{10x-9}$

107. $\sqrt{x^2+3x+7}=x+2$ **108.** $\sqrt{x+2}-\sqrt{x-3}=1$

109. $\sqrt[3]{5x-1}=\sqrt[3]{3x-2}$ **110.** $\sqrt[3]{1-2x}-\sqrt[3]{-x-13}=0$

111. $\sqrt[4]{x-1}+2=0$ **112.** $\sqrt[4]{x+7}=\sqrt[4]{2x}$

Carpenters stabilize wall frames with a diagonal brace, as shown in the figure. The length of the brace is given by

$$L=\sqrt{H^2+W^2}.$$

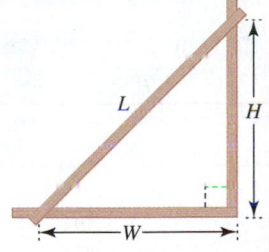

113. Solve this formula for H.

114. If the bottom of the brace is attached 9 ft from the corner and the brace is 12 ft long, how far up the corner post should it be nailed? Give the answer to the nearest tenth of a foot.

10.7 *Write each number as a product of a real number and i. Simplify.*

115. $\sqrt{-16}$ **116.** $\sqrt{-200}$

Perform the indicated operations. Give answers in standard form.

117. $(-2+5i)+(-8-7i)$ **118.** $(5+4i)-(-9-3i)$

119. $\sqrt{-5}\cdot\sqrt{-7}$ **120.** $\sqrt{-25}\cdot\sqrt{-81}$

121. $\dfrac{\sqrt{-72}}{\sqrt{-8}}$ **122.** $(2+3i)(1-i)$

123. $(6-2i)^2$ **124.** $\dfrac{3-i}{2+i}$

Find each power of i.

125. i^{11} **126.** i^{36} **127.** i^{-10} **128.** i^{-8}

Chapter 10 | Mixed Review Exercises

Simplify. Assume that all variables represent positive real numbers.

1. $-\sqrt[4]{256}$

2. $1000^{-2/3}$

3. $\dfrac{z^{-1/5} \cdot z^{3/10}}{z^{7/10}}$

4. $\sqrt[3]{54z^9t^8}$

5. $\sqrt{-49}$

6. $\dfrac{-1}{\sqrt{12}}$

7. $\sqrt[3]{\dfrac{12}{25}}$

8. i^{-1000}

9. $-5\sqrt{18} + 12\sqrt{72}$

10. $(4 - 9i) + (-1 + 2i)$

11. $\dfrac{\sqrt{50}}{\sqrt{-2}}$

12. $\dfrac{3 + \sqrt{54}}{6}$

13. $(3 + 2i)^2$

14. $8\sqrt[3]{x^3y^2} - 2x\sqrt[3]{y^2}$

Solve each equation.

15. $\sqrt{x + 4} = x - 2$

16. $\sqrt[3]{2x - 9} = \sqrt[3]{5x + 3}$

17. $\sqrt{6 + 2x} - 1 = \sqrt{7 - 2x}$

18. $\sqrt{7x + 11} - 5 = 0$

19. $\sqrt{6x + 2} - \sqrt{5x + 3} = 0$

20. $\sqrt{3 + 5x} - \sqrt{x + 11} = 0$

21. $\sqrt{11 + 2x} + 1 = \sqrt{5x + 1}$

22. $\sqrt{5x + 6} - \sqrt{x + 3} = 3$

Answers (left margin):

1. -4
2. $\dfrac{1}{100}$
3. $\dfrac{1}{z^{3/5}}$
4. $3z^3t^2\sqrt[3]{2t^2}$
5. $7i$
6. $-\dfrac{\sqrt{3}}{6}$
7. $\dfrac{\sqrt[3]{60}}{5}$
8. 1
9. $57\sqrt{2}$
10. $3 - 7i$
11. $-5i$
12. $\dfrac{1 + \sqrt{6}}{2}$
13. $5 + 12i$
14. $6x\sqrt[3]{y^2}$
15. $\{5\}$
16. $\{-4\}$
17. $\left\{\dfrac{3}{2}\right\}$
18. $\{2\}$
19. $\{1\}$
20. $\{2\}$
21. $\{7\}$
22. $\{6\}$

Chapter 10 | Test

FOR EXTRA HELP *Step-by-step test solutions are found on the Chapter Test Prep Videos available in* MyMathLab® *or on* YouTube.

▶ *View the complete solutions to all Chapter Test exercises in MyMathLab.*

Evaluate.

1. $-\sqrt{841}$

2. $\sqrt[3]{-512}$

3. $125^{1/3}$

Find a decimal approximation for each radical. Round answers to three decimal places.

4. $\sqrt{478}$

5. $\sqrt[3]{-832}$

6. Graph the function $f(x) = \sqrt{x + 6}$, and give the domain and range.

Simplify. Assume that all variables represent positive real numbers.

7. $\left(\dfrac{16}{25}\right)^{-3/2}$

8. $(-64)^{-4/3}$

9. $\dfrac{3^{2/5}x^{-1/4}y^{2/5}}{3^{-8/5}x^{7/4}y^{1/10}}$

10. $\left(\dfrac{x^{-4}y^{-6}}{x^{-2}y^3}\right)^{-2/3}$

11. $7^{3/4} \cdot 7^{-1/4}$

12. $\sqrt[3]{a^4} \cdot \sqrt[3]{a^7}$

13. Find the exact length of side b in the figure shown at the right.

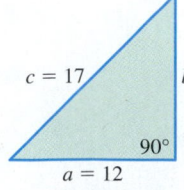

$c = 17$, b, $90°$, $a = 12$

14. Find the distance between the points $(-4, 2)$ and $(2, 10)$.

15. Write an equation of a circle with center $(-4, 6)$ and radius 5.

16. Graph the circle $(x - 2)^2 + (y + 3)^2 = 16$. Identify the center.

Simplify. Assume that all variables represent positive real numbers.

17. $\sqrt{54x^5y^6}$

18. $\sqrt[4]{32a^7b^{13}}$

Answers (left margin):

[10.1]
1. -29
2. -8

[10.2]
3. 5

[10.1]
4. 21.863
5. -9.405
6. 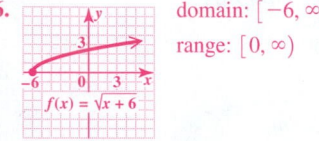 domain: $[-6, \infty)$; range: $[0, \infty)$

$f(x) = \sqrt{x + 6}$

[10.2]
7. $\dfrac{125}{64}$
8. $\dfrac{1}{256}$
9. $\dfrac{9y^{3/10}}{x^2}$
10. $x^{4/3}y^6$
11. $7^{1/2}$, or $\sqrt{7}$

[10.3]
12. $a^3\sqrt[3]{a^2}$, or $a^{11/3}$
13. $\sqrt{145}$
14. 10
15. $(x + 4)^2 + (y - 6)^2 = 25$

16.

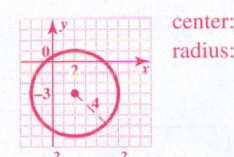

center: $(2, -3)$;
radius: 4

$(x - 2)^2 + (y + 3)^2 = 16$

17. $3x^2 y^3 \sqrt{6x}$ **18.** $2ab^3 \sqrt[4]{2a^3 b}$

19. $\sqrt[6]{200}$

[10.4]

20. $26\sqrt{5}$ **21.** $(2ts - 3t^2)\sqrt[3]{2s^2}$

[10.5]

22. $66 + \sqrt{5}$ **23.** $23 - 4\sqrt{15}$

24. $-\dfrac{\sqrt{10}}{4}$ **25.** $\dfrac{2\sqrt[3]{25}}{5}$

26. $-2(\sqrt{7} - \sqrt{5})$

27. $3 + \sqrt{6}$

[10.6]

28. (a) 59.8

  (b) $T = \dfrac{V_0{}^2 - V^2}{-V^2 k}$, or

  $T = \dfrac{V^2 - V_0{}^2}{V^2 k}$

29. $\{-1\}$ **30.** $\{3\}$

31. $\{-3\}$

[10.7]

32. $-5 - 8i$ **33.** $-2 + 16i$

34. $3 + 4i$ **35.** i

36. (a) true **(b)** true

  (c) false **(d)** true

---

19. $\sqrt{2} \cdot \sqrt[3]{5}$ (Express as a radical.) **20.** $3\sqrt{20} - 5\sqrt{80} + 4\sqrt{500}$

21. $\sqrt[3]{16t^3 s^5} - \sqrt[3]{54t^6 s^2}$ **22.** $(7\sqrt{5} + 4)(2\sqrt{5} - 1)$

23. $(\sqrt{3} - 2\sqrt{5})^2$ **24.** $\dfrac{-5}{\sqrt{40}}$ **25.** $\dfrac{2}{\sqrt[3]{5}}$ **26.** $\dfrac{-4}{\sqrt{7} + \sqrt{5}}$

27. Write $\dfrac{6 + \sqrt{24}}{2}$ in lowest terms.

28. The following formula from physics relates the velocity V of sound to the temperature T.

$$V = \frac{V_0}{\sqrt{1 - kT}}$$

  (a) Approximate V to the nearest tenth if $V_0 = 50$, $k = 0.01$, and $T = 30$.

  (b) Solve the formula for T.

Solve each equation.

29. $\sqrt[3]{5x} = \sqrt[3]{2x - 3}$ **30.** $\sqrt{x + 6} = 9 - 2x$

31. $\sqrt{x + 4} - \sqrt{1 - x} = -1$

In Exercises 32–34, perform the indicated operations. Give the answers in standard form.

32. $(-2 + 5i) - (3 + 6i) - 7i$ **33.** $(1 + 5i)(3 + i)$ **34.** $\dfrac{7 + i}{1 - i}$

35. Simplify i^{37}.

36. Answer *true* or *false* to each of the following.

  (a) $i^2 = -1$ **(b)** $i = \sqrt{-1}$ **(c)** $i = -1$ **(d)** $\sqrt{-3} = i\sqrt{3}$

Chapters R–10 — Cumulative Review Exercises

[1.3–1.5]

1. 1 **2.** $-\dfrac{14}{9}$

[2.3]

3. $\{-4\}$ **4.** $\{-12\}$ **5.** $\{6\}$

[2.8, 8.1]

6. $(-6, \infty)$

[2.7]

7. 36 nickels; **8.** $2\dfrac{2}{39}$ L
  64 quarters

[3.2, 7.1]

9.

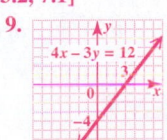

$4x - 3y = 12$

[3.3–3.5, 7.2]

10. $-\dfrac{3}{2}$; $y = -\dfrac{3}{2}x$

Evaluate each expression for $a = -3$, $b = 5$, and $c = -4$.

1. $|2a^2 - 3b + c|$ **2.** $\dfrac{(a + b)(a + c)}{3b - 6}$

Solve each equation or inequality.

3. $3(x + 2) - 4(2x + 3) = -3x + 2$ **4.** $\dfrac{1}{3}x + \dfrac{1}{4}(x + 8) = x + 7$

5. $0.04x + 0.06(100 - x) = 5.88$ **6.** $-5 - 3(x - 2) < 11 - 2(x + 2)$

Solve each problem.

7. A piggy bank has 100 coins, all of which are nickels and quarters. The total value of the money is \$17.80. How many of each denomination are there in the bank?

8. How many liters of pure alcohol must be mixed with 40 L of 18% alcohol to obtain a 22% alcohol solution?

9. Graph the equation $4x - 3y = 12$.

10. Find the slope of the line passing through the points $(-4, 6)$ and $(2, -3)$. Then find the equation of the line and write it in the form $y = mx + b$.

[4.4]
11. $-k^3 - 3k^2 - 8k - 9$

[4.5]
12. $8x^2 + 17x - 21$

[4.7]
13. $3y^3 - 3y^2 + 4y + 1 + \dfrac{-10}{2y + 1}$

[5.2–5.4]
14. $(2p - 3q)(p - q)$
15. $(3k^2 + 4)(k - 1)(k + 1)$
16. $(x + 8)(x^2 - 8x + 64)$

[5.5]
17. $\left\{-3, -\dfrac{5}{2}\right\}$ **18.** $\left\{-\dfrac{2}{5}, 1\right\}$

[6.1]   **[6.2]**
19. $x \neq \pm 3$ **20.** $\dfrac{y}{y + 5}$

[6.4]
21. $\dfrac{4x + 2y}{(x + y)(x - y)}$

[6.5]
22. $-\dfrac{9}{4}$ **23.** $\dfrac{1}{xy - 1}$

[6.6]   **[7.4, 7.5]**
24. $\varnothing$ **25.** $\{(7, -2)\}$

[7.6]
26. $\{(-1, 1, 1)\}$

[7.7]
27. 2-oz letter: $0.61;
   3-oz letter: $0.78

[8.2]
28. $(2, 3)$
29. $(-\infty, 2) \cup (3, \infty)$

[8.3]
30. $\left\{-\dfrac{10}{3}, 1\right\}$

31. $(-\infty, -2] \cup [7, \infty)$

[9.2]
32. -37

[10.2]   **[10.3]**
33. $\dfrac{1}{9}$ **34.** $2x\sqrt[3]{6x^2y^2}$

[10.4]   **[10.3]**
35. $7\sqrt{2}$ **36.** $\sqrt{29}$

[10.6]   **[10.7]**
37. $\{3, 4\}$ **38.** $4 + 2i$

Perform the indicated operations.

11. $(3k^3 - 5k^2 + 8k - 2) - (4k^3 + 11k + 7) + (2k^2 - 5k)$

12. $(8x - 7)(x + 3)$

13. $\dfrac{6y^4 - 3y^3 + 5y^2 + 6y - 9}{2y + 1}$

Factor each polynomial completely.

14. $2p^2 - 5pq + 3q^2$ **15.** $3k^4 + k^2 - 4$ **16.** $x^3 + 512$

Solve by the zero-factor property.

17. $2x^2 + 11x + 15 = 0$ **18.** $5x(x - 1) = 2(1 - x)$

19. For what values of the variable is the rational expression $\dfrac{4}{x^2 - 9}$ undefined?

Perform each operation and express the answer in lowest terms.

20. $\dfrac{y^2 + y - 12}{y^3 + 9y^2 + 20y} \div \dfrac{y^2 - 9}{y^3 + 3y^2}$ **21.** $\dfrac{1}{x + y} + \dfrac{3}{x - y}$

Simplify.

22. $\dfrac{\dfrac{-6}{x - 2}}{\dfrac{8}{3x - 6}}$ **23.** $\dfrac{x^{-1}}{y - x^{-1}}$

24. Solve the equation $\dfrac{x + 1}{x - 3} = \dfrac{4}{x - 3} + 6$.

Solve each system.

25. $3x - y = 23$
      $2x + 3y = 8$

26. $x + y + z = 1$
      $x - y - z = -3$
      $x + y - z = -1$

27. Several years ago, sending five 2-oz letters and three 3-oz letters by first-class mail would have cost $5.39. Sending three 2-oz letters and five 3-oz letters would have cost $5.73. What was the rate for one 2-oz letter and one 3-oz letter? (*Source:* U.S. Postal Service.)

Solve each equation or inequality.

28. $2x + 4 < 10$ and $3x - 1 > 5$ **29.** $2x + 4 > 10$ or $3x - 1 < 5$

30. $|6x + 7| = 13$ **31.** $|2p - 5| \geq 9$

32. If $f(x) = 3x - 7$, find $f(-10)$.

Write each expression in simplest form, using only positive exponents. Assume that all variables represent positive real numbers.

33. $27^{-2/3}$ **34.** $\sqrt[3]{16x^2y} \cdot \sqrt[3]{3x^3y}$ **35.** $\sqrt{50} + \sqrt{8}$

36. Find the distance between the points $(-4, 4)$ and $(-2, 9)$.

37. Solve the equation $\sqrt{3x - 8} = x - 2$.   **38.** Express $\dfrac{6 - 2i}{1 - i}$ in standard form.

11

Quadratic Equations, Inequalities, and Functions

Quadratic functions, one of the topics of this chapter, have graphs that are *parabolas.* Cross sections of telescopes, satellite dishes, and automobile headlights form parabolas, as do the cables that support suspension bridges.

11.1 Solving Quadratic Equations by the Square Root Property

VOCABULARY

☐ quadratic equation
☐ second-degree equation

OBJECTIVE 1   Review the zero-factor property.

Recall from **Section 5.5** that a *quadratic equation* is defined as follows.

> **Quadratic Equation**
>
> A **quadratic equation** (in x here) can be written in the form
>
> $$ax^2 + bx + c = 0,$$
>
> where a, b, and c are real numbers and $a \neq 0$. The given form is called **standard form.**
>
> *Examples:* $4x^2 + 4x - 5 = 0$ and $3x^2 = 4x - 8$
>
> Quadratic equations (The first equation is in standard form.)

A quadratic equation is a **second-degree equation,** that is, an equation with a squared variable term and no terms of greater degree.

In **Section 5.5** we used the zero-factor property to solve quadratic equations.

> **Zero-Factor Property**
>
> **If a and b are real numbers and if $ab = 0$, then $a = 0$ or $b = 0$.**
>
> That is, if the product of two numbers is 0, then at least one of the numbers must be 0. One number must be 0, but both *may* be 0.

NOW TRY EXERCISE 1

Solve each equation using the zero-factor property.

(a) $x^2 - x - 20 = 0$

(b) $x^2 = 36$

EXAMPLE 1   Solving Quadratic Equations Using the Zero-Factor Property

Solve each equation using the zero-factor property.

(a) $x^2 + 4x + 3 = 0$

        $(x + 3)(x + 1) = 0$ Factor.

$x + 3 = 0$ or $x + 1 = 0$ Zero-factor property

$x = -3$ or $x = -1$ Solve each equation.

The solution set is $\{-3, -1\}$.

(b) $x^2 = 9$

         $x^2 - 9 = 0$ Subtract 9.

    $(x + 3)(x - 3) = 0$ Factor.

$x + 3 = 0$ or $x - 3 = 0$ Zero-factor property

$x = -3$ or $x = 3$ Solve each equation.

The solution set is $\{-3, 3\}$.

NOW TRY

OBJECTIVE 2   Solve equations of the form $x^2 = k$, where $k > 0$.

In **Example 1(b),** we might also have solved $x^2 = 9$ by noticing that x must be a number whose square is 9. Thus, $x = \sqrt{9} = 3$ or $x = -\sqrt{9} = -3$.

This is generalized as the **square root property.**

NOW TRY ANSWERS

1. (a) $\{-4, 5\}$ (b) $\{-6, 6\}$

Square Root Property

If x and k are complex numbers and $x^2 = k$, then

$$x = \sqrt{k} \quad \text{or} \quad x = -\sqrt{k}.$$

The solution set is $\left\{-\sqrt{k}, \sqrt{k}\right\}$, which can be written $\left\{\pm\sqrt{k}\right\}$. (The symbol $\pm$ is read "positive or negative" or "plus or minus.")

NOTE When we solve an equation, we must find *all* values of the variable that satisfy the equation. Therefore, we want both the positive and negative square roots of k.

EXAMPLE 2 Solving Quadratic Equations of the Form $x^2 = k$

Solve each equation. Write radicals in simplified form.

(a) $x^2 = 16$

By the square root property, if $x^2 = 16$, then

$$x = \sqrt{16} = 4 \quad \text{or} \quad x = -\sqrt{16} = -4.$$

Check each solution by substituting it for x in the original equation. The solution set is

$$\{4, -4\}, \quad \text{or} \quad \{\pm 4\}.$$

> This notation indicates *two* solutions, one **positive** and one **negative**.

(b) $x^2 = 5$

By the square root property, if $x^2 = 5$, then

$$x = \sqrt{5} \quad \text{or} \quad x = -\sqrt{5}.$$

> Don't forget the negative solution.

The solutions set is $\left\{\sqrt{5}, -\sqrt{5}\right\}$, or $\left\{\pm\sqrt{5}\right\}$.

(c)

$$5m^2 - 40 = 0$$

$$5m^2 = 40 \qquad \text{Add 40.}$$

$$m^2 = 8 \qquad \text{Divide by 5.}$$

> Don't stop here. Simplify the radicals.

$$m = \sqrt{8} \quad \text{or} \quad m = -\sqrt{8} \qquad \text{Square root property}$$

$$m = 2\sqrt{2} \quad \text{or} \quad m = -2\sqrt{2} \qquad \sqrt{8} = \sqrt{4} \cdot \sqrt{2} = 2\sqrt{2}$$

CHECK Substitute each value in the original equation.

$$5m^2 - 40 = 0 \qquad\qquad 5m^2 - 40 = 0$$

$$5(2\sqrt{2})^2 - 40 \overset{?}{=} 0 \quad \text{Let } m = 2\sqrt{2}. \qquad 5(-2\sqrt{2})^2 - 40 \overset{?}{=} 0 \quad \text{Let } m = -2\sqrt{2}.$$

> $(2\sqrt{2})^2$
> $= 2^2 \cdot (\sqrt{2})^2$
> $= 4 \cdot 2$
> $= 8$

$$5(8) - 40 \overset{?}{=} 0 \qquad\qquad 5(8) - 40 \overset{?}{=} 0$$

$$40 - 40 \overset{?}{=} 0 \qquad\qquad 40 - 40 \overset{?}{=} 0$$

$$0 = 0 \ \checkmark \ \text{True} \qquad\qquad 0 = 0 \ \checkmark \ \text{True}$$

The solution set is $\left\{2\sqrt{2}, -2\sqrt{2}\right\}$, or $\left\{\pm 2\sqrt{2}\right\}$.

NOW TRY
EXERCISE 2

Solve each equation. Write radicals in simplified form.

(a) $t^2 = 25$ **(b)** $x^2 = 13$

(c) $3x^2 - 54 = 0$

(d) $2x^2 - 5 = 35$

NOW TRY
EXERCISE 3

Tim is dropping roofing nails from the top of a roof 25 ft high into a large bucket on the ground. Use the formula in **Example 3** to determine how long it will take a nail dropped from 25 ft to hit the bottom of the bucket.

(d)
$$3x^2 + 5 = 11$$
$$3x^2 = 6 \qquad \text{Subtract 5.}$$
$$x^2 = 2 \qquad \text{Divide by 3.}$$
$$x = \sqrt{2} \quad \text{or} \quad x = -\sqrt{2} \qquad \text{Square root property}$$

The solution set is $\{\sqrt{2}, -\sqrt{2}\}$, or $\{\pm\sqrt{2}\}$.

NOW TRY

EXAMPLE 3 Using the Square Root Property in an Application

Galileo Galilei developed a formula for freely falling objects described by

$$d = 16t^2,$$

where d is the distance in feet that an object falls (disregarding air resistance) in t seconds, regardless of weight.

   If the Leaning Tower of Pisa is about 180 ft tall, use Galileo's formula to determine how long it would take an object dropped from the top of the tower to fall to the ground. (*Source:* www.brittanica.com)

Galileo Galilei (1564–1642)

$$d = 16t^2 \qquad \text{Galileo's formula}$$
$$180 = 16t^2 \qquad \text{Let } d = 180.$$
$$11.25 = t^2 \qquad \text{Divide by 16.}$$
$$t = \sqrt{11.25} \quad \text{or} \quad t = -\sqrt{11.25} \qquad \text{Square root property}$$

Time cannot be negative, so we discard $-\sqrt{11.25}$. Using a calculator,

$$\sqrt{11.25} = 3.4 \quad \text{so} \quad t = 3.4. \qquad \text{Round to the nearest tenth.}$$

The object would fall to the ground in 3.4 sec.

NOW TRY

OBJECTIVE 3 Solve equations of the form $(ax + b)^2 = k$, where $k > 0$.

In each equation so far, the exponent 2 appeared with a single variable as its base. We can extend the square root property to solve equations in which the base is a binomial.

EXAMPLE 4 Solving Quadratic Equations of the Form $(x + b)^2 = k$

Solve each equation, and check the solutions.

(a) Use $(x - 3)$ as the base. $\quad (x - 3)^2 = 16$

$$x - 3 = \sqrt{16} \quad \text{or} \quad x - 3 = -\sqrt{16} \qquad \text{Square root property}$$
$$x - 3 = 4 \quad \text{or} \quad x - 3 = -4 \qquad \sqrt{16} = 4$$
$$x = 7 \quad \text{or} \quad x = -1 \qquad \text{Add 3.}$$

CHECK Substitute each value in the original equation.

$(x - 3)^2 = 16$	$(x - 3)^2 = 16$
$(7 - 3)^2 \overset{?}{=} 16 \qquad \text{Let } x = 7.$	$(-1 - 3)^2 \overset{?}{=} 16 \qquad \text{Let } x = -1.$
$4^2 \overset{?}{=} 16 \qquad \text{Subtract.}$	$(-4)^2 \overset{?}{=} 16 \qquad \text{Subtract.}$
$16 = 16 \checkmark \quad \text{True}$	$16 = 16 \checkmark \quad \text{True}$

The solution set is $\{-1, 7\}$.

NOW TRY ANSWERS
2. **(a)** $\{\pm 5\}$ **(b)** $\{\pm\sqrt{13}\}$
   (c) $\{\pm 3\sqrt{2}\}$ **(d)** $\{\pm 2\sqrt{5}\}$
3. 1.25 sec

**NOW TRY
EXERCISE 4**

Solve $(x - 2)^2 = 32$, and check the solutions.

(b) $$(x - 1)^2 = 6$$

$x - 1 = \sqrt{6}$ or $x - 1 = -\sqrt{6}$ Square root property

$x = 1 + \sqrt{6}$ or $x = 1 - \sqrt{6}$ Add 1.

CHECK Substitute each value in the original equation.

$(x - 1)^2 = 6$ $(x - 1)^2 = 6$

$(1 + \sqrt{6} - 1)^2 \overset{?}{=} 6$ Let $x = 1 + \sqrt{6}$. $(1 - \sqrt{6} - 1)^2 \overset{?}{=} 6$ Let $x = 1 - \sqrt{6}$.

$(\sqrt{6})^2 \overset{?}{=} 6$ Simplify. $(-\sqrt{6})^2 \overset{?}{=} 6$ Simplify.

$6 = 6$ ✓   True $6 = 6$ ✓   True

The solution set is $\{1 + \sqrt{6}, 1 - \sqrt{6}\}$, or $\{1 \pm \sqrt{6}\}$.   NOW TRY

**NOW TRY
EXERCISE 5**

Solve $(2t - 4)^2 = 50$. Check the solutions.

EXAMPLE 5 Solving a Quadratic Equation of the Form $(ax + b)^2 = k$

Solve $(3r - 2)^2 = 27$. Check the solutions.

$$(3r - 2)^2 = 27$$

$3r - 2 = \sqrt{27}$ or $3r - 2 = -\sqrt{27}$ Square root property

$3r - 2 = 3\sqrt{3}$ or $3r - 2 = -3\sqrt{3}$ $\sqrt{27} = \sqrt{9} \cdot \sqrt{3} = 3\sqrt{3}$

$3r = 2 + 3\sqrt{3}$ or $3r = 2 - 3\sqrt{3}$ Add 2.

$r = \dfrac{2 + 3\sqrt{3}}{3}$ or $r = \dfrac{2 - 3\sqrt{3}}{3}$ Divide by 3. The fractions cannot be simplified because 3 is *not* a factor in the numerator.

CHECK $$(3r - 2)^2 = 27$$

$$\left(3 \cdot \frac{2 + 3\sqrt{3}}{3} - 2\right)^2 \overset{?}{=} 27 \qquad \text{Let } r = \frac{2 + 3\sqrt{3}}{3}.$$

$$(2 + 3\sqrt{3} - 2)^2 \overset{?}{=} 27 \qquad \text{Multiply.}$$

$$(ab)^2 = a^2b^2 \qquad (3\sqrt{3})^2 \overset{?}{=} 27 \qquad \text{Subtract.}$$

$$27 = 27 \checkmark \quad \text{True}$$

The check of the other value is similar. The solution set is $\left\{\dfrac{2 \pm 3\sqrt{3}}{3}\right\}$.   NOW TRY

OBJECTIVE 4 Solve quadratic equations with solutions that are not real numbers.

If $k < 0$ in the equation $x^2 = k$, then there will be two nonreal complex solutions.

EXAMPLE 6 Solving Quadratic Equations (Nonreal Complex Solutions)

Solve each equation.

(a) $$x^2 = -15$$

$x = \sqrt{-15}$ or $x = -\sqrt{-15}$ Square root property

$x = i\sqrt{15}$ or $x = -i\sqrt{15}$ $\sqrt{-a} = i\sqrt{a}$

The solution set is $\{i\sqrt{15}, -i\sqrt{15}\}$, or $\{\pm i\sqrt{15}\}$.

NOW TRY ANSWERS

4. $\{2 \pm 4\sqrt{2}\}$

5. $\left\{\dfrac{4 \pm 5\sqrt{2}}{2}\right\}$

 **NOW TRY
EXERCISE 6**

Solve each equation.

(a) $t^2 = -24$

(b) $(x + 4)^2 = -36$

NOW TRY ANSWERS

6. (a) $\{\pm 2i\sqrt{6}\}$

  (b) $\{-4 \pm 6i\}$

(b) $(x + 2)^2 = -16$

$x + 2 = \sqrt{-16}$ or $x + 2 = -\sqrt{-16}$ Square root property

$x + 2 = 4i$ or $x + 2 = -4i$ $\sqrt{-16} = 4i$

$x = -2 + 4i$ or $x = -2 - 4i$ Add -2.

The solution set is $\{-2 + 4i, -2 - 4i\}$, or $\{-2 \pm 4i\}$.   **NOW TRY**

11.1 Exercises

 FOR EXTRA HELP   MyMathLab®

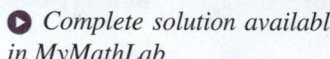 *Complete solution available
in MyMathLab*

Concept Check Answer each question.

1. Which of the following are quadratic equations?

   A. $x + 2 = 0$ **B.** $x^2 - 8x + 16 = 0$ **C.** $2t^2 - 5t = 3$ **D.** $x^3 + x^2 + 4 = 0$

2. Which quadratic equation identified in **Exercise 1** is in standard form?

3. A student incorrectly solved $x^2 - x - 2 = 5$ as follows. **WHAT WENT WRONG?**

$$x^2 - x - 2 = 5$$
$$(x - 2)(x + 1) = 5 \qquad \text{Factor.}$$
$$x - 2 = 5 \quad \text{or} \quad x + 1 = 5 \qquad \text{Zero-factor property}$$
$$x = 7 \quad \text{or} \qquad x = 4 \qquad \text{Solve each equation.}$$

4. A student was asked to solve the quadratic equation $x^2 = 16$ and did not get full credit for the solution set $\{4\}$. **WHAT WENT WRONG?**

Solve each equation using the zero-factor property. See Example 1.

5. $x^2 - x - 56 = 0$ 6. $x^2 - 2x - 99 = 0$ 7. $x^2 = 121$

8. $x^2 = 144$ 9. $3x^2 - 13x = 30$ 10. $5x^2 - 14x = 3$

Solve each equation using the square root property. Simplify all radicals. See Example 2.

11. $x^2 = 81$ 12. $z^2 = 169$ 13. $x^2 = 100$ 14. $m^2 = 64$

15. $x^2 = 14$ 16. $m^2 = 22$ 17. $t^2 = 48$ 18. $x^2 = 54$

19. $x^2 = \dfrac{25}{4}$ 20. $m^2 = \dfrac{36}{121}$ 21. $x^2 = 2.25$ 22. $w^2 = 56.25$

23. $r^2 - 3 = 0$ 24. $x^2 - 13 = 0$ 25. $7x^2 = 4$

26. $2x^2 = 9$ 27. $3n^2 - 72 = 0$ 28. $2x^2 - 80 = 0$

29. $5x^2 + 4 = 8$ 30. $7p^2 - 5 = 11$ 31. $2t^2 + 7 = 61$

32. $3x^2 + 8 = 80$ 33. $-8x^2 = -64$ 34. $-12x^2 = -144$

Solve each equation using the square root property. Simplify all radicals. See Examples 4 and 5.

35. $(x - 3)^2 = 25$ 36. $(x - 7)^2 = 16$ 37. $(x - 4)^2 = 3$

38. $(x + 3)^2 = 11$ 39. $(x - 8)^2 = 27$ 40. $(p - 5)^2 = 40$

41. $(3x + 2)^2 = 49$ 42. $(5t + 3)^2 = 36$ 43. $(4x - 3)^2 = 9$

44. $(7z - 5)^2 = 25$ 45. $(5 - 2x)^2 = 30$ 46. $(3 - 2x)^2 = 70$

▶ **47.** $(3k + 1)^2 = 18$ **48.** $(5z + 6)^2 = 75$ **49.** $\left(\dfrac{1}{2}x + 5\right)^2 = 12$

50. $\left(\dfrac{1}{3}m + 4\right)^2 = 27$ **51.** $\left(x - \dfrac{1}{8}\right)^2 = \dfrac{1}{64}$ **52.** $\left(x - \dfrac{1}{9}\right)^2 = \dfrac{1}{81}$

53. $\left(x - \dfrac{1}{3}\right)^2 = \dfrac{4}{9}$ **54.** $\left(x - \dfrac{1}{5}\right)^2 = \dfrac{16}{25}$ **55.** $\left(x + \dfrac{1}{4}\right)^2 = \dfrac{3}{16}$

56. $\left(x + \dfrac{1}{7}\right)^2 = \dfrac{11}{49}$ **57.** $(4x - 1)^2 - 48 = 0$ **58.** $(2x - 5)^2 - 180 = 0$

Use a calculator to solve each equation. Round answers to the nearest hundredth.

59. $(k + 2.14)^2 = 5.46$ **60.** $(r - 3.91)^2 = 9.28$

61. $(2.11p + 3.42)^2 = 9.58$ **62.** $(1.71m - 6.20)^2 = 5.41$

Find the nonreal complex solutions of each equation. See Example 6.

63. $x^2 = -26$ **64.** $x^2 = -21$ ▶ **65.** $x^2 = -12$ **66.** $x^2 = -18$

67. $(r - 5)^2 = -4$ **68.** $(t + 6)^2 = -9$ **69.** $(6x - 1)^2 = -8$ **70.** $(4m - 7)^2 = -27$

Use Galileo's formula $d = 16t^2$ to solve each problem. Round answers to the nearest tenth. See Example 3.

▶ **71.** The sculpture of American presidents at Mount Rushmore National Memorial is 500 ft above the valley floor. How long would it take a rock dropped from the top of the sculpture to fall to the ground? (*Source:* www.travelsd.com)

72. The Gateway Arch in St. Louis, Missouri, is 630 ft tall. How long would it take an object dropped from the top of the arch to fall to the ground? (*Source:* www.gatewayarch.com)

Solve each problem. See Example 3.

▶ **73.** The area $\mathcal{A}$ of a circle with radius r is given by the formula

$$\mathcal{A} = \pi r^2.$$

If a circle has area 81π in.^{2}, what is its radius?

74. The surface area S of a sphere with radius r is given by the formula

$$S = 4\pi r^2.$$

If a sphere has surface area 36π ft^2, what is its radius?

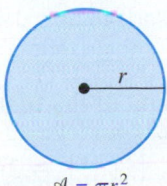

$\mathcal{A} = \pi r^2$

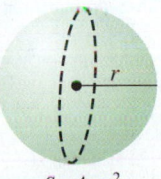

$S = 4\pi r^2$

11.2 Solving Quadratic Equations by Completing the Square

OBJECTIVES

1 Solve quadratic equations by completing the square when the coefficient of the second-degree term is 1.

2 Solve quadratic equations by completing the square when the coefficient of the second-degree term is not 1.

3 Simplify the terms of an equation before solving.

OBJECTIVE 1 Solve quadratic equations by completing the square when the coefficient of the second-degree term is 1.

The methods we have studied so far are not enough to solve an equation such as

$$x^2 + 6x + 7 = 0.$$

We can use the square root property to solve *any* quadratic equation by writing it in the form

Square of a binomial $\rightarrow (x + k)^2 = n.$ ← Constant

That is, we must write the left side of the equation as a perfect square trinomial that can be factored as $(x + k)^2$, the square of a binomial, and the right side must be a constant.

Recall that the perfect square trinomial

$$x^2 + 6x + 9 \quad \text{can be factored as} \quad (x + 3)^2.$$

In the trinomial, the coefficient of x (the first-degree term) is 6 and the constant term is 9. If we take half of 6 and square it, we obtain the constant term, 9.

Coefficient of x Constant

$$\left[\frac{1}{2}(6)\right]^2 = 3^2 = 9$$

Similarly, in $\quad x^2 + 12x + 36, \quad \left[\frac{1}{2}(12)\right]^2 = 6^2 = 36,$

and in $\quad m^2 - 10m + 25, \quad \left[\frac{1}{2}(-10)\right]^2 = (-5)^2 = 25.$

This relationship is true in general and is the idea behind writing a quadratic equation so that the square root property can be applied.

**NOW TRY
EXERCISE 1**
Solve $x^2 + 6x - 2 = 0.$

EXAMPLE 1   Rewriting an Equation to use the Square Root Property

Solve $x^2 + 6x + 7 = 0.$

The trinomial on the left is nonfactorable, so this quadratic equation cannot be solved using the zero-factor property. It is not in the correct form to solve using the square root property. To obtain this form, we need a perfect square trinomial on the left side of the equation.

$$x^2 + 6x + 7 = 0 \qquad \text{Original equation}$$

Only terms with variables remain on the left side. $\quad x^2 + 6x = -7 \qquad \text{Subtract 7.}$

$$x^2 + 6x + \underline{\ ?\ } = -7 \qquad \text{We must add a constant.}$$

Needs to be a perfect square trinomial

Take half the coefficient of the first-degree term, $6x$, and square the result.

$$\left[\frac{1}{2}(6)\right]^2 = 3^2 = 9 \longleftarrow \text{Desired constant}$$

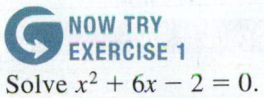

NOW TRY EXERCISE 1

Solve $x^2 + 6x - 2 = 0$.

If we add 9 to each side of $x^2 + 6x = -7$, the equation will have a perfect square trinomial on the left side, as needed.

$$x^2 + 6x = -7 \qquad \text{Transformed equation}$$

This is a key step. $\quad x^2 + 6x + 9 = -7 + 9 \qquad$ Add 9.

$$(x + 3)^2 = 2 \qquad \text{Factor on the left. Add on the right.}$$

Now use the square root property to complete the solution.

$$x + 3 = \sqrt{2} \qquad \text{or} \quad x + 3 = -\sqrt{2}$$

$$x = -3 + \sqrt{2} \quad \text{or} \qquad x = -3 - \sqrt{2}$$

CHECK $\qquad\qquad\qquad\qquad x^2 + 6x + 7 = 0 \qquad$ Original equation

Remember the middle term when squaring $-3 + \sqrt{2}.$

$$\left(-3 + \sqrt{2}\right)^2 + 6\left(-3 + \sqrt{2}\right) + 7 \stackrel{?}{=} 0 \qquad \text{Let } x = -3 + \sqrt{2}.$$

$$9 - 6\sqrt{2} + 2 - 18 + 6\sqrt{2} + 7 \stackrel{?}{=} 0$$

$$0 = 0 \quad \checkmark \quad \text{True}$$

The check of $-3 - \sqrt{2}$ is similar. The solution set is

$$\left\{-3 + \sqrt{2}, -3 - \sqrt{2}\right\}, \quad \text{or} \quad \left\{-3 \pm \sqrt{2}\right\}. \qquad \text{NOW TRY} \,$$

Completing the square is the process of changing the form of the equation in **Example 1** from

$$x^2 + 6x + 7 = 0 \quad \text{to} \quad (x + 3)^2 = 2.$$

Completing the square changes only the form of the equation. To see this, multiply out the left side of $(x + 3)^2 = 2$ and combine like terms. Then subtract 2 from each side to see that the result is $x^2 + 6x + 7 = 0$.

NOW TRY EXERCISE 2

Solve $x^2 - 6x = 9$.

EXAMPLE 2 Completing the Square to Solve a Quadratic Equation

Solve $x^2 - 8x = 5$.

To complete the square on $x^2 - 8x$, take half the coefficient of x and square it.

$$\frac{1}{2}(-8) = -4 \quad \text{and} \quad (-4)^2 = 16$$

Coefficient of x

Add the result, 16, to each side of the equation.

$$x^2 - 8x = 5 \qquad\qquad \text{Given equation}$$

$$x^2 - 8x + 16 = 5 + 16 \qquad \text{Add 16.}$$

$$(x - 4)^2 = 21 \qquad\qquad \text{Factor on the left. Add on the right.}$$

$$x - 4 = \sqrt{21} \qquad \text{or} \quad x - 4 = -\sqrt{21} \qquad \text{Square root property}$$

$$x = 4 + \sqrt{21} \quad \text{or} \qquad x = 4 - \sqrt{21} \qquad \text{Add 4.}$$

NOW TRY ANSWERS
1. $\left\{-3 \pm \sqrt{11}\right\}$
2. $\left\{3 \pm 3\sqrt{2}\right\}$

A check confirms that the solution set is $\left\{4 \pm \sqrt{21}\right\}$. $\qquad$ **NOW TRY**

> **Completing the Square to Solve** $ax^2 + bx + c = 0$ (Where $a \neq 0$)
>
> **Step 1**   **Be sure the second-degree term has coefficient 1.**
> - If the coefficient of the second-degree term is 1, go to Step 2.
> - If it is not 1, but some other nonzero number a, divide each side of the equation by a.
>
> **Step 2**   **Write the equation in correct form.** Make sure that all variable terms are on one side of the equality symbol and the constant term is on the other side.
>
> **Step 3**   **Complete the square.**
> - Take half the coefficient of the first-degree term, and square it.
> - Add the square to each side of the equation.
> - Factor the variable side, which should be a perfect square trinomial, as the square of a binomial. Combine terms on the other side.
>
> **Step 4**   **Solve** the equation using the square root property.

NOW TRY
EXERCISE 3
Solve $x^2 + x - 3 = 0$.

EXAMPLE 3   Solving a Quadratic Equation by Completing the Square ($a = 1$)

Solve $x^2 + 5x - 1 = 0$.

The coefficient of the second-degree term is 1, so we begin with Step 2.

Step 2 $x^2 + 5x = 1$ Add 1 to each side.

Step 3   Take half the coefficient of the first-degree term and square the result.

$$\left[\tfrac{1}{2}(5)\right]^2 = \left(\tfrac{5}{2}\right)^2 = \tfrac{25}{4}$$

$$x^2 + 5x + \frac{25}{4} = 1 + \frac{25}{4}$$ Add the square to each side of the equation.

$$\left(x + \frac{5}{2}\right)^2 = \frac{29}{4}$$ Factor on the left. Add on the right.

Step 4 $x + \dfrac{5}{2} = \sqrt{\dfrac{29}{4}}$ or $x + \dfrac{5}{2} = -\sqrt{\dfrac{29}{4}}$ Square root property

$x + \dfrac{5}{2} = \dfrac{\sqrt{29}}{2}$ or $x + \dfrac{5}{2} = -\dfrac{\sqrt{29}}{2}$ $\sqrt{\dfrac{a}{b}} = \dfrac{\sqrt{a}}{\sqrt{b}}$

$x = -\dfrac{5}{2} + \dfrac{\sqrt{29}}{2}$ or $x = -\dfrac{5}{2} - \dfrac{\sqrt{29}}{2}$ Add $-\dfrac{5}{2}$.

$x = \dfrac{-5 + \sqrt{29}}{2}$ or $x = \dfrac{-5 - \sqrt{29}}{2}$ $\dfrac{a}{c} \pm \dfrac{b}{c} = \dfrac{a \pm b}{c}$

The solution set is $\left\{\dfrac{-5 \pm \sqrt{29}}{2}\right\}$.

NOW TRY

OBJECTIVE 2   Solve quadratic equations by completing the square when the coefficient of the second-degree term is not 1.

NOW TRY ANSWER

3. $\left\{\dfrac{-1 \pm \sqrt{13}}{2}\right\}$

If a quadratic equation has the form

$$ax^2 + bx + c = 0, \quad \text{where} \quad a \neq 1,$$

we obtain 1 as the coefficient of x^2 by dividing each side of the equation by a.

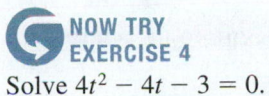

NOW TRY
EXERCISE 4
Solve $4t^2 - 4t - 3 = 0$.

EXAMPLE 4   Solving a Quadratic Equation by Completing the Square ($a \neq 1$)

Solve $4x^2 + 16x - 9 = 0$.

Step 1   **Before completing the square, the coefficient of x^2 must be 1,** not 4. We obtain 1 as the coefficient of x^2 here by dividing each side by 4.

$$4x^2 + 16x - 9 = 0 \qquad \text{Given equation}$$

The coefficient of x^2 must be 1.
$$x^2 + 4x - \frac{9}{4} = 0 \qquad \text{Divide by 4.}$$

Step 2   Write the equation so that all variable terms are on one side of the equation and all constant terms are on the other side.

$$x^2 + 4x = \frac{9}{4} \qquad \text{Add } \tfrac{9}{4}.$$

Step 3   Complete the square by taking half the coefficient of x, and squaring it.

$$\tfrac{1}{2}(4) = 2 \quad \text{and} \quad 2^2 = 4$$

We add the result, 4, to each side of the equation.

$$x^2 + 4x + 4 = \frac{9}{4} + 4 \qquad \text{Add 4.}$$

$$(x + 2)^2 = \frac{25}{4} \qquad \text{Factor; } \tfrac{9}{4} + 4 = \tfrac{9}{4} + \tfrac{16}{4} = \tfrac{25}{4}.$$

Step 4   Solve the equation using the square root property.

$$x + 2 = \sqrt{\frac{25}{4}} \qquad \text{or} \qquad x + 2 = -\sqrt{\frac{25}{4}} \qquad \text{Square root property}$$

$$x + 2 = \frac{5}{2} \qquad \text{or} \qquad x + 2 = -\frac{5}{2} \qquad \text{Take square roots.}$$

$$x = -2 + \frac{5}{2} \qquad \text{or} \qquad x = -2 - \frac{5}{2} \qquad \text{Subtract 2.}$$

$$x = \frac{1}{2} \qquad \text{or} \qquad x = -\frac{9}{2} \qquad -2 = -\tfrac{4}{2}$$

CHECK

$$4x^2 + 16x - 9 = 0 \qquad\qquad\qquad 4x^2 + 16x - 9 = 0$$

$$4\left(\frac{1}{2}\right)^2 + 16\left(\frac{1}{2}\right) - 9 \overset{?}{=} 0 \quad \text{Let } x = \tfrac{1}{2}. \qquad 4\left(-\frac{9}{2}\right)^2 + 16\left(-\frac{9}{2}\right) - 9 \overset{?}{=} 0 \quad \text{Let } x = -\tfrac{9}{2}.$$

$$4\left(\frac{1}{4}\right) + 8 - 9 \overset{?}{=} 0 \qquad\qquad\qquad 4\left(\frac{81}{4}\right) - 72 - 9 \overset{?}{=} 0$$

$$1 + 8 - 9 \overset{?}{=} 0 \qquad\qquad\qquad\qquad 81 - 72 - 9 \overset{?}{=} 0$$

$$0 = 0 \ \checkmark \ \text{True} \qquad\qquad\qquad\qquad 0 = 0 \ \checkmark \ \text{True}$$

NOW TRY ANSWER
4. $\left\{-\frac{1}{2}, \frac{3}{2}\right\}$

The two values, $\frac{1}{2}$ and $-\frac{9}{2}$, check, so the solution set is $\left\{-\frac{9}{2}, \frac{1}{2}\right\}$.   **NOW TRY**

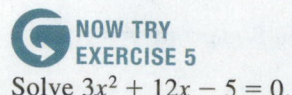

NOW TRY
EXERCISE 5
Solve $3x^2 + 12x - 5 = 0$.

EXAMPLE 5 Solving a Quadratic Equation by Completing the Square ($a \neq 1$)

Solve $2x^2 - 4x - 5 = 0$.

Divide each side by 2 to obtain 1 as the coefficient of the second-degree term.

$$x^2 - 2x - \frac{5}{2} = 0 \qquad \text{Divide by 2. (Step 1)}$$

$$x^2 - 2x = \frac{5}{2} \qquad \text{Add } \tfrac{5}{2}. \text{ (Step 2)}$$

$$\left[\tfrac{1}{2}(-2)\right]^2 = (-1)^2 = 1 \qquad \text{Complete the square. (Step 3)}$$

$$x^2 - 2x + 1 = \frac{5}{2} + 1 \qquad \text{Add 1 to each side.}$$

$$(x - 1)^2 = \frac{7}{2} \qquad \begin{array}{l}\text{Factor on the left.}\\\text{Add on the right.}\end{array}$$

$$x - 1 = \sqrt{\frac{7}{2}} \qquad \text{or} \quad x - 1 = -\sqrt{\frac{7}{2}} \qquad \text{Square root property (Step 4)}$$

$$x = 1 + \sqrt{\frac{7}{2}} \qquad \text{or} \qquad x = 1 - \sqrt{\frac{7}{2}} \qquad \text{Add 1.}$$

$$x = 1 + \frac{\sqrt{14}}{2} \qquad \text{or} \qquad x = 1 - \frac{\sqrt{14}}{2} \qquad \sqrt{\frac{7}{2}} = \frac{\sqrt{7}}{\sqrt{2}} = \frac{\sqrt{7}}{\sqrt{2}} \cdot \frac{\sqrt{2}}{\sqrt{2}} = \frac{\sqrt{14}}{2}$$

Add the two terms in each solution as follows.

$$1 + \frac{\sqrt{14}}{2} = \frac{2}{2} + \frac{\sqrt{14}}{2} = \frac{2 + \sqrt{14}}{2}$$

$$1 = \tfrac{2}{2}$$

$$1 - \frac{\sqrt{14}}{2} = \frac{2}{2} - \frac{\sqrt{14}}{2} = \frac{2 - \sqrt{14}}{2}$$

The solution set is $\left\{\dfrac{2 \pm \sqrt{14}}{2}\right\}$.

NOW TRY

EXAMPLE 6 Solving a Quadratic Equation (Nonreal Complex Solutions)

Solve $4p^2 + 8p + 5 = 0$.

$$4p^2 + 8p + 5 = 0$$

The coefficient of the second-degree term must be 1.

$$p^2 + 2p + \frac{5}{4} = 0 \qquad \text{Divide by 4. (Step 1)}$$

$$p^2 + 2p = -\frac{5}{4} \qquad \text{Subtract } \tfrac{5}{4} \text{ from each side. (Step 2)}$$

The coefficient of p is 2. Take half of 2, square the result, and add it to each side.

$$p^2 + 2p + 1 = -\frac{5}{4} + 1 \qquad \left[\tfrac{1}{2}(2)\right]^2 = 1^2 = 1; \text{ Add 1. (Step 3)}$$

$$(p + 1)^2 = -\frac{1}{4} \qquad \begin{array}{l}\text{Factor on the left.}\\\text{Add on the right.}\end{array}$$

NOW TRY ANSWER

5. $\left\{\dfrac{-6 \pm \sqrt{51}}{3}\right\}$

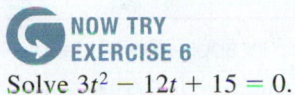

**NOW TRY
EXERCISE 6**
Solve $3t^2 - 12t + 15 = 0$.

$$p + 1 = \sqrt{-\frac{1}{4}} \quad \text{or} \quad p + 1 = -\sqrt{-\frac{1}{4}} \qquad \text{Square root property (Step 4)}$$

$$p + 1 = \frac{1}{2}i \quad \text{or} \quad p + 1 = -\frac{1}{2}i \qquad \sqrt{-\frac{1}{4}} = \frac{1}{2}i$$

$$p = -1 + \frac{1}{2}i \quad \text{or} \quad p = -1 - \frac{1}{2}i \qquad \text{Add } -1.$$

The solution set is $\left\{-1 \pm \frac{1}{2}i\right\}$.

 NOW TRY

OBJECTIVE 3 Simplify the terms of an equation before solving.

**NOW TRY
EXERCISE 7**
Solve $x(x + 5) = 3$

EXAMPLE 7 Simplifying the Terms of an Equation before Solving

Solve $x(x + 7) = 2$.

$$x(x + 7) = 2$$

$$x^2 + 7x = 2 \qquad \text{Multiply.}$$

$$x^2 + 7x + \frac{49}{4} = 2 + \frac{49}{4} \qquad \begin{array}{l}\text{To complete the square, add} \\ \left(\frac{1}{2} \cdot 7\right)^2 = \frac{49}{4} \text{ to each side.}\end{array}$$

$$\left(x + \frac{7}{2}\right)^2 = \frac{57}{4} \qquad \text{Factor; } 2 + \frac{49}{4} = \frac{8}{4} + \frac{49}{4} = \frac{57}{4}.$$

$$x + \frac{7}{2} = \sqrt{\frac{57}{4}} \quad \text{or} \quad x + \frac{7}{2} = -\sqrt{\frac{57}{4}} \qquad \text{Square root property}$$

$$x = -\frac{7}{2} + \frac{\sqrt{57}}{2} \quad \text{or} \quad x = -\frac{7}{2} - \frac{\sqrt{57}}{2} \qquad \begin{array}{l}\text{Subtract } \frac{7}{2}; \\ \text{quotient rule for radicals}\end{array}$$

$$x = \frac{-7 + \sqrt{57}}{2} \quad \text{or} \quad x = \frac{-7 - \sqrt{57}}{2} \qquad \begin{array}{l}\text{Add and subtract} \\ \text{the fractions.}\end{array}$$

The solution set is $\left\{\frac{-7 \pm \sqrt{57}}{2}\right\}$.

 NOW TRY

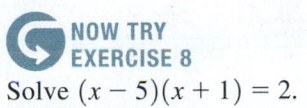

**NOW TRY
EXERCISE 8**
Solve $(x - 5)(x + 1) = 2$.

EXAMPLE 8 Simplifying the Terms of an Equation before Solving

Solve $(x + 3)(x - 1) = 2$.

$$(x + 3)(x - 1) = 2$$

$$x^2 + 2x - 3 = 2 \qquad \text{Multiply using the FOIL method.}$$

$$x^2 + 2x = 5 \qquad \text{Add 3.}$$

$$x^2 + 2x + 1 = 5 + 1 \qquad \text{Add } \left[\frac{1}{2}(2)\right]^2 = 1^2 = 1.$$

$$(x + 1)^2 = 6 \qquad \text{Factor on the left. Add on the right.}$$

NOW TRY ANSWERS

6. $\{2 \pm i\}$

7. $\left\{\dfrac{-5 \pm \sqrt{37}}{2}\right\}$

8. $\{2 \pm \sqrt{11}\}$

$$x + 1 = \sqrt{6} \quad \text{or} \quad x + 1 = -\sqrt{6} \qquad \text{Square root property}$$

$$x = -1 + \sqrt{6} \quad \text{or} \quad x = -1 - \sqrt{6} \qquad \text{Subtract 1.}$$

The solution set is $\left\{-1 \pm \sqrt{6}\right\}$.

 NOW TRY

NOTE The solutions $-1 \pm \sqrt{6}$ given in **Example 8** are *exact*. In applications, decimal solutions are often required. Using the square root key of a calculator yields $\sqrt{6} \approx 2.449$. Approximating the two solutions gives

$$x \approx 1.449 \quad \text{and} \quad x \approx -3.449.$$

11.2 Exercises

     MyMathLab®

▶ *Complete solution available in MyMathLab*

1. *Concept Check* Which one of the two equations

$$(2x + 1)^2 = 5 \quad \text{and} \quad x^2 + 4x = 12,$$

is more suitable for solving by the square root property? Which one is more suitable for solving by completing the square?

2. *Concept Check* Which step is an appropriate way to begin solving the quadratic equation $2x^2 - 4x = 9$ by completing the square?

A. Add 4 to each side of the equation.   **B.** Factor the left side as $2x(x - 2)$.

C. Factor the left side as $x(2x - 4)$.   **D.** Divide each side by 2.

Complete each trinomial so that it is a perfect square. Then factor the trinomial. **See Objective 1.**

▶ **3.** $x^2 + 10x +$ _____   **4.** $x^2 + 16x +$ _____   ▶ **5.** $z^2 - 20z +$ _____

6. $a^2 - 32a +$ _____   **7.** $x^2 + 2x +$ _____   **8.** $m^2 - 2m +$ _____

9. $p^2 - 5p +$ _____   **10.** $x^2 + 3x +$ _____   **11.** $x^2 + \frac{1}{2}x +$ _____

12. $x^2 + \frac{1}{3}x +$ _____   **13.** $x^2 - 0.4x +$ _____   **14.** $x^2 - 0.8x +$ _____

Determine the number that will complete the square to solve each equation, after the constant term has been written on the right side and the coefficient of the second-degree term is 1. Do not actually solve. **See Examples 6–8.**

15. $x^2 + 4x - 2 = 0$ **16.** $t^2 + 2t - 1 = 0$ **17.** $x^2 + 10x + 18 = 0$

18. $x^2 + 8x + 11 = 0$ **19.** $3w^2 - w - 24 = 0$ **20.** $4z^2 - z - 39 = 0$

Concept Check *Solve each equation by completing the square.*

21. $x^2 + 4x = 1$

Take half the coefficient of x and square it.

$$\frac{1}{2} \cdot \underline{\hspace{1cm}} = 2, \quad \text{and} \quad \underline{\hspace{1cm}}^2 = 4.$$

Add _____ to each side of the equation.

$$x^2 + 4x + \underline{\hspace{1cm}} = 1 + 4$$

Factor and add.

$$\underline{\hspace{3cm}}$$

Complete the solution.

22. $3x^2 + 5x - 2 = 0$

Divide each side by _____ to obtain

$$\underline{\hspace{2cm}}. \text{ Add } \frac{2}{3} \text{ to each side}$$

to obtain _____.

Take half the coefficient of x and square it. Add _____ to each side of the equation.

$$x^2 + \frac{5}{3}x + \underline{\hspace{1cm}} = \frac{2}{3} + \frac{25}{36}$$

Factor and add.

$$(\underline{\hspace{1cm}})^2 = \underline{\hspace{1cm}}$$

Complete the solution.

Solve each equation by completing the square. **See Examples 1–3.**

▶ 23. $x^2 - 4x = -3$ **24.** $p^2 - 2p = 8$ **▶ 25.** $x^2 + 2x - 5 = 0$

26. $r^2 + 4r + 1 = 0$ **▶ 27.** $x^2 + 4x - 2 = 0$ **28.** $t^2 + 2t - 1 = 0$

29. $x^2 + 10x + 18 = 0$ **30.** $x^2 + 8x + 11 = 0$ **31.** $x^2 - 8x = -4$

32. $m^2 - 4m = 14$ **▶ 33.** $x^2 + 7x - 1 = 0$ **34.** $x^2 + 13x - 3 = 0$

Solve each equation by completing the square. **See Examples 4–6.**

▶ 35. $4x^2 + 4x = 3$ **36.** $9x^2 + 3x = 2$ **37.** $2k^2 + 5k - 2 = 0$

38. $3r^2 + 2r - 2 = 0$ **▶ 39.** $5x^2 - 10x + 2 = 0$ **40.** $2x^2 - 16x + 25 = 0$

41. $3x^2 - 9x + 5 = 0$ **42.** $6x^2 - 8x - 3 = 0$ **43.** $3x^2 + 7x = 4$

44. $2x^2 + 5x = 1$ **45.** $-x^2 + 2x = -5$ **46.** $-x^2 + 4x = 1$

47. $-3x^2 + 11x + 42 = 0$ **48.** $-9x^2 - 20x + 21 = 0$

49. $0.1x^2 - 0.2x - 0.1 = 0$ **50.** $0.1p^2 - 0.4p + 0.1 = 0$
(*Hint:* First clear the decimals.)   (*Hint:* First clear the decimals.)

51. $x(x - 3) = 1$ **52.** $x(x - 5) = 2$ **53.** $x(x + 3) = -1$

54. $x(x + 7) = -2$ **▶ 55.** $(x + 3)(x - 1) = 5$ **56.** $(x - 8)(x + 2) = 24$

57. $(r - 3)(r - 5) = 2$ **58.** $(x - 1)(x - 7) = 1$

Solve each equation by completing the square. Give (**a**) *exact solutions and* (**b**) *solutions rounded to the nearest thousandth.* **See the Note following Example 8.**

59. $3r^2 - 2 = 6r + 3$ **60.** $4p + 3 = 2p^2 + 2p$

61. $(x + 1)(x + 3) = 2$ **62.** $(x - 3)(x + 1) = 1$

Find the nonreal complex solutions of each equation. **See Example 6.**

63. $m^2 + 4m + 13 = 0$ **64.** $t^2 + 6t + 10 = 0$ **65.** $x^2 + 2x + 7 = 0$

66. $x^2 + 8x + 21 = 0$ **67.** $3r^2 + 4r + 4 = 0$ **68.** $4x^2 + 5x + 5 = 0$

69. $-m^2 - 6m - 12 = 0$ **70.** $-x^2 - 5x - 10 = 0$

Extending Skills   *Solve for x. Assume that a and b represent positive real numbers.*

71. $x^2 - b = 0$ **72.** $x^2 = 4b$ **73.** $4x^2 = b^2 + 16$

74. $9x^2 - 25a = 0$ **75.** $(5x - 2b)^2 = 3a$ **76.** $x^2 - a^2 - 36 = 0$

RELATING CONCEPTS For Individual or Group Work (Exercises 77–82)

The Greeks had a method of completing the square geometrically in which they literally changed a figure into a square. For example, to complete the square for $x^2 + 6x$, we begin with a square of side x, as in the figure on the top. We add three rectangles of width 1 to the right side and the bottom to get a region with area $x^2 + 6x$. To fill in the corner (complete the square), we must add nine 1-by-1 squares as shown.

Work Exercises 77–82 in order.

77. What is the area of the original square?

78. What is the area of each strip?

79. What is the total area of the six strips?

80. What is the area of each small square in the corner of the second figure?

81. What is the total area of the small squares?

82. What is the area of the new "complete" square?

(11.3) Solving Quadratic Equations by the Quadratic Formula

OBJECTIVES

1 Derive the quadratic formula.

2 Solve quadratic equations using the quadratic formula.

3 Use the discriminant to determine number and type of solutions.

VOCABULARY

☐ quadratic formula
☐ discriminant

In this section, we complete the square to solve the general quadratic equation

$$ax^2 + bx + c = 0,$$

where a, b, and c are complex numbers and $a \neq 0$. The solution of this general equation gives a formula for finding the solution of *any* specific quadratic equation.

OBJECTIVE 1 Derive the quadratic formula.

We solve $ax^2 + bx + c = 0$ by completing the square (where $a > 0$) as follows.

$$ax^2 + bx + c = 0 \qquad \text{Use the steps given in Section 11.2.}$$

$$x^2 + \frac{b}{a}x + \frac{c}{a} = 0 \qquad \text{Divide by } a. \text{ (Step 1)}$$

$$x^2 + \frac{b}{a}x = -\frac{c}{a} \qquad \text{Subtract } \tfrac{c}{a}. \text{ (Step 2)}$$

$$\left[\frac{1}{2}\left(\frac{b}{a}\right)\right]^2 = \left(\frac{b}{2a}\right)^2 = \frac{b^2}{4a^2} \qquad \text{Complete the square. (Step 3)}$$

$$x^2 + \frac{b}{a}x + \frac{b^2}{4a^2} = -\frac{c}{a} + \frac{b^2}{4a^2} \qquad \text{Add } \tfrac{b^2}{4a^2} \text{ to each side.}$$

$$\left(x + \frac{b}{2a}\right)^2 = \frac{b^2}{4a^2} + \frac{-c}{a} \qquad \begin{array}{l}\text{Factor on the left.} \\ \text{Rearrange the terms on the right.}\end{array}$$

$$\left(x + \frac{b}{2a}\right)^2 = \frac{b^2}{4a^2} + \frac{-4ac}{4a^2} \qquad \text{Write with a common denominator.}$$

$$\left(x + \frac{b}{2a}\right)^2 = \frac{b^2 - 4ac}{4a^2} \qquad \text{Add fractions.}$$

$$x + \frac{b}{2a} = \sqrt{\frac{b^2 - 4ac}{4a^2}} \quad \text{or} \quad x + \frac{b}{2a} = -\sqrt{\frac{b^2 - 4ac}{4a^2}} \qquad \text{Square root property (Step 4)}$$

We can simplify $\sqrt{\dfrac{b^2 - 4ac}{4a^2}}$ as $\dfrac{\sqrt{b^2 - 4ac}}{\sqrt{4a^2}},$ or $\dfrac{\sqrt{b^2 - 4ac}}{2a}.$

The right side of each equation can be expressed as follows.

$$x + \frac{b}{2a} = \frac{\sqrt{b^2 - 4ac}}{2a} \quad \text{or} \quad x + \frac{b}{2a} = \frac{-\sqrt{b^2 - 4ac}}{2a}$$

$$x = \frac{-b}{2a} + \frac{\sqrt{b^2 - 4ac}}{2a} \quad \text{or} \quad x = \frac{-b}{2a} - \frac{\sqrt{b^2 - 4ac}}{2a}$$

If $a < 0$, the same two solutions are obtained.

$$x = \frac{-b + \sqrt{b^2 - 4ac}}{2a} \quad \text{or} \quad x = \frac{-b - \sqrt{b^2 - 4ac}}{2a}$$

This result is the **quadratic formula,** which is abbreviated as follows.

Quadratic Formula

The solutions of the equation $ax^2 + bx + c = 0$ (where $a \neq 0$) are given by

$$x = \frac{-b \pm \sqrt{b^2 - 4ac}}{2a}.$$

⚠ **CAUTION** *In the quadratic formula, the square root is added to or subtracted from the value of $-b$ before dividing by 2a.*

OBJECTIVE 2 Solve quadratic equations using the quadratic formula.

EXAMPLE 1 Using the Quadratic Formula (Rational Solutions)

Solve $6x^2 - 5x - 4 = 0$.

This equation is in standard form, so we identify the values of a, b, and c. Here a, the coefficient of the second-degree term, is 6, and b, the coefficient of the first-degree term, is -5. The constant c is -4. Now substitute into the quadratic formula.

$$x = \frac{-b \pm \sqrt{b^2 - 4ac}}{2a} \qquad \text{Quadratic formula}$$

$$x = \frac{-(-5) \pm \sqrt{(-5)^2 - 4(6)(-4)}}{2(6)} \qquad a = 6, b = -5, c = -4$$

Use parentheses and substitute carefully to avoid errors.

$$x = \frac{5 \pm \sqrt{25 + 96}}{12}$$

$$x = \frac{5 \pm \sqrt{121}}{12} \qquad \text{Simplify the radical.}$$

$$x = \frac{5 \pm 11}{12} \qquad \text{Take the square root.}$$

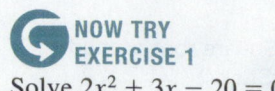

**NOW TRY
EXERCISE 1**
Solve $2x^2 + 3x - 20 = 0$.

There are two values represented, one from the $+$ sign and one from the $-$ sign.

$$x = \frac{5 + 11}{12} = \frac{16}{12} = \frac{4}{3} \quad \text{or} \quad x = \frac{5 - 11}{12} = \frac{-6}{12} = -\frac{1}{2}$$

Check each value in the original equation. The solution set is $\left\{ -\frac{1}{2}, \frac{4}{3} \right\}$.

 NOW TRY

NOTE We could have factored the trinomial and then used the zero-factor property to solve the equation in **Example 1.**

$$6x^2 - 5x - 4 = 0 \qquad \text{See Example 1.}$$

$$(3x - 4)(2x + 1) = 0 \qquad \text{Factor.}$$

$$3x - 4 = 0 \quad \text{or} \quad 2x + 1 = 0 \qquad \text{Zero-factor property}$$

$$3x = 4 \quad \text{or} \qquad 2x = -1 \qquad \text{Solve each equation.}$$

$$x = \frac{4}{3} \quad \text{or} \qquad x = -\frac{1}{2} \qquad \text{Same solutions as in Example 1}$$

When solving a quadratic equation, it is a good idea to try to factor the trinomial first. If it can be factored, then apply the zero-factor property. If it cannot be factored or if factoring is difficult, then use the quadratic formula.

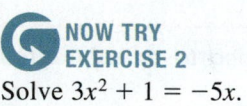

**NOW TRY
EXERCISE 2**
Solve $3x^2 + 1 = -5x$.

EXAMPLE 2 Using the Quadratic Formula (Irrational Solutions)

Solve $4x^2 = 8x - 1$.

Write the equation in standard form as $4x^2 - 8x + 1 = 0$. ◄ This is a key step.

$$x = \frac{-b \pm \sqrt{b^2 - 4ac}}{2a} \qquad \text{Quadratic formula}$$

$$x = \frac{-(-8) \pm \sqrt{(-8)^2 - 4(4)(1)}}{2(4)} \qquad a = 4, b = -8, c = 1$$

$$x = \frac{8 \pm \sqrt{64 - 16}}{8} \qquad \begin{array}{l}\text{Simplify in the numerator}\\ \text{and denominator.}\end{array}$$

$$x = \frac{8 \pm \sqrt{48}}{8} \qquad \text{Subtract under the radical.}$$

$$x = \frac{8 \pm 4\sqrt{3}}{8} \qquad \sqrt{48} = \sqrt{16} \cdot \sqrt{3} = 4\sqrt{3}$$

$$x = \frac{4(2 \pm \sqrt{3})}{4(2)} \qquad \text{Factor.}$$

$$x = \frac{2 \pm \sqrt{3}}{2} \qquad \begin{array}{l}\text{Divide out the common factor 4}\\ \text{to write in lowest terms.}\end{array}$$

The solution set is $\left\{ \frac{2 \pm \sqrt{3}}{2} \right\}$.

 NOW TRY

NOW TRY ANSWERS

1. $\left\{ -4, \frac{5}{2} \right\}$

2. $\left\{ \frac{-5 \pm \sqrt{13}}{6} \right\}$

> ⚠️ **CAUTION**
>
> 1. *Before solving, every quadratic equation must be expressed in standard form* $ax^2 + bx + c = 0$, whether we use the zero-factor property or the quadratic formula.
>
> 2. *When writing solutions in lowest terms, be sure to factor first. Then divide out the common factor.* See the last two steps in **Example 2**.

NOW TRY EXERCISE 3

Solve $(x + 5)(x - 1) = -18$.

EXAMPLE 3 Using the Quadratic Formula (Nonreal Complex Solutions)

Solve $(9x + 3)(x - 1) = -8$.

This is a quadratic equation—when the first terms $9x$ and x are multiplied, we get a second-degree term, $9x^2$. We must write the equation in standard form.

$$(9x + 3)(x - 1) = -8$$

$$9x^2 - 6x - 3 = -8 \qquad \text{Multiply.}$$

$$\text{Standard form} \rightarrow 9x^2 - 6x + 5 = 0 \qquad \text{Add 8.}$$

From the equation $9x^2 - 6x + 5 = 0$, we identify $a = 9$, $b = -6$, and $c = 5$.

$$x = \frac{-b \pm \sqrt{b^2 - 4ac}}{2a} \qquad \text{Quadratic formula}$$

$$x = \frac{-(-6) \pm \sqrt{(-6)^2 - 4(9)(5)}}{2(9)} \qquad \text{Substitute.}$$

$$x = \frac{6 \pm \sqrt{-144}}{18} \qquad \text{Simplify.}$$

$$x = \frac{6 \pm 12i}{18} \qquad \sqrt{-144} = 12i$$

$$x = \frac{6(1 \pm 2i)}{6(3)} \qquad \text{Factor.}$$

$$x = \frac{1 \pm 2i}{3} \qquad \begin{array}{l}\text{Divide out the common factor}\\ \text{6 to write in lowest terms.}\end{array}$$

$$x = \frac{1}{3} \pm \frac{2}{3}i \qquad \begin{array}{l}\text{Standard form } a + bi \text{ for a}\\ \text{complex number}\end{array}$$

The solution set is $\left\{\frac{1}{3} \pm \frac{2}{3}i\right\}$.

NOW TRY

OBJECTIVE 3 Use the discriminant to determine number and type of solutions.

The solutions of the quadratic equation $ax^2 + bx + c = 0$ are given by

$$x = \frac{-b \pm \overbrace{\sqrt{b^2 - 4ac}}^{\text{Discriminant}}}{2a}.$$

The expression under the radical symbol, $b^2 - 4ac$, is called the **discriminant** because it distinguishes among the number of solutions—one or two—and the type of solutions—rational, irrational, or nonreal complex—of a quadratic equation.

NOW TRY ANSWER

3. $\{-2 \pm 3i\}$

Using the Discriminant

If a, b, and c are integers in a quadratic equation $ax^2 + bx + c = 0$, then the discriminant $b^2 - 4ac$ can be used to determine the number and type of solutions of the equation as follows.

Discriminant	Number and Type of Solutions
Positive, and the square of an integer	Two rational solutions
Positive, but not the square of an integer	Two irrational solutions
Zero	One rational solution
Negative	Two nonreal complex solutions

We can also use the discriminant to help decide how to solve a quadratic equation.

> *If a, b, and c are integers and the discriminant is a perfect square (including 0), then the equation can be solved using the zero-factor property. Otherwise, the quadratic formula should be used.*

EXAMPLE 4   Using the Discriminant

Find the discriminant. Use it to predict the number and type of solutions for each equation. Tell whether the equation can be solved using the zero-factor property or whether the quadratic formula should be used.

(a) $6x^2 - x - 15 = 0$

First identify the values of a, b, and c. Because $-x = -1x$, the value of b is -1. We find the discriminant by evaluating $b^2 - 4ac$.

$$b^2 - 4ac$$

Use parentheses and substitute carefully.

$$= (-1)^2 - 4(6)(-15) \qquad a = 6, b = -1, c = -15 \text{ (all integers)}$$
$$= 1 + 360 \qquad \text{Apply the exponent. Multiply.}$$
$$= 361 \qquad \text{Add.}$$
$$= 19^2, \quad \text{which is a perfect square.}$$

The discriminant 361 is a perfect square, so referring to the table we see that there will be two rational solutions. We can solve using the zero-factor property.

(b) $3x^2 - 4x = 5$

Write in standard form as $3x^2 - 4x - 5 = 0$.

$$b^2 - 4ac \qquad \text{Discriminant}$$
$$= (-4)^2 - 4(3)(-5) \qquad a = 3, b = -4, c = -5 \text{ (all integers)}$$
$$= 16 + 60 \qquad \text{Apply the exponent. Multiply.}$$
$$= 76 \qquad \text{Add.}$$

Because 76 is positive but *not* the square of an integer, the equation will have two irrational solutions. We solve using the quadratic formula.

NOW TRY EXERCISE 4

Find the discriminant. Use it to predict the number and type of solutions for each equation. Tell whether the equation can be solved using the zero-factor property or whether the quadratic formula should be used.

(a) $8x^2 - 6x - 5 = 0$

(b) $9x^2 = 24x - 16$

(c) $3x^2 + 2x = -1$

NOW TRY ANSWERS

4. **(a)** 196; two rational solutions; zero-factor property
   (b) 0; one rational solution; zero-factor property
   (c) -8; two nonreal complex solutions; quadratic formula

(c) $4x^2 + x + 1 = 0$

$x = 1x$, so $b = 1$.

$b^2 - 4ac$	Discriminant
$= 1^2 - 4(4)(1)$	$a = 4, b = 1, c = 1$ (all integers)
$= 1 - 16$	Apply the exponent. Multiply.
$= -15$	Subtract.

Because the discriminant is negative, there will be two nonreal complex solutions. We solve using the quadratic formula.

(d) $4x^2 + 9 = 12x$ Write in standard form as $4x^2 - 12x + 9 = 0$.

$b^2 - 4ac$	Discriminant
$= (-12)^2 - 4(4)(9)$	$a = 4, b = -12, c = 9$ (all integers)
$= 144 - 144$	Apply the exponent. Multiply.
$= 0$	Subtract.

The discriminant is 0, so there is only one rational solution. We solve using the zero-factor property.

NOW TRY

11.3 Exercises

FOR EXTRA HELP   ▶ MyMathLab®

▶ *Complete solution available in MyMathLab*

Concept Check *Answer each question.*

1. The documentation for an early version of Microsoft *Word* for Windows used the following for the quadratic formula. Was this correct? If not, correct it.

$$x = -b \pm \frac{\sqrt{b^2 - 4ac}}{2a}$$ Correct or incorrect?

2. One patron wrote the quadratic formula, as shown here, on a wall at the Cadillac Bar in Houston, Texas. Was this correct? If not, correct it.

$$x = \frac{-b\sqrt{b^2 - 4ac}}{2a}$$ Correct or incorrect?

3. A student solved $5x^2 - 5x + 1 = 0$ incorrectly as follows. ***WHAT WENT WRONG?***

$$x = \frac{-(-5) \pm \sqrt{(-5)^2 - 4(5)(1)}}{2(5)}$$

$$x = \frac{5 \pm \sqrt{5}}{10}$$

$$x = \frac{1}{2} \pm \sqrt{5}$$ Solution set: $\left\{ \frac{1}{2} \pm \sqrt{5} \right\}$

4. A student incorrectly claimed that the equation $2x^2 - 5 = 0$ cannot be solved using the quadratic formula because there is no first-degree x-term. ***WHAT WENT WRONG?*** Give the values of a, b, and c for this equation.

*Use the quadratic formula to solve each equation. (All solutions for these equations are real numbers.) See **Examples 1 and 2**.*

▶ **5.** $x^2 - 8x + 15 = 0$ **6.** $x^2 + 3x - 28 = 0$ **7.** $2x^2 + 4x + 1 = 0$

8. $2x^2 + 3x - 1 = 0$

9. $2x^2 - 2x = 1$

10. $9x^2 + 6x = 1$

11. $x^2 + 18 = 10x$

12. $x^2 - 4 = 2x$

13. $4x^2 + 4x - 1 = 0$

14. $4r^2 - 4r - 19 = 0$

15. $2 - 2x = 3x^2$

16. $26r - 2 = 3r^2$

17. $\dfrac{x^2}{4} - \dfrac{x}{2} = 1$

18. $p^2 + \dfrac{p}{3} = \dfrac{1}{6}$

19. $-2t(t + 2) = -3$

20. $-3x(x + 2) = -4$

21. $(r - 3)(r + 5) = 2$

22. $(x + 1)(x - 7) = 1$

23. $(x + 2)(x - 3) = 1$

24. $(x - 5)(x + 2) = 6$

25. $p = \dfrac{5(5 - p)}{3(p + 1)}$

26. $x = \dfrac{2(x + 3)}{x + 5}$

27. $(2x + 1)^2 = x + 4$

28. $(2x - 1)^2 = x + 2$

Use the quadratic formula to solve each equation. (All solutions for these equations are nonreal complex numbers.) ***See Example 3.***

29. $x^2 - 3x + 6 = 0$

30. $x^2 - 5x + 20 = 0$

31. $r^2 - 6r + 14 = 0$

32. $t^2 + 4t + 11 = 0$

33. $4x^2 - 4x = -7$

34. $9x^2 - 6x = -7$

35. $x(3x + 4) = -2$

36. $z(2z + 3) = -2$

37. $(2x - 1)(8x - 4) = -1$

38. $(x - 1)(9x - 3) = -2$

Find the discriminant. Use it to determine whether the solutions for each equation are

    A. *two rational numbers*   **B.** *one rational number*

    C. *two irrational numbers*   **D.** *two nonreal complex numbers.*

Tell whether the equation can be solved using the zero-factor property or whether the quadratic formula should be used. Do not actually solve. ***See Example 4.***

39. $25x^2 + 70x + 49 = 0$

40. $4x^2 - 28x + 49 = 0$

41. $x^2 + 4x + 2 = 0$

42. $9x^2 - 12x - 1 = 0$

43. $3x^2 = 5x + 2$

44. $4x^2 = 4x + 3$

45. $3m^2 - 10m + 15 = 0$

46. $18x^2 + 60x + 82 = 0$

47. Find the discriminant for each quadratic equation. Use it to tell whether the equation can be solved using the zero-factor property or whether the quadratic formula should be used. Then solve each equation.

    (a) $3x^2 + 13x = -12$ **(b)** $2x^2 + 19 = 14x$

48. Based on the answers in **Exercises 39–46,** solve the equation given in each exercise.

    (a) Exercise 39   **(b) Exercise 40**   **(c) Exercise 43**   **(d) Exercise 44**

Extending Skills *Find the value of a, b, or c so that each equation will have exactly one rational solution. (Hint: The discriminant must equal 0 for an equation to have one rational solution.)*

49. $p^2 + bp + 25 = 0$

50. $r^2 - br + 49 = 0$

51. $am^2 + 8m + 1 = 0$

52. $at^2 + 24t + 16 = 0$

53. $9x^2 - 30x + c = 0$

54. $4m^2 + 12m + c = 0$

55. One solution of $4x^2 + bx - 3 = 0$ is $-\dfrac{5}{2}$. Find b and the other solution.

56. One solution of $3x^2 - 7x + c = 0$ is $\dfrac{1}{3}$. Find c and the other solution.

11.4 Solving Equations Quadratic in Form

OBJECTIVES

1. Solve an equation with fractions by writing it in quadratic form.
2. Use quadratic equations to solve applied problems.
3. Solve an equation with radicals by writing it in quadratic form.
4. Solve an equation that is quadratic in form by substitution.

VOCABULARY
☐ quadratic in form

 **NOW TRY EXERCISE 1**

Solve $\dfrac{2}{x} + \dfrac{3}{x+2} = 1$.

NOW TRY ANSWER
1. $\{-1, 4\}$

OBJECTIVE 1 Solve an equation with fractions by writing it in quadratic form.

A variety of nonquadratic equations can be written in the form of a quadratic equation and solved using the methods of this chapter.

EXAMPLE 1 Solving a Rational Equation That Leads to a Quadratic Equation

Solve $\dfrac{1}{x} + \dfrac{1}{x-1} = \dfrac{7}{12}$.

Clear fractions by multiplying each side by the least common denominator, $12x(x-1)$. (The domain is $\{x \mid x \text{ is a real number}, x \neq 0, 1\}$.)

$$12x(x-1)\left(\frac{1}{x} + \frac{1}{x-1}\right) = 12x(x-1)\left(\frac{7}{12}\right) \qquad \text{Multiply by the LCD.}$$

$$12x(x-1)\,\frac{1}{x} + 12x(x-1)\,\frac{1}{x-1} = 12x(x-1)\,\frac{7}{12} \qquad \text{Distributive property}$$

$$12(x-1) + 12x = 7x(x-1) \qquad \text{Multiply.}$$

$$12x - 12 + 12x = 7x^2 - 7x \qquad \text{Distributive property}$$

$$24x - 12 = 7x^2 - 7x \qquad \text{Combine like terms.}$$

This trinomial is factorable. →
$$7x^2 - 31x + 12 = 0 \qquad \text{Standard form}$$

$$(7x - 3)(x - 4) = 0 \qquad \text{Factor.}$$

$$7x - 3 = 0 \quad \text{or} \quad x - 4 = 0 \qquad \text{Zero-factor property}$$

$$7x = 3 \quad \text{or} \quad x = 4 \qquad \text{Solve each equation.}$$

$$x = \frac{3}{7}$$

These values are in the domain. Check them in the original equation. The solution set is $\left\{\frac{3}{7}, 4\right\}$.

NOW TRY

OBJECTIVE 2 Use quadratic equations to solve applied problems.

Some distance-rate-time (or motion) problems lead to quadratic equations.

EXAMPLE 2 Solving a Motion Problem

A riverboat for tourists averages 12 mph in still water. It takes the boat 1 hr, 4 min to travel 6 mi upstream and return. Find the rate of the current.

Step 1 **Read** the problem carefully.

Step 2 **Assign a variable.** Let $x =$ the rate of the current.

The current slows down the boat as it travels upstream, so the rate of the boat traveling upstream is its rate in still water *less* the rate of the current, or $(12 - x)$ mph. See **FIGURE 1** on the next page.

NOW TRY
EXERCISE 2

A small fishing boat averages 18 mph in still water. It takes the boat $\frac{9}{10}$ hr to travel 8 mi upstream and return. Find the rate of the current.

Current

Riverboat traveling upstream—the current slows it down.

FIGURE 1

Similarly, the current speeds up the boat as it travels downstream, so its rate downstream is $(12 + x)$ mph. Thus,

$$12 - x = \text{the rate upstream in miles per hour,}$$

and

$$12 + x = \text{the rate downstream in miles per hour.}$$

	d	r	t
Upstream	6	$12 - x$	$\dfrac{6}{12 - x}$
Downstream	6	$12 + x$	$\dfrac{6}{12 + x}$

Complete a table. Use the distance formula, $d = rt$, solved for time t, $t = \frac{d}{r}$, to write expressions for t.

Step 3 **Write an equation.** We use the total time, 1 hr, 4 min, written as a fraction.

$$1 + \frac{4}{60} = 1 + \frac{1}{15} = \frac{16}{15} \text{ hr} \qquad \text{Total time}$$

The time upstream plus the time downstream equals $\frac{16}{15}$ hr.

Time upstream	+	Time downstream	=	Total time
$\downarrow$		$\downarrow$		$\downarrow$
$\dfrac{6}{12 - x}$	+	$\dfrac{6}{12 + x}$	=	$\dfrac{16}{15}$

Step 4 **Solve** the equation. The LCD is $15(12 - x)(12 + x)$.

$$15(12 - x)(12 + x)\left(\frac{6}{12 - x} + \frac{6}{12 + x}\right)$$

$$= 15(12 - x)(12 + x)\left(\frac{16}{15}\right)$$

Multiply by the LCD.

$$15(12 + x) \cdot 6 + 15(12 - x) \cdot 6 = 16(12 - x)(12 + x)$$

Distributive property; Multiply.

$$90(12 + x) + 90(12 - x) = 16(144 - x^2) \qquad \text{Multiply.}$$

$$1080 + 90x + 1080 - 90x = 2304 - 16x^2 \qquad \text{Distributive property}$$

$$2160 = 2304 - 16x^2 \qquad \text{Combine like terms.}$$

$$16x^2 = 144 \qquad \text{Add } 16x^2. \text{ Subtract } 2160.$$

$$x^2 = 9 \qquad \text{Divide by 16.}$$

$$x = 3 \quad \text{or} \quad x = -3 \qquad \text{Square root property}$$

Step 5 **State the answer.** The current rate cannot be -3, so the answer is 3 mph.

Step 6 **Check** that this value satisfies the original problem.

NOW TRY

PROBLEM-SOLVING HINT Recall from **Section 6.7** that a person's work rate is $\frac{1}{t}$ part of the job per hour, where t is the time in hours required to do the complete job. Thus, the part of the job the person will do in x hours is $\frac{1}{t}x$.

NOW TRY ANSWER
2. 2 mph

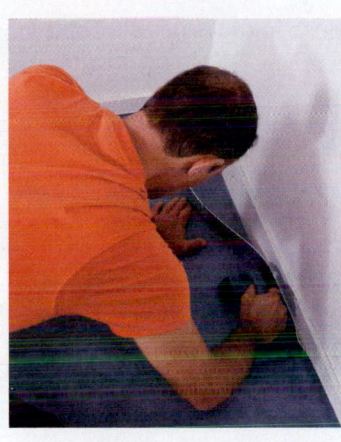

EXAMPLE 3 Solving a Work Problem

It takes two carpet layers 4 hr to carpet a room. If each worked alone, one of them could do the job in 1 hr less time than the other. How long would it take each carpet layer to complete the job alone?

Step 1 **Read** the problem again. There will be two answers.

Step 2 **Assign a variable.**

Let x = the number of hours for the slower carpet layer to complete the job.

Then $x - 1$ = the number of hours for the faster carpet layer to complete the job.

The slower worker's rate is $\frac{1}{x}$, and the faster worker's rate is $\frac{1}{x-1}$. Together they can do the job in 4 hr. Complete a table as shown.

	Rate	Time Working Together	Fractional Part of the Job Done	
Slower Worker	$\frac{1}{x}$	4	$\frac{1}{x}(4)$	Sum is 1 whole job.
Faster Worker	$\frac{1}{x-1}$	4	$\frac{1}{x-1}(4)$	

Step 3 **Write an equation.**

Part done by slower worker + Part done by faster worker = 1 whole job

$$\frac{4}{x} \qquad + \qquad \frac{4}{x-1} \qquad = \qquad 1$$

Step 4 **Solve** the equation from Step 3.

$$x(x-1)\left(\frac{4}{x} + \frac{4}{x-1}\right) = x(x-1)(1) \qquad \text{Multiply by the LCD, } x(x-1).$$

$$4(x-1) + 4x = x(x-1) \qquad \text{Distributive property}$$

$$4x - 4 + 4x = x^2 - x \qquad \text{Distributive property}$$

$$x^2 - 9x + 4 = 0 \qquad \text{Standard form}$$

The trinomial on the left cannot be factored, so the equation cannot be solved using the zero-factor property. We use the quadratic formula.

$$x = \frac{-b \pm \sqrt{b^2 - 4ac}}{2a} \qquad \text{Quadratic formula}$$

$$x = \frac{-(-9) \pm \sqrt{(-9)^2 - 4(1)(4)}}{2(1)} \qquad a = 1, b = -9, c = 4$$

$$x = \frac{9 \pm \sqrt{65}}{2} \qquad \text{Simplify.}$$

$$x = \frac{9 + \sqrt{65}}{2} = 8.5 \quad \text{or} \quad x = \frac{9 - \sqrt{65}}{2} = 0.5 \qquad \text{Use a calculator. Round to the nearest tenth.}$$

**NOW TRY
EXERCISE 3**

Two electricians are running wire to finish a basement. One electrician could finish the job in 2 hr less time than the other. Together, they complete the job in 6 hr. How long (to the nearest tenth) would it take the slower electrician to complete the job alone?

**NOW TRY
EXERCISE 4**

Solve each equation.

(a) $x = \sqrt{9x - 20}$

(b) $x + \sqrt{x} = 20$

Step 5 **State the answer.** Only the solution 8.5 makes sense in the original problem.

If $x = 0.5$, then $x - 1 = 0.5 - 1 = -0.5$,   Time cannot be negative.

which cannot represent the time for the faster worker. The slower worker could do the job in 8.5 hr and the faster in $8.5 - 1 = 7.5$ hr.

Step 6 **Check** that these results satisfy the original problem.   NOW TRY

> **OBJECTIVE 3** Solve an equation with radicals by writing it in quadratic form.

> **EXAMPLE 4** Solving Radical Equations That Lead to Quadratic Equations

Solve each equation.

(a) $x = \sqrt{6x - 8}$

This equation is not quadratic. However, squaring each side of the equation gives a quadratic equation that can be solved using the zero-factor property.

$$x^2 = \left(\sqrt{6x - 8}\right)^2 \qquad \text{Square each side.}$$
$$x^2 = 6x - 8 \qquad \left(\sqrt{a}\right)^2 = a$$

This trinomial is factorable. $\longrightarrow x^2 - 6x + 8 = 0 \qquad \text{Standard form}$

$$(x - 4)(x - 2) = 0 \qquad \text{Factor.}$$
$$x - 4 = 0 \quad \text{or} \quad x - 2 = 0 \qquad \text{Zero-factor property}$$
$$x = 4 \quad \text{or} \qquad x = 2 \qquad \text{Proposed solutions}$$

Squaring each side of a radical equation can introduce extraneous solutions. ***All proposed solutions must be checked in the original (not the squared) equation.***

CHECK $x = \sqrt{6x - 8}$ $x = \sqrt{6x - 8}$

$4 \overset{?}{=} \sqrt{6(4) - 8}$ Let $x = 4$. $2 \overset{?}{=} \sqrt{6(2) - 8}$ Let $x = 2$.

$4 \overset{?}{=} \sqrt{16}$ $2 \overset{?}{=} \sqrt{4}$

$4 = 4$ ✓   True $2 = 2$ ✓   True

Both solutions check, so the solution set is $\{2, 4\}$.

(b) $x + \sqrt{x} = 6$

$$\sqrt{x} = 6 - x \qquad \text{Isolate the radical on one side.}$$
$$\left(\sqrt{x}\right)^2 = (6 - x)^2 \qquad \text{Square each side.}$$
$$x = 36 - 12x + x^2 \qquad (a - b)^2 = a^2 - 2ab + b^2$$
$$x^2 - 13x + 36 = 0 \qquad \text{Standard form}$$
$$(x - 4)(x - 9) = 0 \qquad \text{Factor.}$$
$$x - 4 = 0 \quad \text{or} \quad x - 9 = 0 \qquad \text{Zero-factor property}$$
$$x = 4 \quad \text{or} \qquad x = 9 \qquad \text{Proposed solutions}$$

CHECK $x + \sqrt{x} = 6$ $x + \sqrt{x} = 6$

$4 + \sqrt{4} \overset{?}{=} 6$ Let $x = 4$. $9 + \sqrt{9} \overset{?}{=} 6$ Let $x = 9$.

$6 = 6$ ✓   True $12 = 6$ False

NOW TRY ANSWERS
3. 13.1 hr
4. (a) $\{4, 5\}$ **(b)** $\{16\}$

Only the solution 4 checks, so the solution set is $\{4\}$.   NOW TRY

OBJECTIVE 4 Solve an equation that is quadratic in form by substitution.

A nonquadratic equation that can be written in the form

$$au^2 + bu + c = 0,$$

for $a \neq 0$ and an algebraic expression u, is **quadratic in form.**

Many equations that are quadratic in form can be solved more easily by defining and substituting a "temporary" variable u for an expression involving the variable in the original equation.

NOW TRY EXERCISE 5

Define a variable u in terms of x, and write each equation in the quadratic form $au^2 + bu + c = 0$.

(a) $x^4 - 10x^2 + 9 = 0$

(b) $6(x + 2)^2$
   $- 11(x + 2) + 4 = 0$

EXAMPLE 5 Defining Substitution Variables

Define a variable u in terms of x, and write each equation in the quadratic form $au^2 + bu + c = 0$.

(a) $x^4 - 13x^2 + 36 = 0$

Look at the two terms involving the variable x, ignoring their coefficients. Try to find one variable expression that is the square of the other. Here $x^4 = (x^2)^2$, so we can define $u = x^2$, and rewrite the original equation as a quadratic equation in u.

$$u^2 - 13u + 36 = 0 \qquad \text{Here, } u = x^2.$$

(b) $2(4x - 3)^2 + 7(4x - 3) + 5 = 0$

Because this equation involves both $(4x - 3)^2$ and $(4x - 3)$, we let $u = 4x - 3$.

$$2u^2 + 7u + 5 = 0 \qquad \text{Here, } u = 4x - 3.$$

(c) $2x^{2/3} - 11x^{1/3} + 12 = 0$

We apply a power rule for exponents, $(a^m)^n = a^{mn}$ **(Section 4.1).** Because $(x^{1/3})^2 = x^{2/3}$, we define $u = x^{1/3}$ and write the original equation as follows.

$$2u^2 - 11u + 12 = 0 \qquad \text{Here, } u = x^{1/3}.$$ **NOW TRY**

EXAMPLE 6 Solving Equations That Are Quadratic in Form

Solve each equation.

(a)
$$x^4 - 13x^2 + 36 = 0 \qquad \text{See Example 5(a).}$$

$$(x^2)^2 - 13x^2 + 36 = 0 \qquad x^4 = (x^2)^2$$

$$\boxed{\text{Quadratic in form}} \quad u^2 - 13u + 36 = 0 \qquad \text{Let } u = x^2.$$

$$(u - 4)(u - 9) = 0 \qquad \text{Factor.}$$

$$u - 4 = 0 \quad \text{or} \quad u - 9 = 0 \qquad \text{Zero-factor property}$$

$$\boxed{\text{Don't stop here.}} \quad u = 4 \quad \text{or} \quad u = 9 \qquad \text{Solve.}$$

$$x^2 = 4 \quad \text{or} \quad x^2 = 9 \qquad \text{Substitute } x^2 \text{ for } u.$$

$$x = \pm 2 \quad \text{or} \quad x = \pm 3 \qquad \text{Square root property}$$

Each value can be verified by substituting it into the original equation for x. The equation $x^4 - 13x^2 + 36 = 0$ is a fourth-degree equation and has four solutions, $-3, -2, 2, 3$.* The solution set is abbreviated $\{\pm 2, \pm 3\}$.

NOW TRY ANSWERS

5. (a) $u = x^2$;
   $u^2 - 10u + 9 = 0$
   (b) $u = x + 2$;
   $6u^2 - 11u + 4 = 0$

---

*In general, an equation in which an nth-degree polynomial equals 0 has n complex solutions, although some of them may be repeated.

**NOW TRY
EXERCISE 6**

Solve each equation.

(a) $x^4 - 17x^2 + 16 = 0$

(b) $x^4 + 4 = 8x^2$

(b)

$$4x^4 + 1 = 5x^2$$

$$4(x^2)^2 + 1 = 5x^2 \qquad x^4 = (x^2)^2$$

$$4u^2 + 1 = 5u \qquad \text{Let } u = x^2.$$

$$4u^2 - 5u + 1 = 0 \qquad \text{Standard form}$$

$$(4u - 1)(u - 1) = 0 \qquad \text{Factor.}$$

$$4u - 1 = 0 \quad \text{or} \quad u - 1 = 0 \qquad \text{Zero-factor property}$$

$$u = \frac{1}{4} \quad \text{or} \quad u = 1 \qquad \text{Solve.}$$

> This is a key step. →

$$x^2 = \frac{1}{4} \quad \text{or} \quad x^2 = 1 \qquad \text{Substitute } x^2 \text{ for } u.$$

$$x = \pm\frac{1}{2} \quad \text{or} \quad x = \pm 1 \qquad \text{Square root property}$$

Check that the solution set is $\left\{ \pm\frac{1}{2}, \pm 1 \right\}$.

(c)

$$x^4 = 6x^2 - 3$$

$$x^4 - 6x^2 + 3 = 0 \qquad \text{Standard form}$$

$$(x^2)^2 - 6x^2 + 3 = 0 \qquad x^4 = (x^2)^2$$

$$u^2 - 6u + 3 = 0 \qquad \text{Let } u = x^2.$$

The trinomial on the left is nonfactorable, so we cannot solve the equation using the zero-factor property. To solve, we use the quadratic formula.

$$u = \frac{-(-6) \pm \sqrt{(-6)^2 - 4(1)(3)}}{2(1)} \qquad a = 1, b = -6, c = 3$$

$$u = \frac{6 \pm \sqrt{24}}{2} \qquad \text{Simplify.}$$

$$u = \frac{6 \pm 2\sqrt{6}}{2} \qquad \sqrt{24} = \sqrt{4} \cdot \sqrt{6} = 2\sqrt{6}$$

$$u = \frac{2(3 \pm \sqrt{6})}{2} \qquad \text{Factor.}$$

$$u = 3 \pm \sqrt{6} \qquad \text{Divide out the common factor 2.}$$

> Find *both* square roots in each case. →

$$x^2 = 3 + \sqrt{6} \quad \text{or} \quad x^2 = 3 - \sqrt{6} \qquad \text{Substitute } x^2 \text{ for } u.$$

$$x = \pm\sqrt{3 + \sqrt{6}} \quad \text{or} \quad x = \pm\sqrt{3 - \sqrt{6}} \qquad \text{Square root property}$$

The solution set contains four numbers and is written

$$\left\{ \pm\sqrt{3 + \sqrt{6}}, \pm\sqrt{3 - \sqrt{6}} \right\}.$$

NOW TRY ANSWERS

6. (a) $\{ \pm 1, \pm 4 \}$

(b) $\left\{ \pm\sqrt{4 + 2\sqrt{3}}, \pm\sqrt{4 - 2\sqrt{3}} \right\}$

NOW TRY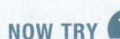

NOTE The quadratic expressions in equations like those in **Examples 6(a) and (b)** can be factored directly.

$$x^4 - 13x^2 + 36 = 0 \qquad \text{Example 6(a) equation}$$

$$(x^2 - 9)(x^2 - 4) = 0 \qquad \text{Factor.}$$

$$(x + 3)(x - 3)(x + 2)(x - 2) = 0 \qquad \text{Factor again.}$$

Using the zero-factor property gives the same solutions that we obtained in **Example 6(a).** Equations that include nonfactorable quadratic expressions (as in **Example 6(c)**) must be solved using substitution and the quadratic formula.

Solving an Equation That Is Quadratic in Form by Substitution

Step 1 **Define a temporary variable u,** based on the relationship between the variable expressions in the given equation. Substitute u in the original equation and rewrite the equation in the form $au^2 + bu + c = 0$.

Step 2 **Solve the quadratic equation obtained in Step 1** either by factoring the trinomial and applying the zero-factor property or by using the quadratic formula.

Step 3 **Replace u with the expression it defined in Step 1.**

Step 4 **Solve the resulting equations for the original variable.**

Step 5 **Check** all solutions by substituting them in the original equation.

EXAMPLE 7 Solving Equations That Are Quadratic in Form

Solve each equation.

(a) $2(4x - 3)^2 + 7(4x - 3) + 5 = 0$

Step 1 Because of the repeated quantity $4x - 3$, substitute u for $4x - 3$.

$$2(4x - 3)^2 + 7(4x - 3) + 5 = 0 \qquad \text{See Example 5(b).}$$

$$2u^2 + 7u + 5 = 0 \qquad \text{Let } u = 4x - 3.$$

Step 2 $$(2u + 5)(u + 1) = 0 \qquad \text{Factor.}$$

$$2u + 5 = 0 \quad \text{or} \quad u + 1 = 0 \qquad \text{Zero-factor property}$$

Don't stop here. $\quad u = -\dfrac{5}{2} \quad \text{or} \qquad u = -1 \qquad \text{Solve for } u.$

Step 3 $$4x - 3 = -\frac{5}{2} \quad \text{or} \quad 4x - 3 = -1 \qquad \text{Substitute } 4x - 3 \text{ for } u.$$

Step 4 $$4x = \frac{1}{2} \quad \text{or} \quad 4x = 2 \qquad \text{Solve for } x.$$

$$x = \frac{1}{8} \quad \text{or} \quad x = \frac{1}{2}$$

Step 5 Check that the solution set of the original equation is $\left\{\frac{1}{8}, \frac{1}{2}\right\}$.

**NOW TRY
EXERCISE 7**
Solve each equation.
(a) $6(x-4)^2 + 11(x-4) - 10 = 0$

(b) $2x^{2/3} - 7x^{1/3} + 3 = 0$

(b) $2x^{2/3} - 11x^{1/3} + 12 = 0$

Step 1 Because $x^{2/3} = (x^{1/3})^2$, we substitute u for $x^{1/3}$.

$$2x^{2/3} - 11x^{1/3} + 12 = 0 \qquad \text{See Example 5(c).}$$

$$2(x^{1/3})^2 - 11x^{1/3} + 12 = 0 \qquad x^{2/3} = (x^{1/3})^2$$

$$2u^2 - 11u + 12 = 0 \qquad \text{Let } u = x^{1/3}.$$

Step 2
$$(2u - 3)(u - 4) = 0 \qquad \text{Factor.}$$

$$2u - 3 = 0 \qquad \text{or} \qquad u - 4 = 0 \qquad \text{Zero-factor property}$$

$$u = \frac{3}{2} \qquad \text{or} \qquad u = 4 \qquad \text{Solve for } u.$$

Step 3
$$x^{1/3} = \frac{3}{2} \qquad \text{or} \qquad x^{1/3} = 4 \qquad \text{Substitute } x^{1/3} \text{ for } u.$$

Step 4
$$(x^{1/3})^3 = \left(\frac{3}{2}\right)^3 \qquad \text{or} \qquad (x^{1/3})^3 = 4^3 \qquad \text{Cube each side.}$$

$$x = \frac{27}{8} \qquad \text{or} \qquad x = 64 \qquad \text{Apply the exponents.}$$

Step 5 Because the original equation involves variables with rational exponents, check that neither of these solutions is extraneous. The solution set is $\left\{\frac{27}{8}, 64\right\}$.

NOW TRY

❗ **CAUTION**   A common error when solving problems like those in **Examples 6 and 7** is to stop too soon. ***Once we have solved for u, we must remember to substitute and solve for the values of the original variable.***

NOW TRY ANSWERS

7. (a) $\left\{\frac{3}{2}, \frac{14}{3}\right\}$ **(b)** $\left\{\frac{1}{8}, 27\right\}$

11.4 Exercises

FOR EXTRA HELP

▶ **MyMathLab®**

▶ *Complete solution available in MyMathLab*

Concept Check   *Based on the discussion and examples of this section, give the first step to solve each equation. Do not actually solve.*

1. $\dfrac{14}{x} = x - 5$

2. $\sqrt{1 + x} + x = 5$

3. $(x^2 + x)^2 - 8(x^2 + x) + 12 = 0$

4. $3x = \sqrt{16 - 10x}$

5. *Concept Check* Study this incorrect "solution." ***WHAT WENT WRONG?***

$$x = \sqrt{3x + 4} \qquad \text{Square}$$
$$x^2 = 3x + 4 \qquad \text{each side.}$$
$$x^2 - 3x - 4 = 0$$
$$(x - 4)(x + 1) = 0$$
$$x - 4 = 0 \quad \text{or} \quad x + 1 = 0$$
$$x = 4 \quad \text{or} \qquad x = -1$$

Solution set: $\{4, -1\}$

6. *Concept Check* Study this incorrect "solution." ***WHAT WENT WRONG?***

$$2(x - 1)^2 - 3(x - 1) + 1 = 0 \qquad \text{Let}$$
$$2u^2 - 3u + 1 = 0 \qquad u = x - 1.$$
$$(2u - 1)(u - 1) = 0$$
$$2u - 1 = 0 \quad \text{or} \quad u - 1 = 0$$
$$u = \frac{1}{2} \quad \text{or} \qquad u = 1$$

Solution set: $\left\{\frac{1}{2}, 1\right\}$

Solve each equation. Check the solutions. See Example 1.

7. $\dfrac{14}{x} = x - 5$

8. $\dfrac{-12}{x} = x + 8$

9. $1 - \dfrac{3}{x} - \dfrac{28}{x^2} = 0$

10. $4 - \dfrac{7}{r} - \dfrac{2}{r^2} = 0$

11. $3 - \dfrac{1}{t} = \dfrac{2}{t^2}$

12. $1 + \dfrac{2}{x} = \dfrac{3}{x^2}$

13. $\dfrac{1}{x} + \dfrac{2}{x+2} = \dfrac{17}{35}$

14. $\dfrac{2}{m} + \dfrac{3}{m+9} = \dfrac{11}{4}$

15. $\dfrac{2}{x+1} + \dfrac{3}{x+2} = \dfrac{7}{2}$

16. $\dfrac{4}{3-p} + \dfrac{2}{5-p} = \dfrac{26}{15}$

17. $\dfrac{3}{2x} - \dfrac{1}{2(x+2)} = 1$

18. $\dfrac{4}{3x} - \dfrac{1}{2(x+1)} = 1$

19. $3 = \dfrac{1}{t+2} + \dfrac{2}{(t+2)^2}$

20. $1 + \dfrac{2}{3z+2} = \dfrac{15}{(3z+2)^2}$

21. $\dfrac{6}{p} = 2 + \dfrac{p}{p+1}$

22. $\dfrac{x}{2-x} + \dfrac{2}{x} = 5$

23. $1 - \dfrac{1}{2x+1} - \dfrac{1}{(2x+1)^2} = 0$

24. $1 - \dfrac{1}{3x-2} - \dfrac{1}{(3x-2)^2} = 0$

Concept Check   *Answer each question.*

25. A boat travels 20 mph in still water, and the rate of the current is t mph.

   (a) What is the rate of the boat when it travels upstream?

   (b) What is the rate of the boat when it travels downstream?

26. It takes m hours to grade a set of papers.

   (a) What is the grader's rate (in job per hour)?

   (b) How much of the job will the grader do in 2 hr?

Solve each problem. Round answers to the nearest tenth. See Examples 2 and 3.

27. In 4 hr, Kerrie can travel 15 mi upriver and come back. The rate of the current is 5 mph. Find the rate of her boat in still water.

Let $x = $ _____.

The rate traveling upriver (*against the current*) is ____ mph.

The rate traveling back downriver (*with the current*) is ____ mph.

Complete the table.

	d	r	t
Up			
Down			

Write an equation, and complete the solution.

28. Carlos can complete a certain lab test in 2 hr less time than Jaime can. If they can finish the job together in 2 hr, how long would it take each of them working alone?

Let $x = $ Jaime's time alone (in hours).

Then ____ = Carlos' time alone (in hours).

Complete the table.

	Rate	Time Working Together	Fractional Part of the Job Done
Carlos			
Jaime			

Write an equation, and complete the solution.

▶ **29.** On a windy day William found that he could travel 16 mi downstream and then 4 mi back upstream at top speed in a total of 48 min. What was the top speed of William's boat if the rate of the current was 15 mph? (Let x represent the rate of the boat in still water.)

	d	r	t
Upstream	4	$x - 15$	
Downstream	16		

30. Vera flew for 6 hr at a constant rate. She traveled 810 mi with the wind, then turned around and traveled 720 mi against the wind. The wind speed was a constant 15 mph. Find the rate of the plane.

	d	r	t
With Wind	810		
Against Wind	720		

31. The distance from Jackson to Lodi is about 40 mi, as is the distance from Lodi to Manteca. Adrian drove from Jackson to Lodi, stopped in Lodi for a high-energy drink, and then drove on to Manteca at 10 mph faster. Driving time for the entire trip was 88 min. Find her rate from Jackson to Lodi. (*Source: State Farm Road Atlas.*)

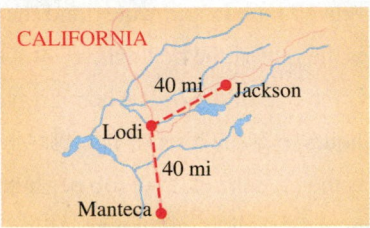

32. Medicine Hat and Cranbrook are 300 km apart. Steve rides his Harley 20 km per hr faster than Mohammad rides his Yamaha. Find Steve's average rate if he travels from Cranbrook to Medicine Hat in $1\frac{1}{4}$ hr less time than Mohammad.

(*Source: State Farm Road Atlas.*)

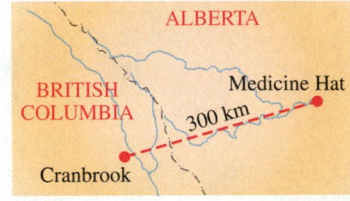

33. Working together, two people can cut a large lawn in 2 hr. One person can do the job alone in 1 hr less time than the other. How long (to the nearest tenth) would it take the faster worker to do the job? (Let x represent the time of the faster worker.)

	Rate	Time Working Together	Fractional Part of the Job Done
Faster Worker	$\frac{1}{x}$	2	
Slower Worker		2	

34. Working together, two people can clean an office building in 5 hr. One person takes 2 hr longer than the other to clean the building alone. How long (to the nearest tenth) would it take the slower worker to clean the building alone? (Let x represent the time of the slower worker.)

	Rate	Time Working Together	Fractional Part of the Job Done
Faster Worker			
Slower Worker	$\frac{1}{x}$		

▶ **35.** Rusty and Nancy are planting flowers. Working alone, Rusty would take 2 hr longer than Nancy to plant the flowers. Working together, they do the job in 12 hr. How long (to the nearest tenth) would it have taken each person working alone?

36. Joel can work through a stack of invoices in 1 hr less time than Noel can. Working together they take $1\frac{1}{2}$ hr. How long (to the nearest tenth) would it take each person working alone?

37. Two pipes together can fill a tank in 2 hr. One of the pipes, used alone, takes 3 hr longer than the other to fill the tank. How long would each pipe take to fill the tank alone?

38. A washing machine can be filled in 6 min if both the hot and cold water taps are fully opened. Filling the washer with hot water alone takes 9 min longer than filling it with cold water alone. How long does it take to fill the washer with cold water?

Solve each equation. Check the solutions. **See Example 4.**

39. $x = \sqrt{7x - 10}$ **40.** $z = \sqrt{5z - 4}$ **41.** $2x = \sqrt{11x + 3}$

42. $4x = \sqrt{6x + 1}$ **43.** $3x = \sqrt{16 - 10x}$ **44.** $4t = \sqrt{8t + 3}$

45. $t + \sqrt{t} = 12$ **46.** $p - 2\sqrt{p} = 8$ **47.** $x = \sqrt{\dfrac{6 - 13x}{5}}$

48. $r = \sqrt{\dfrac{20 - 19r}{6}}$ **49.** $-x = \sqrt{\dfrac{8 - 2x}{3}}$ **50.** $-x = \sqrt{\dfrac{3x + 7}{4}}$

Solve each equation. Check the solutions. **See Examples 5–7.**

51. $x^4 - 29x^2 + 100 = 0$ **52.** $x^4 - 37x^2 + 36 = 0$ **53.** $4q^4 - 13q^2 + 9 = 0$

54. $9x^4 - 25x^2 + 16 = 0$ **55.** $x^4 + 48 = 16x^2$ **56.** $z^4 + 72 = 17z^2$

57. $(x + 3)^2 + 5(x + 3) + 6 = 0$ **58.** $(x - 4)^2 + (x - 4) - 20 = 0$

59. $3(m + 4)^2 - 8 = 2(m + 4)$ **60.** $(t + 5)^2 + 6 = 7(t + 5)$

61. $x^{2/3} + x^{1/3} - 2 = 0$ **62.** $x^{2/3} - 2x^{1/3} - 3 = 0$

63. $r^{2/3} + r^{1/3} - 12 = 0$ **64.** $3x^{2/3} - x^{1/3} - 24 = 0$

65. $4x^{4/3} - 13x^{2/3} + 9 = 0$ **66.** $9t^{4/3} - 25t^{2/3} + 16 = 0$

67. $2 + \dfrac{5}{3x - 1} = \dfrac{-2}{(3x - 1)^2}$ **68.** $3 - \dfrac{7}{2p + 2} = \dfrac{6}{(2p + 2)^2}$

69. $2 - 6(z - 1)^{-2} = (z - 1)^{-1}$ **70.** $3 - 2(x - 1)^{-1} = (x - 1)^{-2}$

The following exercises are not grouped by type. Solve each equation. (Exercises 83 and 84 require knowledge of complex numbers.) **See Examples 1 and 4–7.**

71. $12x^4 - 11x^2 + 2 = 0$ **72.** $\left(x - \dfrac{1}{2}\right)^2 + 5\left(x - \dfrac{1}{2}\right) - 4 = 0$

73. $\sqrt{2x + 3} = 2 + \sqrt{x - 2}$ **74.** $\sqrt{m + 1} = -1 + \sqrt{2m}$

75. $2(1 + \sqrt{r})^2 = 13(1 + \sqrt{r}) - 6$ **76.** $(x^2 + x)^2 + 12 = 8(x^2 + x)$

77. $2m^6 + 11m^3 + 5 = 0$ **78.** $8x^6 + 513x^3 + 64 = 0$

79. $6 = 7(2w - 3)^{-1} + 3(2w - 3)^{-2}$ **80.** $x^6 - 10x^3 = -9$

81. $2x^4 - 9x^2 = -2$ **82.** $8x^4 + 1 = 11x^2$

83. $2x^4 + x^2 - 3 = 0$ **84.** $4x^4 + 5x^2 + 1 = 0$

SUMMARY EXERCISES   Applying Methods for Solving Quadratic Equations

We have introduced four methods for solving quadratic equations written in standard form $ax^2 + bx + c = 0$.

Method	Advantages	Disadvantages
Zero-factor property	This is usually the fastest method.	Not all trinomials are factorable. Some factorable trinomials are difficult to factor.
Square root property	This is the simplest method for solving equations of the form $(ax + b)^2 = c$.	Few equations are given in this form.
Completing the square	This method can always be used, although most people prefer the quadratic formula.	It requires more steps than other methods.
Quadratic formula	This method can always be used.	Sign errors are common when evaluating $\sqrt{b^2 - 4ac}$.

1. square root property
2. zero-factor property
3. quadratic formula
4. quadratic formula
5. zero-factor property
6. square root property

7. $\{\pm\sqrt{7}\}$ **8.** $\left\{-\dfrac{3}{2}, \dfrac{5}{3}\right\}$

9. $\{-3 \pm \sqrt{5}\}$ **10.** $\{-2, 8\}$

11. $\left\{-\dfrac{3}{2}, 4\right\}$ **12.** $\left\{-3, \dfrac{1}{3}\right\}$

13. $\left\{\dfrac{2 \pm \sqrt{2}}{2}\right\}$ **14.** $\{\pm 2i\sqrt{3}\}$

15. $\left\{\dfrac{1}{2}, 2\right\}$ **16.** $\{\pm 1, \pm 3\}$

17. $\left\{\dfrac{-3 \pm 2\sqrt{2}}{2}\right\}$

18. $\left\{\dfrac{4}{5}, 3\right\}$

19. $\{\pm\sqrt{2}, \pm\sqrt{7}\}$

20. $\left\{\dfrac{1 \pm \sqrt{5}}{4}\right\}$

21. $\left\{-\dfrac{1}{2} \pm \dfrac{\sqrt{3}}{2}i\right\}$

22. $\left\{-\dfrac{\sqrt[3]{175}}{5}, 1\right\}$

23. $\left\{\dfrac{3}{2}\right\}$ **24.** $\left\{\dfrac{2}{3}\right\}$

25. $\{\pm 6\sqrt{2}\}$ **26.** $\left\{-\dfrac{2}{3}, 2\right\}$

27. $\{-4, 9\}$ **28.** $\{\pm 13\}$

29. $\left\{1 \pm \dfrac{\sqrt{3}}{3}i\right\}$ **30.** $\{3\}$

31. $\left\{\dfrac{1}{6} \pm \dfrac{\sqrt{47}}{6}i\right\}$

32. $\left\{-\dfrac{1}{3}, \dfrac{1}{6}\right\}$

Concept Check   *Decide whether the* zero-factor property, *the* square root property, *or the* quadratic formula *is most appropriate for solving each quadratic equation. Do not actually solve.*

1. $(2x + 3)^2 = 4$ **2.** $4x^2 - 3x = 1$ **3.** $x^2 + 5x - 8 = 0$

4. $2x^2 + 3x = 1$ **5.** $3x^2 = 2 - 5x$ **6.** $x^2 = 5$

Solve each quadratic equation by the method of your choice.

7. $p^2 = 7$ **8.** $6x^2 - x - 15 = 0$ **9.** $n^2 + 6n + 4 = 0$

10. $(x - 3)^2 = 25$ **11.** $\dfrac{5}{x} + \dfrac{12}{x^2} = 2$ **12.** $3x^2 = 3 - 8x$

13. $2r^2 - 4r + 1 = 0$ ***14.** $x^2 = -12$ **15.** $x\sqrt{2} = \sqrt{5x - 2}$

16. $x^4 - 10x^2 + 9 = 0$ **17.** $(2x + 3)^2 = 8$ **18.** $\dfrac{2}{x} + \dfrac{1}{x - 2} = \dfrac{5}{3}$

19. $t^4 + 14 = 9t^2$ **20.** $8x^2 - 4x = 2$ ***21.** $z^2 + z + 1 = 0$

22. $5x^6 + 2x^3 - 7 = 0$ **23.** $4t^2 - 12t + 9 = 0$ **24.** $x\sqrt{3} = \sqrt{2 - x}$

25. $r^2 - 72 = 0$ **26.** $-3x^2 + 4x = -4$ **27.** $x^2 - 5x - 36 = 0$

28. $w^2 = 169$ ***29.** $3p^2 = 6p - 4$ **30.** $z = \sqrt{\dfrac{5z + 3}{2}}$

***31.** $\dfrac{4}{r^2} + 3 = \dfrac{1}{r}$ **32.** $2(3x - 1)^2 + 5(3x - 1) = -2$

---

*This exercise requires knowledge of complex numbers.

11.5 Formulas and Further Applications

NOW TRY EXERCISE 1

Solve each formula for the specified variable. Keep $\pm$ in the answer in part (a).

(a) $n = \dfrac{ab}{E^2}$ for E

(b) $S = \sqrt{\dfrac{pq}{n}}$ for p

OBJECTIVE 1 Solve formulas involving squares and square roots for specified variables.

EXAMPLE 1 Solving for Specified Variables

Solve each formula for the specified variable. Keep $\pm$ in the answer in part (a).

(a) $w = \dfrac{kFr}{v^2}$ for v

$$w = \dfrac{kFr}{v^2} \quad \text{The goal is to isolate } v \text{ on one side.}$$

$$v^2 w = kFr \qquad \text{Multiply by } v^2.$$

$$v^2 = \dfrac{kFr}{w} \qquad \text{Divide by } w.$$

$$v = \pm\sqrt{\dfrac{kFr}{w}} \qquad \text{Square root property}$$

Include both positive and negative roots.

$$v = \dfrac{\pm\sqrt{kFr}}{\sqrt{w}} \cdot \dfrac{\sqrt{w}}{\sqrt{w}} \qquad \text{Rationalize the denominator.}$$

$$v = \dfrac{\pm\sqrt{kFrw}}{w} \qquad \sqrt{a}\cdot\sqrt{b} = \sqrt{ab};\ \sqrt{a}\cdot\sqrt{a} = a$$

(b) $d = \sqrt{\dfrac{4\mathscr{A}}{\pi}}$ for $\mathscr{A}$

$$d = \sqrt{\dfrac{4\mathscr{A}}{\pi}} \quad \text{The goal is to isolate } \mathscr{A} \text{ on one side.}$$

$$d^2 = \dfrac{4\mathscr{A}}{\pi} \qquad \text{Square each side.}$$

$$\pi d^2 = 4\mathscr{A} \qquad \text{Multiply by } \pi.$$

$$\dfrac{\pi d^2}{4} = \mathscr{A}, \quad \text{or} \quad \mathscr{A} = \dfrac{\pi d^2}{4} \qquad \begin{array}{l}\text{Divide by 4.}\\ \text{Interchange sides.}\end{array}$$ **NOW TRY**

EXAMPLE 2 Solving for a Specified Variable

Solve $s = 2t^2 + kt$ for t.

Because the given equation has terms with t^2 and t, write it in standard form $ax^2 + bx + c = 0$, with t as the variable instead of x.

$$s = 2t^2 + kt$$

$$0 = 2t^2 + kt - s \qquad \text{Subtract } s.$$

$$2t^2 + kt - s = 0 \qquad \text{Standard form}$$

NOW TRY ANSWERS

1. **(a)** $E = \dfrac{\pm\sqrt{abn}}{n}$

   (b) $p = \dfrac{nS^2}{q}$

NOW TRY
EXERCISE 2
Solve for r.
$$r^2 + 9r = -c$$

To solve $2t^2 + kt - s = 0$, use the quadratic formula with $a = 2$, $b = k$, and $c = -s$.

$$t = \frac{-k \pm \sqrt{k^2 - 4(2)(-s)}}{2(2)} \qquad \text{Substitute.}$$

$$t = \frac{-k \pm \sqrt{k^2 + 8s}}{4} \qquad \text{Simplify.}$$

The solutions are $t = \dfrac{-k + \sqrt{k^2 + 8s}}{4}$ and $t = \dfrac{-k - \sqrt{k^2 + 8s}}{4}$.   **NOW TRY**

OBJECTIVE 2 Solve applied problems using the Pythagorean theorem.

The Pythagorean theorem, represented by the equation

$$a^2 + b^2 = c^2, \quad \text{See FIGURE 2.}$$

was introduced in **Sections 5.6 and 10.3.** It is used to solve applications involving right triangles.

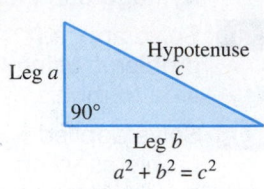

$a^2 + b^2 = c^2$
Pythagorean theorem

FIGURE 2

NOW TRY
EXERCISE 3
Matt is building a new barn, with length 10 ft more than width. While determining the footprint of the barn, he measured the diagonal as 50 ft. What will be the dimensions of the barn?

EXAMPLE 3   Using the Pythagorean Theorem

Two cars left an intersection at the same time, one heading due north, the other due west. Some time later, they were exactly 100 mi apart. The car headed north had gone 20 mi farther than the car headed west. How far had each car traveled?

Step 1   **Read** the problem carefully.

Step 2   **Assign a variable.**

   Let $x = $ the distance traveled by the car headed west.

   Then $x + 20 = $ the distance traveled by the car headed north.

   See **FIGURE 3**. The cars are 100 mi apart, so the hypotenuse of the right triangle equals 100.

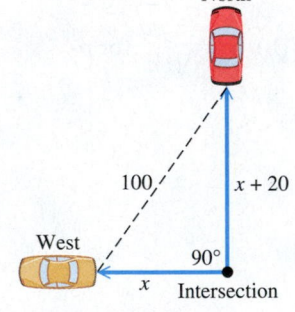

FIGURE 3

Step 3   **Write an equation.** Use the Pythagorean theorem.

$$a^2 + b^2 = c^2 \qquad \text{Pythagorean theorem}$$

$$x^2 + (x + 20)^2 = 100^2 \qquad \text{See FIGURE 3.}$$

$(x + y)^2 = x^2 + 2xy + y^2$

Step 4   **Solve.** $x^2 + x^2 + 40x + 400 = 10{,}000$ Square the binomial.

$$2x^2 + 40x - 9600 = 0 \qquad \text{Standard form}$$

$$x^2 + 20x - 4800 = 0 \qquad \text{Divide by 2.}$$

$$(x + 80)(x - 60) = 0 \qquad \text{Factor.}$$

$$x + 80 = 0 \qquad \text{or} \quad x - 60 = 0 \qquad \text{Zero-factor property}$$

$$x = -80 \quad \text{or} \qquad x = 60 \qquad \text{Solve for } x.$$

NOW TRY ANSWERS

2. $r = \dfrac{-9 \pm \sqrt{81 - 4c}}{2}$

3. 30 ft by 40 ft

Step 5   **State the answer.** Distance cannot be negative, so discard the negative solution. The required distances are 60 mi and $60 + 20 = 80$ mi.

Step 6   **Check.** Here $60^2 + 80^2 = 100^2$, so the answer is correct.   **NOW TRY**

OBJECTIVE 3   Solve applied problems using area formulas.

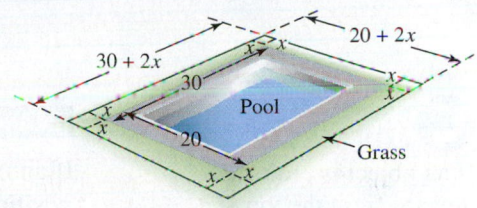

NOW TRY EXERCISE 4

A football practice field is 30 yd wide and 40 yd long. A strip of grass sod of uniform width is to be placed around the perimeter of the practice field. There is enough money budgeted for 296 sq yd of sod. How wide will the strip be?

EXAMPLE 4   Solving an Area Problem

A rectangular reflecting pool in a park is 20 ft wide and 30 ft long. The gardener wants to plant a strip of grass of uniform width around the edge of the pool. She has enough seed to cover 336 ft². How wide will the strip be?

Step 1   **Read** the problem carefully.

FIGURE 4

Step 2   **Assign a variable.** The pool is shown in **FIGURE 4**.

Let x = the unknown width of the grass strip.

Then $20 + 2x$ = the width of the large rectangle (the width of the pool plus two grass strips),

and $30 + 2x$ = the length of the large rectangle.

Step 3   **Write an equation.** Refer to **FIGURE 4**.

$(30 + 2x)(20 + 2x)$ Area of large rectangle (length · width)

$30 \cdot 20$, or 600 Area of pool (in square feet)

The area of the large rectangle minus the area of the pool should equal 336 ft², the area of the grass strip.

Area of large rectangle	−	Area of pool	=	Area of grass
↓		↓		↓

$$(30 + 2x)(20 + 2x) - 600 = 336$$

Step 4   **Solve.**

$600 + 60x + 40x + 4x^2 - 600 = 336$ Multiply.

$4x^2 + 100x - 336 = 0$ Standard form

$x^2 + 25x - 84 = 0$ Divide by 4.

$(x + 28)(x - 3) = 0$ Factor.

$x + 28 = 0$ or $x - 3 = 0$ Zero-factor property

$x = -28$ or $x = 3$ Solve for x.

Step 5   **State the answer.** The width cannot be -28 ft, so the grass strip should be 3 ft wide.

Step 6   **Check.** If $x = 3$, we can find the area of the large rectangle (which includes the grass strip).

$(30 + 2 \cdot 3)(20 + 2 \cdot 3) = 36 \cdot 26 = 936$ ft²   Area of pool and strip

The area of the pool is $30 \cdot 20 = 600$ ft². So, the area of the grass strip is

$936 - 600 = 336$ ft²,   as required.   **NOW TRY**

NOW TRY ANSWER
4. 2 yd

OBJECTIVE 4 Solve applied problems using quadratic functions as models.

Some applied problems can be modeled by *quadratic functions*, which for real numbers a, b, and c, can be written in the form

$$f(x) = ax^2 + bx + c \quad \text{(where } a \neq 0\text{).}$$

NOW TRY EXERCISE 5

If an object is projected upward from the top of a 120-ft building at 60 ft per sec, its position (in feet above the ground) is given by

$$s(t) = -16t^2 + 60t + 120,$$

where t is time in seconds after it was projected. When does it hit the ground (to the nearest tenth)?

EXAMPLE 5 Solving an Applied Problem Using a Quadratic Function

If an object is projected upward from the top of a 144-ft building at 112 ft per sec, its position (in feet above the ground) is given by

$$s(t) = -16t^2 + 112t + 144,$$

where t is time in seconds after it was projected. When does it hit the ground?

When the object hits the ground, its distance above the ground is 0. We must find the value of t that makes $s(t) = 0$.

$$0 = -16t^2 + 112t + 144 \qquad \text{Let } s(t) = 0.$$

$$0 = t^2 - 7t - 9 \qquad \text{Divide by } -16.$$

$$t = \frac{-(-7) \pm \sqrt{(-7)^2 - 4(1)(-9)}}{2(1)} \qquad \begin{array}{l}\text{Substitute into the quadratic formula.}\\ \text{Let } a = 1, b = -7, \text{ and } c = -9.\end{array}$$

$$t = \frac{7 \pm \sqrt{85}}{2} = \frac{7 \pm 9.2}{2} \qquad \begin{array}{l}\text{Use a calculator. Round}\\ \text{to the nearest tenth.}\end{array}$$

The two solutions are

$$t = 8.1 \quad \text{or} \quad t = -1.1.$$

Time cannot be negative, so we discard the negative solution. The object hits the ground 8.1 sec after it is projected.

NOW TRY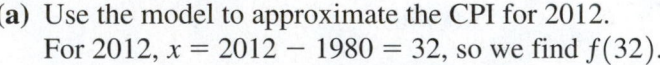

EXAMPLE 6 Using a Quadratic Function to Model the CPI

The Consumer Price Index (CPI) is used to measure trends in prices for a "basket" of goods purchased by typical American families. This index uses a base period of 1982–1984, which means that the index number for that period is 100. The quadratic function

$$f(x) = -0.0003x^2 + 4.55x + 83.7$$

approximates the CPI for the years 1980–2012, where x is the number of years that have elapsed since 1980. (*Source:* Bureau of Labor Statistics.)

(a) Use the model to approximate the CPI for 2012.
For 2012, $x = 2012 - 1980 = 32$, so we find $f(32)$.

$$f(x) = -0.0003x^2 + 4.55x + 83.7 \qquad \text{Given model}$$

$$f(32) = -0.0003(32)^2 + 4.55(32) + 83.7 \qquad \text{Let } x = 32.$$

$$f(32) = 229 \qquad \text{Nearest whole number}$$

NOW TRY ANSWER
5. 5.2 sec after it is projected

According to the model, the CPI for 2012 was 229.

NOW TRY EXERCISE 6

Refer to **Example 6.**

(a) Use the model to approximate the CPI for 2010, to the nearest whole number.

(b) In what year did the CPI reach 175? (Round down for the year.)

(b) In what year did the CPI reach 200?

Find the value of x that makes $f(x) = 200$.

$$f(x) = -0.0003x^2 + 4.55x + 83.7 \qquad \text{Given model}$$

$$200 = -0.0003x^2 + 4.55x + 83.7 \qquad \text{Let } f(x) = 200.$$

$$0 = -0.0003x^2 + 4.55x - 116.3 \qquad \text{Subtract 200.}$$

$$x = \frac{-4.55 \pm \sqrt{4.55^2 - 4(-0.0003)(-116.3)}}{2(-0.0003)} \qquad \begin{array}{l}\text{Use } a = -0.0003, b = 4.55,\\ \text{and } c = -116.3 \text{ in the}\\ \text{quadratic formula.}\end{array}$$

$$x = 25.6 \quad \text{or} \quad x = 15{,}141.1 \qquad \begin{array}{l}\text{Use a calculator. Round}\\ \text{to the nearest tenth.}\end{array}$$

Rounding the first solution 25.6 down, the CPI first reached 200 in

$$1980 + 25 = 2005.$$

NOW TRY ANSWERS

6. (a) 220 **(b)** 2000

(Reject the solution $x = 15{,}141.1$, as this corresponds to a totally unreasonable year.)

NOW TRY

11.5 Exercises

FOR EXTRA HELP

MyMathLab®

▶ *Complete solution available in MyMathLab*

Concept Check *Answer each question.*

1. In solving a formula that has the specified variable in the denominator, what is the first step?

2. What is the first step in solving a formula like $gw^2 = 2r$ for w?

3. What is the first step in solving a formula like $gw^2 = kw + 24$ for w?

4. Why is it particularly important to check all proposed solutions to an applied problem against the information in the original problem?

For each triangle, solve for m in terms of the other variables (where m > 0).

5.

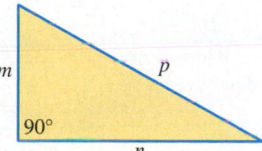

6.

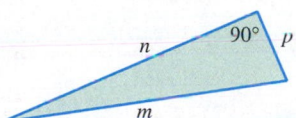

Solve each formula for the specified variable. (Leave $\pm$ in the answers.) **See Examples 1 and 2.**

7. $d = kt^2$ for t

8. $S = 6e^2$ for e

9. $S = 4\pi r^2$ for r

10. $s = kwd^2$ for d

▶ **11.** $I = \dfrac{ks}{d^2}$ for d

12. $R = \dfrac{k}{d^2}$ for d

13. $F = \dfrac{kA}{v^2}$ for v

14. $L = \dfrac{kd^4}{h^2}$ for h

15. $V = \dfrac{1}{3}\pi r^2 h$ for r

16. $V = \pi r^2 h$ for r

▶ **17.** $At^2 + Bt = -C$ for t

18. $S = 2\pi rh + \pi r^2$ for r

19. $D = \sqrt{kh}$ for h

20. $F = \dfrac{k}{\sqrt{d}}$ for d

21. $p = \sqrt{\dfrac{k\ell}{g}}$ for ℓ

22. $p = \sqrt{\dfrac{k\ell}{g}}$ for g

Extending Skills   *Solve each equation for the specified variable. (Leave ± in the answers.)*

23. $p = \dfrac{E^2 R}{(r + R)^2}$ for R (where $E > 0$)

24. $S(6S - t) = t^2$ for S

25. $10p^2 c^2 + 7pcr = 12r^2$ for r

26. $S = vt + \dfrac{1}{2}gt^2$ for t

27. $LI^2 + RI + \dfrac{1}{c} = 0$ for I

28. $P = EI - RI^2$ for I

Solve each problem. When appropriate, round answers to the nearest tenth. ***See Example 3.***

29. Find the lengths of the sides of the triangle.

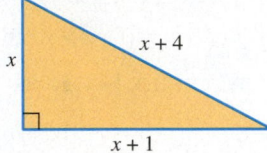

30. Find the lengths of the sides of the triangle.

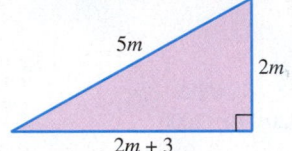

▶ **31.** Two ships leave port at the same time, one heading due south and the other heading due east. Several hours later, they are 170 mi apart. If the ship traveling south traveled 70 mi farther than the other ship, how many miles did they each travel?

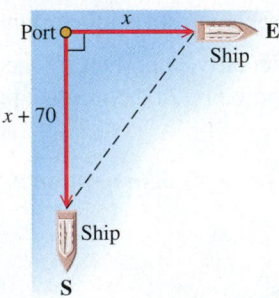

32. Deborah is flying a kite that is 30 ft farther above her hand than its horizontal distance from her. The string from her hand to the kite is 150 ft long. How high is the kite?

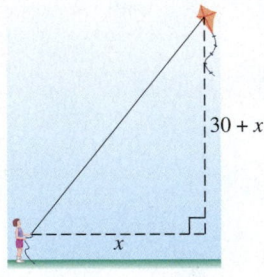

33. A game board is in the shape of a right triangle. The hypotenuse is 2 in. longer than the longer leg, and the longer leg is 1 in. less than twice as long as the shorter leg. How long is each side of the game board?

34. Manuel is planting a vegetable garden in the shape of a right triangle. The longer leg is 3 ft longer than the shorter leg, and the hypotenuse is 3 ft longer than the longer leg. Find the lengths of the three sides of the garden.

35. The diagonal of a rectangular rug measures 26 ft, and the length is 4 ft more than twice the width. Find the length and width of the rug.

36. A 13-ft ladder is leaning against a house. The distance from the bottom of the ladder to the house is 7 ft less than the distance from the top of the ladder to the ground. How far is the bottom of the ladder from the house?

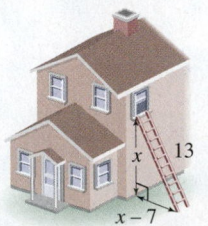

Solve each problem. See Example 4.

▶ **37.** A club swimming pool is 30 ft wide and 40 ft long. The club members want an exposed aggregate border in a strip of uniform width around the pool. They have enough material for 296 ft². How wide can the strip be?

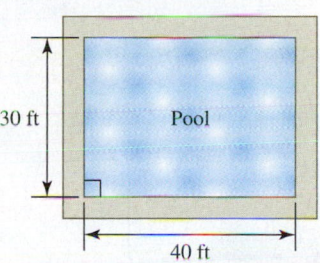

38. Lyudmila wants to buy a rug for a room that is 20 ft long and 15 ft wide. She wants to leave an even strip of flooring uncovered around the edges of the room. How wide a strip will she have if she buys a rug with an area of 234 ft²?

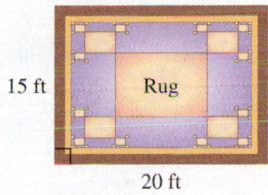

39. A rectangle has a length 2 m less than twice its width. When 5 m are added to the width, the resulting figure is a square with an area of 144 m². Find the dimensions of the original rectangle.

40. Mariana's backyard measures 20 m by 30 m. She wants to put a flower garden in the middle of the yard, leaving a strip of grass of uniform width around the flower garden. Mariana must have 184 m² of grass. Under these conditions, what will the length and width of the garden be?

41. A rectangular piece of sheet metal has a length that is 4 in. less than twice the width. A square piece 2 in. on a side is cut from each corner. The sides are then turned up to form an uncovered box of volume 256 in.³. Find the length and width of the original piece of metal.

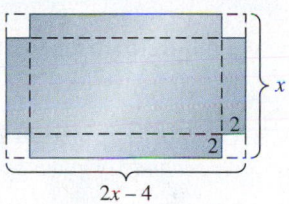

42. A rectangular piece of cardboard is 2 in. longer than it is wide. A square piece 3 in. on a side is cut from each corner. The sides are then turned up to form an uncovered box of volume 765 in.³. Find the dimensions of the original piece of cardboard.

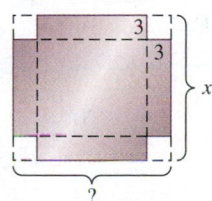

Solve each problem. When appropriate, round answers to the nearest tenth. See Example 5.

▶ **43.** An object is projected directly upward from the ground. After t seconds its distance in feet above the ground is

$$s(t) = 144t - 16t^2.$$

After how many seconds will the object be 128 ft above the ground? (*Hint:* Look for a common factor before solving the equation.)

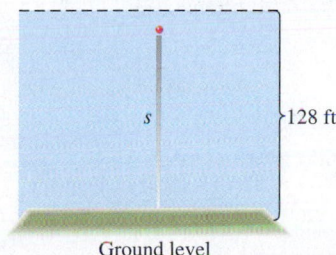

44. When does the object in **Exercise 43** strike the ground?

45. A ball is projected upward from the ground. Its distance in feet from the ground in t seconds is given by

$$s(t) = -16t^2 + 128t.$$

At what times will the ball be 213 ft from the ground?

46. A toy rocket is launched from ground level. Its distance in feet from the ground in t seconds is given by

$$s(t) = -16t^2 + 208t.$$

At what times will the rocket be 550 ft from the ground?

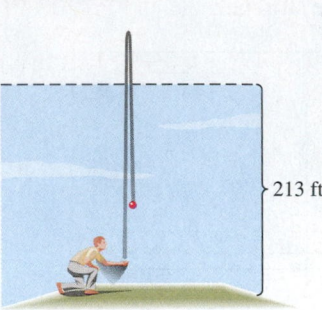

213 ft

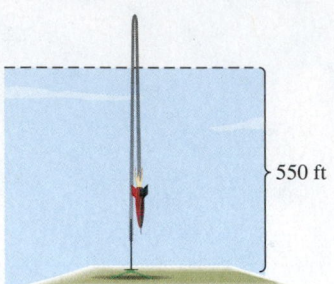

550 ft

47. The following function gives the distance in feet a car going approximately 68 mph will skid in t seconds.

$$D(t) = 13t^2 - 100t$$

Find the time it would take for the car to skid 180 ft.

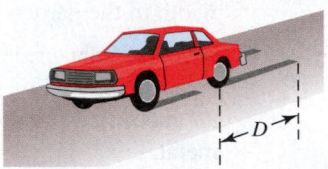

48. Refer to the function in **Exercise 47.** Find the time it would take for the car to skid 500 ft.

A ball is projected upward from ground level, and its distance in feet from the ground in t seconds is given by

$$s(t) = -16t^2 + 160t.$$

49. After how many seconds does the ball reach a height of 400 ft? Describe in words its position at this height.

50. After how many seconds does the ball reach a height of 425 ft? Interpret the mathematical result here.

Extending Skills *Solve each problem using a quadratic equation.*

51. A certain bakery has found that the daily demand for blueberry muffins is $\dfrac{6000}{p}$, where p is the price of a muffin in cents. The daily supply is $3p - 410$. Find the price at which supply and demand are equal.

52. In one area the demand for Blu-ray discs is $\dfrac{1900}{P}$ per day, where P is the price in dollars per disc. The supply is $5P - 1$ per day. At what price, to the nearest cent, does supply equal demand?

53. The formula $A = P(1 + r)^2$ gives the amount A in dollars that P dollars will grow to in 2 yr at interest rate r (where r is given as a decimal), using compound interest. What interest rate will cause \$2000 to grow to \$2142.45 in 2 yr?

54. Use the formula $A = P(1 + r)^2$ to find the interest rate r at which a principal P of \$10,000 will increase to \$10,920.25 in 2 yr.

William Froude was a 19th century naval architect who used the following expression, known as the **Froude number,** *in shipbuilding.*

$$\frac{v^2}{g\ell}$$

This expression was also used by R. McNeill Alexander in his research on dinosaurs. (Source: "How Dinosaurs Ran," Scientific American.)

Use this expression to find the value of v (in meters per second), given g = 9.8 m per sec². (Round to the nearest tenth.)

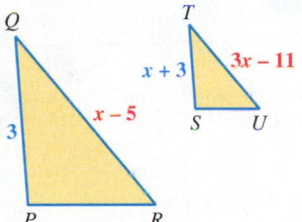

55. Rhinoceros: $\ell = 1.2$;

Froude number = 2.57

56. Triceratops: $\ell = 2.8$;

Froude number = 0.16

Recall that corresponding sides of similar triangles are proportional. Use this fact to find the lengths of the indicated sides of each pair of similar triangles. Check all possible solutions in both triangles. Sides of a triangle cannot be negative (and are not drawn to scale here).

57. Side AC

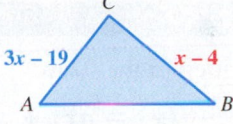
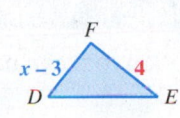

58. Side RQ

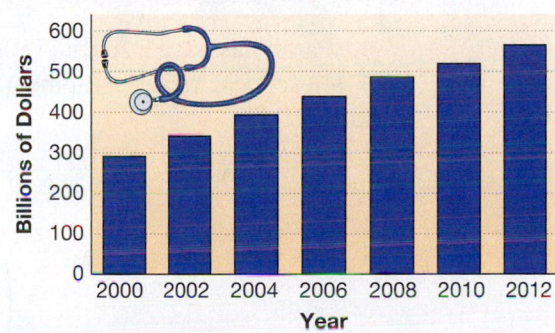

Total spending (in billions of dollars) in the United States from all sources on physician and clinical services for the years 2000–2012 are shown in the bar graph and can be modeled by the quadratic function

$$f(x) = -0.2901x^2 + 25.90x + 291.6.$$

Here, x = 0 represents 2000, x = 2 represents 2002, and so on. Use the graph and the model to work Exercises 59–62. **See Example 6.**

Spending on Physician and Clinical Services

Source: Centers for Medicare and Medicaid Services.

59. Approximate spending on physician and clinical services in 2012 to the nearest $10 billion using **(a)** the graph and **(b)** the model. How do the two approximations compare?

60. Repeat **Exercise 59** for the year 2008.

61. According to the model, in what year did spending on physician and clinical services first exceed $500 billion? (Round down for the year.)

62. Repeat **Exercise 61** for $400 billion.

11.6 Graphs of Quadratic Functions

VOCABULARY

☐ parabola
☐ vertex
☐ axis of symmetry (axis)
☐ quadratic function

OBJECTIVE 1 Graph a quadratic function.

FIGURE 5 gives a graph of the simplest *quadratic function* $y = x^2$. This graph is a **parabola.** The point $(0, 0)$, the lowest point on the curve, is the **vertex** of the parabola. The vertical line through the vertex is the **axis of symmetry,** or simply the **axis,** of the parabola. Here its equation is $x = 0$. A parabola is **symmetric about its axis**—that is, if the graph were folded along the axis, the two portions of the curve would coincide.

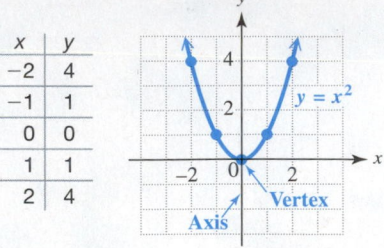

x	y
−2	4
−1	1
0	0
1	1
2	4

FIGURE 5

As **FIGURE 5** suggests, x can be any real number, so the domain of the function $y = x^2$, written in interval notation, is $(-\infty, \infty)$. Values of y are always nonnegative, so the range is $[0, \infty)$.

Quadratic Function

A function that can be written in the form

$$f(x) = ax^2 + bx + c$$

for real numbers a, b, and c, where $a \neq 0$, is a **quadratic function.**

The graph of any quadratic function is a parabola with a vertical axis.

NOTE We use the variable y and function notation $f(x)$ interchangeably. Although we use the letter f most often to name quadratic functions, other letters can be used. We use the capital letter F to distinguish between different parabolas graphed on the same coordinate axes.

Parabolas have a special reflecting property that makes them useful in the design of telescopes, radar equipment, solar furnaces, and automobile headlights. (See **FIGURE 6.**)

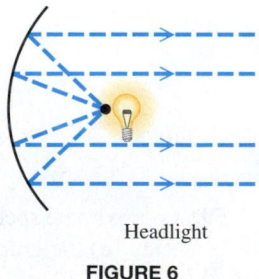

Headlight

FIGURE 6

OBJECTIVE 2 Graph parabolas with horizontal and vertical shifts.

Parabolas need not have their vertices at the origin, as does the graph of $f(x) = x^2$.

NOW TRY
EXERCISE 1

Graph $f(x) = x^2 - 3$.
Give the vertex, axis,
domain, and range.

EXAMPLE 1 Graphing a Parabola (Vertical Shift)

Graph $F(x) = x^2 - 2$.

The graph of $F(x) = x^2 - 2$ has the same shape as that of $f(x) = x^2$ but is *shifted*, or *translated*, 2 units down, with vertex $(0, -2)$. Every function value is 2 less than the corresponding function value of $f(x) = x^2$. Plotting points on both sides of the vertex gives the graph in **FIGURE 7**.

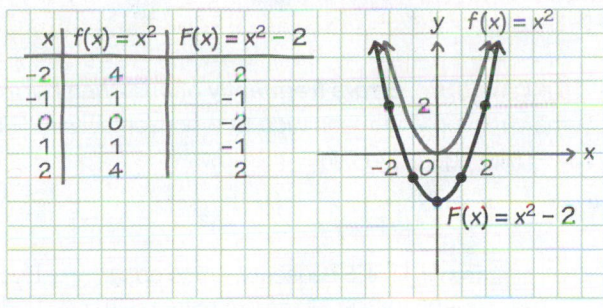

$F(x) = x^2 - 2$
Vertex: $(0, -2)$
Axis of symmetry: $x = 0$
Domain: $(-\infty, \infty)$
Range: $[-2, \infty)$
The graph of $f(x) = x^2$
is shown for comparison.

FIGURE 7

This parabola is symmetric about its axis $x = 0$, so the plotted points are "mirror images" of each other. Because x can be any real number, the domain is $(-\infty, \infty)$. The value of y (or $F(x)$) is always greater than or equal to -2, so the range is $[-2, \infty)$.

NOW TRY ⟳

Vertical Shift of a Parabola

The graph of $\boldsymbol{F(x) = x^2 + k}$ is a parabola.

- The graph has the same shape as the graph of $f(x) = x^2$.
- The parabola is shifted k units up if $k > 0$ and $|k|$ units down if $k < 0$.
- The vertex of the parabola is $(0, k)$.

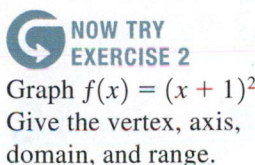
NOW TRY
EXERCISE 2

Graph $f(x) = (x + 1)^2$.
Give the vertex, axis,
domain, and range.

EXAMPLE 2 Graphing a Parabola (Horizontal Shift)

Graph $F(x) = (x - 2)^2$.

If $x = 2$, then $F(x) = 0$, giving the vertex $(2, 0)$. The graph of $F(x) = (x - 2)^2$ has the same shape as that of $f(x) = x^2$ but is shifted 2 units to the right. Plotting points on one side of the vertex, and using symmetry about the axis $x = 2$ to find corresponding points on the other side, gives the graph in **FIGURE 8**.

NOW TRY ANSWERS

1.   2.

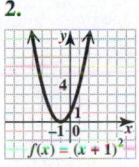

$f(x) = x^2 - 3$

vertex: $(0, -3)$; vertex: $(-1, 0)$;
axis: $x = 0$;   axis: $x = -1$;
domain: $(-\infty, \infty)$; domain: $(-\infty, \infty)$;
range: $[-3, \infty)$ range: $[0, \infty)$

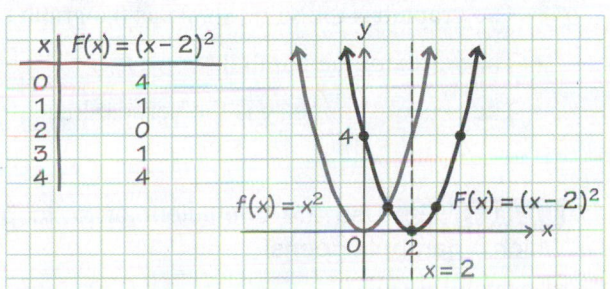

$F(x) = (x - 2)^2$
Vertex: $(2, 0)$
Axis of symmetry: $x = 2$
Domain: $(-\infty, \infty)$
Range: $[0, \infty)$

FIGURE 8

NOW TRY

Horizontal Shift of a Parabola

The graph of $F(x) = (x - h)^2$ is a parabola.

- The graph has the same shape as the graph of $f(x) = x^2$.

- The parabola is shifted h units to the right if $h > 0$ and $|h|$ units to the left if $h < 0$.

- The vertex of the parabola is $(h, 0)$.

🛑 **CAUTION** *Errors frequently occur when horizontal shifts are involved.* To determine the direction and magnitude of a horizontal shift, find the value that causes the expression $x - h$ to equal 0, as shown below.

$$F(x) = (x - 5)^2 \qquad \qquad F(x) = (x + 5)^2$$

Because **+5** causes $x - 5$ to equal 0, the graph of $F(x)$ illustrates a shift of

5 units to the right.

Because **−5** causes $x + 5$ to equal 0, the graph of $F(x)$ illustrates a shift of

5 units to the left.

🔄 **NOW TRY EXERCISE 3**

Graph $f(x) = (x + 1)^2 - 2$. Give the vertex, axis, domain, and range.

EXAMPLE 3 Graphing a Parabola (Horizontal and Vertical Shifts)

Graph $F(x) = (x + 3)^2 - 2$.

This graph has the same shape as that of $f(x) = x^2$, but is shifted 3 units to the left (because $x + 3 = 0$ if $x = -3$) and 2 units down (because of the negative sign in -2). See **FIGURE 9**.

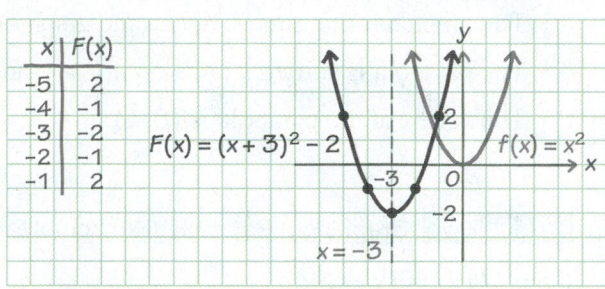

x	F(x)
-5	2
-4	-1
-3	-2
-2	-1
-1	2

$F(x) = (x + 3)^2 - 2$
Vertex: $(-3, -2)$
Axis of symmetry: $x = -3$
Domain: $(-\infty, \infty)$
Range: $[-2, \infty)$

FIGURE 9

NOW TRY

Vertex and Axis of a Parabola

The graph of $F(x) = (x - h)^2 + k$ is a parabola.

- The graph has the same shape as the graph of $f(x) = x^2$.

- The vertex of the parabola is (h, k).

- The axis of symmetry is the vertical line $x = h$.

NOW TRY ANSWER

3. $f(x) = (x + 1)^2 - 2$

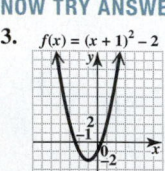

vertex: $(-1, -2)$; axis: $x = -1$; domain: $(-\infty, \infty)$; range: $[-2, \infty)$

OBJECTIVE 3 Use the coefficient of x^2 to predict the shape and direction in which a parabola opens.

Not all parabolas open up, and not all parabolas have the same shape as the graph of $f(x) = x^2$.

**NOW TRY
EXERCISE 4**

Graph $f(x) = -3x^2$.
Give the vertex, axis,
domain, and range.

EXAMPLE 4 Graphing a Parabola That Opens Down

Graph $f(x) = -\frac{1}{2}x^2$.

This parabola is shown in **FIGURE 10**. The coefficient $-\frac{1}{2}$ affects the shape of the graph—the $\frac{1}{2}$ makes the parabola wider $\left(\text{because the values of } \frac{1}{2}x^2 \text{ increase more slowly than those of } x^2\right)$, and the negative sign makes the parabola open down. The graph is not shifted in any direction. Unlike the parabolas graphed in **Examples 1–3**, the vertex $(0, 0)$ has the *greatest* function value of any point on the graph.

x	$f(x)$
-2	-2
-1	$-\frac{1}{2}$
0	0
1	$-\frac{1}{2}$
2	-2

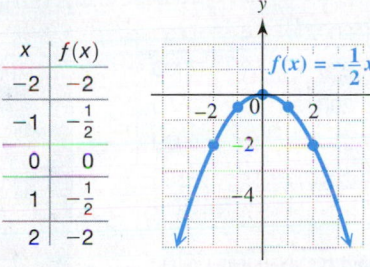

$f(x) = -\frac{1}{2}x^2$

Vertex: $(0, 0)$

Axis of symmetry: $x = 0$

Domain: $(-\infty, \infty)$

Range: $(-\infty, 0]$

FIGURE 10

 NOW TRY

General Characteristics of $F(x) = a(x - h)^2 + k$ (Where $a \neq 0$)

1. The graph of the quadratic function

$$F(x) = a(x - h)^2 + k \quad (\text{where } a \neq 0)$$

   is a parabola with vertex (h, k) and vertical line $x = h$ as axis of symmetry.

2. The graph opens up if $a > 0$ and down if $a < 0$.

3. The graph is wider than that of $f(x) = x^2$ if $0 < |a| < 1$.

   The graph is narrower than that of $f(x) = x^2$ if $|a| > 1$.

**NOW TRY
EXERCISE 5**

Graph $f(x) = 2(x - 1)^2 + 2$.
Give the vertex, axis, domain,
and range.

EXAMPLE 5 Using the General Characteristics to Graph a Parabola

Graph $F(x) = -2(x + 3)^2 + 4$.

The parabola opens down (because $a < 0$) and is narrower than the graph of $f(x) = x^2$ because $|-2| = 2$ and $2 > 1$. This causes values of $F(x)$ to decrease more quickly than those of $f(x) = -x^2$. This parabola has vertex $(-3, 4)$, as shown in **FIGURE 11**. To complete the graph, we plotted the ordered pairs $(-4, 2)$ and, by symmetry, $(-2, 2)$. Symmetry can be used to find additional ordered pairs that satisfy the equation.

NOW TRY ANSWERS

4. **5.**

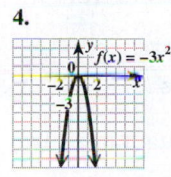

vertex: $(0, 0)$;
axis: $x = 0$;
domain: $(-\infty, \infty)$;
range: $(-\infty, 0]$

$f(x) = 2(x - 1)^2 + 2$
vertex: $(1, 2)$;
axis: $x = 1$;
domain: $(-\infty, \infty)$;
range: $[2, \infty)$

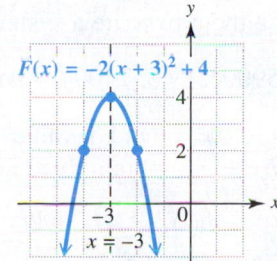

$F(x) = -2(x + 3)^2 + 4$

Vertex: $(-3, 4)$

Axis of symmetry: $x = -3$

Domain: $(-\infty, \infty)$

Range: $(-\infty, 4]$

FIGURE 11

 NOW TRY

OBJECTIVE 4   Find a quadratic function to model data.

EXAMPLE 6   Modeling the Number of Multiple Births

The number of higher-order multiple births (triplets or more) in the United States is shown in the table. Here, x represents the number of years since 1996 and y represents the number of higher-order multiple births (to the nearest hundred).

Year	x	y
1996	0	5900
2000	4	7300
2002	6	7400
2004	8	7300
2006	10	6500
2008	12	6300
2010	14	5500
2012	16	4900

Source: National Center for Health Statistics.

Find a quadratic function that models the data.

A scatter diagram of the ordered pairs (x, y) is shown in **FIGURE 12**. The general shape suggested by the scatter diagram indicates that a parabola should approximate these points, as shown by the dashed curve in **FIGURE 13**. The equation for such a parabola would have a negative coefficient for x^2 because the graph opens down.

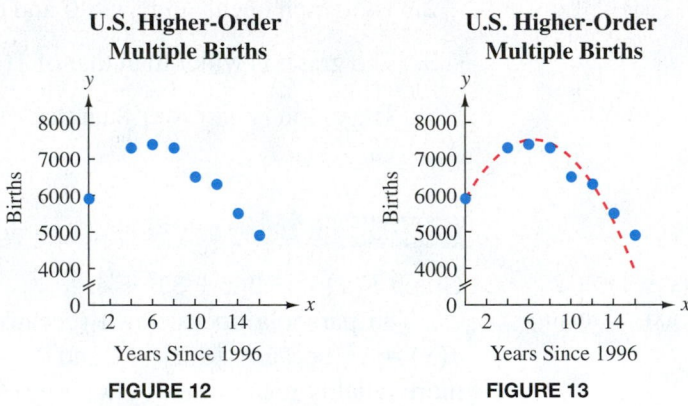

FIGURE 12   **FIGURE 13**

To find a quadratic function of the form

$$y = ax^2 + bx + c$$

that models, or *fits*, these data, we choose three representative ordered pairs from the table and use them to write a system of three equations.

$$(0, 5900), \quad (6, 7400), \quad \text{and} \quad (12, 6300) \qquad \text{Three ordered pairs } (x, y)$$

We substitute the x- and y-values from each ordered pair into the quadratic form $y = ax^2 + bx + c$ to obtain three equations.

$$a(0)^2 + b(0) + c = 5900 \quad \xrightarrow{\text{Simplify.}} \qquad\qquad c = 5900 \qquad (1)$$

$$a(6)^2 + b(6) + c = 7400 \quad \longrightarrow \qquad 36a + 6b + c = 7400 \qquad (2)$$

$$a(12)^2 + b(12) + c = 6300 \quad \longrightarrow \quad 144a + 12b + c = 6300 \qquad (3)$$

NOW TRY EXERCISE 6
Using the points $(0, 5900)$, $(4, 7300)$, and $(12, 6300)$, find another quadratic model for the data on higher-order multiple births in **Example 6.** (Round values of a and b to the nearest tenth.)

To find the values of a, b, and c, we solve this system of three equations in three variables using the methods of **Section 7.6.** From equation (1), $c = 5900$, so we substitute 5900 for c in equations (2) and (3) to obtain two equations in two variables.

$$36a + 6b + 5900 = 7400 \xrightarrow{\text{Subtract 5900.}} 36a + 6b = 1500 \quad (4)$$
$$144a + 12b + 5900 = 6300 \longrightarrow 144a + 12b = 400 \quad (5)$$

We eliminate b from this system of equations in two variables by multiplying equation (4) by -2 and adding the results to equation (5).

$$
\begin{array}{ll}
-72a - 12b = -3000 & \text{Multiply equation (4) by } -2. \\
\underline{144a + 12b = 400} & \text{(5)} \\
72a = -2600 & \text{Add.} \\
a = -36.1 & \text{Use a calculator. Round to} \\
& \text{one decimal place.}
\end{array}
$$

We substitute -36.1 for a in equation (4) to find that $b = 466.6$. (Substituting in equation (5) will give $b = 466.5$ due to rounding procedures.) Using the values we found for a, b, and c, the model is

$$
\begin{array}{ccc}
a & b & c \\
\downarrow & \downarrow & \downarrow
\end{array}
$$
$$y = -36.1x^2 + 466.6x + 5900.$$

NOW TRY

NOW TRY ANSWER
6. $y = -39.6x^2 + 508.4x + 5900$ (Answers may vary slightly due to rounding.)

NOTE If we had chosen three different ordered pairs of data in **Example 6,** a slightly different, though similar, model would result. (See **Now Try Exercise 6.**)

The *quadratic regression* feature on a graphing calculator can also be used to generate the quadratic model that best fits given data. See your owner's manual for details.

11.6 Exercises

FOR EXTRA HELP MyMathLab®

 Complete solution available in MyMathLab

Concept Check *Match each quadratic function in parts (a)–(d) with its graph from choices A–D.*

1. (a) $f(x) = (x + 2)^2 - 1$

(b) $f(x) = (x + 2)^2 + 1$

(c) $f(x) = (x - 2)^2 - 1$

(d) $f(x) = (x - 2)^2 + 1$

A.
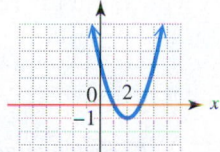

B.
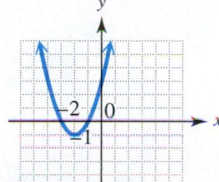

C.
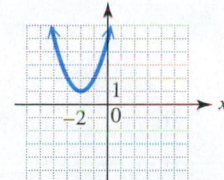

D.
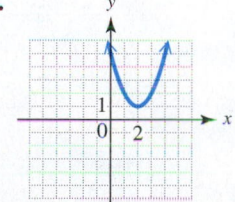

2. (a) $f(x) = -x^2 + 2$

(b) $f(x) = -x^2 - 2$

(c) $f(x) = -(x + 2)^2$

(d) $f(x) = -(x - 2)^2$

A.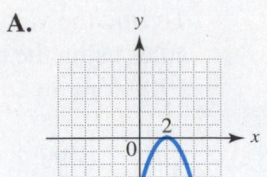

B.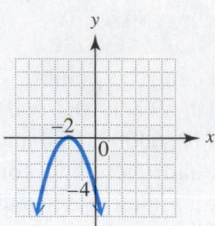

C.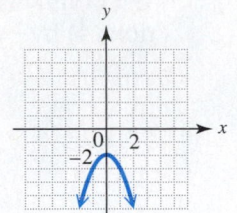

D.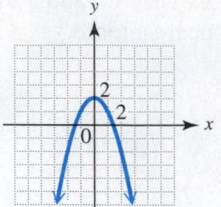

3. *Concept Check*　Match each quadratic function with the description of the parabola that is its graph.

(a) $f(x) = (x - 4)^2 - 2$　　**A.** Vertex $(2, -4)$, opens down

(b) $f(x) = (x - 2)^2 - 4$　　**B.** Vertex $(2, -4)$, opens up

(c) $f(x) = -(x - 4)^2 - 2$　　**C.** Vertex $(4, -2)$, opens down

(d) $f(x) = -(x - 2)^2 - 4$　　**D.** Vertex $(4, -2)$, opens up

4. *Concept Check*　For $f(x) = a(x - h)^2 + k$, in what quadrant is the vertex if the values of h and k are as follows?

(a) $h > 0, k > 0$　　**(b)** $h > 0, k < 0$　　**(c)** $h < 0, k > 0$　　**(d)** $h < 0, k < 0$

Consider the value of a, and make the correct choice.

(e) If $|a| > 1$, then the graph is (*narrower / wider*) than the graph of $f(x) = x^2$.

(f) If $0 < |a| < 1$, then the graph is (*narrower / wider*) than the graph of $f(x) = x^2$.

(g) If $a > 0$, then the graph opens (*up / down*).

(h) If $a < 0$, then the graph opens (*up / down*).

Identify the vertex of each parabola. See Examples 1–4.

5. $f(x) = -3x^2$　　**6.** $f(x) = \dfrac{1}{2}x^2$　　**7.** $f(x) = x^2 + 4$　　**8.** $f(x) = x^2 - 4$

9. $f(x) = (x - 1)^2$　　**10.** $f(x) = (x + 3)^2$　　**11.** $f(x) = (x + 3)^2 - 4$

12. $f(x) = (x + 5)^2 - 8$　　**13.** $f(x) = -(x - 5)^2 + 6$　　**14.** $f(x) = -(x - 2)^2 + 1$

For each quadratic function, tell whether the graph opens up or down and whether the graph is wider, narrower, or the same shape as the graph of $f(x) = x^2$. See Examples 4 and 5.

15. $f(x) = -\dfrac{2}{5}x^2$　　　　　　　**16.** $f(x) = -2x^2$

17. $f(x) = 3x^2 + 1$　　　　　　　**18.** $f(x) = \dfrac{2}{3}x^2 - 4$

19. $f(x) = -4(x + 2)^2 + 5$　　　　　**20.** $f(x) = -\dfrac{1}{3}(x + 6)^2 + 3$

Graph each parabola. Give the vertex, axis of symmetry, domain, and range. See Examples 1–5.

▶ **21.** $f(x) = -2x^2$

22. $f(x) = -\dfrac{1}{3}x^2$

▶ **23.** $f(x) = x^2 - 1$

24. $f(x) = x^2 + 3$

25. $f(x) = -x^2 + 2$

26. $f(x) = -x^2 - 2$

▶ **27.** $f(x) = (x - 4)^2$

28. $f(x) = (x + 1)^2$

▶ **29.** $f(x) = (x + 2)^2 - 1$

30. $f(x) = (x - 1)^2 + 2$

31. $f(x) = 2(x - 2)^2 - 4$

32. $f(x) = 3(x - 2)^2 + 1$

33. $f(x) = -2(x + 3)^2 + 4$

34. $f(x) = -2(x - 2)^2 - 3$

▶ **35.** $f(x) = -\dfrac{1}{2}(x + 1)^2 + 2$

36. $f(x) = -\dfrac{2}{3}(x + 2)^2 + 1$

37. $f(x) = 2(x - 2)^2 - 3$

38. $f(x) = \dfrac{4}{3}(x - 3)^2 - 2$

In Exercises 39–44, decide whether a linear *function* or a *quadratic* function *would be a more appropriate model for each set of graphed data. If linear, tell whether the slope should be* positive *or* negative. *If quadratic, tell whether the coefficient of x^2 should be* positive *or* negative. *See Example 6.*

39. **Time Spent Playing Video Games**

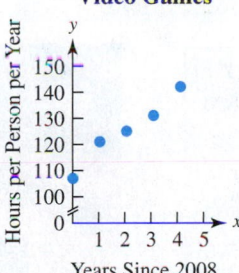

Years Since 2008

Source: www.statisca.com

40. **Average Daily Volume of First-Class Mail**

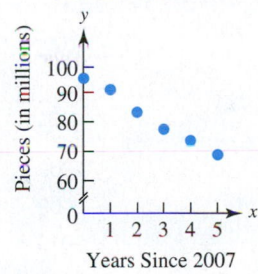

Years Since 2007

Source: U.S. Postal Service.

41. **Food Assistance Spending in Iowa**

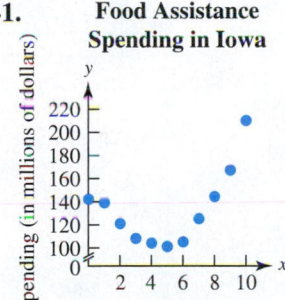

Year

Source: Iowa Department of Human Services.

42. **U.S. Foreign-Born Population**

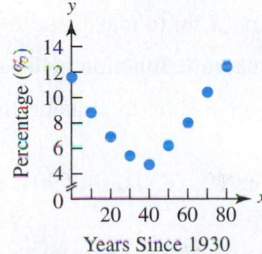

Years Since 1930

Source: U.S. Census Bureau.

43. **High School Students Who Smoke**

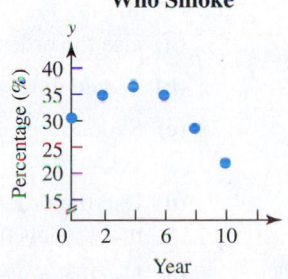

Year

Source: www.cdc.gov

44. **Social Security Assets***

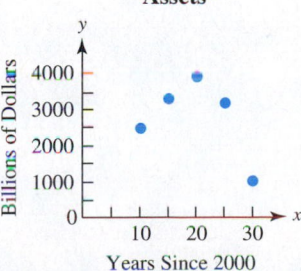

Years Since 2000

*Projected

Source: Social Security Administration.

Solve each problem. See Example 6.

45. Federal student loans (in billions of dollars) are shown in the table, where x represents the number of years since 2008 and y represents total student loans.

Year	x	y
2008	0	76
2009	1	90
2010	2	106
2011	3	112
2012	4	109
2013	5	101

Source: The College Board.

(a) Use the ordered pairs (x, y) to make a scatter diagram of the data.

(b) Would a linear or quadratic function better model the data?

(c) Should the coefficient a of x^2 in a quadratic model $y = ax^2 + bx + c$ be positive or negative?

(d) Use the ordered pairs $(0, 76)$, $(3, 112)$, and $(5, 101)$ to find a quadratic function that models the data.

(e) Use the model from part (d) to approximate total federal student loans (in billions of dollars) for 2010 and 2012. How well does the model approximate the actual data from the table?

46. The number (in thousands) of new, privately owned housing units started in the United States is shown in the table. Here x represents years since 2006 and y represents total housing starts.

Year	x	y
2006	0	1800
2007	1	1360
2008	2	910
2009	3	550
2010	4	590
2011	5	610
2012	6	780

Source: U.S. Census Bureau.

(a) Use the ordered pairs (x, y) to make a scatter diagram of the data.

(b) Would a linear or quadratic function better model the data?

(c) Should the coefficient a of x^2 in a quadratic model $y = ax^2 + bx + c$ be positive or negative?

(d) Use the ordered pairs $(0, 1800)$, $(4, 590)$, and $(6, 780)$ to find a quadratic function that models the data.

(e) Use the model from part (d) to approximate the number of housing starts for 2008 and 2011 to the nearest thousand. How well does the model approximate the actual data from the table?

11.7 More about Parabolas and Their Applications

OBJECTIVE 1 Find the vertex of a vertical parabola.

When the equation of a parabola is given in the form $f(x) = ax^2 + bx + c$, there are two ways to locate the vertex.

1. Complete the square. (See **Examples 1 and 2.**)

2. Use a formula derived by completing the square. (See **Example 3.**)

EXAMPLE 1 Completing the Square to Find the Vertex ($a = 1$)

Find the vertex of the graph of $f(x) = x^2 - 4x + 5$.

We can express $x^2 - 4x + 5$ in the form $(x - h)^2 + k$ by completing the square on $x^2 - 4x$, as in **Section 11.2.** The process is slightly different here because we want to keep $f(x)$ alone on one side of the equation. Instead of adding the appropriate number to each side, we *add and subtract* it on the right.

$$f(x) = x^2 - 4x + 5$$

$$f(x) = (x^2 - 4x \qquad) + 5 \qquad \text{Group the variable terms.}$$

> This is equivalent to adding 0.

$$\left[\tfrac{1}{2}(-4)\right]^2 = (-2)^2 = 4 \qquad \text{Square half the coefficient of the first-degree term.}$$

$$f(x) = (x^2 - 4x + 4 - 4) + 5 \qquad \text{Add and subtract 4.}$$

$$f(x) = (x^2 - 4x + 4) - 4 + 5 \qquad \text{Bring } -4 \text{ outside the parentheses.}$$

$$f(x) = (x - 2)^2 + 1 \qquad \text{Factor. Combine like terms.}$$

The vertex of this parabola is $(2, 1)$.   NOW TRY

NOW TRY EXERCISE 1

Find the vertex of the graph of

$$f(x) = x^2 + 2x - 8.$$

NOW TRY EXERCISE 2

Find the vertex of the graph of

$$f(x) = -4x^2 + 16x - 10.$$

EXAMPLE 2 Completing the Square to Find the Vertex ($a \neq 1$)

Find the vertex of the graph of $f(x) = -3x^2 + 6x - 1$.

Because the x^2-term has a coefficient other than 1, we factor that coefficient out of the first two terms before completing the square.

$$f(x) = -3x^2 + 6x - 1$$

$$f(x) = (-3x^2 + 6x) - 1 \qquad \text{Group the variable terms.}$$

$$f(x) = -3(x^2 - 2x) - 1 \qquad \text{Factor out } -3.$$

$$f(x) = -3(x^2 - 2x \qquad) - 1 \qquad \text{Prepare to complete the square.}$$

$$\left[\tfrac{1}{2}(-2)\right]^2 = (-1)^2 = 1 \qquad \text{Square half the coefficient of the first-degree term.}$$

$$f(x) = -3(x^2 - 2x + 1 - 1) - 1 \qquad \text{Add and subtract 1.}$$

Now bring -1 outside the parentheses. Be sure to multiply it by -3.

> This is a key step.

$$f(x) = -3(x^2 - 2x + 1) + (-3)(-1) - 1 \qquad \text{Distributive property}$$

$$f(x) = -3(x^2 - 2x + 1) + 3 - 1 \qquad \text{Multiply.}$$

$$f(x) = -3(x - 1)^2 + 2 \qquad \text{Factor. Combine like terms.}$$

The vertex is $(1, 2)$.   NOW TRY

NOW TRY ANSWERS
1. $(-1, -9)$ 2. $(2, 6)$

We can complete the square to derive a formula for the vertex of the graph of the quadratic function $f(x) = ax^2 + bx + c$ (where $a \neq 0$).

$f(x) = ax^2 + bx + c$	Standard form
$f(x) = (ax^2 + bx) + c$	Group the terms with x.
$f(x) = a\left(x^2 + \dfrac{b}{a}x \quad\quad\right) + c$	Factor a from the first two terms.
$\left[\dfrac{1}{2}\left(\dfrac{b}{a}\right)\right]^2 = \left(\dfrac{b}{2a}\right)^2 = \dfrac{b^2}{4a^2}$	Square half the coefficient of the first-degree term.
$f(x) = a\left(x^2 + \dfrac{b}{a}x + \dfrac{b^2}{4a^2} - \dfrac{b^2}{4a^2}\right) + c$	Add and subtract $\dfrac{b^2}{4a^2}$.
$f(x) = a\left(x^2 + \dfrac{b}{a}x + \dfrac{b^2}{4a^2}\right) + a\left(-\dfrac{b^2}{4a^2}\right) + c$	Distributive property
$f(x) = a\left(x^2 + \dfrac{b}{a}x + \dfrac{b^2}{4a^2}\right) - \dfrac{b^2}{4a} + c$	$-\dfrac{ab^2}{4a^2} = -\dfrac{b^2}{4a}$
$f(x) = a\left(x + \dfrac{b}{2a}\right)^2 + \dfrac{4ac - b^2}{4a}$	Factor. Rewrite terms with a common denominator.
$f(x) = a\left[x - \left(\dfrac{-b}{2a}\right)\right]^2 + \dfrac{4ac - b^2}{4a}$	$f(x) = a(x - h)^2 + k$ The vertex (h, k) can be expressed in terms of a, b, and c.

$$\underbrace{\phantom{x - \left(\dfrac{-b}{2a}\right)}}_{h} \qquad \underbrace{\phantom{\dfrac{4ac-b^2}{4a}}}_{k}$$

The expression for k can be found by replacing x with $\dfrac{-b}{2a}$. Using function notation, if $y = f(x)$, then the y-value of the vertex is $f\left(\dfrac{-b}{2a}\right)$.

Vertex Formula

The graph of the quadratic function $f(x) = ax^2 + bx + c$ (where $a \neq 0$) has vertex

$$\left(\frac{-b}{2a}, \; f\left(\frac{-b}{2a}\right)\right).$$

The axis of symmetry of the parabola is the line having equation

$$x = \frac{-b}{2a}.$$

EXAMPLE 3   Using the Formula to Find the Vertex

Use the vertex formula to find the vertex of the graph of $f(x) = x^2 - x - 6$.

The x-coordinate of the vertex of the parabola is given by $\dfrac{-b}{2a}$.

$$\frac{-b}{2a} = \frac{-(-1)}{2(1)} = \frac{1}{2} \leftarrow \begin{array}{l} a = 1, b = -1, \text{ and } c = -6. \\ \text{x-coordinate of vertex} \end{array}$$

**NOW TRY
EXERCISE 3**

Use the vertex formula to find the vertex of the graph of

$$f(x) = 3x^2 - 2x + 8.$$

The y-coordinate of the vertex of $f(x) = x^2 - x - 6$ is $f\left(\frac{-b}{2a}\right) = f\left(\frac{1}{2}\right)$.

$$f\left(\frac{1}{2}\right) = \left(\frac{1}{2}\right)^2 - \frac{1}{2} - 6 = \frac{1}{4} - \frac{1}{2} - 6 = -\frac{25}{4} \leftarrow y\text{-coordinate of vertex}$$

The vertex is $\left(\frac{1}{2}, -\frac{25}{4}\right)$.

NOW TRY

OBJECTIVE 2 Graph a quadratic function.

Graphing a Quadratic Function $y = f(x)$

Step 1 **Determine whether the graph opens up or down.**

- If $a > 0$, then the parabola opens up.
- If $a < 0$, then it opens down.

Step 2 **Find the vertex.** Use the vertex formula or complete the square.

Step 3 **Find any intercepts.**

- To find the x-intercepts (if any), solve $f(x) = 0$.
- To find the y-intercept, evaluate $f(0)$.

Step 4 **Complete the graph.** Plot the points found so far. Find and plot additional points as needed, using symmetry about the axis.

EXAMPLE 4 Graphing a Quadratic Function

Graph the quadratic function $f(x) = x^2 - x - 6$.

Step 1 From the equation, $a = 1$, so the graph of the function opens up.

Step 2 The vertex, $\left(\frac{1}{2}, -\frac{25}{4}\right)$, was found in **Example 3** using the vertex formula.

Step 3 Find any intercepts. The vertex, $\left(\frac{1}{2}, -\frac{25}{4}\right)$, is in quadrant IV and the graph opens up, so there will be two x-intercepts. Let $f(x) = 0$ and solve.

$$f(x) = x^2 - x - 6$$

$$0 = x^2 - x - 6 \qquad \text{Let } f(x) = 0.$$

$$0 = (x - 3)(x + 2) \qquad \text{Factor.}$$

$$x - 3 = 0 \quad \text{or} \quad x + 2 = 0 \qquad \text{Zero-factor property}$$

$$x = 3 \quad \text{or} \qquad x = -2 \qquad \text{Solve each equation.}$$

The x-intercepts are $(3, 0)$ and $(-2, 0)$.
Find the y-intercept by evaluating $f(0)$.

$$f(x) = x^2 - x - 6$$

$$f(0) = 0^2 - 0 - 6 \qquad \text{Let } x = 0.$$

$$f(0) = -6 \qquad \text{Apply the exponent. Subtract.}$$

NOW TRY ANSWER
3. $\left(\frac{1}{3}, \frac{23}{3}\right)$

The y-intercept is $(0, -6)$.

**NOW TRY
EXERCISE 4**

Graph the quadratic function

$$f(x) = x^2 + 2x - 3.$$

Give the vertex, axis, domain, and range.

Step 4 Plot the points found so far and additional points as needed using symmetry about the axis, $x = \frac{1}{2}$. The graph is shown in **FIGURE 14**.

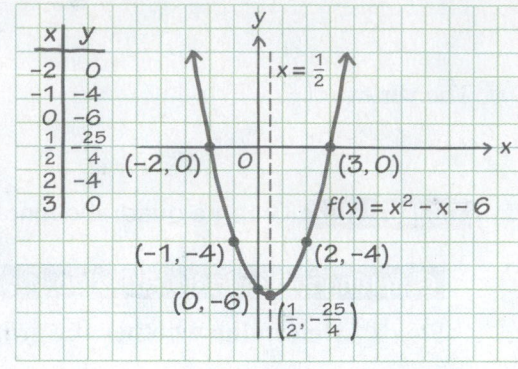

$f(x) = x^2 - x - 6$

Vertex: $\left(\frac{1}{2}, -\frac{25}{4}\right)$

Axis of symmetry: $x = \frac{1}{2}$

Domain: $(-\infty, \infty)$

Range: $\left[-\frac{25}{4}, \infty\right)$

FIGURE 14

 **NOW TRY**

OBJECTIVE 3 Use the discriminant to find the number of *x*-intercepts of a parabola with a vertical axis.

Recall from **Section 11.3** that

$$b^2 - 4ac \qquad \text{Discriminant}$$

is the *discriminant* of the quadratic equation $ax^2 + bx + c = 0$ and that we can use it to determine the number of real solutions of a quadratic equation.

In a similar way, we can use the discriminant of a quadratic *function* to determine the number of *x*-intercepts of its graph. The three possibilities are shown in **FIGURE 15**.

1. If the discriminant is positive, the parabola will have two *x*-intercepts.

2. If the discriminant is 0, there will be only one *x*-intercept, and it will be the vertex of the parabola.

3. If the discriminant is negative, the graph will have no *x*-intercepts.

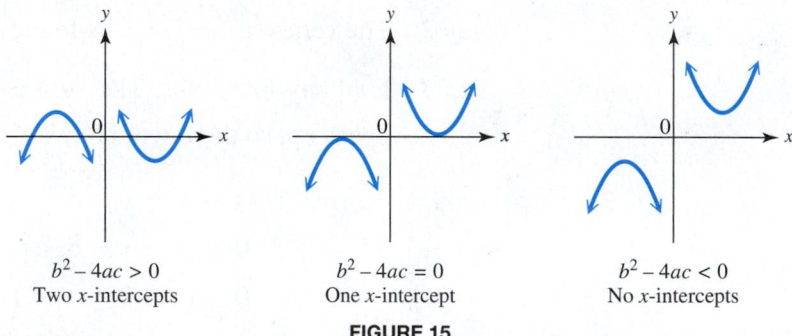

| $b^2 - 4ac > 0$ | $b^2 - 4ac = 0$ | $b^2 - 4ac < 0$ |
| Two *x*-intercepts | One *x*-intercept | No *x*-intercepts |

FIGURE 15

EXAMPLE 5 Using the Discriminant to Determine Number of *x*-Intercepts

Find the discriminant and use it to determine the number of *x*-intercepts of the graph of each quadratic function.

(a) $f(x) = 2x^2 + 3x - 5$

$$b^2 - 4ac \qquad \text{Discriminant}$$

$$= 3^2 - 4(2)(-5) \qquad a = 2, b = 3, c = -5$$

$$= 9 - (-40) \qquad \text{Apply the exponent. Multiply.}$$

$$= 49 \qquad \text{Subtract.}$$

Because the discriminant is positive, the parabola has two *x*-intercepts.

NOW TRY ANSWER

4.
$x = -1$

$(-3, 0)$ $(1, 0)$

$(-1, -4)$ $(0, -3)$

$f(x) = x^2 + 2x - 3$

vertex: $(-1, -4)$; axis: $x = -1$;
domain: $(-\infty, \infty)$; range: $[-4, \infty)$

NOW TRY
EXERCISE 5
Find the discriminant and use it to determine the number of x-intercepts of the graph of each quadratic function.

(a) $f(x) = -2x^2 + 3x - 2$

(b) $f(x) = 3x^2 + 2x - 1$

(c) $f(x) = 4x^2 - 12x + 9$

(b) $f(x) = -3x^2 - 1$

$$b^2 - 4ac \qquad \text{Discriminant}$$
$$= 0^2 - 4(-3)(-1) \qquad a = -3, b = 0, c = -1$$
$$= 0 - 12 \qquad \text{Apply the exponent. Multiply.}$$
$$= -12 \qquad \text{Subtract.}$$

The discriminant is negative, so the graph has no x-intercepts.

(c) $f(x) = 9x^2 + 6x + 1$

$$b^2 - 4ac \qquad \text{Discriminant}$$
$$= 6^2 - 4(9)(1) \qquad a = 9, b = 6, c = 1$$
$$= 36 - 36 \qquad \text{Apply the exponent. Multiply.}$$
$$= 0 \qquad \text{Subtract.}$$

Because the discriminant is 0, the parabola has only one x-intercept (its vertex).

NOW TRY

OBJECTIVE 4 Use quadratic functions to solve problems involving maximum or minimum value.

The vertex of the graph of a quadratic function is either the highest or the lowest point on the parabola. It provides the following information.

1. The y-value of the vertex gives the maximum or minimum value of y.

2. The x-value tells where the maximum or minimum occurs.

PROBLEM-SOLVING HINT In many applied problems we must find the greatest or least value of some quantity. When we can express that quantity in terms of a quadratic function, the value of k in the vertex (h, k) gives that optimum value.

EXAMPLE 6 Finding the Maximum Area of a Rectangular Region

A farmer has 120 ft of fencing to enclose a rectangular area next to a building. (See **FIGURE 16.**) Find the maximum area he can enclose and the dimensions of the field when the area is maximized.

FIGURE 16

Let $x =$ the width of the field.

$$x + x + \text{length} = 120 \qquad \text{Sum of the sides is 120 ft.}$$
$$2x + \text{length} = 120 \qquad \text{Combine like terms.}$$
$$\text{length} = 120 - 2x \qquad \text{Subtract } 2x.$$

NOW TRY ANSWERS
5. (a) -7; none **(b)** 16; two
(c) 0; one

**NOW TRY
EXERCISE 6**
Solve the problem in
Example 6 if the farmer has
only 80 ft of fencing.

The area $\mathcal{A}(x)$ is given by the product of the length and width.

$$\mathcal{A}(x) = (120 - 2x)x \quad \text{Area = length · width}$$
$$\mathcal{A}(x) = 120x - 2x^2 \quad \text{Distributive property}$$

To determine the maximum area, use the vertex formula to find the vertex of the parabola $\mathcal{A}(x) = 120x - 2x^2$. Write the equation in standard form.

$$\mathcal{A}(x) = -2x^2 + 120x \quad a = -2, b = 120, c = 0$$

Then $\quad x = \dfrac{-b}{2a} = \dfrac{-120}{2(-2)} = \dfrac{-120}{-4} = 30,$

and $\quad \mathcal{A}(30) = -2(30)^2 + 120(30) = -2(900) + 3600 = 1800.$

The graph is a parabola that opens down. Its vertex is $(30, 1800)$. The maximum area will be 1800 ft^2 when x, the width, is 30 ft and the length is

$$120 - 2(30) = 60 \text{ ft.} \quad \text{NOW TRY}$$

> **! CAUTION** *Be careful when interpreting the meanings of the coordinates of the* ***vertex.*** The first coordinate, x, gives the value for which the *function value*, y or $f(x)$, is a maximum or a minimum.
>
> Read a problem carefully to determine whether to find the value of the independent variable, the function value, or both.

**NOW TRY
EXERCISE 7**
A stomp rocket is launched
from the ground with an
initial velocity of 48 ft per
sec so that its distance in feet
above the ground after
t seconds is

$s(t) = -16t^2 + 48t.$

Find the maximum height
attained by the rocket and the
number of seconds it takes to
reach that height.

EXAMPLE 7   Finding the Maximum Height Attained by a Projectile

If air resistance is neglected, a projectile on Earth shot straight upward with an initial velocity of 40 m per sec will be at a height s in meters given by

$$s(t) = -4.9t^2 + 40t,$$

where t is the number of seconds elapsed after projection. After how many seconds will it reach its maximum height, and what is this maximum height?

For this function, $a = -4.9$, $b = 40$, and $c = 0$. Use the vertex formula.

$$t = \frac{-b}{2a} = \frac{-40}{2(-4.9)} = 4.1 \quad \text{Use a calculator. Round to the nearest tenth.}$$

This indicates that the maximum height is attained at 4.1 sec. To find this maximum height, calculate $s(4.1)$.

$$s(t) = -4.9t^2 + 40t$$
$$s(4.1) = -4.9(4.1)^2 + 40(4.1) \quad \text{Let } t = 4.1.$$
$$s(4.1) = 81.6 \quad \text{Use a calculator. Round to the nearest tenth.}$$

The projectile will attain a maximum height of 81.6 m at 4.1 sec. $\quad$ **NOW TRY**

OBJECTIVE 5 Graph parabolas with horizontal axes.

If x and y are interchanged in the equation

$$y = ax^2 + bx + c, \quad \text{the equation becomes} \quad x = ay^2 + by + c.$$

Because of the interchange of the roles of x and y, these parabolas are horizontal (with horizontal lines as axes of symmetry).

NOW TRY ANSWERS
6. The field should be 20 ft
   by 40 ft with maximum
   area 800 ft^2.
7. 36 ft; 1.5 sec

Graph of a Horizontal Parabola

The graph of $x = ay^2 + by + c$ or $x = a(y - k)^2 + h$ is a parabola.

- The vertex of the parabola is (h, k).

- The axis of symmetry is the horizontal line $y = k$.

- The graph opens to the right if $a > 0$ and to the left if $a < 0$.

 **NOW TRY
EXERCISE 8**

Graph $x = (y + 2)^2 - 1$.
Give the vertex, axis, domain,
and range.

EXAMPLE 8   Graphing a Horizontal Parabola ($a = 1$)

Graph $x = (y - 2)^2 - 3$. Give the vertex, axis, domain, and range.

This graph has its vertex at $(-3, 2)$ because the roles of x and y are interchanged. It opens to the right (the positive x-direction) because $a = 1$ and $1 > 0$, and has the same shape as $y = x^2$ (but situated horizontally). Plotting a few additional points gives the graph shown in **FIGURE 17**.

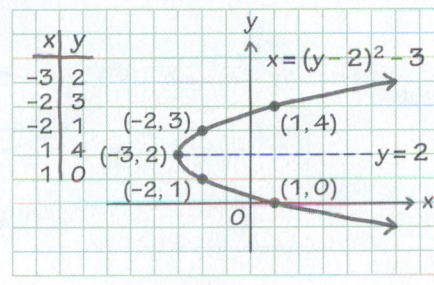

$x = (y - 2)^2 - 3$
Vertex: $(-3, 2)$
Axis of symmetry: $y = 2$
Domain: $[-3, \infty)$
Range: $(-\infty, \infty)$

FIGURE 17   NOW TRY

**NOW TRY
EXERCISE 9**

Graph $x = -3y^2 - 6y - 5$.
Give the vertex, axis, domain,
and range.

EXAMPLE 9   Graphing a Horizontal Parabola ($a \neq 1$)

Graph $x = -2y^2 + 4y - 3$. Give the vertex, axis, domain, and range.

$$x = -2y^2 + 4y - 3$$

$$x = (-2y^2 + 4y) - 3 \qquad \text{Group the variable terms.}$$

$$x = -2(y^2 - 2y) - 3 \qquad \text{Factor out } -2.$$

$$\left[\tfrac{1}{2}(-2)\right]^2 = (-1)^2 = 1$$

$$x = -2(y^2 - 2y + 1 - 1) - 3 \qquad \begin{array}{l}\text{Complete the square within the}\\ \text{parentheses. Add and subtract 1.}\end{array}$$

$$x = -2(y^2 - 2y + 1) + (-2)(-1) - 3 \qquad \text{Distributive property}$$

Be careful here.

$$x = -2(y - 1)^2 - 1 \qquad \text{Factor. Simplify.}$$

Because of the negative coefficient -2 in $x = -2(y - 1)^2 - 1$, the graph opens to the left (the negative x-direction). The graph is narrower than the graph of $y = x^2$ because $|-2| = 2$, and $2 > 1$. See **FIGURE 18**.

NOW TRY ANSWERS

8.   **9.**

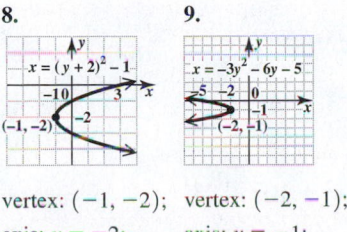

vertex: $(-1, -2)$;   vertex: $(-2, -1)$;
axis: $y = -2$;   axis: $y = -1$;
domain: $[-1, \infty)$;   domain: $(-\infty, -2]$;
range: $(-\infty, \infty)$ range: $(-\infty, \infty)$

$x = -2y^2 + 4y - 3$
Vertex: $(-1, 1)$
Axis of symmetry: $y = 1$
Domain: $(-\infty, -1]$
Range: $(-\infty, \infty)$

FIGURE 18   NOW TRY

! **CAUTION** *Only quadratic equations solved for y (whose graphs are vertical parabolas) are examples of functions.* The horizontal parabolas in **Examples 8 and 9** are *not* graphs of functions because they do not satisfy the conditions of the vertical line test.

▼ Summary of Graphs of Parabolas

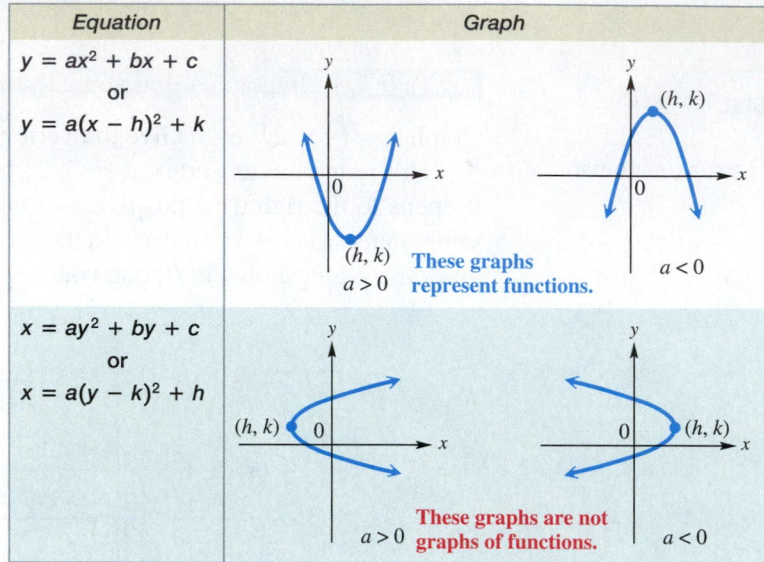

Equation	Graph
$y = ax^2 + bx + c$ or $y = a(x - h)^2 + k$	*These graphs represent functions.*
$x = ay^2 + by + c$ or $x = a(y - k)^2 + h$	*These graphs are not graphs of functions.*

11.7 Exercises

FOR EXTRA HELP

▶ MyMathLab®

▶ *Complete solution available in MyMathLab*

Concept Check   *Answer each question.*

1. How can we determine just by looking at the equation of a parabola whether it has a vertical or a horizontal axis?

2. Why can't the graph of a quadratic function be a parabola with a horizontal axis?

3. How can we determine the number of x-intercepts of the graph of a quadratic function without graphing the function?

4. If the vertex of the graph of a quadratic function is $(1, -3)$, and the graph opens down, how many x-intercepts does the graph have?

5. Which equations have a graph that is a vertical parabola?   A horizontal parabola?

A. $y = -x^2 + 20x + 80$ **B.** $x = 2y^2 + 6y + 5$

C. $x + 1 = (y + 2)^2$ **D.** $f(x) = (x - 4)^2$

6. Which of the equations in **Exercise 5** represent functions?

Find the vertex of each parabola. See Examples 1–3.

▶ **7.** $f(x) = x^2 + 8x + 10$ **8.** $f(x) = x^2 + 10x + 23$

▶ **9.** $f(x) = -2x^2 + 4x - 5$ **10.** $f(x) = -3x^2 + 12x - 8$

▶ **11.** $f(x) = x^2 + x - 7$ **12.** $f(x) = x^2 - x + 5$

Find the vertex of each parabola. For each equation, decide whether the graph opens up, down, to the left, *or* to the right, *and whether it is* wider, narrower, *or the* same shape *as the graph of* $y = x^2$. *If it is a parabola with a vertical axis of symmetry, find the discriminant and use it to determine the number of x-intercepts.* **See Examples 1–3, 5, 8, and 9.**

▶ **13.** $f(x) = 2x^2 + 4x + 5$ **14.** $f(x) = 3x^2 - 6x + 4$ **15.** $f(x) = -x^2 + 5x + 3$

16. $f(x) = -x^2 + 7x + 2$ **17.** $x = \dfrac{1}{3}y^2 + 6y + 24$ **18.** $x = \dfrac{1}{2}y^2 + 10y - 5$

Concept Check   *Match each equation in Exercises 19–24 with its graph in choices A–F.*

19. $y = 2x^2 + 4x - 3$ **20.** $y = -x^2 + 3x + 5$ **21.** $y = -\dfrac{1}{2}x^2 - x + 1$

22. $x = y^2 + 6y + 3$ **23.** $x = -y^2 - 2y + 4$ **24.** $x = 3y^2 + 6y + 5$

A.
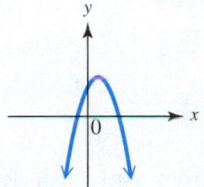

B.
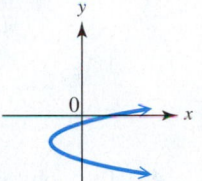

C.
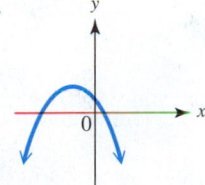

D.
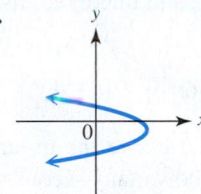

E.
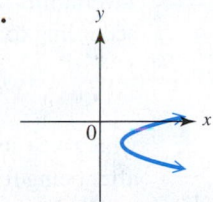

F.
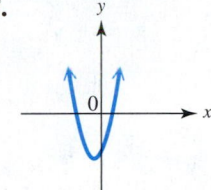

Graph each parabola. (Use the results of **Exercises 7–10** *in Exercises 25, 26, 29, and 30.) Give the vertex, axis of symmetry, domain, and range.* **See Examples 4, 8, and 9.**

▶ **25.** $f(x) = x^2 + 8x + 10$ **26.** $f(x) = x^2 + 10x + 23$

27. $f(x) = x^2 + 2x - 2$ **28.** $f(x) = x^2 + 4x + 3$

29. $f(x) = -2x^2 + 4x - 5$ **30.** $f(x) = -3x^2 + 12x - 8$

▶ **31.** $x = (y + 2)^2 + 1$ **32.** $x = (y + 3)^2 - 2$

33. $x = -(y - 3)^2 - 1$ **34.** $x = -(y - 2)^2 + 4$

35. $x = -\dfrac{1}{5}y^2 + 2y - 4$ **36.** $x = -\dfrac{1}{2}y^2 - 4y - 6$

▶ **37.** $x = 3y^2 + 12y + 5$ **38.** $x = 4y^2 + 16y + 11$

Solve each problem. **See Examples 6 and 7.**

39. Find the pair of numbers whose sum is 40 and whose product is a maximum. (*Hint:* Let x and $40 - x$ represent the two numbers.)

40. Find the pair of numbers whose sum is 60 and whose product is a maximum.

▶ **41.** Polk Community College wants to construct a rectangular parking lot on land bordered on one side by a highway. It has 280 ft of fencing that is to be used to fence off the other three sides. What should be the dimensions of the lot if the enclosed area is to be a maximum? What is the maximum area?

42. Bonnie has 100 ft of fencing material to enclose a rectangular exercise run for her dog. One side of the run will border her house, so she will only need to fence three sides. What dimensions will give the enclosure the maximum area? What is the maximum area?

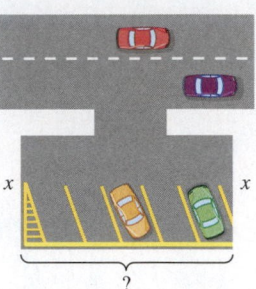

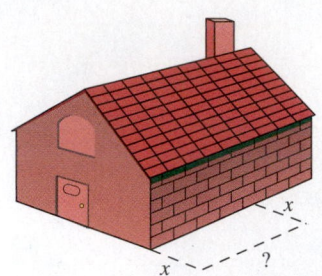

43. Two physics students from American River College find that when a bottle of California wine is shaken several times, held upright, and uncorked, its cork travels according to the function

$$s(t) = -16t^2 + 64t + 1,$$

where s is its height in feet above the ground t seconds after being released. After how many seconds will it reach its maximum height? What is the maximum height?

44. Professor Barbu has found that the number of students attending his intermediate algebra class is approximated by

$$S(x) = -x^2 + 20x + 80,$$

where x is the number of hours that the Campus Center is open daily. Find the number of hours that the center should be open so that the number of students attending class is a maximum. What is this maximum number of students?

45. Klaus has a taco stand. He has found that his daily costs are approximated by

$$C(x) = x^2 - 40x + 610,$$

where $C(x)$ is the cost, in dollars, to sell x units of tacos. Find the number of units of tacos he should sell to minimize his costs. What is the minimum cost?

46. Mohammad has a frozen yogurt cart. His daily costs are approximated by

$$C(x) = x^2 - 70x + 1500,$$

where $C(x)$ is the cost, in dollars, to sell x units of frozen yogurt. Find the number of units of frozen yogurt he must sell to minimize his costs. What is the minimum cost?

47. If an object on Earth is projected upward with an initial velocity of 32 ft per sec, then its height after t seconds is given by

$$s(t) = -16t^2 + 32t.$$

Find the maximum height attained by the object and the number of seconds it takes to hit the ground.

48. A projectile on Earth is fired straight upward so that its distance (in feet) above the ground t seconds after firing is given by

$$s(t) = -16t^2 + 400t.$$

Find the maximum height it reaches and the number of seconds it takes to reach that height.

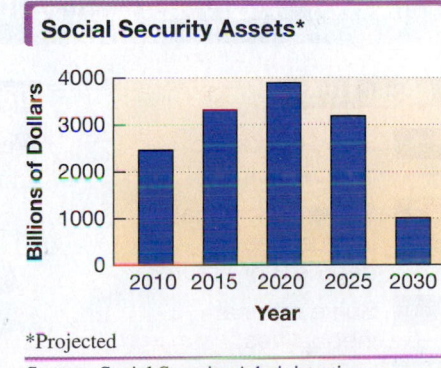

49. The percent of the U.S. population that was foreign-born during the years 1930–2010 can be modeled by the quadratic function

$$f(x) = 0.0043x^2 - 0.3245x + 11.53,$$

where $x = 0$ represents 1930, $x = 10$ represents 1940, and so on. (*Source:* U.S. Census Bureau.)

(a) The coefficient of x^2 in the model is positive, so the graph of this quadratic function is a parabola that opens up. Will the y-value of the vertex of this graph be a maximum or a minimum?

(b) According to the model, in what year during this period was the percent of foreign-born population a minimum? (Round down for the year.) Use the actual x-value of the vertex, to the nearest tenth, to find this percent, also to the nearest tenth.

50. The percent of births in the United States to teenage mothers during the years 2005–2012 can be modeled by the quadratic function

$$f(x) = -0.3661x^2 + 1.565x + 39.21,$$

where $x = 0$ represents 2005, $x = 1$ represents 2006, and so on. (*Source:* CDC.)

(a) The coefficient of x^2 in the model is negative, so the graph of this quadratic function is a parabola that opens down. Will the y-value of the vertex of this graph be a maximum or a minimum?

(b) According to the model, in what year during this period was the percent of births in the United States to teenage mothers a maximum? (Round down for the year.) Use the actual x-value of the vertex, to the nearest tenth, to find this percent, also to the nearest tenth.

The graph shows how Social Security trust fund assets are expected to change, and suggests that a quadratic function would be a good fit to the data. The data are approximated by the function

$$f(x) = -20.57x^2 + 758.9x - 3140.$$

In the model, $x = 10$ represents 2010, $x = 15$ represents 2015, and so on, and $f(x)$ is in billions of dollars.

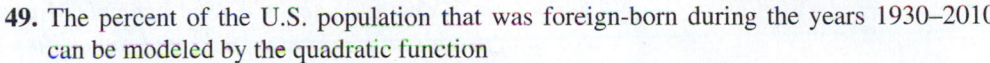

Social Security Assets*

Billions of Dollars

(bar graph with years 2010, 2015, 2020, 2025, 2030 on horizontal axis; vertical axis marked 0, 1000, 2000, 3000, 4000)

Year

*Projected

Source: Social Security Administration.

51. How could we have predicted that this quadratic model would have a negative coefficient for x^2, based only on the graph shown?

52. Algebraically determine the vertex of the graph, with coordinates to four significant digits. Interpret the answer as it applies to this application.

Extending Skills In each problem, find the following.

(a) A function $R(x)$ that describes the total revenue received

(b) The graph of the function from part (a)

(c) The number of unsold seats that will produce the maximum revenue

(d) The maximum revenue

53. A charter flight charges a fare of $200 per person, plus $4 per person for each unsold seat on the plane. The plane holds 100 passengers. Let x represent the number of unsold seats. (*Hint:* To find $R(x)$, multiply the number of people flying, $100 - x$, by the price per ticket, $200 + 4x$.)

54. A charter bus charges a fare of $48 per person, plus $2 per person for each unsold seat on the bus. The bus has 42 seats. Let x represent the number of unsold seats. (*Hint:* To find $R(x)$, multiply the number riding, $42 - x$, by the price per ticket, $48 + 2x$.)

Extending Skills In this section, we completed the square to find the vertex of a parabola in the form $f(x) = ax^2 + bx + c$. (See **Examples 1 and 2**.) Recall from **Section 10.3** that an equation of a circle with center (h, k) and radius r is

$$(x - h)^2 + (y - k)^2 = r^2. \qquad \text{Center-radius form of the equation of a circle}$$

Consider the equation $x^2 + y^2 + 2x + 6y - 15 = 0$, which is also the equation of a circle. We can write this equation in center-radius form by completing the squares on x and y.

$$x^2 + y^2 + 2x + 6y - 15 = 0$$

$$x^2 + y^2 + 2x + 6y = 15 \qquad \text{Transform so that the constant is on the right.}$$

$$(x^2 + 2x \qquad) + (y^2 + 6y \qquad) = 15 \qquad \text{Write in anticipation of completing the square.}$$

$$\left[\tfrac{1}{2}(2)\right]^2 = 1 \qquad \left[\tfrac{1}{2}(6)\right]^2 = 9 \qquad \text{Complete the squares on both } x \text{ and } y. \text{ Add 1 and 9 on } both \text{ sides of the equation.}$$

$$(x^2 + 2x + 1) + (y^2 + 6y + 9) = 15 + 1 + 9$$

$$(x + 1)^2 + (y + 3)^2 = 25 \qquad \text{Factor on the left. Add on the right.}$$

$$[x - (-1)]^2 + [y - (-3)]^2 = 5^2 \qquad \text{Write in center-radius form.}$$

The final equation shows that the circle has center $(-1, -3)$ and radius 5.
Find the center and radius of each circle.

55. $x^2 + y^2 + 4x + 6y + 9 = 0$ **56.** $x^2 + y^2 - 8x - 12y + 3 = 0$

57. $x^2 + y^2 + 10x - 14y - 7 = 0$ **58.** $x^2 + y^2 - 2x + 4y - 4 = 0$

(11.8) Polynomial and Rational Inequalities

OBJECTIVES

1   Solve quadratic inequalities.

2   Solve polynomial inequalities of degree 3 or greater.

3   Solve rational inequalities.

VOCABULARY

☐ quadratic inequality
☐ rational inequality

OBJECTIVE 1 Solve quadratic inequalities.

We can combine methods of solving linear inequalities and methods of solving quadratic equations to solve *quadratic inequalities*.

Quadratic Inequality

A **quadratic inequality** (in x here) can be written in the form

$$ax^2 + bx + c < 0, \qquad ax^2 + bx + c > 0,$$

$$ax^2 + bx + c \leq 0, \quad \text{or} \quad ax^2 + bx + c \geq 0,$$

where a, b, and c are real numbers and $a \neq 0$.

One way to solve a quadratic inequality involves graphing the related quadratic function. This method is justified because the graph is *continuous*—that is, it has no breaks.

**NOW TRY
EXERCISE 1**

Use the graph to solve each quadratic inequality.

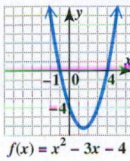

$f(x) = x^2 - 3x - 4$

(a) $x^2 - 3x - 4 > 0$

(b) $x^2 - 3x - 4 < 0$

EXAMPLE 1   **Solving Quadratic Inequalities by Graphing**

Solve each inequality.

(a) $x^2 - x - 12 > 0$

We graph the related quadratic function $f(x) = x^2 - x - 12$. We are particularly interested in the *x*-intercepts, which are found as in **Section 11.7** by letting $f(x) = 0$ and solving the following quadratic equation.

$$x^2 - x - 12 = 0 \qquad \text{Let } f(x) = 0.$$

$$(x - 4)(x + 3) = 0 \qquad \text{Factor.}$$

$$x - 4 = 0 \quad \text{or} \quad x + 3 = 0 \qquad \text{Zero-factor property}$$

$$x = 4 \quad \text{or} \qquad x = -3 \leftarrow \text{The } x\text{-intercepts are } (4, 0) \text{ and } (-3, 0).$$

The graph opens up because the coefficient of x^2 is positive. See **FIGURE 19(a)**. Notice that *x*-values less than -3 or greater than 4 result in *y*-values *greater than* 0. Thus, the solution set of $x^2 - x - 12 > 0$, written in interval notation, is

$$(-\infty, -3) \cup (4, \infty).$$

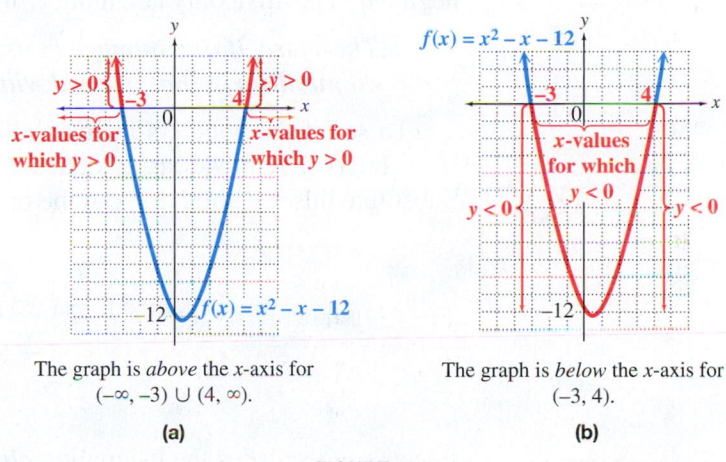

The graph is *above* the *x*-axis for   The graph is *below* the *x*-axis for
$(-\infty, -3) \cup (4, \infty).$ $(-3, 4).$

        (a)   **(b)**

FIGURE 19

(b) $x^2 - x - 12 < 0$

We want values of *y* that are *less than* 0. See **FIGURE 19(b)**. Notice from the graph that *x*-values between -3 and 4 result in *y*-values less than 0. Thus, the solution set of $x^2 - x - 12 < 0$, written in interval notation, is $(-3, 4)$.   **NOW TRY**

NOTE   If the inequalities in **Example 1** had used $\geq$ and $\leq$, the solution sets would have included the *x*-values of the intercepts, which make the quadratic expression equal to 0. They would have been written in interval notation as

$$(-\infty, -3] \cup [4, \infty) \quad \text{and} \quad [-3, 4].$$

Square brackets would indicate that the endpoints -3 and 4 are *included* in the solution sets.

NOW TRY ANSWERS
1. **(a)** $(-\infty, -1) \cup (4, \infty)$
   (b) $(-1, 4)$

Another method for solving a quadratic inequality uses the basic ideas of **Example 1** without actually graphing the related quadratic function.

EXAMPLE 2   Solving a Quadratic Inequality Using Test Values

Solve and graph the solution set of $x^2 - x - 12 > 0$.

Solve the quadratic equation $x^2 - x - 12 = 0$ as in **Example 1(a)**.

$x^2 - x - 12 = 0$	Let $f(x) = 0$.
$(x - 4)(x + 3) = 0$	Factor.
$x - 4 = 0$ or $x + 3 = 0$	Zero-factor property
$x = 4$ or $x = -3$	Solve each equation.

The numbers 4 and -3 divide a number line into Intervals A, B, and C, as shown in **FIGURE 20**. *Be careful to put the lesser number on the left.*

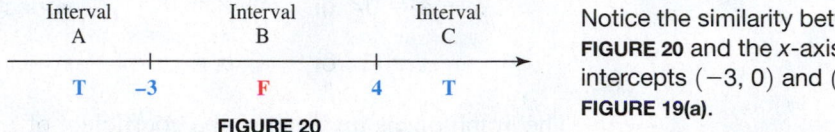

FIGURE 20

Notice the similarity between **FIGURE 20** and the x-axis with intercepts $(-3, 0)$ and $(4, 0)$ in **FIGURE 19(a)**.

The numbers 4 and -3 are the only values that make the quadratic expression $x^2 - x - 12$ equal to 0. All other numbers make the expression either positive or negative. The sign of the expression can change from positive to negative or from negative to positive only at a number that makes it 0.

> *Therefore, if one number in an interval satisfies the inequality, then all numbers in that interval will satisfy the inequality.*

To see if the numbers in Interval A satisfy the inequality, choose any number from Interval A in **FIGURE 20** (that is, any number less than -3). We choose -5. Substitute this test value for x in the original inequality $x^2 - x - 12 > 0$.

$x^2 - x - 12 > 0$	Original inequality
$(-5)^2 - (-5) - 12 \overset{?}{>} 0$	Let $x = -5$.
$25 + 5 - 12 \overset{?}{>} 0$	Simplify.
$18 > 0$ ✓	True

Use parentheses to avoid sign errors.

Because -5 satisfies the inequality, *all* numbers from Interval A are solutions.

Now try 0 from Interval B.

$x^2 - x - 12 > 0$	Original inequality
$0^2 - 0 - 12 \overset{?}{>} 0$	Let $x = 0$.
$-12 > 0$	False

The numbers in Interval B are *not* solutions.

Now try 5 from Interval C.

$x^2 - x - 12 > 0$	Original inequality
$5^2 - 5 - 12 \overset{?}{>} 0$	Let $x = 5$.
$8 > 0$	True

All numbers from Interval C are solutions.

NOW TRY
EXERCISE 2

Solve and graph the solution set.

$$x^2 + 2x - 8 > 0$$

Based on these results (shown by the colored letters in **FIGURE 20**), the solution set includes all numbers in Intervals A and C, as shown in **FIGURE 21**. The solution set is written in interval notation as

$$(-\infty, -3) \cup (4, \infty).$$

−3 and 4 are *not* included because the symbol > does not include equality.

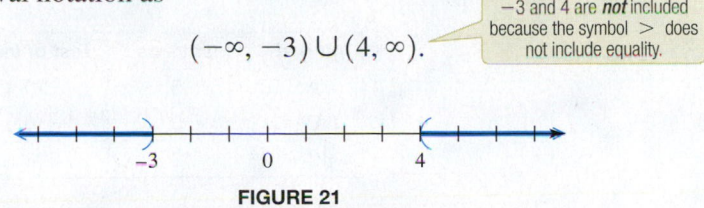

FIGURE 21

This agrees with the solution set found in **Example 1(a).**   **NOW TRY**

Solving a Quadratic Inequality

Step 1   **Write the inequality as an equation and solve it.**

Step 2   **Use the solutions from Step 1 to determine intervals.** Graph the values found in Step 1 on a number line. These values divide the number line into intervals.

Step 3   **Find the intervals that satisfy the inequality.** Substitute a test value from each interval into the original inequality to determine the intervals that satisfy the inequality. All numbers in those intervals are in the solution set. A graph of the solution set will usually look like one of these.

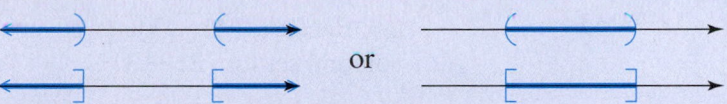

or

Step 4   **Consider the endpoints separately.** The values from Step 1 are included in the solution set if the inequality symbol is ≤ or ≥. They are not included if it is < or >.

EXAMPLE 3   Solving a Quadratic Inequality

Step 1   Solve and graph the solution set of $2x^2 + 5x \leq 12$.

$$2x^2 + 5x = 12 \qquad \text{Related quadratic equation}$$
$$2x^2 + 5x - 12 = 0 \qquad \text{Standard form}$$
$$(2x - 3)(x + 4) = 0 \qquad \text{Factor.}$$
$$2x - 3 = 0 \quad \text{or} \quad x + 4 = 0 \qquad \text{Zero-factor property}$$
$$x = \frac{3}{2} \quad \text{or} \qquad x = -4 \qquad \text{Solve each equation.}$$

Step 2   The numbers $\frac{3}{2}$ and -4 divide a number line into three intervals. See **FIGURE 22.**

NOW TRY ANSWER

2. $(-\infty, -4) \cup (2, \infty)$

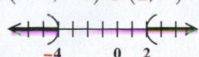

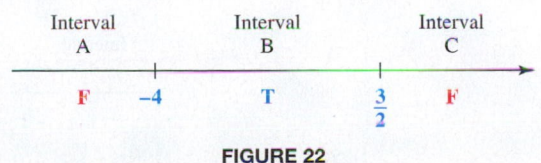

FIGURE 22

**NOW TRY
EXERCISE 3**

Solve and graph the solution set.

$$3x^2 - 11x \le 4$$

Steps 3 and 4 Substitute a test value from each interval in the *original* inequality $2x^2 + 5x \le 12$ to determine which intervals satisfy the inequality.

Interval	Test Value	Test of Inequality	True or False?
A	−5	25 ≤ 12	F
B	0	0 ≤ 12	T
C	2	18 ≤ 12	F

We use a table to organize this information. (Verify it.)

The numbers in Interval B are solutions. See **FIGURE 23**. The solution set is the interval

$$\left[-4, \frac{3}{2}\right].$$ — −4 and $\frac{3}{2}$ are included because the symbol ≤ includes equality.

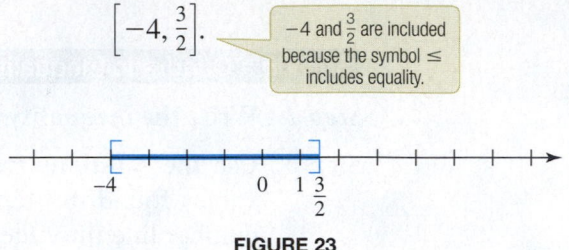

FIGURE 23

NOW TRY

**NOW TRY
EXERCISE 4**

Solve each inequality.

(a) $(4x - 1)^2 > -3$

(b) $(4x - 1)^2 < -3$

EXAMPLE 4   Solving Special Cases

Solve each inequality.

(a) $(2x - 3)^2 > -1$

Because $(2x - 3)^2$ is never negative, it is always greater than −1. Thus, the solution set for $(2x - 3)^2 > -1$ is the set of all real numbers, $(-\infty, \infty)$.

(b) $(2x - 3)^2 < -1$

Using similar reasoning as in part (a), there is no solution for this inequality. The solution set is $\varnothing$.

NOW TRY

OBJECTIVE 2   Solve polynomial inequalities of degree 3 or greater.

EXAMPLE 5   Solving a Third-Degree Polynomial Inequality

Solve and graph the solution set of $(x - 1)(x + 2)(x - 4) \le 0$.

This is a *cubic* (third-degree) inequality rather than a quadratic inequality, but it can be solved using the preceding method by extending the zero-factor property to more than two factors. (Step 1)

$$(x - 1)(x + 2)(x - 4) = 0$$ Set the factored polynomial *equal* to 0.

$x - 1 = 0$ or $x + 2 = 0$ or $x - 4 = 0$ Zero-factor property

$x = 1$ or $x = -2$ or $x = 4$ Solve each equation.

Locate the numbers −2, 1, and 4 on a number line, as in **FIGURE 24**, to determine the Intervals A, B, C, and D. (Step 2)

NOW TRY ANSWERS

3. $\left[-\frac{1}{3}, 4\right]$

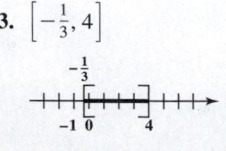

4. (a) $(-\infty, \infty)$ (b) $\varnothing$

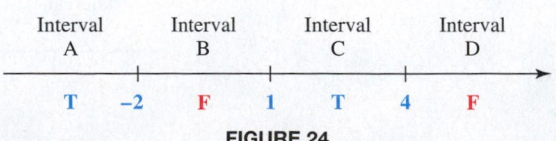

FIGURE 24

**NOW TRY
EXERCISE 5**

Solve and graph the solution set.

$$(x + 4)(x - 3)(2x + 1) \le 0$$

Substitute a test value from each interval in the *original* inequality to determine which intervals satisfy $(x - 1)(x + 2)(x - 4) \le 0$. (Step 3)

Interval	Test Value	Test of Inequality	True or False?
A	−3	$-28 \le 0$	T
B	0	$8 \le 0$	F
C	2	$-8 \le 0$	T
D	5	$28 \le 0$	F

The numbers in Intervals A and C are in the solution set, which is written as the interval $(-\infty, -2] \cup [1, 4]$, and graphed in **FIGURE 25**. The three endpoints are included because the inequality symbol $\le$ includes equality. (Step 4)

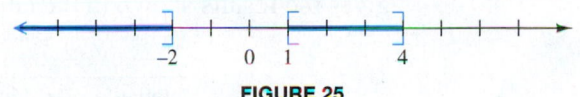

FIGURE 25

NOW TRY

OBJECTIVE 3 Solve rational inequalities.

Rational inequalities involve rational expressions and are solved similarly.

Solving a Rational Inequality

Step 1 **Write the inequality so that 0 is on one side** and there is a single fraction on the other side.

Step 2 **Determine the values that make the numerator or denominator equal to 0.**

Step 3 **Divide a number line into intervals.** Use the values from Step 2.

Step 4 **Find the intervals that satisfy the inequality.** Test a value from each interval by substituting it into the *original* inequality.

Step 5 **Consider the endpoints separately.** Exclude any values that make the denominator 0.

EXAMPLE 6 Solving a Rational Inequality

Solve and graph the solution set of $\dfrac{-1}{x - 3} > 1$.

Write the inequality so that 0 is on one side. (Step 1)

$$\frac{-1}{x - 3} - 1 > 0 \qquad \text{Subtract 1.}$$

$$\frac{-1}{x - 3} - \frac{x - 3}{x - 3} > 0 \qquad \text{Use } x - 3 \text{ as the common denominator.}$$

Be careful with signs.
$$\frac{-1 - x + 3}{x - 3} > 0 \qquad \text{Write the left side as a single fraction.}$$

$$\frac{-x + 2}{x - 3} > 0 \qquad \text{Combine like terms in the numerator.}$$

NOW TRY ANSWER

5. $(-\infty, -4] \cup \left[-\frac{1}{2}, 3\right]$

**NOW TRY
EXERCISE 6**
Solve and graph the solution set.

$$\frac{3}{x+1} > 4$$

The sign of $\frac{-x+2}{x-3}$ will change from positive to negative or negative to positive only at those values that make the numerator or denominator 0. The number 2 makes the numerator 0, and 3 makes the denominator 0. (Step 2) These two numbers, 2 and 3, divide a number line into three intervals. See **FIGURE 26**. (Step 3)

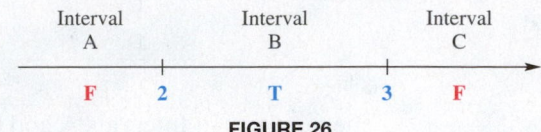

FIGURE 26

Substituting a test value from each interval in the *original* inequality, $\frac{-1}{x-3} > 1$, gives the results shown in the table. (Step 4)

Interval	Test Value	Test of Inequality	True or False?
A	0	$\frac{1}{3} > 1$	F
B	2.5	$2 > 1$	T
C	4	$-1 > 1$	F

The numbers in Interval B are solutions, so the solution set is the interval $(2, 3)$. This interval does not include 3 because it makes the denominator in the original inequality 0. The number 2 is not included either because the inequality symbol $>$ does not include equality. (Step 5) See **FIGURE 27**.

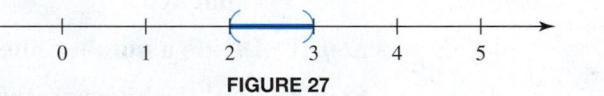

FIGURE 27

NOW TRY

⚠ CAUTION   *When solving a rational inequality, any number that makes the denominator 0 must be excluded from the solution set.*

EXAMPLE 7   Solving a Rational Inequality

Solve and graph the solution set of $\frac{x-2}{x+2} \le 2$.

Write the inequality so that 0 is on one side. (Step 1)

$$\frac{x-2}{x+2} - 2 \le 0 \qquad \text{Subtract 2.}$$

$$\frac{x-2}{x+2} - \frac{2(x+2)}{x+2} \le 0 \qquad \text{Use } x+2 \text{ as the common denominator.}$$

Be careful with signs.

$$\frac{x-2-2x-4}{x+2} \le 0 \qquad \text{Write as a single fraction.}$$

$$\frac{-x-6}{x+2} \le 0 \qquad \text{Combine like terms in the numerator.}$$

NOW TRY ANSWER
6. $\left(-1, -\frac{1}{4}\right)$

NOW TRY
EXERCISE 7

Solve and graph the solution set.

$$\frac{x-3}{x+3} \le 2$$

Because $\frac{-x-6}{x+2} \le 0$, the number -6 makes the numerator 0, and -2 makes the denominator 0. (Step 2) These two numbers determine three intervals on a number line. See **FIGURE 28**. (Step 3)

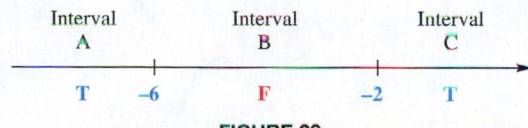

FIGURE 28

Substitute a test value from each interval in the *original* inequality $\frac{x-2}{x+2} \le 2$. (Step 4)

Interval	Test Value	Test of Inequality	True or False?
A	−8	$\frac{5}{3} \le 2$	T
B	−4	$3 \le 2$	F
C	0	$-1 \le 2$	T

The numbers in Intervals A and C are solutions. The solution set is the interval

$$(-\infty, -6] \cup (-2, \infty).$$

The number -6 satisfies the original inequality, but -2 does not because it makes the denominator 0. (Step 5) See **FIGURE 29**.

NOW TRY ANSWER

7. $(-\infty, -9] \cup (-3, \infty)$

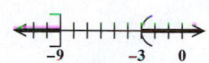

FIGURE 29

NOW TRY

11.8 Exercises

FOR EXTRA HELP ▶ MyMathLab®

▶ *Complete solution available in MyMathLab*

1. Concept Check The solution set of the inequality $x^2 + x - 12 < 0$ is the interval $(-4, 3)$. Without actually performing any work, give the solution set of the inequality

$$x^2 + x - 12 \ge 0.$$

2. Explain how to determine whether to include or exclude endpoints when solving a quadratic or higher-degree inequality.

In each exercise, the graph of a quadratic function f is given. Use the graph to find the solution set of each equation or inequality. **See Example 1.**

▶ **3. (a)** $x^2 - 4x + 3 = 0$

　　(b) $x^2 - 4x + 3 > 0$

　　(c) $x^2 - 4x + 3 < 0$

4. (a) $3x^2 + 10x - 8 = 0$

　　(b) $3x^2 + 10x - 8 \ge 0$

　　(c) $3x^2 + 10x - 8 < 0$

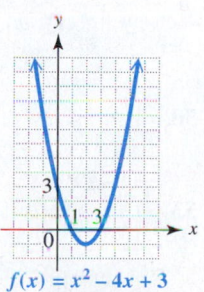

$f(x) = x^2 - 4x + 3$

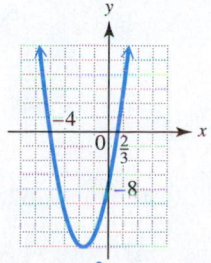

$f(x) = 3x^2 + 10x - 8$

5. (a) $-x^2 + 3x + 10 = 0$

(b) $-x^2 + 3x + 10 \geq 0$

(c) $-x^2 + 3x + 10 \leq 0$

6. (a) $-2x^2 - x + 15 = 0$

(b) $-2x^2 - x + 15 \geq 0$

(c) $-2x^2 - x + 15 \leq 0$

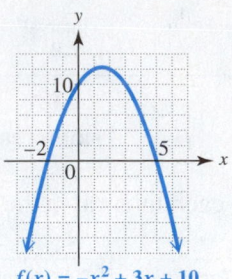

$f(x) = -x^2 + 3x + 10$

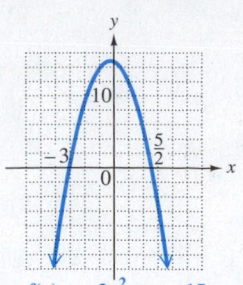

$f(x) = -2x^2 - x + 15$

Solve each inequality, and graph the solution set. **See Examples 2 and 3.** *(Hint: In Exercises 23 and 24, use the quadratic formula.)*

7. $(x + 1)(x - 5) > 0$

8. $(x + 6)(x - 2) > 0$

9. $(x + 4)(x - 6) < 0$

10. $(x + 4)(x - 8) < 0$

▶ 11. $x^2 - 4x + 3 \geq 0$

12. $x^2 - 3x - 10 \geq 0$

13. $10x^2 + 9x \geq 9$

14. $3x^2 + 10x \geq 8$

15. $4x^2 - 9 \leq 0$

16. $9x^2 - 25 \leq 0$

17. $6x^2 + x \geq 1$

18. $4x^2 + 7x \geq -3$

19. $z^2 - 4z \geq 0$

20. $x^2 + 2x < 0$

21. $3x^2 - 5x \leq 0$

22. $2z^2 + 3z > 0$

23. $x^2 - 6x + 6 \geq 0$

24. $3x^2 - 6x + 2 \leq 0$

Solve each inequality. **See Example 4.**

▶ 25. $(4 - 3x)^2 \geq -2$

26. $(7 - 6x)^2 \geq -1$

27. $(3x + 5)^2 \leq -4$

28. $(8x + 5)^2 \leq -5$

29. $(2x + 5)^2 < 0$

30. $(3x - 7)^2 < 0$

31. $(5x - 1)^2 \geq 0$

32. $(4x + 1)^2 \geq 0$

Solve each inequality, and graph the solution set. **See Example 5.**

▶ 33. $(x - 1)(x - 2)(x - 4) < 0$

34. $(2x + 1)(3x - 2)(4x + 7) < 0$

35. $(x - 4)(2x + 3)(3x - 1) \geq 0$

36. $(x + 2)(4x - 3)(2x + 7) \geq 0$

Solve each inequality, and graph the solution set. **See Examples 6 and 7.**

37. $\dfrac{x - 1}{x - 4} > 0$

38. $\dfrac{x + 1}{x - 5} > 0$

39. $\dfrac{2x + 3}{x - 5} \leq 0$

40. $\dfrac{3x + 7}{x - 3} \leq 0$

▶ 41. $\dfrac{8}{x - 2} \geq 2$

42. $\dfrac{20}{x - 1} \geq 1$

43. $\dfrac{3}{2x - 1} < 2$

44. $\dfrac{6}{x - 1} < 1$

45. $\dfrac{x - 3}{x + 2} \geq 2$

46. $\dfrac{m + 4}{m + 5} \geq 2$

▶ 47. $\dfrac{x - 8}{x - 4} < 3$

48. $\dfrac{2t - 3}{t + 1} > 4$

49. $\dfrac{4k}{2k - 1} < k$

50. $\dfrac{r}{r + 2} < 2r$

51. $\dfrac{2x - 3}{x^2 + 1} \geq 0$

52. $\dfrac{9x - 8}{4x^2 + 25} < 0$

53. $\dfrac{(3x - 5)^2}{x + 2} > 0$

54. $\dfrac{(5x - 3)^2}{2x + 1} \leq 0$

Chapter 11 Summary

Key Terms

11.1
quadratic equation
second-degree equation

11.3
quadratic formula
discriminant

11.4
quadratic in form

11.6
parabola
vertex
axis of symmetry (axis)
quadratic function

11.8
quadratic inequality
rational inequality

Test Your Word Power

See how well you have learned the vocabulary in this chapter.

1. The **quadratic formula** is
   A. a formula to find the number of solutions of a quadratic equation
   B. a formula to find the type of solutions of a quadratic equation
   C. the standard form of a quadratic equation
   D. a general formula for solving any quadratic equation.

2. A **quadratic function** is a function that can be written in the form
   A. $f(x) = mx + b$, for real numbers m and b
   B. $f(x) = \frac{P(x)}{Q(x)}$, where $Q(x) \neq 0$
   C. $f(x) = ax^2 + bx + c$, for real numbers a, b, and c ($a \neq 0$)
   D. $f(x) = \sqrt{x}$, for $x \geq 0$.

3. A **parabola** is the graph of
   A. any equation in two variables
   B. a linear equation
   C. an equation of degree 3
   D. a quadratic equation in two variables.

4. The **vertex** of a parabola is
   A. the point where the graph intersects the y-axis
   B. the point where the graph intersects the x-axis
   C. the lowest point on a parabola that opens up or the highest point on a parabola that opens down
   D. the origin.

5. The **axis** of a parabola is
   A. either the x-axis or the y-axis
   B. the vertical line (of a vertical parabola) or the horizontal line (of a horizontal parabola) through the vertex
   C. the lowest or highest point on the graph of a parabola
   D. a line through the origin.

6. A parabola is **symmetric about its axis** because
   A. its graph is near the axis
   B. its graph is identical on each side of the axis
   C. its graph looks different on each side of the axis
   D. its graph intersects the axis.

ANSWERS

1. D; *Example:* The solutions of $ax^2 + bx + c = 0$ ($a \neq 0$) are given by $x = \dfrac{-b \pm \sqrt{b^2 - 4ac}}{2a}$. 2. C; *Examples:* $f(x) = x^2 - 2$, $f(x) = (x + 4)^2 + 1$, $f(x) = x^2 - 4x + 5$ 3. D; *Examples:* See the figures in the Quick Review for **Sections 11.6 and 11.7.** 4. C; *Example:* The graph of $y = (x + 3)^2$ has vertex $(-3, 0)$, which is the lowest point on the graph. 5. B; *Example:* The axis of $y = (x + 3)^2$ is the vertical line $x = -3$. 6. B; *Example:* Because the graph of $y = (x + 3)^2$ is symmetric about its axis $x = -3$, the points $(-2, 1)$ and $(-4, 1)$ are on the graph.

Quick Review

CONCEPTS

EXAMPLES

11.1 Solving Quadratic Equations by the Square Root Property

Square Root Property
If x and k are complex numbers and $x^2 = k$, then
$$x = \sqrt{k} \quad \text{or} \quad x = -\sqrt{k}.$$

Solve $(x - 1)^2 = 8$.
$$x - 1 = \sqrt{8} \quad \text{or} \quad x - 1 = -\sqrt{8}$$
$$x = 1 + 2\sqrt{2} \quad \text{or} \quad x = 1 - 2\sqrt{2}$$
The solution set is $\{1 + 2\sqrt{2}, 1 - 2\sqrt{2}\}$, or $\{1 \pm 2\sqrt{2}\}$.

CONCEPTS	EXAMPLES

11.2 Solving Quadratic Equations by Completing the Square

Completing the Square

Solve $ax^2 + bx + c = 0$ (where $a \neq 0$), as follows.

Step 1 If $a \neq 1$, divide each side by a.

Step 2 Write the equation with the variable terms on one side and the constant on the other.

Step 3 Complete the square.

- Take half the coefficient of x and square it.
- Add the square to each side.
- Factor the perfect square trinomial, and write it as the square of a binomial. Combine terms on the other side.

Step 4 Use the square root property to solve.

Solve $2x^2 - 4x - 18 = 0$.

$$x^2 - 2x - 9 = 0 \qquad \text{Divide by 2.}$$

$$x^2 - 2x = 9 \qquad \text{Add 9.}$$

$$\left[\tfrac{1}{2}(-2)\right]^2 = (-1)^2 = 1$$

$$x^2 - 2x + 1 = 9 + 1 \qquad \text{Add 1.}$$

$$(x - 1)^2 = 10 \qquad \text{Factor. Add.}$$

$x - 1 = \sqrt{10} \qquad$ or $\quad x - 1 = -\sqrt{10}$ Square root property

$x = 1 + \sqrt{10}$ or $x = 1 - \sqrt{10}$

The solution set is $\left\{1 + \sqrt{10},\, 1 - \sqrt{10}\right\}$, or $\left\{1 \pm \sqrt{10}\right\}$.

11.3 The Quadratic Formula

Quadratic Formula

The solutions of $ax^2 + bx + c = 0$ are given by

$$x = \frac{-b \pm \sqrt{b^2 - 4ac}}{2a} \qquad \text{(where } a \neq 0\text{).}$$

The Discriminant

The discriminant $b^2 - 4ac$ of $ax^2 + bx + c = 0$ (where a, b, and c are integers) can be used to determine the number and type of solutions.

Discriminant	Number and Type of Solutions
Positive, and the square of an integer	Two rational solutions
Positive, but not the square of an integer	Two irrational solutions
Zero	One rational solution
Negative	Two nonreal complex solutions

Solve $3x^2 + 5x + 2 = 0$.

$$x = \frac{-5 \pm \sqrt{5^2 - 4(3)(2)}}{2(3)} \qquad a = 3,\, b = 5,\, c = 2$$

$$x = \frac{-5 \pm 1}{6} \qquad \text{Simplify.}$$

$x = -\dfrac{2}{3} \quad$ or $\quad x = -1$ Two solutions, one from $+$ and one from $-$

The solution set is $\left\{-1,\, -\tfrac{2}{3}\right\}$.

For $x^2 + 3x - 10 = 0$, the discriminant is

$$b^2 - 4ac$$

$$= 3^2 - 4(1)(-10) \qquad a = 1,\, b = 3,\, c = -10$$

$$= 49. \qquad \text{Two rational solutions}$$

11.4 Equations Quadratic in Form

A nonquadratic equation that can be written in the form $au^2 + bu + c = 0$, for $a \neq 0$ and an algebraic expression u, is quadratic in form.

Solving an Equation Quadratic in Form (Substitution)

Step 1 Define a temporary variable u.

Step 2 Solve the quadratic equation obtained in Step 1.

Step 3 Replace u with the expression it defined.

Step 4 Solve the resulting equations for the original variable.

Step 5 Check all solutions in the original equation.

Solve $3(x + 5)^2 + 7(x + 5) + 2 = 0$.

$$3u^2 + 7u + 2 = 0 \qquad \text{Let } u = x + 5.$$

$$(3u + 1)(u + 2) = 0 \qquad \text{Factor.}$$

$3u + 1 = 0 \qquad$ or $\quad u + 2 = 0$ Zero-factor property

$u = -\dfrac{1}{3} \quad$ or $\qquad u = -2$ Solve for u.

$x + 5 = -\dfrac{1}{3} \quad$ or $\quad x + 5 = -2$ Replace u with $x + 5$.

$x = -\dfrac{16}{3} \quad$ or $\qquad x = -7$ Subtract 5.

Check that the solution set is $\left\{-7,\, -\tfrac{16}{3}\right\}$.

CONCEPTS	**EXAMPLES**

11.5 Formulas and Further Applications

Solving a Formula for a Squared Variable

Case 1   **If the variable appears only to the second power:** Isolate the squared variable on one side of the equation, and then use the square root property.

Case 2   **If the variable appears to the first and second powers:** Write the equation in standard form, and then use the quadratic formula.

Solve $A = \dfrac{2mp}{r^2}$ for r. (*Case 1*)

$r^2 A = 2mp$ Multiply by r^2.

$r^2 = \dfrac{2mp}{A}$ Divide by A.

$r = \pm\sqrt{\dfrac{2mp}{A}}$ Square root property

$r = \dfrac{\pm\sqrt{2mpA}}{A}$ Rationalize denominator.

Solve $x^2 + rx = t$ for x. (*Case 2*)

$x^2 + rx - t = 0$ Standard form

$x = \dfrac{-r \pm \sqrt{r^2 - 4(1)(-t)}}{2(1)}$

$a = 1, b = r, c = -t$

$x = \dfrac{-r \pm \sqrt{r^2 + 4t}}{2}$ Simplify.

11.6 Graphs of Quadratic Functions

1. The graph of the quadratic function
$$F(x) = a(x - h)^2 + k \quad (\text{where } a \neq 0)$$
is a parabola with vertex (h, k) and the vertical line $x = h$ as axis of symmetry.

2. The graph opens up if $a > 0$ and down if $a < 0$.

3. The graph is wider than the graph of $f(x) = x^2$ if $0 < |a| < 1$ and narrower if $|a| > 1$.

Graph $f(x) = -(x + 3)^2 + 1$.

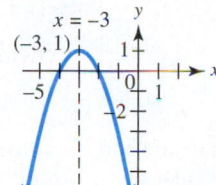

The graph opens down because $a < 0$.

Vertex: $(-3, 1)$

Axis of symmetry: $x = -3$

Domain: $(-\infty, \infty)$

Range: $(-\infty, 1]$

11.7 More about Parabolas and Their Applications

The vertex of the graph of
$$f(x) = ax^2 + bx + c$$
(where $a \neq 0$) may be found by completing the square or using the vertex formula $\left(\dfrac{-b}{2a}, f\left(\dfrac{-b}{2a}\right)\right)$.

Graphing a Quadratic Function

Step 1   Determine whether the graph opens up or down.

Step 2   Find the vertex.

Step 3   Find any intercepts.

Step 4   Find and plot additional points as needed.

Graph $f(x) = x^2 + 4x + 3$.

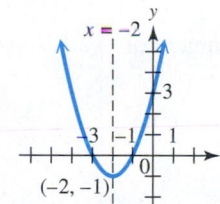

The graph opens up because $a > 0$.

Vertex: $(-2, -1)$

The solutions of $x^2 + 4x + 3 = 0$ are -1 and -3, so the x-intercepts are $(-1, 0)$ and $(-3, 0)$.

$f(0) = 3$, so the y-intercept is $(0, 3)$.

Axis of symmetry: $x = -2$

Domain: $(-\infty, \infty)$

Range: $[-1, \infty)$

Horizontal Parabolas

The graph of
$$x = ay^2 + by + c \quad \text{or} \quad x = a(y - k)^2 + h$$
is a horizontal parabola with vertex (h, k) and the horizontal line $y = k$ as axis of symmetry. The graph opens to the right if $a > 0$ and to the left if $a < 0$.

Horizontal parabolas do not represent functions.

Graph $x = 2y^2 + 6y + 5$.

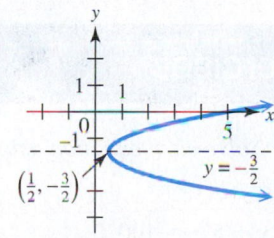

The graph opens to the right because $a > 0$.

Vertex: $\left(\dfrac{1}{2}, -\dfrac{3}{2}\right)$

Axis of symmetry: $y = -\dfrac{3}{2}$

Domain: $\left[\dfrac{1}{2}, \infty\right)$

Range: $(-\infty, \infty)$

CONCEPTS	**EXAMPLES**

11.8 Polynomial and Rational Inequalities

Solving a Quadratic (or Higher-Degree Polynomial) Inequality

Step 1   Write the inequality as an equation and solve.

Step 2   Use the values found in Step 1 to divide a number line into intervals.

Step 3   Substitute a test value from each interval into the *original* inequality to determine the intervals that belong to the solution set.

Step 4   Consider the endpoints separately.

Solve $2x^2 + 5x + 2 < 0$.

$$2x^2 + 5x + 2 = 0 \qquad \text{Related equation}$$

$$(2x + 1)(x + 2) = 0 \qquad \text{Factor.}$$

$$2x + 1 = 0 \quad \text{or} \quad x + 2 = 0 \qquad \text{Zero-factor property}$$

$$x = -\frac{1}{2} \quad \text{or} \qquad x = -2 \qquad \text{Solve each equation.}$$

Intervals: $(-\infty, -2)$, $\left(-2, -\frac{1}{2}\right), \left(-\frac{1}{2}, \infty\right)$

Test values: -3 (Interval A), -1 (Interval B), 0 (Interval C)

$x = -3$ makes the original inequality false, $x = -1$ makes it true, and $x = 0$ makes it false.

The solution set is the interval $\left(-2, -\frac{1}{2}\right)$. The endpoints are not included because the symbol $<$ does not include equality.

Solving a Rational Inequality

Step 1   Write the inequality so that 0 is on one side and there is a single fraction on the other side.

Step 2   Determine the values that make the numerator or denominator 0.

Step 3   Use the values from Step 2 to divide a number line into intervals.

Step 4   Substitute a test value from each interval into the *original* inequality to determine the intervals that belong to the solution set.

Step 5   Consider the endpoints separately.

Solve $\dfrac{x}{x + 2} \geq 4$.

$$\frac{x}{x + 2} - 4 \geq 0 \qquad \text{Subtract 4.}$$

$$\frac{x}{x + 2} - \frac{4(x + 2)}{x + 2} \geq 0 \qquad \text{Write with a common denominator.}$$

$$\frac{-3x - 8}{x + 2} \geq 0 \qquad \text{Subtract fractions.}$$

$-\frac{8}{3}$ makes the numerator 0, and -2 makes the denominator 0.

Intervals: $\left(-\infty, -\frac{8}{3}\right)$, $\left(-\frac{8}{3}, -2\right), (-2, \infty)$

Test values: -4 from Interval A makes the original inequality false, $-\frac{7}{3}$ from Interval B makes it true, and 0 from Interval C makes it false.

The solution set is the interval $\left[-\frac{8}{3}, -2\right)$. The endpoint -2 is not included because it makes the denominator 0.

Chapter 11   Review Exercises

11.1 *Solve each equation using the square root property.*

1. $t^2 = 121$

2. $p^2 = 3$

3. $(2x + 5)^2 = 100$

***4.** $(3x - 2)^2 = -25$

1. $\{\pm 11\}$ 2. $\{\pm \sqrt{3}\}$
3. $\left\{-\dfrac{15}{2}, \dfrac{5}{2}\right\}$ 4. $\left\{\dfrac{2}{3} \pm \dfrac{5}{3}i\right\}$

*This exercise requires knowledge of complex numbers.

5. By the square root property, the first step should be

$$x = \sqrt{12} \quad \text{or} \quad x = -\sqrt{12}.$$

The solution set is $\{\pm 2\sqrt{3}\}$.

6. 5.9 sec

7. $\{-2 \pm \sqrt{19}\}$ **8.** $\left\{\frac{1}{2}, 1\right\}$

9. $\left\{\frac{-4 \pm \sqrt{22}}{2}\right\}$

10. $\left\{\frac{3}{8} \pm \frac{\sqrt{87}}{8}i\right\}$

11. $\left\{-\frac{7}{2}, 3\right\}$

12. $\left\{\frac{-5 \pm \sqrt{53}}{2}\right\}$

13. $\left\{\frac{1 \pm \sqrt{41}}{2}\right\}$

14. $\left\{-\frac{3}{4} \pm \frac{\sqrt{23}}{4}i\right\}$

15. $\left\{\frac{2}{3} \pm \frac{\sqrt{2}}{3}i\right\}$

16. $\left\{\frac{-7 \pm \sqrt{37}}{2}\right\}$

17. (a) 17; C   (b) 64, A

18. (a) −92; D   (b) 0; B

19. $\left\{-\frac{5}{2}, 3\right\}$ **20.** $\left\{-\frac{1}{2}, 1\right\}$

21. $\{-4\}$

22. $\left\{-\frac{11}{6}, -\frac{19}{12}\right\}$

23. $\left\{-\frac{343}{8}, 64\right\}$

24. $\{\pm 1, \pm 3\}$

25. 7 mph   **26.** 40 mph

27. 4.6 hr

28. Zoran: 2.6 hr; Claude: 3.6 hr

---

5. A student gave the following incorrect "solution." **WHAT WENT WRONG?**

$$x^2 = 12$$

$$x = \sqrt{12} \qquad \text{Square root property}$$

$$x = 2\sqrt{3} \qquad \text{Solution set: } \{2\sqrt{3}\}$$

6. The High Roller in Las Vegas, the world's largest observation wheel as of 2014, has a height of about 168 m. Use the metric version of Galileo's formula,

$$d = 4.9t^2 \quad \text{(where } d \text{ is in meters),}$$

to find how long it would take a wallet dropped from the top of the High Roller to reach the ground. Round the answer to the nearest tenth of a second. (*Source:* www.caesars.com)

11.2 *Solve each equation by completing the square.*

7. $x^2 + 4x = 15$ **8.** $2x^2 - 3x = -1$

9. $2z^2 + 8z - 3 = 0$ ***10.** $4x^2 - 3x + 6 = 0$

11.3 *Solve each equation using the quadratic formula.*

11. $2x^2 + x - 21 = 0$ **12.** $x^2 + 5x = 7$ **13.** $(t + 3)(t - 4) = -2$

*14. $2x^2 + 3x + 4 = 0$ ***15.** $3p^2 = 2(2p - 1)$ **16.** $x(2x - 7) = 3x^2 + 3$

Find the discriminant and use it to predict whether the solutions to each equation are

A. *two rational numbers*   **B.** *one rational number*

C. *two irrational numbers*   **D.** *two nonreal complex numbers.*

17. (a) $x^2 + 5x + 2 = 0$ (b) $4t^2 = 3 - 4t$

18. (a) $4x^2 = 6x - 8$ (b) $9z^2 + 30z + 25 = 0$

11.4 *Solve each equation. Check the solutions.*

19. $\dfrac{15}{x} = 2x - 1$ **20.** $\dfrac{1}{n} + \dfrac{2}{n + 1} = 2$

21. $-2r = \sqrt{\dfrac{48 - 20r}{2}}$ **22.** $8(3x + 5)^2 + 2(3x + 5) - 1 = 0$

23. $2x^{2/3} - x^{1/3} - 28 = 0$ **24.** $p^4 - 10p^2 + 9 = 0$

Solve each problem. Round answers to the nearest tenth, as necessary.

25. Bahaa paddled a canoe 20 mi upstream, then paddled back. If the rate of the current was 3 mph and the total trip took 7 hr, what was Bahaa's rate?

26. Carol Ann drove 8 mi to pick up a friend, and then drove 11 mi to a mall at a rate 15 mph faster. If Carol Ann's total travel time was 24 min, what was her rate on the trip to pick up her friend?

27. An old machine processes a batch of checks in 1 hr more time than a new one. How long would it take the old machine to process a batch of checks that the two machines together process in 2 hr?

28. Zoran can process a stack of invoices 1 hr faster than Claude can. Working together, they take 1.5 hr. How long would it take each person working alone?

---

*This exercise requires knowledge of complex numbers.

29. $v = \dfrac{\pm\sqrt{rFkw}}{kw}$

30. $y = \dfrac{6p^2}{z}$

31. $t = \dfrac{3m \pm \sqrt{9m^2 + 24m}}{2m}$

32. 9 ft, 12 ft, 15 ft

33. 12 cm by 20 cm

34. 1 in.

35. 18 in.

36. 3 min

37. $(1, 0)$ **38.** $(3, 7)$

39. $(-4, 3)$ **40.** $\left(\dfrac{2}{3}, -\dfrac{2}{3}\right)$

41. **42.**

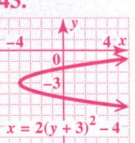
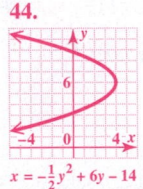

vertex: $(2, -3)$; vertex: $(2, 3)$;
axis: $x = 2$; axis: $x = 2$;
domain: $(-\infty, \infty)$; domain: $(-\infty, \infty)$;
range: $[-3, \infty)$ range: $(-\infty, 3]$

43. **44.**

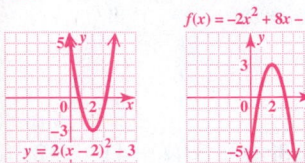

vertex: $(-4, -3)$; vertex: $(4, 6)$;
axis: $y = -3$; axis: $y = 6$;
domain: $[-4, \infty)$; domain: $(-\infty, 4]$;
range: $(-\infty, \infty)$ range: $(-\infty, \infty)$

45. 5 sec; 400 ft

46. length: 50 m; width: 50 m;
maximum area: 2500 m²

47. $\left(-\infty, -\dfrac{3}{2}\right) \cup (4, \infty)$

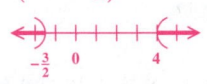

48. $[-4, 3]$

11.5 *Solve each formula for the specified variable. (Give answers with ±.)*

29. $k = \dfrac{rF}{wv^2}$ for v **30.** $p = \sqrt{\dfrac{yz}{6}}$ for y **31.** $mt^2 = 3mt + 6$ for t

Solve each problem. Round answers to the nearest tenth, as necessary.

32. A large machine requires a part in the shape of a right triangle with a hypotenuse 9 ft less than twice the length of the longer leg. The shorter leg must be $\dfrac{3}{4}$ the length of the longer leg. Find the lengths of the three sides of the part.

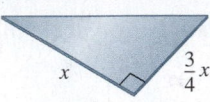

33. A square has an area of 256 cm². If the same amount is removed from one dimension and added to the other, the resulting rectangle has an area 16 cm² less. Find the dimensions of the rectangle.

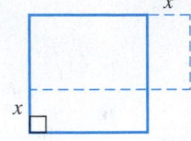

34. Allen wants to buy a mat for a photograph that measures 14 in. by 20 in. He wants to have an even border around the picture when it is mounted on the mat. If the area of the mat he chooses is 352 in.², how wide will the border be?

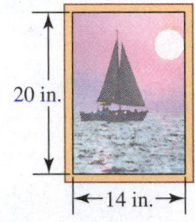

35. If a square piece of cardboard has 3-in. squares cut from its corners and then has the flaps folded up to form an open-top box, the volume of the box is given by the formula

$$V = 3(x - 6)^2,$$

where x is the length of each side of the original piece of cardboard in inches. What original length would yield a box with volume 432 in.³?

36. A searchlight moves horizontally back and forth along a wall. The distance of the light from a starting point at t minutes is given by the quadratic function

$$f(t) = 100t^2 - 300t.$$

How long will it take before the light returns to the starting point?

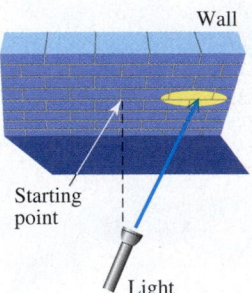

11.6, 11.7 *Identify the vertex of each parabola.*

37. $f(x) = -(x - 1)^2$ **38.** $f(x) = (x - 3)^2 + 7$

39. $x = (y - 3)^2 - 4$ **40.** $y = -3x^2 + 4x - 2$

Graph each parabola. Give the vertex, axis of symmetry, domain, and range.

41. $y = 2(x - 2)^2 - 3$ **42.** $f(x) = -2x^2 + 8x - 5$

43. $x = 2(y + 3)^2 - 4$ **44.** $x = -\dfrac{1}{2}y^2 + 6y - 14$

Solve each problem.

45. The height (in feet) of a projectile t seconds after being fired from Earth into the air is given by

$$f(t) = -16t^2 + 160t.$$

Find the number of seconds required for the projectile to reach maximum height. What is the maximum height?

49. $(-\infty, -5] \cup [-2, 3]$

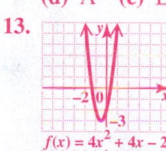

50. $\varnothing$

51. $\left(-\infty, \dfrac{1}{2}\right) \cup (2, \infty)$

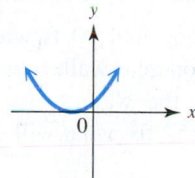

52. $[-3, 2)$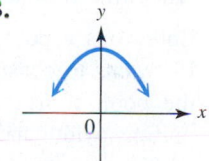

46. Find the length and width of a rectangle having a perimeter of 200 m if the area is to be a maximum. What is the maximum area?

11.8 *Solve each inequality, and graph the solution set.*

47. $(x - 4)(2x + 3) > 0$

48. $x^2 + x \le 12$

49. $(x + 2)(x - 3)(x + 5) \le 0$

50. $(4x + 3)^2 \le -4$

51. $\dfrac{6}{2z - 1} < 2$

52. $\dfrac{3t + 4}{t - 2} \le 1$

Chapter 11   Mixed Review Exercises

1. $R = \dfrac{\pm \sqrt{Vh - r^2 h}}{h}$

2. $\left\{ 1 \pm \dfrac{\sqrt{3}}{3} i \right\}$

3. $\left\{ \dfrac{-11 \pm \sqrt{7}}{3} \right\}$

4. $d = \dfrac{\pm \sqrt{SkI}}{I}$

5. $(-\infty, \infty)$

6. $\{4\}$

7. $\left\{ \pm \sqrt{4 + \sqrt{15}}, \pm \sqrt{4 - \sqrt{15}} \right\}$

8. $\left(-5, -\dfrac{23}{5} \right]$

9. $\left\{ -\dfrac{5}{3}, -\dfrac{3}{2} \right\}$

10. $\{-2, -1, 3, 4\}$

11. $(-\infty, -6) \cup \left(-\dfrac{3}{2}, 1 \right)$

12. (a) F   (b) B   (c) C
    (d) A   (e) E   (f) D

13.

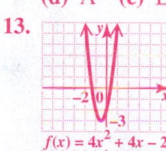

$f(x) = 4x^2 + 4x - 2$

vertex: $\left(-\dfrac{1}{2}, -3 \right)$; axis: $x = -\dfrac{1}{2}$;
domain: $(-\infty, \infty)$; range: $[-3, \infty)$

14. 10 mph

15. length: 2 cm; width: 1.5 cm

Solve each equation or inequality.

1. $V = r^2 + R^2 h$ for R

*2. $3t^2 - 6t = -4$

3. $(3x + 11)^2 = 7$

4. $S = \dfrac{Id^2}{k}$ for d

5. $(8x - 7)^2 \ge -1$

6. $2x - \sqrt{x} = 6$

7. $x^4 - 8x^2 = -1$

8. $\dfrac{-2}{x + 5} \le -5$

9. $6 + \dfrac{15}{s^2} = -\dfrac{19}{s}$

10. $(x^2 - 2x)^2 = 11(x^2 - 2x) - 24$

11. $(r - 1)(2r + 3)(r + 6) < 0$

Work each problem.

12. Match each equation in parts (a)–(f) with the figure that most closely resembles its graph in choices A–F.

(a) $g(x) = x^2 - 5$ (b) $h(x) = -x^2 + 4$ (c) $F(x) = (x - 1)^2$

(d) $G(x) = (x + 1)^2$ (e) $H(x) = (x - 1)^2 + 1$ (f) $K(x) = (x + 1)^2 + 1$

A.   B.   C.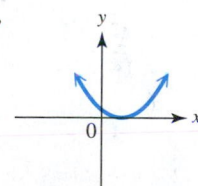

D. 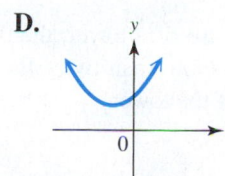   E. 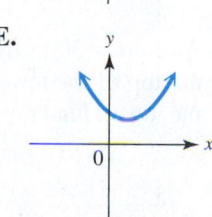   F.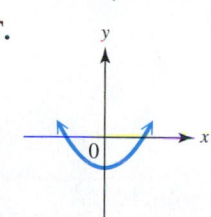

13. Graph $f(x) = 4x^2 + 4x - 2$. Give the vertex, axis of symmetry, domain, and range.

14. In 4 hr, Rajeed can travel 15 mi upriver and come back. The rate of the current is 5 mph. Find the rate of the boat in still water.

15. Two pieces of a large wooden puzzle fit together to form a rectangle with length 1 cm less than twice the width. The diagonal, where the two pieces meet, is 2.5 cm in length. Find the length and width of the rectangle.

---

*This exercise requires knowledge of complex numbers.

Chapter 11 | Test

FOR EXTRA HELP *Step-by-step test solutions are found on the Chapter Test Prep Videos available in* **MyMathLab** *or on* **YouTube**.

▶ *View the complete solutions to all Chapter Test exercises in MyMathLab.*

[11.1, 11.2]

1. $\{\pm 3\sqrt{6}\}$

2. $\left\{-\dfrac{8}{7}, \dfrac{2}{7}\right\}$

3. $\{-1 \pm \sqrt{5}\}$

[11.3]

4. $\left\{\dfrac{3 \pm \sqrt{17}}{4}\right\}$

5. $\left\{\dfrac{2}{3} \pm \dfrac{\sqrt{11}}{3}i\right\}$

[11.1]

6. A

[11.3]

7. discriminant: 88; There are two irrational solutions.

[11.1–11.4]

8. $\left\{\dfrac{2}{3}\right\}$

9. $\left\{-\dfrac{2}{3}, 6\right\}$

10. $\left\{\dfrac{-7 \pm \sqrt{97}}{8}\right\}$

11. $\left\{\pm\dfrac{1}{3}, \pm 2\right\}$ 12. $\left\{-\dfrac{5}{2}, 1\right\}$

[11.5]

13. $r = \dfrac{\pm\sqrt{\pi S}}{2\pi}$

[11.4]

14. Terry: 11.1 hr; Callie: 9.1 hr

15. 7 mph

[11.5]

16. 2 ft   17. 16 m

[11.6, 11.7]

18. A

19.

20.
$f(x) = -x^2 + 4x - 1$

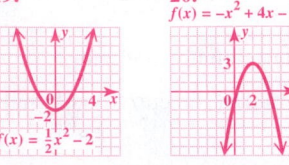
$f(x) = \frac{1}{2}x^2 - 2$

vertex: $(0, -2)$;   vertex: $(2, 3)$;
axis: $x = 0$;   axis: $x = 2$;
domain: $(-\infty, \infty)$; domain: $(-\infty, \infty)$;
range: $[-2, \infty)$ range: $(-\infty, 3]$

Solve each equation using the square root property or completing the square.

1. $t^2 = 54$ 2. $(7x + 3)^2 = 25$ 3. $x^2 + 2x = 4$

Solve each equation using the quadratic formula.

4. $2x^2 - 3x - 1 = 0$ *5. $3t^2 - 4t = -5$

*6. If k is a negative number, then which one of the following equations will have two nonreal complex solutions?

   A. $x^2 = 4k$ B. $x^2 = -4k$ C. $(x + 2)^2 = -k$ D. $x^2 + k = 0$

7. What is the discriminant for $2x^2 - 8x - 3 = 0$? How many and what type of solutions does this equation have? (Do not actually solve.)

Solve each equation by any method.

8. $3x = \sqrt{\dfrac{9x + 2}{2}}$ 9. $3 - \dfrac{16}{x} - \dfrac{12}{x^2} = 0$ 10. $4x^2 + 7x - 3 = 0$

11. $9x^4 + 4 = 37x^2$ 12. $12 = (2n + 1)^2 + (2n + 1)$

13. Solve $S = 4\pi r^2$ for r. (Leave $\pm$ in your answer.)

Solve each problem.

14. Terry and Callie do word processing. For a certain prospectus, Callie can prepare it 2 hr faster than Terry can. If they work together, they can do the entire prospectus in 5 hr. How long will it take each of them working alone to prepare the prospectus? Round answers to the nearest tenth of an hour.

15. Qihong paddled a canoe 10 mi upstream and then paddled back to the starting point. If the rate of the current was 3 mph and the entire trip took $3\frac{1}{2}$ hr, what was Qihong's rate?

16. Endre has a pool 24 ft long and 10 ft wide. He wants to construct a concrete walk around the pool. If he plans for the walk to be of uniform width and cover 152 ft^2, what will the width of the walk be?

17. At a point 30 m from the base of a tower, the distance to the top of the tower is 2 m more than twice the height of the tower. Find the height of the tower.

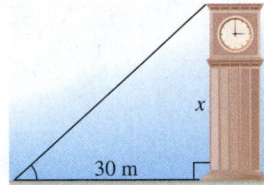

18. Which one of the following figures most closely resembles the graph of $f(x) = a(x - h)^2 + k$ if $a < 0$, $h > 0$, and $k < 0$?

   A.   B.   C.   D.

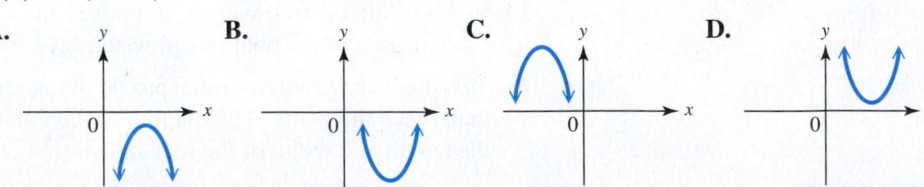

---

*This exercise requires knowledge of complex numbers.

21.

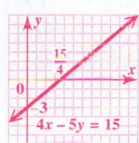

vertex: $(2, 2)$;
axis: $y = 2$;
domain: $(-\infty, 2]$;
range: $(-\infty, \infty)$

$x = -(y - 2)^2 + 2$

22. 160 ft by 320 ft

[11.8]

23. $(-\infty, -5) \cup \left(\dfrac{3}{2}, \infty\right)$

24. $(-\infty, 4) \cup [9, \infty)$

Graph each parabola. Identify the vertex, axis of symmetry, domain, and range.

19. $f(x) = \dfrac{1}{2}x^2 - 2$ **20.** $f(x) = -x^2 + 4x - 1$ **21.** $x = -(y - 2)^2 + 2$

22. Houston Community College is planning to construct a rectangular parking lot on land bordered on one side by a highway. The plan is to use 640 ft of fencing to fence off the other three sides. What should the dimensions of the lot be if the enclosed area is to be a maximum?

Solve each inequality, and graph the solution set.

23. $2x^2 + 7x > 15$ **24.** $\dfrac{5}{t - 4} \le 1$

Chapters R–11 | Cumulative Review Exercises

[1.3, 10.7]

1. (a) $-2, 0, 7$

(b) $-\dfrac{7}{3}, -2, 0, 0.7, 7, \dfrac{32}{3}$

(c) All are real except $\sqrt{-8}$.

(d) All are complex numbers.

[2.3] **[8.3]**

2. $\left\{\dfrac{4}{5}\right\}$ **3.** $\left\{\dfrac{11}{10}, \dfrac{7}{2}\right\}$

[10.6] **[6.6]**

4. $\left\{\dfrac{2}{3}\right\}$ **5.** $\varnothing$

[11.2, 11.3]

6. $\left\{\dfrac{7 \pm \sqrt{177}}{4}\right\}$

[11.4]

7. $\{\pm 1, \pm 2\}$

[2.8, 8.1] **[8.3]**

8. $[1, \infty)$ **9.** $\left[2, \dfrac{8}{3}\right]$

[11.8]

10. $(1, 3)$ **11.** $(-2, 1)$

[3.2, 7.1, 9.1, 9.2]

12.

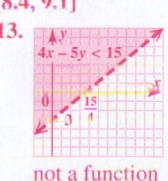

function;
domain: $(-\infty, \infty)$;
range: $(-\infty, \infty)$;

$4x - 5y = 15$

$f(x) = \dfrac{4}{5}x - 3$

[8.4, 9.1]

13.

not a function

---

1. Let $S = \left\{-\dfrac{7}{3}, -2, -\sqrt{3}, 0, 0.7, \sqrt{12}, \sqrt{-8}, 7, \dfrac{32}{3}\right\}$. List the elements of S that are elements of each set.

(a) Integers **(b)** Rational numbers **(c)** Real numbers **(d)** Complex numbers

Solve each equation or inequality.

2. $7 - (4 + 3t) + 2t = -6(t - 2) - 5$ **3.** $|6x - 9| = |-4x + 2|$

4. $2x = \sqrt{\dfrac{5x + 2}{3}}$ **5.** $\dfrac{3}{x - 3} - \dfrac{2}{x - 2} = \dfrac{3}{x^2 - 5x + 6}$

6. $(r - 5)(2r + 3) = 1$ **7.** $x^4 - 5x^2 + 4 = 0$

8. $-2x + 4 \le -x + 3$ **9.** $|3x - 7| \le 1$

10. $x^2 - 4x + 3 < 0$ **11.** $\dfrac{3}{p + 2} > 1$

Graph each relation. Decide whether or not y can be expressed as a function f of x, and if so, give its domain and range, and write using function notation.

12. $4x - 5y = 15$ **13.** $4x - 5y < 15$ **14.** $y = -2(x - 1)^2 + 3$

15. Find the slope and intercepts of the line with equation $-2x + 7y = 16$.

16. Write an equation for the specified line. Express each equation in slope-intercept form.

(a) Through $(2, -3)$ and parallel to the line with equation $5x + 2y = 6$

(b) Through $(-4, 1)$ and perpendicular to the line with equation $5x + 2y = 6$

Solve each system of equations.

17. $2x - 4y = 10$
$9x + 3y = 3$

18. $x + y + 2z = 3$
$-x + y + z = -5$
$2x + 3y - z = -8$

19. Two top-grossing superhero films are *Marvel's The Avengers* and *Spiderman 2*. The two films together grossed $1143 million. *Spiderman 2* grossed $139 million less than *Marvel's The Avengers*. How much did each film gross? (*Source:* Box Office Mojo.)

[11.6]

14.

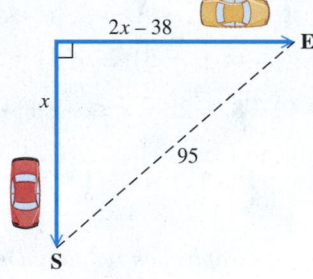

function;
domain: $(-\infty, \infty)$;
range: $(-\infty, 3]$;
$f(x) = -2(x-1)^2 + 3$

[3.2, 3.3, 7.1]

15. $m = \dfrac{2}{7}$; x-intercept: $(-8, 0)$;

y-intercept: $\left(0, \dfrac{16}{7}\right)$

[7.2]

16. (a) $y = -\dfrac{5}{2}x + 2$

     (b) $y = \dfrac{2}{5}x + \dfrac{13}{5}$

[7.3–7.5]   **[7.6]**

17. $\{(1, -2)\}$ **18.** $\{(3, -4, 2)\}$

[7.7]

19. *Marvel's The Avengers:*
    $641 million;
    Spiderman 2: $502 million

[4.1, 4.2]

20. $\dfrac{x^8}{y^4}$ **21.** $\dfrac{4}{xy^2}$

[4.6]

22. $\dfrac{4}{9}t^2 + 12t + 81$

[4.7]

23. $4x^2 - 6x + 11 + \dfrac{4}{x+2}$

[5.1–5.4]

24. $(4m - 3)(6m + 5)$

25. $(2x + 3y)(4x^2 - 6xy + 9y^2)$

26. $(3x - 5y)^2$

[6.2]   **[6.4]**

27. $-\dfrac{5}{18}$ **28.** $-\dfrac{8}{x}$

[6.5]   **[10.3]**

29. $\dfrac{r - s}{r}$ **30.** $\dfrac{3\sqrt[3]{4}}{4}$

[10.5]

31. $\sqrt{7} + \sqrt{5}$

[11.5]

32. southbound car: 57 mi;
    eastbound car: 76 mi

Write with positive exponents only. Assume that variables represent positive real numbers.

20. $\left(\dfrac{x^{-3}y^2}{x^5 y^{-2}}\right)^{-1}$
    21. $\dfrac{(4x^{-2})^2(2y^3)}{8x^{-3}y^5}$

Perform the indicated operations.

22. $\left(\dfrac{2}{3}t + 9\right)^2$
    23. Divide $4x^3 + 2x^2 - x + 26$ by $x + 2$.

Factor completely.

24. $24m^2 + 2m - 15$ **25.** $8x^3 + 27y^3$ **26.** $9x^2 - 30xy + 25y^2$

Simplify. Express each answer in lowest terms. Assume denominators are nonzero.

27. $\dfrac{5x + 2}{-6} \div \dfrac{15x + 6}{5}$ **28.** $\dfrac{3}{2 - x} - \dfrac{5}{x} + \dfrac{6}{x^2 - 2x}$ **29.** $\dfrac{\dfrac{r}{s} - \dfrac{s}{r}}{\dfrac{r}{s} + 1}$

Simplify each radical expression.

30. $\sqrt[3]{\dfrac{27}{16}}$ **31.** $\dfrac{2}{\sqrt{7} - \sqrt{5}}$

32. Two cars left an intersection at the same time, one heading due south and the other due east. Later they were exactly 95 mi apart. The car heading east had gone 38 mi less than twice as far as the car heading south. How far had each car traveled?

12

Inverse, Exponential, and Logarithmic Functions

The magnitudes of earthquakes, intensities of sounds, and population growth and decay are some examples of applications of *exponential* and *logarithmic functions*.

12.1   Inverse Functions

VOCABULARY

☐ one-to-one function
☐ inverse of a function

In this chapter we study two important types of functions, *exponential* and *logarithmic*. These functions are related: They are *inverses* of one another.

OBJECTIVE 1   Decide whether a function is one-to-one and, if it is, find its inverse.

Suppose we define the function

$$G = \{(-2, 2), (-1, 1), (0, 0), (1, 3), (2, 5)\}.$$

We can form another set of ordered pairs from G by interchanging the x- and y-values of each pair in G. We can call this set F, so

$$F = \{(2, -2), (1, -1), (0, 0), (3, 1), (5, 2)\}.$$

To show that these two sets are related as just described, F is called the *inverse* of G. For a function f to have an inverse, f must be a *one-to-one function*.

One-to-One Function

In a **one-to-one function,** each x-value corresponds to only one y-value, and each y-value corresponds to only one x-value.

The function shown in **FIGURE 1(a)** is not one-to-one because the y-value 7 corresponds to *two* x-values, 2 and 3. That is, the ordered pairs $(2, 7)$ and $(3, 7)$ both belong to the function. The function in **FIGURE 1(b)** is one-to-one.

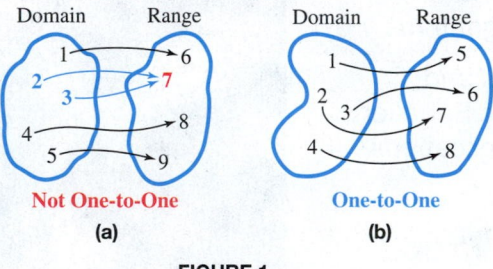

FIGURE 1

The *inverse* of any one-to-one function f is found by interchanging the components of the ordered pairs of f. The inverse of f is written $\boldsymbol{f^{-1}}$. Read f^{-1} as **"the inverse of f"** or **"f-inverse."**

⚠ **CAUTION**   The symbol $f^{-1}(x)$ does **not** represent $\frac{1}{f(x)}$.

Inverse of a Function

The **inverse** of a one-to-one function f, written f^{-1}, is the set of all ordered pairs of the form (y, x), where (x, y) belongs to f. *The inverse is formed by interchanging x and y, so the domain of f becomes the range of f^{-1} and the range of f becomes the domain of f^{-1}.*

For inverses f and f^{-1}, it follows that for all x in their domains,

$$(f \circ f^{-1})(x) = x \quad \text{and} \quad (f^{-1} \circ f)(x) = x.$$

NOW TRY EXERCISE 1

Decide whether each function is one-to-one. If it is, find the inverse.

(a) $F = \{(-1, -2), (0, 0)$
$(1, -2), (2, -8)\}$

(b) $G = \{(0, 0), (1, 1),$
$(4, 2), (9, 3)\}$

(c) A Norwegian physiologist has developed a rule for predicting running times based on the time to run 5 km (5K). An example for one runner is shown here. (*Source:* Stephen Seiler, Agder College, Kristiansand, Norway.)

Distance	Time
1.5K	4:22
3K	9:18
5K	16:00
10K	33:40

EXAMPLE 1 Finding Inverses of One-to-One Functions

Decide whether each function is one-to-one. If it is, find the inverse.

(a) $F = \{(-2, 1), (-1, 0), (0, 1), (1, 2), (2, 2)\}$

Each x-value in f corresponds to just one y-value. However, the y-value 1 corresponds to two x-values, -2 and 0. Also, the y-value 2 corresponds to both 1 and 2. Because some y-values correspond to more than one x-value, F is not one-to-one and does not have an inverse.

(b) $G = \{(3, 1), (0, 2), (2, 3), (4, 0)\}$

Every x-value in G corresponds to only one y-value, and every y-value corresponds to only one x-value, so G is a one-to-one function. The inverse function is found by interchanging the x- and y-values in each ordered pair.

$$G^{-1} = \{(1, 3), (2, 0), (3, 2), (0, 4)\}$$

The domain and range of G become the range and domain, respectively, of G^{-1}.

(c) The table shows the number of days in which the air in the Los Angeles–Long Beach–Santa Ana metropolitan area failed to meet acceptable air-quality standards for the years 2005–2010.

Year	Number of Days Exceeding Standards
2005	29
2006	26
2007	18
2008	31
2009	18
2010	4

Source: U.S. Environmental Protection Agency.

Let f be the function defined in the table, with the years forming the domain and the number of days exceeding the air-quality standards forming the range. Then f is not one-to-one because in two different years (2007 and 2009) the number of days with unacceptable air quality was the same, 18.

NOW TRY

NOW TRY ANSWERS

1. (a) not one-to-one
(b) one-to-one;
$G^{-1} = \{(0, 0), (1, 1),$
$(2, 4), (3, 9)\}$
(c)

Time	Distance
4:22	1.5K
9:18	3K
16:00	5K
33:40	10K

OBJECTIVE 2 Use the horizontal line test to determine whether a function is one-to-one.

By graphing a function and observing the graph, we can use the *horizontal line test* to tell whether the function is one-to-one.

Horizontal Line Test

A function is one-to-one if every horizontal line intersects the graph of the function at most once.

The horizontal line test follows from the definition of a one-to-one function. Any two points that lie on the same horizontal line have the same y-coordinate. No two ordered pairs that belong to a one-to-one function may have the same y-coordinate. Therefore, no horizontal line will intersect the graph of a one-to-one function more than once.

**NOW TRY
EXERCISE 2**

Use the horizontal line test to
determine whether each graph
is the graph of a one-to-one
function.

(a)

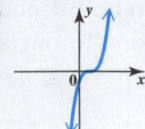

(b)

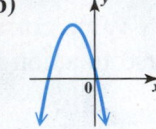

EXAMPLE 2   Using the Horizontal Line Test

Use the horizontal line test to determine whether each graph is the graph of a one-to-one function.

(a)

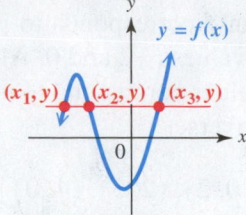

FIGURE 2

(b)

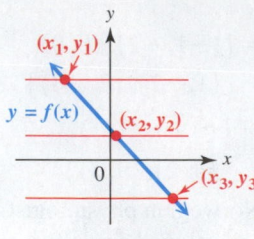

FIGURE 3

Because a horizontal line intersects the graph in more than one point (actually three points), the function in **FIGURE 2** is not one-to-one.

Every horizontal line will intersect the graph in exactly one point. The function in **FIGURE 3** is one-to-one.

NOW TRY

OBJECTIVE 3   Find the equation of the inverse of a function.

The inverse of a one-to-one function is found by interchanging the x- and y-values of each of its ordered pairs. The equation of the inverse of a function $y = f(x)$ is found in the same way.

Finding the Equation of the Inverse of $y = f(x)$

For a one-to-one function f defined by an equation $y = f(x)$, find the defining equation of the inverse as follows.

Step 1   Interchange x and y.

Step 2   Solve for y.

Step 3   Replace y with $f^{-1}(x)$.

EXAMPLE 3   Finding Equations of Inverses

Decide whether each equation defines a one-to-one function. If so, find the equation that defines the inverse.

(a) $f(x) = 2x + 5$

The graph of $y = 2x + 5$ is a nonvertical line, so by the horizontal line test, f is a one-to-one function. To find the inverse, follow the steps.

$$y = 2x + 5 \qquad \text{Let } y = f(x).$$

$$x = 2y + 5 \qquad \text{Interchange } x \text{ and } y. \text{ (Step 1)}$$

$$2y = x - 5 \qquad \text{Solve for } y. \text{ (Step 2)}$$

$$y = \frac{x - 5}{2}$$

$$f^{-1}(x) = \frac{x - 5}{2} \qquad \text{Replace } y \text{ with } f^{-1}(x). \text{ (Step 3)}$$

This equation can be written as follows.

$$f^{-1}(x) = \frac{x}{2} - \frac{5}{2}, \quad \text{or} \quad f^{-1}(x) = \frac{1}{2}x - \frac{5}{2} \qquad \frac{a-b}{c} = \frac{a}{c} - \frac{b}{c}$$

NOW TRY ANSWERS
2. **(a)** one-to-one
   (b) not one-to-one

**NOW TRY
EXERCISE 3**

Decide whether each equation defines a one-to-one function. If so, find the equation that defines the inverse.

(a) $f(x) = 5x - 7$
(b) $f(x) = (x + 1)^2$
(c) $f(x) = x^3 - 4$

Thus, f^{-1} is a linear function. In the function

$$f(x) = 2x + 5,$$

the range value is found by starting with a value of x, multiplying by 2, and adding 5. In the equation for the inverse

$$f^{-1}(x) = \frac{x - 5}{2}, \qquad \text{One form of } f^{-1}(x)$$

we *subtract* 5, and then *divide* by 2. This shows how an inverse is used to "undo" what a function does to the variable x.

(b) $y = x^2 + 2$

This equation has a vertical parabola as its graph, so some horizontal lines will intersect the graph at two points. For example, both $x = 3$ and $x = -3$ correspond to $y = 11$. Because of the x^2-term, there are many pairs of x-values that correspond to the same y-value. This means that the function $y = x^2 + 2$ is not one-to-one and does not have an inverse.

Applying the steps for finding the equation of an inverse leads to the following.

$$y = x^2 + 2$$
$$x = y^2 + 2 \qquad \text{Interchange } x \text{ and } y.$$
$$y^2 = x - 2 \qquad \text{Solve for } y.$$
$$y = \pm\sqrt{x - 2} \qquad \text{Square root property}$$

The last step shows that there are two y-values for each choice of $x > 2$, so we again see that the given function is not one-to-one. It does not have an inverse.

(c) $f(x) = (x - 2)^3$

Because of the cube, each value of x produces a different value of y, so this is a one-to-one function.

$$f(x) = (x - 2)^3$$
$$y = (x - 2)^3 \qquad \text{Replace } f(x) \text{ with } y.$$
$$x = (y - 2)^3 \qquad \text{Interchange } x \text{ and } y.$$
$$\sqrt[3]{x} = \sqrt[3]{(y - 2)^3} \qquad \text{Take the cube root on each side.}$$
$$\sqrt[3]{x} = y - 2 \qquad \sqrt[3]{a^3} = a$$
$$y = \sqrt[3]{x} + 2 \qquad \text{Solve for } y.$$
$$f^{-1}(x) = \sqrt[3]{x} + 2 \qquad \text{Replace } y \text{ with } f^{-1}(x). \qquad \text{NOW TRY}$$

NOW TRY ANSWERS
3. **(a)** one-to-one function;
$f^{-1}(x) = \frac{x + 7}{5}$, or
$f^{-1}(x) = \frac{1}{5}x + \frac{7}{5}$
(b) not a one-to-one function
(c) one-to-one function;
$f^{-1}(x) = \sqrt[3]{x} + 4$

EXAMPLE 4 Using Factoring to Find an Inverse

It is shown in standard college algebra texts that $f(x) = \frac{x + 1}{x - 2}$, $x \neq 2$, is one-to-one. Find $f^{-1}(x)$.

$$f(x) = \frac{x + 1}{x - 2}, \quad x \neq 2 \qquad \text{Given equation}$$
$$y = \frac{x + 1}{x - 2}, \quad x \neq 2 \qquad \text{Replace } f(x) \text{ with } y.$$

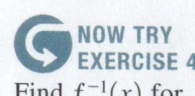

NOW TRY EXERCISE 4

Find $f^{-1}(x)$ for

$$f(x) = \frac{x+3}{x-4}, \quad x \neq 4.$$

Interchange x and y to find the inverse. Then solve for y.

$$x = \frac{y+1}{y-2}, \quad y \neq 2$$

$$x(y-2) = y+1 \qquad \text{Multiply by } y-2.$$

$$xy - 2x = y+1 \qquad \text{Distributive property}$$

$$xy - y = 2x+1 \qquad \text{Subtract } y. \text{ Add } 2x.$$

$$y(x-1) = 2x+1 \qquad \text{Factor out } y.$$

$$y = \frac{2x+1}{x-1} \qquad \text{Divide by } x-1.$$

Replace y with $f^{-1}(x)$ and note the restriction that $x \neq 1$.

$$f^{-1}(x) = \frac{2x+1}{x-1}, \quad x \neq 1$$

NOW TRY

OBJECTIVE 4 Graph f^{-1}, given the graph of f.

One way to graph the inverse of a function f whose equation is given follows.

> **Graphing the Inverse**
>
> **Step 1** Find several ordered pairs that belong to f.
>
> **Step 2** Interchange x and y to obtain ordered pairs that belong to f^{-1}.
>
> **Step 3** Plot those points, and sketch the graph of f^{-1} through them.

We can also select points on the graph of f and *use symmetry* to find corresponding points on the graph of f^{-1}.

For example, suppose the point (a, b) shown in **FIGURE 4** belongs to a one-to-one function f. Then the point (b, a) belongs to f^{-1}. The line segment connecting (a, b) and (b, a) is perpendicular to, and cut in half by, the line $y = x$. The points (a, b) and (b, a) are "mirror images" of each other with respect to $y = x$.

We can find the graph of f^{-1} from the graph of f by locating the mirror image of each point in f with respect to the line $y = x$.

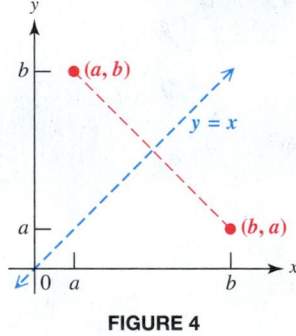

FIGURE 4

EXAMPLE 5 Graphing the Inverse

Use the fact that if (a, b) lies on the graph of f, then (b, a) lies on the graph of f^{-1}.

(a) FIGURE 5(a) on the next page shows the graph of a one-to-one function f. Sketch the graph of f^{-1}.

The points $\left(-1, \frac{1}{2}\right)$, $(0, 1)$, $(1, 2)$, and $(2, 4)$ lie on the graph of f. To find the graph of f^{-1}, plot the points

$$\left(\frac{1}{2}, -1\right), \ (1, 0), \ (2, 1), \text{ and } (4, 2).$$

NOW TRY ANSWER

4. $f^{-1}(x) = \dfrac{4x+3}{x-1}, \quad x \neq 1$

NOW TRY EXERCISE 5

Use the fact that if (a, b) lies on the graph of f, than (b, a) lies on the graph of f^{-1} to graph the inverse of the function f shown.

Join the points with the same shape curve as seen for f. See **FIGURE 5(b)**.

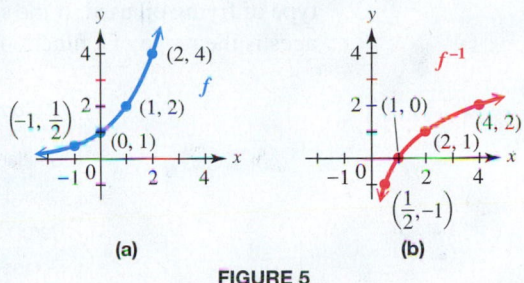

(a)   (b)

FIGURE 5

(b) Graph the inverse of each function f (shown in blue) in **FIGURE 6**.

Each inverse is shown in red. In both cases, the graph of f^{-1} is a reflection of the graph of f across the line $y = x$.

NOW TRY ANSWER

5.

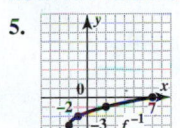

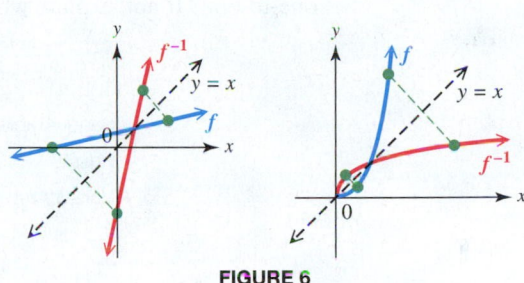

FIGURE 6   **NOW TRY**

12.1 Exercises

FOR EXTRA HELP   ▶ MyMathLab®

▶ *Complete solution available in MyMathLab*

Concept Check *In Exercises 1–4, choose the correct response from the given list.*

1. If a function is made up of ordered pairs in such a way that the same y-value appears in a correspondence with two different x-values, then

   A. the function is one-to-one   **B.** the function is not one-to-one

   C. its graph does not pass the vertical line test   **D.** it has an inverse function associated with it.

2. Which equation defines a one-to-one function? Explain why the others are not, using specific examples.

   A. $f(x) = x$ **B.** $f(x) = x^2$ **C.** $f(x) = |x|$ **D.** $f(x) = -x^2 + 2x - 1$

▶ 3. Only one of the graphs illustrates a one-to-one function. Which one is it? (**See Example 2.**)

   A. 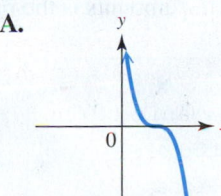   **B.** 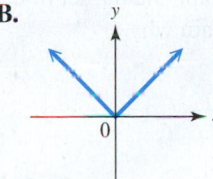   **C.** 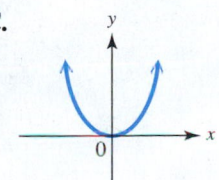   **D.**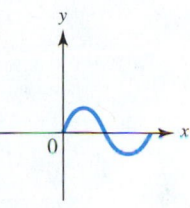

4. If a function f is one-to-one and the point (p, q) lies on the graph of f, then which point *must* lie on the graph of f^{-1}?

   A. $(-p, q)$ **B.** $(-q, -p)$ **C.** $(p, -q)$ **D.** (q, p)

Answer each question.

▶ **5.** The table shows trans fat content in a fast-food product in various countries, based on type of frying oil used. If the set of countries is the domain and the set of trans fat percentages is the range of a function, is it one-to-one? Why or why not?

Country	Percentage of Trans Fat in McDonald's Chicken
Scotland	14
United States	11
Peru	9
Poland	8
Russia	5
Denmark	1

Source: New England Journal of Medicine.

6. The table shows the top world oil exporters in thousands of barrels. If the set of countries is the domain and the set of numbers of barrels exported is the range of a function, is it one-to-one? If not, explain why.

Country	Number of Barrels Exported (in thousands)
Saudi Arabia	8865
Russia	7201
United Arab Emirates	2595
Kuwait	2414
Nigeria	2254
Iraq	2235

Source: U.S. Energy Information Administration.

7. The road mileage between Denver, Colorado, and several selected U.S. cities is shown in the table. If we consider this as a function that pairs each city with a distance, is it a one-to-one function? How could we change the answer to this question by adding 1 mile to one of the distances shown?

City	Distance to Denver (in miles)
Atlanta	1398
Dallas	781
Indianapolis	1058
Kansas City, MO	600
Los Angeles	1059
San Francisco	1235

8. The table lists caffeine amounts in several popular 12-oz sodas, If the set of sodas is the domain and the set of caffeine amounts is the range of a function, is it one-to-one? If not, explain why.

Soda	Caffeine (in mg)
Mountain Dew	55
Diet Coke	45
Dr. Pepper	41
Sunkist Orange Soda	41
Diet Pepsi-Cola	36
Coca-Cola Classic	34

Source: National Soft Drink Association.

If the function is one-to-one, find its inverse. See Examples 1–3.

9. $\{(3, 6), (2, 10), (5, 12)\}$

10. $\left\{(-1, 3), (0, 5), (5, 0), \left(7, -\dfrac{1}{2}\right)\right\}$

11. $\{(-1, 3), (2, 7), (4, 3), (5, 8)\}$

12. $\{(-8, 6), (-4, 3), (0, 6), (5, 10)\}$

13. $f(x) = x + 3$

14. $f(x) = x + 8$

15. $f(x) = -\dfrac{1}{2}x - 2$

16. $f(x) = -\dfrac{1}{4}x - 8$

▶ **17.** $f(x) = 2x + 4$

18. $f(x) = 3x + 1$

19. $f(x) = \sqrt{x - 3}, \quad x \geq 3$

20. $f(x) = \sqrt{x + 2}, \quad x \geq -2$

21. $f(x) = 3x^2 + 2$

22. $f(x) = 4x^2 - 1$

23. $f(x) = x^3 - 4$

24. $f(x) = x^3 + 5$

Each function is one-to-one. Find its inverse. See Example 4.

25. $f(x) = \dfrac{x + 4}{x + 2}, \quad x \neq -2$

26. $f(x) = \dfrac{x + 3}{x + 5}, \quad x \neq -5$

27. $f(x) = \dfrac{4x - 2}{x + 5}, \quad x \neq -5$

28. $f(x) = \dfrac{5x - 10}{x + 4}, \quad x \neq -4$

29. $f(x) = \dfrac{-2x + 1}{2x - 5}, \quad x \neq \dfrac{5}{2}$

30. $f(x) = \dfrac{-3x + 2}{3x - 4}, \quad x \neq \dfrac{4}{3}$

Concept Check *Let $f(x) = 2^x$. We will see in the next section that this function is one-to-one. Find each value, always working part (a) before part (b).*

31. (a) $f(3)$

32. (a) $f(4)$

33. (a) $f(0)$

34. (a) $f(-2)$

(b) $f^{-1}(8)$

(b) $f^{-1}(16)$

(b) $f^{-1}(1)$

(b) $f^{-1}\left(\dfrac{1}{4}\right)$

The graphs of some functions are given in Exercises 35–40.

(a) Use the horizontal line test to determine whether the function is one-to-one.

(b) If the function is one-to-one, then graph the inverse of the function. See Example 5.

▶ **35.**

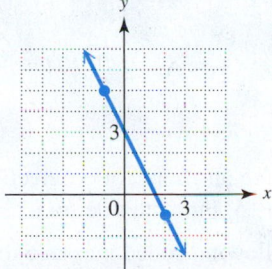

36.

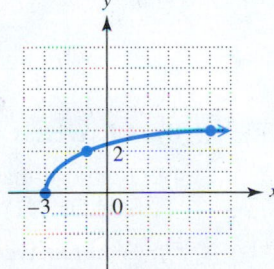

37.

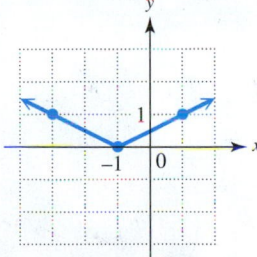

38.

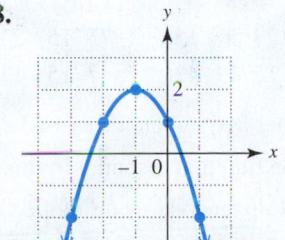

39.

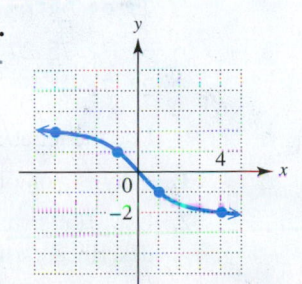

40.

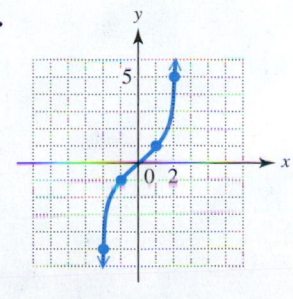

Each function in Exercises 41–48 is one-to-one. Graph the function as a solid line (or curve) and then graph its inverse on the same set of axes as a dashed line (or curve). In Exercises 45–48, complete the table so that graphing the function will be easier. **See Example 5.**

41. $f(x) = 2x - 1$ **42.** $f(x) = 2x + 3$ **43.** $f(x) = -4x$ **44.** $f(x) = -2x$

45. $f(x) = \sqrt{x}$,   **46.** $f(x) = -\sqrt{x}$,   **47.** $f(x) = x^3 - 2$ **48.** $f(x) = x^3 + 3$
$x \geq 0$ $x \geq 0$

x	$f(x)$
0	
1	
4	

x	$f(x)$
0	
1	
4	

x	$f(x)$
-1	
0	
1	
2	

x	$f(x)$
-2	
-1	
0	
1	

RELATING CONCEPTS   For Individual or Group Work   (Exercises 49–52)

Inverse functions can be used to send and receive coded information. A simple example might use the function $f(x) = 2x + 5$. *(Note that it is one-to-one.) Suppose that each letter of the alphabet is assigned a numerical value according to its position, as follows.*

A	1	G	7	L	12	Q	17	V	22
B	2	H	8	M	13	R	18	W	23
C	3	I	9	N	14	S	19	X	24
D	4	J	10	O	15	T	20	Y	25
E	5	K	11	P	16	U	21	Z	26
F	6								

This is an Enigma machine, used by the Germans in World War II to send coded messages.

Using the function, the word ALGEBRA would be encoded as

$$7 \quad 29 \quad 19 \quad 15 \quad 9 \quad 41 \quad 7,$$

because

$$f(A) = f(1) = 2(1) + 5 = 7, \quad f(L) = f(12) = 2(12) + 5 = 29, \quad \text{and so on.}$$

The message would then be decoded by using the inverse of f, which is $f^{-1}(x) = \dfrac{x - 5}{2}$
$\left(\text{or } f^{-1}(x) = \dfrac{1}{2}x - \dfrac{5}{2}\right).$ *For example,*

$$f^{-1}(7) = \frac{7 - 5}{2} = 1 = A, \quad f^{-1}(29) = \frac{29 - 5}{2} = 12 = L, \quad \text{and so on.}$$

Work Exercises 49–52 in order.

49. Suppose that you are an agent for a detective agency. Today's function for your code is $f(x) = 4x - 5$. Find the rule for f^{-1} algebraically.

50. You receive the following coded message today. (Read across from left to right.)

47 95 23 67 −1 59 27 31 51 23 7 −1 43 7 79 43 −1 75 55 67

31 71 75 27 15 23 67 15 −1 75 15 71 75 75 27 31 51

23 71 31 51 7 15 71 43 31 7 15 11 3 67 15 −1 11

Use the letter/number assignment described earlier to decode the message.

51. Why is a one-to-one function essential in this encoding/decoding process?

52. Use $f(x) = x^3 + 4$ to encode your name, using the letter/number assignment described earlier.

12.2 Exponential Functions

OBJECTIVES

1. Use a calculator to find approximations of exponentials.
2. Define and graph exponential functions.
3. Solve exponential equations of the form $a^x = a^k$ for x.
4. Use exponential functions in applications involving growth or decay.

VOCABULARY

☐ exponential function with base a
☐ asymptote
☐ exponential equation

NOW TRY EXERCISE 1

Use a calculator to find an approximation to three decimal places for each exponential expression.

(a) $2^{3.1}$ **(b)** $2^{-1.7}$ **(c)** $2^{1/4}$

OBJECTIVE 1 Use a calculator to find approximations of exponentials.

In **Section 10.2** we showed how to evaluate 2^x for rational values of x.

$$2^3 = 8, \quad 2^{-1} = \frac{1}{2}, \quad 2^{1/2} = \sqrt{2}, \quad \text{and} \quad 2^{3/4} = \sqrt[4]{2^3} = \sqrt[4]{8}$$

Examples of 2^x for rational x

In more advanced courses it is shown that 2^x exists for all real number values of x, both rational and irrational. We can use a calculator to find approximations of exponential expressions that are not easily determined.

EXAMPLE 1 Approximating Using a Calculator

Use a calculator to find an approximation to three decimal places for each exponential expression.

(a) $2^{1.6}$ **(b)** $2^{-1.3}$ **(c)** $2^{1/3}$

FIGURE 7 shows how a TI-84 calculator approximates these values. The display shows more decimal places than we usually need, so we round to three decimal places as directed.

$$2^{1.6} \approx 3.031, \quad 2^{-1.3} \approx 0.406, \quad 2^{1/3} \approx 1.260$$

```
NORMAL FLOAT AUTO REAL RADIAN MP      ▯
2^1.6
                        3.031433133
2^-1.3
                         .4061261982
2^1/3
                          1.25992105
```

FIGURE 7

 **NOW TRY**

OBJECTIVE 2 Define and graph exponential functions.

The following definition of an exponential function assumes that a^x exists for all real numbers x.

Exponential Function

For $a > 0$, $a \neq 1$, and all real numbers x,

$$f(x) = a^x$$

defines the **exponential function with base a**.

NOTE *The two restrictions on the value of a in the definition of an exponential function $f(x) = a^x$ are important.*

1. The restriction $a > 0$ is necessary so that the function can be defined for all real numbers x. Letting a be negative ($a = -2$, for instance) and letting $x = \frac{1}{2}$ would give the expression $(-2)^{1/2}$, which is not real.

2. The restriction $a \neq 1$ is necessary because 1 raised to any power is equal to 1, resulting in the linear function $f(x) = 1$.

NOW TRY ANSWERS
1. (a) 8.574 **(b)** 0.308 **(c)** 1.189

When graphing an exponential function of the form $f(x) = a^x$, pay particular attention to whether $a > 1$ or $0 < a < 1$.

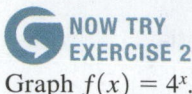

**NOW TRY
EXERCISE 2**

Graph $f(x) = 4^x$.

EXAMPLE 2 Graphing an Exponential Function ($a > 1$)

Graph $f(x) = 2^x$. Then compare it to the graph of $F(x) = 5^x$.

Choose some values of x, and find the corresponding values of $f(x)$. Plotting these points and drawing a smooth curve through them gives the darker graph shown in **FIGURE 8**. This graph is typical of the graph of an exponential function of the form $f(x) = a^x$, where $a > 1$.

The larger the value of a, the faster the graph rises.

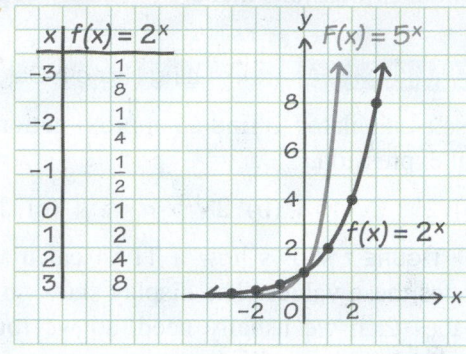

**Exponential function
with base $a > 1$**

Domain: $(-\infty, \infty)$

Range: $(0, \infty)$

y-intercept: $(0, 1)$

The function is one-to-one, and its graph rises from left to right.

FIGURE 8

The vertical line test assures us that the graphs in **FIGURE 8** represent functions. **FIGURE 8** also shows an important characteristic of exponential functions with $a > 1$:

As x gets larger, y increases at a faster and faster rate. **NOW TRY**

🛑 **CAUTION** The graph of this exponential function *approaches* the x-axis, but does **not** touch it.

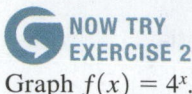

**NOW TRY
EXERCISE 3**

Graph $g(x) = \left(\dfrac{1}{10}\right)^x$.

EXAMPLE 3 Graphing an Exponential Function ($0 < a < 1$)

Graph $g(x) = \left(\dfrac{1}{2}\right)^x$.

Find some points on the graph. The graph in **FIGURE 9** is similar to that of $f(x) = 2^x$ (**FIGURE 8**) except that here *as x gets larger, y decreases.* This graph is typical of the graph of an exponential function of the form $f(x) = a^x$, where $0 < a < 1$.

NOW TRY ANSWERS

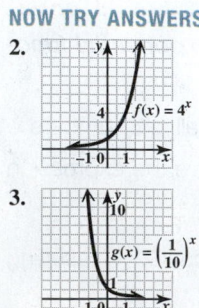

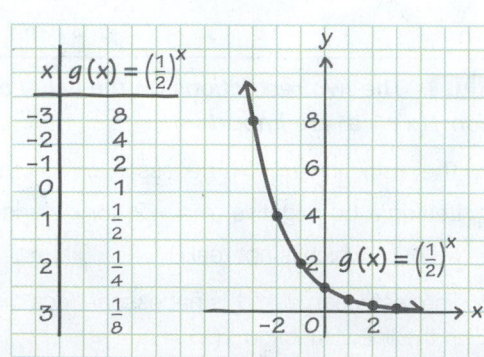

**Exponential function
with base $0 < a < 1$**

Domain: $(-\infty, \infty)$

Range: $(0, \infty)$

y-intercept: $(0, 1)$

The function is one-to-one, and its graph falls from left to right.

FIGURE 9

NOW TRY

Characteristics of the Graph of $f(x) = a^x$

1. The graph contains the point $(0, 1)$, which is its y-intercept.

2. The function is one-to-one.

   • When $a > 1$, the graph will *rise* from left to right. (See **FIGURE 8**.)

   • When $0 < a < 1$, the graph will *fall* from left to right. (See **FIGURE 9**.)

   In both cases, the graph goes from the second quadrant to the first.

3. The graph will approach the x-axis, but never touch it. (Such a line is an **asymptote**.)

4. The domain is $(-\infty, \infty)$, and the range is $(0, \infty)$.

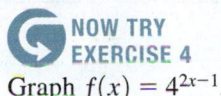

NOW TRY EXERCISE 4

Graph $f(x) = 4^{2x-1}$.

EXAMPLE 4 Graphing a More Complicated Exponential Function

Graph $f(x) = 3^{2x-4}$.

Find some ordered pairs. We let $x = 0$ and $x = 2$ and find values of $f(x)$, or y.

$y = 3^{2(0)-4}$ Let $x = 0$.	$y = 3^{2(2)-4}$ Let $x = 2$.
$y = 3^{-4}$	$y = 3^0$
$y = \dfrac{1}{81}$ $a^{-n} = \frac{1}{a^n}$	$y = 1$ $a^0 = 1$

These ordered pairs, $\left(0, \frac{1}{81}\right)$ and $(2, 1)$, along with the other ordered pairs shown in the table, lead to the graph in **FIGURE 10**.

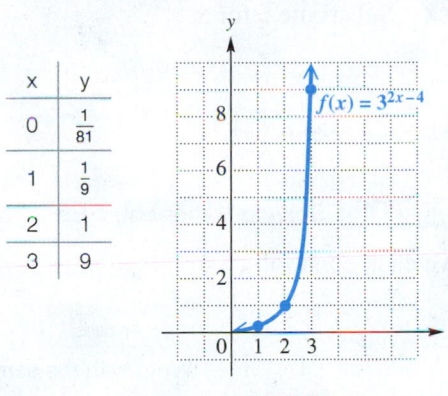

x	y
0	$\frac{1}{81}$
1	$\frac{1}{9}$
2	1
3	9

The graph is similar to the graph of $f(x) = 3^x$ except that it is shifted to the right and rises more rapidly.

FIGURE 10

 NOW TRY

OBJECTIVE 3 Solve exponential equations of the form $a^x = a^k$ for x.

Until this chapter, we have solved only equations that had the variable as a base, like $x^2 = 8$. In these equations, all exponents have been constants. An **exponential equation** is an equation that has a variable in an exponent, such as

$$9^x = 27.$$

We can use the following property to solve certain exponential equations.

NOW TRY ANSWER

4.

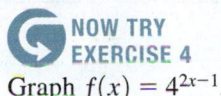

Property for Solving an Exponential Equation

For $a > 0$ and $a \ne 1$, if $a^x = a^y$ then $x = y$.

This property would not necessarily be true if $a = 1$.

Solving an Exponential Equation

Step 1 **Each side must have the same base.** If the two sides of the equation do not have the same base, express each as a power of the same base if possible.

Step 2 **Simplify exponents** if necessary, using the rules of exponents.

Step 3 **Set exponents equal** using the property given in this section.

Step 4 **Solve** the equation obtained in Step 3.

NOTE The steps used in **Examples 5 and 6** cannot be applied to an equation like $3^x = 12$ because Step 1 cannot easily be done. We solve such equations in **Section 12.6**.

NOW TRY EXERCISE 5

Solve the equation.
$$8^x = 16$$

EXAMPLE 5 Solving an Exponential Equation

Solve the equation $9^x = 27$.

$$9^x = 27$$

$$(3^2)^x = 3^3 \qquad \text{Write with the same base;}$$
$$9 = 3^2 \text{ and } 27 = 3^3. \text{ (Step 1)}$$

$$3^{2x} = 3^3 \qquad \text{Power rule for exponents (Step 2)}$$

$$2x = 3 \qquad \text{If } a^x = a^y, \text{ then } x = y. \text{ (Step 3)}$$

$$x = \frac{3}{2} \qquad \text{Solve for } x. \text{ (Step 4)}$$

CHECK Substitute $\frac{3}{2}$ for x.

$$9^x = 9^{3/2} = (9^{1/2})^3 = 3^3 = 27 \quad \checkmark \quad \text{True}$$

The solution set is $\left\{\frac{3}{2}\right\}$.

NOW TRY

EXAMPLE 6 Solving Exponential Equations

Solve each equation.

(a) $4^{3x-1} = 16^{x+2}$

> Be careful multiplying the exponents.

$$4^{3x-1} = (4^2)^{x+2} \qquad \text{Write with the same base; } 16 = 4^2.$$

$$4^{3x-1} = 4^{2x+4} \qquad \text{Power rule for exponents}$$

$$3x - 1 = 2x + 4 \qquad \text{Set exponents equal.}$$

$$x = 5 \qquad \text{Subtract } 2x. \text{ Add 1.}$$

CHECK $4^{3x-1} = 16^{x+2}$

$$4^{3(5)-1} \overset{?}{=} 16^{5+2} \qquad \text{Substitute. Let } x = 5.$$

$$4^{14} \overset{?}{=} 16^7 \qquad \text{Perform the operations in the exponents.}$$

$$4^{14} \overset{?}{=} (4^2)^7 \qquad 16 = 4^2$$

$$4^{14} = 4^{14} \quad \checkmark \quad \text{True}$$

NOW TRY ANSWER
5. $\left\{\frac{4}{3}\right\}$

The solution set is $\{5\}$.

NOW TRY EXERCISE 6

Solve each equation.

(a) $3^{2x-1} = 27^{x+4}$

(b) $5^x = \dfrac{1}{625}$

(c) $\left(\dfrac{2}{7}\right)^x = \dfrac{343}{8}$

(b) $6^x = \dfrac{1}{216}$

      $6^x = \dfrac{1}{6^3}$ $216 = 6^3$

      $6^x = 6^{-3}$ Write with the same base; $\frac{1}{6^3} = 6^{-3}$.

      $x = -3$ Set exponents equal.

CHECK $6^x = 6^{-3} = \dfrac{1}{6^3} = \dfrac{1}{216}$ ✓   Substitute -3 for x. True.

The solution set is $\{-3\}$.

(c) $\left(\dfrac{2}{3}\right)^x = \dfrac{9}{4}$

      $\left(\dfrac{2}{3}\right)^x = \left(\dfrac{4}{9}\right)^{-1}$ $\frac{9}{4} = \left(\frac{4}{9}\right)^{-1}$

      $\left(\dfrac{2}{3}\right)^x = \left[\left(\dfrac{2}{3}\right)^2\right]^{-1}$ Write with the same base.

      $\left(\dfrac{2}{3}\right)^x = \left(\dfrac{2}{3}\right)^{-2}$ Power rule for exponents

      $x = -2$ Set exponents equal.

Check that the solution set is $\{-2\}$.   **NOW TRY**

OBJECTIVE 4 Use exponential functions in applications involving growth or decay.

EXAMPLE 7   Solving an Application Involving Exponential Growth

The graph in **FIGURE 11** shows the concentration of carbon dioxide (in parts per million) in the air. This concentration is increasing exponentially.

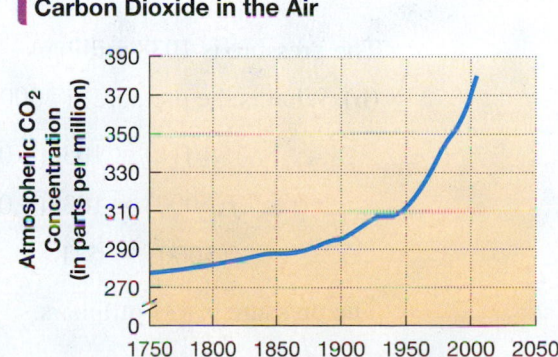

Source: Sacramento Bee; National Oceanic and Atmospheric Administration.

FIGURE 11

NOW TRY
EXERCISE 7
Use the function in **Example 7** to approximate the carbon dioxide concentration in 2000, to the nearest unit.

The data in the preceding graph are approximated by the function

$$f(x) = 266(1.001)^x,$$

where x is number of years since 1750. Use this function and a calculator to approximate the concentration of carbon dioxide in parts per million, to the nearest unit, for each year.

(a) 1900

Because x represents number of years since 1750, $x = 1900 - 1750 = 150$.

$$f(x) = 266(1.001)^x \qquad \text{Given function}$$
$$f(150) = 266(1.001)^{150} \qquad \text{Let } x = 150.$$
$$f(150) \approx 309 \qquad \text{Evaluate with a calculator.}$$

The concentration in 1900 was 309 parts per million.

(b) 1950

$$f(x) = 266(1.001)^x \qquad \text{Given function}$$
$$f(200) = 266(1.001)^{200} \qquad x = 1950 - 1750 = 200$$
$$f(200) \approx 325 \qquad \text{Evaluate with a calculator.}$$

The concentration in 1950 was 325 parts per million. **NOW TRY**

NOW TRY
EXERCISE 8
Use the function in **Example 8** to approximate the pressure at 6000 m, to the nearest unit.

EXAMPLE 8 Applying an Exponential Decay Function

The atmospheric pressure (in millibars) at a given altitude x, in meters, can be approximated by the function

$$f(x) = 1038(1.000134)^{-x}, \quad \text{for values of } x \text{ between 0 and 10,000.}$$

Because the base is greater than 1 and the coefficient of x in the exponent is negative, function values decrease as x increases. This means that as altitude increases, atmospheric pressure decreases. (*Source:* Miller, A. and J. Thompson, *Elements of Meteorology,* Fourth Edition, Charles E. Merrill Publishing Company.)

(a) According to this function, what is the pressure at ground level?

$$f(x) = 1038(1.000134)^{-x} \qquad \text{Given function}$$
$$f(0) = 1038(1.000134)^{-0} \qquad \text{Let } x = 0.$$
$$\text{At ground level, } x = 0. \qquad = 1038(1) \qquad a^0 = 1$$
$$= 1038$$

The pressure is 1038 millibars.

(b) What is the pressure at 5000 m, to the nearest unit?

$$f(x) = 1038(1.000134)^{-x} \qquad \text{Given function}$$
$$f(5000) = 1038(1.000134)^{-5000} \qquad \text{Let } x = 5000.$$
$$f(5000) \approx 531 \qquad \text{Evaluate with a calculator.}$$

The pressure is 531 millibars. **NOW TRY**

NOW TRY ANSWERS
7. 342 parts per million
8. 465 millibars

NOTE The function in **Example 8** is equivalent to

$$f(x) = 1038\left(\frac{1}{1.000134}\right)^x.$$

In this form, the base a satisfies the condition $0 < a < 1$.

12.2 Exercises

  MyMathLab®

▶ *Complete solution available in MyMathLab*

Concept Check   *Provide the answers in Exercises 1–6.*

1. Which point lies on the graph of $f(x) = 3^x$?

   A. $(1, 0)$ B. $(3, 1)$ C. $(0, 1)$ D. $\left(\sqrt{3}, \dfrac{1}{3}\right)$

2. Which statement is true?

   A. The point $\left(\dfrac{1}{2}, \sqrt{5}\right)$ lies on the graph of $f(x) = 5^x$.

   B. $f(x) = 2^x$ is not a one-to-one function.

   C. The y-intercept of the graph of $f(x) = 10^x$ is $(0, 10)$.

   D. The graph of $y = 4^x$ rises at a faster rate than the graph of $y = 10^x$.

3. For an exponential function $f(x) = a^x$, if $a > 1$, then the graph (*rises/falls*) from left to right.

4. For an exponential function $f(x) = a^x$, if $0 < a < 1$, then the graph (*rises/falls*) from left to right.

5. The asymptote of the graph of $f(x) = a^x$

   A. is the x-axis   B. is the y-axis

   C. has equation $x = 1$ D. has equation $y = 1$.

6. Describe the characteristics of the graph of an exponential function. Use the exponential function $f(x) = 3^x$ and the words *asymptote*, *domain*, and *range* in the explanation.

Use a calculator to find an approximation to three decimal places for each exponential expression. See Example 1.

7. $2^{1.9}$ 8. $2^{2.7}$ 9. $2^{-1.54}$ 10. $2^{-1.88}$

11. $10^{0.3}$ 12. $10^{0.5}$ 13. $4^{1/3}$ 14. $6^{1/5}$

15. $\left(\dfrac{1}{3}\right)^{1.5}$ 16. $\left(\dfrac{1}{3}\right)^{2.4}$ 17. $\left(\dfrac{1}{4}\right)^{-3.1}$ 18. $\left(\dfrac{1}{4}\right)^{-1.4}$

Graph each exponential function. See Examples 2–4.

▶ 19. $f(x) = 3^x$ 20. $f(x) = 5^x$ ▶ 21. $g(x) = \left(\dfrac{1}{3}\right)^x$ 22. $g(x) = \left(\dfrac{1}{5}\right)^x$

23. $f(x) = 4^{-x}$ 24. $f(x) = 6^{-x}$ ▶ 25. $f(x) = 2^{2x-2}$ 26. $f(x) = 2^{2x+1}$

Solve each equation. See Examples 5 and 6.

27. $6^x = 36$ 28. $8^x = 64$ ▶ 29. $100^x = 1000$

30. $8^x = 4$ ▶ 31. $16^{2x+1} = 64^{x+3}$ 32. $9^{2x-8} = 27^{x-4}$

33. $5^x = \dfrac{1}{125}$ 34. $3^x = \dfrac{1}{81}$ 35. $5^x = 0.2$

36. $10^x = 0.1$ 37. $\left(\dfrac{3}{2}\right)^x = \dfrac{8}{27}$ 38. $\left(\dfrac{4}{3}\right)^x = \dfrac{27}{64}$

The amount of radioactive material in an ore sample is given by the function

$$A(t) = 100(3.2)^{-0.5t},$$

where A(t) is the amount present, in grams, of the sample t months after the initial measurement.

Use this function for Exercises 39–42.

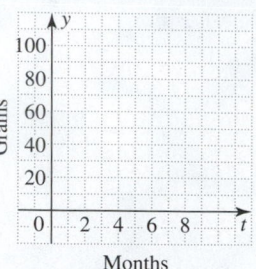

39. How much radioactive material was present at the initial measurement? (*Hint: t* = 0.)

40. How much, to the nearest hundredth, was present 2 months later?

41. How much, to the nearest hundredth, was present 10 months later?

42. Graph the function on the axes as shown.

A major scientific periodical published an article in 1990 dealing with the problem of global warming. The article was accompanied by a graph that illustrated two possible scenarios.

(a) *The warming might be modeled by an exponential function of the form*

$$f(x) = (1.046 \times 10^{-38})(1.0444^x).$$

(b) *The warming might be modeled by a linear function of the form*

$$g(x) = 0.009x - 17.67.$$

In both cases, x represents the year, and the function value represents the increase in degrees Celsius due to the warming. Use these functions to approximate the increase in temperature for each year, to the nearest tenth.

43. 2000   **44.** 2010   **45.** 2020   **46.** 2040

Solve each problem. **See Examples 7 and 8.**

47. The average number of Facebook users in millions at the end of each year from 2004 to 2013 can be modeled by the exponential function

$$f(x) = 2.403(2.394)^x,$$

where $x = 0$ represents 2004, $x = 1$ represents 2005, and so on. Use this model to approximate the number of Facebook users in each year, to the nearest thousandth. (*Source:* http://statista.com)

(a) 2004   **(b)** 2007   **(c)** 2009

48. Based on figures from 1960 through 2010, the municipal solid waste in millions of tons in the United States can be modeled by the exponential function

$$f(x) = 94.78(1.02)^x,$$

where $x = 0$ corresponds to 1960, $x = 10$ corresponds to 1970, and so on. Use this model to approximate the municipal solid waste generated in each year, to the nearest hundredth. (*Source:* U.S. Environmental Protection Agency.)

(a) 1970   **(b)** 1980   **(c)** 1990

(d) In 2005, the actual number was 252.7 million tons. How does this compare to the number that the model gives?

▶ **49.** A small business estimates that the value $V(t)$ of a copy machine is decreasing according to the function

$$V(t) = 5000(2)^{-0.15t},$$

where t is the number of years that have elapsed since the machine was purchased, and $V(t)$ is in dollars.

(a) What was the original value of the machine?

(b) What is the value of the machine 5 yr after purchase, to the nearest dollar?

(c) What is the value of the machine 10 yr after purchase, to the nearest dollar?

(d) Graph the function.

50. Refer to the function in **Exercise 49.**

(a) When will the value of the machine be $2500? (*Hint:* Let $V(t) = 2500$, divide both sides by 5000, and use the method of **Example 5.**)

(b) When will the value of the machine be $1250?

12.3 Logarithmic Functions

OBJECTIVES

1. Define a logarithm.
2. Convert between exponential and logarithmic forms, and evaluate logarithms.
3. Solve logarithmic equations of the form $\log_a b = k$ for a, b, or k.
4. Use the definition of logarithm to simplify logarithmic expressions.
5. Define and graph logarithmic functions.
6. Use logarithmic functions in applications involving growth or decay.

VOCABULARY

☐ logarithm
☐ logarithmic equation
☐ logarithmic function with base a

OBJECTIVE 1 Define a logarithm.

The graph of $y = 2^x$ is the curve shown in blue in **FIGURE 12.** Because $y = 2^x$ defines a one-to-one function, it has an inverse. Interchanging x and y gives

$$x = 2^y, \quad \text{the inverse of} \quad y = 2^x. \qquad \text{Roles of } x \text{ and } y \text{ are interchanged.}$$

As we saw in **Section 12.1,** the graph of the inverse is found by reflecting the graph of $y = 2^x$ across the line $y = x$. The graph of $x = 2^y$ is shown as a red curve in **FIGURE 12.**

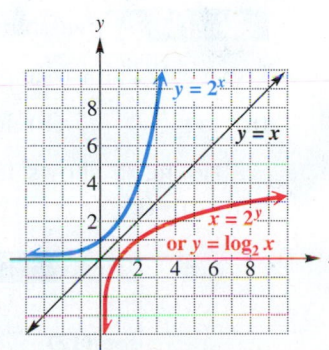

FIGURE 12

We can also write the equation of the red curve using a new notation that involves the concept of *logarithm*.

Logarithm

For all positive numbers a, where $a \neq 1$, and all positive numbers x,

$$y = \log_a x \quad \textbf{means the same as} \quad x = a^y.$$

The abbreviation **log** is used for the word **logarithm.** Read **log$_a$ x** as "the logarithm of x with base a" or "the base a logarithm of x." To remember the location of the base and the exponent in each form, refer to the following diagrams.

Exponent

Logarithmic form: $y = \log_a x$

Base

Exponent

Exponential form: $x = a^y$

Base

Meaning of log$_a$ x

A logarithm is an exponent. *The expression **log$_a$ x** represents the exponent to which the base a must be raised to obtain x.*

OBJECTIVE 2 Convert between exponential and logarithmic forms, and evaluate logarithms.

We can use the definition of logarithm to carry out these conversions.

NOW TRY EXERCISE 1

(a) Write $6^3 = 216$ in logarithmic form.

(b) Write $\log_{64} 4 = \frac{1}{3}$ in exponential form.

EXAMPLE 1   Converting between Exponential and Logarithmic Forms

The table shows several pairs of equivalent forms.

Exponential Form	Logarithmic Form
$3^2 = 9$	$\log_3 9 = 2$
$\left(\frac{1}{5}\right)^{-2} = 25$	$\log_{1/5} 25 = -2$
$10^5 = 100{,}000$	$\log_{10} 100{,}000 = 5$
$4^{-3} = \frac{1}{64}$	$\log_4 \frac{1}{64} = -3$

$y = \log_a x$
means
$x = a^y$.

NOW TRY

EXAMPLE 2   Evaluating Logarithms

Use a calculator to approximate each logarithm to four decimal places.

(a) $\log_2 5$ **(b)** $\log_3 12$ **(c)** $\log_{1/2} 12$ **(d)** $\log_{10} 20$

FIGURE 13 shows how a TI-84 Plus calculator approximates the logarithms in parts (a)–(c).

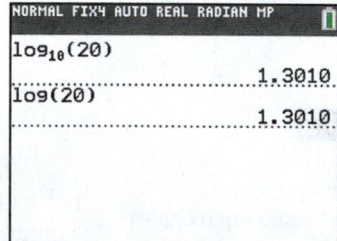

NORMAL FIX4 AUTO REAL RADIAN MP

log$_2$(5)
                    2.3219
log$_3$(12)
                    2.2619
log$_{1/2}$(12)
                    -3.5850

FIGURE 13

NORMAL FIX4 AUTO REAL RADIAN MP

log$_{10}$(20)
                    1.3010
log(20)
                    1.3010

FIGURE 14

NOW TRY ANSWERS

1. (a) $\log_6 216 = 3$

(b) $64^{1/3} = 4$

FIGURE 14 shows the approximation for the expression $\log_{10} 20$ in part (d). Notice that the second display, which indicates log 20 (with no base shown), gives the same result. We shall see in a later section that when no base is indicated, the base is understood to be 10. A base 10 logarithm is a *common logarithm.*

**NOW TRY
EXERCISE 2**

Use a calculator to approximate each logarithm to four decimal places.

(a) $\log_2 7$ **(b)** $\log_5 8$

(c) $\log_{1/3} 12$ **(d)** $\log_{10} 18$

To four decimal places, the logarithms in parts (a)–(d) are as follows.

$$\log_2 5 \approx 2.3219, \qquad \log_3 12 \approx 2.2619,$$
$$\log_{1/2} 12 \approx -3.5850, \qquad \log_{10} 20 \approx 1.3010$$

NOW TRY

NOTE In **Section 12.5** we will introduce another method of calculating logarithms like those in **Example 2(a)–(c)** using the *change-of-base rule*.

OBJECTIVE 3 Solve logarithmic equations of the form $\log_a b = k$ for a, b, or k.

A **logarithmic equation** is an equation with a logarithm in at least one term.

EXAMPLE 3 Solving Logarithmic Equations

Solve each equation.

(a) $\log_4 x = -2$

By the definition of logarithm, $\log_4 x = -2$ is equivalent to $x = 4^{-2}$.

$$x = 4^{-2} = \frac{1}{16}$$

The solution set is $\left\{\frac{1}{16}\right\}$.

(b) $\qquad \log_{1/2}(3x + 1) = 2$

$$3x + 1 = \left(\frac{1}{2}\right)^2 \qquad \text{This is a key step.} \qquad \text{Write in exponential form.}$$

$$3x + 1 = \frac{1}{4} \qquad \text{Apply the exponent.}$$

$$12x + 4 = 1 \qquad \text{Multiply each term by 4.}$$

$$12x = -3 \qquad \text{Subtract 4.}$$

$$x = -\frac{1}{4} \qquad \text{Divide by 12. Write in lowest terms.}$$

CHECK $\qquad \log_{1/2}(3x + 1) = 2$

$$\log_{1/2}\left[3\left(-\frac{1}{4}\right) + 1\right] \overset{?}{=} 2 \qquad \text{Let } x = -\frac{1}{4}.$$

$$\log_{1/2}\frac{1}{4} \overset{?}{=} 2 \qquad \text{Simplify within parentheses.}$$

$$\left(\frac{1}{2}\right)^2 \overset{?}{=} \frac{1}{4} \qquad \text{Write in exponential form.}$$

$$\frac{1}{4} = \frac{1}{4} \quad \checkmark \quad \text{True}$$

NOW TRY ANSWERS

2. (a) 2.8074 **(b)** 1.2920
   (c) −2.2619 **(d)** 1.2553

The solution set is $\left\{-\frac{1}{4}\right\}$.

NOW TRY
EXERCISE 3

Solve each equation.

(a) $\log_2 x = -5$

(b) $\log_{3/2}(2x - 1) = 3$

(c) $\log_x 10 = 2$

(d) $\log_{125} \sqrt[3]{5} = x$

(c) $\log_x 3 = 2$

> Be careful here.
> $-\sqrt{3}$ is extraneous.

$x^2 = 3$ Write in exponential form.

$x = \pm \sqrt{3}$ Take square roots.

Only the *principal* square root satisfies the equation because the base must be a positive number.

CHECK $\log_x 3 = 2$

$\log_{\sqrt{3}} 3 \overset{?}{=} 2$ Let $x = \sqrt{3}$.

$(\sqrt{3})^2 \overset{?}{=} 3$ Write in exponential form.

$3 = 3$ ✓ True

The solution set is $\{\sqrt{3}\}$.

(d) $\log_{49} \sqrt[3]{7} = x$

$49^x = \sqrt[3]{7}$ Write in exponential form.

$(7^2)^x = 7^{1/3}$ Write with the same base.

$7^{2x} = 7^{1/3}$ Power rule for exponents

$2x = \dfrac{1}{3}$ Set exponents equal.

$x = \dfrac{1}{6}$ Divide by 2 $\left(\text{which is the same as multiplying by } \frac{1}{2}\right)$.

Check to verify that the solution set is $\left\{\frac{1}{6}\right\}$.   **NOW TRY**

OBJECTIVE 4 Use the definition of logarithm to simplify logarithmic expressions.

The definition of logarithm allows us to state several special properties.

> **Special Properties of Logarithms**
>
> For any positive real number b, where $b \neq 1$, the following hold true.
>
> $$\log_b b^r = r \qquad b^{\log_b r} = r \quad (\text{where } r > 0)$$
>
> $$\log_b b = 1 \qquad \log_b 1 = 0$$

NOW TRY
EXERCISE 4

Use the special properties to evaluate each expression.

(a) $\log_{10} 10$ **(b)** $\log_8 1$

(c) $\log_{0.1} 1$ **(d)** $\log_3 3^9$

(e) $5^{\log_5 3}$ **(f)** $\log_3 81$

EXAMPLE 4 Using Properties of Logarithms

Use the special properties to evaluate.

(a) $\log_7 7 = 1$ $\log_b b = 1$ **(b)** $\log_{\sqrt{2}} \sqrt{2} = 1$

(c) $\log_9 1 = 0$ $\log_b 1 = 0$ **(d)** $\log_{0.2} 1 = 0$

(e) $\log_2 2^6 = 6$ $\log_b b^r = r$ **(f)** $\log_3 3^{-2.5} = -2.5$

(g) $4^{\log_4 9} = 9$ $b^{\log_b r} = r$ **(h)** $10^{\log_{10} 13} = 13$

(i) $\log_2 32$

We know that $32 = 2^5$, so we replace 32 with 2^5 and use the first property above.

$$\log_2 32 = \log_2 2^5 = 5$$ **NOW TRY**

OBJECTIVE 5 Define and graph logarithmic functions.

> **Logarithmic Function**
>
> If a and x are positive numbers, where $a \neq 1$, then
>
> $$g(x) = \log_a x$$
>
> defines the **logarithmic function with base a.**

NOW TRY
EXERCISE 5

Graph $f(x) = \log_6 x$.

EXAMPLE 5 Graphing a Logarithmic Function ($a > 1$)

Graph $f(x) = \log_2 x$.

By writing $y = f(x) = \log_2 x$ in exponential form as

$$x = 2^y,$$

we can identify ordered pairs that satisfy the equation. It is easier to choose values for y and find the corresponding values of x. Plotting the points in the table of ordered pairs and connecting them with a smooth curve gives the graph in **FIGURE 15**. This graph is typical of logarithmic functions with base $a > 1$.

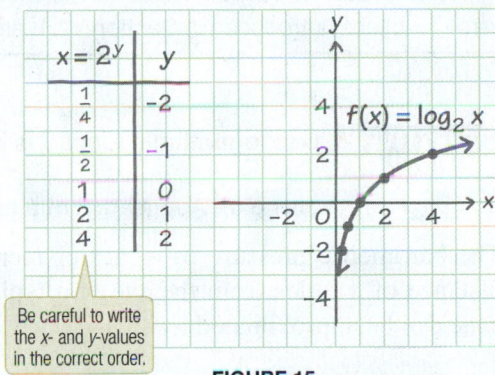

$x = 2^y$	y
$\frac{1}{4}$	-2
$\frac{1}{2}$	-1
1	0
2	1
4	2

Be careful to write the x- and y-values in the correct order.

FIGURE 15

Logarithmic function with base $a > 1$

Domain: $(0, \infty)$

Range: $(-\infty, \infty)$

x-intercept: $(1, 0)$

The function is one-to-one, and its graph rises from left to right.

NOW TRY

NOW TRY
EXERCISE 6

Graph $g(x) = \log_{1/4} x$.

EXAMPLE 6 Graphing a Logarithmic Function ($0 < a < 1$)

Graph $g(x) = \log_{1/2} x$.

We write $y = g(x) = \log_{1/2} x$ in exponential form as

$$x = \left(\frac{1}{2}\right)^y,$$

then choose values for y and find the corresponding values of x. Plotting these points and connecting them with a smooth curve gives the graph in **FIGURE 16**. This graph is typical of logarithmic functions with base $0 < a < 1$.

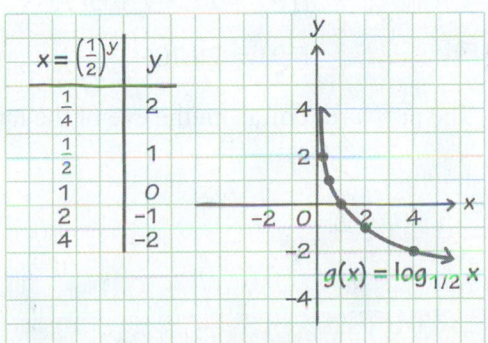

$x = \left(\frac{1}{2}\right)^y$	y
$\frac{1}{4}$	2
$\frac{1}{2}$	1
1	0
2	-1
4	-2

Logarithmic function with base $0 < a < 1$

Domain: $(0, \infty)$

Range: $(-\infty, \infty)$

x-intercept: $(1, 0)$

The function is one-to-one, and its graph falls from left to right.

FIGURE 16

NOW TRY

NOW TRY ANSWERS

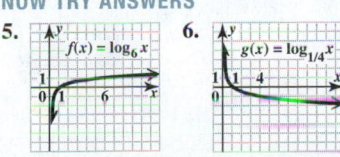

5. $f(x) = \log_6 x$

6. $g(x) = \log_{1/4} x$

Characteristics of the Graph of $g(x) = \log_a x$

1. The graph contains the point $(1, 0)$, which is its x-intercept.

2. The function is one-to-one.

   - When $a > 1$, the graph will *rise* from left to right, from the fourth quadrant to the first. (See **FIGURE 15**.)

   - When $0 < a < 1$, the graph will *fall* from left to right, from the first quadrant to the fourth. (See **FIGURE 16**.)

3. The graph will approach the y-axis, but never touch it. (The y-axis is an asymptote.)

4. The domain is $(0, \infty)$, and the range is $(-\infty, \infty)$.

NOTE See the similar box titled

"Characteristics of the Graph of $f(x) = a^x$"

in **Section 12.2**. Compare the four characteristics one by one to see how the concepts of inverse functions, introduced in **Section 12.1**, are illustrated by these two classes of functions.

OBJECTIVE 6 Use logarithmic functions in applications involving growth or decay.

NOW TRY EXERCISE 7

Suppose the gross national product (GNP) of a small country (in millions of dollars) is approximated by

$$G(t) = 15.0 + 2.00 \log_{10} t,$$

where t is time in years since 2013. Approximate to the nearest tenth the GNP for each value of t.

(a) $t = 1$ **(b)** $t = 10$

EXAMPLE 7 Solving an Application of a Logarithmic Function

The barometric pressure in inches of mercury at a distance of x miles from the eye of a typical hurricane can be approximated by

$$f(x) = 27 + 1.105 \log_{10}(x + 1).$$

(*Source:* Miller, A. and R. Anthes, *Meteorology*, Fifth Edition, Charles E. Merrill Publishing Company.) Approximate the pressure 9 mi from the eye of the hurricane.

Let $x = 9$, and find $f(9)$.

$$f(x) = 27 + 1.105 \log_{10}(x + 1)$$

$$f(9) = 27 + 1.105 \log_{10}(9 + 1) \qquad \text{Let } x = 9.$$

$$f(9) = 27 + 1.105 \log_{10} 10 \qquad \text{Add inside parentheses.}$$

$$f(9) = 27 + 1.105(1) \qquad \log_{10} 10 = 1$$

$$f(9) = 28.105 \qquad \text{Add.}$$

The pressure 9 mi from the eye of the hurricane is 28.105 in.

NOW TRY

NOW TRY ANSWERS
7. **(a)** \$15.0 million
   (b) \$17.0 million

12.3 Exercises

   MyMathLab®

● *Complete solution available in MyMathLab*

1. Concept Check Match each logarithmic equation in Column I with the corresponding exponential equation in Column II.

I	II
(a) $\log_{1/3} 3 = -1$	**A.** $8^{1/3} = \sqrt[3]{8}$
(b) $\log_5 1 = 0$	**B.** $\left(\dfrac{1}{3}\right)^{-1} = 3$
(c) $\log_2 \sqrt{2} = \dfrac{1}{2}$	**C.** $4^1 = 4$
(d) $\log_{10} 1000 = 3$	**D.** $2^{1/2} = \sqrt{2}$
(e) $\log_8 \sqrt[3]{8} = \dfrac{1}{3}$	**E.** $5^0 = 1$
(f) $\log_4 4 = 1$	**F.** $10^3 = 1000$

2. Concept Check Match each logarithm in Column I with its corresponding value in Column II.

I	II
(a) $\log_4 16$	**A.** -2
(b) $\log_3 81$	**B.** -1
(c) $\log_3 \left(\dfrac{1}{3}\right)$	**C.** 2
(d) $\log_{10} 0.01$	**D.** 0
(e) $\log_5 \sqrt{5}$	**E.** $\dfrac{1}{2}$
(f) $\log_{13} 1$	**F.** 4

3. Concept Check The domain of $f(x) = a^x$ is $(-\infty, \infty)$, while the range is $(0, \infty)$. Therefore because $g(x) = \log_a x$ is the inverse of f, the domain of g is _____, while the range of g is _____.

4. Concept Check The graphs of both $f(x) = 3^x$ and $g(x) = \log_3 x$ rise from left to right. Which one rises at a faster rate?

Write in logarithmic form. **See Example 1.**

5. $4^5 = 1024$

6. $3^6 = 729$

7. $\left(\dfrac{1}{2}\right)^{-3} = 8$

8. $\left(\dfrac{1}{6}\right)^{-3} = 216$

9. $10^{-3} = 0.001$

10. $36^{1/2} = 6$

11. $\sqrt[4]{625} = 5$

12. $\sqrt[3]{343} = 7$

13. $8^{-2/3} = \dfrac{1}{4}$

14. $16^{-3/4} = \dfrac{1}{8}$

15. $5^0 = 1$

16. $7^0 = 1$

Write in exponential form. **See Example 1.**

17. $\log_4 64 = 3$

18. $\log_2 512 = 9$

19. $\log_{10} \dfrac{1}{10,000} = -4$

20. $\log_{100} 100 = 1$

21. $\log_6 1 = 0$

22. $\log_\pi 1 = 0$

23. $\log_9 3 = \dfrac{1}{2}$

24. $\log_{64} 2 = \dfrac{1}{6}$

25. $\log_{1/4} \dfrac{1}{2} = \dfrac{1}{2}$

26. $\log_{1/8} \dfrac{1}{2} = \dfrac{1}{3}$

27. $\log_5 5^{-1} = -1$

28. $\log_{10} 10^{-2} = -2$

29. Concept Check Match each logarithm in Column I with its value in Column II.

I	II
(a) $\log_8 8$	**A.** -1
(b) $\log_{16} 1$	**B.** 0
(c) $\log_{0.3} 1$	**C.** 1
(d) $\log_{\sqrt{7}} \sqrt{7}$	**D.** 0.1

30. Concept Check When a student asked his teacher to explain how to evaluate

$$\log_9 3$$

without showing any work, his teacher told him to "Think radically." Explain what the teacher meant by this hint.

Use a calculator to approximate each logarithm to four decimal places. **See Example 2.**

31. $\log_2 9$

32. $\log_2 15$

33. $\log_5 18$

34. $\log_5 26$

35. $\log_{1/4} 12$

36. $\log_{1/5} 27$

37. $\log_2 \left(\dfrac{1}{3} \right)$

38. $\log_2 \left(\dfrac{1}{7} \right)$

39. $\log_{10} 84$

40. $\log_{10} 126$

41. $\log 50$

42. $\log 90$

Solve each equation. **See Example 3.**

▶ **43.** $x = \log_{27} 3$

44. $x = \log_{125} 5$

45. $\log_x 9 = \dfrac{1}{2}$

46. $\log_x 5 = \dfrac{1}{2}$

47. $\log_x 125 = -3$

48. $\log_x 64 = -6$

49. $\log_{12} x = 0$

50. $\log_4 x = 0$

51. $\log_x x = 1$

52. $\log_x 1 = 0$

53. $\log_x \dfrac{1}{25} = -2$

54. $\log_x \dfrac{1}{10} = -1$

55. $\log_8 32 = x$

56. $\log_{81} 27 = x$

57. $\log_\pi \pi^4 = x$

58. $\log_{\sqrt{2}} \left(\sqrt{2} \right)^9 = x$

59. $\log_6 \sqrt{216} = x$

60. $\log_4 \sqrt{64} = x$

61. $\log_4 (2x + 4) = 3$

62. $\log_3 (2x + 7) = 4$

Use the special properties of logarithms to evaluate each expression. **See Example 4.**

63. $\log_3 3$

64. $\log_8 8$

65. $\log_5 1$

66. $\log_{12} 1$

67. $\log_4 4^9$

68. $\log_5 5^6$

69. $\log_2 2^{-1}$

70. $\log_4 4^{-6}$

71. $6^{\log_6 9}$

72. $12^{\log_{12} 3}$

73. $8^{\log_8 5}$

74. $5^{\log_5 11}$

75. $\log_2 64$

76. $\log_2 128$

77. $\log_3 81$

78. $\log_3 27$

79. $\log_4 \left(\dfrac{1}{4} \right)$

80. $\log_6 \left(\dfrac{1}{6} \right)$

81. $\log_6 \sqrt[3]{6}$

82. $\log_9 \sqrt[3]{9}$

If (p, q) is on the graph of $f(x) = a^x$ (for $a > 0$ and $a \neq 1$), then (q, p) is on the graph of $f^{-1}(x) = \log_a x$. Use this fact, and refer to the graphs required in **Exercises 19–24** *of* **Section 12.2** *to graph each logarithmic function.* **See Examples 5 and 6.**

▶ **83.** $g(x) = \log_3 x$

84. $g(x) = \log_5 x$

▶ **85.** $f(x) = \log_{1/3} x$

86. $f(x) = \log_{1/5} x$

87. $g(x) = \log_{1/4} x$ $\left(\text{Hint: } 4^{-x} = \left(\dfrac{1}{4} \right)^x. \right)$

88. $g(x) = \log_{1/6} x$ $\left(\text{Hint: } 6^{-x} = \left(\dfrac{1}{6} \right)^x. \right)$

89. *Concept Check* Explain why 1 is not allowed as a base for a logarithmic function.

90. *Concept Check* Compare the summary of facts about the graph of $f(x) = a^x$ in **Section 12.2** with the similar summary of facts about the graph of $g(x) = \log_a x$ in this section. Make a list of the facts that reinforce the concept that f and g are inverse functions.

Use the graph at the right to predict the value of $f(t)$ for the given value of t.

91. $t = 0$

92. $t = 10$

93. $t = 60$

94. Show that the points determined in **Exercises 91–93** lie on the graph of

$$f(t) = 8 \log_5 (2t + 5).$$

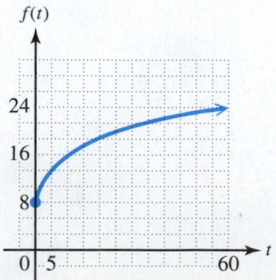

Solve each problem. See Example 7.

95. Sales (in thousands of units) of a new product are approximated by the function

$$S(t) = 100 + 30 \log_3 (2t + 1),$$

where *t* is the number of years after the product is introduced.

   (a) What were the sales, to the nearest unit, after 1 yr?

   (b) What were the sales, to the nearest unit, after 13 yr?

   (c) Graph $y = S(t)$.

96. A study showed that the number of mice in an old abandoned house was approximated by the function

$$M(t) = 6 \log_4 (2t + 4),$$

where *t* is measured in months and $t = 0$ corresponds to January 2014. Find the number of mice in the house for each month.

   (a) January 2014   **(b)** July 2014   **(c)** July 2016   **(d)** Graph $y = M(t)$.

The **Richter scale** *is used to measure the intensity of earthquakes. The Richter scale rating of an earthquake of intensity x is given by*

$$R = \log_{10} \frac{x}{x_0},$$

where x_0 is the intensity of an earthquake of a certain (small) size. The figure here shows Richter scale ratings for selected Southern California earthquakes with magnitudes greater than 4.7.

97. The Northridge earthquake had a Richter scale rating of 6.7. The Landers earthquake had a rating of 7.3. How much more powerful was the Landers quake than the Northridge quake?

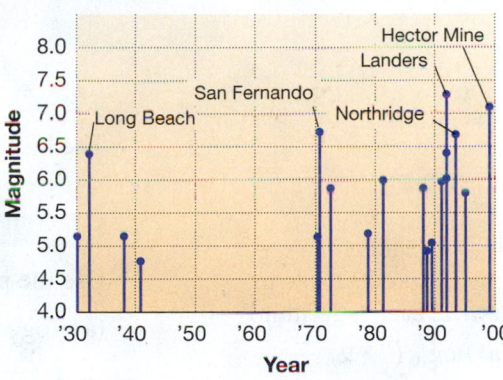

Southern California Earthquakes
(with magnitudes greater than 4.7)

Source: Caltech; U.S. Geological Survey.

98. Compare the smallest rated earthquake in the figure (at 4.8) with the Landers quake (at 7.3). How much more powerful was the Landers quake?

(12.4) Properties of Logarithms

OBJECTIVES

1 Use the product rule for logarithms.

2 Use the quotient rule for logarithms.

3 Use the power rule for logarithms.

4 Use properties to write alternative forms of logarithmic expressions.

Logarithms were used as an aid to numerical calculation for several hundred years. Today the widespread use of calculators has made the use of logarithms for calculation obsolete. However, logarithms are still very important in applications and in further work in mathematics.

OBJECTIVE 1 Use the product rule for logarithms.

One way in which logarithms simplify problems is by changing a problem of multiplication into one of addition. We know that $\log_2 4 = 2$, $\log_2 8 = 3$, and $\log_2 32 = 5$.

$$\log_2 32 = \log_2 4 + \log_2 8 \qquad 5 = 2 + 3$$
$$\log_2 (4 \cdot 8) = \log_2 4 + \log_2 8 \qquad 32 = 4 \cdot 8$$

This is an example of the following rule.

Product Rule for Logarithms

If x, y, and b are positive real numbers, where $b \neq 1$, then the following holds true.

$$\log_b xy = \log_b x + \log_b y$$

That is, the logarithm of a product is the sum of the logarithms of the factors.

NOTE   The word statement of the product rule can be restated by replacing "logarithm" with "exponent." The rule then becomes the familiar rule for multiplying exponential expressions:

The *exponent* of a product is the sum of the *exponents* of the factors.

To prove this rule, let $m = \log_b x$ and $n = \log_b y$, and recall that

$$\log_b x = m \quad \text{means} \quad b^m = x \quad \text{and} \quad \log_b y = n \quad \text{means} \quad b^n = y.$$

Now consider the product xy.

$$xy = b^m \cdot b^n \qquad \text{Substitute.}$$
$$xy = b^{m+n} \qquad \text{Product rule for exponents}$$
$$\log_b xy = m + n \qquad \text{Convert to logarithmic form.}$$
$$\log_b xy = \log_b x + \log_b y \qquad \text{Substitute.}$$

The last statement is the result we wished to prove.

**NOW TRY
EXERCISE 1**

Use the product rule to rewrite each logarithm.

(a) $\log_{10} (7 \cdot 9)$

(b) $\log_5 11 + \log_5 8$

(c) $\log_5 (5x), \quad x > 0$

(d) $\log_2 t^3, \quad t > 0$

EXAMPLE 1   Using the Product Rule

Use the product rule to rewrite each logarithm. Assume $x > 0$.

(a) $\log_5 (6 \cdot 9)$

$= \log_5 6 + \log_5 9 \qquad \text{Product rule}$

(b) $\log_7 8 + \log_7 12$

$= \log_7 (8 \cdot 12) \qquad \text{Product rule}$

$= \log_7 96 \qquad \text{Multiply.}$

(c) $\log_3 (3x)$

$= \log_3 3 + \log_3 x \qquad \text{Product rule}$

$= 1 + \log_3 x \qquad \log_3 3 = 1$

(d) $\log_4 x^3$

$= \log_4 (x \cdot x \cdot x) \qquad x^3 = x \cdot x \cdot x$

$= \log_4 x + \log_4 x + \log_4 x \qquad \text{Product rule}$

$= 3 \log_4 x \qquad \text{Combine like terms.} \qquad$ **NOW TRY**

OBJECTIVE 2   Use the quotient rule for logarithms.

The rule for division is similar to the rule for multiplication.

Quotient Rule for Logarithms

If x, y, and b are positive real numbers, where $b \neq 1$, then the following holds true.

$$\log_b \frac{x}{y} = \log_b x - \log_b y$$

That is, the logarithm of a quotient is the difference of the logarithm of the numerator and the logarithm of the denominator.

NOW TRY ANSWERS

1. (a) $\log_{10} 7 + \log_{10} 9$
  (b) $\log_5 88$
  (c) $1 + \log_5 x$
  (d) $3 \log_2 t$

The proof of this rule is similar to the proof of the product rule.

NOW TRY EXERCISE 2

Use the quotient rule to rewrite each logarithm.

(a) $\log_{10} \dfrac{7}{9}$

(b) $\log_4 x - \log_4 12, \quad x > 0$

(c) $\log_5 \dfrac{25}{27}$

EXAMPLE 2 Using the Quotient Rule

Use the quotient rule to rewrite each logarithm. Assume $x > 0$.

(a) $\log_4 \dfrac{7}{9}$

$= \log_4 7 - \log_4 9$ Quotient rule

(b) $\log_5 6 - \log_5 x$

$= \log_5 \dfrac{6}{x}$ Quotient rule

(c) $\log_3 \dfrac{27}{5}$

$= \log_3 27 - \log_3 5$ Quotient rule

$= 3 - \log_3 5$ $\log_3 27 = 3$

(d) $\log_6 28 - \log_6 7$

$= \log_6 \dfrac{28}{7}$ Quotient rule

$= \log_6 4$ $\frac{28}{7} = 4$ **NOW TRY** ↺

⚠ **CAUTION** *There is no property of logarithms to rewrite the logarithm of a sum or difference.* For example, we *cannot* write $\log_b (x + y)$ in terms of $\log_b x$ and $\log_b y$. Also,

$$\log_b \frac{x}{y} \neq \frac{\log_b x}{\log_b y}.$$

OBJECTIVE 3 Use the power rule for logarithms.

An exponential expression such as

$$2^3 \quad \text{means} \quad 2 \cdot 2 \cdot 2.$$

The base is used as a factor 3 times. Similarly, the product rule can be extended to rewrite the logarithm of a power as the product of the exponent and the logarithm of the base.

$\log_5 2^3$

$= \log_5 (2 \cdot 2 \cdot 2)$

$= \log_5 2 + \log_5 2 + \log_5 2$

$= 3 \log_5 2$

$\log_2 7^4$

$= \log_2 (7 \cdot 7 \cdot 7 \cdot 7)$

$= \log_2 7 + \log_2 7 + \log_2 7 + \log_2 7$

$= 4 \log_2 7$

Furthermore, we saw in **Example 1(d)** that $\log_4 x^3 = 3 \log_4 x$. These examples suggest the following rule.

NOW TRY ANSWERS
2. **(a)** $\log_{10} 7 - \log_{10} 9$
   (b) $\log_4 \frac{x}{12}$
   (c) $2 - \log_5 27$

Power Rule for Logarithms

If x and b are positive real numbers, where $b \neq 1$, and if r is any real number, then the following holds true.

$$\log_b x^r = r \log_b x$$

That is, the logarithm of a number to a power equals the exponent times the logarithm of the number.

Examples: $\log_b m^5 = 5 \log_b m$ and $\log_3 5^4 = 4 \log_3 5$

To prove the power rule, let $\log_b x = m$.

$$b^m = x \qquad \text{Convert to exponential form.}$$

$$(b^m)^r = x^r \qquad \text{Raise to the power } r.$$

$$b^{mr} = x^r \qquad \text{Power rule for exponents}$$

$$\log_b x^r = rm \qquad \text{Convert to logarithmic form; commutative property}$$

$$\log_b x^r = r \log_b x \qquad m = \log_b x \text{ from above}$$

This is the statement to be proved.

As a special case of the power rule, let $r = \dfrac{1}{p}$, so

$$\log_b \sqrt[p]{x} = \log_b x^{1/p} = \frac{1}{p} \log_b x.$$

For example, using this result, with $x > 0$,

$$\log_b \sqrt[5]{x} = \log_b x^{1/5} = \frac{1}{5} \log_b x \quad \text{and} \quad \log_b \sqrt[3]{x^4} = \log_b x^{4/3} = \frac{4}{3} \log_b x.$$

Another special case is as follows.

$$\log_b \frac{1}{x} = \log_b x^{-1} = -\log_b x$$

NOW TRY
EXERCISE 3

Use the power rule to rewrite each logarithm. Assume $a > 0$, $x > 0$, and $a \neq 1$.

(a) $\log_7 5^3$ **(b)** $\log_a \sqrt{10}$

(c) $\log_3 \sqrt[4]{x^3}$ **(d)** $\log_4 \dfrac{1}{x^5}$

EXAMPLE 3   Using the Power Rule

Use the power rule to rewrite each logarithm. Assume $b > 0$, $x > 0$, and $b \neq 1$.

(a) $\log_5 4^2$ **(b)** $\log_b x^5$ **(c)** $\log_b \sqrt{7}$

$\quad = 2 \log_5 4$ $\quad = 5 \log_b x$ $\quad = \log_b 7^{1/2}$ $\sqrt{x} = x^{1/2}$

$\qquad\qquad\qquad\qquad\qquad\qquad\qquad\quad = \dfrac{1}{2} \log_b 7$ Power rule

(d) $\log_2 \sqrt[5]{x^2}$ **(e)** $\log_3 \dfrac{1}{x^4}$

$\quad = \log_2 x^{2/5}$ $\sqrt[5]{x^2} = x^{2/5}$ $\qquad = \log_3 x^{-4}$ Definition of negative exponent

$\quad = \dfrac{2}{5} \log_2 x$ Power rule $\qquad = -4 \log_3 x$ Power rule

NOW TRY

We summarize the properties of logarithms from the previous section and this one.

Properties of Logarithms		

If x, y, and b are positive real numbers, where $b \neq 1$, and r is any real number, then the following hold true.

Special Properties $\quad \log_b b^r = r \qquad b^{\log_b r} = r \quad$ (where $r > 0$)
(from Section 12.3)

Product Rule $\qquad\qquad \log_b xy = \log_b x + \log_b y$

Quotient Rule $\qquad\qquad \log_b \dfrac{x}{y} = \log_b x - \log_b y$

Power Rule $\qquad\qquad\quad \log_b x^r = r \log_b x$

NOW TRY ANSWERS

3. (a) $3 \log_7 5$ **(b)** $\frac{1}{2} \log_a 10$

$\quad$ **(c)** $\frac{3}{4} \log_3 x$ **(d)** $-5 \log_4 x$

OBJECTIVE 4 Use properties to write alternative forms of logarithmic expressions.

NOW TRY
EXERCISE 4

Use properties of logarithms to rewrite each expression if possible. Assume that all variables represent positive real numbers.

(a) $\log_3 9z^4$

(b) $\log_6 \sqrt{\dfrac{n}{3m}}$

(c) $\log_2 x + 3 \log_2 y - \log_2 z$

(d) $\log_5 (x + 10)$
$+ \log_5 (x - 10)$
$- \dfrac{3}{5} \log_5 x, \quad x > 10$

(e) $\log_7 (49 + 2x)$

EXAMPLE 4 Writing Logarithms in Alternative Forms

Use properties of logarithms to rewrite each expression if possible. Assume that all variables represent positive real numbers.

(a) $\log_4 4x^3$

$= \log_4 4 + \log_4 x^3$ Product rule

$= 1 + 3 \log_4 x$ $\log_4 4 = 1$; power rule

(b) $\log_7 \sqrt{\dfrac{m}{n}}$

$= \log_7 \left(\dfrac{m}{n}\right)^{1/2}$ Write the radical expression with a rational exponent.

$= \dfrac{1}{2} \log_7 \dfrac{m}{n}$ Power rule

$= \dfrac{1}{2} (\log_7 m - \log_7 n)$ Quotient rule

(c) $\log_5 \dfrac{a^2}{bc}$

$= \log_5 a^2 - \log_5 bc$ Quotient rule

$= 2 \log_5 a - \log_5 bc$ Power rule

$= 2 \log_5 a - (\log_5 b + \log_5 c)$ Product rule

$= 2 \log_5 a - \log_5 b - \log_5 c$

> Parentheses are necessary here.

(d) $4 \log_b m - \log_b n, \quad b \neq 1$

$= \log_b m^4 - \log_b n$ Power rule

$= \log_b \dfrac{m^4}{n}$ Quotient rule

(e) $\log_b (x + 1) + \log_b (2x + 1) - \dfrac{2}{3} \log_b x, \quad b \neq 1$

$= \log_b (x + 1) + \log_b (2x + 1) - \log_b x^{2/3}$ Power rule

$= \log_b \dfrac{(x + 1)(2x + 1)}{x^{2/3}}$ Product and quotient rules

$= \log_b \dfrac{2x^2 + 3x + 1}{x^{2/3}}$ Multiply in the numerator.

NOW TRY ANSWERS
4. **(a)** $2 + 4 \log_3 z$

(b) $\dfrac{1}{2}(\log_6 n - \log_6 3 - \log_6 m)$

(c) $\log_2 \dfrac{xy^3}{z}$ **(d)** $\log_5 \dfrac{x^2 - 100}{x^{3/5}}$

(e) cannot be rewritten

(f) $\log_8 (2p + 3r)$ cannot be rewritten using the properties of logarithms. *There is no property of logarithms to rewrite the logarithm of a sum.* **NOW TRY**

In the next example, we use numerical values for $\log_2 5$ and $\log_2 3$. While we use the equality symbol to give these values, they are actually approximations because most logarithms of this type are irrational numbers. *We use $=$ with the understanding that the values are correct to four decimal places.*

NOW TRY
EXERCISE 5

Given that $\log_2 7 = 2.8074$ and $\log_2 10 = 3.3219$, use properties of logarithms to evaluate each expression.

(a) $\log_2 70$ **(b)** $\log_2 0.7$

(c) $\log_2 49$

EXAMPLE 5 Using the Properties of Logarithms with Numerical Values

Given that $\log_2 5 = 2.3219$ and $\log_2 3 = 1.5850$, use properties of logarithms to evaluate each expression.

(a) $\log_2 15$

$$= \log_2 (3 \cdot 5) \qquad \text{Factor 15.}$$
$$= \log_2 3 + \log_2 5 \qquad \text{Product rule}$$
$$= 1.5850 + 2.3219 \qquad \text{Substitute the given values.}$$
$$= 3.9069 \qquad \text{Add.}$$

(b) $\log_2 0.6$

$$= \log_2 \frac{3}{5} \qquad 0.6 = \tfrac{6}{10} = \tfrac{3}{5}$$
$$= \log_2 3 - \log_2 5 \qquad \text{Quotient rule}$$
$$= 1.5850 - 2.3219 \qquad \text{Substitute the given values.}$$
$$= -0.7369 \qquad \text{Subtract.}$$

(c) $\log_2 27$

$$= \log_2 3^3 \qquad \text{Write 27 as a power of 3.}$$
$$= 3 \log_2 3 \qquad \text{Power rule}$$
$$= 3(1.5850) \qquad \text{Substitute the given value.}$$
$$= 4.7550 \qquad \text{Multiply.}$$

NOW TRY

NOW TRY
EXERCISE 6

Decide whether each statement is *true* or *false*.

(a) $\log_2 16 + \log_2 16 = \log_2 32$

(b) $(\log_2 4)(\log_3 9) = \log_6 36$

EXAMPLE 6 Deciding Whether Statements about Logarithms Are True

Decide whether each statement is *true* or *false*.

(a) $\log_2 8 - \log_2 4 = \log_2 4$

Evaluate each side.

$\log_2 8 - \log_2 4$	Left side	$\log_2 4$	Right side
$= \log_2 2^3 - \log_2 2^2$	Write 8 and 4 as powers of 2.	$= \log_2 2^2$	Write 4 as a power of 2.
$= 3 - 2$	$\log_a a^x = x$	$= 2$	$\log_a a^x = x$
$= 1$	Subtract.		

The statement is false because $1 \neq 2$.

(b) $\log_3 (\log_2 8) = \dfrac{\log_7 49}{\log_8 64}$

$\log_3 (\log_2 8)$	Left side	$\dfrac{\log_7 49}{\log_8 64}$	Right side
$= \log_3 (\log_2 2^3)$	Write 8 as a power of 2.	$= \dfrac{\log_7 7^2}{\log_8 8^2}$	Write 49 and 64 using exponents.
$= \log_3 3$	$\log_a a^x = x$	$= \dfrac{2}{2}$	$\log_a a^x = x$
$= 1$	$3 = 3^1$	$= 1$	Simplify.

The statement is true because $1 = 1$.

NOW TRY

NOW TRY ANSWERS

5. (a) 6.1293 **(b)** −0.5145
   (c) 5.6148
6. (a) false **(b)** false

12.4 Exercises

 MyMathLab®

▶ *Complete solution available in MyMathLab*

Concept Check *Decide whether each statement of a logarithmic property is* true *or* false. *If it is false, correct it by changing the right side of the equation.*

1. $\log_b x + \log_b y = \log_b (x + y)$

2. $\log_b \dfrac{x}{y} = \log_b x - \log_b y$

3. $\log_b b^x = x$

4. $\log_b x^r = \log_b rx$

5. **Concept Check** A student erroneously wrote $\log_a (x + y) = \log_a x + \log_a y$. When his teacher explained that this was indeed wrong, the student claimed that he had used the distributive property. **WHAT WENT WRONG?**

6. **Concept Check** Consider the following "proof" that $\log_2 16$ does not exist.

$$\log_2 16$$
$$= \log_2 (-4)(-4)$$
$$= \log_2 (-4) + \log_2 (-4)$$

The logarithm of a negative number is not defined, so the final step cannot be evaluated. Thus $\log_2 16$ does not exist. **WHAT WENT WRONG?**

Use the indicated rule of logarithms to complete each equation. See Examples 1–3.

7. $\log_{10} (7 \cdot 8) = $ _____ (product rule)

8. $\log_{10} \dfrac{7}{8}$ = _____ (quotient rule)

9. $3^{\log_3 4}$ = _____ (special property)

10. $\log_{10} 3^6$ = _____ (power rule)

11. $\log_3 3^9$ = _____ (special property)

12. $\log_3 9^2$ = _____ (special property)

Use properties of logarithms to express each logarithm as a sum or difference of logarithms, or as a single number if possible. Assume that all variables represent positive real numbers. See Examples 1–4.

▶ 13. $\log_7 (4 \cdot 5)$

14. $\log_8 (9 \cdot 11)$

▶ 15. $\log_5 \dfrac{8}{3}$

16. $\log_3 \dfrac{7}{5}$

▶ 17. $\log_4 6^2$

18. $\log_5 7^4$

▶ 19. $\log_3 \dfrac{\sqrt[3]{4}}{x^2 y}$

20. $\log_7 \dfrac{\sqrt[3]{13}}{pq^2}$

21. $\log_3 \sqrt{\dfrac{xy}{5}}$

22. $\log_6 \sqrt{\dfrac{pq}{7}}$

23. $\log_2 \dfrac{\sqrt[3]{x} \cdot \sqrt[5]{y}}{r^2}$

24. $\log_4 \dfrac{\sqrt[4]{z} \cdot \sqrt[5]{w}}{s^2}$

Use properties of logarithms to write each expression as a single logarithm. Assume that all variables are defined in such a way that the variable expressions are positive, and bases are positive numbers not equal to 1. See Examples 1–4.

25. $\log_b x + \log_b y$

26. $\log_b w + \log_b z$

27. $\log_a m - \log_a n$

28. $\log_b x - \log_b y$

29. $(\log_a r - \log_a s) + 3 \log_a t$

30. $(\log_a p - \log_a q) + 2 \log_a r$

31. $3 \log_a 5 - 4 \log_a 3$

32. $3 \log_a 5 - \dfrac{1}{2} \log_a 9$

33. $\log_{10}(x+3) + \log_{10}(x+5)$ **34.** $\log_{10}(x+4) + \log_{10}(x+6)$

35. $3 \log_p x + \dfrac{1}{2} \log_p y - \dfrac{3}{2} \log_p z - 3 \log_p a$

36. $\dfrac{1}{3} \log_b x + \dfrac{2}{3} \log_b y - \dfrac{3}{4} \log_b s - \dfrac{2}{3} \log_b t$

To four decimal places, the values of $\log_{10} 2$ *and* $\log_{10} 9$ *are*

$$\log_{10} 2 = 0.3010 \quad \text{and} \quad \log_{10} 9 = 0.9542.$$

Use these values and properties of logarithms to evaluate each expression. DO NOT USE A CALCULATOR. See Example 5.

▶ **37.** $\log_{10} 18$ **38.** $\log_{10} 4$ **39.** $\log_{10} \dfrac{2}{9}$ **40.** $\log_{10} \dfrac{9}{2}$

41. $\log_{10} 36$ **42.** $\log_{10} 162$ **43.** $\log_{10} \sqrt[4]{9}$ **44.** $\log_{10} \sqrt[5]{2}$

45. $\log_{10} 3$ **46.** $\log_{10} \dfrac{1}{9}$ **47.** $\log_{10} 9^5$ **48.** $\log_{10} 2^{19}$

Decide whether each statement is true *or false. See Example 6.*

▶ **49.** $\log_2(8+32) = \log_2 8 + \log_2 32$ **50.** $\log_2(64-16) = \log_2 64 - \log_2 16$

51. $\log_3 7 + \log_3 7^{-1} = 0$ **52.** $\log_3 49 + \log_3 49^{-1} = 0$

53. $\log_6 60 - \log_6 10 = 1$ **54.** $\log_3 8 + \log_3 \dfrac{1}{8} = 0$

55. $\dfrac{\log_{10} 7}{\log_{10} 14} = \dfrac{1}{2}$ **56.** $\dfrac{\log_{10} 10}{\log_{10} 100} = \dfrac{1}{10}$

12.5 Common and Natural Logarithms

VOCABULARY
☐ common logarithm
☐ natural logarithm

Logarithms are important in many applications in biology, engineering, economics, and social science. In this section we find numerical approximations for logarithms. Traditionally, base 10 logarithms were used most often because our number system is base 10. Logarithms to base 10 are **common logarithms,** and

$$\log_{10} x \text{ is abbreviated as } \log x,$$

where the base is understood to be 10.

OBJECTIVE 1 Evaluate common logarithms using a calculator.

In **Example 1,** we give the results of evaluating some common logarithms using a calculator with a (LOG) key. Consult your calculator manual to see how to use this key.

EXAMPLE 1 Evaluating Common Logarithms

Using a calculator, evaluate each logarithm to four decimal places.

(a) $\log 327.1$ **(b)** $\log 437{,}000$

(c) $\log 0.0615$ **(d)** $\log 10^{6.1988}$

**NOW TRY
EXERCISE 1**

Using a calculator, evaluate each logarithm to four decimal places.

(a) log 115

(b) log 539,000

(c) log 0.023

(d) log $10^{12.2139}$

FIGURE 17 shows how a graphing calculator displays these common logarithms to four decimal places.

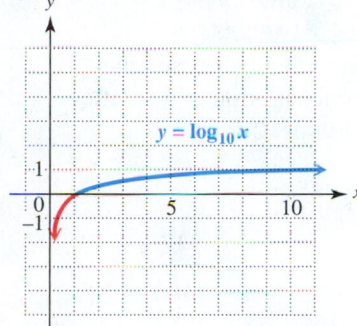

FIGURE 17

NOW TRY

In **Example 1(c),** log 0.0615 $\approx$ −1.2111, which is a negative result. ***The common logarithm of a number between 0 and 1 is always negative*** because the logarithm is the exponent on 10 that produces the number. In this case, we have

$$10^{-1.2111} \approx 0.0615.$$

If the exponent (the logarithm) were positive, the result would be greater than 1 because $10^0 = 1$. The graph in **FIGURE 18** illustrates these concepts.

FIGURE 18

OBJECTIVE 2 Use common logarithms in applications.

In chemistry, pH is a measure of the acidity or alkalinity of a solution. Pure water, for example, has pH 7. In general, acids have pH numbers less than 7, and alkaline solutions have pH values greater than 7, as shown in **FIGURE 19**.

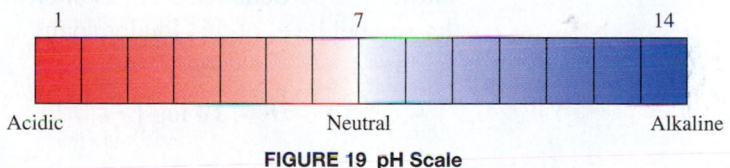

FIGURE 19 pH Scale

The **pH** of a solution is defined as

$$pH = -\log\left[H_3O^+\right],$$

where $\left[H_3O^+\right]$ is the hydronium ion concentration in moles per liter. ***It is customary to round pH values to the nearest tenth.***

EXAMPLE 2 Using pH in an Application

Wetlands are classified as *bogs, fens, marshes,* and *swamps,* on the basis of pH values. A pH value between 6.0 and 7.5, such as that of Summerby Swamp in Michigan's Hiawatha National Forest, indicates that the wetland is a "rich fen." When the pH is between 3.0 and 6.0, the wetland is a "poor fen," and if the pH falls to 3.0 or less, it is a "bog." (*Source:* Mohlenbrock, R., "Summerby Swamp, Michigan," *Natural History.*)

NOW TRY ANSWERS

1. (a) 2.0607   **(b)** 5.7316
   (c) −1.6383   **(d)** 12.2139

**NOW TRY
EXERCISE 2**

Water taken from a wetland
has a hydronium ion
concentration of

$$3.4 \times 10^{-5}.$$

Find the pH value for the water
and classify the wetland as a
rich fen, a poor fen, or a bog.

Suppose that the hydronium ion concentration of a sample of water from a wetland is 6.3×10^{-3}. How would this wetland be classified?

$pH = -\log(6.3 \times 10^{-3})$	$pH = -\log[H_3O^+]$
$pH = -(\log 6.3 + \log 10^{-3})$	Product rule
$pH = -[0.7993 - 3(1)]$	Use a calculator to find log 6.3.
$pH = -0.7993 + 3$	Distributive property
$pH \approx 2.2$	Add.

The pH is less than 3.0, so the wetland is a bog.   **NOW TRY**

**NOW TRY
EXERCISE 3**

Find the hydronium ion
concentration of a solution
with pH 2.6.

EXAMPLE 3   Finding Hydronium Ion Concentration

Find the hydronium ion concentration of drinking water with pH 6.5.

$pH = -\log[H_3O^+]$	
$6.5 = -\log[H_3O^+]$	Let pH = 6.5.
$\log[H_3O^+] = -6.5$	Multiply by -1. Interchange sides.
$[H_3O^+] = 10^{-6.5}$	Write in exponential form, base 10.
$[H_3O^+] \approx 3.2 \times 10^{-7}$	Evaluate with a calculator. **NOW TRY**

The loudness of sound is measured in a unit called a **decibel**, abbreviated **dB**. To measure with this unit, we first assign an intensity of I_0 to a very faint sound, called the **threshold sound**. If a particular sound has intensity I, then the decibel level of this louder sound is

$$D = 10 \log\left(\frac{I}{I_0}\right).$$

Any sound over 85 dB exceeds what hearing experts consider safe. Permanent hearing damage can be suffered at levels above 150 dB.

▼ **Loudness of Common Sounds**

Decibel Level	Example
60	Normal conversation
90	Rush hour traffic, lawn mower
100	Garbage truck, chain saw, pneumatic drill
120	Rock concert, thunderclap
140	Gunshot blast, jet engine
180	Rocket launching pad

Source: Deafness Research Foundation.

**NOW TRY
EXERCISE 4**

Find the decibel level to the
nearest whole number of the
sound from a jet engine with
intensity I of

$$6.312 \times 10^{13}I_0.$$

EXAMPLE 4   Measuring the Loudness of Sound

If music delivered through Bluetooth headphones has intensity I of $3.162 \times 10^{11}I_0$, find the average decibel level. (*Source:* CNET.)

$$D = 10 \log\left(\frac{I}{I_0}\right)$$

$$D = 10 \log\left(\frac{3.162 \times 10^{11}I_0}{I_0}\right)$$ Substitute the given value for *I*.

$$D = 10 \log(3.162 \times 10^{11})$$

$$D \approx 115$$ Evaluate with a calculator. Round to the nearest unit.   **NOW TRY**

NOW TRY ANSWERS
2. 4.5; poor fen
3. 2.5×10^{-3}
4. 138 dB

Leonhard Euler (1707–1783)

The number *e* is named after Euler.

OBJECTIVE 3 Evaluate natural logarithms using a calculator.

Logarithms used in applications are often **natural logarithms,** which have as base the number *e*. The letter *e* was chosen to honor Leonhard Euler, who published extensive results on the number in 1748. It is an irrational number, so its decimal expansion never terminates and never repeats.

One way to see how *e* appears in an exponential situation involves calculating the values of

$$\left(1 + \frac{1}{x}\right)^x \quad \text{as } x \text{ gets larger without bound.}$$

The table shows these values for $x = 1, 10, 100, 1000, 10{,}000,$ and $100{,}000$.

x	$\left(1 + \dfrac{1}{x}\right)^x$
1	2
10	2.59374246
100	2.704813829
1000	2.716923932
10,000	2.718145927
100,000	2.718268237

These approximations are found using a calculator.

It appears that as x gets larger without bound, $\left(1 + \frac{1}{x}\right)^x$ approaches some number. This number is *e*.

> **Approximation for *e***
>
> $$e \approx 2.718281828$$

A scientific or graphing calculator with an $\boxed{e^x}$ key can approximate powers of *e*. See **FIGURE 20**.

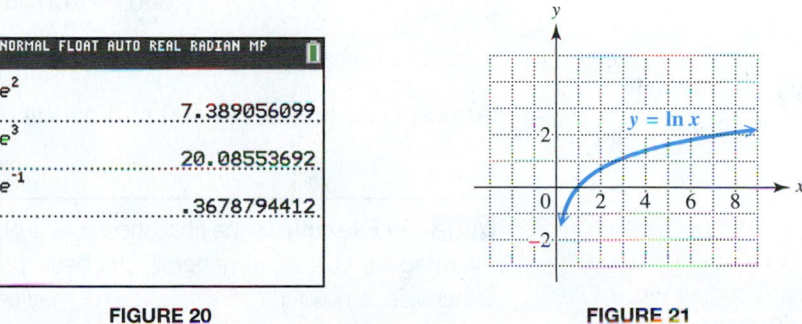

FIGURE 20

FIGURE 21

Logarithms with base *e* are called natural logarithms because they occur in natural situations that involve growth or decay.

The base *e* logarithm of *x* is written **ln *x*** (read "el en *x*").

The graph of $y = \ln x$ is given in **FIGURE 21**.

A calculator key labeled $\boxed{\text{LN}}$ is used to evaluate natural logarithms. Consult your calculator manual to see how to use this key.

NOW TRY
EXERCISE 5

Using a calculator, evaluate each logarithm to four decimal places.

(a) ln 0.26   **(b)** ln 12

(c) ln 150   **(d)** ln $e^{5.8321}$

EXAMPLE 5   Evaluating Natural Logarithms

Using a calculator, evaluate each logarithm to four decimal places.

(a) ln 0.5841 ≈ −0.5377   **(b)** ln 192.7 ≈ 5.2611

(c) ln 10.84 ≈ 2.3832   **(d)** ln $e^{4.6832}$ ≈ 4.6832

FIGURE 22 shows how a graphing calculator displays these natural logarithms to four decimal places. As with common logarithms, *a number between 0 and 1 has a negative natural logarithm.* See part (a), where ln 0.5841 is negative.

```
NORMAL FIX4 AUTO REAL RADIAN MP        🔋
ln(0.5841)
                              -.5377
ln(192.7)
                              5.2611
ln(10.84)
                              2.3832
ln(e^4.6832)
                              4.6832
```

FIGURE 22   NOW TRY

OBJECTIVE 4   Use natural logarithms in applications.

NOW TRY
EXERCISE 6

Use the logarithmic function in **Example 6** to approximate the altitude when atmospheric pressure is 600 millibars. Round to the nearest hundred.

EXAMPLE 6   Applying a Natural Logarithmic Function

The altitude in meters that corresponds to an atmospheric pressure of x millibars is given by the logarithmic function

$$f(x) = 51{,}600 - 7457 \ln x.$$

(*Source:* Miller, A. and J. Thompson, *Elements of Meteorology,* Fourth Edition, Charles E. Merrill Publishing Company.) Use this function to find the altitude when atmospheric pressure is 400 millibars. Round to the nearest hundred.

Let $x = 400$ and substitute in the expression for $f(x)$.

$$f(x) = 51{,}600 - 7457 \ln x$$

$$f(400) = 51{,}600 - 7457 \ln 400 \qquad \text{Let } x = 400.$$

$$f(400) \approx 6900 \qquad\qquad\qquad \text{Evaluate with a calculator.}$$

Atmospheric pressure is 400 millibars at 6900 m.   NOW TRY

NOTE   In **Example 6,** the final answer was obtained using a calculator *without* rounding the intermediate values. In general, it is best to wait until the final step to round the answer. Otherwise, a buildup of round-off error may cause the final answer to have an incorrect final decimal place digit or digits.

OBJECTIVE 5   Use the change-of-base rule.

NOW TRY ANSWERS

5. (a) −1.3471   **(b)** 2.4849
    (c) 5.0106   **(d)** 5.8321
6. 3900 m

In **Example 2** of **Section 12.3** we illustrated how the TI-84 Plus calculator evaluates logarithms for any base. If a calculator does not have this function, we calculate such logarithms using the change-of-base rule, which allows us to convert logarithms from one base to another.

Change-of-Base Rule

If $a > 0$, $a \neq 1$, $b > 0$, $b \neq 1$, and $x > 0$, then the following holds true.

$$\log_a x = \frac{\log_b x}{\log_b a}$$

NOTE Any positive number other than 1 can be used for base b in the change-of-base rule. Usually the only practical bases are e and 10 because calculators generally have dedicated keys for these two bases.

To derive the change-of-base rule, let $\log_a x = m$.

$$\log_a x = m$$

$a^m = x$	Convert to exponential form.
$\log_b a^m = \log_b x$	Take the logarithm on each side.
$m \log_b a = \log_b x$	Power rule
$(\log_a x)(\log_b a) = \log_b x$	Substitute for m.
$\log_a x = \dfrac{\log_b x}{\log_b a}$	Divide by $\log_b a$.

This last statement is the change-of-base rule.

NOW TRY EXERCISE 7

Use a calculator and the change-of-base rule to approximate each logarithm to four decimal places.

(a) $\log_2 7$ **(b)** $\log_5 8$

(c) $\log_{1/3} 12$

EXAMPLE 7 Using the Change-of-Base Rule

Use a calculator and the change-of-base rule to approximate each logarithm to four decimal places.

(a) $\log_2 5$ **(b)** $\log_3 12$ **(c)** $\log_{1/2} 12$

See **FIGURE 23**. To four decimal places, the logarithms in parts (a)–(c) are as follows.

$$\log_2 5 \approx 2.3219, \quad \log_3 12 \approx 2.2619, \quad \log_{1/2} 12 \approx -3.5850$$

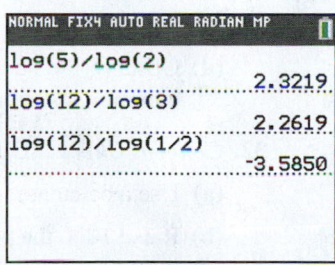

FIGURE 23

Compare these results to **Example 2(a)–(c)** of **Section 12.3.**

NOW TRY

NOTE Either common or natural logarithms can be used when applying the change-of-base rule. Verify that the same results are found in **Example 7** for natural logarithms.

12.5 Exercises

   FOR EXTRA HELP   ▶   MyMathLab®

▶ *Complete solution available in MyMathLab*

Concept Check   *Choose the correct response.*

1. What is the base in the expression log x?

    A. x **B.** 1   **C.** 10   **D.** e

2. What is the base in the expression ln x?

    A. e **B.** 1   **C.** 10   **D.** x

3. Given that $10^0 = 1$ and $10^1 = 10$, between what two consecutive integers is the value of log 6.3?

    A. 6 and 7   **B.** 10 and 11   **C.** 0 and 1   **D.** -1 and 0

4. Given that $e^1 \approx 2.718$ and $e^2 \approx 7.389$, between what two consecutive integers is the value of ln 6.3?

    A. 6 and 7   **B.** 1 and 2   **C.** 2 and 3   **D.** 0 and 1

5. *Concept Check*   Without using a calculator, give the value of log 10^2.

6. *Concept Check*   Without using a calculator, give the value of ln e^2.

You will need a calculator for most of the remaining exercises in this set.

Evaluate each logarithm to four decimal places when appropriate. See Examples 1 and 5.

▶ **7.** log 43   **8.** log 98   **9.** log 328.4

10. log 457.2   **11.** log 0.0326   **12.** log 0.1741

13. log $10^{9.6421}$ **14.** log $10^{3.1112}$ **15.** log (4.76×10^9)

16. log (2.13×10^4) ▶ **17.** ln 7.84   **18.** ln 8.32

19. ln 0.0556   **20.** ln 0.0217   **21.** ln 388.1

22. ln 942.6   **23.** ln $e^{-11.4007}$ **24.** ln $e^{-1.4724}$

25. ln $(8.59 \times e^2)$ **26.** ln $(7.46 \times e^3)$ **27.** ln 10

28. log e **29.** 10 ln e^4 **30.** 15 ln e^3

31. *Concept Check*   Use a calculator to find approximations of each logarithm.

    (a) log 356.8   **(b)** log 35.68   **(c)** log 3.568

    (d) Observe the answers and make a conjecture concerning the decimal values of the common logarithms of numbers greater than 1 that have the same digits.

32. *Concept Check*   Let k represent the number of letters in your last name.

    (a) Use a calculator to find log k.

    (b) Raise 10 to the power indicated by the number found in part (a). What is the result?

    (c) Use the concepts of **Section 12.1** and explain why we obtained the answer found in part (b). Would it matter what number we used for k to observe the same result?

Suppose that water from a wetland area is sampled and found to have the given hydronium ion concentration. Is the wetland a rich fen, a poor fen, or a bog? See Example 2.

33. 3.1×10^{-5} **34.** 2.5×10^{-5} ▶ **35.** 2.5×10^{-2}

36. 3.6×10^{-2} **37.** 2.7×10^{-7} **38.** 2.5×10^{-7}

Find the pH of the substance with the given hydronium ion concentration. ***See Example 2.***

39. Ammonia, 2.5×10^{-12}

40. Sodium bicarbonate, 4.0×10^{-9}

41. Grapes, 5.0×10^{-5}

42. Tuna, 1.3×10^{-6}

Find the hydronium ion concentration of the substance with the given pH. ***See Example 3.***

43. Human blood plasma, 7.4

44. Human gastric contents, 2.0

45. Spinach, 5.4

46. Bananas, 4.6

Solve each problem. ***See Examples 4 and 6.***

47. Managements of sports stadiums and arenas often encourage fans to make as much noise as possible. Find the average decibel level

$$D = 10 \log\left(\frac{I}{I_0}\right)$$

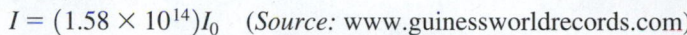

for each venue with the given intensity I.

(a) NFL fans, Kansas City Chiefs at Arrowhead Stadium:

$I = (1.58 \times 10^{14})I_0$ (*Source:* www.guinessworldrecords.com)

(b) NBA fans, Sacramento Kings at Sleep Train Arena:

$I = (3.9 \times 10^{12})I_0$ (*Source:* www.guinessworldrecords.com)

(c) MLB fans, Baltimore Orioles at Camden Yards:

$I = (1.1 \times 10^{12})I_0$ (*Source:* www.baltimoresportsreport.com)

48. The time t in years for an amount increasing at a rate of r (in decimal form) to double is given by

$$t(r) = \frac{\ln 2}{\ln(1 + r)}.$$

This is the **doubling time.** Find the doubling time to the nearest tenth for an investment at each interest rate.

(a) 2% (or 0.02)   **(b)** 5% (or 0.05)   **(c)** 8% (or 0.08)

49. The number of years, $N(x)$, since two independently evolving languages split off from a common ancestral language is approximated by

$$N(x) = -5000 \ln x,$$

where x is the percent of words (in decimal form) from the ancestral language common to both languages now. Find the number of years (to the nearest hundred years) since the split for each percent of common words.

(a) 85% (or 0.85)   **(b)** 35% (or 0.35)   **(c)** 10% (or 0.10)

50. The concentration of a drug injected into the bloodstream decreases with time. The intervals of time T when the drug should be administered are given by

$$T = \frac{1}{k} \ln \frac{C_2}{C_1},$$

where k is a constant determined by the drug in use, C_2 is the concentration at which the drug is harmful, and C_1 is the concentration below which the drug is ineffective. (*Source:* Horelick, B. and S. Koont, "Applications of Calculus to Medicine: Prescribing Safe and Effective Dosage," *UMAP Module 202.*) Thus, if $T = 4$, the drug should be administered every 4 hr. For a certain drug, $k = \frac{1}{3}$, $C_2 = 5$, and $C_1 = 2$. How often should the drug be administered? (*Hint:* Round down.)

51. The growth of outpatient surgeries at hospitals is approximated by

$$f(x) = 10.6049 + 2.3556 \ln x,$$

where x is the number of years since 1990, and $f(x)$ is in millions. (*Source:* American Hospital Association.)

(a) What does this model give for the number of outpatient surgeries in 2011?

(b) According to this model, when did outpatient surgeries reach 17,000,000? (*Hint:* Substitute for $f(x)$, and then write the equation in exponential form to solve it.)

52. In the central Sierra Nevada of California, the percent of moisture that falls as snow rather than rain is approximated by

$$f(x) = 86.3 \ln x - 680,$$

where x is the altitude in feet.

(a) What percent of the moisture at 5000 ft falls as snow?

(b) What percent at 7500 ft falls as snow?

53. The **cost-benefit equation**

$$T(x) = -0.642 - 189 \ln (1 - x)$$

describes the approximate tax $T(x)$, in dollars per ton, that would result in an $x\%$ (in decimal form) reduction in carbon dioxide emissions.

(a) What tax will reduce emissions 25%?

(b) Explain why the equation is not valid for $x = 0$ or $x = 1$.

54. The age in years of a female blue whale of length x in feet is approximated by

$$f(x) = -2.57 \ln \left(\frac{87 - x}{63} \right).$$

(a) How old is a female blue whale that measures 80 ft?

(b) The equation that defines this function has domain $24 < x < 87$. Explain why.

Use the change-of-base rule (with either common or natural logarithms) to approximate each logarithm to four decimal places. **See Example 7.**

▶ **55.** $\log_3 12$ **56.** $\log_4 18$ **57.** $\log_5 3$

58. $\log_7 4$ **59.** $\log_3 \sqrt{2}$ **60.** $\log_6 \sqrt[3]{5}$

61. $\log_\pi e$ **62.** $\log_\pi 10$ **63.** $\log_e 12$

64. $\log_e 15$ **65.** $\log_{12} 3$ **66.** $\log_{18} 4$

67. *Concept Check*   Multiply the results in **Exercises 55 and 65.** Then do the same for **Exercises 56 and 66.** Make a conjecture about the relationship between

$$\log_a b \quad \text{and} \quad \log_b a.$$

68. *Concept Check*   Apply the power rule to the expression in **Exercise 59.** Write the result, and then verify that it is equal to $\log_3 \sqrt{2}$.

12.6 Exponential and Logarithmic Equations; Further Applications

OBJECTIVES

1. Solve equations involving variables in the exponents.
2. Solve equations involving logarithms.
3. Solve applications of compound interest.
4. Solve applications involving base e exponential growth and decay.

VOCABULARY

☐ compound interest
☐ continuous compounding

 **NOW TRY EXERCISE 1**

Solve the equation. Approximate the solution to three decimal places.

$$5^x = 20$$

We solved exponential and logarithmic equations in **Sections 12.2 and 12.3.** General methods for solving these equations depend on the following properties.

> **Properties for Solving Exponential and Logarithmic Equations**
>
> For all real numbers $b > 0$, $b \neq 1$, and any real numbers x and y, the following hold true.
>
> 1. If $x = y$, then $b^x = b^y$.
> 2. If $b^x = b^y$, then $x = y$. (We used this property in **Section 12.2.**)
> 3. If $x = y$, and $x > 0$, $y > 0$, then $\log_b x = \log_b y$.
> 4. If $x > 0$, $y > 0$, and $\log_b x = \log_b y$, then $x = y$.

OBJECTIVE 1 Solve equations involving variables in the exponents.

In **Examples 1 and 2,** we use Property 3.

EXAMPLE 1 Solving an Exponential Equation

Solve $3^x = 12$. Approximate the solution to three decimal places.

$$3^x = 12$$

$$\log 3^x = \log 12 \qquad \text{Property 3 (common logs)}$$

$$x \log 3 = \log 12 \qquad \text{Power rule}$$

$$\text{Exact solution} \longrightarrow x = \frac{\log 12}{\log 3} \qquad \text{Divide by log 3.}$$

$$\text{Decimal approximation} \longrightarrow x \approx 2.262 \qquad \text{Use a calculator.}$$

CHECK $3^x = 3^{2.262} \approx 12$ ✓   True
Evaluate with a calculator.

The solution set is $\{2.262\}$.   **NOW TRY**

> ⚠ **CAUTION** Be careful: $\dfrac{\log 12}{\log 3}$ is **not** equal to log 4. Check to see that
>
> $$\log 4 \approx 0.6021, \quad \text{but} \quad \frac{\log 12}{\log 3} \approx 2.262.$$

> **NOTE** In **Example 1,** we used Property 3 with common logarithms. We could just as easily have used natural logarithms. Verify that
>
> $$\frac{\ln 12}{\ln 3} = \frac{\log 12}{\log 3} \approx 2.262.$$

NOW TRY ANSWER
1. $\{1.861\}$

When an exponential equation has e as the base, as in the next example, it is easiest to use base e (natural) logarithms.

NOW TRY
EXERCISE 2

Solve $e^{0.12x} = 10$. Approximate the solution to three decimal places.

EXAMPLE 2   Solving an Exponential Equation with Base e

Solve $e^{0.003x} = 40$. Approximate the solution to three decimal places.

$$\ln e^{0.003x} = \ln 40 \qquad \text{Property 3 (natural logs)}$$

$$0.003x \ln e = \ln 40 \qquad \text{Power rule}$$

$$0.003x = \ln 40 \qquad \ln e = \ln e^1 = 1$$

$$x = \frac{\ln 40}{0.003} \qquad \text{Divide by 0.003.}$$

$$x \approx 1229.626 \qquad \text{Evaluate with a calculator.}$$

Check that $e^{0.003(1229.626)} \approx 40$. The solution set is $\{1229.626\}$.   **NOW TRY**

NOW TRY
EXERCISE 3

Solve $3^{-x+5} = 8$. Approximate the solution to three decimal places.

EXAMPLE 3   Solving an Exponential Equation

Solve $7^{-x+4} = 17$. Approximate the solution to three decimal places.

$$7^{-x+4} = 17$$

$$\log 7^{-x+4} = \log 17 \qquad \text{Property 3 (common logs)}$$

$$(-x+4)\log 7 = \log 17 \qquad \text{Power rule}$$

$$-x \log 7 + 4 \log 7 = \log 17 \qquad \text{Distributive property}$$

$$-x \log 7 = \log 17 - 4 \log 7 \qquad \text{Subtract 4 log 7.}$$

$$x = \frac{\log 17 - 4 \log 7}{-\log 7} \qquad \text{Divide by } -\log 7.$$

$$x \approx 2.544 \qquad \text{Evaluate with a calculator.}$$

Check that $7^{-2.544+4} \approx 17$. The solution set is $\{2.544\}$.   **NOW TRY**

General Method for Solving an Exponential Equation

Take logarithms having the same base on both sides and then use the power rule of logarithms or the special property $\log_b b^x = x$. (See **Examples 1–3.**)

As a special case, if both sides can be written as exponentials with the same base, do so, and set the exponents equal. (See **Section 12.2.**)

OBJECTIVE 2   Solve equations involving logarithms.

NOW TRY
EXERCISE 4

Solve $\log_6 (2x + 4) = 2$.

EXAMPLE 4   Solving a Logarithmic Equation

Solve $\log_3 (4x + 1) = 4$.

$$\log_3 (4x + 1) = 4$$

$$4x + 1 = 3^4 \qquad \text{Convert to exponential form.}$$

$$4x + 1 = 81 \qquad 3^4 = 81$$

$$4x = 80 \qquad \text{Subtract 1.}$$

$$x = 20 \qquad \text{Divide by 4.}$$

NOW TRY ANSWERS
2. $\{19.188\}$ **3.** $\{3.107\}$
4. $\{16\}$

Check to see that $\log_3 [4(20) + 1] = \log_3 81 = 4$ is true. The solution set is $\{20\}$.

NOW TRY

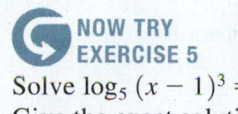

NOW TRY EXERCISE 5

Solve $\log_5 (x - 1)^3 = 2$. Give the exact solution.

EXAMPLE 5 Solving a Logarithmic Equation

Solve $\log_2 (x + 5)^3 = 4$. Give the exact solution.

$$\log_2 (x + 5)^3 = 4$$

$(x + 5)^3 = 2^4$	Convert to exponential form.
$(x + 5)^3 = 16$	$2^4 = 16$
$x + 5 = \sqrt[3]{16}$	Take the cube root on each side.
$x = -5 + \sqrt[3]{16}$	Add -5.
$x = -5 + 2\sqrt[3]{2}$	$\sqrt[3]{16} = \sqrt[3]{8 \cdot 2} = \sqrt[3]{8} \cdot \sqrt[3]{2} = 2\sqrt[3]{2}$

CHECK

$\log_2 (x + 5)^3 = 4$	Original equation
$\log_2 \left(-5 + 2\sqrt[3]{2} + 5\right)^3 \overset{?}{=} 4$	Let $x = -5 + 2\sqrt[3]{2}$.
$\log_2 \left(2\sqrt[3]{2}\right)^3 \overset{?}{=} 4$	Work inside the parentheses.
$\log_2 16 \overset{?}{=} 4$	$\left(2\sqrt[3]{2}\right)^3 = 2^3(\sqrt[3]{2})^3 = 8 \cdot 2 = 16$
$2^4 \overset{?}{=} 16$	Write in exponential form.
$16 = 16$ ✓	True

A true statement results, so the solution set is $\left\{-5 + 2\sqrt[3]{2}\right\}$.   **NOW TRY**

⚠ **CAUTION**   Recall that the domain of $f(x) = \log_b x$ is $(0, \infty)$. *For this reason, always check that each proposed solution of an equation with logarithms yields only logarithms of positive numbers in the original equation.*

EXAMPLE 6 Solving a Logarithmic Equation

Solve $\log_2 (x + 1) - \log_2 x = \log_2 7$.

$$\log_2 (x + 1) - \log_2 x = \log_2 7$$

Transform the left side to an expression with only one logarithm.

$\log_2 \dfrac{x + 1}{x} = \log_2 7$	Quotient rule
$\dfrac{x + 1}{x} = 7$	Property 4
$x + 1 = 7x$	Multiply by x.
$1 = 6x$	Subtract x.
$\dfrac{1}{6} = x$	Divide by 6.

This proposed solution must be checked.

We cannot take the logarithm of a *nonpositive* number, so both $x + 1$ and x must be positive here. If $x = \frac{1}{6}$, then this condition is satisfied.

NOW TRY ANSWER

5. $\left\{1 + \sqrt[3]{25}\right\}$

NOW TRY
EXERCISE 6

Solve.

$\log_4 (2x + 13) - \log_4 (x + 1)$
$= \log_4 10$

CHECK

$\log_2 (x + 1) - \log_2 x = \log_2 7$	Original equation
$\log_2 \left(\dfrac{1}{6} + 1 \right) - \log_2 \dfrac{1}{6} \stackrel{?}{=} \log_2 7$	Let $x = \frac{1}{6}$.
$\log_2 \dfrac{7}{6} - \log_2 \dfrac{1}{6} \stackrel{?}{=} \log_2 7$	Add.
$\log_2 \dfrac{\frac{7}{6}}{\frac{1}{6}} \stackrel{?}{=} \log_2 7$	Quotient rule
$\log_2 7 = \log_2 7$ ✓	True

$\dfrac{\frac{7}{6}}{\frac{1}{6}} = \dfrac{7}{6} \div \dfrac{1}{6} = \dfrac{7}{6} \cdot \dfrac{6}{1} = 7$

A true statement results, so the solution set is $\left\{ \dfrac{1}{6} \right\}$.

NOW TRY

NOW TRY
EXERCISE 7

Solve.

$\log_4 (x + 2) + \log_4 2x = 2$

EXAMPLE 7 Solving a Logarithmic Equation

Solve $\log x + \log (x - 21) = 2$.

$\log x + \log (x - 21) = 2$	
$\log x(x - 21) = 2$	Product rule
$x(x - 21) = 10^2$	Write in exponential form.
$x^2 - 21x = 100$	Distributive property; Multiply.
$x^2 - 21x - 100 = 0$	Standard form
$(x - 25)(x + 4) = 0$	Factor.
$x - 25 = 0 \quad \text{or} \quad x + 4 = 0$	Zero-factor property
$x = 25 \quad \text{or} \qquad x = -4$	Proposed solutions

(The base is 10.)

The value -4 must be rejected as a solution because it leads to the logarithm of a negative number in the original equation.

$$\log (-4) + \log (-4 - 21) = 2 \qquad \text{The left side is undefined.}$$

Check that the only solution is 25, so the solution set is $\{25\}$.

NOW TRY

> **⚠ CAUTION** *Do not reject a potential solution just because it is nonpositive. Reject any value that leads to the logarithm of a nonpositive number.*

Solving a Logarithmic Equation

Step 1 **Transform the equation so that a single logarithm appears on one side.** Use the product rule or quotient rule of logarithms to do this.

Step 2 **Do one of the following.**

   (a) Use Property 4.

   If $\log_b x = \log_b y$, then $x = y$.   (See **Example 6.**)

   (b) Write the equation in exponential form.

   If $\log_b x = k$, then $x = b^k$.   (See **Examples 5 and 7.**)

NOW TRY ANSWERS

6. $\left\{ \dfrac{3}{8} \right\}$ **7.** $\{2\}$

OBJECTIVE 3 Solve applications of compound interest.

We have solved simple interest problems using the formula $I = prt$. In most cases, interest paid or charged is **compound interest** (interest paid on both principal and interest). The formula for compound interest is an application of exponential functions. *In this book, monetary amounts are given to the nearest cent.*

Compound Interest Formula (for a Finite Number of Periods)

If a principal of P dollars is deposited at an annual rate of interest r compounded (paid) n times per year, then the account will contain

$$A = P\left(1 + \frac{r}{n}\right)^{nt}$$

dollars after t years. (In this formula, r is expressed as a decimal.)

**NOW TRY
EXERCISE 8**

How much money will there be in an account at the end of 10 yr if $10,000 is deposited at 2.5% compounded monthly?

EXAMPLE 8 Solving a Compound Interest Problem for *A*

How much money will there be in an account at the end of 5 yr if $1000 is deposited at 3% compounded quarterly? (Assume no withdrawals are made.)

Because interest is compounded quarterly, $n = 4$.

$$A = P\left(1 + \frac{r}{n}\right)^{nt} \qquad \text{Compound interest formula}$$

$$A = 1000\left(1 + \frac{0.03}{4}\right)^{4 \cdot 5} \qquad \begin{array}{l}\text{Substitute } P = 1000, r = 0.03 \text{ (because} \\ 3\% = 0.03), n = 4, \text{ and } t = 5.\end{array}$$

$$A = 1000(1.0075)^{20} \qquad \text{Simplify.}$$

$$A = 1161.18 \qquad \text{Evaluate with a calculator.}$$

The account will contain $1161.18. **NOW TRY**

**NOW TRY
EXERCISE 9**

Find the number of years, to the nearest hundredth, it will take for money deposited in an account paying 4% interest compounded quarterly to double.

EXAMPLE 9 Solving a Compound Interest Problem for *t*

Suppose inflation is averaging 3% per year. To the nearest hundredth of a year, how long will it take for prices to double? (This is the **doubling time** of the money.)

We want the number of years t for P dollars to grow to $2P$ dollars at 3% per year.

$$A = P\left(1 + \frac{r}{n}\right)^{nt} \qquad \text{Compound interest formula}$$

$$2P = P\left(1 + \frac{0.03}{1}\right)^{1t} \qquad \text{Let } A = 2P, r = 0.03, \text{ and } n = 1.$$

$$2 = (1.03)^t \qquad \text{Divide by } P. \text{ Simplify.}$$

$$\log 2 = \log (1.03)^t \qquad \text{Property 3}$$

$$\log 2 = t \log (1.03) \qquad \text{Power rule}$$

$$t = \frac{\log 2}{\log 1.03} \qquad \text{Interchange sides. Divide by log 1.03.}$$

$$t \approx 23.45 \qquad \text{Evaluate with a calculator.}$$

Prices will double in 23.45 yr. To check, verify that $1.03^{23.45} \approx 2$. **NOW TRY**

NOW TRY ANSWERS
8. $12,836.92
9. 17.42 yr

Interest can be compounded over various time periods per year, including

annually,   semiannually,   quarterly,   daily,   and so on.

The number of compounding periods can get larger and larger. If the value of n increases without bound, we have an example of **continuous compounding.** The formula for continuous compounding is derived in advanced courses, and is an example of exponential growth involving the number e.

Continuous Compound Interest Formula

If a principal of P dollars is deposited at an annual rate of interest r compounded continuously for t years, the final amount A on deposit is given by

$$A = Pe^{rt}.$$

**NOW TRY
EXERCISE 10**

Suppose that $4000 is inverse at 3% interest for 2 yr.

(a) How much will the investment grow to if it is compounded continuously?

(b) How long would it take for the original investment to double? Round to the nearest tenth.

EXAMPLE 10   Solving a Continuous Compound Interest Problem

In **Example 8** we found that $1000 invested for 5 yr at 3% interest compounded quarterly would grow to $1161.18.

(a) How much would this investment grow to if it is compounded continuously?

$$A = Pe^{rt} \qquad \text{Continuous compounding formula}$$

$$A = 1000e^{0.03(5)} \qquad \text{Let } P = 1000, r = 0.03, \text{ and } t = 5.$$

$$A = 1000e^{0.15} \qquad \text{Multiply in the exponent.}$$

$$A = 1161.83 \qquad \text{Evaluate with a calculator.}$$

The investment will grow to $1161.83 (which is $0.65 more than the amount in **Example 8** when interest was compounded quarterly).

(b) How long would it take for the initial investment amount to triple? Round to the nearest tenth.

We must find the value of t that will cause A to be $3(\$1000) = \$3000$.

$$A = Pe^{rt} \qquad \text{Continuous compounding formula}$$

$$3000 = 1000e^{0.03t} \qquad \text{Let } A = 3P = 3000, P = 1000, \text{ and } r = 0.03.$$

$$3 = e^{0.03t} \qquad \text{Divide by 1000.}$$

$$\ln 3 = \ln e^{0.03t} \qquad \text{Take natural logarithms.}$$

$$\ln 3 = 0.03t \qquad \text{ln } e^k = k$$

$$t = \frac{\ln 3}{0.03} \qquad \text{Divide by 0.03. Interchange sides.}$$

$$t \approx 36.6 \qquad \text{Evaluate with a calculator.}$$

It would take 36.6 yr for the original investment to triple.   NOW TRY

NOW TRY ANSWERS
10. (a) $4247.35
   (b) 23.1 yr

OBJECTIVE 4 Solve applications involving base e exponential growth and decay.

When situations involve growth or decay of a population, the amount or number of some quantity present at time t can be approximated by

$$f(t) = y_0 e^{kt}.$$

In this equation, y_0 is the amount or number present at time $t = 0$ and k is a constant. The continuous compounding of money is an example of exponential growth. In **Example 11,** we investigate exponential decay.

NOW TRY EXERCISE 11

Radium 226 decays according to the function

$$f(t) = y_0 e^{-0.00043t},$$

where t is time in years.

(a) If an initial sample contains $y_0 = 4.5$ g of radium 226, how many grams, to the nearest tenth, will be present after 150 yr?

(b) What is the half-life of radium 226? Round to the nearest year.

EXAMPLE 11 Solving an Application Involving Exponential Decay

Carbon 14 is a radioactive form of carbon that is found in all living plants and animals. After death, radioactive carbon 14 disintegrates according to the function

$$f(t) = y_0 e^{-0.000121t},$$

where t is time in years, $f(t)$ is the amount of the sample at time t, and y_0 is the initial amount present at $t = 0$.

(a) If an initial sample contains $y_0 = 10$ g of carbon 14, how many grams, to the nearest hundredth, will be present after 3000 yr?

Let $y_0 = 10$ and $t = 3000$ in the formula, and evaluate with a calculator.

$$f(3000) = 10e^{-0.000121(3000)} \approx 6.96 \text{ g}$$

(b) How long would it take to the nearest year for the initial sample to decay to half of its original amount? (This is the **half-life.**)

$f(t) = y_0 e^{-0.000121t}$	Exponential decay formula
$5 = 10e^{-0.000121t}$	Let $y_0 = 10$ and $f(t) = \frac{1}{2}(10) = 5$.
$\frac{1}{2} = e^{-0.000121t}$	Divide by 10.
$\ln \frac{1}{2} = -0.000121t$	Take natural logarithms; $\ln e^k = k$.
$t = \dfrac{\ln \frac{1}{2}}{-0.000121}$	Divide by -0.000121. Interchange sides.
$t \approx 5728$	Evaluate with a calculator.

The half-life is 5728 yr.

NOW TRY ANSWERS
11. (a) 4.2 g   (b) 1612 yr

NOW TRY

12.6 Exercises

FOR EXTRA HELP

 MyMathLab®

 Complete solution available in MyMathLab

Many of the problems in these exercises require a calculator with logarithm capability.

Concept Check   Tell whether common logarithms or natural logarithms would be a better choice to use for solving each equation. Do not actually solve.

1. $10^{0.0025x} = 75$

2. $10^{3x+1} = 13$

3. $e^{x-2} = 24$

4. $e^{-0.28x} = 30$

Solve each equation. Approximate solutions to three decimal places. **See Examples 1 and 3.**

▶ **5.** $7^x = 5$

6. $4^x = 3$

7. $9^{-x+2} = 13$

8. $6^{-x+1} = 22$

9. $3^{2x} = 14$

10. $5^{3x} = 11$

11. $2^{x+3} = 5^x$

12. $6^{x+3} = 4^x$

13. $2^{x+3} = 3^{x-4}$

14. $4^{x-2} = 5^{3x+2}$

15. $4^{2x+3} = 6^{x-1}$

16. $3^{2x+1} = 5^{x-1}$

Solve each equation. Use natural logarithms. When appropriate, approximate solutions to three decimal places. **See Example 2.**

▶ **17.** $e^{0.012x} = 23$

18. $e^{0.006x} = 30$

19. $e^{-0.205x} = 9$

20. $e^{-0.103x} = 7$

21. $\ln e^{3x} = 9$

22. $\ln e^{2x} = 4$

23. $\ln e^{0.45x} = \sqrt{7}$

24. $\ln e^{0.04x} = \sqrt{3}$

25. $\ln e^{-x} = \pi$

26. $\ln e^{2x} = \pi$

27. $e^{\ln 2x} = e^{\ln(x+1)}$

28. $e^{\ln(6-x)} = e^{\ln(4+2x)}$

Solve each equation. Give exact solutions. **See Examples 4 and 5.**

29. $\log_4(2x + 8) = 2$

30. $\log_5(5x + 10) = 3$

31. $\log_3(6x + 5) = 2$

32. $\log_5(12x - 8) = 3$

33. $\log_2(2x - 1) = 5$

34. $\log_6(4x + 2) = 2$

▶ **35.** $\log_7(x + 1)^3 = 2$

36. $\log_4(x - 3)^3 = 4$

37. $\log_2(x^2 + 7) = 4$

38. $\log_6(x^2 + 11) = 2$

39. *Concept Check*   Suppose that in solving a logarithmic equation having the term $\log(x - 3)$, we obtain the proposed solution 2. We know that our algebraic work is correct, so we give $\{2\}$ as the solution set. *WHAT WENT WRONG?*

40. *Concept Check*   Suppose that in solving a logarithmic equation having the term $\log(3 - x)$, we obtain the proposed solution -4. We know that our algebraic work is correct, so we reject -4 and give $\varnothing$ as the solution set. *WHAT WENT WRONG?*

Solve each equation. Give exact solutions. **See Examples 6 and 7.**

41. $\log(6x + 1) = \log 3$

42. $\log(7 - 2x) = \log 4$

▶ **43.** $\log_5(3t + 2) - \log_5 t = \log_5 4$

44. $\log_2(t + 5) - \log_2(t - 1) = \log_2 3$

45. $\log 4x - \log(x - 3) = \log 2$

46. $\log(-x) + \log 3 = \log(2x - 15)$

▶ **47.** $\log_2 x + \log_2(x - 7) = 3$

48. $\log(2x - 1) + \log 10x = \log 10$

49. $\log 5x - \log(2x - 1) = \log 4$

50. $\log_3 x + \log_3(2x + 5) = 1$

51. $\log_2 x + \log_2(x - 6) = 4$

52. $\log_2 x + \log_2(x + 4) = 5$

Solve each problem. **See Examples 8–10.**

▶ **53.** How much money will there be in an account at the end of 6 yr if $2000 is deposited at 4% compounded quarterly? (Assume no withdrawals are made.) To one decimal place, how long will it take for the account to grow to $3000?

54. How much money will there be in an account at the end of 7 yr if $3000 is deposited at 3.5% compounded quarterly? (Assume no withdrawals are made.) To one decimal place, how long will it take for the account to grow to $5000?

▶ **55.** What will be the amount A in an account with initial principal $4000 if interest is compounded continuously at an annual rate of 3.5% for 6 yr? To one decimal place, how long will it take for the initial amount to double?

56. Refer to the first question in **Exercise 54.** Does the money grow to a greater value under those conditions, or when invested for 7 yr at 3% compounded continuously?

57. Find the amount of money in an account after 12 yr if $5000 is deposited at 7% annual interest compounded as follows.

(a) Annually   **(b)** Semiannually   **(c)** Quarterly

(d) Daily (Use $n = 365$.)   **(e)** Continuously

58. Find the amount of money in an account after 8 yr if $4500 is deposited at 6% annual interest compounded as follows.

(a) Annually   **(b)** Semiannually   **(c)** Quarterly

(d) Daily (Use $n = 365$.)   **(e)** Continuously

59. How much money must be deposited today to amount to $1850 in 40 yr at 6.5% compounded continuously?

60. How much money must be deposited today to amount to $1000 in 10 yr at 5% compounded continuously?

Solve each problem. See Example 11.

61. Revenues of software publishers in the United States for the years 2004–2009 can be modeled by the function

$$S(x) = 114,711e^{0.0467x},$$

where $x = 0$ represents 2004, $x = 1$ represents 2005, and so on, and $S(x)$ is in millions of dollars. Approximate, to the nearest unit, revenue for 2007. (*Source:* U.S. Census Bureau.)

62. Based on selected figures obtained during the years 1970–2011, the total number of bachelor's degrees earned in the United States can be modeled by the function

$$D(x) = 794,383e^{0.0174x},$$

where $x = 0$ corresponds to 1970, $x = 10$ corresponds to 1980, and so on. Approximate, to the nearest unit, the number of bachelor's degrees earned in 2010. (*Source:* U.S. National Center for Education Statistics.)

63. Suppose that the amount, in grams, of plutonium 241 present in a given sample is determined by the function

$$A(t) = 2.00e^{-0.053t},$$

where t is measured in years. Approximate the amount present, to the nearest hundredth, in the sample after the given number of years.

(a) 4   **(b)** 10   **(c)** 20   **(d)** What was the initial amount present?

64. Suppose that the amount, in grams, of radium 226 present in a given sample is determined by the function

$$A(t) = 3.25e^{-0.00043t},$$

where t is measured in years. Approximate the amount present, to the nearest hundredth, in the sample after the given number of years.

(a) 20   **(b)** 100   **(c)** 500   **(d)** What was the initial amount present?

▶ **65.** A sample of 400 g of lead 210 decays to polonium 210 according to the function
$$A(t) = 400e^{-0.032t},$$
where t is time in years. Approximate answers to the nearest hundredth.

(a) How much lead will be left in the sample after 25 yr?

(b) How long will it take the initial sample to decay to half of its original amount?

66. The concentration of a drug in a person's system decreases according to the function
$$C(t) = 2e^{-0.125t},$$
where $C(t)$ is in appropriate units, and t is in hours. Approximate answers to the nearest hundredth.

(a) How much of the drug will be in the system after 1 hr?

(b) How long will it take for the concentration to be half of its original amount?

Chapter 12	Summary

Key Terms

12.1	**12.3**	**12.5**	**12.6**
one-to-one function	logarithm	common logarithm	compound interest
inverse of a function	logarithmic equation	natural logarithm	continuous compounding
	logarithmic function		
12.2	with base a		
exponential function			
with base a			
asymptote			
exponential equation			

New Symbols

$f^{-1}(x)$ inverse of $f(x)$

$\log_a x$ logarithm of x with base a

$\log x$ common (base 10) logarithm of x

$\ln x$ natural (base e) logarithm of x

e a constant, approximately 2.718281828

Test Your Word Power

See how well you have learned the vocabulary in this chapter.

1. In a **one-to-one function**
  A. each x-value corresponds to only one y-value
  B. each x-value corresponds to one or more y-values
  C. each x-value is the same as each y-value
  D. each x-value corresponds to only one y-value and each y-value corresponds to only one x-value.

2. If f is a one-to-one function, then the **inverse** of f is
  A. the set of all solutions of f
  B. the set of all ordered pairs formed by interchanging the coordinates of the ordered pairs of f
  C. the set of all ordered pairs that are the opposite (negative) of the coordinates of the ordered pairs of f
  D. an equation involving an exponential expression.

3. An **exponential function** is a function defined by an expression of the form
  A. $f(x) = ax^2 + bx + c$ for real numbers a, b, c $(a \neq 0)$
  B. $f(x) = \log_a x$ for positive numbers a and x $(a \neq 1)$
  C. $f(x) = a^x$ for all real numbers x $(a > 0, a \neq 1)$
  D. $f(x) = \sqrt{x}$ for $x \geq 0$.

4. An **asymptote** is
   A. a line that a graph intersects just once
   B. a line that the graph of a function more and more closely approaches as the x-values increase or decrease without bound
   C. the x-axis or y-axis
   D. a line about which a graph is symmetric.

5. A **logarithmic function** is a function that is defined by an expression of the form
   A. $f(x) = ax^2 + bx + c$ for real numbers a, b, c ($a \neq 0$)
   B. $f(x) = \log_a x$ for positive numbers a and x ($a \neq 1$)
   C. $f(x) = a^x$ for all real numbers x ($a > 0, a \neq 1$)
   D. $f(x) = \sqrt{x}$ for $x \geq 0$.

6. A **logarithm** is
   A. an exponent
   B. a base
   C. an equation
   D. a polynomial.

ANSWERS

1. D; *Example:* The function $f = \{(0, 2), (1, -1), (3, 5), (-2, 3)\}$ is one-to-one.   **2.** B; *Example:* The inverse of the one-to-one function f defined in Answer 1 is $f^{-1} = \{(2, 0), (-1, 1), (5, 3), (3, -2)\}$.   **3.** C; *Examples:* $f(x) = 4^x$, $g(x) = \left(\frac{1}{2}\right)^x$ **4.** B; *Example:* The graph of $f(x) = 2^x$ has the x-axis ($y = 0$) as an asymptote.   **5.** B; *Examples:* $f(x) = \log_3 x$, $g(x) = \log_{1/3} x$ **6.** A; *Example:* $\log_a x$ is the exponent to which a must be raised to obtain x; $\log_3 9 = 2$ because $3^2 = 9$.

Quick Review

CONCEPTS

EXAMPLES

12.1 Inverse Functions

Horizontal Line Test
A function is one-to-one if every horizontal line intersects the graph of the function at most once.

Find f^{-1} if $f(x) = 2x - 3$.

The graph of f is a nonhorizontal (slanted) straight line, so f is one-to-one by the horizontal line test.

Inverse Functions
For a one-to-one function $y = f(x)$, the equation that defines the inverse function f^{-1} is found as follows.

Step 1 Interchange x and y.

Step 2 Solve for y.

Step 3 Replace y with $f^{-1}(x)$.

Find $f^{-1}(x)$ as follows.

$$f(x) = 2x - 3$$
$$y = 2x - 3 \qquad \text{Let } y = f(x).$$
$$x = 2y - 3 \qquad \text{Interchange } x \text{ and } y.$$
$$x + 3 = 2y \qquad \text{Solve for } y.$$
$$y = \frac{x + 3}{2}$$
$$f^{-1}(x) = \frac{1}{2}x + \frac{3}{2} \qquad \text{Replace } y \text{ with } f^{-1}(x).$$

In general, the graph of f^{-1} is the mirror image of the graph of f with respect to the line $y = x$. If (a, b) lies on the graph of f, then (b, a) lies on the graph of f^{-1}.

The graphs of a function f and its inverse f^{-1} are shown here.

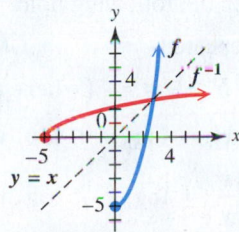

CONCEPTS	EXAMPLES

12.2 Exponential Functions

For $a > 0$, $a \neq 1$, and all real numbers x,

$$f(x) = a^x$$

defines the exponential function with base a.

Graph of $f(x) = a^x$

1. The graph contains the point $(0, 1)$, which is its y-intercept.

2. When $a > 1$, the graph rises from left to right.

   When $0 < a < 1$, the graph falls from left to right.

3. The x-axis is an asymptote.

4. The domain is $(-\infty, \infty)$, and the range is $(0, \infty)$.

$f(x) = 3^x$ defines the exponential function with base 3.

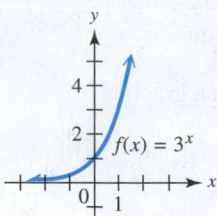

12.3 Logarithmic Functions

For all positive real numbers a, where $a \neq 1$, and all positive real numbers x,

$$y = \log_a x \quad \text{means} \quad x = a^y.$$

Special Properties of Logarithms

For $b > 0$, where $b \neq 1$, the following hold true.

$$\log_b b^r = r \qquad b^{\log_b r} = r \quad \text{(where } r > 0\text{)}$$

$$\log_b b = 1 \qquad \log_b 1 = 0$$

If a and x are positive real numbers, where $a \neq 1$, then

$$g(x) = \log_a x$$

defines the logarithmic function with base a.

Graph of $g(x) = \log_a x$

1. The graph contains the point $(1, 0)$, which is its x-intercept.

2. When $a > 1$, the graph rises from left to right.

   When $0 < a < 1$, the graph falls from left to right.

3. The y-axis is an asymptote.

4. The domain is $(0, \infty)$, and the range is $(-\infty, \infty)$.

$$y = \log_2 x \quad \text{means} \quad x = 2^y.$$

$$\log_5 5^6 = 6 \qquad 4^{\log_4 6} = 6$$

$$\log_3 3 = 1 \qquad \log_5 1 = 0$$

$g(x) = \log_3 x$ defines the logarithmic function with base 3.

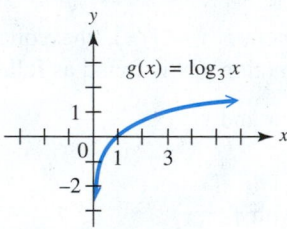

12.4 Properties of Logarithms

If x, y, and b are positive real numbers, where $b \neq 1$, and r is any real number, then the following hold.

Special Properties (repeated)

$$\log_b b^r = r \quad \text{and} \quad b^{\log_b r} = r \quad \text{(where } r > 0\text{)}$$

Product Rule $\quad \log_b xy = \log_b x + \log_b y$	
Quotient Rule $\quad \log_b \dfrac{x}{y} = \log_b x - \log_b y$	
Power Rule $\quad \log_b x^r = r \log_b x$	

$6^{\log_6 10} = 10 \qquad \log_3 3^4 = 4$ Special properties

$\log_2 3m = \log_2 3 + \log_2 m$ Product rule

$\log_5 \dfrac{9}{4} = \log_5 9 - \log_5 4$ Quotient rule

$\log_{10} 2^3 = 3 \log_{10} 2$ Power rule

CONCEPTS	EXAMPLES

12.5 Common and Natural Logarithms

Common logarithms (base 10) are used in applications such as pH, sound level, and intensity of an earthquake. Use the (LOG) key of a calculator to evaluate common logarithms.

Use the formula $\textbf{pH} = -\log [\textbf{H}_3\textbf{O}^+]$ to find the pH (to one decimal place) of grapes with hydronium ion concentration 5.0×10^{-5}.

$$pH = -\log(5.0 \times 10^{-5}) \qquad \text{Substitute.}$$

$$pH = -(\log 5.0 + \log 10^{-5}) \qquad \text{Property of logarithms}$$

$$pH \approx 4.3 \qquad \text{Evaluate with a calculator.}$$

Natural logarithms (base e) are found in formulas for applications of growth and decay, such as continuous compounding of money, decay of chemical compounds, and biological growth. Use the (LN) key or both the (INV) and (e^x) keys to evaluate natural logarithms.

Use the formula for doubling time (in years)

$$t(r) = \frac{\ln 2}{\ln(1+r)}$$

to find doubling time to the nearest tenth of a year, at a rate of 4%.

$$t(0.04) = \frac{\ln 2}{\ln(1+0.04)} \approx 17.7 \qquad \begin{array}{l}\text{Let } r = 0.04, \text{ and evaluate}\\ \text{with a calculator.}\end{array}$$

The doubling time is 17.7 yr.

Change-of-Base Rule

If $a > 0$, $a \neq 1$, $b > 0$, $b \neq 1$, and $x > 0$, then the following holds true.

$$\log_a x = \frac{\log_b x}{\log_b a}$$

Use the change-of-base rule to approximate $\log_3 17$.

$$\log_3 17 = \frac{\ln 17}{\ln 3} = \frac{\log 17}{\log 3} \approx 2.5789$$

12.6 Exponential and Logarithmic Equations; Further Applications

To solve exponential equations, use these properties (where $b > 0$, $b \neq 1$).

1. If $b^x = b^y$, then $x = y$.

Solve. $\qquad 2^{3x} = 2^5$

$$3x = 5 \qquad \text{Set exponents equal.}$$

$$x = \frac{5}{3} \qquad \text{Divide by 3.}$$

The solution set is $\left\{\frac{5}{3}\right\}$.

2. If $x = y$, $x > 0$, $y > 0$, then $\log_b x = \log_b y$.

Solve. $\qquad 5^x = 8$

$$\log 5^x = \log 8 \qquad \text{Take common logarithms.}$$

$$x \log 5 = \log 8 \qquad \text{Power rule}$$

$$x = \frac{\log 8}{\log 5} \approx 1.292 \qquad \text{Divide by log 5.}$$

The solution set is $\{1.292\}$.

To solve logarithmic equations, use these properties, where $b > 0$, $b \neq 1$, $x > 0$, and $y > 0$. First use the properties of **Section 12.4**, if necessary, to write the equation in the proper form.

1. If $\log_b x = \log_b y$, then $x = y$.

Solve. $\qquad \log_3 2x = \log_3(x+1)$

$$2x = x + 1$$

$$x = 1 \qquad \text{Subtract } x.$$

This value checks, so the solution set is $\{1\}$.

CONCEPTS	EXAMPLES
2. If $\log_b x = k$, then $x = b^k$.	Solve. $\log_2 (3x - 1) = 4$
	$3x - 1 = 2^4$ Exponential form
	$3x - 1 = 16$ Apply the exponent.
	$3x = 17$ Add 1.
	$x = \dfrac{17}{3}$ Divide by 3.
Always check any proposed solutions in logarithmic equations.	This value checks, so the solution set is $\left\{\dfrac{17}{3}\right\}$.

Chapter 12   Review Exercises

1. not one-to-one

2. one-to-one

3. $f^{-1}(x) = \dfrac{x - 7}{-3}$, or

$f^{-1}(x) = -\dfrac{1}{3}x + \dfrac{7}{3}$

4. $f^{-1}(x) = \dfrac{x^3 + 4}{6}$

5. not one-to-one

6. This function is not one-to-one because two states in the list have minimum wage $8.00.

7.   **8.**

   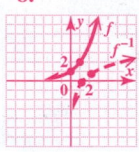

9. 172.466   **10.** 0.034

11. 0.079

12.1 *Determine whether each graph is the graph of a one-to-one function.*

1.   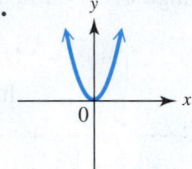   **2.**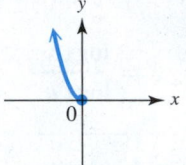

Determine whether each function is one-to-one. If it is, find its inverse.

3. $f(x) = -3x + 7$ **4.** $f(x) = \sqrt[3]{6x - 4}$ **5.** $f(x) = -x^2 + 3$

6. The table lists basic minimum wages in several states. If the set of states is the domain and the set of wage amounts is the range of a function, is it one-to-one? If not, explain why.

State	Minimum Wage (in dollars)
Washington	9.04
Nevada	8.25
California	8.00
Massachusetts	8.00
Ohio	7.70
Iowa	7.25

Source: U.S. Department of Labor.

Each function graphed is one-to-one. Graph its inverse.

7.   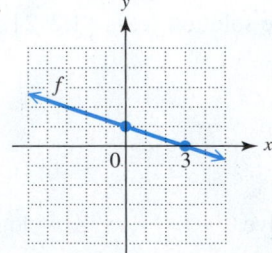   **8.**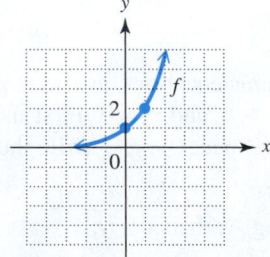

12.2 *Use a calculator to find an approximation to three decimal places for each exponential expression.*

9. $5^{3.2}$ **10.** $\left(\dfrac{1}{5}\right)^{2.1}$ **11.** $8.3^{-1.2}$

12.

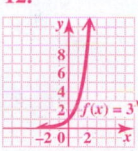

13. **14.**

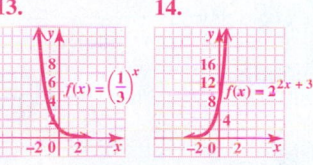

15. $\left\{\dfrac{1}{2}\right\}$ **16.** $\{4\}$

17. $\left\{\dfrac{3}{7}\right\}$

18. (a) 25 million tons
(b) 17 million tons
(c) 11 million tons

19. 2.5850   **20.** 0.5646

21. 1.7404

22.

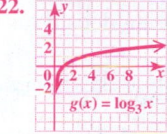

23.

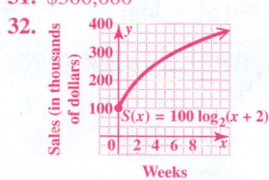

24. (a) 12 **(b)** 13 **(c)** 4

25. $\{2\}$ **26.** $\left\{\dfrac{3}{2}\right\}$

27. $\{7\}$ **28.** $\{8\}$

29. $\{4\}$ **30.** $\left\{\dfrac{1}{36}\right\}$

31. \$300,000

32.

Sales (in thousands of dollars)
$S(x) = 100 \log_2(x + 2)$
Weeks

33. $\log_2 3 + \log_2 x + 2 \log_2 y$

34. $\dfrac{1}{2} \log_4 x + 2 \log_4 w - \log_4 z$

35. $\log_b \dfrac{3x}{y^2}$ **36.** $\log_3 \left(\dfrac{x + 7}{4x + 6}\right)$

37. 1.4609   **38.** -0.5901
39. 3.3638   **40.** -1.3587

Graph each exponential function.

12. $f(x) = 3^x$ **13.** $f(x) = \left(\dfrac{1}{3}\right)^x$ **14.** $f(x) = 2^{2x+3}$

Solve each equation.

15. $5^{2x+1} = 25$ **16.** $4^{3x} = 8^{x+4}$ **17.** $\left(\dfrac{1}{27}\right)^{x-1} = 9^{2x}$

18. Sulfur dioxide emissions in the United States, in millions of tons, from 1980 through 2012 can be approximated by the exponential function

$$f(x) = 30.7032(1.0403)^{-x},$$

where $x = 0$ corresponds to 1980, $x = 5$ to 1985, and so on. Use this function to approximate emissions, to the nearest million tons, for each year. (*Source:* EPA.)

(a) 1985   **(b)** 1995   **(c)** 2005

12.3 *Use a calculator to approximate each logarithm to four decimal places.*

19. $\log_2 6$ **20.** $\log_7 3$ **21.** $\log_{10} 55$

Graph each logarithmic function.

22. $g(x) = \log_3 x$ (*Hint:* See **Exercise 12.**)   **23.** $g(x) = \log_{1/3} x$ (*Hint:* See **Exercise 13.**)

24. Evaluate without using a calculator.

(a) $4^{\log_4 12}$ **(b)** $\log_9 9^{13}$ **(c)** $\log_5 625$

Solve each equation.

25. $\log_8 64 = x$ **26.** $\log_2 \sqrt{8} = x$ **27.** $\log_x \left(\dfrac{1}{49}\right) = -2$

28. $\log_4 x = \dfrac{3}{2}$ **29.** $\log_k 4 = 1$ **30.** $\log_6 x = -2$

A company has found that total sales, in thousands of dollars, are given by the function

$$S(x) = 100 \log_2 (x + 2),$$

where x is the number of weeks after a major advertising campaign was introduced.

31. What were total sales 6 weeks after the campaign was introduced?

32. Graph the function.

12.4 *Use properties of logarithms to express each logarithm as a sum or difference of logarithms. Assume that all variables represent positive real numbers.*

33. $\log_2 3xy^2$ **34.** $\log_4 \dfrac{\sqrt{x} \cdot w^2}{z}$

Use properties of logarithms to write each expression as a single logarithm. Assume that all variables represent positive real numbers, $b \neq 1$.

35. $\log_b 3 + \log_b x - 2 \log_b y$ **36.** $\log_3 (x + 7) - \log_3 (4x + 6)$

12.5 *Evaluate each logarithm to four decimal places.*

37. $\log 28.9$ **38.** $\log 0.257$ **39.** $\ln 28.9$ **40.** $\ln 0.257$

41. 0.9251   **42.** 1.7925

43. 6.4   **44.** 8.4

45. 2.5×10^{-5}

46. Magnitude 1 is about 6.3 times as intense as magnitude 3.

47. **(a)** 18 yr   **(b)** 12 yr
(c) 7 yr   **(d)** 6 yr
(e) Each comparison shows approximately the same number. For example, in part (a) the doubling time is 18 yr (rounded) and $\frac{72}{4} = 18$.

Thus, the formula $t = \frac{72}{100r}$ (called the *rule of 72*) is an excellent approximation of the doubling time formula.

48. $\{2.042\}$

49. $\{4.907\}$ **50.** $\{18.310\}$

51. $\left\{\frac{1}{9}\right\}$ **52.** $\{-6 + \sqrt[3]{25}\}$

53. $\{2\}$ **54.** $\left\{\frac{3}{8}\right\}$

55. $\{4\}$ **56.** $\{1\}$

57. When the power rule was applied in the second step, the domain was changed from $\{x \mid x \neq 0\}$ to $\{x \mid x > 0\}$. The valid solution -10 was "lost." The solution set is $\{\pm 10\}$.

58. D

59. $24,403.80

Use the change-of-base rule (with either common or natural logarithms) to approximate each logarithm to four decimal places.

41. $\log_{16} 13$ **42.** $\log_4 12$

Find the pH of each substance with the given hydronium ion concentration.

43. Milk, 4.0×10^{-7} **44.** Crackers, 3.8×10^{-9}

45. If orange juice has pH 4.6, what is its hydronium ion concentration?

46. The magnitude of a star is given by the equation

$$M = 6 - 2.5 \log \frac{I}{I_0},$$

where I_0 is the measure of the faintest star and I is the actual intensity of the star being measured. The dimmest stars are of magnitude 6, and the brightest are of magnitude 1. Determine the ratio of intensities between stars of magnitude 1 and 3.

47. **Section 12.5, Exercise 48** introduced the doubling function

$$t(r) = \frac{\ln 2}{\ln (1 + r)},$$

that gives the number of years required to double money when it is invested at interest rate r (in decimal form) compounded annually. To the nearest year, how long does it take to double money at each rate?

(a) 4%   **(b)** 6%   **(c)** 10%   **(d)** 12%

(e) Compare each answer in parts (a)–(d) with the following numbers. Explain.

$$\frac{72}{4}, \quad \frac{72}{6}, \quad \frac{72}{10}, \quad \frac{72}{12}$$

12.6 *Solve each equation. Approximate solutions to three decimal places.*

48. $3^x = 9.42$ **49.** $2^{x-1} = 15$ **50.** $e^{0.06x} = 3$

Solve each equation. Give exact solutions.

51. $\log_3 (9x + 8) = 2$ **52.** $\log_5 (x + 6)^3 = 2$

53. $\log_3 (x + 2) - \log_3 x = \log_3 2$ **54.** $\log (2x + 3) = 1 + \log x$

55. $\log_4 x + \log_4 (8 - x) = 2$ **56.** $\log_2 x + \log_2 (x + 15) = \log_2 16$

57. Consider the following "solution" of the equation $\log x^2 = 2$. **WHAT WENT WRONG?** Give the correct solution set.

$\log x^2 = 2$	Original equation
$2 \log x = 2$	Power rule for logarithms
$\log x = 1$	Divide each side by 2.
$x = 10^1$	Write in exponential form.
$x = 10$	$10^1 = 10$

Solution set: $\{10\}$

58. Which is *not* a representation of the solution of $7^x = 23$?

A. $\dfrac{\log 23}{\log 7}$ **B.** $\dfrac{\ln 23}{\ln 7}$ **C.** $\log_7 23$ **D.** $\log_{23} 7$

Solve each problem. Use a calculator as necessary.

59. If $20,000 is deposited at 4% annual interest compounded quarterly, how much will be in the account after 5 yr, assuming no withdrawals are made?

60. How much will $10,000 compounded continuously at 3.75% annual interest amount to in 3 yr?

61. Which is a better plan?

> *Plan A:* Invest $1000 at 4% compounded quarterly for 3 yr
>
> *Plan B:* Invest $1000 at 3.9% compounded monthly for 3 yr

62. What is the half-life of a radioactive substance that decays according to the function

$$Q(t) = A_0 e^{-0.05t}, \quad \text{where } t \text{ is in days, to the nearest tenth?}$$

*A machine purchased for business use **depreciates**, or loses value, over a period of years. The value of the machine at the end of its useful life is its **scrap value**. By one method of depreciation, the scrap value, S, is given by*

$$S = C(1 - r)^n,$$

where C is original cost, n is useful life in years, and r is the constant percent of depreciation.

63. Find the scrap value of a machine costing $30,000, having a useful life of 12 yr and a constant annual rate of depreciation of 15%. Round to the nearest dollar.

64. A machine has a "half-life" of 6 yr. Find the constant annual rate of depreciation, to the nearest percent.

60. $11,190.72
61. Plan A is better because it would pay $2.92 more.
62. 13.9 days
63. $4267   **64.** 11%

Chapter 12 | Mixed Review Exercises

Evaluate.

1. $\log_2 128$ **2.** $5^{\log_5 36}$ **3.** $e^{\ln 4}$

4. $10^{\log e}$ **5.** $\log_3 3^{-5}$ **6.** $\ln e^{5.4}$

Solve each equation.

7. $\log_3 (x + 9) = 4$ **8.** $\ln e^x = 3$ **9.** $\log_x \dfrac{1}{81} = 2$

10. $27^x = 81$ **11.** $2^{2x-3} = 8$ **12.** $8^{3x} = 5^{x+1}$

13. $5^{x+2} = 25^{2x+1}$ **14.** $\log_3 (x + 1) - \log_3 x = 2$

15. $\log (3x - 1) = \log 10$ **16.** $\ln (x^2 + 3x + 4) = \ln 2$

1. 7   **2.** 36
3. 4   **4.** e
5. −5   **6.** 5.4

7. {72}   **8.** {3}
9. $\left\{\dfrac{1}{9}\right\}$ **10.** $\left\{\dfrac{4}{3}\right\}$
11. {3}   **12.** {0.348}
13. {0}   **14.** $\left\{\dfrac{1}{8}\right\}$
15. $\left\{\dfrac{11}{3}\right\}$ **16.** {−2, −1}

Solve each problem.

17. Recall from **Exercise 49** in **Section 12.5** that the number of years, $N(x)$, since two independently evolving languages split off from a common ancestral language is approximated by

$$N(x) = -5000 \ln x,$$

where x is the percent of words from the ancestral language common to both languages now. Find x to the nearest percent if the split occurred 2000 yr ago.

18. To solve the equation $5^x = 7$, we must find the exponent to which 5 must be raised in order to obtain 7. This is $\log_5 7$.

(a) Use the change-of-base rule and a calculator to find $\log_5 7$.

(b) Raise 5 to the number found in part (a). What is the result?

(c) Using as many decimal places as the calculator gives, write the solution set of $5^x = 7$.

19. Let m be the number of letters in your first name, and let n be the number of letters in your last name.

(a) Explain what $\log_m n$ means.   (b) Use a calculator to find $\log_m n$.

(c) Raise m to the power indicated by the number found in part (b). What is the result?

17. 67%
18. (a) 1.209061955
   (b) 7
   (c) {1.209061955}
19. Answers will vary. Suppose the name is Jeffery Cole, with $m = 7$ and $n = 4$.
   (a) $\log_7 4$ is the exponent to which 7 must be raised to obtain 4.
   (b) 0.7124143742
   (c) 4

20. (a) 0.325   **(b)** 0.673

20. One measure of the diversity of the species in an ecological community is the **index of diversity,** the logarithmic expression

$$-(p_1 \ln p_1 + p_2 \ln p_2 + \cdots + p_n \ln p_n),$$

where $p_1, p_2, \ldots, p_n$ are the proportions of a sample belonging to each of n species in the sample. (*Source:* Ludwig, J. and J. Reynolds, *Statistical Ecology: A Primer on Methods and Computing,* New York, John Wiley and Sons.) Approximate the index of diversity to the nearest thousandth if a sample of 100 from a community produces the following numbers.

(a) 90 of one species, 10 of another   **(b)** 60 of one species, 40 of another

---

Chapter 12   Test

FOR EXTRA HELP   *Step-by-step test solutions are found on the Chapter Test Prep Videos available in* MyMathLab® *or on* You Tube.

▶ *View the complete solutions to all Chapter Test exercises in MyMathLab.*

[12.1]

1. (a) not one-to-one
   (b) one-to-one
2. $f^{-1}(x) = x^3 - 7$
3.

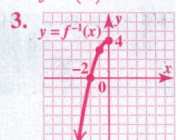

[12.2]

4.

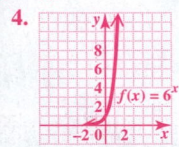

[12.3]

5.

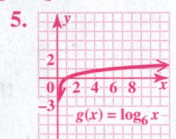

6. 9
7. 6   **8.** 0

[12.2]

9. $\{-4\}$ **10.** $\left\{-\dfrac{13}{3}\right\}$

11. (a) 775 millibars
    (b) 265 millibars

[12.3]
12. $\log_4 0.0625 = -2$
13. $7^2 = 49$ **14.** $\{32\}$
15. $\left\{\dfrac{1}{2}\right\}$ **16.** $\{2\}$

17. 5; 2; fifth; 32

[12.4]
18. $2 \log_3 x + \log_3 y$

19. $\dfrac{1}{2} \log_5 x - \log_5 y - \log_5 z$

---

1. Decide whether each function is one-to-one.

(a) $f(x) = x^2 + 9$

(b)

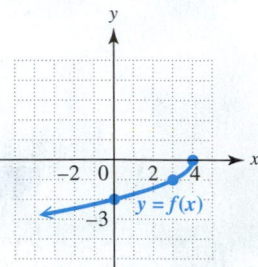

2. Find $f^{-1}(x)$ for the one-to-one function

$$f(x) = \sqrt[3]{x + 7}.$$

3. Graph the inverse given the graph of $y = f(x)$.

Graph each function.

4. $f(x) = 6^x$ **5.** $g(x) = \log_6 x$

Use the special properties of logarithms to evaluate each expression.

6. $7^{\log_7 9}$ **7.** $\log_3 3^6$ **8.** $\log_5 1$

Solve each equation. Give exact solutions.

9. $5^x = \dfrac{1}{625}$ **10.** $2^{3x-7} = 8^{2x+2}$

11. The atmospheric pressure (in millibars) at a given altitude x (in meters) is approximated by

$$f(x) = 1013e^{-0.0001341x}.$$

Use this function to approximate the atmospheric pressure at each altitude.

(a) 2000 m   **(b)** 10,000 m

12. Write in logarithmic form: $4^{-2} = 0.0625.$

13. Write in exponential form: $\log_7 49 = 2.$

Solve each equation.

14. $\log_{1/2} x = -5$ **15.** $x = \log_9 3$ **16.** $\log_x 16 = 4$

17. Fill in the blanks with the correct responses: The value of $\log_2 32$ is _____. This means that if we raise _____ to the _____ power, the result is _____.

Use properties of logarithms to write each expression as a sum or difference of logarithms. Assume that variables represent positive real numbers.

18. $\log_3 x^2 y$ **19.** $\log_5 \left(\dfrac{\sqrt{x}}{yz}\right)$

20. $\log_b \dfrac{s^3}{t}$

21. $\log_b \dfrac{r^{1/4}s^2}{t^{2/3}}$

Use properties of logarithms to write each expression as a single logarithm. Assume that variables represent positive real numbers, $b \neq 1$.

20. $3 \log_b s - \log_b t$

21. $\dfrac{1}{4} \log_b r + 2 \log_b s - \dfrac{2}{3} \log_b t$

[12.5]

22. 1.3636

23. −0.1985

24. (a) $\dfrac{\log 19}{\log 3}$ **(b)** $\dfrac{\ln 19}{\ln 3}$

(c) 2.6801

Evaluate each logarithm to four decimal places.

22. $\log 23.1$ **23.** $\ln 0.82$

24. Use the change-of-base rule to express $\log_3 19$ as described.

(a) in terms of common logarithms **(b)** in terms of natural logarithms

(c) approximated to four decimal places

[12.6]

25. {3.966} **26.** {3}

27. \$12,507.51

28. (a) \$19,260.38 **(b)** 13.9 yr

25. Solve $3^x = 78$. Approximate the solution to three decimal places.

26. Solve $\log_8 (x + 5) + \log_8 (x - 2) = 1$.

27. Suppose that \$10,000 is invested at 4.5% annual interest, compounded quarterly. How much will be in the account in 5 yr if no money is withdrawn?

28. Suppose that \$15,000 is invested at 5% annual interest, compounded continuously.

(a) How much will be in the account in 5 yr if no money is withdrawn?

(b) To one decimal place, how long will it take for the initial principal to double?

Chapters R–12 | Cumulative Review Exercises

[1.3]

1. $-2, 0, 6, \dfrac{30}{3}$ (or 10)

2. $-\dfrac{9}{4}, -2, 0, 0.6, 6, \dfrac{30}{3}$ (or 10)

3. $-\sqrt{2}, \sqrt{11}$

Let $S = \left\{ -\dfrac{9}{4}, -2, -\sqrt{2}, 0, 0.6, \sqrt{11}, \sqrt{-8}, 6, \dfrac{30}{3} \right\}$. List the elements of S that are members of each set.

1. Integers **2.** Rational numbers **3.** Irrational numbers

[1.3–1.5]

4. 16 **5.** −39

Simplify each expression.

4. $|-8| + 6 - |-2| - (-6 + 2)$ **5.** $2(-5) + (-8)(4) - (-3)$

[2.3] **[2.8, 8.1]**

6. $\left\{ -\dfrac{2}{3} \right\}$ **7.** $[1, \infty)$

Solve each equation or inequality.

6. $7 - (3 + 4x) + 2x = -5(x - 1) - 3$ **7.** $2x + 2 \leq 5x - 1$

[8.3]

8. {−2, 7}

9. $(-\infty, -3) \cup (2, \infty)$

8. $|2x - 5| = 9$ **9.** $|4x + 2| > 10$

[10.6] **[11.2, 11.3]**

10. {0, 4} **11.** $\left\{ \dfrac{1 \pm \sqrt{13}}{6} \right\}$

10. $\sqrt{2x + 1} - \sqrt{x} = 1$ **11.** $3x^2 - x - 1 = 0$

[11.8]

12. $(-\infty, -4) \cup (2, \infty)$

12. $x^2 + 2x - 8 > 0$ **13.** $x^4 - 5x^2 + 4 = 0$ **14.** $5^{x+3} = \left(\dfrac{1}{25} \right)^{3x+2}$

[11.4] **[12.2]**

13. $\{ \pm 1, \pm 2 \}$ **14.** {−1}

Perform the indicated operations.

15. $(2p + 3)(3p - 1)$ **16.** $(4k - 3)^2$

[4.5] **[4.6]**

15. $6p^2 + 7p - 3$ **16.** $16k^2 - 24k + 9$

17. $(3m^3 + 2m^2 - 5m) - (8m^3 + 2m - 4)$ **18.** Divide $15x^3 - x^2 + 22x + 8$ by $3x + 1$.

[4.4]

17. $-5m^3 + 2m^2 - 7m + 4$

Factor.

19. $8x + x^3$ **20.** $24y^2 - 7y - 6$ **21.** $5z^3 - 19z^2 - 4z$

[4.7] **[5.1]**

18. $5x^2 - 2x + 8$ **19.** $x(8 + x^2)$

22. $16a^2 - 25b^4$ **23.** $8c^3 + d^3$ **24.** $16r^2 + 56rq + 49q^2$

[5.2, 5.3]
20. $(3y - 2)(8y + 3)$
21. $z(5z + 1)(z - 4)$

[5.4]
22. $(4a + 5b^2)(4a - 5b^2)$
23. $(2c + d)(4c^2 - 2cd + d^2)$
24. $(4r + 7q)^2$

[4.1, 4.2]   **[6.2]**
25. $-\dfrac{1875p^{13}}{8}$ **26.** $\dfrac{x+5}{x+4}$

[6.4]
27. $\dfrac{-3k - 19}{(k+3)(k-2)}$

[3.3, 7.1, 9.1]
28. **(a)** yes
  (b) 2192;
   The number of travelers
   increased by an average of
   2192 thousand per year
   during these years.

[7.2]
29. $y = \dfrac{3}{4}x - \dfrac{19}{4}$

[7.3–7.5]   **[7.6]**
30. $\{(4, 2)\}$ **31.** $\{(1, -1, 4)\}$

[7.7]
32. 6 lb

[3.2, 7.1]   **[8.4]**
33.   **34.**

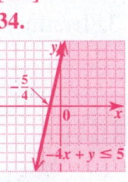

[11.6]   **[12.2]**
35.   **36.**

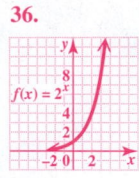

[12.3]   **[10.3]**
37.   **38.** $12\sqrt{2}$

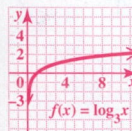

[10.4]   **[10.7]**
39. $-27\sqrt{2}$ **40.** 41

[12.4]
41. $3 \log x + \dfrac{1}{2} \log y - \log z$

[12.6]
42. **(a)** 25,000 **(b)** 30,500
  (c) 37,300
  (d) in 3.5 hr, or at 3:30 P.M.

Perform the indicated operations.

25. $\dfrac{(5p^3)^4(-3p^7)}{2p^2(4p^4)}$

26. $\dfrac{x^2 - 9}{x^2 + 7x + 12} \div \dfrac{x - 3}{x + 5}$

27. $\dfrac{2}{k + 3} - \dfrac{5}{k - 2}$

Solve each problem.

28. The graph indicates the number of international travelers to the United States from 2006 through 2010.

  (a) Is this the graph of a function?

  (b) What is the slope of the line in the graph? Interpret the slope in the context of international travelers to the United States.

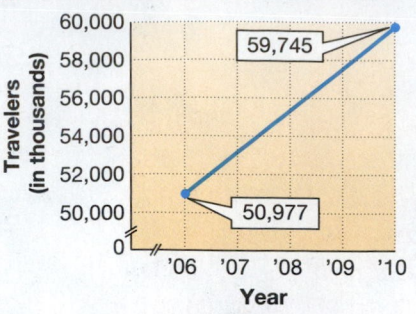

Source: U.S. Department of Commerce.

29. Find an equation of the line passing through $(5, -1)$ and parallel to the line with equation $3x - 4y = 12$. Write the equation in slope-intercept form.

Solve each system.

30. $5x - 3y = 14$
  $2x + 5y = 18$

31. $x + 2y + 3z = 11$
  $3x - y + z = 8$
  $2x + 2y - 3z = -12$

32. Candy worth \$1.00 per lb is to be mixed with 10 lb of candy worth \$1.96 per lb to obtain a mixture that will be sold for \$1.60 per lb. How many pounds of the \$1.00 candy should be used?

Number of Pounds	Price per Pound (in dollars)	Value (in dollars)
x	1.00	$1x$
	1.60	

Graph.

33. $5x + 2y = 10$

34. $-4x + y \le 5$

35. $f(x) = \dfrac{1}{3}(x - 1)^2 + 2$

36. $f(x) = 2^x$

37. $f(x) = \log_3 x$

Simplify.

38. $\sqrt{288}$

39. $2\sqrt{32} - 5\sqrt{98}$

40. Multiply $(5 + 4i)(5 - 4i)$.

41. Use properties of logarithms to write the following as a sum or difference of logarithms. Assume that variables represent positive real numbers.

$$\log \frac{x^3 \sqrt{y}}{z}$$

42. Let the number of bacteria present in a certain culture be given by

$$B(t) = 25{,}000 e^{0.2t},$$

where t is time measured in hours, and $t = 0$ corresponds to noon. Approximate, to the nearest hundred, the number of bacteria present at each time.

  (a) noon   **(b)** 1 P.M.   **(c)** 2 P.M.   **(d)** When will the population double?

APPENDIX A

Review of Exponents, Polynomials, and Factoring

OBJECTIVES

1. Review the basic rules for exponents.
2. Review operations with polynomials.
3. Review factoring techniques.

(Transition from Beginning to Intermediate Algebra)

OBJECTIVE 1 Review the basic rules for exponents.

In **Sections 4.1 and 4.2,** we introduced the following.

Definitions and Rules for Exponents

If no denominators are 0, the following hold true for any integers m and n.

		Examples
Product rule	$a^m \cdot a^n = a^{m+n}$	$7^4 \cdot 7^5 = 7^9$
Zero exponent	$a^0 = 1 \quad (a \neq 0)$	$(-3)^0 = 1$
Negative exponent	$a^{-n} = \dfrac{1}{a^n}$	$5^{-3} = \dfrac{1}{5^3}$
Quotient rule	$\dfrac{a^m}{a^n} = a^{m-n}$	$\dfrac{2^2}{2^5} = 2^{-3} = \dfrac{1}{2^3}$
Power rules (a)	$(a^m)^n = a^{mn}$	$(4^2)^3 = 4^6$
(b)	$(ab)^m = a^m b^m$	$(3k)^4 = 3^4 k^4$
(c)	$\left(\dfrac{a}{b}\right)^m = \dfrac{a^m}{b^m}$	$\left(\dfrac{2}{3}\right)^2 = \dfrac{2^2}{3^2}$
Negative-to-positive rules	$\dfrac{a^{-m}}{b^{-n}} = \dfrac{b^n}{a^m}$	$\dfrac{2^{-4}}{5^{-3}} = \dfrac{5^3}{2^4}$
	$\left(\dfrac{a}{b}\right)^{-m} = \left(\dfrac{b}{a}\right)^m$	$\left(\dfrac{4}{7}\right)^{-2} = \left(\dfrac{7}{4}\right)^2$

EXAMPLE 1 Applying Definitions and Rules for Exponents

Simplify. Write answers using only positive exponents. Assume that all variables represent nonzero real numbers.

(a) $(x^2 y^{-3})(x^{-5} y^7)$

$= (x^{2+(-5)})(y^{-3+7})$ Product rule

$= x^{-3} y^4$ Add exponents.

$= \dfrac{1}{x^3} y^4$ Definition of negative exponent

$= \dfrac{y^4}{x^3}$ $\dfrac{1}{x^3} y^4 = \dfrac{1}{x^3} \cdot \dfrac{y^4}{1} = \dfrac{y^4}{x^3}$

**NOW TRY
EXERCISE 1**

Simplify. Write answers using only positive exponents. Assume that all variables represent nonzero real numbers.

(a) $(m^{-8}n^4)(m^4n^{-3})$

(b) $-8^0 + 8^0$

(c) $\dfrac{(p^{-3}q)^4}{(p^2q^5)^2}$

(d) $\left(\dfrac{2x^{-2}y}{x^2y^{-4}}\right)^{-4}$

(b) $(-5)^0 + (-5^0)$

$= 1 + (-1)$ $\quad (-5^0) = -1 \cdot 5^0 = -1 \cdot 1 = -1$

$= 0$ $\qquad\qquad$ Add.

(c) $\dfrac{(t^5s^{-4})^2}{(t^{-3}s^5)^3}$

$= \dfrac{t^{10}s^{-8}}{t^{-9}s^{15}}$ Power rule (b)

$= \dfrac{t^{10}t^9}{s^{15}s^8}$ Negative-to-positive rule

$= \dfrac{t^{19}}{s^{23}}$ Product rule

(d) $\left(\dfrac{-3x^{-4}y}{x^5y^{-4}}\right)^{-2}$

$= \left(\dfrac{x^5y^{-4}}{-3x^{-4}y}\right)^2$ Negative-to-positive rule

$= \dfrac{x^{10}y^{-8}}{9x^{-8}y^2}$ Power rules (b) and (c)

$= \dfrac{x^{18}}{9y^{10}}$ Quotient rule

(e) $(2x^2y^3z)^2(x^4y^2)^3$

$= (4x^4y^6z^2)(x^{12}y^6)$ Power rule (b)

$= 4x^{16}y^{12}z^2$ Product rule   **NOW TRY**

OBJECTIVE 2   Review operations with polynomials.

These arithmetic operations with polynomials were covered in **Sections 4.4–4.6.**

> **Adding and Subtracting Polynomials**
>
> To add polynomials, add like terms.
>
> To subtract polynomials, change all signs in the second polynomial and add the result to the first polynomial.

**NOW TRY
EXERCISE 2**

Add or subtract as indicated.

(a) $(3x^3 + x^2 - 5x - 6) + (-6x^3 + 2x^2 + 4x - 1)$

(b) Subtract.

$\quad 4x^2 + 7x - 5$
$\underline{-5x^2 - 2x + 3}$

NOW TRY ANSWERS

1. (a) $\dfrac{n}{m^4}$ **(b)** 0

$\quad$ **(c)** $\dfrac{1}{p^{16}q^6}$ **(d)** $\dfrac{x^{16}}{16y^{20}}$

2. (a) $-3x^3 + 3x^2 - x - 7$

$\quad$ **(b)** $9x^2 + 9x - 8$

EXAMPLE 2   Adding and Subtracting Polynomials

Add or subtract as indicated.

(a) $(-4x^3 + 3x^2 - 8x + 2) + (5x^3 - 8x^2 + 12x - 3)$

$= (-4x^3 + 5x^3) + (3x^2 - 8x^2) + (-8x + 12x) + (2 - 3)$

$\qquad$ Commutative and associative properties

$= (-4 + 5)x^3 + (3 - 8)x^2 + (-8 + 12)x + (2 - 3)$

$\qquad$ Distributive property

$= x^3 - 5x^2 + 4x - 1$ Simplify.

(b) $-4(x^2 + 3x - 6) - (2x^2 - 3x + 7)$

$= -4x^2 - 12x + 24 - 2x^2 + 3x - 7$ Distributive property; definition of subtraction

$= -6x^2 - 9x + 17$ Combine like terms.

(c) Subtract.

$\quad 2t^2 - 3t - 4$ Change the sign $\quad 2t^2 - 3t - 4$
$\underline{-8t^2 + 4t - 1}$ of each term in $\underline{8t^2 - 4t + 1}$ Change signs.
$\qquad\qquad\qquad$ $-8t^2 + 4t - 1$, $\quad 10t^2 - 7t - 3$ Add.   **NOW TRY**
$\qquad\qquad\qquad$ and add.

Multiplying Polynomials

To multiply two polynomials, multiply each term of the second polynomial by each term of the first polynomial and add the products.

To multiply two binomials, use the FOIL method. (See **Section 4.5.**)

The special product rules are useful when multiplying binomials.

Special Product Rules

For x and y, the following hold true.

$$\left.\begin{array}{l}(x + y)^2 = x^2 + 2xy + y^2 \\ (x - y)^2 = x^2 - 2xy + y^2\end{array}\right\} \quad \text{Square of a binomial}$$

$$(x + y)(x - y) = x^2 - y^2 \quad \begin{array}{l}\text{Product of a sum and} \\ \text{difference of two terms}\end{array}$$

**NOW TRY
EXERCISE 3**

Find each product.

(a) $(6x - 5)(2x - 3)$

(b) $(4m - 3n)(4m + 3n)$

(c) $(7z + 1)^2$

(d) $(r + 3)(r^2 - 3r + 9)$

EXAMPLE 3 Multiplying Polynomials

Find each product.

(a) $(4y - 1)(3y + 2)$

$$\underbrace{\text{First}}_{\text{terms}} \quad \underbrace{\text{Outer}}_{\text{terms}} \quad \underbrace{\text{Inner}}_{\text{terms}} \quad \underbrace{\text{Last}}_{\text{terms}}$$

$= 4y(3y) + 4y(2) - 1(3y) - 1(2)$ FOIL method

$= 12y^2 + 8y - 3y - 2$ Multiply.

$= 12y^2 + 5y - 2$ Combine like terms.

(b) $(3x + 5y)(3x - 5y)$ $(ab)^2 = a^2b^2$, **not** ab^2.

$= (3x)^2 - (5y)^2$ $(x + y)(x - y) = x^2 - y^2$

$= 9x^2 - 25y^2$ Power rule (b)

(c) $(2t + 3)^2$

$= (2t)^2 + 2(2t)(3) + 3^2$ $(x + y)^2 = x^2 + 2xy + y^2$

$= 4t^2 + 12t + 9$ Remember the middle term.

(d) $(5x - 1)^2$

$= (5x)^2 - 2(5x)(1) + 1^2$ $(x - y)^2 = x^2 - 2xy + y^2$

$= 25x^2 - 10x + 1$ $(5x)^2 = 5^2x^2 = 25x^2$

(e) $(3x + 2)(9x^2 - 6x + 4)$

$$\begin{array}{r} 9x^2 - 6x + 4 \\ 3x + 2 \\ \hline 18x^2 - 12x + 8 \\ 27x^3 - 18x^2 + 12x \\ \hline 27x^3 + 8 \end{array}$$

Multiply vertically.

$\longleftarrow 2(9x^2 - 6x + 4)$

Be sure to write like terms in columns. $\longleftarrow 3x(9x^2 - 6x + 4)$

Add.

The product is a sum of cubes, $27x^3 + 8$.

NOW TRY ANSWERS

3. (a) $12x^2 - 28x + 15$

(b) $16m^2 - 9n^2$

(c) $49z^2 + 14z + 1$

(d) $r^3 + 27$

NOW TRY

OBJECTIVE 3 Review factoring techniques.

Factoring involves writing a polynomial as a product. See **Chapter 5.**

Factoring a Polynomial

Question 1 **Is there a common factor other than 1?** If so, factor it out.

Question 2 **How many terms are in the polynomial?**

Two terms: Is it a difference of squares or a sum or difference of cubes? If so, factor as in **Section 5.4.**

$$x^2 - y^2 = (x + y)(x - y) \qquad \text{Difference of squares}$$

$$x^3 - y^3 = (x - y)(x^2 + xy + y^2) \qquad \text{Difference of cubes}$$

$$x^3 + y^3 = (x + y)(x^2 - xy + y^2) \qquad \text{Sum of cubes}$$

Three terms: Is it a perfect square trinomial?

$$x^2 + 2xy + y^2 = (x + y)^2$$
$$x^2 - 2xy + y^2 = (x - y)^2 \qquad \text{Perfect square trinomials}$$

If the trinomial is not a perfect square trinomial, what is the coefficient of the second-degree term?

• If it is 1, use the method of **Section 5.2.**

• If it is not 1, use the general factoring methods of **Section 5.3.**

Four terms: Try to factor by grouping, as in **Section 5.1.**

Question 3 **Can any factors be factored further?** If so, factor them.

EXAMPLE 4 Factoring Polynomials

Factor each polynomial completely.

(a) $6x^2y^3 - 12x^3y^2$

$$= 6x^2y^2 \cdot y - 6x^2y^2 \cdot 2x \qquad 6x^2y^2 \text{ is the greatest common factor.}$$

$$= 6x^2y^2(y - 2x) \qquad \text{Distributive property}$$

(b) $3x^2 - x - 2$

To find the factors, find two terms that multiply to give $3x^2$ (here $3x$ and x) and two terms that multiply to give -2 (here $+2$ and -1). Make sure that the sum of the outer and inner products in the factored form is the middle term of the trinomial, $-x$.

$$3x^2 - x - 2 \quad \text{factors as} \quad (3x + 2)(x - 1).$$

CHECK To check, multiply the factors using the FOIL method.

(c) $3x^2 - 27x + 42$

$$= 3(x^2 - 9x + 14) \qquad \text{Factor out the common factor.}$$

$$= 3(x - 7)(x - 2) \qquad \text{Factor the trinomial.}$$

(d) $100t^2 - 81$

$$= (10t)^2 - 9^2 \qquad \text{Difference of squares}$$

$$= (10t + 9)(10t - 9) \qquad x^2 - y^2 = (x + y)(x - y)$$

NOW TRY
EXERCISE 4
Factor each polynomial completely.

(a) $5x^2 - 20x - 60$

(b) $10t^2 + 13t - 3$

(c) $49x^2 + 42x + 9$

(d) $mn - 2n + 5m - 10$

(e) $27x^3 - 1000$

(e) $4x^2 + 20xy + 25y^2$

The terms $4x^2$ and $25y^2$ are both perfect squares, so this trinomial might factor as a perfect square trinomial.

$$\text{Try to factor}\quad 4x^2 + 20xy + 25y^2 \quad \text{as} \quad (2x + 5y)^2.$$

CHECK   Take twice the product of the two terms in the squared binomial.

$$2 \cdot 2x \cdot 5y = 20xy \longleftarrow \text{Middle term of } 4x^2 + 20xy + 25y^2$$

Twice —— | —— Last term
First term

Because $20xy$ is the middle term of the trinomial, the trinomial is a perfect square.

$$4x^2 + 20xy + 25y^2 \quad \text{factors as} \quad (2x + 5y)^2.$$

(f) $1000x^3 - 27$

$$= (10x)^3 - 3^3 \qquad\qquad \text{Difference of cubes}$$

$$= (10x - 3)\big[(10x)^2 + 10x(3) + 3^2\big] \qquad x^3 - y^3 = (x - y)(x^2 + xy + y^2)$$

$$= (10x - 3)(100x^2 + 30x + 9) \qquad (10x)^2 = 10^2x^2$$

(g) $6xy - 3x + 4y - 2$

Because there are four terms, try factoring by grouping.

$$6xy - 3x + 4y - 2$$

$$= (6xy - 3x) + (4y - 2) \qquad \text{Group the terms.}$$

$$= 3x(2y - 1) + 2(2y - 1) \qquad \text{Factor each group.}$$

$$= (2y - 1)(3x + 2) \qquad \text{Factor out } 2y - 1.$$

In the final step, factor out the greatest common factor, the binomial $2y - 1$.

NOW TRY

NOW TRY ANSWERS
4. (a) $5(x - 6)(x + 2)$
(b) $(5t - 1)(2t + 3)$
(c) $(7x + 3)^2$
(d) $(n + 5)(m - 2)$
(e) $(3x - 10)(9x^2 + 30x + 100)$

A Exercises

FOR
EXTRA   MyMathLab®
HELP

Simplify each expression. Write the answers using only positive exponents. Assume that all variables represent positive real numbers. **See Example 1.**

1. $(a^4b^{-3})(a^{-6}b^2)$

2. $(t^{-3}s^{-5})(t^8s^{-2})$

3. $(5x^{-2}y)^2(2xy^4)^2$

4. $(7x^{-3}y^4)^3(2x^{-1}y^{-4})^2$

5. $-6^0 + (-6)^0$

6. $(-12)^0 - 12^0$

7. $\dfrac{(2w^{-1}x^2y^{-1})^3}{(4w^5x^{-2}y)^2}$

8. $\dfrac{(5p^{-3}q^2r^{-4})^2}{(10p^4q^{-1}r^5)^{-1}}$

9. $\left(\dfrac{-4a^{-2}b^4}{a^3b^{-1}}\right)^{-3}$

10. $\left(\dfrac{r^{-3}s^{-8}}{-6r^2s^{-4}}\right)^{-2}$

11. $(7x^{-4}y^2z^{-2})^{-2}(7x^4y^{-1}z^3)^2$

12. $(3m^{-5}n^2p^{-4})^3(3m^4n^{-3}p^5)^{-2}$

Add or subtract as indicated. **See Example 2.**

13. $(2a^4 + 3a^3 - 6a^2 + 5a - 12) + (-8a^4 + 8a^3 - 14a^2 + 21a - 3)$

14. $(-6r^4 - 3r^3 + 12r^2 - 9r + 9) + (8r^4 - 13r^3 - 14r^2 - 10r - 3)$

15. $(6x^3 - 12x^2 + 3x - 4) - (-2x^3 + 6x^2 - 3x + 12)$

16. $(10y^3 - 4y^2 + 8y + 7) - (7y^3 + 5y^2 - 2y - 13)$

17. Add.
$$5x^2y + 2xy^2 + y^3$$
$$\underline{-4x^2y - 3xy^2 + 5y^3}$$

18. Add.
$$6ab^3 - 2a^2b^2 + 3b^5$$
$$\underline{8ab^3 + 12a^2b^2 - 8b^5}$$

19. $3(5x^2 - 12x + 4) - 2(9x^2 + 13x - 10)$

20. $-4(2t^3 - 3t^2 + 4t - 1) - 3(-8t^3 + 3t^2 - 2t + 9)$

21. Subtract.
$$6x^3 - 2x^2 + 3x - 1$$
$$\underline{-4x^3 + 2x^2 - 6x + 3}$$

22. Subtract.
$$-9y^3 - 2y^2 + 3y - 8$$
$$\underline{-8y^3 + 4y^2 + 3y + 1}$$

Find each product. **See Example 3.**

23. $(3x + 1)(2x - 7)$ **24.** $(5z + 3)(2z - 3)$ **25.** $(4x - 1)(x - 2)$

26. $(7t - 3)(t - 4)$ **27.** $(4t + 3)(4t - 3)$ **28.** $(6x + 1)(6x - 1)$

29. $(2y^2 + 4)(2y^2 - 4)$ **30.** $(3b^3 + 2t)(3b^3 - 2t)$ **31.** $(4x - 3)^2$

32. $(9t + 2)^2$ **33.** $(6r + 5y)^2$ **34.** $(8m - 3n)^2$

35. $(c + 2d)(c^2 - 2cd + 4d^2)$ **36.** $(f + 3g)(f^2 - 3fg + 9g^2)$

37. $(4x - 1)(16x^2 + 4x + 1)$ **38.** $(5r - 2)(25r^2 + 10r + 4)$

39. $(7t + 5s)(2t^2 + 5st - s^2)$ **40.** $(8p + 3q)(2p^2 - 4pq + q^2)$

Factor each polynomial completely. **See Example 4.**

41. $8x^3y^4 + 12x^2y^3 + 36xy^4$ **42.** $10m^5n + 4m^2n^3 + 18m^3n^2$ **43.** $x^2 - 2x - 15$

44. $x^2 + x - 12$ **45.** $2x^2 - 9x - 18$ **46.** $3x^2 + 2x - 8$

47. $36t^2 - 25$ **48.** $49r^2 - 9$ **49.** $16t^2 + 24t + 9$

50. $25t^2 + 90t + 81$ **51.** $4m^2p - 12mnp + 9n^2p$ **52.** $16p^2r - 40pqr + 25q^2r$

53. $x^3 + 1$ **54.** $x^3 + 27$

55. $8t^3 + 125$ **56.** $27s^3 + 64$

57. $t^6 - 125$ **58.** $w^6 - 27$

59. $5xt + 15xr + 2yt + 6yr$ **60.** $3am + 18mb + 2an + 12nb$

61. $6ar + 12br - 5as - 10bs$ **62.** $7mt + 35ms - 2nt - 10ns$

63. $t^4 - 1$ **64.** $r^4 - 81$

65. $4x^2 + 12xy + 9y^2 - 1$ **66.** $81t^2 + 36ty + 4y^2 - 9$

67. $4x^2 - 28x + 40$ **68.** $2x^2 - 18x + 36$

Synthetic Division

OBJECTIVE 1 Use synthetic division to divide by a polynomial of the form $x - k$.

If a polynomial in x is divided by a binomial of the form $x - k$, a shortcut method called **synthetic division** can be used. Consider the following.

Polynomial Division	Synthetic Division

$$
\begin{array}{r}
3x^2 + 9x + 25 \\
x - 3\overline{)3x^3 + 0x^2 - 2x + 5} \\
\underline{3x^3 - 9x^2} \\
9x^2 - 2x \\
\underline{9x^2 - 27x} \\
25x + 5 \\
\underline{25x - 75} \\
80
\end{array}
$$

$$
\begin{array}{r}
3 \quad 9 \quad 25 \\
1 - 3\overline{)3 \quad 0 \quad -2 \quad 5} \\
\underline{3 \ -9} \\
9 \ -2 \\
\underline{9 \ -27} \\
25 \quad 5 \\
\underline{25 \ -75} \\
80
\end{array}
$$

On the right above, exactly the same division is shown written without the variables. This is why it is *essential* to use 0 as a placeholder in synthetic division. All the numbers in color on the right are repetitions of the numbers directly above them, so we omit them, as shown on the left below.

$$
\begin{array}{r}
3 \quad 9 \quad 25 \\
1 - 3\overline{)3 \quad 0 \quad -2 \quad 5} \\
\underline{-9} \\
9 \ -2 \\
\underline{-27} \\
25 \quad 5 \\
\underline{-75} \\
80
\end{array}
$$

$$
\begin{array}{r}
3 \quad 9 \quad 25 \\
1 - 3\overline{)3 \quad 0 \quad -2 \quad 5} \\
\underline{-9} \\
9 \\
\underline{-27} \\
25 \\
\underline{-75} \\
80
\end{array}
$$

The numbers in color on the left are again repetitions of the numbers directly above them. They too are omitted, as shown on the right above. If we bring the 3 in the dividend down to the beginning of the bottom row, the top row can be omitted because it duplicates the bottom row.

$$
\begin{array}{r}
1 - 3\overline{)3 \quad 0 \quad -2 \quad 5} \\
\underline{-9 \ -27 \ -75} \\
3 \quad 9 \quad 25 \quad 80
\end{array}
$$

We omit the 1 at the upper left—it represents $1x$, which will always be the first term in the divisor. To simplify the arithmetic, we replace subtraction in the second row by addition. To compensate, we change the -3 at the upper left to its additive inverse, 3.

$$\text{Additive inverse} \longrightarrow 3\overline{)3 \quad\; 0 \quad -2 \quad\;\; 5}$$

$$\begin{array}{cccc} & 9 & 27 & 75 \longleftarrow \text{Signs changed} \\ \hline 3 & 9 & 25 & 80 \longleftarrow \text{Remainder} \end{array}$$

The quotient is read from the bottom row.

$$3x^2 + 9x + 25 + \frac{80}{x - 3}$$

The first three numbers in the bottom row are the coefficients of the quotient polynomial with degree 1 less than the degree of the dividend. The last number gives the remainder.

> ***Synthetic division is used only when dividing a polynomial by a binomial of the form $x - k$.***

**NOW TRY
EXERCISE 1**

Use synthetic division to divide $4x^3 + 18x^2 + 19x + 7$ by $x + 3$.

EXAMPLE 1   Using Synthetic Division

Use synthetic division to divide $5x^2 + 16x + 15$ by $x + 2$.

We change $x + 2$ into the form $x - k$ by writing it as

$$x + 2 = x - (-2), \quad \text{where } k = -2.$$

Now we write the coefficients of $5x^2 + 16x + 15$, placing -2 to the left.

$$x + 2 \text{ leads to } -2. \longrightarrow -2\overline{)5 \quad 16 \quad 15} \longleftarrow \text{Coefficients}$$

$$-2\overline{)5 \quad\;\; 16 \quad\;\; 15}$$
$$ \;\; {-10}$$
$$5$$

Bring down the 5, and multiply: $-2 \cdot 5 = -10.$

$$-2\overline{)5 \quad\;\; 16 \quad\;\; 15}$$
$$\;\; {-10} \quad {-12}$$
$$5 \quad\;\; 6$$

Add 16 and -10, getting 6, and multiply 6 and -2 to get -12.

$$-2\overline{)5 \quad\;\; 16 \quad\;\; 15}$$
$$\;\; {-10} \quad {-12}$$

Add 15 and -12, getting 3.

Read the result from the bottom row. $\longrightarrow$

$$5 \quad\;\; 6 \quad\;\;\; 3 \longleftarrow \text{Remainder}$$

Remember that a fraction bar means division.

$$\frac{5x^2 + 16x + 15}{x + 2} = 5x + 6 + \frac{3}{x + 2}$$

NOW TRY

**NOW TRY
EXERCISE 2**

Use synthetic division to divide $-3x^4 + 13x^3 - 6x^2 + 31$ by $x - 4$.

EXAMPLE 2   Using Synthetic Division with a Missing Term

Use synthetic division to divide $-4x^5 + x^4 + 6x^3 + 2x^2 + 50$ by $x - 2$.

$$2\overline{)-4 \quad\;\; 1 \quad\;\; 6 \quad\;\; 2 \quad\;\;\; 0 \quad\;\; 50}$$
$$\;\; {-8} \quad {-14} \quad {-16} \quad {-28} \quad {-56}$$
$${-4} \quad {-7} \quad {-8} \quad {-14} \quad {-28} \quad {-6}$$

Use the steps given above, first inserting a 0 for the missing x-term.

Read the result from the bottom row.

NOW TRY ANSWERS

1. $4x^2 + 6x + 1 + \frac{4}{x + 3}$

2. $-3x^3 + x^2 - 2x - 8 + \frac{-1}{x - 4}$

$$\frac{-4x^5 + x^4 + 6x^3 + 2x^2 + 50}{x - 2} = -4x^4 - 7x^3 - 8x^2 - 14x - 28 + \frac{-6}{x - 2}$$

NOW TRY

OBJECTIVE 2 Use the remainder theorem to evaluate a polynomial.

We can use synthetic division to evaluate polynomials. In the synthetic division of **Example 2,** where the polynomial was divided by $x - 2$, the remainder was -6. Replacing x in the polynomial with 2 gives the following.

$$-4x^5 + x^4 + 6x^3 + 2x^2 + 50$$

$$= -4 \cdot 2^5 + 2^4 + 6 \cdot 2^3 + 2 \cdot 2^2 + 50 \qquad \text{Replace } x \text{ with 2.}$$

$$= -4 \cdot 32 + 16 + 6 \cdot 8 + 2 \cdot 4 + 50 \qquad \text{Evaluate the powers.}$$

$$= -128 + 16 + 48 + 8 + 50 \qquad \text{Multiply.}$$

$$= -6 \qquad \text{Add.}$$

This number, -6, is the same number as the remainder. Dividing by $x - 2$ produced a remainder equal to the result when x is replaced with 2. This always happens, as the following **remainder theorem** states. This result is proved in more advanced courses.

> **Remainder Theorem**
>
> If the polynomial $P(x)$ is divided by $x - k$, then the remainder is equal to $P(k)$.

**NOW TRY
EXERCISE 3**

Let $P(x) = 3x^3 - 2x^2 + 5x + 30$. Use the remainder theorem to evaluate $P(-2)$.

EXAMPLE 3 Using the Remainder Theorem

Let $P(x) = 2x^3 - 5x^2 - 3x + 11$. Use the remainder theorem to evaluate $P(-2)$.

Divide $P(x)$ by $x - (-2)$, using synthetic division.

$$\text{Value of } k \longrightarrow -2)\overline{\begin{array}{rrrr} 2 & -5 & -3 & 11 \\ & -4 & 18 & -30 \\ \hline 2 & -9 & 15 & -19 \end{array}} \longleftarrow \text{Remainder}$$

Thus, $P(-2) = -19$.

NOW TRY

OBJECTIVE 3 Use the remainder theorem to decide whether a given number is a solution of an equation.

**NOW TRY
EXERCISE 4**

Use the remainder theorem to decide whether -4 is a solution of the equation.

$$5x^3 + 19x^2 - 2x + 8 = 0$$

EXAMPLE 4 Using the Remainder Theorem

Use the remainder theorem to decide whether -5 is a solution of the equation.

$$2x^4 + 12x^3 + 6x^2 - 5x + 75 = 0$$

If synthetic division gives a remainder of 0, then -5 is a solution. Otherwise, it is not.

$$\text{Proposed solution} \longrightarrow -5)\overline{\begin{array}{rrrrr} 2 & 12 & 6 & -5 & 75 \\ & -10 & -10 & 20 & -75 \\ \hline 2 & 2 & -4 & 15 & 0 \end{array}} \longleftarrow \text{Remainder}$$

Because the remainder is 0, the polynomial has value 0 when $k = -5$. So -5 is a solution of the given equation.

NOW TRY

The synthetic division in **Example 4** shows that $x - (-5)$ divides the polynomial with 0 remainder. Thus $x - (-5) = x + 5$ is a *factor* of the polynomial.

NOW TRY ANSWERS
3. -12 **4.** yes

$$2x^4 + 12x^3 + 6x^2 - 5x + 75 \quad \text{factors as} \quad (x + 5)(2x^3 + 2x^2 - 4x + 15).$$

The second factor is the quotient polynomial in the last row of the synthetic division.

B Exercises

FOR EXTRA HELP   MyMathLab®

Concept Check   *Complete the statement.*

1. The factored form of $2x^3 + x^2 - x + 10$ is $(x + 2)(2x^2 - 3x + 5)$. The remainder theorem assures us that when $2x^3 + x^2 - x + 10$ is divided by $x + 2$, the remainder is _____.

2. See **Exercise 1.** Given $P(x) = 2x^3 + x^2 - x + 10$, we can determine that $P(-2) =$ _____.

Use synthetic division to find each quotient. **See Examples 1 and 2.**

3. $\dfrac{x^2 - 6x + 5}{x - 1}$

4. $\dfrac{x^2 - 4x - 21}{x + 3}$

5. $\dfrac{4m^2 + 19m - 5}{m + 5}$

6. $\dfrac{3x^2 - 5x - 12}{x - 3}$

7. $\dfrac{2a^2 + 8a + 13}{a + 2}$

8. $\dfrac{4y^2 - 5y - 20}{y - 4}$

9. $(p^2 - 3p + 5) \div (p + 1)$

10. $(z^2 + 4z - 6) \div (z - 5)$

11. $\dfrac{4a^3 - 3a^2 + 2a - 3}{a - 1}$

12. $\dfrac{5p^3 - 6p^2 + 3p + 14}{p + 1}$

13. $(x^5 - 2x^3 + 3x^2 - 4x - 2) \div (x - 2)$

14. $(2y^5 - 5y^4 - 3y^2 - 6y - 23) \div (y - 3)$

15. $(-4r^6 - 3r^5 - 3r^4 + 5r^3 - 6r^2 + 3r + 3) \div (r - 1)$

16. $(2t^6 - 3t^5 + 2t^4 - 5t^3 + 6t^2 - 3t - 2) \div (t - 2)$

17. $(-3y^5 + 2y^4 - 5y^3 - 6y^2 - 1) \div (y + 2)$

18. $(m^6 + 2m^4 - 5m + 11) \div (m - 2)$

Use the remainder theorem to evaluate $P(k)$. **See Example 3.**

19. $P(x) = 2x^3 - 4x^2 + 5x - 3;$ $k = 2$

20. $P(x) = x^3 + 3x^2 - x + 5;$ $k = -1$

21. $P(x) = -x^3 - 5x^2 - 4x - 2;$ $k = -4$

22. $P(x) = -x^3 + 5x^2 - 3x + 4;$ $k = 3$

23. $P(x) = 2x^3 - 4x^2 + 5x - 33;$ $k = 3$

24. $P(x) = x^3 - 3x^2 + 4x - 4;$ $k = 2$

Use the remainder theorem to decide whether the given number is a solution of the equation. **See Example 4.**

25. $x^3 - 2x^2 - 3x + 10 = 0;$ $x = -2$

26. $x^3 - 3x^2 - x + 10 = 0;$ $x = -2$

27. $3x^3 + 2x^2 - 2x + 11 = 0;$ $x = -2$

28. $3x^3 + 10x^2 + 3x - 9 = 0;$ $x = -2$

29. $2x^3 - x^2 - 13x + 24 = 0;$ $x = -3$

30. $5x^3 + 22x^2 + x - 28 = 0;$ $x = -4$

31. $x^4 + 2x^3 - 3x^2 + 8x = 8;$ $x = -2$

32. $x^4 - x^3 - 6x^2 + 5x = -10;$ $x = -2$

RELATING CONCEPTS   For Individual or Group Work   (Exercises 33–38)

We can show a connection between dividing one polynomial by another and factoring the first polynomial. Let $P(x) = 2x^2 + 5x - 12$. **Work Exercises 33–38 in order.**

33. Factor $P(x)$.   34. Solve $P(x) = 0$.   35. Evaluate $P(-4)$.   36. Evaluate $P\left(\frac{3}{2}\right)$.

37. Complete the sentence: If $P(a) = 0$, then $x -$ _____ is a factor of $P(x)$.

38. Use the conclusion reached in **Exercise 37** to decide whether $x - 3$ is a factor of $Q(x) = 3x^3 - 4x^2 - 17x + 6$. Factor $Q(x)$ completely.

ANSWERS TO SELECTED EXERCISES

In this section we provide the answers that we think most students will obtain when they work the exercises using the methods explained in the text. If your answer does not look exactly like the one given here, it is not necessarily wrong. In many cases, there are equivalent forms of the answer that are correct. For example, if the answer section shows $\frac{3}{4}$ and your answer is 0.75, you have obtained the right answer, but written it in a different (yet equivalent) form. Unless the directions specify otherwise, 0.75 is just as valid an answer as $\frac{3}{4}$.

In general, if your answer does not agree with the one given in the text, see whether it can be transformed into the other form. If it can, then it is the correct answer. If you still have doubts, talk with your instructor.

R   PREALGEBRA REVIEW

Section R.1 (pages 12–16)

1. true   **3.** false; This is an improper fraction. Its value is 1.
5. false; The fraction $\frac{13}{39}$ is written in lowest terms as $\frac{1}{3}$.
7. false; *Product* refers to multiplication, so the product of 10 and 2 is 20.   **9.** C   **11.** A   **13.** prime   **15.** composite; $2 \cdot 3 \cdot 5$
17. composite; $2 \cdot 2 \cdot 2 \cdot 2 \cdot 2 \cdot 2$ **19.** neither   **21.** composite; $3 \cdot 19$
23. prime   **25.** composite; $2 \cdot 2 \cdot 31$ **27.** composite; $2 \cdot 2 \cdot 5 \cdot 5 \cdot 5$
29. composite; $2 \cdot 7 \cdot 13 \cdot 19$ **31.** $\frac{1}{2}$ **33.** $\frac{5}{6}$ **35.** $\frac{16}{25}$ **37.** $\frac{1}{5}$ **39.** $\frac{6}{5}$
41. $1\frac{5}{7}$ **43.** $6\frac{5}{12}$ **45.** $7\frac{6}{11}$ **47.** $\frac{13}{5}$ **49.** $\frac{83}{8}$ **51.** $\frac{51}{4}$ **53.** $\frac{24}{35}$
55. $\frac{5}{8}$ **57.** $\frac{6}{25}$ **59.** $\frac{6}{5}$, or $1\frac{1}{5}$ **61.** 9   **63.** $\frac{65}{12}$, or $5\frac{5}{12}$ **65.** $\frac{38}{5}$, or $7\frac{3}{5}$
67. $\frac{10}{3}$, or $3\frac{1}{3}$ **69.** 12   **71.** $\frac{1}{16}$ **73.** 10   **75.** 18   **77.** $\frac{35}{24}$, or $1\frac{11}{24}$
79. $\frac{84}{47}$, or $1\frac{37}{47}$ **81.** $\frac{11}{15}$ **83.** $\frac{2}{3}$ **85.** $\frac{8}{9}$ **87.** $\frac{29}{24}$, or $1\frac{5}{24}$ **89.** $\frac{43}{8}$, or $5\frac{3}{8}$
91. $\frac{101}{20}$, or $5\frac{1}{20}$ **93.** $\frac{5}{9}$ **95.** $\frac{2}{3}$ **97.** $\frac{1}{4}$ **99.** $\frac{17}{36}$ **101.** $\frac{67}{20}$, or $3\frac{7}{20}$
103. $\frac{11}{12}$ **105.** (a) $\frac{1}{2}$ (b) $\frac{1}{4}$ (c) $\frac{1}{3}$ (d) $\frac{1}{6}$ **107.** 6 cups   **109.** $1\frac{1}{8}$ in.
111. $\frac{9}{16}$ in.   **113.** $618\frac{3}{4}$ ft   **115.** $5\frac{5}{24}$ in.   **117.** 8 cakes (There will be some sugar left over.)   **119.** $16\frac{5}{8}$ yd   **121.** $3\frac{3}{8}$ in.   **123.** $\frac{3}{50}$
125. $4\frac{4}{5}$ million, or 4,800,000   **127.** C

Section R.2 (pages 23–25)

1. (a) 6   (b) 9   (c) 1   (d) 7   (e) 4   **3.** (a) 46.25   (b) 46.2
(c) 46   (d) 50   **5.** $\frac{4}{10}$ **7.** $\frac{64}{100}$ **9.** $\frac{138}{1000}$ **11.** $\frac{43}{1000}$ **13.** $\frac{3805}{1000}$
15. 143.094   **17.** 25.61   **19.** 15.33   **21.** 21.77   **23.** 81.716
25. 15.211   **27.** 116.48   **29.** 0.006   **31.** 7.15   **33.** 2.05
35. 5711.6   **37.** 94   **39.** 0.162   **41.** 1.2403   **43.** 1%   **45.** $\frac{1}{20}$
47. $12\frac{1}{2}$%, or 12.5%   **49.** 0.25; 25%   **51.** $\frac{1}{2}$; 0.5   **53.** $\frac{3}{4}$; 75%
55. 0.375   **57.** 1.25   **59.** $0.\overline{5}$; 0.556   **61.** $0.1\overline{6}$; 0.167   **63.** 0.54
65. 0.07   **67.** 1.17   **69.** 0.024   **71.** 0.0625   **73.** 0.008   **75.** 79%
77. 2%   **79.** 0.4%   **81.** 128%   **83.** 40%   **85.** 600%   **87.** $\frac{51}{100}$

---

89. $\frac{3}{20}$ **91.** $\frac{1}{50}$ **93.** $\frac{7}{5}$, or $1\frac{2}{5}$ **95.** $\frac{3}{40}$ **97.** 80%   **99.** 14%
101. $18.\overline{18}$%   **103.** 225%   **105.** $216.\overline{6}$%   **107.** 160   **109.** 4.8
111. 109.2   **113.** $17.80; $106.80   **115.** $119.25; $675.75
117. 19.8 million   **119.** 13%

1   THE REAL NUMBER SYSTEM

Section 1.1 (pages 33–35)

1. false; $3^2 = 3 \cdot 3 = 9$ **3.** false; A number raised to the first power is that number, so $3^1 = 3$.   **5.** false; $4 + 3(8 - 2)$ means $4 + 3(6)$, which simplifies to $4 + 18$, or 22. The common error leading to 42 is adding 4 to 3 and then multiplying by 6. One must follow the rules for order of operations.   **7.** ①, ②   **9.** ①, ③, ②   **11.** ②, ④, ③, ①   **13.** 9
15. 49   **17.** 144   **19.** 64   **21.** 1000   **23.** 81   **25.** 1024   **27.** $\frac{1}{36}$
29. $\frac{16}{81}$ **31.** 0.064   **33.** 32   **35.** 58   **37.** 22.2   **39.** $\frac{49}{30}$, or $1\frac{19}{30}$
41. 12   **43.** 13   **45.** 26   **47.** 4   **49.** 42   **51.** 5   **53.** 41
55. 95   **57.** 90   **59.** 14   **61.** 9   **63.** $3 \cdot (6 + 4) \cdot 2$
65. $10 - (7 - 3)$ **67.** $(8 + 2)^2$ **69.** $16 \leq 16$; true
71. $61 \leq 60$; false   **73.** $0 \geq 0$; true   **75.** $45 \geq 46$; false
77. $66 > 72$; false   **79.** $2 \geq 3$; false   **81.** $3 \geq 3$; true
83. Five is less than seventeen; true   **85.** Five is not equal to eight; true
87. Seven is greater than or equal to fourteen; false   **89.** Fifteen is less than or equal to fifteen; true   **91.** One-third is equal to three-tenths; false
93. Two and five-tenths is greater than two and fifty-hundredths; false
95. $15 = 5 + 10$ **97.** $9 > 5 - 4$ **99.** $16 \neq 19$ **101.** $\frac{1}{2} \leq \frac{2}{4}$
103. $20 > 5$ **105.** $\frac{3}{4} < \frac{4}{5}$ **107.** $1.3 \leq 2.5$ **109.** (a) $14.7 - 40 \cdot 0.13$
(b) 9.5   (c) 8.075; walking (5 mph)   (d) $14.7 - 55 \cdot 0.11$;
8.65; 7.3525, swimming   **111.** Alaska, Texas, California, Idaho
113. Alaska, Texas, California, Idaho, Missouri

Section 1.2 (pages 40–42)

1. B   **3.** A   **5.** $2x^3 = 2 \cdot x \cdot x \cdot x$, while $2x \cdot 2x \cdot 2x = (2x)^3$.
7. The exponent 2 applies only to its base, which is x.   **9.** (a) 11
(b) 13   **11.** (a) 16   (b) 24   **13.** (a) 64   (b) 144   **15.** (a) $\frac{5}{3}$
(b) $\frac{7}{3}$ **17.** (a) $\frac{7}{8}$ (b) $\frac{13}{12}$ **19.** (a) 52   (b) 114   **21.** (a) 25.836
(b) 38.754   **23.** (a) 24   (b) 28   **25.** (a) 12   (b) 33   **27.** (a) 6
(b) $\frac{9}{5}$ **29.** (a) $\frac{4}{3}$ (b) $\frac{13}{6}$ **31.** (a) $\frac{2}{7}$ (b) $\frac{16}{27}$ **33.** (a) 12   (b) 55
35. (a) 1   (b) $\frac{28}{17}$ **37.** (a) 3.684   (b) 8.841   **39.** $12x$ **41.** $x + 9$
43. $x - 4$ **45.** $7 - x$ **47.** $x - 8$ **49.** $\frac{18}{x}$ **51.** $6(x - 4)$
53. yes   **55.** no   **57.** yes   **59.** yes   **61.** yes   **63.** no
65. $x + 8 = 18$; 10   **67.** $2x + 1 = 5$; 2   **69.** $16 - \frac{3}{4}x = 13$; 4
71. $3x = 2x + 8$; 8   **73.** expression   **75.** equation   **77.** equation
79. 70 yr   **81.** 76 yr

Section 1.3 (pages 49–51)

1. 0   **3.** positive   **5.** quotient; denominator   **7. (a)** A   **(b)** A   **(c)** B
(d) B   **9.** This is not true. The absolute value of 0 is 0, and 0 is not positive. We could say that *absolute value is never negative,* or *absolute value
is always nonnegative.*   **11.** 4   **13.** 0   **15.** One example is $\sqrt{13}$.
There are others.   **17.** true   **19.** true   **21.** false

In Exercises 23–27, answers will vary.

23. $\frac{1}{2}, \frac{5}{8}, 1\frac{3}{4}$ **25.** $-3\frac{1}{2}, -\frac{2}{3}, \frac{3}{7}$ **27.** $\sqrt{5}, \pi, -\sqrt{3}$
29. 2,259,105   **31.** -3424 **33.** 46.77
35.   **37.**

39.   **41. (a)** 3, 7   **(b)** 0, 3, 7

(c) $-9, 0, 3, 7$ **(d)** $-9, -1\frac{1}{4}, -\frac{3}{5}, 0, 0.\overline{1}, 3, 5.9, 7$ **(e)** $-\sqrt{7}, \sqrt{5}$
(f) All are real numbers.   **43. (a)** 7   **(b)** 7   **45. (a)** -8 **(b)** 8
47. (a) $\frac{3}{4}$ **(b)** $\frac{3}{4}$ **49. (a)** -5.6 **(b)** 5.6   **51.** 6   **53.** -12
55. $-\frac{2}{3}$ **57.** 3   **59.** -3 **61.** -11 **63.** -7 **65.** 4
67. $|-3.5|$, or 3.5   **69.** $-|-6|$, or -6 **71.** $|5-3|$, or 2   **73.** true
75. true   **77.** true   **79.** false   **81.** true   **83.** false
85. Public transportation, 2010 to 2011
87. Communication, 2009 to 2010

Section 1.4 (pages 60–65)

1. negative   **3.** negative

5. $-8; -6; 2$ **7.** positive   **9.** negative   **11.** -8 **13.** -12
15. 2   **17.** -2 **19.** -9 **21.** 0   **23.** $-\frac{3}{5}$ **25.** $\frac{1}{2}$ **27.** $-\frac{19}{24}$
29. $-\frac{3}{4}$ **31.** 8.9   **33.** -6.01 **35.** 12   **37.** 5   **39.** 2   **41.** -9
43. 0   **45.** -7.7 **47.** -8 **49.** 0   **51.** -20 **53.** -3 **55.** -4
57. -8 **59.** -14 **61.** 9   **63.** -4 **65.** 4   **67.** $\frac{3}{4}$ **69.** $-\frac{11}{8}$, or $-1\frac{3}{8}$
71. $\frac{15}{8}$, or $1\frac{7}{8}$ **73.** 11.6   **75.** -9.9 **77.** 10   **79.** -5 **81.** 11
83. -10 **85.** 22   **87.** -2 **89.** $-\frac{17}{8}$, or $-2\frac{1}{8}$ **91.** $-\frac{1}{4}$, or -0.25
93. -6 **95.** -12 **97.** -5.891 **99.** $-5 + 12 + 6; 13$
101. $[-19 + (-4)] + 14; -9$ **103.** $[-4 + (-10)] + 12; -2$
105. $\left[\frac{5}{7} + \left(-\frac{9}{7}\right)\right] + \frac{2}{7}; -\frac{2}{7}$ **107.** $4 - (-8); 12$
109. $-2 - 8; -10$ **111.** $[9 + (-4)] - 7; -2$
113. $[8 - (-5)] - 12; 1$ **115.** -10 **117.** $+10$ **119.** -12
121. -184 m   **123.** $120°$F   **125.** $-69°$F   **127.** 17
129. (a) 4.9%   **(b)** Americans spent more money than they earned,
which means they had to dip into savings or increase borrowing.
131. $5540   **133.** $1045.55   **135.** $323.83   **137.** -11.03%
139. 13.8%   **141.** 50,395 ft   **143.** 1345 ft   **145.** 136 ft

Section 1.5 (pages 74–77)

1. greater than 0   **3.** less than 0   **5.** greater than 0   **7.** equal to 0
9. undefined; 0; Examples include $\frac{1}{0}$, which is undefined, and $\frac{0}{1}$, which
equals 0.   **11.** -30 **13.** 30   **15.** 120   **17.** -33 **19.** 0   **21.** -2.38

23. $\frac{5}{12}$ **25.** $-\frac{1}{6}$ **27.** 6   **29.** $-32, -16, -8, -4, -2, -1, 1, 2, 4, 8, 16, 32$
31. $-40, -20, -10, -8, -5, -4, -2, -1, 1, 2, 4, 5, 8, 10, 20, 40$
33. $-31, -1, 1, 31$ **35.** 3   **37.** -7 **39.** 8   **41.** -6
43. -4 **45.** $\frac{32}{3}$, or $10\frac{2}{3}$ **47.** $-\frac{15}{16}$ **49.** 0   **51.** undefined
53. -11 **55.** -2 **57.** 35   **59.** 13   **61.** -22 **63.** 6   **65.** -18
67. 67   **69.** -8 **71.** 3   **73.** 7   **75.** 4   **77.** -1 **79.** 4   **81.** -3
83. 29   **85.** 47   **87.** 72   **89.** $-\frac{78}{25}$ **91.** 0   **93.** -23
95. undefined   **97.** $9 + (-9)(2); -9$ **99.** $-4 - 2(-1)(6); 8$
101. $(1.5)(-3.2) - 9; -13.8$ **103.** $12[9 - (-8)]; 204$
105. $\frac{-12}{-5 + (-1)}; 2$ **107.** $\frac{15 + (-3)}{4(-3)}; -1$ **109.** $\frac{2}{3}[8 - (-1)]; 6$
111. $0.20(-5 \cdot 6); -6$ **113.** $\left(\frac{1}{2} + \frac{5}{8}\right)\left(\frac{3}{5} - \frac{1}{3}\right); \frac{3}{10}$ **115.** $\frac{-\frac{1}{2}\left(\frac{3}{4}\right)}{-\frac{2}{3}}; \frac{9}{16}$
117. $\frac{x}{3} = -3; -9$ **119.** $x - 6 = 4; 10$ **121.** $x + 5 = -5; -10$
123. $8\frac{2}{5}$ **125.** 4   **127.** 2   **129. (a)** 6 is divisible by 2.
(b) 9 is not divisible by 2.   **131. (a)** 64 is divisible by 4.
(b) 35 is not divisible by 4.   **133. (a)** 2 is divisible by 2 and
$1 + 5 + 2 + 4 + 8 + 2 + 2 = 24$ is divisible by 3.   **(b)** Although 0 is
divisible by 2, $2 + 8 + 7 + 3 + 5 + 9 + 0 = 34$ is not divisible by 3.
135. (a) $4 + 1 + 1 + 4 + 1 + 0 + 7 = 18$ is divisible by 9.
(b) $2 + 2 + 8 + 7 + 3 + 2 + 1 = 25$ is not divisible by 9.

Section 1.6 (pages 85–87)

1. (a) B   **(b)** F   **(c)** C   **(d)** I   **(e)** B   **(f)** D, F   **(g)** B   **(h)** A
(i) G   **(j)** H   **3.** yes   **5.** no   **7.** no   **9.** (foreign sales) clerk;
foreign (sales clerk)   **11.** Subtraction is not associative.
13. row 1: $-5, \frac{1}{5}$; row 2: 10, $-\frac{1}{10}$; row 3: $\frac{1}{2}$, -2; row 4: $-\frac{3}{8}, \frac{8}{3}$; row 5: $-x$,
$\frac{1}{x}$; row 6: y, $-\frac{1}{y}$; opposite; the same   **15.** -15; commutative property
17. 3; commutative property   **19.** 6; associative property
21. 7; associative property   **23.** commutative property
25. associative property   **27.** associative property
29. inverse property   **31.** inverse property   **33.** identity property
35. commutative property   **37.** distributive property
39. identity property   **41.** distributive property   **43.** 150   **45.** 2010
47. 400   **49.** 1400   **51.** 470   **53.** -9300 **55.** 11   **57.** 0
59. -0.38 **61.** 1   **63.** The student made a sign error. The expression following the first equality symbol should be $-3(4) - 3(-6)$.
$-3(4 - 6)$ means $-3(4) - 3(-6)$, which simplifies to $-12 + 18$, or 6.
65. We must multiply $\frac{3}{4}$ by 1 in the form of a fraction, $\frac{3}{3}$; $\frac{3}{4} \cdot \frac{3}{3} = \frac{9}{12}$
67. 85   **69.** $4t + 12$ **71.** $7z - 56$ **73.** $-8r - 24$ **75.** $-2x - \frac{3}{4}$
77. $-5y + 20$ **79.** $12x + 10$ **81.** $-6x + 15$ **83.** $-48x - 6$
85. $-16y - 20z$ **87.** $24r + 32s - 40y$ **89.** $-24x - 9y - 12z$
91. $-4t - 3m$ **93.** $5c + 4d$ **95.** $q - 5r + 8s$

Section 1.7 (pages 91–93)

1. B   **3.** C   **5.** The student made a sign error when applying the
distributive property: $7x - 2(3 - 2x)$ means $7x - 2(3) - 2(-2x)$, which
simplifies to $7x - 6 + 4x$, or $11x - 6$.   **7.** $4r + 11$ **9.** $21x - 28y$
11. $5 + 2x - 6y$ **13.** $-7 + 3p$ **15.** $2 - 3x$ **17.** -12 **19.** 3

21. 1   **23.** -1 **25.** $\frac{1}{2}$ **27.** $\frac{2}{5}$ **29.** -0.5 **31.** 10   **33.** like

35. unlike   **37.** like   **39.** unlike   **41.** $13y$ **43.** $-9x$ **45.** $13b$

47. $7k + 15$ **49.** $-4y$ **51.** $2x + 6$ **53.** $14 - 7m$ **55.** $-17 + x$

57. $23x$ **59.** $-\frac{28}{3} - \frac{1}{3}t$ **61.** $9y^2$ **63.** $-14p^3 + 5p^2$ **65.** $8x + 15$

67. $22 - 4y$ **69.** $-19p + 16$ **71.** $-t + 3$ **73.** $5x + 15$

75. $15 - 9x$ **77.** $-16y + 63$ **79.** $4r + 15$ **81.** $12k - 5$

83. $-\frac{3}{2}y + 16$ **85.** $-2x + 4$ **87.** $-\frac{14}{3}x - \frac{22}{3}$ **89.** $-23.7y - 12.6$

91. $-2k - 3$ **93.** $4x - 7$ **95.** $(4x + 8) + (3x - 2)$; $7x + 6$

97. $(5x + 1) - (x - 7)$; $4x + 8$ **99.** $(x + 3) + 5x$; $6x + 3$

101. $(13 + 6x) - (-7x)$; $13 + 13x$ **103.** $2(3x + 4) - (-4 + 6x)$; 12

105. $1000 + 5x$ (dollars)   **106.** $750 + 3y$ (dollars)

107. $1000 + 5x + 750 + 3y$ (dollars)   **108.** $1750 + 5x + 3y$ (dollars)

---

2   LINEAR EQUATIONS AND INEQUALITIES IN ONE VARIABLE

Section 2.1 (pages 109–110)

1. equation; expression   **3.** equivalent equations   **5.** (a) expression;
$x + 15$ (b) expression; $y + 7$ (c) equation; $\{-1\}$ (d) equation;
$\{-17\}$ **7.** A, B   **9.** $\{12\}$ **11.** $\{31\}$ **13.** $\{-3\}$ **15.** $\{4\}$

17. $\{-9\}$ **19.** $\left\{-\frac{3}{4}\right\}$ **21.** $\{-10\}$ **23.** $\{-13\}$ **25.** $\{10\}$

27. $\left\{\frac{4}{15}\right\}$ **29.** $\{6.3\}$ **31.** $\{-16.9\}$ **33.** $\{7\}$ **35.** $\{-4\}$

37. $\{-3\}$ **39.** $\{2\}$ **41.** $\{-6\}$ **43.** $\{-5\}$ **45.** $\{-2\}$ **47.** $\{3\}$

49. $\{0\}$ **51.** $\{-7\}$ **53.** $\{-3\}$ **55.** $\{0\}$ **57.** $\{2\}$ **59.** $\{-16\}$

61. $\{0\}$ **63.** $\{2\}$ **65.** $\{13\}$ **67.** $\{-4\}$ **69.** $\{0\}$ **71.** $\left\{\frac{7}{15}\right\}$

73. $\{7\}$ **75.** $\{-4\}$ **77.** $\{13\}$ **79.** $\{29\}$ **81.** $\{18\}$ **83.** $\{12\}$

85. Answers will vary. One example is $x - 6 = -8$.   **87.** $3x = 2x + 17$;
$\{17\}$ **89.** $7x - 6x = -9$; $\{-9\}$

Section 2.2 (pages 115–117)

1. (a) multiplication property of equality   (b) addition property of
equality   (c) multiplication property of equality   (d) addition property
of equality   **3.** B   **5.** $\frac{5}{4}$ **7.** 10   **9.** $-\frac{2}{9}$ **11.** -1 **13.** 6   **15.** -4

17. 0.12   **19.** -1 **21.** $\{6\}$ **23.** $\left\{\frac{13}{2}\right\}$ **25.** $\{-5\}$ **27.** $\{-4\}$

29. $\left\{-\frac{18}{5}\right\}$, or $\{-3.6\}$ **31.** $\{12\}$ **33.** $\{0\}$ **35.** $\{-12\}$ **37.** $\left\{\frac{3}{4}\right\}$

39. $\{40\}$ **41.** $\{-30\}$ **43.** $\{-2.4\}$ **45.** $\{3.5\}$ **47.** $\{-12.2\}$

49. $\{-48\}$ **51.** $\{72\}$ **53.** $\{-35\}$ **55.** $\{14\}$ **57.** $\{18\}$

59. $\left\{-\frac{27}{35}\right\}$ **61.** $\{3\}$ **63.** $\{-5\}$ **65.** $\{20\}$ **67.** $\{7\}$ **69.** $\{0\}$

71. $\left\{-\frac{3}{5}\right\}$ **73.** $\{18\}$ **75.** $\{-6\}$ **77.** Answers will vary. One
example is $\frac{3}{2}x = -6$.   **79.** $4x = 6$; $\left\{\frac{3}{2}\right\}$ **81.** $\frac{x}{-5} = 2$; $\{-10\}$

Section 2.3 (pages 125–126)

1. Use the addition property of equality to subtract 8 from each side.
3. Clear parentheses by using the distributive property.   **5.** Clear
fractions by multiplying by the LCD, 6.   **7.** (a) identity; B
(b) conditional; A   (c) contradiction; C   **9.** $\{4\}$ **11.** $\{-5\}$ **13.** $\left\{\frac{5}{2}\right\}$

15. $\{-1\}$ **17.** $\left\{-\frac{1}{2}\right\}$ **19.** $\{-3\}$ **21.** $\{5\}$ **23.** $\{0\}$ **25.** $\left\{\frac{4}{3}\right\}$

27. $\left\{-\frac{5}{3}\right\}$ **29.** $\varnothing$ **31.** $\{5\}$ **33.** $\{0\}$ **35.** $\{$all real numbers$\}$

37. $\varnothing$ **39.** $\{5\}$ **41.** $\{12\}$ **43.** $\{11\}$ **45.** $\{0\}$ **47.** $\{18\}$

49. $\{3\}$ **51.** $\left\{\frac{5}{4}\right\}$ **53.** $\{120\}$ **55.** $\{6\}$ **57.** $\{15,000\}$ **59.** $\{8\}$

61. $\{0\}$ **63.** $\{$all real numbers$\}$ **65.** $\{4\}$ **67.** $\{20\}$ **69.** $\varnothing$

71. $11 - q$ **73.** $\frac{9}{x}$ **75.** $65 - h$ **77.** $x + 15$; $x - 5$ **79.** $25r$

81. $\frac{t}{5}$ **83.** $3x + 2y$

Section 2.4 (pages 137–141)

1. D; There cannot be a fractional number of cars.   **3.** A; Distance
cannot be negative.   **5.** $x + 9 = -26$; -35 **7.** $8(x + 6) = 104$; 7
9. $5x + 2 = 4x + 5$; 3   **11.** $3(x - 2) = x + 6$; 6   **13.** $\frac{3}{4}x + 6 = x - 4$; 40

15. $3x + (x + 7) = -11 - 2x$; -3 **17.** *Step 1:* Republicans;
Step 2: $x - 4$; Democrats; *Step 3:* $(x - 4) + x = 150$; Democrats: 73,
Republicans: 77   **19.** New York: 29 screens; Ohio: 27 screens

21. Democrats: 52; Republicans: 46   **23.** Madonna: $228 million;
Springsteen: $199 million   **25.** wins: 66; losses: 16   **27.** orange: 97 mg;
pineapple: 25 mg   **29.** 112 DVDs   **31.** onions: 81.3 kg; grilled steak:
536.3 kg   **33.** 1950 Denver nickel: $16.00; 1945 Philadelphia nickel: $8.00
35. whole wheat: 25.6 oz; rye: 6.4 oz   **37.** American: 18; United: 11;
Southwest: 26   **39.** shortest piece: 15 in.; middle piece: 20 in.; longest
piece: 24 in.   **41.** gold: 46; silver: 29; bronze: 29   **43.** 36 million mi
45. *A* and *B*: 40°; *C*: 100°   **47.** 68, 69   **49.** 101, 102   **51.** 10, 12
53. 17, 19   **55.** 10, 11   **57.** 18   **59.** 15, 17, 19   **61.** 18°   **63.** 20°
65. 39°   **67.** 50°

Section 2.5 (pages 148–152)

1. area   **3.** perimeter   **5.** area   **7.** area   **9.** $P = 26$ **11.** $\mathscr{A} = 64$
13. $b = 4$ **15.** $t = 5.6$ **17.** $B = 14$ **19.** $r = 2.6$ **21.** $r = 10$

23. $\mathscr{A} = 50.24$ **25.** $r = 6$ **27.** $V = 150$ **29.** $V = 52$

31. $V = 7234.56$ **33.** $I = \$600$ **35.** $p = \$550$ **37.** $t = 1.5$ yr
39. length: 18 in.; width: 9 in.   **41.** length: 14 m; width: 4 m
43. shortest: 5 in.; medium: 7 in.; longest: 8 in.   **45.** two equal sides: 7 m;
third side: 10 m   **47.** perimeter: 5.4 m; area: 1.8 m²   **49.** 10 ft
51. 154,000 ft²   **53.** 194.48 ft²; 49.42 ft   **55.** 23,800.10 ft²
57. length: 36 in.; volume: 11,664 in.³   **59.** 48°, 132°   **61.** 55°, 35°
63. 51°, 51°   **65.** 105°, 105°

We give one answer for Exercises 67–103. There are other correct forms.

67. $t = \dfrac{d}{r}$ **69.** $b = \dfrac{\mathscr{A}}{h}$ **71.** $d = \dfrac{C}{\pi}$ **73.** $H = \dfrac{V}{LW}$ **75.** $r = \dfrac{I}{pt}$

77. $h = \dfrac{2\mathscr{A}}{b}$ **79.** $h = \dfrac{3V}{\pi r^2}$ **81.** $b = P - a - c$ **83.** $W = \dfrac{P - 2L}{2}$

85. $m = \dfrac{y - b}{x}$ **87.** $y = \dfrac{C - Ax}{B}$ **89.** $r = \dfrac{M - C}{C}$ **91.** $a = \dfrac{P - 2b}{2}$

93. $b = 2S - a - c$ **95.** $F = \dfrac{9C + 160}{5}$ **97.** $y = -6x + 4$

99. $y = 5x - 2$ **101.** $y = \dfrac{3}{5}x - 3$ **103.** $y = \dfrac{1}{3}x - 4$

Section 2.6 (pages 157–162)

1. (a) C **(b)** D **(c)** B **(d)** A **3.** $\frac{4}{3}$ **5.** $\frac{4}{3}$ **7.** $\frac{15}{2}$ **9.** $\frac{1}{5}$ **11.** $\frac{5}{6}$
13. 10 lb; \$0.749 **15.** 64 oz; \$0.047 **17.** 32 oz; \$0.531
19. 32 oz; \$0.056 **21.** 263 oz; \$0.076 **23.** true **25.** false
27. true **29.** $\{35\}$ **31.** $\{7\}$ **33.** $\left\{\frac{45}{2}\right\}$ **35.** $\{2\}$ **37.** $\{-1\}$
39. $\{5\}$ **41.** $\left\{-\frac{31}{5}\right\}$ **43.** $\{-2\}$ **45.** \$30.00 **47.** \$8.75
49. \$67.50 **51.** \$56.40 **53.** 50,000 fish **55.** 4 ft **57.** 2.7 in.
59. 2.0 in. **61.** $2\frac{5}{8}$ cups **63.** \$404.76 **65.** $x = 4$ **67.** $x = 8$
69. $x = 22.5$; $y = 25.5$

71. (a) 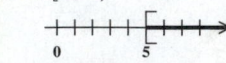   **(b)** 54 ft **73.** \$239 **75.** \$280

77. (a) 2625 mg

(b) $\dfrac{125 \text{ mg}}{5 \text{ mL}} = \dfrac{2625 \text{ mg}}{x \text{ mL}}$

(c) 105 mL

79. 140.4 **81.** 700 **83.** 425 **85.** 8% **87.** 120% **89.** 80%
91. 28% **93.** 32% **95.** \$3000

Section 2.7 (pages 169–173)

1. A **3.** C **5.** D **7.** A **9.** 45 L **11.** \$750 **13.** \$17.50
15. 160 L **17.** $13\frac{1}{3}$ L **19.** 4 L **21.** 20 mL **23.** 4 L **25.** \$2100
at 5%; \$900 at 4% **27.** \$2500 at 6%; \$13,500 at 5% **29.** \$1700 at 8%;
\$800 at 2% **31.** 10 nickels **33.** 46-cent stamps: 25; 20-cent stamps: 20
35. Arabian Mocha: 7 lb; Colombian Decaf: 3.5 lb **37.** 530 mi
39. 2.668 hr **41.** 9.14 m per sec **43.** 8.51 m per sec **45.** $7\frac{1}{2}$ hr
47. 5 hr **49.** $1\frac{3}{4}$ hr **51.** eastbound: 300 mph; westbound: 450 mph
53. slower car: 40 mph; faster car: 60 mph **55.** Bob: 7 yr old; Kevin:
21 yr old **57.** width: 3 ft; length: 9 ft **59.** \$650

Section 2.8 (pages 184–187)

1. $>, <$ (or $<, >$); $\geq, \leq$ (or $\leq, \geq$) **3.** $(0, \infty)$ **5.** $x > -4$ **7.** $x \leq 4$
9. $(-\infty, 4]$ **11.** $(-\infty, -3)$

13. $(4, \infty)$ **15.** $(-\infty, 0]$

17. $\left[-\frac{1}{2}, \infty\right)$ **19.** $[1, \infty)$

21. $[5, \infty)$ **23.** $(-\infty, -11)$

25. It must be reversed when multiplying or dividing by a negative
number.

27. $(-\infty, 6)$ **29.** $[-10, \infty)$

31. $(-\infty, -3)$ **33.** $(-\infty, 0]$

35. $(20, \infty)$ **37.** $[-3, \infty)$

39. $(-\infty, -3]$ **41.** $(-1, \infty)$

43. $[-5, \infty)$ **45.** $(-\infty, 2]$

47. $\left(-\infty, \frac{3}{2}\right)$ **49.** $(-\infty, 1)$

51. $(-\infty, 0]$ **53.** $\left(-\frac{1}{2}, \infty\right)$

55. $[4, \infty)$ **57.** $[2, \infty)$

59. $(-\infty, 32)$ **61.** $(-\infty, 6]$

63. $\left[\frac{5}{12}, \infty\right)$ **65.** $(-21, \infty)$

67. $x \geq 18$ **69.** $x > 5$ **71.** $x \leq 20$ **73.** 83 or greater **75.** more
than 3.8 in. **77.** all numbers greater than 16 **79.** It is never less
than $-13°$F. **81.** 32 or greater **83.** 12 min **85.** $5x - 100$
87. $(5x - 100) - (125 + 4x)$, which simplifies to $x - 225$; $x > 225$
89. $-1 < x < 2$ **91.** $-1 < x \leq 2$

93. $[8, 10]$ **95.** $(0, 10)$

97. $(-3, 4)$ **99.** $(-2, 1]$

101. $[-1, 6]$ **103.** $(1, 3)$

105. $(-6, 2)$ **107.** $\left(-\frac{11}{6}, -\frac{2}{3}\right)$

109. $[3, 7)$

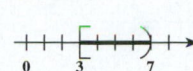

111. $[-26, 6]$

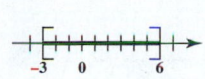

113. $[-3, 6]$

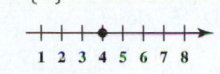

115. $\left[-\frac{24}{5}, 0\right]$

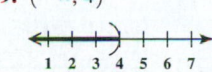

117. $\{4\}$

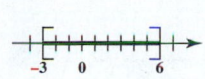

118. $(4, \infty)$

119. $(-\infty, 4)$

120. The graph would be the set of all real numbers.

(c)

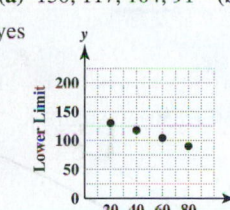

(d) The points lie approximately in a linear pattern. Rates at which 2-year college students complete a degree within 3 years were decreasing.

83. (a) 130; 117; 104; 91 **(b)** (20, 130), (40, 117), (60, 104), (80, 91)

(c) yes

85. between 130 and 170 beats per minute; between 117 and 153 beats per minute

---

3 LINEAR EQUATIONS AND INEQUALITIES IN TWO VARIABLES

Section 3.1 (pages 207–210)

1. does; do not **3.** II **5.** 3 **7.** negative; negative **9.** positive; negative **11.** If $xy < 0$, then either $x < 0$ and $y > 0$ or $x > 0$ and $y < 0$. If $x < 0$ and $y > 0$, then the point lies in quadrant II. If $x > 0$ and $y < 0$, then the point lies in quadrant IV. **13.** between 2010 and 2011 and 2011 and 2012 **15.** 2011: 9%; 2012: 8%; decline: 1%

17. yes **19.** yes **21.** no **23.** yes **25.** yes **27.** no **29.** 17 **31.** -5 **33.** -1 **35.** -7 **37.** 8; 6; 3; (0, 8); (6, 0); (3, 4) **39.** -9; 4; 9; $(-9, 0)$; (0, 4); (9, 8) **41.** 12; 12; 12; (12, 3); (12, 8); (12, 0) **43.** -10; -10; -10; $(4, -10)$; $(0, -10)$; $(-4, -10)$ **45.** -2; -2; -2; $(9, -2)$; $(2, -2)$; $(0, -2)$ **47.** 4; 4; 4; (4, 4); (4, 0); $(4, -4)$ **49.** No, the ordered pair (3, 4) represents the point 3 units to the right of the origin and 4 units up from the x-axis. The ordered pair (4, 3) represents the point 4 units to the right of the origin and 3 units up from the x-axis. **51.** (2, 4); I **53.** $(-5, 4)$; II **55.** (3, 0); no quadrant **57.** $(4, -4)$; IV

59.–69.

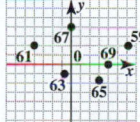

71. -3; 6; -2; 4

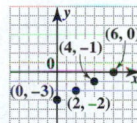

73. -3; 4; -6; $-\frac{4}{3}$

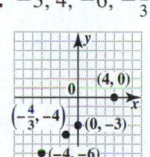

75. -4; -4; -4; -4

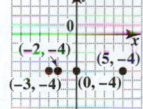

77. The points in each graph appear to lie on a straight line.

79. (a) (5, 45) **(b)** (6, 50)

81. (a) (2008, 29.3), (2009, 28.3), (2010, 28.0), (2011, 26.9), (2012, 25.4), (2013, 22.5) **(b)** (2013, 22.5) means that 22.5 percent of 2-year college students in 2013 received a degree within 3 years.

Section 3.2 (pages 220–223)

1. By; C; 0 **3. (a)** A **(b)** C **(c)** D **(d)** B **5.** x-intercept: (4, 0); y-intercept: $(0, -4)$ **7.** x-intercept: $(-2, 0)$; y-intercept: $(0, -3)$ **9. (a)** D **(b)** C **(c)** B **(d)** A

11. 5; 5; 3

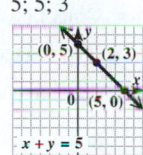

13. 1; 3; -1

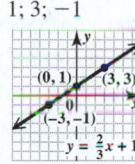

15. -6; -2; -5

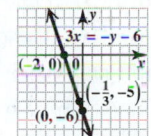

17. $(8, 0)$; $(0, -8)$ **19.** $(4, 0)$; $(0, -10)$ **21.** $(0, 0)$; $(0, 0)$ **23.** $(2, 0)$; $(0, 4)$ **25.** $(6, 0)$; $(0, -2)$ **27.** $(0, 0)$; $(0, 0)$ **29.** $(4, 0)$; no y-intercept **31.** no x-intercept; $(0, 2.5)$

33.

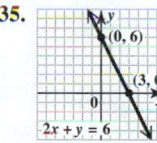

35.

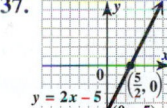

37.

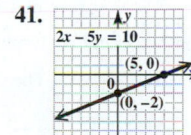

39.

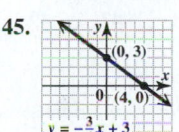

41.

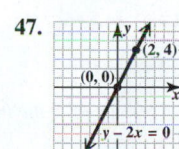

43.

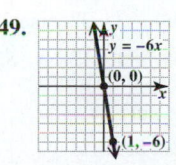

45.

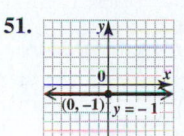

47.

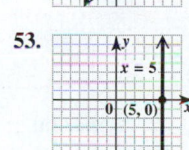

49.

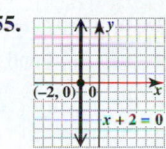

51.

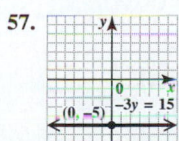

53.

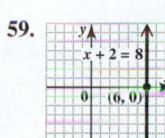

55.

57.

59.

In Exercises 61–67, descriptions may vary.

61. The graph is a line with x-intercept $(-3, 0)$ and y-intercept $(0, 9)$.

63. The graph is a vertical line with x-intercept $(11, 0)$.

65. The graph is a horizontal line with y-intercept $(0, -2)$.

67. The graph has x- and y-intercepts $(0, 0)$. It passes through the points $(2, 1)$ and $(4, 2)$.

69. 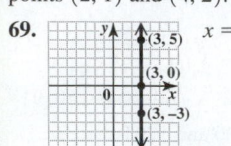 **71.**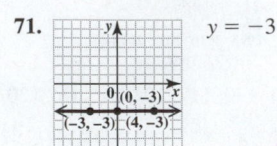

73. (a) 121 lb, 143 lb, 176 lb
(b) $(62, 121), (66, 143),$
$(72, 176)$

75. (a) $62.50; $100   **(b)** 200
(c) $(50, 62.50), (100, 100),$
$(200, 175)$

(c) 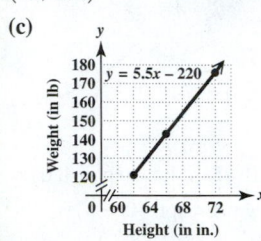   **(d)**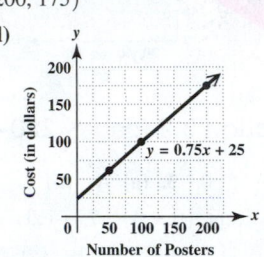

(d) 68 in.; 68 in.

77. (a) $30,000   **(b)** $15,000   **(c)** $5000   **(d)** After 5 yr, the SUV has a value of $5000.   **79. (a)** 2000: 30.1 lb; 2008: 32.6 lb; 2012: 33.8 lb
(b) 2000: 29.8 lb; 2008: 32.8 lb; 2012: 33.5 lb   **(c)** The values are quite close.   **(d)** 36.9 lb; It is very close to the USDA projection.

Section 3.3 (pages 232–236)

1. steepness; vertical; horizontal   **3. (a)** 6   **(b)** 4   **(c)** $\frac{6}{4}$, or $\frac{3}{2}$; slope of the line   **5. (a)** C   **(b)** A   **(c)** D   **(d)** B

In Exercises 7 and 9, sketches will vary.

7. The line must fall from left to right.   **9.** The line must be vertical.
11. (a) negative   **(b)** zero   **13. (a)** positive   **(b)** negative
15. (a) zero   **(b)** negative   **17.** Because he found the difference $3 - 5 = -2$ in the numerator, he should have subtracted in the same order in the denominator to obtain $-1 - 2 = -3$. The correct slope is $\frac{-2}{-3} = \frac{2}{3}$.
19. $\frac{8}{27}$ **21.** $-\frac{2}{3}$ **23.** 4   **25.** $-\frac{1}{2}$ **27.** 0   **29.** $\frac{5}{4}$ **31.** $\frac{3}{2}$ **33.** 0
35. -3 **37.** undefined   **39.** $\frac{1}{4}$ **41.** $-\frac{1}{2}$ **43.** 5   **45.** $\frac{1}{4}$ **47.** $\frac{3}{2}$ **49.** $\frac{3}{2}$
51. -1 **53.** 0   **55.** undefined

In part (a) of Exercises 57 and 59, we used the intercepts. Other points can be used.

57. (a) $(5, 0)$ and $(0, 10)$; -2 **(b)** $y = -2x + 10$; -2
59. (a) $(3, 0)$ and $(0, -5)$; $\frac{5}{3}$ **(b)** $y = \frac{5}{3}x - 5$; $\frac{5}{3}$
61. (a) 1   **(b)** $(-4, 0); (0, 4)$ **63. (a)** $-\frac{1}{3}$ **(b)** $(-6, 0); (0, -2)$

(c) 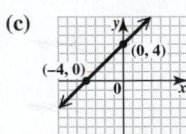   **(c)**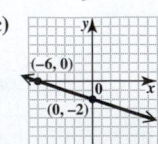

65. $-3; \frac{1}{3}$ **67.** A   **69.** $-\frac{2}{5}; -\frac{2}{5}$; parallel   **71.** $\frac{8}{9}; -\frac{4}{3}$; neither
73. $\frac{3}{2}; -\frac{2}{3}$; perpendicular   **75.** $5; \frac{1}{5}$; neither
77. (a) $(2004, 817), (2012, 1661)$ **(b)** 105.5   **(c)** Music purchases increased by 844 million units in 8 yr, or 105.5 million units per year.
79. 0.5   **80.** positive; increased   **81.** 0.5%   **82.** -0.2 **83.** negative; decreased   **84.** 0.2%

Section 3.4 (pages 242–245)

1. $m; (0, b)$ **3. (a)** C   **(b)** B   **(c)** A   **(d)** D   **5.** slope: $\frac{5}{2}$; y-intercept: $(0, -4)$ **7.** slope: -1; y-intercept: $(0, 9)$ **9.** slope: $\frac{1}{5}$; y-intercept: $\left(0, -\frac{3}{10}\right)$

11. 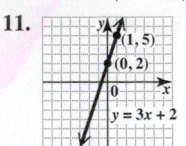   **13.** 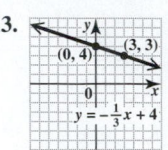   **15.**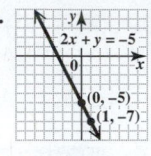

17. 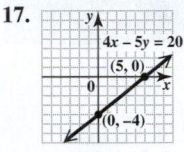   **19.** 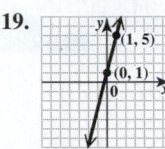   **21.**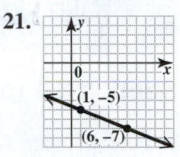

23. 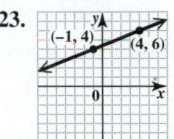   **25.** 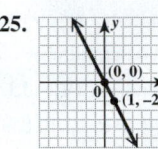   **27.**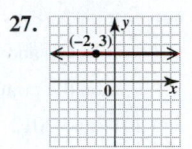

29. 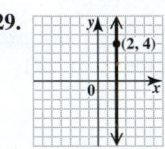   **31.**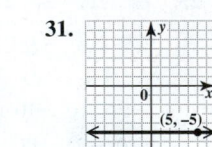

33. y-axis   **35.** $y = 3x - 3$ **37.** $y = -x + 3$ **39.** $y = -\frac{1}{2}x + 2$
41. $y = 4x - 3$ **43.** $y = -x - 7$ **45.** $y = 2x - 7$ **47.** $y = -4x - 1$
49. $y = x - 6$ **51.** $y = \frac{3}{4}x + 4$ **53.** $y = 3$ **55.** $x = 2$
57. $x = 0$ **59.** $y = -6$
61. (a) 2   **(b)** $(0, -1)$ **63. (a)** $-\frac{1}{3}$ **(b)** $(0, -2)$
(c) $y = 2x - 1$ **(c)** $y = -\frac{1}{3}x - 2$
(d) 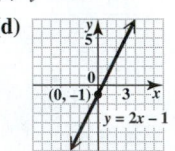   **(d)**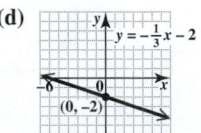

65. (a) 0.05; commission rate   **(b)** $(0; 2000)$; base salary per month
(c) $2500   **(d)** $30,000   **67. (a)** $400   **(b)** $0.25
(c) $y = 0.25x + 400$ **(d)** $425   **(e)** 1500   **69.** $y = -\frac{A}{B}x + \frac{C}{B}$
70. (a) $-\frac{2}{3}$ **(b)** 2   **(c)** $\frac{3}{7}$ **71.** $\left(0, \frac{C}{B}\right)$ **72. (a)** $(0, 6)$ **(b)** $\left(0, \frac{1}{2}\right)$
(c) $(0, -3)$

Section 3.5 (pages 249–252)

1. (a) D   (b) C   (c) B   (d) E   (e) A   **3.** A, B, D   **5.** $y = 5x + 2$

7. $y = x - 9$ **9.** $y = -3x - 4$ **11.** $y = -x + 1$ **13.** $y = \frac{2}{3}x + \frac{19}{3}$

15. $y = -\frac{4}{5}x + \frac{9}{5}$ **17.** (a) $y = x + 6$ (b) $x - y = -6$

19. (a) $y = \frac{1}{2}x + 2$ (b) $x - 2y = -4$ **21.** (a) $y = -\frac{3}{5}x - \frac{11}{5}$

(b) $3x + 5y = -11$ **23.** (a) $y = -\frac{1}{3}x + \frac{22}{9}$ (b) $3x + 9y = 22$

25. (a) $y = 3x - 9$ (b) $3x - y = 9$ **27.** (a) $y = -\frac{2}{3}x + \frac{4}{3}$

(b) $2x + 3y = 4$ **29.** $y = -2x - 3$ **31.** $y = 4x - 5$ **33.** $y = \frac{3}{4}x - \frac{9}{2}$

35. (a) $(1, 2530), (2, 2790), (3, 2940), (4, 3070), (5, 3220)$

(b) yes

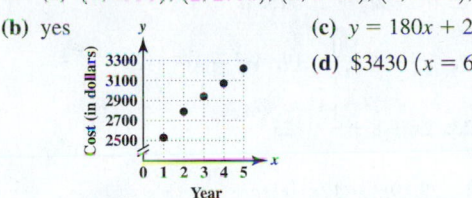

(c) $y = 180x + 2350$

(d) $\$3430 \ (x = 6)$

37. $y = 15x + 37$ **39.** $y = 0.2375x + 59.7$

---

4   EXPONENTS AND POLYNOMIALS

Section 4.1 (pages 268–270)

1. false; $3^3 = 3 \cdot 3 \cdot 3 = 27$ **3.** true   **5.** w^6 **7.** $\left(\frac{1}{2}\right)^6$ **9.** $(-4)^4$

11. $(-7y)^4$ **13.** In $(-3)^4$, -3 is the base. In -3^4, 3 is the base.
$(-3)^4 = 81; -3^4 = -81$ **15.** base: 3; exponent: 5; 243

17. base: -3; exponent: 5; -243 **19.** base: $-6x$; exponent: 4

21. base: x; exponent: 4   **23.** $2; 5; 8^7$ **25.** 5^8 **27.** 4^{12} **29.** $(-7)^9$

31. t^{24} **33.** $-56r^7$ **35.** $42p^{10}$ **37.** $-30x^9$ **39.** The product rule
does not apply.   **41.** The product rule does not apply.   **43.** 4^6 **45.** t^{20}

47. $7^3 r^3$ **49.** $5^5 x^5 y^5$ **51.** 5^{12} **53.** -8^{15} **55.** $8q^3 r^3$ **57.** $\dfrac{9^8}{5^8}$

59. $\dfrac{1}{2^3}$ **61.** $\dfrac{a^3}{b^3}$ **63.** $\dfrac{x^3}{2^3}$ **65.** $-\dfrac{2^5 x^5}{y^5}$ **67.** $\dfrac{5^5}{2^5}$ **69.** $\dfrac{9^5}{8^3}$ **71.** $2^{12} x^{12}$

73. $-6^5 p^5$ **75.** $6^5 x^{10} y^{15}$ **77.** x^{21} **79.** $4w^4 x^{26} y^7$ **81.** $-r^{18} s^{17}$

83. $\dfrac{5^3 a^6 b^{15}}{c^{18}}$, or $\dfrac{125 a^6 b^{15}}{c^{18}}$ **85.** Using the product rule, the expression
should be simplified as follows: $(10^2)^3 = 10^{2 \cdot 3} = 10^6$.   **87.** $12x^5$

89. $6p^7$ **91.** $125x^6$ **93.** $\$304.16$ **95.** $\$1640.16$

Section 4.2 (pages 276–278)

1. negative   **3.** negative   **5.** positive   **7.** 0   **9.** 1   **11.** 1   **13.** -1

15. -1 **17.** 0   **19.** 0   **21.** 0   **23.** 0   **25.** (a) B   (b) C   (c) D

(d) B   (e) E   (f) B   **27.** 2   **29.** $\dfrac{1}{64}$ **31.** 16   **33.** $\dfrac{49}{36}$ **35.** $\dfrac{1}{81}$

37. $\dfrac{8}{15}$ **39.** $-\dfrac{7}{18}$ **41.** $\dfrac{7}{2}$ **43.** 11; 8; 3; 125   **45.** 125   **47.** $\dfrac{125}{9}$ **49.** 25

51. x^{15} **53.** 216   **55.** $2r^4$ **57.** $\dfrac{25}{64}$ **59.** $\dfrac{p^5}{q^8}$ **61.** r^9 **63.** $\dfrac{yz^2}{4x^3}$

65. $a + b$ **67.** $(x + 2y)^2$ **69.** 343   **71.** $\dfrac{1}{x^2}$ **73.** $\dfrac{64x}{9}$ **75.** $\dfrac{x^2 z^4}{y^2}$

77. $6x$ **79.** $\dfrac{1}{m^{10} n^5}$ **81.** $\dfrac{1}{xyz}$ **83.** $x^3 y^9$ **85.** $\dfrac{1}{2r}$ **87.** $\dfrac{1}{-4y}$

---

89. The student attempted to use the quotient rule with unequal bases.
The correct way to simplify is $\dfrac{16^3}{2^2} = \dfrac{(2^4)^3}{2^2} = \dfrac{2^{12}}{2^2} = 2^{10} = 1024$.

91. $\dfrac{a^{11}}{2b^5}$ **93.** $\dfrac{108}{y^5 z^3}$ **95.** $\dfrac{9z^2}{400x^3}$

Section 4.3 (pages 283–287)

1. (a) C   (b) A   (c) B   (d) D   **3.** in scientific notation   **5.** not in
scientific notation; 5.6×10^6 **7.** not in scientific notation; 8×10^1

9. not in scientific notation; 4×10^{-3} **11.** (a) 6; 4; 6.3; 4   (b) 5; 2;
5.71; -2 **13.** 5.876×10^9 **15.** 8.235×10^4 **17.** 7×10^{-6}

19. 2.03×10^{-3} **21.** -1.3×10^7 **23.** -6×10^{-3} **25.** 750,000

27. 5,677,000,000,000   **29.** 1,000,000,000,000   **31.** 6.21   **33.** 0.00078

35. 0.000000005134   **37.** -0.004 **39.** $-810,000$ **41.** (a) 6×10^{11}

(b) 600,000,000,000   **43.** (a) 1.5×10^7 (b) 15,000,000

45. (a) -6×10^4 (b) $-60,000$ **47.** (a) 2.4×10^2 (b) 240

49. (a) 6.3×10^{-2} (b) 0.063   **51.** (a) 3×10^{-4} (b) 0.0003

53. (a) -4×10 (b) -40 **55.** (a) 1.3×10^{-5} (b) 0.000013

57. (a) 5×10^2 (b) 500   **59.** (a) -3×10^6 (b) $-3,000,000$

61. (a) 2×10^{-7} (b) 0.0000002   **63.** 4.7E-7 **65.** 2E7   **67.** 1E1

69. 1.04×10^8 **71.** 9.2×10^{-3} **73.** 6×10^9 **75.** 0.000002

77. 4.2×10^{42} **79.** 1.5×10^{17} mi   **81.** $\$3186$ **83.** $\$53,185$

85. 3.59×10^2, or 359 sec   **87.** $\$98.28$ **89.** 6×10^{17}, or
600,000,000,000,000,000; 3.6×10^{19}, or 36,000,000,000,000,000,000

91. The Chile earthquake was 10 times as intense as the Southern
Sumatra earthquake.   **92.** The Obihoro earthquake was 10 times as
intense as the Hindu Kush earthquake.   **93.** The Kamchatka earthquake
was about 3.16 times as intense as the Southern Sumatra earthquake.

94. The Chile earthquake would be 100 times as intense.

Section 4.4 (pages 294–297)

1. 4; 6   **3.** 9   **5.** 19   **7.** 0   **9.** one; 6   **11.** one; 1   **13.** two; $-19, -1$

15. three; 1, 8, 5   **17.** $2m^5$ **19.** $-r^5$ **21.** It cannot be simplified.

23. $-5x^5$ **25.** $5p^9 + 4p^7$ **27.** 0   **29.** $-2xy^2$ **31.** $-\dfrac{22}{15} tu^7$

33. already simplified; 4; binomial   **35.** $11m^4 - 7m^3 - 3m^2$; 4; trinomial

37. x^4; 4; monomial   **39.** 7; 0; monomial   **41.** $-13ab$; 2; monomial

43. (a) -3 (b) 0   **45.** (a) 14   (b) -19 **47.** (a) 36   (b) -12

49. $5x^2 - 2x$ **51.** $5m^2 + 3m + 2$ **53.** $\dfrac{7}{6}x^2 - \dfrac{2}{15}x + \dfrac{5}{6}$

55. $6m^3 + m^2 + 4m - 14$ **57.** $3y^3 - 11y^2$ **59.** $4x^4 - 4x^2 + 4x$

61. $15m^3 - 13m^2 + 8m + 11$ **63.** $5m^2 - 14m + 6$ **65.** $4x^3 + 2x^2 + 5x$

67. $-11y^4 + 8y^2 + y$ **69.** $a^4 - a^2 + 1$ **71.** $5m^2 + 8m - 10$

73. $-6x^2 - 12x + 12$ **75.** -10 **77.** $4b - 5c$ **79.** $6x - xy - 7$

81. $-3x^2 y - 15xy - 3xy^2$ **83.** $8x^2 + 8x + 6$ **85.** $2x^2 + 8x$

87. $8t^2 + 8t + 13$ **89.** (a) $23y + 5t$ (b) $25°, 67°, 88°$ **91.** $-7x - 1$

93. $0, -3, -4, -3, 0$ **95.** $7, 1, -1, 1, 7$ **97.** $0, 3, 4, 3, 0$

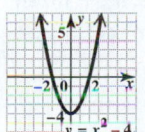
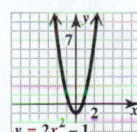
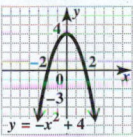

99. 4, 1, 0, 1, 4

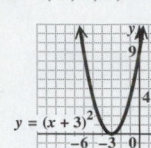

101. 63; If a dog is 9 in dog years, then it is 63 in human years. **102. (a)** 37 **(b)** 68 **(c)** 80 **103.** 2.5; 130 **104.** 6; $27

77. $\frac{1}{2}m^2 - 2n^2$ **79.** $9a^2 - 4$ **81.** $\pi x^2 + 4\pi x + 4\pi$
83. $x^3 + 6x^2 + 12x + 8$ **85.** $(a + b)^2$ **86.** a^2 **87.** $2ab$ **88.** b^2
89. $a^2 + 2ab + b^2$ **90.** They both represent the area of the entire large square. **91.** 1225 **92.** $30^2 + 2(30)(5) + 5^2$ **93.** 1225
94. They are equal.

Section 4.5 (pages 302–305)

1. (a) B **(b)** D **(c)** A **(d)** C **3.** distributive **5.** one
7. $15y^{11}$ **9.** $30a^9$ **11.** $15pq^2$ **13.** $-18m^3n^2$ **15.** $9y^{10}$
17. $-8x^{10}$ **19.** $6m^2 + 4m$ **21.** $-6p^4 + 12p^3$
23. $-16z^2 - 24z^3 - 24z^4$ **25.** $6y^3 + 4y^4 + 10y^7$
27. $28r^5 - 32r^4 + 36r^3$ **29.** $6a^4 - 12a^3b + 15a^2b^2$
31. $3m^2$; $2mn$; $-n^3$; $21m^5n^2 + 14m^4n^3 - 7m^3n^5$
33. $12x^3 + 26x^2 + 10x + 1$ **35.** $72y^3 - 70y^2 + 21y - 2$
37. $20m^4 - m^3 - 8m^2 - 17m - 15$
39. $6x^6 - 3x^5 - 4x^4 + 4x^3 - 5x^2 + 8x - 3$ **41.** $5x^4 - 13x^3 + 20x^2 + 7x + 5$ **43.** $3x^5 + 18x^4 - 2x^3 - 8x^2 + 24x$ **45.** first row: x^2, $4x$; second row: $3x$, 12; Product: $x^2 + 7x + 12$ **47.** first row: $2x^3$, $6x^2$, $4x$; second row: x^2, $3x$, 2; Product: $2x^3 + 7x^2 + 7x + 2$
49. (a) $2p$; $3p$; $6p^2$ **(b)** $2p$; 7; $14p$ **(c)** -5; $3p$; $-15p$ **(d)** -5; 7; -35 **(e)** $6p^2 - p - 35$ **51.** $m^2 + 12m + 35$ **53.** $n^2 + 3n - 4$
55. $12x^2 + 10x - 12$ **57.** $81 - t^2$ **59.** $9x^2 - 12x + 4$
61. $10a^2 + 37a + 7$ **63.** $12 + 8m - 15m^2$ **65.** $20 - 7x - 3x^2$
67. $3t^2 + 5st - 12s^2$ **69.** $8xy - 4x + 6y - 3$ **71.** $15x^2 + xy - 6y^2$
73. $6y^5 - 21y^4 - 45y^3$ **75.** $-200r^7 + 32r^3$ **77. (a)** $3y^2 + 10y + 7$
(b) $8y + 16$ **79.** $x^2 + 14x + 49$ **81.** $a^2 - 16$ **83.** $4p^2 - 20p + 25$
85. $25k^2 + 30kq + 9q^2$ **87.** $m^3 - 15m^2 + 75m - 125$
89. $8a^3 + 12a^2 + 6a + 1$ **91.** $-9a^3 + 33a^2 + 12a$
93. $56m^2 - 14m - 21$ **95.** $81r^4 - 216r^3s + 216r^2s^2 - 96rs^3 + 16s^4$
97. $6p^8 + 15p^7 + 12p^6 + 36p^5 + 15p^4$ **99.** $-24x^8 - 28x^7 + 32x^6 + 20x^5$ **101.** $6p^4 - \frac{5}{2}p^2q - \frac{25}{12}q^2$ **103.** $14x + 49$
105. $\pi x^2 - 9$ **107.** $30x + 60$ **108.** $30x + 60 = 600$; $\{18\}$
109. 10 ft by 60 ft **110.** 140 ft **111.** $450 **112.** $2870

Section 4.6 (pages 308–310)

1. (a) $4x$; $16x^2$ **(b)** $4x$; 3; $24x$ **(c)** 3; 9 **(d)** $16x^2 + 24x + 9$
3. $m^2 + 4m + 4$ **5.** $r^2 - 6r + 9$ **7.** $x^2 + 4xy + 4y^2$
9. $25p^2 + 20pq + 4q^2$ **11.** $16x^2 - 24x + 9$ **13.** $16a^2 + 40ab + 25b^2$
15. $36m^2 - \frac{48}{5}mn + \frac{16}{25}n^2$ **17.** $\frac{1}{4}x^2 + \frac{1}{3}x + \frac{1}{9}$ **19.** $2x^2 + 24x + 72$
21. $9t^3 - 6t^2 + t$ **23.** $48t^3 + 24t^2 + 3t$ **25.** $-16r^2 + 16r - 4$
27. $k^2 - 25$ **29.** $16 - 9t^2$ **31.** $25x^2 - 4$ **33.** $25y^2 - 9x^2$
35. $100x^2 - 9y^2$ **37.** $4x^4 - 25$ **39.** $\frac{9}{16} - x^2$ **41.** $81y^2 - \frac{4}{9}$
43. $25q^3 - q$ **45.** $-5a^2 + 5b^6$ **47.** $2k^2 - \frac{1}{2}$ **49.** $-x^2 + 1$
51. $x^3 + 3x^2 + 3x + 1$ **53.** $t^3 - 9t^2 + 27t - 27$ **55.** $r^3 + 15r^2 + 75r + 125$ **57.** $8a^3 + 12a^2 + 6a + 1$ **59.** $256x^4 - 256x^3 + 96x^2 - 16x + 1$ **61.** $81r^4 - 216r^3t + 216r^2t^2 - 96rt^3 + 16t^4$
63. $2x^4 + 6x^3 + 6x^2 + 2x$ **65.** $-4t^4 - 36t^3 - 108t^2 - 108t$
67. $x^4 - 2x^2y^2 + y^4$ **69.** no **71.** 9999 **73.** 39,999 **75.** $399\frac{3}{4}$

Section 4.7 (pages 316–319)

1. $10x^2 + 8$; 2; $5x^2 + 4$ **3.** $5x^2 + 4$; 2 (These may be reversed.); $10x^2 + 8$ **5.** $6p^4$; $18p^7$; $2p^2 + 6p^5$ **7.** $30x^3 - 10x + 5$
9. $4m^3 - 2m^2 + 1$ **11.** $4t^4 - 2t^2 + 2t$ **13.** $a^4 - a + \frac{2}{a}$
15. $-3p^2 - 2 + \frac{1}{p}$ **17.** $7r^2 - 6 + \frac{1}{r}$ **19.** $4x^3 - 3x^2 + 2x$
21. $-9x^2 + 5x + 1$ **23.** $2x + 8 + \frac{12}{x}$ **25.** $\frac{4x^2}{3} + x + \frac{2}{3x}$
27. $-27x^3 + 10x^2 + 4$ **29.** $9r^3 - 12r^2 + 2r + \frac{26}{3} - \frac{2}{3r}$
31. $-m^2 + 3m - \frac{4}{m}$ **33.** $-3a + 4 + \frac{5}{a}$ **35.** $\frac{12}{x} - \frac{6}{x^2} + \frac{14}{x^3} - \frac{10}{x^4}$
37. $6x^4y^2 - 4xy + 2xy^2 - x^4y$ **39.** $x + 2$ **41.** $2y - 5$
43. $p - 4 + \frac{44}{p + 6}$ **45.** $6m - 1$ **47.** $2a - 14 + \frac{74}{2a + 3}$
49. $4x^2 - 7x + 3$ **51.** $4k^3 - k + 2$ **53.** $5y^3 + 2y - 3 + \frac{-5}{y + 1}$
55. $3k^2 + 2k - 2 + \frac{6}{k - 2}$ **57.** $2p^3 - 6p^2 + 7p - 4 + \frac{14}{3p + 1}$
59. $x^2 + 3x + 3$ **61.** $2x^2 - 6x + 19 + \frac{-55}{x + 3}$ **63.** $r^2 + 2 + \frac{13}{r^2 - 4}$
65. $3x^2 + 3x - 1 + \frac{1}{x - 1}$ **67.** $y^2 - 3y + 9$ **69.** $a^2 + 5$
71. $x^2 - 4x + 2 + \frac{9x - 4}{x^2 + 3}$ **73.** $x^3 + 3x^2 - x + 5$
75. $\frac{3}{2}a - 10 + \frac{77}{2a + 6}$ **77.** $x^2 + \frac{8}{3}x - \frac{1}{3} + \frac{4}{3x - 3}$ **79.** $x^2 + x - 3$
81. $48m^2 + 96m + 24$ **83.** $5x^2 - 11x + 14$

5 · FACTORING AND APPLICATIONS

Section 5.1 (pages 336–338)

1. product; multiplying **3.** 4 **5.** 4 **7.** 6 **9.** 1 **11.** 8 **13.** $10x^3$
15. xy^2 **17.** 6 **19.** $6m^3n^2$ **21.** factored **23.** not factored
25. The correct factored form is $18x^3y^2 + 9xy = 9xy(2x^2y + 1)$. If a polynomial has two terms, then the product of the factors must have two terms. $9xy(2x^2y) = 18x^3y^2$ is just one term. **27.** $3m^2$ **29.** $2z^4$
31. $2mn^4$ **33.** $y + 2$ **35.** $a - 2$ **37.** $2 + 3xy$ **39.** $x(x - 4)$
41. $3t(2t + 5)$ **43.** $9m(3m^2 - 1)$ **45.** $m^2(m - 1)$ **47.** $8z^2(2z^2 + 3)$
49. $-6x^2(2x + 1)$ **51.** $5y^6(13y^4 + 7)$ **53.** no common factor (except 1) **55.** $8mn^3(1 + 3m)$ **57.** $13y^2(y^6 + 2y^2 - 3)$
59. $-2x(2x^2 - 5x + 3)$ **61.** $9p^3q(4p^3 + 5p^2q^3 + 9q)$
63. $a^3(a^2 + 2b^2 - 3a^2b^2 + 4ab^3)$ **65.** $(x + 2)(c - d)$

67. $(m + 2n)(m + n)$ **69.** $(p - 4)(q^2 + 1)$ **71.** not in factored form; $(7t + 4)(8 + x)$ **73.** in factored form **75.** not in factored form
77. The student should factor out -2, instead of 2, in the second step to obtain $x^2(x + 4) - 2(x + 4)$, which can be factored as $(x + 4)(x^2 - 2)$.
79. $(p + 4)(p + q)$ **81.** $(a - 2)(a + b)$ **83.** $(z + 2)(7z - a)$
85. $(3r + 2y)(6r - x)$ **87.** $(a^2 + b^2)(3a + 2b)$ **89.** $(3 - a)(4 - b)$
91. $(4m - p^2)(4m^2 - p)$ **93.** $(y + 3)(y + x)$ **95.** $(5 - 2p)(m + 3)$
97. $(3r + 2y)(6r - t)$ **99.** $(1 + 2b)(a^5 - 3)$
101. $8(2a + 5b^2)(a + b)$ **103.** $2p(p + q^2)(p - q)$

Section 5.2 (pages 342–344)

1. a and b must have different signs, one positive and one negative.
3. C **5.** $a^2 + 13a + 36$ **7.** 1 and 48, -1 and -48, 2 and 24, -2 and -24, 3 and 16, -3 and -16, 4 and 12, -4 and -12, 6 and 8, -6 and -8; The pair with a sum of -19 is -3 and -16. **9.** 1 and -24, -1 and 24, 2 and -12, -2 and 12, 3 and -8, -3 and 8, 4 and -6, -4 and 6; The pair with a sum of -5 is 3 and -8. **11.** 20; 12; table entries: 2, 2, 12, 4, 4, 9; 10 and 2; $(y + 10)(y + 2)$ **13.** $p + 6$ **15.** $x + 11$ **17.** $x - 8$
19. $y - 5$ **21.** $x + 11$ **23.** $y - 9$ **25.** $(y + 8)(y + 1)$
27. $(b + 3)(b + 5)$ **29.** $(m + 5)(m - 4)$ **31.** $(y - 5)(y - 3)$
33. prime **35.** $(z - 7)(z - 8)$ **37.** $(r - 6)(r + 5)$
39. $(a + 4)(a - 12)$ **41.** prime **43.** $(x + 16)(x - 2)$
45. $(r + 2a)(r + a)$ **47.** $(x + y)(x + 3y)$ **49.** $(t + 2z)(t - 3z)$
51. $(v - 5w)(v - 6w)$ **53.** $(m + 6n)(m - 2n)$ **55.** $(a - 6b)(a - 3b)$
57. $4(x + 5)(x - 2)$ **59.** $2t(t + 1)(t + 3)$ **61.** $2x^4(x - 3)(x + 7)$
63. $6z^2(z - 3)(z - 1)$ **65.** $5m^2(m^3 - 5m^2 + 8)$
67. $x(x - 4y)(x - 3y)$ **69.** $a^3(a + 4b)(a - b)$
71. $z^8(z - 7y)(z + 3y)$ **73.** $mn(m - 6n)(m - 4n)$
75. $yz(y + 3z)(y - 2z)$ **77.** $(a + b)(x + 4)(x - 3)$
79. $(2p + q)(r - 9)(r - 3)$

Section 5.3 (pages 350–352)

1. $(2t + 1)(5t + 2)$ **3.** $(3z - 2)(5z - 3)$ **5.** $(2s - t)(4s + 3t)$
7. (a) 2, 12, 24, 11 (b) 3, 8 (Order is irrelevant.) (c) $3m, 8m$
(d) $2m^2 + 3m + 8m + 12$ (e) $(2m + 3)(m + 4)$
(f) $(2m + 3)(m + 4) = 2m^2 + 8m + 3m + 12$, and combining like terms gives the original trinomial $2m^2 + 11m + 12$. **9.** B **11.** B **13.** A
15. $(4x + 4)$ cannot be a factor because its terms have a common factor of 4, but those of the polynomial do not. The correct factored form is $(4x - 3)(3x + 4)$. **17.** $2a + 5b$ **19.** $x^2 + 3x - 4$; $x + 4, x - 1$, or $x - 1, x + 4$ **21.** $2z^2 - 5z - 3$; $2z + 1, z - 3$, or $z - 3, 2z + 1$
23. $(3a + 7)(a + 1)$ **25.** $(2y + 3)(y + 2)$ **27.** $(3m - 1)(5m + 2)$
29. $(3s - 1)(4s + 5)$ **31.** $(5m - 4)(2m - 3)$ **33.** $(4w - 1)(2w - 3)$
35. $(4y + 1)(5y - 11)$ **37.** prime **39.** $2(5x + 3)(2x + 1)$
41. $3(4x - 1)(2x - 3)$ **43.** $-q(5m + 2)(8m - 3)$
45. $3n^2(5n - 3)(n - 2)$ **47.** $y^2(5x - 4)(3x + 1)$
49. $(5a + 3b)(a - 2b)$ **51.** $(4s + 5t)(3s - t)$
53. $m^4n(3m + 2n)(2m + n)$ **55.** prime **57.** $(3x + 4)(x + 4)$

59. $-5x(2x + 7)(x - 4)$ **61.** prime **63.** $(24y + 7x)(y - 2x)$
65. $(18x^2 - 5y)(2x^2 - 3y)$ **67.** $2(24a + b)(a - 2b)$
69. $x^2y^5(10x - 1)(x + 4)$ **71.** $4ab^2(9a + 1)(a - 3)$
73. $(12x - 5)(2x - 3)$ **75.** $(8x^2 - 3)(3x^2 + 8)$
77. $(4x + 3y)(6x + 5y)$ **79.** $-1(x + 7)(x - 3)$
81. $-1(3x + 4)(x - 1)$ **83.** $-1(a + 2b)(2a + b)$
85. $(m + 1)^3(5q - 2)(5q + 1)$ **87.** $(r + 3)^3(3x + 2y)^2$
89. $-4, 4$ **91.** $-11, -7, 7, 11$

Section 5.4 (pages 359–362)

1. 1; 4; 9; 16; 25; 36; 49; 64; 81; 100; 121; 144; 169; 196; 225; 256; 289; 324; 361; 400 **3.** A, D **5.** The binomial $4x^2 + 16$ can be factored as $4(x^2 + 4)$. *After* any common factor is removed, a sum of squares (like $x^2 + 4$ here) *cannot* be factored. **7.** $(y + 5)(y - 5)$
9. $(x + 12)(x - 12)$ **11.** prime **13.** prime **15.** $4(m^2 + 4)$
17. $(3r + 2)(3r - 2)$ **19.** $4(3x + 2)(3x - 2)$
21. $(14p + 15)(14p - 15)$ **23.** $(4r + 5a)(4r - 5a)$
25. $(9x + 7y)(9x - 7y)$ **27.** $6(3x + y)(3x - y)$ **29.** prime
31. $(2 + x)(2 - x)$ **33.** $(6 + 5t)(6 - 5t)$ **35.** $x(x^2 + 4)$
37. $x^2(x + 1)(x - 1)$ **39.** $(p^2 + 7)(p^2 - 7)$
41. $(x^2 + 1)(x + 1)(x - 1)$ **43.** $(p^2 + 16)(p + 4)(p - 4)$
45. B, C **47.** 10 **49.** 9 **51.** $(w + 1)^2$ **53.** $(x - 4)^2$ **55.** prime
57. $2(x + 6)^2$ **59.** $(2x + 3)^2$ **61.** $(4x - 5)^2$ **63.** $(7x - 2y)^2$
65. $(8x + 3y)^2$ **67.** $2(5h - 2y)^2$ **69.** $k(4k^2 - 4k + 9)$
71. $z^2(25z^2 + 5z + 1)$ **73.** 1; 8; 27; 64; 125; 216; 343; 512; 729; 1000
75. C, D **77.** (a) neither of these (b) perfect cube (c) perfect square (d) perfect square (e) both of these (f) perfect cube
79. $(a - 1)(a^2 + a + 1)$ **81.** $(m + 2)(m^2 - 2m + 4)$
83. $(y - 6)(y^2 + 6y + 36)$ **85.** $(k + 10)(k^2 - 10k + 100)$
87. $(3x - 4)(9x^2 + 12x + 16)$ **89.** $6(p + 1)(p^2 - p + 1)$
91. $5(x + 2)(x^2 - 2x + 4)$ **93.** $(y - 2x)(y^2 + 2yx + 4x^2)$
95. $2(x - 2y)(x^2 + 2xy + 4y^2)$ **97.** $(2p + 9q)(4p^2 - 18pq + 81q^2)$
99. $(3a + 4b)(9a^2 - 12ab + 16b^2)$
101. $(5t + 2s)(25t^2 - 10ts + 4s^2)$
103. $(2x - 5y^2)(4x^2 + 10xy^2 + 25y^4)$
105. $(3m^2 + 2n)(9m^4 - 6m^2n + 4n^2)$
107. $(x + y)(x^2 - xy + y^2)(x^6 - x^3y^3 + y^6)$ **109.** $\left(p + \frac{1}{3}\right)\left(p - \frac{1}{3}\right)$
111. $\left(6m + \frac{4}{5}\right)\left(6m - \frac{4}{5}\right)$ **113.** $(x + 0.8)(x - 0.8)$
115. $\left(t + \frac{1}{2}\right)^2$ **117.** $(x - 0.9)^2$ **119.** $\left(x + \frac{1}{2}\right)\left(x^2 - \frac{1}{2}x + \frac{1}{4}\right)$
121. $4mn$ **123.** $(m - p + 2)(m + p)$

Section 5.5 (pages 371–373)

1. $ax^2 + bx + c$ **3.** 0; zero-factor **5.** the term with greatest degree is greater than two (It is *cubic*.) **7.** (a) linear (b) quadratic
(c) quadratic (d) linear **9.** The variable x is another factor to set equal to 0, so the solution set is $\left\{0, \frac{7}{7}\right\}$. **11.** $\{-5, 2\}$ **13.** $\left\{3, \frac{7}{2}\right\}$

15. $\left\{-\frac{1}{2}, \frac{1}{6}\right\}$ **17.** $\left\{-\frac{5}{6}, 0\right\}$ **19.** $\left\{0, \frac{4}{3}\right\}$ **21.** $\{6\}$ **23.** $\{-2, -1\}$
25. $\{1, 2\}$ **27.** $\{-8, 3\}$ **29.** $\{-1, 3\}$ **31.** $\{-2, -1\}$ **33.** $\{-4\}$
35. $\left\{-2, \frac{1}{3}\right\}$ **37.** $\left\{-\frac{4}{3}, \frac{1}{2}\right\}$ **39.** $\left\{-\frac{2}{3}\right\}$ **41.** $\{-3, 3\}$ **43.** $\left\{-\frac{7}{4}, \frac{7}{4}\right\}$
45. $\{-11, 11\}$ **47.** $\{-6, 0\}$ **49.** $\{0, 7\}$ **51.** $\left\{0, \frac{1}{2}\right\}$ **53.** $\{2, 5\}$
55. $\left\{-4, \frac{1}{2}\right\}$ **57.** $\{-17, 4\}$ **59.** $\{-4, 12\}$ **61.** $\{-9, -2\}$
63. $\left\{-\frac{7}{3}, 0, \frac{7}{3}\right\}$ **65.** $\{-2, 0, 4\}$ **67.** $\{-5, 0, 4\}$ **69.** $\left\{0, \frac{1}{2}, 4\right\}$
71. $\{-3, 0, 5\}$ **73.** $\left\{-\frac{5}{2}, \frac{1}{3}, 5\right\}$ **75.** $\left\{-\frac{7}{2}, -3, 1\right\}$ **77.** $\{-1, 3\}$
79. $\{-1, 3\}$ **81.** $\{3\}$ **83.** $\left\{-\frac{2}{3}, 4\right\}$ **85.** $\left\{-\frac{4}{3}, -1, \frac{1}{2}\right\}$
87. (a) 64; 144; 4; 6   **(b)** No time has elapsed, so the object hasn't
fallen (been released) yet.

Section 5.6 (pages 378–384)

1. Read; variable; equation; Solve; answer; Check, original
3. $\mathcal{A} = bh$; *Step 3:* $45 = (2x + 1)(x + 1)$; *Step 4:* $x = 4$ or $x = -\frac{11}{2}$;
Step 5: base: 9 units; height: 5 units; *Step 6:* $9 \cdot 5 = 45$
5. $\mathcal{A} = LW$; *Step 3:* $80 = (x + 8)(x - 8)$; *Step 4:* $x = 12$ or $x = -12$;
Step 5: length: 20 units; width: 4 units; *Step 6:* $20 \cdot 4 = 80$
7. length: 14 cm; width: 12 cm   **9.** base: 12 in.; height: 5 in.
11. height: 13 in.; width: 10 in.   **13.** length: 15 in.; width: 12 in.
15. mirror: 7 ft; painting: 9 ft   **17.** 20, 21   **19.** 0, 1, 2 or 7, 8, 9
21. $-3, -1$ or 7, 9   **23.** 7, 9, 11   **25.** $-2, 0, 2$ or 6, 8, 10   **27.** 12 cm
29. length: 20 in.; width: 15 in.; diagonal: 25 in.   **31.** 12 mi   **33.** 8 ft
35. 112 ft   **37.** 256 ft   **39. (a)** 1 sec   **(b)** $\frac{1}{2}$ sec and $1\frac{1}{2}$ sec   **(c)** 3 sec
(d) The negative solution, -1, does not make sense because t represents
time, which cannot be negative.   **41. (a)** 10   **(b)** 111 million; The
result obtained from the model is more than 109 million, the actual
number for 2000.   **(c)** 22   **(d)** 343 million; The result is more than
326 million, the actual number.   **(e)** 24   **(f)** 391 million

| 6 | **RATIONAL EXPRESSIONS AND APPLICATIONS** |

Section 6.1 (pages 400–403)

1. 3; 3; -5 **3.** B, D   **5.** B   **7.** $-3; -3; -3; -6; \frac{3}{5}$
9. (a) $\frac{7}{10}$ **(b)** $\frac{8}{15}$ **11. (a)** 0   **(b)** -1 **13. (a)** $-\frac{64}{15}$ **(b)** undefined
15. (a) undefined   **(b)** $\frac{8}{25}$ **17. (a)** 0   **(b)** 0   **19. (a)** 0
(b) undefined   **21.** $x \neq 0$ **23.** $y \neq 0$ **25.** $x \neq 6$ **27.** $x \neq -\frac{5}{3}$
29. $m \neq -3, m \neq 2$ **31.** It is never undefined.   **33.** It is never
undefined.   **35.** numerator: $x^2, 4x$; denominator: $x, 4$ **37.** $3r^2$ **39.** $\frac{2}{5}$
41. $\frac{x - 1}{x + 1}$ **43.** $\frac{7}{5}$ **45.** $\frac{6}{7}$ **47.** $m - n$ **49.** $\frac{2}{t - 3}$ **51.** $\frac{3(2m + 1)}{4}$
53. $\frac{3m}{5}$ **55.** $\frac{3r - 2s}{3}$ **57.** $\frac{1}{x + 6}$ **59.** $\frac{x + 3}{x - 3}$ **61.** $\frac{13x}{7}$ **63.** $k - 3$
65. $\frac{x - 3}{x + 1}$ **67.** $\frac{x + 1}{x - 1}$ **69.** $\frac{x + 2}{x - 4}$ **71.** $-\frac{3}{7t}$ **73.** $\frac{z - 3}{z + 5}$
75. $\frac{r + s}{r - s}$ **77.** $\frac{a + b}{a - b}$ **79.** $\frac{m + n}{2}$ **81.** $\frac{x^2 + 1}{x}$ **83.** $1 - p + p^2$

85. $x^2 + 3x + 9$ **87.** $-\frac{b^2 + ba + a^2}{a + b}$ **89.** $\frac{k^2 - 2k + 4}{k - 2}$ **91.** $\frac{z + 3}{z}$
93. $\frac{1 - 2r}{2}$ **95.** -1 **97.** $-(m + 1)$ **99.** -1 **101.** It is already in
lowest terms.   **103.** -2

Answers may vary in Exercises 105, 107, and 109.
105. $\frac{-(x + 4)}{x - 3}, \frac{-x - 4}{x - 3}, \frac{x + 4}{-(x - 3)}, \frac{x + 4}{-x + 3}$ **107.** $\frac{-(2x - 3)}{x + 3},$
$\frac{-2x + 3}{x + 3}, \frac{2x - 3}{-(x + 3)}, \frac{2x - 3}{-x - 3}$ **109.** $\frac{-(3x - 1)}{5x - 6}, \frac{-3x + 1}{5x - 6}, \frac{3x - 1}{-(5x - 6)},$
$\frac{3x - 1}{-5x + 6}$ **111.** $x^2 + 3$ **113. (a)** 0   **(b)** 1.6   **(c)** 4.1
(d) The number of vehicles waiting also increases.   **115.** Both yield
$2x + 3$.   **116.** Both yield $2x + 1$.   **117.** Both yield $x^2 + 1$.
118. Both yield $x^2 + 2$.

Section 6.2 (pages 408–410)

1. (a) B   **(b)** D   **(c)** C   **(d)** A   **3.** $\frac{3a}{2}$ **5.** $-\frac{4x^4}{3}$ **7.** $\frac{2}{c + d}$
9. $4(x - y)$ **11.** $\frac{t^2}{2}$ **13.** $\frac{x + 3}{2x}$ **15.** $x - 2$; 3; $x - 2$; 5; $\frac{3}{4}$ **17.** 5
19. $-\frac{3}{2t^4}$ **21.** $\frac{1}{4}$ **23.** $-\frac{35}{8}$ **25.** $\frac{2(x + 2)}{x(x - 1)}$ **27.** $\frac{x(x - 3)}{6}$
29. $\frac{5(x - 4)}{x^2(x + 4)}$ **31.** $\frac{-4t(t + 1)}{t - 1}$ **33.** $\frac{10}{9}$ **35.** $-\frac{3}{4}$ **37.** $-\frac{9}{2}$
39. $\frac{p + 4}{p + 2}$ **41.** -1 **43.** $-\frac{m + 2}{m + 1}$ **45.** $\frac{(2x - 1)(x + 2)}{x - 1}$
47. $\frac{(k - 1)^2}{(k + 1)(2k - 1)}$ **49.** $\frac{4k - 1}{3k - 2}$ **51.** $\frac{m + 4p}{m + p}$ **53.** $\frac{m + 6}{m + 3}$
55. $\frac{y + 3}{y + 4}$ **57.** $\frac{m}{m + 5}$ **59.** $\frac{r + 6s}{r + s}$ **61.** $\frac{(q - 3)^2(q + 2)^2}{q + 1}$
63. $\frac{3 - a - b}{2a - b}$ **65.** $-\frac{(x + y)^2(x^2 - xy + y^2)}{3y(y - x)(x - y)}$, or
$\frac{(x + y)^2(x^2 - xy + y^2)}{3y(x - y)^2}$ **67.** $\frac{x + 10}{10}$ **69.** $\frac{5xy^2}{4q}$

Section 6.3 (pages 413–416)

1. C   **3.** C   **5.** 5; 5; one; 5; 2; 5; 50   **7.** 60   **9.** 1800   **11.** x^5
13. $30p$ **15.** $180y^4$ **17.** $84r^5$ **19.** $15a^5b^3$ **21.** r^9t^3
23. $(x + 1)(x - 1)$ **25.** $12p(p - 2)$ **27.** $28m^2(3m - 5)$
29. $30(b - 2)$ **31.** $18(r - 2)$ **33.** $c - d$ or $d - c$ **35.** $m - 3$ or $3 - m$
37. $p - q$ or $q - p$ **39.** $2(x + 1)(x - 1)$ **41.** $3(x - 4)^2$
43. $12p(p + 5)^2$ **45.** $8(y + 2)(y + 1)$ **47.** $k(k + 5)(k - 2)$
49. $a(a + 6)(a - 3)$ **51.** $(p + 3)(p + 5)(p - 6)$
53. $(k + 3)(k - 5)(k + 7)(k + 8)$ **55.** $\frac{20}{55}$ **57.** $\frac{-45}{9k}$
59. $\frac{60m^2k^3}{32k^4}$ **61.** $\frac{57z}{6z - 18}$ **63.** $\frac{-4a}{18a - 36}$ **65.** $\frac{6(k + 1)}{k(k - 4)(k + 1)}$
67. $\frac{(t - r)(4r - t)}{t^3 - r^3}$ **69.** $\frac{2y(z - y)(y - z)}{y^4 - z^3y}$, or $\frac{-2y(y - z)^2}{y^4 - z^3y}$

71. $\dfrac{36r(r+1)}{(r-3)(r+2)(r+1)}$ **73.** $\dfrac{ab(a+2b)}{2a^3b+a^2b^2-ab^3}$ **75.** 7

76. 1   **77.** identity property of multiplication   **78.** 7   **79.** 1

80. identity property of multiplication

Section 6.4 (pages 422–425)

1. E   **3.** C   **5.** B   **7.** G   **9.** 2; 2; 5; 5; 10x; $\dfrac{57}{10x}$ **11.** $\dfrac{11}{m}$

13. $\dfrac{4}{y+4}$ **15.** 1   **17.** $\dfrac{m-1}{m+1}$ **19.** b **21.** x **23.** $y-6$ **25.** $\dfrac{1}{x-3}$

27. -1 **29.** $\dfrac{3z+5}{15}$ **31.** $\dfrac{10-7r}{14}$ **33.** $\dfrac{-3x-2}{4x}$ **35.** $\dfrac{61}{28t}$

37. $\dfrac{x+1}{2}$ **39.** $\dfrac{5x+9}{6x}$ **41.** $\dfrac{7-6p}{3p^2}$ **43.** $\dfrac{-k-8}{k(k+4)}$ **45.** $\dfrac{x+4}{x+2}$

47. $\dfrac{6m^2+23m-2}{(m+2)(m+1)(m+5)}$ **49.** $\dfrac{4y^2-y+5}{(y+1)^2(y-1)}$ **51.** $\dfrac{3}{t}$

53. $m-2$; $2-m$ **55.** $\dfrac{-2}{x-5}$, or $\dfrac{2}{5-x}$ **57.** -4 **59.** $\dfrac{-5}{x-y^2}$, or

$\dfrac{5}{y^2-x}$ **61.** $\dfrac{x+y}{5x-3y}$, or $\dfrac{-x-y}{3y-5x}$ **63.** $\dfrac{-6}{4p-5}$, or $\dfrac{6}{5-4p}$

65. 3   **67.** $\dfrac{-m-n}{2(m-n)}$ **69.** $\dfrac{-x^2+6x+11}{(x+3)(x-3)(x+1)}$

71. $\dfrac{-5q^2-13q+7}{(3q-2)(q+4)(2q-3)}$ **73.** $y-7$ **75.** $\dfrac{7x+31}{x+4}$

77. $\dfrac{-5x+13}{4x}$ **79.** $\dfrac{2x^2+6x}{(x-7)(x-3)}$, or $\dfrac{2x(x+3)}{(x-7)(x-3)}$

81. $\dfrac{2a+21}{3(a-2)}$ **83.** $\dfrac{x-8}{2(x-3)}$, or $\dfrac{8-x}{2(3-x)}$ **85.** $\dfrac{1}{x-2}$

87. $\dfrac{9r+2}{r(r+2)(r-1)}$ **89.** $\dfrac{2(x^2+3xy+4y^2)}{(x+y)(x+y)(x+3y)}$, or

$\dfrac{2(x^2+3xy+4y^2)}{(x+y)^2(x+3y)}$ **91.** $\dfrac{15r^2+10ry-y^2}{(3r+2y)(6r-y)(6r+y)}$

93. (a) $\dfrac{9k^2+6k+26}{5(3k+1)}$ (b) $\dfrac{1}{4}$ **95.** $\dfrac{10x}{49(101-x)}$

Section 6.5 (pages 432–434)

1. (a) 6; $\dfrac{1}{6}$ (b) 12; $\dfrac{3}{4}$ (c) $\dfrac{1}{6}\div\dfrac{3}{4}$ (d) $\dfrac{2}{9}$ **3.** Choice D is correct,

because every sign has been changed in the fraction. This means it

was multiplied by $\dfrac{-1}{-1}=1$.   **5.** $\dfrac{1}{20}$ **7.** -6 **9.** $\dfrac{1}{xy}$ **11.** $\dfrac{2a^2b}{3}$

13. $\dfrac{m(m+2)}{3(m-4)}$ **15.** $\dfrac{2}{x}$ **17.** $\dfrac{8}{x}$ **19.** $\dfrac{a^2-5}{a^2+1}$ **21.** $\dfrac{31}{50}$ **23.** $\dfrac{y^2+x^2}{xy(y-x)}$

25. $\dfrac{40-12p}{85p}$, or $\dfrac{4(10-3p)}{85p}$ **27.** $\dfrac{5y-2x}{3+4xy}$ **29.** $\dfrac{a-2}{2a}$ **31.** $\dfrac{z-5}{4}$

33. $\dfrac{-m}{m+2}$ **35.** $\dfrac{x+8}{-x+7}$ **37.** $\dfrac{x^2+y^2}{x^2-y^2}$, or $\dfrac{x^2+y^2}{(x+y)(x-y)}$

39. $\dfrac{ab}{a+b}$ **41.** $\dfrac{3m(m-3)}{(m-1)(m-8)}$ **43.** $\dfrac{2x-7}{3x+1}$ **45.** $\dfrac{y+4}{y-8}$

47. $\dfrac{30}{(a+b)(a-b)}$ **49.** $\dfrac{x+y}{x^2+xy+y^2}$ **51.** $\dfrac{x^2y^2}{y^2+x^2}$

53. $\dfrac{y^2+x^2}{xy^2+x^2y}$, or $\dfrac{y^2+x^2}{xy(y+x)}$ **55.** $\dfrac{p^2+k}{p^2-3k}$ **57.** $\dfrac{1}{2xy}$ **59.** $\dfrac{5}{3}$ **61.** $\dfrac{13}{2}$

63. $\dfrac{19r}{15}$ **65.** $\dfrac{\frac{3}{8}+\frac{5}{6}}{2}$ **66.** $\dfrac{29}{48}$ **67.** $\dfrac{29}{48}$ **68.** Answers will vary.

Section 6.6 (pages 442–445)

1. 12   **3.** xyz **5.** Yes, it is acceptable because $\dfrac{1}{3-x}$ is equivalent to

$\dfrac{-1}{x-3}$.   **7.** expression; $\dfrac{43}{40}x$ **9.** equation; $\left\{\dfrac{40}{43}\right\}$ **11.** expression; $-\dfrac{1}{10}x$

13. equation; $\{-10\}$ **15.** equation; $\{0\}$ **17.** $x\ne -2,0$

19. $x\ne -3,4,-\dfrac{1}{2}$ **21.** $x\ne -9,1,-2,2$ **23.** $\left\{\dfrac{1}{4}\right\}$ **25.** $\left\{-\dfrac{3}{4}\right\}$

27. $\{-15\}$ **29.** $\{7\}$ **31.** $\{-15\}$ **33.** $\{-5\}$ **35.** $\{-6\}$ **37.** $\varnothing$

39. $\{5\}$ **41.** $\{4\}$ **43.** $\{5\}$ **45.** $\left\{x\,\middle|\,x\ne +\dfrac{4}{3}\right\}$ **47.** $\{1\}$ **49.** $\{4\}$

51. $\{5\}$ **53.** $\{-4\}$ **55.** $\{-2,12\}$ **57.** $\varnothing$ **59.** $\{3\}$ **61.** $\{3\}$

63. $\{-3\}$ **65.** $\left\{-\dfrac{1}{5},3\right\}$ **67.** $\left\{-\dfrac{1}{2},5\right\}$ **69.** $\{3\}$ **71.** $\left\{-\dfrac{1}{3},3\right\}$

73. $\{-1\}$ **75.** $\{-6\}$ **77.** $\left\{-6,\dfrac{1}{2}\right\}$ **79.** $\{6\}$ **81.** $\{0\}$

83. $F=\dfrac{ma}{k}$ **85.** $a=\dfrac{kF}{m}$ **87.** $R=\dfrac{E-Ir}{I}$, or $R=\dfrac{E}{I}-r$

89. $\mathscr{A}=\dfrac{h(B+b)}{2}$ **91.** $a=\dfrac{2S-ndL}{nd}$, or $a=\dfrac{2S}{nd}-L$ **93.** $y=\dfrac{xz}{x+z}$

95. $t=\dfrac{rs}{rs-2s-3r}$, or $t=\dfrac{-rs}{-rs+2s+3r}$ **97.** $z=\dfrac{3y}{5-9xy}$, or

$z=\dfrac{-3y}{9xy-5}$ **99.** $t=\dfrac{2x-1}{x+1}$, or $t=\dfrac{-2x+1}{-x-1}$ **101.** (a) -3 (b) -1

(c) $-3,-1$ **102.** $\dfrac{15}{2x}$ **103.** If $x=0$, the divisor R is equal to 0, and

division by 0 is undefined.   **104.** $(x+3)(x+1)$

105. $\dfrac{7}{x+1}$ **106.** $\dfrac{11x+21}{4x}$ **107.** $\varnothing$ **108.** We know that -3 is not

allowed, because P and R are undefined for $x=-3$.

Section 6.7 (pages 452–455)

1. into a headwind: $(m-5)$ mph; with a tailwind: $(m+5)$ mph

3. $\dfrac{1}{10}$ job per hr   **5.** (a) the amount   (b) $5+x$ (c) $\dfrac{5+x}{6}=\dfrac{13}{3}$

7. x represents the original numerator; $\dfrac{x+3}{(x-4)+3}=\dfrac{3}{2}$; $\dfrac{9}{5}$

9. x represents the original numerator; $\dfrac{x+2}{3x-2}=1$; $\dfrac{2}{6}$

11. x represents the number; $\dfrac{1}{6}x=x+5$; -6

13. x represents the quantity; $x+\dfrac{3}{4}x+\dfrac{1}{2}x+\dfrac{1}{3}x=93$; 36

15. 18.809 min   **17.** 331.763 m per min   **19.** 3.565 hr

21. $\dfrac{500}{x-10}=\dfrac{600}{x+10}$ **23.** 8 mph   **25.** 32 mph   **27.** 165 mph

29. 3 mph   **31.** 18.5 mph   **33.** $\dfrac{1}{8}t+\dfrac{1}{6}t=1$, or $\dfrac{1}{8}+\dfrac{1}{6}=\dfrac{1}{t}$

35. $2\dfrac{2}{5}$ hr   **37.** $5\dfrac{5}{11}$ hr   **39.** 3 hr   **41.** $2\dfrac{7}{10}$ hr   **43.** $9\dfrac{1}{11}$ min

7   GRAPHS, LINEAR EQUATIONS, AND SYSTEMS

Section 7.1 (pages 479–484)

1. (a) I **(b)** III **(c)** II **(d)** IV **(e)** none **(f)** none

3. (a) $-3; 3; 2; -1$ **5. (a)** $-4; 5; -\frac{12}{5}; \frac{5}{4}$ **7.** $(6, 0); (0, 4)$

(b)
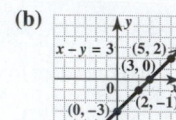
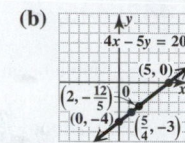
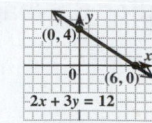

9. $(6, 0); (0, -2)$ **11.** $(-2, 0); \left(0, -\frac{5}{3}\right)$ **13.** none; $(0, 5)$

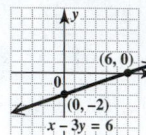
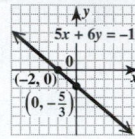
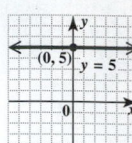

15. $(-4, 0)$; none **17.** $(0, 0); (0, 0)$ **19.** $(0, 0); (0, 0)$

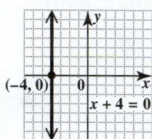
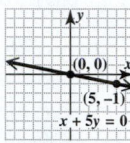
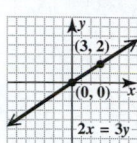

21. (a) $(-2, 0); (0, 3)$ **23. (a)** C **(b)** D **(c)** B **(d)** A

(b) B **25.** $(-5, -1)$ **27.** $\left(\frac{9}{2}, -\frac{3}{2}\right)$

(c) **29.** $\left(0, \frac{11}{2}\right)$ **31.** $(2.1, 0.9)$ **33.** A, B, D, F

35. (a) 2 **(b)** 0 **(c)** undefined

(d) $-\frac{1}{3}$ **(e)** 1 **(f)** -4 **(g)** -1 **(h)** $\frac{7}{4}$

37. (a) 8 **(b)** rises **39. (a)** $\frac{5}{6}$ **(b)** rises

41. (a) 0 **(b)** horizontal **43. (a)** $-\frac{1}{2}$ **(b)** falls **45. (a)** undefined

(b) vertical **47. (a)** -1 **(b)** falls **49.** -2 **51.** $\frac{4}{3}$ **53.** $-\frac{5}{2}$

55. undefined

In part (a) of Exercises 57 and 59, we used the intercepts. Other points can be used.

57. (a) $(4, 0)$ and $(0, -8)$; 2 **(b)** $y = 2x - 8$; 2 **(c)** $A = 2, B = -1$; 2

59. (a) $(-3, 0)$ and $(0, -3)$; -1 **(b)** $y = -x - 3$; -1

(c) $A = 1, B = 1$; -1

61. $-\frac{1}{2}$ **63.** $\frac{5}{2}$ **65.** 4

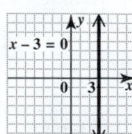
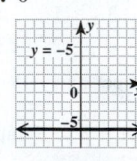
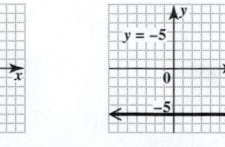

67. undefined **69.** 0 **71.**

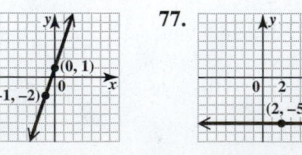

73. **75.** **77.**

79.
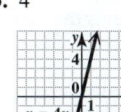

81. parallel **83.** perpendicular **85.** neither
87. parallel **89.** neither **91.** perpendicular
93. $-\$4000$ per year; The value of the machine is decreasing \$4000 per year during these years.
95. 0% per year (or no change); The percent of pay raise is not changing—it is 3% per year during these years. **97. (a)** In 2012, there were 326 million wireless subscriber connections in the U.S. **(b)** 14.2 **(c)** The number of subscribers increased by an average of 14.2 million per year from 2007 to 2012. **99. (a)** -7 theaters per yr **(b)** The negative slope means that the number of drive-in theaters decreased by an average of 7 per year from 2005 to 2012. **101.** \$0.08 per year; The price of a gallon of gasoline increased by an average of \$0.08 per year from 1980 to 2012.
103. -1859 thousand cameras per year; The number of digital cameras sold decreased by an average of 1859 thousand per year from 2010 to 2013. **105.** Because the slopes of both pairs of opposite sides are equal, the figure is a parallelogram. **107.** $\frac{1}{3}$ **108.** $\frac{1}{3}$ **109.** $\frac{1}{3}$ **110.** $\frac{1}{3} = \frac{1}{3} = \frac{1}{3}$ is true. **111.** collinear **112.** not collinear

Section 7.2 (pages 493–497)

1. A **3.** A **5.** $3x + y = 10$ **7.** A **9.** C **11.** H **13.** B

15. $y = 5x + 15$ **17.** $y = -\frac{2}{3}x + \frac{4}{5}$ **19.** $y = x - 1$ **21.** $y = \frac{2}{5}x + 5$

23. $y = \frac{2}{3}x + 1$ **25.** $y = -x - 2$ **27.** $y = 2x - 4$ **29.** $y = -\frac{3}{5}x + 3$

31. (a) $y = x + 4$ **(b)** 1 **33. (a)** $y = -\frac{6}{5}x + 6$ **(b)** $-\frac{6}{5}$

(c) $(0, 4)$ **(c)** $(0, 6)$

(d)
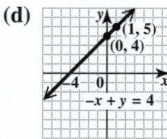

(d)
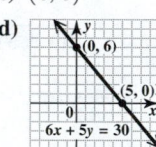

35. (a) $y = \frac{4}{5}x - 4$ **(b)** $\frac{4}{5}$ **37. (a)** $y = -\frac{1}{2}x - 2$ **(b)** $-\frac{1}{2}$

(c) $(0, -4)$ **(c)** $(0, -2)$

(d) **(d)**

39. (a) $y = -2x + 18$ **(b)** $2x + y = 18$ **41. (a)** $y = -\frac{3}{4}x + \frac{5}{2}$

(b) $3x + 4y = 10$ **43. (a)** $y = \frac{1}{2}x + \frac{13}{2}$ **(b)** $x - 2y = -13$

45. (a) $y = 4x - 12$ **(b)** $4x - y = 12$ **47. (a)** $y = 1.4x + 4$

(b) $7x - 5y = -20$ **49.** $2x - y = 2$ **51.** $x + 2y = 8$

53. $y = 5$ **55.** $x = 7$ **57.** $y = -3$ **59.** $2x - 13y = -6$

61. $y = 5$ **63.** $x = 9$ **65.** $y = -\frac{3}{2}$ **67.** $y = 8$ **69.** $x = 0.5$

71. (a) $y = 3x - 19$ **(b)** $3x - y = 19$ **73. (a)** $y = \frac{1}{2}x - 1$

(b) $x - 2y = 2$ **75. (a)** $y = -\frac{1}{2}x + 9$ **(b)** $x + 2y = 18$

77. (a) $y = 7$ **(b)** $y = 7$ **79.** $y = 45x$; $(0, 0), (5, 225), (10, 450)$

81. $y = 3.75x$; $(0, 0), (5, 18.75), (10, 37.50)$ **83.** $y = 140x$; $(0, 0),$ $(5, 700), (10, 1400)$ **85. (a)** $y = 149x + 15$ **(b)** $(5, 760)$; The cost for 5 tickets and a parking pass is \$760. **(c)** \$313 **87. (a)** $y = 41x + 99$

(b) $(5, 304)$; The cost for a 5-month membership is \$304. **(c)** \$591

89. (a) $y = 95x + 36$ **(b)** $(5, 511)$; The cost of the plan for 5 months is \$511. **(c)** \$2316 **91. (a)** $y = 6x + 30$ **(b)** $(5, 60)$; It costs \$60 to rent the saw for 5 days. **(c)** 18 days **93. (a)** $y = -1357x + 7030$; Sales of portable media/MP3 players in the United States decreased by \$1357 million per year from 2010 to 2013. **(b)** \$5673 million **95. (a)** $y = 3.875x + 31.3$ **(b)** \$73.9 billion; It is very close to the actual value. **97.** 32; 212 **98.** $(0, 32)$ and $(100, 212)$ **99.** $\frac{9}{5}$ **100.** $F = \frac{9}{5}C + 32$ **101.** $C = \frac{5}{9}(F - 32)$ **102.** 86° **103.** 10° **104.** −40°

Section 7.3 (pages 503–506)

1. system of linear equations; same **3.** inconsistent; no; independent **5.** dependent; consistent; infinitely many **7.** It is not a solution of the system because it is not a solution of the second equation, $2x + y = 4$. **9.** A; The ordered-pair solution must be in quadrant II, and $(-4, -4)$ is in quadrant III. **11. (a)** B **(b)** C **(c)** D **(d)** A **13.** no **15.** yes **17.** yes **19.** no **21.** yes **23.** $\{(4, 2)\}$ **25.** $\{(0, 4)\}$ **27.** $\{(4, -1)\}$

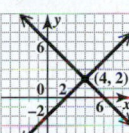 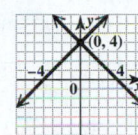 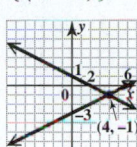

In Exercises 29–41, we do not show the graphs.
29. $\{(1, 3)\}$ **31.** $\{(x, y) \mid 3x + y = 5\}$ (dependent equations) **33.** $\{(0, 2)\}$ **35.** $\varnothing$ (inconsistent system) **37.** $\{(x, y) \mid 3x - y = -6\}$ (dependent equations) **39.** $\{(4, -3)\}$ **41.** $\varnothing$ (inconsistent system) **43. (a)** neither **(b)** intersecting lines **(c)** one solution **45. (a)** dependent **(b)** one line **(c)** infinite number of solutions **47. (a)** neither **(b)** intersecting lines **(c)** one solution **49. (a)** inconsistent **(b)** parallel lines **(c)** no solution **51. (a)** 1980–2000 **(b)** 2001; about 750 newspapers **(c)** $(2001, 750)$ **53.** 40; 30 **55.** Supply exceeds demand.

Section 7.4 (pages 512–513)

1. The student must find the value of y and write the solution as an ordered pair. The solution set is $\{(3, 0)\}$. **3.** A false statement, such as $0 = 3$, occurs. **5.** $\{(3, 9)\}$ **7.** $\{(7, 3)\}$ **9.** $\{(-4, 8)\}$ **11.** $\{(3, -2)\}$ **13.** $\{(0, 5)\}$ **15.** $\{(x, y) \mid 3x - y = 5\}$ **17.** $\left\{\left(\frac{1}{4}, -\frac{1}{2}\right)\right\}$ **19.** $\varnothing$ **21.** $\{(x, y) \mid 2x - y = -12\}$ **23.** $\{(1, 5)\}$ **25.** $\varnothing$ **27.** $\{(0, 0)\}$ **29.** $\{(2, 6)\}$ **31.** $\{(2, -4)\}$ **33.** $\{(-2, 1)\}$ **35.** $\{(x, y) \mid x + 2y = 48\}$ **37.** $\{(10, 4)\}$ **39.** $\{(4, -9)\}$ **41.** To find the total cost, multiply the number of bicycles (x) by the cost per bicycle (\$400), and add the fixed cost (\$5000). Thus, $y_1 = 400x + 5000$ gives this total cost (in dollars). **42.** $y_2 = 600x$ **43.** $y_1 = 400x + 5000$, $y_2 = 600x$; solution set: $\{(25, 15{,}000)\}$ **44.** 25; 15,000; 15,000

Section 7.5 (pages 518–519)

1. false; The solution set is $\varnothing$. **3.** $\{(4, 6)\}$ **5.** $\{(-1, -3)\}$ **7.** $\{(-2, 3)\}$ **9.** $\left\{\left(\frac{1}{2}, 4\right)\right\}$ **11.** $\{(3, -6)\}$ **13.** $\{(0, 4)\}$ **15.** $\{(0, 0)\}$ **17.** $\{(7, 4)\}$ **19.** $\{(0, 3)\}$ **21.** $\{(3, 0)\}$ **23.** $\{(x, y) \mid x - 3y = -4\}$ **25.** $\varnothing$ **27.** $\{(-3, 2)\}$ **29.** $\{(11, 15)\}$ **31.** $\left\{\left(13, -\frac{7}{5}\right)\right\}$ **33.** $\{(6, -4)\}$ **35.** $\{(x, y) \mid x + 3y = 6\}$ **37.** $\varnothing$ **39.** $\left\{\left(-\frac{5}{7}, -\frac{2}{7}\right)\right\}$ **41.** $\left\{\left(\frac{1}{8}, -\frac{5}{6}\right)\right\}$

Section 7.6 (pages 526–528)

1. Answers will vary. Some possible answers are **(a)** two perpendicular walls and the ceiling in a normal room, **(b)** the floors of three different levels of an office building, and **(c)** three pages of a book (because they intersect in the spine). **3.** The statement means that when -1 is substituted for x, 2 is substituted for y, and 3 is substituted for z in the three equations, the resulting three statements are true. **5.** 4 **7.** $\{(3, 2, 1)\}$ **9.** $\{(1, 4, -3)\}$ **11.** $\{(0, 2, -5)\}$ **13.** $\{(1, 0, 3)\}$ **15.** $\left\{\left(1, \frac{3}{10}, \frac{2}{5}\right)\right\}$ **17.** $\left\{\left(-\frac{7}{3}, \frac{22}{3}, 7\right)\right\}$ **19.** $\{(-12, 18, 0)\}$ **21.** $\{(0.8, -1.5, 2.3)\}$ **23.** $\{(4, 5, 3)\}$ **25.** $\{(2, 2, 2)\}$ **27.** $\left\{\left(\frac{8}{3}, \frac{2}{3}, 3\right)\right\}$ **29.** $\{(-1, 0, 0)\}$ **31.** $\{(-4, 6, 2)\}$ **33.** $\{(-3, 5, -6)\}$ **35.** $\varnothing$; inconsistent system **37.** $\{(x, y, z) \mid x - y + 4z = 8\}$; dependent equations **39.** $\{(3, 0, 2)\}$ **41.** $\{(x, y, z) \mid 2x + y - z = 6\}$; dependent equations **43.** $\{(0, 0, 0)\}$ **45.** $\varnothing$; inconsistent system **47.** $\{(2, 1, 5, 3)\}$ **49.** $\{(-2, 0, 1, 4)\}$

Section 7.7 (pages 539–545)

1. (a) 6 oz **(b)** 15 oz **(c)** 24 oz **(d)** 30 oz **3.** \$1.89x **5. (a)** $(10 - x)$ mph **(b)** $(10 + x)$ mph **7.** wins: 92; losses: 70 **9.** length: 78 ft; width: 36 ft **11.** AT&T: \$126.4 billion; Verizon: \$115.8 billion **13.** $x = 40$ and $y = 50$, so the angles measure 40° and 50°. **15.** hockey: \$354.84; basketball: \$315.66 **17.** ribeye: \$17.29; salmon: \$14.99 **19.** DVD: \$12.96; Blu-ray: \$19.49 **21.** 6 gal of 25%; 14 gal of 35% **23.** pure acid: 6 L; 10% acid: 48 L **25.** nuts: 14 kg; cereal: 16 kg **27.** \$1000 at 2%; \$2000 at 4% **29.** train: 60 km per hr; plane: 160 km per hr **31.** scooter: 25 mph; bicycle: 10 mph **33.** boat: 21 mph; current: 3 mph **35.** \$0.75-per-lb candy: 5.22 lb; \$1.25-per-lb candy: 3.78 lb **37.** general admission: 76; with student ID: 108 **39.** 8 for a citron; 5 for a wood apple **41.** $x + y + z = 180$; angle measures: 70°, 30°, 80° **43.** first: 20°; second: 70°; third: 90° **45.** shortest: 12 cm; middle: 25 cm; longest: 33 cm **47.** gold: 13; silver: 11; bronze: 9 **49.** general admission: 1170; courtside: 985; bench: 130 **51.** bookstore A: 140; bookstore B: 280; bookstore C: 380 **53.** first chemical: 50 kg; second chemical: 400 kg; third chemical: 300 kg **55.** wins: 30; losses: 12; overtime losses: 6 **57.** box of fish: 8 oz; box of bugs: 2 oz; box of worms: 5 oz

8 INEQUALITIES AND ABSOLUTE VALUE

Section 8.1 (pages 566–567)

1. D **3.** B **5.** F **7.** (a) $x < 100$ (b) $100 \le x \le 129$
(c) $130 \le x \le 159$ (d) $160 \le x \le 189$ (e) $x \ge 190$

9. $[16, \infty)$

11. $(7, \infty)$

13. $(-\infty, -4)$

15. $(-\infty, -40]$

17. $(-\infty, 4]$

19. $(-\infty, -10]$

21. $(-\infty, 14)$

23. $\left(-\infty, -\dfrac{15}{2}\right)$

25. $\left[\dfrac{1}{2}, \infty\right)$

27. $[2, \infty)$

29. $(3, \infty)$

31. $(-\infty, 4)$

33. $\left(-\infty, \dfrac{23}{6}\right]$

35. $\left(-\infty, \dfrac{76}{11}\right)$

37. $(-\infty, \infty)$

39. $\varnothing$

41. A; B

43. $(1, 11)$

45. $[-14, 10]$

47. $[-5, 6]$

49. $\left[-\dfrac{14}{3}, 2\right]$

51. $\left(-\dfrac{1}{3}, \dfrac{1}{9}\right]$

53. $\left[-1, \dfrac{5}{2}\right]$

55. $\left[-\dfrac{1}{2}, \dfrac{35}{2}\right]$

Section 8.2 (pages 573–575)

1. true **3.** false; The union is $(-\infty, 7) \cup (7, \infty)$. **5.** false;
The intersection is $\varnothing$. **7.** $\{1, 3, 5\}$, or B **9.** $\{4\}$, or D **11.** $\varnothing$

13. $\{1, 2, 3, 4, 5, 6\}$, or A

15.

17.

19. $(-3, 2)$

21. $(-\infty, 2]$

23. $\varnothing$

25. $[5, 9]$

27. $(-3, -1)$

29. $(-\infty, 4]$

31.

33.

35. $(-\infty, 8]$

37. $[-2, \infty)$

39. $(-\infty, \infty)$

41. $(-\infty, -5) \cup (5, \infty)$

43. $(-\infty, -1) \cup (2, \infty)$

45. $(-\infty, \infty)$

47. $[-4, -1]$ **49.** $[-9, -6]$
51. $(-\infty, 3)$ **53.** $[3, 9]$

55. intersection; $(-5, -1)$

57. union; $(-\infty, 4)$

59. union; $(-\infty, 0] \cup [2, \infty)$ **61.** intersection; $[4, 12]$

63. {Tuition and fees} **65.** {Tuition and fees, Board rates, Dormitory
charges} **67.** Maria, Joe **68.** none of them **69.** none of them
70. Luigi, Than **71.** Maria, Joe **72.** all of them

Section 8.3 (pages 583–586)

1. E; C; D; B; A **3.** (a) one (b) two (c) none **5.** $\{-12, 12\}$
7. $\{-5, 5\}$ **9.** $\{-6, 12\}$ **11.** $\{-5, 6\}$ **13.** $\left\{-3, \dfrac{11}{2}\right\}$
15. $\left\{-\dfrac{19}{2}, \dfrac{9}{2}\right\}$ **17.** $\left\{\dfrac{7}{3}, 3\right\}$ **19.** $\{12, 36\}$ **21.** $\{-12, 12\}$
23. $\{-10, -2\}$ **25.** $\left\{-\dfrac{32}{3}, 8\right\}$ **27.** $\{-75, 175\}$
29. $(-\infty, -3) \cup (3, \infty)$ **31.** $(-\infty, -4] \cup [4, \infty)$

33. $(-\infty, -25] \cup [15, \infty)$ **35.** $\left(-\infty, -\dfrac{12}{5}\right) \cup \left(\dfrac{8}{5}, \infty\right)$

37. $(-\infty, -2) \cup (8, \infty)$ **39.** $\left(-\infty, -\dfrac{9}{5}\right] \cup [3, \infty)$

41. (a) [number line] **(b)** [number line]

43. $[-3, 3]$ [number line]   **45.** $(-4, 4)$ [number line]

47. $(-25, 15)$ [number line]   **49.** $\left[-\frac{12}{5}, \frac{8}{5}\right]$ [number line]

51. $[-2, 8]$ [number line]   **53.** $\left(-\frac{9}{5}, 3\right)$ [number line]

55. $(-\infty, -5) \cup (13, \infty)$ [number line]   **57.** $\left(-\frac{13}{3}, 3\right)$ [number line]

59. $\{-6, -1\}$ [number line]   **61.** $\left[-\frac{10}{3}, 4\right]$ [number line]

63. $\left[-\frac{7}{6}, -\frac{5}{6}\right]$ [number line]   **65.** $[3, 13]$ [number line]

67. $\left(-\infty, \frac{1}{2}\right] \cup \left[\frac{7}{6}, \infty\right)$ **69.** $\left\{-\frac{5}{3}, \frac{11}{3}\right\}$ **71.** $(-\infty, -20) \cup (40, \infty)$

73. $\{-5, 1\}$ **75.** $\{3, 9\}$ **77.** $\{0, 20\}$ **79.** $\{-5, 5\}$

81. $\{-5, -3\}$ **83.** $(-\infty, -3) \cup (2, \infty)$ **85.** $[-10, 0]$

87. $(-\infty, 20] \cup [30, \infty)$ **89.** $\left\{-\frac{5}{3}, \frac{1}{3}\right\}$ **91.** $\{-1, 3\}$ **93.** $\left\{-3, \frac{5}{3}\right\}$

95. $\left\{-\frac{1}{3}, -\frac{1}{15}\right\}$ **97.** $\left\{-\frac{5}{4}\right\}$ **99.** $(-\infty, \infty)$ **101.** $\varnothing$ **103.** $\left\{-\frac{1}{4}\right\}$

105. $\varnothing$ **107.** $(-\infty, \infty)$ **109.** $\left\{-\frac{3}{7}\right\}$ **111.** $\left\{\frac{2}{5}\right\}$ **113.** $(-\infty, \infty)$

115. $\varnothing$ **117.** between 30.72 and 33.28 oz, inclusive **119.** between

31.2 and 32.8 oz, inclusive **121.** $(-0.05, 0.05)$

123. $(2.74975, 2.75025)$ **125.** between 6.8 and 9.8 lb

127. $|x - 1000| \le 100$; $900 \le x \le 1100$ **129.** 810.5 ft

130. Bank of America Center, Texaco Heritage Plaza **131.** Williams

Tower, Bank of America Center, Texaco Heritage Plaza, Enterprise

Plaza, Centerpoint Energy Plaza, Continental Center I, Fulbright Tower

132. (a) $|x - 810.5| \ge 95$ **(b)** $x \ge 905.5$ or $x \le 715.5$

(c) JPMorgan Chase Tower, Wells Fargo Plaza, One Shell Plaza

(d) It makes sense because it includes all buildings *not* listed in the

answer to **Exercise 131.**

Section 8.4 (pages 593–595)

1. (a) yes **(b)** yes **(c)** no **(d)** yes **3. (a)** no **(b)** no **(c)** no

(d) yes **5.** solid; below **7.** dashed; above **9.** $\le$ **11.** $>$

13. 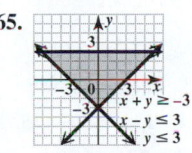 **15.** 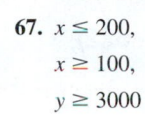 **17.**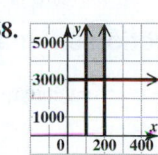

19. 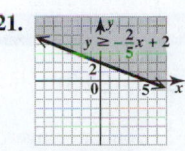 **21.** 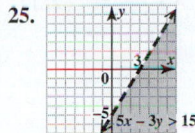 **23.**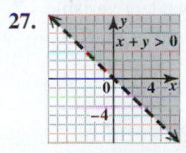

25. 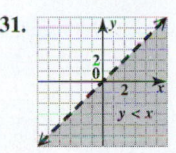 **27.** 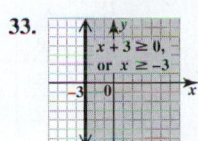 **29.**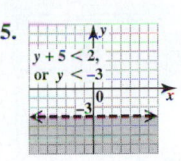

31. [graph] **33.** [graph] **35.** [graph]

37. 2; $(0, -4)$; $2x - 4$; solid; above; $\ge$; $\ge 2x - 4$ **39.** C **41.** B

43. (a) no **(b)** yes **(c)** no **45.** [graph]

47. [graph] **49.** [graph] **51.** [graph]

53. [graph] **55.** [graph] **57.** [graph]

59. [graph] **61.** [graph] **63.** [graph]

65. [graph] **67.** $x \le 200$, $x \ge 100$, $y \ge 3000$ **68.** [graph]

69. $C = 50x + 100y$ **70.** Some examples are $(100, 5000)$, $(150, 3000)$,

and $(150, 5000)$. The corner points are $(100, 3000)$ and $(200, 3000)$.

71. The least value occurs when $x = 100$ and $y = 3000$. **72.** The

company should use 100 workers and manufacture 3000 units to achieve

the least possible cost.

9 | RELATIONS AND FUNCTIONS

Section 9.1 (pages 611–613)

1. relation; ordered pairs   **3.** domain; range

5. independent variable; dependent variable   **7.** $\{(2, -2), (2, 0), (2, 1)\}$

9. $\{(1960, 0.76), (1980, 2.69), (2000, 5.39), (2013, 8.38)\}$

11. $\{(A, 4), (B, 3), (C, 2), (D, 1), (F, 0)\}$

In Exercises 13 and 15, answers will vary.

13. 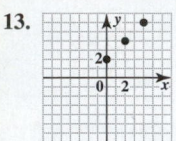   **15.**

x	y
-3	-4
-3	1
2	0

17. function; domain: $\{5, 3, 4, 7\}$; range: $\{1, 2, 9, 6\}$ **19.** not a function; domain: $\{2, 0\}$; range: $\{4, 2, 5\}$ **21.** function; domain: $\{-3, 4, -2\}$; range: $\{1, 7\}$ **23.** not a function; domain: $\{1, 0, 2\}$; range: $\{1, -1, 0, 4, -4\}$ **25.** function; domain: $\{2, 5, 11, 17, 3\}$; range: $\{1, 7, 20\}$ **27.** not a function; domain: $\{1\}$; range: $\{5, 2, -1, -4\}$ **29.** function; domain: $\{4, 2, 0, -2\}$; range: $\{-3\}$ **31.** function; domain: $\{-2, 0, 3\}$; range: $\{2, 3\}$ **33.** function; domain: $(-\infty, \infty)$; range: $(-\infty, \infty)$ **35.** not a function; domain: $\{-2\}$; range: $(-\infty, \infty)$ **37.** not a function; domain: $(-\infty, 0]$; range: $(-\infty, \infty)$ **39.** function; domain: $(-\infty, \infty)$; range: $(-\infty, 4]$ **41.** not a function; domain: $[-4, 4]$; range: $[-3, 3]$ **43.** not a function; domain: $(-\infty, \infty)$; range: $[2, \infty)$ **45.** function; $(-\infty, \infty)$ **47.** function; $(-\infty, \infty)$ **49.** function; $(-\infty, \infty)$ **51.** not a function; $[0, \infty)$ **53.** not a function; $(-\infty, \infty)$ **55.** function; $(-\infty, \infty)$ **57.** function; $(-\infty, \infty)$ **59.** function; $(-\infty, 0) \cup (0, \infty)$ **61.** function; $(-\infty, 4) \cup (4, \infty)$ **63.** function; $\left(-\infty, -\frac{1}{2}\right) \cup \left(-\frac{1}{2}, \infty\right)$ **65.** not a function; $[1, \infty)$ **67.** function; $(-\infty, 0) \cup (0, \infty)$ **69. (a)** yes   **(b)** domain: $\{2009, 2010, 2011, 2012, 2013\}$; range: $\{44.0, 43.4, 43.1, 42.9\}$ **(c)** 42.9; 2010   **(d)** Answers will vary. Two possible answers are $(2010, 43.4)$ and $(2013, 43.1)$.

Section 9.2 (pages 618–622)

1. $f(x)$; function; domain; x; f of x (or "f at x")   **3.** line; -2; linear; $-2x + 4$; -2; 3; -2 **5.** 4   **7.** 13   **9.** -11 **11.** 4   **13.** -296 **15.** 3   **17.** 2.75   **19.** $-3p + 4$ **21.** $3x + 4$ **23.** $-3x - 2$ **25.** $-6t + 1$ **27.** $-\pi^2 + 4\pi + 1$ **29.** $-3x - 3h + 4$ **31.** $-\frac{p^2}{9} + \frac{4p}{3} + 1$ **33. (a)** -1 **(b)** -1 **35. (a)** 2   **(b)** 3   **37. (a)** 15   **(b)** 10   **39. (a)** 4   **(b)** 1   **41. (a)** 3   **(b)** -3 **43. (a)** -3 **(b)** 2   **45. (a)** 2   **(b)** 0   **(c)** -1 **47. (a)** $f(x) = -\frac{1}{3}x + 4$ **(b)** 3   **49. (a)** $f(x) = 3 - 2x^2$ **(b)** -15 **51. (a)** $f(x) = \frac{4}{3}x - \frac{8}{3}$ **(b)** $\frac{4}{3}$ **53.** domain: $(-\infty, \infty)$;   **55.** domain: $(-\infty, \infty)$;   **57.** domain: $(-\infty, \infty)$;   range: $(-\infty, \infty)$ range: $(-\infty, \infty)$ range: $(-\infty, \infty)$

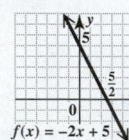   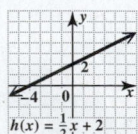   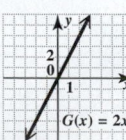

59. domain: $(-\infty, \infty)$;   **61.** domain: $(-\infty, \infty)$;   range: $\{-4\}$ range: $\{0\}$

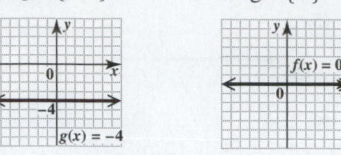

63. x-axis   **65. (a)** 11.25 (dollars)   **(b)** 3 is the value of the independent variable, which represents a package weight of 3 lb. $f(3)$ is the value of the dependent variable, which represents the cost to mail a 3-lb package.   **(c)** \$18.75; $f(5) = 18.75$ **67. (a)** $f(x) = 12x + 100$ **(b)** 1600; The cost to print 125 t-shirts is \$1600.   **(c)** 75; $f(75) = 1000$; The cost to print 75 t-shirts is \$1000.   **69. (a)** 1.1   **(b)** 5   **(c)** -1.2 **(d)** $(0, 3.5)$ **(e)** $f(x) = -1.2x + 3.5$ **71. (a)** $[0, 100]$; $[0, 3000]$ **(b)** 25 hr; 25 hr   **(c)** 2000 gal   **(d)** $f(0) = 0$; The pool is empty at time 0.   **(e)** $f(25) = 3000$; After 25 hr, there are 3000 gal of water in the pool.   **73. (a)** 194.53 cm   **(b)** 177.29 cm   **(c)** 177.41 cm   **(d)** 163.65 cm   **75.** Because it falls from left to right, the slope is negative.

76. $-\frac{3}{2}$ **77.** $-\frac{3}{2}$; $\frac{2}{3}$ **78.** $\left(\frac{7}{3}, 0\right)$ **79.** $\left(0, \frac{7}{2}\right)$ **80.** $f(x) = -\frac{3}{2}x + \frac{7}{2}$ **81.** $-\frac{17}{2}$ **82.** $\frac{23}{3}$

Section 9.3 (pages 629–632)

1. polynomial; one; terms; powers   **3. (a)** -10 **(b)** 8   **(c)** -4 **5. (a)** 8   **(b)** -10 **(c)** 0   **7. (a)** 8   **(b)** 2   **(c)** 4   **9. (a)** 7   **(b)** 1   **(c)** 1   **11. (a)** 8   **(b)** 74   **(c)** 6   **13. (a)** -11 **(b)** 4   **(c)** -8 **15. (a)** $8x - 3$ **(b)** $2x - 17$ **17. (a)** $-x^2 + 12x - 12$ **(b)** $9x^2 + 4x + 6$ **19.** $f(x)$ and $g(x)$ can be any two polynomials that have a sum of $3x^3 - x + 3$, such as $f(x) = 3x^3 + 1$ and $g(x) = -x + 2$.   **21.** $x^2 + 2x - 9$ **23.** 6   **25.** $x^2 - x - 6$ **27.** 6   **29.** -33 **31.** 0   **33.** $-\frac{9}{4}$ **35.** $-\frac{9}{2}$ **37. (a)** $P(x) = 8.49x - 50$ **(b)** \$799   **39.** $10x^2 - 2x$ **41.** $2x^2 - x - 3$ **43.** $8x^3 - 27$ **45.** $2x^3 - 18x$ **47.** -20 **49.** $2x^2 - 6x$ **51.** 36   **53.** $\frac{35}{4}$ **55.** $\frac{1859}{64}$ **57.** $5x - 1$; 0   **59.** $2x - 3$; -1 **61.** $4x^2 + 6x + 9$; $\frac{3}{2}$ **63.** $\frac{x^2 - 9}{2x}$, $x \neq 0$ **65.** $-\frac{5}{4}$ **67.** $\frac{x - 3}{2x}$, $x \neq 0$ **69.** 0   **71.** $-\frac{35}{4}$ **73.** $\frac{7}{2}$ **75.** B   **77.** A   **79.** 6   **81.** 83   **83.** 53   **85.** 13   **87.** $2x^2 + 11$ **89.** $2x - 2$ **91.** $\frac{97}{4}$ **93.** 8   **95.** 1   **97.** 9   **99.** 1   **101.** $(f \circ g)(x) = 63{,}360x$; It computes the number of inches in x miles.   **103. (a)** $s = \frac{x}{4}$ **(b)** $y = \frac{x^2}{16}$ **(c)** 2.25   **105.** $(\mathcal{A} \circ r)(t) = 4\pi t^2$; This is the area of the circular layer as a function of time.

Section 9.4 (pages 638–641)

1. increases; decreases   **3.** direct   **5.** direct   **7.** inverse   **9.** inverse   **11.** inverse   **13.** direct   **15.** joint   **17.** combined   **19.** The perimeter of a square varies directly as the length of its side.   **21.** The surface area of a sphere varies directly as the square of its radius.   **23.** The area of a triangle varies jointly as the lengths of its base and height.   **25.** 4; 2; 4π; $\frac{4}{3}\pi$; $\frac{1}{2}$; $\frac{1}{3}\pi$ **27.** $A = kb$ **29.** $h = \frac{k}{t}$ **31.** $M = kd^2$

33. $I = kgh$ **35.** 36   **37.** $\frac{16}{9}$ **39.** 0.625   **41.** $\frac{16}{5}$ **43.** $222\frac{2}{9}$
45. \$3.92   **47.** 8 lb   **49.** 450 cm³   **51.** 256 ft   **53.** $106\frac{2}{3}$ mph
55. 100 cycles per sec   **57.** $21\frac{1}{3}$ foot-candles   **59.** \$420
61. 11.8 lb   **63.** 448.1 lb   **65.** 68,600 calls   **67.** Answers will vary.

10   ROOTS, RADICALS, AND ROOT FUNCTIONS

Section 10.1 (pages 658–661)

1. true   **3.** false; Zero has only one square root.   **5.** true   **7.** a must be
positive.   **9.** a must be negative.   **11.** $-3, 3$ **13.** $-8, 8$ **15.** $-13, 13$
17. $-\frac{5}{14}, \frac{5}{14}$ **19.** $-30, 30$ **21.** 1   **23.** 7   **25.** 10   **27.** -4
29. -16 **31.** $-\frac{12}{11}$ **33.** 0.8   **35.** It is not a real number.   **37.** It is not
a real number.   **39.** 19   **41.** 19   **43.** $\frac{2}{3}$ **45.** $3x^2 + 4$ **47.** rational; 5
49. irrational; 5.385   **51.** rational; -8 **53.** irrational; -17.321
55. It is not a real number.   **57.** irrational; 34.641   **59.** 9 and 10
61. 7 and 8   **63.** -7 and -6 **65.** 4 and 5   **67.** 1; 8; 27; 64; 125; 216;
343; 512; 729; 1000   **69. (a)** E   **(b)** F   **(c)** D   **(d)** B   **(e)** C   **(f)** E
71. -9 **73.** 6   **75.** -4 **77.** -8 **79.** 6   **81.** -2 **83.** It is not a
real number.   **85.** 2   **87.** It is not a real number.   **89.** $\frac{8}{9}$ **91.** $\frac{4}{3}$
93. $-\frac{1}{2}$ **95.** 3   **97.** 0.5   **99.** -0.7 **101.** 0.1

In Exercises 103–109, we give the domain and then the range.

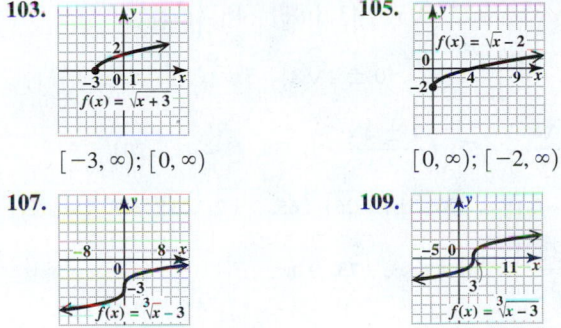

103. $[-3, \infty); [0, \infty)$ **105.** $[0, \infty); [-2, \infty)$
107. $(-\infty, \infty); (-\infty, \infty)$ **109.** $(-\infty, \infty); (-\infty, \infty)$

111. 12   **113.** 10   **115.** 2   **117.** -9 **119.** -5 **121.** $|x|$ **123.** $|z|$
125. x **127.** x^5 **129.** $|x|^5$ (or $|x^5|$)   **131.** C   **133.** 97.381
135. 16.863   **137.** -9.055 **139.** 7.507   **141.** 3.162   **143.** 1.885
145. 1,183,000 cycles per sec   **147.** 10 mi   **149.** 392,000 mi²
151. (a) 1.732 amps   **(b)** 2.236 amps

Section 10.2 (pages 668–670)

1. C   **3.** A   **5.** H   **7.** B   **9.** D   **11.** 13   **13.** 9   **15.** 2   **17.** $\frac{8}{9}$
19. -3 **21.** It is not a real number.   **23.** 1000   **25.** 27   **27.** -1024
29. 16   **31.** $\frac{1}{8}$ **33.** $\frac{1}{512}$ **35.** $\frac{9}{25}$ **37.** $\frac{27}{8}$ **39.** $\sqrt{10}$ **41.** $\left(\sqrt[4]{8}\right)^3$
43. $\left(\sqrt[8]{9q}\right)^5 - \left(\sqrt[3]{2x}\right)^2$ **45.** $\dfrac{1}{\left(\sqrt{2m}\right)^3}$ **47.** $\left(\sqrt[3]{2y+x}\right)^2$
49. $\dfrac{1}{\left(\sqrt[3]{3m^4 + 2k^2}\right)^2}$ **51.** 64   **53.** 64   **55.** x^{10} **57.** $\sqrt[6]{x^5}$
59. $\sqrt[15]{t^8}$ **61.** 9   **63.** 4   **65.** y **67.** $x^{5/12}$ **69.** $k^{2/3}$
71. x^3y^8 **73.** $\dfrac{1}{x^{10/3}}$ **75.** $\dfrac{1}{m^{1/4}n^{3/4}}$ **77.** p^2 **79.** $\dfrac{c^{11/3}}{b^{11/4}}$ **81.** $\dfrac{q^{5/3}}{9p^{7/2}}$

83. $p + 2p^2$ **85.** $k^{7/4} - k^{3/4}$ **87.** $6 + 18a$ **89.** $-5x^2 + 5x$
91. $x^{17/20}$ **93.** $\dfrac{1}{x^{3/2}}$ **95.** $y^{5/6}z^{1/3}$ **97.** $m^{1/12}$ **99.** $x^{1/8}$ **101.** $x^{1/24}$
103. $\sqrt{a^2 + b^2} = \sqrt{3^2 + 4^2} = 5; a + b = 3 + 4 = 7; 5 \neq 7$
105. 4.5 hr   **107.** 19.0°; The table gives 19°.
109. 4.2°; The table gives 4°.

Section 10.3 (pages 678–682)

1. D   **3.** B   **5.** D   **7.** Because there are only two factors of
$\sqrt[3]{x}$, $\sqrt[3]{x} \cdot \sqrt[3]{x} = \left(\sqrt[3]{x}\right)^2$, or $\sqrt[3]{x^2}$.   **9.** $\sqrt{9}$, or 3   **11.** $\sqrt{36}$, or 6
13. $\sqrt{30}$ **15.** $\sqrt{14x}$ **17.** $\sqrt{42pqr}$ **19.** $\sqrt[3]{10}$ **21.** $\sqrt[3]{14xy}$
23. $\sqrt[4]{33}$ **25.** $\sqrt[4]{6x^3}$ **27.** This expression cannot be simplified by the
product rule.   **29.** $\frac{8}{11}$ **31.** $\dfrac{\sqrt{3}}{5}$ **33.** $\dfrac{\sqrt{x}}{5}$ **35.** $\dfrac{p^3}{9}$ **37.** $-\frac{3}{4}$
39. $\dfrac{\sqrt[3]{r^2}}{2}$ **41.** $-\dfrac{3}{x}$ **43.** $\dfrac{1}{x^3}$ **45.** $2\sqrt{3}$ **47.** $12\sqrt{2}$ **49.** $-4\sqrt{2}$
51. $-2\sqrt{7}$ **53.** This radical cannot be simplified further.   **55.** $4\sqrt[3]{2}$
57. $-2\sqrt[3]{2}$ **59.** $2\sqrt[3]{5}$ **61.** $-4\sqrt[4]{2}$ **63.** $2\sqrt[5]{2}$ **65.** $-3\sqrt[5]{2}$
67. $2\sqrt[6]{2}$ **69.** $6k\sqrt{2}$ **71.** $12xy^4\sqrt{xy}$ **73.** $11x^3$ **75.** $-3t^4$
77. $-10m^4z^2$ **79.** $5a^2b^3c^4$ **81.** $\frac{1}{2}r^2t^5$ **83.** $5x\sqrt{2x}$ **85.** $-10r^5\sqrt{5r}$
87. $x^3y^4\sqrt{13x}$ **89.** $2z^2w^3$ **91.** $-2zt^2\sqrt[3]{2z^2t}$ **93.** $3x^3y^4$
95. $-3r^3s^2\sqrt[4]{2r^3s^2}$ **97.** $\dfrac{y^5\sqrt{y}}{6}$ **99.** $\dfrac{x^5\sqrt[3]{x}}{3}$ **101.** $4\sqrt{3}$ **103.** $\sqrt{5}$
105. $x^2\sqrt{x}$ **107.** $\sqrt[6]{432}$ **109.** $\sqrt[12]{6912}$ **111.** $\sqrt[6]{x^5}$ **113.** 5
115. $8\sqrt{2}$ **117.** $2\sqrt{14}$ **119.** 13   **121.** $9\sqrt{2}$ **123.** $\sqrt{17}$ **125.** 5
127. $6\sqrt{2}$ **129.** $\sqrt{5y^2 - 2xy + x^2}$ **131. (a)** B   **(b)** C   **(c)** D
(d) A   **133.** $x^2 + y^2 = 144$ **135.** $(x + 4)^2 + (y - 3)^2 = 4$
137. $(x + 8)^2 + (y + 5)^2 = 5$

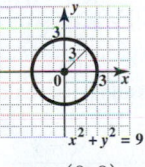 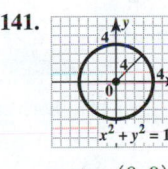 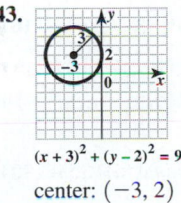

139. $x^2 + y^2 = 9$ center: $(0, 0)$ **141.** $x^2 + y^2 = 16$ center: $(0, 0)$
143. $(x + 3)^2 + (y - 2)^2 = 9$ center: $(-3, 2)$

145. $(x - 2)^2 + (y - 3)^2 = 4$ center: $(2, 3)$
147. $2\sqrt{106} + 4\sqrt{2}$
149. (a) $d = 1.224\sqrt{h}$ **(b)** 15.3 mi
151. 27.0 in.   **153.** 581

Section 10.4 (pages 685–687)

1. B   **3.** 15; Each radical expression simplifies to a whole number.
5. -4 **7.** $7\sqrt{3}$ **9.** $14\sqrt[3]{2}$ **11.** $5\sqrt[4]{2}$ **13.** $24\sqrt{2}$
15. The expression cannot be simplified further.   **17.** $20\sqrt{5}$
19. $4\sqrt{2x}$ **21.** $-11m\sqrt{2}$ **23.** $7\sqrt[3]{2}$ **25.** $2\sqrt[3]{x}$ **27.** $-7\sqrt[3]{x^2y}$
29. $-x\sqrt[3]{xy^2}$ **31.** $19\sqrt[4]{2}$ **33.** $x\sqrt[4]{xy}$ **35.** $9\sqrt[4]{2a^3}$
37. $(4 + 3xy)\sqrt[3]{xy^2}$ **39.** $4t\sqrt[3]{3st} - 3s\sqrt{3st}$ **41.** $4x\sqrt[3]{x} + 6x\sqrt[4]{x}$

43. $2\sqrt{2} - 2$ **45.** $\dfrac{5\sqrt{5}}{6}$ **47.** $\dfrac{7\sqrt{2}}{6}$ **49.** $\dfrac{5\sqrt{2}}{3}$ **51.** $5\sqrt{2} + 4$

53. $\dfrac{5 + 3x}{x^4}$ **55.** $\dfrac{m\sqrt[3]{m^2}}{2}$ **57.** $\dfrac{3x\sqrt[3]{2} - 4\sqrt[3]{5}}{x^3}$ **59. (a)** $\sqrt{7}$

(b) 2.645751311   **(c)** 2.645751311   **(d)** equal   **61.** A; 42 m

63. $\left(12\sqrt{5} + 5\sqrt{3}\right)$ in.   **65.** $\left(24\sqrt{2} + 12\sqrt{3}\right)$ in.

Section 10.5 (pages 694–697)

1. E   **3.** A   **5.** D   **7.** $3\sqrt{6} + 2\sqrt{3}$ **9.** $20\sqrt{2}$ **11.** -2

13. -1 **15.** 6   **17.** $\sqrt{6} - \sqrt{2} + \sqrt{3} - 1$

19. $\sqrt{22} + \sqrt{55} - \sqrt{14} - \sqrt{35}$ **21.** $8 - \sqrt{15}$ **23.** $9 + 4\sqrt{5}$

25. $26 - 2\sqrt{105}$ **27.** $4 - \sqrt[3]{36}$ **29.** 10

31. $6x + 3\sqrt{x} - 2\sqrt{5x} - \sqrt{5}$ **33.** $9r - s$

35. $4\sqrt[3]{4y^2} - 19\sqrt[3]{2y} - 5$ **37.** $3x - 4$ **39.** $4x - y$ **41.** $2\sqrt{6} - 1$

43. $\sqrt{7}$ **45.** $5\sqrt{3}$ **47.** $\dfrac{\sqrt{6}}{2}$ **49.** $\dfrac{9\sqrt{15}}{5}$ **51.** $-\dfrac{7\sqrt{3}}{12}$ **53.** $\dfrac{\sqrt{14}}{2}$

55. $-\dfrac{\sqrt{14}}{10}$ **57.** $\dfrac{2\sqrt{6x}}{x}$ **59.** $\dfrac{-8\sqrt{3k}}{k}$ **61.** $\dfrac{-5m^2\sqrt{6mn}}{n^2}$

63. $\dfrac{12x^3\sqrt{2xy}}{y^5}$ **65.** $\dfrac{5\sqrt{2my}}{y^2}$ **67.** $-\dfrac{4k\sqrt{3z}}{z}$ **69.** $\dfrac{\sqrt[3]{18}}{3}$ **71.** $\dfrac{\sqrt[3]{12}}{3}$

73. $\dfrac{\sqrt[3]{18}}{4}$ **75.** $-\dfrac{\sqrt[3]{2pr}}{r}$ **77.** $\dfrac{x^2\sqrt[3]{y^2}}{y}$ **79.** $\dfrac{2\sqrt[4]{x^3}}{x}$ **81.** $\dfrac{\sqrt[4]{2yz^3}}{z}$

83. $\dfrac{3\left(4 - \sqrt{5}\right)}{11}$ **85.** $\dfrac{6\sqrt{2} + 4}{7}$ **87.** $\dfrac{2\left(3\sqrt{5} - 2\sqrt{3}\right)}{33}$

89. $2\sqrt{3} + \sqrt{10} - 3\sqrt{2} - \sqrt{15}$ **91.** $\sqrt{m} - 2$

93. $\dfrac{4\left(\sqrt{x} + 2\sqrt{y}\right)}{x - 4y}$ **95.** $\dfrac{x - 2\sqrt{xy} + y}{x - y}$ **97.** $\dfrac{5\sqrt{k}\left(2\sqrt{k} - \sqrt{q}\right)}{4k - q}$

99. $3 - 2\sqrt{6}$ **101.** $1 - \sqrt{5}$ **103.** $\dfrac{4 - 2\sqrt{2}}{3}$ **105.** $\dfrac{6 + 2\sqrt{6p}}{3}$

107. $\dfrac{3\sqrt{x + y}}{x + y}$ **109.** $\dfrac{p\sqrt{p + 2}}{p + 2}$ **111.** Each expression is approximately

equal to 0.2588190451.   **113.** $\dfrac{33}{8\left(6 + \sqrt{3}\right)}$ **115.** $\dfrac{4x - y}{3x\left(2\sqrt{x} + \sqrt{y}\right)}$

Section 10.6 (pages 703–705)

1. (a) yes   **(b)** no   **3. (a)** yes   **(b)** no   **5.** No. There is no solution.
The radical expression, which is nonnegative, cannot equal a negative
number.   **7.** $\{11\}$ **9.** $\left\{\frac{1}{3}\right\}$ **11.** $\varnothing$ **13.** $\{5\}$ **15.** $\{18\}$ **17.** $\{5\}$
19. $\{4\}$ **21.** $\{17\}$ **23.** $\{5\}$ **25.** $\varnothing$ **27.** $\{0\}$ **29.** $\{0\}$ **31.** $\varnothing$
33. $\{1\}$ **35.** It is incorrect to just square each term. The right side
should be $(8 - x)^2 = 64 - 16x + x^2$. The correct first step is $3x + 4 =$
$64 - 16x + x^2$, and the solution set is $\{4\}$.   **37.** $\{1\}$ **39.** $\{-1\}$
41. $\{14\}$ **43.** $\{8\}$ **45.** $\{0\}$ **47.** $\varnothing$ **49.** $\{7\}$ **51.** $\{7\}$
53. $\{4, 20\}$ **55.** $\varnothing$ **57.** $\left\{\frac{5}{4}\right\}$ **59.** $\{9, 17\}$ **61.** $\left\{\frac{1}{4}, 1\right\}$

63. $L = CZ^2$ **65.** $K = \dfrac{V^2m}{2}$ **67.** $M = \dfrac{r^2F}{m}$ **69.** $r = \dfrac{a}{4\pi^2N^2}$

Section 10.7 (pages 710–712)

1. nonreal complex, complex   **3.** real, complex   **5.** pure imaginary,
nonreal complex, complex   **7.** i **9.** -1 **11.** $-i$ **13.** $13i$ **15.** $-12i$
17. $i\sqrt{5}$ **19.** $4i\sqrt{3}$ **21.** $-\sqrt{105}$ **23.** -10 **25.** $i\sqrt{33}$ **27.** $\sqrt{3}$
29. $5i$ **31.** -2 **33.** $-1 + 7i$ **35.** 0 **37.** $7 + 3i$ **39.** -2
41. $1 + 13i$ **43.** $6 + 6i$ **45.** $4 + 2i$ **47.** -81 **49.** -16
51. $-10 - 30i$ **53.** $10 - 5i$ **55.** $-9 + 40i$ **57.** $-16 + 30i$ **59.** 153
61. 97   **63.** 4   **65. (a)** $a - bi$ **(b)** $a^2; b^2$ **67.** $1 + i$ **69.** $2 + 2i$
71. $-1 + 2i$ **73.** $-\frac{5}{13} - \frac{12}{13}i$ **75.** $1 - 3i$ **77.** $1 + 3i$ **79.** -1
81. i **83.** -1 **85.** $-i$ **87.** $-i$ **89.** $\frac{1}{2} + \frac{1}{2}i$

11	**QUADRATIC EQUATIONS, INEQUALITIES, AND FUNCTIONS**

Section 11.1 (pages 728–729)

1. B, C   **3.** The zero-factor property requires a product equal to 0. The
first step should have been to rewrite the equation with 0 on one side.
5. $\{-7, 8\}$ **7.** $\{\pm 11\}$ **9.** $\left\{-\frac{5}{3}, 6\right\}$ **11.** $\{\pm 9\}$ **13.** $\{\pm 10\}$
15. $\{\pm\sqrt{14}\}$ **17.** $\{\pm 4\sqrt{3}\}$ **19.** $\left\{\pm\frac{5}{2}\right\}$ **21.** $\{\pm 1.5\}$
23. $\{\pm\sqrt{3}\}$ **25.** $\left\{\pm\dfrac{2\sqrt{7}}{7}\right\}$ **27.** $\{\pm 2\sqrt{6}\}$ **29.** $\left\{\pm\dfrac{2\sqrt{5}}{5}\right\}$
31. $\{\pm 3\sqrt{3}\}$ **33.** $\{\pm 2\sqrt{2}\}$ **35.** $\{-2, 8\}$ **37.** $\{4 \pm \sqrt{3}\}$
39. $\{8 \pm 3\sqrt{3}\}$ **41.** $\left\{-3, \frac{5}{3}\right\}$ **43.** $\left\{0, \frac{3}{2}\right\}$ **45.** $\left\{\dfrac{5 \pm \sqrt{30}}{2}\right\}$
47. $\left\{\dfrac{-1 \pm 3\sqrt{2}}{3}\right\}$ **49.** $\{-10 \pm 4\sqrt{3}\}$ **51.** $\left\{0, \frac{1}{4}\right\}$ **53.** $\left\{-\frac{1}{3}, 1\right\}$
55. $\left\{\dfrac{-1 \pm \sqrt{3}}{4}\right\}$ **57.** $\left\{\dfrac{1 \pm 4\sqrt{3}}{4}\right\}$ **59.** $\{-4.48, 0.20\}$
61. $\{-3.09, -0.15\}$ **63.** $\{\pm i\sqrt{26}\}$ **65.** $\{\pm 2i\sqrt{3}\}$ **67.** $\{5 \pm 2i\}$
69. $\left\{\dfrac{1}{6} \pm \dfrac{\sqrt{2}}{3}i\right\}$ **71.** 5.6 sec   **73.** 9 in.

Section 11.2 (pages 736–738)

1. Solve $(2x + 1)^2 = 5$ by the square root property. Solve $x^2 + 4x = 12$
by completing the square.   **3.** 25; $(x + 5)^2$ **5.** 100; $(z - 10)^2$
7. 1; $(x + 1)^2$ **9.** $\frac{25}{4}; \left(p - \frac{5}{2}\right)^2$ **11.** $\frac{1}{16}; \left(x + \frac{1}{4}\right)^2$ **13.** 0.04; $(x - 0.2)^2$
15. 4   **17.** 25   **19.** $\frac{1}{36}$ **21.** 4; 2; 4; 4; $(x + 2)^2 = 5$; $\{-2 \pm \sqrt{5}\}$
23. $\{1, 3\}$ **25.** $\{-1 \pm \sqrt{6}\}$ **27.** $\{-2 \pm \sqrt{6}\}$ **29.** $\{-5 \pm \sqrt{7}\}$
31. $\{4 \pm 2\sqrt{3}\}$ **33.** $\left\{\dfrac{-7 \pm \sqrt{53}}{2}\right\}$ **35.** $\left\{-\frac{3}{2}, \frac{1}{2}\right\}$
37. $\left\{\dfrac{-5 \pm \sqrt{41}}{4}\right\}$ **39.** $\left\{\dfrac{5 \pm \sqrt{15}}{5}\right\}$ **41.** $\left\{\dfrac{9 \pm \sqrt{21}}{6}\right\}$
43. $\left\{\dfrac{-7 \pm \sqrt{97}}{6}\right\}$ **45.** $\{1 \pm \sqrt{6}\}$ **47.** $\left\{-\frac{7}{3}, 6\right\}$ **49.** $\{1 \pm \sqrt{2}\}$
51. $\left\{\dfrac{3 \pm \sqrt{13}}{2}\right\}$ **53.** $\left\{\dfrac{-3 \pm \sqrt{5}}{2}\right\}$ **55.** $\{-4, 2\}$
57. $\{4 \pm \sqrt{3}\}$ **59. (a)** $\left\{\dfrac{3 \pm 2\sqrt{6}}{3}\right\}$ **(b)** $\{-0.633, 2.633\}$

61. (a) $\{-2 \pm \sqrt{3}\}$ **(b)** $\{-3.732, -0.268\}$ **63.** $\{-2 \pm 3i\}$

65. $\{-1 \pm i\sqrt{6}\}$ **67.** $\left\{-\dfrac{2}{3} \pm \dfrac{2\sqrt{2}}{3}i\right\}$ **69.** $\{-3 \pm i\sqrt{3}\}$

71. $\{\pm\sqrt{b}\}$ **73.** $\left\{\dfrac{\pm\sqrt{b^2+16}}{2}\right\}$ **75.** $\left\{\dfrac{2b \pm \sqrt{3a}}{5}\right\}$ **77.** x^2

78. x **79.** $6x$ **80.** 1 **81.** 9 **82.** $(x+3)^2$, or $x^2 + 6x + 9$

Section 11.3 (pages 743–744)

1. No. The fraction bar should extend under the term $-b$. The correct

formula is $x = \dfrac{-b \pm \sqrt{b^2-4ac}}{2a}$.   **3.** The last step is wrong. Because

5 is not a common factor in the numerator, the fraction cannot be simpli-

fied. The solution set is $\left\{\dfrac{5 \pm \sqrt{5}}{10}\right\}$.   **5.** $\{3,5\}$ **7.** $\left\{\dfrac{-2 \pm \sqrt{2}}{2}\right\}$

9. $\left\{\dfrac{1 \pm \sqrt{3}}{2}\right\}$ **11.** $\{5 \pm \sqrt{7}\}$ **13.** $\left\{\dfrac{-1 \pm \sqrt{2}}{2}\right\}$

15. $\left\{\dfrac{-1 \pm \sqrt{7}}{3}\right\}$ **17.** $\{1 \pm \sqrt{5}\}$ **19.** $\left\{\dfrac{-2 \pm \sqrt{10}}{2}\right\}$

21. $\{-1 \pm 3\sqrt{2}\}$ **23.** $\left\{\dfrac{1 \pm \sqrt{29}}{2}\right\}$ **25.** $\left\{\dfrac{-4 \pm \sqrt{91}}{3}\right\}$

27. $\left\{\dfrac{-3 \pm \sqrt{57}}{8}\right\}$ **29.** $\left\{\dfrac{3}{2} \pm \dfrac{\sqrt{15}}{2}i\right\}$ **31.** $\{3 \pm i\sqrt{5}\}$

33. $\left\{\dfrac{1}{2} \pm \dfrac{\sqrt{6}}{2}i\right\}$ **35.** $\left\{-\dfrac{2}{3} \pm \dfrac{\sqrt{2}}{3}i\right\}$ **37.** $\left\{\dfrac{1}{2} \pm \dfrac{1}{4}i\right\}$

39. 0; B; zero-factor property   **41.** 8; C; quadratic formula   **43.** 49; A;

zero-factor property   **45.** -80; D; quadratic formula   **47. (a)** 25; zero-

factor property; $\left\{-3, -\dfrac{4}{3}\right\}$ **(b)** 44; quadratic formula; $\left\{\dfrac{7 \pm \sqrt{11}}{2}\right\}$

49. -10 or 10 **51.** 16   **53.** 25   **55.** $b = \dfrac{44}{5}; \dfrac{3}{10}$

Section 11.4 (pages 752–755)

1. Multiply by the LCD, x.   **3.** Substitute a variable for $x^2 + x$.

5. The proposed solution -1 does not check. The solution set is $\{4\}$.

7. $\{-2, 7\}$ **9.** $\{-4, 7\}$ **11.** $\left\{-\dfrac{2}{3}, 1\right\}$ **13.** $\left\{-\dfrac{14}{17}, 5\right\}$

15. $\left\{-\dfrac{11}{7}, 0\right\}$ **17.** $\left\{\dfrac{-1 \pm \sqrt{13}}{2}\right\}$ **19.** $\left\{-\dfrac{8}{3}, -1\right\}$

21. $\left\{\dfrac{2 \pm \sqrt{22}}{3}\right\}$ **23.** $\left\{\dfrac{-1 \pm \sqrt{5}}{4}\right\}$ **25. (a)** $(20 - t)$ mph

(b) $(20 + t)$ mph   **27.** the rate of her boat in still water;

$x - 5$; $x + 5$; row 1 of table: 15, $x - 5$, $\dfrac{15}{x-5}$; row 2 of table: 15,

$x + 5$, $\dfrac{15}{x+5}$; $\dfrac{15}{x-5} + \dfrac{15}{x+5} = 4$; 10 mph   **29.** 25 mph   **31.** 50 mph

33. 3.6 hr   **35.** Rusty: 25.0 hr; Nancy: 23.0 hr   **37.** 3 hr; 6 hr

39. $\{2, 5\}$ **41.** $\{3\}$ **43.** $\left\{\dfrac{8}{9}\right\}$ **45.** $\{9\}$ **47.** $\left\{\dfrac{2}{5}\right\}$ **49.** $\{-2\}$

51. $\{\pm 2, \pm 5\}$ **53.** $\left\{\pm 1, \pm \dfrac{3}{2}\right\}$ **55.** $\{\pm 2, \pm 2\sqrt{3}\}$

57. $\{-6, -5\}$ **59.** $\left\{-\dfrac{16}{3}, -2\right\}$ **61.** $\{-8, 1\}$ **63.** $\{-64, 27\}$

65. $\left\{\pm 1, \pm \dfrac{27}{8}\right\}$ **67.** $\left\{-\dfrac{1}{3}, \dfrac{1}{6}\right\}$ **69.** $\left\{-\dfrac{1}{2}, 3\right\}$ **71.** $\left\{\pm \dfrac{\sqrt{6}}{3}, \pm \dfrac{1}{2}\right\}$

73. $\{3, 11\}$ **75.** $\{25\}$ **77.** $\left\{-\sqrt[3]{5}, -\dfrac{\sqrt[3]{4}}{2}\right\}$ **79.** $\left\{\dfrac{4}{3}, \dfrac{9}{4}\right\}$

81. $\left\{\pm \dfrac{\sqrt{9 + \sqrt{65}}}{2}, \pm \dfrac{\sqrt{9 - \sqrt{65}}}{2}\right\}$ **83.** $\left\{\pm 1, \pm \dfrac{\sqrt{6}}{2}i\right\}$

Section 11.5 (pages 761–765)

1. Find a common denominator, and then multiply both sides by the

common denominator.   **3.** Write it in standard form (with 0 on one side,

in decreasing powers of w).   **5.** $m = \sqrt{p^2 - n^2}$ **7.** $t = \dfrac{\pm\sqrt{dk}}{k}$

9. $r = \dfrac{\pm\sqrt{S\pi}}{2\pi}$ **11.** $d = \dfrac{\pm\sqrt{skI}}{I}$ **13.** $v = \dfrac{\pm\sqrt{kAF}}{F}$

15. $r = \dfrac{\pm\sqrt{3\pi Vh}}{\pi h}$ **17.** $t = \dfrac{-B \pm \sqrt{B^2 - 4AC}}{2A}$ **19.** $h = \dfrac{D^2}{k}$

21. $\ell = \dfrac{p^2 g}{k}$ **23.** $R = \dfrac{E^2 - 2pr \pm E\sqrt{E^2 - 4pr}}{2p}$

25. $r = \dfrac{5pc}{4}$ or $r = -\dfrac{2pc}{3}$ **27.** $I = \dfrac{-cR \pm \sqrt{c^2 R^2 - 4cL}}{2cL}$

29. 7.9, 8.9, 11.9   **31.** eastbound ship: 80 mi; southbound ship: 150 mi

33. 8 in., 15 in., 17 in.   **35.** length: 24 ft; width: 10 ft   **37.** 2 ft

39. 7 m by 12 m   **41.** 20 in. by 12 in.   **43.** 1 sec and 8 sec

45. 2.4 sec and 5.6 sec   **47.** 9.2 sec   **49.** It reaches its *maximum* height

at 5 sec because this is the only time it reaches 400 ft.   **51.** \$1.50

53. 0.035, or 3.5%   **55.** 5.5 m per sec   **57.** 5 or 14   **59. (a)** \$560 billion

(b) \$560 billion; They are the same.   **61.** 2008

Section 11.6 (pages 771–774)

1. (a) B   **(b)** C   **(c)** A   **(d)** D   **3. (a)** D   **(b)** B   **(c)** C   **(d)** A

5. $(0, 0)$ **7.** $(0, 4)$ **9.** $(1, 0)$ **11.** $(-3, -4)$ **13.** $(5, 6)$

15. down; wider   **17.** up; narrower   **19.** down; narrower

21.

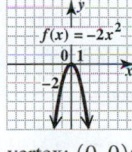

vertex: $(0, 0)$;

axis: $x = 0$;

domain: $(-\infty, \infty)$;

range: $(-\infty, 0]$

23.

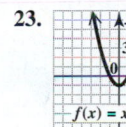

vertex: $(0, -1)$;

axis: $x = 0$;

domain: $(-\infty, \infty)$;

range: $[-1, \infty)$

25.

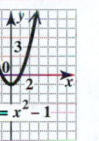

vertex: $(0, 2)$;

axis: $x = 0$;

domain: $(-\infty, \infty)$;

range: $(-\infty, 2]$

27.

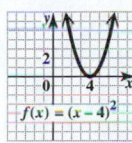

vertex: $(4, 0)$;

axis: $x = 4$;

domain: $(-\infty, \infty)$;

range: $[0, \infty)$

29.

vertex: $(-2, -1)$;

axis: $x = -2$;

domain: $(-\infty, \infty)$;

range: $[-1, \infty)$

31.

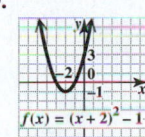

vertex: $(2, -4)$;

axis: $x = 2$;

domain: $(-\infty, \infty)$;

range: $[-4, \infty)$

33. $f(x) = -2(x+3)^2 + 4$

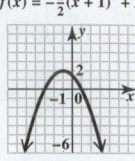

vertex: $(-3, 4)$;
axis: $x = -3$;
domain: $(-\infty, \infty)$;
range: $(-\infty, 4]$

35. $f(x) = -\frac{1}{2}(x+1)^2 + 2$

vertex: $(-1, 2)$;
axis: $x = -1$;
domain: $(-\infty, \infty)$;
range: $(-\infty, 2]$

37.

$f(x) = 2(x-2)^2 - 3$

vertex: $(2, -3)$;
axis: $x = 2$;
domain: $(-\infty, \infty)$;
range: $[-3, \infty)$

39. linear function; positive **41.** quadratic function; positive

43. quadratic function; negative

45. (a)

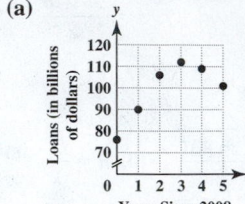

Years Since 2008

(b) quadratic function
(c) negative
(d) $y = -3.5x^2 + 22.5x + 76$
(e) 2010: \$107 billion;
2012: \$110 billion; The model
approximates the data very well.

Section 11.7 (pages 782–786)

1. If x is squared, it has a vertical axis. If y is squared, it has a horizontal
axis. **3.** Find the discriminant of the function. If it is positive, there
are two x-intercepts. If it is 0, there is one x-intercept (at the vertex), and
if it is negative, there is no x-intercept. **5.** A, D are vertical parabolas.
B, C are horizontal parabolas. **7.** $(-4, -6)$ **9.** $(1, -3)$
11. $\left(-\frac{1}{2}, -\frac{29}{4}\right)$ **13.** $(-1, 3)$; up; narrower; no x-intercepts
15. $\left(\frac{5}{2}, \frac{37}{4}\right)$; down; same shape; two x-intercepts **17.** $(-3, -9)$;
to the right; wider **19.** F **21.** C **23.** D

25. $f(x) = x^2 + 8x + 10$

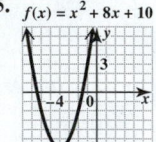

vertex: $(-4, -6)$;
axis: $x = -4$;
domain: $(-\infty, \infty)$;
range: $[-6, \infty)$

27.

$f(x) = x^2 + 2x - 2$

vertex: $(-1, -3)$;
axis: $x = -1$;
domain: $(-\infty, \infty)$;
range: $[-3, \infty)$

29.

$f(x) = -2x^2 + 4x - 5$

vertex: $(1, -3)$;
axis: $x = 1$;
domain: $(-\infty, \infty)$;
range: $(-\infty, -3]$

31.

$x = (y+2)^2 + 1$

vertex: $(1, -2)$;
axis: $y = -2$;
domain: $[1, \infty)$;
range: $(-\infty, \infty)$

33.

$x = -(y-3)^2 - 1$

vertex: $(-1, 3)$;
axis: $y = 3$;
domain: $(-\infty, -1]$;
range: $(-\infty, \infty)$

35.

$x = -\frac{1}{5}y^2 + 2y - 4$

vertex: $(1, 5)$;
axis: $y = 5$;
domain: $(-\infty, 1]$;
range: $(-\infty, \infty)$

37.

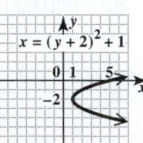

$x = 3y^2 + 12y + 5$

vertex: $(-7, -2)$;
axis: $y = -2$;
domain: $[-7, \infty)$;
range: $(-\infty, \infty)$

39. 20 and 20
41. 140 ft by 70 ft; 9800 ft^2
43. 2 sec; 65 ft
45. 20 units; \$210

47. 16 ft; 2 sec **49. (a)** minimum **(b)** 1967; 5.4%

51. The coefficient of x^2 is negative because a parabola that models the
data must open down.

53. (a) $R(x) = (100 - x)(200 + 4x)$
$= 20{,}000 + 200x - 4x^2$

(b)

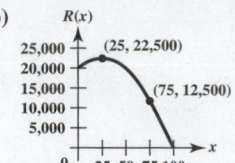

(c) 25 **(d)** \$22,500

55. center: $(-2, -3)$; $r = 2$

57. center: $(-5, 7)$; $r = 9$

Section 11.8 (pages 793–794)

1. $(-\infty, -4] \cup [3, \infty)$ **3. (a)** $\{1, 3\}$ **(b)** $(-\infty, 1) \cup (3, \infty)$
(c) $(1, 3)$ **5. (a)** $\{-2, 5\}$ **(b)** $[-2, 5]$ **(c)** $(-\infty, -2] \cup [5, \infty)$
7. $(-\infty, -1) \cup (5, \infty)$ **9.** $(-4, 6)$

11. $(-\infty, 1] \cup [3, \infty)$ **13.** $\left(-\infty, -\frac{3}{2}\right] \cup \left[\frac{3}{5}, \infty\right)$

15. $\left[-\frac{3}{2}, \frac{3}{2}\right]$ **17.** $\left(-\infty, -\frac{1}{2}\right] \cup \left[\frac{1}{3}, \infty\right)$

19. $(-\infty, 0] \cup [4, \infty)$ **21.** $\left[0, \frac{5}{3}\right]$

23. $\left(-\infty, 3 - \sqrt{3}\right] \cup \left[3 + \sqrt{3}, \infty\right)$ **25.** $(-\infty, \infty)$ **27.** $\varnothing$
29. $\varnothing$ **31.** $(-\infty, \infty)$

33. $(-\infty, 1) \cup (2, 4)$ **35.** $\left[-\frac{3}{2}, \frac{1}{3}\right] \cup [4, \infty)$

37. $(-\infty, 1) \cup (4, \infty)$ **39.** $\left[-\frac{3}{2}, 5\right]$

41. $(2, 6]$ **43.** $\left(-\infty, \frac{1}{2}\right) \cup \left(\frac{5}{4}, \infty\right)$

45. $[-7, -2)$ **47.** $(-\infty, 2) \cup (4, \infty)$

49. $\left(0, \frac{1}{2}\right) \cup \left(\frac{5}{2}, \infty\right)$ **51.** $\left[\frac{3}{2}, \infty\right)$

53. $\left(-2, \frac{5}{3}\right) \cup \left(\frac{5}{3}, \infty\right)$

12 INVERSE, EXPONENTIAL, AND LOGARITHMIC FUNCTIONS

Section 12.1 (pages 811–814)

1. B **3.** A **5.** This function is one-to-one. **7.** Yes, it is one-to-one. By adding 1 to 1058, two distances would be the same, so the function would not be one-to-one. **9.** $\{(6, 3), (10, 2), (12, 5)\}$

11. not one-to-one **13.** $f^{-1}(x) = x - 3$ **15.** $f^{-1}(x) = -2x - 4$

17. $f^{-1}(x) = \dfrac{x - 4}{2}$, or $f^{-1}(x) = \dfrac{1}{2}x - 2$ **19.** $f^{-1}(x) = x^2 + 3, x \geq 0$

21. not one-to-one **23.** $f^{-1}(x) = \sqrt[3]{x + 4}$

25. $f^{-1}(x) = \dfrac{-2x + 4}{x - 1}, x \neq 1$ **27.** $f^{-1}(x) = \dfrac{-5x - 2}{x - 4}, x \neq 4$

29. $f^{-1}(x) = \dfrac{5x + 1}{2x + 2}, x \neq -1$ **31.** (a) 8 (b) 3

33. (a) 1 (b) 0

35. (a) one-to-one **37.** (a) not one-to-one **39.** (a) one-to-one

(b) 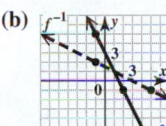 (b)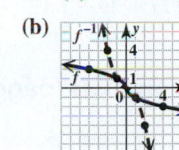

41. 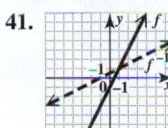 **43.** 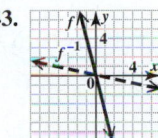 **45.**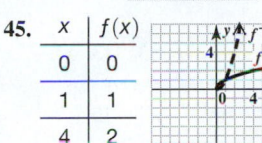

x	f(x)
0	0
1	1
4	2

47.

x	f(x)
−1	−3
0	−2
1	−1
2	6

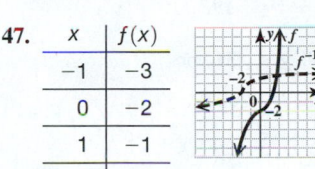

49. $f^{-1}(x) = \dfrac{x + 5}{4}$, or $f^{-1}(x) = \dfrac{1}{4}x + \dfrac{5}{4}$

50. MY GRAPHING CALCULATOR IS THE GREATEST THING SINCE SLICED BREAD. **51.** If the function were not one-to-one, there would be ambiguity in some of the characters, as they could represent more than one letter. **52.** Answers will vary. For example, Jane Doe is 1004 5 2748 129 68 3379 129.

Section 12.2 (pages 821–823)

1. C **3.** rises **5.** A **7.** 3.732 **9.** 0.344 **11.** 1.995 **13.** 1.587
15. 0.192 **17.** 73.517 **19.** 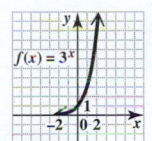 **21.**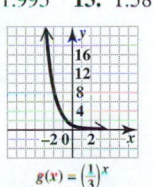

23. 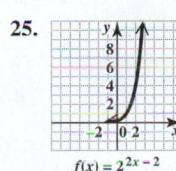 **25.** 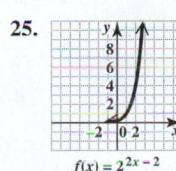 **27.** $\{2\}$ **29.** $\left\{\frac{3}{2}\right\}$
31. $\{7\}$ **33.** $\{-3\}$
35. $\{-1\}$ **37.** $\{-3\}$
39. 100 g **41.** 0.30 g

43. (a) 0.6°C (b) 0.3°C **45.** (a) 1.4°C (b) 0.5°C
47. (a) 2.403 million (b) 32.971 million (c) 188.962 million
49. (a) $5000 (b) $2973 (c) $1768
(d)
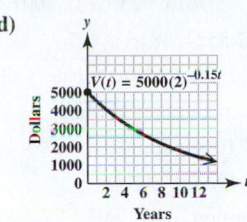

Section 12.3 (pages 829–831)

1. (a) B (b) E (c) D (d) F (e) A (f) C **3.** $(0, \infty); (-\infty, \infty)$
5. $\log_4 1024 = 5$ **7.** $\log_{1/2} 8 = -3$ **9.** $\log_{10} 0.001 = -3$
11. $\log_{625} 5 = \frac{1}{4}$ **13.** $\log_8 \frac{1}{4} = -\frac{2}{3}$ **15.** $\log_5 1 = 0$ **17.** $4^3 = 64$
19. $10^{-4} = \frac{1}{10,000}$ **21.** $6^0 = 1$ **23.** $9^{1/2} = 3$ **25.** $\left(\frac{1}{4}\right)^{1/2} = \frac{1}{2}$
27. $5^{-1} = 5^{-1}$ **29.** (a) C (b) B (c) B (d) C **31.** 3.1699
33. 1.7959 **35.** −1.7925 **37.** −1.5850 **39.** 1.9243 **41.** 1.6990
43. $\left\{\frac{1}{3}\right\}$ **45.** $\{81\}$ **47.** $\left\{\frac{1}{5}\right\}$ **49.** $\{1\}$ **51.** $\{x \mid x > 0, x \neq 1\}$
53. $\{5\}$ **55.** $\left\{\frac{5}{3}\right\}$ **57.** $\{4\}$ **59.** $\left\{\frac{3}{2}\right\}$ **61.** $\{30\}$ **63.** 1 **65.** 0
67. 9 **69.** −1 **71.** 9 **73.** 5 **75.** 6 **77.** 4 **79.** −1 **81.** $\frac{1}{3}$
83. 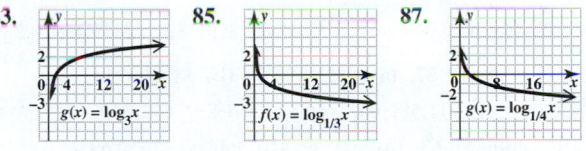 **85.** **87.**

89. Every power of 1 is equal to 1, and thus it cannot be used as a base.
91. 8 **93.** 24 **95.** (a) 130 thousand units (b) 190 thousand units
(c) 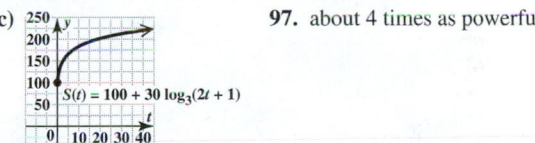 **97.** about 4 times as powerful

Section 12.4 (pages 837–838)

1. false; $\log_b x + \log_b y = \log_b xy$ **3.** true **5.** In the notation $\log_a (x + y)$, the parentheses do not indicate multiplication. They indicate that $x + y$ is the result of raising a to some power. **7.** $\log_{10} 7 + \log_{10} 8$
9. 4 **11.** 9 **13.** $\log_7 4 + \log_7 5$ **15.** $\log_5 8 - \log_5 3$ **17.** $2 \log_4 6$
19. $\frac{1}{3} \log_3 4 - 2 \log_3 x - \log_3 y$ **21.** $\frac{1}{2} \log_3 x + \frac{1}{2} \log_3 y - \frac{1}{2} \log_3 5$
23. $\frac{1}{3} \log_2 x + \frac{1}{5} \log_2 y - 2 \log_2 r$ **25.** $\log_b xy$ **27.** $\log_a \dfrac{m}{n}$
29. $\log_a \dfrac{rt^3}{s}$ **31.** $\log_a \dfrac{125}{81}$ **33.** $\log_{10} (x^2 + 8x + 15)$ **35.** $\log_p \dfrac{x^3 y^{1/2}}{z^{3/2} a^3}$
37. 1.2552 **39.** −0.6532 **41.** 1.5562 **43.** 0.2386 **45.** 0.4771
47. 4.7710 **49.** false **51.** true **53.** true **55.** false

Section 12.5 (pages 844–846)

1. C **3.** C **5.** 2 **7.** 1.6335 **9.** 2.5164 **11.** -1.4868 **13.** 9.6421
15. 9.6776 **17.** 2.0592 **19.** -2.8896 **21.** 5.9613 **23.** -11.4007
25. 4.1506 **27.** 2.3026 **29.** 40 **31. (a)** 2.552424846
(b) 1.552424846 **(c)** 0.552424846 **(d)** The whole number parts will
vary, but the decimal parts are the same. **33.** poor fen **35.** bog
37. rich fen **39.** 11.6 **41.** 4.3 **43.** 4.0×10^{-8} **45.** 4.0×10^{-6}
47. (a) 142 dB **(b)** 126 dB **(c)** 120 dB **49. (a)** 800 yr
(b) 5200 yr **(c)** 11,500 yr **51. (a)** 17.8 million **(b)** 2005
53. (a) \$54 per ton **(b)** If $x = 0$, then $\ln(1 - x) = \ln 1 = 0$, so $T(x)$
would be negative. If $x = 1$, then $\ln(1 - x) = \ln 0$, but the domain of
$\ln x$ is $(0, \infty)$. **55.** 2.2619 **57.** 0.6826 **59.** 0.3155 **61.** 0.8736
63. 2.4849 **65.** 0.4421 **67.** In each case, the product is 1. $\log_a b$ and
$\log_b a$ are reciprocals.

Section 12.6 (pages 853–856)

1. common logarithms **3.** natural logarithms **5.** $\{0.827\}$
7. $\{0.833\}$ **9.** $\{1.201\}$ **11.** $\{2.269\}$ **13.** $\{15.967\}$ **15.** $\{-6.067\}$
17. $\{261.291\}$ **19.** $\{-10.718\}$ **21.** $\{3\}$ **23.** $\{5.879\}$
25. $\{-\pi\}$, or $\{-3.142\}$ **27.** $\{1\}$ **29.** $\{4\}$ **31.** $\left\{\frac{2}{3}\right\}$ **33.** $\left\{\frac{33}{2}\right\}$
35. $\{-1 + \sqrt[3]{49}\}$ **37.** $\{\pm 3\}$ **39.** 2 cannot be a solution because
$\log(2 - 3) = \log(-1)$, and -1 is not in the domain of $\log x$. **41.** $\left\{\frac{1}{3}\right\}$
43. $\{2\}$ **45.** $\varnothing$ **47.** $\{8\}$ **49.** $\left\{\frac{4}{3}\right\}$ **51.** $\{8\}$ **53.** \$2539.47; 10.2 yr
55. \$4934.71; 19.8 yr **57. (a)** \$11,260.96 **(b)** \$11,416.64
(c) \$11,497.99 **(d)** \$11,580.90 **(e)** \$11,581.83 **59.** \$137.41
61. \$131,962 million **63. (a)** 1.62 g **(b)** 1.18 g **(c)** 0.69 g
(d) 2.00 g **65. (a)** 179.73 g **(b)** 21.66 yr

Appendix A (pages 871–872)

1. $\dfrac{1}{a^2 b}$ **3.** $\dfrac{100y^{10}}{x^2}$ **5.** 0 **7.** $\dfrac{x^{10}}{2w^{13}y^5}$ **9.** $-\dfrac{a^{15}}{64b^{15}}$ **11.** $\dfrac{x^{16}z^{10}}{y^6}$
13. $-6a^4 + 11a^3 - 20a^2 + 26a - 15$ **15.** $8x^3 - 18x^2 + 6x - 16$
17. $x^2y - xy^2 + 6y^3$ **19.** $-3x^2 - 62x + 32$ **21.** $10x^3 - 4x^2 + 9x - 4$
23. $6x^2 - 19x - 7$ **25.** $4x^2 - 9x + 2$ **27.** $16t^2 - 9$ **29.** $4y^4 - 16$
31. $16x^2 - 24x + 9$ **33.** $36r^2 + 60ry + 25y^2$ **35.** $c^3 + 8d^3$
37. $64x^3 - 1$ **39.** $14t^3 + 45st^2 + 18s^2t - 5s^3$
41. $4xy^3(2x^2y + 3x + 9y)$ **43.** $(x + 3)(x - 5)$ **45.** $(2x + 3)(x - 6)$
47. $(6t + 5)(6t - 5)$ **49.** $(4t + 3)^2$ **51.** $p(2m - 3n)^2$
53. $(x + 1)(x^2 - x + 1)$ **55.** $(2t + 5)(4t^2 - 10t + 25)$
57. $(t^2 - 5)(t^4 + 5t^2 + 25)$ **59.** $(5x + 2y)(t + 3r)$
61. $(6r - 5s)(a + 2b)$ **63.** $(t^2 + 1)(t + 1)(t - 1)$
65. $(2x + 3y - 1)(2x + 3y + 1)$ **67.** $4(x - 5)(x - 2)$

Appendix B (page 876)

1. 0 **3.** $x - 5$ **5.** $4m - 1$ **7.** $2a + 4 + \dfrac{5}{a + 2}$ **9.** $p - 4 + \dfrac{9}{p + 1}$
11. $4a^2 + a + 3$ **13.** $x^4 + 2x^3 + 2x^2 + 7x + 10 + \dfrac{18}{x - 2}$
15. $-4r^5 - 7r^4 - 10r^3 - 5r^2 - 11r - 8 + \dfrac{-5}{r - 1}$
17. $-3y^4 + 8y^3 - 21y^2 + 36y - 72 + \dfrac{143}{y + 2}$ **19.** 7
21. -2 **23.** 0 **25.** yes **27.** no **29.** yes **31.** no
33. $(2x - 3)(x + 4)$ **34.** $\left\{-4, \frac{3}{2}\right\}$ **35.** 0 **36.** 0 **37.** a
38. Yes, $x - 3$ is a factor; $Q(x) = (x - 3)(3x - 1)(x + 2)$

GLOSSARY

For a more complete discussion, see the section(s) in parentheses.

A

absolute value The absolute value of a number is the distance between 0 and the number on a number line. (Section 1.3)

absolute value equation An absolute value equation is an equation that involves the absolute value of a variable expression. (Section 8.3)

absolute value inequality An absolute value inequality is an inequality that involves the absolute value of a variable expression. (Section 8.3)

addends In the addition $x + y$, the addends are x and y. (Section 1.4)

additive inverse (opposite) The additive inverse of a number x, symbolized $-x$, is the number that is the same distance from 0 on the number line as x, but on the opposite side of 0. The number 0 is its own additive inverse. For all real numbers x, $x + (-x) = (-x) + x = 0$. (Section 1.3)

algebraic expression An algebraic expression is a sequence of constants, variables, operation symbols, and/or grouping symbols (such as parentheses) formed according to the rules of algebra. (Section 1.2)

area The area of a plane (two-dimensional) geometric figure is the measure (in square units) of the surface covered by the figure. (Section 2.5)

asymptote A line that a graph more and more closely approaches as the graph gets farther away from the origin is an asymptote of the graph. (Section 12.2)

axis of symmetry (axis) The axis of symmetry of a parabola is the vertical line or horizontal line (depending on the orientation of the graph) through the vertex of the parabola. (Sections 4.4, 11.6)

B

base (in an exponential expression) The base in an exponential expression is the expression that is the repeated factor. In b^x, b is the base. (Sections 1.1, 4.1)

base (in percents) The base in a percent equation is the whole amount being considered. (Section 2.6)

binomial A binomial is a polynomial consisting of exactly two terms. (Section 4.4)

boundary line In the graph of a linear inequality, the boundary line separates the region that satisfies the inequality from the region that does not satisfy the inequality. (Section 8.4)

C

center The fixed point that is a fixed distance from all the points that form a circle is the center of the circle. (Section 10.3)

circle A circle is the set of all points in a plane that lie a fixed distance from a fixed point. (Section 10.3)

circle graph (pie chart) A circle graph (or pie chart) is a circle divided into sectors, or wedges, whose sizes show the relative magnitudes of the categories of data being represented. (Section R.1)

coefficient (See **numerical coefficient.**)

combined variation A relationship among variables that involves both direct and inverse variation is combined variation. (Section 9.4)

common factor An integer that is a factor of two or more integers is a common factor of those integers. (Section 5.1)

common logarithm A common logarithm is a logarithm having base 10. (Section 12.5)

complementary angles (complements) Complementary angles are two angles whose measures have a sum of 90°. (Section 2.4)

complex conjugates The complex numbers $a + bi$ and $a - bi$ are complex conjugates. (Section 10.7)

complex fraction A complex fraction is a quotient with one or more fractions in the numerator, denominator, or both. (Section 6.5)

complex number A complex number is any number that can be written in the form $a + bi$, where a and b are real numbers and i is the imaginary unit. (Section 10.7)

components In an ordered pair (x, y), x and y are the components. (Section 7.1)

composite function If g is a function of x, and f is a function of $g(x)$, then $f(g(x))$ defines the composite function of f and g. It is symbolized $(f \circ g)(x)$. (Section 9.3)

composite number A natural number greater than 1 that is not prime is a composite number. It is composed of prime factors represented in one and only one way. (Section R.1)

composition of functions The process of finding a composite function is called composition of functions. (Section 9.3)

compound inequality A compound inequality consists of two inequalities linked by a connective word such as *and* or *or*. (Section 8.2)

compound interest Compound interest is paid or charged on both principal and previous interest. (Section 12.6)

conditional equation A conditional equation is only true under certain conditions. (Section 2.3)

conjugate The conjugate of $a + b$ is $a - b$. (Sections 4.6, 10.5)

consecutive even (or odd) integers Two consecutive even integers, such as 4 and 6, differ by 2. Two consecutive odd integers, such as 3 and 5, also differ by 2. (Sections 2.4, 5.6)

consecutive integers Two integers that differ by 1 are consecutive integers. (Sections 2.4, 5.6)

consistent system A system of equations with a solution is a consistent system. (Section 7.3)

constant A fixed, unchanging number is a constant. (Section 1.2)

constant function A constant function is a linear function whose graph is a horizontal line. It has the form $f(x) = b$, where b is a real number. (Section 9.2)

constant of variation In the variation equations $y = kx$, $y = \frac{k}{x}$, and $y = kxz$, the real number k is the constant of variation. (Section 9.4)

continuous compounding Continuous compounding of interest involves compounding where the number of compounding periods is allowed to increase without bound, leading to a formula with the number e. (Section 12.6)

contradiction A contradiction is an equation that has no solution. (Section 2.3)

coordinate on a number line   Every point on a number line is associated with a unique real number, which is the coordinate of the point. (Section 1.3)

coordinates of a point in a plane   The numbers in an ordered pair are called the coordinates of the corresponding point in the plane. (Sections 3.1, 7.1)

cross products   The cross products in the proportion $\frac{a}{b} = \frac{c}{d}$ are ad and bc. (Section 2.6)

cube root   A number b is a cube root of a if $b^3 = a$ is true. (Section 10.1)

cube root function   The function $f(x) = \sqrt[3]{x}$ is the cube root function. (Section 10.1)

D

decimal   A number written with place values as powers of ten, often using a decimal point, is a decimal number. (Section R.2)

decimal places   In a decimal number, each digit occupies a decimal place that is a power of ten. (Section R.2)

degree   A degree is a basic unit of measure for angles in which one degree (1°) is $\frac{1}{360}$ of a complete revolution. (Section 2.4)

degree of a polynomial   The degree of a polynomial is the greatest degree of any of the terms in the polynomial. (Section 4.4)

degree of a term   The degree of a term is the sum of the exponents on the variables in the term. (Section 4.4)

denominator   The number below the fraction bar in a fraction is the denominator. It indicates the number of equal parts in a whole. (Section R.1)

dependent equations   Equations of a system that have the same graph (because they are different forms of the same equation) are dependent equations. (Section 7.3)

dependent variable   In an equation relating x and y, if the value of the variable y depends on the value of the variable x, then y is the dependent variable. (Section 9.1)

descending powers   A polynomial in one variable is written in descending powers of the variable if the exponents on the variables of the terms of the polynomial decrease from left to right. (Section 4.4)

difference   The result of subtracting two numbers is their difference. (Sections R.1, 1.4)

direct variation y varies directly as x if there exists a real number k such that $y = kx$. (Section 9.4)

discriminant   The discriminant of the quadratic equation $ax^2 + bx + c = 0$ is the quantity $b^2 - 4ac$ under the radical in the quadratic formula. (Section 11.3)

dividend   In the quotient $\frac{a}{b}$, the dividend is a. (Sections R.1, R.2, 1.5)

divisor   In the quotient $\frac{a}{b}$, the divisor is b. (Sections R.1, R.2, 1.5)

domain   The set of all first components (x-values) in the ordered pairs of a relation is the domain. (Section 9.1)

double solution   A double solution of a quadratic equation is a solution that appears twice in the solution process but represents only one distinct value. (Section 5.5)

E

elements   The elements of a set are the objects that belong to the set. (Section 1.2)

empty set (null set)   The empty set, denoted by { } or ∅, is the set containing no elements. (Section 2.3)

equation   An equation is a statement that two algebraic expressions are equal. (Sections 1.2, 2.1)

equivalent equations   Equivalent equations are equations that have the same solution set. (Section 2.1)

exponent (power)   An exponent, or power, is a number that indicates how many times its base is used as a factor. In b^x, x is the exponent (power). (Sections 1.1, 4.1)

exponential equation   An exponential equation is an equation that has a variable in at least one exponent. (Section 12.2)

exponential expression   A number or letter (variable) written with an exponent is an exponential expression. (Sections 1.1, 4.1)

exponential function with base a   An exponential function with base a is a function of the form $f(x) = a^x$, where $a > 0$ and $a \neq 1$ for all real numbers x. (Section 12.2)

extraneous solution   A proposed solution to an equation that does not satisfy the original equation is an extraneous solution. (Sections 6.6, 10.6)

extremes of a proportion   In the proportion $\frac{a}{b} = \frac{c}{d}$, the a- and d-terms are the extremes. (Section 2.6)

F

factor   If a, b, and c represent numbers and $a \cdot b = c$, then a and b are factors of c. (Sections R.1, 1.5, 5.1)

factored form   An expression is in factored form when it is written as a product. (Section 5.1)

first-degree equation   A first-degree (linear) equation has no term with the variable to a power greater than 1. (Section 7.1)

focus variable   In the elimination method for solving a system of equations in three variables, the first variable to be eliminated is the focus variable. (Section 7.6)

FOIL method   FOIL is a mnemonic device that represents a method for multiplying two binomials $(a + b)(c + d)$. Multiply **F**irst terms ac, **O**uter terms ad, **I**nner terms bc, and **L**ast terms bd. Then combine like terms. (Section 4.5)

formula   A formula is an equation in which variables are used to describe a relationship among several quantities. (Section 2.5)

fourth root   A number b is a fourth root of a if $b^4 = a$ is true. (Section 10.1)

fraction   A number expressed as a quotient of integers in the form $\frac{a}{b}$ is a fraction. (Section R.1)

function   A function is a relation in which, for each distinct value of the first component of the ordered pairs, there is exactly one value of the second component. (Section 9.1)

G

graph of an equation   The graph of an equation in two variables is the set of all points that correspond to all of the ordered pairs that satisfy the equation. (Sections 3.2, 7.1)

graph of a number   The point on a number line that corresponds to a number is its graph. (Section 1.3)

greatest common factor (GCF)   The greatest common factor of a list of integers is the largest factor of all those integers. The greatest common factor of the terms of a polynomial is the largest factor of all the terms in the polynomial. (Sections R.1, 5.1)

H

horizontal line   In a plane, a horizontal line is parallel to the y-axis and has no x-intercept. (Section 3.2)

hypotenuse   The side opposite the right angle in a right triangle is the longest side and is the hypotenuse. (Sections 5.6, 10.3)

I

identity An identity is an equation that is true for all valid replacements of the variable. It has an infinite number of solutions. (Section 2.3)

identity element for addition For all real numbers a, $a + 0 = 0 + a = a$. The number 0 is the identity element for addition. (Section 1.6)

identity element for multiplication For all real numbers a, $a \cdot 1 = 1 \cdot a = a$. The number 1 is the identity element for multiplication. (Section 1.6)

imaginary part The imaginary part of the complex number $a + bi$ is b. (Section 10.7)

improper fraction A fraction with numerator greater than or equal to denominator is an improper fraction. Its value is greater than or equal to 1. (Section R.1)

inconsistent system An inconsistent system of equations is a system with no solution. (Section 7.3)

independent equations Equations of a system that have different graphs are independent equations. (Section 7.3)

independent variable In an equation relating x and y, if the value of the variable y depends on the value of the variable x, then x is the independent variable. (Section 9.1)

index (order) In a radical of the form $\sqrt[n]{a}$, n is the index or order. (Section 10.1)

inequality An inequality is a statement that two expressions may not be equal. (Sections 1.1, 2.8, 8.1)

inner product When using the FOIL method to multiply two binomials $(a + b)(c + d)$, the inner product is bc. (Section 4.5)

integers $\{\ldots, -2, -1, 0, 1, 2, \ldots\}$ is the set of integers. (Section 1.3)

intersection The intersection of two sets A and B, symbolized $A \cap B$, is the set of elements that belong to *both A and B*. (Section 8.2)

interval An interval is a portion of a number line. (Sections 2.8, 8.1)

inverse of a function f The inverse of a one-to-one function f, written f^{-1}, is the set of all ordered pairs of the form (y, x), where (x, y) belongs to f. (Section 12.1)

inverse variation y varies inversely as x if there exists a real number k such that $y = \frac{k}{x}$. (Section 9.4)

irrational number An irrational number cannot be written as the quotient of two integers, but can be represented by a point on a number line. (Sections 1.3, 10.1)

J

joint variation y varies jointly as x and z if there exists a real number k such that $y = kxz$. (Section 9.4)

L

leading term When a polynomial is written in descending powers of the variable, the greatest-degree term is written first and is the leading term of the polynomial. (Section 4.4)

least common denominator (LCD) Given several denominators, the least multiple that is divisible by all the denominators is the least common denominator. (Sections R.1, 6.3)

legs (of a right triangle) The two shorter perpendicular sides of a right triangle are the legs. (Sections 5.6, 10.3)

like terms Terms with exactly the same variables raised to exactly the same powers are like terms. (Sections 1.7, 4.4)

line graph A line graph is a series of line segments in two dimensions that connect points representing data. (Section 3.1)

linear equation in one variable A linear equation in one variable (here x) can be written in the form $Ax + B = C$, where A, B, and C are real numbers and $A \neq 0$. (Section 2.1)

linear equation in two variables A linear equation in two variables (here x and y) is an equation that can be written in the form $Ax + By = C$, where A, B, and C are real numbers and A and B are not both 0. (Sections 3.1, 7.1)

linear function A function f that can be written in the form $f(x) = ax + b$, where a and b are real numbers, is a linear function. The value of a is the slope m of the graph of the function. (Section 9.2)

linear inequality in one variable A linear inequality in one variable (here x) can be written in the form $Ax + B < C$, $Ax + B \leq C$, $Ax + B > C$, or $Ax + B \geq C$, where A, B, and C are real numbers and $A \neq 0$. (Sections 2.8, 8.1)

linear inequality in two variables A linear inequality in two variables (here x and y) can be written in the form $Ax + By < C$, $Ax + By \leq C$, $Ax + By > C$, or $Ax + By \geq C$, where A, B, and C are real numbers and A and B are not both 0. (Section 8.4)

logarithm A logarithm is an exponent. The expression $\log_a x$ represents the exponent to which the base a must be raised to obtain x. (Section 12.3)

logarithmic equation A logarithmic equation is an equation with a logarithm of a variable expression in at least one term. (Section 12.3)

logarithmic function with base a A logarithmic function with base a is a function of the form $g(x) = \log_a x$, where $a > 0$ and $a \neq 1$ for all positive real numbers x. (Section 12.3)

lowest terms A fraction is in lowest terms if the greatest common factor of the numerator and denominator is 1. (Sections R.1, 6.1)

M

means of a proportion In the proportion $\frac{a}{b} = \frac{c}{d}$, the b- and c-terms are the means. (Section 2.6)

minuend In the subtraction $x - y$, the minuend is x. (Section 1.4)

mixed number A mixed number includes a whole number and a fraction written together and is understood to be the sum of the whole number and the fraction. (Section R.1)

monomial A monomial is a polynomial consisting of exactly one term. (Section 4.4)

multiplicative inverse (reciprocal) The multiplicative inverse (reciprocal) of a nonzero number x, symbolized $\frac{1}{x}$, is the real number which has the property that the product of the two numbers is 1. For all nonzero real numbers x, $\frac{1}{x} \cdot x = x \cdot \frac{1}{x} = 1$. (Sections R.1, 1.5)

N

natural (counting) numbers The set of natural numbers is the set of numbers $\{1, 2, 3, 4, \ldots\}$. (Sections R.1, 1.3)

natural logarithm A natural logarithm is a logarithm having base e. (Section 12.5)

negative number A negative number is located to the left of 0 on a number line. (Section 1.3)

negative square root For even indexes, the symbols $-\sqrt{}$, $-\sqrt[4]{}$, $-\sqrt[6]{}$, $\ldots$, $-\sqrt[n]{}$ are used for negative roots. (Section 10.1)

nonreal complex number If $a + bi$ has $b \neq 0$, then it is a nonreal complex number. (Section 10.7)

number line   A line that has a point designated to correspond to the real number 0, and a standard unit chosen to represent the distance between 0 and 1, is a number line. All real numbers correspond to one and only one number on such a line. (Section 1.3)

numerator   The number above the fraction bar in a fraction is the numerator. It shows how many of the equivalent parts are being considered. (Section R.1)

numerical coefficient (coefficient)   The numerical factor in a term is the numerical coefficient, or simply the coefficient. (Sections 1.7, 4.4)

O

one-to-one function   In a one-to-one function each x-value corresponds to only one y-value, and each y-value corresponds to only one x-value. (Section 12.1)

order   (See **index**.)

ordered pair   An ordered pair is a pair of numbers written within parentheses in the form (x, y). (Sections 3.1, 7.1)

ordered triple   An ordered triple is a triple of numbers written within parentheses in the form (x, y, z). (Section 7.6)

origin   The point at which the x-axis and y-axis of a rectangular coordinate system intersect is the origin. (Sections 3.1, 7.1)

outer product   When using the FOIL method to multiply two binomials $(a + b)(c + d)$, the outer product is ad. (Section 4.5)

P

parabola   The graph of a second-degree (quadratic) equation in two variables is a parabola. (Sections 4.4, 11.6)

parallel lines   Parallel lines are two lines in the same plane that never intersect. (Section 3.3)

percent   Percent, written with the symbol %, means per one hundred. (Sections R.2, 2.6)

percentage   A percentage is a part of a whole. (Section 2.6)

perfect cube   A perfect cube is a number with a rational cube root. (Sections 5.4, 10.1)

perfect fourth power   A perfect fourth power is a number with a rational fourth root. (Section 10.1)

perfect square   A perfect square is a number with a rational square root. (Sections 5.4, 10.1)

perfect square trinomial   A perfect square trinomial is a trinomial that can be factored as the square of a binomial. (Section 5.4)

perimeter   The perimeter of a plane (two-dimensional) geometric figure is the measure of the outer boundary of the figure. For a polygon (e.g., a rectangle, square, or triangle), it is the sum of the lengths of the sides. (Section 2.5)

perpendicular lines   Perpendicular lines are two lines that intersect to form a right (90°) angle. (Section 3.3)

plane   In a rectangular coordinate system, the x- and y-axes form a plane—a flat surface illustrated by a sheet of paper. (Section 3.1)

plot   To plot an ordered pair is to locate it on a rectangular coordinate system. (Sections 3.1, 7.1)

polynomial   A polynomial is a term or a finite sum of terms in which all coefficients are real, all variables have whole number exponents, and no variables appear in denominators. (Section 4.4)

polynomial function   A function defined by a polynomial in one variable, consisting of one or more terms, is a polynomial function. (Section 9.3)

positive number   A positive number is located to the right of 0 on a number line. (Section 1.3)

positive (principal) square root   For even indexes, the symbols $\sqrt{\ }$, $\sqrt[4]{\ }$, $\sqrt[6]{\ }, \ldots, \sqrt[n]{\ }$ are used for positive roots, which are the principal roots. (Section 10.1)

prime number   A natural number greater than 1 is prime if it has only 1 and itself as factors. (Section R.1)

prime polynomial   A prime polynomial is a polynomial that cannot be factored into factors having only integer coefficients. (Section 5.2)

product   The result of multiplying two numbers is their product. (Sections R.1, 1.5)

proper fraction   A fraction with numerator less than denominator is a proper fraction. Its value is less than 1. (Section R.1)

proportion   A proportion is a statement that two ratios are equal. (Section 2.6)

proposed solution   A value that appears as an apparent solution after a rational or radical equation has been solved according to standard methods is a proposed solution for the original equation. It may or may not be an actual solution and must be checked. (Sections 6.6, 10.6)

pure imaginary number   A complex number $a + bi$ where $a = 0$ and $b \neq 0$ is a pure imaginary number. (Section 10.7)

Q

quadrant   A quadrant is one of the four regions in the plane determined by the axes in a rectangular coordinate system. (Sections 3.1, 7.1)

quadratic equation   A quadratic equation (in x here) is an equation that can be written in the form $ax^2 + bx + c = 0$, where a, b, and c are real numbers, with $a \neq 0$. (Sections 5.5, 11.1)

quadratic formula   The quadratic formula is a general formula used to solve a quadratic equation of the form $ax^2 + bx + c = 0$, where $a \neq 0$. It is $x = \dfrac{-b \pm \sqrt{b^2 - 4ac}}{2a}$. (Section 11.3)

quadratic function   A function that can be written in the form $f(x) = ax^2 + bx + c$, for real numbers a, b, and c, where $a \neq 0$, is a quadratic function. (Section 11.6)

quadratic inequality   A quadratic inequality (in x here) can be written in the form $ax^2 + bx + c < 0$ or $ax^2 + bx + c > 0$ (or with $\leq$ or $\geq$), where a, b, and c are real numbers and $a \neq 0$. (Section 11.7)

quadratic in form   A nonquadratic equation that can be written in the form $au^2 + bu + c = 0$, for $a \neq 0$ and an algebraic expression u, is quadratic in form. (Section 11.4)

quotient   The result of dividing two numbers is their quotient. (Sections R.1, R.2, 1.5)

R

radical   An expression consisting of a radical symbol, root index, and radicand is a radical. (Section 10.1)

radical equation   A radical equation is an equation with a variable in at least one radicand. (Section 10.6)

radical expression   A radical expression is an algebraic expression that contains a radical. (Section 10.1)

radicand   The number or expression under a radical symbol is the radicand. (Section 10.1)

radius   The radius of a circle is the fixed distance between the center and any point on the circle. (Section 10.3)

range   The set of all second components (y-values) in the ordered pairs of a relation is the range. (Section 9.1)

ratio   A ratio is a comparison of two quantities using a quotient. (Section 2.6)

rational expression   The quotient of two polynomials with denominator not 0 is a rational expression. (Section 6.1)

rational inequality An inequality that involves rational expressions is a rational inequality. (Section 11.8)

rational numbers Rational numbers can be written as the quotient of two integers, with denominator not 0. (Section 1.3)

real numbers Real numbers include all numbers that can be represented by points on a number line—that is, all rational and irrational numbers. (Section 1.3)

real part The real part of the complex number $a + bi$ is a. (Section 10.7)

reciprocal (See **multiplicative inverse.**)

rectangular (Cartesian) coordinate system The x-axis and y-axis placed at a right angle at their zero points form a rectangular coordinate system. It is also called the Cartesian coordinate system. (Sections 3.1, 7.1)

relation A relation is any set of ordered pairs. (Section 9.1)

repeating decimal A decimal number that does not have a final digit (such as 0.333 . . .) is a repeating decimal. (Section R.2)

right angle A right angle measures 90°. (Section 2.4)

rise Rise refers to the vertical change between two points on a line—that is, the change in y-values. (Sections 3.3, 7.1)

run Run refers to the horizontal change between two points on a line—that is, the change in x-values. (Sections 3.3, 7.1)

S

scatter diagram A scatter diagram is a graph of ordered pairs of data. (Section 3.1)

scientific notation A number is written in scientific notation when it is expressed in the form $a \times 10^n$, where $1 \leq |a| < 10$ and n is an integer. (Section 4.3)

second-degree equation A second-degree (quadratic) equation has a squared variable term and no terms of greater degree. (Section 11.1)

set A set is a collection of objects. (Section 1.2)

signed numbers Signed numbers are numbers that can be written with a positive or negative sign. (Section 1.3)

slope The ratio of the vertical change in y to the horizontal change in x for any two points on a line is the slope of the line. (Sections 3.3, 7.1)

solution of an equation A solution of an equation is any replacement for the variable that makes the equation true. (Sections 1.2, 2.1)

solution of a system A solution of a system of equations in two variables is an ordered pair (x, y) that makes all equations true at the same time. (Section 7.3)

solution set The set of all solutions of an equation is the solution set. (Section 2.1)

solution set of a system of linear equations The set of all ordered pairs that satisfy all equations of a linear system at the same time is the solution set. (Section 7.3)

solution set of a system of linear inequalities The set of all ordered pairs that satisfy all inequalities of a linear system at the same time is the solution set. (Section 8.4)

square root The inverse of squaring a number is taking its square root. That is, a number a is a square root of k if $a^2 = k$ is true. (Section 10.1)

square root function The root function $f(x) = \sqrt{x}$, where $x \geq 0$, is the square root function. (Section 10.1)

standard notation If a decimal number is written in its usual expanded form showing all decimal places (in contrast to *scientific notation*), it is in standard form. (Section 4.3)

straight angle A straight angle measures 180°. (Section 2.4)

subtrahend In the subtraction $x - y$, the subtrahend is y. (Section 1.4)

sum The result of adding two numbers is their sum. (Sections R.1, 1.4)

supplementary angles (supplements) Supplementary angles are two angles whose measures have a sum of 180°. (Section 2.4)

system of linear equations (linear system) A system of linear equations consists of two or more linear equations to be solved at the same time. (Section 7.3)

system of linear inequalities A system of linear inequalities consists of two or more linear inequalities to be solved at the same time. (Section 8.4)

T

table of ordered pairs (values) A table of ordered pairs (values) is an organized way of displaying ordered pairs. (Sections 3.1, 7.1)

term A term is a number, a variable, or the product or quotient of a number and one or more variables raised to powers. (Sections 1.7, 4.4)

terminating decimal A decimal number that has a final digit (such as 5 in 0.25) is a terminating decimal. (Section R.2)

terms of a proportion The terms of the proportion $\frac{a}{b} = \frac{c}{d}$ are a, b, c, and d. (Section 2.6)

three-part inequality An inequality that says that one number is between two other numbers is a three-part inequality. (Sections 2.8, 8.1)

trinomial A trinomial is a polynomial consisting of exactly three terms. (Section 4.4)

U

union The union of two sets A and B, symbolized $A \cup B$, is the set of elements that belong to *either A or B* (or both). (Section 8.2)

unlike terms Unlike terms are terms that do not have the same variable, or terms with the same variables but whose variables are not raised to the same powers. (Sections 1.7, 4.4)

V

variable A variable is a symbol, usually a letter, used to represent an unknown number. (Section 1.2)

vertex The point on a parabola that has the least y-value (if the parabola opens up) or the greatest y-value (if the parabola opens down) is the vertex of the parabola. (Sections 4.4, 11.6)

vertical angles When two intersecting lines are drawn, the angles that lie opposite each other have the same measure and are vertical angles. (Section 2.5)

vertical line In a plane, a vertical line is parallel to the x-axis and has no y-intercept. (Section 3.2)

volume The volume (in cubic units) of a three-dimensional figure is a measure of the space occupied by the figure. (Section 2.5)

W

whole numbers The set of whole numbers is $\{0, 1, 2, 3, 4, \ldots\}$. (Sections R.1, 1.3)

working equation In the elimination method for solving a system of equations in three variables, the working equation is used twice to eliminate the focus variable. (Section 7.6)

X

x-axis The horizontal number line in a rectangular coordinate system is the x-axis. (Sections 3.1, 7.1)

x-intercept A point where a graph intersects the x-axis is an x-intercept. (Sections 3.2, 7.1)

Y

y-axis The vertical number line in a rectangular coordinate system is the y-axis. (Sections 3.1, 7.1)

y-intercept A point where a graph intersects the y-axis is a y-intercept. (Sections 3.2, 7.1)

PHOTO CREDITS

FRONT MATTER
p. vT Margaret L. Lial; **p. vB** John Hornsby

CHAPTER R
p. 11 Cristovao/Shutterstock; **p. 15** Ryan McVay/Stockbyte/Getty Images; **p. 26** WavebreakMediaMicro/Fotolia

CHAPTER 1
p. 27 Michael Flippo/Fotolia; **p. 35** Monkey Business/Fotolia; **p. 42** Monkey Business/Fotolia; **p. 50** Kenishirotie/Fotolia; **p. 51** Gstockstudio/Fotolia; **p. 62** Orhan Cam/Shutterstock; **p. 63** Shefkate/Fotolia; **p. 65** GaryLHampton/iStockPhoto; **p. 75** Anjelagr/Fotolia; **p. 80** Terry McGinnis; **p. 94** James Thew/Fotolia; **p. 102** Sascha Burkard/Fotolia

CHAPTER 2
p. 103 Huaxiadragon/Fotolia; **p. 111** Alex_SK/Fotolia; **p. 130** PCN Photography/Alamy; **p. 138** Emkaplin/Fotolia; **p. 139T** DenisNata/Fotolia; **p. 139B** Les Cunliffe/Fotolia; **p. 153** Jaimie Duplass/Fotolia; **p. 159T** Xtr2007/Fotolia; **p. 159BL** Chokkicx/iStock/360/Getty Images; **p. 159BR** Petro Feketa/Fotolia; **p. 161** Sumire8/Fotolia; **p. 163** Stuart Jenner/Shutterstock; **p. 170** Wavebreakmedia/Shutterstock; **p. 172** Christopher Nolan/Fotolia; **p. 181** Kali9/Getty Images; **p. 185** Micromonkey/Fotolia; **p. 186** Rob/Fotolia; **p. 188** Studio DMM Photography, Designs & Art/Shutterstock

CHAPTER 3
p. 199 Eppic/Fotolia; **p. 200** ArenaCreative/Fotolia; **p. 204** KingPhoto/Fotolia; **p. 206** Felix Mizioznikov/Fotolia; **p. 210** Andresr/Shutterstock; **p. 219** Iofoto/Shutterstock; **p. 248** Karen Roach/Fotolia; **p. 251** Armadillo Stock/Shutterstock

CHAPTER 4
p. 261 Sara Piaseczynski; **p. 270** Goodluz/Fotolia; **p. 286T** Romario Ien/Fotolia; **p. 286BL** Mystique/Fotolia; **p. 286BR** Rafael Ramirez Lee/Shutterstock; **p. 297** Brodetskaya Elena/Fotolia; **p. 326** James Thew/Fotolia

CHAPTER 5
p. 329 Image Asset Management Ltd./Alamy; **p. 365** Nickolae/Fotolia; **p. 373** 4745052183/Shutterstock; **p. 378** Gino Santa Maria/Fotolia; **p. 383** Philipp Nemenz/Getty Images

CHAPTER 6
p. 393 SMI/Newscom; **p. 425** Denis Tabler/Fotolia; **p. 450** Rafael Ben-Ari/Fotolia; **p. 451** Johnny Habell/Shutterstock; **p. 453** Alma_sacra/Fotolia

CHAPTER 7
p. 467 Filtv/Fotolia; **p. 478** Joanna Zielinska/Fotolia; **p. 492** Photastic/Shutterstock; **p. 495** Holbox/Shutterstock; **p. 496** Wrangler/Shutterstock; **p. 497** Barry Blackburn/Shutterstock; **p. 513** Vadim Ponomarenko/Fotolia; **p. 529T** Photoestelar/Fotolia; **p. 529B** Rob Marmion/Shutterstock; **p. 536** Morgan Lane Photography/Shutterstock; **p. 540T** Nadia Zagainova/Shutterstock; **p. 540B** Joe Gough/Fotolia; **p. 543L** Hieronymus Ukkel/Fotolia; **p. 543R** Dwphotos/Shutterstock; **p. 544T** Stephen Coburn/Shutterstock; **p. 544B** Shutterstock

CHAPTER 8
p. 559 Degtiarova Viktoriia/Shutterstock; **p. 573** Faraways/Shutterstock; **p. 575** Michaeljung/Fotolia; **p. 586** Lunamarina/Fotolia

CHAPTER 9
p. 603 Konstantin Yolshin/Fotolia; **p. 620T** C Squared Studios/Stockbyte/Getty Images; **p. 620B** Comstock/Stockbyte/Getty Images; **p. 622** Neelsky/Shutterstock; **p. 627** Gorbelabda/Shutterstock; **p. 638** Avava/Shutterstock; **p. 641** John Hornsby

INDEX

A

Absolute value
  distance definition of, 576–577
  evaluating, 48
  explanation of, 47, 576
  simplifying square roots using, 656
Absolute value equations, 576–582
Absolute value inequalities, 576–582
Addition
  associative property of, 79, 84
  commutative property of, 78–79, 84
  of complex numbers, 708
  of decimals, 17–18
  of fractions, 6–8
  with grouping symbols, 56–57
  identity element for, 80
  identity property of, 80–81, 84
  inverse for, 81
  of multivariable polynomials, 293
  of negative numbers, 52–53
  on a number line, 52
  in order of operations, 29
  of polynomial functions, 625
  of polynomials, 291, 293, 868
  of radical expressions, 683–685
  of rational expressions, 416–419
  of real numbers, 52–54, 56–58
  of signed numbers, 52–54
  summary of properties of, 84
  of terms, 298
  word phrases for, 57–58
Addition property
  of equality, 104–108
  of inequality, 176–177, 561
Additive identity element, 80
Additive inverse
  explanation of, 47, 81
  finding for real numbers, 46–47, 55
  symbol for, 56
Agreement on domain, 610
Algebraic expressions
  distinguishing from equations, 40
  evaluating, 37
  explanation of, 36
  from word phrases, 38, 91, 124
  simplifying, 88–91
Angles
  complementary, 135–136
  measure of, 135–136, 144–145
  right, 135
  straight, 135, 144–145
  supplementary, 135–136, 540
  vertical, 144–145, 540
Apogee, 636

Approximately equal symbol, 657
Area
  of a circle, 148
  of a rectangle, 529–530, 759
  rules for exponents for, 267
  of a trapezoid, 142, 148
  of a triangle, 148
Associative properties
  distinguishing from commutative, 79–80
  explanation of, 79, 84
Average, 76
Average rate of change, 478–479
Axis
  of a coordinate system, 204, 468
  of a parabola, 293–294, 766, 768, 781
  of symmetry, 293–294

B

Base
  of an exponential expression, 28, 262
  in a percentage discussion, 156
Basic principle of fractions, 3
Binomials
  conjugates of, 692
  explanation of, 290
  factoring, 353
  finding greater powers of, 307–308
  finding product of the sum and difference of two terms, 306–307
  multiplication by FOIL method, 300–301
  multiplication of, 688–689
  squares of, 305–306
  steps to multiply by FOIL method, 300
Boundary line, 588
Braces
  explanation of, 29
  as a grouping symbol, 29
  for set notation, 39
Brackets as grouping symbol, 29–31
Business production problems, 537–539

C

Calculator graphing
  for approximation of roots, 657–658
  for generating quadratic models, 771
Cartesian coordinate system
  explanation of, 204, 468
  plotting points on, 205
Celsius-Fahrenheit relationship, 497
Center-radius form of a circle, 678
Change-of-base rule, 842–843
Characteristic impedance, 703

Circle(s)
  area of, 148
  center-radius form of, 678
  circumference of, 148
  equation of, 677–678
  graph, 11
  graphing, 677–678
Circumference of a circle, 148
Coefficient(s), 88, 288
Collinear points, 484
Combined variation, 637–638
Common denominators, 6–7
Common factors, 3, 330
Common logarithms
  applications of, 839–840
  evaluating, 838–839
  explanation of, 838
Commutative properties
  distinguishing from associative, 79–80
  explanation of, 78–79, 84
Complementary angles, 135–136
Completing the square method
  for finding the vertex, 775–777
  for solving quadratic equations, 730–736
Complex conjugates, 709
Complex fractions
  explanation of, 425
  simplifying, 426–430
  steps to simplify, 426, 428
Complex numbers
  addition of, 708
  conjugates of, 709
  division of, 709–710
  explanation of, 707
  imaginary part of, 707
  multiplication of, 708–709
  nonreal, 707
  real part of, 707
  standard form of, 707
  subtraction of, 708
Components, 468
Composite function
  explanation of, 627
  finding, 627–629
Composite numbers, 2
Composition of functions, 626–629
Compound inequalities
  with *and*, 568–570
  explanation of, 568
  with *or*, 571–572
Compound interest
  continuous, 852
  formula for, 270, 764, 851
Concours d'elegance, 425